ENVIRONMENTAL CHEMISTRY

ENVIRONMENTAL CHEMISTRY

(SECOND EDITION)

P S Sindhu
MSc., PhD
Formerly
Associate Professor in Chemistry
Hans Raj College, University of Delhi
Delhi

Publishing Globally

NEW AGE INTERNATIONAL (P) LIMITED, PUBLISHERS
LONDON • NEW DELHI • NAIROBI
Bangalore • Chennai • Cochin • Guwahati • Hyderabad • Kolkata • Lucknow • Mumbai
Visit us at **www.newagepublishers.com**

Published by New Age International (P) Ltd., Publishers
First Edition: 2002
Second Edition: 2010
Reprint: 2017

GLOBAL OFFICES

- **New Delhi** **NEW AGE INTERNATIONAL (P) LIMITED, PUBLISHERS**
  7/30 A, Daryaganj, New Delhi-110002, (INDIA)
  Tel.: (011) 23253771, 23253472, **Telefax:** 23267437, 43551305
  E-mail: contactus@newagepublishers.com • Visit us at www.newagepublishers.com
- **London** **NEW AGE INTERNATIONAL (UK) LTD.**
  27 Old Gloucester Street, London, WC1N 3AX, UK
  E-mail: info@newacademicscience.co.uk • Visit us at www.newacademicscience.co.uk
- **Nairobi** **NEW AGE GOLDEN (EAST AFRICA) LTD.**
  Ground Floor, Westlands Arcade, Chiromo Road (Next to Naivas Supermarket)
  Westlands, Nairobi, KENYA, **Tel.:** 00-254-713848772, 00-254-725700286
  E-mail: kenya@newagepublishers.com

BRANCHES

- **Bangalore** 37/10, 8th Cross (Near Hanuman Temple), Azad Nagar, Chamarajpet, Bangalore- 560 018
  Tel.: (080) 26756823, **Telefax:** 26756820, **E-mail: bangalore@newagepublishers.com**
- **Chennai** 26, Damodaran Street, T. Nagar, Chennai-600 017, **Tel.:** (044) 24353401, **Telefax:** 24351463
  E-mail: chennai@newagepublishers.com
- **Cochin** CC-39/1016, Carrier Station Road, Ernakulam South, Cochin-682 016, **Tel.:** (0484) 2377303, **Telefax:** 4051304
  E-mail: cochin@newagepublishers.com
- **Guwahati** Hemsen Complex, Mohd. Shah Road, Paltan Bazar, Near Starline Hotel, Guwahati-781 008,
  Tel.: (0361) 2513881, **Telefax:** 2543669, **E-mail: guwahati@newagepublishers.com**
- **Hyderabad** 105, 1st Floor, Madhiray Kaveri Tower, 3-2-19, Azam Jahi Road, Near Kumar Theater, Nimboliadda
  Kachiguda, Hyderabad-500 027, **Tel.:** (040) 24652456, **Telefax:** 24652457
  E-mail: hyderabad@newagepublishers.com
- **Kolkata** RDB Chambers (Formerly Lotus Cinema) 106A, 1st Floor, S N Banerjee Road, Kolkata-700 014
  Tel.: (033) 22273773, **Telefax:** 22275247, **E-mail: kolkata@newagepublishers.com**
- **Lucknow** 16-A, Jopling Road, Lucknow-226 001, **Tel.:** (0522) 2209578, 4045297, **Telefax:** 2204098
  E-mail: lucknow@newagepublishers.com
- **Mumbai** 142C, Victor House, Ground Floor, N.M. Joshi Marg, Lower Parel, Mumbai-400 013
  Tel.: (022) 24927869, **Telefax:** 24915415, **E-mail: mumbai@newagepublishers.com**
- **New Delhi** 22, Golden House, Daryaganj, New Delhi-110 002, **Tel.:** (011) 23262368, 23262370, **Telefax:** 43551305
  E-mail: sales@newagepublishers.com

ISBN: 978-81-224-2731-8
C-16-12-9933

Printed in India at Saras Graphics, Rai, Haryana.
Typeset at Pagitek Graphics, Delhi.

NEW AGE INTERNATIONAL (P) LIMITED, PUBLISHERS
7/30 A, Daryaganj, New Delhi-110002
Visit us at **www.newagepublishers.com**
(CIN: U74899DL1966PTC004618)

Dedicated to

I. COPENHAGEN CLIMATE COUNCIL, DENMARK
Working to combat global warming

II. VATAVARAN
Working for a pollution-free environment

Preface to the Second Edition

The primary purpose in reviewing the first edition of this volume was to bring it up-to-date. The organization is essentially the same and it covers as did the first edition: all phases of structure, properties and origin of pollution. Environmental Chemistry is now a mature, viable discipline. It is an exciting area that combines chemical principles to the biggest challenge facing human kind, the maintenance and enhancement of environmental quality. Man has generated maximum wealth in the last two centuries, but that has degraded the environment. The complex interactions that occur among terrestrial, atmospheric, aquatic, living and anthropological environments make Environmental Science. Since all these interactions involve chemical reactions thus very much part of chemistry though these reactions affect all the living being, yet it is more of chemistry than any thing else. It is the understanding of the reasons that explain every fact in terms of chemical principles and effects on our environment that has made it a viable discipline. This is the approach that has been followed in this book. Understanding of environment this way will help us to explain the mechanism and implications of any problem.

Scientist involved in different fields of study considers them best suited to guard the various aspects of our environment. A photo chemist is best suited for atmosphere and reactions taking place in it, a biologist investigating water and water pollution, soil scientist are concerned with soil, soil erosion and pollution, an entomologist primary interest lies in pesticides, nuclear scientist are engaged in the study of radiochemistry, polymer scientists study polymer and plastics, medical doctors are mainly worried about the effects of various metal ions on life. In view of such diverse nature of scientists, a comprehensive treatise on environment is not available. Here is an effort to fulfil such a requirement.

Since environmental pollution is a multidisciplinary subject hence quite complex and development of a convincing effect is bound to be lengthy but if developed this way will convince any reader to its consequence. There are quite a large number of books available that state facts but this is the only book that gives a detailed but lucid appreciation of the subject for students and teachers alike. Even in this edition the basic structure of the book has not been disturbed. I have tried to make many reasoning simpler but explicit in this revised edition. To facilitate the readers to understand the problems, large number of solved problems are included. At the end of Chapter 13 Part I and Chapter 10 Part II large number of chapterwise problems have been included. These may be taken as question bank that will help examiners in paper setting. Question banks are followed by new appendix. It is my conviction that student must spend some time on computer internet, because use of internet improves memory and power to concentrate on any job. In a way that will improve you as human being. Further computer is the todays need and will never be reversed. Keeping these things in mind Appendix-I in both part I and II certain wonderful internet sites on each chapter relevant to the part has been added. Glossary of various terms is given at the end of the book. Some printing mistakes have also been corrected and many facts and their explanations have been upgraded.

I will like to take this opportunity to acknowledge my thanks and appreciation to New Age International Publishers Limited for undertaking to publish second edition of this book. I must acknowledge my thanks to Dr. Naresh K.Mehra, Professor of Zoology, Delhi University, who expired sometime back, but had gone through Part II of the book and offered numerous suggestions, all of which have been included. I must also thank my wife Dr. Jagjit Kaur Sindhu who restrained from distracting me whenever I was working on this book.

P.S. SINDHU

Preface to the First Edition

Scientists involved in different fields of study consider themselves best suited to guard the various aspects of our environment. A photochemist is best suited for atmosphere and reactions taking place in it, a biologist investigates water and water pollution, soil scientists are concerned with soil, soil erosion and pollution, an entomologists primary interest lies in pesticides, nuclear scientists are engaged in the study of radiochemistry, polymer scientists study polymers and plastics, medical doctors are mainly worried about the effects of various metal ions on life. In view of such diverse nature of scientists a comprehensive treatise on environment is not yet available. Here is an effort to fulfil such a requirement.

All the processes that take place in the universe are either physical or chemical in nature. Changes in our environment are predominantly chemical processes. So much so that life is reduced to a chain of chemical reactions. When some reaction in this chain gets altered or stops, it may lead to death. The present nature of life has developed in a certain environment. As pollution is changing the chemical composition of our environment, the nature of chemical reaction and rate of reaction may change and thus will affect life and behaviour. A better understanding of nature and structure of molecules involved in our environmental changes is therefore necessary. Such an understanding is within the purview of a good physical chemist. This is another reason why this book was written.

Reactions associated with atmospheric pollution are primarily photochemical in nature which are entirely different from pollution reactions taking place in water or earth or other related states. So as a matter of convenience the book is divided into two independent parts. Part I deals with Atmospheric Chemistry and Part 11 deals with pollution reactions taking place in liquid or solid state or their interfaces.

To facilitate the readers to understand the problems, large number of solved problems are included. At the end of Chapter 13 Part I and Chapter 10 Part II, large number of chapterwise problems have been included. This may be taken as question bank that will help examiners in paper setting. Glossary of various terms is given at the end of the book.

Part I

Atmosphere of earth is a fairly complex and various changes in it are primarily photochemical in nature which require an understanding of photochemistry of gases and thus of Atomic and Molecular Spectroscopy, States of Molecules and Reaction Kinetics. These topics are introduced in Chapters 2 and 3 as required for treatment in subsequent chapters. This basic information was used to explain molecular spectra of almost all molecules which have been related as to how dissociation and formation of complex radicals take place in the atmosphere. Along with the mechanism of these reactions, electronic states of reactants, products and reaction rate constant have also been given. A better understanding of nature and structure of involved molecules give a better understanding of environmental pollution. All these make, in a way, a quantitative treatment which dispels any vague talking.

Earth can be taken as a black body radiator and how air pollution leads to its warming up has been discussed in Chapter 6. Further a quantitative treatment will definitely help us to appreciate better the processes taking place in smog formation, ozone hole at Antarctica, acid rains and increased particulate matter in atmosphere and thus their various effects. Different pollutants have different origin and as to how to control air pollution has been discussed in Chapter 11.

Air pollutants are dispersed before going through a change. This dispersion is dependent on stability and turbulent structure of atmosphere and thus controlled by prevailing meteorological conditions which have been discussed in Chapter 1 and 10. The implication of this on climate change has been discussed in Chapter 6. Here how the El-Nino-Southern oscillation can be used to our benefit has also been discussed. Chapter 12 deals with some specific cases of air pollution in India.

Noise is known to have a non-auditory effect on physiology of man and could be dangerous to his health. This is an important menace in the Indian context where making noise in all forms in our habit and has been discussed in Chapter 13.

The various concentration units used in gaseous reactions, changing the units of light absorption coefficient and some common conversion factors for gas phase reactions are given in Appendix III, IV and V respectively.

Part II

Man has invented many compounds for its use and convenience. However, during the course of manufacturing those compounds or after their use are such that they end up as pollutants in different states. In every chapter, structure and mechanism of their effects have been emphasized. It is my feeling that there still prevails some confusion about the identification of pollution and their solutions. Even then an attempt has been made to present the mechanism at every stage.

An environmentalist is one who has a concern for all types of life. Such a concern becomes stronger when one understands his environment. Since chemistry is a multidisciplinary subject hence a little complex but it is intended to give a lucid appreciation of subject for students and teachers alike.

I have assumed that the user of this book has an understanding of science at the undergraduate level. Each part and each chapter has been developed almost independent of any other but interrelationships are well presented. Almost at every stage, why and how of chemistry of our environmental reactions have been discussed.

I would like to take this opportunity to acknowledge my thanks and appreciation to many publishers and authors whose data and illustrations have been used in this book with appropriate references. I am also thankful to my publisher New Age International (P) Limited for undertaking the publication of this book with the hope that the book will find wide circulation.

I must acknowledge my thanks to Dr. Naresh K. Mehra, Professor of Zoology, Delhi University who went through Part II of this book and offered numerous suggestions, incorporation of which changed the complexion of some of the section. I must also thank my wife Dr. Jagjit Kaur Sindhu who restrained from distracting me whenever I was working on this book.

P.S. SINDHU

Contents

Preface to the Second Edition vii

Preface to the First Edition ix

PART–I
Chemistry of Atmosphere

1. Earth's Atmosphere **1–25**

1.1 Evolution of Life and Earth 1
1.2 Mechanism of Reactions 6
1.3 Extent of Atmosphere 9
1.4 Temperature Lapse Rates and Atmosphere Stability 11
1.5 Atmospheric Stability 12
1.6 Circulation of the Atmospheric Air 13
1.7 Angular Momentum of Atmosphere 16
1.8 Coriolis Effect 17
1.8a The Geostrophic Wind 18
1.8b Gradient Wind 19
1.8c The Effect of Friction 20
1.8d Vertical Motion-Divergence 21
1.9 Local Wind Systems 21
References 25

2. Principles of Thermodynamics and Gas Kinetics **26–44**

2.1 Introduction 26
2.2 Material Balance 27
2.3 Energy Laws 33
2.3a Zeroth Law of Thermodynamics 33
2.3b First Law of Thermodynamics 33

2.3c Second Law of Thermodynamics 36
2.3d Thermodynamic Efficiency of Energy Conversion Processes 37
2.3e Third Law of Thermodynamics 38
2.4 Kinetics of Gas Reactions 39
2.4a Reaction Rates 40
2.4b Reaction Rate Laws and Reaction Rate Constants 41
2.4c Half Lives and Life Times of Reaction 42
References 44

3. Energy States and Photochemistry 45–74

3.1 Energy Levels and Atomic Spectra 45
3.1a Selection Rule 47
3.2 Energy Levels and Molecular Spectra 48
3.2a Electromagnetic Radiation 48
3.3 Intensity of Spectral Lines 49
3.4 Width of Spectral Lines 50
3.5 Molecular Vibrations 51
3.6 Vibrations in Polyatomic Molecules 55
3.7 Electronic States in Diatomic Molecules 58
3.8 Electronic Transition in Diatomic Molecules 60
3.9 Intensity of Spectral Lines 63
3.10 Dissociation Energy and Dissociation Products 65
3.11 Predissociation 66
3.12 Electronic States of Polyatomic Molecules 68
3.13 Absorption of Light and Photochemistry 68
3.13a Grottus-Draper Law 69
3.13b Stark-Einstein Law 69
3.13c Beer Lambert Law 70
3.14 Quantum Yield 71
3.15 Possible Fates of an Electronically Excited Molecule 71
References 74

4. Photochemistry of Atmosphere 75–104

4.1 Introduction 75
4.2 Oxygen, O_2 76
4.3 Ozone, O_3 80
4.4 Water, H_2O 82
4.5 Ammonia, NH_3 84
4.6 Nitrogen, N_2 84
4.7 Nitrogen Oxides in the Absence of Hydrocarbons 85
4.7a Nitric Oxide, NO 85
4.7b Nitrogen Dioxide, NO_2 87
4.7c Nitrate Radical, NO_3 91
4.7d Nitrous Acid, HONO 92

4.7e Nitric Acid, HNO_3 93
4.7f Peroxynitric Acid, HO_2NO_2 94
4.8 Methane, CH_4 95
4.9 Organic Nitrogenous Compounds 95
4.10 Calculated Tropospheric Lifetimes of Some Representative Organics 97
References 104

5. Chemistry of Stratospheric Ozone (A Fragile Earth's Shield) **105–123**

5.1 Significance of Atmospheric Ozone 105
5.2 Natural Control of Stratospheric Ozone 106
5.3 Photochemistry of Water Vapour in Stratosphere 108
5.4 Role of ClO_X in Stratosphere 108
5.5 Stratospheric Pollution and the SST 109
5.6 UV Radiation and Skin Cancer 110
5.7 Chlorofluoro Carbons in Atmosphere 111
5.8 Other Halogen Containing Compounds 118
5.9 Probing the Ozone Shield 119
5.10 Ozone Hole 120
5.11 The Sun 121
References 122

6. Carbon Dioxide and Climate **124–162**

6.1 Carbon Monoxide, CO 124
6.1a Role of NO in O_3 Production and CO Oxidation 125
6.2 Carbon Dioxide, CO_2 127
6.3 Grow More Trees 127
6.3a Ancient Phoenicia and Slope Farming 127
6.3b Dead Cities of Syria 128
6.3c Rajasthan 128
6.3d South Korea 128
6.4 Energy Balance of the Earth 129
6.5 Green House Effect 133
6.6 Quantitative Green House Effect 137
6.7 Ozone Loss may Reduce Global Warming 139
6.8 Climate Change Variability 139
6.9 Implication of Climate Change 143
6.10 The Challenge of Climate Change and Climate Prediction—International Efforts 146
6.11 The El Nino: Southern Oscillation Phenomenon 148
6.12 Small Scale Climate Effect 152
6.13 Ecological Consequences 153
6.14 Why the Temperature of Delhi is Rising Year after Year? 155
6.15 Climate Changes in India due to Green House Effect 160
References 161

7. Acid Deposition **163–177**

7.1 Oxidation of SO_2 in Troposphere 163
  7.1a Hydroxyl Radical 165
  7.1b Ozone + Olefin Reaction 166
  7.1c Ground State Oxygen Atom, O (^3P) 167
  7.1d Homogenous Aqueous Phase Reactions 167
  7.1e Heterogenous Reactions on Surfaces 170
7.2 Oxidation of NO_2 to Nitric Acid 171
  7.2a Homogeneous Gas Phase Reactions 171
  7.2b Aqueous Phase Reactions 172
7.3 Sulphuric and Nitric Acids: Comparison and Contrast 173
7.4 Deposition Processes 173
7.5 Impact of Acid Deposition 174
  7.5a Injury to Vegetation 174
  7.5b Aluminium Concentrations 174
  7.5c Mercury Poisoning 174
  7.5d General Effects of Acid Deposition 175
7.6 Acid Fog 176
7.7 Hydrogen Sulphide 176
7.8 Future Projections 177
References 177

8. Photochemical Smog **178–197**

8.1 Temperature Inversion 178
8.2 General Nature of Smog 182
  8.2a London Type Smog 183
  8.2b Photochemical Smog 183
8.3 Photochemical Smog 184
  8.3a Basic Photochemical Cycle of NO, NO_2 and Ozone 184
  8.3b Catalytic Cycle of NO $\rightarrow$ NO_2 Conversion 185
  8.3c Significance of the Hydrocarbon/NO_x Ratio 187
  8.3d Peroxy Acetyl Nitrate, PAN 188
8.4 Particulate Organics 190
8.5 Inhibition of Photochemical Smog 192
8.6 Adverse Effects of Photochemical Smog 193
8.7 Air Quality Standards 195
References 197

9. Particulate Matter in Atmosphere **198–221**

9.1 Introduction 198
9.2 Brownian Diffusion 200
9.3 Size Distribution 201
9.4 Light Scattering and Visibility Reduction 203
9.5 Mechanisms of Aerosol Formation in Air 206

9.6 Chemical Composition of Particulate Matter 208
9.7 Particles as Pollutant 211
9.7a Diesel—Particulate Matter 212
9.8 Airborne Polycyclic Aromatic Hydrocarbons and Their Derivatives 212
9.9 Lead in Fine Particles 214
9.10 Fly Ash Problem 216
9.11 Particulate Matter Effects on Plant Growth 217
9.12 Smoker—A Polluter 218
9.12a How a Smoker is a Polluter? 219
9.12b How to Stop Smoking? 219
9.13 Space Garbage: A Space Pollutant 220
References 220

10. Atmospheric Dispersion **222–240**

10.1 Introduction 222
10.2 The Point Source Gaussian Plume Model 223
10.3 Plume Rise 230
10.4 Downwind Concentration under a Temperature Inversion 232
10.5 A Line Source Dispersion Model 235
10.6 Deposition from Aerosols 236
10.7 Area Source Models 237
10.8 Emission Controls 239
References 240

11. Air Pollution Control **241–270**

11.1 Introduction 241
11.2 Removal of Dry Particulate Matter 241
11.2a Filters 241
11.2b Reverse Pulse 242
11.2c Reverse Flow 242
11.2d Pressure Drop 244
11.3 Electrostatic Precipitators 245
11.4 Cyclone Separators 247
11.4a Pressure Drop 249
11.4b Velocity Ratio 250
11.5 Scrubbers 253
11.5a Chemical Additions 254
11.5b Other Systems 254
11.6 Gaseous Pollutants and Their Removal 255
11.7 Controls for Specific Gaseous Air Pollutants 256
11.7a Sulphur Dioxide 256
11.7b Hydrogen Sulphide 259
11.7c Ammonia, Amines and Pyridine Bases 260
11.7d Nitrogen Oxides 260
11.7e Fluorine and Fluorides 260

11.8 Odours 261
11.9 The Automobile Fuel 262
11.10 Emission from Two Stroke Engines 264
11.11 Emission from Four Stroke Engines 265
11.12 Emission from Diesel Vehicles 267
11.13 Catalytic Converters 267
11.14 Gas Turbines and Jet Engines 269
References 270

12. Air Pollution in India **271–284**

12.1 Introduction 271
12.2 Chemical Industries 276
12.3 The Bhopal Gas Tragedy 276
12.4 State of Some Metro Cities 281
12.5 The Polluted Air and Dyeing 283
12.6 Clean Production & Improve Productivity 283
References 284

13. Noise Pollution **285–296**

13.1 Introduction 285
13.2 Sound Pressure and Power Levels 285
13.3 Effects of Noise and Vibrations on Man 287
13.4 Effect of Noise on Performance 290
13.5 Health Hazards of Noise Pollution 291
13.6 Prevention of Hearing Loss in Industry 292
13.7 Noise Pollution in Delhi, Chennai and Lucknow 293
13.8 Control of Noise Pollution 295
References 295

Chapterwise Problems **297–322**

Appendices **323–340**

Appendix I : Some Very Good Internet Sites on Our Atmosphere 323
Appendix II : 2009 United Nations Climate Change Conference (COP15) Copenhagen 324
Appendix III : Approximate Bond Dissociation Energies 332
Appendix IV : Concentration Units 336
Appendix V : Conversion Factor for Changing Light Absorption Coefficient from One Unit to Another 338
Appendix VI : Some Common Conversion Factors for Gas Phase Reactions 339

PART–II
Water, Earth and Environment

1. Water and Earth **3–35**

1.1 Evolution of the Earth and Life 3
1.2 Mechanisms of Reactions 8
1.3 Extent of Atmosphere 11
1.4 Water 12
1.5 The Arrangement of H_2O Molecules in Water 15
1.6 Ionic Water 20
1.7 Aqueous Electrolytes 24
1.8 Clathrate Type of Hydrates 27
1.9 Hydrologic Cycle 29
1.10 The Geologic Cycle 31
References 35

2. Ecology **36–45**

2.1 Introduction 36
2.2 Food Chains and Food Webs 37
2.3 Natural Selection 39
2.4 Response to Physical Factors 40
2.5 Thermal Pollution 42
2.6 Control of Thermal Pollution 44
References 45

3. Water Pollution **46–105**

3.1 Introduction 46
3.2 Pollution of Lakes 47
3.3 Pollution in Rivers 48
3.4 Effluents 51
3.5 Oxygen in Water 52
3.6 Carbon Dioxide in Water 57
3.7 Nutrients 64
3.8 Organic and Inorganic Compounds 67
3.9 Analytical Examination of Sewage 70
3.10 Indian Rivers 77
3.11 Industrial Management of Water 80
3.12 The Chemistry of Sewage Purification 82
3.13 Industrial Effluents 90
3.14 Treatment and Disposal of Industrial Wastes 92
3.15 Some Havoc Causing Cases of Untreated Industrial Effluents 94
3.16 Some Methods of Purifying Industrial Effluents 95
3.17 Chemistry and Biochemistry of Sewage 96
References 100

4. Soils and the Environment **102–171**

4.1 Introduction 102
4.2 Clay Mineral Content 105
4.3 Crystal Structure of Clay Minerals 106
4.4 Identification of Clay Minerals in Soils 115
4.5 Iron and Aluminium Hydroxides 121
4.6 Crystal Structure of Feldspars 122
4.7 Hydrolysis of Feldspars 124
4.8 Acid Character of Clay Minerals 127
4.9 Adsorption of Anions by Clay Minerals 131
4.10 Soil Organic Matter 138
4.11 Reactions of HA's and FA's with Metals and Minerals 141
4.12 Decomposition of Soil Minerals by Organic Matter 143
4.13 The Role of Clay Fraction in the Particle Arrangement and Stability of Soil Aggregates 144
4.14 Wind Erosion of Soil 147
4.15 Soil Acidification 148
4.16 Effects of Soil Acidity on Plants 153
4.17 Liming Acid Soils 154
4.18 Nitrate, its Sources and as a Health Hazard 157
4.19 Pesticides in Soils 158
4.20 Heavy Metals in Soil (See Also Section 5.6) 165
4.21 Water Management for Soils in India 166
References 169

5. Environmental Chemistry of Some Elements **172–212**

5.1 Introduction 172
5.2 The Carbon Cycle, CO_2 and Carbonate 175
5.3 The Nitrogen Cycle 177
5.4 The Phosphorus Cycle 183
5.5 Sulphur and the Sulphur Cycle 186
5.6 Biologically Important Metals and Trace Element 190
5.6a Introduction 190
5.6b Beryllium 192
5.6c Boron 192
5.6d Magnesium and Calcium 193
5.6e Chromium 194
5.6f Manganese 194
5.6g Iron 195
5.6h Copper 199
5.6i Zinc 199
5.6j Arsenic 202
5.6k Selenium 204
5.6l Cadmium 205

5.6m Mercury 206
5.6n Lead 209
5.6o The Halogens 210
References 212

6. Polymers and Plastics **213–233**

6.1 Introduction 213
6.2 Polymer Synthesis 214
6.3 Polymers Degradation 218
6.4 Photochemical Degradation 220
6.5 Biodegradation of Naturally Occurring Polymeric Substances 224
6.6 Disposable Synthetic Polymers 226
6.7 Polymer Recycling 228
6.8 Carry Bags—A Menace 232
References 233

7. Fossil Fuels and Petroleum **234–269**

7.1 Nature of Petroleum 234
7.2 The Automobile Fuel 234
7.3 Emissions from Two-stroke Engine 239
7.4 Emission from Diesel Vehicles 241
7.5 Catalytic Converters 244
7.6 Alternative Fuels 246
7.7 Compressed Natural Gas (CNG) 248
7.8 Propane 249
7.9 Methanol and Ethanol 250
7.10 Biofuel 250
7.11 Hydrocarbons in our Environment 251
7.12 Environmental Reactions of Petroleum and Petroleum Products 253
7.12a Reaction with Ground State Molecular Oxygen $^3\Sigma_g$ 254
7.12b Reaction with Singlet Molecular Oxygen ($^1\Delta_g$) 256
7.12c Reaction with Ozone 257
7.13 Biodegradation of Hydrocarbons 257
7.14 Crude Oil and Marine Life 259
7.15 Coal 261
7.16 Detergents 262
7.17 The Builders 265
7.18 Environmental Effects of Biodegradable Organic Compounds 267
References 268

8. Pesticides **270–309**

8.1 Introduction 270
8.2 Chemistry of Chlorinated Organic Compounds 271
8.3 Polychlorinated Biphenyls (PCB) 273

8.4 Insecticides 274
8.5 Pesticides Interfering with Respiration 275
8.6 Second Generation Pesticides 276
8.7 Lindane 279
8.8 Polychlorinated Cyclopentadiene Derivatives 281
8.9 Auxins Herbicides 284
8.10 Long-term Effects of Organochloro Compounds 286
8.11 Organophosphorus Pesticides 288
8.12 The Carbamates 295
8.13 Third-generation Controller: Pheromones 297
8.14 Use of Pheromones and Attractants in Insect Control 300
8.15 Juvenile and Moulting Hormones 301
8.16 Fourth Generation Insect Controller, Antijuvenile Hormone from Plants 303
8.17 Insect Sterilisation 304
8.18 Current Position of Pesticides in India 304
References 308

9. Nuclear Chemistry and the Environment **310–350**

9.1 Introduction 310
9.2 The Kinetics of Nuclear Disintegration 312
9.3 Kinetics of Disintegration of Mixtures of Radionuclides 315
9.4 Radiation Units 319
9.5 Radiation Chemistry 325
9.6 Biological Effects of Radiations 326
9.7 The Interaction of Ionizing Radiations with Cells 328
9.8 Manmade and Natural Radiation 330
9.9 Maximum Permissible Doses 333
9.10 Nuclear Fission and Nuclear Reactor 337
9.11 Radiation Hazard from Reactors 342
9.12 Radioactive Waste Management 345
References 350

10. Garbage Management **351–363**

10.1 Garbage 351
10.2 Hospital Wastes 353
10.3 Non-biodegradable but Recyclable Wastes 355
10.4 Non-biodegradable Non-recyclable Garbage 356
10.5 Some Toxic Chemical Garbage 357
10.6 Bio-degradable Garbage 358
10.7 Oil Sludge and Oil Zapper 359
10.8 Biogas 360
10.9 Thermal Power Production and Fly Ash 360
References 363

Chapterwise Problems 365–376

Appendices 377–381

Appendix I : Some Very Good Internet Sites on Environmental Pollution 377
Appendix II : Some Relationships between Commonly used Units 378
Units 378
Physical Constants 380
Periodic Table 381

Glossary 383–395

Index of Part–I 397–400

Index of Part–II 401–404

PART–I

Chemistry of Atmosphere

1 Earth's Atmosphere

1.1 EVOLUTION OF LIFE AND EARTH

It is hypothesized that evolution of earth took place approximately five billion years ago when a cloud in the interstellar gas, known as solar nebula, collapsed forming protostars and planetary systems. Out of these clouds of cosmic dust came out earth, sphere of molten rocks. Gases bubbled out of the hot interior forming reducing atmosphere of methane, ammonia, hydrogen cyanide, water vapours, helium and hydrogen, etc., that surrounded the earth. As these hot gases moved upward, from high to low pressure, which must have led to fall in temperature and, thus, condensation of gases which fell back upon the surface of the infant earth in the form of rain.

The cooling in the upper strata of the atmosphere caused torrents of rain, and the process continued for many million years, for the rains immediately vapourized in the beginning, long before they even reached the surface. Gradually, as the earth cooled and rains could reach the surface, shallow oceans resulted. The rains fell not for days, not for months, nor for years but perhaps for centuries, until the world was covered with water. The ceaseless downpour washed away millions of tons of minerals, some mountains, and slashed out huge valleys across the surface of the earth. The rains finally slowed and stopped. The lowest region of earth's surface had been filled with a chemical soup in which the first form of sea and life developed.

The earth at that time appeared to have been rocked by explosions so gigantic and violent, that a thousand hydrogen bombs exploded together could not possibly match them. Mountains arose up out of the earth, and deep oceans were dug up. In the beginning, these first mountains were eroded by the torrential rains, and new ones were subsequently formed.

Lightning flashed through the atmosphere. The composition of the atmosphere has been deduced by spectroscopy and telemetry of allied planets. Space probes have helped a lot in this regard. Earth's primitive atmosphere consisted of methane, ammonia, hydrogen cyanide and water vapours. Hydrogen and helium had already escaped from the gravitational pull of earth. These gases were subjected to the strong ionising radiation that dissociated these molecules into positive and negative ions. Water vapours condense more readily on particles carrying negative charge than those carrying a positive charge. Thus the negative electric charges were brought down into the lower part of the cloud, while the positive charge were left in the upper part of it, thereby producing sufficient high differences in the electrical potential for lightning to strike between the clouds. These almost continuous electric discharges in the atmosphere for many thousand years made these molecule to unite to form various amino acids. The atmosphere was still the same reducing atmosphere. The amino acid produced in atmosphere continuously precipitated down to ground forming concentrated solution in ocean water.

The bulk of evidence indicates that primitive atmosphere of earth was reducing and that present oxidising atmosphere and hydrosphere are relatively recent, and are in fact, the product of biological activity. Bacterial decomposition as well as oxidation was not a problem in the prebiotic world. This led to reversal of equilibria and thus to structural stabilization which lead to polymerization of amino acids. Once polymerisation started, the danger of dissolution diminishes, for the polymers may organise into a stable configuration. Hydrophobic bonding comes into play and the polymer is surrounded by a hydration sheath which in effect protects it from the solvent powers of the bulk water. These polymers transformed into a highly organised configuration by the electrolytes in the sea water.

Oparin[7, 8] demonstrated in his experiments that suspended proteinoid microspheres assume a coacervate structure upon repeated hydration and dehydration, this might have occurred in rocky crevices along the sea edges. It was at this level that the mechanisms of evolution and natural selection came into operation. Coacervates are more stable and more successful than their fellow molecules in the competition for nutrient material and so survived, while the weaker contemporaries perished and their substance was devoured. One important structural feature separates the coacervate droplet from the cell—the cell has a membrane of lipids which have one hydrophobic and one hydrophilic end, and tend to pack in sheets, and such sheets or membranes made possible the isolation of the interior of a coacervate from its surroundings.

It is claimed that there are six stages of molecular complexity to the evolution of life. The scientists represent this hierarchy of polymer complexity firstly by the simple polypeptide chain, then the α-helix, followed by the folded coil, then linked folded coils as in *myoglobin* and finally the linking of different folded coils together with a ribonucleic acid chain as in the *tobacco mosaic virus*—one of the simplest living things.

The complexity of living molecules renders the probability of their formation small. But there were immense number of atoms and molecules colliding continuously with each other for few hundred million years. In a long span of time even such a small probability may lead to the formation of complex organic molecules *viz.*, nucleic acids that led to the appearance of life. Somehow, at a certain time, or more likely on many occasions in the long past, a strange and wonderful thing happened. One of these nucleic acids, perhaps with the aid of a protein, which acted as an enzyme, was able to reproduce itself. Now there were two of these remarkable groups of chemicals. The two floated their separate ways, each gathering in new chemicals from the soupy sea. In a short time each became two, and the two became four and the four became eight, soon the seas were filled with many of these "prelife" forms. Could these be the beginning of Life ?

Ultraviolet and cosmic rays continued to stream down upon the earth. Often these cosmic rays struck one of the "prelife" forms. Every now and then, this caused a change in the nucleic acid—a mutation. Most of the time, the changes were not very important or, perhaps, they even hurt the tiny "prelife" forms, and made it less able to survive in the soupy seas. Once in a while, the change was helpful and made the small 'prelife' form better. Perhaps it could reproduce more rapidly, and it was able to capture new chemicals better than its neighbours, perhaps it formed a thicker covering around itself to protect the chemicals inside.

Life did not appear suddenly. Instead, it came gradually, with one small step at a time. Perhaps it took millions of years, or may be tens of millions—we do not know. Some of the creatures that are known today, such as viruses, seem to be living at times and non-living at other times. Finally, a creature was formed, which all would recognise as life. It could live by itself, made its own food, or gobble it down from the seas, and reproduce itself. Is this the way life actually began on earth? No one can say for sure, but scientists do know from their experiments in the laboratory, that this is the way it could have happened, but there is no eye witness accounts at all of how actually life began.

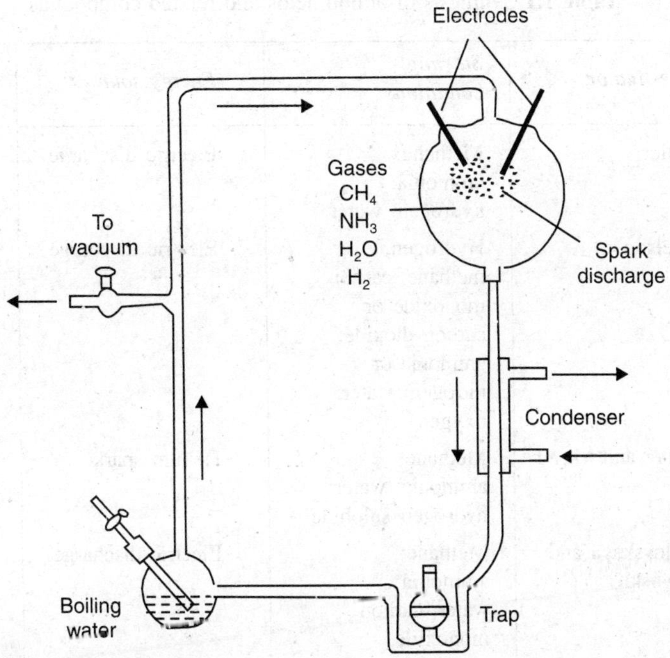

Fig. 1.1 Miller's experimental apparatus (one of several models). Most chemical reactions take place in the spark chamber, and the reaction products are condensed. The boiling flask of water keeps the system circulating in the direction shown by the arrows. The amino acids collected at trap were detected by chromatography.

In the 1950's some scientists thought of ways to test ideas on the origin of life right in the laboratory. An American researcher,[5, 6] Stanley Miller, recreated in their laboratory at the University of Chicago, what was thought to be the atmosphere of the ancient earth. For nearly a week they sent electric sparks through this atmosphere. At the end of this time, they were surprised to find some of the simple chemicals of life in their apparatus. Among them were amino acids—the building blocks of proteins. A summary of some primitive earth experiments is given in Table 1.1.

Scientists throughout the world became very excited at this discovery. They not only repeated the Miller's experiment, but they made changes of their own. One scientist added a little phosphate—the kind one finds in the rocks of earth, and the amino acids that were formed, joined with each other to form larger compounds. According to all the chemical tests, they were proteins. But, were they like the proteins found in living things? The scientists fed them to bacteria, who took them as though they were real proteins. Still other researchers made chemicals like the nucleic acid residues porphyrins, which are related to the haemoglobin of our blood, and chlorophyll of the plants. The porphyrins can absorb solar energy and can use it for later reactions.

Another complicated chemical made thus, and which could have been found on the ancient earth was **adenosine triphosphate (ATP)**. This is a very important chemical of life, for it is used as a storehouse of energy by almost all living things. ATP is the basic currency for the exchange of energy inside the cell. Without ATP life could not exist. So, the scientists showed in the laboratory that almost all the chemicals needed for life could be formed from the non-living chemicals present in the strange atmosphere and the primeval broth.

Table 1.1 Synthesis of amino acids and related compounds

Date	*Investigators*	*Starting conditions*	*Energy sources*	*Significant products*
1953	Miller	Methane; ammonia; hydrogen; water	Electric discharge	Glycine, alanine; β-alanine, glutamic acid
1955–56	Abelson	Hydrogen, methane, carbon monoxide or carbon dioxide; ammonia or nitrogen; water; oxygen	Electric discharge	Alanine, glycine, sarcosine, serine, aspartic acid (only when conditions are non-oxidising)
1957	Walter and Mayer	Methane; ammonia; water; hydrogen sulphide	Helium spark	NH_4SCN (ammonium thiocyanate)
1959	Pavlovskaya and Pasynskii	Methane; ammonia; water; carbon monoxide	Electric discharge	Common amino acids (depletion of hydrogen from system enhances amino acid synthesis); aminobutyric acid; lysine
1960	Groth and Von Weysenhoff	Methane; ethane; ammonia; water	Ultraviolet light; 407 nm; 219.6 nm	Glycine; alanine; aminobutyric acid
1959–65	Becker, *et al.*; Hanafusa *et al.*; Ponnamperuma *et al.;* Steinman, *et al.*	*n*-carboxyamino acid anhydrides; aminoacetonitrile sulphate; kaolin; dicyandiamide; HCl and corresponding amino acids	Heat to 100°C	Aspartic acid, threonine, serine, glutamic acid; proline; glycine; alanine; valine; isoleucine; leucine; tyrosine; phenylalanine.
1965	Grossenbacher; Knight	Methane; ammonia bubbled over ammonium hydroxide	—	Several common amino acids plus threonine; serine; glutamic acid; isoleucine
1965		Methane; ammonia; water	1300°K heat	Aspartic acid; threonine; serine; alloisoleucine; isoleucine; phenylalanine; β-alanine; glutamic acid; etc.

An important question still remained—these strange chemicals could have been formed, but could they have come together to produce life by themselves? Life is not merely self reproduction of nucleic acids, it demands controlled growth following nutrition in a closed membrane bound system, the cell. This means that metabolism is an essential part of life.

The Russian researcher, Oparin[7, 8] tried a new experiment. He mixed solutions of two substances—gelatin and gum arabic together, and tiny droplets formed. A mixture of these droplets with enzymes helped chemicals to react much more quickly than they did in an open system, when the enzymes were just dissolved in water with no droplets present. The findings of this experiments were just like in the case of life.

Oparin's[7, 8] droplets were the beginning of an answer. But in many ways they were not like living cells at all, for his droplets were of many different sizes and shapes, and they easily broke apart. Even more important, Oparin's droplets of gelatin and gum arabic were made from chemicals that had been produced by living things. Could non-living chemicals come together to form such droplets ?

Researchers continued the search. They took the amino acids that were made in a simulated atmosphere like that of the ancient earth, heated them and formed proteins. Then they added water, as could have occurred on the surface of the ancient earth when it rained or when the tides of the ocean lapped at volcanic shores, and an amazing thing happened. The proteins formed tiny balls and these balls were rather like Oparin's droplets, but there was an important difference, they were all about the size of bacterial cells.

The scientists looked at these tiny protein balls under the powerful microscopes. They appeared rather like bacteria. They had a thickened coat around them just like the bacteria. And this coat, like that of a bacterium, seemed able to let certain chemicals pass through it while it stopped others. More tests were run, and it was found that the proteins of these balls, or microspheres as they are called, acted like enzymes. Then it was discovered that these protein microspheres could reproduce if they were kept in the solution in which they were originally formed, for a week or so, they began to form small buds.

These researchers grew excited because there are living things such as yeasts and certain bacteria that reproduce by forming buds. What would happen if the buds were separated from their parent microspheres? It turned out to be easy to separate buds, an electric shock or even a sudden bump would do it. These buds were then left quietly in the solution to grow. They kept growing until they reached the size of their parents. Then they stopped growing. The researchers left the daughter microspheres in the solution and soon these protein balls began to form buds of their own! A new generation of protein microspheres were being formed.

The little protein microspheres formed entirely from non-living chemicals, appeared almost like life. This is not life as we know it today, for they do not contain nucleic acids. Scientists believe that the two self-reproducing systems, the nucleic acids and the proteins, could have come together in the seas of the ancient earth and could have integrated to form a whole, a primitive cell, somewhat similar to the primitive bacterial cell without a well organised nucleus. At this stage, it seems more logical to assume that life has originated as many independent systems, which have later got integrated together to evolve into a simple cell.

At the earliest stage of evolution of life on earth no direct sunlight could penetrate the surface of earth. The primitive micro-organism must have been able to exist in this damp darkness, and in the absence of oxygen, must have used the dissolved organic substances as their food, whereas others grew accustomed to purely organic food. This second class of organisms obtained their energy through the oxidation of sulphur and iron compounds. However, some of the primitive organisms may not have been satisfied with purely vegetable diet. These organisms began to eat each other and the necessity of catching prey or running away from pursuit led to the development of an ability to move.

As time passed, the surface of earth grew cooler, the heavy clouds hiding the sun gradually thinned out. Sun's radiation produced atmospheric oxygen proportionately as a result of two interrelated process—

(*i*) Dissociation of atmospheric water vapours to hydrogen and oxygen by solar radiation, and
(*ii*) Loss of hydrogen, from the earth's gravitation field.

Ammonia is a strong absorber of ultraviolet radiation. The presence of ammonia in the primitive atmosphere probably inhibited and slowed the dissociation of water vapour for the first several million years. As the abundance of free oxygen in primitive atmosphere slowly rose from 0.001 to 0.1 of its present amount, it had an important effect on the total composition of atmosphere. An amount of free oxygen as large as 0.1 of the present day value is sufficient to oxidize methane and ammonia to carbon dioxide and nitrogen. Thus slow dissociation of water vapours may have changed the total composition of the atmosphere through its oxidizing effect. As the level of ammonia in atmosphere decreased, the rate of oxygen production by photodissociation increased. With the presence of CO_2 photosynthetic plant life became possible. Photosynthesis in turn, released much more additional oxygen by metabolic breakdown of water and carbon dioxide. It is this plant-produced oxygen that has slowly accumulated to provide the abundant oxygen in our planet's present atmosphere.

Under the intense radiation present in the upper atmosphere, oxygen formed a layer of ozone (O_3), which absorbed most of the sun's UV radiation and prevented it from reaching the lower atmosphere and surface of earth. Without this O_3 screen, life would be possible under rocks, in deep water or in other places where direct sunlight could not penetrate. The bacteria and blue green algae of the early, oxygen poor atmosphere, may have lived in such restricted environments. Only after the formation of ozone screen, existence of life became possible first in shallow water and ultimately on the land surface. In this way, life on earth began about three billion years ago, since then a multitude of different types of organisms have emerged, prospered and died out, leaving their fossils on earth for the history to be reconstructed.

The development of man from his not-so-good looking and not-so-intelligent ancestors has taken place during the last two million years. Probably the earliest stage of that branch of organic evolution originated in Africa. In 1959, the remains of *Zinjanthropus*, who may have been the great-great-great ... grand daddy of us all, were found in East Africa by British archeologist, Dr. Louis Leakey. *Zinjanthropus*—the name means, East African man, lived 17,50,000 years ago and is the earliest known maker of stone age.

1.2 MECHANISM OF REACTIONS

It may be recalled that modern biochemical pathways are the products of millions of years of slow evolution. It is unlikely that anything resembling a biochemical pathway today was present or even possible among the mixture of chemicals proposed in the primeval broth, or in the atmosphere. With few exceptions the simulated conditions employed by most researchers are harsh, unsafe to living systems of today, and destructive to polymers. It is possible, however, that the mechanisms involved in the synthesis of organic compounds on the primitive earth were harsh, that the first "prelife" forms retained these mechanisms, and that evolution of milder chemistry took place primarily in these structures.

Miller[5, 6] considered two possible mechanisms: (*i*) Hydrogen cyanide, aldehydes, nitriles, and amines, and part of the polymers were synthesized in the electric discharge and the amino, hydroxy–aliphatic acids were formed by hydrolysis of the respective nitriles in the solution. (*ii*) All the products identified were synthesized in the gas phase from free radicals and ions, formed in the electric discharges. His tests on both these alternatives, seem to favour the first of these.

HCN is an early product. It may have several origins. It can form directly from the interaction between methane and ammonia, both constituents of the earlier atmosphere.

$$CH_4 + NH_4 \longrightarrow HCN + 3H_2$$

Other hydrocarbons as they form can also give rise to HCN on reacting with ammonia. Carbon monoxide is still another source for HCN.

$$CO + NH_3 \longrightarrow HCN + H_2O$$

Synthesis of organic compounds like ethane with carbon-carbon bonds may have formed in the electric spark, yielding two-carbon compounds, e.g., C_2H_2, C_2H_4, C_2H_6. These compounds show addition reactions with HCN

$$C_2H_4 + HCN \longrightarrow CH_3CH_2CN \text{ (Propionitrile)}$$

$$C_2H_2 + 2HCN \longrightarrow NCCH_2CH_2CN \text{ (Succinodinitrile)}$$

The corresponding acids are obtained when the nitriles are hydrolyzed in aqueous phase.

$$CH_3CH_2CN + 2H_2O \longrightarrow CH_3CH_2COOH + NH_3 \text{ (Propionic acids)}$$

$$NCCH_2CH_2CN + 4H_2O \longrightarrow HOOCCH_2CH_2COOH + 2NH_3 \text{ (Succinic acid)}$$

With HCN, aldehydes form the corresponding hydroxy nitriles, which on hydrolysis will yield the hydroxy acids.

$$\underset{\text{Acetaldehyde}}{CH_3CHO} + HCN \longrightarrow CH_3CHOHCN \xrightarrow{2H_2O} \underset{\text{Lactic acid}}{CH_3CHOHCOOH} + NH_3$$

$$\underset{\text{Propionaldehyde}}{CH_3CH_2CHO} + HCN \longrightarrow CH_3\,CH_2\,CHOHCN$$

$$\downarrow 2H_2O$$

$$\underset{\alpha\text{-Hydroxy butyric acid}}{CH_3CH_2CHOHCOOH} + NH_3$$

With ammonia, the hydroxynitriles can form the corresponding aminonitriles from which the amino acids may be obtained by the hydrolysis.

$$CH_3CHOHCN \xrightarrow{NH_3} CH_3CHNH_2CN + H_2O$$

$$\downarrow 2H_2O$$

$$\underset{\text{Alanine}}{CH_3CHNH_2COOH} + NH_3$$

$$CH_3CH_2CHOHCN \xrightarrow{NH_3} CH_3CH_2CHNH_2CN + H_2O$$

$$\downarrow 2H_2O$$

$$\underset{\alpha\text{-Aminobutyric acid}}{CH_3CH_2CHNH_2COOH} + NH_3$$

The use of heat in simulated prebiological synthesis has revealed some interesting facts. Vacillating temperatures can bring about, in sequence, the synthesis of amino acids, the polymerization of these into protein and the formation of microspheres. The ease with which amino acids will polymerise into polypeptides, especially in the presence of an excess of aspartic and glutamic acids, makes it possible to visualise the natural conditions under which such events might have taken place. Ionizing radiations, very high temperatures, and electric spark create high energy species. Under the impact of sufficient energy, compounds dissociate and give rise to free radicals and ions. The results of these interactions can be very complex and unpredictable. Recombination of free radicals of carbon will form predominantly branched chain compounds. It suggested that the formation of acetylene from carbon radicals as an intermediate in the abiological synthesis of styrene which, in turn, would give rise to the aromatic amino acids: phenylalanine and tyrosine.

One of the most interesting mechanisms has to do with the reactions of activated HCN. Hydrogen cyanide appears among the first products of the gases in mixtures used in experiments of the abiological synthesis of organic compounds. Adenine among the other reaction products was detected from a mixture of HCN, ammonia, and water. This was confirmed in subsequent work and the probable mechanism suggested is given below:

(1) 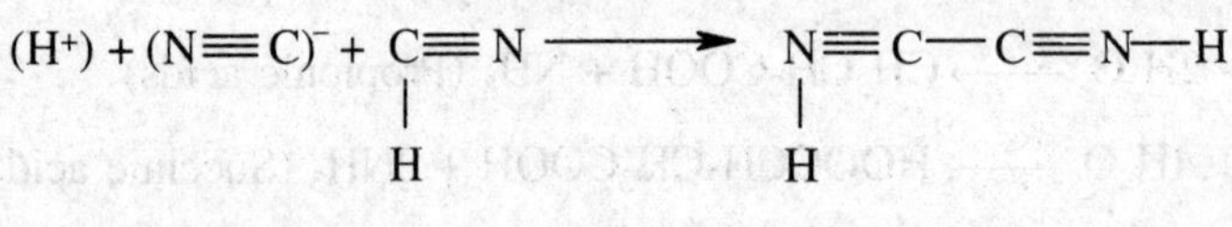

(2) 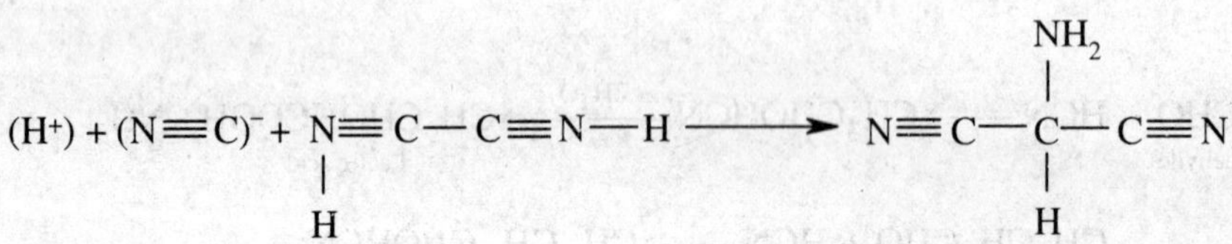

(3) 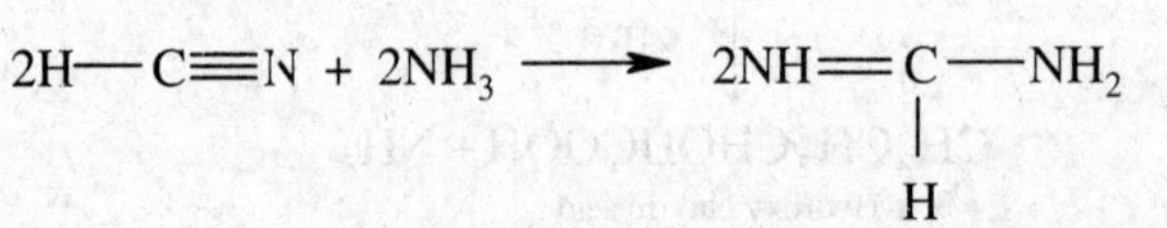

(4) 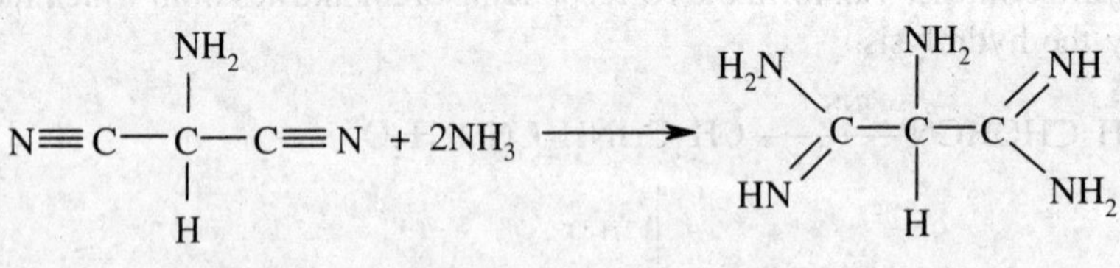

(5) 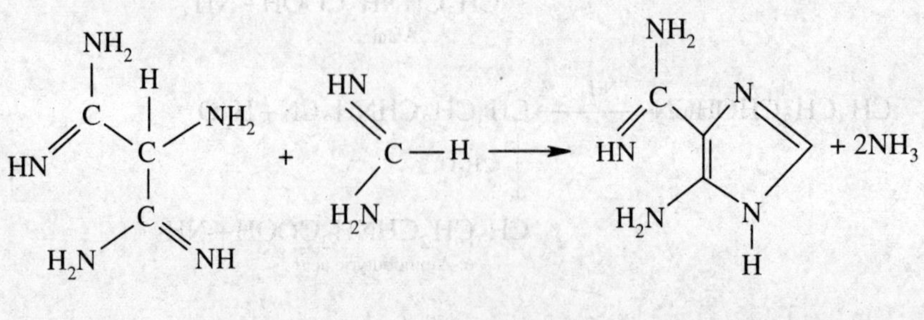

(6) 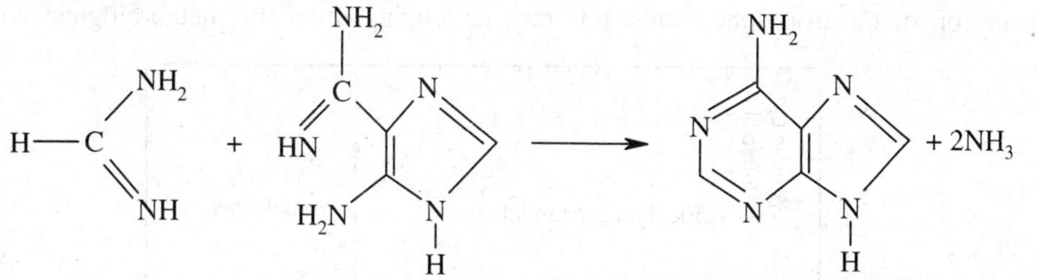

Overall reaction: **5HCN = Adenine**

Similar theories, as above, have been suggested for the synthesis of other purines such as guanine, xanthine, etc. Most of these ideas have been tested experimentally and found to be correct, when appropriate experimental conditions prevailed.

It has already been pointed out that available evidence favour that the prebiological atmosphere had composition comprising mainly carbon monoxide, carbon dioxide, nitrogen and hydrogen, and the seas attained a pH of 8-9. People have often asked, 'what kind of prebiological chemistry can occur at ordinary temperatures in dilute solutions at pH 8. Calvin[4] has doubts, because dehydration reactions are not thermodynamically feasible in excess water, and the primeval broth had lots of it. At high or low pH, HCN will not polymerise, but at these pH it readily does. Hydrolysis of a tetramer of HCN yields Glycine, and still others yield alanine, serine, aspartic acid, and glutamic acid. Uptill now, so many different views have resulted, that any attempt to summarise all would easily fill a treatise, and therefore only the most prevalent ones have been discussed.

1.3 EXTENT OF ATMOSPHERE

A vast ocean of air surrounds earth as a thin transparent veil. One quarter of the atmosphere is below 2 km, one half below the altitude of 5 km and three quarters of it is below 9 km. But highly rarefied fringes of atmosphere extend hundreds of kilometers above earth's surface, and it is difficult to say where our atmosphere ends. The total mass of atmosphere is 5.1×10^{18} kg which is only 0.33% of the total mass of water in oceans. The entire atmosphere is distributed over the 5.1×10^{14} m^2 of the earth's surface. This implies that atmospheric pressure at sea level is about 10^4 kg m^{-2} or 10^3 g cm^{-2}. This is approximately equivalent to one atmosphere (1013 m bar)**. The earth's atmosphere comprises 78% molecular nitrogen (N_2), 21% molecular oxygen (O_2), 0.93% argon (Ar), 0.03% carbon dioxide CO_2. These gases are uniformly mixed in troposphere and stratosphere except at great heights.

Starting from the earth's surface, the atmosphere can be divided into several distinct regions based on the temperature-altitude profile, as shown in Fig. 1.2.

The temperature initially falls upto –63°C with increasing altitude at the rate of 5.3° C per km. This region is called troposphere but the temperature change above 15 km is reversed. In the stratosphere, from 15 to 50 km the temperature increases from –63°C to 15°C. From 50 to ~ 85 km is mesosphere, the temperature drops to –88°C until ~85 km where thermosphere or ionosphere begins. Here temperature increases with altitude.

The decrease in temperature with increasing altitude in troposphere is the result of a strong heating effect at the earth's surface from the absorption of visible and ultraviolet solar radiation. Accompanying

**10^3 m bar ≡ 10^3 g cm^{-2} which when multiplied with acceleration due to gravity gives pressure in Nm^{-2}.

this, there is strong vertical mixing so that particles and gaseous air pollutants can move from the earth's surface to the top of the troposphere in a few day, depending upon the meteorological conditions.

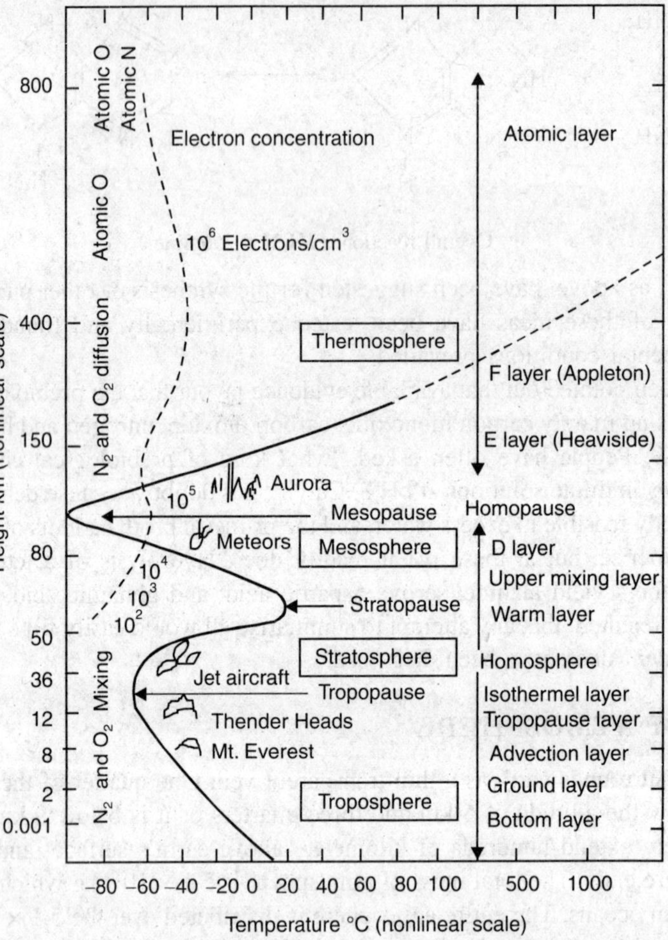

Fig. 1.2 Structure of the earth's atmosphere.

Essentially all the water vapours, clouds, precipitate in the earth's atmosphere. Storms, hurricanes, tornadoes and also typhoons also originate here; as a result, removal of pollutants by precipitation scavenging can be an important process.

In the stratosphere, a series of photochemical reactions involve ozone and molecular oxygen occur. Ozone strongly absorbs solar radiation in the range 210 to 290 nm whereas oxygen absorbs at ≤ 200 nm. The absorption of light, primarily by ozone, is a major factor responsible for the increase in temperature with altitude in the stratosphere. This initiates a series of reactions in which ozone is formed and destroyed leading to steady state concentration of ozone. This concentration in the stratosphere is essential for life, because it absorbs light of wavelength ≤ 290 nm. As a result light of wavelength longer than 290 nm can reach troposphere. This short wavelength cut off sets limits on troposphere photochemistry. Relatively less vertical mixing occurs in stratosphere, and no precipitation can take place in this region. As a result, massive injection of particles, such as from volcanic eruption, often produce layers of particles in the stratosphere which persist for long periods of time.

In the mesosphere from ~ 50 to 85 km the temperature again falls with altitude and vertical mixing within the region occurs. This temperature trend is due to the decrease in the O_3 concentration with altitude in this region; this, in turn, reduces the heat release through absorption and decomposition of ozone.

At ~ 85 km the temperature starts rising again because of increased absorption of solar radiation of wavelength ≤ 200 nm by oxygen and nitrogen as well as other atomic species. This region is known as Thermosphere. The stratum of the atmosphere lying between 20 and 110 km is known as chemosphere. The transition zones between the various shells of the atmosphere are known as tropopause, stratopause and mesopause.

The temperature profile in Fig. 1.2 is generalized one; not only some segments of curve are highly variable (bottom layer temperature range from −50 to + 80°C, upper layer mixing from −80 to + 70°C, F-layer from + 60 to + 1000°C), but the overall shape of the curve is strongly dependent on latitude and terrestrial and solar condition.

1.4 TEMPERATURE LAPSE RATES AND ATMOSPHERE STABILITY

Air pollutants emitted from various anthropogenic sources has to be first transported and diluted in the atmosphere before these undergo various physical and chemical changes. Effective dispersion of pollutants in the atmosphere depends primarily on the stability and turbulent structure of atmosphere. In the broadest sense dispersion is controlled by meteorological conditions prevailing in the atmosphere.

In the day time, the ground is heated by solar radiation and the air near the ground is heated by conduction from the ground. Assuming that air behaves as an ideal gas so that

$$P = \rho RT$$

where ρ is density of air. This packet of air will rise until it reaches a place in troposphere where surrounding has the same density *i.e.* same temperature and pressure. As air moves from surface of earth, upward there is expansion and thus fall in pressure. Expansion of gas means doing work against intermolecular attraction for which energy is taken from the air itself. This decreases energy of air and hence the temperature decreases. The expansion can be assumed to be adiabatic as no heat flows into or out of the packet during the process. Derivation of the equation for temperature and pressure changes in reversible adiabatic expansion of a gas is given in most Physical chemistry textbooks. The result is

$$\frac{P_1}{P_2} = \left[\frac{T_1}{T_2}\right]^{C_p/R}$$

where C_p is the heat capacity at constant pressure of the gas and subscripts 1 and 2 refer to properties of the gas before and after expansion, respectively. Qualitatively this expression shows that if gas is expanding, the temperature of the gas will drop. This equation, together with the known decrease of pressure with height, allows us to calculate the adiabatic lapse rate.

Three different situations arise.

1. Environmental lapse rate, $\left(\frac{dT}{dz}\right)_{\text{envir}} = 0.65°\text{C}/100\ \text{m}$. This is followed from ground level up to an altitude of about 2 km where there is considerable variation. This is due to the character of underlying surface and the rate of radiation emanating from it. On sunny days because of strong solar heating from the ground the environmental lapse rate often exceeds 1°C/100 m, although this is usually confined to the lowest 200 metres.

2. Adiabatic lapse rate of dry air $\Gamma = \left(\frac{dT}{dz}\right)_{adia} = -\frac{g}{C_P} = -\frac{9.806\ ms^{-2}}{1005\ J\ kg^{-1}\ °C^{-1}} \cdot \frac{1J}{kg\ m^2 s^2}$ $= 0.00976°C\ cm^{-1} \cong 1°C/100\ m$. As dry air moves upward and expands slowly without exchange of heat, then such parcel at any height is in equilibrium, and value of this lapse rate is 1°C/100 m. If the actual air temperature decreases faster with increasing elevation than this lapse rate, the air will be unstable and rapid mixing and dilution of pollutants will occur. Conversely if the actual air temperature drops more slowly than the adiabatic lapse rate, the air will be stable and pollutant will concentrate. The relation is derived on the assumptions that air parcel could be treated as ideal gas that could be moved without heat transfer between it and surrounding. Both these assumptions are very good. It also assumes that air was dry. This may or may not be a good assumption.

3. Adiabatic lapse rate of wet air $\left(\frac{dT}{dz}\right)_{wet.\ air} < 1°C/100\ m$. If air has some water in it, C_p changes slightly from the value assumed, but not enough to warrant a correction. On the other hand, if enough water vapour is present then condensation occurs when the parcel is raised and cooled, latent heat will be released. The added heat means air parcel will not cool as rapidly as dry air. Unlike dry air lapse rate, wet air adiabatic lapse rate is not constant. This is because the amount of moisture that air can hold before reaching saturation and beginning of condensation is a function of temperature. A reasonable average wet lapse rate is about 0.6°C/100 m.

1.5 ATMOSPHERIC STABILITY

When the environment lapse rate and dry adiabatic lapse rate are exactly the same, a rising parcel of air will have the same pressure, temperature and density of the surroundings and would experience no buoyant force. Such an atmosphere is said to be neutrally stable, where a displaced mass of air neither tends to return to its original position nor tends to continue its displacement. When $\left(\frac{dT}{dz}\right)_{envir.} > \left(\frac{dT}{dz}\right)_{adia.}$ the atmosphere is said to be super adiabatic. A rising parcel of air, cooling at the adiabatic rate, will be warmer and less dense than the surrounding environment. As a result, it becomes more buoyant and tends to continue its upward motion. Since vertical motion is enhanced by buoyancy, such an atmosphere is called unstable. In the unstable atmosphere the air from different altitudes mix thoroughly and disperse the effluents rapidly, hence desirable for preventing pollution.

When $\left(\frac{dT}{dz}\right)_{envir.} < \left(\frac{dT}{dz}\right)_{adia.}$, a rising air parcel becomes cooler and thus more dense than its surrounding and tends to fall back to its original position. Such atmospheric condition is quite stable and the lapse rate is called subadiabatic. As vertical mixing of pollutants do not take place, their level rapidly build up. A similar reasoning may be applied to the case of wet lapse rate and dry lapse rate.

The extreme case of a stable atmosphere occur when temperature increases with altitude. Such a lapse rate is known as negative lapse rate. Under such conditions of temperature inversion, the atmosphere is very stable.

Temperature Inversion

Temperature inversion represent the extreme case of atmospheric stability. There are several causes of inversion but three important cases will be discussed.

(*i*) **Radiation Inversions:** The surface of the earth cools down at night by radiating energy toward space. On a cloudy night, this radiation tends to be absorbed by water vapour in atmosphere which in turn reradiates back some of that energy back to the ground (see detail in Section 8.1). On a clear winter night as the ground emits radiation, it directly goes to space, thereby cooling the surface by radiation cooling. This cools the air near the surface of earth than air above it creating an inversion. Radiation inversion occur close to the ground mostly during winter and last for only a few hours. It often begins as the traffic builds up in the evening, which traps auto exhaust at ground level, particularly CO. In the morning as the sun warms the ground and the inversion begins to break up, pollutants that have been trapped in the stable air are brought back to earth by a process known as *fumigation.* Valleys and low lying areas are particularly affected by this type of inversion because denser and colder air tends to sink down beneath the warmer air.

(*ii*) **Advective Inversion:** It is formed when warm air moves over a cold surface or cold air horizontally. The inversion can be ground based in the former or elevated in the later case. An example of an elevated advective inversion occurs when a hill range, forces warm land breeze not to flow at high levels but a cool sea breeze flows at low levels in the opposite direction.

(*iii*) **Subsidence Inversions:** These are associated with high pressure weather systems, known as *anticyclones.* It may last for months, occurs at high elevation and are more common in summer than winter. Imagine atmosphere made up of three layers: topmost as subsiding air, followed by inversion layer and third as surface layer. Subsidence inversion are associated with high-pressure weather systems, known as *anticyclones.* Air in the middle of high pressure zone is descending while on the edges, it is rising Air near the ground moves outward from the center, while air aloft moves toward the center from the edges. The result is a massive vertical circulation system. Air in the inversion layer experience greater pressure and is compressed and thus heated. If its temperature at an elevation z_1 is T_1, then as it falls to elevation z_2 it will be heated adiabatically to $T_2 = T_1 + \Gamma\ (z_2 - z_1)$. Thus heating warms the descending air to a higher temperature than the air below whose temperature is dictated by ground conditions. The result is an inversion, located anywhere from several hundred meters above surface to several thousand meters that persist as long as the high pressure weather system persist.

Since subsiding air is getting warmer, it can hold more water. Without sources of new moisture, its relative humidity drops, thus there is little chance of cloud formation. The result is high pressure zone creates clear, dry weather with lots of sunshine during the day and clear sky at night. Clear skies allow solar warming of the earth's surface. This helps to create super-adiabatic conditions under the inversion during the day time and hence good mixing. At night the surface can cool quickly by radiation, which may result in a radiation inversion located under subsidence inversion as shown in Fig. 1.3. These semipermanent highs are the result of general atmospheric circulation pattern shown in Fig. 1.4.

1.6 CIRCULATION OF THE ATMOSPHERIC AIR

The behaviour of the atmosphere determine weather, climate and destination of pollutants within the global ecosystem. In order to appreciate the consequences of human activity, one must understand the characteristic properties of atmosphere. Earth is unique planet, which has a climate propitious for life. However the weather and the climate of earth depend upon the uneven heating of earth by sun due to two reasons.

(*i*) The solar rays strike at the equator at much smaller surface area thus receive much more solar heat per unit area than they receive at higher latitudes.

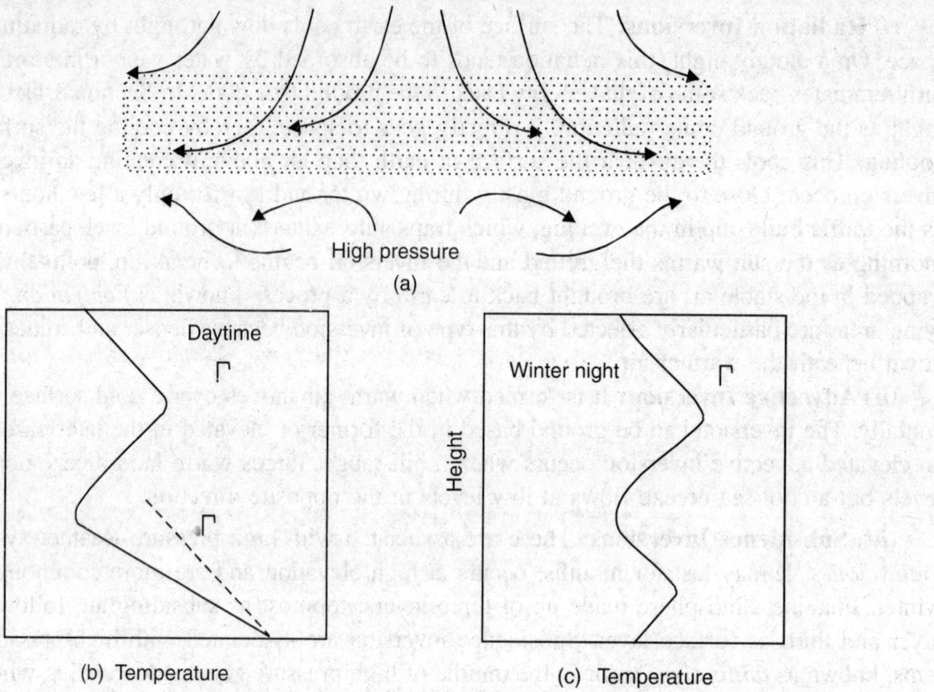

Fig. 1.3 Subsidence inversion (*a*) As air descends it is warmed adiabatically, creating clear skies and an upper level inversion (dots); (*b*) during the day, air under the inversion may be unstable due to solar warming of surface; (*c*) radiation inversions may form at night under the subsidence inversion, especially in winter.

(*ii*) At the equator sunlight pass through the shortest distance of atmosphere while at higher latitudes, it passes through greater quantities of air which absorbs greater sunlight. Hence more sunlight reaches the equatorial surface than reaches the polar surface.

The atmosphere and ground radiate heat to outer space day and night. However, it turns out that on an annual basis polar regions radiate more heat to space than they receive from sun. This fact accounts for temperature difference between the warm equatorial and cold polar regions of the earth. The temperature difference between equator and the poles creates a thermodynamic engine.

Hot air is less dense than cold, and tend to rise near the equator and flow towards the poles at some elevation above the surface of the earth. The colder air from the polar region is then expected to flow towards the equator nearer to the surface of the earth. This model of circulation of air does not take into account (*i*) the unequal heating of land and sea, (*ii*) variation of solar radiation at various latitudes at different times of year, and (*iii*) earth's rotation about its axis.

If earth was not rotating, a better model can assume there are three such separate circulation systems in each hemisphere, as shown in Fig. 1.4 follows:

1. The Polar zone, extending from each pole to about 60 degree north or south latitude. Here the cold air moves from east to west forming to giant freezing whirlpools which characterise the Arctic and Antarctic regions.
2. The Temperate zone, between 60th and 30th north or south latitudes with prevailing westerly winds.

3. The Subtropical zone forming a belt 60 degree wide around the equator where warm winds blow mostly from east to west, called trade winds.

The equator is directly under the sun and here air will be heated to maximum. As this air approaches the tropopause, it begins to turn, some heading towards north and some towards south. It descends at around latitude of 30° and then returns to the equator, forming what is called Hadley cell. In a similar fashion, though not as distinct, there are Hadley cells linked as in a chain between 30° and 60° latitudes, and between 60° and poles. The descending air at 30° creates a persistant high pressure zone with corresponding lack of clouds or rainfall, that is a contributing factor to the creation of the world's great deserts. The desert of southern California, the south western United States, the Sahara, the Children desert, the Kalahari in South Africa, and the Great deserts of Australia are all located at roughly 30° latitudes, for the subsidence inversion taking place. Conversely, the rising air near the equator and near 60° latitude tends to be moist after passing over oceans. As it rises and cools, the moisture condenses, causing clouds and rainfall. For this reason some of the wettest regions of the world are located in those bands of latitude.

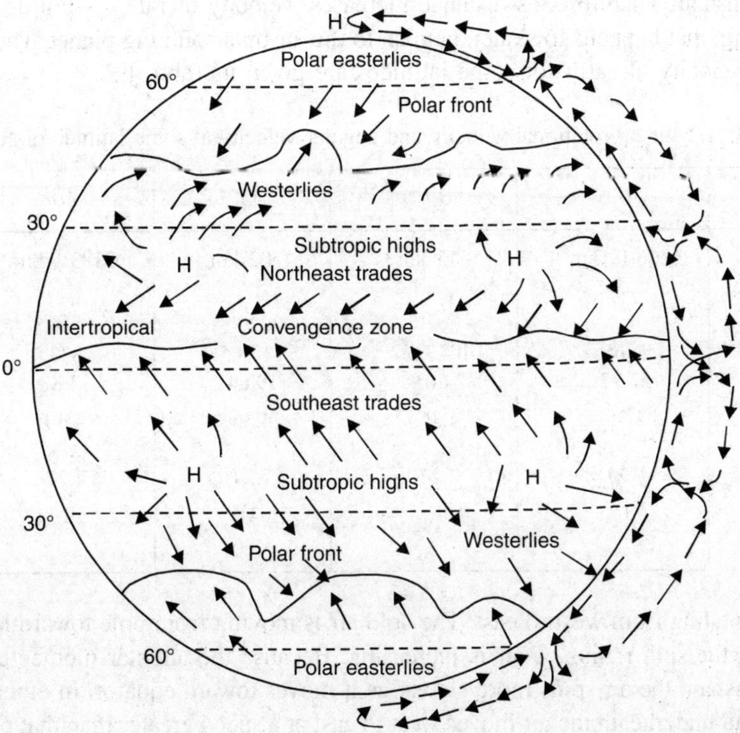

Fig. 1.4 General atmospheric circulation patterns for the rotating earth. High pressure zones are common around 30° latitude, producing clear skies and subsidence inversion (from G.O.P. Obasi, 1996).

The air that moves along the surface of the earth close to Hadley cell is affected by Coriolis forces (Section 1.7), causing it to curve toward the west if it is moving toward the equator or towards the east if it is moving towards poles. The resulting winds between 30° and 60° are known as westerlies and between 30° and equator are the trade winds. Near the equatorial, there is little wind since the air is mostly rising. That band is called equatorial doldrums. Similarly, the surface air is relatively calm around 30° latitude forming a band called the horse latitudes.

While within each zone the winds are rather steady and strong, the boundaries between the zones, where the air moves mostly upward or downward, are characterized by almost windless, conditions. The boundaries between polar zone and temperate zone are known as polar fronts. This is an unstable front. The boundaries between temperate and subtropical air circulation zones are known as the horse latitudes. No good sailing winds blow along the equator, in the belt known to seamen as the doldrums.

1.7 ANGULAR MOMENTUM OF ATMOSPHERE

When an object is rotary, it has an angular momentum, which is equal to the product of the object's mass, velocity and distance from the axis of rotation. An ice skater who is spinning, has an angular momentum. An object in motion continues in motion unless acted on by an external force. The skater spinning at a particular rate has a certain angular momentum. When the skater extends his arms, the rate of rotation slows down. By pulling in the arms, the skater increases the rate of rotation. This phenomenon has been described as conservation of momentum. When the skater's arms are extended, the average distance of the skater's mass from the spin axis is increased. However, the angular momentum of the skater remains constant. Therefore, as distance increases, velocity or rate of spin decreases.

The same thing, that happens to skater, happen to the air on a spinning planet. The linear rotational velocity, angular velocity of earth for some latitudes are given in Table 1.2.

Table 1.2 Linear rotational velocity and angular velocity at some latitude of earth

Latitude θ	*0*	*30*	*45*	*60*	*90*
Radius at respective latitude	6366.19 km	5513 km	4502 km	3183.1 km	Zero
Linear rotational velocity in kmh^{-1}	1666	1443.3	1178.5	833	Zero
in km min^{-1}	27.77	24.05	19.64	13.83	Zero
Angular velocity in $min^{-1} \times 10^3$	4.36	4.36	4.36	4.36	Zero
in $sec^{-1} \times 10^5$	7.27	7.27	7.27	7.27	Zero
Relative angular velocity	1	1	1	1	Zero

The earth is rotating from west to east. The cold air is moving from pole toward latitudes (a north-south movement), the spin radius of air is increasing. Because the angular momentum of the atmosphere remains constant, the air spins more slowly as it moves toward equator. In other words, at lower latitudes the ground underneath the air moves west to east at a speed greater than that of air. This slower spinning of the air relative to the spinning ground produces for an observer, at the air ground interfere, the westward flow of polar air masses that are moving equatorward. These flows are called the polar Easterlies, because the air is moving from east.

Near the equator, the air also is moving equator wards from 30 degree north to south latitude the north-east and south-east trade winds are the result of the same westward flow effect described above. However, air starting at 30° latitude moving towards 60° latitude is reducing its spin radius. As a result, the rotational speed of the air increases. Therefore, this poleward moving air in the middle latitudes moves from the west more rapidly than the Earth's surface, and the observer on the surface detects west-to-east movement of air over the surface. The west-to-east moving air masses are called the prevailing westerlies and occur at the middle latitudes. The polar easterlies, prevailing westerlies, north-east and south-east trade winds are shown in Fig. 1.4.

1.8 CORIOLIS EFFECT

Three forces of importance to horizontal motion of air are pressure gradient, gravity and friction. Primarily because of horizontal temperature gradients, there are horizontal density gradients and, consequently, horizontal pressure gradient. The direction of the pressure difference and thus movement of air is from high to low pressure.

Surface temperature differences makes air move poleward or equator ward, and changes in angular velocity of the air eastward or westward relative to earth surface. If the earth were not rotating, the wind would blow from high to low pressure. Close to the earth, it would be slowed by friction between the atmosphere and the earth's surface, but would maintain the same direction with height. However, since the earth undergoes rotation, there is an apparent force acting upon horizontal atmospheric motions when examined from a point of reference on the earth's surface. This effect is called the Coriolis effects and it causes the wind in the northern hemisphere to be deflected to right of a straight path and in the southern hemisphere to the left. In other words, the Coriolis effect gives a twist to the air movement.

This phenomenon can best be understood by considering what would happen to a rocket that is aimed straight to the south from Delhi A rocket fired from Delhi towards Bangalore would hit the earth far west of Bangalore. Hence looking along into path from Delhi toward Bangalore, one would observe that rocket seems to be deflected to the right. This deflection occurs because the linear rotational velocity is greater at low latitudes than at high latitudes. The linear rotational velocity of the surface is 1666 km h^{-1} at the equator and velocity is zero km h^{-1} at each pole. The linear rotational velocity difference between Bangalore and Delhi is about 174 km h^{-1}. So while the rocket is enroute, Bangalore moves out from its reach. The rocket should be aimed east of Bangalore in order to impact at Bangalore.

Another example, consider a wind of velocity 10 $ms^{-1} \equiv 36$ kmh^{-1} blowing in the direction of the 0° longitude meridian across the north pole. The wind in an absolute sense continues to blow in this direction for 1 hr, and a parcel of air starting at the pole travel 36 km in this period. However, since the earth turns 360° every 24 hr, or 15° per hour, it has rotated by 15° in an hour and 15° meridian is now beneath the wind vector. As viewed from space the flow has continued in a straight line. However, as viewed from the earth, the flow has undergone an apparent deflection to the right. The force required to produce this apparent deflection is the Coriolis force per unit mass and is equal to $F_c = vf$ where v is wind speed and f is the Coriolis parameter equals $2\,\Omega \sin\theta$. Here Ω is the angular speed of the earth's rotation $= 7.27 \times 10^{-5}\,s^{-1}$ and θ is the latitude. It is seen that f is maximum at the poles ($\theta = 90°$) and zero at equator ($\theta = 0°$). For the above example, the deflecting force at poles is 1.45×10^{-3} m s^{-1} and at $\theta = 90°$, $F_c = 0$. The value of f at various latitudes are as follow:

θ	0	10	20	30	40	50	60	70	80	90
$f \times 10^5$ sec^{-1}	0	2.52	4.97	7.27	9.35	11.14	12.6	13.66	14.32	14.54

Thus at 30° latitude with a wind speed of 10 ms^{-1}, Coriolis force would be $7.27 \times 10^{-5} \times 1000 =$ 0.0727 cms^{-2}, while with the same wind speed at 60 degree latitude the Coriolis force would equal 0.126 cms^{-2}.

The Coriolis force is hardly a real force but rather the effect of earth's rotation on a moving body. This deflecting force (*i*) is always directed at right angles to the direction of air flow, (*ii*) affects only wind direction, not wind speed, (*iii*) is affected by wind speed *i.e.* stronger the wind the greater the deflecting force and (*iv*) is strongest at poles and weakest on equator where it eventually becomes non existent.

The forth condition is the most difficult to understand. It is a consequence of the extent of reorientation of the surface over which the wind blows and is dependent upon latitude. For example, at the poles where the land surface is perpendicular to the earth's axis, the daily rotation causes the horizon to make one complete turn about a vertical axis each day. Restated more simply, the surface over which the wind blows makes one complete rotation each day. At the equator the earth's surface is parallel to the earth's axis, consequently it has got rotation about an axis oriented vertically to this surface. Therefore, at the equator the surface over which the air blows does not rotate in a horizontal sense. To summarise, the magnitude of the Coriolis force is dependent upon latitude, it is strongest at the poles and weakest equatorward where it eventually becomes non existent.

1.8a The Geostrophic Wind

Over the earth's surface variations in air pressure are determined from barometric readings taken at hundred weather stations. These pressure data are shown on a weather map by means of isobars, which are lines connecting places of equal air pressure. The spacing of the isobars indicates the amount of pressure change occurring over a given distance and is expressed as the pressure gradient. Thus, the wind speeds should continuously accelerate for as long as this imbalance exists. But winds do not get faster indefinitely hence some other force or forces must be in operation. These forces are

(*i*) **Frictional Forces:** Significant for surface winds upto 4-5 km height.
(*ii*) **Coriolis Force:** Above 5 km, where frictional force is insignificant, it becomes significant.

To understand the operation of these forces, imagine a non-moving parcel of air at the starting points in Fig. 1.5. Since this parcel of air has no motion, the Coriolis force exerts no influence, only pressure gradient force can act. Pressure gradient force is always directed perpendicular to the isobars. The parcel begins to accelerate directly towards the area of low pressure. As soon as the flow begins, the Coriolis force cause a deflection to the right for winds in Northern Hemisphere.

As the parcel continues to accelerate, the Coriolis force intensifies. Thus increased speed results in a further deflection. Eventually, the wind will be turned so that it is flowing parallel to the isobars, with the pressure gradient force directed towards the area of low pressure and opposed by the Coriolis force

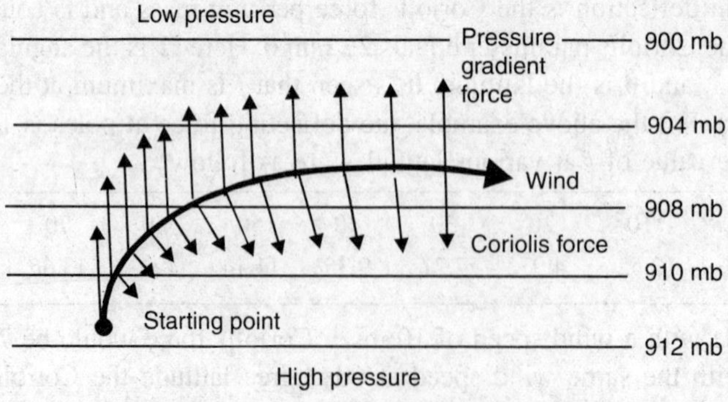

Fig. 1.5 The geostrophic wind. Upper-level winds are deflected by the Coriolis force until the Coriolis force just balances the pressure gradient force. Above 600 m, where friction is negligible, these winds will flow nearly parallel to the isobars and are called geostrophic winds.

which is directed towards the region of high pressure (Fig. 1.5). When these two opposing forces are equal, the wind continue to flow parallel to the isobars at a constant speed. Since the Coriolis force is proportional to wind speed, it will remain constant and a balance between these forces will be maintained and air will flow along isobars the winds generated by this balance are called geostrophic winds. The simple relationship between wind direction and pressure distribution was first formulated by the Dutch metrologist Buys Ballott, in 1857. Essentially the *Buy Ballott's law* states: *In the Northern Hemisphere if you stand with your back to the wind, low pressure will be to your left and high pressure to your right. In the Southern Hemisphere the situation is reversed since the Coriolis deflection is to the left.*

The importance of the idealised geostrophic flow lies in the fact that it gives a good approximation of determining both the wind direction and speed by measuring pressure field orientation and spacing of isobars. The pressure gradient force per unit mass is given by expression

$$F_p = \frac{1}{\rho}\frac{\Delta p}{\Delta d}$$

where F_p is pressure gradient force per unit mass, ρ is density of air, Δp is the pressure difference obtained from isobar spacing and Δd is distance between isobars. Using this formula and standard sea level density of 1.29×10^{-3} g cm^{-3} pressure gradient force for any isobaric surface can be calculated. For example, suppose 4 millibar $\equiv 4 \times 981 = 3924$ Nm^{-2} isobar are spaced 200 km apart. Then

$$F_p = \frac{1 \times 3.924 \times 10^3}{1.29 \times 10^{-3} \times 2 \times 10^7} = 0.152 \text{ cm s}^{-2}$$

Comparing this value with that obtained for the Coriolis force calculated earlier suggests that a balance between forces is possible at reasonable wind speeds.

1.8b Gradient Wind

When the isobars are curved, an additional force centrifugal force act outward from the centre of curvature around the low pressure. The resultant balance of force occurs when the pressure gradient force balance the sum of the Coriolis and centrifugal force (Fig. 1.6a), and wind continues parallel to the isobars. In the case of curvature around, high pressure a balance of force occurs when the sum of the pressure gradient and centrifugal forces equals the Coriolis force (Fig. 1.6b). To maintain a given

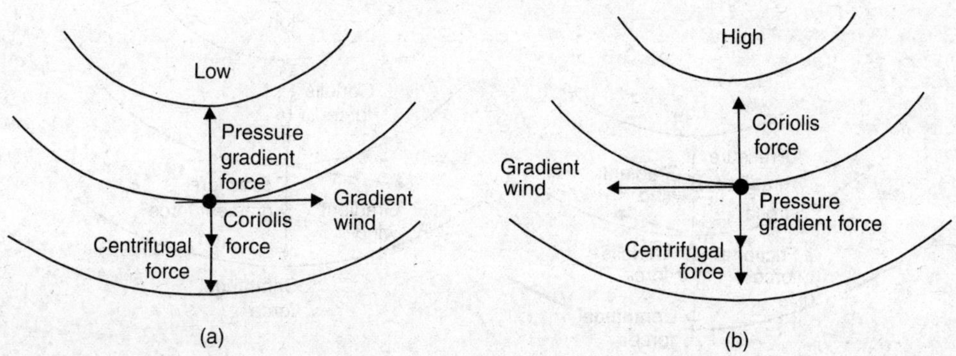

Fig. 1.6 Balance of forces resulting in gradient wind around (*a*) low pressure (*b*) high pressure. Note that the wind speed is greater for a given pressure gradient force than that around low pressure.

gradient wind speed, a greater pressure gradient force (tighter spacing of the isobars) is required in the flow around low pressure systems than in the flow around high pressure system.

1.8c The Effect of Friction

The frictional effect of the earth's surface on the atmosphere increases as the earth's surface is approached from aloft. Assuming that we start with geostrophic balance aloft, consider what happens to the wind as we move down, toward the earth. The effect of friction is to slow the wind velocity, which, in turn, decrease the Coriolis force. The wind then turns toward low pressure until the resultant vector of the frictional force and Coriolis force balances the pressure gradient force (Fig. 1.7). The greater the friction, the slower the wind and greater the amount of turning toward low pressure. The turning of the wind from the surface through the friction layer is called the Ekman spiral. A radial plot or hodograph of the winds through the friction layer is shown diagrammatically in Fig. 1.7. Note that this frictional effect will cause pollutants released at two different heights to tend to move in different directions.

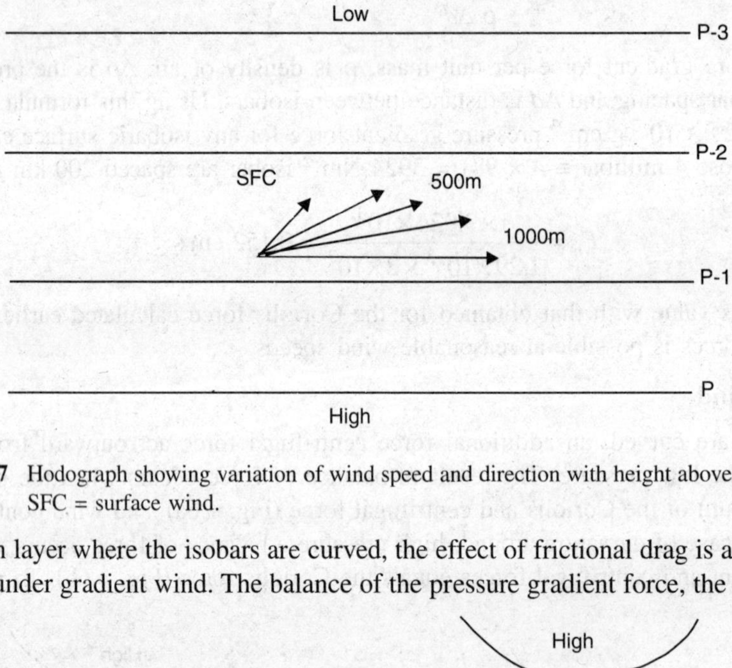

Fig. 1.7 Hodograph showing variation of wind speed and direction with height above ground SFC = surface wind.

In the friction layer where the isobars are curved, the effect of frictional drag is added to the forces discussed above under gradient wind. The balance of the pressure gradient force, the Coriolis deviating

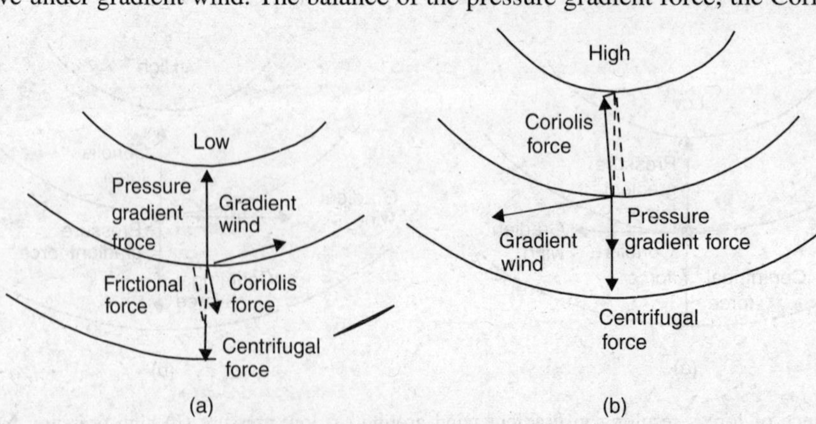

Fig. 1.8 Effect of friction upon gradient wind around (a) low pressure (b) high pressure.

force, the centrifugal force, and the frictional drag in the circuit of the curved isobars results in wind flow around low pressure and high pressure in the Northern Hemisphere, as shown in Fig. 1.8.

1.8d Vertical Motion-Divergence

So far in discussing motion in the atmosphere, only horizontal motions have been emphasized. Although of much smaller magnitude than horizontal motions, vertical motions are important both to daily weather formation and to the transport and dispersion of pollutants.

Persistent vertical motions are linked to the horizontal motions. If there is divergence (spreading) of the horizontal flow, there is sinking (downward, vertical motion) of air from above to compensate, and perhaps precipitation from moist air masses, as well as transportation of air and its pollutants from near the surface to higher altitudes.

1.9 LOCAL WIND SYSTEMS

Frequently, local-wind systems are superimposed upon the large-scale wind systems (discussed earlier). These local flows are especially important and may dominate when the larger-scale flow becomes minor and indefinite. Local wind systems are usually quite significant in terms of the transport and dispersion of air pollutants.

Sea and Land Breezes: The sea breeze is a result of the differential heating a land and water surfaces by incoming solar radiation. Since solar radiation penetrates several meters into body of water, it warms very slowly. In contrast, only upper few centimeters of land surfaces are heated and warming occurs rapidly in response to solar heating. Therefore, especially on clear summer days, the land surface heats up rapidly, warming the air near the surface and decreasing its density. This causes the air to rise over the land, decreasing the atmospheric pressure near the surface relative to the pressure at the same altitude over the water surface. The rising air decreases the pressure over the land relative to that above the water at altitudes of approximately 100–200 m. The air that rises over the land surface is replaced by cooler air from over the water surface. This air, in turn, is replaced by subsiding air from somewhat higher layers of the atmosphere over the water. Air from the higher pressure zone several hundred meters above the surface then flow from the land surface to over the water, completing a circular flow. Any general flow due to large-scale pressure systems will be super imposed upon the breeze and may either reinforce or inhibit it. Ignoring the large-scale influences, the strength of the sea breeze will generally be a function of the temperature excess of the air above the land surface over that above the water surface.

Just as heating in the day time occurs more quickly over land than over water, at night radiational cooling occurs more quickly over land. The pressure pattern tends to be the reverse of that in the day time. The warmer air tends to rise over the water, which is replaced by the land breeze from land to water, with the reverse flow (water to land) completing the circular flow at altitudes some what aloft. Frequently at night, the temperature differences between land and water are smaller than those during the day time, and therefore the land breeze has a lower speed.

Mount and Valley Winds: Solar heating and radiation cooling influence local flows in terrain situations. Consider midday heating of a south-facing mountain sides. As the slope heats, the air adjacent to the slope warms, its density is decreased and air attempts to ascend (Fig. 1.9a). Near the top of the slope, the air tends to rise vertically. Along each portion of the slope farther down the mountain, it is easier for each rising parcel of air to move upslope, replacing the parcel a head of it rather than rising vertically. This upslope flow is valley winds.

At night when radiation cooling occurs on slopes the cool dense air near the surface descends along the slope (Fig. 1.9b). This is downslope wind. To compensate for this descending air, air farther from the slope is cooled very little, hence warmer air relative to the descending air rises. This frequently result in a closed circular path. Where the downslope winds occur on opposite slopes of a valley, the cold air can accumulate on the valley floor. If there is any slope to the valley floor, this pool of cold air can move down the valley, resulting in a drainage or canyon wind.

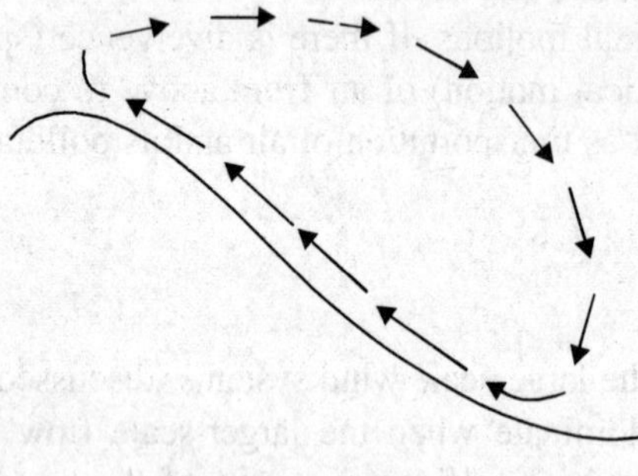

Fig. 1.9 (*a*) Upslope wind (daytime) due to greater solar heating on the valley's side than in its centre.

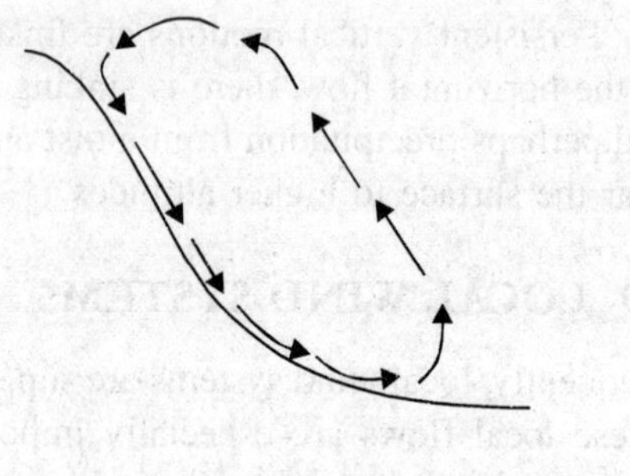

Fig. 1.9 (*b*) Downslope wind (night) more rapid radiational cooling on the valley's slope than in its centre.

Different combinations of valley and mountain slope, especially with same slopes nearly perpendicular to the incoming radiation and others in deep shadow, lead to many combinations of wind patterns, many nearly unique. Also, each local flow may be modified by the regional wind, that result from the prevailing pressure patterns. Table 1.3 gives characteristics of eight different situations, depending on the orientation of the ridgeline and valley with respect to the sun, wind direction perpendicular or parallel to the ridgeline, and time of day. Figure 1.10 shows examples of some of the mountain and valley winds listed in Table 1.3.

Table 1.3 Generalized mesoscale wind flow pattern associated with different combinations of wind direction and ridgeline orientation

Wind direction relative to ridgeline	*Time of day*	*Ridge line orientation*	
		East-West	*North-South*
Parallel	Day	South-Facing slope is heated-single helix (Fig. 1.10a)	Upslope flow on both heated slopes-double helix. (Fig. 1.10b)
	Night	Downslope flow on both slopes-double helix	Downslope flow on both slopes-double helix
Perpendicular	Day	South-Facing slope is heated. Northwind-stationary eddy fills the valley (Fig. 1.10c) south wind-eddy suppressed, flow without separation (Fig. 1.10e)	Upslope on both heated slopes-stationary eddy on the one side of valley (Fig. 1.10d)
	Night	Indefinite flow-extreme stagnation in valley bottom	Indefinite flow-extreme stagnation in valley bottom (Fig. 1.10f)

These are rather idealised circulations compared to the observed flows at any one time.

The effect of solar radiation is different with valley orientation. An east-west valley has only one slope that is significantly heated—the south facing slope may be near normal with midday shine. A north-south valley will have both slopes heated at mid-day. The effect of flow in relation to valley orientation is such that wind flows perpendicular to valleys, tend to form circular eddies and encourage local flows. On the other hand wind flow parallel to valley tend to discourage local flow and sweep clean the valley, with stronger wind speeds.

Keeping in mind that the resultant flow is the combination of the general and local flows; the lighter the general flow, the greater is the opportunity for generation of local flows.

Complicated terrain such as a major canyon with numerous side canyons, will produce complicated unique flow, especially when side canyon drainage flows reinforce the drainage flow in the main valley.

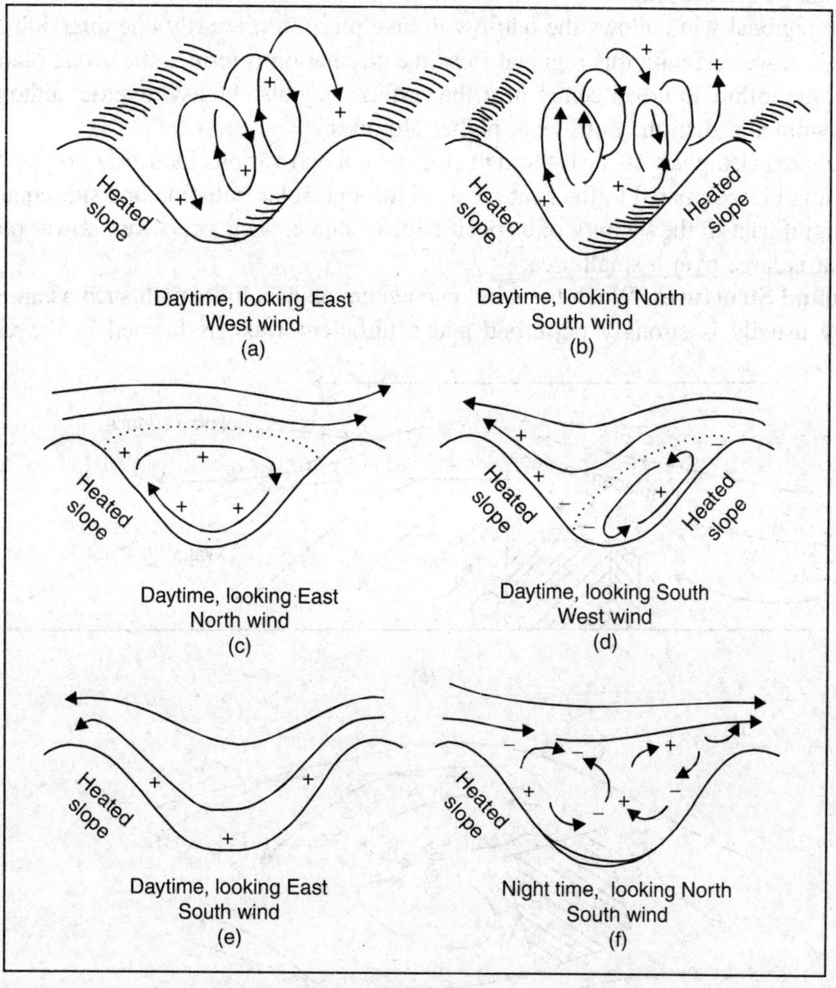

Fig. 1.10 Local valley-ridge flow patterns, Ref. Table 1.3.

Urban-Rural Circulation: Urban areas have different roughness and thermal characteristics from their rural surrounding. Although the increased roughness affect both the vertical wind profile and the vertical temperature profile, but the effects due to thermal features are more dominant. The aspalt, concrete, and steel of urban areas gets heated up quickly, and have a high heat storing capacity compared to the soil and vegetation of rural area. Also, some surfaces of buildings are normal to the sun's rays just after sunrise and also before sunset, allowing warming throughout the day. The result is that the urban areas become warmer than its surroundings during the day and stores sufficient heat to reradiate heat during the night that keeps the urban atmosphere considerably warmer than its rural surroundings.

Under the lightest winds, the air rises over the warmest part of the urban core, drawing cooler air from all sides, subsidence will replace this air in rural areas, and a closed torus (doughnut)-shaped circulation occurs with an outflow above the urban area. The circulation is referred to as the urban heat island. The strength of the resulting flow is dependant upon the difference in temperature between the urban centre and it surroundings.

When the regional wind allows the outflow to take place in primarily one direction and the rising warm urban air moves off with this regional flow, the circulation is termed the urban plume, under this circumstance, the inflow to urban centre near the surface may also be asymmetric, although it is more likely to be symmetric than the outflow at higher altitudes.

The urban area also gives off heat through combustion and various industrial processes though this heat is quite small as compared to the heat received through solar radiation and subsequently released. It can be of significant in the vicinity of a specific local source, such as a steam power plant where the release of heat is large over a small area.

Flow around Structures: When the wind encounters objects in its path such as an isolated structure, the flow usually is strongly perturbed and a turbulent wake is formed in the vicinity of the

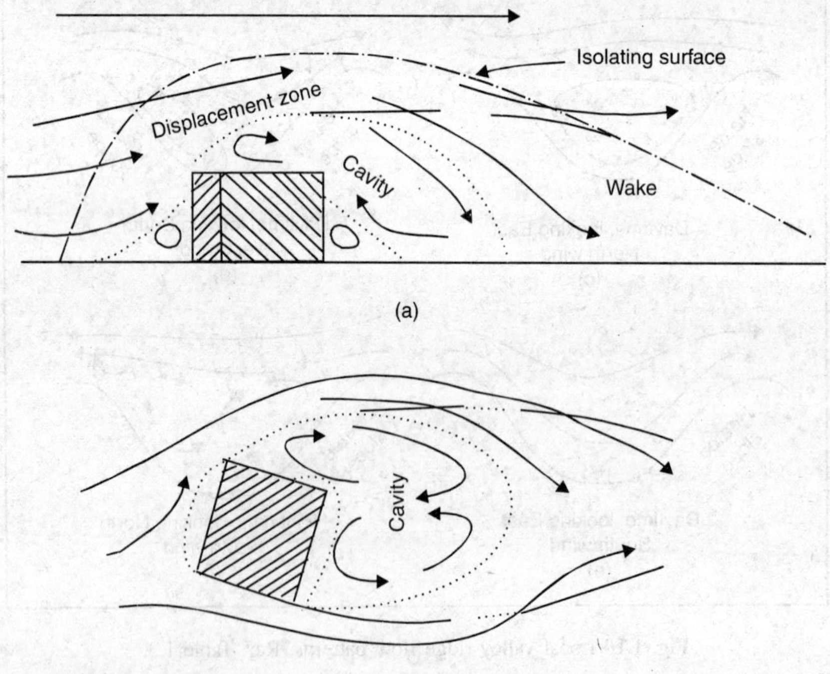

Fig. 1.11 Aerodynamic flow around a cube: (*a*) side view, (*b*) plan view.

structure, especially down wind of it. If the structure is semistreamlined in shape, the flow may move around it with little disturbance. Since most structure have edges and corners the generation of a turbulent wake is quite common. Figure 1.11 shows schematically the heat flow in the vicinity of cubical structure. The disturbed flow consist of a cavity with strong turbulence and mixing, a wake extending downwind from the cavity to a distance equivalent to a number of structure side length, a displacement zone where flow is displaced before entering the wake, and a region of flow that is displaced away from the structure but is not caught in the wake. Wind tunnels, as well as water channels are extremely useful in moderating the building wake effects. Trees can act as wind tunnels while various water bodies can act as water channels.

REFERENCES

1. Barbara, J., Finlayson Pitts, J.N. Pitts Jr, *Atmospheric Chemistry, Fundamentals and Experimental Techniques,* John Wiley & Sons, New York (1986).
2. Hamptreys, W.J., *Physics of Air,* 3rd Edition, McGraw-Hill, New York (1940), Reprinted by Door, New York, (1964).
3. Palmer, E. and C.W. Newton, *Atmospheric Circulation System, Their Structures and Physical Interpretation,* Academic Press, New York (1969).
4. Calvin, M., *Chemical Evolution,* Oxford Univ. Press, London.
5. Miller, S.L., and Orgel L.E., *The Origin of Life,* Englewood Cliff's Prentice Hall, New York (1958).
6. Miller, S.L., Production of Some Amino Acids under Possible Primitive Earth. *J. Am. Chem. Soc.,* 77, 2351-2361 (1955).
7. Oparin, A.I., The Origin of Life and Origin of Enzyme, *Adv. Enzymology,* 27, 347-80 (1965).
8. Oparin, A.I., *Genesis and Evolutionary Development in Life,* Academic Press, New York (1968).
9. Orgil, L.E., *The Origin of Life Molecules and Nature Selection,* John Wiley, New York.
10. Prof. Gadwin, O.P. Obasi, Rajiv Gandhi Institute for Contemporary Studies (RGICS, Paper 36, "*Climate, Climate Change Variability and Predictability*", New York (1996).

2

Principles of Thermodynamics and Gas Kinetics

2.1 INTRODUCTION

All activities in the universe require energy. Nothing moves, changes, grow or decays without the expenditure of energy. Energy is essential for all form of life. Energy flows downhill from high potential sources like stars with millions of degrees temperature to the cosmic cold of outerspace at a few degrees above absolute zero. The planet Earth intercepts some of this energy flux from the sun, captures, transforms and stores some of it and reradiates the rest back to outer space. During much of the 5 billion years since the evolution of earth, the natural system of it evolved under the impelling forces of radiative energy flow. Life evolved, atmosphere composition and structure changed and millions of organisms developed during this time with this energy flux, which was and is necessary for naturally developed system of Earth.

During millions of years of birth, growth, death and decay of organisms, a very small fraction of the solar energy was left behind as inanimate organic compounds to form reservoirs of coal, oil and gas. With the release of this energy and the creative consequence of an inventive mind, man constructed a complex society of machines, buildings, chemicals etc. All this gave man enormous mobility and tremendous consumption of energy which upsets the energy balance of natural system through release of poisonous chemicals and thereby polluting the atmosphere. If there had been an infinite supply of energy as compared to consumption, it would have lasted indefinitely. However, we have reserves of coal, oil, gas and uranium, implying the life to extinction with their exhaustion.

Society has been near steady state during most of the human history; only during the last two centuries has there been an enormous growth of human society. Our economic system has been predicted to be on continued growth state. As energy becomes more costly, the net energy gains become smaller, and the mineral resources grow more scarce as population densities increase. In order to survive and optimize the use of energy, it is necessary that rate of growth of population be reduced to 0.5%. In other words, India should go in a big way for Family Planning.

Energy is ability to do work. Work is defined as what happens when a force acts through a distance. On pushing a cart, work is done by exerting force on cart and move it through a distance. On walking, certain force is applied to push the earth backward and thus some work is done by walking a certain distance. In driving an automobile, certain energy is consumed by burning some fuel to make automobile move.

The different form of energy are mechanical, electrical, chemical, magnetic, nuclear and heat. The energy contained in gasoline is chemical energy, but when exploded in the engine cylinder, it is rapidly converted to mechanical energy as the expanding vapour drives the piston and thus automobiles.

Some of the terms referred above are defined along with the units are as follows. Only SI units have been used throughout the book.

1. Force = Mass × Acceleration

$$\text{Unit: Newton (N)} = \text{Kilogram} \times \text{metre sec}^{-2}$$

$$N = kg \times ms^{-2}$$

2. Work = Force × Distance = Energy

$$\text{Joule (J)} = \text{Newton} \times \text{meter}$$

$$J = Nm$$

3. Power = Energy × Time^{-1}

$$\text{Watt} = \text{Joule} \times \text{sec}^{-1}$$

$$W = Js^{-1}$$

Because power is energy expended per unit time, energy can be expressed

$$\text{Energy} = \text{Power} \times \text{Time}$$

$$\text{Watthour} = \text{Watt} \times \text{Time}$$

As large quantities of energy and fuel are used in society, it requires too many powers often to express energy consumed which is expressed in tons of coal equivalent, barrels of oil or quads, 1 metric ton (10^6 g) of coal (mtce)

$$= 29.3 \times 10^9 \text{ J}$$

1 barrel (bbl) of oil (42 gallons) $= 6.12 \times 10^9 \text{ J} = 194$ WYr

1 cubic metre (m^3) of natural gas

$$= 37.08 \times 10^9 \text{ J} = 1.165 \text{ WYr}$$

1 metric ton of oil (mton)

$$= 1.43 \text{ mtce}$$
$$= 41.9 \times 10^9 \text{ J}$$

1 quad $= 172 \times 10^6$ bbl of oil

$$= 36.0 \times 10^6 \text{ metric ton of coal}$$
$$= 2.83 \times 10^{10} \text{ m}^3 \text{ of gas}$$

There is another unit of ton that is also used which is called short ton.

1 short ton = 907.2 kg

= 0.9072 metric ton

2.2 MATERIAL BALANCE

When a chemical reaction takes place, matter is neither created nor destroyed. This concept allows us to track material pollutants from one place to another with mass balance equations. Mass balance may involve three situations.

Situation 1 : Steady State Conditions

Define the particular region in space that is to be analyzed. The region may include a simple chemical mixing tank, coal fired plant, a lake, a river stream, an air basin over a city or the globe itself. In this region a steady state or equilibrium condition prevails, so that a substance that enter it, leaves unchanged.

So that

$$\text{Input rate} = \text{Output rate} \quad \text{...(2.1)}$$

This may be explained with a river into which a polluted stream falls. The input river water flows at rate Q_s (volume/time) with pollutant concentration C_s (mass/volume). The polluted stream input has a flow rate Q_w with pollutant concentration C_w. If steady state condition prevails then mass balance based on (2.1) can be written as

$$C_s Q_s + C_w Q_w = C_m Q_m \quad \text{...(2.2)}$$

Situation 2 : Steady State with Non-conservative Pollutant

A substance that enters a region may not leave the region unchanged. Some of it may be converted to some other substance. For example N_2 may be oxidized to NO when fuel burns. Many pollutants undergo chemical, biological or nuclear reaction at a rate sufficient to treat them as non-conservative system. Then,

$$\text{Input rate} = \text{Output rate} + \text{Decay rate} \quad \text{...(2.3)}$$

The decay of nonconservative substance is frequently modeled as a first order reaction i.e. rate of loss of the substance is proportional to concentration

$$\frac{dc}{dt} = -kc \quad \text{...(2.4)}$$

where k is reaction rate constant with dimension time^{-1} and negative sign implies loss of substance with time and c is pollutant's concentration. This equation can be integrated as

$$\int_{c_0}^{c} \frac{dc}{c} = \int_{0}^{t} -kdt$$

which yield

$$\ln \frac{C}{C_0} = -kt$$

or

$$\frac{C}{C_0} = e^{-kt} \quad \text{...(2.5)}$$

where C_0 is the initial concentration. If difference is uniformly distributed throughout a volume V, then total amount of substance is CV and total rate of decay of the substance is $\frac{d(CV)}{dt} = \frac{Vdc}{dt}$ and decay rate $= kCV$, so that

$$\text{Input rate} = \text{Output rate} = kCV \quad \text{...(2.6)}$$

This equation implies complete mixing up of reactants all the time. These two situations will be explained with two examples.

Example 2.1: A river is flowing at 10^{-2} Ls^{-1} has a tributary feeding into it with a flow 5×10^{-3} Ls^{-1}. The concentrations of chlorides upstream junction are 20 mg L^{-1} and 40 mg L^{-1} in river and tributary respectively. Calculate the down stream chloride concentration.

Solution: This may be treated as conservative system with complete mixing so that by Eq. 2.2.

$$C_sQ_s + C_wQ_w = C_mQ_m = C_m(Q_s + Q_w)$$

or

$$C_m = \frac{C_sQ_s + C_wQ_w}{Q_s + Q_w}$$

$$= \frac{20\times10^{-2} + 40\times5\times10^{-3}}{10^{-2} + 5\times10^{-3}}$$

$$= 26.7 \text{ mg L}^{-1}.$$

Down stream chloride concentration = 26.7 mg L^{-1}.

Example 2.2: Consider a lake with volume 10^4 litres that is fed by a stream with flow rate 5×10^{-3} Ls^{-1}, and pollutant concentration 10 mg L^{-1}. There is also a sewage outfall that discharges 5×10^{-4} Ls^{-1} of the same pollutant into the lake with concentration 100 mg L^{-1}. The sewage decomposition reaction rate constant of pollutant is 2.31×10^{-6} s^{-1}. Calculate the steady state concentration of stream.

Solution: This is a case of non-conservative system

$$\text{Input rate} = C_sQ_s + C_wQ_w$$

$$= 5\times10^{-3}\times10 + 5\times10^{-4}\times100 = 0.1 \text{ mg s}^{-1}$$

$$\text{Output rate} = C_m(Q_s + Q_w)$$

$$= C(5\times10^{-3} + 5\times10^{-4}) = 5.5\times10^{-3}C$$

$$\text{Decay rate} = k\text{CV} = 2.31\times10^{-6}\times\text{C}\times10^4$$

$$= 23.1\times10^{-3}\times\text{C mg s}^{-1}$$

Now using equation 2.6

$$\text{Input rate} = \text{Output rate} + \text{Decay rate}$$

$$0.1 = 5.5\times10^{-3}\text{C} + 23.1\times10^{-3}\text{C}$$

$$= 28.6\times10^{-3}\text{C}$$

or

$$\text{C} = \frac{0.10}{28.6\times10^{-3}} = 3.5 \text{ mg L}^{-1}$$

Steady state concentration of stream = 3.5 mg L^{-1}.

Situation 3 : Step Function Response

Quite often, we are interested in how concentration will change with time when there is sudden change in amount of pollutant entering the system. This is known as step function response of the system.

The system can be modeled as a box of volume V, that has equal flow into and output as Q. Contents of box are completely mixed at all times so that pollutant concentration C in the box is the same as the concentration leaving the box. The total mass of pollutant in the box is VC and rate of increase of pollutant in the box is $\frac{VdC}{dt}$. Let the total rate at which pollutant enters the box as S. If pollutants are nonconservative they will be modeled with a first order reaction rate coefficient k so that

Input rate = Output rate + Accumulation rate + Decay rate

or Accumulation rate = Input rate – Output rate – Decay rate

or
$$\frac{VdC}{dt} = S - QC - kCV \quad ...(2.7)$$

As a steady state approximation at $t = \infty$, $\frac{dC}{dt} = 0$ so that

$$C(\infty) = \frac{S}{Q + kV} \quad ...(2.8)$$

so that
$$\frac{dC}{dt} = \frac{Q + kV}{V}\left(C - \frac{S}{Q + kV}\right) \quad ...(2.9)$$

One way to solve this differential equation is to change variables. If we assume

$$y = C - \frac{S}{Q + kV} \quad ...(2.10)$$

then
$$\frac{dy}{dt} = \frac{dC}{dt}$$

Equation 2.9 becomes

$$\frac{dy}{dt} = -\left(\frac{kV + S}{V}\right)y \quad ...(2.11)$$

which is like differential Equation 2.4. The solution of which is

$$y = y_0 e^{-(k+Q/v)t} \quad ...(2.12)$$

where y_0 is value y at $t = 0$. If C_0 is concentration in the box at time $t = 0$ from 2.10

$$y_0 = C_0 - \frac{S}{Q + kV} \quad ...(2.13)$$

substituting 2.10 and 2.13 into 2.12 yields

$$C - \frac{S}{Q+kV} = \left(C_0 - \frac{S}{Q+kV}\right) e^{-\left(k+\frac{Q}{V}\right)t}$$

or

$$C_t = \left(C_0 - \frac{S}{Q+kV}\right) e^{-\left(k+\frac{Q}{V}\right)t} + \frac{S}{Q+kV} \qquad ...(2.14)$$

Equation 2.14 can be made to look slightly simpler by inserting the steady state solution from (2.8)

$$C_t = \left[C_0 - C_\infty\right] e^{-\left(k+\frac{Q}{V}\right)t} + C_\infty \qquad ...(2.15)$$

This equation should make sense.

Example 2.3: Consider the 10^4 L lake analyzed in Example 2.2. Suppose the lake conditions is deemed inacceptable and to solve the problem it is decided to completely divert the sewage outfall around lake eliminating as a source of pollution. The incoming stream still has flow $Q_s = 5 \times 10^{-3}$ Ls^{-1} and conc. $C_s = 10$ mg L^{-1}. Assuming complete mixing conditions find the concentration of pollutants in the lake after a week of diversion.

Solution: For this situation

$C_0 = 3.5$ mg L^{-1} $\qquad V = 10^4$ L

$k = 2.31 \times 10^{-6}$ s^{-1} $\qquad Q = 5.0$ ms^{-1}

The total rate at which pollutant is entering the lake from incoming stream is

$$S = Q_s C_s = 5 \times 10^{-3} \times 10 \text{ mg s}^{-1} = 5 \times 10^2 \text{ mg s}^{-1}$$

$$C_\infty = \frac{S}{Q+kV} = \frac{5}{28.1}$$

$$= \frac{5\times10^{-2}}{5\times10^{-3} + 2.31\times10^{-6}\times10^{4}} = \frac{50\times10^{-3}}{28.1\times10^{-3}}$$

$$= 1.779 \text{ mg L}^{-1}$$

Using Equation 2.15

$$C(7 \text{ days}) = (3.5 - 1.779)\exp - \left(2.31\times10^{-6} + \frac{5\times10^{-3}}{10^{-4}}\right) 7\times24\times3600 + 1.779$$

$$= 1.721\exp^{-1.7} + 1.779$$

$$= 2.09 \text{ mg L}^{-1}.$$

Pollutant concentration in lake after 7 days of diversion = 2.09 mg L^{-1}.

Example 2.4: Global precipitation has been estimated to average about 1 m of water per year across the globe. If surface area of earth is 5.10×10^{14} m^2 find the energy required to cause that much of water to evaporate each year. Compare your answer to world energy consumption of 3.3×10^{17} kJ and to the average rate at which sunlight strike earth surface 167 Wm^{-2}.

Solution: Energy to vaporise 1 kg of water = 2499 kJ/kg

Total volume of water vaporised = $1 \times 5.10 \times 10^{14}$ m^3

As an approximation density of water = 10^3 kg/m^3

Total mass of water vaporised = $5.10 \times 10^{14} \times 10^3$ kg

Total energy required to vaporise this much water

$$= 5.10 \times 10^{17} \times 2499 = 1.25 \times 10^{21} \text{ kJ/year.}$$

This is about 4000 times the world energy consumption.

Averaged over the globe, this energy rate is

$$= \frac{1.25 \times 10^{24}}{365 \times 24 \times 3600 \times 5.10 \times 10^{14}} = 78 \text{ Wm}^{-2}$$

which is about half of the 167 Wm^{-2} of incoming sunlight striking earth's surface.

Example 2.5: Typical coal based power plants has an energy content of 24 kJ/g and an average carbon content of about 62%. Suppose the average plant burn fuel with 2% sulphur and 10% ash. Assume that this plant with 3 units of heat energy delivers 1 unit of electrical energy.

(*a*) Per kilowatt hour of electrical energy produced, find the emission of SO_2, particulate matter and carbon.

(*b*) If we assume all of the particulate matter is made up of fly ash, how efficient must the particulate control system be to meet the particulate emission limit. The clean air standard in 130 g of elemental sulphur per 10^6 kJ and particulate emission to 13 g/10^6 kJ.

Solution: Since 3 kW hr of heat is required to produce 1 kWhr of electricity.

Heat input = 3 kW hr = 3×3600 = 10800 kJ

Sulphur emission are thus restricted to

$$\text{S. emission} = \frac{130}{10^6} \times 10800 = 1.40 \text{ g of S/ kWhr}$$

As 1 gm of S ≡ 2 gm of SO_2 so 2.8 g SO_2/kW hr will be emitted. The particulate emission would be

$$\text{Particulate emission} = \frac{13 \text{ g}}{10^6 \text{ kJ}} \times 10800 = 0.14 \text{ g/kW hr.}$$

To find carbon emission let us first find amount of coal burned/kW hr.

$$\text{Coal input} = \frac{10800 \text{ kJ/kW hr}}{24 \text{ kJ/g}} = 450 \text{ g coal/kW hr}$$

Since coal is 62% carbon.

$$\text{Carbon emission} = 0.62 \times 450 = 280 \text{ g C/kW hr}$$

Burning 450 g coal containing 2% sulphur will release $0.02 \times 450 = 9$ g of sulphur. Since allowable sulphur is 1.4 g, the removal efficiency must be

$$1 - \frac{1.4}{9.0} = 0.85 = 85\%$$

Since 10% of coal is ash, 70% of that is fly ash, the total fly ash generated will be

$$= 0.70 \times 0.10 \times 450 \text{ g Coal/kW hr} = 31.5 \text{ g fly ash}$$

The allowable particulate matter is restricted to 0.14 g/kW hr so that

$$\text{Particulate removal efficiency} = 1 - \frac{0.14}{31.4} = 0.995 = 99.5\%.$$

2.3 ENERGY LAWS

The quantitative relation of conversion of energy into useful work are given by zeroth, first, second and third laws of thermodynamics.

2.3a Zeroth Law of Thermodynamics

"If two systems, A and B, are in equilibrium with a third system, C, then A and B are also in equilibrium with each other". In general, all spontaneous processes tend towards equilibrium. Once the system is at equilibrium no exchange of energy can take place thus no useful work can be obtained from it.

2.3b First Law of Thermodynamics

This law is a statement of the law of conservation of energy which states, "*Energy can neither be created nor destroyed, only it can be transformed.*" This law appears sometimes as

$$(U_2 - U_1) = \Delta U = q - w = q - PdV \qquad ...(2.16)$$

where ΔU is the change in the internal energy of the system being studied when heat q flows in or out and work w is done on or by the system. The system being studied can be at any part of the universe—a city, a forest, a person, a nuclear power plant—within certain limits that will not be discussed here. When ΔU is positive, the internal energy of the system is increasing; when it is negative, the internal energy of the system is decreasing. When q is positive, heat is absorbed by the system, when it is negative, heat is evolved by the system. When w is positive, work is done by the system and when w is negative work is done on the system.

The heat released during various chemical reactions, for example, the combustion of coal or oil at constant pressure

$$q = \Delta H = H_2 - H_1 \qquad ...(2.17)$$

where ΔH is enthalpy change during the reaction. If a chemical reaction takes place in a system, the temperature of the system, immediately after the reaction is generally different from the temperature before the reaction. To restore the system to its initial temperature, heat must flow either to or from the surroundings. If the system is hotter after the reaction than before, heat must flow to the surrounding to bring the system to the initial temperature. In this case, the reaction is exothermic, by the convention for heat flow, the heat of the reaction is negative. If the system is colder after the reaction than before, heat must flow from the surroundings to restore the system to its initial temperature. In this case, the reaction is endothermic and heat of reaction is positive.

The change in state of a system produced by a specified chemical reaction is definite. Since enthalpy change is a function of the state, the corresponding enthalpy change is definite. Thus if a specified set of reactants are transformed to a specified set of products by more than one sequence of reactions, the total enthalpy change must be same for every sequence. Suppose sodium chloride is synthesized by two different sequences as follow:

Sequence 1

$$Na(s) + H_2O(l) \longrightarrow NaOH(s) + \frac{1}{2}H_2(g) \qquad \Delta H_1 = -139.78 \text{ kJ mol}^{-1}$$

$$\frac{1}{2}H_2(g) + \frac{1}{2}Cl_2(g) \longrightarrow HCl(g) \qquad \Delta H_2 = -92.3 \text{ kJ mol}^{-1}$$

$$HCl(g) + NaOH(s) \longrightarrow NaCl(s) + H_2O \qquad \Delta H_3 = -179.06 \text{ kJ mol}^{-1}$$

$$\text{Net change } Na(s) + \frac{1}{2}Cl_2(g) \longrightarrow NaCl(s) \qquad \Delta H = ?$$

$$\Delta H = \Delta H_1 + \Delta H_2 + \Delta H_3 = -411.15 \text{ kJ mol}^{-1}$$

Sequence 2

$$\frac{1}{2}H_2(g) + \frac{1}{2}Cl_2(g) \longrightarrow HCl(g) \qquad \Delta H_2 = -92.31 \text{ kJ mol}^{-1}$$

$$Na(s) + HCl(g) \longrightarrow NaCl(s) + \frac{1}{2}H_2 \qquad \Delta H_5 = -318.84 \text{ kJ mol}^{-1}$$

$$\text{Net change } Na(s) + \frac{1}{2}Cl(g) \longrightarrow NaCl(s) \qquad \Delta H = ?$$

$$\Delta H = \Delta H_2 + \Delta H_5 = -411.15 \text{ kJ mol}^{-1}$$

The net change is obtained by adding together all the reactions in the sequence. In either case total enthalpy change for the same reaction is same, though net products have been obtained in different ways. Such functions like enthalpy are state functions and independent of the path of the process.

The above results can be further simplified. The increase in enthalpy in a reaction is the heat of reaction or enthalpy of formation of the compound as for example

$$2Fe(s) + \frac{3}{2}O_2(g) \longrightarrow Fe_2O_3(s)$$

$$\frac{1}{2}N_2(g) + 2H_2(g) + \frac{1}{2}Cl(g) \longrightarrow NH_4Cl(s)$$

If ΔH for these reactions is written in terms of molar enthalpies $(\bar{H})$

$$\Delta H_f(Fe_2O_3, s) = \bar{H}(Fe_2O_3, s) - 2\bar{H}(Fe, s) - \frac{3}{2}\bar{H}(O_2, g)$$

$$\Delta H_f(NH_4Cl, s) = \bar{H}(NH_4Cl, s) - \frac{1}{2}\bar{H}(N_2, g) - 2\bar{H}(H_2, g) - \frac{1}{2}(Cl_2, g)$$

The enthalpy of every element in its standard state of aggregation, at 1 atm pressure and at 298.15 K is assigned equal to zero. Hence iron, oxygen (g), nitrogen (g), hydrogen (g), chlorine (g) have zero enthalpy

$$\Delta H_f^\circ\left[Fe_2O_3(s)\right] = \bar{H}^\circ\left[Fe_2O_3(s)\right]$$

and
$$\Delta H_f^\circ\left[NH_4Cl(s)\right] = \bar{H}^\circ\left[NH_4Cl(s)\right]$$

For elementary solids that exist in more than one crystalline form, the form that is more stable at 298 K and 1 atm, is assigned $\bar{H}^\circ = 0$. For example this assignment goes to rhombic sulphur, graphite carbon, and not to monoclinic sulphur or diamond. Enthalpy of formation of various compounds under standard state conditions are available in any standard book on Thermodynamics. Some of these values are given in Table 2.1.

Table 2.1 Standard molar enthalpies of formation at 298 K

Compound	ΔH_f° *in kJ mol*$^{-1}$	*Compound*	ΔH_f° *in kJ mol*$^{-1}$
CH_4 (*g*)	– 74.81	N_2O	+ 81.51
C_2H_4 (*g*)	+ 52.3	NO	+ 90.2
C_2H_6 (*g*)	– 84.64	NO_2	+ 33.2
C_9H_{20}	– 1148.4	N_2O_4	+ 9.66
CH_3CHO	– 166.3	SO_2	– 296.8
HCHO	– 156.3	H_2S	– 20.6
H_2O (*g*)	– 241.8	CO_2	– 393.5
NH_3 (*g*)	– 46.1		
HNO_2	– 400.4	CO	– 110.5
HNO_3	– 174.1	OH	+ 4.11
HCl	– 92.3	HO_2	– 244.72
HO_2NO_2	+ 593.37	ClNO	+ 467.87
$CH_3\underset{\underset{O}{\|}}{C}-OONO_2$ (PAN)	+ 1216	O_3	– 142.7

These values help to evaluate enthalpy of various reactions without carrying out.

Consider the rate of energy released when gasoline is burnt, assuming that it is made up of C_9H_{20} alkane only

$$C_9H_{20} + 14O_2 \longrightarrow 9CO_2 + 10H_2O$$

$$\Delta H_f^\circ\left[C_9H_{20}\right] = -1148.4 \text{ kJ mol}^{-1}, \qquad \Delta H_f^\circ\left[CO_2\right] = -393.5 \text{ kJ mol}^{-1}$$

$$\Delta H_f^\circ\left[H_2O\right] = -241.8 \text{ kJ mol}^{-1}$$

$$\Delta H = 9\Delta H_f^\circ[CO_2] + 10\Delta H_f^\circ[H_2O] \quad - \Delta H_f^\circ[C_9H_{20}]$$

$$= 9\times -393.5 + 10\times -241.8 + 1148.4$$

$$= -4361.1 \text{ kJ mol}^{-1}$$

The enthalpy of this reaction is –4361.1 kJ mol^{-1}

2.3c Second Law of Thermodynamics

If gasoline, consisting of C_9H_{20} alkane, used in an automobile, which is moving at a speed of 50 km per hour, then entire energy of –4361.1 kJ $mole^{-1}$ is not converted to work. About 75% of the energy is dissipated as heat, only 25% is converted to useful work *i.e.* –3279.9 kJ mol^{-1} is dissipated as heat while –1093.3 kJ mol^{-1} is converted to useful work. In order to discuss the inevitable, an availability of some of the heat to do some useful work under various circumstances, second law was introduced.

The second law of thermodynamics is a quantitative statement that most occurrences in this universe are irreversible. For example, a hot cup of coffee inevitably cools to room temperature, but a cold cup of coffee has never been observed to extract heat from a cold room in order to become hot coffee. Surprisingly the transfer of heat from the room to the cup of coffee in order to heat it above room temperature would not violate first law of thermodynamics even though this kind of heat transfer never occurs in real life. One statement is very specific in this regards "*No process is possible whose sole effect is to remove heat from an object at lower temperature and to supply this heat to an object at higher temperature.*" This is a very definite but a qualitative statement.

Another statement of second law is "*No process involving an energy transformation will occur spontaneously, unless there is a degradation of the energy from a concentrated form into a dispersed form*".

Because some energy is always dispersed into unavailable form, no spontaneous transformation of energy say light into potential energy (protoplasma) is 100% efficient. Entropy is a measure of unavailable energy resulting from transformation: the term is also used as a general index of the disorder associated with energy degradation. A gas for example has higher entropy than a crystalline solid, partly because the gas molecules can have many positions and velocities that are constantly changing at a fixed temperature and pressure. The molecules in a crystalline solid however are fixed in space in a crystalline lattice, although they vibrate about their fixed positions; the crystal has fewer accessible states than the gas and therefore lower entropy. One may also consider that the crystal is more ordered than gas and therefore has lower entropy.

During the discussions of heat and energy definition of entropy in terms of heat is particularly useful. The entropy change of a system during some process, ΔS, may be defined as

$$\Delta S = \int dq_{rev}/T \qquad ...(2.18)$$

if temperature during the process is constant

$$\Delta S = q_{rev}/T \qquad ...(2.19)$$

where q_{rev} refers to the heat that would have been absorbed or given off during the process, if it had proceeded in a reversible process. However for an irreversible process

$$\Delta S > \int dq_{rev}/T \qquad ...(2.20)$$

In terms of entropy changes, the second law states "*No process will ever occur in which the entropy of the universe decrease*". So much so that any system of man or nature that does not conform to this law is indeed doomed.

As in the case of a hot cup of coffee in a cold room will eventually cool down to room temperature. This is a general observation; any hot objects will heat a cold object so that the temperature of the hot object will decrease and temperature of the cold object will increase until temperature of both objects is same. If the room is large, increase in temperature of air may be negligible and thus total entropy has increased.

When light is absorbed by some object, it becomes warmer as light has been transformed into another kind of energy, heat energy. Heat energy comprising vibration and motion of molecules, increases and thus disorder or entropy increases. The absorption of sunlight by land and water creates hot and cold areas, leading air to flow, which may drive wind mills and perform work to produce electricity. The energy that is consumed is not actually used up. Rather it is converted from a state of high utility to a state of low utility, gasoline in an automobile tank is indeed used up as gasoline, however the energy in it is not destroyed but converted into forms no longer usable by automobile.

Sunlight reaches the biosphere at a rate of 96 kJ m^{-2} min^{-1} but is attenuated exponentially as it passes through the atmosphere, at most 67% (56.38 kJ m^{-2} min^{-1}) may reach the earth's surface at noon in a clear summer day. Solar radiation is further attenuated and the spectral distribution of its energy is greatly altered as it passes through cloud cover, water and vegetation. The daily input of sunlight to the autotropic layer of an ecosystem varies between 12600-16800 kJ m^{-2}. The 24-hour flush of heat is considerably less than the incoming solar radiation.

Radiation energy reaching the surface of the earth is about 10% ultraviolet, 45% visible and 45% infrared. Visible light is least attenuated as it passes through the atmosphere. The cool deep shade of forest is due to absorption of visible and far infrared radiation by the folliage over head. Chlorophyll absorbs the blue and red lights, and water in leaves and water vapours absorbs the infra red heat energy. Thus blue and red lights are the most useful for photosynthesis.

2.3d Thermodynamic Efficiency of Energy Conversion Processes

In an automobile, chemical energy from the combustion of gasoline appears as heat energy at about 1000K in the combustion chamber above the pistons. Some of the energy appears as work as the piston moves and drives the transmission generator etc. As stated earlier only, 25% of the input energy appears as useful work during the operation of car.

Such an upper limit may be calculated by assuming that engine operates in a cyclic way as a reversible engine operating between a highest temperature T_1 (in this case 1000 K) and a lowest temperature T_2 (in this case not lower than 350 K) somewhat below the boiling point of water. Work is done in a cycle called Carnot Cycle and is discussed at length in any Physical Chemistry textbook. The efficiency of such an engine is found to be as the network output divided by heat input at high temperature and if the engine is reversible then efficiency η

$$\eta_{rev}(\text{Carnot cycle}) = \frac{T_1 - T_2}{T_2} \quad ...(2.21)$$

$$\eta_{irrev} < \eta_{rev} \quad ...(2.22)$$

where T_1 and T_2 must be the absolute temperature. Thus the efficiency of reversible engine depends on the temperature between which it operates, it becomes more efficient as T_1 and T_2 move further apart.

Thus for an automobile with $T_1 = 1000$ K, $T_2 = 350$ K

$$\eta_{rev} = \frac{1000-350}{1000} = 0.65$$

The maximum efficiency of an engine is 65% of energy that would be converted into useful work.

Many automobile engines operate in a Otto cycle in which more than two temperatures need to be considered. The maximum efficiency of a reversible Otto engine can be expressed as

$$\eta_{rev}\,(\text{Otto engine}) = \left(1 - \frac{1}{r^{\gamma-1}}\right) \qquad ...(2.23)$$

where, r is the compression ratio of engine and γ is the ratio of heat capacity at constant pressure to heat capacity at constant volume ($\gamma = C_p/C_v$) of the gas in engine ($\gamma = 1.4$ for air). For Otto engine with $r = 9$ and $\gamma = 1.4$ the maximum efficiency would be 0.58 or 58%. The actual efficiency in a car, is 25% only, very far from maximum possible efficiency calculated from Equations 2.21 to 2.23.

Not all engines operate as far from their maximum efficiencies as gasoline engines in cars, the best of the engines used in steam electric plants have actual efficiency of 40% in contrast to maximum theoretical efficiencies of about 60%. Thus there is always a lot of waste of heat energy which is inevitable.

Energy conversion processes that do not involve heat engine may have no theoretical limitations to the efficiency of the conversion. For example, the direct conversion of solar energy to electrical energy and conversion of chemical energy directly to electrical energy.

As mentioned earlier, the waste heat from various engines is usually released into the environment and cause "Thermal Pollution". This makes cities warmer than their surrounding countrysides partly because heat released from all of the machines are mainly used in the cities.

2.3e Third Law of Thermodynamics

It states "*The entropy of a pure, perfectly crystalline substance is zero at absolute zero of temperature*". Consider the constant pressure transformation of a solid from the absolute zero of temperature to some temperature T below it melting point

$$\text{Solid}\,(0K,\,p) \longrightarrow \text{Solid}\,(T,\,p)$$

$$\Delta S = S_T - S_0 = \int_0^T \frac{C_p}{T}\,dT \qquad ...(2.24)$$

or

$$S_T = S_0 + \int_0^T \frac{C_p}{T}\,dT$$

By third law of thermodynamics $S_0 = 0$, so that absolute entropy at any temperature can be calculated to be equal to

$$S_T = \int_0^T \frac{C_p}{T}\,dT \qquad ...(2.25)$$

The most rigorous definition of entropy is in terms of the number of states that are possible for a system under condition of temperature, pressure, composition etc. So that

$$S_T = k \ln \Omega \qquad \text{...(2.26)}$$

where Ω is the number of states possible in the system and k is the Boltzman constant. The state of the system involves the position, energy of every atom and molecule in the system. The more states that are possible for a system, the less is known about which state the system is actually in; therefore one can think of entropy as a measure of our ignorance. However, for all practical purposes entropy may be taken as a measure of disorder.

2.4 KINETICS OF GAS REACTIONS

Those reactions which cannot be broken down into two or more simpler reactions are called Elementary reactions. Generally they consist of one, two or three reactant species and are referred to as unimolecular, bimolecular or intermolecular processes respectively. The role of third molecule is to act as an inert gas which stabilizes the energy rich product from a highly exothermic bimolecular reaction by siphoning off the excess energy and thus preventing dissociation back into reactants. In such cases, one need not be specific of the colliding third body, for which the symbol "M" is used. Examples of these three classes of gas phase reactions are

Unimolecular: The thermal decomposition of PAN (peroxyacetyl nitrate)

$$CH_3\underset{\underset{O}{\|}}{C}—OONO_2 \longrightarrow CH_3\underset{\underset{O}{\|}}{C}—OO + NO_2 \qquad \text{...(2.27)}$$

Bimolecular: Formation of nitrogen dioxide

$$O_3 + NO \longrightarrow NO_2 + O_2 \qquad \text{...(2.28)}$$

Trimolecular: The formation of ozone by reaction of a ground state oxygen atom, $O(^3P)$ with O_2

$$O(^3P) + O_2 + M \longrightarrow O_3 + M \qquad \text{...(2.29)}$$

An overall reaction includes two or more elementary reactions. Thus if a single reaction has four or more reactant, it cannot be an elementary process. Oxidation in air of methane by OH radicals in the presence of nitric oxide is a good example. The major stable products, formaldehyde and NO_2 are formed in a sequence of five elementary reactions

$$CH_4 + \dot{O}H \longrightarrow \dot{C}H_3 + H_2O \qquad \text{...(2.30)}$$

$$\dot{C}H_3 + O_2 \longrightarrow CH_3\dot{O}_2 \qquad \text{...(2.31)}$$

$$CH_3\dot{O}_2 + NO \longrightarrow CH_3\dot{O} + NO_2 \qquad \text{...(2.32)}$$

$$CH_3\dot{O} + O_2 \longrightarrow HCHO + H\dot{O}_2 \qquad \text{...(2.33)}$$

$$H\dot{O}_2 + NO \longrightarrow NO_2 + H\dot{O} \qquad \text{...(2.34)}$$

In fact, this is an example of chain reactions with the initiating step (2.30) and chain carrying steps which regenerate OH radicals. However, this reaction sequence does not go on recycling OH radicals indefinitely, and thus cannot be called as catalyst. The OH can combine with NO_2 to give nitric acid, and HO_2 can self combine to form hydrogen peroxide and O_2.

$$\dot{H}O + NO_2 \xrightarrow{M} HONO_2 \quad ...(2.35)$$

$$2H\dot{O}_2 \xrightarrow{M} H_2O_2 + O_2 \quad ...(2.36)$$

These are elementary chain termination processes because the HO and HO_2 radicals form molecular products that are relatively stable in troposphere. A mechanism of a reaction is the sequence of elementary reaction leading to the formation of products. Thus the reaction sequence shown above 2.30 to 2.34, 2.35 and 2.36 is a mechanism for the oxidation of CH_4 in troposphere.

2.4a Reaction Rates

The rates of reaction is defined as the change in concentration of a reactant or products with time. For simple reactions occurring with unit stoichiometry, the rate expressed in terms of reactant disappearance is the same as the rate in terms of product formation as for reaction (2.28)

$$NO + O_3 \longrightarrow NO_2 + O_2 \quad ...(2.28)$$

the rate is defined as

$$\text{rate} = -\frac{d[O_3]}{dt} = -\frac{d[NO]}{dt} = \frac{d[NO_2]}{dt} = \frac{d[O_2]}{dt}$$

i.e. rate = rate of disappearance of O_3 = rate of disappearance of NO
= rate of formation of NO_2 = rate of formation of O_2

For reaction of more general form

$$pP + qQ \longrightarrow rR + sS \quad ...(2.37)$$

where stoichiometric coefficients p, q, r and s are not all unity, the rate in terms of disappearance of P may not be equal to the disappearance of Q or appearance of R or S. To take such difference into account, the rates of generalised reaction is defined as

$$\text{rate} = -\frac{1}{p}\frac{d[P]}{dt} = -\frac{1}{q}\frac{d[Q]}{dt} = \frac{1}{r}\frac{d[R]}{dt} = \frac{1}{s}\frac{d[S]}{dt} \quad ...(2.38)$$

For example thermal oxidation of NO by oxygen

$$2NO + O_2 \longrightarrow 2NO_2 \quad ...(2.39)$$

$$\text{rate} = -\frac{1}{2}\frac{d[NO]}{dt} = -\frac{d[O_2]}{dt} = \frac{1}{2}\frac{d[NO_2]}{dt}$$

The reaction rate constants are given as k_{pqr} which is reaction rate constant for reaction *p.qr* where p represent the chapter number so that qr is reaction number in chapter p. The reaction rate constant k_{-pqr} is rate constant for reverse reaction *p.qr*.

2.4b Reaction Rate Laws and Reaction Rate Constants

The importance of distinguishing between elementary and overall reactions comes in formulating reaction rate laws. For elementary reactions only, the rate law may be written directly from stoichiometry equation. Thus for the general elementary gas phase reaction

$$pP + qQ \longrightarrow rR + sS \qquad ...(2.40)$$

$$\text{rate} = k[P]^p [Q]^q$$

where $p + q \leq 3$ by definition of elementary reaction. For example, the rate expression for the elementary reaction (2.28) is given by

$$NO + O_3 \longrightarrow NO_2 + O_2 \qquad ...(2.28)$$

$$\text{rate} = k_{228} [O_3] [NO]$$

where k_{228} is called reaction rate constant which has a specific value for a particular reaction at a particular temperature. The order of a reaction is defined as the sum of the exponents in the rate law. Thus reaction 2.28 has an order of 1 + 1 = 2 *i.e.* second order. The order with respect to each species appearing in the rate law is the exponent of concentration of that species; thus reaction 2.28 is first order in both O_3 and NO.

Similarly the thermal oxidation of NO (reaction 2.39) by oxygen molecule is a third order reaction

$$2NO + O_2 \longrightarrow 2NO_2 \qquad ...(2.39)$$

of which the rate equation from stoichiometry or molecularly can be written as

$$\text{rate} = k_{240} [O_2][NO]^2 \qquad ...(2.41)$$

However, in troposphere, the O_2 concentration is always so large relative to NO that it is effectively a constant and thus can be incorporated in reaction rate constant. The rate law can now be written as

$$\text{rate} = k_{241} [NO_2]^2$$

$$k_{241} = k_{240} [O_2] \qquad ...(2.42)$$

The reaction will be referred to as a pseudo-second order reaction. These rate equation have a significance if the reaction is an elementary reaction. The general overall rate law may be of the form

$$\text{rate} = k[P]^l [Q]^m [R]^n [S]^o$$

It is important to stress that in contrast to elementary reaction, in overall reaction, the exponents of rate laws, *l, m, n, o* do not necessarily bear a relationship to the stoichiometric coefficients of the reaction *p, q, r, s*. Since units of rate are concentration × time^{-1}, the units of reaction rate constant must be concentration $[1 - (l + m + n + o)]$ time^{-1}. For usual gas phase reaction *n* and *o* are zero so that

$$\text{Reaction Rate constant} = k = \frac{\text{rate}}{[P]^l [Q]^m}$$

units of k are conc$^{[(1 - l + m)]}$ time^{-1}.

Thus for first order reaction $\qquad m + l = 1$

second order reaction $m + l = 2$

third order reaction $m + l = 3$

so that units of Reaction Rate constants, k, for

First order reaction s^{-1}

Second order reaction $cm^3\ molecule^{-1}\ s^{-1}$

Third order reaction $cm^6\ molecules^{-2}\ s^{-1}$

Tables 2.2 and 2.3 give some of the common conversion factors for gas phase reactions constants.

Table 2.2 Conversion between units of concentration assuming one atmospheric pressure (760 Torr) and 298 K

Part per	*Unit*	*Molecules, Atoms or Radicals per cm³*
10^6	1 ppm	$2.46 \times 10^{13}\ cm^{-3}$ molecule
10^8	1 pphm	$2.46 \times 10^{11}\ cm^{-3}$ molecule
10^9	1 ppb	$2.46 \times 10^{10}\ cm^{-3}$ molecule
10^{12}	1ppt	$2.46 \times 10^{7}\ cm^{-3}$ molecule
$1\ \mu gm^{-1}$	= ppm 40.9 × molar mass	

Table 2.3 Some common conversion factors for liquid and gas phase reactions

Concentration	
	$1\ mole\ L^{-1} = 6.02 \times 10^{20}\ molecule\ cm^{-3}$
	$1\ ppm = 2.46 \times 10^{13}\ molecule\ cm^{-3}$
	$1\ atm = 760\ Torr = 4.09 \times 10^{-2}\ mole\ L^{-1} = 2.46 \times 10^{19}\ molecule\ cm^{-3}$
Second order rate constant	
	$cm^3\ molecules^{-1}\ s^{-1} \times 6.02 \times 10^{26} = L\ mole^{-1}\ s^{-1}$
	$ppm^{-1}\ m^{-1} \times 4.08 \times 10^5 = L\ mole^{-1}\ s^{-1}$
	$ppm^{-1}\ m^{-1} \times 6.77 \times 10^{16} = cm^3\ molecule^{-1}\ s^{-1}$
	$atm^{-1}\ s^{-1} \times 4.06 \times 10^{-20} = cm^3\ molecule^{-1}\ s^{-1}$
Third order reaction	
	$cm^6\ molecule^{-2}\ s^{-1} \times 3.63 \times 10^{41} = L^2\ mole^{-2}\ s^{-1}$
	$ppm^{-2}\ min^{-1} \times 9.97 \times 10^{12} = L^2\ mole^{-2}\ s^{-1}$
	$ppm^{-2}\ min^{-1} \times 2.75 \times 10^{29} = cm^6\ molecule^{-2}\ s^{-1}$

Solution Phase Reactions: Concentration units are moles L^{-1} giving corresponding rate constants of $L\ mole^{-1}\ s^{-1}$ and $L^2\ mole^{-2}\ s^{-1}$ for second and third order reactions.

In this book, a fast, second or pseudo-second order gas phase reaction is one with a rate constant greater than $10^{-13}\ cm^3\ molecule^{-1}s^{-1}$ and a slow reaction with reaction rate constant $10^{-18}\ cm^3\ molecule^{-1}\ s^{-1}$.

2.4c Half Lives and Life Times of Reaction

A rate constant is a quantitative measure of how fast reactions proceed and therefore is an indicator of how long a given set of reactants will survive in atmosphere under a particular set of reactant concentrations. However, reaction rate constant is not a parameter, which by itself is readily related to the average length of time a species will survive in atmosphere before reacting. A more intuitively meaningful parameter is half life time $T_{1/2}$ or the natural life time τ of a pollutant with respect to reaction.

Table 2.4 Relationship between the Reaction Rate Constant, half lives and life time for First, Second and Third order Reactions

Reaction order	*Reaction*	*Half life time of A*	*Life time of A*
First	$A \xrightarrow{k_1} \text{product}$	$T_{1/2}^{A} = \dfrac{0.693}{k_1}$	$\tau^{A} = \dfrac{1}{k_1}$
Second	$A + B \xrightarrow{k_2} \text{product}$	$T_{1/2}^{A} = \dfrac{0.693}{k_2[B]}$	$\tau^{A} = \dfrac{1}{k_2[B]}$
Third	$A + B + C \xrightarrow{k_3} \text{product}$	$T_{1/2}^{A} = \dfrac{0.693}{k_3[B][C]}$	$\tau^{A} = \dfrac{1}{k_3[B][C]}$

The half life time is defined as time required for the concentration of a reactant to fall to one half of its initial value whereas life time is defined as the time it takes for reactant concentration to fall to 1/2 of its initial value. Both $T_{1/2}$ and τ are directly related to the reaction rate constant and to the concentration. These relationship are given in Table 2.4.

Let us use these to calculate the natural life time of some species in three examples at 298 K, P = 1 atm and average HO concentration 5 × 10^5 and 5 × 10^7 HO radical molecule cm^{-3} in clean air or one polluted with smog.

Example 2.6: Thermal decomposition of PAN

$$CH_3\overset{\overset{\large O}{\|}}{C}OONO_2 \xrightarrow{k_1} CH_3\overset{\overset{\large O}{\|}}{C}\text{—}OO + NO_2 \qquad k_1 = 3.6\times10^{-4}\,s^{-1}$$

$$T_{1/2}^{PAN} = \frac{0.693}{k_1} = 1926 \text{ sec} = 32 \text{ min}.$$

$$\tau^{PAN} = \frac{1}{k_1} = \frac{1}{3.6\times10^{-4}} = 2778 \text{ sec} = 46.3 \text{ min}.$$

Example 2.7: Second order reaction–oxidation of methane by OH radical

$$CH_4 + \dot{O}H \xrightarrow{k_2} \dot{C}H_3 + H_2O \qquad k_2 = 8.4\times10^{-15}\,cm^3 molecule^{-1} s^{-1}$$

For OH concentration = 5 × 10^5 molecule cm^{-3}

$$\tau_{OH}^{CH_4} = \frac{1}{k_2[OH]} = \frac{1}{8.4\times10^{-15}\times5\times10^5} = 2.38\times10^8 \text{ s}$$

$$= 7.55 \text{ years}$$

For OH concentration = 5×10^7 molecule cm^{-3}

$$\tau_{OH}^{CH_4} = \frac{1}{8.4\times10^{-15}\times5\times10^7} = 2.38\times10^6 \text{ s} = 27.5 \text{ days}$$

Example 2.8: Third order reaction–Formation of O_3

$$O + O_2 \xrightarrow{k_3} O_3 \qquad k_3 = 6.0\times10^{-34} \text{ cm}^6\text{molecule}^{-2}\text{s}^{-1}$$

As at 1 atm pressure this reaction is still in the low concentration limit may be taken as second order reaction. This effective second order reaction rate constant $k_3' = k_3 \times 1$ atm = $k_3 \times 2.46 \times 10^{19}$ molecule cm^{-3} (Table 2.3) where 1 atm of air is taken as the third body so that k_3' = 1.476×10^{-14} cm^3 mol^{-1} s^{-1}. Thus for $[O_2] = 5.2 \times 10^{18}$ cm^3 molecule.

$$\tau_{O_2}^{O} = \frac{1}{k_3'[O_2]} = \frac{1}{1.476\times10^{-14}\times5.2\times10^{18}} = 1.3\times10^{-5} \text{ s}$$

$$= 13 \mu\text{s}$$

Three points should be made about these calculations

(*a*) They are valid only for the specified reactions,

(*b*) For second and third order reactions $T_{1/2}$ and τ depends on concentration of other reactants, and

(*c*) In this book, a fast, second order or pseudo second order gas phase reaction is one with a rate constant greater than ~ 10^{-13} cm^3 molecules^{-1} s^{-1} and a slow reaction with reaction rate constant less than 10^{-18} cm^3 molecule^{-1} s^{-1}.

REFERENCES

1. Atkins, P.W., *Physical Chemistry,* Oxford University Press, London (1982).
2. Benson, S.W., *The Foundations of Chemical Kinetics*, McGraw-Hill, New York (1960).
3. Castellan, G.W., *Physical Chemistry*, Addison-Wesley/Narosa, New Delhi (1995).
4. Gardiner, W.C., *Rates and Mechanism of Chemical Reactions,* Benjamin-Cummings, Menlo Park, CA (1969).
5. Koltz, I., *Energy Changes is Biochemical Reactions,* Academic Press, New York (1967).
6. Koltz, I., and Rosenberg R.M., *Chemical Thermodynamics,* Benjamin, New York (1972).
7. Kaplus, M. and Porter R.N., *Atoms and Molecules,* W.A. Benjamin, New York (1970).
8. Laidler, K.J. and Meiser J.H., *Physical Chemistry,* Benjamin-Cummings, Menlo Park, CA (1982).

3

Energy States and Photochemistry

3.1 ENERGY LEVELS AND ATOMIC SPECTRA

Most of the chemical reactions taking place in atmosphere are photochemical in nature. For a reaction to be photochemically active, electromagnetic radiation must interact with the reactant, thus show absorption, and hence involvement of spectroscopy. In other words to understand photochemistry, it is essential that one should know spectroscopy.

Since in spectroscopy atomic or molecular transition between various energy levels take place, understanding of these energy states is also a must for understanding photochemistry. Only brief principles relevant to us will be explained here. For detailed understanding of energy states, one should consult references[4, 5, 6, 8], given at end of this chapter. However, a very lucid treatment of it is given by the author in his book on *Molecular Spectroscopy.*

All energy levels in atoms and molecules are quantized in nature. Wave mechanics, in principle, can be extended to untangle the complications of atomic and molecular structure. It, however, proves difficult in practice. In view of this, an explanation based on the interaction of angular momentum vector of various electrons in an atom proved capable of untangling mere complication. For this reason "Vector model of atom" is an accepted model to explain the complex atomic spectra.

It assumes that the outer electrons move in a central field with inner closed shell being spherically symmetrical with force and field for the outermost electrons. The inner closed shells of atoms do not contribute to the complexity of the spectrum, but only to the absolute position of spectral lines. Thus all alkali metals have similar spectra, but the position of lines only get shifted to longer wavelength.

The magnitude of angular momentum of a quantized system can have limited values of $\sqrt{l(l+1)}\,\hbar$, where l is azimuthal quantum number of atom and half integral values $\left(0, \frac{1}{2}, 1, ...\right)$ for spin angular momentum. As angular momentum, vector along *x, y, z* direction do not commute, only one of the components of angular momentum can be known precisely in a system. As a convention, it is taken to be the *z*-component and the only possible values of the component are $m\,\hbar$ where $-l \le m \le l$.

Coupling of two or more vectors is the cause of complex atomic spectra. This concept of coupling is called, Russell-Saunders coupling or L-S coupling, and will be introduced in a rather empirical fashion. The general scheme of coupling is as follows.

1. First take the l's of each electron outside the closed shell and form them into angular momentum vector L. The magnitude of L can be equal to $(l_1 + l_2)$, $(l_1 + l_2 - 1)$,...., $(l_1 - l_2)$. Each L can have $(2L + 1)$ various values. A notation grew up in which different L values are described according to the symbols

L	0	1	2	3	4	5	6	7
Symbol	S	p	D	F	G	H	I	K

2. The spins of each electrons outside the closed shell are coupled to give resultant spin S.
3. Each L and S vector are coupled to get resultant J vector.

The particular state is said to be a multiplet, and can have a multiplicity equal to the number of J values.

For example, if $S = \frac{1}{2}$ then $J = L + \frac{1}{2}$ and $L - \frac{1}{2}$. Thus the multiplicity will be $2S + 1$ provided $L > S$. If $L < S$ there is only one possible value of J although $2S + 1 > 1$, *e.g.* for $l = 0$, $s = \frac{1}{2}$; $L = 0$, $S = \frac{1}{2}$ or $J = \frac{1}{2}$. Thus the state is singlet even though $2S + 1 = 2$. So if $L > S$, the multiplicity will be given by $2L + 1$. The energy state with quantum number L, S. J is indicated by a code symbol which is called "Term Symbol" as

$$^{2S+1}L_J \quad \textbf{Term Symbol} \qquad ...(3.1)$$

For example, $^4D_{7/2}$ means $L = 2$ for D state, $S = 3/2$ or $2S + 1 = 4$ and $J = 2+\frac{3}{2}=\frac{7}{2}$. These states, so derived by L-S coupling, follow Hund's rule.

In order to determine the terms resulting from two non-equivalent electrons, say in a case of a p-electron and a d-electron, various possible L values are calculated as

p-electron $\quad l_1 = 1$

d-electron $\quad l_2 = 2$

$L = 2 + 1, 2 + 1 - 1, 2 + 1 - 2$,...., $2 - 1 = 3, 2, 1$, *i.e.*, the two electrons can be either parallel or anti parallel. S is, therefore, 1 or 0 which means that triplet as well singlet terms result S. In all there will be six terms ^1P, ^1D, ^1F, ^3P, ^3D, ^3F. Similarly, non-equivalent p-electrons p^2 or p^4 not all the terms derived for two non-equivalent electrons are possible. While considering non-equivalent electrons, Pauli's exclusion principle need not be taken into account. This assumption is justified, since, when the electron have different n or l values, the Pauli's exclusion principle is already satisfied. When, however the two electrons have equal n and equal l, they must at least differ in their values of m_l or m_s. As in p^2 case m_l is same for both electron (+1, 0, –1) and, therefore, the two electrons cannot both have $m_s = +\frac{1}{2}$ or both $-\frac{1}{2}$. That is, their spins can only be antiparallel for a D-term, giving ^1D only and ^3D is not possible. Further considerations show that two equivalent *p*-electrons give only the terms ^1S, ^3P, ^1D and three equivalent *p*-electrons give ^4S, ^2P, ^2D states. There are same additional examples for different configurations given in Table 3.1.

Table 3.1 Terms of some equivalent and non-equivalent atomic configurations

Equivalent electrons		*Non-equivalent electrons*	
Electron configuration	*Term*	*Electron configuration*	*Term*
s^2	^1S	ss	^1S, ^3S
p^2	^1S, ^1D, ^3P	sp	^1P, ^3D
p^3	^2P, ^2D, ^4S	sd	^1D, ^3D
p^4	^1S, ^1D, ^3P	pp	^1S, ^1P, ^1D, ^3S, ^3P, ^3D
p^5	^2P	pd	^1P, ^1D, ^1F, ^3P, ^3D, ^3F
p^6	^1S	dd	^1S, ^1P, ^1D, ^1F, ^1G
			^3S, ^3P, ^3D, ^3F, ^3G
d^2	^1S, ^1D, ^1G, ^3P, ^3F	sss	^2S, ^2S, ^4S
d^3	^2p, ^2D, ^2F, ^2G, ^2H, ^4P, ^4F	ssp	^2P, ^2P, ^4P
d^4	^1S, ^1D, ^1F, ^1G, ^1I, ^3P	spd	^2P, ^2D, ^2F, ^2P, ^2D, ^2F
	^3D, ^3F, ^3G		^4P, ^4D, ^4F
	^3H, ^5D	ppp	^2S, ^2P, ^2D, ^2F, ^4S, ^4P, ^4D, ^4F
d^5	^2S, ^2P, ^2D, ^2F, ^2G, ^2H,	ppd	^2S, ^2P, ^2D, ^2F, ^2G,
	^2I, ^4P, ^4D, ^4F, ^4G, ^6S		^4S, ^4P, ^4D, ^4F, ^4G

A spectral line result from the transition of atom from one energy level to another

$$\varepsilon_2 - \varepsilon_1 = h\nu \qquad ...(3.2)$$

According to international agreement the upper state always come first. However, if atom or molecule is going from lower energy to higher energy state, it is called absorption spectra, while if it is going from higher energy state to lower energy state, it is called an emission spectra.

3.1a Selection Rule

In place of energy states, term symbols are used. The selection rules for the total angular momentum is $\Delta J = 0, \pm 1$ with the restriction that

$$J = 0 \not\leftrightarrow J = 0 \ (\not\leftrightarrow \text{ means cannot combine with})$$

Under the same conditions, the selection rule for orbital angular momentum is $\Delta L = 0, \pm 1$ and $\Delta S = 0$. The selection rule under consideration may be true to a first approximation. A transition may be forbidden as a dipole radiation, but may be allowed as quadruple radiation or magnetic dipole radiation and therefore, may appear, even though very weakly.

In addition to these, atomic states are divided into the even and odd states, called gerade and ungerade denoted by '*g*' and '*u*'. For example, ground state of oxygen atom is $2s^2\ 2p^4$ is an even state and is written as ^3P_g. The ground state of Nitrogen is $2s^2\ 2p^3$, $^4S_{3/2}$ is an odd state written as $^4S_{3/2u}$.

Laporte Selection Rule. The German words gerade (even) and ungerade (odd) are added as labels to states and wave function to denote their behaviour under inversion (their parity).

It states that electric dipole transitions allowed are those involving a change of parity; thus

$$g \not\leftrightarrow g \quad \& \quad u \not\leftrightarrow u \quad \text{are forbidden transition}$$

but

$$g \leftrightarrow u \quad \text{transition is allowed.}$$

3.2 ENERGY LEVELS AND MOLECULAR SPECTRA

3.2a Electromagnetic Radiation

Light has both wave like and particle like properties. As a wave, it is combination of oscillating electric and magnetic fields perpendicular to each other and to the direction of propagation. The distance between consecutive peaks is the wavelength λ, and the number of complete cycles passing a fixed point in one second is the frequency ν. They are inversely proportional through a relationship

$$\nu = c/\lambda, \qquad ...(3.3)$$

where, c is velocity of light in vacuum, 3×10^8 m s^{-1}. A red radiation with wavelength $\lambda = 6.8 \times 10^{-7}$ m the corresponding frequency is

$$\nu = \frac{c}{\lambda} = \frac{3 \times 10^8}{6.8 \times 10^{-7}} = 4.542 \times 10^{14} \text{ Hz}$$

Such an inconvenient large numerical value forced scientists to introduce a third quantity called wave number, $\bar{\nu}$, which is simply given by the relation

$$\bar{\nu} = \frac{1}{\lambda} \text{ cm}^{-1} \text{ with } \lambda \text{ is in cm.} \qquad ...(3.4)$$

Thus for red light, wave number is

$$\bar{\nu} = \frac{1}{6.6 \times 10^{-7}} = 15151.51 \text{ cm}^{-1}$$

Consider light of wavelength λ as a photon particle, the energy of a quantum of

$$\varepsilon = h\nu = \frac{hc}{\lambda}$$

where h is Planck's constant 6.62×10^{-34} J s. Since chemists generally deal experimentally with moles of material a convenient unit is, a mole of photons defined as 1 Einstein. The energy of 1 Einstein of wavelength λ in nanometers (nm) is

$$E = N\varepsilon = \frac{Nhc}{\lambda} = \frac{6.02 \times 10^{23} hc}{\lambda} = \frac{1.196 \times 10^5}{\lambda} \text{ kJ/Einstein}^{-1} \qquad ...(3.5)$$

Figure 3.1 illustrates the regions into which electromagnetic radiation has been divided. The boundaries between the regions are by no means precise, although the molecular processes associated with each region are quite different.

The region of most direct interest to troposphere photochemistry which ranges from visible at 700 nm to near ultraviolet at ~290 nm.

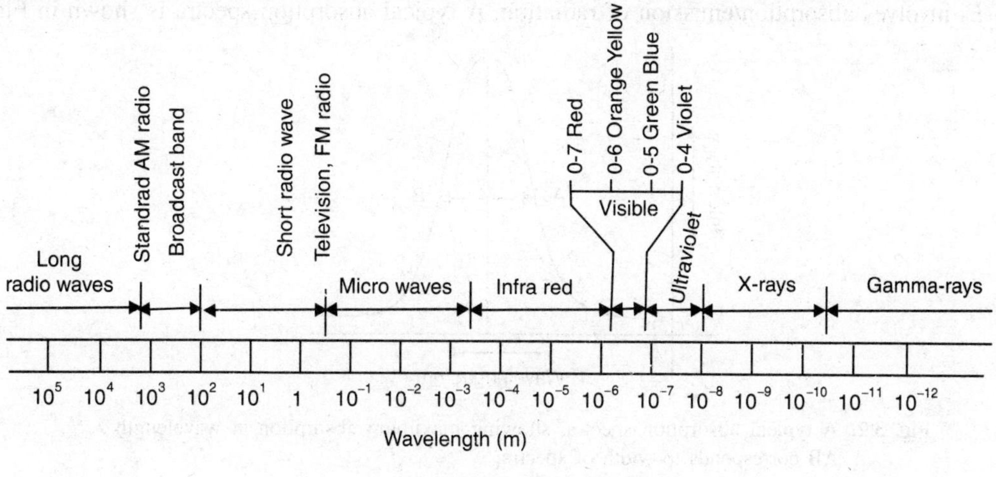

Fig. 3.1 The Electromagnetic spectrum.

Table 3.2 Typical wavelength, frequencies, wave number and energies of visible region

Name	*Wavelength* λ *in nm*	*Frequency* ν, *in Hz*	*Wave number* $\bar{\nu}$ *in* cm^{-1}	*Energy in kJ einstein*$^{-1}$
Red	660	4.55×10^{14}	1.51×10^4	1.81×10^2
Orange	620	4.8×10^{14}	1.6×10^4	1.9×10^9
Yellow	580	5.2×10^{14}	1.7×10^4	2.1×10^2
Green	530	5.7×10^{14}	1.9×10^4	2.3×10^2
Blue	470	6.4×10^{14}	2.1×10^4	2.5×10^2
Violet	420	7.1×10^{14}	2.4×10^4	2.8×10^4
Near Ultraviolet	400–200	$(7.5 - 15.0) \times 10^{14})$	$(2.5 - 5) \times 10^4$	$(3.0 - 6)\ 10^2$

Our simplified treatment of molecular spectra is based on the approximation that:

(*i*) Only regions of interest for Environmental Chemistry are vibrational and electronic spectra only, rest of the types of spectra will not be dealt with here. For all practical purposes, it is assumed that molecular spectra is made up of vibrational and electronic spectra only. Vibrational spectra is observed in infrared region while electronic spectra is observed in visible and ultraviolet region.

(*ii*) Molecules may be assumed to behave as if only vibrational and electronic energy levels are there. These energies of molecules are independent of each other, *i.e.*, $E_{\text{total}} = E_{\text{vib}} + E_{\text{elec}}$ and our discussion follow that order.

3.3 INTENSITY OF SPECTRAL LINES

The spectral lines are usually presented as a graph in which extinction co-efficient or intensity of absorbance (log I_0/I) are in the ordinate and wavelength or frequency are in absissa. Absorption intensity is dependent on concentration and has been discussed in Section 3.13. Transition of molecules from

$E_1 \rightarrow E_2$ involves absorption/emission of radiation. A typical absorption spectra is shown in Fig. 3.2.

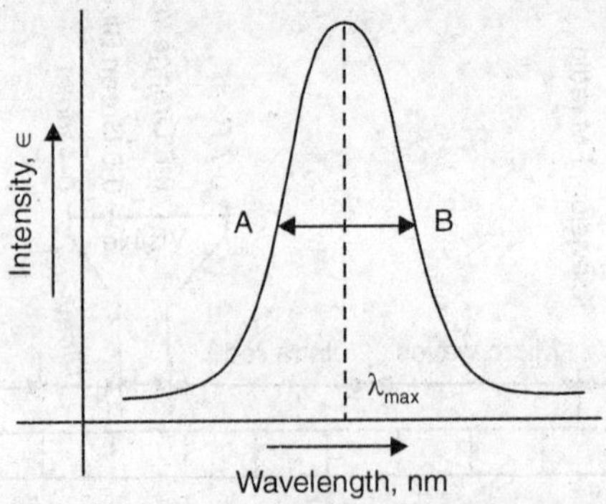

Fig. 3.2. A typical absorption spectra, showing maximum absorption at wavelength λ_{max}. AB corresponds to width of spectra.

There are two factors with which intensity of spectral lines are affected (*i*) Transition probability (*ii*) Population of state.

(*i*) **Transition Probability:** The intensity of spectral lines arising from a molecular transition between a pair of states *i* and *j* with wave function ψ_i and ψ_j representing the states, on which transition probability R_{ij} depends. It is defined as

$$R_{ij} \propto \int \psi_i \; M \; \psi_j \; d\tau \qquad ...(3.5a)$$

Larger the value of R_{ij} more intense be spectral lines. The condition under which the integral fail to vanish for various possible pairs of states of a given system are known as selection rule values. A transition between a pair of energy levels for which R_{ij} vanishes is said to be forbidden transition and appear as weak spectral line. Absorption band with ε_{max} value > 10^3 m^2 mol^{-1} are considered to be of high intensity or strong bands where as those with ε_{max} values < 10^2 m^2 mol^{-1} are known as low intensity or weak bands.

(*ii*) **Population in State:** If transition from 2nd energy level to a third energy level are equally probable, then the most intense spectral line will arise from the level, which has the greater population. Boltzmann's distribution gives this population distribution. If n_i is number of molecules with energy E_i, then the number of molecules, n_j with energy E_j is

$$n_j = n_i e^{-(E_j - E_i)/kT} \qquad ...(3.6)$$

k is Boltzmann constant. Obviously, before molecules are excited from state ψ_i this state must be populated. Larger the population of the state ψ_i more intense will be transition. However certain transitions are not allowed by selection rule and invariably of low intensity though have high population in the ground state.

3.4 WIDTH OF SPECTRAL LINES

Spectral transitions are function of measured position in terms of frequency, wavelength or wave numbers. The spread of spectra over a range of frequency is called width of spectra. The width ($\Delta\nu$) of a spectral band is usually defined in terms of the half band width, that is width of band at half height. It depends on two factors.

Doppler Broadening: Doppler effect shifts the wavelength of light emitted by a moving molecule. If the gas molecules are moving and radiation is viewed in x direction, the wavelength of light emitted by molecule with a component of velocity V_x in x direction is

$$\lambda = \lambda_0 \left(1 \pm \frac{V_x}{c}\right) \quad ...(3.7)$$

The plus sign corresponds to a molecule receding from the observer and minus to a molecule approaching the observer. Molecule move towards, as well as away from observer in a continuous range of velocities, thus even a perfectly monochromatic radiation moving in this way would appear to emit a range of frequencies. This is called Doppler broadening of spectral lines.

Pressure Broadening: The molecules absorbing radiation in a gas undergo collision. In each collision there is certain possibility that a molecule initially in excited state will make a radiationless transition to a lower state, so that the life time of excited state will be decreased. This will add certain frequencies to a monochromatic beam thereby broadening the spectra. The number of collision per second depends on the pressure of gas, and this effect is therefore called Pressure broadening, is also called as *collision broadening.*

3.5 MOLECULAR VIBRATIONS

The bond between two atoms in a diatomic molecule can be viewed as a vibrating spring in which, as the inter-nuclear distance changes from the equilibrium value r_{eq}, the atoms tend to restore them to equilibrium position, Fig. 3.4. In a real diatomic molecule, potential energy function is described by Morse Potential, *i.e.*

$$U(q) = D_e\left(1 - e^{\beta q}\right) \quad ...(3.8)$$

where, $q = | x_2 - x_1 |$ measures the distortion of bond from the equilibrium length, *i.e.*, change in bond length, as shown in Fig. 3.3, and β, a constant $= 1.2177 \times 10^7 \; \nu_{osc}\sqrt{\dfrac{\mu}{D_e}}$ where, μ is the reduced mass of molecule, ν_{osc} the oscillating frequency and D_e the dissociation energy. If one substitutes this potential energy of this oscillator into Schrödinger wave equation, one finds that quantized vibrational energies are given by

$$E_V = h\nu_{vib}\left(V + \frac{1}{2}\right) - h\nu_{vib}\, x_e\left(V + \frac{1}{2}\right)^2 \quad ...(3.9)$$

where V is vibrational quantum number with allowed values of 0, 1, 2, ... ; x_e is the anharmonicity constant, characteristic of molecule. It should be noted that x_e is different from $|x_2 - x_1|$ = change in bond length. In order to obtain the allowed energy level, E_V a plot of $U(q)$ versus $q = |x_2 - x_1|$ change in bond length (Fig. 3.3) gives an asymmetric curve what is called the Morse potential curve. The Morse Potential curve for the NO molecule is shown in Fig. 3.3 with potential energy $U(q)$ in electron volts. It is conventional to express vibration modes and energy levels in cm^{-1} as follow.

$$\frac{E_V}{hc} = G(V) = \bar{\nu}_{vib}\left(V + \frac{1}{2}\right) - \bar{\nu}_{vib}x_e\left(V + \frac{1}{2}\right)^2 \quad ...(3.9a)$$

The selection rule for vibrational spectra is

$$\Delta V = \pm 1, \pm 2, \pm 3$$

The minus sign correspond to vibrational emission spectra and plus to vibrational absorption spectra.

The fundamental absorption band, also called the normal band correspond to $\Delta V = 1$; $V = 0 \rightarrow 1$ transition, *i.e.*

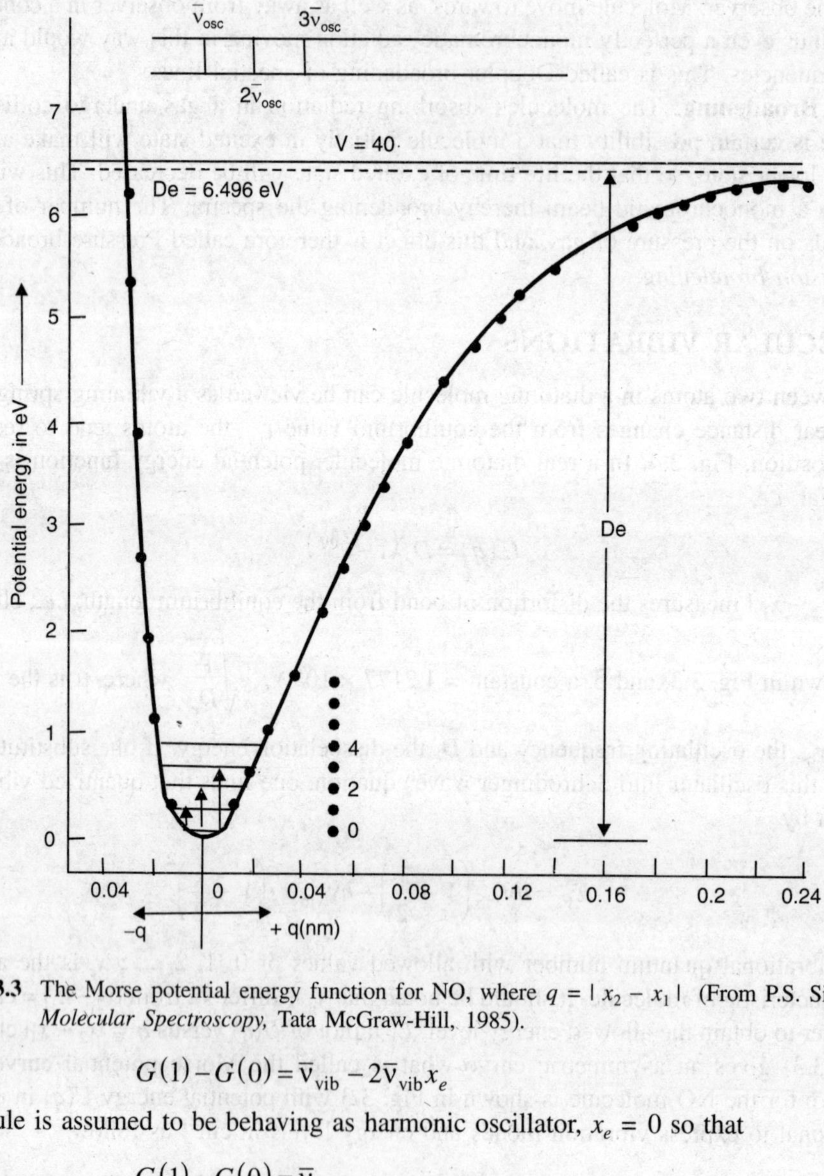

Fig. 3.3 The Morse potential energy function for NO, where $q = | x_2 - x_1 |$ (From P.S. Sindhu: *Molecular Spectroscopy*, Tata McGraw-Hill, 1985).

$$G(1) - G(0) = \bar{\nu}_{vib} - 2\bar{\nu}_{vib} x_e \qquad ...(3.10)$$

If molecule is assumed to be behaving as harmonic oscillator, $x_e = 0$ so that

$$G(1) - G(0) = \bar{\nu}_{vib} \qquad ...(3.11)$$

and this vibrational frequency is given by

$$\bar{\nu}_{vib} = \frac{1}{2\pi c}\sqrt{\frac{k}{\mu}} \qquad ...(3.12)$$

where k = force constant and has a specific value for a particular bond, and μ is reduced mass of the molecule. Thus for NO as a harmonic oscillator, $x_e = 0$, $\bar{\nu}_{vib} = 1904\ cm^{-1}$. This value is a little larger than the value, if it were anharmonic oscillator (Table 3.3). From this value of 1904 cm^{-1}, force constant can be calculated for NO molecule from equation 3.12, with

$$\mu(NO) = 12.495 \times 10^{-27}\ kg$$

$$k = 4\pi^2 \mu \bar{\nu}_{vib}^2 c^2$$

$$= 4\pi^2 2 \times 12.495 \times 10^{-27} \times 1904 \times 1904 \times 3 \times 10^{10} \times 3 \times 10^{10}$$

$$= 1609.4\ Nm^{-1}$$

Bands corresponding to $\Delta V = +2, +3$ and transitions $V = 0 \rightarrow 2$ and $V = 0 \rightarrow 3$ one called first and second overtone given by

$$G(2) - G(0) = 2\bar{\nu}_{vib} - 6\bar{\nu}_{vib} x_e \qquad ...(3.13)$$

$$G(3) - G(0) = 3\bar{\nu}_{vib} - 12\bar{\nu}_{vib} x_e \qquad ...(3.14)$$

Since x_e, the anharmonicity constant is quite small, the difference between first hot band, second hot band and fundamental frequency is quite small but not negligible. This has been shown for NO molecule in Table 3.3. Energy level for NO molecule are given by

$$G(V) = 1904.4\left(V + \frac{1}{2}\right) - 1904.4 \times 7.43 \times 10^{-3} \times \left(V + \frac{1}{2}\right)^3$$

$$= 1904.4\left(V + \frac{1}{2}\right) - 14.187\left(V + \frac{1}{2}\right)^2 \qquad ...(3.15)$$

Table 3.3 Vibrational spectra data for nitric oxide

Energy levels G(V) in cm^{-1}	*G(0) 948.6* *G(1) 2824.7*	*G(2) 4672.7* *G(3) 6492.6*	*G(4) 8284.3*
Vibrational Transition	Fundamental G(1) – G(0)	First hot band G(2) – G(1)	Second hot band G(3) – G(2)
Absorption band in cm^{-1}	1876.1	1847.9	1819.9
Vibrational Transition	Third hot band G(4) – G(3)	First Overtone G(2) – G(0)	Second Overtone G(3) – G(0)
Absorption band in cm^{-1}	1791.7	3724.0	5543.9

Molecules are continuously in a state of vibration and have the vibrational energy levels. However, when electromagnetic radiation interacts with molecules, transition from one vibrational state to another takes place. **In order that such interaction may take place, it is essential that molecular vibration should show a change in dipole moment.** Since homonuclear diatomic molecules are symmetrically stretched or compressed equally on both sides, no change in dipole moment is expected, and thus will not show a vibrational spectra. However, molecules, like NO, CO, are not equally stretched or compressed on both sides, hence a change in dipole moment, and thus show vibrational spectra. This has a vivid consequence. Had the atmosphere been made of N_2, O_2 gases only, it should have been transparent to infrared radiation, the radiation that lead to vibrational transitions. Introduction of CO, NO, CO_2, water vapours as pollutants in air decreases its transparency, and thus Green House effect comes in.

A diatomic molecule may be considered as made up of two point masses m_1 and m_2 joined by chemical bond which behaves as a spring. The only forces are those due to interaction potential, U. The point masses m_1 and m_2 move only along the bond axis with displacement x_1 and x_2, as shown in Fig. 3.4.

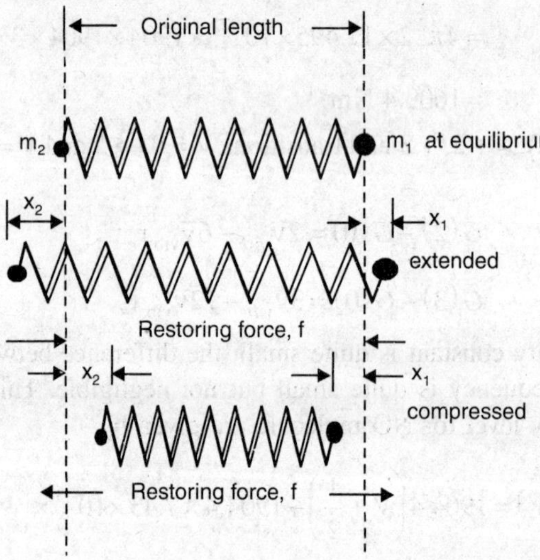

Fig. 3.4 The vibrating system of two particles m_1, m_2 and a massless spring showing the displacement x_1 and x_2 and the restoring force, *f.* (From P.S. Sindhu: *Fundamentals of Molecular Spectroscopy*, New Age International (P) Ltd., 2005).

The two point masses are stretched from equilibrium position by x_1 and x_2 distance when restoring force tries to compress the molecule. The point masses m_1 and m_2 move from stretched state to equilibrium state and then bond gets compressed. The restoring force which acts in opposite direction to the direction in which point masses move to restore the molecule to equilibrium position and then to stretched state. In this way, the movement of atoms from equilibrium position to stretched followed by compressed state, and then again to equilibrium position completes one vibration. During the course of vibration, the force of stretching is not equal to the force for compression. This means for the same force the work done for compression is less than work done for extension. That is why the potential energy diagram (Fig. 3.3) is asymmetric in nature. Every time when the molecule is in equilibrium position, the entire energy of the molecule is the potential energy so that

$$U = E_v$$

or

$$D_e\left(1-e^{-\beta q}\right)^2 = h\nu_{\text{vib}}\left(V+\frac{1}{2}\right) - h\nu_{\text{vib}}x_e\left(V+\frac{1}{2}\right)^2 \quad \text{...(3.16)}$$

This follows that as vibrational quantum number, V increases, $q = |x_2 - x_1|$ goes on increasing and energy levels can be drawn over the potential energy diagram (Fig. 3.3). As V increases, anharmonicity term increasingly make a dominant contribution, for example

$$G(39) = 1904.4 \times 39.5 - 14.187 \times 39.5 \times 39.5$$

$$= 75221.8 - 22135.2 = 53088.5 \text{ cm}^{-1}$$

$$G(39) - G(38) = 797.8 \text{ cm}^{-1}$$

In other words, energy levels are not equally spread but difference decreases with the increase in vibrational quantum number. The absorption band for NO for V = 38 → 39 is 797.8 cm^{-1} which is a decrease by 58.1% of fundamental vibration. These energy levels go on converging till at the dissociation limit this function becomes continuous function. Vibrational energy level as given by Equation 3.9(a)

$$G(V) = \left(V + \frac{1}{2}\right)\bar{\nu}_{\text{vib}} - x_e\bar{\nu}_{\text{vib}}\left(V + \frac{1}{2}\right)^2$$

so that energy difference between successive energy level can be written as

$$\frac{\Delta E}{hc} = G(V+1) - G(V) = \bar{\nu}_{\text{vib}}\left[1 - 2x_e(V+1)\right] \quad \text{...(3.17)}$$

The separation ΔE goes on decreasing with increasing V till a continuum appears when $\Delta E \to 0$ so that

$$V_{\max} = \frac{1}{2x_e} - 1 \quad \text{...(3.18)}$$

This $V_{\max}$ corresponds to distortion to be so large that molecule is not restored to its normal state but dissociates. On substitution this value of $V_{\max}$ after rounding to the nearest number, in Equation 3.8 the energy corresponding to the dissociation of molecule can be calculated. As in NO molecule

$$V_{\max} = \frac{1}{2 \times 7.43 \times 10^{-3}} - 1$$

$$= 66.2 \approx 67$$

substituting this value in Equation 3.13

$G(67)$ = Dissociation Energy of NO

= 639075 cm^{-1}

= 764.05 kJ mole^{-1}

= 7.92 eV

Experimental Value = 52393.3 cm^{-1} = 626.3 kJ mole^{-1}

= 6.49 eV

which is in very good agreement.

3.6 VIBRATIONS IN POLYATOMIC MOLECULES

As we go from diatomic to polyatomic molecule with N atoms, the number of vibrations possible are

$3N - 6$, if molecule is nonlinear and $3N - 5$ if molecule is linear. Of the $3N - 6$ normal modes of vibrations, $(N - 1)$ may be described as bond stretching vibrations. The remaining $(2N - 5)$ modes (in nonlinear molecule), $2N - 4$ modes (in linear molecule) are angle bending vibrations.

The nuclei may be regarded as mass points in a potential field due to the bonding. When atoms are slightly displaced from their equilibrium positions and then released, molecules perform vibrations which in general have complicated forms. Any such vibration is a superposition of a limited number of so called normal vibrations, in general with different amplitudes and phases. Each normal mode is such that in the absence of other normal modes, every nuclei performs simple harmonic oscillations about its equilibrium positions and all these oscillations are in phase. Thus each normal vibration may be regarded as simple harmonic oscillation and normal coordinates so constructed as to express all the individual displacements of the nuclei. Each normal vibration has its own frequency. Two (or three) of these frequencies may be identical when the modes are said to show twofold (or threefold) degeneracy.

At any one time, all the atoms will be vibrating about say their mean position. If we visualize this picture of the vibrations of all the nuclei, it must be very complicated into $3N - 6$ vibrations are imposed on whole molecule. This complicated motion of molecule is resolved into $3N - 6$ vibrations by a process called Normal Coordinate Analysis, and each of these vibration is called normal vibration when every atom performs a simple harmonic motion with same characteristic frequency. All atoms move in phase with one another but with different amplitudes. However those motions which lead to a change of dipole moment interact with electromagnetic radiation to go from one vibrational state to other and thus show an absorption peak in infrared spectra. Those vibrations which do not involve a change of dipole moment do not interact with electromagnetic radiation and thus do not show infrared spectra. However the existence of these vibrations are identified with the help of Raman Spectra.

Water Molecule

As a general rule of Thumb, "Fundamental vibration with highest value is assigned to asymmetric stretching and with minimum value corresponds to bending mode." This rule applies to all types of molecules.

Water has a triatomic symmetrical bent structure with a dipole moment. It should have $3N - 6 = 3 \times 3 - 6 = 3$ fundamental vibrations, two $(N - 1 = 2)$ of which are bond stretching vibrations and $2N - 5 = 1$ as angle bending modes. The water vapours show three sharp absorption peaks in its IR-spectra at 3755, 3650 and 1595 cm^{-1}.

The one with highest value, 3755 cm^{-1} corresponds to asymmetrical stretching and one with lowest value 1595 as bending mode. The third correspond to symmetrical stretching vibrations. These vibrations may be represented as follow.

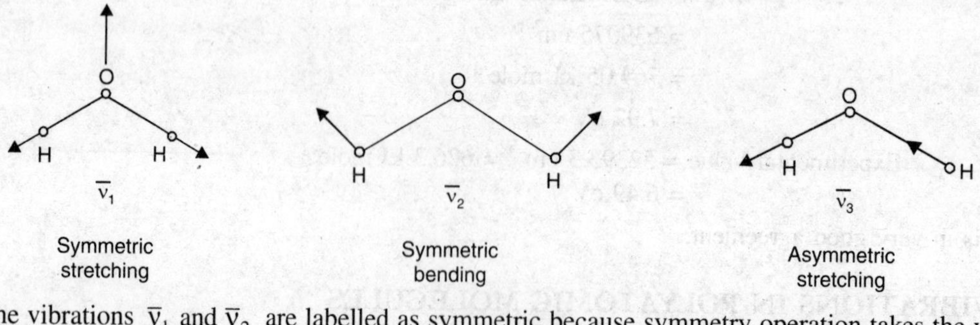

The vibrations $\bar{\nu}_1$ and $\bar{\nu}_2$ are labelled as symmetric because symmetry operation takes the vibrating molecules to an equivalent configuration. Similar reasoning makes $\bar{\nu}_3$ as asymmetric stretching. As all the three vibrations are associated with change of dipole moment, hence they are IR-active. Some other molecules with structure similar to water, and present in air show IR-spectra as given below.

Molecule	$\bar{\nu}_1$	$\bar{\nu}_2$	$\bar{\nu}_3$
NO_2	1320	749	1618
NO_2^-	1328	828	1661
SO_2	1151	518	1360
O_3	1043	710	1740

Carbondioxide Molecule

Carbondioxide is another molecule of interest which has a linear symmetrical structure with zero dipole moment. It should show $3N - 5 = 4$, four fundamental vibrations. Out of these $N - 1 = 2$, two are bond stretching modes and $2N - 4 = 2$ *i.e.*, one double degenerate angle bending modes as shown below.

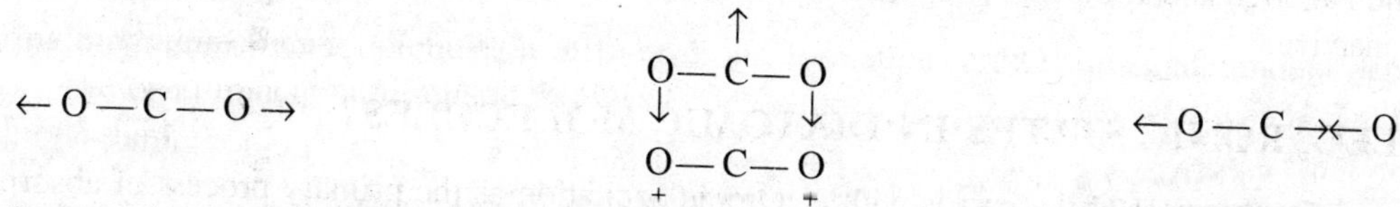

Symmetrical stretching $\bar{\nu}_1 = 1340\ cm^{-1}$ IR-inactive

Bending mode $\bar{\nu}_2 = 667\ cm^{-1}$ IR-active

Asymmetrical stretching $\bar{\nu}_3 = 2349\ cm^{-1}$ IR-active.

These three vibrational modes 1340, 667, 2349 cm^{-1} were assigned to $\bar{\nu}_1, \bar{\nu}_2, \bar{\nu}_3$. As a rule of thumb the highest value, $\bar{\nu}_3 = 2349\ cm^{-1}$ is assigned to asymmetrical stretching and lowest $\bar{\nu}_2 = 667\ cm^{-1}$ to double degenerate bending mode. The symmetrical stretching $\bar{\nu}_1 = 1340\ cm^{-1}$ is IR inactive because both carbon oxygen bonds are extended equally but in opposite direction so dipole moment remains zero. The $\bar{\nu}_3$ is IR active in which one bond is extended and other compressed and vice versa leading to a change in dipole moment. The $\bar{\nu}_2$ vibrations involve angle bending vibration, one in the plane of paper and other in a plane perpendicular to this plane. In this vibration molecule goes from linear to bent structure and thus develop dipole moment and thus IR active. This vibrational mode is double degenerate. Fundamental vibrational frequencies of some other molecules found in atmosphere having similar geometry are given below.

Molecule	$\bar{\nu}_1$ *IR-inactive*	ν_2 *IR-active*	ν_3 *IR-active*
CS_2	658	397	1533
NO_2^+	1400	667	2360

Nitrous Oxide

Nitrous oxide is an asymmetrical linear molecule. It should show four fundamental vibrations as shown below.

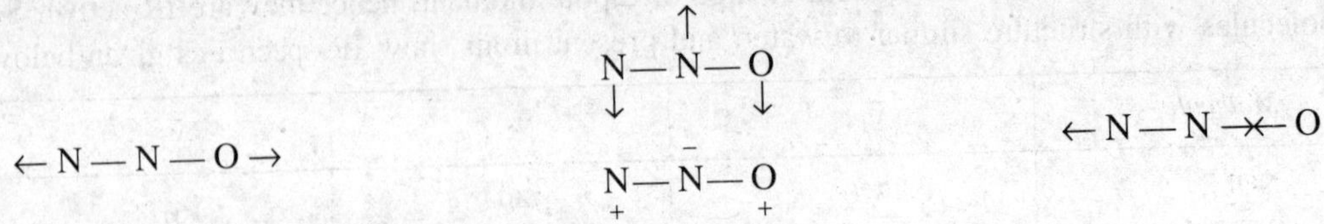

Symmetrical stretching $\bar{\nu}_1 = 1285\ cm^{-1}$

Bending mode $\bar{\nu}_2 = 589\ cm^{-1}$

Asymmetrical stretching $\bar{\nu}_3 = 2224\ cm^{-1}$

All three involve change of dipole moment and thus IR-active.

Methane

In methane $3N - 6 = 9$, nine fundamental vibrations are observed. Out of which only two fundamental absorption bands are observed in infrared region. Both the vibrations $\bar{\nu}_3$ and $\bar{\nu}_4$ are triplydegenerate observation at 3020 and 1306 cm^{-1}. Other three vibrations do not involve change in dipole moment and thus IR-inactive.

3.7 ELECTRONIC STATES IN DIATOMIC MOLECULES

Our concern for the most of the part is with electronic excitation as the primary process of absorption leads to photochemical transformations. Absorption in different spectral regions may lead to different electronic states of the absorbing molecule and electronically different intermediate species, as well as to different products of the photochemical reactions. It is, therefore, important to understand the designation of electronic states of molecules in so far as they relate to possible electronic transitions. Electronic states of atoms have been dealt in Section 3.1. The following is a brief summary of the methods used in arriving at molecular state descriptions, for more details the reader can refer to standard textbooks, of quantum mechanics[4, 5, 6, 7, 8], molecular spectroscopy, and the mathematical formulation of these principles.

For diatomic (or linear polyatomic molecules), it is possible to designate the electronic states by a set of quantum numbers analogous to atoms. Thus, electrons with atomic orbital angular momentum quantum number l can have quantized molecular values λ from 0 to l and $\Lambda = |\Sigma\lambda_i|, \Sigma\lambda_i - 1, |\Sigma\lambda_i - 2 \ldots 0$ and these combine in a manner analogous to polyelectronic atoms to give a quantum number Λ representing the total orbital angular momentum along the inter-nuclear axis. Λ can have values 0, 1, 2, 3, 4... L (where L is the quantum number for the resultant orbital angular momentum for all of the electrons in the molecule), and are designated, respectively, Σ, Π, Δ, Φ, Γ states, analogous to *s, p, d* for the hydrogen atom and to *S, P, D*... for polyelectronic atoms for $\Lambda = 0, 1, 2, 3, 4$. Thus for π electrons $\lambda_1 = 1$, $\lambda_2 = 1$; $\Lambda = 2$, 0 or Σ & Δ states. Similarly, the total molecular spin quantum number is S and the component of S along the inter-nuclear axis is S with possible values S, $(S - 1)$ ...0... $- (S - 1)$, $-S$. The sum of these two quantum numbers is a quantum number analogous to J, $\Omega = (\Lambda + S)$, $(\Lambda + S - 1)$, ... $(\Lambda - S)$, and the total term symbol for a specific diatomic or linear polyatomic electronic state is

$$\text{State of molecule} = {}^{2S+1}\Lambda_{\Omega} \qquad \textbf{Term Symbol}$$

In addition to these quantum numbers Λ, S and Ω there are two other properties of homonuclear diatomic species dealing with the symmetry of its wave function, properties that will be shown later to be important in determining the probability of a molecule absorbing a photon of light, and hence of being excited to a specific electronic state.

(*a*) If inversion of all electrons through a centre of symmetry of the molecule leads to no change in sign of electronic wave function, the state is *even,* or gerade (*g*); if a change of sign results, the state is *odd,* or ungerade (*u*). The symbol *g* or *u* is included in the molecular term symbol as a second subscript to Λ.

(*b*) If reflection in a plane of symmetry passing through the nuclei (containing the inter-nuclear axis of symmetry) leads to no change in sign of the electronic wave function, the state is *positive* (+); if it changes sign, the state is *negative* (–) in a non degenerate state. The symbol + or – is written as a second superscript to Λ. This element of symmetry applies to $\Lambda = 0$ or Σ states.

Molecular oxygen provides a practical, atmospheric, photochemical example for illustrating these state designation factors. The electronic configuration for ground state O_2 (16 electrons) is

$$(1s\sigma_g)^2(1s\sigma_u^*)^2(2s\sigma_g)^2(2s\sigma_u^*)^2(2p_x\sigma_g)^2$$

$$(2p_y\pi_u)^2(2p_zp_u)^2(2p_y\pi_g^*)^1(2p_z\pi_g^*)^1$$

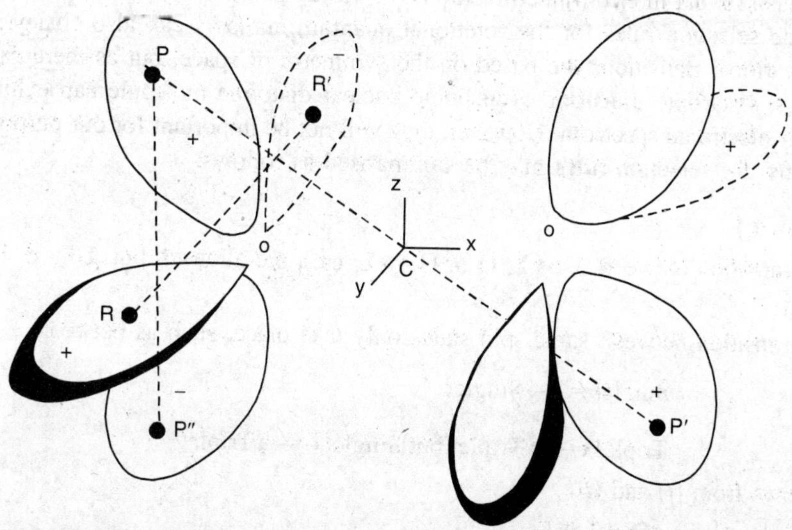

Fig. 3.5 Schematic drawing of the $(2p_y\pi_g^*)$ and $(2p_x\pi_g^*)$ orbitals of O_2. The $(2p_y\pi_g^*)$ orbital is in the *x-y* plane, and the $(2p_z\pi_g^*)$ orbital is on the *x-z* plane. Inversion through the center of symmetry *C* from *P* to *P′* does not change sign, whereas inversion through a plane of symmetry passing through the two nuclei such as the *x-y* plane (from *P* to *P″*) or the *x-z* plane (from *R* to *R′*) does lead to a change in sign.

The (l*s* σ_g), etc. designate molecular orbitals, the asterisk (*) refers to as antibonding orbitals (i.e., the density of electrons between two oxygen atoms in this type of orbital is less than that for the two free atoms, and, therefore, there is a repulsive force between them), and the superscript gives the number of electrons in the molecular orbital. Since there are two unpaired electrons in the equivalent (degenerate) $(2p_y\pi_g^*)$ and $(2p_z\pi_g^*)$ orbitals, by Hund's rule of maximum multiplicity the total spin is

1 and the ground state is a triplet. The quantum number Λ can be 2 or 0; however, if $\Lambda = 2$ both electrons would have the same $\lambda = \pm 1$ value which is forbidden by the Pauli exclusion principle for electrons in equivalent orbitals also with the same spins, and therefore only the $\Lambda = 0$ or Σ state is allowed for $S = 1$. Figure 3.4 shows the wave functions for the two electrons in the $2p_y\pi_g^*$ and $2p_z\pi_g^*$ orbitals for the lowest energy configuration from which it is seen that interchange through the centre of symmetry does not lead to a change in sign, whereas interchange through the plane of symmetry does. The ground state of oxygen is, therefore, *gerade* and *negative* and the term symbols is $^3\Sigma_g^-$.

While the ground electronic state of O_2 is $^3\Sigma_g^-$ state, there are two other possible states (of some what higher energies) for the electrons in these same two equivalent orbitals in which the total spin is zero (multiplicity = 1). In this, either + 1 or −1, leads to $\Lambda = 2$ and a $^1\Delta$ state or different $\lambda = 0$ and hence $a^1\,\Sigma$. From the same symmetry considerations based on Fig. 3.4 that were used to obtain the term symbol for the two electronically excited states are $^1\Delta_g$ and $^1\Sigma_g^+$. On the other hand, if an electron in a lower energy orbital (such as the $p_z\pi_u^*$ state) is excited to the $[(2p_z\pi_u)$ and $(2p_z\pi_g^*)]$ are singly occupied; the Pauli's exclusion principle no longer applies since the two unpaired electrons occupy separate orbitals, and six states are possible $^1\Sigma_u^+$, $^1\Sigma_u^-$, $^1\Delta_u$, $^3\Sigma_u^+$, $^3\Sigma_u^-$ and $^3\Delta_u$.

3.8 ELECTRONIC TRANSITION IN DIATOMIC MOLECULES

It may be emphasised that in electronic transitions the vibrational and rotational quantum numbers can also change. The selection rules for the rotational quantum numbers can also change. The selection rules for the rotational transitions are based on the symmetry of space, but as there can be a shift of charge during an electronic transition even homo-nuclear diatomic molecule can exhibit a rotational structure in their electronic spectrum. However, they will not be important for our purpose. For electric dipole transitions, the selection rules may be summarised as follows:

(*i*) $\Delta\Lambda = 0, \pm 1$

Thus transition between $\Sigma \leftrightarrow \Sigma$, Π or $\Pi \leftrightarrow \Sigma$, etc., are allowed, but $\Sigma \leftrightarrow \Delta$ is not allowed.

(*ii*) $\Delta S = 0$

Thus transition between same spin states only take place, such as between

$$\text{Singlet} \longleftrightarrow \text{Singlet}$$

$$\text{Triplet} \longleftrightarrow \text{Triplet but Singlet} \not\longleftrightarrow \text{Triplet}$$

(*iii*) It follows from (*i*) and (*ii*)

$$\Delta\Omega = 0, \pm 1$$

(*iv*) There are also restrictions on symmetry changes

$$+ \longleftrightarrow + \qquad - \longleftrightarrow - \qquad + \not\longleftrightarrow -$$

$$g \not\longleftrightarrow g \qquad u \not\longleftrightarrow u \qquad g \longleftrightarrow u$$

Such a direction is used with spectroscopic notations, transitions not allowed by the above selection rules, are called forbidden transition but may appear only as very weak absorption bands. Such transitions are result of magnetic dipole transitions, as in O_2 at 7600 Å, due to a $^3\Sigma_g^+ \rightarrow {}^1\Sigma_g^+$ transition or due to quadruple transitions or due to the presence of external fields or collision with other molecules. Table 3.4 summarise the various allowed transitions.

In spectroscopic notation, the ground state of a molecule is always designated *X*, and is taken to be of zero energy. Excited singlet states are designated *A*, *B*, *C*... in order of increasing energy. Excited triplet states are likewise designated *a*, *b*, *c*... in order of increasing energy. According to international agreement, the upper states always comes first and the direction of arrows correspond to spectra as follows:

$\rightarrow$ emission spectra
$\leftarrow$ absorption spectra
$\leftrightarrow$ emission and absorption spectra.

Such a direction is used with spectroscopic notations.

Table 3.4 Allowed electronic transitions of diatomic molecules having light nuclei

Homo nuclear diatomic molecules	*Hetero nuclear diatomic molecules*
$\Sigma_g^+ \leftrightarrow \Sigma_u^+$	$\Sigma^+ \leftrightarrow \Sigma^+$
$\Sigma_g^- \leftrightarrow \Sigma_u^-$	$\Sigma^- \leftrightarrow \Sigma^-$
$\Pi_g \leftrightarrow \Sigma_u^+$	$\Pi \leftrightarrow \Sigma^-$
$\Pi_u \leftrightarrow \Sigma_g^-$	
$\Pi_g \leftrightarrow \Pi_u$	$\Pi \leftrightarrow \Pi$
$\Pi_g \leftrightarrow \Delta_u$	$\Pi \leftrightarrow \Delta$
$\Pi_u \leftrightarrow \Delta_g$	
$\Delta_g \leftrightarrow \Delta_u$	$\Delta \leftrightarrow \Delta$

Some of the low lying electronic transitions of carbon monoxide CO, as representative molecule are given in Table 3.5. It is assumed that ground state of CO is zero. The table shows various transitions, change in angular momentum in terms of term value and in terms of spectroscopic notations.

The ground state of oxygen molecule is $X, {}^3\Sigma_g^-$, a triplet state, as a result only transition to upper triplet states are spin allowed. The transition from the ground state to the $A, {}^3\Sigma_u^-$ state is also not theoretically forbidden because it involves a (+) $\rightarrow$ (–), however this transition

$$X, {}^3\Sigma_g^- \rightarrow A, {}^3\Sigma_u^- \qquad 35007 \text{ cm}^{-1} \text{ (285 nm)}$$

Table 3.5 Electronic spectra of CO molecule

Angular momentum change in term value	*Singlet ↔ Singlet*		*Singlet ↔ Triplet*		*Triplet ↔ Triplet*	
	Spectroscopic notation	*cm⁻¹*	*Spectroscopic notation*	*cm⁻¹*	*Spectroscopic notation*	*cm⁻¹*
$^3\Pi_r \leftrightarrow {}^1\Sigma^+$			$a \leftrightarrow X$	48473		-
$^3\Sigma^+ \leftarrow {}^1\Sigma^+$		-	$a' \leftarrow X$	55355		-
$^3\Sigma^+ \rightarrow {}^3\Pi_r$		-		-	$a' \rightarrow a$	6882.4
$^3\Delta_1 \leftarrow {}^1\Sigma^+$		-	$d \rightarrow X$	60621		-
$^3\Delta_1 \rightarrow {}^3\pi_x$				-	$d \rightarrow a$	12148.7
$^3\Sigma^- \leftarrow {}^1\Sigma^+$		-	$e \leftarrow X$	63705		-
$^3\Sigma^- \rightarrow {}^3\Pi_r$				-	$e \rightarrow a$	15232
$^1\Pi \rightarrow {}^1\Sigma^+$	$A \leftrightarrow X$	64748		-		-
$^1\Sigma^- \rightarrow {}^1\Sigma^+$	$I \leftarrow X$	64546		-		-
$^1\Delta \leftarrow {}^1\Sigma^+$	$D \leftarrow X$	65391		-		-
$^3\Sigma^+ \leftarrow {}^1\Sigma^+$		-	$b \leftrightarrow X$	83831		-
$^3\Sigma^+ \rightarrow {}^3\Pi_r$		-		-	$b \rightarrow a$	35358
$^1\Sigma^+ \leftrightarrow {}^1\Sigma^+$	$B \leftrightarrow X$	86916		-		-
$^1\Sigma^+ \leftrightarrow {}^1\Pi$	$B \rightarrow A$	22171		-		-
$^3\Sigma^+ \leftrightarrow {}^1\Sigma^+$		-	$j \leftarrow X$	90988		-
$^1\Sigma^+ \leftrightarrow {}^1\Sigma^+$	$C \leftrightarrow X$	91919		-		-
$^3\Pi \leftarrow {}^1\Sigma^+$		-	$c \leftarrow X$	92097		-
$^3\Pi \rightarrow {}^3\Pi_r$		-		-	$c \rightarrow a$	43604
$^1\Pi \leftrightarrow {}^1\Sigma^+$	$E \leftarrow X$	92930		-		-
$^1\Pi \leftarrow {}^1\Pi^+$	$A \rightarrow E$	28185		-		-

occurs weakly resulting in weak absorption band ($\sigma \leq 10^{-23}$ cm^2 molecule^{-1}) known as the Herzberg continuum.

The transition

$$X\ ^3\Sigma_g^- \rightarrow B,\ ^3\Sigma_u^- \qquad 49358 \text{ cm}^{-1} \text{ (202 nm)}$$

is allowed shows a banded structure and so correspond to transition from $V'' = 0$ and $V'' = 1$ (*i.e.*, hot band) of the ground state to different vibrational levels of upper state. The upper $^3\Sigma_g^-$ state is crossed by repulsive $^3\Pi_u$ state at $V = 4$ providing a mechanism for the production of two ground state oxygen atoms, O (^{3}P) atoms, from *B*, $^3\Sigma_u^-$ state of oxygen molecule. Triplet–singlet transition are also the forbidden transitions according to the selection rule. The first transition in infrared region, correspond to transition to ground vibrational state of $^1\Delta_g$ state

$$X^3\Sigma_g^- \rightarrow a,\ ^1\Delta_g \qquad (\text{IR light})\ 7882\ \text{cm}^{-1} \quad (1268\ \text{nm})$$

$$X^3\Sigma_g^- \rightarrow b,\ ^1\Sigma_u^+ \qquad (\text{Red light})\ 3120\ \text{cm}^{-1}\ (762\ \text{nm})$$

The absorption spectrum of oxygen becomes continuous at 175 nm with a strong absorption down to 130 nm. This continuum is believed to be due to dissociation of the $B\,^3\Sigma_u^-$ to $O(^3P)+O(^1D)$. Below 130 nm, a banded absorption again appears at a wavelength 133.2 nm to produce O (^1S) atom.

3.9 INTENSITY OF SPECTRAL LINES

Intensities of various lines in electronic transitions are explained with the help of the Franck-Condon principle. It states "The time required for molecular vibrations ($\approx 10^{-12}$ s) is much longer than for that required for an electronic transition ($\approx 10^{-15}$ s) so that the electronic transition in a molecule take place so rapidly in comparison to vibrational motion that immediately afterwards the nuclei still have very much the same relative position and velocity as before the transition. The consequence of it will be illustrated with three typical cases:

$$r_e' = r_e'' \ (\text{Fig. 3.6a}), \qquad r_e' > r_e'' \ (\text{Fig. 3.6b}) \ \text{ and } \ r_e' >> r_e''(\text{Fig. 3.6c})$$

Single dash correspond to excited state and double dash to ground state.

Case I: The potential energy minima (Fig. 3.6a) lies very nearly one above the outer as in the near ultraviolet spectrum of the (O_2) molecule $\left(X \leftarrow a \text{ or } ^1\Delta_g \leftrightarrow {}^3\Sigma_g^-\right)$ transition. If zero point energy is disregarded and the molecule is initially at the minimum of the lower curve, then a transition to the minimum of the upper curve (0—0 band) follows the **Franck-Condon principle**. On the other hand, a transition into a higher vibrational state *CD* would be possible only when, at the moment of the electronic jump, either the nuclear position or velocity or both alter to an appreciable extent. At point *E*, the molecule has kinetic energy *EB*. Only at the turning points, the kinetic energy is same as that in initial state. Thus, such a transition is forbidden or at least improbable. Therefore 0—0 and 1—0 bands show a relative band intensity of 11 : 1 for O_2 molecule with the following values of the equilibrium bond lengths and vibration frequencies for the 0—0 band:

$$r_e' = 0.1216\ \text{nm} \qquad \bar{\nu}' = 1483\ \text{cm}^{-1}$$

$$r_e'' = 0.12075\ \text{nm} \qquad \bar{\nu}'' = 1580\ \text{cm}^{-1}$$

showing that the potential energy curves are indeed similar, at least in the neighbourhood of the minima, $\bar{\nu}''V$ and $\bar{\nu}'V$ are vibrational absorption frequencies in ground and excited states.

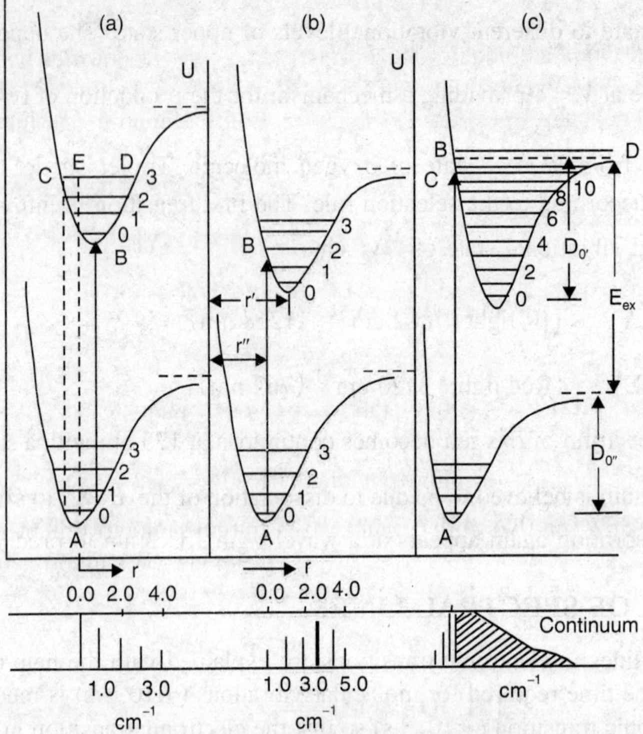

Fig. 3.6 Potential energy curves explaining the intensity distribution in the absorption spectra as stated by Franck-Condon Principle (From P.S. Sindhu: *Fundamentals of Molecular Spectroscopy,* New Age International (P) Ltd., New Delhi, 2005).

Case II: The minimum upper potential energy lies at a moderately greater r_e' value (0.01 – 0.02 nm, Fig. 3.6b) than the lower as in the case of CO spectra ($X \leftrightarrow A$ or $^1\Pi \leftarrow {}^1\Sigma^+$ transition). Therefore the 0—0 band in CO is not the most probable, but the vertical upward transition from *A* to *B* involving no change in the inter-nuclear distance and velocity is 4 – 0 band. The 2 – 0 and 3 – 0 bands are the strongest whose left turning points lie in the neighbourhood of *B*. This band system gives the values

$$r_e = 0.1235 \text{ nm} \qquad \bar{\nu}_e' = 1515.6 \text{ cm}^{-1}$$

$$r_c'' = 0.1128 \text{ nm} \qquad \bar{\nu}_e'' = 2170.2 \text{ cm}^{-1}$$

Case III: The potential energy minima in two curves (Fig. 3.6c) lies at a considerable internuclear distance $r_e' >> r_e''$ corresponding to transition $X \leftrightarrow B$ or $^3\Pi \leftarrow {}^1\Sigma_g^+$ for I_2 molecule. The transition point in the upper curve at *B* lies asymptotic to this curve, therefore corresponding to the continuous spectrum. This leads to dissociation of the molecule without any vibration and since the atoms which are formed may take up any value of kinetic energy, the transitions are not quantized and a continuum results. As in the case of I_2, the 0—0 band is absent in the absorption of spectra. The energy of the 0—0 band is determined by combining absorption spectra with fluorescence spectrum and it gives values:

$$r_e' = 0.3024 \text{ nm} \qquad \bar{\nu}_e' = 126 \text{ cm}^{-1}$$

$$r_e'' = 0.2667 \text{ nm} \qquad \bar{\nu}_e'' = 215 \text{ cm}^{-1}$$

Emission Spectra

In emission spectra, the transition may be from any upper vibrational state to a lower state. The variation of intensity between progressions and successions based upon different V' will depend upon the mode of excitation.

3.10 DISSOCIATION ENERGY AND DISSOCIATION PRODUCTS

Dissociation energy D_0 corresponds to the heat of reaction

$$AB \rightleftarrows A + B$$

It corresponds to the work done to dissociate the molecule from the lowest level (V″ = 0) of the ground electronic state into normal atoms. The spectroscopic determination of D_0 involves two steps: (*a*) a dissociation energy to the upper state products, the dissociation limit, and (*b*) information on the electronic states of the products.

Consider the initial state of the molecule to be in the ground vibration state with the potential energy curve, as given in Fig. 3.6c. The dotted lines represent the dissociation of the normal and excited molecule into atoms, the dissociation energies being D_0'' from $V'' = 0$ and D_0' from $V' = 0$. The total energy of the dissociation products from the upper state is greater by an amount E_x called the excitation energy of product so that

$$\bar{\nu}(\text{continuum}) = D_0'' + E_x \text{ cm}^{-1} \qquad \text{...(3.19)}$$

Thus knowing E, the dissociation energy can be calculated. The ultraviolet absorption bands of oxygen molecule for the transition.

$$X\,^3\Sigma_g^- \rightarrow B\,^3\Sigma_u^- \qquad 202 \text{ nm} = 6.138 \text{ eV}$$

These bands are found to converge to a very clear limit and are followed by a continuous spectrum. The convergence limit lies at 202 nm, corresponding to 6.138 eV.

The low-lying states of oxygen atoms are ^3P ground state and ^1D, ^1S which lie at 1.967 and 4.188 eV above ^3P. Higher states need not be considered. The conceivable combinations, the dissociation products to give the O_2 molecule are : $^3P + {}^3P$, $^3P + {}^1D$, $^3P + {}^1S$, $^1D + {}^1D$, $^1D + {}^1S$, $^1S + {}^1S$. Of these combinations, only $^3P + {}^1D$ and $^3P + {}^1S$ can result to a $^3\Sigma_u^-$ state and thus be considered. The dissociation energy from these combinations is either 6.138 – 1.967 = 4.17 eV or 6.138 – 4.188 = 1.95 eV. The latter value is smaller than the energy of the highest observed vibrational level in the ground state which lies 3.4 eV above the lowest level $V = 0$. Thus there remains only the value $D'' = 4.17$ eV = 402 kJ mol^{-1} with a dissociation product in the ^3P and ^1D states.

Another method is from Morse potential as described in Section 3.5. In many molecules, the internuclear distance in the lower and upper energy states are such that the transition near dissociation is of negligible probability. In such cases vibrational results are used. These energy levels go on converging near the dissociation limit and may be considered as a continuous function of V'' without much error. V_{max} will be given by

$$V_{max} = \frac{1}{2x_e} - 1 \qquad ...(3.18)$$

On substituting this value of V_{max} after rounding to the next lowest number in Equation 3.9, the energy corresponding to the dissociation of a molecule can be calculated. This energy will be equal to the dissociation energy. The O_2^+ molecule in the normal state is a good example of the vibrational levels $\bar{\nu}$ = 1876 cm^{-1} ; $\bar{\nu}\, x_e$ = 16.53 cm^{-1}, so that V_{max} = 56.0. With

$$G(V) = 1876.4\,(V + 1/2) - 16.53(V + 1/2)^2 \text{ cm}^{-1} \text{ for } O_2^+ \text{ molecule}$$

This gives the dissociation energy of O_2^+ = 53248 cm^{-1} = 6.60 eV. The experimental dissociation energy of O_2^+ = 6.48 eV, which is in very good agreement to the calculated value. This method has an added advantage that dissociation energy can be conveniently calculated for the excited state.

3.11 PREDISSOCIATION

Predissociation is a phenomenon involving three electronic states: ground state *X* and two excited states *A* and *B*. If the potential energy curve corresponding to a state *B* to which direct transition is not possible crosses a state *A* to which transition from the ground state *X* is possible, and if at the level of the crossing point the nuclear have sufficient potential energy to dissociate in the state *B*, then a radiationless transition from state *A* to state *B* may result in the dissociation of the molecule. This is known as predissociation and may be used to fix a limit to the dissociation energy.

Consider the NO molecule with potential energy curves *X* and *B* only (Fig. 3.7) of two $^2\Pi$ states of this molecules, the ground state and upper state of the so-called β bands which lies 45500 cm^{-1} above the ground state. Since the heat of dissociation for the ground state amounts to only 42730 cm^{-1} a

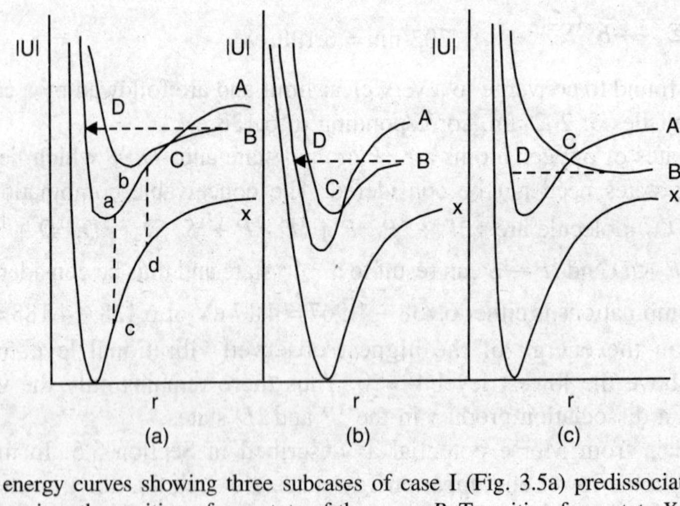

Fig. 3.7 Potential energy curves showing three subcases of case I (Fig. 3.5a) predissociation. In each case of broken line gives the position of asymtote of the curve *B*. Transition from state *X* take place to the left of the curve *A*. (From P.S. Sindhu: *Fundamentals of Molecular Spectroscopy*, New Age International (P) Ltd., New Delhi, 2005).

predissociation of the upper $^2\Pi$ state into the lower in terms of energy possible is for all vibrational and rotational levels of the upper state.

However, no predissociation is observed because it violates the Franck-Condon principle.

If, in the example of NO a radiationless transition to the lower state were to take place during the vibrational motion *ab* in the upper state, the amount of energy *ac* or *bd* or some intermediate amount would have to be supplied instantaneously by the amount *ac* or *bd* or some intermediate amount which is not allowed. Hence, this transition will not lead to the radiationless transition, unless there are three electronic states *X, A, B,* and is in agreement with the observation.

Corresponding to the three forms of energy of the molecule (electronic, vibrational, rotational), three kinds of overlapping of molecular energy levels occur, each leading to three cases of predissociation.

Case I: Overlapping of a certain vibrational or rotational level of an electronic state by the dissociation continuum belonging to another molecular electronic state will lead to radiationless transition into this other dissociated molecule.

Case II: Overlapping of the higher vibrational levels of molecular electronic state by a dissociation continuum joining onto a lower dissociation limit of the same electronic state will lead to radiationless splitting off of the particular atom or group of atoms (predissociation by vibration).

Case III: Overlapping of the higher rotational levels of a given vibrational level of a diatomic molecule by the dissociation continuum belonging to the same electronic state radiationless dissociation without a change of the electron state (predissociation by rotation).

The Franck-Condon principle viewed classically implies that a radiationless transition between two electronic states can only take place if the two potential energy curves intersect, and the wave mechanical viewpoint gives essentially the same conclusions, but the regions in which transitions are possible are broadened. When the predissociations of case I are interpreted in terms of three potential-energy curves (Figs. 3.6 a, b, c), transition by absorption may take place from the ground state *X* to the left hand ascending branch of curve *A*; direct transition to *B* by absorption of radiation does not take place. If a vibrational level of state *A* above *D* is reached, predissociation will take place. The spectrum will show the disappearance or diffuseness of the band structure at this place, and by a weakening or disappearance of the band corresponding to the dissociation energy in the emission spectra. In subcases (*a*) and (*b*), the point of interaction lies on or below the dissociation limit of state *B* (Figs. 3.6 a, b). Thus in these cases the energy equivalent to the beginning of diffuseness in the band structure or to the breaking off in emission, known as the predissociation limit, is the energy required to dissociate the molecule in state *B*. In the third subcase (*c*), the point of intersection lies above the dissociation limit of *B*. Here the predissociation limit does not coincide with the dissociation limit but lies higher.

Predissociation is governed not only by the intersection of the potential energy curves but by the selection rules which specify the types of state between which transition may take place. Since predissociation is nothing but a special case of the perturbations, exactly the same selection rules hold as for perturbations. These selection rules are:

1. Both states must have the same total angular momentum, *i.e.*, $\Delta J = 0$.
2. Both states must have the same spin multiplicity, *i.e.*, $\Delta S = 0$.
3. Both states must be positive or both must be negative, *i.e.*, $+ \not\leftrightarrow -$.
4. The Λ or Ω values as the case may be of the two states may differ only by 0 or ±1, *i.e.*, $\Delta\Lambda$ or $\Delta\Omega = 0, \pm 1$.
5. For identical nuclei, both states must have the same symmetry in the nuclei, *i.e.*, $s \not\leftrightarrow a$.

The selection rules 1, 3 and 5 are perfectly rigorous. The second rule, $\Delta \Lambda = \pm 1$ are called heterogeneous perturbations. Homogeneous perturbations may occur even in an approximation in which the finer interaction of rotation and electronic motion is neglected.

The use of predissociation to determine dissociation energies is complicated by the fact that there are other effects which may cause anomalies in the structure of bands and which may be mistaken for predissociations. When two energy levels of different electronic states nearly coincide, and if transition between them is allowed by the selection rules and the Franck-Condon principle, they interact. The energy levels, are perturbed and effects are produced which may be mistaken for accidental predissociation, if not for a predissociation. Occasionally, it is difficult to decide whether the alleged predissociation is real or not. CO is an example of a molecule where such a question has been controversial. On the other hand, the diatomic phosphorus molecule provides a good example for the determination of a dissociation energy unambiguously by means for predissociation.

3.12 ELECTRONIC STATES OF POLYATOMIC MOLECULES

Electronic spectra analysis of polyatomic molecules is much more complex than that of diatomic molecule. The simplest approach is to use symmetry consideration for its understanding details of various approaches to the designation of various electronic states, and how to interpret the spectra are given in the author's books on molecular spectroscopy. If a symmetry operation is carried out over the wave equation, no change in the energy level is expected. As a consequence, the electronic wave function for the degenerate state transform like the one irreducible representation, IR of the molecule. The IR can only be symmetric or antisymmetric for each of the symmetry operation permitted by the symmetry of the module.

For degenerate states, the wave function may change into a linear combination of the two or more degenerate wave functions. Different wave functions may behave differently with respect to various symmetry operations of a given point group but since in general not all the symmetry elements of a point group are interdependent on one another, only certain combinations of behaviour of the eigen functions with regard to symmetry operations are possible. Such combinations of wave functions, degenerate and transform according to the IR of the point group considered. Each wave function and therefore each electronic state belongs to one of the IR of the point group.

The molecule H_2O, H_2CO_3, SO_2 having the symmetry corresponding to C_{2v} will have electronic states of type A_1, A_2, B_1, B_2, the only IR's of the point group C_{2v}. The coupling of the spin with the orbital motion may lead to splitting of molecular electronic states into $2S + 1$ component where S is the total electronic spin of the molecule. This multiplicity is written as a superscript in front of the symbol. Thus for $S = 0$ gives states ${}^1A_1$, ${}^1B_1$, ${}^1B_2$; $S = \frac{1}{2}$ gives ${}^2A_1$, ${}^2A_2$, ${}^2B_1$, ${}^1B_2$ an so on. Whenever there is an electronic angular momentum, magnetic moment will be present. As all degenerate states of axial molecules have an electronic angular momentum and thus a fairly strong magnetic field that can orient the spin, fairly large multiplet splitting is expected except for light atom molecules.

In non degenerate states there is no electronic angular momentum.

3.13 ABSORPTION OF LIGHT AND PHOTOCHEMISTRY

Photochemistry concerns itself with the study of the effect of radiant energy on chemical reactions, and with the rates and mechanisms by which reactions initiated by light proceed. With photochemical reactions the rate can be controlled by varying the intensity of the light used for irradiation. However, any increase in the rate of photochemical reaction with temperature is due to thermal reaction and not to photochemical reaction. In spontaneous reactions, light acts to speed up the thermal reactions, *i.e.*, it acts as an catalyst. In non-spontaneous reactions, the radiant energy may increase the free energy of the reactants sufficiently to make the ΔG negative. An outstanding example of such a process is photosynthesis. Under the action of sunlight by chlorophyll, the green coloring matter of vegetation, carbon dioxide and water combine in plant complex to form carbohydrate material and oxygen. On removal of the light the products oxidize slowly back to carbon dioxide and water, releasing at the same time the energy accumulated from the sun's radiation. Other examples of such reversals of thermodynamically spontaneous process are photochemical conversion of oxygen to ozone and decomposition of HCl.

3.12a Grottus-Draper Law

The first law of photochemistry, called the Grottus-Draper law, states: "*Only the light which is absorbed by a molecule can be effective in producing photochemical changes in the molecule*". When light falls over a system one of the following three change may happen.

1. The system is perfectly transparent to light of wavelength λ, means this light does not interact with molecules and cannot lead to photochemical reaction.
2. The system is translucent to light of wavelength λ, means this light cannot pass through the medium, rather it is scattered, thus interaction between light and molecules is quite weak, so photochemical reactions may not be possible.
3. The system absorbs the light of wavelength λ, means this light interacts with molecules and thus a photochemical reaction may take place. Therefore, to assess the potential for photochemically induced changes, it is essential to know the absorption spectra of reactants. In order to fully characterise the reaction mechanism, one must also know the absorption spectra of the intermediate reaction products.

3.12b Stark-Einstein Law

A second important law of photochemistry, known as the Stark-Einstein Law, follows directly from the particle behaviour of electromagnetic radiations.

> "*Absorption of radiation is a one photon process. Absorption of one photon excites one atom or molecule is primary (initiating) step and all subsequent physical and chemical reactions follow from this excited species*" *i.e.*,

$$\varepsilon = h\nu = \frac{hc}{\lambda}$$

where $h\nu$ is a quantum of light called photon and h is Planck's constant 6.62×10^{-34} Js. Since chemists generally deal experimentally with moles of material, a convenient unit is a mole of photons defined as 1 einstein. The energy of 1 einstein of wavelength λ in nm is

$$E = 6.02\times10^{23}\ h\nu = \frac{1.196\times10^{5}}{\lambda}\ \text{kJ einstein}^{-1} \quad \text{...(3.5)}$$

if λ is in Angstrom $$E = \frac{1.196\times10^{6}}{\lambda}\ \text{kJ einstein}^{-1}$$

Another unit used in photochemistry to express the energy is electron volt; 1 eV = 96.46 kJ mole^{-1} and λ in nm thus

$$E = \frac{1.24 \times 10^3}{\lambda} \text{ eV}$$

Referring to the dissociation of NO molecule with bond dissociation energy of 6.496 eV correspond to wavelength of 191 nm (Equation 3.5).

$$\lambda = \frac{Nhc}{E} = \frac{1.24 \times 10^3}{6.496} = 190.88 \text{ nm}$$

Thus absorption of one photon of light of wavelength 190.88 nm by an nitric oxide molecule, NO will give it sufficient energy to break a part or dissociate.

However, flash lamps of extremely high intensity or photons from pulsed lasers, simultaneous absorption of two photons is possible. In such situations it is said that this law is not strictly obeyed.

3.12c Beer Lambert Law

The intensity of absorption can be expressed as transmittance (T) which is defined as the ratio of intensity of radiation transmitted I from the sample to that of the radiation incident on the sample (I_0) *i.e.*,

$$T = I/I_0$$

Intensity of absorption is more conveniently expressed in terms of absorbance (A) *i.e.*

$$A = \log (1/T) = \log_{10} I_0/I$$

Absorbance of a band is related to the sample thickness (l) and concentration (c) of absorbing species. The relation is expressed in the form of Beer-Lambert Law as

$$A = \log_{10} I_0/I = \varepsilon cl \qquad ...(3.20)$$

Absorbance is a dimensionless quantity, concentration is usually expressed in mol dm^{-3} and path length in cm, hence ε, extinction coefficient has the units of dm^3 mol^{-1} cm^{-1}. If we use SI units of mol dm^{-3} for concentration and m for path length, the units of ε will be m^2 mol^{-1}. We can obtain the value of ε in m^{2}mol^{-1} from those in dm^3 mol^{-1} cm^{-1} units as follow

$$\varepsilon = \text{dm}^3 \text{ mol}^{-1} \text{ cm}^{-1} = 10^{-3}\text{m}^3(10^{-2}\text{m})^{-1} = 10^{-1} \text{ m}^2 \text{ mol}^{-1}$$

λ_{max} value: The value of wavelength at which absorption maximum occur is called λ_{max}.

ε_{max} value: It is known as molar absorptive or molar extinction coefficient is a measure of intensity of absorption and is characteristic of a particular compound at a given wavelength.

The most common units used in gas phase photochemistry for concentration are number of molecules cm^{-3} symbolised by N with natural logarithms. The gas phase absorption coefficient, designated σ, is then in units of cm^2 molecule^{-1}; σ is also known as absorption cross-section. The Beer-Lamberts Law becomes

$$\ln \frac{I_0}{I} = \sigma N l \qquad ...(3.21)$$

A second widely used set of unit for gaseous species is concentrations as pressure, in atmosphere, so that

$$\ln \frac{I_0}{I} = kpl \qquad ...(3.22)$$

Since pressure depends upon temperature, a reference temperature 270 or 298 K must be specified.

For a mixture of absorbing molecules where each molecule has an extinction coefficient ε at concentration c_i (i = 1, 2, ...) the Beer-Lambert law becomes

$$\frac{I}{I_0} = 10^{-(\varepsilon_1 c_1 + \varepsilon_2 c_2 + \varepsilon_3 c_3 ...)l} \qquad ...(3.23)$$

3.13 QUANTUM YIELD

The second law of photochemistry requires that each quantum that is absorbed brings about a change in one molecule. The overall number of molecules reacting chemically as a consequence of this absorption of a single photon may be virtually any value ranging from zero to very large number (10^6). This latter is interpreted as arising from subsequent chain reactions involving reactive intermediates.

The efficiency of a photochemical reaction is quantum yield, which is the number of molecules undergoing a change per photon absorbed. Generally, the quantum yield relates to a specific species; thus for the generalised reaction.

$$pP + qQ \longrightarrow rR + sS$$

$$\Phi_P = \frac{\text{number of } P \text{ molecules reacted}}{\text{number of photons absorbed}}$$

$$= \frac{\text{number of moles of } P \text{ reacted}}{\text{number of einstein absorbed}} \qquad ...(3.24)$$

The quantum yield Φ represent overall efficiency of the reaction initiated by the absorption of light. It involves changes not only brought about in the initial or primary act of light absorption but also changes occurring in subsequent thermal reactions by molecules, not directly associated with light absorption. In contrast to this, the primary quantum yield Φ is the fraction of the absorbing molecules decomposed in the kinetically simple primary process in which the photon of light is absorbed,

$$P \xrightarrow{h\nu} \text{photo products}$$

$$\Phi = \frac{\text{number of moles of } P \text{ reacted}}{\text{number of einstein absorbed}}$$

$$= \frac{\text{rate of reaction}}{I_a}$$

so that the rate of a primary process is ΦI_a. Since light absorption in the primary process is one for one event, the maximum value for Φ is unity; physical decay may lead to $\Phi < 1$.

3.14 POSSIBLE FATES OF AN ELECTRONICALLY EXCITED MOLECULE

Once a molecule is excited into an electronically excited state by absorption of a photon, it can undergo a number of primary process. Photochemical processes are those in which the excited species dissociates, isomerises, rearranges or react with another molecule. Photophysical processes include radiative transition in which excited molecule emits light in the form fluorescence or phosphorescence and returns to the ground state, intramolecular non radiative transitions in which some of all the energy of the absorbed photons ultimately converted to heat.

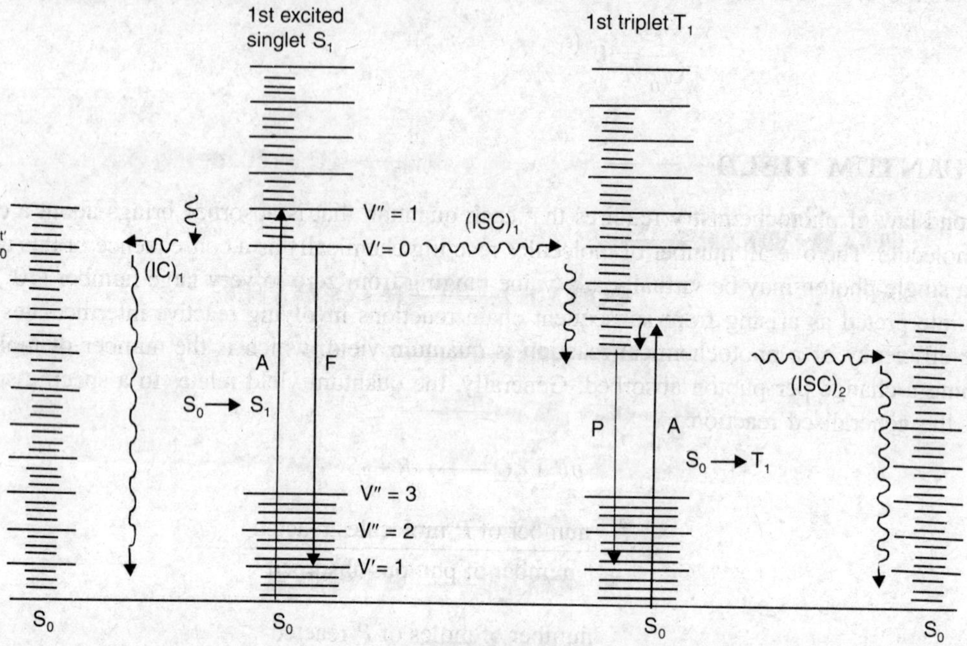

Fig. 3.8. Jablonski diagram illustrating photophysical radiative and non-radiative transitions. S_0 = ground singlet state, S_1 = 1st excited singlet state. T_l = 1st triplet state. A = absorption of light, F = fluorescence, P = phosphorescence, IC = internal conversion. ISC = intersystem crossing. Radiative transitions are shown by solid lines, non-radiative by wavy lines. Photochemical processes are not indicated. (From Barbara J Finlayson Pitts *et al.*: *Atmospheric Chemistry,* John Wiley & Sons, New York, 1986).

These photophysical processes are often displayed in the form of the Jablonski type energy level diagram shown in Fig. 3.8. Common singlet states are labelled as S_0, S_1, S_2 and so on, and triplets are labelled as T_1, T_2, T_3, and so on in order of increasing energy. Vibrational states are shown as being approximately equally spaced horizontal lines. Radiative transitions for example fluorescence (F) and phosphorescence (P) are shown as solid lines and non-radiative transitions as wavy lines. Vertical distances between the vibrational levels of the singlet ground state S_0 and the two electronically excited states, the first excited singlet S_1, and its triplet T_1 corresponding to their energy gaps.

The excited molecule can undergo the following process in addition to direct photoionization.

(*a*) If the excited state is completely repulsive, *i.e.*, there are no attractive forces strong enough to lead to a potential energy minimum, then dissociation into two or molecular fragments, free radicals ions will occur within approximately one vibration ($10^{-12} - 10^{-13}$).

(*b*) Even if the excited electronic state is 'bound' *i.e.* it possesses a potential energy minimum, the absorbed photon may be of sufficient energy to raise the vibrational energy to a dissociative limit. This type of direct photodissociation may also be extremely rapid for diatomic molecules. An example of photodissociation is a spin allowed dissociation of O_3 at $\lambda < 320$ nm

$$O_3(B_2) + h\nu(\lambda < 320 \text{ nm}) \rightarrow O(^1D) + O_2(^1\Lambda_g)$$

Both O_2 and O are in their electronic excited states. With more complex species, the large number itself may make its dissociative lifetime very large and either of the competitive processes, de-excitation or chemical reaction may take place.

(*c*) Fluorescence (F) is defined as emission of light due to a transition between states of like multiplicity, *i.e.*, $S_1 \rightarrow S_0 + h\nu$. The molecule may return to the stable ground state by radiative emission of light from the lowest vibrational level. This is an allowed transition and hence the lifetime of upper state with respect to fluorescence is usually short, typically 10^{-6}–10^{-9} s. However, this emission will be of lower energy than the excitation energy for molecule because of the nonradiative dissipation of vibrational energy in the excited state (F). For example, the fluorescence life of OH in the electronically excited $A\ ^2\Sigma^+$ state is 0.7 μs.

(*d*) Phosphorescence (P) is defined as emission light due to a transition between states of different spin multiplicities. Because this is not an allowed transition for an ideal unperturbed molecule, phosphorescence lifetimes tend to be relatively long 10^{-3} – 10^{-2} s.

(*e*) Inter-system crossing (ISC) is the intramolecular crossing from one state to another of different multiplicity without emission of radiation. In Fig. 3.8 (ISC), shows the transfer from first excited singlet state S_1 to the first excited triplet state. Since the process is horizontal, the total energy remains the same and the molecule initially produced in upper vibrational level of T_1 from which it is deactivated as shown by the vertical wavy line. Since this is a non-radiative process, it is not limited by the $\Delta S = 0$. The resulting new excitation can no longer undergo a rapid radiative decay to ground state because it is forbidden. Similarly, $(ISC)_2$ shows the intersystems crossing from T_1 to upper vibrational states of the ground state, S_0 from which vibrational deactivation to $V'' = 0$ then occurs.

(*f*) Internal conversion (IC) process is the intramolecular crossing of an excited molecule from one state to another of the same multiplicity without the emission of radiation. As seen in Fig. 3.8, the horizontal wavy line (IC) represent the internal conversion of $S_1 \rightarrow S_0$ this is generally followed by vibrational deactivation to $V'' = 0$ but no chemical change. Such process is quite rare but found in aromatic molecules and cynanine dyes.

(*g*) The excited (donor *D*) molecule may be deactivated by radiationless energy transfer to another (acceptor *A*) molecule. Because of the forbidden character of triplet state in phosphorescence, triplet state is long lived than the excited singlet state. It can, therefore, readily undergo collisional energy transfer at normal encounter of distances by (the spin allowed reaction.

$$D \text{ (Triplet)} + A \text{ (Singlet)} \longrightarrow D \text{ (Singlet)} + A \text{ (Triplet)}$$

provided the transfer is exothermic ($\Delta E_{acceptor} < \Delta E_{donor}$) or at least less than approximately 10 kJ endothermic. Triplet states are quenched in most fluid solutions by this type of energy transfer to dissolved oxygen and solvent impurities, even in the absence of added quenchers.

The relative efficiency of various photophysical and photochemical primary process are described in terms of quantum yield. For example, HCHO absorbs light from 250 to 370 nm decomposing either by a free radical process or by intramolecular process producing stable molecules.

$$HCHO + h\nu \xrightarrow{a} \dot{H} + H\dot{C}O$$

$$\xrightarrow{b} H_2 + CO$$

Reaction (*a*) is a significant source of HO_2 and, hence, ultimately OH in troposphere because both H atom and HCO radical react with oxygen in air to form HO_2. The overall quantum yield of H_2 and CO in the reaction is defined as

$$\Phi_{H_2} = \frac{\text{number of molecules of } H_2 \text{ formed}}{\text{number of photons absorbed by HCHO}}$$

and

$$\Phi_{CO} = \frac{\text{number of molecules of CO formed}}{\text{number of photons absorbed by HCHO}}$$

Photolysis of pure HCHO vapours does not produce H_2 and CO directly from the primary molecular decomposition, *b*, but also from the subsequent secondary reactions of the reactive species HCHO formed in a reaction.

$$H + HCO \longrightarrow H_2 + CO$$

$$2H \xrightarrow{M} H_2$$

$$2HCO \xrightarrow[a]{} 2CO + H_2$$

$$\xrightarrow[b]{} CO + HCHO$$

The magnitude of quantum yield gives valuable indication of the relative importance of photophysical and photochemical processes and in some cases of the mechanism of a photochemical reaction. For example small quantum yield $\Phi < 1$ indicates photophysical process must be important. On the other hand quantum yield greater than one ($\Phi > 1$) indicates that a chain reaction is occuring. For example, in the photolysis of a mixture of Cl_2 and H_2, Φ can be as high as 10^6.

REFERENCES

1. Calvert, J.G. and Pitts J.N., *Photochemistry,* Wiley, New York (1966).
2. Clayton, R.K., *Light and Living Matter,* The Physical Part Vol. 1, The Biology Part Vol. 2. McGraw-Hill, New York (1971).
3. Herzberg, G., *Atomic Spectra and Atomic Structure,* 2nd Ed., Dover, New York (1944).
4. Sindhu, P.S., *Molecular Spectroscopy,* Tata McGraw-Hill, New Delhi (1985).
5. Sindhu, P.S., *Fundamentals of Molecular Spectroscopy*, New Age International (P) Ltd., New Delhi (2005).
6. Herzberg, G., *Spectra of Diatomic Molecules,* Van Nostrand, Princeton, New Jersey (1944).
7. Herzberg, G., *Electronic Spectra and Electronic Structure of Polyatomic Molecules,* Van Nostrand, Princeton, New Jersey (1966).
8. Simsons, J.P., *Photochemistry and Spectroscopy,* Wiley, London (1971).
9. Wells, C.H. J., *Photochemistry,* Chapman and Halls, London (1972).
10. Barbara, J. Finlays on Pitts, James N. Pitts (Jr), *Atmospheric Chemistry,* John Wiley & Sons, New York (1986).
11. Sindhu, P.S., *Elements of Molecular Spectroscopy,* New Age International (P) Ltd., New Delhi (2007).

4 Photochemistry of Atmosphere

4.1 INTRODUCTION

The source of energy of all chemical reactions in atmosphere is sunlight which excite many stable molecules to form radicals/molecules in their excited states. These products initiate many reactions, which most of time are not photochemical reactions. At a particular latitude, maximum time for which sunlight is available is 10 to 12 hours. In a clean atmosphere, sunlight initiate many reaction which reach to completion or equilibrium state in the night. The initiator molecules reach an equilibrium concentration or are eliminated. So that the next day sunlight and atmosphere starts a fresh. In this way, a cyclic process in favour of life on earth had been in existence for quite long time. However introduction of various new gaseous products into atmosphere, due to industrialisation and change in our life style produce new radicals/molecules which have been producing new reaction products, thereby breaking and changing the cycle that existed previously. Pollution of the atmosphere may go against survival of life on earth. For example, emission of nitrogen oxides, hydrocarbons by internal combustion engine drastically affects the quality of our environment contributing to photochemical processes which were not there previously.

The total intensity of sunlight outside the earth's atmosphere is characterised by solar constant defined as, the total amount of light received per unit area, normal to the direction of propagation of light, the mean value is 1372 W m^{-2}. Outside the atmosphere, the solar flux approximates black body emission at 6000 K. However, light absorption or scattering by atmospheric constituents modifies the spectral distribution. The presence of atomic and molecular oxygen, nitrogen and ozone as the absorbing species in the upper atmosphere modifies the spectral distribution so much that only wavelength of $\lambda > 290$ nm is available for photochemical reactions in troposphere. The intensity and wavelength distribution of sunlight at the earth's surface depends on both the nature and concentration of gases and path length through which the light passes (Beer-Lambert's Law). The path length, that is, distance from the outer end of atmospheric to an observer on the earth's surface is a function of the angle of the earth and hence time of day, latitude and season. The presence of clouds and reflection of light from earth's surface also alters the light intensity from time to time. All effects of scattering and absorption of light by gas molecules and particulate matter make it much more complex. Consequently these effects will not be taken up here. For this, one may refer to classic monograph of Leighton[3]. More recently, Peterson[6] recalculated these solar fluxes from 290 to 700 run using radiative transfer model developed by Dave[1]. Demergian *et al.*[2] then applied them to the photolysis of some important species. Pitts and Pitts[5] book on atmospheric chemistry is the latest edition in the series.

REACTIONS IN ATMOSPHERE

4.2 OXYGEN, O_2

The most abundant photochemical reaction in atmosphere, is the photolysis of oxygen molecule. Figure 4.1 shows the potential energy curves for the some of the important electronic states of O_2. Table 4.1 gives some of the electronic states of O_2 molecules and various observed electronic transitions in those states. Oxygen is a diatomic paramagnetic molecule with bond dissociation energy of 5.1 eV

Table 4.1 Electronic spectra of O_2 molecule

Angular momentum change in term value	*Triplet ↔ Triplet*		*Triplet ↔ Singlet*		*Singlet ↔ Singlet*	
	Spectroscopic notation	*cm⁻¹*	*Spectroscopic notation*	*cm⁻¹*	*Spectroscopic notation*	*cm⁻¹*
$^1\Delta_g \leftrightarrow {}^3\Sigma_g^-$	–		$a \leftrightarrow X$	7882	–	
$^1\Sigma_g^+ \leftrightarrow {}^3\Sigma_g^-$	–		$b \leftrightarrow X$	13121	–	
$^1\Sigma_g^- \to {}^1\Delta_g$	–		–		$b \to a$	5238.5
$^1\Sigma_u^- \leftrightarrow {}^3\Sigma_g^-$	–		$c \leftrightarrow X$	32664	–	
$^1\Sigma_u^- \to {}^1\Delta_g$	–		–		$c \to a$	24782
$^3\Delta_u \leftrightarrow {}^3\Sigma_g^-$	$A' \leftarrow X$	34320	–		–	
$^3\Delta_4 \to {}^1\Delta_g$	–		$A' \to a$	26440		
$^3\Sigma_u^+ \leftrightarrow {}^3\Sigma_g^-$	$A \leftrightarrow X$	35007	–		–	
$^3\Sigma_u^+ \to {}^1\Sigma_g^+$			$A \to b$	22886		
$^3\Sigma_u^+ \to {}^1\Delta_g$			$A \to a$	27125		
$^3\Sigma_u^- \leftrightarrow {}^3\Sigma_g^-$	$B \leftrightarrow X$ Schumann Runge.	49358	–		–	
$^3\Pi_g \leftarrow {}^3\Sigma_g^-$	$C \leftarrow X$	65670	–		–	
$^1\Pi_g \leftarrow {}^3\Sigma_g^-$	–		–		$d \leftarrow X$	64320
$^3\Sigma_u^- \leftarrow {}^3\Sigma_g^-$	$D \leftarrow X$	75450	–		–	

corresponding to a photon of wavelength 243 nm. Oxygen begins to absorb just below this wavelength in a spectral region shown as Herzberg continuum in the region 200 to 300 nm. Absorption is weak ($\sigma < 10^{-3}$ cm^2 molecule^{-1}) at 200 nm, but the fact that the spectrum is continuum in this region indicates that dissociation occurs with quantum yield = 1. Actually the excitation is to the weakly bound $^3\Sigma_u^+$ state which readily dissociates into two ground state $^3P_2$ oxygen atoms.

$$O_2\left(X,\ ^3\Sigma_g^-\right) \xrightarrow[\lambda=200\text{ nm}]{h\nu} O_2\left(A,\ ^3\Sigma_u^+\right) \longrightarrow 2O\left(^3P_2\right) \quad ...(4.1)$$

This transition violates the selection rule because the $- \rightarrow +$ is a forbidden transition consequently show weak absorption spectra.

$$O_2\left(X,\ ^3\Sigma_g^-\right) \xrightarrow[\lambda=130-200\text{ nm}]{h\nu} O_2\left(B,\ ^3\Sigma_u^-\right) \quad ...(4.2)$$

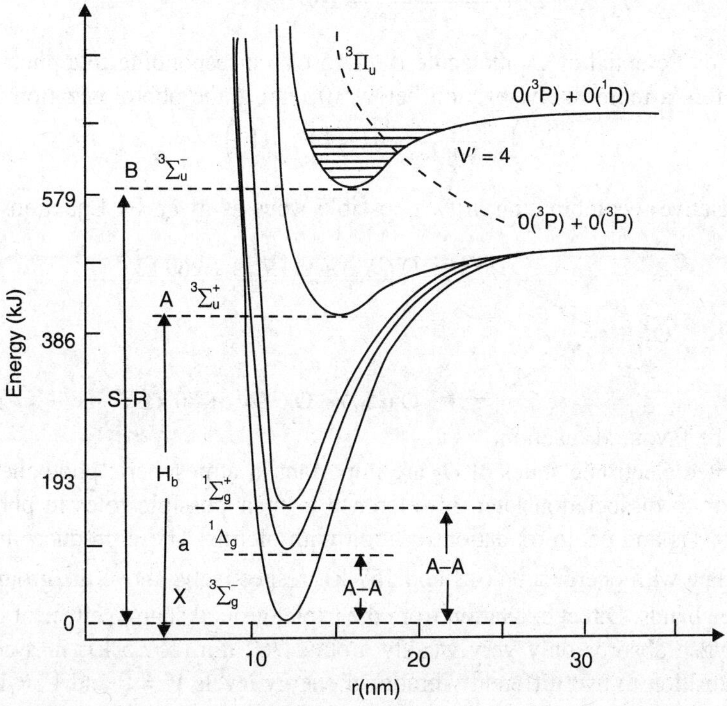

Fig. 4.1 Potential energy curves for ground and first four excited states of O_2 : S–R = Schumann-Runge system, H_b = Herzberg continuum, A–A = atmospheric bands. (From Gaydon, 10).

The transition (Eq. 4.2) is allowed transition and results in absorption in the 130–200 nm region known as Schumaan Runge-range. A banded structure is seen from about 175–200 nm which corresponds to transition from $V'' = 0$ and $V'' = 1$ of ground state to different vibrational levels of the upper excited electronic state. These bands get closer together so the wavelength decreases, and at 176 nm the bands converge into Schumann-Runge range which reaches maximum absorption at approximately 147 nm ($\sigma = 0.00203 \times 10^{-20}$ cm^2 molecule^{-1}). Excitation in this region is to the bound state $^3\Sigma_u^-$, an allowed transition, hence large absorption coefficient. The $^3\Sigma_u^-$ is crossed by the repulsive state Π_u state

(Fig. 4.1) $V' = 4$ without any potential minima providing a mechanism for the production of one excited and one ground state oxygen atoms as

$$O_2\left(X,\ ^3\Sigma_g^-\right) \xrightarrow{147\ \text{nm}} O_2\left(B,\ ^3\Sigma_u^-\right) \longrightarrow O\left(^3P_2\right) + O\left(^1D_2\right) \qquad \text{...(4.3)}$$

Around 132 nm, oxygen molecule goes to $^3\Sigma_u^+$ state which has sufficient energy to produce oxygen atoms in even higher electronic states, and some dissociate to the $^1S_0$ state

$$O_2\left(X,\ ^3\Sigma_g^-\right) \xrightarrow[\lambda<130\ \text{nm}]{h\nu} O_2\left(D,\ ^3\Sigma_u^-\right) \longrightarrow O\left(^3P_2\right) + O\left(^1S_0\right) \qquad \text{...(4.4)}$$

Betweeen 125–100 nm additional discrete and diffuse bands are observed but have been partly assigned and may belong to Rydberg series converging to the first ionization potential.

Below 92.3 nm, two excited atoms are produced

$$O_2\left(X,\ ^3\Sigma_g^-\right) \xrightarrow[\lambda<92.3\ \text{nm}]{h\nu} 2O\left(^1S_0\right) \qquad \text{...(4.5)}$$

As the Ionization Potential of O_2 molecule is 12.15 eV corresponding to a photon wavelength λ = 102 nm, and therefore a more likely reaction below 102 nm is the photoionization

$$O_2\left(X,\ ^3\Sigma_g^-\right) \xrightarrow[\lambda<102\ \text{nm}]{h\nu} O_2^+\left(X,\ ^2\Pi_g\right) + e \qquad \text{...(4.6)}$$

followed by dissociative recombination in two possible ways as given by Equations 4.7 and 4.8.

$$O_2^+ + e \longrightarrow O\,(^1S_0) + O\,(^3P_2) + 269\ \text{kJ} \qquad \text{...(4.7)}$$

$$O_2^+ + e \longrightarrow O\,(^1S_0) + O\,(^1D_2) + 80\ \text{kJ} \qquad \text{...(4.8)}$$

with 4.7 found to be favoured reaction.

Two other excited electronic states of O_2 are important in atmospheric photochemistry because of the specific reaction to dissociation later, in reference to their possible roles in photochemical smog cycle, ozone photolysis and photo oxidation reaction with olefins. These produce singlet excited state $a,\ ^1\Delta_g$ and $b,\ ^1\Sigma_g^+$ state with energies 94.1 kJ and 157 kJ, respectively, above the ground state known as atmospheric oxygen bands. Direct excitation from the triplet ground state to either of these states is spin forbidden, and oxygen absorbs only very weakly around 762 nm (157.5 kJ) in a completely banded structure due to transition to two different vibrational energy levels $V' = 0$ and $V' = 1$ within the $a,\ ^1\Delta_g$ state. The reverse radiative process is of course also spin-forbidden and these states, when produced are therefore relatively stable in the atmosphere, a factor contributing to their importance; the radiative life-time of the $^1\Sigma_g^+$ is 12 seconds while that of $^1\Delta_g$ is 65 seconds. Furthermore, collisional quenching is also inefficient, rate constant of deactivation by ground state O_2 being 5×10^{-16} and 2×10^{-18} cm^3 mole-cule^{-1}s^{-1} for the $^1\Sigma_g^+$ and $^1\Delta_g$ states respectively. In the absence of collisional deactivation, the emission is slow from this, normally forbidden radiative transitions

$$O_2\left(b,\ ^1\Sigma_g^+\right) \longrightarrow O_2\left(X,\ ^3\Sigma_g^-\right) + h\nu\ \left(13121\ \text{cm}^{-1}\right) \qquad \text{...(4.9)}$$

The 762 nm band is due to $b,\ ^1\Sigma_g^+$ state O_2 which is often observed in system containing the $a,\ ^1\Delta_g$ state because of the energy pooling of reaction.

$$O_2\left(a,{}^1\Delta_g^+\right)+O_2\left(a,{}^1\Delta_g^+\right)\longrightarrow O_2\left(X,{}^3\Sigma_g^-\right)+O_2\left(b,{}^1\Sigma_g^+\right) \quad ...(4.10)$$

$$O_2\left(a,{}^1\Delta_g^+\right)\longrightarrow O_2\left(X,{}^3\Sigma_g^-\right)+h\nu \ \left(7882\ \text{cm}^{-1}\right) \quad ...(4.11)$$

The radiation emitted by reaction 4.9 + 4.11 lie in the visible region and is the cause of *Night Glow* observed at night in the sky.

Ground state, ${}^3P_2$ oxygen atoms might be expected to combine directly to form O_2. However, simultaneous conversion of energy and momentum required the presence of a third body in the non-radiation process.

$$2O\left({}^3P_2\right)+M\longrightarrow O_2\left(X,{}^3\Sigma_g^-\right)+M \quad ...(4.12)$$

where M can be another oxygen atom or molecular nitrogen molecule. Reaction 4.12 is considered kinetically simple and trimolecular:

$$\text{Rate}=k_{412}[O]^2[M] \qquad k_{412}=3\times10^{-33}\,\text{cm}^6\text{molecule}^{-2}\text{s}^{-1}$$

The rate constant is very nearly temperature independent. In fact, it has a slightly inverse temperature behaviour, the rate decreases with increasing temperature implying a negative activation energy which, of course, is impossible physically if activation energy is considered to be the height of a potential barrier. The reason for the negative activation energy is that oxygen atoms form an unstable intermediate complex with M

$$O+M \rightleftharpoons OM$$

$$OM+O\longrightarrow O_2+M \quad ...(4.13)$$

As the temperature is raised, OM becomes more unstable so that its concentration is decreased, thereby decreasing the rate of reaction 4.12, assuming equilibrium between O, M and OM leads to the same rate law as that for the trimolecular reaction. For reaction 4.12; reaction rate constant is of the order of magnitude of collisional frequency for triple collisions, indicating that recombination occurs essentially at every encounter. Even so, at the very low pressures encountered in the upper atmosphere (10^{-6} atm at 100 km), the rate of reaction (4.12) is small as compared to rate of photodissociation and therefore above 100 km the primary oxygen species present is atomic oxygen.

It has been seen that excited singlet (${}^1D_2$ and ${}^1S_0$) atoms are formed from the vacuum UV photodissociation of molecular oxygen. These species also relax to triplet ground state oxygen with the emission of light

$$O\left({}^1D_2\right)\longrightarrow O\left({}^3P_2\right)+h\nu \ \left(15873.0\ \text{cm}^{-1}\right)$$

$$O\left({}^1S_0\right)\longrightarrow O\left({}^3P_2\right)+h\nu \ \left(17310\ \text{cm}^{-1}\right)$$

Again both are forbidden transitions and, therefore, show only weak emission contributing to the air glow and aurora. Below 140 km, however, collisional deactivation with any atomic or molecular species M may become the dominant reaction, particularly with O (${}^1D_2$)

$$O\left({}^1D_2\right)+M\longrightarrow O\left({}^3P_2\right)+M \quad ...(4.14)$$

$$k_{414} = 2.9 \times 10^{-11} \text{ cm}^3 \text{ molecule}^{-1} \text{ s}^{-1}$$

If $M = O_2$ excitation is to the ${}^1\Sigma_g^+$ state which decays collisionally to the state ${}^1\Delta_g$ state

$$O(^1D_2) + O_2(X, \Sigma_g^-) \longrightarrow O(^3P_2) + O_2(b, {}^1\Sigma_g^+) \xrightarrow{M} O_2(a, {}^1\Delta_g) \quad \text{...(4.15)}$$

An important aspect of this energy transfer is that spin is conserved, that is, oxygen atom undergoes a singlet-triplet transition while the reverse triplet-singlet reaction occurs with molecular oxygen, so that total reactant and product spin are the same. This is a very efficient process under exothermic conditions, probably occurring within an order of magnitude of collisional frequency, and therefore reaction 4.15 is one of the major sources of singlet oxygen $(a, {}^1\Delta_g)$ in the atmosphere.

4.3 OZONE, O_3

Ozone forms only 10^{-5} % of entire atmosphere but plays an extremely important role to the life on earth. The extremely small amount of O_3 causes the absorption of solar radiation with $\lambda < 300$ nm, which would have harmful effect on living organism. About 100 km above earths surface, three body recombination reaction 4.12 becomes important because of high concentration of oxygen atoms at this atmospheric height. However, another three body reaction possible is

$$O(^3P_2) + O_2(X, {}^3\Sigma_g^-) + M \longrightarrow O_3 + M \quad \text{...(4.16)}$$

$$k_{416} = 5.6 \times 10^{-34} \text{ cm}^6 \text{ molecule}^{-2}\text{s}^{-1}$$

The rate constant is considerable less than that of the atom combination. This reaction is the sole source of ozone in the stratosphere and mesosphere as well as being a major source of destruction of oxygen atoms. Ozone is one of the most amazing important and remarkable minor constituents in our atmosphere against the damaging UV radiation, which it absorbs. In this way it becomes the main energy reservoir in the upper atmosphere and hence a factor in climate regulation. On the other hand, ozone is one of the most toxic inorganic chemicals known, and its presence at relatively high concentrations in lower atmospheric polluted air makes it a potentially dangerous substance for man and organisms.

The bond dissociation energy of Ozone is 1.05 eV (101.64 kJ mol^{-1}) corresponding to a photon of wavelength λ = 1180 nm and the quantum yield of dissociation is unity at all wavelength below 1180 nm. Actually, this value is due to predissociation obtained from thermochemical data. However, ozone absorbs only weakly in the visible region (Chappuis bands) leading to ground state species

$$O_3 \xrightarrow{h\nu} O_2(X, {}^3\Sigma_g^-) + O(^3P_2) \quad \text{...(4.17)}$$

Reactions 4.3, 4.4, 4.16 and 4.17 together lead to formation of Ozone. Reaction 4.17, however, does not destroy ozone as it recombines immediately according to Equation 4.16. If there were no other reactions than these four reactions, it can be conceived that eventually all atmospheric oxygen would had been converted into ozone and whole life would have been wiped out.

The wavelength of maximum absorption in visible region is at 600 nm with $\sigma = 24.4 \times 10^{-20}$ cm^2

molecule^{-1}. The major absorption of ozone is in the ultraviolet (Hartley Continuum) commencing at 334 nm with λ_{max} = 254 nm; σ = 568.4 × 10^{-20} cm^2 molecule^{-1}, as shown in Fig. 4.2.

This ultraviolet photolysis is quite complex involving extensive secondary reactions. At $\lambda \geq 320$ nm, the predominant dissociative reactions is

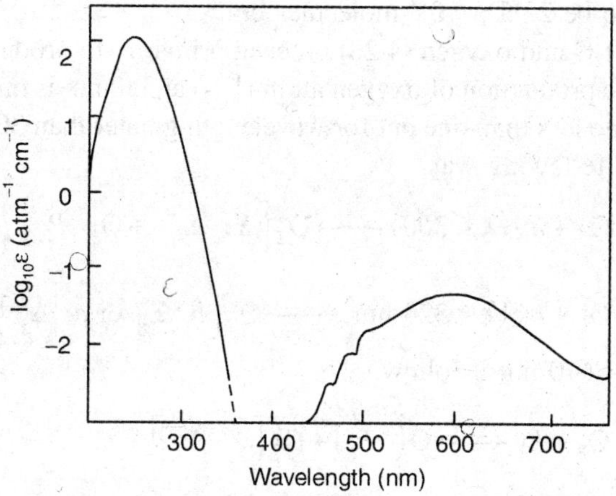

Fig. 4.2 Absorption spectrum of O_3. Note logarithmic scale for the extinction coefficient.

$$O_3 \xrightarrow[\lambda \geq 320]{h\nu} O(^3P_2) + O_2(a,\ ^1\Delta_g \text{ or } b,\ ^1\Sigma_g^+) \quad ...(4.18)$$

but below 320 nm it is

$$O_3 \xrightarrow{h\nu} O(^1D_2) + O_2(a,\ ^1\Delta_g) \quad ...(4.19)$$

The most important aspect of O_3 photochemistry for troposphere is the yield and wavelength dependence in the 4.19 reaction since it is a source of OH radical via its reaction with water

$$O(^1D_2) + H_2O \longrightarrow 2\dot{O}H \quad ...(4.20)$$

$$k_{420} = 2.2 \times 10^{-10}\,\text{cm}^3\text{molecule}^{-1}\text{s}^{-1}$$

This very fast reaction occurs in competition with the deactivation of O ($^1D_2$) by air

$$O(^1D_2) + M \longrightarrow O(^3P_2) + M \quad ...(4.21)$$

$$k_{414} = 2.9 \times 10^{-11}\ \text{cm}^3\ \text{molecule}^{-1}\ \text{s}^{-1}$$

In atmosphere of air at 50% relative humidity and at 298 K, approximately 10% of the O ($^1D_2$) produced by 4.19, reacts with water to form OH radical. Thus the rate equation for 4.20 and 4.21 reactions

$$\frac{-d[O(^1D_2)]}{dt} = k_{420}[H_2O][O(^1D_2)] = k_{414}[M][O(^1D_2)] \quad ...(4.22)$$

Using reaction rate constants k_{420} and k_{414} from 4.20 and 4.21 one obtains

$$\frac{k_{421}}{k_{420}} = \frac{[H_2O]}{[M]} = 0.132 \qquad ...(4.23)$$

If water concentration is 3.9×10^{17} molecule cm^{-3}, concentration of M from Eq. 4.23 can be calculated which comes to be 2.95×10^{18} molecules cm^{-3}.

Thus the reaction of H_2O and oxygen (4.20) proceed primarily to produce OH group. The quantum yield of the reaction for the production of oxygen atom ($^1D_2$) at 350 nm is one. However, for wavelength less than 300 nm it is a little less than one but for wavelength greater than 300 nm, it drops very rapidly. Photolysis proceed in the following way

$$O_3 + h\nu(\lambda < 300) \longrightarrow O_2\left[X, {}^3\Sigma_g^-\right] + O\left[{}^3P_2\right] \qquad ...(4.24)$$

$$O_3 + h\nu(\lambda < 320 \text{ nm}) \longrightarrow O_2\left[b, {}^1\Sigma_g^+ \text{ or } a, {}^1\Delta_g\right] + O\left[{}^3P_2\right] \qquad ...(4.18)$$

In Chappius region, 550 – 610 nm it follow

$$O_3 + h\nu \longrightarrow O\left({}^3P_2\right) + O_2\left[X, {}^3\Sigma_g^-\right] \qquad ...(4.25)$$

However, in urban and suburban areas sufficient NO_2 is present, the production of O ($^3P_2$) from photolysis will be much less than that from NO_2. In addition, the oxygen atom will combine with O_2 molecule to produce O_3 with no net loss of ozone, as a result photolysis at $\lambda > 320$ nm is not of significance in troposphere chemistry. Further reactions will be considered in connection with oxides of nitrogen. Electronic spectra of O_3 is as given in Table 4.2.

Table 4.2 Electronic spectra of ozone molecule

Transition	$\tilde{A} \leftrightarrow X$	$\tilde{B} \leftrightarrow X$	$\tilde{C} \leftrightarrow X$	$\tilde{D} \leftrightarrow X$
Region in nm (Fig. 4.2)	1000-700 Diffuse band	610-550 Chappius band diffuse	374-300 Huggins band diffuse	300-220 Hartly band extremely intense $\lambda_{max} = 254$

The absorption cross section of ozone in the Hartley maximum at 254 nm is 6.09×10^{-18} cm^2 $molecule^{-1}$ *i.e.*, a layer of 0.007 cm thickness, at atmospheric pressure, absorbs 60% of the incident light. This is the reasons why atmosphere with an ozone layer of 0.3 cm atom is completely opaque to solar and stellar radiation in the Hartley band region. On the other hand, the absorption coefficient in the Huggins bands varies from about $2.03 = 10^{-19}$ cm^2 $molecule^{-1}$ at 300 nm to 4.06×10^{-22} cm^2 $molecule^{-1}$ at 343 nm and to 1.28×10^{-25} cm^2 $molecule^{-1}$ at 365 nm. For comparison, the absorption coefficient for the strongest of the Chappius band is 2.436×10^{-21} cm^2 $molecule^{-1}$.

4.4 WATER, H_2O

Water absorbs radiation below 186 nm, the absorption spectrum consisting of a continuum ($\lambda_{max} = 165$ nm) and a banded region between 125 and 143 nm superimposed on a second continuum. Some of the energy states and transition of water molecule are given in Table 4.3. The presence of two continuums indicates that two different excited states are involved in photochemistry of water, the probable reactions are

$$H_2O \xrightarrow[\lambda<186\text{ nm}]{h\nu} H(^2S_{1/2}) + HO(X,\ ^2\Pi) \qquad ...(4.26)$$

$$H_2O \xrightarrow[\lambda<124\text{ nm}]{h\nu} O(^1D) + H_2(X,\ ^1\Sigma_g^+) \qquad ...(4.27)$$

Table 4.3 Some of the energy states and transition of water molecule in the range 111-196 nm

Energy state	*Point group*	*Electronic configuration*	*Transition and range of transition energy in cm^{-1}*		*Remarks*
$X\ (^1A_1)$	C_{2v}	$(^1b_2)^2\ (3a_1)^2\ (1b_1)^2$	–		
$A\ (^1B_1)$	C_{2v}	$...(^1b_2)^2\ (3a_1)^2\ (1b_1)^2\ (4a_1)$	53763-68965	$A \leftarrow X$	
$B\ (^1A_1)$	$D_{\infty h}$	$...(3a_1)^2\ (1b_1)^2\ (4a_1)$	70922-80000	$B \leftarrow X$	Progression of very diffuse bands
$C\ (^1B_1)$	C_{2v}	$...(3a_2)^2\ (1b_1)^2\ (5a_1)$	80580-83752	$C \leftarrow X$	
$D(^1A_1)$	C_{2v}	$...(3a_1)^2\ (1b_1)^2\ (2b_1)$	81967-85324	$D \leftarrow X$	Diffuse band
$E(^1B_1)$	C_{2v}	$...(3a_1)^2\ (1b_1)^2\ (4sa_1)$	88652	$E \leftarrow X$	Single Bond
F	C_{2v}	$...(3a_1)^2\ (1b_1)^2\ (3d)$	89686	$F \leftarrow X$	Single Bond

Bond dissociation energy of water molecule for unsymmetrical dissociation is 493.3 kJ mol^{-1} or 5.113 eV which correspond to a photon of wavelength 242 nm (41320 cm^{-1}). However, there is no experimental evidence corresponding to such a primary process because water does not show absorption around this wavelength.

In water molecule the highest filled orbital in ground state is a nonbonding orbital $1b_1$ while the lowest unfilled antibonding orbital is $4a_1$. In the first excited state corresponding to removal of an electron from the former to the latter orbital, may weaken the bonding. If the equilibrium configuration in this excited state is symmetrical, an unsymmetrical dissociation into HO + H is unlikely to occur while for symmetrical dissociation into H + O + H, the energy is not sufficient. Indeed the anti-bonding orbital $4a_1$ may favour bonding between the two hydrogen atoms. The H – O – H angle is likely to be reduced in excited state and this may favour into O + H_2. On account of the spin conservation rule, the oxygen atom must be in the ^{1}D state upon dissociation. Hydrogen molecule is formed from two hydrogen atoms not bound to oxygen atom. Such a dissociation process thus leads to internal combination. The probable mechanism is

$$H-O-H \xrightarrow[125\text{ nm}]{h\nu} H+O+H \longrightarrow O+H_2$$

The precise mechanism is not clear for the formation of oxygen which is major product in water photolysis. However for H_2O, photochemical evidence shows that at 185 nm dissociation into H + OH predominates, suggesting that upper state of the continuum is either repulsive or has an unsymmetrical equilibrium position. The likely mechanics appears to be

$$H + HO \longrightarrow H_2 + O(^3P_2) \qquad ...(4.28)$$

followed by reaction 4.12 when oxygen atom in the presence of a third body combines to form oxygen molecule, although reaction of two hydroxyl radicals has also been proposed

$$\dot{O}H + H\dot{O} \longrightarrow H_2 + O_2 \qquad ...(4.29)$$

4.5 AMMONIA, NH_3

Ammonia is produced by biological process and hence exist in both clear and polluted atmosphere. The dissociation energy of NH bond, D (NH_2—H) is 4.3 eV ≡ 414.8 kJ mol^{-1} which correspond to photon wavelength 288 nm. As this wavelength lies in UV region it cannot be photolyzed. It reacts with OH radicals

$$NH_3 + \dot{O}H \longrightarrow \dot{N}H_2 + H_2O \qquad ...(4.72)$$

$$k_{472} = 1.6 \times 10^{-13} \text{ cm}^{-3}, \text{ molecule}^{-1}\text{s}^{-1}$$

At an OH concentration of 1×10^6 cm^{-3}, the lifetime of NH_3, with respect to above reaction would be 72 days. However NH_3 and thus NH_4^+ plays a significant role in atmosphere. It is the only common base which is highly soluble, essential for the formation of amino acids, proteins and for the evolution of life.

Ammonia show very strong UV absorption in the region 220–105 nm. Ammonia was one of the major constituent in the primitive atmosphere. This absorption is so strong, it is believed that it inhibited the photodissociation of water vapours. As slowly the abundance of free oxygen in primitive atmosphere slowly rose from 0.02% to 2%, methane and ammonia got oxidized to carbon dioxide and nitrogen and may be some N_2O also, which inhibited, from the beginning the complete conversion of entire oxygen into ozone.

4.6 NITROGEN, N_2

Nitrogen is not so important photochemically because of its large bond energy 7.374 eV ≡ 707.7 kJ mol^{-1} corresponding to photon wavelength of 169 nm, which limits its photochemical activity above the ozone layer. Nitrogen molecule absorbs weakly between 169 and 200 nm, absorption in this region leading to spin forbidden excitation from ground state

$$N_2\left(X, {}^1\Sigma_g^+\right) \longrightarrow N_2\left(A, {}^1\Sigma_u^+\right) \qquad ...(4.30)$$

which is a source of energy of highly excited oxygen atom by collisional energy transfer

$$N_2\left(A, {}^3\Sigma_u^+\right) + O\left({}^3P_2\right) \longrightarrow N_2\left(X, {}^1\Sigma_g^+\right) + O\left({}^1S_0\right)$$

Below 100 nm photodissociation may occur through photoionization and dissociative recombination involving O_2 and producing nitric oxide.

$$N_2\left(X, {}^1\Sigma_g^+\right) \xrightarrow[\lambda<100\,\text{nm}]{h\nu} N_2^+ + e \qquad ...(4.31)$$

$$N_2^+ + O_2\left(X, {}^1\Sigma_g^-\right) \longrightarrow NO^+ + NO \qquad ...(4.32)$$

$$NO^+ + e \longrightarrow N\left({}^2D_{3/2}\right) + O \qquad ...(4.33)$$

Deactivation of energetically excited nitrogen atom is a source of singlet oxygen

$$N\left({}^2D_{3/2}\right)+O_2\left(X,\ {}^3\Sigma_g^-\right)\longrightarrow N\left({}^4S_{3/2}\right)+O_2\left(a,{}^1\Delta_g\right) \quad ...(4.34)$$

Formation of Nitric oxide may also take place as a result of high energy trimolecular collision.

$$N\left({}^4S_{3/2}\right)+O\left({}^3P_2\right)+M\longrightarrow NO+M \quad ...(4.35)$$

4.7 NITROGEN OXIDES IN THE ABSENCE OF HYDROCARBONS

4.7a Nitric Oxide, NO

Nitric oxide, NO is important non-toxic, primary pollutant emitted by both mobile and stationary sources along with small quantity of NO_2. In ambient air NO get converted to NO_2. As NO_2 is emitted directly as well as formed from chemical reaction of NO, it is a primary and secondary pollutant. A part of electronic spectra of NO is given in Table 4.4. As NO does not absorb at $\lambda > 230$ nm and therefore it can not be the primary initiator of photochemical reaction in a polluted lower atmosphere. The bond dissociation energy of NO is 6.496 eV $\equiv$ 626.8 kJ mol^{-1} corresponding to a photon of wavelength

Table 4.4 Electronic spectra of NO molecule

Angular momentum change in term-value	*Spectroscopic notation*	*cm^{-1}*
$\frac{3}{2}\leftarrow\frac{1}{2}$	–	119.73
${}^4\Pi_i\rightarrow{}^2\Pi_r$	$a\rightarrow x$	38000
${}^2\Sigma^+\leftrightarrow{}^2\Pi_r$	$A\leftrightarrow x$	44080
	γ-bands	44200
${}^2\Pi_r\leftrightarrow{}^2\Pi_r$	$B\leftrightarrow x$	45392
	β-bands	45481
${}^4\Sigma^-\rightarrow{}^4\Pi_i$	$b\rightarrow a$	10326
${}^2\Pi_r\rightarrow{}^2\Pi_r$	$C\leftrightarrow X$	52251
${}^2\Pi_r\rightarrow{}^2\Sigma^+$	$C\leftrightarrow A$	8172
${}^2\Sigma^+\leftrightarrow{}^2\Pi_r$	$D\leftrightarrow X$	53172
${}^2\Sigma^+\rightarrow{}^2\Sigma^+$	$D\rightarrow A$	9092
${}^2\Delta_i\leftrightarrow{}^2\Pi_r$	$B'\leftrightarrow X$	59900
${}^2\Delta_i\rightarrow{}^2\Pi_r$	$B'\rightarrow B$	14508
${}^2\Delta_i\rightarrow{}^2\Pi_r$	$B'\rightarrow C$	
${}^2\Sigma^+\leftarrow{}^2\Pi_r$	$E\leftarrow X$	60863

190.9 nm which is not of relevance to lower atmosphere. The absorbance observed at 226 nm correspond to the transition

$$NO(X,\ ^2\Pi_r) \longleftrightarrow NO(A,\ ^2\Sigma^+)$$

A major difficulty until recently with this system has been to explain the very rapid oxidation of NO to NO_2. Thermal oxidation is thermodynamically feasible but if the reaction is kinetically simple, it must be trimolecular

$$2NO + O_2 \longrightarrow 2NO_2 \quad ...(4.36)$$

$$k_{436} = 2.1\times10^{-38}\ \text{cm}^6\ \text{molecule}^{-2}\ \text{s}^{-1}$$

However, when one lowers the NO concentration from high concentration to ppb-ppm level, the speed of the oxidation drops to the point where its rate is very small. For example at 100 Torr, NO (~1.3 × 10^5 ppm), approximately 85% of NO is oxidized in 15 s. However, at 0.1 ppm, approximately 226 days would be required to achieve the same net oxidation. Thus this thermal oxidation of NO could not explain the rapid conversion of NO to NO_2.

The problem was solved with the discovery of hydroxy radical (HO_2) radicals involved in the oxidation of not only NO but also CO with following reactions:

$$\dot{O}H + CO \longrightarrow \dot{H} + CO_2 \quad ...(4.37)$$

$$k_{437} = 1.7\times10^{-13}\ \text{cm}^3\ \text{molecule}^{-1}\ \text{s}^{-1}$$

$$\dot{H} + O_2 \longrightarrow H\dot{O}_2 \quad ...(4.38)$$

$$k_{438} = 2\times10^{-37}\ \text{cm}^6\ \text{molecule}^{-1}\ \text{s}^{-1}$$

$$H\dot{O}_2 + NO \longrightarrow \dot{O}H + NO_2 \quad ...(4.39)$$

$$k_{439} = 8.3\times10^{-12}\ \text{cm}^3\ \text{molecule}^{-1}\text{s}^{-1}$$

These reaction sequences have, indeed, proven important in clearing atmosphere, leaving very small amounts of NO. In polluted atmosphere organics play a role similar to CO in carrying on chain reactions. In this cycle, propane, say organics is oxidized to propionaldehyde C_2H_5CHO, two molecules of NO are converted to NO_2 and OH is regenerated. Of course, the propionaldehyde can itself then be attacked by OH and a similar cycle occurs leading to further NO oxidation. The chain reactions are eventually terminated by such reactions as

$$H\dot{O}_2 + H\dot{O}_2 \longrightarrow H_2O_2 + O_2 \quad ...(4.40)$$

$$R\dot{O}_2 + H\dot{O}_2 \longrightarrow ROOH + O_2 \quad ...(4.41)$$

$$\dot{O}H + NO_2 \xrightarrow{M} HNO_3 \quad ...(4.42)$$

A major source of OH in both clean and polluted air is photolysis of O_3 in sunlight to produce electronically excited oxygen atom, $O(^1D_2)$.

$$O_3 \xrightarrow[\lambda<320]{h\nu} O(^1D_2) + O_2(a,\ ^1\Delta_g) \quad ...(4.19)$$

$$O(^1D_2) + H_2O \longrightarrow 2\dot{O}H \qquad ...(4.20)$$

NO also reacts with O_3

$$NO + O_3 \longrightarrow NO_2 + O_2 \qquad ...(4.43)$$

$$k_{443} = 1.8 \times 10^{-14} \text{ cm}^3 \text{ molecule}^{-1}\text{s}^{-1}$$

Thus significant concentration of NO and O_3 are not found simultaneously in the same air mass. Reaction of NO with OH and alkoxy radicals produces nitrous acid and in part organic nitrites.

$$NO + \dot{O}H \xrightarrow{M} HONO \qquad ...(4.44)$$

$$k_{444} = 6.5 \times 10^{-12} \text{cm}^3\text{molecule}^{-1}\text{s}^{-1}$$

$$NO + R\dot{O} \longrightarrow RONO \qquad ...(4.45)$$

$$k_{445} = 3 \times 10^{-11} \text{cm}^3\text{molecule}^{-1}\text{s}^{-1}$$

These two reactions do not permanently remove NO from the atmosphere because of photo-dissociation of HONO and RONO at $\lambda < 400$ nm give back OH radicals.

NO reacts very rapidly with NO_3

$$NO + NO_3 \longrightarrow 2NO_2 \qquad ...(4.46)$$

$$k_{446} = 3 \times 10^{-11} \text{s}^{-1}.$$

In summary, NO can react in ambient air with HO_2, RO_2 as well as O_3, OH, RO and NO_3. The reaction with HO_2 and RO_2 are central part of the chain reactions in which organics are oxidized and NO is converted to NO_2 in both clean and polluted atmospheres. The reaction of NO and O_3 controls the development of O_3 peak in polluted urban areas. The reaction of NO and OH and in part with RO radicals as temporary night time storage for NO, since the major products HONO and RONO rapidly photolyze at dawn. Finally, the reaction with NO_3 controls to a great extent the concentration of NO_2 in night time air masses.

4.7b Nitrogen Dioxide, NO_2

Nitrogen dioxide is a orange gas with a broad intense absorption band over most of the visible and a ultraviolet regions with a maxima at 400 nm (σ = 30.86 × 10^{-20} cm^2 molecule^{-1}) as shown in Fig. 4.3. Some of the energy states and electronic transitions observed are given in Table 4.5. As it absorbs over a range of visible and *UV* region, it is an important molecule for photochemical reaction. Undoubtedly, this gas is a major atmospheric absorber leading to photochemical smog. The dissociation energy of NO_2, D(N–O) = 3.114 eV ≡ 300.4 kJ mol^{-1} and D (NO_2) = 4.50 eV ≡ 434.6 kJ mol corresponding to a photon wavelengths 398.2 nm and 275.2 nm and absorption below these wavelength leads to almost complete photodissociation with quantum yield equal to one.

$$NO_2\,[X, {}^2A_1] \xrightarrow[\text{visible}]{h\nu} NO\,(^2\Pi) + O\,(^3P_2) \qquad ...(4.47a)$$

$$NO_2\,[X, {}^2A_1] \xrightarrow[UV]{h\nu} NO\,(^2\Pi) + O\,(^1D_2) \qquad ...(4.47b)$$

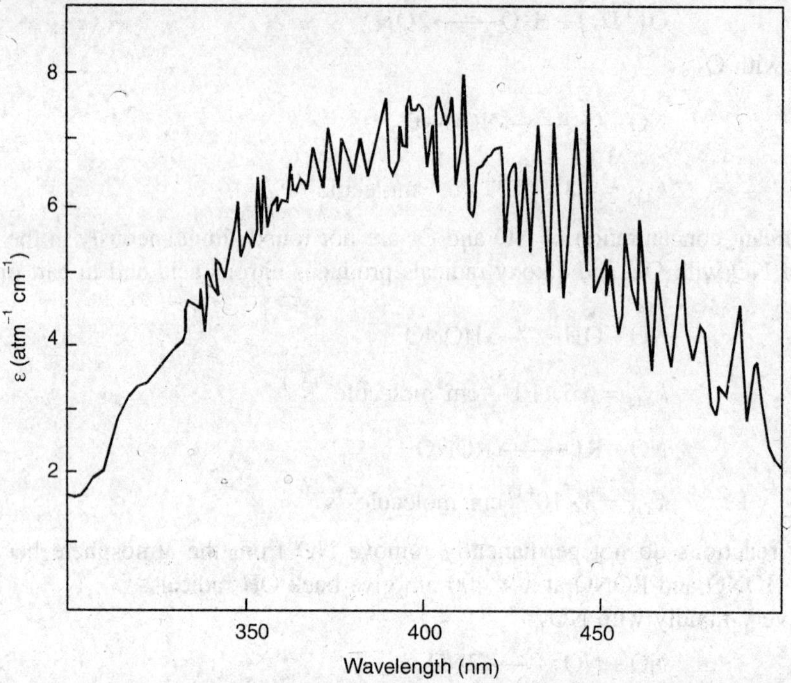

Fig. 4.3 Absorption spectrum of NO_2 at 25°C.

Table 4.5 Some of the energy states and transition of nitrogen dioxide molecule

Various energy states	*Energy in cm⁻¹*	*Observed transition*	*Remarks*
$X(^2A_1)$	0		
$\tilde{A}(^2B_1)$	15000 (666 nm)	$\tilde{X} \leftrightarrow \tilde{A}$ 1000–320 nm	Predissociation limit at 397.9 nm.
$\tilde{B}(^2B_2)$	40126 (249 nm)	$\tilde{X} \leftrightarrow \tilde{B}$ 258–235 nm	Predissociation limit at 248 nm.
$\tilde{C}$	42500 (235 nm)	$\tilde{C} \leftrightarrow \tilde{X}$ 235–200 nm	
$\tilde{D}$	50000 (200 nm)	$\tilde{D} \leftrightarrow \tilde{X}$ 200–165 nm	Continuum with maximum at 170 nm and diffuse bands.
$\tilde{E}(^2\Sigma_u^+)$	58309 (171 nm)	$\tilde{E} \leftrightarrow \tilde{X}$ 165–135 nm	
$\tilde{F}$	76711 (130 nm)	$\tilde{F} \leftrightarrow \tilde{X}$ 130–125 nm	Progression of three bands.

There are a number of potential fates in atmosphere including photolysis as well as reaction with OH, O_3, NO_3, O (^3P), HO_2, RO_2. The photolysis is the only known anthropogenic source of O_3 via 4.47a followed by 4.16, *i.e.*,

$$O(^3P_2) + O_2 \xrightarrow{M} O_3 \qquad ...(4.16)$$

The third order rate constant for this reaction is 5.6 × 10^{-34} cm^6 $molecule^{-2}$ s^{-1}. Using $[O_2]$ = 0.21 atm = 5.2 × 10^{18} molecules cm^{-3} and $[M] = [N_2]$ = 20.8 × 10^{18} molecule cm^{-3}, the lifetime of oxygen atom formed by NO_2 at the earth's surface.

$$\tau(O) = \frac{1}{k_{416}[O_2][M]} = \frac{1}{5.6\times10^{-34}\times5.2\times10^{18}\times20\times10^{18}}$$

$$= 16.5 \text{ ms}$$

The photolytic rate constant obviously depends on solar energy and spectral distribution and hence on factors such as latitude, season, solar angle to earth and degree of cloud cover. However, the rate constant for reaction (4.47a) is $k_{447a} = 9\times10^{-2}\,s^{-1}$ corresponding to NO_2 lifetime greater than 0.185 min.

Reaction 4.48 with OH is important in forming nitric acid in atmosphere and hence in contributing to acid rain and fogs

$$NO_2 + OH \xrightarrow{M} HONO_2 \qquad ...(4.48)$$

This reaction occurs primarily during day light hours since OH is produced by photochemical reactions. This second order reaction has a rate constant k_{448} $[M]$ = 1.1 × 10^{-11} cm^3 $molecule^{-1}$ s^{-1}. At an OH concentration of 5 × 10^6 molecule cm^{-3}, the lifetime of NO_2 with respect this reaction is 5 hours. Thus, photolysis is faster than reaction with OH, but is a major source of nitric acid in atmosphere.

Reaction 4.49 of NO_2 with O_3 is another important reaction in troposphere where O_3 and NO_2 are present in significant concentrations.

$$NO_2 + O_3 \xrightarrow{a} NO_3 + O_2$$

$$\xrightarrow{b} NO + 2O_2 \qquad ...(4.49)$$

Thus, it is the source of the nitrate radical which is now believed to play a significant role in the oxidation of organics and in the formation of HNO_3 in troposphere. Since these reaction do not require light, contribute to the night time chemistry of nitrogen oxides. The reaction rate constant for (a) path is 3.2 × 10^{-17} cm^{-3} $molecule^{-1}$ s^{-1} with lifetime of NO_2 at an O_3 concentration of 0.08 ppm is 4.4 hours. A small portion of the reaction go by the path (b).

Reaction 4.50 of NO_3 with NO_2 can also occur at night

$$NO_2 + NO_3 \overset{M}{\rightleftharpoons} N_2O_5 \qquad ...(4.50)$$

$$\longrightarrow NO + NO_2 + O_2 \qquad ...(4.51)$$

where NO_3 is formed and not removed rapidly by other processes such as reactions with NO. The N_2O_5 formed decompose back to NO_2 and NO_3 so that reaction 4.50 is an equilibrium reaction with equilibrium constant value 2.44 × 10^{-1} cm^3 $molecule^{-1}$ at 298 K with rate constant for forward reaction at low pressure, k_{450} = 4.5 × 10^{-30} cm^6 $molecule^{-2}$ s^{-1}. The rate constant of reverse reaction can be calculated from equilibrium constant and this rate constant *i.e.* $k_{-450} = Keq/k_{450}$ = 1.844 × 10^{-19} cm^3 $molecule^{-1}$ s^{-1}.

Reaction of NO_2 with NO_3 can also produce NO, NO_2 and O_2 (4.51) with reaction rate constant $k_{451} = 4.0 \times 10^{-16}$ cm^3 molecule^{-1} s^{-1}. This is four order slower in magnitude than that giving N_2O_5 and is relatively unimportant. It does, however, place an upper limit on the lifetime of NO_3 with nitrogen dioxide concentrations being 2.46×10^{12} to 19.68×10^{12} molecule cm^{-3} in the morning and afternoon. The upper limit for which NO_3 stays in air is 18 min and 2.12 min in the morning and afternoon, respectively.

Reaction 4.52 of NO_2 with ground state oxygen atom has a two-body route producing NO and O_2 and a three-body both producing NO_3

$$NO_2 + O(^3P) \xrightarrow{a} NO + O_2 \quad \text{...(4.52)}$$

$$\xrightarrow[a]{M} NO_3 \quad \text{...(4.52a)}$$

The rate constant of the biomolecular reaction k_{452} is 9.3×10^{-12} cm^3 molecule^{-1} s^{-1} while that of intermolecular reaction $k_{452a} = 2 \times 10^{-12}$ cm^2 molecule^{-1} s^{-1}. Although the concentration of O (^3P) has not been measured in ambient air, it is expected to peak 8×10^4 cm^{-3} molecule in an urban heavily polluted atmosphere as that of Delhi. The lifetime of NO_2 with respect to reaction 4.52 at this concentration is 13 days, too slow to be of significance to troposphere.

Reaction 4.53 of NO_2 with HO_2 forms peroxy nitric acid

$$NO_2 + H\dot{O}_2 \xrightarrow{M} HO_2NO_2 \quad \text{...(4.53)}$$

However this compound is thermally unstable and rapidly decompose back to HO_2 and NO_2. The rate constants are

$$k_{453} = 1.4 \times 10^{-12}\ \text{cm}^3\text{molecule}^{-1}\text{s}^{-1} \qquad k_{-453} = 0.085\ \text{s}^{-1}$$

The reverse reaction gives the lifetime of $HO_2NO_2 \approx 12$ s making it to fall apart as rapidly as it is formed. Modeling calculations suggest HO_2 concentration in highly polluted city to be 2×10^9 cm^{-3} molecule thereby giving the lifetime of NO_2 for the forward reaction to be ~ 6 min. This also shows that reverse reaction has no consequence in terms of removal of NO_2 from atmosphere.

Reaction 4.54 of alkylperoxy radical, RO_2 with NO_2

$$NO_2 + R\dot{O}_2 \overset{M}{\rightleftharpoons} RO_2NO_2 \quad \text{...(4.54)}$$

form alkyl peroxynitrate and like HO_2NO_2 it is also thermally unstable and decomposes back to reactants with lifetime of $RO_2NO_2 \leq 1$ sec so that this reaction can not be a permanent sink for NO_2.

NO_2 can also react with alkoxy radicals to form alkyl nitrate or if an abstractable hydrogen is available, nitrous acid and carbonyl compound may be formed.

$$NO_2 + R\dot{O} \xrightarrow[a]{M} RONO_2$$

$$\xrightarrow{b} HONO + R_1R_2CO \quad \text{...(4.55)}$$

The bimolecular rate constant at 1 atm is 1.5×10^{-11} cm^3 molecule^{-1} s^{-1}. At an alkoxy concentrations of 10^4 cm^{-3} molecule, for a polluted atmosphere, lifetime of NO_2 for reaction 4.55 is 77 days. Thus this reaction is too slow to contribute significantly to the removal of NO_2 from troposphere.

A possible non-photochemical reaction of NO_2 is the reaction with NaCl on the surface of ocean water

$$2NO_2 + NaCl \longrightarrow ClNO + NaNO_3 \quad \text{...(4.56)}$$

In an clean atmosphere containing NO, NO_2 and air, the reaction controlling the concentration of NO and NO_2 are

$$NO_2 \xrightarrow[\lambda=430\text{ nm}]{h\nu} NO + O(^3P_2) \qquad ...(4.47a)$$

$$O(^3P_2) + O_2 \xrightarrow{M} O_3 \qquad ...(4.16)$$

$$O_3 + NO \longrightarrow NO_2 + O_2 \qquad ...(4.43)$$

$$\frac{d[O_3]}{dt} = 0 = k_{416}[O][O_2][M] - k_{443}[NO][O_3]$$

$$[O_3] = \frac{k_{416}[O][O_2][M]}{k_{443}[NO]} \qquad ...(4.57)$$

If the rate of change of oxygen atom concentration is considered to be zero in a steady state, then

$$\frac{d[O]}{dt} = 0 = k_{446}[NO_2] - k_{416}[O][O_2][M]$$

or

$$[O] = \frac{k_{446}[NO_2]}{k_{416}[O_2][M]} \qquad ...(4.58)$$

substituting from Equation 4.58 to 4.57 one obtains

$$[O_3] = \frac{k_{446}[NO_2]}{k_{443}[NO]}$$

which on rearranging gives

$$\frac{k_{446}}{k_{443}} = \frac{[O_3][NO]}{[NO_2]} \qquad ...(4.59)$$

$$= \frac{3\times10^{-11}}{1.8\times10^{-14}} = 1666.7 \text{ molecule cm}^{-3} \text{ at 298 K}$$

This relationship (4.59) is known as **Leighton relationship** and reactions 4.46, 4.16 and 4.43 are the basis of it. It has a practical application that instead of carrying out computer programming to obtain $[O_3]$, [NO] and $[NO_2]$ separately, if two of the three concentrations are known third can be calculated. At sun set and sunrise deviation from this relation is expected. In general the Leighton relationship is expected to hold when sufficient NO is present, that the reaction 4.44 implies major loss O_3, for example at relatively low hydrocarbon to NO ratios.

4.7c Nitrate Radical, NO_3

The nitrate radical is formed from the reaction of NO_2 with O_3

$$NO_2 + O_3 \longrightarrow NO_3 + O_2 \qquad ...(4.49)$$

It is found in polluted atmosphere with a concentration of 350 ppt, the upper limit for which it stays in atmosphere is 18 min. Since it is not a photochemical product, it plays a key role in the night time, sink for troposphere chemistry of olefins, phenols and cresols and is one which get transformed to nitric acid.

Once formed, NO_3 reacts with NO_2

$$NO_2 + NO_3 \overset{M}{\rightleftharpoons} N_2O_5 \qquad ...(4.50)$$

$$\longrightarrow NO + NO_2 + O_2 \qquad ...(4.51)$$

or with NO_3

$$NO + NO_3 \longrightarrow 2NO_2 \qquad ...(4.46)$$

$$k_{446} = 3.0 \times 10^{-11}\ cm^3 molecule^{-1} s^{-1}.$$

NO_3 radical shows an absorption spectrum in the range 570 to 670 nm with maximum absorption at 672 nm, thus an active product for photolysis.

Photodissociation of NO_3 gives two possible sets of products

$$NO_3 + h\nu \xrightarrow{a} NO + O_2 \qquad ...(4.60)$$

$$\xrightarrow{b} NO_2 + O(^3P_2) \qquad ...(4.61)$$

The reaction constants for these reactions are $k_{460} = 0.022\ s^{-1}$ and $k_{461} = 0.18\ s^{-1}$ respectively in the atmosphere for an overhead sum for wavelength $470 < \lambda < 650$ nm. Due to these reactions it is rapidly photolysed at day time but at night when O_3 and NO_2 are present, the detectable concentration of NO_3 is reached. Thermal decomposition of it has also been suggested

$$NO_3 \xrightarrow{M} NO + O_2 \qquad ...(4.62)$$

4.7d Nitrous Acid, HONO

The probable mechanisms for HONO formation in atmosphere is the reaction between OH radical and NO molecule

$$NO + \dot{O}H \xrightarrow{M} HONO \qquad ...(4.44)$$

$$k_{444} = 6.5 \times 10^{-12}\ cm^3 molecule^{-1} s^{-1}$$

with an observed concentration of 8 ppb. This reaction produces a small steady state HONO concentration because of rapid photolysis of HONO. Photolysis of nitrous acid is a major source of OH radical, particularly at dawn, in polluted urban area

$$HONO \xrightarrow[\lambda<400]{h\nu} \dot{O}H + NO \qquad ...(4.63)$$

$$\dot{O}H + HONO \longrightarrow H_2O + NO_2 \qquad ...(4.64)$$

$$k_{464} = 6.6 \times 10^{-12}\ cm^3 molecule^{-1} s^{-1}$$

Although OH reacts rapidly with HONO, significant concentrations of OH are only present when photolysis of HONO take place. For OH concentration of 5 × 10^5 and 5 × 10^6 molecule cm^{-3}, the lifetime of HONO as well as NO with respect to reactions 4.64 and 4.44 are 3.5 days and 8.4 hours, respectively.

The alkaline hydrolysis of peroxyacetyl nitrate, PAN has been known for years to produce nitrite ion

$$CH_3\underset{\underset{O}{\|}}{C}-OONO_2 + 2OH^{-1} \longrightarrow CH_3\underset{\underset{O}{\|}}{C}O + O_2 + NO_2^- + H_2O \quad ...(4.65)$$

However, there is no clear evidence at the present time that reaction 4.65 does occur to an appreciable extent in atmospheric droplets, because of the acidic nature of atmospheric particles and fogs. In short, HONO is clearly a key species in troposphere through its photolysis to produce OH radical. This is an area which require more work.

4.7e Nitric Acid, HNO_3

During high temperature, combustion of fuels in air, formation of NO and small amount of NO_2 take place. The predominant atmospheric fate of the NO emitted is oxidation to NO_2 which can then react with OH during

$$HO + NO_2 \xrightarrow{M} HNO_3 \quad ...(4.48)$$

This secondary reaction is relatively rapid with k_{448} = 1.1 × 10^{-11} cm^3 $molecule^{-1}$ s^{-1} and approximately ten times that of SO_2–OH reaction. For a typical tropospheric place, OH concentration of 1 × 10^6 molecule cm^{-3}, the lifetime of NO_2 is 1 day compared to 13 days for SO_2–OH reaction. This relatively rapid oxidation of NO_2 to HNO_3 implies that HNO_3 formation should occur closer to the source than sulphuric acid formation.

A second potential important source of HNO_3 involves the oxidation of NO_2 with O_3 to nitrate radical

$$O_3 + NO_2 \longrightarrow NO_3 + O_2 \quad ...(4.49a)$$

$$k_{449a} = 3.2 \times 10^{-17}\ cm^3 molecules^{-1} s^{-1}$$

$$NO_3 + NO_2 \rightleftarrows N_2O_5 \quad ...(4.50)$$

$$NO_3 + RH \longrightarrow R + HNO_3 \quad ...(4.66)$$

$$NO_3 + RCHO \longrightarrow R + HNO_3 \quad ...(4.67)$$

In polluted urban atmosphere, alkane concentrations excluding CH_4, are typically 100 *ppb* (2.4 × 10^{12} molecules cm^{-3}), formaldehyde and acetaldehyde concentrations are in the range 5 × 10^{11} and 2.5 × 10^{11} molecules cm^{-3}, respectively. The calculated night time rates of HNO_3 formation for these compounds are 1.25 × 10^9, 1.5 × 10^9 and 4.5 × 10^9 molecules cm^{-3} h^{-1}, respectively, leading to a net rate of HNO_3 formation of 7.5 × 10^{10} molecules cm^{-3} h^{-1} from hydrogen abstraction by NO_3. This can be compared to the daytime rate of formation of HNO_3 by reaction with OH radicals. At NO_2, OH concentration of 1.25 × 10^{12} and 1 × 10^6 molecules cm^{-3}, respectively, the day time acid formation rate come to be 5 × 10^{10} molecule cm^{-3} h^{-1}. Thus in polluted atmosphere, as a crude estimate 15% of HNO_3

formation may be due to hydrogen abstraction by NO_3 radicals.

i.e., $$\frac{d[HNO_3]}{dt} = k_{466}[NO_3][RH]$$

$$= 5\times10^{-17}\times2.5\times10^{9}\times2.4\times10^{12}$$

$$= 3.1\times10^{5}\text{ molecules cm}^{-3}\text{ s}^{-1}$$

$$= 10.8\times10^{8}\text{ molecules cm}^{-3}\text{ hour}^{-1}$$

Apparently, gaseous HNO_3 has a long chemical life and as a result can act as a terminator for tropospheric chain reactions.

It absorbs light, weakly in shorter wavelength of ultraviolet. However, photolysis of HNO_3 produces OH and NO_2

$$HNO_3 \xrightarrow[(200<\lambda<320\text{ nm})]{hv} OH + NO_2 \qquad ...(4.68)$$

Organic nitrates have an absorption band that tails into the *UV* region with λ = 290 nm. The three primary process proposed for organic nitrates are:

$$RCH_2ONO_2 + hv \xrightarrow{a} RCH_2O + NO_2 \qquad ...(4.69a)$$

$$\xrightarrow{b} R-\overset{\overset{\displaystyle O}{\|}}{C}-H + HONO \qquad ...(4.69b)$$

$$\xrightarrow{c} RCH_2ONO + O \qquad ...(4.69c)$$

It is believed that 4.69a is the most important decomposition route. The most important organic nitrate in trophosphere chemistry is peroxy acetyl nitrate (PAN).

However, HNO_3 undergoes rapid decomposition at earth's surface as well as absorption into water droplets which contribute to a shorter overall lifetime in atmosphere than expected from purely chemical processes.

The OH radical reacts with HNO_3

$$\dot{O}H + HNO_3 \longrightarrow H_2O + NO_3 \qquad ...(4.70)$$

It can react with gaseous ammonia to produce ammonium nitrate, a deliquescent solid. At 25°C and 62% relative humidity, it exists as solid, but at relative humidity > 62% it exist as hydrated solid. The relationship between dissociation constant of NH_4NO_3 and relative humidity does not change significantly with pH in acidic region. Most of the homogenous gas phase reactions of HNO_3 in troposphere are slow, HNO_3 is removed from atmosphere primarily by absorption in to cloud or fog water or rain water or deposition at the earth's surface. The high solubility of HNO_3 ensures that the absorption processes are relatively fast which contributes to the observation of significant concentration of HNO_3 and H_2SO_4 in acid rain and fogs.

4.7f Peroxynitric Acid, HO_2NO_2

The only reaction known to form HO_2NO_2 is the reaction 4.53 of HO_2, but its thermal decomposition is fast making it an equilibrium reaction

$$HO_2NO_2 \underset{k_{453}=0.085\ s^{-1}}{\overset{M}{\rightleftharpoons}} HO_2 + NO_2 \quad ...(4.53)$$

with lifetime of HO_2NO_2 due to thermal decomposition of only 12 sec., much shorter than that for photolysis reaction with OH

$$OH + HO_2NO_2 \longrightarrow H_2O + O_2 + NO_2 \quad ...(4.71)$$

$$k_{471} = 5.2 \times 10^{-12}\ cm^3\ molecule^{-1}s^{-1}$$

with OH concentration of 1×10^6 molecule cm^{-3}, the lifetime of HO_2NO_2 with respect to the above reaction is found to be ~ 2 days.

4.8 METHANE, CH_4

Methane is a symmetrical tetrahedral molecule with bond length 10.94 nm with bond dissociation energy of D (CH_3—H) = 4.40 eV (424.53 kJ mol^{-1}), which corresponds to a photon of wavelength 282 nm. As this radiation is not available in the troposphere hence methane is photochemically inactive.

Absorption of gaseous methane begins at 145.5 nm rather suddenly reaching absorption coefficient k = 70 cm^{-1} at 137.0 nm and rising to a maximum of 1500 cm^{-1} at about 93 nm with maximum photoionization cross section = 1173.9 cm^{-1} at 77 nm. It is seen that below 77 nm all absorption is due to photoionization but at longer wavelength part of above 94.6 nm all of the absorption must be associated with dissociation processes. All excited (unstable) states of CH_4 are of Rydberg type continua because these Rydberg states are reached in a region of their potential surfaces above the lowest dissociation limit or because they are intersected by one of the states arising from CH_3 + H, CH_2 + H_2 or CH_2 + H + H or CH + H_2 + H. As a consequence such a strong predissociation results that the spectrum is continuous.

If one of the electron is removed bond dissociation energy is greatly reduced for

$$CH_4^+ \longrightarrow CH_3^+ + H \quad ...(4.72)$$

$$D = 1.34\ eV.$$

The same small stability towards dissociation into CH_3 + H must be expected for Rydberg states of CH_4. As a consequence equilibrium in these states will deviate considerably from that of the ground state and will have much larger value. Therefore in absorption from ground state, the unstable part of the potential surface is reached and only a continuous absorption is expected in agreement with observation. In fact no discrete electronic spectrum of CH_4 has been observed. The methane formed from various sources in troposphere is converted to carbon dioxide by the 5.15 and 5.16 reactions.

4.9 ORGANIC NITROGENOUS COMPOUNDS

The formation of **peroxyacetyl nitrate** (PAN), takes place by the reaction of acetyl radicals with O_2 and NO_2. There are a variety of sources of the acetyl radicals.

$$CH_3CHO + \dot{O}H \longrightarrow CH_3\dot{C}{=}O + H_2O \quad ...(4.73)$$

$$CH_3CHO + NO_3 \longrightarrow CH_3\dot{C}{=}O + HNO_3 \quad ...(4.67)$$

$$CH_3\dot{C}O + O_2 \longrightarrow CH_3\overset{\overset{\displaystyle O}{\|}}{C}{-}O\dot{O} \quad ...(4.74)$$

$$CH_3\overset{O}{\overset{\|}{C}}O\dot{O} + NO_2 \rightleftharpoons CH_3\overset{O}{\overset{\|}{C}}{-}OONO_2 \qquad \text{...(4.75)}$$

Photolysis of biacetyl also produces acetyl radicals

$$CH_3 - \overset{O}{\overset{\|}{C}} - \overset{O}{\overset{\|}{C}} - CH_3 \xrightarrow{h\nu} 2CH_3 - \overset{O}{\overset{\|}{C}}\bullet$$

Similarly, the oxidation of 2 butene by OH radical produces acetaldehyde as a product so that the photo oxidation of alkane in the presence of NO_x gives PAN. The atmospheric oxidation of ethane also has been a significant source of PAN in troposphere and lower stratosphere through following reactions

$$C_2H_6 + \dot{O}H \longrightarrow \dot{C}_2H_5 + H_2O$$

$$\dot{C}_2H_5 + O_2 \longrightarrow C_2H_5\dot{O}_2$$

$$C_2H_5\dot{O}_2 + NO \longrightarrow C_2H_5\dot{O} + NO_2$$

$$C_2H_5\dot{O} + O_2 \longrightarrow CH_3CHO + H\dot{O}_2$$

An equilibrium exists between the formation and decomposition of PAN.

$$CH_3\overset{O}{\overset{\|}{C}}O\dot{O} + NO_2 \rightleftharpoons CH_3\overset{O}{\overset{\|}{C}}OONO_2$$

The rate constant for decomposition of PAN in above reaction is $3.57 \times 10^{-4}\ s^{-1}$ with lifetime of 46.68 min.

PAN acts as a storage reservoir of NO_x. When NO_2/NO ratio is high and/or the temperature is low the decomposition reaction is inhibited and hence PAN concentration can be built up. However, when fresh NO is introduced into the atmosphere or temperature rises, PAN decomposes and releases $CH_3\overset{O}{\overset{\|}{C}}O\dot{O} + NO_2$. This rate is a non-photolytic free radical source and, hence, can be important in night time chemistry.

$CH_3\overset{O}{\overset{\|}{C}}O\dot{O}$ released can react in a sequence, leading to oxidation of NO to NO_2 as follows:

$$CH_3\overset{O}{\overset{\|}{C}}O\dot{O} + NO \longrightarrow CH_3\overset{O}{\overset{\|}{C}}\dot{O} + NO_2$$

$$CH_3\overset{O}{\overset{\|}{C}}\dot{O} \longrightarrow \dot{C}H_3 + CO_2$$

$$\mathrm{CH_3 + O_2 \longrightarrow CH_3\dot{O}_2}$$

$$\mathrm{CH_3\dot{O}_2 + NO \longrightarrow CH_3\dot{O} + NO_2}$$

$$\mathrm{CH_3\dot{O} + O_2 \longrightarrow HCHO + H\dot{O}_2}$$

$$\mathrm{H\dot{O}_2 + NO \longrightarrow \dot{O}H + NO_2}$$

$$\mathrm{\dot{O}H + NO \xrightarrow{M} HONO} \quad ...(4.44)$$

$$\mathrm{\dot{O}H + NO_2 \xrightarrow{M} HNO_3} \quad ...(4.48)$$

$$\mathrm{\dot{O}H + Organic \longrightarrow HCHO — CH_3CHO\ etc.}$$

This sequence forms photochemically active species such as HCHO, HONOs which can photolyze at dawn accelerating photochemical smog formation. It may be pointed out that thermal decomposition of PAN is believed to present the only sink for PAN in atmosphere. Its reactions with OH and O_3 are very slow.

PAN may serve as a useful indicator of photochemical air pollution caused by long range transport. While elevated O_3 levels are often used for this purpose, the existence of highly variable source of O_3, such as stratospheric injection and photooxidation in clean troposphere makes assessment of the contribution of anthropogenic sources, peroxy benzoyl nitrate (PBzN). It is thought to be about two order of magnitude stronger than PAN in terms of eye irritation. The upper limit of its concentration is 0.07 ppb.

4.10 CALCULATED TROPOSPHERIC LIFETIMES OF SOME REPRESENTATIVE ORGANICS

The troposphere contains, in addition to O_3, a variety of trace but highly reactive, oxidizing species, which participate in the day time photooxidation of organics. Labile species include the hydroxyl ($\dot{O}H$), hydroperoxyl ($H\dot{O}_2$), alkylperoxy ($R\dot{O}_2$), radicals, ground state $O(^3P)$ and electronically excited $O\ (^1D)$ oxygen atoms, electronically excited singlet oxygen $O_2\ [^1\Delta_g]$ and nitrate radical (NO_3).

Adding to the complexity of the atmospheric system is the fact that there are hundreds, if not thousands, of organics present in the natural and polluted trophosphere including alkanes, alkenes, alkynes and aromatics and their derivatives containing oxygen, nitrogen sulphur and halogen atoms. Further, depending on the physical state of the organic, for example gaseous and absorbed on the surface of an aerosol particle, the reaction may be homogenous or heterogenous in nature. Clearly given this vast array of possible reactions, the determination of those specific processes that may have significant impact on the physical, chemical and biological properties of a given air parcel is a formidable task. However, our knowledge of kinetics and mechanisms of elementary reactions of the labile oxidizing species with organics and inorganics has increased dramatically. It has been found that various chemical reactions can be reduced to the reactions of bonds. In other words, various chemical reactions represent making and breaking of bonds. Similar bonds will give rise to similar reactions, consequently compound is not so important. For example, rate constants for gas phase reactions of O_3, OH, Oxygen atoms, NO_3 are becoming available. Furthermore, for OH reactions understanding of structure-reactivity relationships for various classes of organics is emerging. Such information now permits one to estimate with some confidence, the relative contributions of each of the oxidizing species to the net rate of removal of most organics and inorganics from atmosphere by homogenous gas phase processes. This, of course, narrows down significantly the number of different types of reactions that should be considered while evaluating the chemistry of a specific air mass.

This point can be illustrated by calculating the tropospheric lifetime (τ) of some model organics. For second order reaction between an organic, *A*, and an oxidizing species, Ox

$$A + \text{Ox} \xrightarrow{k_{ax}} \text{products}$$

$$\tau = \frac{1}{k_{ax}\,[\text{Ox}]}$$

While the choices of the concentrations of various oxidants are somewhat arbitrary, but these are of very high concentrations, representative of the peaks which in most cases will not exist for substantial period of time. Hence these values are not even the averages, even in highly polluted area. These concentrations are given in Table 4.6.

These calculations merely illustrate that some reactions in the matrix of organics and oxidizing species in the atmosphere are sufficiently slow so can be omitted from further considerations. The organics and their lifetimes of chemical reactions are cited in Table 4.6.

Some of observed results are as follows:

1. The oxidants attack an alkanes in hydrogen atom abstraction reaction. The C—H bond dissociation energies of primary, secondary and tertiary are 404, 387 and 379 kJ mol^{-1}, respectively. Since, C—H bond is being broken, the reaction is fastest for the weaker, tertiary C—H bonds and slowest for primary C—H bonds.

Table 4.6 Calculated minimum tropospheric lifetimes of some organics reactions with oxidizing agent at certain peak concentrations of O_3, OH, HO_2, O (^3P), NO_3

	Oxidizing species					*Nature of reaction*
Concentration molecules cm^{-3}	*O_3 5 × 10^{12} (200 ppb)*	*$\dot{O}H$ 1 × 10^7 (0.4 ppt)*	*$H\dot{O}_2$ 2 × 10^9 (81 ppt)*	*O (^3P) 8 × 10^4 (0.003 ppt)*	*NO_3 1.1 × 10^{10} (430 ppt)*	
*n*Butane						RH + *Ox* → *R*
τ	≥ 6.5 × 10^2 yr	11 h	~ 10^3 yr	18 yr	29 days	Hydrogen
k	(≤ 9.8 × 10^{-24})	(25 × 10^{-12})	1.6 × 10^{-20}	2.2 × 10^{-12}	3.6 × 10^{17}	abstraction
trans-2-Butene						
τ	17min	26min	≥ 4 yr	6.3 days	4 min	
k	(2.0 × 10^{-16})	(6.5 × 10^{-11})	(≤ 4 × 10^{-18})	2.3 × 10^{-11}	3.8 × 10^{-13}	
Formaldehyde						
τ	≥ 3.2 × 10^3 yr	3.1 h	1.9 h	2.5 yr	1.8 days	
k	(≤ 2 × 10^{-24})	(9.0 × 10^{-12})	(7.5 × 10^{-14})	(1.6 × 10^{-13})	(5.8 × 10^{-16})	
Phenol						
τ	> 4.6 days	1 h	nd	nd	24s	
k	(5 × 10^{19})	(2.8 × 19^{-11})			(3.8 × 10^{-12})	

2. The day-time concentrations of $\dot{O}H$ and night-time concentrations of NO_3 radicals vary over wide ranges, but one can assume the ratio $[NO_3] / [OH]$ is 10^3. Recalling that relative atmospheric lifetimes of an organic to attack by $\dot{O}H$ or $\dot{N}O_3$ radicals τ_{OH}/τ_{NO_3} is given by the ratio kNO_3/kOH. For the compounds referred in Table 4.6, n-butane, trans-2 Butene, formaldehyde and phenol, there ratio are equal to 0.0144, 5.8, 0.064 and 136, respectively. These small values show that attack by NO_3 is responsible for only a small portion of the total removal of organic. For the simple straight chain alkanes like n-butane, which react more slowly with NO_3, their removal can be attributed to almost entirely to reaction with OH.
3. The bond dissociation energy for allylic C—H bond is 330 kJ mol^{-1} compared to ~ 370–410 kJ mol^{-1} for most other bonds. Thus for long-chain alkenes containing weak secondary and tertiary allylic C—H bonds, hydrogen atom abstraction in competition with addition has been suggested.

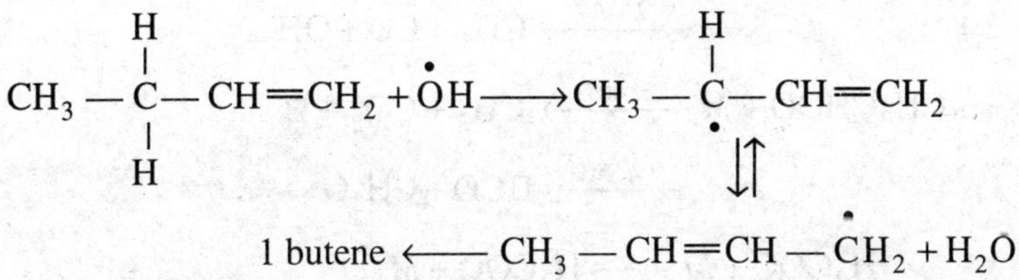

...(4.76)

At room temperature, the result of a low pressure study suggest that as much as 20% of the reaction with 1 butene proceeds via abstraction. However, recent works suggest that this abstraction is negligible ($\leq$ 10% of reaction) and can thus be ignored. Rather OH-alkene reactions are now agreed to be addition reaction at double bond in accordance to Markonikov Rule. Such addition product radicals help to oxidize NO to NO_2.

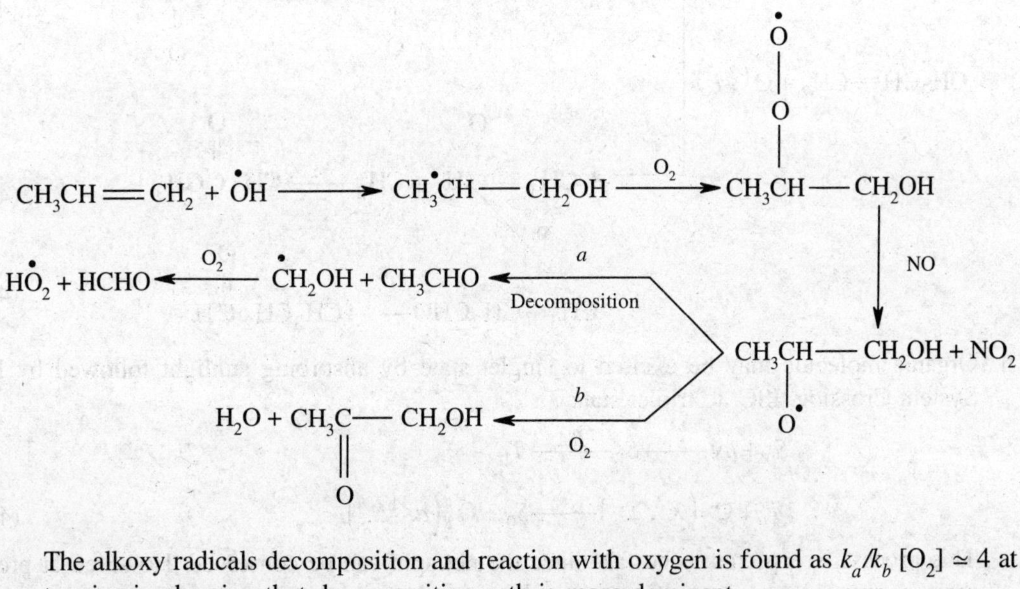

The alkoxy radicals decomposition and reaction with oxygen is found as $k_a/k_b\,[O_2] \simeq 4$ at 700 torr in air, showing that decomposition path is more dominant.

4. Ozone does not act as a good hydrogen abstractor and thus can be neglected. Gas phase reaction with alkenes has been known for years to produce aldehydes, ketones, acids as well as CO, CO_2 and H_2O.

$$CH_3CH{=}CH_2 + O_3 \longrightarrow CH_3\underset{|}{C}H{-}\underset{|}{C}H_2 \ (\text{O–O–O ozonide ring}) \xrightarrow{a} CH_3CHOO + HCHO \qquad ...(4.77)$$

$$\xrightarrow{b} CH_3CHO + H_2COO \qquad ...(4.78)$$

$$CH_3CHOO + M \xrightarrow{40\%} CH_3CHOO$$

$$\xrightarrow{a,\ 12\%} CH_4 + CO_2$$

$$\xrightarrow{b,\ 19\%} CH_3 + CO + OH$$

$$CH_3CHOO + M \xrightarrow{c,\ 40\%} CH_3 + CO_2 + H \qquad ...(4.79)$$

$$\xrightarrow{d,\ 5\%} HCO + CH_3O$$

$$H_2COO + M \longrightarrow H_2COO + M$$

$$\longrightarrow CO + H_2O \qquad ...(4.80)$$

5. O(^3P) can add to double bond at either end of it as

$$CH_3CH{=}CH_2 + O(^3P) \longrightarrow CH_3{-}CH{-}\overset{H}{\underset{O}{C}}H \longrightarrow CH_3CH_2\underset{\underset{O}{\|}}{C}H \qquad ...(4.81)$$

$$CH_3CH{=}CH_2 + O(^3P) \longrightarrow CH_3{-}\overset{O}{\overset{|}{C}}H{-}CH_2 \xrightarrow{a} CH_3\overset{\overset{O}{\|}}{C}CH_3 \qquad ...(4.82)$$

$$\xrightarrow{b} CH_3 + CH_2CHO \longrightarrow CH_3CH_2\overset{\overset{O}{\|}}{C}H \qquad ...(4.83)$$

6. Organic molecule may be excited to singlet state by absorbing sunlight followed by Inter System Crossing, ISC, to triplet state

$$S_0 + h\nu \longrightarrow S_1 \xrightarrow{ISC} T_1$$

$$T_1 + O_2\left(X^3, \Sigma_g\right) \longrightarrow S_0 + O_2\left(a, {}^1\Delta_g\right) \qquad ...(4.84)$$

This process is expected to be fast since the net spin can be conserved. However, at present there appears to be no significant gas phase exothermic chemical reaction which produced substantial concentrations of O_2 ($^1\Delta_g$) in troposphere. Under moderately polluted atmosphere its concentration is 10^8 molecules cm^{-3}. Its reaction with ethene is as follows:

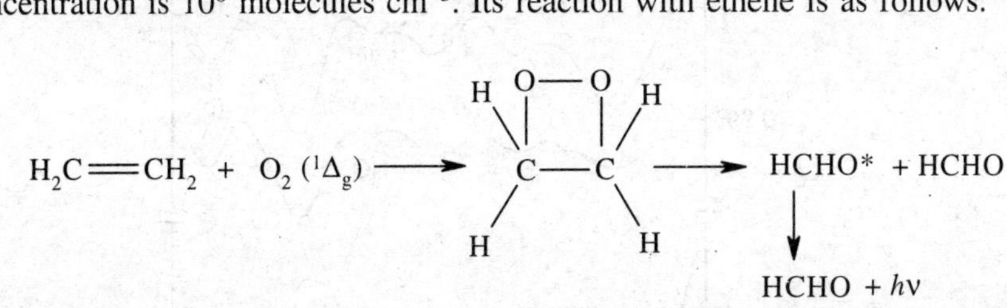

...(4.85)

Thus formaldehyde and chemiluminescence has been observed from the singlet O_2 with ethene as well as methyl, ethyl, n-butyl-vinyl ethers. In summary $O_2(^1\Delta_g)$ does not participate directly to a significant extent in the overall oxidation of hydrocarbons in photochemical air pollution. However, because of the nature of the products of its reaction, the role of $O_2(^1\Delta_g)$ in photosensitized oxidations particularly in biological systems, remains of considerable interest. Formaldehyde is an important pollutant in atmosphere. Its bond dissociation energies D (H—HCO) is 3.56 eV ≡ 343.5 kJ mol^{-1} corresponding to photon wavelength of 348 nm while for D (2H + CO) = 4.53 eV = 437.1 kJ mol^{-1} corresponding to photon wavelength of 274 nm. The absorption spectra of formaldehyde is in sharp contrast to continuous spectra for higher aldehydes is as shown in Fig. 4.4.

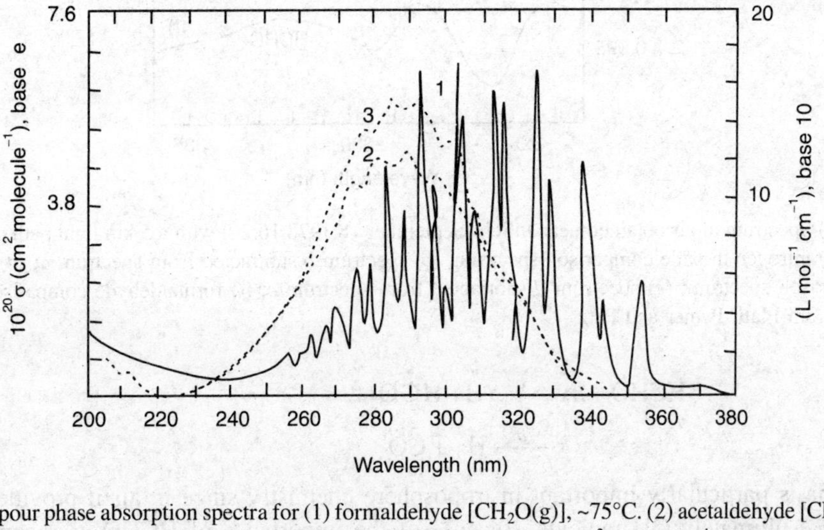

Fig. 4.4 Vapour phase absorption spectra for (1) formaldehyde [$CH_2O(g)$], ~75°C. (2) acetaldehyde [$CH_3CHO(g)$], 25°C; (3) propionaldehyde [$C_2H_5CHO(g)$], 25°C. Left axis σ (cm^2 $molecule^{-1}$), base *e*; right axis is *c* (*L* $mole^{-1}$ cm^{-1}), base 10. Note the loss of structure as the size of molecule increases. (From Calvert and Pitts, *Photochemistry,* John Wiley & Sons, New York[8].

Formaldehyde spectra shows a banded structure which allows the application of Differential Optical Absorption Spectrophotometry, DOAS. The details of this technique can be found in Pitts and Pitts[5] book entitled *Atmospheric Chemistry* (p. 338). This technique helps to identify the presence of HCHO in air, is self explanatory in Fig. 4.5. Absorption spectra of HCHO extends to ~ 370 nm whereas the higher aldehydes only absorb to 345 nm. Formaldehyde has two photodissociation paths

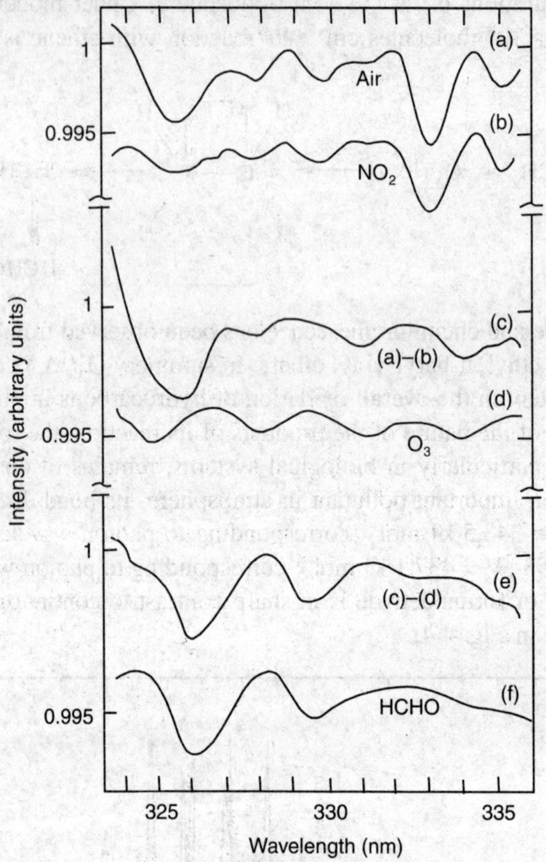

Fig. 4.5 (*a*) Spectrum of air obtained near Julich (September 18,1978,16:20) with a 5 km light path using DOAS; (*b*) nitrogen dioxide comparison spectrum; (*c*) spectrum *b* subtracted from spectrum *a*; (*d*) ozone comparison spectrum; (*e*) spectrum *d* subtracted from spectrum *c*; (*f*) formaldehyde comparison spectrum. (From Platt; Pemer and Patz[9].

$$HCHO + h\nu \xrightarrow{a} \dot{H} + H\dot{C}O$$
$$\xrightarrow{b} H_2 + CO \qquad ...(4.85)$$

Path 4.85a is particularly important in troposphere chemistry since in air it provides a source of H_2O and hence ultimately OH radicals. Because of the importance of HCHO as a source of OH in troposphere, a great deal of work has been carried out on the absorption cross-section and primary quantum yields. For further details of the electronic states in electronic spectra, the reader can refer to book by Sindhu[7]. The product energy distribution involved in HCHO the reader can refer to the *review* by Moore and Weisshan[4].

Like the photolysis reaction, formaldehyde reacts with OH radical to generate HCO radical as

$$HCHO + H\dot{O} \longrightarrow H\dot{C}O + H_2O \qquad ...(4.86)$$

$$k_{486} = 9.0\times10^{-12}\ cm^{-3} molecule\ s^{-1}$$

HCO formed either way generates O_2H and thus HO radical with following mechanism.

$$H\dot{C}O + O_2 \longrightarrow H\dot{O}_2 + CO \quad ...(4.87)$$

$$H\dot{O}_2 + NO \longrightarrow \dot{O}H + NO_2 \quad ...(4.88)$$

$$H\dot{O}_2 + HCHO \rightleftharpoons HOOCH_2\dot{O} \rightleftharpoons \dot{O}OCH_2OH \quad ...(4.89)$$

$$k_{489} \approx 7.5 \times 10^{-14} \text{ cm}^{-3} \text{ molecule s}^{-1}\ k_{-489} = 1.5 \text{ s}^{-1}.$$

This HO_2 reaction is quite rapid and is competitive with the reaction 4.86 in which formaldehyde is attacked by OH radicals. The vector proceeds via the initial addition of HO_2 to the double bond to form alkoxy radical which can rapidly isomerize to a peroxy radical. The peroxy radical is reasonable stable as shown by the formation of peroxy nitrate, $HOCH_2COONO_2$. What is particularly interesting is that same peroxy radical should be formed by addition of O_2 to the radical CH_2OH. However, there is a great deal of evidence that the $O_2 + CH_2OH$ reaction gives HO_2 + HCHO. In this reaction the adduct containing the 136 kJ mol^{-1} excess energy from addition reaction decomposes rapidly compared to collisional deactivation.

$$O_2 + CH_2OH \longrightarrow [OOCH_2OH]^* \xrightarrow{a} HO_2 + HCHO \xrightarrow{b} OOCH_2OH \quad ...(4.90)$$

This peroxy radical may react with NO and then with O_2 ultimately forming formic acid

$$\dot{O}OCH_2OH + NO \longrightarrow \dot{O}CH_2OH + NO_2 \quad ...(4.91)$$

$$\dot{O}CH_2OH \xrightarrow{O_2} HCOOH + H\dot{O}_2 \quad ...(4.92)$$

$$k_{492} \approx 3.5 \times 10^{-14} \text{ molecule cm}^{-3}\text{s}^{-1}.$$

At low NO concentration, peroxy radical will react with HO_2 or RO_2 or NO_3 also leading to abstraction of H atom from HCHO in a manner analogous to OH radical

$$NO_3 + HCHO \longrightarrow H\dot{C}O + HNO_3 \quad ...(4.93)$$

$$k_{493} \approx 5.6 \times 10^{-16} \text{ molecule cm}^{-3}\text{s}^{-1}$$

The four possible fates of aldehydes in the atmosphere are thus reaction with OH, HO_2 and NO_3 and photolysis. Since formaldehyde is the major aldehydic component of auto exhaust and is to be likely the major aldehyde in ambient air, let us compare its loss by reaction with OH, HO_2, NO_3 to that by photolysis

Table 4.7 compares the lifetime of HCHO with respect to photolysis and reaction with OH, HO_2 and NO_3 in a relatively clean and in a moderately polluted atmosphere using the rate constant given earlier and given in the table. Comparing these lifetimes to those from photolysis one can see that during day light hours, photolysis will predominate in the loss of HCHO in relatively clean atmosphere. In polluted atmosphere, reaction with HO_2 and OH will be comparable to photolysis. At night when the OH and HO_2 concentration are low, photolysis does not occur and NO_3 can contribute to the aldehyde loss. However, this is slow compared to the day-time loss by reaction with HO and HO_2.

Table 4.7 Lifetimes of HCHO with respect to photolysis and reaction with OH, HO_2 and NO_3 in typical relatively clean and polluted atmosphere

Loss process	*Reaction rate constant*	*Estimated lifetime*
Photolysis[a]	$k = 22.3 \times 10^{-6}\ s^{-1}$	8.6 *h*
Noon, January 1		
Noon, July 1	$k = 50.6 \times 10^{-6}\ s^{-1}$	3.8 *h*
Near Sunrise/Sunset July 1	$(\theta = 86°) k = 110.6 \times 10^{-6}\ s^{-1}$	174 *h*
Reaction with OH	$k_{486} = 9 \times 10^{-12}$ molecule $cm^{-3} s^{-1}$	
Relatively clean conditions		2.57
Moderately polluted conditions		6.17
Reaction with HO_2	$k_{498} = 7.5 \times 10^{-14}$ molecule $cm^{-3} s^{-1}$	
Relatively clean conditions		1.54 days
Moderately polluted conditions		3.7 *h*
Reaction with NO_3	$k_{493} = 5.6 \times 10^{-16}$ molecule $cm^{-3} s^{-1}$	
Relatively clean conditions		82.67 days
Relatively polluted condition		8.3 days

*Concentration under relatively clean conditions were taken to be as follow:
$[OH] = 5 \times 10^5$; $[HO_2] = 1 \times 10^8$; $[NO_3] = 2.5 \times 10^5\ cm^{-1}$. Under moderately polluted conditions each concentration was taken to be a factor of 10 higher.

REFERENCES

1. Dave, J.V., Development of Programs for Computing Characteristics of Ultraviolet Radiation. Final report under contract NAS 5-21680 NASA Report. CR-139134 National Aeronautics and Space Administration, Goddard Space Fight Centre, Greenbeld, Maryland NTIS No. NT5-10746/6SL (1972).
2. Demerjian, K.L., Schere and Peterson J.T., Theoretical Estimates of Actinic (Spherically Integrated) Flux and Photolytic Rate Constants of Atmospheric Species in the Lower Troposphere, *Adv. Environ. Sc. Technol.* 10, 369 (1980).
3. Leighton, P.A., *Photochemistry of Air Pollution,* Academic Press, New York (1961).
4. Moore, C.B., Weisshan J.C., Formaldehyde Photochemistry, *Ann. Rev. Phys. Chem.* 34, 525 (1983).
5. Pitts, B.J.E. and Pitts James N. (Jr.), *Atmospheric Chemistry,* John Wiley & Sons, New York (1986).
6. Peterson, J .T., Calculated Acinic Fluxes 240-700 nm for Air Polluted Photochemistry Application, US. Environmental Protection Agency Report EPA 600/4-76-025 June (1976).
7. Sindhu, P.S., *Molecular Spectroscopy,* Tata McGraw-Hill, New Delhi (1985).
8. Calvert, J.G. and Pitts J.N., *Photochemistry,* John Wiley & Sons, New York (1966).
9. Platt, U., Perner and Patz H.W., Simultaneous Measurement of Atmospheric CH_2O, O_3 and NO_2 by Differential Optical Absorption, *Nature*, 285, 312 (1980a).
10. Gaydon, A.G., *Dissociation Energies and Spectra of Diatomic Molecules,* Chapman and Hall, New York (1968).

5

Chemistry of Stratospheric Ozone

(A FRAGILE EARTH'S SHIELD)

5.1 SIGNIFICANCE OF ATMOSPHERIC OZONE

Common observation of rainbows demonstrates that sunlight contains all the visible colours, from red to violet. Besides, the solar spectrum contains variable radiation—low energy infrared and high energy ultraviolet (*UV*) radiation. Although all incandescent bodies emit all these radiations, it was known for a century that *UV* radiation from the sun is plentiful only to a wavelength of about 290 nm, it is completely absent at shorter wavelengths. Numerous observations of light from a variety of stars subsequently, were found to show the same sharp *UV* cut off at 290 nm that is characteristic of sunlight. This cut off is not some special property of the sun and stars but rather a property of the Earth's atmosphere, through which all celestial light must pass to reach the ground.

In 1880s, British chemist W.N. Harbley observed that Ozone, O_3 has exactly the same light absorbing spectra and he deduced that ozone in the upper atmosphere is indeed the atmospheric material that prevents exceedingly energetic *UV* radiation at wavelength shorter than 290 nm from reaching the earth's surface. It forms 10^{-5} % of entire atmosphere. On the average, Ozone is more plentiful near the poles than at the Equator and more abundant in winter than in summer. Despite its low concentration, continuation of the Ozone shield in approximately in its present form is exceedingly important for mankind. An average Ozone concentration variation with height is shown in Fig. 5.1. It shows that maximum concentration of Ozone is around a height of 25 km.

The Chemistry of Oxygen species O, O_2 and O_3 in upper atmosphere was first given by Chapman known as Chapman cycle. When oxygen molecule absorbs *UV* radiation of $\lambda < 220$ nm, the molecule dissociate into reactive oxygen atoms (Equation 4.1).

$$O_2\left(X,\ ^3\Sigma_g^-\right) \xrightarrow[\lambda=200\text{ nm}]{h\nu} 2O\left(^3P_2\right) \qquad ...(4.1)$$

Atoms of oxygen then can attach themselves to other molecules of O_2 and form ozone molecule (Equation 4.16)

$$O\left(^3P_2\right) + O_2\left(X,\ ^3\Sigma_g^-\right) \xrightarrow{M} O_3 + M \qquad ...(4.16)$$

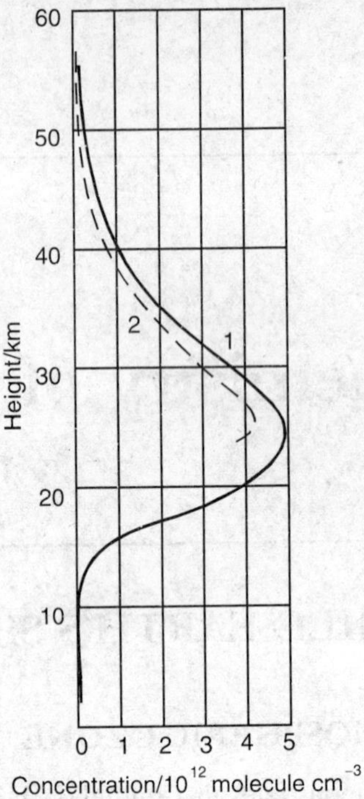

Fig. 5.1 Absolute ozone concentration, calculated and actual, as a function of altitude 1. Calculated ozone concentration, 2 actual ozone concentration.

This formation of ozone molecule in a highly excited state, of which extra energy is transmitted to other molecules (M) and converted to heat plays an important role in creating part of the atmosphere known as the stratosphere. Ozone is itself a very reactive molecule and occasionally intercepts atomic oxygen transforming both into molecular oxygen.

$$O_3 + O \longrightarrow 2O_2\left(X, {}^3\Sigma_g^-\right)$$

Whereas UV radiation at wavelength longer than 242 nm cannot be absorbed by O_2, it is absorbed by O_3 causing the molecule to split into one atom and one molecule of oxygen.

$$O_3 \xrightarrow[\lambda > 310\ \text{nm}]{h\nu} O\left({}^3P_2\right) + O_2\left(a, {}^1\Delta_g \text{ or } b\,{}^1\Sigma_g^+\right) \qquad ...(4.18)$$

In summary, Ozone is constantly created by solar radiation photolysis on earth's atmosphere and is constantly removed by reaction with atomic oxygen and radiation.

5.2 NATURAL CONTROL OF STRATOSPHERIC OZONE

Such large removal process was uncovered about 1970 by Paul Crutzen[10, 11]. The atmosphere becomes thinner and colder with increasing attitude for the first nine to sixteen km above ground. Almost all the water vapours are removed at lower altitudes. In the troposphere, that part of the atmosphere between the surface and about nine to sixteen km altitude, the gases undergo rapid vertical mixing and are cleansed periodically by rain storms. Consequently, chemicals released in the troposphere can be removed rapidly and long term pollution is relatively unlikely. The same nitrogen oxides found in the stratospheric zone acting as catalytic cycle are also found near the earth's surface do not reach stratosphere because they are removed by tropospheric weather processes.

Those NO and NO_2 molecules found in the stratospheric are formed there from another oxide of nitrogen, nitrous oxide, N_2O. Approximately three molecules of every ten million in the atmosphere are N_2O, their presence is resulting from bacterial denitrification or the destruction of nitrogen containing residues in the soil and in the ocean, returns nitrogen compounds to the atmosphere, whereas most of this nitrogen is in the form of N_2 a small fraction is released as N_2O; this process maintains atmospheric N_2O at its steady level.

N_2O is a linear assymetrical molecule. It shows absorption starting at 306.5 nm with a flat maxima near 290 nm. It is followed by second and third continuum starting at 282 and 260 nm with maximum at 273 and 182 nm respectively. These three continuum with various dissociation products

$$N_2O \longrightarrow N_2(^1\Sigma) + O(^3P)$$

$$\longrightarrow N_2(^1\Sigma) + O(^1D)$$

$$\longrightarrow NO(^2\Pi) + N(^4S)$$

$$\longrightarrow N_2(^1\Sigma) + O(^1S)$$

The bond dissociation energy of N_2O are D (N_2—O) = 1.677 eV = 161.8 kJ mol^{-1} and D (NO—N) = 4.93 eV = 475.64 kJ mol^{-1} corresponding to photons of wavelength 739 and 251.5 nm respectively.

N_2O molecule is not a very reactive molecule, neither is it removed from the atmosphere by rainfall. Hence molecule of N_2O can survive unchanged in the atmosphere for decades, long enough to mix upward in the high stratosphere above 32 km. Here solar *UV* radiation causes many otherwise inert molecules to become chemically reactive. At this level most of the N_2O is converted to N_2, but about 5% become NO which then participate in the NO_x catalytic chain

$$O(^1D) + N_2O \xrightarrow{a} N_2 + O_2 \quad \text{.. (5.1)}$$

$$\xrightarrow{b} 2NO \quad \text{...(5.2)}$$

NO attacks Ozone very efficiently to produce molecular oxygen and NO_2 which then reacts with atomic oxygen to form NO again plus another molecule of O_2. After both reactions have occurred,

$$NO + O_3 \longrightarrow NO_2 + O_2 \quad \text{...(4.43)}$$

$$NO_2 + O \longrightarrow NO + O_2 \quad \text{...(4.52)}$$

the NO molecule is still present, although one molecule of O_3 and one oxygen atom have been removed. The NO—NO_2 catalytic system often abbreviated as NO_X can remove hundreds or thousands of ozone molecules before being terminated by reaction with another free radicals as in

$$OH + NO_2 \longrightarrow HNO_3 \text{ (nitric acid)} \quad \text{...(4.48)}$$

The ozone concentration of the natural stratosphere is primarily controlled by these two chemical cycles the solar *UV* formation of O_3 from O_2 and the biological nitrogen cycle that keeps the atmosphere supplied with N_2O. The formation of ozone by the action of solar *UV* radiation on atmosphere O_2

represents the expenditure of more energy than that consumed by all man's activities combined together. The possibility of directly interfering with or supplementing solar ozone production is therefore far beyond man's present capabilities.

5.3 PHOTOCHEMISTRY OF WATER VAPOUR IN STRATOSPHERE

Although the water vapour content of the upper stratosphere is exceedingly small, it exerts a strong effect on ozone content. Vertical mixing is very slow in stratosphere due to the extreme dryness in this region, whereas the mixing ratio of water vapour is 10^{-2} at ground level and it has a value of 3×10^{-6} in stratosphere. The presence of water vapour and species such as OH in stratosphere lowers the amount of ozone, as the latter is destroyed during the reactions

$$O_3 + \dot{O}H \longrightarrow H\dot{O}_2 + O_2$$

$$O_3 + \dot{O}H \longrightarrow 2O_2 + \dot{H} \qquad \text{...(5.3)}$$

From water vapours H, H_2 and OH are formed under the effect of solar of radiation according to the equation

$$H_2O + h\nu \longrightarrow H\left({}^2S_{1/2}\right) + OH\left(X,\ {}^2\Pi\right) \text{ or } OH\left(A,\ {}^2\Sigma^+\right) \qquad \text{...(4.54)}$$

$$H_2O + h\nu \longrightarrow H_2 + O\left({}^3P_2\right) \qquad \text{...(4.28)}$$

Other reaction can also take place in HO_x system as, for example

$$O_3 + H\dot{O}_2 \longrightarrow \dot{O}H + O_2 + O_2$$

The changes brought by the moisture in the stratosphere by the exhaust gases of high flying aircraft clearly states that water acts as a pollutant in the stratosphere.

5.4 ROLE OF ClO_x IN STRATOSPHERE

NO_x is closely coupled to ClO_x reactions. Thus ClO formed from Chloro Fluoro Carbon (CFC) reaction (5.11 and 5.12a) reacts with NO_2 to produce chlorine nitrate in essentially 100% yield

$$ClO + NO_2 \longrightarrow ClONO_2 \qquad \text{...(5.4)}$$

which acts as a temporary reservoir for chlorine at night. At dawn $ClONO_2$ photolyzes forming chlorine atoms

$$ClONO_2 + h\nu \longrightarrow Cl + NO_3 \qquad \text{...(5.5)}$$

In addition to these interactions between ClO_x and NO_x chemistry there are also important ClO_x—HO_X Cl—CH_4 interactions, in which chlorine atom is converted to HCl.

$$Cl + O_3 \longrightarrow ClO + O_2 \qquad \text{...(5.6)}$$

$$ClO + O \longrightarrow Cl + O_2 \qquad \text{...(5.7)}$$

$$CH_4 + Cl \longrightarrow CH_3 + HCl \qquad \text{...(5.8)}$$

$$H\dot{O}_2 + \dot{C}l \longrightarrow HCl + O_2 \qquad ...(5.9)$$

ClO reacts with HO_2 to form HOCl which along with HCl and $ClONO_2$ is another temporary reservoir of chlorine atoms

$$Cl\dot{O} + H\dot{O}_2 \longrightarrow HOCl + O_2 \qquad ...(5.10)$$

However, like $ClONO_2$, HOCl photolyzes to form chlorine atom. All these ozone removing long chain reactions, fall within man's current technological reach. Introduction of such reactive free radicals as NO and Cl in stratosphere may continue to remove O_3 for long time.

5.5 STRATOSPHERIC POLLUTION AND THE SST

Pollutants can enter the stratosphere through two routes.

1. Direct release of free radicals or molecules into stratosphere which are easily transformed into pollutant.
2. Compounds like N_2O are released at the surface of earth that can only get converted into NO—NO_2 molecules in stratosphere.

The second route has already been discussed. However jet aircraft capable of operating at high altitude can release NO_x molecules directly into stratosphere. In the late 1960's, as plans took shape for building the Anglo-French Concorde supersonic transport (SST), the Soviet Tupolev-144 SST and the US Boeing SST, questions were raised about possible environmental effects. The potential stratospheric ozone problems with such aircraft were not a direct result of the supersonic velocities at which they fly but rather of the high altitudes for which their flight was designed. Current subsonic jets generally fly at altitudes of 9-11 km and the Concorde and Tu-144 at 16 km. In general, the higher the aircraft flies, the longer its released pollutants take to drift down through the stratosphere and thus they have more time to catalyze the removal of ozone.

When air is heated to high temperatures by passage through a jet engine some N_2 and O_2 reacts to form NO molecules, ready to participate in the NO_x catalytic chain. In early 1971, physical chemist Harold Johnston warned that NO directly released in stratosphere from SST engines represented a real danger to ozone layer, if the proposed fleet of 500 or more Boeing SST's were actually built and flown. In the aftermath of the US Senate decision in 1971, primarily on economic grounds to halt the development of the Boeing SST, the US government funded an intensive, three year study of the environment effects from SST—the "Climatic Impact Assessment Programme, CIAP" (1972-74). The examination of the stratosphere carried out during CIAP have provided more information about the stratosphere than that acquired from all prior scientific efforts.

During CIAP, high altitude measurements with balloons and aircraft confirmed that NO and NO_2 are actually present in stratosphere as expected, and that the NO_x catalytic cycle was indeed the chief ingredient missing from the Chapman mechanisms. Stratospheric model calculations indicated that projected depletion of ozone was strongly dependent upon the altitude of flight of the aircraft with highest ozone losses for the proposed 500 Boeing plane fleet. Much smaller effects were calculated for Concorde and Tupolev SST and still smaller for the current fleets of subsonic aircraft. Present calculations have indicated negligible changes in ozone level for flight altitudes below ~ 20 km, even for large fleets of aircrafts.

Although the current phase of SST building has subsided without the construction of 500 aircraft fleets, the potential stratospheric environmental problems have been outlined and will undoubtedly require thorough consideration as the second generation of SST comes into discussion during the next 10–15 years. The advantage of higher altitude and higher speed needs to be weighted against the problems accompanying higher release altitudes for such pollutants as NO_x and water vapours.

5.6 UV RADIATION AND SKIN CANCER

During the ClAP study, investigators also considered the various environmental consequences of stratospheric ozone loss. Two physical changes to the atmosphere appeared predominant: increased penetration of *UV* radiation to the Earth's surface and alteration in the temperature structure of the stratosphere itself. The change in stratospheric structure could influence the world's climate, whereas increased *UV* radiations at the surface would affect many biological systems.

Solar *UV* radiation can be conveniently divided into three categories *UV-A* with wavelength of 320-400 nm; *UV-B* with wavelengths of 280-320 nm which includes the ozone cutoff region and *UV-C* with wavelengths shorter than 280 nm, all of which are absorbed in stratosphere by either O_3 or O_2 as shown in Fig. 5.2.

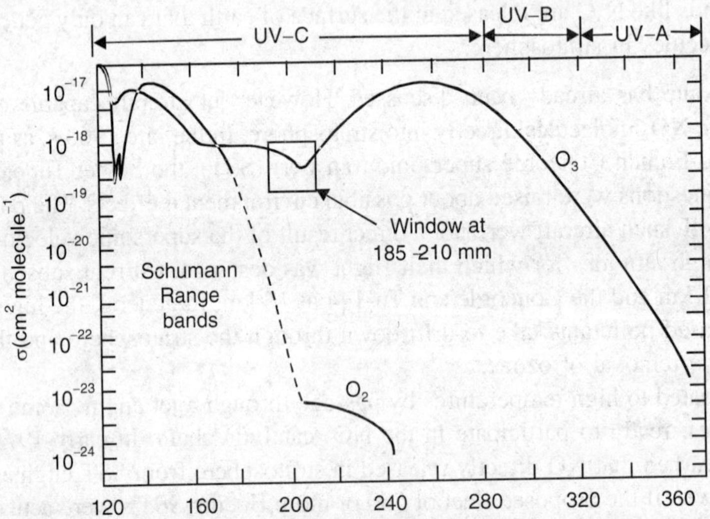

Fig. 5.2 Absorption cross-sections for O_2 and O_3 from 120 to 360 nm, showing the window at ~185 to 210 nm.

Most of the *UV-A* radiation currently reaches the surface of earth and its passage through atmosphere has very little or no affect. Evolution of all biological species including humans, has taken place under the light of intense *UV-A* radiation of small light exposure and would not have survived if they were affected by it. On the other hand, absorption of *UV-C* by O_3 and O_2 molecules in the stratosphere is so strong that no *UV-C* would penetrate to the surface even after a major loss of stratospheric ozone. The major effect of change in ozone concentration would be on the transmission of *UV-B* radiation near the ozone cut off. A decrease in ozone concentration in stratosphere will increase the fraction of *UV-B* penetrating to the surface with consequent increase in biological effects.

One obvious effect of *UV-B* radiation on lightskinned humans is reddening of the skin. A prolonged, exposure may lead to sunburn. Controlled experiments demonstrated that even more prolonged exposure to *UV-B* induces skin cancer in laboratory animals, whereas such experiments have not been carried out with human beings. Natural *UV-B* exposure varies sufficiently from one geographical location to another so that semi-controlled experiments have been performed by choices of living site. The average concentration of ozone is lowest near the Equator. Thus people living in the equatorial latitudes are exposed to more *UV-B* radiation than those living in temperate on polar regions. In addition, atmospheric absorption of *UV-B* is also dependent upon zenith angle of the sun. As the sun rises further above the horizon, penetration of *UV-B* radiation increases. All regions receive maximum *UV-B* radiation during the four hours around noon. Exposure to *UV-B* in temperate regions is also much stronger in summer than in winter.

Careful studies by the US National Cancer Institute showed that the incidence of human skin cancer is several times greater in the southern part of the US than in the northern part. These investigation have shown that light skinned people are more sensitive to skin cancer than dark skinned population having the presence of the light absorbing pigment melanin in the skin. The incidence of skin cancer is particularly high in Queensland, Australia where a predominantly white population lives in the equatorial region.

Skin cancer occurs in three principal forms. Two are relatively mild forms identified by the layers of cells in which they grow, basal or squamous. These are slow-growing cancer that can be readily treated medically and are primarily a source of discomfort and disfigurement but rarely can become a cause of death, results from prolonged, cumulative exposure to *UV-B*, are commonly found among older people. The third form of skin cancer, malignant melanoma, grows rapidly, frequently spreads to other organs. Malignant melanoma is also often found among younger people. Studies have indicated that exposure to solar *UV-B* is also a prime contributing factor to its occurrence.

Records in the Scandinavian countries are particularly comprehensive and clearly show that the incidence of malignant melanoma has increased there by a factor of four to five between 1950 and 1975. Undoubtedly this increase is associated with the tendency for these fair skinned individuals to expose more of the body to direct sunlight for longer amount of time.

Predictions of increased incidence of skin cancer by loss of atmospheric ozone, require understanding of two separate factors. First, how large will be the average increase in *UV-B* radiation resulting from the loss of ozone? Second, how much increase in the rate of skin cancer is expected from the anticipated increase in solar *UV-B*-radiation ? Based on careful calculations, the increase in *UV-B* radiation has been placed at approximately 20% for a 10% loss in ozone. Finally, the estimate of the increase in skin cancer as a function of the increase in *UV-B* is basically derived from the record of skin cancer rates observed for different latitudes and is less certain, but it would result in a 40-50% increase in basel and squamous-cell skin cancer and about 30% increase in malignant melanoma.

Since all biological systems have evolved under the protection of the ozone shield, it is quite unlikely that man is the only species sensitive to an increase in *UV-B* radiation. Special concern have been expressed for very small species such as oceanic plankton, the entire organism and not just the skin will be exposed to *UV-B* radiation. Changes in growth characteristics of plants, including agricultural crops; and effects on the behaviour of insects is anticipated. Increased exposure to *UV* radiation in laboratory reveals inhibition of plant (chloroplast) growth, exhibiting abnormal swelling of chloroplast membranes. Though such effects, cannot be extended to atmospheric ozone loss, yet significant effect will be there.

5.7 CHLOROFLUORO CARBONS IN ATMOSPHERE

Chlorofluoro carbons (CFCs) that is compounds containing chlorine, fluorine atoms first came into use for refrigeration around 1930, because of their non-flammability and least chemical reactivity. Later on when low pressure valves were developed, CFCs became standard propellants for dispensing aerosol. By 1950 CFCs were widely used for refrigeration, air conditioning, aerosol propellants, cleaning solvents, plastic foams, in fast food packing, in dry cleaning, for sterilizing surgical instruments and for cleaning electronic equipments. A list of these chemicals are given in Table 5.1.

Table 5.1 Uses of chlorofluoro carbons

CFC's		*Uses*
Methane Based		
CCl_3F	F—11	Air conditioning, Refrigeration, Cleaning forms
CCl_2F_2	F—12	Aerosols, Air conditioning, Refrigeration, Sterilizer for surgical instruments
$CHClF_2$	F—22	Aerosol propellants, cleaning solvent
Ethane Based		
$CCl_2F—CClF_2$	F—113	Precision cleaning foams for electronics Industry, Refrigeration.
$CClF_2—CClF_2$	F—114	Aerosol, Refrigeration, Cleaning agent, Foams
$CClF_2—CF_3$	F—115	- do - and airconditioning
Halons		
CF_2BrCl	F—1211	Portable fire extinguisher.
CF_3Br	F—1301	Total flood fire extinguisher
C_2F_4 Br	F—2402	Fire extinguisher

Chlorofluoro carbons containing carbon and possibly hydrogen have been used extensively in the industrialised nations in the past decades primarily as propellants in aerosol spray cans, as refrigerants, and as blowing agents e.g. for producing polyurethane foam. Their chemical characteristics have made them ideally suited for such uses as they are generally non toxic and chemically inert. Thus, they can be used around open flames, and leaks in refrigerator units do not represent a health hazard, as older units operated on coolants such as SO_2.

The principal chlorofluoro carbons in use are CCl_3F, CCl_2F_2 and $CHClF_2$. These are often referred to as F—11, F—12, F—22 after the Dupont trade name Freon. The number following this reflect number of hydrogen, fluorine atoms respectively. The first number is the number of hydrogen atoms plus one, second gives the number of fluorine atoms. For chlorofluoro carbon containing two or more carbon atoms, a three digit numbering system is used, the first digit corresponds to the number of carbon atoms minus one, the second to the number of hydrogen plus one and third to the number of fluorine atoms. Thus CCl_2FCClF_2 is F—113 and $CClF_2CClF_2$ as F—114. To determine the chemical formula from a CFC trade name, simply add 90 to the number and interpret the three digits as follow. The right most digit is the number of fluorine atoms, the middle digit is the number of hydrogen atoms and left digit is the number of carbons. Some examples will make the point clear.

		C	*H*	*F*	*Cl*	*Formula*
F—12	12 + 90 = 102	1	0	2	2	CF_2Cl_2
F—22	22 + 90 = 112	1	1	2	1	CF_2HCl
F—115	115 + 90 = 205	2	0	5	1	C_2F_5Cl
F—124	124 + 90 = 214	2	1	4	1	C_2F_4HCl

The trade name of CFC can be calculated in a reverse way as in $C_3F_6Cl_2$ has three carbons, no hydrogen, six fluorines so its number is 306, subtracting 90 from 306 gives 216. The trade name of $C_3F_6Cl_2$–F—216. However, this procedure cannot be applied to Halons. For Halons containing chlorine, bromine, hydrogen and fluorine, the first number denotes number of carbons, second to number of fluorine, third to number of chlorine and fourth to number of bromines. Thus CF_2BrCl F—1211, CF_3Br is F—1301 and $C_2F_4Br_2$ is F—2402.

The chlorofluoro carbons have very long lifetime in stratosphere. This is consequence of the fact that they do not absorb light of wavelength above 290 nm and do not react at significant rates with O_3 and OH. They have a insignificant reaction with NO_3 also. In addition to the lack of chemical sinks, they do not appear to be substantial physical sinks because they are not very soluble in water and hence are not removed rapidly by rainouts.

As a result, CFC's reside in troposphere for years, slowly diffusing up across the troposphere into stratosphere. Calculations, based on mass balance approach show that lifetime of $CFCl_3$ (F—11) and CF_2Cl_2 (F—12) in the global atmosphere, that is, the time to diffuse to stratosphere and undergo photolysis appear in the range of 50-150 years. Since the wavelength distribution increase with increasing

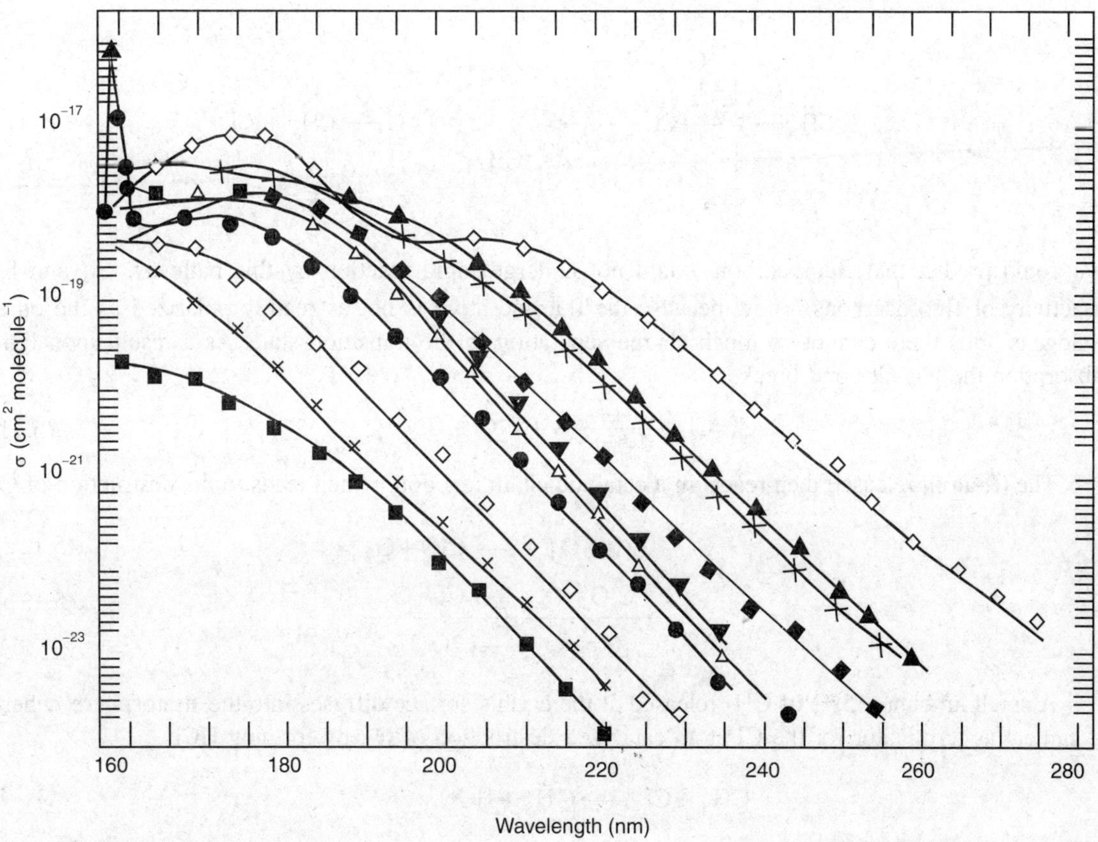

Fig. 5.3 Semi logarithmic plot of the absorption cross sections of the halogenated methanes at 298K : +, $CHCl_3$; Δ, $CHCl_2$ F; ×, $CHClF_2$; ◆, CH_2Cl_2; ◇, CH_2ClF; • CH_3Cl; □, CCl_4; ▲, CCl_3F (F-11); ▼, CCl_2F_2 (F-12); ■ $CClF_3$ (from Hubrich and Stuhl, 1980).

altitude, the CFC's eventually become exposed to the wavelength of light, which they can absorb. CFC's show an absorption maxima in the range 170-220 nm.

Figure 5.3 shows the absorption cross section of some halogenated methanes from 160 to 280 nm. The absorption becomes very weak beyond 240 nm in the case of F—11, and 220 nm in the case of F—12. As shown in Fig. 5.2 both O_2 and O_3 absorb in the *UV* region from 120–360 nm, there is a window in the overlapping O_2 and O_3 absorption from 185 to 210 nm, that is, a region where the total light absorption is in a shallow minimum. By coincidence this is region where the CFC absorbs light (Fig. 5.3).

The C—Cl has the lowest bond dissociation energy in CFC is 3.172 eV = 305 kJ mol^{-1} whereas that for the strong C—F bond, the dissociation energy is 4.784 eV = 460 kJ mol^{-1}. These correspond to photon of wavelength 392 and 260 nm. As a result, upon light absorption, the weaker C—Cl bond breaks first.

The stability of the Freons and other CFC's is due to the special characteristics of the C—F bond, which is much stronger than C—Cl bond. Also the C—F bond distance (13-14 nm) is much shorter than C—Cl bond distance (~18 nm). Since bond energies provide a rough estimate about the rate of S_{N_1} reaction.

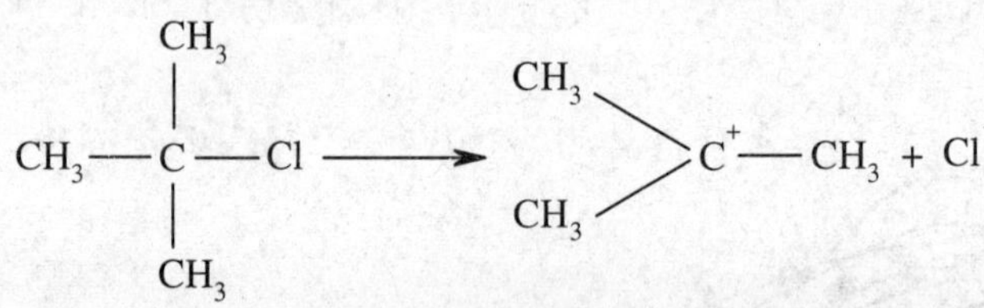

we could predict that fluorocarbon would not undergo rapid reaction by this pathway. S_{N_1} and E2 reactivity of fluorocarbons is low because the fluorine atom is not as readily polarized as the other halogens, thus there cannot be much charge separation in the transition state. As a result upon light absorption the C—Cl bond breaks

$$CF_2Cl_2 \xrightarrow[(UV\text{–}C)]{h\nu} CF_2Cl + Cl \qquad \text{...(5.11)}$$

The Cl-atom released then reacts in a catalytic chain reaction, which leads to the destruction of O_3

$$Cl + O_3 \longrightarrow ClO + O_2 \qquad \text{...(5.12 a)}$$

$$ClO + O \longrightarrow Cl + O_2 \qquad \text{...(5.12 b)}$$

$$\text{Net} \quad O_3 + O \longrightarrow 2O_2$$

A small amount (15%) of CH_4 released at the earth's surface diffuses into the stratosphere, where it can act as terminator for the Cl atom catalyzed destruction of O_3, by forming HCl.

$$CH_4 + Cl \longrightarrow \dot{C}H_3 + HCl \qquad \text{...(5.13)}$$

Although the HCl can recycle to produce Cl-atom via reaction with OH.

$$HCl + \dot{O}H \longrightarrow H_2O + \dot{C}l \qquad \text{...(5.14)}$$

But some of it may diffuse down through the tropopause, get dissolved by rains, thereby may act as a permanent chlorine sink. The observed increase in methane in troposphere may therefore lower the stratospheric chlorine atom concentration and hence decreases the destruction of ozone layer. Methane might also contribute to ozone production in the stratosphere, independent of CFC's cycle via the CH_4—OH—NO_x chemistry.

Methane oxidation in natural troposphere occurs by OH attack

$$CH_4 + \dot{O}H \longrightarrow \dot{C}H_3 + H_2O$$

$$\dot{C}H_3 + O_2 \longrightarrow CH_3\dot{O}_2$$

Again the fate of peroxy radical depends on the ambient concentration of NO. In the presence of sufficient NO, CH_3O_2 reacts with NO

$$CH_3\dot{O}_2 + NO \longrightarrow CH_3\dot{O} + NO_2 \quad ...(5.15)$$

$$CH_3\dot{O} + O_2 \longrightarrow HCHO + H\dot{O}_2$$

HO_2 so produced oxidizes another NO to NO_2

$$H\dot{O}_2 + NO \longrightarrow H\dot{O} + NO_2$$

Thus, on further oxidation of HCHO, two molecules of NO have been oxidized to NO_2 and hence two molecules of O_3 can result from photolysis of the NO_2 via reaction (4.47a and 4.16).

However, if the NO level is low, reaction (5.15) may be sufficiently slow so that reaction (5.16) with HO_2 may compete:

$$CH_3\dot{O}_2 + H\dot{O}_2 \longrightarrow CH_3OOH + O_2 \quad ...(5.16)$$

In this case, O_3 is not formed as in the CH_4 oxidation with $k_{515} = 7.6 \times 10^{-12}$ cm^3 molecule^{-1} s^{-1} and $k_{516} = 3 \times 10^{-12}$ cm^3 molecule^{-1} s^{-1}, removal of CH_3O_2 by 5.15 is equal to that by 5.16 *i.e.*

$$[CH_3O_2][NO]k_{515} = k_{516}[CH_2O_2][O_2H]$$

or

$$[NO] = \frac{k_{516}}{k_{515}}[O_2H]$$

Thus for HO_2 concentration of 1×10^8 molecules cm^{-3}, NO concentration be calculated:

$$[NO] = \frac{3 \times 10^{-12}}{7.6 \times 10^{-12}} \times 1 \times 10^8 = 0.3947 \times 10^8 \text{ molecules cm}^{-1}$$

$$= 1.6 \text{ ppt}$$

It is thus expected that hydroperoxides should be formed in the clean troposphere if NO level are very low.

As discussed at the beginning of this chapter, the effect of NO_X emissions on the O_3 concentrations depends critically on the altitude of emission. Thus injection in the mid and upper stratosphere will lead to net O_3 destruction via reaction 4.43 and 4.52. However, the opposite effect, that is a net increase in O_3 is predicted when NO is injected into the lower stratosphere and upper trophosphere. This seemingly paradoxical result is due to competing processes, for NO_2 loss formed in 4.43. In the mid and upper stratosphere, reaction 4.52 dominates the NO_2 loss so that the 'odd oxygen' species—O_3 and O are destroyed in the cycle consisting of 4.43 and 4.52

$$NO + O_3 \longrightarrow NO_2 + O_2 \quad ...(4.43)$$

$$NO_2 + O \longrightarrow NO + O_2 \quad ...(4.52)$$

$$\text{Net} \quad O_3 + O \longrightarrow 2O_2$$

However in the lower stratosphere and upper troposphere, the O atom concentration are lower and photolysis of NO_2 to form O, which then produce O_3 become competitive.

$$NO_2 + h\nu \longrightarrow NO + O \quad ...(4.47)$$

$$O + O_2 \xrightarrow{M} O_3 \quad ...(4.16)$$

Figure 5.4 schematically shows the interactions between these species in the troposphere and stratosphere.

Nitrous oxide N_2O is emitted in biological processes at the earth's surface, and its emission rate may be modified by anthropogenic activities by the increasing use of fertilizers. It has no known significant tropospheric reaction, except possibly decomposition when adsorbed on surfaces such as sands, however, it appears to be too slow to be a significant loss process. Hence like CFC, N_2O slowly diffuses into stratosphere. Here it is either photolyzed to N_2 and O_2 or is converted to NO via reactions.

$$N_2O \longrightarrow N_2 + O_2 \quad ...(5.1)$$

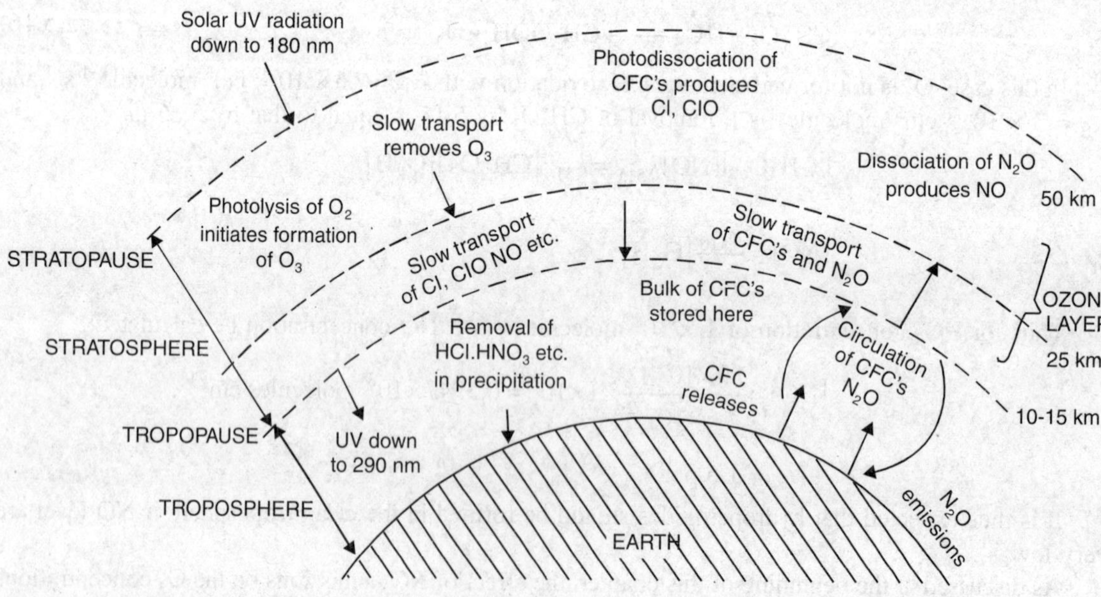

Fig. 5.4 Schematic diagram of the processes determining the concentration of ozone in the stratosphere.

$$N_2O + O(^1D) \longrightarrow 2NO \qquad ...(5.2)$$

The NO produced in 5.2 then reacts via 4.43 and 4.52 above to destroy oxygen atom. Since the production of NO in the reaction 5.2 occurs primarily between 20 and 40 km, the net effect is decrease in O_3.

NO_x is closely coupled to ClO_x reaction as has been discussed in Section 5.4. Because of the close coupling of various chemical cycles, substantial change is predicted in column O_3. To illustrate the coupling between chemical cycles and how this alters the model prediction, let us take one of the more significant changes—the change in HO_2—NO rate and examine why this changed the predicted column abundance of O_3 in reaction 4.39.

Prior to 1973, at room temperature the rate constant reported for NO and HO_2 reaction was 2×10^{-15} cm^3 $molecule^{-1}$. However, measurement carried from 1973 to 1977, the increasing values with the largest being 8.8×10^{-12} cm^3 $molecule^{-1}$ s^{-1} have been reported and have been confirmed by number of other research groups. When this revised rate was used as a model input, an increase in loss of column O_3 due to CFC but a decrease due to NO resulted. This occurs because of the rapid removal of NO by reaction with HO_2, shifts the fate of NO away from its reaction 4.43 which destroys O_3. It has also changed NO_2/NO ratio in favour of NO_2

$$HO_2 + NO \longrightarrow NO_2 + OH \qquad ...(4.39)$$

Part of NO_2 then photolyzes ultimately to produce O_3

$$NO_2 + O_2 \xrightarrow{h\nu} NO_3 + O \qquad ...(4.47)$$

$$O + O_2 \longrightarrow O_3 \qquad ...(4.16)$$

On the other hand, the increased OH production via reaction 4.39 increases the rate of recycling of the HCl back to the catalytically active Cl atom

$$OH + HCl \longrightarrow H_2O + Cl \qquad ...(5.14)$$

Thus a change in this rate constant in HO_X—NO_X cycle had a dramatic effect on the predicted changes in the total O_3 concentrations illustrating the importance of coupled reaction chemistry.

CO_2 can interact with the HO_X—NO_X—NO_X—ClO_X cycles through its effect on the radiation balance. Although CO_2 acts to warm the troposphere by absorbing infrared-radiation (green house effect) and convert it to heat due to a complicated balance between emission and absorption, it has the opposite effect in stratosphere.

Increase in CO_2 due to fossil fuel combustion are thus expected to warm the troposphere, but cool the stratosphere. This cooling effect changes the rates of the chemical reaction in stratosphere. The net effect on an increase in CO_2 on the rates is expected to result in an increase in the O_3 concentration. However, the absorption of light by O_3 in stratosphere produces heat; this increased heating from increased O_3 may partially offset the decreased temperature due to CO_2.

It is important to recognize that the time scale for altering the O_3 column is quite different for different species. For example, the long time required by CFC's to diffuse from earth's surface into stratosphere means that CFC released today will have their effect on the O_3 many years later. On the other hand, NO_x released from aircraft directly into the upper atmosphere causes effects that are essentially immediate. This difference presents difficult control strategy in the case of CFC's.

The calculated loss in ozone from the CFC's already in atmosphere is about 1%, and the eventual ozone loss would be more than 10% for long term yearly release of 800,000 loss of F—12 and F—11.

Synthesized more than 40 years ago as refrigerant, F—12 has been highly successful in this application.

5.8 OTHER HALOGEN CONTAINING COMPOUNDS

In addition to CFCs, a number of other halogen containing organic compounds are emitted into the troposphere. These include compounds such as carbon tetrachloride CCl_4, Methyl chloroform CH_3CCl_3, dichloro methane CH_2Cl_2, perchloroethane C_2Cl_6, trichloroethene $C_2H_3Cl_3$, Methylchloride CH_3Cl, Methyl Bromide CH_3Br.

An important difference between CFCs and other compounds is the presence of abstractable hydrogen atoms and/or reactive C=C double bonds. Thus OH attack on such species is much faster than for F—11 and F—12 and as a result, their tropospheric lifetimes are much shorter. Table 5.2 gives the room temperature rate constants for OH reaction with these molecules as well as the corresponding lifetimes for an OH concentration of 1×10^6 radicals cm^{-3} during the day. Some of these compounds may be removed in the troposphere, balance of which being transported into the stratosphere, where they too may produce chlorine atoms on photolysis.

Calculations indicate that some of these halocarbons have already contributed to the depletion of O_3 layer and others might play important role in future.

Although the emphasis in this chapter has been on chlorine containing compounds, brominated compounds emitted in troposphere may also deplete ozone layer. Anthropogenic sources of organic bromine compounds include their use as soil fumigants CH_3Br, gasoline additives $C_2H_4Br_2$ flame retardants $CBrF_3$ and CBr F_2CBrF_2.

Table 5.2 Annual Emission rates of some Halocarbons, their reaction rate constants with OH reaction (eq. 5.14) at 298K, their lifetimes under relatively clean conditions

Halocarbon	*1979 Emission rate (10^6 kg)*	k *cm^{-3} molecule^{-1} s^{-1}*	$\tau = \frac{1}{k[OH]}$
$CFCl_3$ (F-11)	283		—
CF_2Cl_2 (F-12)	340.0	—	—
CCl_4	118.0	—	—
CH_3Cl	—	4.3×10^{-14}	270 days
CH_3CCl_3	522.3	1.2×10^{-14}	2.6 yr.
C_2HCl_3	459.1	2.2×10^{-12}	5 days
C_2Cl_4	667.0	1.7×10^{-13}	68 days
CH_2Cl_2	356.0	1.4×10^{-13}	83 days

The troposphere chemistry of bromine compounds is similar to that of chlorine containing compounds. Thus CH_3Br with its abstractable hydrogen atom is attacked by OH in the troposphere with a mean lifetime of ~ 1 year for an OH concentration 1×10^6 cm^{-3}. Alternatively, it can diffuse into the stratosphere where it not only continues to react with OH, but also photolyzes to produce Br atoms which participate in catalytic ozone destruction via reaction analogous to reaction 5.6 and 5.7.

Some exceptions of bromine containing compounds are there, for example, Br does not abstract hydrogen from CH_4. Bromine involves primarily reaction with HO_2 only, which is slower than the corresponding HO_2 + Cl reaction. In addition, BrO photolyzes much faster than ClO and hence photolysis can recycle BrO into Br atom. As a result of these differences compared to ClO, Br is relatively more efficient than Cl as ozone destruction spends more time in catalytically atomic Br form and less in the form of HBr.

A particular interesting aspect of stratospheric bromine chemistry is the possibility of synergistic interactions between the bromine and chlorine cycles via reaction

$$BrO + ClO \longrightarrow Br + Cl + O_2 \qquad ...(5.17)$$

In this way ClO_x—BrO_x interaction could lead to enhanced O_3 destruction in the lower stratosphere.

5.9 PROBING THE OZONE SHIELD

Direct measurements of ozone loss from man's activities are very difficult because of the natural variability of ozone concentration at any single location. Continuous measurement is needed at each monitoring station to gain averages over the days, months, seasons and years, and many stations are needed globally. Measurements at the few stations where continuous records are available since 1930s show year-to-year natural variation that some investigators have associated with the well known 11-year cycle of solar activity. With such natural variations, good world wide measurements over a 40–50 years period at many stations probably would be needed to detect an ozone change of 1–2% introduced by man. Such records do not exist, and probably the best that can be done currently is the detection of an additional man made change of 3% persisting for at least five year. Because waiting many years for a change of the magnitude would entail severe risk.

The increase in ozone levels during the 1960s has not been satisfactorily explained, although both the solar sunspot maximum in 1969 and the almost total cessation of nuclear testing in atmosphere at the beginning of the decade have been widely discussed as contributing causes. The decrease since 1970 as the sun emerged from a solar sunspot minimum and has increased sunspot activity with a maximum in 1980-81. This only leads us to conclude that sunspot activity do not affect ozone layer. It is only affected by sun flame as discussed in Section 5.11.

Continuous ozone measurement by satellite is currently in progress, but the technique requires further refinement. Since no satellite data are available prior to 1970, long-term trends from such instruments will not be available for many years.

Although the basic processes that maintain natural levels of atmospheric ozone are reasonably well understood, the deviation in concentration from year-to-year requires much more precise analysis. In certain cases, however, short term changes have been explained successfully. For instance, in August 1972 the sun erupted violently with long flaming prominences and its magnetic disturbances were felt on earth. The cataclysm also spewed out huge numbers of energetic hydrogen nuclei or protons. This "solar proton event" —the largest observed during the two decades in which they have been recorded—caused extensive ionization in the high atmosphere at polar latitudes. The ionized species so created reacted rapidly to form NO, and this sudden appearance of catalytic material caused an ozone reduction that was recorded in satellite ozone measurements. The amount of data returned by such satellites is so voluminous, however that the immediate atmospheric effects of this solar proton event remained hidden until atmospheric calculations were performed in 1975. A subsequent search is still unexamined satellite data for 1972 showed a decrease within one day of about 16% in ozone concentration above an altitude of 25 miles and above latitude 70°N, no effect was observed at lower latitudes where the Earth's magnetic field turned away the solar protons far above the stratosphere, preventing the formation of NO. This solar proton event is the best example providing actual observation of the removal of stratospheric ozone by the NO_x catalytic effect.

The protection of the ozone layer has been promoted in international frameworks such as the Vienna convention for the protection of the ozone layer and in Montreal protocol, which were concluded in 1985 and 1987, respectively. The protocol stipulated that the production and consumption of controlled CFCs (F-11, -12, -113, -114, -115) in 1998 should be half that of 1986. It was recognised internationally however, that the agreement alone would not be sufficient to protect ozone layer but be implemented with stronger restrictions.

5.10 OZONE HOLE

Ozone is a photochemical product of oxygen which is formed by solar radiation of short wavelength. The existence of ozone belt around the earth is mainly responsible for filtering out some of the lethal ultraviolet radiation from reaching the earth, there by acting as a protective shield for sustaining life of plants, animals and humans on earth.

For most of the year the stratosphere over Antarctica has fairly high concentration of ozone. The ozone molecules formed over the tropics level along with chlorofluro carbons moves to polar regions via atmospheric motion. This traps the ozone over the south pole for several months except winter. It is within this vortex the substantial ozone loss was first detected in 1980. The major ozone damage occurs during winter when tiny ice particles are formed in polar stratosphere. These particles act as catalyst to enhance, directly or indirectly, the conversion of chlorine into its ozone destroying form, chlorine monoxide. The entire mechanism of ozone depletion is quite complicated but derived from TOMS version 6 on Nimbas 7 TOMS (1979–92) and on Meteor 3 (1993–1994) reported by NASA.

In the course observations conducted at Japan's Syowa station in the Antarctica, Chubachi Shigeru, a senior researcher of the Meteorological Research Institute, Japan Meteorological Agency, found that there was a hole in the ozone, 15 to 25 km above earth. Knowing that there would be no sun over the Antarctica from May to August, and that it would be impossible to perform ozone observation using sunlight, he devised a way to use moonlight to monitor the observations throughout the year. A highly sensitive electronic equipment capable of functioning in relatively dim light was incorporated into a Dobson spectrophotometer, making it possible to continue to obtain data on ozone by moonlight. September 4, 1980, was the coldest day on record. On that day, the data recorded by Chubachi's ozone observation team was unusual. It happened to be the first day of sunlight observation in four months. The data showed that the amount of ozone was significantly less than it had been on the day before, when observations were made in moonlight everyday thereafter, the data continued to show a decrease in ozone layer, not reversing until October 28, when the data went back. Chubachi and his colleagues had stumbled across the discovery that ozone layer was indeed being depleted. Similar observation on the ozone hole was recorded by the National Aeronautic and Space Administration (NASA). Ozone hole was first observed in October 1980 which became thinner and thinner year after year. This is vividly clear from the increase in red colour intensity in the Figures of 1980, 1985, 1989, 1991, 1993 and 1994 shown here. Halogen atoms, particular chlorine atoms were mainly responsible for the depletion of ozone layer. So it was anticipated that concentration of ClO in Antarctic stratosphere should be quite high. However, if NO_x be present in significant concentration, it would interfere in the ozone depletion. In other words, rates of depletion should have been quite low which is not the case. Hence the concentration of NO_x should decrease towards the hole concentration and ClO should be quite high.

These anticipations were confirmed in 1987 when a significant decrease in NO concentrations and increase in ClO concentration were observed near the ozone hole, hence confirmed what was anticipated. These results confirm the existence of hole over Antarctica due to the ozone loss.

Reactions are particularly intense within stratosphere clouds that forms over Antarctica in the extremely cold night of southern hemisphere winter. Reactions which occur on the surfaces of ice particles within the clouds release chlorine and bromine from CFC and Halons in active form that accumulate through winter. When sun rises in the spring, the clouds break up to release active chlorine and bromine which rapidly destroys ozone. The result is the ozone hole, an area of sharp decline in ozone concentration over most of the Antarctica for about two months during the southern hemisphere spring.

The discovery of the Antarctic ozone hole raised the possibility that a similar hole could exist over Arctic region. Early results from a series of measurements in US inter 1982–83 suggested that ozone loss over Arctic region also exist. However, this disappeared in subsequent years but has reappeared in 1983 and 1994 as shown in photograph high red colour. This shows that ozone depletion over north pole is there but not to the degree it was over Antarctica. However the mechanisms of depletion may be same.

Stratospheric air above Arctic is generally warmer and less confined than Antarctica and fewer clouds form there. Arctic depletion is therefore less severe, though in recent years it has proved worse than expected. Polar ozone depletion is accelerated by atmospheric circulation on which moves CFC's in stratosphere away from the tropics towards both poles.

Ozone depletion will make earth susceptible to ultraviolet radiation. DNA molecule show the most intense absorption of *UV* light at 265 nm. The long wavelength extreme of this absorption band extends to just above 300 nm. Proteins have absorption maxima typically around 275 to 285 nm and these absorption bands also extends above 300 nm. Plants are sensitive to *UV* radiation below 300 nm. Damage caused to DNA, protein and plants by *UV-B* and *UV-C* radiation will be cumulative and it is the integrated dose over many years that will produce skin cancer, may change the genetic codes or cause mutation in plants. It will damage DNA molecules which could cause development of strains of microorganisms against which man has no resistance or strains resistant to antibiotics. Ultraviolet radiation will increase skin cancer, sunburns and eye damage to human beings. It may impair certain crop plants and may adversely affect a wide variety of organisms from bacteria to vertebrates. It will also lead to faster deterioration of paints, fabrics, plastics and other such material, magnitude of which is uncertain but significant.

Ozone depletion may also affect the climate though not clearly understood. Ozone absorbs and reradiates radiation in the infrared and ultraviolet region. With its depletion, spectrum of incoming radiation will change and hence may produce new molecular species. These molecules may change the radiation emitting and absorbing properties of atmosphere. In the absence of ozone layer the radiations will directly flow to the earth surface and hence change the vertical distribution of molecules in the troposphere and stratosphere, which in turn change the rate of vertical circulation. All these may influence the climate change, but how, that can not be anticipated.

5.11 THE SUN

The impassive face that the sun offers us is only an appearance. It is a ball of gas–hydrogen, helium and traces of heavy elements–bought to incandescence and ionised into plasma. This sphere not only lights but also heats the universe. Every moment it is hurling into space, out to Earth and other planets, shreds of gaseous matter with the power of several billion thermonuclear bombs. This continuous torrent of charged particles—protons, electrons and heavy ions—has gradually reduced the rocks on the surface of Mercury and the Moon to powder. Our planet is protected by a magnetic shield that moves the charged particles towards the poles or shunts them aside. Periodically (roughly every 11 years) this solar wind is unleashed and blows in violent gusts. The consequences can be spectacular, for example it affects the ozone layer around the earth, as was discussed in Section 5.10. Another example, the solar wind tripped the entire electric grid of the province of Quebec on March 13, 1989, disabling it for six hours.

The phenomenon at the source of these disturbances can now be observed. Two images captured by the American and European space craft SOHO (Solar and Heliospheric Observatory) on September 14, 2000 and January 18, 2001 bear witness to the violent ejection of matter in the chromosphere, the gaseous layer 200 km thick that envelops the visible surface of the Sun. Two gigantic plasma protuberances, which could each contain several tens of Earths are clearly visible at the top right of the first image and at the lower left in the second. They shoot up to several hundreds of thousands of kilometers above the surface, equivalent to the distance between Earth and the Moon. Some of these arcs of matter remain suspended for several days, even months, on end. Others more unstable explode and release a torrent of charged gas in a solar eruption. These flares produce rays similar to Cosmic rays in composition 90% protons and about 10% α-particles. In the process the cosmic rays may increase in intensity thousand folds. These large flares occur on the average every 4–5 years. Apparently if flares are in the direction of Earth can let loose electromagnetic storms with the consequences described above which include destruction of Ozone layer on earth. It may be repeated that it is these flares and not the spots with which electromagnetic perturbation on earth is associated. Emission of electrically charged particles which when hit earth atmosphere produce perturbation of the terrestrial magnetic field and polar aurorae. The latter manifest themselves in field fluctuation and vibration of magnetic pointer of compass. These oscillations give rise to sudden radio flashes thus noises in radio and telegraphic communications.

These images, captured in extreme ultraviolet (here, waves emitted by ionised iron), show among other things that the disk itself is far from homogenous. The coolest zones (6000K) appear in dark shades; the hottest are lighter. Within the hot zones the temperature climbs to 60,000K. The visible structure in these images are linked to solar magnetic fields. The Sun's magnetic field is ten thousand times more powerful that Earth's, but, above all, much stable. Its polarity shifts every 22 years, and in that interval it acquires a very complex structure. Instead of emerging from the Sun at the two poles, the magnetic field forms multiple loops that burst out and then dive back below the surface of the sun in areas of dark sunspots (not seen here). The movement of ionized gas in the chromosphere follows these trajectories. Much research will be necessary, however, to better understand the formation of these structures and to predict, even if it is only hours in advance, eruptions directed towards our planet, that would allow us to protect the most vulnerable electronic equipment from magnetic storms and to prevent airplanes from flying over the polar regions during the strongest gusts.

REFERENCES

1. Chang, J.S. and Duewer W.H., Modeling Chemical Processes in Stratosphere, *Ann. Rev. Physi. Chem.* 30, 443 (1979).

2. DeMore, W.B., Molina M.J., Watson R.T., Golden D.M., Hampson R.F., Kurylo M.J., Howard C.J. and Ravishankara A.R., *Chemical Kinetics and Photochemical Data for Use in Stratospheric Modeling, Evaluation No. 6,* JPL Publ. No. 83-62 September 15, (1983).

3. DeMore, W.B., Margitan J.J., Molina M.J., Watson R.T., Golden D.M., Hampson R.F., Kurylo M.J., Howard C.J. and Ravishakara A.R., *Chemical Kinetics and Photochemical Data for Use in Stratospheric Modeling, Evaluation No. 7*, JPL Publ. No. 85-37 (1985).
4. Turco, R.P., *Stratospheric Ozone Perturbations in Ozone in the Free Atmosphere,* R.C. Whitten and S.S. Prasad Eds. Van Nostrand, Reinhold, New York, Chap 5, (1985).
5. Turco, R.P., *Photochemistry of the Stratosphere in the Photochemistry of Atmosphere, Earth, the other Planets and Comments,* J.S. Levine Eds., Academic Press, New York (1985b).
6. Wayne, R.P., *Chemistry of Atmosphere,* Clarendon Press, Oxford (1985).
7. Whitten, R.P. and Prasad S.S., Eds *Ozone in the Free Atmosphere*, Van Nostrand, Reinhold, New York (1985).
8. National Aeronautics and Space, *Vertical, Present State of Knowledge in the Upper Atmosphere,* January, (1984).
9. National Research Council, *Causes and Effects of Changes in Stratospheric Ozone:*. Committee on causes and effects of changes in stratospheric ozone, update 1983, Environmental Studies Board, Commission on Physical Sciences, Mathematics and Resources, National Academy Press, Washington DC (1984).
10. Crutzen, P.J., Chlorofluoromethanes: Threats to the Ozone Layer, *Rev. Geophysics Space Phys*., 17, 1824 (1979).
11. Paul J. Crutzen "Influence of Nitrogen Oxides on Atmospheric Ozone Content", *Q.J.R. Meteorol. Soc.*, 96, 320 (1970).
12. Harold, S. Johnston, "Reduction of Stratospheric Ozone by Nitrogen Oxide Catalyst from Supersonic Transport Exhaust", *Science*, 173, 517 (1971).
13. Howard, C.J. and Evenson K.M., "Kinetics of the Reaction with HO_2 and NO", *Geophys. Res. Lett.* 4,437 (1977).
14. Profit and 11 others, (1989). A Chemical Definition of the Boundary of the Antarctic Ozone Hole.; I. *Geophys. Res.,* 94 (D9) 11437-11448 (1989).
15. Hubrich and Stuble, "The Ultraviolet Spectra of Some Halogenated Methanes and Ethanes of Atmospheric Interest", *Journal of Photochemistry and Photobiology* 12,93-107 (1980).

6

Carbon Dioxide and Climate

6.1 CARBON MONOXIDE, CO

Carbon monoxide has a bond length of 11.28 nm with singlet ground state as $^1\Sigma^+$. Bond dissociation energy of CO is D (C—O) = 11.09 eV = 1070 kJ mol^{-1} corresponding to a photon of wavelength 111.8 nm. Some of the various electronic transition of CO are given in Table 3.6. The lowest transition of CO is $^3\Pi_r \leftrightarrow {}^1\Sigma^+$ corresponding to an absorption at (206 nm) 48473 cm^{-1} which lies in near *UV* and thus is of not much interest to the photochemical reaction.

Carbon monoxide is highly toxic gas with a very high concentration in urban atmosphere with no warning property such as odour. Up till now we do not know about any cycle by which CO can be removed from our atmosphere.

With increasing concentration the usual sequence of warning signal and symptoms of injury are headache, dizziness, flickering of the eyes, ringing in the ears, vomiting, palpitations, pressure on chest, difficulty in breathing, unconsciousness and death. Haemoglobin in our blood acts as an oxygen carrier from lungs to tissues as

$$Hb + O_2 \longrightarrow HbO_2$$

Carbon monoxide have two nonbonding pairs of electrons. One of this pair of electron has the highest energy and thus most reactive and lies over carbon atom. This pair of electrons form coordinate bond with metal to give metal carbonyls as M–C–O. Similar Fe–CO bond formation take place in haemoglobin. Carbon monoxide combines with haemoglobin 210 times readily than oxygen thus prevents it from performing its important function of transporting the oxygen from lungs to tissues

$$HbO_2 + CO \longrightarrow HbCO + O_2$$

This reaction is almost irreversible. Not only is the oxygen displaced from haemoglobin molecule but remaining oxygen tend to be bound somewhat more firmly, thus further decreasing the oxygen delivery capacity of haemoglobin. In the initial stages for each 1 ppm of CO with which body is in equilibrium, 0.16% of body haemoglobin will be inactivated. At normal breathing rate, 7-8 hours are required to reach a blood saturation at CO air level from 50 to 100 ppm.

6.1a Role of NO in O_3 Production and CO Oxidation

In the natural troposphere, NO_x and CO are produced by natural processes as well as by industrial units. The sole known oxidant for CO in atmosphere is the hydroxyl radical. The major source of OH is the photolysis of O_3 followed by the reaction of electronically excited O (^{1}D) atom with water. The OH—CO reaction sets off a sequence which produces HO_2

$$CO + \dot{O}H \longrightarrow \dot{H} + CO_2 \quad ...(4.37)$$

$$\dot{H} + O_2 \xrightarrow{M} H\dot{O}_2 \quad ...(4.38)$$

The fate of HO_2, however, depends on the ambient concentration of NO. If sufficient HO_2 is present, HO_2 converts NO to NO_2 which is then photolyzed to produce O_3,

$$H\dot{O}_2 + NO \longrightarrow \dot{O}H + NO_2 \quad ...(4.39)$$

$$NO_2 \xrightarrow[\lambda \leq 430\text{ nm}]{h\nu} NO + O \quad ...(4.47b)$$

$$O + O_2 \xrightarrow{M} O_3 \quad ...(4.16)$$

The sequence 4.37, 4.47b and 4.16 adds upto the following stoichiometry

$$CO + 2O_2 \longrightarrow CO_2 + O_3$$

That is one O_3 molecule is produced for each CO oxidised.

On the other hand, if the NO concentration is very low, reaction 4.39 will be very fast and hence either the self reaction 4.37 of the two radicals or the reaction 6.2 of HO_2 with O_3 can begin to compete with 4.39.

$$H\dot{O}_2 + H\dot{O}_2 \longrightarrow H_2O_2 + O_2 \quad ...(6.1)$$

$$H\dot{O}_2 + O_3 \longrightarrow \dot{O}H + 2O_2 \quad ...(6.2)$$

Reactions 4.37, 4.38, 6.1 and 6.3,

$$H_2O_2 + h\nu \longrightarrow 2\dot{O}H \quad ...(6.3)$$

taken together add upto the following net reactions

$$2CO + O_2 \longrightarrow 2CO_2$$

Thus O_3 is neither destroyed nor formed in the oxidation of CO. Finally, if $H\dot{O}_2$ reacts with O_3 instead, then the sum of reactions 4.37, 4.38 and 6.2, is

$$CO + O_3 \longrightarrow CO_2 + O_2$$

that is one molecule of ozone is destroyed in the oxidation of CO molecule.

The NO concentration at which these various paths become equal can be calculated from a knowledge of the rate constants k_{439}, k_{61}, k_{62} for these $H\dot{O}_2$ reactions, for some typical concentrations of $H\dot{O}_2$ and O_3. The values of these rate constants at 298 K are

$k_{439} = 8.3 \times 10^{-12}$ cm^3 molecule s^{-1}

$k_{61} = 5.6 \times 10^{-12}$ cm^3 molecule s^{-1}

$k_{62} = 2.0 \times 10^{-15}$ cm^3 molecule s^{-1}

In order for the self-reaction 6.1 of $H\dot{O}_2$ take equal to be reaction 4.39 of HO_2 with NO

$$k_{439}[\mathrm{NO}] = k_{61}\left[H\dot{O}_2\right]$$

Taking $H\dot{O}_2$ concentration of 1×10^8 cm^{-3} as being typical of natural troposphere during daylight hours, then

$$[\mathrm{NO}] = \frac{k_{61}}{k_{439}}\left[H\dot{O}_2\right] = \frac{5.6\times10^{-12}}{8.3\times10^{-12}}\times1\times10^8 = 6.75\times10^7 \text{ molecule cm}^{-3}$$

$$= 2.74 \text{ ppt}$$

Similarly, for reaction 6.2 of $H\dot{O}_2$ with O_3 to compete with reaction 4.39 of $H\dot{O}_2$ with NO

$$k_{439}[\mathrm{NO}] = k_{62}[O_3]$$

Taking a typical O_3 concentration of 40 ppt (1×10^{12} cm^{-3}), then

$$[\mathrm{NO}] = \frac{k_{62}}{k_{439}}[O_3] = \frac{2.0\times10^{-15}}{8.3\times10^{-12}}\times1\times10^{12} \text{ cm}^{-3}$$

$$= 2.4\times10^8 \text{ molecule cm}^{-3} = 10 \text{ ppt}$$

In short, the NO concentration must be ~ 3–10 ppt or less in order for reactions of $H\dot{O}_2$, other than that with NO, to be significant. While such small concentrations of NO have been measured in some locations, it is not clear in how much of the global troposphere this is the case.

Table 6.1 summarises the possible fates of $H\dot{O}_2$ and the resulting impact on O_3 production or destruction during CO oxidation.

Table 6.1 Possible fates of HO_2 in the natural troposphere and resulting impact on O_3 during CO oxidation

Fate of HO_2		*Impact of O_3 during oxidation of one CO molecule*	*Approximate conc. of NO at which rate of HO_2 removal (4.39) equals that in 6.1 or 6.2*
$H\dot{O}_2 + NO \longrightarrow \dot{O}H + NO_2$	4.39	One O_3 formed	–
$H\dot{O}_2 + H\dot{O}_2 \longrightarrow H_2O_2 + O_2$	6.1	No O_3 formed or destroyed	3 ppt
$H\dot{O}_2 + O_3 \longrightarrow \dot{O}H + 2O_2$	6.2	One O_3 destroyed	10 ppt

The above results are with the assumption $[H\dot{O}_2] = 1 \times 10^{12}\ cm^{-3}$, $[O_3] = 1 \times 10^{12}\ cm^{-3}$ and 50% relative humidity.

6.2 CARBON DIOXIDE, CO_2

Carbon dioxide is a linear symmetrical molecule with C—O bond length of 11.62 nm. The ground state of this linear molecule is singlet state $^1\Sigma_g^+$. It does not show any absorption in visible and near ultraviolet region, rather is completely transparent at least up to 210 nm. The first known absorption of CO_2 starts gradually at 175 nm reaches a maxima at 147.5 nm showing diffuse bands which are overlapped by a continuum. The intensities and spacing of the band are quite irregular. In the excited state the molecule is bent (122° ± 2°) transition from the minimum of the upper state leads to very highly excited vibrational levels in the visible and near ultraviolet region. The excited electronic state has been identified as a $^1B_2$ state derived from a $^1\Delta_2$ state of a linear molecule.

The second absorption region lies in 139–122 nm is somewhat sharper and more regular with an intensity maxima near 133 nm.

Bond dissociation energy of D (O—CO) = 5.453 eV = 526.16 kJ mol^{-1}, which corresponds to a photon of wavelength 227.3 nm. The precise mechanism of this dissociation is not clear.

As all the absorption bands are observed in near ultraviolet region and beyond, free carbon dioxide molecule is of not much importance to photochemical reactions in atmosphere.

6.3 GROW MORE TREES

Carbon dioxide is transformed from its gaseous state into oxygen, carbohydrates and other organic materials by photosynthesis. Carbon dioxide gets tagged to chlorophyll material of plants, which undergoes photosynthesis. Respiration by all organisms returns carbon dioxide to atmosphere. This simple CO_2 cycle and role of CO_2 in water as buffer has been discussed very explicitly in another publication of author. This CO_2 concentration is increasing at a rate in excess of 1 ppm per year as a result of the burning of fossil fuels, the manufacture of cement and the cutting of forests. This increase in CO_2 content is significantly affecting the earth's climate.

Prior to industrialisation, CO_2 in and out of atmosphere might have been in equilibrium at least for short period of time. It is now estimated that the atmospheric concentration of carbon dioxide prior to the industrial revolution might have been about 268 ppm. There is evidence that the concentration was as low as 150 ppm about 5,000 years ago. This equilibria had been maintained by the plants and forests. The consequences of deforestation, overgrazing, intensive burning and land scarring has brought change in local climates, declining water tables and erosion of soil. This increase, in local concentration of CO_2, which reduced atmospheric humidity and cloud cover, increased heat reflectivity and lowering of rainfall. It has been shown that forests help to maintain the level of rainfall for their own existence, and that deforestation results not only in an immediate lowering of ground water level but also lowers the rainfall. Thus, deforestation increased the CO_2 concentration in air and made the atmosphere dry and erosion of soil converted it to desert. Some of the historical evidences will make the arguments amply clear.

6.3a Ancient Phoenicia and Slope Farming

About 5,300 years ago, the Phoenicians migrated from desert to settle along the eastern coast of the Mediterranean sea and established the coastal towns of Tyre and Sidon, Beyrouth and the port of Byblos (present day Jbail). The land is mountainous with a relief of about 3,000 meters and, at that time, was heavily forested with the famous cedars of Lebanon. These trees became the timber supply for the alluvial plains of the Nile and Mesopotamia. As the limited, flat land along the coast was populated to the carrying capacity, people moved to the slopes and as the slopes were cleared and cultivated, they were subject to soil erosion. Today there are a large number of terrace walls in various states of repair, indicating that the ancient Phoenician farmers attempted to control erosion with rock wall across the slope as many as 40 to 50 centuries ago.

The cedars of Lebanon retreated under the axe, and today very little of the original forest of approximately 2,600 sq km is left. Today the bare limestone slopes strewn with remnants of farmers terrace walls are testimony to the results of erosion and the decline and loss of the resources of a country.

6.3b Dead Cities of Syria

There exist in northern Syria a number of formerly prosperous cities that are now for all practical purposes dead. Before the invasion by Persians and Arabs, cities such as El Bare prospered from the conversion of forest land to farm land, which resulted in export of olive oil and wine to Rome. After the invasion, which destroyed their agriculture, as much as 2 meters of soil eroded from the slopes, and today the 13 centuries of neglect can be seen in the nearly complete destruction of the land. What is left of the formerly productive land an artificial desert generally lacking vegetation, water and soil.

6.3c Rajasthan

The Great Thar desert of western India has increased its size by 60,000 acres in the last 100 years. Dr. A. Krishnan of the Central Arid Zone Research Institute, Jodhpur, said in 1973 that archaeological evidence and recent carbon dating studies had established beyond doubt that the Rajasthan desert was largely caused by deforestation, overgrazing and deterioration of soil, which means that this desert is primarily a man-made desert.

Apparently, industrialisation or no industrialisation, forest are a must for the survival of urban man. So grow more trees to survive.

6.3d South Korea

A revolution of growing more trees took place in South Korea. During the early 1970's, cutting of trees in South Korea for fuel was not an unusual phenomena. Even the fallen leaves and twigs, which build soil fertility, were collected and burnt. The hill slopes were rendered barren. No one realised the seriousness of the problem, until the Han River overflowed drowning 672 in the capital Seoul and leaving over 3 lakhs homeless.

The catastrophe shocked the people of South Korea and brought them into action. A ten-year plan to replant approximately one million hectares of eroding forest was proposed. The work was enormous and so was enthusiasm of the people. About 20,000 local forestry associations came up, and their two million members devoted a month in each spring to planting trees. They returned each autumn to remove the deadly weeds. The government provided seedlings, fertiliser and technical assistance. The land owners were guaranteed 10 per cent of the yield from forest products.

The project was concluded six months ahead of schedule. Once again the countryside became green and more productive. The annual harvest of chestnut and pine mushrooms increased many folds. The replanted hillsides and watersheds cannot cork up rainfall and resist erosion. The risk of floods and landslides have been reduced. That's is green revolution!

6.4 ENERGY BALANCE OF THE EARTH

The sun can be considered a spherical light source of diameter 1.4×10^6 km located at 1.5×10^8 km from the earths surface. The discussion on the general circulation of the atmosphere in Chapter 1 proposed that most of the heating of the earth's surface comes from solar radiation. The solar energy reaching the surface of the earth each year is about 3.06×10^{24} kW. Heat generated by radioactive processes in the earth and conduction from the core contribute 8.76×10^{18} kW and man's activities contribute about 4×10^{18} kW per year. This means that less than 0.1 per cent of the total energy reaching the earth's surface each year comes from processes other than direct solar radiation. Therefore, the energy input to the earth is as if it all came from the sun.

The sun is almost a black body radiator (with superimposed line spectra). A perfect black body is one that absorbs all radiation hitting it and hence is also a perfect emitter. The energy density (energy per unit volume) per wavelength increment emitted by a black body is given by Planck's equation

$$\frac{dU(\lambda)}{d\lambda} = \frac{8\pi hc\left[\exp(-hc/\lambda kT)\right]}{\lambda^5\left[1-\exp(-hc/\lambda kT)\right]} \quad ...(6.4)$$

where $k = 1.38 \times 10^{-23}$ J molecule^{-1} K^{-1} is Boltzmann constant, h is Planck's constant 6.63×10^{-34} J s, c is the speed of light in vacuum 3×10^8 m s^{-1}; λ is the wavelength of interest in meters; and T is the absolute temperature.

Since the sun is emitting radiation mostly from its surface and can be considered as a black body radiator. The total intensity of radiation, or radiant flux over all wavelengths M, from a unit area of a black body at any temperature is given by the Stefan-Boltzmann Law

$$M = \sigma T^4 \quad ...(6.5)$$

where $\sigma = 5.672 \times 10^{-8}$ Wm^{-2} deg^{-1} or 34.39×10^{-11} Jm^{-2} s^{-1}K^{-4} is called Stefan-Boltzmann constant. Radiation from a black body ceases at a temperature of absolute zero, 0 K.

A black body is one which is capable of emitting and absorbing all frequencies of radiation without favour. A very good approximation to it is a pinhole in a container because the radiation leaking out of the hole has been reflected around inside so many times that it has come to thermal equilibrium with the walls. The origin of radiation from such a heated body is by the rapid vibrations of particles. These quantized oscillator emit radiant energy in the form of electromagnetic waves.

This emission of radiation spectra is solely dependent on the equilibrium temperature of black body and independent of its composition. The average temperature of earth at any time is close to 290 K and that of sun 6000 K. The energy density variation for these two bodies are shown in Fig. 6.1. It is seen that there is a wavelength of maximum emission λ_{max} which shifts to a lower wavelength as temperature is increased. The wavelength at which maximum emission of radiation occurs at any temperature is given by Wien's displacement law

$$\lambda_m T = 2.898 \times 10^{-3}\,\text{m K} \quad ...(6.6)$$

From the maxima of intensity of any black body, using this relation average temperature can be calculated. As in sun (Fig. 6.1) λ_m = 483 nm, temperature of sun comes to 6000 K.

At 290 K the entire range of radiation emitted lies in the infrared region while at 6000 K (λ_{max} = 483 nm) 90% of solar radiation is in visible (420 – 700 nm) and infrared (700 – 0.03 nm). The radiation intensity of the sun varies somewhat due to many fluctuating factors like nuclear reaction taking place in it. This is particularly the case for ultraviolet region and extremely variable in the X-ray region of the spectrum.

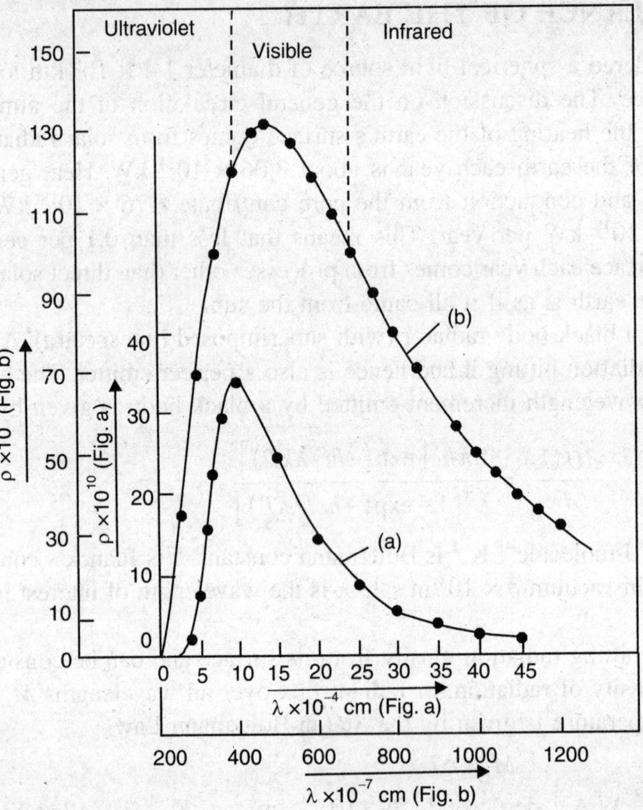

Fig. 6.1 Distribution of energy on (*a*) Earth 290 K (*b*) Sun 6000 K as a black body radiator.

Table 6.2 Fundamental vibrational spectral lines of some important gases

Molecule	*Vibrational frequencies in* cm^{-1}			
	v_1 *Symmetrical stretching*	v_2 *Bending mode*	v_3 *Asymmetrical stretching*	v_4
CO	1743.4 (5.739)			
CO_2	1340 (7.2) IR-inactive	667 (14.99) IR-active	2349 (4.26) IR-active	
H_2O	3650 (2.73)	1595 (6.27)	3755 (2.66)	
NO	2376 (4.208)			
NO_2	1220 (7.58)	749 (13.15)	1618 (6.18)	
N_2O	1285 (7.78)	588 (17.0)	2224 (4.5)	
CH_4	2916 (3.43) 1R-inactive	1533 (6.52) 1R-inactive	3020 (3.31) 1R-active	1306 (7.66) 1R-active
SO_2	1151 (8.68)	518 (19.3)	1360 (7.34)	
O_3	1043 (9.587)	710 (0.1408)	1740 (5.747)	

Note: Values given in brackets corresponds to wavelength in micrometers for reference to Fig. 6.2.

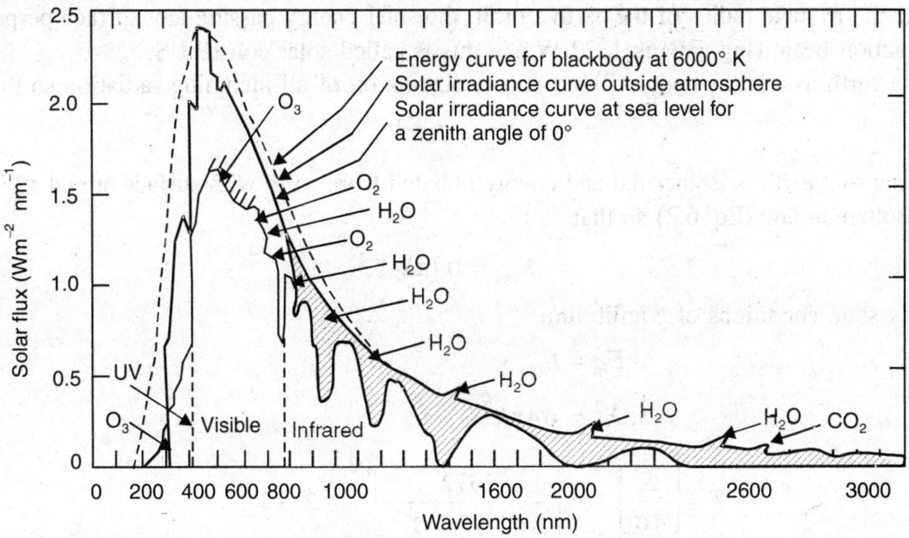

Fig. 6.2 Solar flux outside the atmosphere and at sea level, respectively. The emission of a blackbody at 6000 K is also shown for comparison. The species responsible for light absorption in the various regions (O_3, H_2O, etc.) are also shown.

The solar radiation that actually penetrates to surface of earth below the atmosphere no longer has the spectral distribution of black body at 6000 K. From the making and breaking of ozone molecule in stratosphere where almost whole of ultraviolet and other high energy radiations are removed. A large portion of sun's infrared radiation is absorbed by water vapours and carbondioxide in the atmosphere with numerical values given in Table 6.2. It is apparent from Fig. 6.2 that this absorption is complete at some wavelengths. This leaves only visible light to reach earth's surface.

Solar radiation is emitted in all directions from sun and very little of this reaches earth. In fact earth is so far from the sun that it picks up only 2×10^{-9} of the total solar radiation. At a distance from the

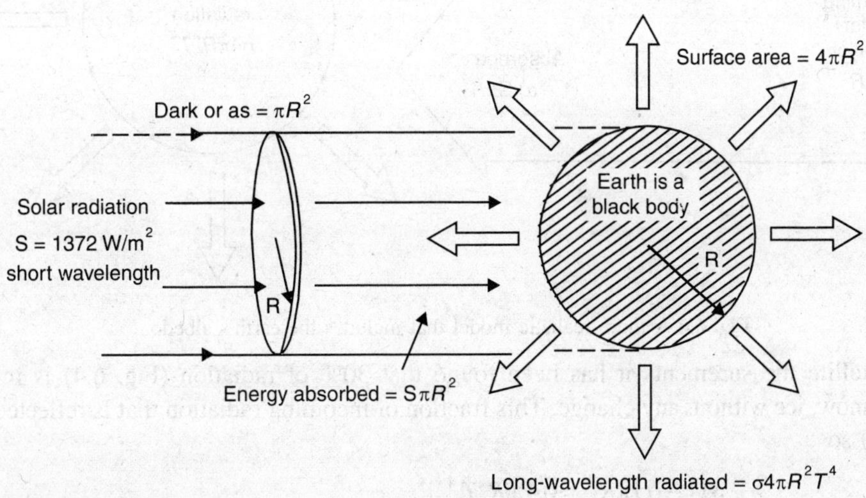

Fig. 6.3 In this simplest model, the earth is treated as a black body, absorbing all radiation impinging upon it, and radiating an equal amount.

sun equal to the average radius of the earth's orbit, the solar energy passing any surface perpendicular to solar radiation beam (Fig. 6.3) is 1372 Wm^{-2}; this is called solar constant S.

Treating earth as a black body will amount to absorption of all incoming radiation so that

$$E_{in} = S\pi R^2$$

Assuming that earth is isothermal and energy radiated from earth with surface area 4 πR^2 is given by Stefan-Boltzman law (Eq. 6.2) so that

$$E_{out} = \sigma 4\pi R^2 T^4$$

under steady state conditions of equilibrium

$$E_{in} = E_{out}$$

or

$$S\pi R^2 = \sigma 4\pi R^2 T^4$$

or

$$T = \left[\frac{S}{4\sigma}\right]^{1/4} = \left[\frac{1372}{4\times 5.67\times 10^{-8}}\right]^{1/4} = 279 \text{ K.}$$

This answer is remarkably close to the actual global average temperature of 290 K. Unfortunately the apparent accuracy is to a large extent coincidental rather than the result of this being particularly good model.

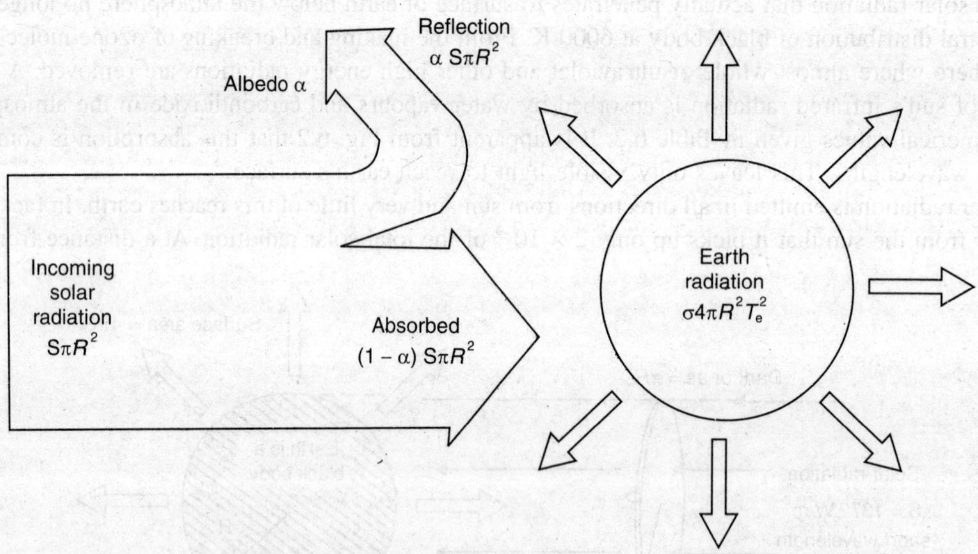

Fig. 6.4 A more realistic model that includes the earth's albedo.

From satellite measurements it has been found that 30% of radiation (Fig. 6.4) is immediately reflected by snow, ice without any change. This fraction of incoming radiation that is reflected is known as albedo (α) so

$$S(1-\alpha)\pi R^2 = \sigma 4\pi R^2 T_e^4$$

$$T_e = \left[\frac{S(1-\alpha)}{4\sigma}\right]^{1/4} = \left[\frac{1372(1-0.7)}{4\times 5.67\times 10^{-8}}\right]^{1/4}$$

$$= 255 \text{ K.}$$

The correction that has been applied to account for albedo has unfortunately made our estimate 255 K worse, with an error of 12%. We need to find an explanation for the earth is not that cold. The key factor that makes this model differ so much from reality is that it has not taken into account the interaction between atmosphere and radiation emitted from earth surface. That is, it has not included the Green House Effect.

The incoming solar radiation E_{in} per year can be calculated as

$$E_{\text{in}} = S\times 365\times 24\times 60\times 60\times 4\pi R^2$$

$$= 1372\times 365\times 24\times 60\times 60\times 4\times 3.141\times (75\times 10^6)^2$$

$$= 3.06\times 10^{24} \text{ kW per year.}$$

It is 3.06×10^{24} kW of solar radiation that reach earth and it is necessary that 3.06×10^{24} kW per year leave earth. If less energy than that leaves the earth, it will become hotter. This statement holds good if atmosphere was not there. Interaction of radiation with atmosphere traps certain quantity of heat so that $E_{\text{in}} \neq E_{\text{out}}$, that is the Green house effect, leads to warming up of earth.

6.5 GREEN HOUSE EFFECT

Warming up of earth due to pollution is due to green house effect. A related phenomenon is observed in a glass house in Agricultural Universities called Green House, in which plants are placed. As glass is translucent to infrared radiation, it allows only 10% of this radiation to escape, the remainder 90% of radiation is trapped which increases its temperature. The temperature of the green house stabilize at a higher level than that of the outer air. The same effect occurs in a closed automobile that is left out in sun. In both cases energy retention in green house is caused by lack of convection i.e. lack of mixing of the interior air with the surrounding atmosphere. Both in green house and the car interior may thus become considerably warmer than the temperature of the surrounding atmosphere. A similar but more complex interactions occur between solar radiation and the gases that surround earth.

The entire range of radiation emitted by earth falls in the infrared region (4 – 20 μm). The interaction of infrared radiation with molecules make them go from one vibrational energy level to another vibrational energy level. The essential condition for the radiation to interact for these transitions is that vibrational mode should be associated with a change in dipole moment. However, homonuclear gas molecules N_2 and O_2 do have vibrational energy levels, but are symmetrically stretched or compressed, thus involve no change in dipole moment and thus do not interact with infrared radiation. As a result homonuclear molecules like N_2, O_2 are transparent to infrared radiation. Molecular vibrations of carbon dioxide, methane, chlorofloro carbons, NO_2 and H_2O and small amounts of other gases involve change in dipole moment hence do absorb infrared radiation but the most prominent is CO_2 because of its higher comparative concentration. Vibrational spectral lines of some of the gases making contribution to green house effect are given in Table 6.2.

The earth's surface radiate at an average temperature (290 K), and this radiation can be absorbed almost completely by atmospheric water vapour and carbon dioxide except in a few wavelength regions, the so called "atmospheric window". The main atmospheric window is from 8.5-11 μm with partially opaque "window" from 7-8.5 μm. The region below 7 and above 14 μm is completely opaque to infrared radiation. When the sky is clear, terrestrial radiation in the wavelength region of the "window" 7–14 μm escapes directly to space, outside the atmosphere. Hence on clear nights, especially in winter, radiation escape through the atmospheric window that leads to cooling of earth. Clouds when present, reflect this radiation back to the earth's surface and does not allow it to cool.

CO_2, H_2O, NO_2 and CH_4 do absorb the incoming solar radiation, as is apparent from Table 6.2 and Fig. 6.2. However, there are small regions in which no absorption takes place. The CO_2 absorption bands centered at 14.99 and 4.26 μm are responsible for much of this return radiation as is also the absorption for water vapour centred at 2.7 and 6.27 μm and continues to a very long wavelength. In addition to water vapour, nitrogen peroxide and methane absorb with band centres at 6.18, 7.58, 13.35 and 3.31, 3.43, 6.5, 7.6 μm, respectively, also contribute to the radiation returned to the ground surface though contribution for methane and NO_2 are not shown in Fig. 6.5. This figure shows how absorption bands associated with CO_2 and H_2O vapours interfere with the free passage of radiation emitted by earth to outer space. CO_2, H_2O vapours absorb part of the outgoing stream of infrared radiation from the ground. Then these excited molecules reradiate this energy both outward towards space and back downward towards the ground. This returns some of the radiant flux of the thermal energy to the ground, thereby raising the surface temperature and contributing to the warmer climate. The spectral characteristics of the radiation returned to the ground from the atmosphere are shown in Fig. 6.5.

Most of the infrared interactions occur when the frequency of the radiation closely matches that of the vibrational spectral lines of the molecule. The width of a spectral line (as discussed in section 3.3 and 3.4) is an important element in understanding the importance for the absorption of radiation. In the Earth's atmosphere these spectral widths are primarily determined by "pressure broadening", which is the distortion of the spectrum due to the collision with another molecule. Most of the infrared absorption in the atmosphere can be thought of as occurring when two molecules are colliding. The absorption due to a photon interacting with a lone molecule is relatively small. This three-body aspect of the problem, one photon and two molecules, makes direct quantum mechanical computation for molecules of interest more challenging. Careful laboratory spectroscopic measurements rather than *ab initio* quantum mechanical computations provide the basis for most of the radiative transfer calculations used in studies of the atmosphere.

As discussed in Chapter 3, section 3.5 and 3.6, infrared radiation interacts with a vibration only if it involves a change of dipole moment, not otherwise. The major constituent of the atmosphere: oxygen (O_2), nitrogen (N_2); do not interact with infrared radiation. The oxygen and nitrogen molecules do vibrate symmetrically, without any change in dipole moment hence can neither absorb nor emit infrared radiation. In the Earth's atmosphere, the dominant infrared absorbing gases are carbon dioxide, water vapours and ozone (O_3). The same molecules are also the dominant infrared emitting molecules. Carbon dioxide is a linear molecule, but its bending and asymmetrical stretching vibrational modes involve change of dipole moment thus carbon dioxide molecules can absorb infrared radiation in which the molecule bends with the carbon in the middle moving one way and the oxygens in the end moving the other way, creating some charge separation, a *dipole moment,* thus carbon dioxide molecules can absorb IR radiation. However asymmetric stretch is at too high a frequency to be effectively excited by atmospheric temperature collisions, although it does contribute to absorption of IR radiation. Collisions will immediately transfer this energy to heating the surrounding gas. On the other hand, other CO_2 molecules will be vibrationally excited by collisions. Roughly 5% of CO_2 molecules are vibrationally excited at room temperature and it is this 5% that radiates. A substantial part of the greenhouse effect due to carbon dioxide exists because this vibration is easily excited by infrared radiation. The symmetric stretching vibrational mode does not involve change of dipole moment thus do not interact with infrared radiation. The vibrational modes of water are at too high energies to effectively radiate, but do absorb higher frequency IR radiation. Water vapour has a bent shape. All the three vibrational modes involve change of dipole moment, which means that IR light can be emitted and absorbed. These transitions can

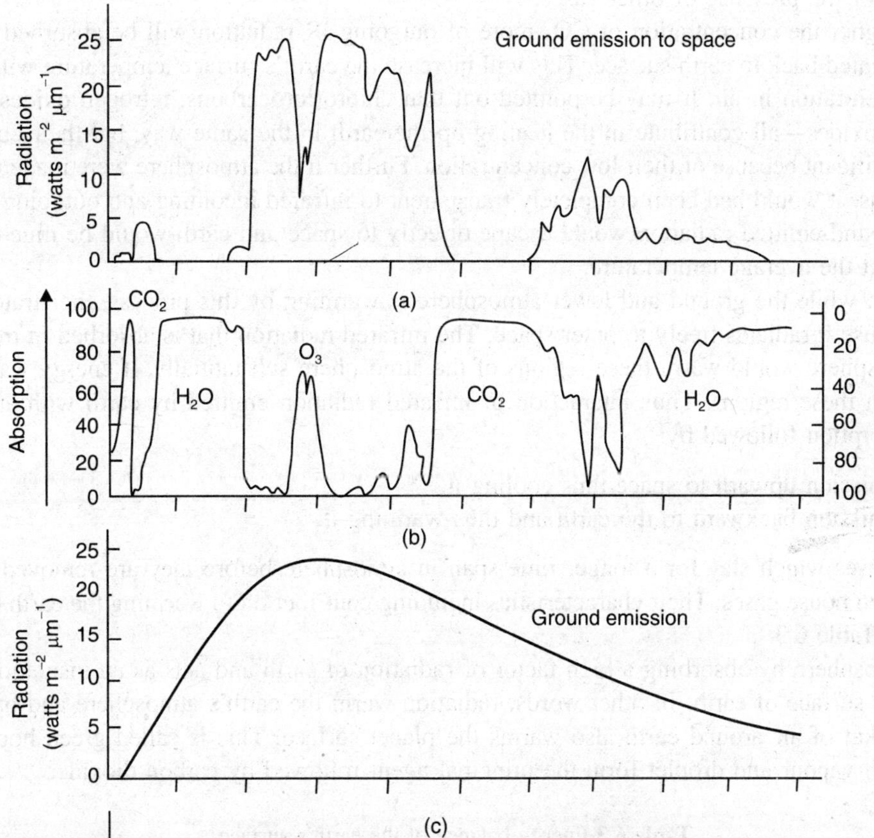

Fig. 6.5 The infrared spectral distribution of thermal radiation and atmospheric transmission. (*a*) Radiation from the ground that reaches outer space; (*b*) Transmission (and absorption) by atmospheric gases; (*c*) The "blackbody" radiation emitted by the ground surface; and (*d*) Radiation received at the ground from the atmosphere.

also be produced by collisional energy transfer. Clouds are also very important infrared absorbers. Therefore, water has multiple effects on infrared radiation, through its vapour phase and through its condensed phases. Other absorbers of significance include methane, nitrous oxide and the chlorofluorocarbons. Pressure broadening increases the overlap between the spectral lines due to different gases, introducing some confusion. As a result, the absorption due to one gas cannot be thought of as independent of the presence of other gases.

Thus, higher the concentration of CO_2 more of outgoing IR radiation will be absorbed and more will be reradiated back to earth surface. This will increase the earth's surface temperature with increase in CO_2 concentration in air. It may be pointed out that chloroflorocarbons, nitrogen oxides, methane and sulphur oxides—all contribute in the heating up the earth in the same way, but their contribution is not so significant because of their low concentration. Further if the atmosphere were made up of only N_2 and O_2 gas, it would had been completely transparent to infrared incoming and outgoing radiation, all of the ground-emitted radiation would escape directly to space and earth would be much colder at 255 K than at the average temperature.

However, while the ground and lower atmosphere is warming by this process, the stratosphere is cooling because it radiates freely to outer space. The infrared radiation that is absorbed in troposphere and in stratosphere would warm these regions of the atmosphere substantially, if energy was not also radiated from these regions. Thus interaction of infrared radiation emitted by earth with atmosphere leads to absorption followed by:

(*i*) reemission upward to space thus cooling it.
(*ii*) reemission backward to the earth and thus warming it.

Those gases which stay for a longer time span in atmosphere before they are removed are more effective green house gases. Their characteristics including contribution to warming the earth have been compiled in Table 6.3.

The atmosphere by absorbing a high factor of radiation of earth and acts as an insulation to keep heat near thé surface of earth. In other words, radiation warm the earth's atmosphere and by creating a warm blanket of air around earth also warms the planet surface. This is called green house effect. Water in both vapour and droplet form the principal agent followed by carbon dioxide.

Table 6.3 Energy balance of the earth's surface

Type of energy	*Quantity J cm^{-2} min^{-1}*
Solar radiation absorbed	+ 1.0608
Back radiation from atmosphere	+ 1.883
Long wavelength radiation (emitted)	– 2.333
Heat transfer to atmosphere by conduction	– 0.20
Heat transfer to atmosphere by evaporation	– 0.4108
Net energy absorbed	0

A green house coefficient may be defined as the ratio of the surface temperature to the absolute radiation temperature of the whole system. For the earth, this coefficient is 290/255 = 1.137, while for the planet Venus it appears to be 700 K/265 K = 2.60. Venus, thus, has a much larger green-house effect than earth, this is presently believed to be caused by high concentrations of sulphuric acid in the Venus' atmosphere. Earth has the possibility of varying its green house effect by means of changes in carbon dioxide content of the atmosphere, since this compound is a prime absorber of the infrared radiation emitted from the earth's surface.

6.6 QUANTITATIVE GREEN HOUSE EFFECT

The processes which determine the temperature of earth and atmosphere at any location is quite complicated. These processes include absorption and reflection of radiation by various bodies.

Let us add some quantitative information to the simple green house effect. Consider earth, its atmosphere and outer space as three separate regions. We will normalize energy flows between these regions by expressing them in terms of rates per unit surface area of earth. The total amount of radiation striking earth is $S\pi R^2$. Distributed over the entire surface of earth, the incoming solar radiation is equal to

$$\frac{S\pi R^2}{4\pi R^2} = \frac{S}{4} = \frac{1372}{4} = 343 \text{ Wm}^{-2}$$

The average albedo of earth and atmosphere varies from 30 to 50% with average value equal to 30%. Since albedo is 30 percent, incoming radiation reflected = 0.3 × 343 = 103 Wm^{-2}. Of this 103 Wm^{-2}, it is estimated that 89 W/m^2 are reflected off the atmosphere itself while the 14 Wm^2 are reflected off the earth surface.

The radiation absorbed by atmosphere and earth = 0.7 × 343 = 240 W^{-2}, of this 79 Wm^{-2} is absorbed by the green house gases of atmosphere. The remaining 161 Wm^{-2} reaches earth surface where it is absorbed, only this radiation may have given earth a temperature of 231 K.

Treating the earth as a black body, with Stefan-Boltzmann relation the rate at which energy is radiated from the earth's surface towards atmosphere can be estimated. At 290 K the global average temperature

$$\frac{\text{Energy radiated by surface}}{\text{Surface area}} = \frac{\sigma 4\pi R^2 T^4}{4\pi R^2} = \sigma T^4$$

$$= 5.67 \times 10^{-8} \times (290)^4 = 401 \text{ Wm}^{-2}$$

Of this 401, only 20 Wm^{-2} passes directly through atmosphere, mostly through the atmospheric windows. The remaining 381 Wm^{-2} is absorbed by green house gases in atmosphere. The atmosphere then radiates 353 Wm^{-2} back to surface, as pollution increases this amount goes on increasing and thus warming up of earth take place. All these energy flow are shown in Fig. 6.6.

If this model is internally self consistent, the rate of energy gain should be equal to the rate or energy loss in each of the three regions, space atmosphere and earth's surface.

	Energy gain		*Energy loss*
Earth's surface	161 + 358	=	94 + 19 +20 + 381
Atmosphere	79 + 94 + 19 + 381	=	220 + 353
Space	103 + 20 + 220	=	343

The model shows the necessary balances and shows that atmosphere has a definite role in the warming up of the earth. Change in any one number will disturb the entire equilibria that is prevailing on earth that will change weathering pattern.

A number of attempts have been made[3, 4, 5] to model the combined effects on global temperature by CO_2 alongwith other green house gases. The equilibrium temperature increase ΔT associated with these combination is given by

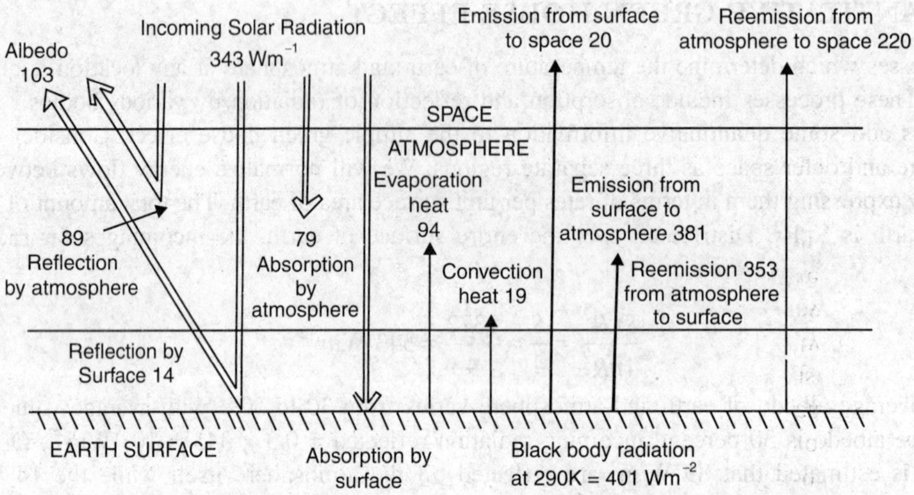

Fig. 6.6 Global average energy flows between space, the atmosphere, and the earth's surface. Units are watts per square meter of surface area.

$$\Delta T = 9.968 \log \frac{(CO_2)}{(CO_2)_0} + 0.057\left[(N_2O)^{0.5} - (N_2O)_0^{0.5}\right]$$

$$+0.019\left[(CH_4)^{0.5} - (CH_4)_0^{0.5}\right]$$

$$+0.14\left[(CFC-11) - (CFC-11)_0\right] + 0.16\left[(CFC-12) - (CFC-12)_0\right]$$

Since this relation gives us a temperature change, each compound must have some reference concentration, which the subscript 0 designates. Using the data from the pre-industrial concentration, and 1985 concentration, the combined equilibrium temperature change and concentration in 2030 are compiled in Table 6.4.

Table 6.4 Temperature changes caused by individual green house gases

Gas	*1850*	*1985*	*ΔT 1985-1850*	*2030*	*ΔT 2030-1850*	*% of total temperature change*
CO_2	280 ppm	345	0.904	446 ppm	2.013	71.2
CH_4	1150 ppb	1790	0.159	2807 ppb	0.362	12.8
N_2O	285 ppb	305	0.033	282 ppb	0.152	5.4
CFC-11	0	0.24 ppb	0.034	0.7392 ppb	0.103	3.64
CFC-12	0	0.4 ppb	0.064	1.232 ppb	0.197	6.96
			1.2°C		2.83°C	

The above example suggest that approximately 71% increase would be caused by CO_2 and 29% by rest of the gases. This change in temperature is for certain concentrations of gases can be calculated for any other concentrations. This perspective emphasizes the important conclusion that future temperature control requires control on CO_2. Also as the concentration of green house gases increase, more heat is transferred to the earth's surface and thus increase in temperature is expected.

6.7 OZONE LOSS MAY REDUCE GLOBAL WARMING

In a study it has been found that penetration of UV-B radiation to troposphere will disintegrate ozone molecule there. The radiation will dissociate water molecule to produce hydroxyl molecule.

$$H_2O \xrightarrow[\lambda<186\,\text{nm}]{h\nu} H\left({}^2S_{1/2}\right) + HO\left(X, {}^2\Pi\right) \quad ...(4.26)$$

These radicals will interact with SO_2, provided we pollute the atmosphere that much, and produce SO_4^{2-}

$$H\dot{O} + SO_2 \xrightarrow{M} HOS\dot{O}_2 \quad ...(7.9)$$

$$HOS\dot{O}_2 + O_2 \longrightarrow H\dot{O}_2 + SO_3 \quad ...(7.14)$$

$$SO_3 + H_2O \longrightarrow H_2SO_4 \quad ...(7.16)$$

While the kinetics of these elementary reactions are reasonably well defined and it is known that a significant fraction of $HOSO_2$ radical formed ultimately winds up as H_2SO_4 aerosol. These sulphate ions form cloud condensation nuclei. Thus increase in UV radiation would enhance cloud formation which could counter warming effect.

Further studies have shown that hydroxyl radical formation in troposphere is sensitive to the thickness of ozone layer. Therefore marked reduction in global ozone that has been observed through 80's and particularly in the early 90's are likely to have led to the increase in hydroxyl radical concentration by 3%.

Also a doubling of CO_2 concentration would increase water concentration by 30% which would increase both OH concentration and thus cloud forming nuclei by 15%. This hydroxyl ions are the main sinks for methane and CO, which will lead to decrease in growth rates of both CH_4 and CO as observed in 1991 and 1992 when there was decrease in concentration of ozone in stratosphere. However such an increase in rain may be associated with acid rains which will affect plant, animal life of all sorts and degrade soils. How we will adjust to the increased OH concentration, a strong oxidizing agent, is still not well understood.

6.8 CLIMATE CHANGE VARIABILITY

The earth's climate is mankind's one of the greatest natural resources. The very existence of man depends fundamentally on a benign climate. Right climate helps us to produce the crops that feed us, the right amount of rainfall occurring at the right times, provide the water resources which are fundamental to our existence. Tropical cyclones, floods, droughts extreme cold spells or extreme heat waves have a damaging effect. Sustenance of humankind, biosphere, plants and animal, all depend on the right seasonal changes in temperature and rainfall. However, human activities today have reached the stage where they are producing a discernible change in climate. The term climate change, often refers to changes in modern climate, 90 to 95% of which have been caused by human action. Consequently the term *anthropogenic* climate change is frequently adopted; this phenomenon is also referred to in the main stream media as global warming. In some cases, the term is also used with a presumption of human causation, as in the United Nations Framework Convention on Climate Change (UNFCCC). The UNFCCC uses "climate variability" for non human caused variations.

The climate system consists of the atmosphere, the hydrosphere, the cryosphere, the lithosphere and the biosphere. Disturbance in anyone will affect the others.

Cooling of Earth and Green House Effect

From the time of inception of earth, 4 to 5 billion years ago, it has been surrounded by gases. These gases had contributed in a big way to the cooling of the earth. The cooling in the upper strata of atmosphere caused torrent of rains that continued for centuries, removed lot of water vapours. Thinning down of water vapours, a green house gas, in atmosphere helped in cooling of earth. This must have been associated with lot of lightning discharges passing through atmosphere for centuries, that led to the formation of various amino acids and thus removal of NH_3, HCN and CH_4, that also led to cooling of earth.

Gamma and ultraviolet rays of sun dissociated water to oxygen and hydrogen molecules, removing more of water vapours, again must have been associated with the cooling. Oxygen so released, slowly changed the character of atmosphere from reducing to oxidising which was associated with replacement of methane by carbon dioxide. Since methane is more effective green house gas than carbon dioxide, its replacement must have been associated with cooling.

Precipitation of various minerals such as carbonates in sea and on the surface of earth must have led to removal of lot of carbon dioxide and thus further cooling.

Oxygen in today's atmosphere resulted largely by photosynthesis by plants As photosynthesis gradually increased, CO_2 concentration must have decreased that must have enhanced its cooling. As oxygen level in atmosphere increased, ozone formation increased, eliminating the UV radiation to reach earth, that again led to cooling of earth.

In the past two million years, roughly the time that man have inhabited the earth, there have been about 20 glacial and interglacial periods involving temperature changes as a result of non-anthropogenic forces, though exact causes are still not understood.

Long Time-Scale Climate Variability

The earth's climate has never been static. Climate is a dynamic regime subject to changes at all time scales ranging from decades to millennia to millions of years. Many scientific clues can be found in glacial records which indicate that past climates have been quite different from the present. From all the available evidences it is found that global temperature variation in the past are shown in Fig. 6.7. Prominent are the glacial-interglacial cycles at about 1,00,000 years intervals when climate was mostly cooler than at present. Global mean surface temperature typically varied by 4–5°C through these ice age cycles with large changes in ice volume and sea level, and temperature variation as great as 10–15°C in some middle and high latitude regions of the northern hemisphere. Since the beginning of the current interglacial epoch, about 10,000 years before present (Fig. 6.7b) global temperature had fluctuated within a much smaller range. Some fluctuation have nevertheless lasted several centuries, including in "Little Ice Age" from the sixteenth to nineteenth centuries which was global in extent.

Causes of Significant Longer-Term Climate Variability

One important factor in the repeated glaciation on earth is related to alternations of the earth's orbital parameters—the "Milankovitch hypothesis". These alternations take the form of a recurrent variation of the eccentricity of earth's elliptical orbit around the sun within a period of about 1,00,000 years (resulting 111 small changes in the distance of the earth from the sun in given seasons) and slow changes in the angle of tilt of the earth's axis, which affects the amplitude of the seasonal insolation. This phenomenon causes a variation of only about 0.2 per cent in the amount of solar radiation intercepted by the earth. Such a small change in the solar radiation flux must be in some manner amplified by the earth's climate system in a considerable amount to induce the profound changes in climatic

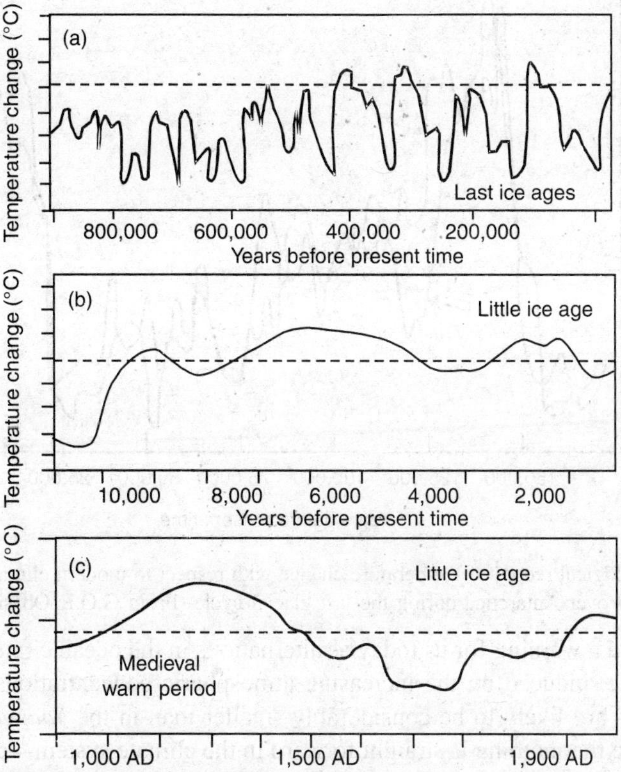

Fig. 6.7 Schematic diagrams of global temperature variations since the Pleistocene on three time-scales: (*a*) The last million years (*b*) The last ten thousand years, and (*c*) the last thousand years. The dotted line nominally represents conditions near the beginning of the century. (From G.O.P. Obasi, 1996)

conditions, such as those which have occurred in the transition to ice age. The ice sheets of Antarctica and Greenland preserve the climatological history of the last ice age. The composition of deep ice cores provides a basis for estimating the air temperature over many millennia. Small air bubbles are trapped in the ice sheet when it was formed and the composition of these air bubbles have been analysed to give an estimate of variation of the atmospheric composition paralleling the temperature record (Fig. 6.8).

This shows that variations in carbon dioxide during ice age cycles are also important factors serving to modify and perhaps to amplify other forcing effects.

Ocean-induced Changes

There are also rapid changes in climate on time scale of about a century which cannot be directly related to orbital effects or to changes in atmospheric composition. A notable example is the *Younger-Dryas cold episode*, which was manifested by an abrupt reversal of the general warming trend in progress about 10,500 years before the last episode of continental glaciation came to an end. There is as yet no consensus on the reasons for this reversal which lasted about 500 years and ended very suddenly. However, since the change was strongest around the North Atlantic Ocean, it has been suggested that this climatic reversal had its physical origins in a marked decrease in the surface temperature of this ocean, following large-scale melting of the ice-sheet and consequential changes in the oceanic circulation.

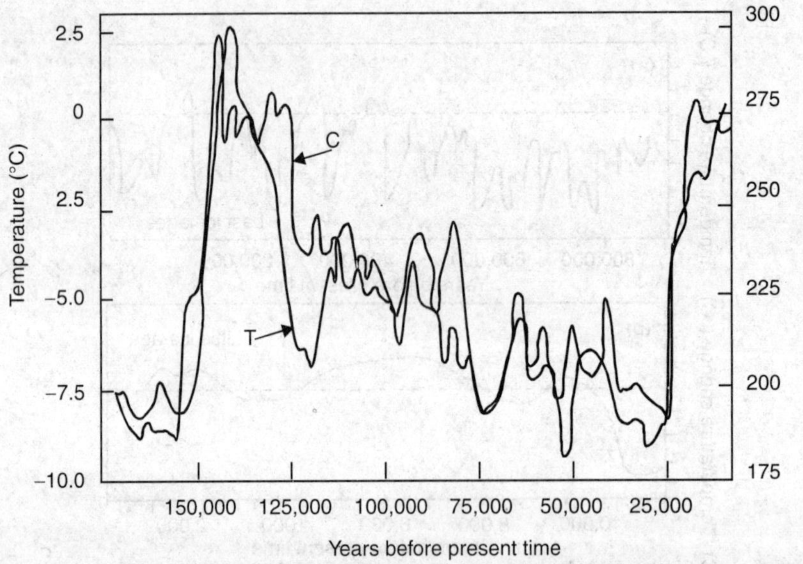

Fig. 6.8 Paleoclimatological record of temperature change with respect to modern climate and atmospheric CO_2 concentration over Antarctica during the last glacial cycle (From G.O.P. Obasi, 1996).

These changes hold a warning for us today as alternations in the ocean circulation could result from modification of climate, induced by the increasing atmospheric concentrations of green house gases, though the alternations are likely to be considerably smaller than in the *Younger-Dryas* event. This is brought out to illustrate that nothing is straight forward in the climate system—changes may sometimes be very unexpected. Great care is needed in studying the climate system and to take into account all the possible degrees of freedom and interactions between the different components of the climate system.

The period since the end of the last glaciation (about 10,500 years ago) has been characterised by smaller changes in the global average temperature with a range of probably about 2°C, though it is not clear whether all the fluctuations were truly global. However, large regional changes in hydrological conditions have occurred, particularly in the tropics. Wetter conditions in the Sahara from 12,000 to 4,000 years ago, enabled man to survive by hunting and fishing, on what is today among the most arid regions on earth. Pollen sequences from lake beds of North-West India suggest that periods with subdued monsoon activity existed during the last glacial maximum, but in the epoch 8000 to 2500 years ago, there was a humid climate with frequent floods. The late 10th to the early 13th centuries (AD 950–1250) appear to have been exceptionally warm in western Europe, Iceland and Greenland—a period that is known in Europe as the "Medieval Climatic Optimum". On the other hand, it seems that China might have been cold in this period (mainly in the winters), although the southern part of Japan was warm. This episode of generally widespread warmth is notable, in which there was no evidence of any increase in greenhouse gases.

A striking indicator of climate changes are the advances and retreats of glaciers in mountainous regions of the world. In the most recent cooler period, the "Little Ice Age", glaciers advanced markedly in almost all high mountainous regions, and glaciers were much more extensive 100–200 years ago than now. Since 1850, most glaciers have been retreating rapidly.

Warming up of earth will make the ice at poles to be increasingly converted to water, resulting in rise in world sea level. As climate warms, the extent of snow and ice over the earth will tend to decrease. As snow and ice are good reflectors of incoming radiation, therefore a decrease in snow and ice will increase radiation absorbed and enhance the warming process. Also as climate warms up, the atmosphere can hold more water vapours, but water vapour a strong green-house gas and so this effect also automatically tend to reinforce a warming trend. As the temperature increases, soil becomes dry and dust particles goes into atmosphere conveniently. Dust particles may also be emitted into the atmosphere as a result of volcanic eruptions. In cold places, these particles increase smog formation which makes day appear shorter. In hot places, dust particles scatter sunlight by decreasing the extent of sunlight received. In either case photochemical conversion of CO_2 to O_2 will decrease. As a consequence, the lung efficiency of man will decrease, which will have serious effects on health.

Changes in earth's albedo may also be implicated in climate changes. Although large-scale cultivation, irrigation, damming of rivers to form lakes and cutting down of forests all result into a minor change in the earth's albedo. Since there are many unknown factors, it would be better if man could leave the composition of atmosphere unchanged.

The present rate of increase of green-house gases in the atmosphere increase global mean temperature by anywhere in the range 1 to 3.5°C by the end of this century. This rate of increase of temperature would be highest known in the past 10,000 years.

Variation in weather can be large, but analysis of meteorological and other data over large areas and over decades or more, have provided evidence for some important systematic changes but no definite order has been found.

Regional changes are also evident. For example, recent warming has been greatest over the mid-latitude continents in winter and spring with a few areas of cooling, such as the North Atlantic Ocean. Precipitation has increased over land in high latitudes of the Northern Hemisphere especially during the cold season.

The global sea level has risen by between 10 to 25 cm over the past 100 years and much of the rise may be related to the increase in global mean temperature.

There are inadequate data to determine whether consistent global changes in climate variability or weather extremes have occurred over the 20th century. On regional scales, there is clear evidence of changes in some extremes and climate variability indicators *e.g.*, more forests in several widespread areas; a decrease in the proportion of rainfall in India.

In 1990's persistent warm-phase of the El Nino-Southern Oscillation (which causes droughts and floods in many areas) was unusual in the context of the last 120 years.

6.9 IMPLICATION OF CLIMATE CHANGE

Extreme Weather Events

Among the dramatic occurrences of extreme weather phenomenon and their direct impact on human, the most significant are tropical cyclones, floods, droughts, tornadoes, lightning, hail and heavy snowfall.

In an average year, about 80 tropical cyclones form over warm ocean water in various parts of the tropics and affect 50 different countries. On an average about 20,000 people lose their lives each year and the damage caused by the associated violent winds, floods and tidal surges can cost many thousands of millions of US dollars. In some years, individual cyclones can have even more tragic consequences. In 1970 and 1991, for example, severe cyclones in the Bay of Bengal resulted in the loss of some 3,00,000 and 1,38,000 lives respectively. There was a superclone in 1999 in Orissa where winds blew with more than 221 kmpH velocity. Most of the 250 coastal villages were wiped off, more than 20000 people, many thousand cattle heads and birds died. Mangrove forests and sand dunes along the coastal belt had been acting as a natural barrier against cyclones or tidal surgers, have been wiped. These forest have been destroyed in the recent years and Orissa has no disaster management plan. So misery mounted the need to provide accurate and timely warnings of such weather disasters has become important. However, there are hopeful signs that this is being achieved in many parts of world.

The intensity of hurricane are most difficult to predict. Hurricane can be modeled as an idealized heat engines running between warm heat reservoir (the sea), at around 300-330 K and a cold reservoir 15-18 km up in the tropical troposphere at about 200 K. The strongest wind speed or lowest sea level pressure is proportional to the difference in absolute temperature between these two reservoir.

This may bring about of local increases in precipitation intensity and the probability of extreme rainfall events and the risk of severe floods in some regions. On the other hand, intensified droughts are probable in other areas. Furthermore, although current in the occurrence, intensity of geographical distribution of severe storms, such as tropical cyclones, recent monitoring has shown records in the frequencies (*e.g.*, in 1993, Philippines) and increased intensity (*e.g.*, Hurricane Andrew in 1992, USA) of these storms and previously unknown tracks (*e.g.*, tropical cyclone across Southern Africa in January 1996 from the Indian Ocean to the Atlantic Ocean).

On August 28/29, 2005 Hurricane Katrina struck USA Gulf cost states Louisiana, Mississipi and Albama. It started as a tornado followed by a storm surge hurricane, wind blowing at 230 km/hour. The sea waves nearly 3 meters high splashed on the shores blowing away houses, flooding cities which remained under 1.5 meters of water for more than 5 days, killing about 25000 people, 50000 injured and damage to properties worth $ 100 billion. Toxic chemicals in New Orleans city floods have made it unsafe for human habitation for a decade.

Agriculture and Forests

Agriculture is a basic area of human activity that is particularly sensitive to climate. The production, collection, transportation, storage, processing and distribution of agricultural produce are all greatly influenced by climatic factors. It frequently happens that if a particular area experiences unfavourable seasons climatically, major or even virtually total loss of crops can occur. In many tropical countries dependency on specific annual rainfall, events, such as, of course, the monsoon, losses on an average may be 30 per cent or more of the annual agricultural production if the rains are exceedingly delayed, or fall short of the normal value. A significant proportion of the agricultural production losses that result, could be avoided by the better use of appropriate information on climate that is already available with the National Meteorological and Hydrological Services. The benefit from the application of this information far surpasses its cost. Furthermore, climate information and weather forecasts can be used to reduce crop and livestock losses, anticipate when and where irrigation would be needed, when pesticides should be applied, the choice of the best harvest time and so on. Prevalence of particularly dry or particularly wet conditions can affect the incidence of pests. An example of this is the major problem of devastation caused by locust swarms in some parts of the world. Locusts can begin to breed a few days after the onset of rain in an otherwise dry environment. If we can predict this rainfall, action can be taken to prevent swarms from developing and thus to avert significant crop losses.

All these studies indicate that climate change will certainly have important consequences on agriculture. Crop yields and changes in productivity will vary considerably from place to place, with marked alterations in patterns of production. Productivity would be expected to decrease in regions of high present day vulnerability that are least able to adjust, especially in the tropics and sub-tropics, including parts of south-east Asia, the Sahel region of Africa and parts of South America. On the other hand, potential productivity in some high and mid-latitude regions may increase because of a long-growing season, and limited new areas for production could open up, but this would be confined to the northern hemisphere. Increasing dryness and summer heat in certain present temperate mid-latitude areas could reduce grain and horticultural production.

A substantial fraction of the earth's existing forest area could be expected to undergo major changes. The rate of climate change will be rapid compared to that of the growth of forest species, and forest zones could decline in a climate to which they are increasingly ill-acclimatised. Generally, flora and fauna will not be able to adapt rapidly to climatic zones which could be expected to move up to several hundred kilometers poleward over the next 50 years. Some species will be lost as a result of the increased climatic stress, leading to a reduction in global biospheric diversity. On the other hand, there could be improved efficiency of plant growth in a higher carbon dioxide (CO_2) environment. Unfortunately, there is already a major problem of deforestation existing in many developed and developing countries of the world. This is not only having a negative impact on plant and animal life, on soil conservation and watershed protection, but is a contributor to global warming by reducing the number of trees that can retain CO_2, a green-house gas.

Water Resources

Another basic concern to humankind is the availability of fresh water resources. Meeting the water demands in quality and quantity of increasing populations and in support of agriculture are currently vital problems in many parts of the world. Projections indicate a significant decrease in per capita freshwater availability. In order to develop and improve the administration of water resources, detailed knowledge of rainfall that can be normally expected and the variations that can occur both in space and time are essential for all countries. This knowledge also underpins the need for planning national responses to droughts and floods. Again, predictions of anomalous wet and dry periods, a season or more in advance, would be of enormous benefit in managing fresh water resources.

Over exploitation or mismanagement of water resources may deplete its resources, which can be restored through water harvesting. This can be done by recharging of well that raises its water level. Another aspect of water harvesting involve building check dams, there by creating a water reservoir, that slows down run off water. Those rivers over which such check dams are made have water throughout the year to meet the irrigation needs. This also recharges the ground water through water sheet intervention.

Desertification

The anticipated changes in the global hydrological cycle, as a result of global warming, would have major impact on regional water resources, while some parts of the globe would have increased rainfall, droughts would likely intensify in others. Even relatively small increases in temperature or reduction in precipitation, may aggravate the already serious problems of quantity and quality of water supplies in many parts of the world. Existing deserts would likely become hotter but not significantly wetter, limiting still further any productive use of the marginal land. The risk of desertification will be increased as the environment becomes drier, and the soil becomes further degraded through erosion and compaction as well as human activities. In this connection, the world community should take appropriate action, through the negotiation of an international convention to combat desertification. This is already evident in Delhi which is heading towards desert. The only way to retard it is by growing more trees.

Sea-level Rise

Another serious consequence of global warming is the rise in sea level, mainly through the thermal expansion of sea-water, with considerable potential socio-economic impacts. Current estimates range from 15 to 95 cm. A 50 cm sea-level rise would threaten low islands and coastal zones. A 95 cm rise would render some island countries uninhabitable, displace tens of millions of people, create menace in many low lying urban areas, flood productive lands, contaminate fresh water supplies and change coastlines. The major low-lying deltas in the Bay of Bengal and many small flat islands in the Indian Ocean could be threatened. These impacts would be exacerbated if floods and storms became more severe. The additional coastal protection required will involve major extra costs.

Impacts on Other Activities

In case of significant warming, the global area, extent and volume of the terrestrial cryosphere (comprising seasonal snow cover, near-surface layers of permafrost and ice masses such as the ice caps and glaciers) will be substantially reduced. In particular, between one-third and one-half of the existing mountain glacier mass could disappear over the next century. Glacial recession and loss of ice from ice sheets will add to sea-level rise.

Many other activities are fundamentally dependent on climate. For instance, major economic benefits can be derived from the ability to foresee and plan energy consumption, such as the high levels of energy used in unusually cold spells, or nowadays in hot spells as the use of air conditioning increases. In the area of marine activities, some nations rely to a considerable extent on food from the sea. Marine climate information is needed in supporting the management and expended in exploitation of coastal and deep-sea fisheries. The safety of all types of crafts at sea and air requires accurate weather forecasts and warnings. In the building industry, good architectural design, taking advantage of information on the likely range of temperature humidity and wind speed can contribute significantly to energy conservation.

Climate change will also exert impact on human health, leisure and tourism industries. Certain climate conditions very much influence the human body. Cardiovascular diseases are known to be more likely in people subject to severe climatic stress, whether extreme heat or extreme cold. Many tropical areas are prone to diseases which depend to a varying extent on heat and humidity. In other regions, respiratory illnesses are aggravated by cold, damp weather, or by pollution. Finally, it should be noted that tourism and leisure industries are now a major source of income in many countries and most of these activities depend on climate. Clearly, any changes in climate, which upset current weather patterns, or produce greater extreme conditions, would impact adversely on these activities.

6.10 THE CHALLENGE OF CLIMATE CHANGE AND CLIMATE PREDICTION—INTERNATIONAL EFFORTS

A search to explanation of impacts of climate change that could be induced by human activities, confident answers, explanations and clarification are still awaited. Understanding climate and thus detecting climate change require data collection and the need for the long, accurate and consistent data records are crucial to the problem.

The longer the data record, the more is the likelihood of detecting changes. The greater the accuracy of the data set, the lesser liable it is to conflicting interpretations, which goes to the heart of the question of uncertainties that are so commonly associated with evaluation of the scientific assessment of climate change. Consistency is vital so that we are sure that we can distinguish the detail in various elements in the data.

Monitoring, Assessment, Research and Prediction Programmes

The history of internationally coordinated research efforts goes back to the latter part of the last century and involved the predecessor of World Meteorological Organization, (WMO), the International Meteorological Organisation, IMO. Among more recent and large scale programmes were the International Geophysical Year (IGY-1957-58) and the Global Atmospheric Research Programme (GARP), spanning 15 years from 1967 to 1982 with the close cooperation of WMO and International Council of Scientific Unions (ICSU). The focal points of GARP were the major field experiments, such as the First GARP Global Experiment (FGGE) which was carried out between December 1978 to November 1979. Following the first World Climate Conference in 1979, the WMO established the World Climate Programme (WCP) to address the full range of climate change issues. These earlier programmes, which involved both data collection and research, and which accumulated a vast amount of experience, led to our understanding of the climate system and lay the foundation for the present-day atmospheric research programmes.

Another very important activity, jointly sponsored by WMO and the United Nations Environment Programme (UNEP), is the activity of Intergovernmental Panel on Climate Change (IPCC), which carries out periodic assessments of the science, impacts and the economics of climate change, and of the response options available to manage the climate change. Through the IPCC assessments, gaps in data and knowledge are identified. The technical and scientific findings of IPCC play a major role in the negotiation of the United Nations Framework Convention on Climate Change, which came into effect in March 1994. There is a close linkage between IPCC, WCRP and the data-related WMO activities.

Prediction Methodology

The basic tool to predict climate or climate change is the model of the earth's general circulation (Fig. 1.3). Models of this type are based on the laws of physics and use descriptions in simplified physical terms, for instance, the effects of clouds and their influence on the incoming solar radiation and radiation emitted from the earth's surface, or mixing in the oceans. The most advanced general circulation models now cover the entire earth, and include representations of the atmosphere, ocean, cryosphere and land surfaces. There is no need to go into details concerning the construction of these models but suffice to say they are extremely complex, particularly when accurate reproduction of the important processes in climate, the interactions between different components of the climate system, such as ocean, atmosphere, land surface and the cryosphere have to be taken into account is the need of the hour. Models for the earth's climate system, have been developed with the help of most powerful of modern computers. The concept of such models has been known for many years but it is only comparatively recent that their use for climate studies has become practical. Scientific organisations which possess super computers, such as the Indian Meteorological Department, are playing a special role in this regard.

It should be recognised here that the earth's climate system is in essence a vast and complex generator of random natural variations, it is, in fact, a classical example of a chaotic system. Chaotic or random variability occurs on all-time scales and can obscure predictable climate signals and climate change patterns. Nevertheless, some variations in the natural climate system do recur with recognisable patterns, and may be associated with specific mechanisms which can be definitely predicted for a finite period of time, before being overwhelmed by chaotic dynamics. On the other hand, in the prediction of long-term climate change as a result of, for example, increasing quantities of atmospheric carbon dioxide, we have to compare the statistics of the model's representation of the existing climate with the statistics of the model's representation of the changed climate. If the model is good and includes all the important climate processes, it will be very much like those of the real atmosphere of today. The differences between these statistics provide an estimate of the climate change we can expect (for example, differences in mean temperature or inter-annual variability). It is worth repeating that, because of the basic chaotic nature of the climate system, there is no possibility of predicting in detail the climate for years ahead. The best we can hope for on these time scales is a quantitative estimate of how long-term climate statistics may change.

6.11 THE EL NINO: SOUTHERN OSCILLATION PHENOMENON

The El Nino: Southern Oscillation (ENSO) phenomenon (see Fig. 6.9), is the first predictable variation that has been clearly identified, and represents the first major breakthrough in climate prediction. Pacific ocean water move to the west and conditions common in west appears in east. This brings heavy rains to South America and caused drought in south-eastern India, Australia and South Africa.

Normally westward blowing trade winds caused by the rotation of earth and conditions in the tropics push surface water across the Pacifics towards Asia. The coast of Peru is washed by the cold water from Antarctica which is responsible for its desert like character. This cold current supports one of the world great fisheries and makes Peru the world largest exporter of fish meal. The warm water piles along the coast of Indonesia, Australia and Philippines pushing sea level more than 30 cm above those on the South American side of the Pacifics. This gives rains for asian countries and dry coastal area of Ecauador, Chile and Peru where comparatively cold water is left.

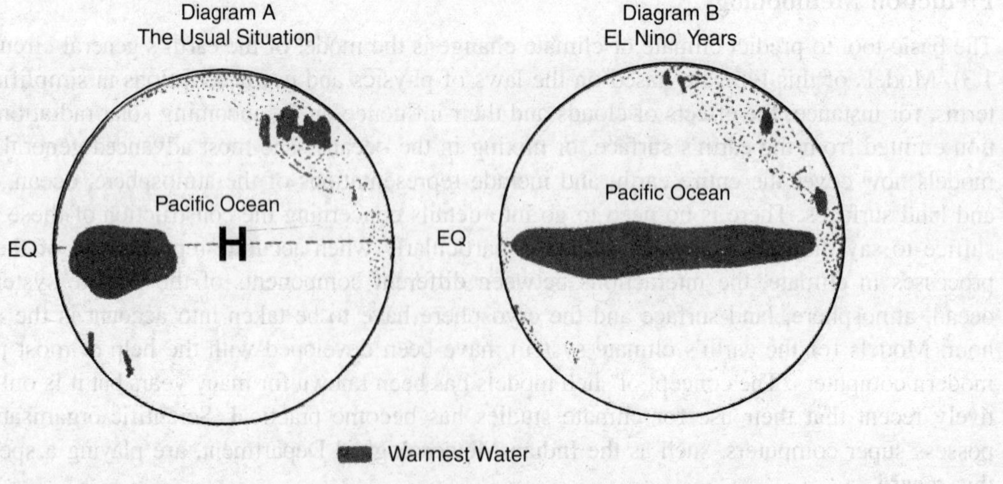

Fig. 6.9 El Nino and the Southern Oscillation (From G.O.P. Obasi, 1996).

This process is reversed in El Nino and its periodic occurrence has been known for many years. The term El Nino (Spanish for "Christ child") was originally used by fishermen along the coasts of Ecuador and Peru to refer to unusually warm offshore ocean conditions that typically appeared near Christmas and lasted for several months. Fish were less abundant near the South American coast. Most of them raised water temperatures along the coast as well as in a belt stretching 8000 km, across the equatorial Pacific. The weaker of these events raised sea temperatures by only about 1°C and had only minor impacts on South American fisheries. However, strong events, such as that of 1982–83 (see Fig. 6.10), left an imprint not only on local weather and marine life, but on climatic conditions around the earth.

During period of El Nino phenomenon, however easterly trade winds collapse or even reverse. As the slight weakening of winds caused a modest change in sea surface temperature and wind pressure increases. The warm water of the western Pacific flows back eastward and surface temperature increases significantly on the western coasts of South America. As this happen, the wet weather conditions normally present in the eastern pacific move to the west and arid conditions common to west appear in east. This brings heavy rains to South America and cause drought in south eastern Asia, India, Australia and South Africa.

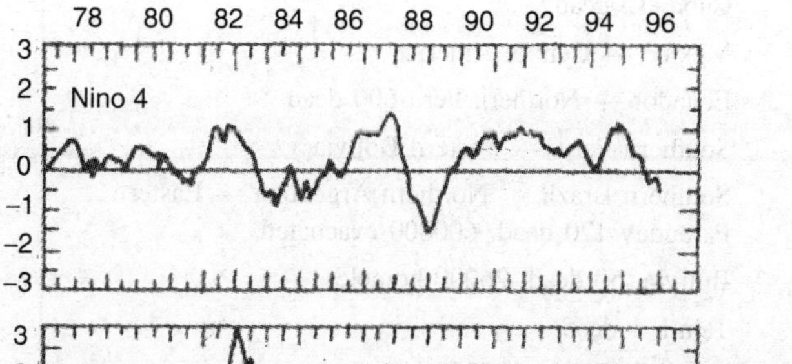

Fig. 6.10 Equatorial Pacific sea surface temperature anomaly (°C) for the areas indicated in the figure Nino 1 + 2 is the average over the Nino 1 and Nino 2 regions. Anomalies are departures from the adjusted climatology (Reynolds and Smith, 1995, 1 Climate 8, 1571-1583).

Reacting to the changes in wind speed and direction associated with El Nino, the sea level at Christmas Island in the mid-Pacific rose several centimetres, and by October 1982, sea-level rose up to 30 cm had spread 10,000 km, east to Ecuador. The sea level rise in the east were accompanied by simultaneous drops in the western pacific, exposing and destroying the upper layers of fragile coral reefs thereby seriously affecting marine life. By the time conditions returned to normal in the mid-1983, 25 per cent of the year's fur seal and sea lion adults and all of the pups had died. Fishermen in Ecuador and Peru suffered heavily when the Anchovy harvest failed and Sardines unexpectedly moved south.

Table 6.5 Effects related to 1982-83 El Nino

Phenomenon	*Victims*	*Damage (in $)*
Storms	US Mountain and Pacific States 45 dead	1.1 b
Hooding	US Gulf States 50 dead	1.1 b
Huricane	Hawaii, I dead	230M
Storms	Northeastern US 66 dead	–
Flooding	Cuba, 15 dead	170M
Drought	Mexico — Central America	600M
Flooding	Ecuador — Northern Peru 600 dead	650M
Drought	Southern Peru — Western Bolivia	240M
Hooding	Southern Brazil — Northern Argentina — Eastern Paraguay 170 dead, 600000 evacuated	3b
Hooding	Bolivia, 50 dead, 26000 homeless	300M
Hurricane	Tahiti, 1 dead	50M
Drought, Fires	Australia 71 dead, 8000 homeless	2.5b
Drought	Indonesia 340 dead	500M
Drought	Philippines	450M
Wet Weather	Southern China 600 dead	600M
Drought	Southern Indian — Sri Lanka	150M
Cold	Middle East (Lebanon) 65 dead	50M
Disease, Starvation	Southern Africa	lb
Drought	Siberian Peninsula — N. Africa	200M
Flooding	Western Europe 25 dead	200M

On land there were equally dramatic effects [see Table 6.5]. In Ecuador and northern Peru, upto 250 cm of rain fell during a six-month period. The abnormal wind patterns steered tropical cyclones away from their usual tracks to islands such as Hawaii and Tahiti, unaccustomed to such severe weather. Heavy rain fell over the central instead of over the western Pacific, resulting in severe droughts, and disastrous forest fires in Indonesia and Australia. In 1998, fires destroyed some two million hectares of Indonesia's forest. The drought, along with an economic crisis left about five million Indonesian in desperate need of food and water. These conditions helped to set the stage for riots that led no the downfall of President Suharto. El Nino also took the blame for blistering temperature of 38°C in Texas for a record 30 days. Winter storms battered Southern California and caused widespread flooding across the southern United States. Overall the loss to the world economy as a result of this El Nino is estimated to have amounted to over US $8 billion.

Where El Nino turns typical weather pattern upside down—drying out lush rain forest and turning arid area into wet lands—La Nina (Spanish for—little girl) generally exaggerates the conditions considered normal. With El Nino departure the south western America would return to drier conditions. Add La Nina, and dry may turn to drought. El Nino are not always followed by La Nina, nor does the strength of one reflects that of other. 1998 El Nino may have been the most powerful of the 20th Century. India experienced the worst summer in 50 years which left 3000 dead, millions drenched in sweat. The south-eastern wind from Bay of Bengal which blows across in May were missing this time. These winds normally nullify the effect of heat waves brought by the surface winds from the north-western plains of the country causing thunder storms and preventing the temperature from rising. These unchecked heat waves were due to deforestation and El Nino phenomenon, resulted from climate change.

The Tropical Ocean and Global Atmosphere Programme

This 1982-83 event was a tremendous spurt to studying ocean-atmosphere interactions in the tropical Pacific and the climatic impacts around the world. The study, called the "Tropical Ocean and Global Atmosphere" (TOGA) programme, was conducted as part of the World Climate Research Programme. The process was as follows: The normal distribution of atmospheric mass along the equator results in a pattern of high and low pressure such that low pressure is over the western ocean and the high pressure over the eastern ocean (see Fig. 6.9). In certain years, the pattern reverses itself, and then gradually reverts back to normal.

The changes in atmospheric pressure patterns are linked to changes in the tropical oceans. Under normal atmospheric conditions, the sea surface in the eastern Pacific is relatively cool. In El Nino years, atmospheric pressure is unusually high in the western Pacific and dry weather conditions prevail (see Fig. 6.11). The sea surface in the central and eastern Pacific tends to be unusually warm with lower atmospheric pressure. Moreover, these changes in the Pacific are now known to be associated with other unusual and sometimes catastrophic changes around the world.

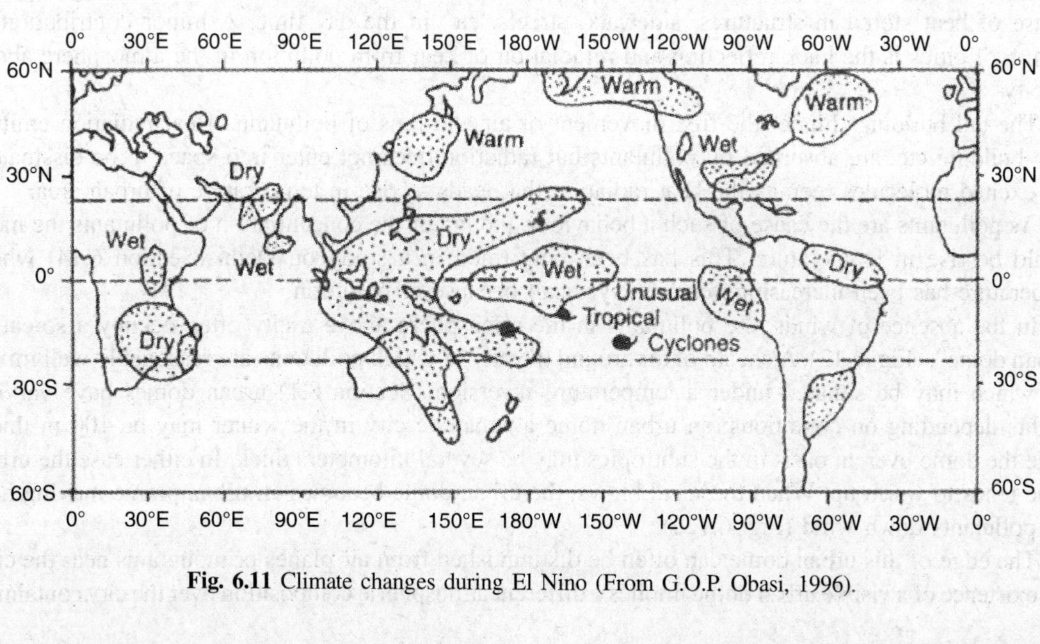

Fig. 6.11 Climate changes during El Nino (From G.O.P. Obasi, 1996).

Using these measurements in models, it is possible to predict El Nino events in the tropical Pacific and the consequent changes expected round the world in terms of rainfall, drought, and regions of heat or cold up to a year in advance. So much so that these models can reproduce the characteristics of a typical event and their effects on weather patterns throughout the world. It is important to point out that, while these major strides have been made, the relationship between the El Nino, the Southern Oscillation and the monsoons have not yet been clearly defined. Progress in this area would be the key to major advances in the prediction of monsoons.

Immediate Benefits of El Nino Predictions

Peru, among other countries provides, an excellent example of how valuable El Nino predictions can be, for as soon as the prediction for the coming season is made, farmers' representatives and government officials met to decide on the appropriate combination of crops to sow in order to maximize the overall yield. Other countries are taking similar steps, including Australia, Brazil, Ethiopia, India and Indonesia. It is the tropical regions where the impacts of El Nino are most marked and the accuracy of the predictions is greatest. The phase of ENSO has also been used to predict with significant success, the number of tropical cyclones in the North Atlantic well in advance. For instance in 1995, the number of tropical cyclones for the year was predicted to jump suddenly from below average, as it had been for over a decade, to well above normal in 1995. In fact, the actual number was well in excess of the prediction. One can visualise the enormous socio-economic benefits as officials were able to gear up for an intense hurricane season.

6.12 SMALL SCALE CLIMATE EFFECT

Man's effect on small-scale climate changes are known and worth discussing for that reason. Every town, reservoir, and ploughed field changes the local climate to some extent. On an average, urban climate are warmer and have more precipitation than the surrounding countryside.

Urban heat is generally assumed to arise from three causes. The most obvious source is the heat from combustion and the general energy use in the cities. A major cause of the warmer nights is the release of heat stored in structures, sidewalk, streets, etc. in the day time. A minor contribution to warmth of cities is the back reflection and reradiation of heat from pollution in the atmosphere above the city.

The tall buildings hinder the free movement of air and thus of pollutants. Heat radiation emitted from building etc. are absorbed by pollutants but radiation does not enter into space to be dissipated. The excited molecules reemits the heat radiation that leads to rise in temperature of urban area.

As pollutants are the cause of such a behaviour, the larger the concentration of pollutants the more should be rise in temperature. This has been confirmed in a study on Delhi (Section 6.14) where temperature has been increasing year after year all because of pollution.

In the absence of winds, the pollutants in the atmosphere above a city often occupy a so-called "urban dome", (Fig. 6.12). A sheath of air around the city in which pollutants are reasonably well mixed and which may be situated under a temperature inversion (Section 8.1) urban domes have varying heights depending on conditions; an urban dome a subarctic city in the winter may be 100 m thick, while the dome over an oasis in the subtropics may be several kilometers thick. In either case the urban dome tends to warm up. When the wind blows, the urban dome becomes an urban plume moving heat and pollutants down wind (Fig. 6.12b).

The edge of this urban dome can often be distinguished from air planes or mountains near the city. The existence of a visible urban dome implies a different atmospheric composition over the city containing smoke, dust, SO_2 and so on in much higher concentrations than country air. All these reduce the intensity of incoming solar radiation.

The lower wind velocity in cities probably is caused by many obstacles (buildings that are in the path of many wind in a city). The lack of standing water and the fast drainage of precipitation into storm drains result in a decrease of local evaporation and thus in the observed decrease of humidity in cities. In such situation, it is necessary that city should have large number of fountains in the city.

However, when a city contains many cooling towers and there is generally a high water vapour output from burning of fossil fuels and industry, this, together with extra condensation nuclei produced by the city, may cause an increased frequency of showers and thunderstorms down winds from the city and a weak drizzle with in a city that lowers the temperature.

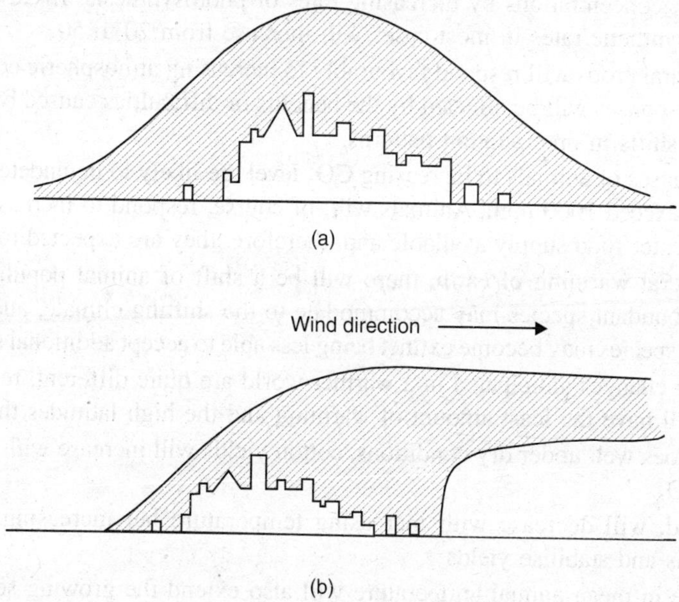

Fig. 6.12 The urban (*a*) dome (*b*) plume.

The other alternative is to grow more trees. Carbon dioxide will get converted to oxygen photochemically and thus reduce the pollutants and urban dome will break to allow the radiation to escape which will decrease the temperature of city.

6.13 ECOLOGICAL CONSEQUENCES

A warmer world in some regions and drier in other will cause dislocations that are beyond control of any modern society or has ever experienced. Our knowledge concerning expected responses by flora and fauna are very limited. Some of what we know are as follows:

1. The greatest air temperature increases are expected to occur in the northern hemisphere winter at high latitudes as a result of a substantial reduction in the amount of sea ice and snow cover on the land because of the warming. This increased melting of ice during summer will delay the formation of ice during autumn and that in turn reduces the thermal insulation of the sea surface during the early winter. Normal sea ice acts as a thermal blanket between the water and

the air. The lack of sea ice allows the heat stored in the ocean to warm air above it and thereby to increase the winter air temperature.

2. Since India lies at low latitudes, change in seasons will be insignificant.
3. Naturally, all coastal dwellers should be much concerned with sea level rise or fall. Low lying area of Kolkata and Dhaka will be flooded: Major coastal cities such as New York, London, Leningrad, Amsterdam, Washington DC, Stockholm, Venice, Hamburg, Bangkok, Singapore and Jakarta may be flooded. Higher sea level will mean coastal loading, increased storm erosion and disappearance of many small islands. The mean sea level has risen by 12 cm during the past century, but has been rising more rapidly during the past 10 years.
4. All plants require CO_2 for photosynthesis and growth. Plants will respond favourably to increased CO_2 concentrations by increasing rates of photosynthesis. If CO_2 concentration doubles, photosynthetic rates in most cases will increase from 20 to 50.
5. All agricultural crops will respond favourably to increasing atmospheric concentrations of CO_2 and these responses will be affected by the benefits or difficulties caused by increased temperatures or by shifts in precipitation patterns.
6. Direct responses of animals to increasing CO_2 level are likely to be undetectable except if concentrations exceed 1000 ppm. Animals will, of course, respond to increased plant biomass by having a greater food supply available and, therefore, they are expected to increase in number.
7. With a general warming of earth, there will be a shift of animal populations toward higher latitudes. Abundant species may accommodate to the shifting climate, and dominate, whereas endangered species may become extinct being less able to accept additional stress or competition.
8. The climate changes anticipated in a warmer world are quite different, region by region. Low latitudes will have the least amount of warming and the high latitudes the most.
9. As cotton does well under dry conditions, cotton yields will increase with increase in temperature and CO_2.
10. Wheat yield, will decrease with increasing temperature but increasing levels of CO_2 may mitigate this and stabilise yields.
11. An increase in mean annual temperature will also extend the growing season for most crops and this could be beneficial. However, it will also extent the time when insect pests can breed, and may make it possible for some species of insects to have another generation during the year. An important ecological law is that numbers of insects increase exponentially with the number of generations per season. Increased moisture in some regions may increase the number of pests and pathogens and increased dryness may reduce the numbers.
12. Rivers, streams and lakes will be affected in many ways by increasing levels of CO_2. More flooding of tropical watersheds may lead to an increase of Schistosomias which is a chronic parasitic disease potentially fatal to human and other mammals. It is spread by snails that act as hosts for the larvae of the parasite. The larvae become free, swimming in stagnant water. The parasites enter the skin of humans bathing in or using the water and lay eggs in the liver. It leads to weight loss, restlessness and liver disorders ensue.
13. Increased rainfall in tropical and subtropical regions producing more stagnant water will be the breeding areas of mosquito and thus increase in malaria. An increase of flowing water, will provide a reproductive habitat for the black fly and the dreadful *River Blindness* disease known as chocerciasis, may increase in tropical countries. The black fly is a purveyor of this parasite.

6.14 WHY THE TEMPERATURE OF DELHI IS RISING YEAR AFTER YEAR?

Industries, motor vehicles, air-conditioners consume energy to produce pollution and heat in cities. However, how much energy is used is a variable factor. Its variation will change the temperature of a city. Hence, study of variation of temperature over a time period will let us know how polluted a city is. Such a study was a carried out by the author.

Daily maximum temperature and maximum mass of water in air in New Delhi from May to July for 1993, 1994, 1995 was collected. Both maximum temperature and maximum mass of water vapours per cubic metre of air was plotted against dates. However, because of the many fluctuating factors operating in atmosphere, sudden rise and fall temperature as well as mass of water vapours was observed. This disturbs the value even of subsequent days. Keeping this in mind most probable lines were drawn which showed not the exact values but a definite trend. These trends are shown in Fig. 6.13 and 6.14. Some prominent observations are very apparent from these figures.

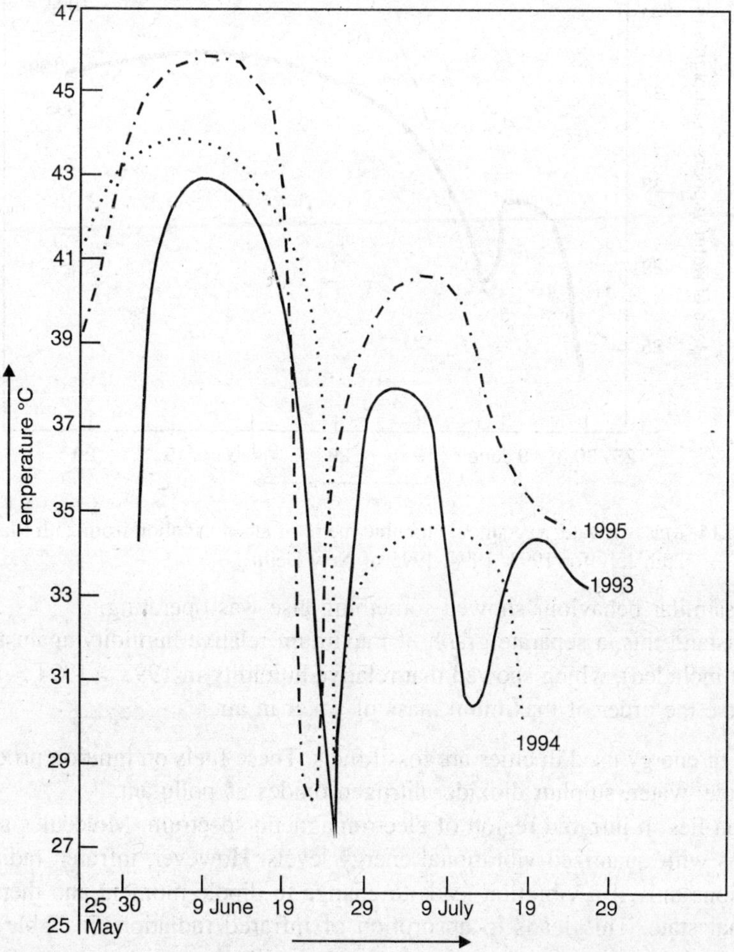

Fig. 6.13 Temperature variation from 25th May to 25th July for 1993, 1994, 1995 of New Delhi.

1. The temperature of Delhi in summer, 1995 was higher than in summer of 1994 which was higher than summer of 1993. However, how much temperature increase has taken place on a particular day was a variable factor. Similar variation is expected throughout the year.
2. Mass of water in air in 1995 summer was more than 1994 summer which was more than 1992 summers. Similar variation is expected rest of the year.
3. The variation in mass water in air is not analogous to temperature variation. Had the Fig. 6.13, and 6.14 or curves in Fig. 6.14 showed the analogous behaviour, cause could have been the same.

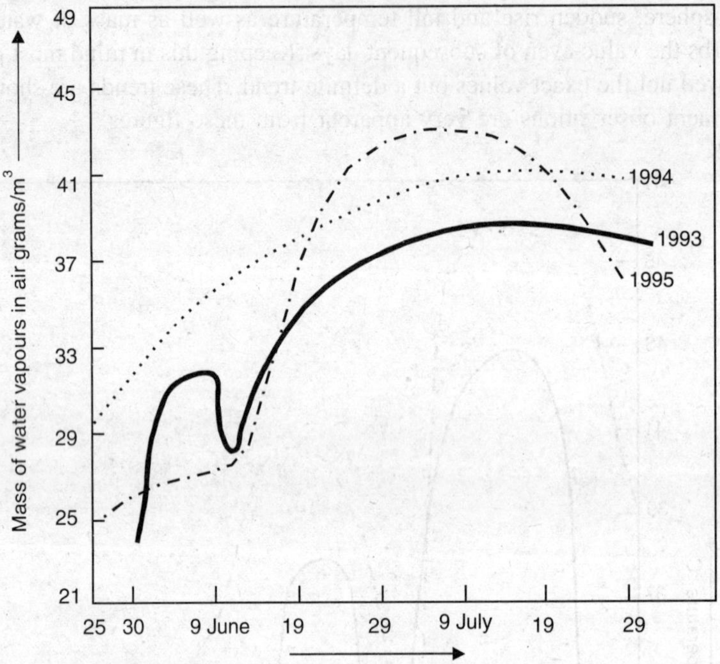

Fig. 6.14 Mass of water vapours in air cubic metre of air in variation from 25th May to 25th July for 1993, 1994, 1995 of New Delhi.

4. Their dissimilar behaviour showed something else was operating.
5. To understand this, a separate graph of maximum relative humidity against dates was plotted (Fig. not included), which showed that relative humidity in 1993 > 1994 > 1995. This order is the reverse the order of maximum mass of water in air.

Main source of energy used in cities are fossil fuels. These fuels on ignition produce heat radiation and carbon dioxide, water, sulphur dioxide, nitrogen oxides as pollutant.

Heat radiation lies in infrared region of electromagnetic spectrum. Molecules are in a continuous state of vibrations with quantized vibrational energy levels. However, infrared radiation interact with molecular vibrations only if a vibration leads to change in dipole moment and then molecule goes to excited vibrational state. This leads to absorption of infrared radiation. In Table 6.2 are given the frequencies at which some molecules of interest show IR absorptions. Nitrogen and oxygen are homomolecular diatomic molecules involving no change in dipole moment in its vibration and hence do not show any infrared absorption line. Had the atmosphere containing absorption, nitrogen and oxygen only, then heat radiation emitted by the earth would not had interacted with atmosphere and thus enters back to space to be dissipated. This means no rise in temperature of urban area. This is what happens in rural area. However, as air become polluted with carbon dioxide, water, sulphur-dioxide, etc., heat radiation is absorbed by these molecules *i.e.*, the radiation does not enter into the space to be dissipated. The excited molecules re-emit the heat radiation that lead to rise in temperature.

As the tall buildings hinder the free movement of air and thus pollutants, air movement is blocked. Such polluted air form a urban dome over the city. It does not allow heat radiation to escape and hence temperature of urban areas is always higher than rural area. As the pollutants are the cause of such a behaviour, the larger the concentration of pollutant, the more should be rise in temperature.

As in Figure 6.13 temperature behaviour in three years is similar, but year after year, the rise in temperature shows that urban dome has more pollutant and consequently year after year temperature of Delhi is increasing. *Hence, pollution in Delhi in* 1994 *was more than* 1993 *and* 1995 was *more polluted than* 1994.

Every man required 13.5 kg of air, 2 kg of water and 1.2 kg of food everyday. Since largest amount is of air, it should be free of pollutants, the pollutant sources cannot be removed. The alternative solution of pollution is dilution which means removal of carbon dioxide from air. By planting large number of trees. Carbon dioxide will get converted to oxygen photochemically thus dilute the pollutants. It may be pointed out that catalytic converters change nitrogen oxides, carbon monoxide to nitrogen and carbon dioxide, hence major exhaust gas carbon dioxide stay in the atmosphere though nitrogen oxides and carbon monoxide are completely eliminated. Carbon dioxide can only be regulated by planting more trees.

Burning of fossil fuels, produce maximum carbon dioxide and water vapour. This means water vapour concentration should increase, which is corroborated by our observation (Fig. 6.13). However, as the behaviour of water vapour concentration variation in three years is not the same as that of temperature, it means that water vapour source is not only fossil fuel but evaporation from water surface also. The latter source may have been changing over the period of time. As the relative humidity is in the order 1993 > 1994 > 1995 means that mass of water in air has increased due to increased burning of fossil fuels but water coming from evaporation has decreased.

The urban heat island as in Delhi should induce an upward motion of air which would result in water condensation to form clouds and thus to small showers, in a polluted city. Such diminished atmospheric stability should lead to fall of temperature. Rather non existence of clouds and almost no such showers in Delhi show that atmosphere here is highly stable. This stability is a consequence of relative humidity which was greater in 1993 and least in 1995. This means Delhi is becoming more and more dry for the prevailing temperature thus should have tendency to more water evaporation. In crude way it can also be said that before cloud are formed, they are evaporated.

The various sources of water evaporation are:

(*i*) Evaporation from our body skin;

(*ii*) Evaporation from river Yamuna and open drains. Evaporation from Yamuna must have increased because of increase in temperature of Delhi. However, over the years open drains have decreased hence evaporation from drains must have decreased;

(*iii*) Evaporation from soil: Most of the high rising buildings that have come up have covered the entire area with stone flooring. Footpaths have already been covered with stone so as to keep the surrounding clean. Such steps means less water available for evaporation. There is an overall decrease in water coming from evaporation, and (*iv*) Evaporation from tree leaves. As trees in Delhi have been decreasing, evaporation of water from tree tops must have decreased. If trees are not there, microorganisms may die and dry soil will be left. This means Delhi will be heading towards a desert.

There is an overall decrease in water coming from evaporation. The consequence of such dryness will be more evaporation from existing source, which will effect the Delhi citizens as follows:

(*i*) Our skin is made up of protein molecules held together by water molecules. If more water molecules are removed by evaporation, cross linking between protein molecules will take place which will lead to appearance of early old age;

(*ii*) More evaporation from Yamuna means acute shortage of water;

(*iii*) If soil surface is covered, it will amount to less water leaching down the ground, which will lower the water table of Delhi. Dry surface means hot winds which may cause further soil erosion.

(*iv*) The roots of trees keep the soil particles together and wet. If trees are not there, dry soil will be left with no granular structure, which will increase particulate matter in air in Delhi.

All these factors will lead to desertification of Delhi.

In order that Delhi's atmosphere so stabilises that no change be possible, it is better we should go for planting more trees.

It will increase open soil surface and thus fall in water table will be retarded. The roots of trees will keep soil wet and thus decrease in particulate matter in air. More are trees, more evaporation of water and thus may alter the drying up of Delhi. Evaporation will bring down the temperature near trees, but temperature under sun will be higher. This will introduce lot of temperature fluctuations within a small area and thus induce air movement which will break the urban dome of Delhi and temperature will fall. This will restore the water content of air to normal. Ridge forest that has already been destroyed may be difficult to reclaim. The alternative is to ask all high rising building house owners to plant trees. They should behave in a more responsible way and thus plant and maintain trees. All along the innerside of the boundary walls of these buildings, trees should be planted separated by a distance of 6 feet. The space between the trees should be filled by planting shrubs. Such plantations will change the complexation of environment. So "*Grow more Trees*" to save cities from pollution.

The total forest area of world was estimated to be 7000 mh (million hectare) in 1900. By 2000, it was reduced to 2370 mh and if the trend of reduction continues then we are heading towards desert. Forests in India are being lost at the rate of about 1.6 mh per annum, with present forest cover of 64 mh, *i.e.*, 19.44 per cent of the total land mass, which is much less than the standard 33 per cent as recommended in the National Forest Policy. The maximum deforestation was in Delhi and Punjab followed by Rajasthan, Gujarat and Haryana. In the early fifties, more than fifty per cent of Delhi was forest and now only 1.5 per cent is forest. The actual forest cover of various states in India is given in Table 6.6.

India was once densely covered with extremely rich and productive natural forest. It is imaginable that today man has turned it into a near desert.

Before ending the Chapter, I will like to quote an historic incident. In northern India, the magnificent city of Fatehpur Sikri, just 23 miles west of Agra was built in the 16th century, only to be abandoned within two decades. When this site was selected by Akbar, it was a thick forest, but deforestation changed the climate. Looking at Fatehpur Sikri today in the midst of oven-like desert, it is hard to envision the productive and well watered environment which gave birth to it 400 years ago. Once wells dry, there is no going back. Same thing can happen to Delhi, within a Decade or so.

Table 6.6 Actual forest cover of 1991

State/UTs	*Area in sq kms*		*Actual forest cover as % of geographical area*
	Geographical area	*Actual forest cover*	
Andhra Pradesh	275068	47290	17.2
Arunachal Pradesh	83743	68757	82.1
Assam	78438	24751	31.5
Andaman & Nicobar	8249	7622	92.4
Bihar	173877	26668	15.3
Goa	3702	1253	33.8
Gujarat	196024	11907	6.1
Chandigarh	114	5	4.5
Delhi	1493	22	1.5
Haryana	44212	513	1.2
Himachal Pradesh	55673	11780	21.2
Jammu & Kashmir	222235	20064	9.0
Karnataka	191791	32199	16.8
Kerala	38863	10292	26.5
Manipur	22.327	17.685	79.2
Maharashtra	307690	44044	14.3
Mizoram	21081	18853	89.4
Punjab	50362	1343	2.7
Rajasthan	342239	12835	3.7
Sikkim	7096	3033	42.7
Tamil Nadu	130058	17713	13.6
Tripura	10486	5535	52.8
Uttar Pradesh	294411	33609	11.4
Meghalaya	22429	15875	70.8
Nagaland	16579	14321	86.4
Orissa	155707	47205	30.3
Dadra and Nagar Haveli	491	206	42
Daman and Diu	112	2	1.8
Lakshadeep	32	-	-
Pondicherry	492	-	-
Madhya Pradesh	443446	135785	30.6
Total	3287263	639182	19.44

Source: The State of Forest Report, 1991, GOI.

6.15 CLIMATE CHANGES IN INDIA DUE TO GREEN HOUSE EFFECT

India is paying for the sins committed by Industrialized nations, especially USA and Japan, responsible for almost 80% of the Green House Gases emitted in the world and 90% of the disaster that has and that may take place in India. Here are some facts:

1. By 2050 India will be at least 3°C hotter, many of its rivers would have dried up, floods and droughts would be regular, cyclones more intense associated with all this, severe food and water shortage.
2. Delhi's winter months are shrinking regularly. Last to last year it arrived in late October, last year in late November.
3. Glaciers are receding year after year. The Gangotri glacier that feeds rivers like Ganges has already lost one third of its 24 km length in the last 50 years. It is losing 18 meters every year. Pindari glaciers, a favourite among the trekkers, melt over 130 meters every year. The Chota Shign glacier in Himachal Pradesh loses 6 meters annually. By 2035 most glaciers in India would have disappeared.
4. The average rainfall in India has remained unchanged but several aberrations have surfaced. There has been increase of 40-80% rainfall in Haryana, Punjab, Delhi and floods in the arid state of Rajasthan. Kerala, North-eastern states with heavy rainfall have shown a consistent drop in volume.
5. The rice yield would be substantially less in next 50 years because of rapid warming.
6. In the little village of Pubong Fatak lives a 97 years old man, Phul Bahadur, 13 km from Darjeeling in West Bengal. This old man is living a chronicle of time in the famed hill station. "When I migrated from Nepal in 1926, snow used to cover the thick forests of Darjeeling for more than three months", he remembers. "Now there is no snow and no trees to shelter me from the harsh winds, I feel like my house is getting blown away" he fears. Most senior residents have similar stories.
7. Some years ago the maximum temperature recorded in the Darjeeling were 14.89°C. There has been consistent increase in temperature, which has touched a record high of 28°C in summer 2000. Fans running on full speed have become a common sight now in summer, in Darjeeling as well as Gangtok, the capital of Sikkim. Barely 5-6 years ago people use to come to office in suits even in summers, now one cannot sit without fan. Teenagers walk around in sleeveless blouses.
8. Mild winter drizzle was earlier accompanied by frost in Darjeeling. Leaves would dry up during December and January. The winter winds use to blow the tree fruits that would decompose due to moisture from the winter rain. Lightning during monsoon would fix nitrogen in the soil helping the seedling to germinate. By June the plants would be big enough to resist heavy rainfall. In the past few years' winters are dry, free from frost, that inhibits soil decomposition. Monsoon rains arrive early, hampering the growth and germination of seedling and eventually affecting flowering.
9. Seasonal flowers are showing unusual signs. Scizanthus usually bloom between March and May. In past few years they bloom either a month late or a month early. Anthurium, an ornamental plants used to be conformable below 1220 m now thriving at 2285 m. Similar is the case with many other plants and trees. Honey from Darjeeling has a very characteristic flavour, thickness and purity, but this year there has been no honey.
10. In past 5-6 years quality of cabbage has deteriorated. They become flaccid and open up much earlier than they used to, which ruins the taste of vegetable. Cabbages are called cold crop and are

usually grown when the temperature drops. The cell system in the cabbage head remains compact in winter and tastiest cabbage comes from high altitudes.

11. While cabbage fares badly, cucumber and tomato, which grow better in warmer conditions, are doing well even in the Batasia region at an altitude of 1980 m where there was hardly any cultivation of tomato earlier.
12. Among the unique features of Kashmir valley is that it had five seasons cycle. The fifth season, called Sont has disappeared now. It used to last for 30-45 days, from end of February to Baisakhi, upto mid April. In this season there used to be precipitation comprising light snowfall, sleet (snow and rain) and sunshine that culminated into torrential rains by the end of March. The average temperature used to be 8 °C. Almond, the first tree to blossom would be just about to flower.
13. Saifuddin Soz, former Union Minister of Environment and Forest, who lives in Srinagar, reminisces, "During my childhood, the Srinagar used to be snow bound and the courtyards had almost 7 feet of snow till May, not to speak of mountains buried under 35 feet of snow. He regrets that snow has decreased sharply in past 5-6 years, because it has become warm in early February, now. Kashmir the land of abundant water has suffered a drought every year since 1997. The river Jhelum fed by glaciers melting from March to May, now runs dry in the peak of summer as melting take place in January and February.
14. Abdul Salaam Bhatt, owner of houseboat Salaam Palace, in Dal lake, says when he was 15; he used to wear two vests, a kurta, two sweaters and a jacket. Now he is 55, it is much warmer now and he needs just one jacket. He recalls lake use to freeze in December and January down to a depth of one foot and vehicles used to ply on the ice sheet. It hardly freezes now.
15. There are three types of rainfalls in Sikkim. January brings with it light rains and strong winds due to western disturbances. During March and April comes heavy cyclonic showers called Norwester rains. This is followed by monsoon from June to September. The rain patterns have changed. The intensity of Norwester winds has increased in the last 3-4 years leading to greater number of drain bursts. The monsoon rainy days have declined though concentration and intensity has increased. The amount of rain that would have fallen in ten days, falls in two days.
16. As pollination disappeared in Sikkim, mosquitoes are making an appearance particularly in summer. Patients with illness common to tropical areas — diarrhoea, fever, jaundice has increased in past 3-4 years. Viruses that thrive in this type of climate have also increased. Many children complain of skin diseases and infections. Incidence of Malaria has increased in Kashmir over the years, which was free from this disease.
17. Orange needs 20°C for normal growth. Now the temperature has reached 28°C, which has declined orange output. Rabbi crops, which used to grow at 1500 m, now have to be grown at 2000 m. If it is orange in Sikkim, in Kashmir it is apple. Despite increasing area of cultivation, apple production in Kashmir has remained stagnant for the past few years due to rise in temperature and dry weather. They used to be attractive red in colour, now they look dull. Even maize is hard to be cultivated now.

REFERENCES

1. Campbell, T., *Do It Yourself,* Whether Book Oxmoor House, Birmingham (1979).
2. Lutgens, F.K. and Tarbuck E.J., *The Atmosphere,* Prentice Hall, Englewood, Cliff (1979).
3. Robert, W.O. and Lansford H., *The Climate Mandate,* W.H. Free-man, San Francisco (1979).
4. Sindhu, P.S., "Electrochemical Science–A Non-polluting Technology for 2000 AD", *Chemistry Education.* October Dec., 15-21 (1991).

5. Sindhu, P.S., *Fundamentals of Molecular Spectroscopy*, New Age International (P) Ltd., New Delhi (2005).
6. Sauer, C.O., *Land and Life A Selection from the Writings of Carl Leightly* (*ed*), Berkely, University of California Press (1967).
7. Kriohnan, A., *Calcutta,* Statesman, page 6, Sept. 15, (1973) .
8. Lowder Milk W.C. (1960) "Lessons from the Old World to the American in Land Use" Smithsonian Report for pp. 413-28, (1943).
9. Gadwin, O.P. Obasi, Rajiv Gandhi Memorial Institute for Contemporary Studies (RGICS, Paper 36, 1996), "Climate, Climate Change Variability and Predictability".

7

Acid Deposition

Acid rains are common in Europe and USA. Fogs with pH values of 2 or less also have been observed. Relevant questions about acid deposition include the following: What acids are present? In what concentration? What are there precursors? How are they formed? Does oxidation occur in the gas phase, in aerosol fog cloud, or rain droplets or on the surface of solids or in all of these systems? In this chapter, the wide range of chemical reactions are proposed to explain the conversion of gaseous SO_2 to sulphuric acid aerosols and sulphate, gaseous NO_x to nitric acid and nitrates have been dealt with. Within last several years, rapid decline in forest has been observed in some areas of Europe. The total forest area affected in West Germany is 8 per cent in 1982, 34 per cent in 1983 and 50 per cent in 1984 thereby eroding a wide variety of trees. The cause is acid deposition, but many of the effects may not be due to a combination of air pollutants acting in concert with acids plus O_3.

7.1 OXIDATION OF SO_2 IN TROPOSPHERE

The combustion of sulphur present in fossil fuels (solid, liquid and gas) leads to emission of SO_2, which is subsequently oxidized to sulphuric acid.

Sulphur dioxide is a triatomic molecule with bent symmetrical structure like water with +4 as the oxidation state of sulphur. The absorption spectrum of SO_2 consist of a weak absorption band in near ultraviolet with λ_{max} = 374 nm and $\sigma = 0.016 \times 10^{-20}$ cm^{-2} $molecule^{-1}$ and a strong ultraviolet band starting at 337 nm with λ_{max} = 294 nm and $\sigma = 40.6 \times 10^{-20}$ cm^2 $molecule^{-1}$. This absorption peak at 294 nm forms two singlet excited states of SO_2

$$SO_2\left(X,\,^1A_1\right) \xrightarrow[240<\lambda<330\text{ nm}]{h\nu} SO_2\left(^1A_2,\,^1B_1\right) \qquad ...(7.1)$$

while the weak absorption involves a spin forbidden transition to triplet state ($^3B_1$).

$$SO_2\left(X,\,^1A_1\right) \xrightarrow[340<\lambda<400\text{ nm}]{h\nu} SO_2\left(^3B_1\right) \qquad ...(7.2)$$

The bond dissociation energy of SO_2, D (OS – O) is 5.61 eV = 541.3 kJ mol^{-1} which corresponds to photon wavelength of 220 nm. SO_2 can dissociate to SO + O only at wavelength below 220 nm.

$$SO_2\left(X,\,^1A_1\right) \xrightarrow[\lambda<220\text{ nm}]{h\nu} SO\left(^3\Sigma_u^-\right) + O\left(^3P\right)$$

This reactions cannot occur in troposphere where only wavelength of 290 nm and above are present. This suggest that SO_2 is photochemically inactive with reactions of this type. Therefore, light in either band is not available for absorption and photodissociation to SO and O is not possible. The weak band at 374 nm region show strong Zeeman splitting suggesting this absorption band correspond to spin forbidden radiative transition to the lowest lying triplet state, the ground state being singlet.

$$SO_2\left(X^1, A_1\right) \xrightarrow{hv} SO_2\left({}^3B_1\right) \quad ...(7.2)$$

The occurrence of this forbidden component of transition is not caused by simple vibronic interaction with another triplet state of type, ${}^3B_1$, but by vibrationally induced spin orbit coupling. This coupling can mix the A_2 component of the ${}^3B_1$ state with the ${}^1B_1$ state of ν_3 (b_2) excited by an odd number of quanta and thus can enable the 1 – 0, 3 – 0 bands of the ${}^3B_1 - {}^1A_1$ transition from the neighbours $\tilde{X}$, ${}^1A_1\,\tilde{A}$, ${}^1B_1$ transition.

The absorption at the 294 nm band is due to the first excited singlet state corresponding to ${}^1A \rightarrow {}^1B_1$ transition as shown by Equation 7.1. However, this excited singlet state may also get converted to triplet state either by inter-system crossing

$$SO_2\left({}^1B_1\right) \xrightarrow{ISC} SO_2\left({}^3B_1\right) \quad ...(7.3)$$

$$k_{73} \approx 10^3\,s^{-1}$$

or by collisional induction with SO_2

$$SO_2\left({}^1B_1\right) + SO_2\left({}^1A_1\right) \longrightarrow SO_2\left({}^3B_1\right) + SO_2\left({}^1A_1\right)$$

$$k_{74} = 1.7 \times 10^{-12}\ cm^3\ molecule^{-1}s^{-1} \quad ...(7.4)$$

which occurs at every collision. Triplet SO_2 has a very long lifetime, approximately 10^3 s, and presumably it is this state that is responsible for much sulphur dioxide photochemistry and homogenous oxidation. For example, SO_3 is a product of the reaction of SO_2 (${}^3B_1$) with ground state SO_2 (${}^1A_1$)

$$SO_2\left({}^3B_1\right) + SO_2\left({}^1A_1\right) \longrightarrow SO_3 + SO \quad ...(7.5)$$

with SO also giving SO_3

$$SO + SO_2 \longrightarrow SO_3 + S \quad ...(7.6)$$

The photo oxidation of SO_2 is much faster in the presence of hydrocarbons and oxides of nitrogen than in a pollution free atmosphere. The triplet SO_2 is another potential oxidant of nitric oxide

$$SO_2\left({}^3B_1\right) + NO \longrightarrow SO + NO_2$$

and

$$NO_2 + SO_2 \longrightarrow SO_3 + NO \quad ...(7.7)$$

with the net reaction being the NO catalyzed oxidation of triplet SO_2 from ground singlet state SO_2.

$$SO_2\left({}^3B_1\right) + SO_2\left({}^1A_1\right) \longrightarrow SO_3 + SO \quad ...(7.8)$$

The reaction comparable to Equation 7.8 does not occur between ground state SO_2 molecules. This is another example of how a molecule in different electronic states react differently, and suggest a potentially important photochemical role of SO_2 in reaction 7.8 that will contribute to the observed rapid removal of SO_2 in photochemical smog systems.

7.1a Hydroxyl Radical

There are a number of candid reactions for the oxidation of ground state SO_2 with O_3 or with labile species as OH, HO_2, CH_3O_2 and NO_3 radicals and O (^3P). However, all available evidence suggest the only gas phase process that is fast and efficient enough to account for most of the sulphuric acid aerosol formed by gas phase processes is reaction of SO_2 with the OH radical

$$\dot{O}H + SO_2 \xrightarrow{M} HOS\dot{O}_2 \qquad ...(7.9)$$

While the kinetics of the elementary reaction 7.9 are reasonably well defined, and it is known that a significant fraction of the $HOSO_2$ radical formed ultimately winds up as H_2SO_4 aerosol. In polluted atmosphere containing O_2, NO, NO_2, peroxy radicals as HO_2, RO_2 and water, the reactions are not well understood.

Davis *et al.*[4] proposed that in troposphere and lower stratosphere the predominant fate of $HOSO_2$ is reaction with O_2 as

$$HOS\dot{O}_2 + O_2 \xrightarrow{M} HOSO_2\,\dot{O}_2 \qquad ...(7.10a)$$

The $HOSO_2O_2$ radical is relatively stable and will react with atmospheric water vapours to form hydrated radical adduct

$$HOSO_2\,\dot{O}_2 + H_2O \longrightarrow HOSO_2O_2\,(H_2O) \qquad ...(7.10b)$$

Reaction 7.10b is followed by further addition of H_2O molecules until hydrated cluster of the form $HOSO_2O_2\,(H_2O)_x$ are produced. These clusters may react with NO, SO_2, HO_2 radical by quasi-heterogenous processes.

The products could then lead to aerosol species, including H_2SO_4, by both condensation and coagulation.

The adduct $HOSO_2O_2$ has been proposed to react in a manner similar to the alkoxy radicals for example to oxidize NO to NO_2

$$HOSO_2\,\dot{O}_2 + NO \longrightarrow HOSO_2\,\dot{O} + NO_2 \qquad ...(7.11)$$

However, if the concentration of NO are very low radical-radical reaction may take place as discussed above and hydration of $HOSO_2O_2$ may precede such reactions in atmosphere. Formation of H_2SO_4 from $HOSO_2O_2$ radical can be envisaged via a number of reactions, including hydrogen abstraction from hydrocarbons.

$$HOSO_2\,\dot{O} + -\overset{H}{\underset{|}{\overset{|}{C}}}- \longrightarrow HOSO_2OH + -\underset{|}{\dot{C}}- \qquad ...(7.12)$$

or perhaps with NO or NO_2 followed by hydrolysis

$$HOSO_2\,\dot{O} \xrightarrow{NO} HOSO_2ONO \xrightarrow{H_2O} HOSO_2OH + HONO \qquad ...(7.13)$$

$$HOSO_2\,\dot{O} \xrightarrow{NO_2} HOSO_2ONO_2 \xrightarrow{H_2O} HOSO_2OH + HNO_3$$

An alternate pathway for reaction 7.10a is

$$HOS\dot{O}_2 + O_2 \xrightarrow{M} H\dot{O}_2 + SO_3 \qquad ...(7.14)$$

$$k_{714} \approx 4 \pm 2 \times 10^{-13} \ cm^3 mol^{-1} s^{-1}$$

$$HO_2 + NO \longrightarrow \dot{O}H + NO_2 \qquad ...(7.15)$$

The sulphur trioxide formed in 7.14 would react with water vapours to form H_2SO_4.

$$SO_3 + H_2O \longrightarrow H_2SO_4 \qquad ...(7.16)$$

$$k_{716} \approx 10^{-12} \ cm^3 mol^{-1} s^{-1}$$

Meagher and co-worker[6] concluded, based on smog chamber studies of irradiated SO_2-propene-butane-NO_x—H_2O mixtures, that the OH—SO_2 reaction leads 98.5% of the time to HO_2 formation. In order to explain their O_3 yields, they also suggested that one of the intermediates in the OH—SO_2 oxidation must be O_3.

However, the problem with above mechanism is that reaction 7.14 is estimated to be −24.72 kJ mol^{-1} endothermic. However, this problem may be overcome if adduct $HOSO_2$ becomes hydrated before it reacts with O_2 with overall reaction written as

$$H\dot{O} + SO_2 + (O_2, H_2O) \longrightarrow H_2SO_4 + H\dot{O}_2 \qquad ...(7.17)$$

One notes that concept of hydration is present in both the mechanisms, except that in the former one hydrates the $HOSO_2O_2$ and in the latter the $HOSO_2$ radical.

7.1b Ozone + Olefin Reaction

A second possible oxidant of gaseous SO_2 in polluted air, is the stabilised Criegee biradical RCHOO formed in ozone-olefin reaction 4.77

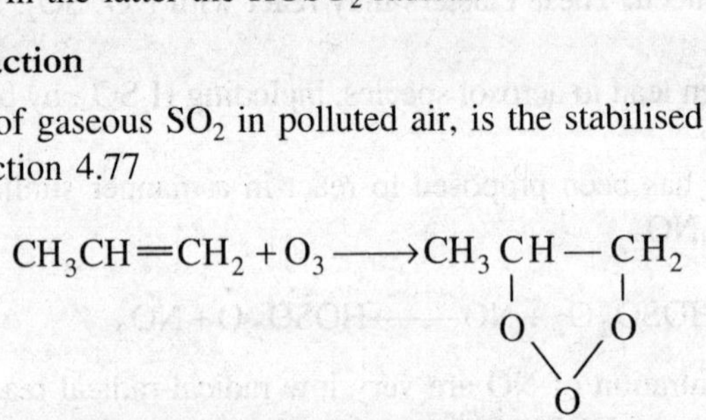

$$\xrightarrow{a} CH_3CHOO + HCHO \qquad ...(4.77)$$

$$\xrightarrow{b} CH_3CHO + H_2COO \qquad ...(4.78)$$

Since in gas phase O_3 itself react very slowly with SO_2 ($k < 8 \times 10^{-24}$ cm^3 $molecule^{-1}$ s^{-1}), the oxidation must be attributed to intermediates produced in ozone-olefin reaction.

$$CH_3CHOO + SO_2 \longrightarrow CH_3CHO + SO_3 \qquad ...(7.18)$$

$$k_{718} = 3 \times 10^{-14} cm^3 mol^{-1} s^{-1}$$

Water has been shown to react with ozone-olefin intermediates but addition of water inhibits the SO_2 oxidation to SO_3. The rate constant of similar reaction with NO and NO_2 are 7×10^{-12} and 7×10^{-13} cm^3 $molecule^{-1}s^{-1}$. Since these rate constants are more than k_{718}, most of the intermediates will be used by nitrogen oxides as a result 7.18 cannot be a major rate for SO_2 oxidation.

7.1c Ground State Oxygen Atom, O (^{3}P)

The reaction of O (^{3}P) atom with SO_2

$$SO_2 + O(^3P) \xrightarrow{M} SO_3 \qquad ...(7.19)$$

$$k_{719} = 3.3 \times 10^{-14}\ cm^3 mol^{-1} s^{-1}$$

The concentration of O (^3P) atoms in ambient air is expected to be only ≤ 10^5 atoms cm^{-3} so that lifetime of SO_2 in reaction 7.19 is more than 9.6 years making an insignificant contribution to the oxidation of SO_2. However, if relatively high NO_2 concentration are present in sunlight, significant concentration of O (^3P) may be produced. This may lead to a measurable rate of the SO_2 oxidation.

7.1d Homogenous Aqueous Phase Reactions

The presence of aqueous droplets in the form of aerosols, clouds, fogs and rain in the troposphere offers another phase in which SO_2 oxidation may occur. Thus, sulphur dioxide gas dissolves in water to set up equilibria

$$SO_2(g) + H_2O \rightleftharpoons SO_2 H_2O \qquad k_{720a} = 1.24\ M\ atm^{-1} \qquad ...(7.20a)$$

$$SO_2H_2O \rightleftharpoons HSO_3^- + H^+ \qquad k_{720b} = 1.32 \times 10^{-2} M \qquad ...(7.20b)$$

$$2HSO_3^- \rightleftharpoons SO_3^{2-} + H^+HSO_3^- \qquad k_{721} = 6.42 \times 10^{-8} M \qquad ...(7.21)$$

As a result, dissolved SO_2 really includes three chemical species, Hydrated SO_2, bisulphite ion (HSO_3^-) and the sulphite ion (SO_3^{2-}). The predominant form depends on the acidity of the solution. The half time of any of equilibria to shift to right or left of the equilibria reaction is few milli second or less, it is immaterial which of the three species. SO_2H_2O, HSO_3^- or SO_3^{2-}, is the actual reactant in any particular oxidation, the equilibria will be re-established relatively rapidly.

In the gas phase SO_2 concentration range 0.2–200 ppb, and over a pH range of 0–6 concentration from ~ 10^{-9} to 10^{-3} mol L^{-1} are anticipated. As expected the concentration fall as pH falls. Because of the different forms in which dissolved SO_2 exists in solution, the oxidation state (*i.e.*, + 4) often used to denote all these forms of SO_2 taken together, as

$$S(IV) = SO_2 \cdot H_2O + HSO_3^- + SO_3^{2-}$$

As discussed in more details below, the rates of various aqueous phase reactions show different dependencies on pH. Some have rate coefficients that increase with increasing pH (*e.g.*, O_3) while others (*e.g.*, H_2O_2) show the opposite trend.

In the first case, both the kinetics and solubility of S(IV) vary with pH in the same direction. As a result, the overall rate of production of S(VI) by such reactions typically shows a strong pH dependence, so that it is often only at higher pH values that they represent important oxidation pathways of S(IV). Such oxidative pathways may initially contribute significantly to acid formation in a relatively neutral droplet; however as more and more of acid is formed, the rate of formations of S(VI) by these reactions will decrease. This self-quenching effect may lead either to a lowered net oxidation of SO_2 in the atmosphere or to other pathways that do not show this pH dependence becoming relatively more important.

On the other hand, if the rate coefficient decreases with increasing pH, then the reaction rate and solubility of S(IV) work in opposite directions. The net result in terms of S(VI) production may be, as in the case of H_2O_2, a relatively small dependence on pH. Such oxidation may thus contribute in a relatively constant fashion over the entire range of pH values of atmospheric interest. It may be pointed out that kinetics of oxidation in only one portion of a sequence of processes leads SO_2 to aqueous phase sulphate. The sequence of steps, depicted in Fig. 7.1 is as follows:

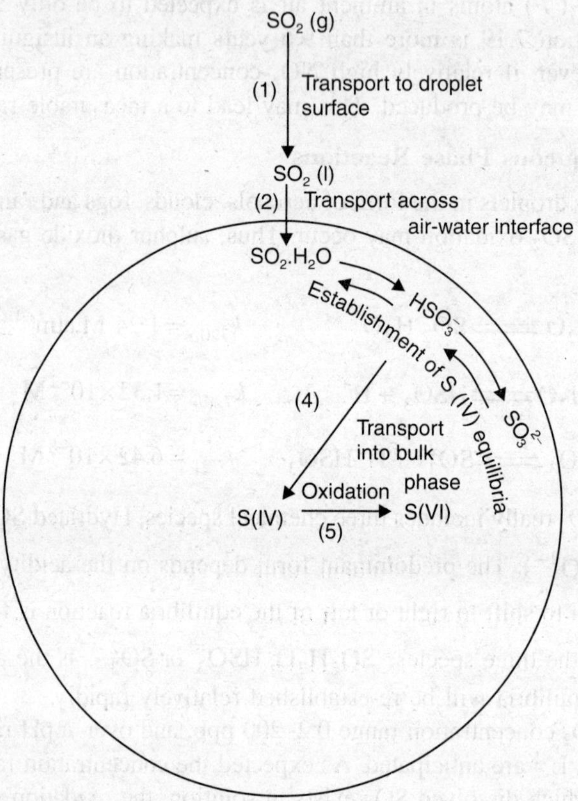

Fig. 7.1 Scheme involved in the transfer of SO_2 from the gas phase to the aqueous phase of an atmospheric water droplet and its oxidation in the liquid phase SO_2 (g) = gas phase SO_2, SO_2(l) = SO_2 at water-gas interface.

1. Transport of gas to the surface of the droplet. Time for gas phase diffusion to the surface are in the range 10^{-10} to 10^{-4} s for droplets with radii from 10^{-5} to 10^{-2} cm.
2. Transfer of gas across the air-liquid interface time to establish an equilibria for gases of interest over a droplet, pH range 2-6 is less than 10^{-8} to 10^{-1} s.
3. Formation of aqueous phase equilibria of the dissolved species, through equilibria 7.20-7.21 in the case of SO_2. This occurs on a time scale of milliseconds or less.
4. Transport of dissolved species within the aqueous phase by molecular diffusion, takes ~ 10^{-6} to 1s for droplets with radii from 10^{-5} to 10^{-2} cm.
5. Reactions in the droplets.

The major oxidising species in water are O_3 and H_2O_2. There are potential sources of O_3 and H_2O_2 in aqueous droplets, one might anticipate both chemistry and photochemistry of these highly reactive species to occurs. Photochemistry of O_3 has been discussed in Chapter IV, particularly reaction 4.19 and 4.20 as

$$O_3 \xrightarrow[\lambda \leq 320\text{ nm}]{hv} O + O_2 \quad \text{...(4.19 \& 4.20)}$$

The photolysis of H_2O_2 gives hydroxyl free radical

$$H_2O_2 \xrightarrow[\lambda \leq 360\text{ nm}]{hv} 2\dot{O}H \quad \text{...(7.22)}$$

The major loss process for O_3 and H_2O_2 in droplets will depend on the presence and concentration of other species in solution. However, importance of each reaction is not being assessed quantitatively since this will depend on particular conditions.

O_3 can react with OH^- and HO_2^- to produce superoxide ion

$$O_3 + OH^- \longrightarrow H\dot{O}_2 + O_2^- \quad \text{...(7.23)}$$

$$O_3 + HO_2^- \longrightarrow H\dot{O} + O_2^- + O_2 \quad \text{...(7.24)}$$

$$O_3 + \dot{O}H \longrightarrow H\dot{O}_2 + O_2 \qquad k_{725} = 1.1 \times 10^8\ \text{l mol}^{-1}\text{s}^{-1} \quad \text{...(7.25)}$$

$$O_3 + H\dot{O}_2 \longrightarrow \dot{O}H + 2O_2 \qquad k_{726} = 4.0 \times 10^6\ \text{l mol}^{-1}\text{s}^{-1} \quad \text{...(7.26)}$$

In urban area, nitric acid and S(IV) will also be present and O_3 can react with these

$$O_3 + NO_2^- \longrightarrow NO_3^- + O_2 \qquad k_{727} = 1.6 \times 10^5\ \text{l mol}^{-1}\text{s}^{-1} \quad \text{...(7.27)}$$

$$O_3 + HSO_3^- + OH^- \longrightarrow SO_4^{2-} + H_2O + O_2 \quad \text{...(7.28)}$$

From pH 1 to 3, rate constant increases by one order in magnitude. The rate expression over the pH range 1.0 – 6.2 can be represented as

$$\text{rate}\left(\text{mol L}^{-1}\text{s}^{-1}\right) = \frac{d[O_3]}{dt} = \left(k_a + k_b[OH^-][HSO_3^-][O]\right)$$

where $k_a = 3.8 \times 10^5\ M^{-1}s^{-1}$ and $k_b = 1.05 \times 10^{16}\ M^2s^{-1}$ at 25°C

However, these reactions are usually not included in models of reactions in atmospheric droplets, presumably because the rate constant are low to moderate and low organic concentration together yield low overall rates compared to other possible fates. While photolyzed solution of H_2O_2 and reaction 7.30 and 7.31 have been suggested to predominate over photolysis under many conditions.

$$O_2 + HCO_3^- \longrightarrow H\dot{O}_2^- + CO_3^- \quad \text{...(7.29)}$$

$$H_2O_2 + \dot{C}O_3^- \longrightarrow HCO_3^- + H\dot{O}_2 \qquad k_{730} = 8.0 \times 10^5\ \text{l mol}^{-2}\text{s}^{-1} \quad \text{...(7.30)}$$

$$H_2O_2 + \dot{O}H \longrightarrow H\dot{O}_2 + H_2O \qquad k_{731} = 2.7 \times 10^7\ \text{l mol}^{-1}\text{s}^{-1} \quad \text{...(7.31)}$$

$$H_2O_2 + HSO_3^- \longrightarrow HSO_4^- + H_2O \quad \text{...(7.32)}$$

$$H_2O_2 + HONO \xrightarrow{H_3O^+} HNO_3 + H_2O \quad k_{733} = 4.6\times10^3[H^+]1\ mol^{-1}s^{-1} \qquad ...(7.33)$$

Radical reactions similar to those in gas phase might be expected, *e.g.*, hydrogen abstraction from alkanes by OH and NO_3 followed by alkyl peroxy radicals that can oxidize NO as in the gas phase.

$$RH + NO_3 \longrightarrow \dot{R} + HNO_3 \qquad ...(7.34)$$

$$\dot{R} + O_2 \longrightarrow R\dot{O}_2 \qquad ...(7.35)$$

$$R\dot{O}_2 + NO \longrightarrow R\dot{O} + NO_2 \qquad ...(7.36)$$

$$R\dot{O}_2 + H\dot{O}_2 \longrightarrow ROOH + O_2 \qquad ...(7.37)$$

$$ROOH + HSO_3^- \longrightarrow HSO_4^- + ROH \qquad ...(7.38)$$

Oxidation of S(IV) by O_3 has been suggested to occur via free radical pathways since O_3 can produce free radical, such as OH in aqueous solutions which may oxidise SO_2 as follows :

$$HSO_3^- + OH \longrightarrow H_2O + SO_3^- \qquad k_{739} = 9.5\times10^9\ 1\ mol^{-1}\ s^{-1} \qquad ...(7.39)$$

$$SO_3^- + O_2 \longrightarrow SO_5^- \qquad k_{740} = 1\times10^8\ 1\ mol^{-1}s^{-1} \qquad ...(7.40)$$

$$2SO_5^- \longrightarrow \dot{S}O_4^- + SO_4^{2-} + O_2 \qquad k_{741} = 1.0\times10^{11}\ 1\ mol^{-1}s^{-1} \qquad ...(7.41)$$

$$\dot{S}O_4^- + HCO_3^- \xrightarrow{H_2O} SO_4^{2-} + H_3O^+ + C\dot{O}_3^- \qquad k_{742} = 9\times10^6\ 1\ mol^{-1}s^{-1} \qquad ...(7.42)$$

$$\dot{S}O_4^- + H_2O_2 \longrightarrow HSO_4^- + H\dot{O}_2 \qquad k_{743} = 1.2\times10^7\ 1\ mol^{-1}s^{-1} \qquad ...(7.43)$$

$$\dot{S}O_4^- + Cl^- \longrightarrow SO_4^{2-} + Cl \qquad k_{744} = 2\times10^8\ 1\ mol^{-1}s^{-1} \qquad ...(7.44)$$

Oxides of nitrogen in dissolved state oxidise S(IV) as follows:

$$2NO_2 + HSO_3^- \xrightarrow{4H_2O} 3H_3O^+ + 2NO_2^- + SO_4^{2-}$$

7.1e Heterogenous Reactions on Surfaces

Adsorption of SO_2 on the surface of solids, followed by its oxidation on the surface, may provide a third route for the formation of sulphuric acid. Various surfaces may be provided by carbon particles, fly ash, dust, MgO, Fe_2O_3, V_2O_5 and MnO_2, which are known to oxidize S(IV) to S(VI). The rate of removal depends on the nature of particular surface, the presence of co-pollutants such as NO_2 as in the case carbon surface, as well as on the relative humidity. The increase with increasing water vapour suggest that oxidation of the SO_2 may occur in a thin film of water on the surface of the solids. However, the potential contribution of such surface photochemical reactions is intriguing and clearly an area for further study. These photochemical surface reaction do oxidize S(IV) to S(VI).

7.2 OXIDATION OF NO_2 TO NITRIC ACID

Oxidation of oxides of sulphur and nitrogen are so interrelated that they cannot be separated. Some of the reactions will be repetition but that cannot be helped.

7.2a Homogeneous Gas Phase Reactions

The predominant atmospheric fate of NO emitted is oxidation to NO_2 which can then react with OH during the day

$$NO_2 + \dot{O}H \xrightarrow{M} HNO_3 \qquad k_{448} = 1.1\times10^{-11}\,cm^3mol^{-1}s^{-1} \qquad ...(4.48)$$

This reaction is relatively rapid, approximately ten times than that for SO_2—OH 7.9 reaction. For a typical tropospheric OH concentration of 1×10^6 radicals cm^{-3}, the lifetime of NO_2 with respect to 4.48 is only ~ 1 day compared to 13 days for SO_2 under the same conditions. This relatively rapid oxidation of NO_2 to HNO_3 implies that nitric acid formation should occur closer to the source than sulphuric acid formation. However, if significant concentration of NO_2 are tied up in the form PAN, the effective lifetime of NO_2 may be extended.

A second potentially important source of HNO_3 involves reactions of the nitrate radical with some organics.

$$\dot{N}O_3 + RH \longrightarrow \dot{R} + HNO_3 \qquad k_{745} = 5\times10^{-17}\,cm^3mol^{-1}s^{-1} \qquad ...(7.45)$$

$$\dot{N}O_3 + RCHO \longrightarrow R\dot{C}O + HNO_3 \qquad ...(7.46)$$

$$k_{HCHO} = 5.6\times10^{-16} \ \& \ k_{CH_3CHO} = 1.6\times10^{-15}\ cm^3mol^{-1}s^{-1}$$

In polluted urban atmosphere, alkane concentration (excluding CH_4) are typically 2.5×10^{12} mol cm^{-3} and that of HCHO and CH_3CHO are 5×10^{11} and 2.5×10^{11} mol cm^{-3}, respectively, and NO_3 conc. of 2.5×10^9 mol cm^{-3}, the calculated night time rates of HNO_3 formation from these compounds are 3.1×10^5; 7×10^5; 10×10^5 mol cm^{-3} s^{-1} or 0.045, 0.1008 and 0.144 ppb h^{-1} respectively, leading to a net rate of HNO_3 formation of ~ 0.3 ppb h^{-1}.

This can be compared to the daytime rate of formation of HNO_3 by reaction, 4.48. At an NO_2 conc. of 1.25×10^{12} mol cm^{-3} and OH average day time concentrations of 1.6×10^6 mol cm^{-3}, rate of HNO_3 formation at daytime comes to be 2.2×10^7 mol cm^{-3} s^{-1} = 1.68 ppb h^{-1}. Thus, in a polluted atmosphere, as a crude estimate, ~ 9.46% of HNO_3 is formed be due to hydrogen abstraction reactions of nitrate radical at night.

In the presence of NO_2 and O_3, the nitrate radical is formed which is in equilibrium with N_2O_5

$$O_2 + NO_2 \xrightarrow{a} NO_3 + O_2 \qquad k_{449a} = 3.2\times10^{-17}\ cm^3\ mol^{-1}s^{-1} \qquad ...(4.49a)$$

$$NO_3 + NO_2 \rightleftarrows N_2O_5 \qquad ...(4.50)$$

Hydrolysis of N_2O_5 is a potential source of nitric acid

$$N_2O_5 + H_2O \longrightarrow 2HNO_3 \qquad k_{747} \le 1.3\times10^{-21}\ cm^3mol^{-1}s^{-1} \qquad ...(7.47)$$

Because reaction 7.47 appears to be so slow, the reaction sequence 4.49a, 4.50 and 7.47 has often been ruled out as a significant source of nitric acid, but it does contribute to the formation of HNO_3. Thus Tuazon and co-workers[9] calculated that under typical conditions of Los Angeles during the night

(NO_2 at 3 ppb, NO_3 100 ppt), reactions 4.49a, 4.50 and 7.47 would result in an HNO_3 formation rate of 0.3 ppb h^{-1} at 50 per cent relative humidity. These rates are comparable to those from the nitrate radical reaction. The upper limit to k_{747} refer to the reaction of N_2O_5 with gaseous water vapour; the dissolution of N_2O_5 into water droplet followed by reaction with the liquid water or the reaction of N_2O_5 with adsorbed water on surfaces may be significantly faster. Figure 7.2 shows the rates of nitric acid production in air parcel predicted by Russell and co-worker[7] by reaction 4.49a, 4.50, 7.47, 4.48 and 7.47 during one 24 hr day for conditions typical of smog episode in Los Angeles.

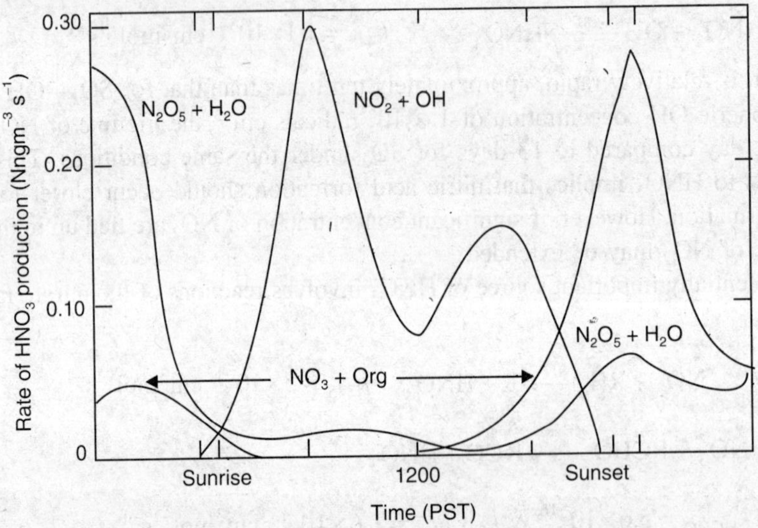

Fig. 7.2 Diurnal variation in production of HNO_3, by various reaction paths in ngm^{-3} s^{-1} as nitrogen, N.

7.2b Aqueous Phase Reactions

Reaction of NO_2 with water has been suggested in the past as a source of nitric acid as

$$2NO_2(g) + H_2O(l) \rightleftharpoons 2H^+ + NO_2^- + NO_3^- \quad ...(7.48)$$

This reaction is now known to be too slow due to two factors:

(*a*) low solubility of NO_2;

(*b*) the second order dependence of the rate on NO_2 concentrations. Thus while 7.48 can be rapid at high NO_2 partial pressures, the rate rapidly drops as the concentration approach ambient levels.

N_2O_5 hydrolysis discussed above may occur not only in the gas phase but also in the liquid phase, and N_2O_5 can be a precursor to nitric acid. Another source of nitric acid can be nitrate radical in water may abstract Hydrogen atom from water or alkene or aldehyde present in it to produce nitric acid.

Results of some cloud chamber studies suggest that H_2O_2 may oxidise NO_2 rapidly to nitrate in cloud water droplets. However, reaction rate is too slow to be of significance with rate constant of 10 1 mol^{-1} s^{-1}. In conclusion, some measurement of the chemical composition of precipitation and the surrounding air mass suggest that significant amounts of nitric acid may be generated in the aqueous phase in the atmosphere.

7.3 SULPHURIC AND NITRIC ACIDS: COMPARISON AND CONTRAST

Once formed in atmosphere, sulphuric acid and nitric acid show quite different behaviour both physically and chemically.

1. Nitric acid is more volatile and thus exists in significant concentrations in gas phase while sulphuric acid has a very low vapour pressure under ambient conditions and hence exist in the form of particles.
2. Both nitric acid and sulphuric acid can react with bases present in atmosphere to form salts.

$$HNO_3 + NH_3 \rightleftharpoons NH_4NO_3$$

   It is a highly soluble salt but the dissociation constant is a sensitive function of temperature. This makes the HNO_3 to easily revolatilise even after forming the ammonium salt. No analogous physical and chemical changes exist for sulphuric acid and $(NH_4)_2SO_4$.
3. If deposition of SO_2 on surfaces is occurring at a constant rate in competition with its oxidation, more of the SO_2 will end up deposited and less transformed into sulphuric acid in the atmosphere, of course, if deposited SO_2 forms acid on the surface, the net result in terms of effects may be same. However, no such deposition of NO_x is observed.
4. Hydrocarbon—NO_x—SO_x. Chemistry not only controls the gas-phase processes leading to acid formation in atmosphere, but also liquid phase processes.
5. The physical and chemical properties of nitric acid and sulphuric acids are thus different. However, their chemistry are inter-twined via the OH radical and other oxidant involved in converting NO_x and SO_2 into acids. For example where significant concentrations of NO_x and SO_2 are present, increasing NO_x will decrease the gas-phase rate of formation of sulphuric acid, because increasing NO_x increases the rate of reaction 4.48 thus lowering the concentration of gas-phase OH which result in rate of oxidation of SO_2. At high gas-phase concentration of NO, little H_2O_2 would be formed because of the reaction

$$H\dot{O}_2 + NO \longrightarrow \dot{O}H + NO_2 \quad \text{...(7.49)}$$

7.4 DEPOSITION PROCESSES

Once acids are formed, these acids can be deposited at the earths surface in two ways:

(*i*) **Wet Deposition:** Pollutants may be dissolved in clouds, fog, rain or snow when these water droplets falls on the surface of earth (which includes not only soil, but grass, trees, buildings, etc.) is termed wet deposition. Because of the highly variable nature of precipitation events, quantitative estimate of wet deposition is difficult.

(*ii*) **Dry Deposition:** Pollutants in the form of either gases or small particles can also be transported to ground level and absorbed and/or adsorbed by materials there without first being dissolved in atmospheric water droplets is called dry deposition. It should be noted that the surface itself may be wet or dry; the term dry deposition only refers to the mechanism of transport to the surface and not to the nature of surface.

The ratio of dry to wet removal of SO_2 has been estimated to be 2 : 1, particularly in dry, arid and semiarid regions as in Delhi, Rajasthan and eastern Uttar Pradesh where dry deposition is clearly important. Dry deposition drops a lot of acidic material close to the industrial source, throughout central

Europe where the deposition rate was four times the rate of acid precipitation in rain and snow in the month of April 1974 rainfall pH 2.4 was reported at Pitlochy, 2.7 on Scotland, 2.9 on the west coast of Norway and 3.5 in Iceland. The lowest annual pH recorded was at De Bilt, Netherlands in 1967, giving a value of 3.78.

Acid deposition whether wet or dry affects all components of changing landscape. Calcium carbonate neutralize acids in solution as

$$CaCO_3 + H_2SO_4 \longrightarrow Ca^{+2} + SO_4^{2-} + H_2O + CO_2$$

The ions released, Ca^{2+}, and SO_4^{2-} neutralize one another and therefore buffer the system against acidification. Acid deposition increases the rate of corrosion of metals, thereby decreasing there lifetime.

7.5 IMPACT OF ACID DEPOSITION

7.5a Injury to Vegetation

Sulphur dioxide injures the vegetation by initial disruption of cellular integrity which usually appears in the spongy parenchymal cells. Subsequently, the palisade layer immediately above it is affected. These areas first appear water soaked, then become dry and papery and usually bleach to a light ivory or tan colour. The final effect is a pattern of light coloured blotches, mainly interveinal. The veins may remain green and intact, tracing a green pattern across the blotches which seem to have run together. With time and/or severe injury, the veins may also become affected and bleach out. Continuous chronic exposure to low level of sulphur dioxide will produce a diffuse chlorosis in the leaves of many broad-leaved plants and trees.

Grasses have veins which run in parallel rows without branching. However, the basic blotching pattern of collapse with sulphur dioxide still holds true. The final bleached pattern gives a streaked effect, often only the principal central vein remains intact. The grasses have no palisade layer, so the collapse is generally uniform through the thickness of the blade.

Sulphur dioxide is thought to enter the leaf as a gas and to act directly on the functional cells near the stomata. There is a definite atmospheric concentration response threshold for acute injury that depends upon the variety of plant and on the age of the leaf. Apparently the cell has capacity to detoxify the sulphur dioxide or sulphite, by conversion to sulphate at a certain rate. If the concentration of sulphur dioxide is so large that it exceeds the rate of conversion to sulphate, the sulphite builds up, the water relation of the cell are disrupted, followed by plasmolysis and then collapse.

7.5b Aluminium Concentrations

Waters affected by acid deposition increase aluminium concentration. Aluminium hydroxide is common in soils, solubility of which increases with the increase in hydrogen ion concentration. The Al^{3+}, Fe^{3+} and Mg^{2+} cations neutralize negative charge carried by organic compounds. The result is that both the aluminium and the organic material precipitate out in the soil. However as more organic matter forms at the surface, acid deposition must go deeper into the soil to find additional aluminium to interact with. When a soil becomes very acidic as a result of acid deposition, Al^{3+} cations are released from the soil, enter the water percolating into streams and lakes. These Al^{3+} ions are highly toxic to fish particularly when it crosses a concentration of 90 to 270 ppm.

7.5c Mercury Poisoning

Acid deposition releases mercury from the watershed or lake sediments. Inorganic mercury is converted to methyl mercury by anaerobic microbial activity and once produced accumulates rapidly in the aquatic food web and in the tissues of higher organisms. It is also toxic to humans. Many fatalities occured in Japan at Minamata and Nigata between 1955 and 1962 when residents ate fish and shell fish contaminated by mercury released by industries not by acid deposition.

7.5d General Effects of Acid Deposition

The potential effects of acid deposition are many:

1. Acidification of a lake or stream results in disruption of the food web. The species diversity of the lake or stream decreases, organism behaviour is affected and organism may become weak and die. Acidophillic algae, mosses or fungi grow abundantly and litter accumulates in the bottom of the lake or stream.
2. Demineralisation of soil occurs. Basic cations, such as Ca^{2+}, Mg^{2+}, K^+ and Na^+ are leached away, replaced by acid cation such as Mg^{2+}, Al^{3+} and hydroxy aluminium and pH of soil declines. However, soils are limed to keep them basic. In other words, comparatively more lime has to be added to soils so that acid deposition does not affect the crops.
3. Nutrient cycles are disrupted and as a result soil bacteria and fungi die. Litter accumulates because decomposers are absent, as a result there is accumulation of organic acids in soils. Nitrifying bacteria's capacity to fix nitrogen diminishes rapidly below pH 6 and an increase in the abundance of fungi may result. Average number of nodules, containing nitrogen fixing bacteria on beans and soyabeans decreases, with the increasing concentration of SO_2.
4. Damage to crops and forests occur. Potential impact of acid deposition on plants include (*i*) damage to the leaf cuticle, (*ii*) interference with normal stomata opening and closing, (*iii*) poisoning of plant cells after diffusion of acid compounds through the stomata or cuticle, (*iv*) interference with normal metabolism or growth process, (*v*) interference with flowering, germination and other reproductive processes, (*vi*) interference with nitrogen fixation by inhibiting microbial or fungal activities, and (*vii*) possible synergistic interactions with other environmental stress factors.
5. Acid deposition falling on plant canopies of forests is modified during its transport from canopy to soil. Organic compounds such as hormones, amino acids, carbohydrates are leached from leaves through fall water. The nature of leachate or through fall depends on tree characteristic, leaf structure and morphology, stand density, age and composition and rate of precipitation. An enrichment of inorganic ions in the water fall after contact with vegetation take place, but as the hydrogen ion concentration increases, cations loss increases.
6. Fresh water lakes are fairly alkaline with Ca^{2+}, Mg^{2+} and HCO_3^- as the dominant ions. Acid neutralising capacity of water depend upon the HCO_3^- concentration. Many invertebrates are affected by falling pH, as also are species of phytoplankton and zooplankton. Even the reproduction of amphibians is impaired. Many species of fish feed on plankton. As the food supply for fish is altered, fish population decline. Thousands of lakes in Sweden are devoid of fish, hundreds of lakes in the United States are without fish, in Southern Norway 2033 lakes were surveyed in 1975 and 741 were reported to be without fish.
7. Deterioration of limestone monuments, bridges, highways and buildings takes place as sulphate ion replace carbonate ion in calcium carbonate thereby forming water soluble compounds. Throughout Europe many famous buildings, monuments, and art treasures have deteriorated at an alarming rate because of the erosional effects of acid deposition.

7.6 ACID FOG

The processes that lead to formation of sulphuric and nitric acids in precipitation also occur in fogs. The prime difference between fog and clouds, which is relevant to the chemistry, in fogs, the liquid water content is of the order 0.1 g m^3, whereas in clouds, it is approximately an order of larger magnitude. This smaller liquid water content in fogs, combined with the relatively rapid condensation and evaporation processes which occur as fog forms and dissipates, can result in urban area that are very acidic and concentrations of SO_4^{2-}, NO_3^- and NH_4^+ also much higher. Fogs are observed after days of dense haze suggest that condensation and evaporation of water vapour on pre-existing aerosol particles were the major processes controlling the water concentrations.

High acid concentration in fog can be a possible cause of health and/or plant effects. The net contribution of acid fogs to the total deposition of acids at the earth's surface may be relatively small. Thus evaporation of the fog may occur prior to deposition leaving an aerosol similar to that on which the water vapour originally condensed. In addition the smaller particle size in fogs compared to clouds and rain drops results in a much smaller total volume of solution in the form of fogs and in slower settling rates. However, given their potential for adverse effects on health, vegetation, and materials, the relative-importance of acid fogs may be rather high.

7.7 HYDROGEN SULPHIDE

Wet muds can produce H_2S by decay processes analogous to the one that produce methane, thus mobilising the sulphur in biogenic organic material. An even more important source of H_2S especially in marine wet lands where the supply of sulphate ion is ample is the action of sulphate reducing bacteria. Volcanic emission contain mostly SO_2 and small amounts of H_2S.

H_2S is a bent molecule like H_2O with bond dissociation energy D (H—SH) = 3.26 eV which correspond to photon wavelength of –380 nm. Electronic spectra of it shows dissociation continuum at 379 nm and 358 nm by the following reaction

$$H_2S(^1A_1) \longrightarrow SH(^2\Pi) + H(^2S) \qquad \lambda = 379 \text{ nm}$$

$$\longrightarrow S(^1D) + H_2(^1\Sigma_g^+) \qquad \lambda = 358 \text{ nm}$$

Since these dissociations correspond to predissociation, H_2S is not photochemically active.

If oxygen is available, H_2S will use that powerful oxidant for the metabolic oxidation but in the absence of oxygen, H_2S will resort to nitrate (denitrification), and when nitrate (or nitrite) is exhausted, it will use, as last resort, sulphate as an oxidant. The H_2S thereby produced can escape, by oxidising both in waters and in the atmosphere and added to be atmospheric SO_2 burden.

Rural areas release most of their sulphur in summer in the form of H_2S, whereas urban areas add more of their sulphur to the atmosphere in the winter heating season as SO_2. The soil is an important sink of sulphur as well as a source; if the sulphur content of soil is low there will be a net uptake of sulphur while if it is high, there will be a net loss to atmosphere. Yet just how the H_2S gets oxidised to SO_2 and SO_2 to SO_3 is as follow: The direct oxidation of H_2S by ozone in gas phase is slow, but in the presence of water, such as in fog and cloud droplets the reaction may be fast. However, the reaction mechanism is not understood. The small tropospheric concentration of atomic oxygen from the photolysis of O_3 and NO_2 may oxidise

$$H_2S + O \longrightarrow OH + HS$$

and in the presence of a third body specially in polluted air to SO_2. Rest is the chemistry of SO_2.

7.8 FUTURE PROJECTIONS

The sources of air pollution contributing to acid deposition are many, including electric utilities, industrial and commercial power plants and boilers, automobiles, trucks and other vehicles. In addition there are natural sources of sulphur and nitrogen compounds entering the atmosphere. Indeed, the weight of evidence at present strongly supports the overwhelming contribution of human activities to acid deposition in these industrialised areas.

The chemical complexity involved in acid deposition has a direct bearing on the issue of control of acid rain. Thus in the parliament of India, a bill to control acid rain in the country through control of upwind SO_2 emission should be proposed. A major uncertainty associated with the enactment of such legislation may have a quantitative effects on acid deposition of reducing the precursor emission. Specially, if SO_2 emissions are reduced by some percentage, acid deposition will at a particular location decrease by a corresponding amount if everything else remain constant. One of the major source of SO_2 is petrol. The presence of sulphur in petrol enhances the knocking action of vehicles which also enhances the emission of more of nitrogens oxides and carbon dioxide. So sulphur free gasoline and diesel should be used, when knocking action of petrol will improve, oxides of nitrogen will be reduced. A detailed discussion is done in Section 11.7.

REFERENCES

1. Benson, S.W., "Thermochemistry and Kinetics of Sulphur Containing Molecules and Radicals", *Chem. Rev.* 78, 23 (1978).
2. Teasley, J.I., Ed. *SO_2, NO and NO_2 Oxidation Mechanisms, Atmospheric Consideration, Acid Precipitation Series*, Butterworth, Baston (1984).
   Calvert, J.G. and W.R. Stockwell, "The Mechanism and Rates of the Gas Phase Oxidation of Sulphur Dioxide and the Nitrogen Oxides in the Atmosphere", pp. 1-60.
3. Hoffmann, M.R. and Jocob, D.J., "Kinetics and Mechanism of Catalytic Oxidation of Dissolved Sulphur Dioxide in Aqueous Solution: An Application to Night time Fog Water Chemistry" in Vol. 3, (2i), Page 101-172.
4. Davis, D.D., Ravishankara, A.R. and Fischer, S., "SO_2 Oxidation via Hydroxyl Radical: Atmospheric Fate of HSO_x Radicals", Geophys Res. Lett. 6, 113 (1979).
5. Lee, Y.N. and Schwartz, S.E., "Kinetics of Oxidation of Aqueous S(IV) by Nitrogen dioxide, in Precipitation Scavenging, Dry Deposition and Resuspension", Vol. 1, HR. Prappacher, R.G. Semonin and WGN Slinn Eds Elsevier, New York, 453-470 (1983).
6. Meaghor, J.F., Olszyna, K.J. & Luna, M., "The Effect of SO_2 Gas Phase Oxidation of Hydroxyl Smog Chemistry", Atmos Environ 18, 2095 (1984).
7. Russell, A.G., McRae, G.J. and Cass, G.R., "*Acid Deposition of Photochemical Oxidation Products, A Study Using a Lagrangian Trayectory Model in Air Pollution Modelling and Its Application III*", C. De Wispelaere Ed. Plenem Press, New York, 539-564 (1984).
8. Sein Feld, J.H., *Atmospheric Chemistry and Physics of Air Pollution,* Wiley, New York (1986).
9. Tuazon, E.C., Atkinson, C.N., Plam, Winer A.M, Pitts J.N., Jr., "The Reaction of Gas Phase N_2O_5 with Water Vapour" Geophys. Res. Lett., 10, 953 (1983).
10. Von Nieding, G. Wagner H.M., Wivbungen von photoxidantiesauf Mensch and Tier, in Luft qualit at skriterien fur photo chemische Oxidation Report 5/83, pp. 133-203, Umlvelt bundesamt, E. Schmidt Verlag, Berlin (1983).
11. *The Handbook of Environmental Chemistry,* Hans Gusten "Formation, Transport and Control of Photochemical Smog", Springer Verlag, Berlin.

8 Photochemical Smog

8.1 TEMPERATURE INVERSION

In the day time, the ground is heated by solar radiation, and the air near the ground is heated by conduction from the ground. Assume that air behaves as an ideal gas with n mol so that

$$V = \frac{nRT}{P} \qquad ...(8.1)$$

V must increase, and density n/V must decrease on heating. Consequently, this packet of air will rise until it reaches a place in troposphere where the surrounding air has the same density, *i.e.*, same temperature and pressure. As air moves from surface of earth, upward, there is fall in pressure. This leads to expansion of gas, which means doing work against intermolecular attraction for which energy is taken from the air itself. This decreases energy of air and hence the temperature. The expansion can be assumed to be adiabatic as no heat flows into or out of the packet during the process. Derivation of the equation for temperature and pressure changes in reversible adiabatic expansion of an ideal gas are given in most of the physical chemistry and many general chemistry textbooks. The result is

$$\frac{P_1}{P_2} = [T_1/T_2]^{C_P/R} \qquad ...(8.2)$$

where C_P is the heat capacity at constant pressure of the gas and subscripts 1 and 2 refer properties of the gas before and after expansion, respectively. Qualitatively this expression shows that if the gas is expanding, that is, if $P_2 < P_1$ then $T_2 < T_1$, and the temperature of the gas will drop. This equation, together with the known decrease of pressure with height, allows calculation of the adiabatic lapse rate of dry air.

The rate of decrease of temperature with increase in height is called the Lapse rate. Since the air temperature below stratosphere on an average decreases with height, the average environment lapse rate in the troposphere is positive. Two significant lapse rates are:

(*i*) The adiabatic lapse rate for dry air. A dry air moves upward and expands slowly without exchange of heat, then such parcels at any height are in equilibrium and value of lapse rate is 1°C/100 m.

(*ii*) The adiabatic lapse rate for water-saturated air is known as the saturation lapse rate—this lapse rate is smaller than dry adiabatic lapse rate because of the release of latent heat as the air parcel ascends and cools. The saturation lapse rate varies with temperature and height.

The rising air cools faster than the temperature lapse rate of the environment, which involves a decrease of only 0.65°C per 100 m. Rising dry air will eventually reach a height where temperature and, therefore, the density matches that of the surrounding air. If the rising air contains a lot of moisture that condenses as the air rises and becomes cooler, then heat is given off during this condensation and the air cools more slowly with height, *i.e.*, its lapse rate is less than the dry air. This moist air will, therefore, rise higher than dry air, and will probably form clouds as it rises.

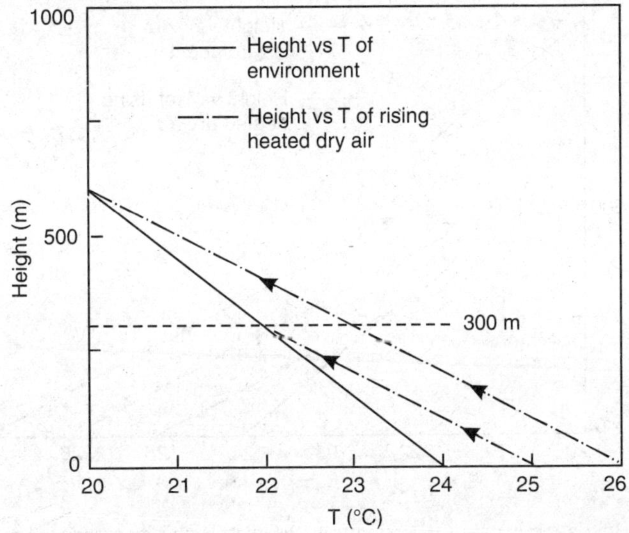

Fig. 8.1 The fate of dry air heated by the earth's surface when the temperature lapse rate of the environment is normal.

Figure 8.1 shows what happens to heated air when the troposphere is normal. The solid line shows the temperature of the lower atmosphere with earth surface temperature as 24°C and there is no temperature inversion in the atmosphere. If the sun is shining the surface of the earth, it absorbs heat and becomes warmer which will heat the air just above it by conduction. The two-dash lines in Fig. 8.1 refer to two different parcels of air heated by the surface of earth to 25° and 26°C, respectively. Both parcels of air will begin to rise as discussed above; they rise with the temperature lapse rate of rising dry air. Since the temperature of these two parcels of air are decreasing faster, 1°C/100 m, the temperature of the air through which they are rising, 0.65°C/100 m each parcel will eventually reach a height at which its temperature equals that of the air already present. Its pressure and density are then also equal to that of the surrounding atmosphere, and the parcel will stop rising. Figure 8.1 shows that the parcel that was heated to 25°C will rise to a height just below 300 m, while the parcel that was heated to 26°C by the surface of the earth will rise to a height just below 600 m. Hotter air rises to a higher altitude.

If a temperature inversion exists in the atmosphere at an altitude below which the heated air would normally rise, the heated air may be trapped below the inversion along with any pollutants that were present in the heated air. Figure 8.2 shows how heated air may be trapped when there is a temperature inversion in the troposphere. In this figure, the solid line showing the variation of atmospheric temperature with altitude is not linear; it has a region in which temperature increases with increasing altitude between approximately 400 and 600 m altitude. This region of increasing temperature is known as temperature inversion. It is in this region of temperature inversion, formation of smog as well as photochemical smog takes place. Above and below this region, air exhibits its normal lapse rate of 0.65°C/100 m. If the sun shines and the earth's surface is heated, the resulting ground-level heated air rises, as shown by the dashed lines in Fig. 8.2. Three parcels of air are shown, one heated to 25°C, the second to 26°C and third to 27°C. The parcel of air heated to 25°C rises to same altitude, just below 300 m, as the similar parcel in Fig. 8.1. This happens because temperature inversion occurs above this altitude and can effect only air that is attempting to rise through the inversion.

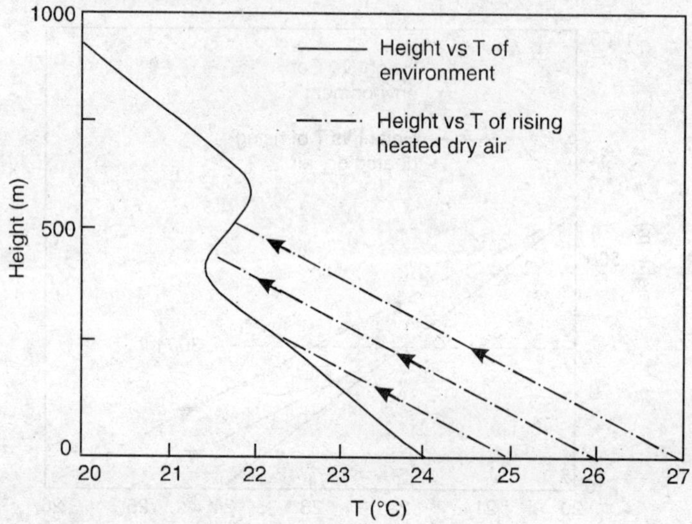

Fig. 8.2 How a temperature inversion traps heated air and its associated pollutants.

Both the 26°C and 27°C parcels of air are trapped by inversion each reaches an altitude at which its temperature equals that of the air already present with in the temperature inversion zone. The 26°C parcel of air rises to less than 500 m, considerably lower than the 600 m it would ordinarily reach.

These projections can be on a moment-to-moment basis, the result would not be worth that much effort. In practice, what is usually done is to obtain a single daily temperature profile, usually at night or in the early morning, using a balloon equipped to transmit the necessary data. The dry adiabatic temperature line is drawn from the surface, usually with a starting point equal to the average maximum surface temperature for the month. The elevation where these lines cross is known as the maximum mixing Fig. 8.3 depth.

The product of the maximum mixing depth and the average wind speed within the mixing depth is sometimes used as an indicator of the atmosphere's dispersive capability. This product is known as the ventilation coefficient (m^2s^{-1}), values of ventilation coefficient less than about 6000 m^2s^{-1} are considered indicative of high air pollution potential.

Wind speed generally increase with height and it is essential to know wind speed at various heights. The following power law expression is frequently used for elevation less than a few hundred meters above the ground

$$\frac{u_1}{u_2} = \left(\frac{Z_1}{Z_2}\right)^p \qquad \text{...(8.2}a\text{)}$$

where u_1 and u_2 wind speed at higher and lower elevation

Z_1, Z_2 = higher and lower elevation, respectively

p = a dimension less parameter that varies with atmospheric stability. The values for various terrains is given in Table 8.1.

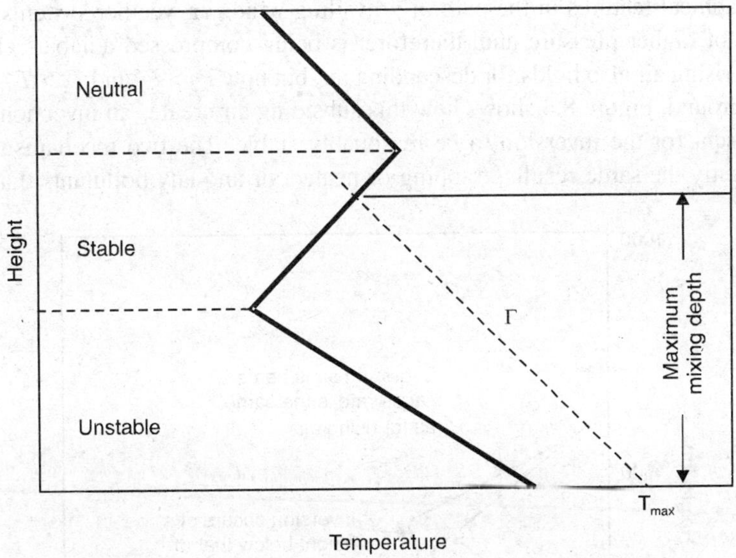

Fig. 8.3 The maximum mixing depth is obtained by projecting the dry adiabatic lapse rate to the point of intersection with the atmospheric temperature profile. Notice that the mixing depth does not necessarily correspond to the layer of unstable air.

Table 8.1 Wind profile exponent p for rough terrain

Stability class	*Description*	*Exponent, p*
A	Very stable	0.15
B	Moderately stable	0.15
C	Slightly unstable	0.20
D	Neutral	0.25
E	Slightly stable	0.40
F	Stable	0.60

For smooth terrain, multiply p by 0.6; see Table 10.1 for further description of the stability classification used here.

These temperature inversions that trap rising air are naturally occurring phenomena that may have two different causes, *i.e.* radiation inversion and subsidence inversion—explained in the following paragraphs. London tends to have radiation inversions in the winter, one of these trapped a highly polluted fog in 1952, producing the notorious London killer fog which was considered responsible for over 3,000 deaths. Los Angeles tends to have subsidence inversion that trap pollutants in the summer and fall.

A radiation inversion generally occurs at night, when air near the ground is cooled by conduction from the cooled surface, especially on a clear night, when radiation cooling occurs. Higher up in the troposphere, the air is not cooled appreciably, so, in the absence of winds, there is colder air underneath warmer air, an inversion by definition.

A subsidence inversion occurs in a region where high altitude air is pushed down nearer the ground because of topographical features in the path of prevailing winds or weather systems. This air is being pushed to regions of higher pressure and, therefore, is being compressed adiabatically. The same Eq. 8.2 that is held for rising air also holds for descending air, but now $P_2 > P_1$ and $T_2 > T_1$, *i.e.*, the air warms up as it nears the ground. Figure 8.4 shows how this subsiding air creates an inversion; again horizontal winds must be absent for the inversion to be reasonably stable. The two mechanisms for forming an inversion have exactly the same results, trapping of heated air and any pollutants that may be present.

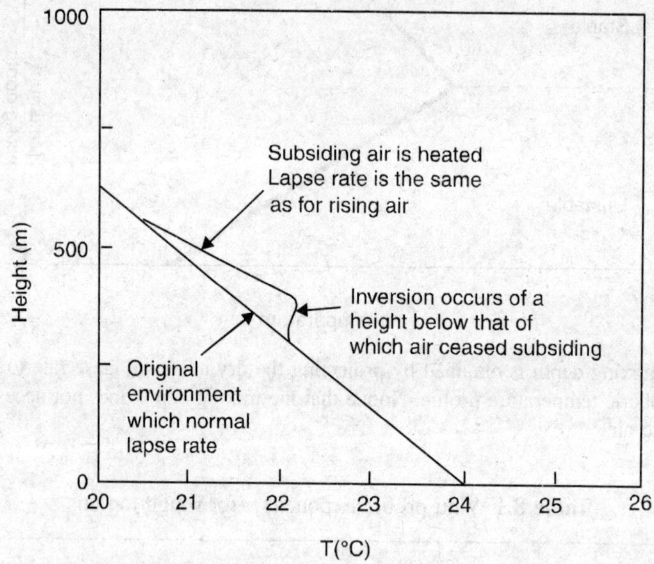

Fig. 8.4 Mechanism by which a subsidence inversion occurs.

8.2 GENERAL NATURE OF SMOG

The overall oxygen and ozone balance at lower altitude may be contributed by higher concentrations of minor constituents. The minor constituents may provide, intermediate free radicals that can participate in catalytic destruction of free oxygen atoms and ozone. As a consequence, a number of reactions may occur producing a variety of eye-and-throat irritants aerosols and reduced visibility, and other irritating or destructive species manifesting themselves in the general category of photochemical smog. Table 8.2 lists some of the characteristics of London and photochemical smogs.

Smog is made of two words: smoke plus fog. Smog word is misleading since smoke and fog are not the key component unlike the notorious "killer smogs" that were a part of London life, as well as that of many other highly industrial urban areas, for centuries before the mass clean up efforts were undertaken in the last few decades, photochemical smog is of recent origin. It was first detected through plant damage in Los Angeles in 1944, but has now spread to so many metropolitan cities like Delhi, Mumbai that it must be considered one of the most pressing actual or potential worldwide problem with global aspect transcending national boundaries, particularly in the Northern hemisphere in a belt between 30° and 60° latitude.

Table 8.2 Characteristics of sulphurous, "London" smog and "photochemical" smog

Characteristics	*Sulphurous, London smogs*	*Photochemical smogs*
First occurrence	Centuries old, 4 days smog with 3000 death in London in 1952.	Los Angeles (1944)
Primary pollutant	SO_2 + particulate matter	Organics, NO_x
Secondary pollutant	H_2SO_4 + aerosols, sulphates, sulphonic acid	Ozone, PAN, HNO_3, aldehydes particulate, nitrate, sulphates
Temperature	Cool (≤ 3°C)	Hot (≥ 25°C)
Relative humidity	High, usually Fogging	Low, usually hot and dry
Cause of conversion	Radiation (inversion)	Subsistence (inversion)
Principal sources	Industrial and House hold fuel combustion (Coal & Petroleum)	Motor vehicle fuel combustion
Effect on humans	Lung and Throat irritation	Eye irritation
Nature of compounds	Reducing	Oxidizing
Time of occurrence of episodes	Winter Months (espccially yearly in morning)	Around midday of summer months

8.2a London Type Smog

It occurs mainly in the early morning hours in winter where there is relatively high humidity. Air near the ground is cooled by conduction from cooled surface, especially on a clear night when radiation cooling occurs. Higher up in the troposphere, the air is not cooled appreciably, so, in the absence of winds, there is colder air underneath warmer air when temperature inversion takes place. In this belt of stable air, chemical reaction takes place that lead to formation of London type smog.

Sulphur dioxide is the major contaminant, and this creates a chemically reducing atmosphere because of the ease with which sulphur dioxide can be oxidised. The chemical reactions in this type of smog are not photochemical in nature. Throat irritation is felt in this smog.

8.3 PHOTOCHEMICAL SMOG

On the other hand, photochemical smog occurs on warm sunny days (summer and winter) with initially clear skies and low humidity. A subsidence inversion of temperature in a heated polluted air between reasonably stable horizontal layer.

On sunny days when mobile traffic levels is high and eye irritation is common complaint, following events take place in succession:

- Nitric oxide (NO) is converted to nitrogen dioxide (NO_2);
- When all of the NO has disappeared, ozone starts to appear;

  The hydrocarbons are oxidised and disappear, and
- Reaction products such as aldehydes and PAN are formed.

There is a dramatic difference in visibility that can occur within only a few hours of smog generation. While the peak smog effects may be generated around noon time, they may not show up until several hours later "down wind".

Automobile engines and power plants are the major man made sources of nitrogen oxides and hydrocarbons. There are also natural sources of these materials, and on a global basis these latter sources may lead to far greater emissions than from the man made sources. The importance of artificial sources is that they generate sufficiently high concentrations of the pollutants in an urban environment with appropriate climatic, geographic and solar radiation conditions to undergo the photochemical reactions leading to photochemical smog.

NO is the dominant species generally comprising over 95 per cent of the total NO_x. NO_2 is formed from NO in the exhaust gases by a termolecular reaction.

$$2NO + O_2 \longrightarrow 2NO_2$$

The presence of even a small amount of NO_2 in the atmosphere is sufficient to trigger the complex series of reaction to produce photochemical smog through the photolysis of NO_2 by solar light in the 300-420 nm wavelength range.

$$NO_2 \xrightarrow[\lambda<420\text{ nm}]{h\nu} NO(^2\Pi) + O[^3P_2] \qquad ...(8.3)$$

The quantum yield for NO_2 photolysis is close to unity at 366 nm. As the wavelength of light is increased to 398 nm, quantum yield decreases with a sharp cutoff for wavelength greater than 400 nm.

8.3a Basic Photochemical Cycle of NO, NO_2 and Ozone

The rate of photodissociation of NO_2 in Eq. 8.3 is very fast and is sufficiently fast to trigger the production of ozone in the lower troposphere.

$$O[^3P_2] + O_2 \xrightarrow{M} O_3 \qquad ...(8.4)$$

The lifetime of oxygen atom formed by NO_2 at earths surface is 16.5 μ sec. The rate of formation of ozone has been widely used as a measure of smog intensity. A fraction of the NO emitted in the air parcel reacts with ozone to regenerate NO_2.

$$O_3 + NO \longrightarrow NO_2 + O_2 \qquad k_{85} = 1.8\times10^{-14}\ \text{cm}^3\text{mol}^{-1}\text{s}^{-1} \qquad ...(8.5)$$

This reaction is also quite fast and in sunlight the three species NO, NO_2 and O_3 attain a quasi-steady state so that

$$\frac{k_{446}}{k_{85}} = \frac{k_{446}}{k_{443}} = \frac{[NO][O_3]}{[NO_2]} = \text{mol cm}^{-3} = 1667\ \text{mol cm}^{-3} \qquad ...(4.59)$$

This relationship is called the Leighton relationship and enables a rapid convenient prediction of ozone level in a sunlight irradiated air. If either O_3 or NO is present in ambient air at a high level, the other must necessarily be present at a rather low level. This relation is not valid at night when NO will react with ozone until either NO or ozone is depleted. A substantial amount of NO_2 generated at night will participate in the next day's photosmog chemistry a typical urban diurnal cycle of NO, NO_2 and ozone is shown in Fig. 8.5.

The morning rush hour adds more NO and reactive hydrocarbons to the atmospheric mixture. In the early morning NO is converted rapidly to NO_2. As a consequence of photochemical reaction 8.3, peak values of O_3 are reached in the early afternoon. The-time lag between maximum precursor emission of NO_x and maximum ozone concentration is highly variable and depends on the atmospheric turbulence and other metrological parameters *i.e.*, hot sunny weather, under low humidity conditions. The characteristic symptoms of the smog are the brown haze in atmosphere, reduced visibility, eye irritation, respiratory distress and plant damage.

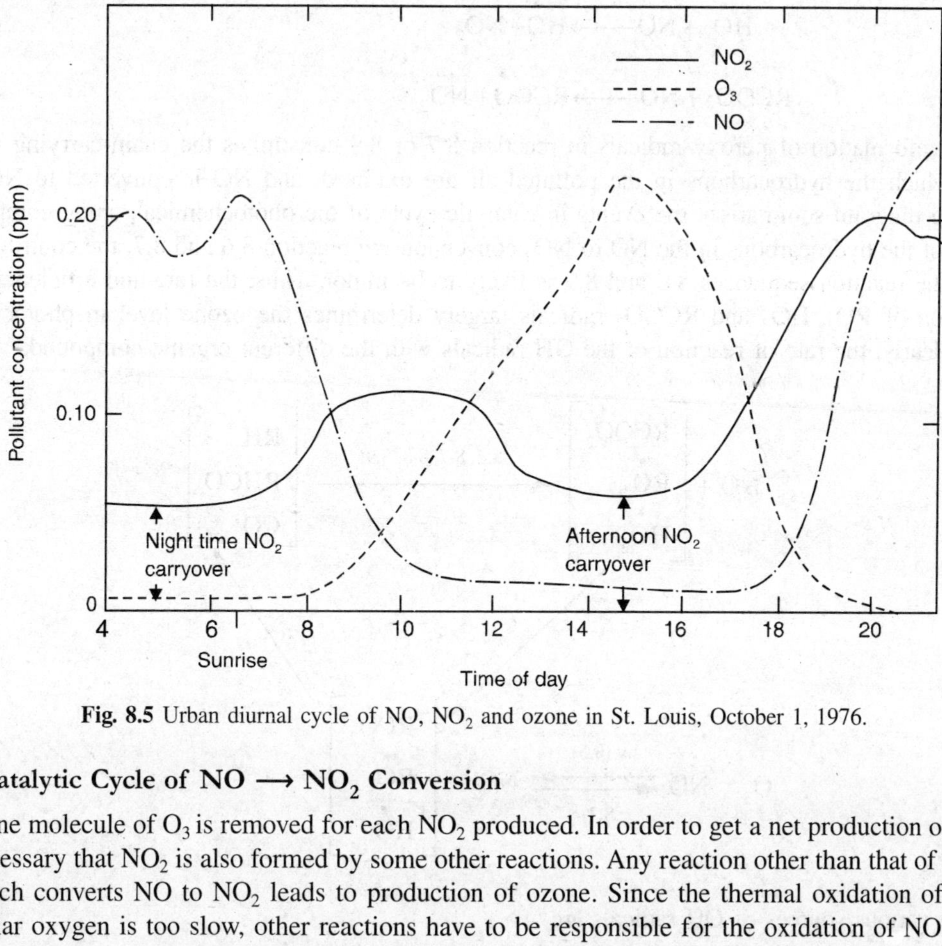

Fig. 8.5 Urban diurnal cycle of NO, NO_2 and ozone in St. Louis, October 1, 1976.

8.3b Catalytic Cycle of NO ⟶ NO_2 Conversion

Since one molecule of O_3 is removed for each NO_2 produced. In order to get a net production of ozone, it is necessary that NO_2 is also formed by some other reactions. Any reaction other than that of reaction 8.5 which converts NO to NO_2 leads to production of ozone. Since the thermal oxidation of NO by molecular oxygen is too slow, other reactions have to be responsible for the oxidation of NO.

In polluted air, a variety of highly reactive transient radicals are formed, most promising are HO, RO, hydroperoxide (HO_2), alkyl peroxide RO_2 and alkyl peroxy radicals ($RCOO_2$).

The OH radical and alkyloxy radicals react with hydrocarbons, aldehydes or carbon monoxide present in the polluted air in which they efficiently generate HO_2 and RO_2 radicals

$$\dot{O}H + RH \longrightarrow \dot{R} + H_2O \qquad \text{...(8.6)}$$

$$\dot{R} + O_2 \longrightarrow R\dot{O}_2 \qquad \text{...(8.7)}$$

$$\dot{O}H + CO \longrightarrow CO_2 + \dot{H} \qquad \text{...(8.8)}$$

$$H + O_2 \xrightarrow{M} H\dot{O}_2 \qquad ...(8.9)$$

$$R\dot{O} + O_2 \longrightarrow H\dot{O} + \underset{\text{(aldehyde \& ketone)}}{RCHO} \qquad ...(8.10)$$

$$R\dot{O}_2 + NO \longrightarrow R\dot{O} + NO_2 \qquad ...(8.11)$$

$$H\dot{O}_2 + NO \longrightarrow H\dot{O} + NO_2 \qquad ...(8.12)$$

$$RCO\dot{O}_2 + NO \longrightarrow RCO\dot{O} + NO_2 \qquad ...(8.13)$$

The reformation of peroxy radicals in reaction 8.7 or 8.9 constitutes the chain-carrying reaction during which the hydrocarbons in the polluted air are oxidised, and NO is converted to NO_2. The following diagram summarises the events in catalytic cycle of the photochemical, smog compared to the role of the hydrocarbons in the NO to NO_2 conversion *via* reaction 8.6 and 8.7, the contribution of CO_2 in the reaction sequences 8.8 and 8.9 is likely to be minor. Thus, the rate and efficiency of the production of RO_2, HO_2 and $RCOO_2$ radicals largely determines the ozone level in photochemical smog. Clearly, the rate of reaction of the OH radicals with the different organic compounds.

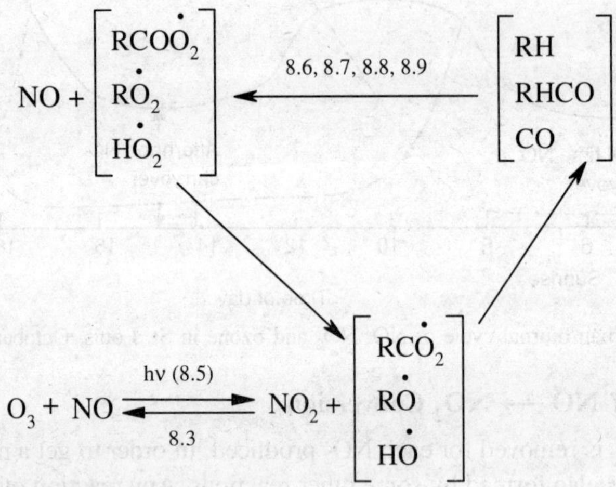

The various sources of OH radicals are

$$H_2O + O\left[{}^1D_2\right] \longrightarrow 2\dot{O}H \qquad ...(4.20)$$

$$HONO \xrightarrow[\lambda<400\text{ nm}]{h\nu} H\dot{O} + NO \qquad ...(4.64)$$

$$H_2O_2 \xrightarrow[\lambda<370\text{ nm}]{h\nu} 2\,H\dot{O} \qquad ...(7.22)$$

$$HCHO \xrightarrow[\lambda<370\text{ nm}]{h\nu} H\dot{C}O + H$$

$$\longrightarrow H_2 + CO \qquad ...(4.85)$$

$$RCHO \xrightarrow[\lambda<350]{h\nu} \dot{R} + \dot{C}HO$$

In air, the formyl radical HCO as well as the hydrogen atom produced in reaction 4.85 subsequently may form HO_2 radicals

$$\dot{C}HO + O_2 \longrightarrow H\dot{O}_2 + CO$$

$$H + O_2 \longrightarrow H\dot{O}_2 \qquad ...(4.87)$$

In Table 8.3, the most important radicals sources for photochemical generation of the active OH and HO_2 radicals in the catalytic NO to NO_2 conversion are summarized.

Table 8.3 Photochemical radical sources in the atmosphere, wavelength and photodissociation rate constant for Central Europe at summer solistice [1]

Photochemical process, consecutive reactions numbers		*Wavelength range [nm]*	*k [s⁻¹]*
$NO_2 \xrightarrow{h\nu} NO + O$	Reaction 4.47a	400 - 290	0.0078
$HONO \xrightarrow{h\nu} OH + NO$	Reaction 4.64	390 - 290	ca 0.001
$HCHO \xrightarrow{h\nu} H + CHO$	Reaction 4.85	330 - 290	1.5×10^{-5}
$H_3CCHO \xrightarrow{h\nu} CH_3 + CHO$		320 - 290	3.4×10^{-6}
$O_3 \xrightarrow{h\nu} O + O_2$	Reaction 4.18	320 - 290	2×10^{-5}

8.3c Significance of the Hydrocarbon/NO_x Ratio

Under different ambient conditions, the rate of ozone formation can be limited by the availability of either NO_x or hydrocarbon in the polluted air. All hydrocarbons with the exception of methane are summarized as NMHC, non-methane hydrocarbons. The significance of NMHC/NO_x ratio in the ozone formation is shown in Fig. 8.6. This is a diagram of the so-called isopleths of maximum afternoon O_3 concentration as a function of the early morning NMHC & NO concentration. It is derived from a chemical kinetics model of photochemical smog, which was tested in smog chamber studies by irradiating automobile exhaust mixtures with changing NMHC/NO_x ratio at constant light intensity. The following conclusion can be drawn from it:

— Reduction of the NMHC emission will yield ozone reduction. In most urban areas where the ambient NMHC/NO_x ratio is less than about 10 : 1, NMHC reduction are most effective in reducing ozone level. In remote areas where the ambient NMHC/NO_x ratio is about 30 : 1 or above, increases or decreases in ambient NMHC concentrations may have an insignificant effect on O_3 concentrations.

— Reduction of NO_x can lead to ozone reductions and increase depends on the NMHC/NO_x ratio. At low NMHC/NO_x ratio typical of an urban area, allowing NO_x to increase, can lower the ozone concentration locally through reaction 8.5. The downward transport of NO can increase the ozone concentrations. At high $NMHC/NO_x$ ratios, reducing NO_x will be most effective in reducing the ozone level.

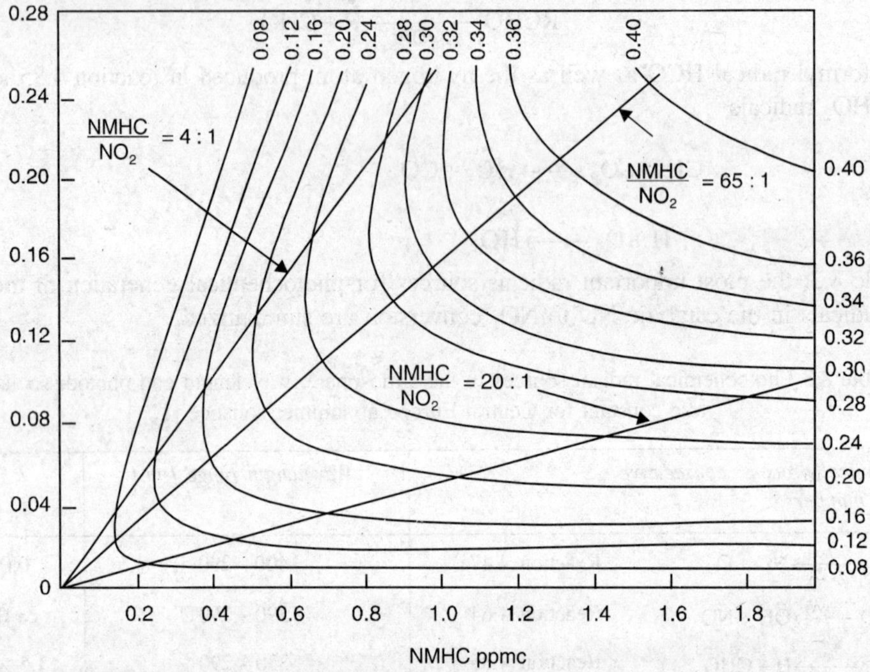

Fig. 8.6 Isopleth plot of the dependency of maximum afternoon ozone concentration to morning precursor concentrations $NMHC/NO_2$. (From Hans Gusten, *Handbook of Environmental Chemistry,* Springer Verlag, Berlin.)

Methane is a natural component of atmosphere and a possible chain mechanism with CH_3O_2 as a chain carrier also leading to NO oxidation is

$$H\dot{O} + CH_4 \longrightarrow \dot{C}H_3 + H_2O \quad \text{...(8.14)}$$

$$\dot{C}H_3 + O_2 + M \longrightarrow CH_3\dot{O}_2 + M \quad \text{...(8.15)}$$

$$CH_3\dot{O}_2 + NO \longrightarrow CH_3\dot{O} + NO_2 \quad \text{...(8.16)}$$

$$CH_3\dot{O} + O_2 \longrightarrow HCHO + H\dot{O}_2 \quad \text{...(8.17)}$$

Reaction 8.17 is a source of formaldehyde as well as generating the hydroperoxyl radical. It is now believed that the direct photolysis of formaldehyde and other aldehydes are important initiators of photochemical smog.

8.3d Peroxy Acetyl Nitrate, PAN

Peroxy acetyl nitrate is the principal member of nitrogenous compound produced in photochemical smog. The formation of PAN takes place by the reaction of acetyl radical with O_2 and NO_2. There are a variety of sources of acetyl radicals.

$$CH_3CHO + \dot{O}H \longrightarrow CH_3\dot{C}{=}O + H_2O \qquad ...(4.73)$$

$$CH_3CHO + \dot{N}O_3 \longrightarrow CH_3\dot{C}{=}O + HNO_3 \qquad ...(4.67)$$

$$CH_3\dot{C}{=}O + O_2 \longrightarrow CH_3\overset{\dot{O}_2}{\overset{|}{C}}{=}O \qquad ...(4.74)$$

$$CH_3\overset{O}{\overset{\|}{C}}O\dot{O} + NO_2 \longrightarrow CH_3\overset{O}{\overset{\|}{C}}{-}OONO_2 \qquad ...(4.75)$$

The atmospheric oxidation of ethane also has been a significant source of PAN in troposphere through following reactions

$$C_2H_6 + \dot{O}H \longrightarrow \dot{C}_2H_5 + H_2O \qquad ...(8.18)$$

$$\dot{C}_2H_5 + O_2 \longrightarrow C_2H_5\dot{O}_2 \qquad ...(8.19)$$

$$C_2H_5\dot{O}_2 + NO \longrightarrow C_2H_5\dot{O} + NO_2 \qquad ...(8.20)$$

$$C_2H_5\dot{O} + O_2 \longrightarrow CH_3CHO + H\dot{O}_2 \qquad ...(8.21)$$

An equilibrium exists between the formation and decomposition of PAN.

$$CH_3\overset{O}{\overset{\|}{C}}OONO_2 \longrightarrow CH_3\overset{O}{\overset{\|}{C}}O\dot{O} + NO_2 \qquad ...(8.22)$$

$$k_{822} = 3.57 \times 10^{-4} \qquad \tau = 46.68 \text{ min}$$

PAN acts as a storage reservoir of NO_x. When NO_2/NO ratio is high and/or the temperature is low the decomposition reaction is inhibited and hence PAN concentration can build up. When fresh NO is introduced into atmosphere or temperature rise, acetyperoxy (CH_3COO_2) and NO_2 are released through decomposition of PAN and thermal decomposition of PAN is the only sink in atmosphere. Its reaction with OH and O_3 are very slow.

Another peroxy nitrate of particular interest is peroxy benzoyl nitrate (PBzN). It is thought to be about two orders of magnitude stronger than PAN in terms of eye irritation. It is formed in an analogous manner to PAN.

$$C_6H_5CHO + \dot{N}O_3 \longrightarrow HNO_3 + C_6H_5\overset{O}{\overset{\|}{C}}\bullet$$

$$C_6H_5\overset{O}{\overset{\|}{C}}\bullet \xrightarrow{O_2} C_6H_5\overset{O}{\overset{\|}{C}}{-}O{-}O\bullet \qquad ...(8.23)$$

$$\underset{\text{PBzN}}{C_6H_5\overset{O}{\overset{\|}{C}}{-}OONO_2} \xleftarrow{NO_2} C_6H_5\overset{O}{\overset{\|}{C}}{-}O{-}O\bullet$$

PAN and PBzN are strong eye irritants and have been suggested a possible etiological agent in the growing incidences of skin cancer. PAN strongly react with sulphydryl groups of enzymes and with other sulphur containing compounds such as amino acids. It is believed that these reactions explain the eye irritating properties of PAN as well as its toxic effect on plants.

8.4 PARTICULATE ORGANICS

In a photochemical oxidizing atmosphere such as that of Los Angels, difunctional substituted alkane derivatives of the type

$$X-(CH_2)_n-X \text{ and } X-\underset{}{\overset{\displaystyle CH_3\atop |}{(CH_2)_{n-1}}}-X$$

with n = 1–5 have been observed in a number of studies. The substituent X can be —COOH, CHO —CH_2OH, —CH_2, ONO, COONO or — $COONO_2$. For example, Table 8.4 shows some of the difunctional species identified in submicron fraction of particles, suggesting there secondary nature, that is, these compounds are formed from chemical reactions in the atmosphere.

Table 8.4 Some difunctionally substituted alkane derivatives found in submicron ambient particles in urban air

Compound	*n*
HOOC $(CH_2)_n$ COOH	1–5
HOOC $(CH_2)_n$ CHO, HOOC $(CH_2)_n$ CH_2OH	3–5
HOOC $(CH_2)_n$ CH_2ONO or CHO $(CH_2)_n CH_2O\dot{N}O_2$	3–5
$CHO(CH_2)_n$ CHO, $CHO(CH_2)_n$ CH_2OH	3–5
$HOOC(CH_2)_n$ COONO or $CHO(CH_2)_n$ $COONO_2$	3–5
$CHO(CH_2)_n$ COONO	3, 4
$HOOC(CH_2)_n$ $COONO_2$	4, 5

Subsequently, even larger branched chain dicarboxylic acids (Table 8.5) were observed in Los Angeles aerosols.

Monosubstituted alkanes of the type CH_3 — $(CH_2)_n$ — X were not observed, which was attributed to their high vapour pressure so that they exist mainly in the gas phase.

Table 8.5 Some alkyl dicarboxylic acids identified in ambient particles in urban air

$HOOC(CH_2)_n$ COOH	$(n = 1-8)$
$HOOC-\underset{}{\overset{\displaystyle CH_3\atop |}{CH}}-COOH$	$HOOCCH_2\overset{\displaystyle CH_3\atop |}{CH}CH_2COOH$

(Contd.)

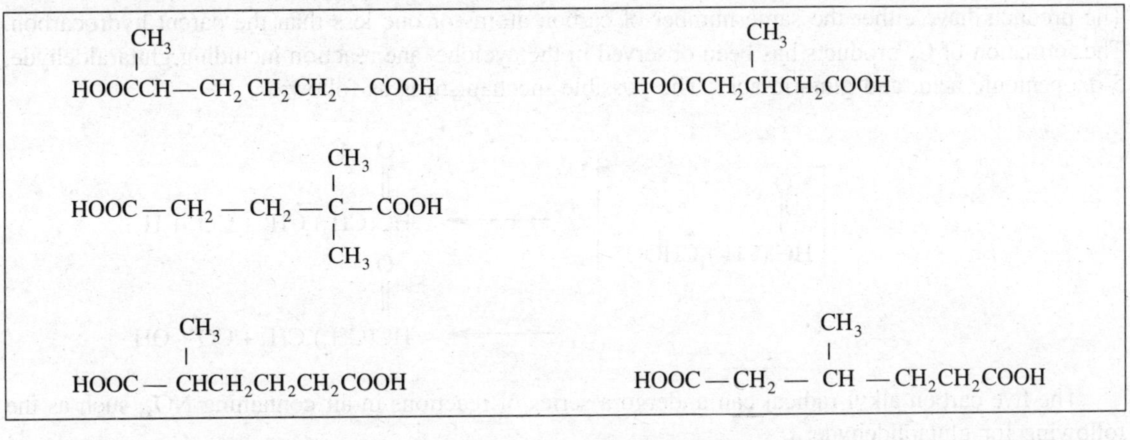

Beside these secondary organics from gas to particle conversion, there are primary carbonaceous particulates, emitted directly from sources. The most likely precursors appear to be cyclic olefins and perhaps dialkenes which can be oxidized to give stable species containing two substituted groups in the same molecule. The reaction of O_3 with cyclohexene is expected to lead to the excited biradicals.

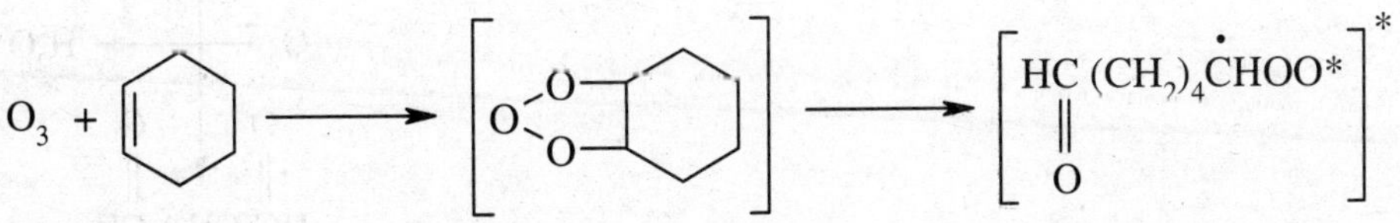

The fraction of the excited biradical stabilized under tropospheric condition is uncertain even for relatively simple alkenes. In addition the decomposition paths of the biradical are unknown. However, one can envision reactions, such as the following leading from excited biradical to stable multi-functional compounds.

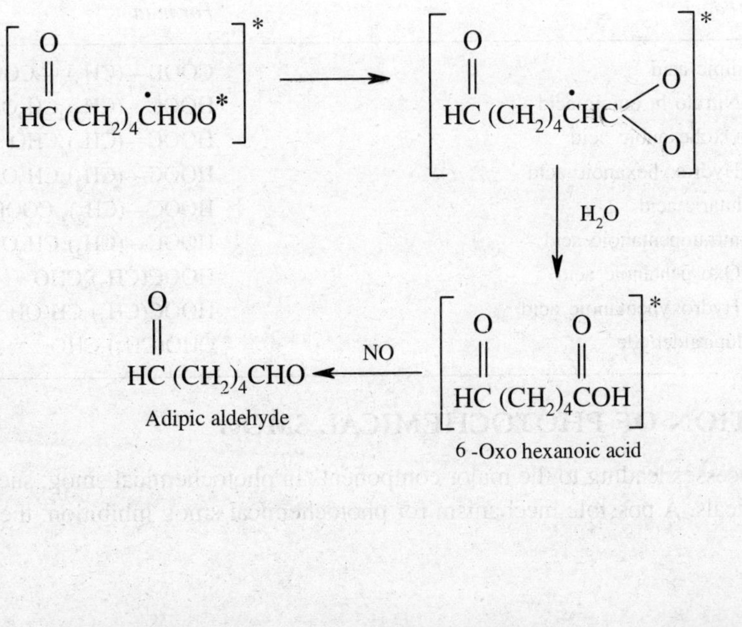

The products have either the same number of carbon atoms or one less than the parent hydrocarbon. The formation of C_5 products has been observed in the cyclohexane reaction including glutaraldehyde, 5-oxopentonic acid, and glutaric acid. The possible mechanism is as follow:

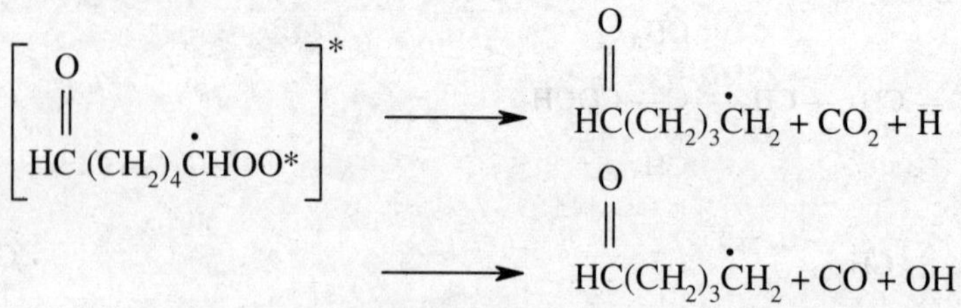

The five carbon alkyl radical can undergo a series of reactions in air containing NO_x, such as the following for glutaraldehyde.

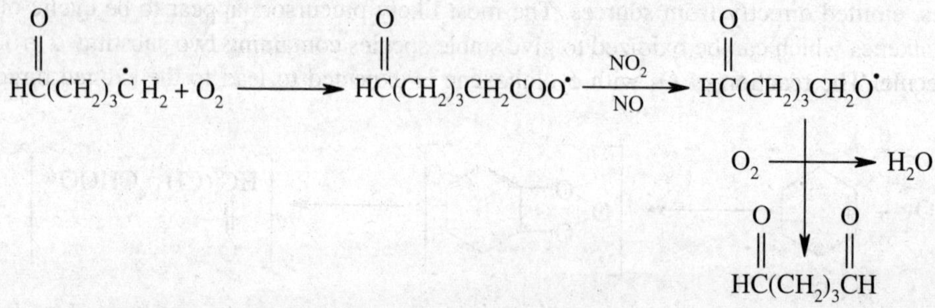

The glutaraldehyde presumably acts as the source of the C_5 compounds containing carboxylic acid groups (Table 8.6) through its further oxidation. Table 8.6 shows the product identified in the particles formed in the cyclohexene-NO-ambient air system.

Table 8.6 Compounds identified in particles from the cyclohexene NO_x-ambient air oxidation

Product	*Formula*
Adipic acid	$COOH—(CH_2)_4—COOH$
6-Nitrato hexanoic acid	$HOOC—(CH_2)_4CH_2ONO_2$
6-Oxohexanoic acid	$HOOC—(CH_2)_4CHO$
6-Hydroxyhexanoic acid	$HOOC—(CH_2)_4CH_2OH$
Glutaric acid	$HOOC—(CH_2)_3\ COOH$
5-nitratopentanoic acid	$HOOC—(CH_2)_3CH_2ONO_2$
5-Oxo pentanoic acid	$HOOC(CH_2)_3CHO$
5-Hydroxypentanoic acid	$HOOC(CH_2)_3CH_2OH$
Glutaraldehyde	$CHO(CH_2)_3CHO$

8.5 INHIBITION OF PHOTOCHEMICAL SMOG

Many of the processes leading to the major components in photochemical smog, such as O_3, NO_2, PAN involve free radicals. A possible mechanism for photochemical smog inhibition, therefore, might be to add free radical scavengers to the atmosphere, thereby terminating or breaking the chain processes leading to photochemical oxidants. One particularly promising free radical scavenger is diethyl hydroxyl amine, $(C_2H_5)_2NOH$, or DEHA. Addition of only 3 m Torr of this compound to a propylene-NO–O_2, mixture with a total pressure of 100 Torr, completely inhibits the oxidation of NO to NO_2, while addition of 2.5 m Torr results in 50 per cent inhibition. Apparently inhibition results from radical abstraction of hydrogen.

$$(C_2H_5)_2\,NOH + OH \longrightarrow H_2O + (C_2H_5)_2\,NO \qquad ...(8.24)$$

The compound DEHA is a clear, volatile, liquid with only mild odour. So far it has proven to be a non-toxic and non-mutagenic, although some urinary animal tests suggest that it may be metabolized into a mutagen. Presumably, if it is found to be completely safe, it will be tested by dispersing it into the atmosphere in the early morning hours either by evaporation from cannisters or by spraying it from moving vehicles or airplanes.

Several arguments can be used against such a scheme for inhibiting photochemical smog until much more is known about its long term effects on humans and on the environment. Pitts has pointed out that DEHA is similar to diethyl nitroso amine $(C_2H_5)_2N$—NO a potent carcinogen. It may also be that introduction of a free radical scavenger at one point will simply delay smog formation until later in the day when the polluted atmosphere has moved down wind, thereby giving the problem to suburb rural areas or to another city.

8.6 ADVERSE EFFECTS OF PHOTOCHEMICAL SMOG

While discussing the measures to control the photochemical oxidants, one has to consider the damage to the human population, animals, vegetation and to materials. Since ozone is the main secondary pollutants of photochemical smog, in general, the human symptoms and vegetation injuries caused by ozone is considered. Table 8.7 summarises the effects of ozone, which led to the ambient air quality criteria for ozone.

Table 8.7 Effects or ozone to life and standards

Concentration of Ozone in air (ppm)	*Symptoms in humans, and threshold of damage to vegetation*
1.0	Cough, extreme fatigue, lack of coordination; increased airway resistance; decreased forced expiratory volume.
0.5	Chest contraction, impaired carbon monoxide diffusion capacity; decrease in lung function without exercise.
0.3	Headache, chest discomfort sufficient to prevent completion of exercise; decrease in lung function exercising subjects.
0.25	Increase in incidence and severity of asthma attacks; moderate eye irritation.
0.15	In sensitive individuals, reduction in pulmonary lung function; chest discomfit, irritation of the respiratory tract, coughing and wheezing. Threshold for injury to vegetation.
0.12	The US National Primary and Secondary Ambient Air Quality Standard attained when the expected number of days per calender year with maximum hourly concentration above 0.12 ppm is equal or less than 1, as determined in a specified manner

All oxidizing compounds in smog with oxidizing power greater than oxygen have been termed as oxidant. Eye irritation is the first symptom of smog injury to human tissue, but it appears to be temporary injury. Eye irritation is noticeable when oxidant concentration in smog exceeds 0.25 ppm.

Respiratory irritation and difficulty in breathing is common, and the correlation of oxidant concentration with its irritating action has implications beyond the simple annoyance factor. Near 0.15–0.1 ppm oxidant, impairment of pulmonary function in emphysema patients, impaired performance of athletes, and aggravation of asthama is observed, sulphur oxides further aggravate the bronchial asthma.

Smog may be able to damage the lungs by the formation of compounds that harden the tissue. Nitrogen dioxide and ozone caused stiffening of the delicate molecular structure of the proteins collagen and elastin. There are materials in lungs tissue that give it elasticity, hence the effect was hardening of the lungs and, thus, respiration problems as damages to the membranes surrounding pulmonary are caused. Capillaries and water passes from blood vessels to lung alveoli causing pulmonary oedema.

Ozone may also cause damage to fibers, dyes, paints and other organic materials. Mechanical strength of natural rubber, synthetic polymers deteriorates by the action of ozone. Ozone reacts with double bonds in organic material, in the process the material cracks and become brittle. Regarding the degradation of rubber and some elastomers by ozone, the prevention or retardation of these process by means of antiozonant additives is of great interest. Photochemical oxidants also damages exterior household paints, pigments and are major hazard to work of art.

Ozone, nitrogen dioxide and PAN are the identified toxicants of photochemical smog, their effect on vegetation is as follow:

Ozone: The initial effect of ozone is in the palisade layer and involves scattered groups of cells. The cells lose water, plasmolyze, the cell contents become disrupted and disintegrate with the production of a dark pigment in some cases (stipple) or its absence (fleck). With continued exposure or high level exposure, these, upper surface dead areas, will enlarge, coalesce, and eventually injure the spongy cells to form necrotic lesions through the leaf and eventually collapsed area. In certain severe cases, ozone injury may appear very similar to sulphur dioxide.

Ozone enters the leaf through the stoma, but in leaves with two types of cells, the gas preferentially attacks the palisade cells. The reason for this is unknown, nor is it clear as to whether ozone reaches the palisade cells or first causes some metabolic imbalances with the possible production of metabolic poisons—which then affects the palisade cells.

PAN: PAN has been accepted as the primary phytotoxicant which causes the smog-type injury. The initial collapse is in the spongy parenchyma cells surrounding the air space into which a stoma opens. In some cases, the effect is limited more or less to the cells nearest the lower epidermis. This produces the characteristic under surface silvering, glazing, or bronzing. In other cases, such as petunia and tobaccos, the collapse may be through the entire thickness of the leaf. Since the cells of these leaves are sensitive only for a short period during development, a cross-leaf banding associated with sequential cell maturation along the leaf blade from tip to base develop. In grasses this effect is quite common.

While PAN injury may affect the entire thickness of the leaf and cause the collapsed area to lose its green colour, this type of injury should not be confused with SO_2 damage. Collapse due to PAN usually does not cause as marked a difference in leaf thickness as that due to SO_2. The general PAN damage pattern is often more of a cross-leaf banding rather than of the blotching or streaking associated with SO_2. Even when severe damage occurs, the more typical symptoms can normally be seen around the edges of the tissue collapse area.

Nitrogen Oxides: Neither nitric oxide nor nitrogen dioxide have been considered of major concern in vegetation damage since several studies suggest that levels sufficient to injure vegetation would be for above any known ambient levels. However, exposing plant to 0.5 ppm of NO_2 for longtime, leads to decline in growth of plants.

8.7 AIR QUALITY STANDARDS

A number of pollutants have documented effects on people, plants or material concentration, or approaching those in polluted air. In the United States, seven of them are known as "criteria pollutants" and National Ambient Air Quality Standards (NAAQS) have been set for them to protect public health and welfare. Tables 8.8 and 8.9 list these compounds. NAAQS standards and recommended values and limits set by some other countries and the World Health Organization are given in Table 8.9.

Table 8.8 National Ambient Air Quality Standards (NAAQS) set by the United States Federal Government

	Primary		*Secondary*	
Pollutants	*Concentration*	*Time*	*Concentration*	*Time*
CO	9.0 ppm	8h	9.0 ppm	8h
	35.0 ppm	1h	35.0 ppm	1h
SO_2	0.03 ppm	Annual Arithmatic mean	—	—
	0.14 ppm	24h	0.5 ppm	3h
O_3	0.12 ppm	1h	0.12 ppm	1h
NO_2	0.05 ppm	Annual Arithmatic mean	0.05 ppm	Annual Arithmatic mean
NMHC	0.24	Average 6 to 9 am	0.24	Average 6 to 9 am
Total Suspended	75 μg m^{-3}	Average Geometric mean	60 μg m^{-3}	Annual Geometric mean
	260 μg m^{-3}	24h	150 μg m^{-3}	24h
Lead	1.5 μg m^{-3}	Quarterly average	1.5 μg m^{-3}	Quarterly average

Ambient air quality standard has two components: a concentration and a time. For example, the US NAAQS for O_3 is 0.12 ppm for 1 h. This means that to avoid deleterious effects, the O_3 concentration should not exceed 0.12 ppm for more than 1 hour. An additional restriction is that the average number of days per year above the standard should not be more than one. The United States sets two types of NAAQS—primary and secondary. They are based on information contained in air quality criteria documents (US Environmental Protection Agency, 1978, 1979, 1982, a, b 1984)[9]. These contain a wealth of information on all aspects of the criteria pollutants, as do the documents on all aspects of the criteria pollutants, as do the documents by WHO and should be consulted for detailed information and references regarding pollutant sources, ambient levels, chemical transformation effect and so on.

Note that the definition of a US primary ambient air quality standard is designed "to protect the public health" with "an adequate margin of safety". These standards are set to protect even the most susceptible groups in the population, including those with cardiac and respiratory disease, and new-born infants whose defence system are not well developed.

Table 8.9 Selected recommended values or limits for various air pollutants in some countries around the world and by World Health Organisation

Pollutant	*Country*	*Concentration*	*Time*
CO	WHO	25 ppm	24 h
		30 ppm	8 h
		100 ppm	1 h
	Canada	13 ppm	8 h
	Japan	20 ppm	8 h
	CIS	1.3 ppm	24 h
	Germany	26 ppm	0.5 h
SO_2	WHO	38-57 ppb	24 h
		15-23 ppb	Annual A.M.
	Sweden	0.29 ppm	1 h
		0.12 ppm	24 h
	Netherlands	0.05 ppm	24 h
	CIS	0.02 ppm	24 h
	Germany	0.06 ppm	24 h
O_3	WHO	0.10 ppm	1 h
	Japan	0.06 ppm	1 h
NO_2	Japan	0.04-0.06 ppm	24 h
	Germany	0.05 ppm	2-12 mon
	CIS	0.05 ppm	24 h
NMHC	Canada	0.24 ppm	
Total suspended particulates	WHO	60-90 $\mu g\ m^{-3}$	Annual average
	Canada	120 $\mu g\ m^{-3}$	24 h
		70 $\mu g\ m^{-3}$	1 year
	Sweden	260 $\mu g\ m^{-3}$	24 h
		75 $\mu g\ m^{-3}$	1 yr
	Japan	200 $\mu g\ m^{-3}$	1 h
	CIS	150 $\mu g\ m^{-3}$	24 h
Lead	CIS	0.7 $\mu g\ m^{-3}$	24 h

Source: Sevenson *et al.*, (1983) *World Health Organization* 1972, (1978-79).

Secondary, NAAQS are set to protect public welfare. This includes economic losses due to damage, including visibility degradation, and personal discomfort. For most pollutants, the primary and secondary NAAQS are same. However, for the particulates they are not; the secondary guidelines are more stringent than the primary standards.

It should be noted that WHO standards in Table 8.9 are short-term goals; more stringent long-term goals for various pollutants have also been set. These long-term goals at which no ill effects would be anticipated are concentrations falling between the short-term ones in Table 8.8 and the natural background of pollutants.

REFERENCES

1. Becker, K.H., Lobel, J. Schurathu: Bildung, Transport and Kontrolle von photoxidatioen in: Luft qualitats kristerien fur photochemis che oxidation Report 5/83, pp. 3–132 univelt bundes amp E. Schmidt Verlag, Berlin (1983).
2. Demerjian, K.L., Kerr, J.A. Calvert, J.G., "The Mechanism of Photochemical Smog" *Adva. Environ Sci: Technol.*, 4, 1–256 (1974).
3. Leighton, P.A., *Photochemistry of Air Pollution,* Academic Press, New York (1961).
4. Niki, H., Daby, E.E. Weinstok, B., "Mechanism of Smog Reactions" Chapter 2 in "Photochemical Smog and Ozone Reactions", *Advances Chem Series* 113, 16–57 (1972).
5. National Primary and Secondary Ambient Air Quality Standards, Federal Register 36, 8186 (1971).
6. Stephen, E.R., "The Formation, Reactions and Properties of Peroxyacetyl Nitrates in Photochemical Air Pollution", *Advan. Environ. Sci: Technol.*, 1,119–146 (1969).
7. Sevenson, G. *et al.*, "Motor Vehicles and Cleaner Air", Report of Swedish Government Committee on Automative Air Pollution, Stockholm (1983).
8. World Health Organization "Photochemical Oxidant", Environmental Health Criteria No. 7, Geneva, Switzer land (1978).
9. US Environmental Protection Agency, "Air Quality Criteria for Particulate and Sulphur Oxides", December, 1982a. Air Quality Criteria for Oxides of Nitrogen Sept. 1982b, Air Quality Criteria for Ozone and other Photochemical Oxidant, April 1978, Air Quality Criteria for Carbon monoxide, October 1979, Addendum Review of the NAAQS for carbon monoxide, 1983; Reassessment of Scientific and Technical Information, 1984 Revised Evaluation of Health Effects Associated with Carbon monoxide exposure, 1984, Air Quality Criteria for Lead in Review (1984).
10. Von Nieding, G., Wagner, H.M., Wivbungen von photooxidantiesauf Mensch and Tier, in Luft qualit at skriterien fur photo chemische Oxidation Report 5/83, pp. 133–203, Umlvelt bundesamt, E. Schmidt Verlag, Berlin (1983).
11. *The Handbook of Environmental Chemistry,* Hans Gusten, "*Formation, Transport and Control of Photochemical Smog*", Springer Verlag, Berlin.

9

Particulate Matter in Atmosphere

9.1 INTRODUCTION

Particles, or particulate matter, may be solid or liquid with diameter between ~ 0.002 μm and ~ 100 μm. The lower limit of size cannot be defined because there is no accepted criterion as to when a cluster of molecules become particles. However, particle with diameter of ~ 0.002 μm have been measured, and this is the smallest size detectable by condensation nuclei counter. The upper limit corresponds to the size of fine drizzle or very fine sand. These particles are so large that they quickly fall out of the atmosphere and, hence, do not remain suspended for significant periods of time. Tiny particles, solid or liquid, in air are called Aerosol.

Particles may be either emitted into the atmosphere directly or formed there by chemical reaction referred as primary and secondary particles. Formation of particles in photochemical smog is an example of secondary particles. In addition to particle number, concentration, their mass, size, chemical composition and optical properties are important, but size is the most important.

Particles usually referred to as having a radius or diameter, implying the biggest sphere it envelops around it if the particle is rotated about its axis. The diameter of this sphere is called the diameter of particle.

However, many particles in the atmosphere have quite irregular shapes for which geometrical diameter is not meaningful. So in practice the size of such irregular shaped particles is expressed in terms of some sort of equivalent or effective diameter which depends on a physical rather than a geometrical, property. The most commonly used is the aerodynamic diameter, D_a, which is defined as the diameter of a sphere of unit density (1g cm^{-3}) which has the same terminal falling speed in air as the particle under consideration. This effective diameter is particularly useful because it can determine residence time in the air, it reflects the various regions of the respiratory system in which particles of different size become deposited, D_a is given by relation

$$D_a = D_g k \sqrt{\frac{\rho_p}{\rho_0}}$$

D_g is geometrical diameter, ρ_p is density of particle neglecting the buoyancy effects of air, ρ_0 is reference density and k is a shape factor which is equal to 1 in the case of sphere. Particle densities are often less than the bulk densities of pure substance due to pores, voids, and cracks in the particles. Throughout this chapter we use the term diameter, with understanding that it is the aerodynamic diameter of particle unless stated otherwise.

The diameter of the particle can be calculated by applying the Strokes law. The settling velocity occurs when the frictional and gravitational forces are balanced, *i.e.*,

$$F_{\text{Gravity}} = mg$$

$$= \frac{\pi D}{6} \rho_p g = F \qquad \text{...(9.1)}$$

The particles move relative to the gas and, hence, is subjected to a resistance force acting over the particles sphere of diameter D due to laminar flow of air over them.

$$F = 3\,\pi\,\eta\,VD \qquad \text{...(9.1a)}$$

where ρ density of particles, V is velocity of particles and η is viscosity of gas so that

$$3\pi\eta\,V D = \frac{\pi D^3}{6}\rho_p g \quad \text{or} \quad V = \frac{\rho_p g D^2}{18\eta} \qquad \text{...(9.2)}$$

or

$$D = \sqrt{\frac{18\eta V}{\rho_p g}} \qquad \text{...(9.2a)}$$

The viscosity of air at 25°C is 186.3×10^{-7} kg m^{-1} s^{-1}, $g = 9.81$ ms^{-2}, velocity in ms^{-1}, ρ_p in kg m^{-3} will give diameter in meters.

However as the particle becomes smaller, the air molecules appear less as a continuous fluid and more as discrete molecules separated by space through which the particles can slip. The net effect is that the particles can move faster than predicted by equation 9.2 due to this slipping between gas molecules. To correct for this a correction factor must be applied to the resistance force predicted by Stroke law, equation. The correction factor, C is a number greater than 1. Thus equation (9.1a) is modified to

$$F_G = \frac{3\pi\eta VD}{C}$$

and the settling velocity becomes

$$V = \frac{D^2 \rho_p g C}{18\eta} \qquad \text{...(9.3)}$$

The values of C called Cunningham correction factor, for different diameter values is as follow:

D, Diameter in μm	2	1.5	1.0	0.5	0.2	0.1	0.01	0.001
Values of C	1	1.11	1.166	1.33	1.88	2.89	22.45	219.28

It is seen that particles with diameters of the order of 10 μm or greater have settling velocity ≥ 0.1 cms^{-1} and hence will settle out of the atmosphere relatively rapidly. However, those particles with diameters ≤ 1 μm will remain suspended for long periods of time and hence can participate in atmospheric transformation.

9.2 BROWNIAN DIFFUSION

In the absence of convection, particles ≤ 1 μm diameter are transported primarily by Brownian diffusion; this is primarily responsible for the rapid coagulation of particles in the Aitken nuclei range.

The rates of Brownian diffusion can be quantified by reference to Fig. 9.1.

In the left section (A) of the box of cross sectional area 1 cm^2 there are N particles cm^{-3} with the right section (B) is initially empty. Particles (or gas molecules) will always diffuse from a region of higher concentration to one of lower concentrations. The rate at which they diffuse depend upon the

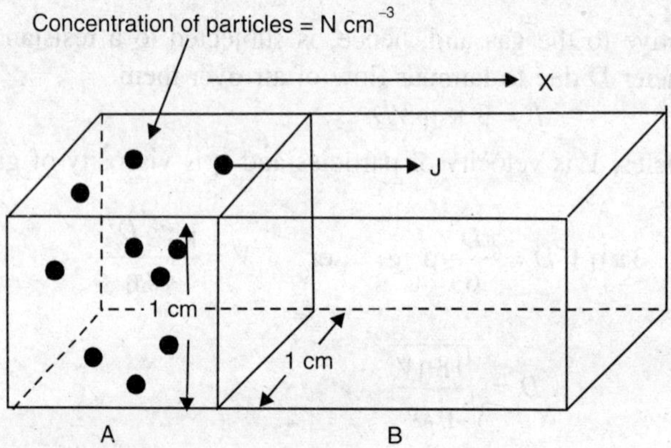

Fig. 9.1 Schematic diagram showing diffusion of particles from a region of high concentration (A) to a region of low concentration (B). J = flux of particles passing through the plane dividing A and B.

concentration gradient, $\frac{dN}{dx}$; larger the gradient, faster the rate of diffusion. This is the basis of the well known Fick's first Law of diffusion.

$$J = -\mathcal{D}\frac{dN}{dx}$$

The constant $\mathcal{D}$ is known as the diffusion coefficient and is proportionality constant relating the flux to the concentration gradient.

$$\mathcal{D} = \frac{kTC}{3\pi\eta D}$$

k is Boltzmann constant, T temperature in °K, C the Cunningham correction factor, η viscosity of the gas in which particles are suspended, and $\mathcal{D}$ is the particle diameter.

The relative importance of Brownian diffusion and gravitational settling in the deposition of particles may be seen by calculating the total deposition of particles onto a horizontal surface by these two processes in a given period of time under certain conditions.

Table 9.1 show the results of such a calculation for particle diameter from 0.001 to 10 μm.

At a diameter of ~ 0.2 μm the two mechanisms become equal, with greatly exceeding gravitational settling for particles in the Aitken nuclei range. However, if calculation for the mass of particles deposited were made, result would be quite different.

Table 9.1 Cumulative deposition of unit density particles on to a horizontal surface from unit aerosol concentration during 100s by diffusion and gravitational settling

D, Diameter μm	*C*	*Cumulative Deposition*		*Ratio Diffusion/ Settling*
		Diffusion number cm^{-2}	*Settling number cm*$^{-2}$	
0.001	219.28	2.5	6.5×10^{-5}	3.8×10^{4}
0.01	22.45	0.26	6.7×10^{-4}	390
0.1	2.89	2.9×10^{-2}	8.5×10^{-3}	3.4
0.2	1.88	—	—	—
0.5	1.33	—	—	—
1.0	1.166	5.9×10^{-3}	0.35	1.7×10^{-2}
2.0	1	—	–	—
10	1	1.7×10^{-3}	31	5.5×10^{-5}
100	1	5.5×10^{-4}	200	2.2×10^{-7}

It should be kept in mind that these calculated rates of diffusion and gravitational settling are only applicable to still air. In fact, in the atmosphere the air is rarely still and is usually undergoing some degree of turbulent motion. In this case the transport of particles become more complex and faster due to the velocity gradient and converted pattern of air flow.

9.3 SIZE DISTRIBUTION

Particulates are expressed in terms of mass of particulate matter per unit volume of air, restricting to only those particles with diameter ≤ 10 μm which are small enough to penetrate beyond the average adult's nose when inhaled. Particles larger than 10 μm are invariably ignored. Based on particles size, other parameters which are important for aerosol are mass of particle, surface area and volume. Taken all these functions together, atmospheric aerosol can be divided into three district groups.

1. Coarse Particles — diameter ≥ 2.5 μm
2. Fine Particles — diameter < 2.5 μm
3. $2i$ = Accumulation range — diameter ≤ 2–0.08 μm
4. $2u$ = Aitken nuclei range — diameter ≤ 0.08 μm

Figure 9.2 summarises these three ranges as well as the major sources and removal for each one.

Coarse Particles $\mathcal{D}$ > 2.5 μm

The particles are usually produced by mechanical processes such as grinding, wind or erosion. As a result they are large and hence settle out of the atmosphere by sedimentation. In the atmosphere, transport of coarse particles over long distance can occur. These particles are predominantly inorganic such as sand, sea salt and so on in this range. Because the source and sink are different from the smaller modes, the occurrence of particles in this mode tends to be only weakly associated with fine particle mode. One notes that the majority of biological particles, spores, pollens, and so on tend to be in the group of coarse particles.

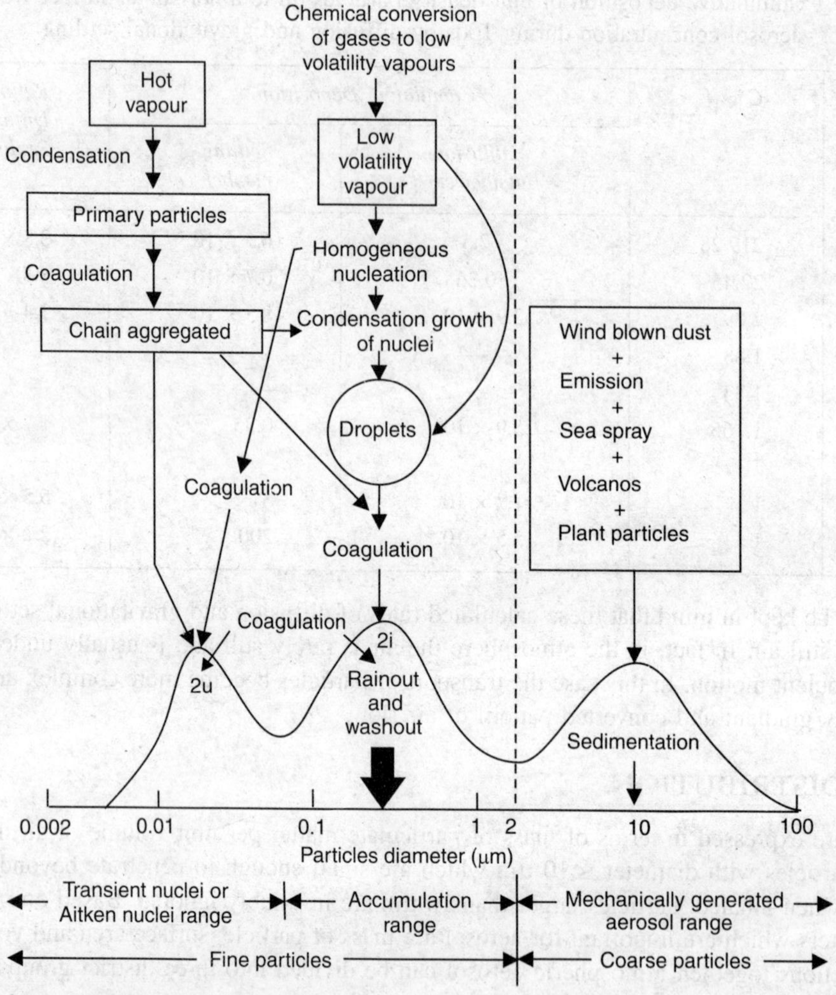

Fig. 9.2 Schematic diagram of an atmospheric aerosol size distribution showing the three modes, the main source of mass of each mode, and the principal processes involved in inserting mass into and removing mass from each mode.

Accumulation Range Particles $\mathcal{D} \leq 2$ - 0.08 μm

These particle's diameter range lies between 0.08 to 1–2 μm and arise from the condensation vapours of low volatility or from coagulation of smaller particles in the 2*u* nuclei range either with themselves or with the larger particles in the Accumulation range. As the mobility of small particles is very high, the rates of self coagulation of smaller particles is very high. These particles contain far more organics but also contains small amount of soluble inorganics such as NH_4^+, NO_3^-, SO_4^{2-} and so on.

Particles in the Accumulation range represent 5 per cent of the total particle number but make up a significant portion (e.g. 50 per cent) of the aerosol mass. They are removed relatively slowly by incorporation into cloud droplets followed by rainout, or by washout during precipitation. Alternatively they may be carried to surfaces by diffusion and advection and undergo dry deposition.

Aitken Nuclei $\mathcal{D} < 0.08$ μm

These particles arise from ambient temperature, gas to particle conversion and combustion processes in which hot, supersaturated vapours are formed and subsequently undergo condensation, and are primarily observed in significant concentration near such sources. These particles act as nuclei for the condensation of low vapour pressure gaseous species, causing them to grow towards the Accumulation range, alternatively these nuclei may grow larger by coagulation. The life time of these particles is short, sometimes on the order of minutes.

Combustion process (power plants and diesel engines) produce particles in Aitken range as well as in Accumulation range, the number of each depends on the nature of engine, fuel and operating conditions. Most particles produced by a light duty diesel engine are initially in the Aitken nuclei range but bigger particles grow rapidly. The number concentration drops dramatically as coagulation proceeds.

Under roadway condition, a typical 10^3 dilution factor operates with atmospheric dispersion, which dilutes the aerosol.

9.4 LIGHT SCATTERING AND VISIBILITY REDUCTION

Air pollution reduces visibility because liquid and solid airborne materials both absorb and scatter light and is related to size, concentration and physical characteristics of the particulate pollutants present when the contrast between the object and the background approaches zero, eye cannot distinguish the contrast between the object and background, the object cannot be seen, and it is said to be beyond the limit of visibility.

This alteration of contrast or light intensity is due to the absorption and scattering of light by the atmosphere. If a light path from an object to an observer path of length x is illuminated by a light beam of intensity I, this light is passing through incremental distance dx, is reduced by absorption and scattering by an amount dI, proportional to the intensity I. This may be written as the following expression.

$$dI = -\sigma I \, dx \qquad \text{...(9.4)}$$

where σ is the extinction or attenuation coefficient and minus sign indicates that intensity is being reduced by the effect of atmosphere. By integration over the path length 9.4 takes the form of Beer-Lambert's Law Eq. (2.34)

$$I/I_0 = e^{-\sigma x} \qquad \text{...(9.5)}$$

It can also be shown that apparent contrast C at a distance x can be expressed in a similar manner as

$$C = C_0 e^{-\sigma x} \qquad \text{...(9.6)}$$

where C_0 is the actual contrast of the object relative to its background.

The attenuation coefficient σ includes the effect of both scattering and absorption. It is sometimes useful to consider these two processes separately in such case Eq. 9.5 becomes

$$I = I_0 e^{-(b+k)x} \qquad \text{...(9.7)}$$

where b is the attenuation coefficient due to scattering and k is due to the absorption.

The scattering of light by particles depend on the size of the particles relative to the wavelength of light:

(*i*) particle diameter << λ, is known as Rayleigh scattering;

(*ii*) $\mathcal{D} \cong \lambda$, known as Mie scattering, and

(*iii*) $\mathcal{D} >> \lambda$.

Since we are concerned here only with visible and near ultraviolet light in the aitken region, that is $290 \leq \lambda \leq 750$ nm the first case correspond to particles with $\mathcal{D} \leq 0.03$ μm and third to particles with ≥ 10 μm. Particles with size between these two extremes fall in the second category, where $\mathcal{D} \cong \lambda$, this is the most important size regions for atmospheric particles.

The Rayleigh scattering is of importance in air pollution work. If the Rayleigh scattering were the only contributor to the extinction of light, a visibility of more than 240 km could be calculated. It is predominant for particle in atmosphere which are much smaller in size than the wavelength of incident light.

A common convention, when discussing light scattering as a function of particle size, is to define a dimensionless size parameter α, which is the ratio of circumference of the particle to the wavelength of the incident light

$$\alpha = \pi\mathcal{D}/\lambda$$

Very small particles ($\alpha << 3$) behave like gaseous molecules and, hence, produce Rayleigh scattering. Because the particles or molecules undergoing this type of scattering are small relative to the incident wavelength, the entire species is subjected at any instant of time to what appears to be uniform magnetic field; this creates dipole that oscillates with changing electromagnetic field of the light wave and reradiates the energy in all directions. Thus, Rayleigh scattering is symmetric in the forward and backward directions relative to the incident light beam.

Very large particles, on the other hand, ($\mathcal{D} >> \lambda$ ($\alpha >> 3$)) undergo geometric scattering, where the light beam refracted through the particle can be treated using classical optics. Between these two regimes where $\mathcal{D} = \lambda$, $\alpha \cong 3$, much more complex light scattering occur called the Mie scattering. Because particles undergoing the Mie scattering have dimensions of the same order as the wavelength of the incident light, the electromagnetic field of the light wave is not uniform over the entire particle at any instant of time, and a three-dimensional charge distribution is set up in the scattering particle. In 1908, Mie developed the solution for scattering of light of wavelength λ by a homogenous sphere of diameter *d*. As shown in Fig. 9.3, light is considered to be incident on the sphere and to be scattered at

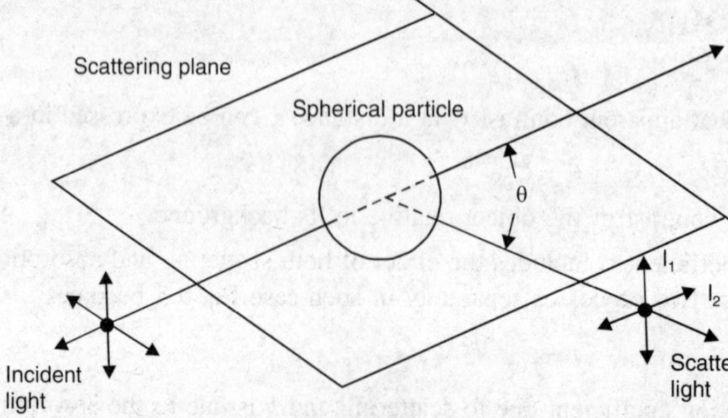

Fig. 9.3 Diagram showing scattering angle, scattering plane, and the polarized components of scattered light.

various angles θ to the direction of the unscattered beam. The incident and scattered beams are shown as the combination of two independent polarized beams; one I_1 has its electric vector perpendicular to the scattering plane defined by the incident and scattered beams, and the other I_2 is parallel to it. The intensity of the light at a distance R and a scattering angle θ from the particle is given by

$$I(\theta, R) = \frac{I_0 \lambda^2 (L_1 + L_2)}{8\pi^2 R^2} \quad ...(9.8)$$

where L_1 and L_2 are known as the Mie intensity parameters for the perpendicular polarized and parallel polarized components of the scattering light and is a complex function of refractive index (m) of the scatterer, the size parameter α and scattering angle θ.

Unlike the Rayleigh scattering, the Mie scattering takes place predominantly in the forward direction. This can be seen in Fig. 9.4 which shows the Mie intensity parameters L_1 and L_2 as a function of the scattering angle θ for three different values of the size parameters α, assuming the droplets are composed entirely of liquid water (m = 1.33). For α = 2.0 and 10 both L_1 and L_2 fall from a maximum value at θ = 0 as θ increases. For smaller particles with α = 0.8, L_2 initially falls as θ increases, but then rises again as θ approaches 180° corresponding to back scattering; for these particles, L_1 decreases slightly from 0° to 180°. However, in all cases L_1 and L_2 and hence scattering intensity show their maximum values at θ = 0°, corresponding to forward light scattering.

The variation of scattering light intensity with θ (Fig. 9.4) clearly becomes more complex as the particle size increases with sharp oscillations seen at α = 10.

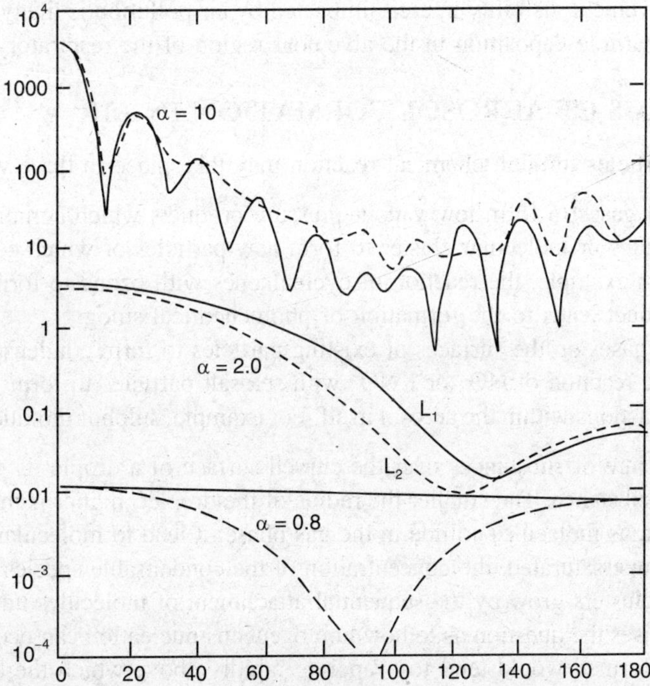

Fig. 9.4 The Mie intensity parameters versus scattering angle for water droplets (m = 1.333) having α = 0.8, 2.0 and 10.0. Solid lines are L_1, and dashed lines are L_2 (From H.C. Hinds, *Aerosol Technology*, John Wiley and Sons Inc., 1982).

Absorption: Although many common air pollutants, such as submicron sized wood smoke, oil mist and other smokes, produce light extinction primarily by scattering of smoke, as open hearth emission and soot, absorption becomes an important factor in the light extinction calculations. The absorption coefficient of haze is more or less equivalent to scattering coefficient. The absorption factor is particularly important when close to the ground and then decreases with altitude.

The applicability of this information should be readily apparent to the air pollution investigator, who is concerned with the alleviation of visibility restrictions. If a significant improvement in visibility has to be brought about, problem of controlling the emission of atmospheric generation of submicron particles has to be solved. The fact that emission control for small particles is much more expensive when control of large sized particles makes the problem different to tackle.

The reduction of visibility depends on aerosol particulate size and size distribution. The visibility (in km) V, varies inversely with concentration, M (μg m^{-3}) according to

$$V = K/M$$

The value of the K, has been found to vary between 900 and 3600 km μg m^{-3} with average value 1800 km μg m^{-3}. So on an average with a particle burden of 100 μg m^{-3}, visibility would be about 18 km but three times this burden the visibility is reduced to less than 6 km.

Humidity also affects the production of aerosols and thus affects the visibility. For a relative humidity 80%, it has been found in Los Angeles that reduction in visibility from 6 to 3 km occurs. The visibility further gets reduced to 2 km as relative humidity increases to 90%. However, the exact effect of humidity on visibility has been a function of the air pollutants present, which will differ for each city. In any case an improvement visibility in areas impacted by air pollution is likely to be accompanied by a reduction in total particle deposition in the alveoloar region of the respiratory system as well.

9.5 MECHANISMS OF AEROSOL FORMATION IN AIR

The formation of particles through chemical reaction may take place in three ways:

- Reactions of gases to form low vapour pressure products which combine with other condensable molecules or molecular cluster to form new particles or which condense on preexisting particles. For example, the reaction of cycloalkenes with ozone to form multifunctional oxygenated product leads to the formation of photochemical smog.
- Reaction of gases on the surfaces of existing particles to form condensed phase products. For example, the reaction of NO_2 or KNO_3 with sea salt particles to form $NaNO_3$.
- Chemical reactions within the aerosol itself. For example, sulphur dioxide oxidation to sulphate.

The vapour pressure of substances over the curved surface of a droplet is greater than over a flat surface of the same substance. The smaller the radius of the droplet, higher is the vapour pressure over the surface. However, as molecules collide in the gas phase, it lead to molecular cluster formation. By making the system supersaturated, the concentration of the condensable species increases, as does that of the clusters. The clusters grow by the sequential attachment of molecules until they reach a critical diameter D*. This raises the question as to how homogeneous nucleation can occur at all, since the first very small droplets formed would tend to evaporate rapidly above which the droplets are stable and grow, and below which they evaporate. The critical diameter is given by

$$D^* = \frac{4\gamma V}{kT \ln S} \qquad ...(9.9)$$

where γ is surface tension of the chemical forming the particle, V is the molecular volume and S is saturation ratio defined as the ratio of the actual vapour pressure to the equilibrium vapour pressure at temperature T.

In experiments, to simulate photochemical smog formation, particle free air was used. Figure 9.5 shows the concentration of nuclei as a function of time when smoggy ambient air, which has been filtered to remove particles, is irradiated with sunlight. It is seen that the number of particles rises quickly and dramatically, by about four orders of magnitude, peaks and then falls. The initial sharp rise is due to nucleation of low vapour pressure products of gas phase reactions. However, as the number of nuclei increases, two processes set in, which tend to limit the formation of new nuclei. These are scavenging of the molecular clusters by the nuclei before they have reached the critical diameter D^* needed to form a stable nucleus and coagulation of nuclei to form larger particles. When the rates of these heterogenous processes become equal to the rate of nuclei formation, the number of nuclei remains constant. Subsequently, the total number of particles decrease as coagulation occurs.

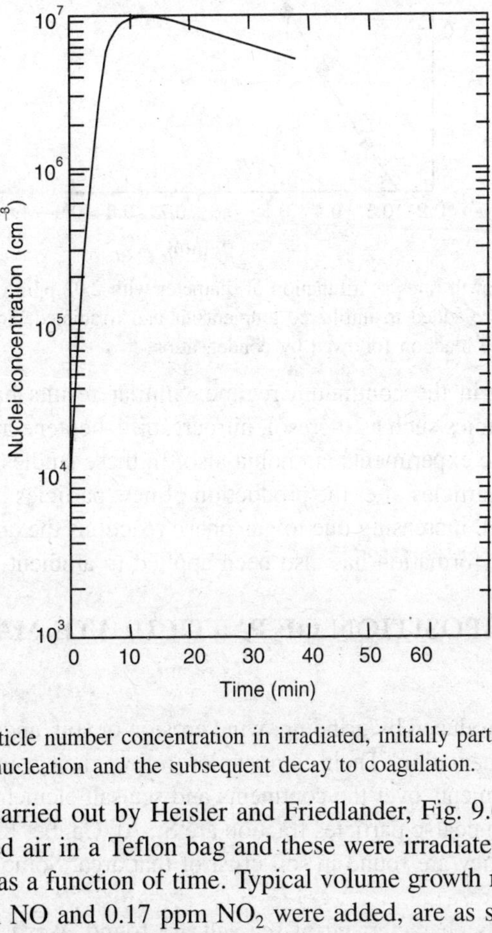

Fig. 9.5 Change of total particle number concentration in irradiated, initially particle free air. Initial concentration rise is due to nucleation and the subsequent decay to coagulation.

In another experiment carried out by Heisler and Friedlander, Fig. 9.6 in which hydrocarbon and NO_x were added to unfiltered air in a Teflon bag and these were irradiated with sunlight. The particle growth rates were followed as a function of time. Typical volume growth rates for the case where 2.02 ppm cyclohexene, 0.34 ppm NO and 0.17 ppm NO_2 were added, are as shown in Fig. 9.6. The mean free path is ~ 0.07 μm in air at room temperature and atmospheric pressure, so that for the case of gas phase reactions, the conditions are in continuum-transition regime. The solid line is the best least square fit for the case where aerosol growth is due to gas phase reactions, followed by condensation of low vapour pressure products, the growth rate varies directly with the particle diameter when the diameter

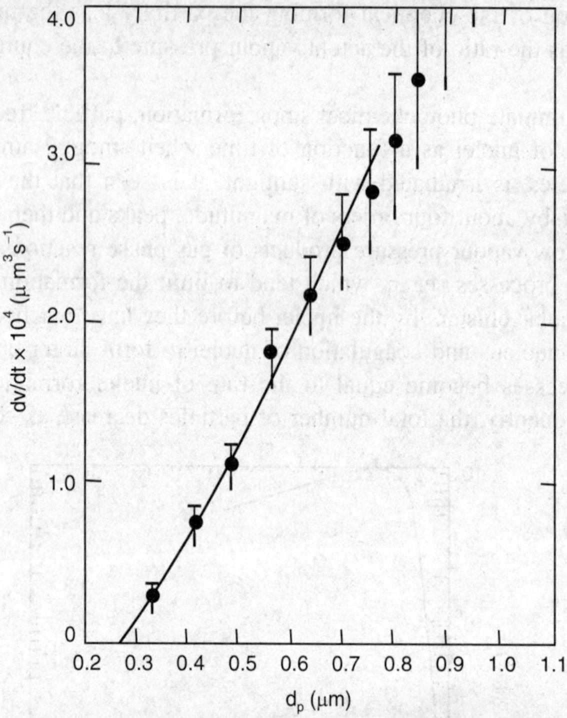

Fig. 9.6 Particle volume growth rates as a function of diameter with 2.02 ppm cyclohexene, 0.34 ppm NO and 0.17 ppm NO_2 were added to unfiltered ambient air and irradiated. Solid line is best least square fit assuming gas phase reaction followed by condensation.

are sufficiently large to be in the continuum regime. Similar results have been found in studies of aerosol formation from organics such as *o*-cresol, nitrocresol, 1-heptene and methyl sulphide, irradiated in air with NO_x and in some experiments ammonia also. In these studies, evidence was also found for evaporation of the smaller particles after the production of new particles had stopped, but while aerosol mass concentration were still increasing due to gas phase reactions the approach to elucidating mechanisms of secondary aerosol formation has also been applied to ambient air giving similar results.

9.6 CHEMICAL COMPOSITION OF PARTICULATE MATTER

Coarse Particles

Such particles are usually produced by grinding, wind erosion of soil, and are usually large, hence settle out in atmosphere by sedimentation. Their composition reflect their source. Coarse particles are generally comprised of soil elements over the continents and sea salt elements over ocean. Thus over land the major elements found in coarse particles fraction are Si, Al, Ca, Fe, K, Ti, Mn and Sr, and these are often in the same ratio as they are found in soil crust of that area. Some of biological coarse particles spores, pollens are also there.

In marine area, aerosols characteristic of sea salt are found. Wave action entrains air and forms bubbles which rise to surface. As they rise, dissolved organics may become adsorbed on them. The bubbles burst on reaching the surface, producing small droplets that are ejected into air. The particles with a wide range from less than 0.1 μm to greater than 100 μm are formed. As evaporation of water in the droplet into the surrounding air mass occurs, the salt becomes more concentrated. If the relative humidity is below deliquescence point of NaCl (75%), evaporation may occur leaving a solid salt particle. Along with the salt is an organic material originally associated with the bubble. Thus marine-derived salt contains sterols, fatty acids salts, fatty alcohols, so is the microlayer (the first 150 μm) of sea. This micro-layer is involved in the aerosol formation process from the sea, adsorbed organics are expected on the surface of such sea salt particle. Most of the elements found in coarse particles is volatile and are relatively chemically inert.

Fine Particles

Fine particles generally arise from conversions of gas to particle. Components of fine particles are sulphate, nitrate, organic and elemental carbons and ammonium ions. In the urban region, fine particles are enriched in elements such as As, Cd, Ca, Mo, Pb, Sb. The relative proportions of these species depend on the nature of emissions into air, photochemical activity and relative humidity. The carbon containing compounds found in fine particles are long chain hydrocarbons, polycyclic aromatic hydrocarbons, with or without polar groups like carboxylic acids, nitrates or sulphates. These long chain organics with one or more polar group act as surfactants to condensed phase, suggests that these may act as surfactant in aqueous atmospheric aerosols forming an organic film over the surface of aerosol. The role of organic films on droplets may have some or all of the following effects:

(*a*) Reduction of the rate of evaporation of water from droplets.
(*b*) Inhibition of the transport of stable molecules of highly reactive free radicals such as OH and HO_2 from thc gas phase into the droplets, thereby reducing their roles in atmospheric aqueous oxidations.
(*c*) Reduction of the efficiency with which the particles are scavenged by larger clouds and rain droplets.

Thus, the presence of organic films may increase the lifetime of such particles in atmosphere compared to as water droplets.

Particulate Organics

In a photochemical oxidizing atmosphere such as that of Los Angeles the difunctional substituted alkane derivatives of the type

$$X-(CH_2)_n-X \text{ and } X-\underset{|}{\overset{CH_3}{}}\!\!(CH_2)_{n-1}-X$$

where n = 1–5 have been observed in a number of studies. The substituent X can be —COOH, CHO. Tables 8.3 and 8.4 show some of the difunctional species identified in submicron fraction of particles, suggesting they were secondary in nature, that is, were formed from chemical reactions in atmosphere. These difunctional species are solid at room temperature, consequently on condensation give secondary fine particle. The most likely precursors appear to be cyclic olefins and perhaps dialkenes which can be oxidized to give stable species containing two substituted groups in the same molecule. The reaction of O_3 with cyclohexene is expected to lead to the excited biradicals.

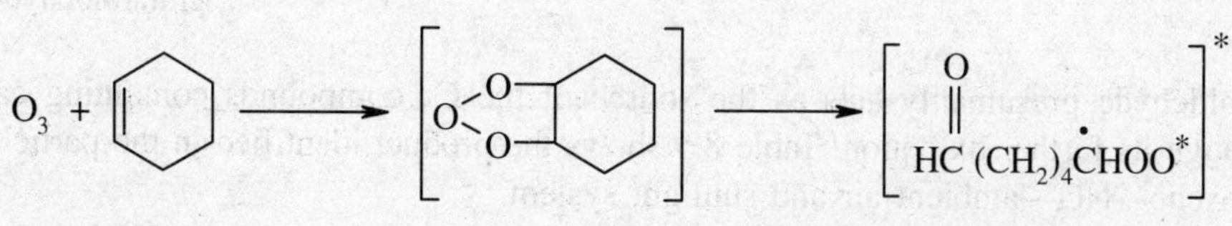

The fraction of excited biradical stabilized under tropospheric conditions is uncertain even for relatively simple alkenes. Besides, the decomposition paths of the biradical are unknown. However, one can envision reactions such as the following leading from excited biradical to stable multifunctional compounds.

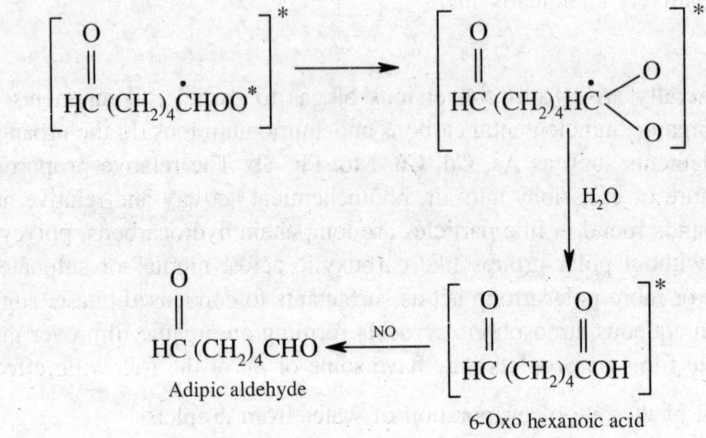

The formation of C_5 products has been observed in the cyclohexane reaction.

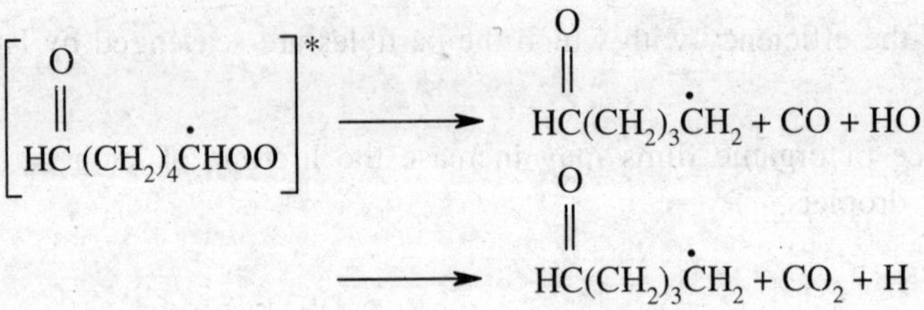

The five carbon alkyl radicals can undergo a series of reaction in air containing NO_x such as the following glutaraldehyde

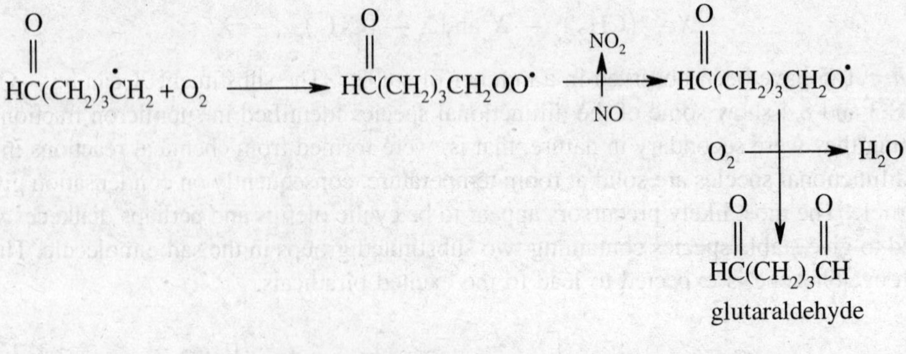

The glutaraldehyde presumably acts as the source of the C_5 compounds containing carboxlic acid groups through its further oxidation. Table 8.5 shows the product identified in the particles formed in the cyclohexene—NO—ambient air and sunlight system.

In short, the same type of aerosol organic products have been identified both in model systems and in polluted urban ambient air. While the reaction mechanisms leading to the observed products are, to a great extent, speculative at present, they can, in general, be rationalised based on the oxidation of analogous compounds.

Beside organic carbon, element carbon can also be an important component of aerosol especially in areas where significant quantity of wood is burned or use of diesel automobile is quite prominant.

9.7 PARTICLES AS POLLUTANT

Air quality standards for particulate matter are written in terms of total suspended particulate (TSP), *i.e.*, non size fractionated particles. The rationale for moving toward a size-related air quality standard based on the health effects will be apparent from an examination of Fig. 9.7, a diagram of human respiratory tract. Larger particles, if inhaled, are removed in head or upper respiratory tract. The respiratory system from the nose through the tracheobronichial region is covered by a layer of mucus secreting epithelium and serves to condition the air before it reaches the alveoli. Air is warmed to body temperature, is saturated with moisture. Mucus is continuously moving upwards. The epithelial lining cells are ciliated, that is they have hair like protoplasmic appendages that project into the passage ways and by continuous coordinated rhythmic action, work the mucus and entrap foreign material upward to the throat where it is either swallowed or expelled. In the air free from particulate matter, mucus is white in colour but a polluted air makes it back. How black is mucus reflects extent of air pollution.

Both the mucus secreting ability and anciliary action of the epithelium have been sensible to adversities and excess exposure to smoke, dust and toxic substances impair their functioning causing an increase in susceptibility to various respiratory disorder. On the other hand, finer particles fall in the repairable size range. These particles can reach the alveolar region which is not coated with a protective mucus layer. Here the clearance time for deposited particles is much greater than in the upper respiratory tract, and hence the potential for health effects in much greater.

Particulate pollutants have a bearing on penetration beyond the respiratory passages into the lungs. Nasal passage prevents coarser particulates bigger than 5 microns from entering into the respiratory system. Particles with a size of about 1 micron enter into lungs easily and rapidly. Actually aerosols with particle less than 1 micron size may reach the alveoli of lungs and damage lung tissues (Figs. 9.8-9.15).

This electron microscope photograph shows carbon particles deposited in lungs, which act as small walls which inhibit the air movement in the lungs. Since lungs are always working, oxygen in it is converted to CO_2 but can not leave lungs. The air of these chambers, is renewed by diffusion, a slow process not by any ebbing and flowing of air. This CO_2 and particulate matter leads to host of respiratory diseases, cancer, asthma, bronchitis. Particulate matter concentration is so high in metro cities like Delhi, that the number of people suffering from respiratory diseases has increased exponentially. In 1998 more than 10,000 persons died due to air borne diseases and 2 out of every 10 children suffer from asthma. So much so that those suffering from asthma when visit Delhi their condition deteriorate. These respiratory diseases cost country an expenditure of Rs. 4550 crore per year.

Soluble particles will be absorbed into the blood from alveoli while insoluble are carried to lymphatic streams and get deposited in pulmonary lymphatic deposit point or lymph glands where they create toxicity problems. Nature of toxicity and problems associated depend on the composition of particulate matter.

9.7a Diesel—Particulate Matter

The trend of using diesel in motor cars has grown rapidly over the last few years. The sole factor responsible for this is the low running cost. Diesel price is low to help the agriculture sector and support public transport. But this is a myth, because the biggest benefactor of its low cost are urban consumers.

Diesel engines are primarily responsible for particulate matter emission, 95 per cent of NO_x and 96 per cent of SO_2. Ozone is a secondary pollutant produced as a result of interaction of NO_x, hydrocarbons, particulate matter and sunlight leading to smog formation. Changes needed in the diesel engine to reduce NO_x emission tend to increase particulate matter emission or vice versa. Also particulate matter of diesel is coated with polycyclic aromatic hydrocarbons, 1,8-dinitropyrene which are highly carcinogenic. The greater is the number of particles in the emission, the smaller is their average size and more deeply they are able to penetrate the respiratory track. Moreover lung cells react first with components on the surface of particles. This makes the ultra fine particles dangerous as they have a large surface area where they absorbs a greater amount of toxic chemicals than the larger particles. Switching from regular to ultra low sulphur fuel also reduce average particle size significantly. In other words, cleaner the diesel, tinier the particulate matter, increasing its toxic potential.

To reduce particulate matter, a device called a particulate trap is needed, but they are still in the experimental stage. Thus neither improving diesel fuel quality nor tightening the engine quality helps anyway. Thus it is necessary to ban the registration of new diesel driven vehicular fleet. This ban should stay in place till a way to control the deadly particle emission from diesel exhaust has been found out.

9.8 AIRBORNE POLYCYCLIC AROMATIC HYDROCARBONS AND THEIR DERIVATIVES

In this section, a brief treatment of the role of polycyclic aromatic hydrocarbons (PAH), because of the carcinogenic nature of several representative molecules present in respirable ambient particles will be given. These topics have been treated in details in classics on *Polycyclic Hydrocarbons* by[3, 4, 5, 8, 9].

In the late 19th century, usually high rates of skin cancer were reported for workers in paraffin refinery, shale oil and coal industries. The chemical carcinogen(s) responsible for this activity of coal tar and pitch is documented by Phillips (1983) in 'Fifty Years of Benzo (*a*) pyrene (BaP)″ which was identified and synthesised in 1933. The structure of BaP and Benzo(e)pyrene (BeP) are shown below:

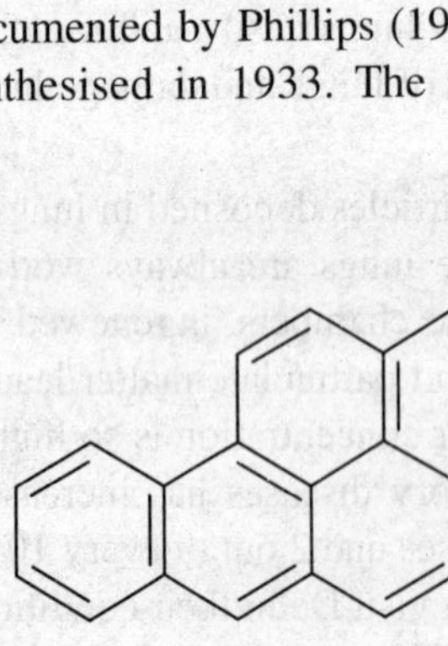

Benzo(a) pyrene
(BaP)

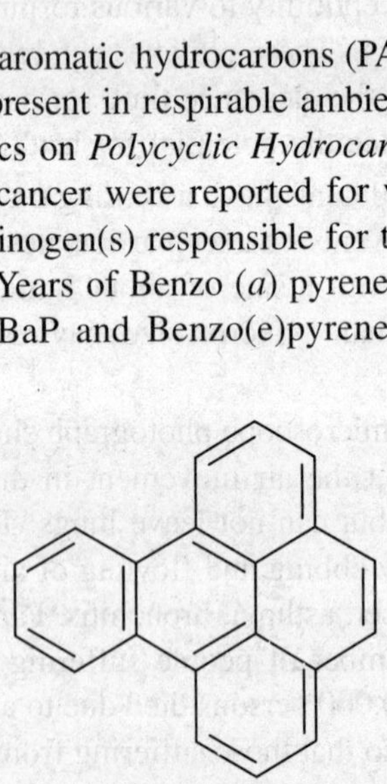

Benzo(e) pyrene
(BeP)

By 1970, it was well recognised that particles and Primary Particulate Organic matter (POM) collected from motor vehicle exhaust, smoke and fly ash were associated with PAH including their nitrogen analogues. Diesel received special emphasis because of much greater emission of soot than for a gasoline fuel.

PAH and their heteroatom analogues are formed during the incomplete combustion of organic matter. The mechanism of formation of PAH involves production of free radicals by pyrolysis at ~ 500-800°C of fuel hydrocarbons in the chemically reducing zone of a flame burning with an insufficient supply of oxygen. The C_2 fragments as well as C_1 and higher radicals, combine rapidly in reducing atmosphere to form partially condensed aromatic molecules. On cooling the reaction mixture, these PAH condense from the vapour phase into existing particulate substrates with a product distribution that generally reflects their thermodynamic stabilities in the oxygen deficient flame. PAH observed in atmosphere range from bicyclic species such as naphthalene present in gas phase to PAH containing seven or more fused rings such as coronene which are present solely adsorbed on particles.

An obviously-important physical property determining the distribution of PAH in air-water-soil environment is the vapour pressure. Table 9.2 gives the vapour pressures at 25°C of some representative PAH. PAH with five or six rings are predominantly adsorbed on particulate matter.

Table 9.2 Carcinogenic activities, vapour pressure and solubilities in water at 25°C of some environmental relevant PAH

Compound	*Molar mass*	*Carcinogenic activity*	*Vapour pressure mm of Hg*[(12)]	*Solubility* $\mu g\ l^{-1}$ [(9)]
Naphthalene	128	0	7.8×10^{-2}	31700
Acenaphthylene	152	0	6.7×10^{-3}	—
Fluorene	166	0	6×10^{-4}	1980
Phenanthrene	178	0	1.2×10^{-4}	1290
Anthracene	178	0	6.0×10^{-6}	73
Fluoranthene	202	+	9.2×10^{-6}	260
Pyrene	202	0	4.5×10^{-6}	135
Benz (*a*) anthracene	228	+	2.1×10^{-7}	14
Chrysene	228	+	6.4×10^{-9}	2.0
(BaP) Benzo (*a*) Pyrene	252	++	5.6×10^{-9}	0.05
(BeP) Benzo (e) Pyrene	252	0/+	—	3.8
Benzo (ghi) perylene	276	+	—	0.3
Coronene	300	0/+	—	0.1

The vapour pressure of these compounds can be significantly reduced by their adsorption on various types of substrates. Table 9.2 also show that PAH have very low solubilities in pure water. This has major implications when one considers the distribution of PAH and their derivative through equilibrias: air ↔ water ↔ soil in its environment.

A wide variety of PAH are released to atmosphere, adsorbed on surfaces of combustion-generated respirable particles. These combustion generated particles are released into urban atmosphere containing gaseous copollutants such O_3, NO_2, SO_2, HNO_3, PAN and so on, are concurrently exposed to sunlight and molecular oxygen, to produce a very large number of compounds and thus large number of pollutants.

Given this reaction system, it is clear why quantitative answers to basic questions on rates, products and mechanisms of the atmospheric reactions of particle-bound PAH suspended in urban air have been, and continue to be elusive. Significant progress has been made by laboratory studies on rates and products of reactions of PAH deposited on various specific substrates and exposed in dark and in light to simulated polluted atmosphere. However, there remains the major problem of extrapolating the results to conditions that prevail in ambient atmosphere.

It is now well established that certain PAH, deposited on a variety of substrates and exposed to a range of concentrations of gaseous NO_2 + HNO_3 from ppm to ambient level react to form mono and dinitro PAH. For example, in the first product study in a simulated atmosphere, BaP deposited on a GF filter was exposed for 8 hours to air containing 0.25 ppm of NO_2 and traces of gaseous HNO_3. About 20 per cent of the BaP was converted to 6–NO_2–BaP and small amount of 1– and 3–NO_2 isomers. The presence of small amount of gaseous HNO_3 alongwith NO_2 is essential for efficient nitration. Adsorbed PAH, order of reactivity may be similar to that in solution, but the rates and mechanism of reaction may not necessarily be correlated to solution reactions.

It has been found that polluted air contain significant (upto ~ 430 ppt) of NO_3 radical, which together is in equilibrium with gaseous N_2O_5, while NO_3 had a low reactivity but N_2O_5 is a potent nitrating agent for certain PAH. Its significance is the night time gas phase reaction with N_2O_5 that leads to formation of NO_2–PAN.

The other possible fate of certain NO_2 –PAH is their photodecomposition, ultimately into quinones and possibly phenolic derivatives. 6–NO_2–BaP deposited on silica gel when photolyzed gave the expected isomer quinones of BaP.

PAH exposed to the gaseous portion of filtered photochemical smog as well as low level of O_3 in pure air led to ozonolysis which led to formation of host of oxy-PAH formed as a result of substitution and ring opening reaction. Exact mechanism in ozone with PAH is still not clear.

9.9 LEAD IN FINE PARTICLES

Table 9.3 shows the estimated emission of lead to atmosphere on a global scale; clearly gasoline combustion in motor vehicles is by far the greatest source of lead pollution.

Table 9.3 Estimated global atmospheric lead emissions

Source category	*Emission meteric tons/year*	*Percentage of total*
Oil and gasoline combustion	273,000	60.8
Coal combustion	14,000	3.1
Waste incineration	8,900	2.0
Wood combustion	4,500	1.0
Iron & steel production	50,000	11.1
Primary copper smelting	27,000	6.6
Primary lead smelting	31,000	6.9
Primary zinc smelting	16,000	3.5
Primary Nickel smelting	2,500	0.5
Industrial application	7,400	1.6
Miscellaneous	14,920	3.3
Total	449,200	100%

In high compression engines, the hydrocarbons of simple petroleum distillates preignite before the top stroke of compression cycle. Preignition decreases the engine power because it results in an increase in the pressure on the piston before it has completed the upward stroke of the compression cycle. The preignition usually occurs explosively (knocking), which jams the engines and hasten its eventual demise. It was discovered that addition of 3 ml of "lead" per gallon prevents preignition. The lead that is added to gasoline is a mixture of 60 per cent tetraethyl methy lead $Pb(C_2H_5)_4$ and/or tetramethy lead $Pb(CH_3)_4$, about 35–40 per cent ethylene dibromide $BrCH_2$—CH_2Br, ethylene dichloride $ClCH_2$—CH_2Cl, 2 per cent dye, solvent and stabiliser. Lead prevents preignition and uneven burning by its conversion to lead oxide, a compound that quenches the free radicals, reactive intermediates with unpaired electrons formed in the combustion process. This quenching decreases the combustion rate and prevents detonation and preignition. The ethylene dibromide and ethylene dichloride facilitate the removal of the lead from the cylinder by its conversion to volatile lead halides that are thrown out from the engine into atmosphere.

Of the lead used in gasoline, only 25 per cent is retained in automobiles, the remainder being emitted in the form of particles. These particles are mainly PbBrCl, the volatile lead halide. In atmosphere, these compounds can react to form lead carbonates such as $2PbCO_3$. $Pb(OH)_2$ and oxide particulate. The range of lead aerosol particle sizes appears to be 0.2 to 0.9μ and these particles are even found embedded in the bark of trees near highway soils, particularly those rich in clay and organic matter, bind lead very strongly with a small amount of unburnt tetraethyl lead. One-third of the mass is in the fine particles form which when inhaled goes to our system.

The current standard of safe blood lead level is twenty μg per decilitre. The blood does not need lead, yet everyone absorbs it in varying amounts. Lead is not excreted but is stored for many years in tissue, chiefly in the bone marrow from which it is released back into blood stream to wreak cellular havoc. For middle-aged man, the risk of hypertension rises, for pregnant women, it means greater chance of miscarriage, premature delivery and still birth. Children are susceptible to a range of phychological, neurological, kidney and blood abnormalities, including partial learning loss, hyperactivity, lower IQ scores, impaired ability to absorb iron and calcium, reduced ability to metabolize vitamin D, disturbance in the formation and maintenance of red blood cells, decreased muscle bone growth, interference with the creation and function of certain important enzymes and amino acids which may cause cancer. Biologically, inert lead also accumulates in teeth, hair, nails. Lead can interact with sulfuydryl groups (–SH) in biomaterials, thus rendering them inaccessible to perform their biochemical function and interfering with cellular metabolism. Exposure to external Pb over an extended period of time can result in chronic nephritis or scarring and shrinking of kidney tissues.

Anyone can be affected by lead present in air or in fine particulate that one may inhale lead, that may wreak cellular havoc in any of the above process.

One theory (2) goes that the down fall of the Roman Empire was not because of barbarians nor was it because of moral corruption, but was due to lead poisoning. In those days, lead was extensively used in utensils, weapons, cosmetics, wine vessels and water pipes which may have been responsible for imperial madness and for infertility and miscarriage rates that kept the ruling classes from replacing themselves. The process of distillation was not known in those days. So after preparing grape wine in lead wine vessels, a small amount of acid was added as a result leaded wine was used for drinking. Thus lead was the true cause of madness of the Caesar and that brought the empire to ruin. Bones from Roman graves do show extreme concentration of the lead. And that was only in Rome. Now it is the world whole.

There remains a serious problem with respect to leaded gasoline—the Pb poisons the catalytic systems designed to remove noxious automobile exhaust emissions. If lead is removed, increased wear and tear and increased exhaust emission result thereby increasing vehicle operating costs and decreasing vehicle life expectancy. Certain aromatic antiknock compounds such as benzene, toluene, can be substituted for lead but aromatics and the products of their partial oxidation, such as phenols are known carcinogens. Lead removal will increase the cost of gasoline by an estimated 10%. Unleaded gasoline also results in greater reduction of atmospheric clarity and more soiling. Clearly then the problem of leaded gasoline is not a simple one.

Sulphur is one of the important constituent of gasoline. It interferes with the antibiotic action which leads to emission of lot of nitrogen oxides and carbon monoxide along with sulphur dioxide. Thus various pollutants emitted on gasoline combustion are

(*i*) Lead salt particles
(*ii*) Sulphur dioxide
(*iii*) TEL vapours
(*iv*) Hydrocarbon vapours
(*v*) Nitrogen oxides
(*vi*) Carbon monoxide
(*vii*) Carbon dioxide
(*viii*) Water vapour
(*ix*) Carbon particles

(*i*), (*ii*) and (*iii*) can be removed with reformation of fuel and improvement of engines. Use of catalytic converter can convert (*iv*), (*v*), (*vi*) to CO_2, H_2, N_2.

9.10 FLY ASH PROBLEM

Thermal power station are a source of major power supply in India. These stations also rank among the worst air polluters in the country. The main effluents from these plants are fly ash and SO_2. 75 per cent of coal produced, goes into power generation. Coal of India has low sulphur contents generally less than 1 per cent but a high ash content, 25 to 40 per cent. High ash coal mean more wear and tear of plant and machinery, low thermal efficiency of the boiler, slagging, choking and scaling of furnaces and tubing and most serious of all generation of a large amount of slag and fly ash. As a rule of thumb, for every MW of installed capacity approximately one acre of land is required for disposal of ash generated, the material accumulating to a height of 8 to 10 meters. Most of the Indian thermal power plants having little land space, simply let the solid waste run into nearby river, choking the natural drainage system and despoiling the land down stream. Between 85 to 90 per cent of solid waste generated in the form of bottom ash, but 10-15 per cent is lost as fall out from chimneys. The fall out increases the suspended particles matter in the air and adversely affect the agriculture, land flora and vegetation around.

In every thermal power station, chimney should have a working electric precipitator to remove the fly ash leaving the chimney. It may be added that using electric precipitator attached to chimney in no way decreases the amount of fly ash coming out of a thermal power station. It removes fly ash from air and gives a control over the fly ash so as to be disposed off properly. This fly ash along with bottom ash should be mixed with gypsum, lime and packed near the power station itself. This mixture on mixing with water will work like cement, and public should be encouraged to purchase it. This can be an extra source of income for thermal power plant.

Fly ash is generally grey in colour, abrasive, acidic, refractory in nature and has a fineness of 4000 to 8000 $cm^2\ g^{-1}$ and possesses pozzolonic characteristics. Its particles ranging in size from as much as 120 to less than 5 microns in equivalent diameters. As these particles are quite light, get airborne very fast and pollute the atmosphere. Long inhalation of fly ash causes, silicosis, fibroses of lungs, bronchitis and pnemonitis. It corrodes structural surfaces and its deposition affects horticulture. Fly ash disposal in sea/rivers causes damage to aquatic life cycles. Slurry disposal lagoons/setting tanks become breeding grounds for mosquitoes and bacteria. Crops on thousand of acres in the vicinity of power plants are damaged, nasal and respiratory disorder increases.

Fly ash today is a national problem. Existing 75 thermal power plants alone need about 50,000 acres of land for disposal of fly ash during their life span of 30 years. The requirement of land, for disposal of fly ash slurry would increase with the addition of more thermal power plants. The annual expenditure on road transportation just for dumping the fly ash is about Rs. 50 crores. Considering 10 per cent annual growth generation through thermal power plant, the annual fly ash generation is expected to exceed 100 million tons by 2010 A.D.

9.11 PARTICULATE MATTER EFFECTS ON PLANT GROWTH

Plant growth is a photochemical process. Plants detect the onset of different seasons of the year and respond to it in ways that we have come to accept as normal. As the winter season draws to a close and the warmer weather of spring arrives seeds germinate and produce new plants, new leaves appear from tubers and bulbs of herbaceous perennials and the dormant winter buds of trees and shrubs swell open to form new leaves. Some plants flower in spring, other in mid summer and still other in the autumn.

How do plants detect the season and what mechanism do they use to respond to it. The relative lengths of the day and night control the flowering and dormancy of plants. This phenomenon is called photoperiodicity. By controlling the lengths of the daily light and dark periods with artificial lights, plants that normally flower at widely different times of the year can be brought to flower simultaneously.

Plants apparently measure time because they react differently to the long days of summer and short days of autumn. Plants actually measure the duration of darkness and main function of the light period is to operate the photosynthetic machinery. If a long dark period is interrupted with a brief interval of light, summer flowering plants such as barley, flower as though they were receiving long days and short nights.

In order to be photochemically effective in producing a physiological response, it must be absorbed by a special kind of chemical compound within the plant. Those colours that are absorbed most readily by this compound are usually the most effective in producing the response. This compound is called "Phytochrome" a soluble protein molecule that has been detected in wide variety of plant species and appear to be of universal occurrence. Spectrum of this protein show a sharp absorbance maximum to red light at 660 μm, we call it P660.

Far red light changes phytochrome to its red sensitive form, P660 while red light changes phytochrome to its far red sensitive form which show an absorption at 735 μm we call it P735. However, red light germinate seeds and plant grows but if far red is given growth of plants is inhibited. Red light produces a small change in phytochrome P660 that enables it to combine with a receptor site of an enzyme. This allows the enzyme to initiate a synthetic pathway that lead to plant growth.

Physiological experiment have shown that the far red absorbing form of phytochrome P735 was slowly converted back into red absorbing form, P660 in darkness. The dark conversion is an enzymatic process, which is markedly slow·-d by lowering of temperature and does not occur at all in the absence of oxygen, available only at night. Every plant germinates in a particular season. Throughout the year an equilibria

$$P660 \rightleftharpoons P735 \text{ with } K = C_{P735}/C_{P660}$$

For different seasons the value of K is different. Different plants have different minimum value of K to germinate. When that value is attained, it is the proper season for the plant to germinate. Next day, the germination starts. In other words, this minimum ratio and thus K value decides the germination season for plant. The dark conversion of phytochrome provides the plants with a means of sensing time and is probably the basis by which plant distinguishes long nights from short nights.

Plants seldom grow well under constant temperature conditions. They need rhythmic or cyclic temperature changes daily or over longer periods; this is called thermoperiodicity. Even after seeds germination, plants also show a daily thermoperiodicity already provides an answer to geographical distribution of plants, their genetic constitution determines what temperature they require and whether these should change daily or seasonally. This genetic constitution consist of the inherent properties, which are mostly located in chromosomes and which determines the reactions of a plant toward environmental factors such as temperatures.

As particulate matter increase in atmosphere the intensity of light reaching the surface decrease. The temperature of earth but may be offset by green house effect, and thus disturb the K value, consequently plant growth through thermoperiodicity may be affected. Also particulate matter scatter the light changing the spectrum of visible light, particularly the red light, reaching the surface of earth. In the process, intensity of red light in it will decrease which will affect K value and thus the germination and flowering of various plants. The overall effect may be to retard the plant growth and thus will disturb equilibria. This should be evident from the fact that nothing grows in the shade of tree where there is no red light.

Each plant of a genetic constitution is associated with a set of pests in soils. Acceleration or retardation in their growth will disturb the pest equilibria in soil which may affect the plant and environment in the ways that cannot be predicted at this stage.

Particulate matter deposited on plant leaves block the stomata of plant thus inhibiting the rate of transpiration of minerals from soil. Deposited particulate restricts the absorption of CO_2 thereby reducing the rate of photosynthesis. It thus will retard the growth of plants and crop production. Dust mixed with mist or rain form a thick crust on the upper leaf surface which shields the sunlight necessary for carbon assimilation photochemical reactions.

9.12 SMOKER—A POLLUTER

Smoking produce about one thousand very reactive type of molecules, including free radicals, the structure of most of these molecules is established. These reactive molecules increase the pulse rate, raises blood pressure. Many of these compounds cause clotting which lead to angina and heart attack. Carbon monoxide in smoke combines with haemoglobin in red blood cell and impair their capacity to transport oxygen to vital organs. In 1964, it was established that smoking leads to lung cancer. Medical evidence pointed out a close link between smoking and enlargement of air sacs of the lungs causing breathlessness, heart diseases, bronchitis, paralytic stroke, and birth of low weight babies. Highly reactive molecules produced by smokers also lead to death of nearly 3,000 nonsmoker in the USA each year from passive smoking. All these factors indicate that cigarette and *beedi* smoking is a big menace to mankind that reduces man's life span by over ten years and thus a pollutant.

9.12a How a Smoker is a Polluter?

With time, a smoker loses his inner strength and develop many different diseases. These diseases may not be due to smoking but because resistance in his body has decreased. Say the smoker develops tuberculosis (TB). Now when he smokes, air goes into his lungs but what he exhales is not only CO_2 and N_2 but also TB bacteria. These bacteria are inhaled by his family members, his office colleagues, who may develop TB with no fault of their. Since TB in people around him has been caused because of him, A *Smoker* is a *Polluter*.

Many non-smokers die because of smokers every year. So much so in Europe and in USA many firms do not give job to smokers, which has reduced the consumption of cigarettes there. Strange enough, in India cigarette business is flourishing.

It has been observed that smoking is tempting in poverty-ridden Third World countries. Tobacco consumption in less developed countries is estimated to be rising at 2.7 per cent annually, whereas smoking habits are declining by 1.8 per cent every year in industrialized nations. India raked Rs. 2,700 crores in 1992-93; Rs. 3,000 crores in 1993-94 and Rs. 6,000 crores in 1997-98 from the tobacco industry. China raked in a revenue of Rs. 4.8 billion in 1992, though it will turnout to be a health hazard.

India is the third largest tobacco producer after China and USA, where in 1992, 34 per cent of cancer deaths were due to tobacco-related cancer. It does provide a livelihood to 25 million Indians, including labourers and tobacco farmers. Its farming is heavily dependent on pesticides. It consume considerable amount of paper for packing—about 6 km of paper per hour on modern packing machines—and spews out deadly tobacco that can kill not only smoker but his innocent office colleagues as well.

Apparently it is quite a complex problem. A legislation might take some wind off its sails, but ultimately saving citizens boils down to it individual responsibility. Each smoker has to decide that he has not to smoke.

9.12b How to Stop Smoking?

The Heart Care Foundation of India has formulated on how to quit smoking:

1. According to the guidelines, the four cigarettes which are habit forming and difficult to leave should be the ones to be given up first. These are the cigarettes smoked with the morning tea, breakfast, lunch and dinner. Most smokers when trying to quit, give up the in-between but retain these crucial four, with the result they can never stop.
2. The smokers first cut down their consumption of tea and coffee as this contain caffeine which increases the urge to smoke. Studies have revealed that chlonidin tablets help in curbing this urge.
3. If you feel an urge to smoke, pour a couple of drops of clove oil on the inner portion of your tongue. Brush your teeth three times a day. Nicotine has a tendency to accumulate on the surface of the teeth, and brushing helps to remove it.
4. Never smoke in the house—always go out. Consumption of fruits and vegetables will suppress the urge. Avoid alcoholic drinks, and in any case do not smoke while you drink.
5. Do not keep cigarettes at home to offer to your guests. Remember smoking is a slow poison and will kill sooner than later.

6. In early stages of giving up smoking, you will be irritable and tense. Meditation for about 20 minutes twice a day is a very effective counter-technique. It will calm you and help you to relax.

So make your move now. Do not let a 7.5 cm tube of tobacco dictate your life or cause your death.

9.13 SPACE GARBAGE: A SPACE POLLUTANT

Artificial debris in space, called space garbage, is growing at a catastrophic rate, and may pose a major hazard to future space operations, aerospace experts have warned. The most serious debris source is the fragmentation of satellite payloads and rockets. Solar radiation and extreme thermal environment in space almost vacuum space can breakdown the bonding agent in paints coating the surface of space crafts to release thousands of tiny flakes of paint that could orbit earth for years.

The drag effect by the earth's atmosphere does remove some of the debris but atmosphere drag is highly inefficient at altitudes above 750 km. A growing volume of traffic to earth orbit coupled with an increase in debris deposition rate may lead to catastropic growth within 50 years. C.I. Mcinnes, an aerospace scientist at the University of Glasgow, UK, has called for new policies to postpone such a catastrophic growth by a careful control of debris deposition at the most susceptible altitudes.

The accumulation of these debris is creating a band like structure around earth. Scientist have categorized over 18,000 pieces of space debris and believe million more tiny in orbit and invisible from earth. They include old satellites, rocket fragments, space mechanism and even flake of paints. The breakup of US seven Delton second stage rocket is said to account for nearly 30 per cent of debris in orbit.

At the current growth rate, these debris will be presenting a serious hazard to future operations in space. A 0.2 mm diameter paint flake reportedly hit the window of space shuttle during its seventh space mission causing a tiny crater. Researcher believe that a particle of 1 cm radius could cause serious damage and endanger the lives of astronauts. On the other hand, radars on earth cannot detect objects smaller than 10 cm radius above an altitude of 1000 km.

C.I. Macinnes says that number of debris is of less importance than the policies that control the altitudes at which debris is deposited. If inactive pay load are transferred from the 1000 km band to a local minimum density distribution, the collisional growth could be substantially delayed.

REFERENCES

1. Barbara, J., Finlayson-Pitts & James, N. Pitts Jr., *Atmospheric Chemistry,* John Wiley & Sons (1985).
2. Jagjit Kaur Sindhu, "Lead Poisoning" *Chemistry Education*, pp. 47-48 (1992).
3. Clar, E., *Polycyclic Hydrocarbons,* 2 Vols., Academic Press, , pp. 974, New York (1964).
4. Bgorseth, A. and Ramdahl, T., Eds. *Handbook of Polycyclic Aromatic Hydrocarbons,* Vol. 2, Emission Sources and Recent Advances in Analytical Chemistry, Dekker, New York (1985).
5. Grimmer, G. and Pott, F., "Occurrence of PAH in Environmental Carcinogens: Polycyclic Aromatic Hydrocarbons" G. Grimmer Eds. CRC Press, Boca Raton FL, pp. 61-129 (1983).
6. Husar, R.B. and Whitby, K.T., "Growth Mechanisms and Size Spectra of Photochemical Aerosols", *Environ. Sc. Technol.*, 7,241 (1973).
7. Heisler, S.L. and Friedlander, S.K., "Gas to Particle conversion in Photochemical Smog: Aerosol Growth Laws and Mechanism for Organics", *Atoms Environ.*, 11, 157 (1977).

8. Hoffman, D. and Wynder, E.L., "Organic Particulate Pollutant-Chemical Analysis and Bioassas for Carcinogenicity in Air Pollution", Vol. II 3rd Ed. A.C. Stern Ed., 361-455, Academic Press, New York (1977).
9. Lee, M.L., Novotony, M.V. & Bartle, K.D., *Analytical Chemistry of Polycyclic Aromatic Compounds,* Academic Press, New York (1981).
10. Philips, D.H., "Fifty Years of Benzo(a) Pyrene", *Nature*, 303,468 (1983).
11. Hinds, W.C., *Aerosol Technology,* John Wiley & Sons Inc., New York (1982).
12. Sonnefeld, W.J., *et al.*, "Dynamic Coupled Column Liquid Chromatography Determination of Ambient Temperature Vapour Pressure PAH", *Anal. Chem.* 55, 275 (1983).

10 Atmospheric Dispersion

10.1 INTRODUCTION

The clean air act specifies new source emission standards as well as ambient air quality standards. The connecting link between the two is atmosphere. How do pollutants behave once they have been emitted? How their concentration in air can be predicted? How can the limitations imposed by Pollution Control Board be shown to be satisfied for new sources in an attainment area? How can the improvement in air quality be predicted and attained when certain restrictions are imposed?

The answer can be given by the computer models that use information such as predicted emissions, smokestack height, wind data, atmospheric temperature profiles, ambient temperature, solar insolation, and local terrain features.

The atmospheric temperature profile affects the dispersion of pollutants from a smoke stack, are shown in Fig. 10.1.

Figure 10.1(*a*). If pollutants are emitted from a smokestack into a neutral stable atmosphere such a symmetrical plume is expected. The term used to describe this plume is *coning*.

Figure 10.1(*b*). If atmosphere is very unstable, and there is a rapid vertical air movement, both up and down, producing a *looping* plume.

Figure 10.1(*c*). A stable atmosphere greatly restricts the dispersion of the plume in the vertical direction, though it spreads horizontally. The result is called *fanning* plume.

Figure 10.1(*d*). Emissions from a smoke stack that is under inversion layer head downward much more easily than upward. The resulting *fumigation* can lead to greatly elevated down wind, ground level concentrations.

Figure 10.1(*e*). The stake here is above an inversion layer where mixing in the upward direction is uninhibited, but downward motion is greatly restricted by the inversion's stable air. Such plumes called *lofting* are helpful to the people at the ground level.

A common approach to air pollution control is to build taller and taller stacks so as to emit pollutants above inversion. The flow around an isolated building is characterized by the formation of a cavity, (as shown in Fig. 1.10) behind the building and a wake further down stream. So the height of stack should be at least two and a half times the height of the surrounding building is an empirical rule-of-thumb. If stack height is below this value the plume is trapped in the wake zone of building because of the poor mixing between the cavity and the main stream. An unfortunate consequence of this approach has been that pollutants released from tall stacks are then able to travel great distances, so that effects such as acid deposition can be felt hundreds of miles from the source.

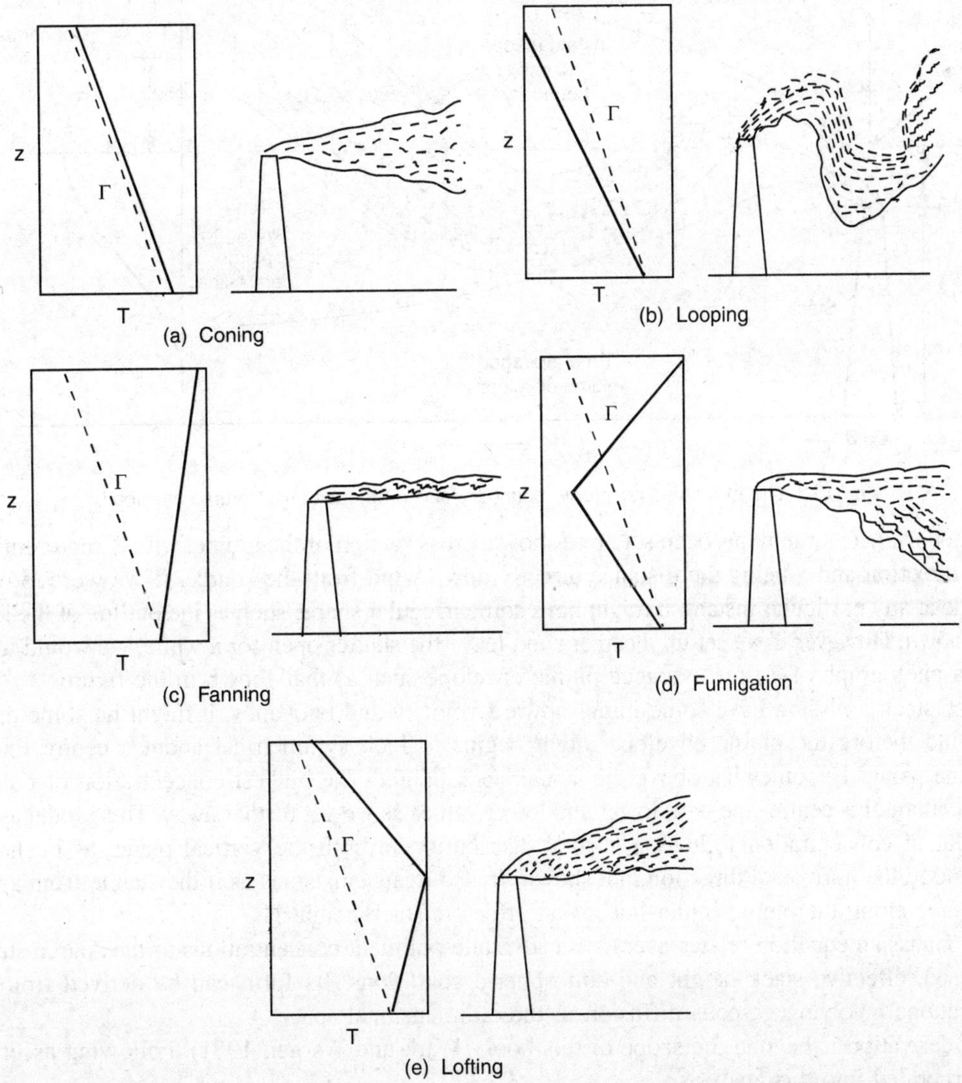

Fig. 10.1 Effect of atmospheric lapse rates and stack heights on plume behaviour. The dashed line is the dry adiabatic lapse rate for reference (*Source:* Williamson, S.J. Fundamentals of Air Pollution, 1973 Addison-Wesley Publication and Co.)

10.2 THE POINT SOURCE GAUSSIAN PLUME MODEL

Every computer program tries to relate emission to air quality and tries to calculate the time averaged pollutant concentration down wind from a source by using a Gaussian distribution curve. The basic Gaussian model applies to single point source, such as smokestack but can be modified to account for line sources (such as emission from motor vehicles along a highway) or area sources. To begin consider just a single point source such as shown in Fig. 10.2.

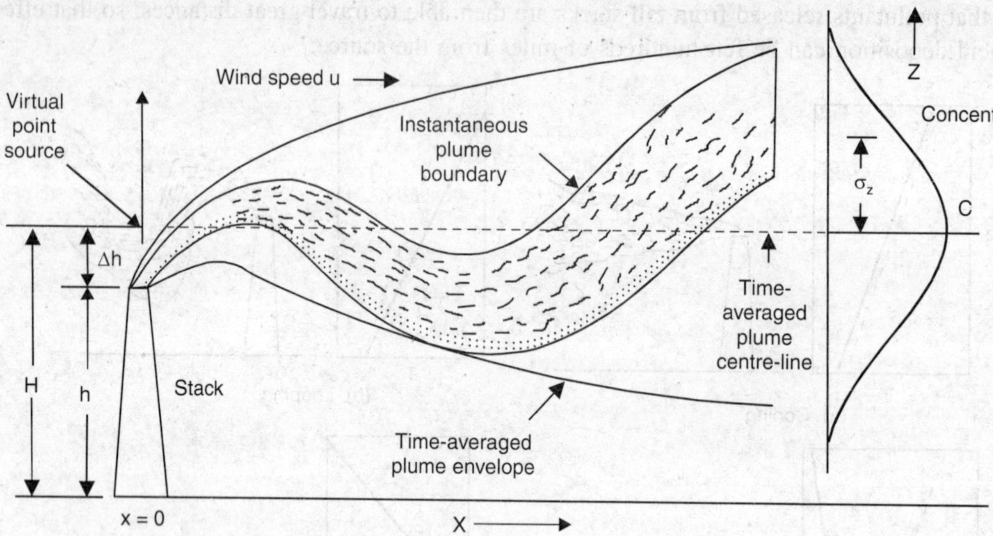

Fig. 10.2 The instantaneous plume boundary and a time averaged plume envelope.

The coordinate system has been set up to show a cross section of the plume, with Z representing the vertical direction and *x* being the distance directly down wind from the source. If we were to observe the plume at any particular instant, it might have some irregular shape, such as the outline of the looping plume shown. However if we set up a camera and leave the shutter open for a while, we would imagine getting a photograph of a time averaged plume envelope such as that shown in the figure.

Since stack emission have some initial upward velocity and buoyancy, it might be some distance down wind, before the plume envelope might begins to look symmetrical about a centre line. The centre-line would be somewhat above the actual stack height. The highest concentration of pollutants would be along this centre-line with lower and lower values as we get further away. This model assumes that pollutant concentration follows a normal distribution in both the vertical plane, as is shown in figure, and in the horizontal direction, not shown. It also treats emissions as if they came from a virtual point source along the plume center-line, at an effective stack height H.

The Gaussian equation relates average steady state pollutant concentrations to the source strength, wind speed, effective stack height and atmospheric conditions. Its form can be derived from basic considerations involving gaseous diffusion in three dimensional space.

The derivation is beyond the scope of this book (Wark and Warner, 1981). Following assumption are incorporated into this analysis.

(*i*) The rate of emission from the source is constant.

(*ii*) The wind speed is constant both with time and elevation.

(*iii*) The pollutant is conservative; that is, it is not lost by decay, chemical reaction, or deposition. When it hits the ground, none is absorbed and all is reflected.

(*iv*) The terrain is relatively flat open country.

The three dimensional coordinate system established in Fig. 10.3 has the stack at the origin, with distance directly downwind given by *x*, distance off the downwind axis specified by *y*, and elevation given by *z*. Since our concern is going to be only with receptors (people and ecosystem) at ground level, the form of the Gaussian plume equation given here is less general than it can be, and applies only for $z = 0$.

$$C(x, y) = \frac{Q}{\pi u \sigma_y \sigma_z}\left(\exp\frac{-H}{2\sigma_z^2}\right)\left(\exp\frac{-y^2}{2\sigma_y^2}\right) \quad ...(10.1)$$

where

			units
$C(x, y)$	=	Concentration at ground level at the point (x, y)	$\mu g\ m^{-3}$
x	=	distance directly down wind	m
y	=	horizontal distance from plume centre-line	m
Q	=	emission rate of pollutant	$Mg\ s^{-1}$
H	=	effective stack height	m
		$H = h + \Delta h$ (Fig. 10.2)	
u	=	average wind speed at height H	ms^{-1}
σ_y	=	horizontal dispersion coefficient (standard deviation)	m
σ_z	=	vertical dispersion coefficient (standard deviation)	m

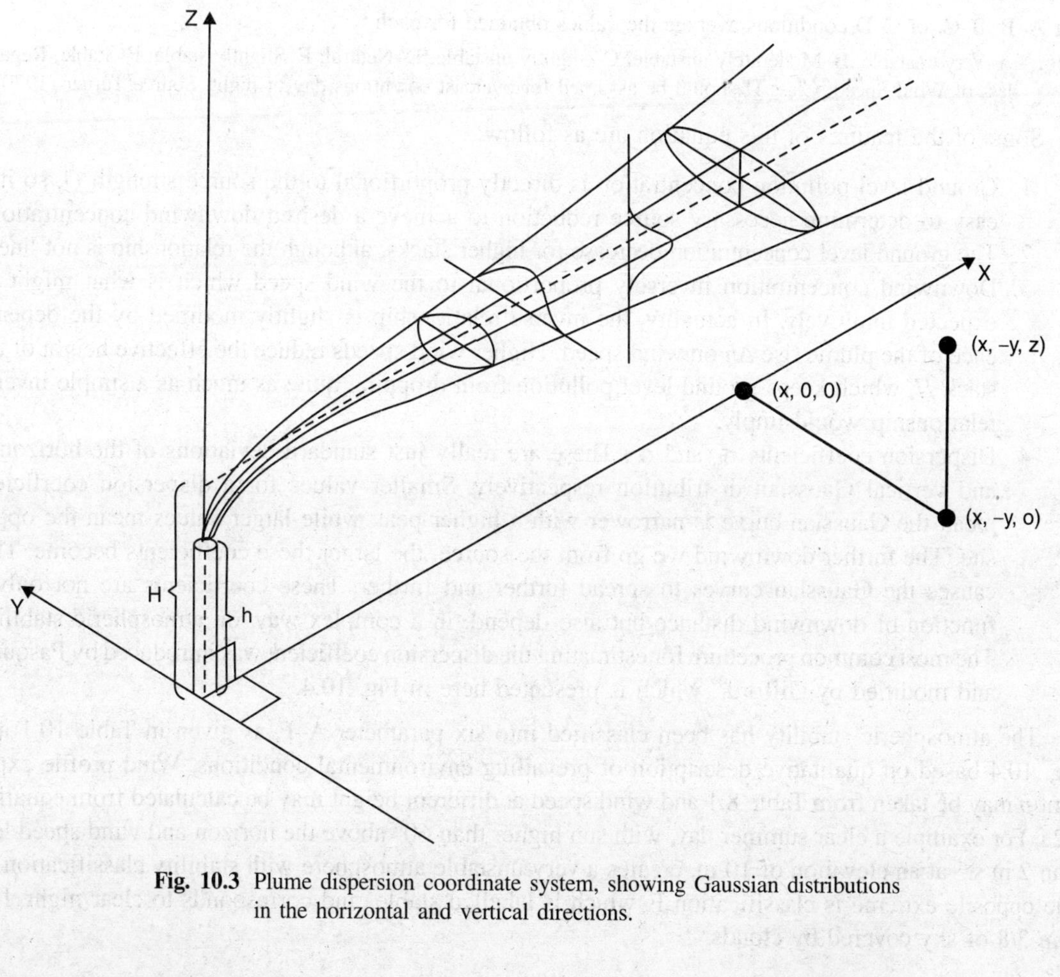

Fig. 10.3 Plume dispersion coordinate system, showing Gaussian distributions in the horizontal and vertical directions.

Table 10.1 Atmospheric stability classification

Surface wind speed[a] ms^{-1}	*Strong[b]*	*Day solar insolation moderate[c]*	*Slight[d]*	*Night cloudiness[e]*	
				Cloudy (≥ 4/8)	*Clear* (≤ 3/8)
< 2	A	A-B	B	E	F
2-3	A-B	B	C	E	F
3-5	B	B-C	C	D	E
5-6	C	C-D	D	D	D
> 6	C	D	D	D	D

[a]Surface wind speed is measured at 10 m above ground.
[b]Corresponds to clear summer day with sun higher than 60° above the horizon.
[c]Corresponds to a summer day with a few broken clouds or clear day with sun 35-60° above the horizon.
[d]Corresponds to a fall afternoon, or a cloudy summer day, or clear summer day with sun 15-35°C above horizon.
[e]Cloudiness is defined as the fraction of sky covered by clouds.
[f]For A–B, B–C, or C–D conditions average the values obtained for each.

Note: A–Very unstable; B–Moderately unstable; C–Slightly unstable; D–Neutral; E–Slightly stable; F–Stable, Regardless of Wind Speed, Class D should be assumed for overcast conditions day or night. Source Turner (1970).

Some of the features of this equation are as follow:

1. Ground level pollution concentration is directly proportional to the source strength Q, so it is easy to determine necessary source reduction to achieve a desired downwind concentration.
2. The ground level concentration decrease for higher stacks, although the relationship is not linear.
3. Downwind concentration inversely proportional to the wind speed which is what might be expected intuitively. In actuality, the inverse relationship is slightly modified by the dependence of the plume rise Δh on wind speed. Higher wind speeds reduce the effective height of the stack H, which keeps ground level pollution from dropping quite as much as a simple inverse relationship would imply.
4. Dispersion coefficients σ_y and σ_z. These are really just standard deviations of the horizontal and vertical Gaussian distribution respectively. Smaller values for a dispersion coefficient mean the Gaussian curve is narrower with a higher peak while larger values mean the opposite. The further down wind we go from the source, the larger these coefficients become. This causes the Gaussian curves to spread further and further. These coefficients are not only a function of downwind distance but also depend, in a complex way, on atmospheric stability. The most common procedure for estimating the dispersion coefficient was introduced by Pasquill[8] and modified by Gifford[9], which is presented here in Fig. 10.4.

The atmospheric stability has been classified into six parameter A–F, as given in Table 10.1 and Fig. 10.4 based on qualitative description of prevailing environmental conditions. Wind profile exponent p may be taken from Table 8.1 and wind speed at different height may be calculated from equation 8.2a. For example a clear summer day, with sun higher than 60° above the horizon and wind speed less than 2 m s^{-1} at an elevation of 10 m, creates a very unstable atmosphere with stability classification A. The opposite extreme is classification F, which is labelled stable, and corresponds to clear night, less than 3/8 of sky covered by clouds.

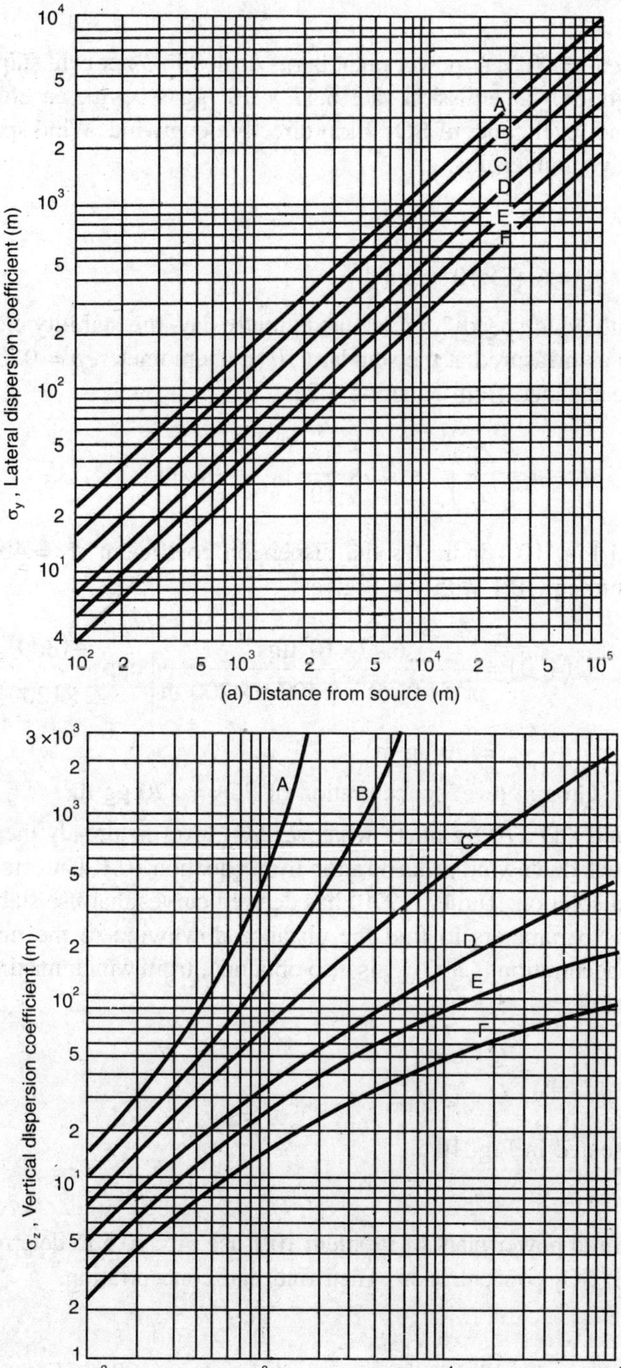

Fig. 10.4 Gaussian dispersion coefficients as a function of distance downwind (*a*) horizontal coefficient σ_y; (*b*) vertical coefficient σ_z.

Example 10.1

A new 40 per cent efficient 1000 MW power plant burns coal with 3 per cent sulphur and a heat content of 12000 Btu/lb. It emits SO_2 at emission rate 6.47×10^8 μg s^{-1} with an effective height equal to 300 m. Predict the ground level conc. of SO_2 4 km directly downwind. Wind speed at 10 m height are 2.5 m s^{-1} and it is cloudy summer day.

Solution

$$Q = 6.47 \times 10^8 \ \mu g\ s^{-1}$$

From Table 10.1 with wind speed 2.5 m s^{-1} and cloudy day, the stability classification is *C*. Table 8.1 uses the wind speed as measured at the standard 10 m anemometer, $p = 0.2$, so that wind speed at anemometer height be calculated from equation 8.2a.

$$u_1 = u_2\left(\frac{Z_1}{Z_2}\right)^{0.2} = 2.5\left(\frac{300}{10}\right)^{0.20} = 4.9\ \mathrm{m\,s^{-1}}$$

At 4 km downwind Fig. 10.4 indicates the dispersion coefficient $\sigma_y = 400$ m and $\sigma_z = 200$ m. Plugging these values into eq. 10.1 with $y = 0$ gives

$$C(4, 0) = \frac{6.47 \times 10^8\ \mu g s^{-1}}{\pi \times 4.9\ ms^{-1} \times 400\,m \times 200\ m}\left[\exp\frac{-(300)^2}{2 \times (200)^2}\right]$$

$$= 170\ \mu g\ m^3.$$

Ground level concentration of SO_2 = 170 μg m^3.

So power plant would add 170 μg m^3 to whatever SO_2 level is already there.

How to determine peak downwind concentration from equation 10.1. Unfortunately it is not possible to derive a mathematical solution. Turner (1970) has derived curves that use stability classification and effective stack height as parameters to find the distance downwind to the maximum concentration (C_{max}). A normalized concentration $(Cu/Q)_{max}$ is also obtained, from which maximum concentration can be found using the relation

$$C_{max} = \frac{Q}{U}\left(\frac{Cu}{Q}\right)_{max}.$$

The curves are presented in Fig. 10.5.

Example 10.2

For the 1000 MW coal fired power plant of Problem 10.1 use Fig. 10.5 to determine the distance down wind to reach maximum SO_2 concentration. Then find that concentration.

Solution

The stability classification is *C* and the effective stack height is 300 m. From Fig. 10.5 the distance downwind x_{max} is just 4 km. Reading down from the point on figure corresponding to stability class *C* and effective stack height H = 300 m $(Cu/Q)_{max}$ looks to be about 1.5×10^{-6} m^{-2}. Wind speed the effective height of stack was found in Problem 10.1 to be u = 4.9 m s^{-1} and $Q = 6.47 \times 10^8$ μg/s that

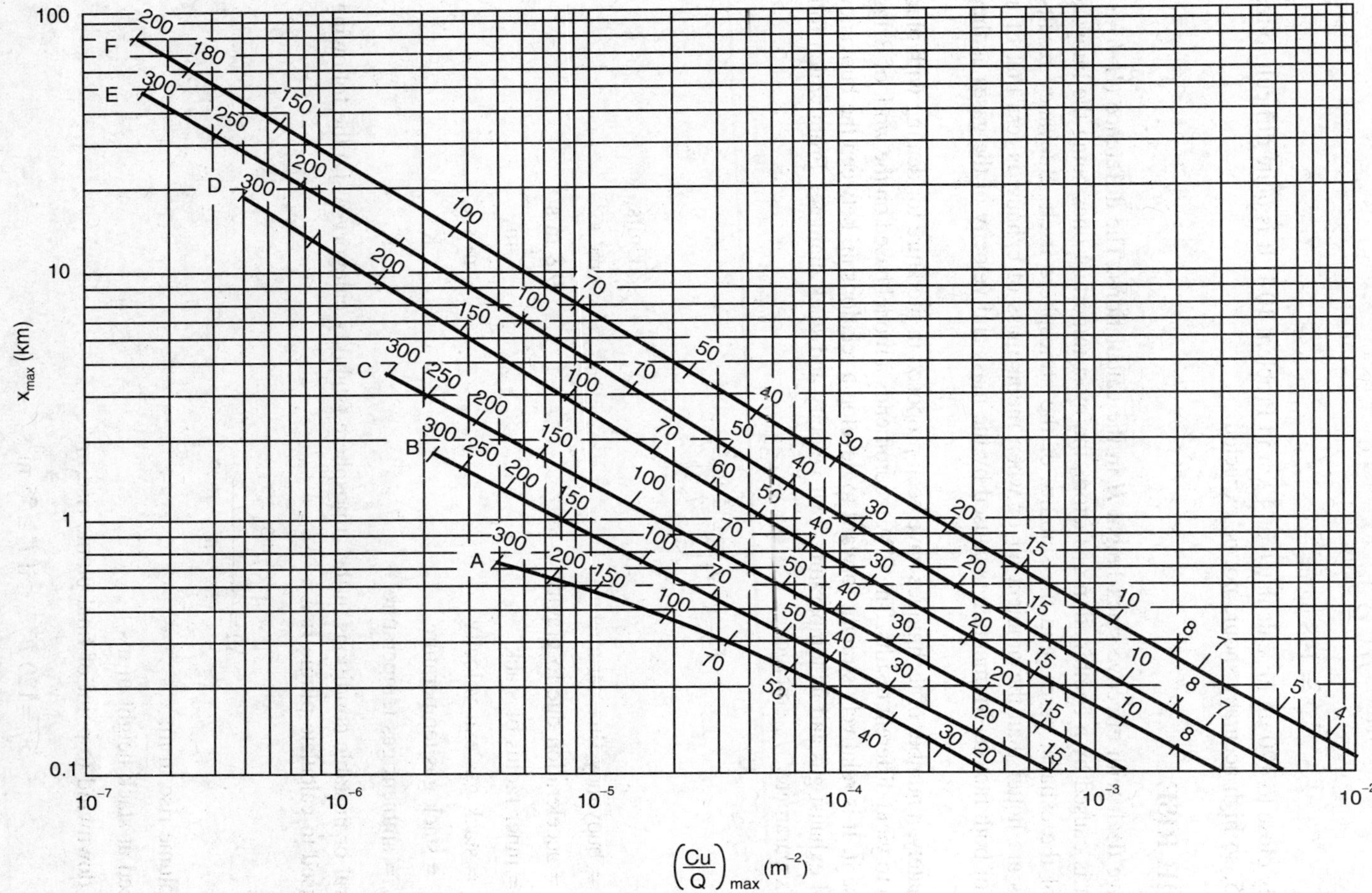

Fig. 10.5 To determine the peak downwind plane concentration, enter the graph at the appropriate stability classification and effective stack height (numbers above the lines, in meters) and then move across to find the distance to the peak, and down, to find a parameter from which the peak concentrations can be found (Turner, 1970).

$$C_{max} = \frac{Q}{U}\left(\frac{Cu}{Q}\right)_{max} = \frac{6.47\times10^{-8}}{4.9} 1.5\times10^{-6}$$

$$= 198\ \mu g\ m^3$$

which is fairly close to 170 μg m^3 calculated at 4 km in Problem 10.1. It is quite difficult to read Fig. 10.4 and 10.5, so high accuracy should not be expected.

10.3 PLUME RISE

So far we have dealt with effective stack height H in the calculations. The difference $(H–h)$ is called plume rise. It is caused by a combination of factors, the most important ones being the buoyancy and momentum of the exhaust gases, and the stability of the atmosphere itself. Buoyancy results when exhaust gases are hotter than the ambient air or when molar mass of exhaust is less that of air or a combination of both factors. Momentum is caused by the mass and velocity of the gases as they leave the stack.

Unfortunately, a number of techniques have been proposed in literature for dealing with plume rise and they tend to yield different results. The EPA recommends a model based on the work of Briggs[1] and so we will use it. It is relatively straight forward to develop a relationship between the buoyancy force on a parcel of exhaust gas and its temperature, exit velocity and stack diameter. From which following buoyancy flux parameter, F, emerges.

$$F = g\,r\,v_s\left(1-\frac{T_a}{T_s}\right) \quad \text{...(10.2)}$$

where	Units
F = buoyancy flux parameter	$m^4\ s^{-3}$
g = acceleration due to gravity	9.8 m s^{-2}
r = inner radius of stack	m
v_s = stack gas exit velocity	m s^{-1}
T_a = stack gas temperature	K
T_b = ambient gas temperature	K

For neutral or unstable conditions in the atmosphere (stability category A–D) the following equation can be used to calculate plume rise.

$$\Delta h = \frac{1.6\,F^{1/3}x_f^{2/3}}{u} \quad \text{...(10.3)}$$

where Δh = plume rise in m.

u = wind speed at stack height in ms^{-1}

x_f = distance downwind to point of final plume rise, m.

Use $\quad x_f = 120\ F^{0.4} \quad$ if $F \geq 55\ m^4s^{-3}$

$\quad x_f = 50\ F^{5/8} \quad$ if $F < 55\ m^4s^{-3}$

For stable windy conditions (stability category E and F) use the following

$$\Delta h = 2.4\left(\frac{F}{uS}\right)^{1/3} \quad ...(10.4)$$

where S is stability parameter with unit s^{-1} given by

$$S = \frac{g}{T_a}\left(\frac{dT_a}{dZ} + \Gamma\right) \quad ...(10.5)$$

where Γ = adiabatic dry air lapse rate = 0.01° C m^{-1}. The derivative $\frac{dT_a}{dZ}$ represent the actual rate of change of ambient temperature with altitude.

Example 10.3

A 750 MW coal fired plant has a 250 m stack with inside radius 4 m. The exit velocity of stack gases is estimated at 15 m s^{-1} at a temperature of 413 K and ambient temperature 298K and wind speed at stack height are estimated to be 5 m s^{-1}. Estimate the effective height of stack if (*a*) atmosphere is stable with temperature increasing at the rate of 2°C/km (*b*) the atmosphere is slightly unstable, Class C.

Solution

First find the buoyancy parameter F by using Eq. 10.2

$$F = gr^2 v_s\left(1 - \frac{T_a}{T_s}\right)$$

$$= 9.81 \times (4)^2 \times 15 \times \left(1 - \frac{298}{413}\right) = 655 \text{ m}^4\text{s}^{-3}$$

(*a*) With stable atmosphere by using Eq. 10.5

$$S = \frac{g}{T_a}\left(\frac{dT_a}{dZ} + \Gamma\right)$$

$$= \frac{9.81}{298}(0.002 + 0.01) = 0.0004 \text{ s}^{-2}$$

By Eq. 10.4 $\quad \Delta h = 2.4\left(\frac{F}{uS}\right)^{1/3}$

$$= 2.4\left(\frac{655}{5 \times 0.0004}\right)^{1/3} = 165$$

so effective height of stack = 250 + 165 = 415 m.

(*b*) With an unstable atmosphere, Class C we use Eq. 10.2. Since $F > 55$ m^4/s^3 the distance down wind to the point of final plume rise that should be used is

$$x_f = 120F^{0.4} = 120 \times (655)^{0.4} = 1600 \text{ m}$$

$$\Delta h = 1.6 F^{1/3} x_f^{2/3} / u = 1.6(655)^{1/3} (1600)^{2/3} / 5 = 380 \text{ m}$$

and the effective stack height = H = 250 + 380 = 630 m.

10.4 DOWNWIND CONCENTRATION UNDER A TEMPERATURE INVERSION

The Gaussian plume equation, as presented thus far, applies to an atmosphere where the temperature profile is a simple straight line. If as is often the case there is an inversion above the effective stack height, then Gaussian equation has to be modified to account for the fact that the vertical dispersion of pollutants is limited by the inversion.

If the pollutants are assumed to reflect off the inversion layer, just as they were assumed to reflect off the ground in the basic Gaussian equation, then an estimate at any point downwind require an analysis of these multiple reflections. That complexity can be avoided if we are willing to restrict our predictions of plume concentration to distances far enough downwind that the summation of these multiple reflections converges into a closed form solution. Beyond that distance, the air is considered to be completely mixed under the inversion, with uniform concentrations from ground level to the bottom of the inversion layer.

Turner[4] suggested the following modified Gaussian equation to estimate concentrations down wind under an inversion. It is based on the assumption that the downwind distance from the source is at least twice the distance to where the plume first interacts with the inversion layer

$$C(x,y) = \frac{Q}{(2\pi)^{1/2} u\sigma_y L}\left(\exp\frac{-y^2}{2\sigma_y^2}\right) \text{ for } x \geq 2X_L \qquad ...(10.6)$$

where L = elevation of the bottom of the inversions layer.

X_L = the distance downwind where the plume first encounters the inversion layer.

The distance X_L occurs at the point where the vertical dispersion coefficient, σ_z is equal to

$$\sigma_z = 0.47(\mathrm{L} - \mathrm{H}) \qquad \text{at}(x = X_L) \qquad ...(10.7)$$

X_L can be estimated by using Fig. 10.4b. For $x \leq X_L$ the standard Gaussian plume equation (10.1) may be used to estimate downwind concentration. For concentration between X_L and 2 X_L concentration can be estimated by interpolating between the values for $x = X_L$ and $x = 2X_L$.

Example 10.4

Consider a stack with effective height 100 m emitting SO_2 at the rate of 200 g s^{-1}. Winds at 10 m elevation are blowing at 5.5 m s^{-1} and at the stack height they are 10 m s^{-1}. It is a clear summer day with the sun nearly over head and an inversion layer starting at 300 m. Estimate the ground level SO_2 concentration at a distance downwind twice that where reflection begins to occur from the inversion.

Solution

Equation 10.7 gives σ_z at the distance X_L

$$\sigma_z = 0.47(L - H) = 0.47(300 - 100) = 94 \text{ m.}$$

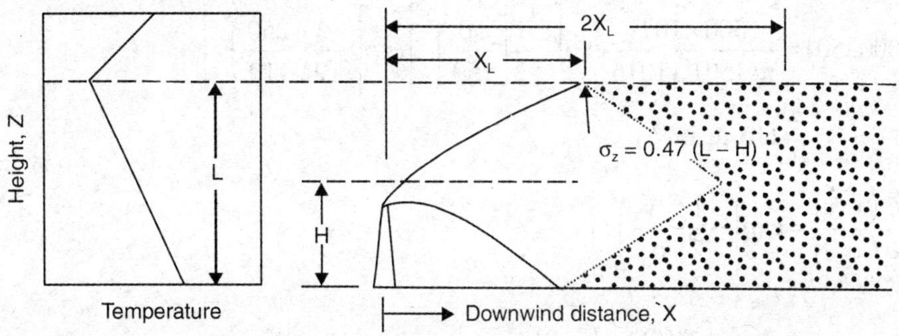

Fig. 10.6 Plume dispersion under an elevated inversion. Equation 10.6 applies for distance greater than 2X down wind, where X_i occurs at the point where σ_i = 0.47 (L – H).

To find X_L we can use Fig. 10.4b. A clear summer day with 5.5 m/s wind speed correspond to stability Class C′ (Table 10.1). Entering Fig. 10.4b with σ_z = 94 m, X_L = 1600 m.

To find the concentration at a distance $x = 2\ X_L = 2 \times 1600 = 3200$ m we can use equation (10.6) but we need to estimate σ_y at that point. Using Fig. 10.4a at x = 3200 and C class, σ_y = 300 m. The concentration directly downwind correspond to y = 0 so

$$C(2X_l, 0) = \frac{Q}{(2\pi)^{1/2} u\sigma_y L} = \frac{200 \text{ g s}^{-1}}{(2\pi)^{1/2} 10 \text{ m s}^{-1} 300 \text{ m } 300}$$

$$= 89 \times 10^{-6} \text{ g m}^{-3} = 89\ \mu\text{g m}^{-3}.$$

Example 10.5

A chimney with a design stack height of 250 m is emitting SO_2 at a rate of 500 g s^{-1} on a sunny day in June with a moderate wind speed 6 m s^{-1} at stack altitude. Estimate the concentration of SO_2 down wind for the following situation.

C (1000, 0); C(1000, 50), if C (1000, *y*, 0) is 100 μg what is the value of *y*.

Solution

On a sunny day in June, the solar radiation will be strong and air very unstable. Table 8.1 show that p for this case be 0.25. Therefore,

$$u_1 = u_2\left(\frac{Z_1}{Z_2}\right)^p = 6\left(\frac{10}{250}\right)^{0.25} = 2.7 \text{ m s}^{-1}$$

Table 10.1 show that this speed corresponds B class stability. The values of σ_y and σ_z can be read from Fig. 10.4a and b. For a distance 1000 m. σ_y = 150 m, σ_z = 110 m with Q = 500 g s, u = 6 m s^{-1}, H = 250 m.

$$(a)\quad C(1000,0) = \frac{500 \times 10^6}{\pi(150)(110)6} \exp\left[-\frac{1}{2}\frac{(250)^2}{(110)^2}\right] = 122\ \mu\text{g m}^{-3}.$$

$$(b)\ \ C(1000,\ 50)=\frac{500\times10^6}{\pi(150)(110)6}\exp\left[-\frac{1}{2}\left[\frac{50}{150}\right]^2\right]\exp\left[-\frac{1}{2}\left(\frac{250}{110}\right)^2\right]$$

$$=115\ \mu\text{g m}^{-3}$$

$$\frac{C(x,y)}{C(x,0)}=\exp\left[-\frac{1}{2}\left(\frac{y}{\sigma_y}\right)^2\right]$$

or
$$y=2^{1/2}\ln\left[\frac{C(SO_2)(x,0)}{C(SO_2)(x,y)}\right]^{1/2}\sigma_y$$

$$=\left[2\ln\left(\frac{122}{100}\right)\right]^{1/2}(150)=94.6\approx 95\text{ m.}$$

Example 10.6

If in above problem the stack diameter is 5 m exit velocity is 13.5 m s^{-1} and gas temperature at exit is 145 °C what is the plume rise for an ambient air of 30°C. Calculate the ground level concentration on the plume centre line at the downwind distance of 1 km. The wind speed is 6 m s^{-1}.

Solution

This is a case of unstable or neutral atom.

$$F=gr^2v\left(1-\frac{T_a}{T_s}\right) \qquad \text{...(10.2)}$$

$$=9.8\times(2.5)^2\,13.5\left(1-\frac{303}{418}\right)=227\text{ m}^4/\text{s}^2$$

$$\Delta h=\frac{1.6F^{1/3}x_f^{2/3}}{u}$$

Since $\quad F\geq55\quad x_f=120\ F^{0.4}=120(227)^{0.4}=1050.9$

$$\Delta h=\frac{1.6F^{1/3}x_f^{2/3}}{u}$$

$$=\frac{1.6(227)^{1/3}(1050.9)^{2/3}}{6}=166\text{ m}$$

Effective height of stack = 250 + 166 = 416 m.

Using the value of σ_y = 150, σ_z = 110 and H = 416

C (1000, 0) can be calculate

$$C(1000,0)=\frac{Q}{\pi\sigma_y\sigma_z u}\exp\left[-\frac{1}{2}\frac{H^2}{\sigma_2^2}\right]$$

$$=\frac{500\times10^6}{\pi\times150\times110\times6}\exp\left[-\frac{(416)^2}{2(110)^2}\right]=1.26\mu g\,m^{-3}$$

There is a significant reduction in ground level concentration as compared to the case for a zero plume where

$$C=\frac{500\times10^6}{\pi\times150\times110\times6}\exp\left[-\frac{1}{2}\left(\frac{250}{110}\right)^2\right]=121\mu g\ m^{-3}.$$

10.5 A LINE SOURCE DISPERSION MODEL

In some circumstances it is appropriate to model sources distributed along a line as if they have formed a continuously emitting, infinite line source. Examples of line sources that might be modelled this way include motor vehicles travelling along a straight section of highway, agricultural burning along the edge of a field or a line of industrial sources on the bank of river. For simplicity consider the case of an infinite length source at ground level, with winds blowing perpendicular to the line. Under these circumstances, the ground level concentration of pollutant at distance x from the line source may be described as follow

$$C(x)=\frac{2Q}{\sqrt{2\pi}\,\sigma_z u} \qquad \text{...(10.8)}$$

where Q is an emission rate per unit distance along the line (g/m-s)

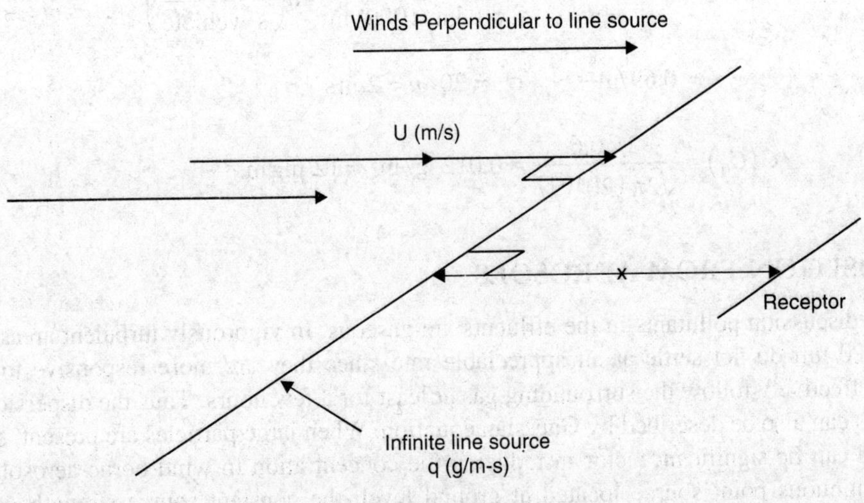

Fig. 10.7 A line source dispersion model.

Example 10.7

Suppose a highway has 10 vehicles passing a given spot per second each emitting 3.4 g/mile of CO. If wind is perpendicular to the highway and blowing at 5 mph (2.2 m s^{-1}) on an overcast day, estimate the ground level CO concentration 200m from free way.

Solution

We need to estimate the CO emission per meter free way

$$Q = 10 \text{ vehicles/s} \times 3.4 \text{ g/veh mile} \times 1 \text{ mile/1609}$$
$$= 0.021 \text{ g/m-s} = 21 \text{ mg/m-s}$$

We need vertical dispersion coefficient σ_z. Since this is an overcast day, the footnotes of Table 10.1 indicate we should use stability classification *D*. Using Fig. 10.4b at a distance 200 m. σ_z is 9 m. Substituting these values into the equation (10.8).

$$C(0.2 \text{ km}) = \frac{2Q}{\sqrt{2\pi}\,\sigma_z u} = \frac{2\times 21}{(2\pi)^{1/3}\; 2.2\times 9} = 0.845 \text{ mg/m}^3$$

The 1 hour CO air quality standard is 40 mg/m^3 so this estimate is well within the air standard.

Example 10.8

The traffic density for a highway is 1200 vehicles and the vehicle speed is 80 km/hr with 40 g/s as the average CO emission per vehicle. Estimate the concentration 500 m downwind of the highway. The wind speed is 2 ms^{-1} and neutral conditions prevail.

Solution

As mentioned in footnote of Table 10.1 atmospheric stability classification correspond to *D*. The graph of Fig. 10.4b show σ_z = 20 for *D* classification and 500 m from source.

$$Q = \frac{1200 \text{ vehicles}}{\text{hr}} \cdot \frac{1}{80} \frac{\text{hr}}{\text{km}} \frac{1}{1000} \frac{\text{km}}{\text{m}} 40\left(\frac{\text{g}}{\text{s}} \frac{1}{\text{vehicle}}\right)$$

$$= 0.69 \text{ /m–s} \qquad \sigma_z = 20 \quad u = 2 \text{ m s}^{-1}$$

$$C(C_0) = \frac{2\times 0.6}{\sqrt{2\pi}(20)(2)} = 0.012 \text{ g/m}^3 = 12 \text{ mg m}^{-3}$$

10.6 DEPOSITION FROM AEROSOLS

In our earlier discussion pollutants in the effluents are gaseous. In vigorously turbulent areas, particles of size upto 20 μm do not settle at an appreciable rate since they are more responsive to the eddy motions and effectively follow the surrounding gas at least for a few hours. Thus the dispersion of such small particles can also be described by Gaussian equation. When large particles are present, deposition on the ground can be significant factor in reducing the concentration in wind borne aerosols.

For a continuous point source located at ground level, the constant source strength *Q* given in Eq. 10.1 is replaced by *Q* (*x*), the effective source strength at the downwind distance *x*. This can be related to the original source strength Q at the origin by the relation.

$$Q(x) = Q\exp\left[\frac{-4v_t\, x}{\sqrt{2\pi}\, u\sigma_z\, n'}\right] \quad ...(10.9)$$

so that

$$C(x, y, z) = \left[\frac{Q}{\pi u\sigma_x\sigma_z}\right]\exp\left[\frac{-4v_t\, x(p+1)}{\sqrt{2\pi}\, u\sigma_z}\right]\exp\left[-\frac{1}{2}\left[\frac{y^2}{\sigma_y^2}+\frac{z^2}{\sigma_z^2}\right]\right] \quad ...(10.9a)$$

Example 10.9

A ground level source is emitting smoke at a rate of 1 g s^{-1}. The wind speed is 5 m s^{-1} and neutral condition prevails. Calculate the ground level concentration on the axis of smoke cloud at a down wind distance of 100 m. Particle deposition velocity v_t = 0.04 m s^{-1}.

Solution

For neutral conditions atmospheric stability correspond to D (Table 10.1) and p = 0.25 (Table 8.1). From Fig. 10.4a & 10.4b at a distance 100 m $\sigma_y = 8$ m and $\sigma_z = 5$ m. Also $u = 5\ \text{m s}^{-1}$, $v_t = 0.04\ \text{m s}^{-1}$, $Q = 1\ \text{g s}^{-1}$.

Substituting these values in Eq. 10.2

$$C(100, 0, 0) = \frac{1}{\pi\times5\times8\times5}\exp\left[\frac{-2(0.04)100\times1.25}{\sqrt{2\pi}\times5\times5\times0.25}\right]$$

$$= 0.938\ \text{mg m}^3.$$

10.7 AREA SOURCE MODELS

For distributed sources, there are a number of approaches that can be taken to estimate pollutants concentrations. If there are a modest number of point sources, it is reasonable to use the point source Gaussian plume equation for each source to predict its individual contribution. Then, by superposition, find the total concentration at a given location by summing the individual contributions. Multiple use of the Gaussian line source equation is another approach. By dividing an area into a series of parallel strips then treating each strip as a line source, the total concentration of any strip can be estimated.

A much simpler, more intuitive approach can be taken to estimate pollutant concentration over an area such as a city, by using the box model concepts. Consider the airshed over an urban area to be represented by a rectangular box, such as shown in Fig. 10.8 with base dimension X, Y, and height Z. The box is oriented so that wind, with speed u is normal to one side of the box. The height of the box is determined by atmospheric conditions and we could consider it to be just the mixing depth. Emissions per unit area will be represented by q_s (g/m^2-s).

Consider the air blowing into the box on the upwind side to have pollutant concentration C_{in} for simplicity assume that no pollutants is lost from the box along the sides parallel to the wind or from the top. We will assume that the pollutants are rapidly and completely mixed in the box, creating a uniform average concentration C_u. Finally, we will treat the pollutants as if they are conservative; that is, they do not react, decay, or fall out of the air stream. All of these restrictions can be modified in more sophisticated version of a box model.

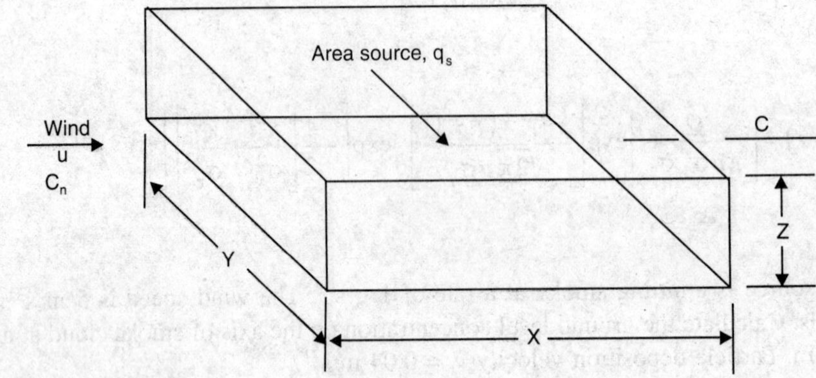

Fig. 10.8 Box model for an airshed over a city. Emissions per unit area are given by q, pollutants are assumed to be uniformly mixed in the box with concentration C, while upwind of the box the concentration is C_{in}.

Working with pollutants mass, the amount of pollution in the box is the volume of the box times the concentrations $XYZC$. The rate at which air is entering and leaving the box is the area of either end times the wind speed YZu, so the rate at which pollution is entering the box is $YZuC_{in}$. The rate at which it leaves the box is $YZuC_u$. If we assume the pollutant is conservative, then we can write the following mass balance for the box.

Rate of change of pollution in the box	=	Rate of pollution entering the box	–	Rate of pollution leaving the box

$$XYZ\frac{dC}{dt} = q_s XY + YZuC_{\text{in}} - YZuC \quad ...(10.10)$$

C_{in} & C = Pollutant concentration is the incoming air and in the airshed respectively

q_s = emission rate per unit area

u = average windspeed against one edge of the box.

The steady state solution to 10.10 can be obtained so that $dC/dt = 0$ and

$$C(\infty) = \frac{q_s X}{uZ} + C_{\text{in}} \quad ...(10.11)$$

If the air entering the box is clean, the steady state concentration is proportional to emission rate and inversely proportional to ventilation coefficient (product mixing depth and wind speed). If it is not clean, we just add the effect of incoming concentration solution to equation 10.10 can be obtained to get time dependent increase in pollutant above city. Let C (0) be the concentration of pollutant in the airshed above the city at time $t = 0$, so that solution becomes

$$C(t) = \left(\frac{q_s X}{uZ} + C_{\text{in}}\right)\left(1 - e^{-ut/X}\right) + C(0)^{-ut/X} \quad ...(10.12)$$

If we assume the incoming wind blows no pollution into the box and if the initial concentration in the box is zero, then Eq. 10.13 simplifies to

$$C(t) = \frac{q_s X}{uZ}\left(1 - e^{-ut/X}\right) \qquad ...(10.13)$$

When $t = X/u$, the exponential function becomes e^{-1} and the concentration reaches 63% of its final value. That value of time has various names. It is called the *time constant*, the *ventilation time* or the *residence time*.

Example 10.10

In a square city of 15 km side, there are 200000 cars on the road each being driven 30 km between 4 pm and 6 pm and each emitting 3 g/km of CO. It is a clear winter evening with a radiation inversion that restricts mixing to estimate 200 m and the wind is bringing clean air in at a steady rate of 1.0 m/s along an edge of the city. Use a box model to estimate the CO concentration at 6 pm if there was no CO in the air at 4 pm and the only source of CO is cars. Assume that CO is conservative and that there is complete and instantaneous mixing in the box.

Solution

The emission per m^2, q_s would be

$$q_s - \frac{200000 \times 300 \text{ km/car} \times 3 \text{ g/km}}{\left(15 \times 10^3\right)^2 \times 3600 \text{ s}/hv \times 2 \text{ hr}} = 1.1 \times 10^{-5} \text{ g m}^2\text{-s}$$

The concentration after 2 hours (7200 s) would be

$$C(t) = \frac{q_s Y}{uZ}\left(1 - e^{-ut/Y}\right)$$

$$= \frac{1.1 \times 10^{-5} \text{ g/m}^2\text{-s} \times 15 \times 10^3 \text{ m}}{1.0 \text{ m/s} \times 20.0 \text{ m}}\left(1 - \exp\frac{1.0 \text{m/s } 7200 \text{ s}}{15000 \text{ m}}\right)$$

$$= 3.2 \times 10^{-3} \text{ g/m}^3 = 3.2 \text{ mg/m}^3.$$

10.8 EMISSION CONTROLS

Since most air pollutants are produced during combustion, one of the most important way to reduce emission is to simply reduce the consumption of fossil fuels. There are two broad approaches that can reduce the fuel consumption.

1. Increase energy efficiency
2. Substitute less polluting energy sources for fossil fuels.

Energy efficiency can be increased by having more efficient appliances and automobiles, improve lighting system in building, efficient manufacturing process, better insulated homes and offices can reduce the energy required to run our society. The second approach is to decrease our reliance on fossil fuels by increasing power production from solar, wind, hydroelectric, geothermal sources to the extent that fossil fuels continue to be used, there are three general approaches that can reduce emission.

1. Reduce the emission potential of fuel itself. *Examples:* Include switching to fuels with less sulphur or nitrogen content in power plants, coal fired plants fuel can be chemically created to remove some of the sulphur or nitrogen before combustion, replacement of coal by fossil gas like propane produced in large volumes otherwise.
2. Reduce emission by improving the combustion process itself. *Examples:* Use burners in power plants that reduce NO_x emissions, new fluidized bed boilers that reduce both NO_x and SO_x, modification in internal combustion engine to decrease.
3. Capture the emission after they have been formed but before they are released. On power plants should have precipitation to remove particulate matter and flue gas desulphurization techniques, used after combustion but before the exhaust stack. On automobiles there should be a catalytic converter put on the exhaust pipe itself.

Low NO_x burners employ a staged combustion process that delays mixing the fuel and air in the boiler. In the first stage of combustion, the fuel starts burning in an air starved environment, causing the fuel bound nitrogen to be released as nitrogen gas rather than NO_x. The following stage introduces more air to allow complete combustion of the fuel to take place. Potential NO_x reduction of 45-60 per cent seem likely.

Flue-gas Desulphurization Technologies

In most wet scrubbers, finely pulverized limestone is mixed with water to create a slurry that is sprayed into the flue gas. When calcium sulphate dihydrate or a calcium sulphate ppt.

The precipitate is removed from the scrubber as a sludge.

$$CaCO_3 + SO_2 + \frac{1}{2}O_2 + 2H_2O \longrightarrow CaSO_4\,2H_2O + CO_2 \qquad \text{90\% efficient}$$

Wet scrubber sometimes use lime when following reaction take place

$$CaO + SO_2 + \frac{1}{2}O_2 + 2H_2O \longrightarrow CaSO_4\,2H_2O \qquad \text{95\% efficient}$$

Particulate matters are removed by cyclones, baghouses or electric precipitators as will be discussed in Chapter 11.

REFERENCES

1. Gul Briggs, G.A., Discussion of Chimney Plumes in Neutral and Stable Surroundings, *Atmospheric Environment*, Vol. 6, p. 507 (1972).
2. Wark, K. and C.F. Warner, *Air Pollution, its Origin and Control*, Harper and Row, New York (1981).
3. Williamson, S.J., *Fundamentals of Air Pollution,* Addison Wesley, Reading MA (1973).
4. Turner, D.B., *Workbook of Atmospheric Dispersion Estimates*, US Environmental Protection Agency, Washington DC (1970).
5. Flagan, R.C. and Seinfeld, J.H., *Fundamentals of Air Pollution Engineering*, Prentice Hall, Englewood Cliffs, NJ (1988).
6. Peterson, W.B., *User's Guide for PAL–A Gaussian Plume Algorithm for Point, Area and Line Sources*, US Environmental Protection Agency, Research Triangle Park, NC (1978).
7. Rao, C.S., *Environmental Pollution Control Engineering*, New Age International (P) Ltd., New Delhi (1990).
8. Pasquill, F., The Estimation of the Dispersion of Wind Born Material, Meteorological Magazine, 90:1063 (1961).
9. Gifford, F.A. Uses of Routine Meteorological Observations for Estimating Atmospheric Dispersion, Nuclear Safety, 2 (4) (1961).

11 Air Pollution Control

11.1 INTRODUCTION

In order to control air pollution, the polluted gas must pass through a device or system, which may collect or destroy the pollutant and release clean gas to the atmosphere. The device selected must be specific for the concerned pollutant. If the pollutant is an aerosol, the device used will, in most cases, be different from the one used for a gaseous pollutant. If the aerosol is a dry solid, a different device has to be used than for liquid droplets.

Once the control is in place, its operation and maintenance become a major concern. Important reasons for an operation and maintenance programme are : (*i*) the necessity of continuously meeting emission regulations; (*ii*) prolonged control equipment life; (*iii*) maintaining productivity of the process served by the control device; (*iv*) reducing operation cost; (*v*) promoting better public relations and avoiding community alienation, and (*vi*) promoting better relations with regulatory officials.

The operation and maintenance program has the following minimum requirements:

(*i*) An equipment and record system with equipment information, warranties, instruction manuals, etc.
(*ii*) Lubrication and cleaning schedules.
(*iii*) Planning and scheduling of preventive maintenance.
(*iv*) A storeroom and inventory system for spare parts and supplies.
(*v*) Listing of maintenance personnel.
(*vi*) Costs and budget for operation and maintenance.
(*vii*) Storage of special tools and equipment.

11.2 REMOVAL OF DRY PARTICULATE MATTER

Dry aerosol, or particulate matter, differ physically, chemically and electrically from carrying gas stream so that their removal should present no major difficulties. Particulate matter can be removed by a filter, wet scrubber, or electrostatic precipitator with equally satisfactory results.

11.2a Filters

A filter removes particulate matter from the carrying gas stream because the particulate impinges upon and then adheres to the filter material. As time passes, the deposit of particulate matter becomes greater and the deposit then acts as a filtering medium. When the deposit becomes so heavy that the pressure necessary to force the gas through the filter becomes excessive, or the flow reduction severely impairs the process, the filter must either be replaced or cleaned.

The filter medium can be fibrous, such as cloth; granular, such as sand; a rigid solid, such as a screen; or a mat, such as a felt pad. It can be in the shape of a tube, sheet, bed, fluidized bed, or any other desired form. The material can be natural or man made fibers, granules, cloth, felt, paper metal, ceramic, glass or plastic. In all cases particulate matter will be removed from the gas stream only if it adheres to the filter surface. Added to the clean filter is the resistance that occurs when the adhering particles form a cake on the filter surface. This cake increases thickness as approximately a linear function of time, and the pressure difference necessary to cause the same gas flow with time also becomes a linear function. Usually, the pressure available at the filter is limited so that as the cake builds up, the flow decreases. Filter cleaning can be based; therefore on: (*i*) increased pressure drop across the filter; (*ii*) decreased volume of gas flow, and (*iii*) time elapsed since the last cleaning.

The industrial filters called bag houses or filter bags are fabricated from cloth, with material and weave selected to fit the specific application. Cotton and synthetic fabrics are used for relatively low temperatures, with glass cloth fabrics being used for elevated temperature, upto 290°C.

The permeability of filter varies with the tightness of the cloth and is determined by passing air through the cloth at a 1.27 cm water pressure drop. Permeability from 3 to 10 $m^3\ min^{-1}\ m^{-2}$ cloth are typical expressed as filter velocity for air to cloth ratio, A/C which is 5 to 17 $cm\ s^{-1}$ or 5 : 1 to 17 : 1 for these clean cloth permeabilities. The entire filter unit is usually designed in such a manner that filtering can still take place in part of the bag house while cleaning is taking place in another part. The various types of baghouses categorised by the type of cleaning method are as follows:

11.2b Reverse Pulse

These units have a short but strong pulse of air of about 100 ms duration. The cleaning air pulses pass through a venturi tube at the top of the bag and into the bag to expand it, as shown in Fig. 11.1. The filter cakes which was built up on the outside of the bag falls into the hopper and is removed. As filtering occur on the outside, each bag is supported by an internal cage. Advantage of pulse jets are longer life and smaller size. Bag service is extended by the low-bag movement during cleaning which causes little wear and tear. Unit size can be smaller because of higher filtering velocities, closer bag placement and no out of service requirement for cleaning. Pulse jets can operate at 3–4 times filtration velocity. Bag movement during cleaning is so small that bags can be placed closer to one another which reduces the size of the unit. The unit can continue to operate during cleaning as the cleaning pulse is only momentary during which no additional filter capacity is required for a continuous operation.

11.2c Reverse Flow

The reverse flow bag houses are cleaned by forcing the air backwards through the bags in a direction opposite to that of normal filtering. Units that use bag filter on the inside of the woven bags, reverse flow air makes the bag collapse inward. When the bag is brought back on line, it expands and the dislodged cake falls into the hopper through the open end at the bottom of the bag. By contrast, each felt envelop filter element is supported by an internal rigid wire mesh frame and separator and filtering occurs on the outside. This is shown in Fig. 11.2. These units operate at 1–3 cms^{-1} velocities. Compartments are isolated periodically and clean air is blown into the open front end of each envelop. It requires about 0.1 to 1 m^3 of air per m^2 of filter area for cleaning.

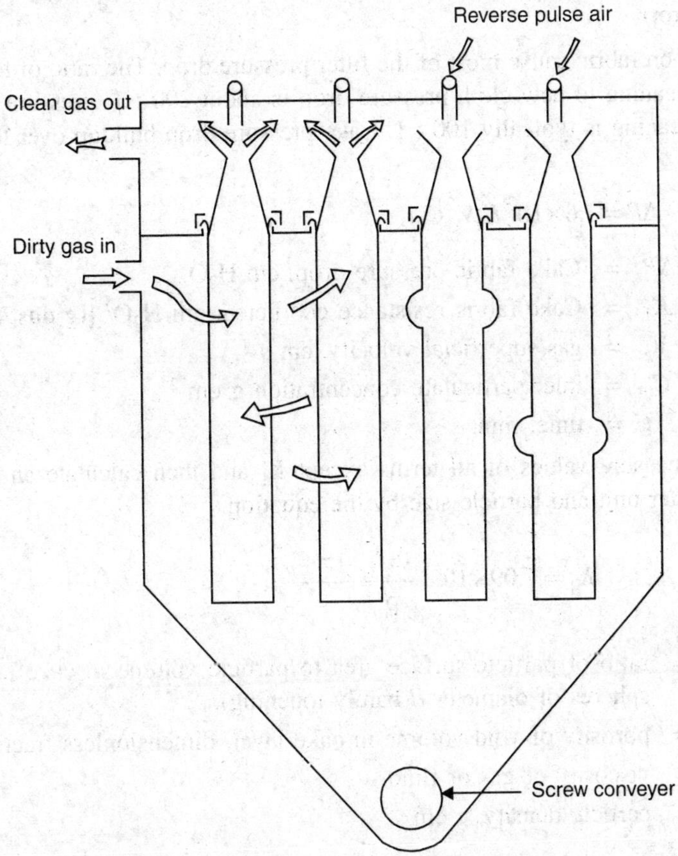

Fig. 11.1 Reverse pulse baghouse.

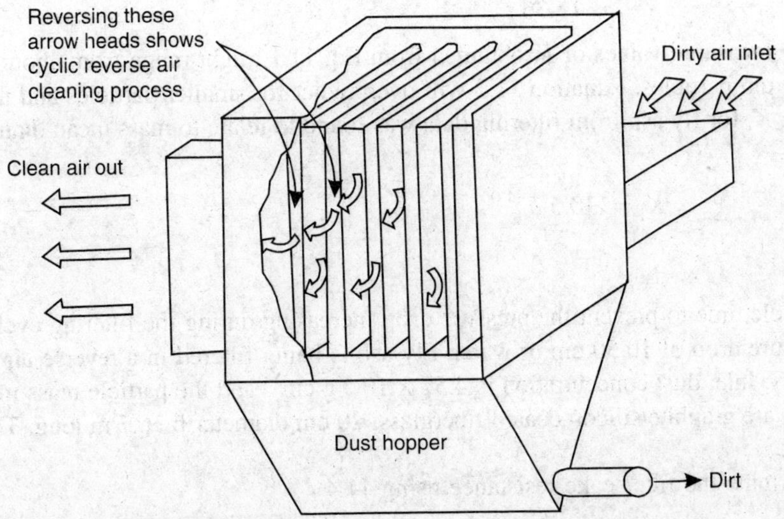

Fig. 11.2 Reverse air unit with felt envelope filters.

11.2d Pressure Drop

Deposited particles on fabric cause most of the filter pressure drop. The ratio of terminal pressure drop of a filter before cleaning to new cloth pressure drop is about 200 : 1, pressure drop ratio for before cleaning to after cleaning is typically 100 : 1. Cake pressure drop buildup over the time interval t can be estimated by

$$\Delta P = 3.6\times10^3 K_1 V_g^2 C_0 t \qquad ...(11.1)$$

where
ΔP = Cake fabric pressure drop, cm H_2O
K_1 = Cake fabric resistance coefficient cm H_2O/ [(g dust/cm^2)(cm/min)].
V_g = gas superficial velocity, cm s^{-1}.
C_0 = inlet particulate concentration g cm^{-2}
t = time, min.

It is better to measure values of all terms except K_1 and then calculate an empirical K_1 for the specific dust and filter unit and particle size by the equation.

$$K_1 = 7.09\times10^{-5}\,\frac{\eta_g S^2}{\rho_p}\,\frac{1-\varepsilon_p}{\varepsilon_p^3} \qquad ...(11.2)$$

where
S = ratio of particle surface area to particle volume in cake layer cm^{-1} ($\cong 6/d$ for spheres of diameter d barely touching).
ε_p = porosity of void volume in cake layer, dimensionless fraction.
η_g = viscosity of gas or fluid.
ρ_p = particle density, g cm^{-3}.

Values for porosity can be estimated for spherical particles of diameter d in micron

$$\varepsilon_p = \frac{\ln d + 614}{15.36} \qquad ...(11.3)$$

Experimentally measured values of K_1 obtained from Eq. 11.1 might range from about 122 for 0.1 μm to 0.03 for 100 μm particles. Equation 11.2 will overpredict for smaller particles and under predict for larger particles. K_1 for fly ash from filtering data was found to relate to mass mean diameter in microns by

$$K_1 = 110.5\, d^{-1.06} \qquad ...(11.4)$$

Example 11.1

Estimate the cycle time to prevent the pressure drop increasing during the filtering cycle from exceedingly low pressure drop of 10.50 cm of water. Fly ash is being filtered in a reverse air filter at 1.4 cm s^{-1} filter velocity. Inlet dust concentration is 4.57×10^{-6} g cm^{-3} and the particle mass mean diameter is 8 μm. The bags are graphite/silicon coated fiberglass, 20 cm diameter by 6.7 m long. The coal has low sulphur content.

Solution: Determine the filter cake resistance using 11.4

$$K_1 = 110.5(8)^{-1.06} = 12.19 \text{ cm } H_2O\left[\left(\text{g dust/cm}^2\right)(\text{cm/min})\right]$$

The cycle time from equation 11.1 is

$$t = \frac{10.5}{(3.6\times10^{3})(12.19)(1.4)^{2}(4.57\times10^{-6})} = 26.71 \text{ min}$$

Most of these systems operate with cycle times of 30 min.

Typical A/C ratios vary with both fabric construction and type of fabric. In general, woven fabrics operate at lower velocities (< 3 cm/s) to prevent driving the dust through the light material. Felts operate at higher velocities because the material is thicker and the dust does not penetrate as easily. Felts provide a greater filtration depth. Recommended A/C ratio filtration velocities may be:

Glass-woven	1 : 1 cm^3/s per cm^2 (1 cm s^{-1})
Nomex-woven	1.5 : 1 cm^3/s per cm^2 (1.5 cm s^{-1})
Nomex-felt	4 : 1 cm^3/s per cm^2 (4 cm s^{-1}).

11.3 ELECTROSTATIC PRECIPITATORS

High voltage electrostatic precipitator (ESP) have been widely used throughout the world for particulate removal. Advantage of ESP are as follow

(*i*) It can easily handle high-temperature gases, which makes it a likely choice for boilers, steel furnaces etc.

(*ii*) It has an extremely small pressure drop, therefore fan costs are minimized.

(*iii*) It has an extremely high collection efficiency. It can handle a wide range of particulate sizes and dust concentrations.

(*iv*) Its operating and maintenance cost is very low. Three disadvantages of ESP are (*i*) Its initial cost is very high (*ii*) a large amount of space is required for installation (*iii*) ESP are not suitable for combustible particles.

Particles in air may have a positive, negative or no charge because air is a poor conductor of electricity. Introduction of a high energy electrical field in a gas result in the formation of large concentration of charged ions. This charge is given to particle in electrically active region (the corona region) around a negative discharge electrode. Electrons are stripped from neutral gas molecules in the corona region resulting in an avalanche of electrons moving toward the collector electrode as shown Fig. 11.3.

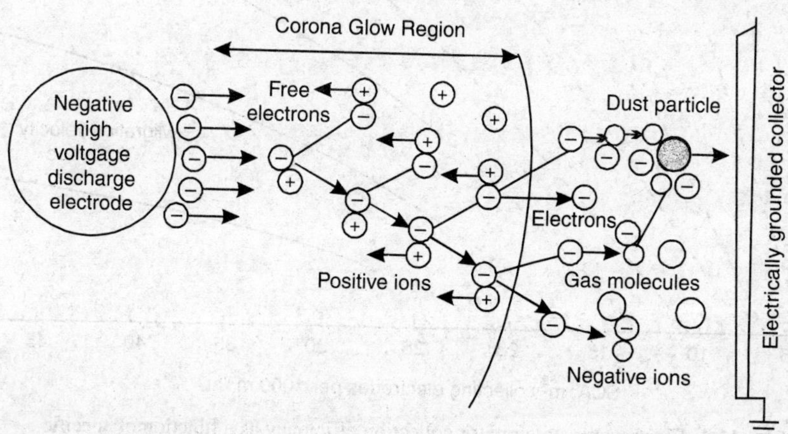

Fig. 11.3 Particle charging process by corona discharge.

The positive ions return to the discharge electrode. Electron moving toward the collector form negative ions from the gas molecules, G. The ions come in contact with dust particles to give them large charges. For dry air this voltage is 10000 volt/cm.

In an electrical field, a charged particle moves toward appropriate electrode with a velocity, V_e

$$V_e = \eta_p \frac{E\,1.7\times10^{-24}}{\eta_g d_p}\left(1+\frac{6.21\times10^{-4}T}{d_p}\right) \quad \text{...(11.5)}$$

where η_p = number of electron charges/particle (from Fig. 11.4)

E = Field strength in volt cm^{-1}.

η_g = viscosity of gas in g cm^{-1} s.

d_p = particle diameter in m and T is temperature in K.

ESPs are usually large and because of this require a careful design. ESP collection efficiency is based on test data obtained for each dust-air system. The overall ESP collection efficiency is given by ε_0 by the equation

$$\varepsilon_0 = 1 - \exp\left[-w\left(\frac{36A}{Q}\right)n\right] \quad \text{...(11.6)}$$

In this form, w represents the corrected migration velocity in cm/s and its value is obtained by test data. The ratio A/Q is collecting electrode area in m^2 over gas flow rate in $m^3\ h^{-1}$ and is known as

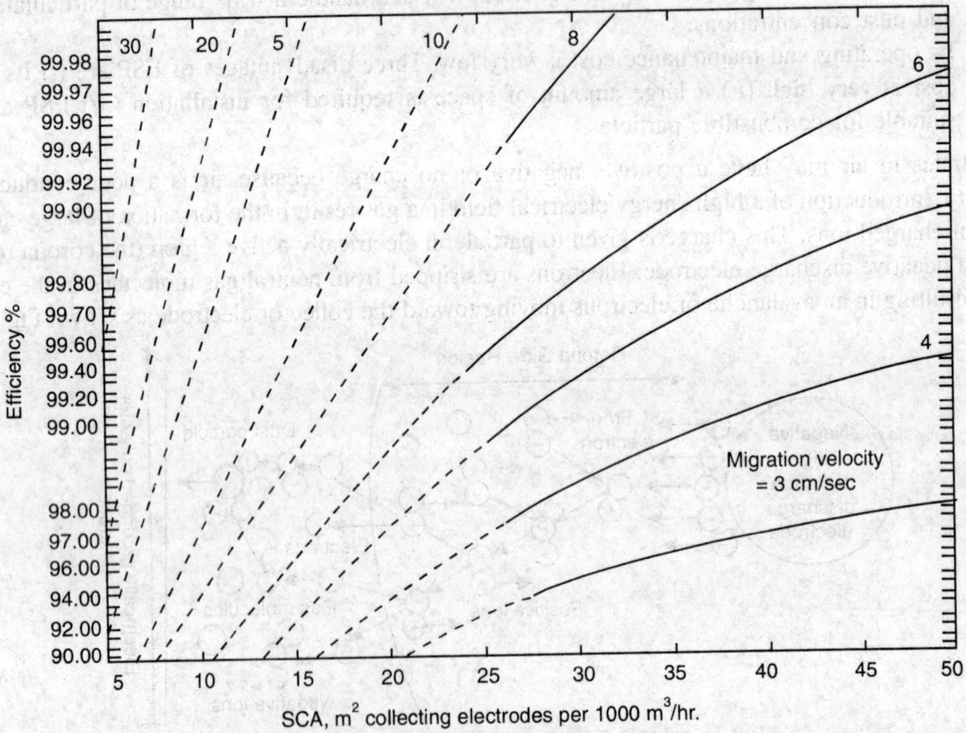

Fig. 11.4 Electrostatic precipitator collection efficiency as a function of specific collecting area and migration velocity.

specific collecting area, SCA : The exponent n vary from about 0.5 to 1 depending on the value of SCA. Units are dimensionalised by the value 36. Equation 11.6 can be written as

$$w = \frac{\ln(1-\varepsilon_0)}{(36A/Q)n} \quad ...(11.7)$$

This value of w is a corrected average to account for dust size, size distribution, gas and dust unevenness within a particular system, charging effectiveness reentrainment, dust resistivity, etc. Values of n close to unity should be used for SCA less than 25 m^2 collecting electrode area per 1000 m^2 h^{-1} gas flow and close to 0.5 above this. This is shown in Fig. 11.4 where the dashed portion of lines is for $n = 1$ and solid portion is for $n = 0.5$, for all values of SCA. Using Fig. 11.4 or Equation 11.6 collection efficiency at various SCA can be estimated. High efficiency ESP is for fly ash usually require SCA of about 25 m^2 per 1000 m^3 per hr indicating that migration velocity is relatively low (*i.e.*, 3 cm s^{-1}) with efficiency 93%. Doubling the SCA from 25 to 50 increases efficiency to 97.5% an increase by 4.5%. On the other hand, if SCA is kept 25, but migration velocity is increased from 3 to 4 cm s^{-1}, efficiency increases from 93 to 96.5%. This raises efficiency w from 3 to 4 or to 6. Some typical values used for fly ash precipitator are summarised in Table 11.1.

Table 11.1 Typical design parameters for fly ash precipitator

Parameter	*Range*
Gas Velocity	75-200 cm s^{-1},
Migration Velocity	1.4-15 cm s^{-1},
Specific Collecting Area	5-45 m^2 per 1000 m^3 h^{-1}
Corona Power	30-300 W per 100,000 m^3 h^{-1}
Corona Current	5-75 nano amp per cm^2
Plate Area	500-7500 m^2 per electrical set
Number of Flight Tension Section in Gas Flow Direction	2-8
Degree of Sectionalisation	0.23-2.35 high tension bus sections per 100000 m^3 h^{-1}

11.4 CYCLONE SEPARATORS

Cyclones obtain their name by similarity to natural storm. They can be highly efficient but usually are of medium-low efficiency, low pressure drop and therefore low operating cost precleaning devices. Their are numerous variations of design ranging from complex arrangements of multiple small cyclones in a common hopper to single large units of various configuration.

In a dry cyclone, dusty inlet gases are forced to swirl around inside the cylindrical body producing a centrifugal force that moves the particles to the wall. This action is depicted in Fig. 11.5. The particles, whether liquid or solid, move out of the small opening at the end of the cone and the cleaned gases leave the other end. The inlet can be either at the top or the bottom, as shown in Fig. 11.5. Liquid sprays could be introduced into any of these design.

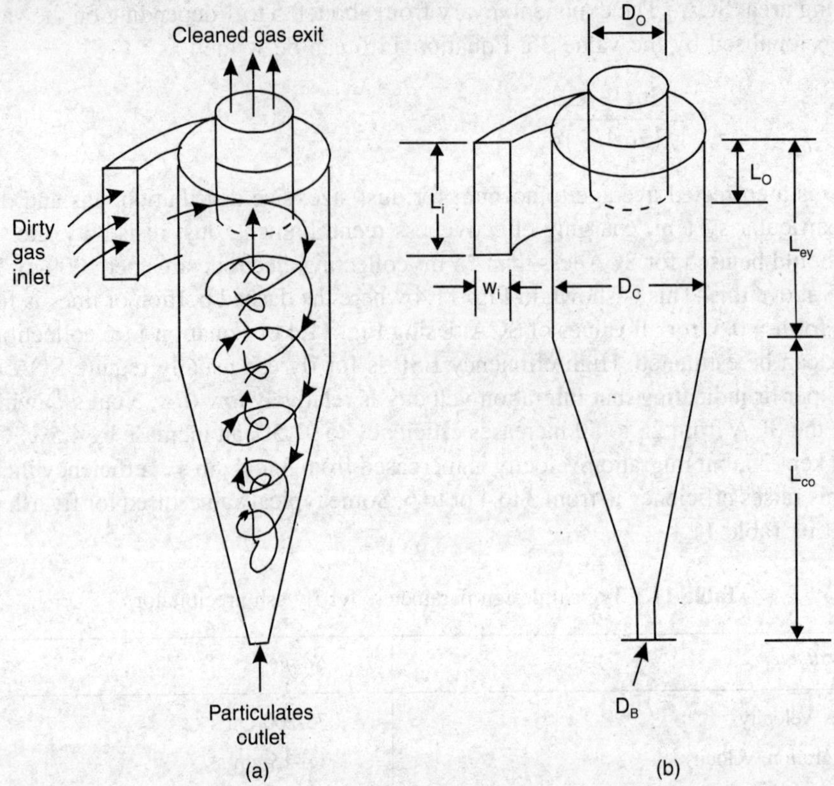

Fig. 11.5 (*a*) Conceptualized cyclone separator action (*b*) Cyclone design configuration.

Table 11.2 Dimensionless design ratios for tangential entry cyclones

Symbols[a]	Nomenclature	*High Efficiency*		*Conventional*	
		Stairmand	*Swift*	*Lapple*	*Swift*
D_c	Body Diameter	1.0	1.0	1.0	1.0
L_i	Inlet Length	0.5	0.44	0.5	0.5
W_i	Inlet Width	0.2	0.21	0.25	0.25
L_o	Outlet Length	0.5	0.5	0.625	0.6
D_o	Outlet Diameter	0.5	0.4	0.5	0.5
L_{ey}	Cylinder Length	1.5	1.4	2.0	1.75
L_{co}	Cone Length	2.5	2.5	2.0	2.0
H_o	Overall Height	4.0	3.9	4.0	3.75
D_B	Bottom Diameter	0.375	0.4	0.25	0.4
l	Natural Length	2.45	2.04	2.30	2.30
G	Configuration Factor[b]	551.3	669.2	402.9	381.8

(*a*) See Figure 11.5b.

(*b*) Dimensionless constants, not ratios.

There are several basic design dimensions that have been standardised by those who study cyclone collection efficiency. These are given in Table 11.2 as dimensionless ratios based on a cyclone body diameter, D_c of 1.0. These high efficiency design configurations are designated by Stairmand (8) and Swift (9) and the conventional designs are by Lapple (5) and Swift (9). Figure 11.5b locates the symbols and depicts a high efficiency unit. Volumetric flow rates are roughly approximated as $175\ D_c^2$, $220\ D_c^2$ and $500\ D_c^2 \text{m}^3\text{h}^{-1}$ for high efficiency, conventional and low efficiency designs.

Usual industrial capacities range from 200–85,000 $m^3\ h^{-1}$ per cyclone. Inlet velocities range for high efficiency to low efficiency lie between from 1500 to 2700 cm s^{-1}. Pressure drops of only 15 cm H_2O maximum are typical for high efficiency units and pressure drops of about 5 cm H_2O are normal. The last two entries in Table 11.2, natural length and configuration factor are not defined in Fig. 11.5b. Natural length, l of a cyclone is considered to be the length required for a complete gas vortex configuration given by

$$l = 23\ D_o \left(D_c^2 / L_i W_i\right)^{1/3} \quad \text{...(11.8)}$$

and should be less than the overall cyclone height minus the gas outlet length, L_o to maintain the full gas vortex. Configuration factor, G is a dimensionless ratio dependent upon design configuration.

11.4a Pressure Drop

Cyclone Pressure drop, ΔP, is dependent mainly on the velocity and can be estimated by

$$\Delta P = 8.19 \times 10^{-3}\ \rho_g V_i^2 \left(\frac{L_i W_i}{D_0^2}\right)\left(\frac{1}{0.0057 C_i + 1}\right) \quad \text{...(11.9)}$$

where ΔP is in cm H_2O, V_i = inlet velocity, cm s^{-1} and C_i = inlet dust load gm^{-3}.

The fractional collection efficiency, ε_i, for specific size particles as

$$\varepsilon_i = 1 - \exp\left\{-2.0\left[\frac{G \tau_i Q(1+\eta_v)}{D_c^3}\right]^{\frac{0.5}{1+\eta_v}}\right\} \quad \text{...(11.10)}$$

where τ_i = relaxation time for the specific size particle given by

$$\tau = \frac{d^2 \rho_p \times 10^4}{18 \eta_g} = \frac{m_p C}{3\pi \eta_g d} \quad \text{...(11.11)}$$

G = cyclone configuration factor, dimensionless (see Table 11.2), Q = gas flowrate, $m^3\ s^{-1}$, η_v = vortex exponent, m = mass of particle, d = diameter of particle, d_p diameter of particle in μm, ρ_g = density of gas or fluid, η_g = viscosity of gas or fluid, t = temperature in °C. The vortex exponent η_v and the Cunningham correction factor, C, are

$$\eta_v = 1 - \left(1 - 0.669 D_c^{0.14}\right)\left(\frac{t + 273}{293}\right)^{0.3} \quad \text{...(11.12)}$$

$$C = 1 + \left(6.21 \times 10^{-4}\right) T / d_p \quad \text{...(11.13)}$$

Overall cyclone particle collection efficiency, ε_o may be estimated for a specific size, size distribution and density of dust, if a sufficient number of size groups are used by

$$\varepsilon_0 = \Sigma m_i \varepsilon_i \qquad ...(11.14)$$

where m_i mass of each individual size fraction used for calculating ε_i.

11.4b Velocity Ratio

The ratio of inlet velocity V_i, to saltation velocity V_d is an important factor in optimising high efficiency cyclone design. The term saltation velocity is the minimum gas velocity required to prevent the dropping or settling out of particles in cm s^{-1} given by

$$V_d = 0.228\, V_a \left[\frac{(W_i/D_c)^{0.4}}{(1 - W_i/D_c)^{1/3}} D_c^{0.067}\, V_i^{2/3} \right] \qquad ...(11.15)$$

where V_i is inlet velocity in cm s^{-1}, V_a is an approximation of gas path velocity in cm s^{-1} given by

$$V_a = 1.10 \left[g\eta_g \frac{(\rho_p - \rho_g)}{\rho_g^2} \right]^{1/3} \qquad ...(11.16)$$

Particle collection efficiency is maximum in cyclones when the ratio of $(V_i/V_d)_b$ equals 1.25 but reentrainment occurs when $V_i/V_d \geq 1.35$. The bracket and subscript b in the first ratio indicate that this value is specified for base conditions, *i.e.*, t = 38°C and particle density of 2.58 g cm^{-3} so that

$$V_i/V_d = \frac{(V_i/V_d)_b}{f_t\, f_p} \qquad ...(11.17)$$

Correction factors derived from data show that temperature correction factor, f_t is

$$f_t = \frac{6.88}{10.52 - \ln t} \qquad ...(11.18)$$

and

$$f_p = \frac{9.932}{3.11 - \ln \rho_p} \qquad ...(11.19)$$

In summary the design constraints are:

1. V_i with in the range 1500–2700 cm s^{-1}.
2. ΔP < 25 cm water gauge.
3. $(V_i/V_d)_b \cong 1.25$ for optimum collection efficiency
4. V_i/V_d < 1.35 to reduce reentrainment
5. $l \leq H_0 - L_0$ to maintain full gas vortex
6. $G/\Delta P$ is maximised.

Higher collection efficiency may be achieved with smaller parallel cyclones to accommodate the total gas flow and obtain the required collection efficiency.

Example 11.2

Design a cyclone precleaner for maximum particle collection efficiency on a 93,690 A m^3 h^{-1} gas stream at 100°C. Consider the air is release at 600 g s^{-1} of dust with mean diameter of particles as 12 μm, ρ_g is 3.6 × 10^{-4} and ρ_p is 1.5 × 10^{-4} g cm^{-3}. Estimate the pressure drop and overall collection efficiency of the designed unit, assuming that V_i = 2100 cm s^{-1}.

Solution: Use the Stairmand high efficiency design Table 10.2. Under these conditions, optimum inlet to saltation velocity ratio from Equations 11.17, 11.18 and 11.19 is

$$V_i/V_d = \frac{1.25}{\dfrac{6.88}{(10.52 - \ln 100)}\left(\dfrac{9.932}{12.32 - \ln 1.5}\right)} = 1.289$$

and it should not exceed V_i/V_d = 1.35.

Proceed by obtaining quantities that do not depend on design size and velocity at 100°C (Molar mass of air = 29)

$$\rho_g = \left(\frac{29\,g}{g\,\text{mol}}\right)\left(\frac{g\ \text{mol}}{22400\ \text{cm}^3}\right)\frac{293}{373} = 1.017\times10^{-3}\ \text{g cm}^{-3}$$

Approximate gas path velocity from equation 11.16 with η_g of air at 100°C = 1.5 (from tables)

$$V_a = 1.10\left[\frac{(980)(2.10\times10^{-4})(1.5)}{(1.017\times10^{-3})^2}\right]^{1/3} = 73.5\ \text{cm s}^{-1}$$

Arbitrarily chose a medium inlet velocity of 2100 cm s^{-1}. If four cyclones in parallel were used, each would receive 93690 A m^3 h^{-1} ÷ 4 = 23490 A m^3 h^{-1}

or
$$Q = \frac{23490}{60\times60} = 6.525\ \text{A m}^3\text{s}^{-1}$$

size of each inlet would then be

$$\left(\frac{6.525\,\text{Am}^3\text{s}^{-1}}{2100\ \text{cm s}^{-1}}\right)\left(\frac{100\ \text{cm}}{\text{m}}\right) = 0.3107\ \text{m}^2$$

Designing the cyclone based on Stairmand design

$$W_i L_i = (0.2\,D_c)(0.5\,D_c) = 0.3107$$

or
$$D_c = \left(\frac{0.3107}{0.2\times0.5}\right)^{1/2} = 1.763\ \text{m}$$

This is an acceptable starting point. Calculate saltation velocity, V_d using Equation 11.15, W_i and D_L values may be taken from Table 11.2 and

$$V_d = (0.228)(73.5)\frac{(0.2)^{0.4}}{(1-0.2)^{1/3}}(1.763)^{0.067}(2100)^{2/3} = 1{,}615\ \text{cm s}^{-1}$$

Actual ratio V_i/V_d = 2100/1615 = 1.30, which is close to optimum, picking a new inlet velocity of 2000 cm s^{-1} gives a new value of D_c of 1.80, V_d 1566 cm s^{-1} and V_i/V_d = 1.28. Interpolating yields an optimum inlet velocity of 2050 cm s^{-1} and D_c of 1.78 m. A unit with no vanes would have a pressure drop from Eq. 11.9

$$C_i = \left(\frac{600\ \text{g}}{\text{s}}\right)\left(\frac{3600\ \text{s}}{\text{hr}}\right)\left(\frac{\text{hr}}{93690\ \text{m}^3}\right) = 23.05\ \text{g m}^{-3}$$

$$\Delta P = (8.19\times10^{-3})(1.017\times10^{-3})(2050)^2\,\frac{(0.5)(0.2)}{(0.5)^2}\left[\frac{1}{(0.0057)(23.05+1)}\right]$$

$$= 12.4 \text{ cm water.}$$

This pressure drop is acceptable and would be reduced to about 6.2 water with a neutral vane. Checking the final constraint using equation 11.8

$$l = (2.3)\left[(0.5)(1.78)\right]\left[\frac{1}{(0.5)(0.2)}\right]^{1/3} = 4.41 \text{ m}$$

and $H_0 - L_0$ is $(4-0.5)(1.78) = 6.23$ m

so $l < H_0 - L_0$

This can be confirmed by the inspection of Table 11.3 but collection efficiency can not be calculated. The dust size distribution is plotted on log probability paper and arbitrarily divided into 10 size fractions, as shown in Table 11.3. Cyclone particle-collection efficiency changes only slightly for the larger particles but rapidly for the smaller particles so smaller size fraction are used for the smaller sizes.

Relaxation times are calculated from equation 11.11 and included in the Table 11.3, for example

$$\tau_{4.1\ \mu\text{m}} = \frac{(4.1\times10^{-4}\ \text{cm})^2\,(1.5\ \text{g cm}^{-3})}{(18)(2.28\times10^{-4}\ \text{g cm}^{-1}\text{s}^{-1})} = 6.144\times10^{-5}\ \text{s}$$

The vortex exponent from equation 11.12 is

$$\eta_v = 1 - \left[1-(0.669)(1.78)^{0.14}\right]\left[\frac{100+273}{293}\right]^{0.3} = 0.705$$

and

$$\varepsilon_{4.1\ \mu\text{m}} = 1 - \exp\left[-2.0\left[\frac{(551.3)(6.144\times10^{-5})(6.525)(1.705)}{(1.78)^3}\right]^{\frac{0.5}{1.705}}\right]$$

$$= 0.5953$$

Grade fractional efficiencies are calculated for this dust and this cyclone under these conditions as listed in Table 11.3. Overall efficiency by Equation 11.14 using Table 11.3 data is 0.7817 or 78.2% for this dust, this cyclone and under these conditions.

Final optimization would consist of varying the number of cyclones and maximizing the $G/\Delta P$ ratio.

Table 11.3 Calculations summarized for example problem values to determine cyclone grade fractional efficiencies

Dust Cumulative Per cent Oversize	*Grade Average Diameter (μm)*	*Dust Mass Fraction (m)*	*Average Diam Relaxation Time (s)*	*Grade Fractional Efficiencies (ε_j)*
> 75	50	0.25	9.137×10^{-3}	0.9801
55-75	19	0.20	1.39×10^{-3}	0.8915
40-55	11.5	0.15	4.834×10^{-4}	0.8089
30-40	7.8	0.10	2.224×10^{-4}	0.7323
25-30	5.5	0.05	1.106×10^{-4}	0.6583
20-25	5.0	0.05	9.137×10^{-5}	0.6377
15-30	4.1	0.05	6.144×10^{-5}	0.5953
10-15	2.3	0.05	1.933×10^{-5}	0.4747
< 5	1.3	0.05	6.177×10^{-6}	0.3691

The dimension of various parts as shown in Fig. 11.5b in accordance to the requirement of the above problem are given in Table 11.4.

Table 11.4 Size of various parts of cyclone in meters accordance to problem 10.2

	D_c	L_i	W_i	L_o	D_o	L_{cy}	L_{co}	H_o	D_B
Dimensionless Stairmand design ratio	1	0.5	0.2	0.5	0.5	1.5	2.5	4.0	0.375
Actual size	1.75	0.875	0.35	0.875	0.875	2.625	4.375	7.0	0.636

11.5 SCRUBBERS

Scrubbers or wet collectors, wet the particles by spraying liquid over an aerosols; which are subsequently removed from air. On spraying a liquid on aerosol the particles are trapped from supporting gaseous medium to the liquid scrubbing medium. The liquid droplet impinge upon particle surface to make it heavier. The smaller the droplet the greater is the collection efficiency. The wetted particles are collected on a surface followed by their removal from the surface. The collecting surface can be in the form of a bed or simply a wetted surface. Increasing either the gas velocity or the liquid droplet in a scrubber will increase the efficiency because of the greater number of collisions per unit time. The ultimate scrubber is the venturi scrubber which operates at extremely high gas and liquid velocities with a very high pressure drop across the venturi throat.

Spinning disks, cups or other related devices are used to physically fracture liquid by discharging it at high velocities from the periphery of a rapidly rotating device. The disks are usually flat or curved plates and either solid or perforated. Rotating arms are not truly spinning disks, but are a combination of nozzles (drilled holes) in pipe that are mechanically rotated to help break up and distribute the spray. The centrifugal force developed by a spinning disk is sufficient to overcome the surface tension of the liquid flowing across the disk causing it to breakup into small droplet, which are flung from the edges of the disk in a 360° arc.

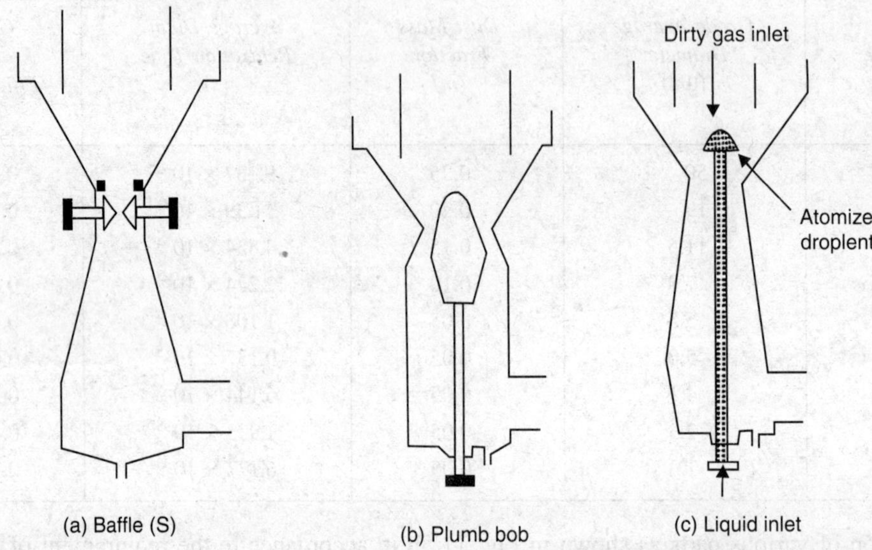

Fig. 11.6 (*a*) Baffle scrubbers (*b*) Plumb bob scrubber (*c*) Throat pneumatic automization for venturi scrubber.

The scrubbers saturate the inlet gases and cool them within a few centimeters from the point of injection of liquid for low temperature gases, but takes a considerable distance for high temperature gases. The water vapours leave with scrubbed gases but the bulk of the scrubbing liquid is entrained as liquid and must be separated from the gas. This mist elimination is also part of the unit. Velocities of the order of 500 cm s^{-1} are necessary for complete atomization, yet velocities of 2500 cm s^{-1} give effective particle removal and high gas absorption. Particle collection efficiency in a scrubber is related to the gas phase pressure drop.

11.5a Chemical Additions

Sodium, magnesium and calcium ions can absorb SO_2 as sulphites and sulphates. This increases SO_2 removal and prevents the liquid phase from becoming a strongly corrosive acid. Particle removal may be improved by use of surfactants. Very small amount of these reduces the surface tension of liquid which make atomization easier, enhances the particle wettability and reduces deposition of solid. Typically nonionic, low-foaming surfactants are best for scrubbers which improves the particle collection efficiency of the unit.

11.5b Other Systems

Many unique systems have been proposed:

(*i*) Ceramic candles, which are thimble shaped, porous acid resistant ceramic tubes. Although efficiency exceeding 98 per cent have been reported, the candles have high maintenance requirements because they are very fragile.

(*ii*) Electric cyclones, which utilises an electrode is the centre of cyclone to establish an electric field within the cyclone body. This device is more efficient than the standard cyclone. It probably is more applicable to mists and droplets than to dry particulate due to possible fire/explosion hazards with compatible dusts.

(*iii*) Sonic agglomerators, which have been used experimentally for sulphuric acid mist and as a mist eliminators. Commercial development is not projected at this time because the energy requirements are considerably greater than those for venturi scrubbers of similar capacity.

11.6 GASEOUS POLLUTANTS AND THEIR REMOVAL

Sulphur dioxide, the oxides of nitrogen and many others are naturally present in the atmosphere in low concentration, play a vital part in natural cycles of growth and decay. Unusually high natural concentration does occur; examples being methane, hydrogen sulphide from geothermal sources, but these are exceptional circumstances. So air pollutants are those in relatively high concentrations which result from the chemical and biological processes used by man.

Air pollutants are generally produced in dilute concentrations, together with large volumes of a carrier gas. One way of controlling air pollutants is to remove the source, either by changing the process or by stopping the process altogether. But this is not possible. The other alternative is to improve the basic processes with stricter pollution controls. Four general methods of separating gaseous pollutants are currently in use. These are (*i*) absorption in liquid, (*ii*) adsorption on a solid surface, (*iii*) condensation to a liquid, and (*iv*) conversion into less polluting or non polluting gas.

Gas absorption is a standard operation, which can be carried out either counter currently, with gas and liquid flowing in opposite directions or co-currently with both fluids flowing in the same directions. When dealing with relatively high concentrations of pollutant gases (of the order of 1% or more), then counter current flow is the common practice, since a reasonable concentration difference, resulting in an optimum driving force, occurs throughout the whole system, the lowest gas stream concentration contracting the weakest liquid concentration (referring to the dilute concentration of gas in the liquid). However, in the removal of pollutant gases already in relatively low concentration in the carrier gas, co-current scrubbers, with special design features giving large contact surfaces for short contact times are often used. These contractors may be scrubbers or intimate contact being achieved by having small bubbles of gas in the liquid or small droplets of adsorbing liquid in the gas.

In general, gaseous pollutants in concentrations down to 0.1% by volume, *i.e.*, 1000 ppm are effectively handled by scrubbers. These include hydrogen sulphide, sulphur dioxide, ammonia, nitrogen oxides and hydrocarbons. The gases can often be recovered, possibly by steam distillation and absorbing liquid recycled.

The process involved in adsorption of a gas on a solid are fundamentally similar to their absorption in a liquid, as the molecules diffuse across a boundary layer to the solid gas interface. At the solid surface, however, the molecules are retained, and in time buildup a concentration such that the concentration difference diffusion driving force is lost, and only an interchange of molecules, rather than not removal from the gas phase, takes place.

The surface available for adsorption may be very large in case of some materials such as activated carbon, silica gel. If the air pollutant is not present in large concentration, as in the case of order molecules, it may take a long time before the entire adsorbent surface is occupied.

Homogeneous gas reactions can also be used for control of some gaseous pollutants and in particular for hydrocarbons. These gases can be completely oxidized to CO_2 and H_2O which are harmless. This requires the gases to be retained in combustion chamber for a reasonable time 0.3–0.5s at flame temperature but it may add to the cost of process.

Catalytic combustion uses heterogeneous catalyst, which can reduce the reaction temperature for oxidation from about 800°C to 400°C. This way lesser amount of fuel is burnt consequently less pollution. Of course, this decreases the rate of the reaction.

11.7 CONTROLS FOR SPECIFIC GASEOUS AIR POLLUTANTS

11.7a Sulphur Dioxide

Sources: Burning of fossil fuels, roasting of sulphide ores of copper, lead and zinc, from the sulphuric acid plants.

Sulphur dioxide, from smelters or from combustion of fossil fuels, is not easily soluble in water, and it is necessary to use alkaline solutions or alkaline solids. Many processes have been suggested, and they can be classified into those that use a cheap reagent which are adsorbed and reacted with sulphur dioxide, is subsequently thrown away, and those that use an expensive reagent which can be recovered, in the latter, the sulphur is also recovered and then used for making either pure sulphur dioxide, sulphuric acid, or if ammonia is used, ammonium sulphate.

Lime stone, lime or dolomite has been added to the gases in the combustion chamber when the formation of sulphate take place

$$CaO + SO_2 + \frac{1}{2}O_2 \longrightarrow CaSO_4$$

The dry sulphate powder can then be taken out by particle recovery, or can be washed out in scrubber. Removal efficiencies of 40–70% have been achieved with dry recovery and over 90% with wet collection. The alternative is using lime to magnesium oxide slurries, the latter producing sulphite

$$MgO + SO_2 \longrightarrow MgSO_3$$

A modification which has led to several successful commercial application is the double alkali process, where a more active alkali such as ammonia solution, is used to remove the sulphur dioxide, and this is thus recovered by using a second alkali, lime to produce the sulphate $CaSO_4$, either for sale or disposal and ammonia for reuse. Ammonia in the vapour state has also been advocated as reacting effectively with SO_2 at 150°C

$$2NH_3 + SO_2 + H_2O \longrightarrow (NH_4)_2SO_3$$

$$NH_3 + SO_2 + H_2O \longrightarrow NH_4HSO_3$$

The gases are then scrubbed with water, and the resultant solution is reacted with NaOH, lime or zinc oxide slurries, to recover ammonia

$$(NH_4)_2SO_3 + 2NaOH \longrightarrow Na_2SO_3 + 2NH_3 + 2H_2O$$

$$NH_4HSO_3 + NaOH \longrightarrow NaHSO_3 + NH_3 + H_2O$$

Such a process may be well over 95% effective in removing sulphur dioxide.

Water itself, while a poor absorbent for sulphur dioxide, can be used in combination with a catalyst of active carbon. Here oxidation takes place on the moist catalyst surface, and dilute (10–15%) sulphuric acid is produced. This is called the Lurgi Sulphuric acid process.

Manganese oxide is also a good adsorbent and is being used in several pilot plants as are sodium sulphite, sodium citrate.

The simplest of all processes is of course the direct oxidation of SO_2 to SO_3

$$SO_2 + \frac{1}{2}O_2 \longrightarrow SO_3$$

Thus, the standard catalytic oxidation in sulphuric acid manufacturing is generally carried out at 450–600°C where conversion is satisfactory and the rate is fast, and is self-supporting if the gases are at 450°C, the reaction temperature. Low concentrations of the order of 0.1–0.4% common in flue gases require special catalysts, which have recently been developed. At lower temperature certain activated carbons are also suitable for the oxidation of sulphur dioxide that can be recovered for subsequent use in an acid plant.

Chemically interesting is the process where after absorption by ammonia, sulphur as well as ammonium sulphate is recovered from ammonium hydrogen sulphate and thiosulphate solution by a disproportionation reaction carried out in a high pressure autoclave

$$2NH_4HSO_3 + (NH_4)_2S_2O_3 \longrightarrow 2(NH_4)_2SO_4 + 2S + H_2O$$

$$T = 170°C, 1.4\ MPa$$

Both the sulphur as well as ammonium sulphate can be used commercially.

A mixture of molten alkali carbonates is also being used for SO_2 recovery, the mixture being 32% Li_2CO_3, 33% Na_2CO_3 and 35% K_2CO_3 by weight. These give a eutectic mixture, melting at 397°C which appears as clear liquid at 450°C. If the alkali atoms are written as *M*, the absorption stage is

$$SO_2 + M_2CO_3 \longrightarrow M_2SO_3 + CO_2$$

The sulphite can be oxidized as

$$M_2SO_3 + \frac{1}{2}O_2 \longrightarrow M_2SO_4$$

After absorption the sulphite and sulphate are reduced with carbon at 800°C.

$$2M_2SO_3 + 3C \longrightarrow 2M_2S + 3CO_2$$

$$M_2SO_4 + 2C \longrightarrow M_2S + 2CO_2$$

Some sulphate and sulphide can also arise from the reaction, and is known to be important at 800°C in the reducer

$$4M_2SO_3 \longrightarrow M_2S + 3M_2SO_4$$

The carbon dioxide is used to regenerate the carbonate with water vapour at 450°C

$$M_2S + CO_2 + H_2O \longrightarrow M_2CO_3 + H_2S$$

The hydrogen sulphide produced is in turn oxidized to sulphur, over a catalyst in a Claus plant

$$H_2S + \frac{1}{2}O \longrightarrow S + H_2O$$

Difficulties with this process are the highly corrosive molten carbonates, which have to be handled at high temperatures. Special stainless steels are used at the lower temperatures, where a protective $LiCrO_2$ film forms, and at the very high temperatures (upto 900°C) the materials of construction are high density alumina. The lithium carbonate is a relatively expensive material and considerably care must be taken to reduce losses and recover the lithium from wastes.

Chemical scrubbants used in venturi and other types of scrubbers during flue gas desulfurization (FGD) contain alkaline cation such sodium, calcium and magnesium, listed in order of decreasing reactivity. Lime and limestone are less expensive, so products produced are usually discarded, while sodium and magnesium chemicals are regenerated.

Specific relationships for absorption of SO_2 from flue gas in venturi scrubbers using calcium and magnesium are presented here to show some of the significant relationships. These equations are specific to certain operating conditions and systems. For MgO scrubbing at a liquid to gas ratio L_G, equal to 5.41 m^{-3}, SO_2 removal efficiency is

$$\varepsilon_0 = 1 - \exp\left[\left(6.86\,\Delta P^{-1.014}\, y^{-3.75+0.271\ln y}\, 10^{6-0.031\,\text{pH}}\right) - 3\right] \quad ...(11.20)$$

where ΔP is in cm H_2O, y is inlet SO_2 concentration in ppm and pH is actual pH of scrubbing slurry. For CaO–SO_2 system, expression is as follows:

$$\varepsilon_0 = 0.307 + 0.0457(S) + 3.75\times10^{-3}(\Delta P) + 0.048 L_G$$
$$+ 0.01516(I) - 0.0258(I^2) - 5.98\times10^{-3}(C_s) \quad ...(11.21)$$

where S = stoichiometry, moles CaO/mole SO_2 input

ΔP = venturi scrubber pressure drop, cm H_2O.

L_G = Liquid to gas ratio 1 m^{-3}

I = Ionic strength of Na^+ present, molarity

C_s = slurry solids concentration, percent age by weight

Example 11.3

Predict SO_2 removal efficiency by FCD in a single-stage MgO venturi scrubber and a two stage CaO scrubber. Pressure drop for each state is 17 cm H_2O. Inlet SO_2 concentration is 1600 ppm, scrubber slurries have a pH of 7, contains 5% by weight solids and a ratio of 5.4 l liquid per m^3 gas. Considering that adequate chemical is available for absorption and that the lime system stoichiometry is 1 mole CaO/mole SO_2 input and no sodium ions are present (Note that normally stoichiometry is expressed as mole CaO/mole SO_2 absorbed).

Solution: Absorption efficiency of SO_2 in the single-stage MgO system using Equation 10.20 is

$$\varepsilon_0 = 1 - \exp\left[(6.86)(17)^{-1.014}\left(1600^{-3.75+0.271\ln 1600}\right)\left(10^{6-0.031(7)}\right) - 3\right] = 0.911$$

Absorption for the first stage of CaO system using Equation 10.21 is

$$\varepsilon_0 = 0.307 + 0.0457(1) + 3.75\times10^{-3}(17) + 0.048\,(5.4) - 5.98\times10^{-3}(5) = 0.646$$

Efficiency for two stages in series under these conditions would be

$$\varepsilon_0 = 0.646 + 0.646(1 - 0.646) = 0.875$$

These data suggest that a single stage of separation with CaO solution is not very effective. Two-stage absorption increases the efficiency to 87.5%, which can be further increased by using more reactive lime-like thiosorbic lime.

11.7b Hydrogen Sulphide

Source: Coal gasification, biological reactions involving sulphide components, and in some chemical reactions.

Removal: Coal gas contain 1% ammonia but that may remove half of the H_2S. However, with additional ammonia, hydrogen sulphide can be effectively removed. The gases need, however, to be further scrubbed to remove entrained ammonia. The product is usually ammonium sulphate and H_2S which is by this time has sufficiently high concentration to be used for sulphur recovery by the Claus process.

An alternative to ammonia, the ethyl and methylamines are used in a range of processes. As the reagent is relatively expensive, it is recovered after use as an absorbent by steam stripping and is then recycled. The first compound to be used was triethanolamine (TEA = $C_2H_4OH)_3N$ but this proved to be difficult to regenerate, and so a 15–20% aqueous solution of monoethanol amine (MEA) was brought into use. Its disadvantage is largely that some reagent is lost because a heat stable compound is formed with carbonyl sulphide (COS) diethanolamine. MEA is, therefore, used for natural gas purification, when virtually no COS impurity is present while diethanolamine (DEA) is used for coal and refinery gas.

Sodium thioarsenate is used as an absorbent of H_2S when the reaction that take place

$$H_2S + Na_4As_2S_2O_2 \rightleftharpoons Na_2As_2S_6O + H_2O$$

which is regenerated using atmosphere oxygen as oxidizing agent

$$2Na_4As_2S_6O + O_2 \rightleftharpoons 2Na_4As_2S_5O_2 + 2S$$

A very widely used process in recent years uses organic dyestuffs in the critical oxidation-reduction stages. The hydrogen sulphide in concentrations of 10–10000 ppm (1%) is first absorbed in an alkaline solution maintained at pH 8.5-9.5 and containing sodium carbonate, sodium vandate anthraquinone 2 : 6 and 2 : 7 disulphonate (ADA). Rochelle salt (potassium sodium tartarate) is added as a sequestering agent to prevent precipitation of the vandate. The reactions are as follows:

Absorption	$H_2S + Na_2CO_3 \longrightarrow NaHS + NaHCO_3$
Sulphur production	$2NaHS + H_2S + 4NaVO_3 \longrightarrow Na_2V_4O_9 + 4NaOH + 2S$
Recovery of vandate	$Na_2V_4O_9 + 2NaOH + H_2O + 2ADA \longrightarrow 4NaVO_3 + 2ADA$
Oxidation of the ADA	$2ADA(\text{reduced}) + O_2 \longrightarrow ADA + 2H_2O$

The process plant, inspite of these complex reactions, is quite simple, a packed column being used for the absorption stage, followed by a delay tank and a sulphur extraction plant. The existing concentrations of hydrogen sulphide are of the order of 1 ppm.

Ferric oxide is the other reagent material that has been widely used for hydrogen sulphide removal. In aqueous suspension the ferric oxide acts largely as a catalyst. Again suspension passing down a packed column is the absorbent. Again the suspension passing down a packed column is absorbent, and sulphur is removed by blowing atmospheric oxygen through the sulphide containing solution. Unfortunately, the sulphur from this process tends to be contaminated with iron oxide, reducing in a finely divided state on peat or fiberous material or in pellets, has long been used in "oxide boxes" for the removal of small amounts of hydrogen sulphide. The oxide may be subsequently recovered by combustion of the carrier material or by solvent extraction, using for example perchloro-ethylene. Here in general terms, the reactions are

absorption $$6H_2S + 2Fe_2O_3 \longrightarrow 2Fe_2S_3 + 6H_2O$$

regeneration $$2Fe_2S_3 + 6O_2 \longrightarrow 2Fe_2O_3 + 3SO_2$$

The regeneration can also be carried out at high temperatures. Hydrogen sulphide, at concentrations of 14 g m^{-3} in gas has been reduced to 0.01–0.03% by passing the gas through a four-stage fluidized bed maintained at 340–360°C. The recovery stage is achieved by heating the ferric sulphide to 800° C in air. The sulphur dioxide driven off can be used in making sulphuric acid. Unfortunately, in coal gas, H_2S is usually accompanied by organic sulphides and thiophene, and the degree of removal is 70–80% for the sulphides and only 30–45% for thiophene. By placing a pretreatment stage into the system, which turns the sulphides and thiophenes to H_2S, using special catalysts (chromic alumina, cobalt-molybdenum, zinc oxide etc.) the overall process can be improved. There are many other processes that are used for H_2S removal.

11.7c Ammonia, Amines and Pyridine Bases

These are all formed in coal carbonization, the first two also found in the waste gases of meat, offal, fish, etc. Ammonia and the amines are very soluble in water and strong oxides, because of the reaction giving ammonium or alkyl ammonium ions. However, the solubility of these compounds in water falls off rapidly with rising temperature. One method is a two-stage system, the first being a cooler of some sort where gases are cooled to 30–50°C followed by a water scrubber. The other system is a single-stage absorber and crystallizer in which sulphuric acid is strayed and ammonium sulphate crystallizing out in the bottom. The pyridine bases are also removed in this process. The final traces of amines can be removed by passing the gas through deep beds of activated charcoal.

11.7d Nitrogen Oxides

There are six stable oxides of nitrogen (N_2O NO, NO_2, N_2O_4, N_2O_3 and N_2O_5), and of these NO is mainly produced by combustion, while NO_2, N_2O_3 and N_2O_5 are given off in some industrial operations such as pickling of metals prior to other treatment such as anodizing. The volumes from the latter processes are comparatively small, and scrubbing the gases with alkaline solution has been tried where maximum of 70% of nitrogen gases are removed. What does appear promising are those systems where the gases are passed rapidly over the liquid surface on special wave formed plates and such systems have reported efficiencies around 95%. Systems where the gases are passed through moist activated carbon also appear to be effective, although removal efficiencies are not quoted.

When nitrogen oxides are produced by combustion, the very large volumes and low concentration prelude their removal by scrubbers. However, it has been found possible by careful attention to burner design and operation, using tangential, instead of opposed firing, two-stage combustion, partial recirculation of combustion gases (upto 25%), and in some cases, derating the boiler, to reduce oxide of nitrogen production from well over 1000 ppm to 42 ppm.

11.7e Fluorine and Fluorides

Very low concentration of fluorides (of the order of pptm) can damage vegetation, and so the removal of fluorides from waste gases of aluminium electrolysis plant, fertiliser plants from glass, etc., is essential. Hydrogen fluoride is very soluble in water and simple counter flow scrubbers wet cell washers consisting of a succession of packed boxes filled with straw and other material and strayed with water, cross flow packed scrubbers and floating bet scrubbers where plastic balls are fluidized in air and water have been found effective.

When fluorine is present a 5–10%, sodium hydroxide solution should be used. In aqueous and very low caustic (less than 2%) solutions, the very poisonous fluorine oxide, OF_2, is formed. In packed lowers, with liquid gas contact times of about 1 minute, the following reaction occur

$$F_2 + 2NaOH \longrightarrow \frac{1}{2}O_2 + H_2O + 2NaF$$

$$HF + NaOH \longrightarrow NaF + H_2O$$

Sodium fluoride is then treated with lime to regenerate the hydroxide

$$2NaF + CaO + H_2O \longrightarrow CaF_2\,(ppt) + 2NaOH$$

A very elegant method of recovering gaseous and particulate fluorides from aluminium smelters has been developed, where the waste gases are passed through a fluidized bed of alumina, which will subsequently be used in the smelter and so return the emitted fluoride. The temperature of absorption is 65–160°C and the bed thickness 50–300 mm. The efficiency of the process is 99.2% or better for gaseous fluorides, and over 90% for particulate fluorides with outlet concentrations being reduced from 100–200 mg m^{-1} to less than 3 mg m^{-3}.

11.8 ODOURS

Unlike the gaseous air pollutants discussed above, odours generally occur in very low concentrations—ppm or less—and must be reduced to much lower concentrations. Either adsorption on activated carbon or combustion in a flame or on a catalyst surface one of the three methods employed. Adsorption of odour on activated carbon is widely used for food and cooking smells, mercaptan, cresols, butyric acid. The adsorption capacity depends on the surface area of carbon (200–1200 $m^2\ g^{-1}$), the temperature of adsorption and material. Usually, active carbon beds last from few weeks to several months before they require renewal.

Flame combustion is widely used for disposal of waste hydrocarbon odour in oil refineries and chemical works, where the fume energy, concentration of the material is such as to produce self supporting combustion "flares" are used, but if these are fed with an aromatic material and air they tend to turn with a smoky flame. This is overcome by injecting steam into the flame which result in the water gas reaction, producing hydrogen and carbon monoxide, both of which burn smokelessly.

Catalytic combustion has the advantage that it occurs at 300–400°C instead of at more than twice this temperature, as in direct flame incineration and so the problems of high temperature heat exchange, and additional energy input in the case of low fume energy are reduced. Unfortunately; the catalysts used are expensive, and need occasional regeneration due to poisoning. This can be due to chlorides from cuprous chloride with copper oxide catalysts, sulphur compounds affect alumina catalysts and organic phosphorus compounds from lubricating oil form phosphoric acid. Heavy metal also deactivate the catalytic action. Thus use of catalytic oxidation should be restricted to those application where there is little possibility of catalyst poisoning as in treating the vapour coffee, roasting, bread-baking varnish cooking.

11.9 THE AUTOMOBILE FUEL

The low boiling fraction of crude petroleum obtained by distillation is called gasoline and is a source of energy in automobile internal combustion engine. The fuel mixture of gasoline vapour and air burns with in the working cylinder. Most of the automobiles are driven by a four-stroke engine (Fig. 11.7).

The gasoline is mixed with air in carburettor and drawn into the cylinder on the first stroke. The air gasoline vapour mixture is compressed on the second stroke and ignited by spark plug. The expanding gases formed by combustion then drive the piston down on the third stroke that provides power to drive the car and combustion product are forced out of cylinder in the fourth stroke. The engine gives maximum efficiency when gasoline-air mixture is highly compressed before firing in cylinder. The modern engines squeeze this mixture to one seventh or even to one-eighth of its normal volume.

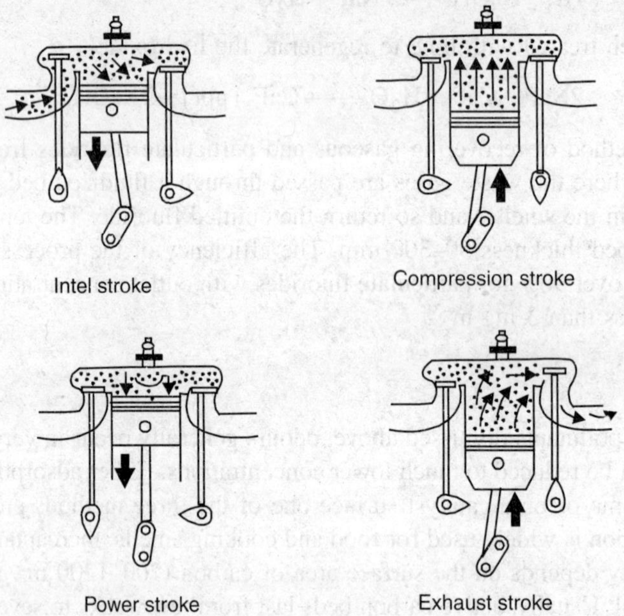

Fig. 11.7 Schematic of a four-stroke cycle engine.

As the compression ratio increases, a point is reached when knocking is observed. On sparking all the fuel does not burn smoothly and the end portion burns with explosive violence giving rise to metallic rattle called knocking. This results in loss of power. The loss of power is given by a rating of a fuel by octane number. Higher the octane number, less is the rattling. Thus pure linear alkane gasoline has an octane number of 68 only. One should understand that whenever heat energy is not being converted to work, some other chemical reaction to utilize this energy has to take place. Thus metallic rattling or knocking leads formation of nitrogen oxides. The other consequence of rattling is that fuel may not burn completely when carbon monoxide, carbon and gasoline vapours will be present in the exhaust.

Addition of tetraethyl lead, TEL to gasoline increases its octane rating from 68 to 98. TEL is a mixture of about 60% tetraethyl lead or tetramethyl lead, 35% ethyl dibromide and ethyl dichloride, 2% dye and rest is solvent and stabiliser. 0.66 g/lt of this TEL is added to gasoline. It prevents preignition and even burning by its conversion to lead oxide, a compound that quenches the free radicals reactive intermediates with unpaired electrons formed in the combustion process. Ethyl dibromide and Ethyl dichloride facilitates the removal of the lead from cylinder by its conversion to volatile lead halides that are exhausted from engine into atmosphere. Lead bromides and chlorides are toxic air pollutant.

Sulphur is one of the important constituent of gasoline. It interferes with the antiknock action of TEL which lead to emission of lot of nitrogen oxides and carbon monoxide along with sulphur dioxide. Thus various pollutants emitted on gasoline combustion are:

(*i*) Lead salt particles, (*ii*) Sulphur dioxide, (*iii*) TEL vapours, (*iv*) Hydrocarbon vapours, (*v*) Nitrogen oxides, (*vi*) Carbon monoxide, (*vii*) Carbon dioxide, (*viii*) Water vapours, (*ix*) Carbon particle.

(*i*), (*ii*) and (*iii*) can be removed with the reformation of fuel and improvement of engine. Use of catalytic converter can convert (*iv*), (*v*), (*vi*) to CO_2, H_2O, N_2.

Lead Pollution: This will be explained with Delhi as a specific example. Gasoline consumed in Delhi in 1985–86 was 2.5844×10^9 litres, in 1989–90 it went upto 3.9657×10^9 litres and in 1994–95 it went upto 4.10664×10^9 litres. In each of these gasoline, 0.15 g/lt of lead is assumed to be added (which is equivalent to 0.66 gilt of TEL), though in the earlier years it was more than double this value. Assume that 50% of this gasoline is burnt outside Delhi, and rest in Delhi. The population of Delhi is taken to be 1,00,00,000.

The safe limit of lead in our system is 25 micrograms per litre. On burning 1000 litre of petrol, 0.1198 kg of lead chloride as particulate matter is thrown into the air. The amounts for the three years is as follow:

1985–86	531 kg/day	≡	53.1 mg/day/person
1989–90	815 kg/day	≡	81.5 mg/day/person
1994–95	843.8 kg/day	≡	84.38 mg/day/person

The amount for 1994–95 is equivalent to 84.38 mg/day/person. If 99.9% of this lead salt just falls over plants or soil and 0.1% of it is inhaled, everyone must have been inhaling 0.0844 mg/day of lead chloride every day in the year 1994–95. This is quite a high amount. Every citizen of Delhi must be having high lead concentration in their bloods. Lead that has been falling over grass, vegetables, grown around Delhi: High lead soils must be having high lead concentration which also enters our systems. As lead salts are mostly insoluble, Delhi cannot get rid of lead that had been falling over it for at least four to five decades. Under such situation we should stop adding lead to petrol immediately.

Reformation of Gasoline

Following steps may be taken in the order of priority:

1. Sulphur should be removed from gasoline for which many processes are there.
2. On passing the crude naphtha vapours over Pt-Alumina as catalyst at 770°C aromatic hydrocarbons are obtained. Sulphur free gasoline is blended with benzene, toulene and xylene (BTX) with octane number 104 which is more than octane rating of lead gasoline. Such a gasoline can be used without TEL, without knocking effect. However exhaust of such a gasoline contains benzene which is toxic.
3. A mixture of sulphur-free gasoline is mixed with minimum benzene, toulene and xylene and 0.013 g/lt of lead is added to raise its octane number to 98. It is this lead free gasoline that is being supplied in Metro cities of India from April 1995.
4. In USA, unleaded gasoline was introduced in 1974 to cater to cars equipped with catalysts, leaded gasoline was banned there from January 1995.
5. Presence of oxygenates, e.g., ethanol, methanol, ethers in gasoline are advantageous as oxygenates help in reducing carbon monoxide emission, especially from older vehicles and they are excellent octane improver.

6. Some of the fuels currently being used are liquified petroleum gas (LPG), liquified propane gas, compressed natural gas (CNG). The supply of some of these fuel is very limited. Other fuel problems involving storage, distribution, and power requirements have to be considered.

11.10 EMISSION FROM TWO STROKE ENGINES

Mostly the Crankcase scavenged two stroke engine is used to power the Indian two and three wheelers and is shown in Fig. 11.8. In a conventional two-stroke engines, the fuel air mixture from the carburettor enters the engine crankcase, where it is compressed by a factor of 1.1 to 1.3 and during scavenging period enters the engine cylinder via transfer ports. The fresh fuel-air mixture in the crankcase, which is slightly compressed pushes the burned gases out of the cylinder. During this process, a portion of the fresh mixture also finds its way to the exhaust port and is lost without being burnt. This phenomenon of fresh mixture loss is sometimes termed as mixture short circuiting. It has been estimated that depending upon the engine operating conditions, 15 to 40% of total fuel supplied to the engine is short circuited. This is the principle cause of high hydrocarbon emissions observed from these engines. This results in a poor fuel economy of the vehicles powered by these engines.

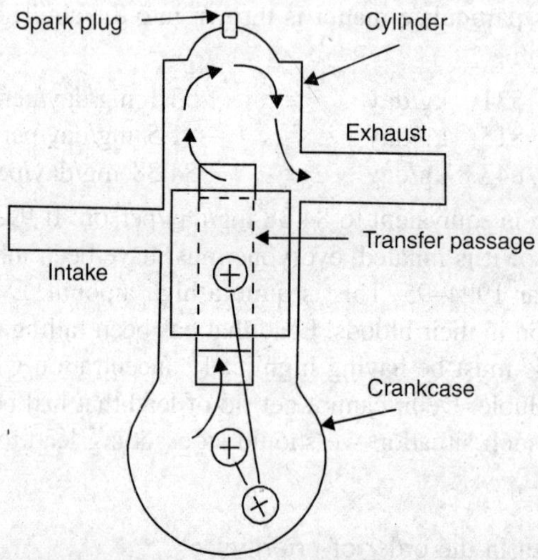

Fig. 11.8 Schematic Diagram of a Conventional Crankcase Scavenged Two-Stroke Engine.

The imperfect scavenging of the two-stroke engine cylinder of burnt gases being retained in the engine cylinder. The presence of a high amount of residual burnt gases in the cylinder results in slow flame propagation and poor combustion. As a result has high hydrocarbon, while NO_x emissions are rather low. The emission depends upon the engine load and speed and may vary considerably. The CO concentration varies in the range of 1 : 1 to 4.8%, while hydrocarbon emission vary from 12000 to 36500 ppm, C, 2000 to 6090 ppm hexane equivalent. The concentration NO_x is less than 200 ppm. Depending upon the design, there are considerable variations in emissions, research and development efforts to reduce scavenged through fuel losses may be discussed into following categories:

(*i*) Improvements in the engine with premised fuel air scavenging.
(*ii*) Scavenging with stratified charging, and
(*iii*) Fuel injections and scavenging by air.

However these changes have not made a significant difference. Scooter is one of the main pollutant on the Indian roads. Reforms to two-stroke engine to reduce pollution is not possible. That is why scooters with four stroke have been introduced on Indian roads. Catalytic converter for this engine, if introduced may help to have pollution free exhaust. It may be added that scooter in any form is not on the roads of Europe and U.S.A.

Unconventional and more complex design have been made to reduce the hydrocarbon emissions. Such emission controlled technology has been well presented in the Indian Institute of Petroleum, Dehradun, Report on Vehicles Emission and Control Perspectives in India[4].

11.11 EMISSION FROM FOUR STROKE ENGINES

All the Indian cars and jeeps are powered with four stroke, spark ignited, gasoline fueled carburetted engines. Some of the cars are equipped with positive crankcase ventilation system (PCV). Some of the Indian models employ exhaust gas recirculation, evaporative emission control technologies.

Gases and vapours that under pressure escape from the combustion chamber past the engine piston and rings and enter the crankcase to be discharged to the atmosphere through draught tube and other vents are known as crankcase blow by. The blow by gases emitted from the crankcase are rich in hydrocarbon and contribute about 20% of total hydrocarbon emission on uncontrolled gasoline vehicles.

The hydrocarbon rich blow by gases are prevented from escaping into atmosphere and are recycled to intake air system through a positive crankcase ventilation (PCV) valve. This valve restricts flow of blow by gases during idling to prevent rough idling and high flow at lighter loads.

A schematic diagram of the system is shown in Fig. 11.9. This is a low cost hardware that has been adopted in India by the car producers. PCV can knock off a substantial portion of the hydrocarbon emissions.

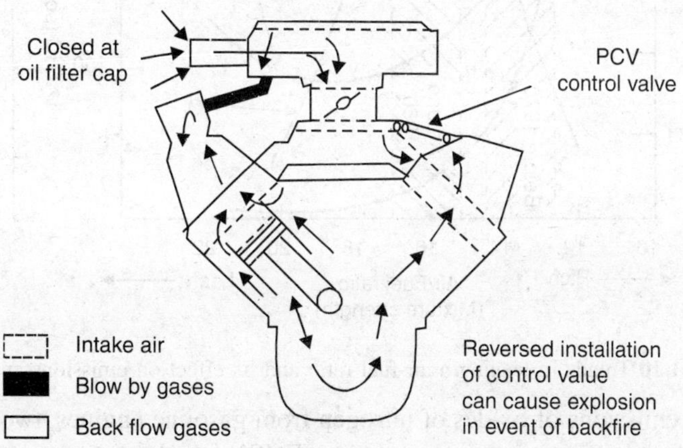

Fig. 11.9 Positive crankcase ventilation system (PCV).

Evaporative emissions are estimated to contribute to another 20% of total hydrocarbons. The two major sources of evaporative emissions are carburettor bowl and fuel tank. There is no standard for evaporative emissions and no evaporation emission control system is presently in use on any of the automotive vehicles. Hydrocarbon emission in the exhaust are primarily caused due to "Wall quenched phenomenon, *i.e.*, quenching or extinction of the flame at the walls of engine combustion chamber due to the heat transfer leaving behind a layer or volume of unburned and partially burned fuel air mixture. High combustion pressure and temperature promote reaction leading to the formation of oxides of nitrogen. Besides, there are several reaction resulting in particulates. Sulphur dioxide and lead emission are also there. Sulphur acts as a knocking agent in the fuel and increases the CO and NO_x content in the exhaust. The air-fuel ratio is one of the most important parameters, which greatly influences the exhaust out emissions. Figure 11.10 shows the trends in learning of air-fuel ratio and its effect on emission and fuel combustion. Reduced compression ratio improved combustion chamber design, lean burn engines, stratified charge engines, etc., help greatly in reducing the exhaust emissions.

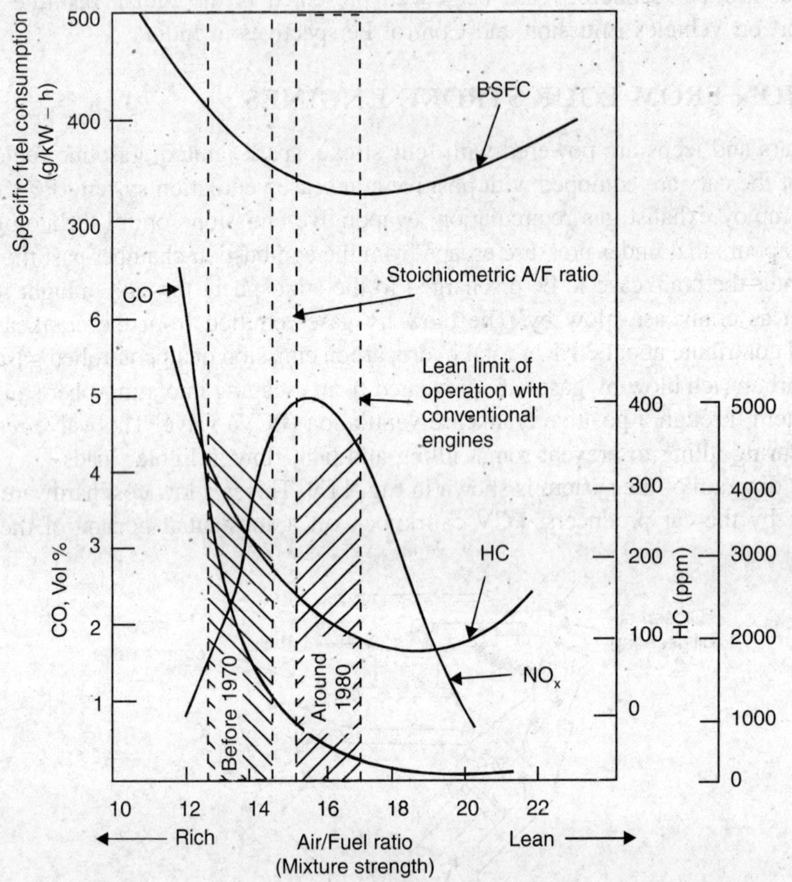

Fig. 11.10 Trends in leaning a air-fuel ratio and its effect on emissions and fuel.

In order to lower emissions of oxides of nitrogen from gasoline engines, two general systems have been developed. The first is exhaust gas recirculation (EGR), which mixes a portion of the exhaust gas with the incoming fuel–air charge, thus reducing temperatures within the combustion chamber. This recirculation is controlled by valving and associated with plumbing and electronics so that it occurs during periods of highest NO_x production when some power reduction can be tolerated: a cruising condition at highway speed. Other alternatives are to use another catalytic converter, in series with the HC/CO converter, which decomposes the oxides of nitrogen to oxygen and nitrogen before the gases are exhausted from the tail pipe or to use a three way converter, which converts all the gases NO, CO, CO_2, HC to N_2, O_2, CO_2 and water vapours.

11.12 EMISSION FROM DIESEL VEHICLES

The diesel cycle is regulated by fuel flow only air flow remaining constant with engine speed. Because the diesel engine is normally operated well on the lean side of the stoichiometric mixture (40 : 1) or more emission of unburned hydrocarbons and carbon monoxide is minimized. Almost all the light and heavy duty commercial vehicles in India are powered by four stroke compression ignited diesel engines.

Most of these vehicles are powered by direct injection diesel engines excepting one make of light commercial vehicles and some passenger cars. Most diesel vehicles are naturally aspirated. Turbo-charging is being employed in some models. Pollutants emitted by diesel engines are CO, NO_x, unburnt hydrocarbons, SO_2, smoke particulates, aldehydes and odour. The exhaust smoke from the diesel is made up of soot particles as the volatility of diesel being very low, not much evaporative losses occur from the fuel tanks. The concentration of CO_2 and unburned hydrocarbon in diesel exhaust are relatively low compared to gasoline engine, but nitrogen oxides are present in high concentration typical emission from a diesel engine under different operation conditions are given in Table 11.5.

Table 11.5 Typical emissions from a 4 stroke diesel engines under different operating conditions

Operation conditions	*CO%*	*HC ppm*	*NO_x ppm*	*Formaldehyde*
Full load rated speed	0.20	29	921	4.3
Half load rate speed	0.03	70	493	6.9
No load rated speed	0.03	90	109	1.8
Idle	0.03	106	119	6.8

The details of Emission control technologies being developed are available in "Vehicles Emission and Control Perspectives in India, A state of the art", report published by Indian Institute of Petroleum (IIP), Dehradun.

11.13 CATALYTIC CONVERTERS

All successful catalytic converters developed so far employ precious metals as catalyst Platinum/ Rhodium. Hydrocarbon are oxidized over the catalyst in two-stroke engine but CO conversion is low as the hydrocarbon are not completely oxidized to CO_2. There is only a partial oxidation to produce CO. Emission tests on moped and motorcycle are given in Table 11.6.

Table 11.6 Emission test results

	1.2 hp Moped		*2.7 hp Moped*		*150 cc Motorcycle*	
Emissions g/km.	*Without catalyst*	*With catalyst*	*Without catalyst*	*With catalyst*	*Without catalyst*	*With catalyst*
CO	1.013	0.015	0.771	0.022	21.7	1.7
HC	2.716	0.090	3.205	0.037	16.9	10.4
NO_x	0.049	0.049	0.09	0.067	0.01	0.04

From Table 11.6, it may be noted that engine design improvements resulted decrease in CO and hydrocarbon emissions. Use of catalytic converters reduced the emission further considerably. Results with fresh and aged catalyst for 15000 km show there is negligible decrease in its efficiency. However, lead is extremely poisonous for precious metal thus has to be eliminated from gasoline. If oil to fuel ratio is higher say more than 2% catalyst, cell blockage occur by carbon and other residues.

Oxidation catalyst for cars were introduced first in the US market in 1975 to meet the CO and HC emission standards. It oxidizes the CO and HC to CO_2 and H_2O and hence called oxidation catalyst. The catalyst is made up of Pt-Pd with active catalyst material amounting to 0.05 tray ounces (1.6 g) per car of platinum and palladium in ratio 5 : 2. Rhodium acts as a promoter and further increases its efficiency.

This is fitted on the exhaust pipes of a car and renders noxious fumes harmless (see Fig. 11.11) Since April 1995 these converters-priced between Rs. 10,000 and Rs. 15,000 each depending on type—have been mandatory for new cars registered in the four metro cities (Mumbai, Delhi, Kolkata and Chennai), the converters only work with unleaded fuel.

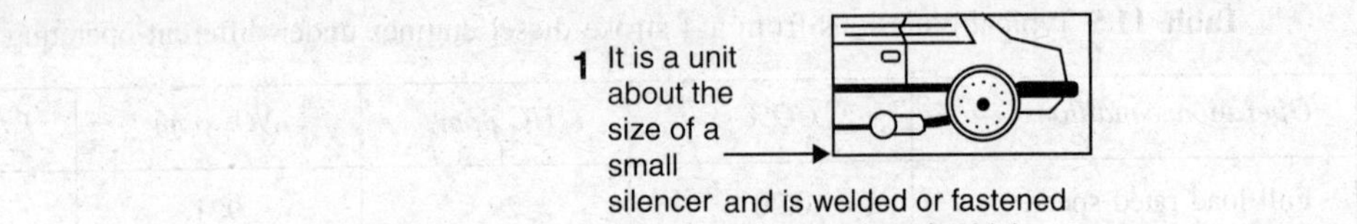

Fig. 11.11 Catalytic converter and how it works.

Further research led to the development of a variety of catalytic converters that can control all the three emissions HC, CO and NO_x simultaneously and hence the three catalysts. This system uses a catalyst system which operates best near the stoichiometric air-fuel ratio. To maintain the exhaust, closed loop air-fuel ratio, control systems are required. If air-fuel ratio is maintained correctly, more than 90% reduction in all the three pollutants is achieved. Various automotive manufactures have developed closed loop-controlled carburettor and fuel injection systems to meet this requirement and also to improve vehicle fuel economy and performance.

The advent of Zircona and/or Titanio exhaust gas oxygen sensor (lambda-sensor) was a great help in designing closed loop feed back A/F control system. It measures the amount of excess oxygen in the exhaust relative to the oxygen in air. When the exhaust goes from rich to lean or *vice-versa,* the oxygen sensor produces a signal. This signal can be used to control the amount of fuel supplied to the engine. Figure 11.12 shows the Ford three-way catalyst/feedback control system in the form of a block diagram.

The components comprising the control system include:

(*a*) Exhaust gas oxygen (EGO) sensor,
(*b*) Electronic control unit (ECU), and
(*c*) Vacuum regulator solenoid (VRS).

The three way catalyst system, while reducing the levels NO_x, hydrocarbons and CO, the draw back that it increases the concentrations of reduced species, such as ammonia, cyanide and organic amines, hydrogen sulphide and organic sulphides.

Diesel oxidation catalyst (DOC) are noble catalyst similar to those used in gasoline. There use on monolithic honey comb as support in recommended. Depending on the type of engine and its exhaust, DOC oxidizes 30 to 80% of HC and 40 to 90% of CO. They do not reduce NO_x as there is always an excess of oxygen present in diesel exhaust.

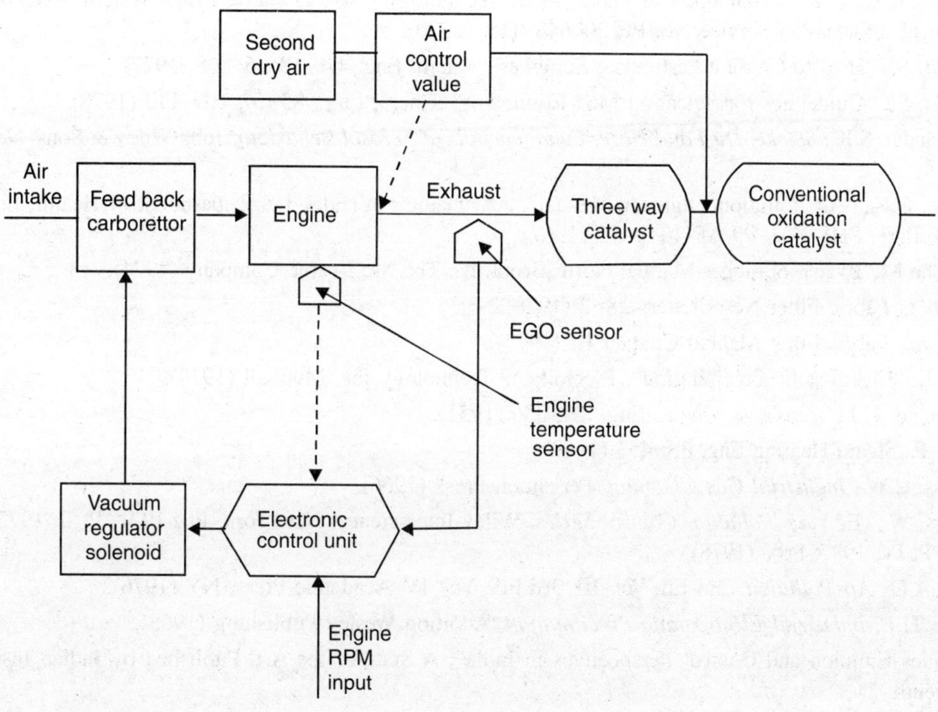

Fig. 11.12 Ford three-way catalyst/feedback control system block diagram.

That is why three-way catalyst are not effective in diesel engines. DOC show very little oxidation of dry soot. These, however, remove 30 to 50% total organic fractions.

11.14 GAS TURBINES AND JET ENGINES

The turbine is designed to produce torque at the output shaft while the jet engine allows most of the hot gases to expand into atmosphere, producing usable thrust. Emissions from both turbines and jet are similar as are their control methods. The emissions are primarily unburned hydrocarbons, unburned carbon which results in the visible exhaust and oxides of nitrogen. By redesigning the fuel spray nozzles and reducing cooling air to the combustion chambers to permit more complete combustion, control of the unburned hydrocarbons and the unburned carbon may be accomplished. The gas turbine engine for automotive or truck use could be either a simple turbine, a regenerative turbine, a free turbine, or any combination.

The atmosphere of the world cannot continue to accept greater and greater amounts of emissions from mobile sources as our transportation systems expand. Probably the ultimate answer to the problem of emission from millions of private automobiles is an alternative transportation system. It must be remembered, however, that even rail systems and bus systems do emit some air pollution. Rail systems are expensive and lack flexibility. Buses offer much greater flexibility at lower cost than rail systems, but in order to operate efficiently and effectively, they would require separate roadways with good workshops to maintain buses and honest people to look after all these systems.

REFERENCES

1. Billings, C.E., *et al.*, "Handbook of Fabric Filter Technology", Vol. 1, Fabric Filter System Study, National Technical Information Service No. PB 200648 (Dec. 1970).
2. Clavert, S., "How to Chase a Particulate Scrubber", Chem. Eng., 84 (18), 54–68 (1977).
3. Clavert, S., "Guidelines for Selecting Mist Eliminator", Chem. Eng., 85 (5), 109–112 (1978).
4. Friedlander, S.K., *Smoke, Dust and Haze, Fundamentals of Aerosol Behaviour,* John Wiley & Sons, New York, (1977).
5. Lapple, C.E., "Air Pollution Engineer Manual", J.A. Danielson (Ed.), US Department of Health, Education and Welfare, PHS No. 999-AP-40, p. 95 (1967).
6. (*i*) The Mc Ilvaine Scrubber Manual North Brook IL, The Mc Ilvaine Company (1974).
   (*ii*) *Ibid.,* Fabric Filter News letters 28: 2 (1978).
   (*iii*) *Ibid.,* Fabric Filter Manual Chapter III (1977).
7. Katz, J., "Electrostatic Precipitation", Precipitator Technology Inc. Munhall (1979).
8. Stairmand, C.J., Trans Inst. Chem. Eng., 29, 356 (1951).
9. Swift, P., Steam Heating Eng. 38: 453 (1969).
10. Strauss, C.W., *Industrial Gas Cleaning,* Pergamon Press (1966).
11. Strauss, W., (Ed.) *Air Pollution Control Series,* Wiley-Interscience, New York, Pt I,1971, Pt II, 1972, Pt III, 1976, Pt IV, 1977, Pt V (1978).
12. Stem, A.C., *Air Pollution,* 2nd Ed, Vol. III, 3rd Ed., Vol. IV, Academic Press, NY (1976).
13. White, H.J., *Industrial Electrostatic Precipitation,* Addition Wesley Publishing (1963).
14. "Vehicles Eminion and Control Perspectives in India", A State of the Art, Published by Indian Institute of Petroleum.

12

Air Pollution in India

12.1 INTRODUCTION

The main products of burning fossil fuels are CO_2, CO, SO_2, NO_x, suspended particulate matter (SPM), hydrocarbons and metallic tracers. India is world's fifth largest producer of these gases. Air quality data in India's major cities indicate that ambient level of NO_x, SO_2 and suspended particulate matter are often higher than WHO and Indian standards. Major contributors include thermal power station, industrial factories, vehicle, and non commercial fuels such as coal briquettes, animal dung and trash. Other specific polluters arise from particular sources; for instance, urea dust from fertiliser factory, fluoride from aluminium production, cement and lime dust from cements plants, radiation from nuclear power station.

Thermal power stations are a source of major power supply in India. These stations also rank among the worst air polluters in the country. The main effluents from these plants are flyash and SO_2. 75 per cent of coal produced, goes into power generation. Metallurgy industries also consume quite a lot of coal and produce similar pollutants. Indian coal has low sulphur contents generally less than 1%, but a high ash content 25 to 40%. High ash coal means more wear and tear of plant and machinery, low thermal efficiency of the boiler, slagging, choking and scaling of furnaces and tubing, and most-serious of all, generation of a large amount of slag and fly ash. As a rule of thumb, for every MW of installed capacity approximately one acre of land is required for disposal of the ash generated, the material accumulating to a height of 8 to 10 meters. Most of the Indian thermal power plants having little land space, simply let the solid waste run into nearby river, choking the natural drainage system and despoiling the land down stream. Between 85 to 90% of solid waste generated in the form of bottom ash, but 10–15% is lost as fallout from chimneys. The fall out increases the suspended particulate matter in the air and adversely affects the agriculture, land, flora and vegetation around.

In every thermal power station, chimney should have a working electric precipitator to remove the fly ash leaving the chimney. It may be added that using electric precipitator attached to chimney in no way decreases the amount of fly ash coming out of a thermal power station. It removes fly ash from air and gives a control over the fly ash so as to dispose it off properly. This fly ash along with bottom ash should be mixed with gypsum, lime and packed near the power station itself. This mixture on mixing with water will work like cement, and public should be encouraged to purchase it. This will be an extra source of income for thermal power plant.

Fly ash is generally grey in colour, abrasive, acidic, refractory in nature and has a fineness of 4000 to 8000 cm^2 g^{-1} and possesses pozzolonic characteristics. Its particles ranging in size from as much as 120 to less than 5 microns in equivalent diameters being light, get airborne very fast and pollute the atmosphere. Long inhalation of fly ash causes, silicosis, fibroses of lungs, bronchitis and pneumonitis. It corrodes structural surfaces and its deposition affects horticulture. Fly ash disposal in sea/rivers causes damage to aquatic life cycles. Slurry disposal lagoons/setting tanks become breeding grounds for mosquitoes and bacteria. In addition fly ash holds the potential to contaminate the underground water resources with traces of toxic metals present in it.

Fly ash today is a national problem. Existing thermal power plants alone need more than 50,000 acres of land for disposal of fly ash, during their life span of 30 years. The requirement of land, for disposal of fly ash slurry would increase with the addition of more thermal power plants. The annual expenditure on road transportation just for dumping the fly ash is more than Rs. 50 crores. Considering 10 per cent annual growth generation through thermal power plants, the annual fly ash generation is expected to exceed 125 million tons by 2025 A.D.

Appreciating the overall concern for environmental problems being created by fly ash, which otherwise is a very useful byproduct of thermal plants, the Technology Information Forecasting and Assessment Council (TIFAC)[3] of the Department of Science and Technology (DST), identified "Safe disposal and gainful utilisation of Fly ash" as a thrust area and commissioned a Techno-Market Survey: "Technologies for disposal of Thermal Power Station Fly ash" during 1990. The report is a bird's eye view of the technologies either already developed or in the process of development. Some of the technologies used are:

(*i*) Demonstration of Roller Compacted Concrete (RCC) dam construction technologies. Okhla Flyover bridge embankment was made from fly ash and was inaugurated on 3rd Jan., 1996.

(*ii*) It can also be used for construction of road pavement as has been done in "Hanuman-Setu" Yamuna Bazar, Delhi.

(*iii*) Soil treatment for dwelling units at abandoned ash ponds, NFL, Panipat has been completed.

Some industries have also come forward to solve this problem. The Delhi-based Ballarpur Industries Limited (BILT), of which the 1991–92 turnover was Rs. 911 crore, produces bricks from fly ash. BILT had entered into venture with Autoclaved Aerated Concrete (AAC) to make bricks and ceiling panel from fly ash at Palwal in Haryana about 70 km from Delhi. The 200 tones of fly ash material everyday come from NTPC Badarpur Thermal Power Station near the capital. Fly ash is calcinated with lime gypsum and water when it forms cement like substance. This is moulded into bricks and panels which are stronger than cement. These bricks are better insulated and more stable building material. BILT will use Ytong of Germany's technology to make reinforced slabs for roofs, floors, building blocks and wall panels. Even TISCO is helping an entrepreneur V.K. Aggarwal to set up the 8-crore Tatanagar Bricks Company, at Jamshedpur in Jharkhand.

Two technocrats from the Institute for Solid Waste Research and Ecological Balance, N. Bhanumathidas and her husband N. Kalidas have succeeded in creating fly ash concrete, which matches portland cement in strength, but costs just half as much. While they themselves are setting up the 18-crore project at Visakhapatnam in Andhra Pradesh alongwith a German Company Sabzgitter Industries.

Initially, the Gujarat Narmada Valley Fertiliser Company (GNFC) used the 400 tones of fly ash it generates everyday to reclaim slushy land. With about one lakh square meters of land in Zadeshwa village in Gujarat back on the map, it has now formed the 300-crore Gujarat Narmada Fly ash Company to convert the fly ash into bricks. So in this way fly ash can be used in a better way.

The smaller cities and rural areas around them where thermal power stations have been located are often badly hit by the air pollution. The 440 MW thermal power plant at Bhatinda in Punjab used to produce about 1200 tones of fly ash; about 1 kg per resident. The city came under a permanent blanket of smoke, fly ash coats every surface and a good deep breath used to wind up in a racking cough. The electrostatic precipitator rarely works.

Crops on thousand of acres in the vicinity of power plant was damaged, nasal and respiratory disorders has increased. A cruel irony was that Bhatinda got no power from this plant. It was with Bhatinda power plant but it can be with any power plant. It is suggested that electric precipitator is a must with every power plant which should be regularly cleaned to keep it working properly. Use of fly ash for making bricks should be encouraged. These bricks will also decrease the use of clay soil used for making bricks which will also retard clay soil erosion.

A typical Indian 200 MW power station using S-coal emits 50 tonnes of SO_2 and even higher amounts of soot. SO_2 can be removed by a method developed by Kawasaki Steel, Chiba Plant in Japan for Desulphurization and Denitrification facility, as shown in Fig. 12.1. Gases from sintering facility are subjected to dry electrical precipitation to remove dust, next to desulphurization using a slaked lime and gypsum method and then wet electrical precipitation to remove the finest particulate matter. The gases processed through an after burner, a heat exchanger and a heating furnace to elevate their temperature to about 350°C. After ammonia is blown into them, denitrification takes place in reactor filled with catalyst using a contact reduction method with ammonia, thereby removing more than 10 per cent of nitrogen oxides. The cleaned gases return through a heat exchanger and are vented from a stack 200 meters high. In this method by-product gypsum, sulphur, sodium sulphite are recovered which can be used in such establishments as pulp plants and glass factories. Desulphurization will also remove any metal/metal vapours that may have formed in the ignition process

The Suspended Particulate Matter (SPM) counts in India are much above the comparative figures of other countries. Dust level in India are also high being a tropical country. Therefore control of SPM generated by human activities is very important. High count of SPM cause respiratory stress and diseases including TB and asthma.

The second biggest polluter of atmosphere in India are fertiliser plants. The main pollutants are fluorine gas, suspended particulate matter, sulphur dioxide and trioxides, hydrocarbons and particulate from nitrogen based plants. Oxides of sulphur emitted in these plants are much more than the thermal power stations. Most of the plants in India have poor pollution control. However, if the outgoing gases are desulphurized, all other gases and SPM will be removed.

Third biggest polluter of air in India are textile mills. The textile mills operate in humid environment and handle a large variety of organic chemicals in large amounts. Their major affluents are cotton dust, smoke and other combustion wastes, including kerosene and naphtha vapours, sulphuric acid, nitrogen oxides, chlorine gas, formaldehyde and chlorine dioxide.

Areas around textile mills are usually covered with a thin deposit of cotton fluff or dust. Workers inside the mills and residents outside have little choice but to inhale it. Continuous inhalation causes biosinosis, weaken's respiratory functions and resistance to lung diseases, especially TB and chronic bronchitis. In addition to general pollution control, the workers who inhale lot of acidic gases while working should be served with a glass of milk before they leave the factory. Factory workers should also be asked to eat more of jaggery (*Gur*). A general method of pollution control for textile mills cannot be given, because it will vary from plant to plant. However, textile mills should not be located in congested areas of city. These mills should have a good cross ventilation of air with large number of exhaust fans.

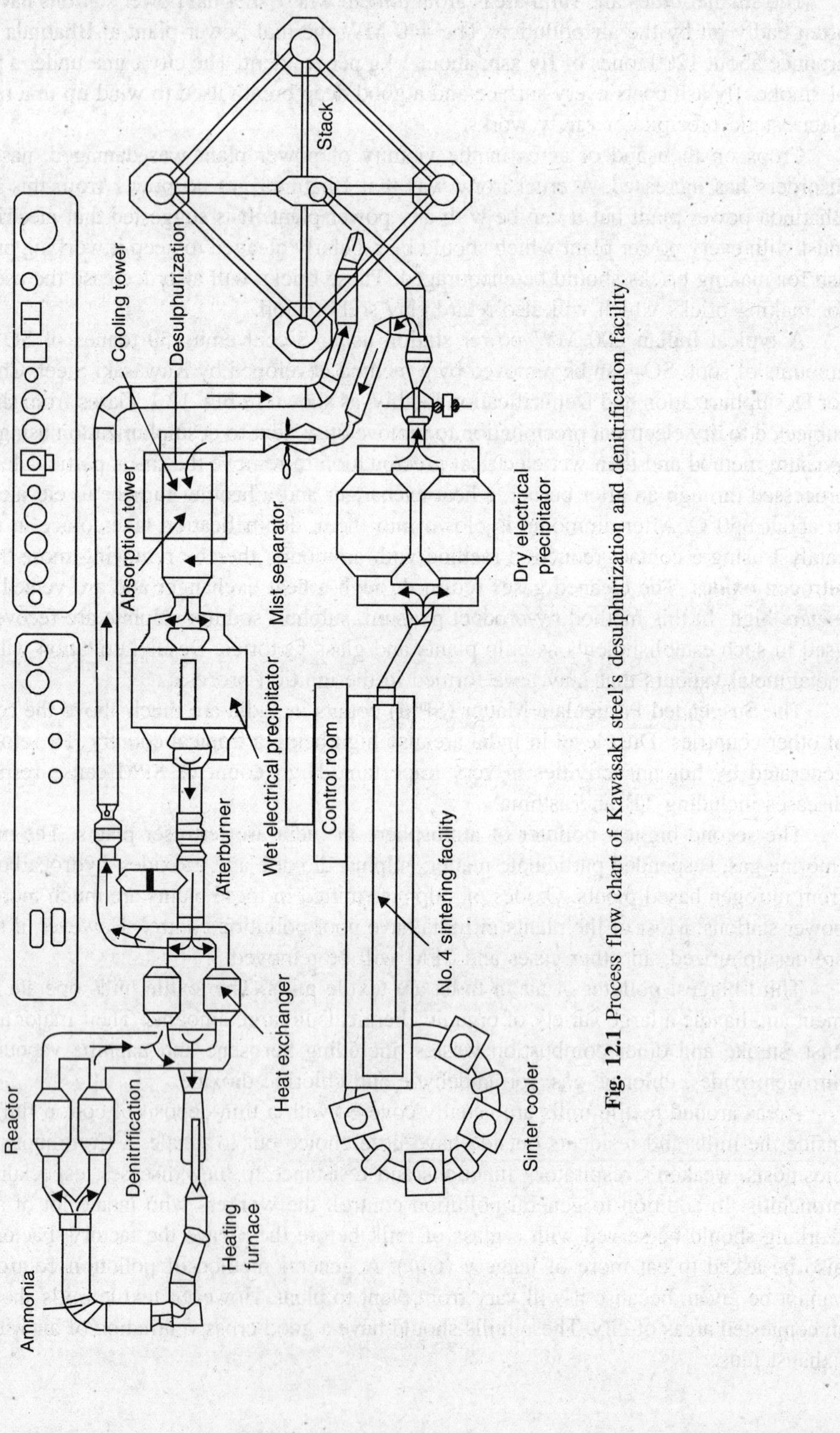

Fig. 12.1 Process flow chart of Kawasaki steel's desulphurization and denitrification facility

Water and air are two major inputs of any industry. We cannot control air movement but water can be controlled and checked. Since in industry, in the course of product-formation, temperature change take place and to control the temperature water is consumed. So if more water is used by an industry, it should be more inefficient with respect to energy consumed and must be throwing out more effluents. In other words, an industry that uses lesser water, lesser is energy consumed, and thus more efficient and consequently will be throwing out less effluent. So if water consumed by an industry can be controlled, industrial pollution can be controlled.

Factories using similar technologies and processes consume widely varying quantities of water. Steel plant use 12–21 m^3 of water per ton of steel, fertiliser plants use 9 to 40 m^3 per ton of paper. The wide variation in specific consumption of water among the 19 paper plants, 9 fertiliser plants adopting the similar technologies reflect difference in operational efficiency and use of outdated technologies. The lowest water consumption figures of our plants are far higher than the international level, shows Indian industry in a poor light. As in the case of steel manufacturers in the USA, Europe and South-East Asia, required as little as 10 per cent of the water consumed by an Indian plant of same size.

Japan, Germany, the USA and many other countries have controlled pollution significantly by controlling water. They made water wastage and consumption more expensive deliberately and it worked well. In 1965, Japan's industrial output was worth \$21 per cubic metre of water which has increased in 25 years to \$77. This lead to water conservation and recycling. In the former West-Germany, similar strategies between 1975–90 increased industrial output by 44 per cent. In the USA, industrial water consumption in 2000 AD was only one-third of what it was in 1977 mainly because of the increased cost of disposing of the waste water imposed by environmental legislation: Indian industries should not be allowed to consume large amounts of water. If it is priced properly, industries will treasure it as priceless commodity. In other words, if we control water management, industrial pollution can be checked. The details of this are given in a report published by the author[2].

The exhaust fumes emitted by Indian traffic contain carbon monoxide, carbon dioxide, oxides of nitrogen, hydrocarbons, aldehydes and lead halides. The black smoke emitted by diesel engines contains more of particulate matter than the exhaust of petrol engines, though other pollutants are less.

About one-fifth of the vehicular population in India is concentrated in the major metropolitan cities. 70 per cent of the vehicles having two and three wheels, which are mostly two-stroke driven engines, while petrol-driven vehicles with four-stroke engines constitute only 14 per cent and diesel driven vehicles make up 8 per cent total. Evidently two wheelers have become the most popular mode of transport. The number of two wheelers have been growing at a rate of about 20 per cent per year, and it is estimated that these vehicles will be around 100 million by 2010 which will constitute more than 80 per cent of the traffic. Even this modest population of vehicular population has become a serious problem in urban area.

60–70 per cent of automobile pollution is through exhaust pipe, 20 per cent occur through blow by from crank case breather and remaining through evaporative emissions from the fuel tank breather, carburettor and spillage losses. Heavy duty diesel-powered vehicles contributes more of NO_x and SPM, while like-powered vehicles and motors are major source of CO and hydrocarbons.

For the prevention and control of vehicular pollution, it is necessary to have a multipronged approach. This should, interalia be addressed to the following tasks:

(*i*) Upgradation of fuel quality and use of cleaner fuel with no sulphur in it.

(*ii*) Enforcement of emission standards for vehicles on the road and at manufacturing stage.

(*iii*) Introduction of Rapid Transport System in all metro cities and in upcoming metro cities.

The quality of fuel used in India is one of the worst in the world. It is, therefore, essential to upgrade the quality of fuel. The fuel reformation involves conversion of aliphatic compound to aromatic compounds by the process called Plat-forming. Such a fuel that has been reformed called blended fuel with benzene, toluene and xylene can be used without TEL. While reforming the fuel, gasoline should be made free from sulphur. Public transport system as well as three wheeler be encouraged to change to CNG or polypropylene. Two stroke engine in two wheelers should be banned in favour of four stroke engine. These changes will eliminate major pollutants particulate matter, oxides of nitrogen, lead and sulphur.

One of the reasons that there is so much congestion on the roads is that government do not provide a good transport in metro cities. The government should introduce some mass rapid transport system in the metro cities, ordinary man will prefer to travel by these transport system and that will decrease pollution.

12.2 CHEMICAL INDUSTRIES

The Indian chemical industries are the second largest in the world, brewing more than 102 million tonnes of chemical every year. Unofficial estimates say that there "are over 11,0000 chemicals being marketed by companies making idyllic premises like Union Carbide Corporation (UCC), Bhopal, did 33 years ago. The industry here is ignorant of properties and noxious potential of the chemicals it produces in increasing quantities. Of the 3350 chemical compounds, for instance used in pesticide sector, the toxicity of only 168 is known. In cosmetic industry of the 3490 chemicals that are in use, toxicity of 400 is only known. Overall toxicity level of 32,400 chemicals out of 36,000 being manufactured in India are still unknown.

National consumption of fertiliser went up from 2,23,000 tonnes in 1959 to 12 million tonnes in 1992. Average production of ammonia went up from 200 tonnes a day to 1,500 tonnes a day in the same period, the production target for phosphorous pentoxide has been set at 5 million tonnes from present production of 1.5 million tonnes. The toxicity of over 60 per cent of the chemicals that go into the making of fertiliser is unknown. The rocketing production graphs have created a huge toxic belt at Ahmedabad-Vapi belt in Gujarat (textiles and basic chemicals), Kota in Rajasthan (dying), Thane-Kalyan in Maharastra (mostly chemical) for drugs and pigments. The much applauded commercial successes of these regions has a dark underside that has swollen over the years, smothering contrary in formation. Ignorance about the exact characteristic of a chemical and its behaviour in varying climate and production conditions is one of the reasons why fear of disaster like MIC cannot be dismissed. The chemical industry is hardly expected to spend millions on production of new chemicals and then get the products banned.

Untrained workers blissfully ignorant of what they are handling in their factories face personal risks, on the one hand, and may cause more accidents and hence prepared for another Bhopal disaster. To make things better and to avoid such disasters our politicians and industrialist class of our country should take more of responsibilities on this account and Ministry of Environment should take a driving seat.

12.3 THE BHOPAL GAS TRAGEDY

Date: December 3, 1984
Venue: Bhopal, Madhya Pradesh, India

Source: Union Carbide Factory, Manufacturer of Carbyl (Carbamate pesticide) that uses methyl isocyanate (MIC)

It was a chilly night of 2-3 December 1984 when the accident that is the worst chemical disaster in history took place. It took a heavy toll of human lives. People started dying with in hours and more than 2000 lives were lost in the first few days. Late Heeresh Chandra, the man who had to deal with it first hand, presented the forensic aspect at the third Indo-Pacific Congress on Legal Medicine held in Madras in December 1989. According to Ivory Doney who reviewed the proceedings, "A silent stunned audience listened with awe the terrible story of 1984, when on the tragic day poisonous fumes killed hundreds of people or maimed thousands of them in the ensuing months. He equated this tragedy to "Pompeii suddenly engulfed in the dust of Vesuvius or Hiroshima when the atom bomb was dropped". He concludes 'that the story should be told at some future international forensic meetings again and again.

Union Carbide previously produced carbaryl without MIC but after 1979 began using MIC because it was cheaper. In the early 1980, the demand for pesticide had fallen, the factory was operating at a loss leading to a series of cost cut measures from around 1982. These measures affected the two interrelated areas: worker's condition and the safety regulations of equipment installed at the plant.

There was a massive leak of MIC stored for a long period in the incriminated Tank 610 of the Pesticide Plant. Extremely violent chemical reactions led to the alarmingly high pressure in the tank that lead to the formations of 21 compounds. All this lead to increase in pressure that ruptured safety disk and lead to escape of MIC. The tank 610 was connected to tank 619 through a safety valve, but no body opened the valve between the two tanks to relieve pressure in 610. Actually the MIC tank alarm had not worked for four years. The gases escaped from an atmospheric vent line 36 meters in the air. The vent gas scrubber was supposed to spray caustic soda on escaping vapours to neutralize them but the scrubber had been out of service for five months before the disaster. To reduce energy cost, refrigeration system designed to inhibit the volatilization of MIC had been left idle-the MIC was kept at 20°C and not at 4.5° C as advised by the manual. As the refrigeration system was not working temperature rose making the reactions more vigorous. Water curtain would have neutralized MIC but was designed to reach to a height of 13 meters, while the MIC vapours were gushing out 36 meters above the ground. The flare tower could not be used because a length of piping was corroded and had not been replaced. The slip blind plates that could have prevented water from pipes being cleaned from leaking into MIC tank via faulty valves were not installed. Their installation had been omitted from cleaning check list. Virtually every relevant safety instrument was either in short supply, malfunctioned or designed improperly. One or more of these factors contributed to the accidental and uncontrolled runaway reaction. It may be pointed out here that the toxicity was not known about any of the compounds, including MIC.

At major hospitals of the city, the patients could be graded symptomatically into four categories: (*i*) Minor eye ailments, throat irritation and cough (*ii*) Severe conjunctivitis, eye ailment eczema keratitis, acute bronchitis and drowsiness (*iii*) Lung membranes were damaged that lead to cardio-respiratory distress (*iv*) Convulsions, followed by cardio respiratory arrest. Intense fatigue and muscular weakness was another common feature. Several thousands survived with a variety of morbidity and permanent disabilities with lingering ailments and incapacity to work. Non-availability of any information about the toxicity of even the parent compound MIC was a great impediment to institute's "detoxication measures" and lay down guidelines for therapeutic intervention and management of the victims.

Heeresh Chandra and his colleagues started performing autopsies with in 72 h of disaster. It was noticed that the usual post-mortem lucidity or cyanosis was not present; instead there was pinkish discolouration over all parts of body. The most important findings were in the lungs. There was a gross increase in the weight of nearly two and a half to three times the normal. The entire respiratory tract showed a series of pathological changes. The lungs were heavily waterlogged and had a distinctive cherry red colour (Fig. 12.2a). The mucous was intensely congested. The trachea and the major division of the bronchi revealed narcotizing or ulcerative changes. The following (Fig. 12.2 and 12.3) striking microscopic findings were noticed by S. Sriramachari[4]. Severe rachitic and bronchitis with denudation of the epithelium was seen in some sections (Fig. 12.2b). There was marked congestion and thickening of the alveolar septa (Fig. 12.2c). The alveoli were filled with eosinophilic albuminous fluid (Fig. 12.3).

Immediate treatment to patients that was given was sodium thiosulphate (NST) treatment. With adequate precautions and statistical appraisal, it was observed that NST was beneficial in relieving the symptomatology of the victims. Thus a major policy decision was taken to use NST as a therapeutic measure. It was shown later on that HCN level of blood victims were more than double compared with living or normal autopsy samples. Thus it was established that there was some degree of HCN toxicity. However because of low HCN concentration in atmosphere and acidic medium in human system, formation of CN^- must have been very very small. Since it is CN^- ion that leads to abrupt death, because of its very low concentration, cannot be the major cause of the deaths.

As the gases escaped from an atmospheric vent line from high pressure to atmospheric pressure there must have been significant fall of temperature. Further, expansion of gas would have decreased the concentration of reactants. These two factors decreased the concentration of reactants. These two factors must have contributed to decrease the rate of reactions taking place in tank to almost stop. As a result, once gases escaped the tank, addition of significant amount of new products may not be there. Also the concentration of 21 compounds formed in tank 610 when dispersed over whole Bhopal would have been very small, making insignificant contribution to the tragedy. This leaves us with MIC as the main cause of the Bhopal tragedy.

Blood haemoglobin (Hb) is made up of globins (94 per cent) and haem (6 per cent). The composition in globins depends on the species from which it is isolated. Globins consist of four-polypeptide chain and in human haemoglobin (Hb) the chains are of two types with terminal acid group, α-chain, valyl-leucyl end group, β-chain valyl-histidyl-leucyl end group. Normal adult Hb consists of α-chain (141 amino acids) and (β-chains (146 amino acids) each. It is the valine present in α and β chains, which is of interest to us.

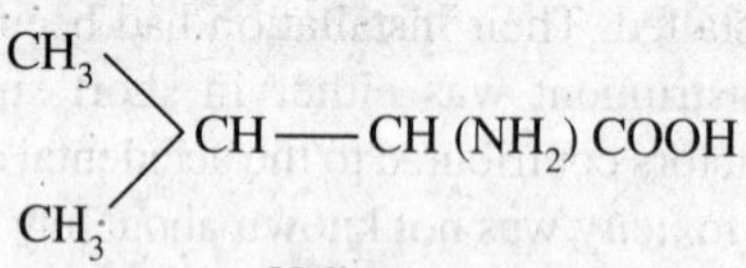

Valine

It was confirmed that end group valine of Hb–molecule reacted with MIC to form carbamoylated end terminal valine residues which de-linked and reorganized to form methyl valine hydantoin (HYH). Figure 12.4 shows the possible stages in the interaction of MIC and terminal valine residues of Hb chain. Carbamoylation of Hb was proof of the MIC crossing the alveolar-capillary barrier and spreading throughout the body as is also shown by autopsy finding, Fig. 12.2b, c and 12.3.

Glutathione consist of glutamic acid, cystine and glycine with following structure. The amino acid valine has two methyl, electron releasing groups and that may have facilitated the formation of N-carbamyolation. This is followed by cyclization that lead to hydantoin formation (Fig. 12.4) with its release from the rest of globulin.

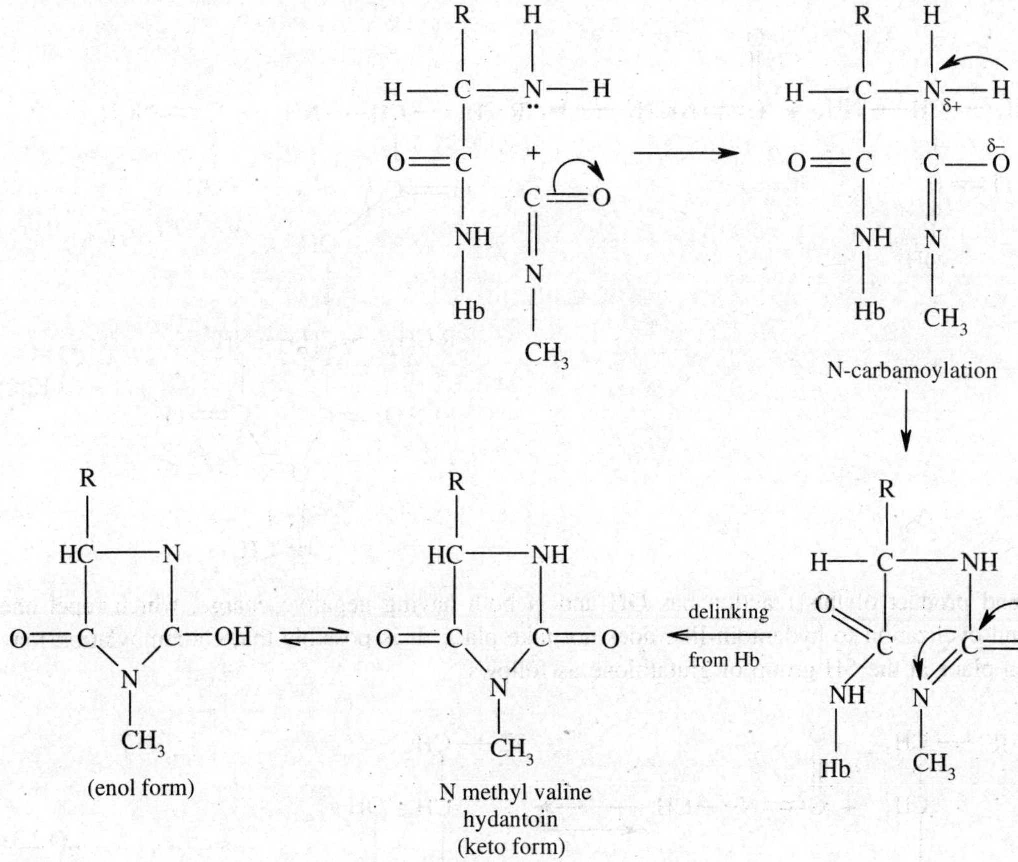

...(12.1)

Fig. 12.4 Chemical pathways of N carbamoylation Hb-part of Globin; R here represent group $(CH_3)_2CH^-$.

However the results of N-carbamoylation did not fully answer the issues of recurrence and repeated episodes of illness in patients. It was demonstrated later that glutathione undergoes S-carbamoylation.

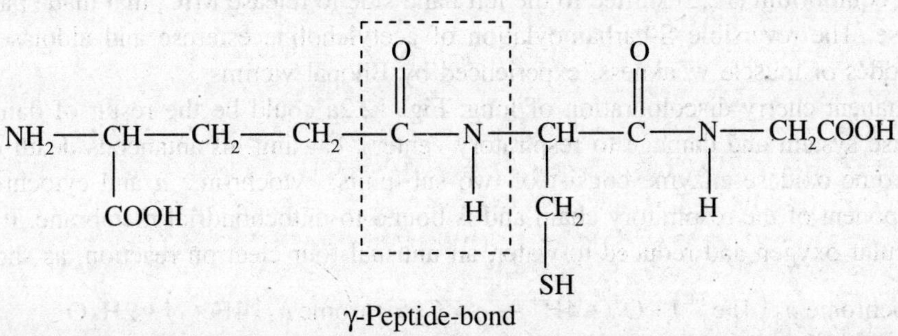

Glutathione probably helps to maintain sulphural groups of protein in the reduced state, iron in ferrous state and acts as a reducing agent. The terminal NH_2 group of glutamate may have undergone carbamoylation by reaction 12.2

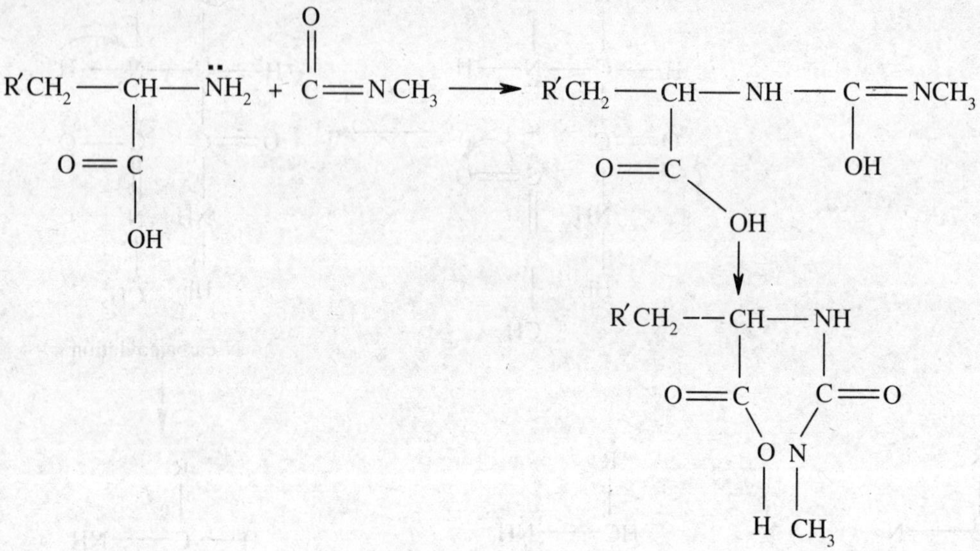

...(12.2)

The end product of this reaction has OH and N both having negative charge, which repel one another and cyclization to hydantoin like does not take place. It is possible that carbamoylation may have taken place at the SH group of glutathione as follows:

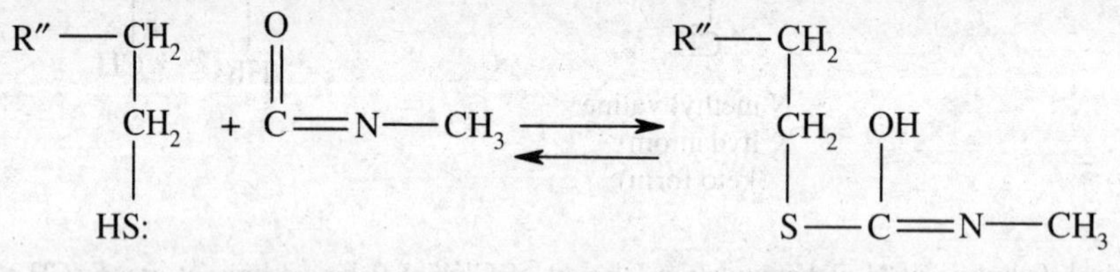

...(12.3)

This carbamoylation is called S-carbamoylation and is an equilibrium reaction. This S-carbamoylation functioned as an exchange pool or reservoir of isocynate in body. As the concentration of MIC in body decreased the equilibrium (12.3) shifted to the left hand side to release MIC, that made patients repeatedly to relapse. The reversible S-carbamoylation of acetylcholine esterase and aldolase account for repeated episodes of muscle weakness, experienced by Bhopal victims.

The permanent cherry discolouration of lung, Fig. 12.2a could be the result of damage to cytochrome oxidase system and damage to respiratory centers, causing instantaneous death of many victims. Cytochrome oxidase enzyme consist of two sub units: cytochrome *a* and cytochrome a_3. It is terminal component of the respiratory chain and is bound to mitochondrial membrane. It carries electron to molecular oxygen and reduced to water, an unusual four electron reaction, as shown below

$$\text{Cytochrome } a_3\left(4Fe^{2+}\right) + O_2 + 4H^+ \longrightarrow \text{Cytochrome } a_3\left(4Fe^{3+}\right) + 2H_2O \qquad ...(12.4)$$

Damage to cytochrome oxidase system means reaction 12.4 cannot take place, thus damage to respiratory centers causing instantaneous death. The subsequent alterations in the blood gases resulted in cherry red colour (Fig. 12.2a).

The flooding of the respiratory system with massive MIC led to rapid stimulation of the respiratory centers and eventual death. The uniqueness of MIC toxicity became apparent with rapid development of chemical binding of MIC to end terminal amino and SH groups. The successful demonstration of N-carbamoylation of blood clearly provided an explanation of the cherry red discolouration. It seems that biochemical lesion of Bhopal disaster may lie between the interplay of N- and S-carbamoylation.

This raises the question, did hydantoin play a role that lead to death of so many person? The answer is Yes. The moment hydantoin formation took place, it got converted to cation as

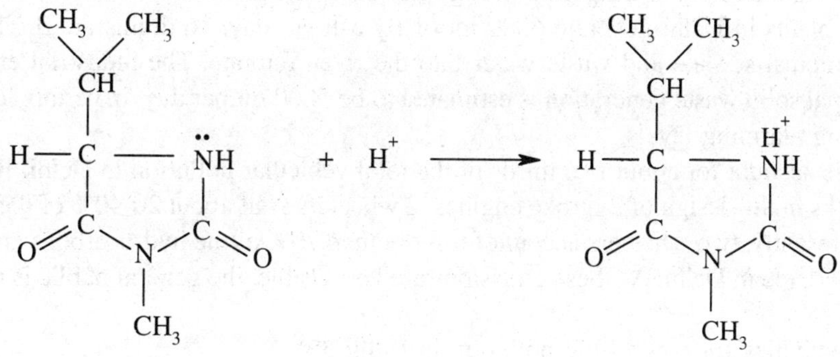

Hydantoin — Hydantoin cation

This must be an irreversible reaction and it decreased the concentration of hydrogen ion and slowly incapacitating glutathione to act as oxygen carrier and that lead to the death.

So overall reaction was MIC lead to N-carbamoylation, which lead to the formation of N-Methylvaline Hydantoin. Subsequently, hydantoin removed protons from the pathway of Glutathione incapacitating to carry oxygen. As the concentration of hydantoin increased, more and more glutathione molecules were incapacitated to carry oxygen and that lead to death. However, when the concentration of MIC decreased in the system, more of it was released from reversal of S-carbamoylation and that lead to relapse of patients. Relapsing of patients is a sign of nonavailability of oxygen and thus corroborates the mechanism.

12.4 STATE OF SOME METRO CITIES

While the large metropolitan cities suffer the worst level of air pollution in scale, small towns are catching up fast, but there is no quantitative data available for such area.

Mumbai: Mumbai predicament is probably the worst only because of city's peculiar geography. The main offenders of air pollution are the Rastriya Chemicals and Fertilizers, Bharat Refineries, Hindustan Petroleum refineries, Tata thermal power plant, Bombay gas company and nearly 450 major industries. Transport and industries contribute maximum to the pollution in Mumbai.

All these big units, with the exception of gas company are located in the Trombay-Chembur area. The 100-year old Bombay gas company is situated in Lalbaug. Also in the Lalbaug-Parel belt are most of Bombay Textile mills whose tall chimneys spew tonnes of pollutants into the atmosphere everyday. Chembur has the highest concentration of chemical industries in the country and is called now Gas-Chembur by the local inhabitants. To make the matter worse, RCF plants and refineries are about to expand or are already in the process of doing so. Tata's have already put in a massive 500 MW unit and have applied for yet another of the same capacity.

The impact of this pollution is evident from a detailed study of 4000 people conducted by R.S. Kamat of Mumbai KEM Hospital and Air Pollution Prevention Cell of Mumbai Municipal Corporation in 1977–78. The health of residents in a clean western suburb like Khar was compared with the residents of Chembur and Lalbaug. The survey revealed that people living in the congested industrial area suffered from a much higher incidence of disease like chronic bronchitis, TB, skin allergy, anaemia, irritation of eyes. The absentee rate of workers was much higher, particular in Lalbaug's textile mills. Also, there was a rise in the number of deaths due to cancer in Lalbaug twice as many as in Chembur or Khar.

Delhi: Vehicular pollution contributes 67% of the total air pollution load (approximately 3000 mt per day) in Delhi. Another 25% of air pollution is generated by industries and coal based thermal plants. The three power plants in Delhi generate 6000 mt of fly ash per day. 16 drains are discharging about 1900 mld of municipal sewage and waste water into the river Yamuna. The industrial effluent load is 320 mld. Municipal solid waste generation is estimated to be 5000 mt per day. In certain localities noise levels are attaining alarming levels.

Two-wheelers account for about two thirds of the total vehicular pollution in Delhi. Because of the inherent drawbacks in the design of 2-stroke engines, 2 wheelers emit about 20-40% of the fuel unburnt/partially burnt. Presently, two wheelers account for more than 70% of the hydrocarbons and nearly 50% of the carbon monoxide in Delhi. As these emissions are less visible, the general public is not concerned about it.

The salient activities for controlling pollution in Delhi are

1. Main emission norms for all vehicles be tightened. All vehicles must comply to Bharat Stage III/Euro III emission norms.
2. All two stroke engines be banned and must be replaced by four stroke engines.
3. Only CNG driven 3 wheelers be allowed to ply in Delhi.
4. The three coal based power plants should switch over to beneficiated coal/superior quality of coal.
5. Provision for dry collection and storage facilities for fly ash should be made in power plants.
6. Polluting industries be shifted from nonconforming to conforming areas.
7. Ensure that pollution control devices are installed in every air polluting industry.

Chennai: Chennai is a little better off. Its major industries include the Madras Refineries Ltd. (MRL) and Madras Fertilizers Ltd. (MFL), Kothari Chemicals and the thermal power station at Ennore and Basin Bridge. MRL processes crude oil with high sulphur content and thermal power station use high sulphur coal to contaminate environment with sulphur dioxide. The MFL unit vents urea dust into air. Large amounts of ammonia leakage from pipelines and storage tanks cause eye irritation and coughing. Fly ash of the power plants is dumped off the coast into sea. Graphite India spews out fine graphite dust day and night which settles in open surface killing plants, flowers, enters human body causing stomach aches and many other problems.

Kolkata: According to Dr. B.B. Suderesan, NEERI emits a total of 1,299 tons of pollutants a day, 60 per cent of the Kolkata residents suffer from respiratory diseases because of air pollution. The effect of such pollution on a person is equivalent to that of smoking 20 cigarettes/day.

The apex courts order confirms increasingly interventionist stance it has taken on the issue of environmental pollution and ecological balance. Earlier in January 1995, it ordered the closure of 30 polluting industries for their flagrant violation of environment norms. Following a flutter in the media and the state government and agitation among the trade unions which feared loss of over 4,000 jobs, the court suspended the order for three months but not before warning that discharge of effluents into the Ganges and release of toxic gases into atmosphere would not be taken kindly. The 30 identified units may by far be the worst pollutants but they are not the only one. There are hundreds of small and medium units that have been contributing to pollution of Kolkata. Unlike the major industrial units, these units have no funds to install safety devices and have been seeking financial assistance from the state, which has so far not come.

12.5 THE POLLUTED AIR AND DYEING

In 1982, the Centre for Science and Environment published a report "The state of India's Environment, 1982". The following is being quoted from it:

People of Nalavgulu village of Chitradurga district in Karnataka face a Hobson's choice: Either they can cling to their ancestral village whose air and water are badly poisoned by the effluents of a chemical factory and die a painful slow death or look for a new place to settle. They had no where to go, however, and nobody was in a hurry to save them.

The culprit of this environment disaster is Harihar Polyfibers–Rs. 25-crore Birla company which was established in 1969. What was once a baby boon for the villagers, providing many jobs, has turned into a virtual nightmare. The fortnightly **India Today** reports that as production began, Harihar Polyfibers tall chimneys started showing soot mingled with chemicals and other toxic byproducts. Now pillars supporting huts, beams holding the roofs of houses almost crumble into powder at the barest touch. A brass' pitcher presented by Harihar Polyfibers to Nalavgulu school two years earlier, crumbled when the magazine photographer tried to lift it.

The variety of diseases reportedly afflicting men, women and children include eruptions, boils, pock marks as in small pox, dry skin, and skin irritation, crashed soles, burning sensation accompanied by cough, yellowing of the eyes as in jaundice and an unknown malady of the intestines". Livestock and other animals were also noticeably suffering from unusual ailments.

Water tanks near the factory are covered with a stream of froth. The froth bubbles keep on popping about and their spray settles down on human eye, ear, nose, besides contaminating exposed food.

Treated effluent waste is dumped in the Tungabhadra river an open drain. "We have put up a Rs. 1.2 crore treatment system and are spending Rs. 30,000 a day to treat the water discharged", said K.I. Philip, Chief Executive of the plant, in an interview "Why should we spend another Rs. 25 lakhs for covered drainage pipes?"

According to Philip, after installation of a Rs. 3.2-core electrostatic precipitator, "The air the Nalavaguluans breath is as pure as mountain air" yet India Today's correspondents reported, "Less than 24 hours exposure to Nalavagulu atmosphere was enough to make us itchy and our eyes watery".

The Nalavagulu residents are desperate. All they want a new 40 acres of land elsewhere with dwellings, but no one provided it.

12.6 CLEAN PRODUCTION & IMPROVE PRODUCTIVITY

Indian industrialisation, like others, is costly to public health and the environment. When end of pipe pollution controls are added to curb industrial discharges, less damage occurs. But these solutions are expensive and are often not optimal from an environmental perspective.

Clean production is a comprehensive, preventive approach to environmental protection. This requires organisations to be creative and investigate all phases of the manufacturing processes and life cycles. Cleaner production thus encompasses action such as energy and raw materials conservation, elimination of toxic substances (as raw materials and as product constituents), and reducing the amount of wastes and pollutants created by products and processes, thereby lowering the amounts emitted to air, land and water.

Reduction of Sulphide in Effluent from Sulphur Black Dyeing at Century Textiles and Industries Ltd., Mumbai—an example.

A study of the available methods of sulphur black colour dyeing and the treatment options was made. Art investigation was made into an alternative to sodium sulphide. It was found that an alkaline solution of glucose could bring about satisfactory conversion of sulphur dyes. However, the high cost of glucose was the main constraint in practice.

A market survey was conducted for procuring an equivalent chemical at a competitive price. This led to the identification of the by-product of the maize starch industry, hydrol, which contained about 50 per cent of reducing sugars. Experiments revealed that 100 parts of Na_2S could be substituted by 65 parts of hydrol plus 25 parts of NaOH. The dissolving and the affinity reactions were carried out in a single stage.

The substitution with hydrol was implemented with a redesigned mixing strategy. The dyeings obtained after substitution were seen to be equivalent to conventional dyeing in depth of shades, fastness properties, etc., plus there were some other improvements in the quality of the dyed product. The process has been in use since April 1990 and has resulted in the reduction in sulphide concentration of the effluent from 30 ppm to 2 ppm. The substitution resulted in a slight increase in the BOD load on plant, but this increase was found not to be critical and was easily manageable with the existing biological treatment system.

Enabling Technology: Theoretical understanding of the chemistry is involved and a search for suitable reagents.

Advantages: (*i*) The reduction of sulphide in effluent; (*ii*) improved settling characteristics in the secondary settling tank of the activated sludge unit; (*iii*) less corrosion in the treatment plant due to reduced sulphide levels in the effluent, and (*iv*) the foul smell of sulphide in the work place was eliminated.

Economic Benefits: No capital expenditure was involved in this substitution and in fact, the operating cost were found to be marginally lower. The saving of not installing an additional effluent treatment facilities was about \$20,000 in capital expenses and about \$3000 in running expenses per annum. The substitution has not incurred any additional operating costs.

REFERENCES

1. "*The State of India's Environment*", A Citizen Report, 1982, published by the Centre for Science and Environment, New Delhi (1982).
2. Sindhu, P.S. and Malik Iqbal, "Control Water – Control Industrial Pollution" A report published by Vatavaran, NGO, New Delhi (1996).
3. TIFAC-report, "Technologies for Disposal of Thermal Power Station Fly Ash", published by Director - Fly Ash Mission, Technology Information, Forecasting and Assessment Council (TIFAC), Department of Science and Technology, Ministry of Science and Technology, Government of India, Technology Bhavan, New Mehrauli Road, New Delhi-110016.
4. Sriramachari, S., "The Bhopal Gas Tragedy: An Environmental Disaster", *Current Science*, Vol. 86, 905-920, (April 2004).

13

Noise Pollution

13.1 INTRODUCTION

Sound waves behave like ripples on a pond. Like the ripples, sound waves move away from the source at a constant speed. Like the ripples on the pond, sound waves simply pass through the atmosphere by exchange of momentum. There is no net transfer of matter from the source, just as in case of the pond there was no net fluid flow. In the pond, one can see a two-dimensional pattern of circular waves spreading out from source. In atmosphere, of course, the pattern is three-dimensional and the sound waves are spherical. The speed of sound in air varies slightly with temperature and humidity, but is independent of frequency. For most purposes, it may be regarded as having a constant value of about 341.4 ms^{-1}. The speed of sound in water is about 1478.3 ms^{-1}.

Most sources of the encountered sound waves are due to some mechanical vibration, such as the vibration of a panel covering a machine. In order to reduce the sound produced, the visibility amplitude must be decreased. This is because sound pressure (P_{rms}) is directly proportional to amplitude (A), as $P_{rms} = A/\sqrt{2}$.

However, in some sources of sound, there are no mechanical vibrations involved. Such sources are found in air flows such as jet flow and pipe or duct flow. Here the sound is caused by forces in fluid itself due to the process of high velocity fluid mixing with fluid moving with lower velocity. Some reduction in the sound generation can be achieved with a proper understanding of the process involved. However, the best approach in these cases is to try to absorb the sound once it has been produced.

13.2 SOUND PRESSURE AND POWER LEVELS

The Decibel: The thing to remember about the decibel is that it represents a relative measurement or ratio. Each quantity measured in decibel is expressed as a ratio relative to a reference, power, pressure, intensity or whatever quantity we are considering. For this reason, whenever we use the decibel we also use the word level... sound power level, sound pressure level, *etc.,* to remind us that we are referring to a reference quantity and that our measurement is relative, not absolute.

The Bel word originated in the 1920s to describe attenuation of signals in telephone cables. One Bel (named after Alexander Graham Bel) is the ratio R given by

$$\log_{10} R = 1 \qquad \therefore \qquad R = 10^1 = 10 \qquad \text{...(13.1)}$$

Because this is a large ratio, a smaller ratio, the decibel is used more. One decibel is the ratio r given by

$$\log_{10} r = 0.1$$

or
$$10 \log_{10} r = 1 \text{ dB} \quad \text{...(13.2)}$$

$$r = 10^{0.1} = 1.26 \quad \text{...(13.3)}$$

Thus, the decibel represents the ratio of 2 numbers equal to 1.26.

Sound Pressure Level (SPL or L_P): This is often abbreviated to SPL or L.

$$\text{SPL} = 10 \log_{10} \frac{\langle P^2 \rangle}{\langle P_{\text{ref}}^2 \rangle} = 20 \log_{10} \frac{P}{P_{\text{ref}}^2} \text{ dB} \quad \text{...(13.4)}$$

where P is the rms (root mean square) sound pressure and P_{ref} is an rms reference pressure = 0.0002 μ bar = 0.00002 Nm^{-2}. Note 1 atmosphere = 14.7 × 10^{-6} psi. The reference pressure 0.0002 μ bar has been chosen as the smallest sound that an average young adult can just hear in quiet surrounding. Figure 13.1 will help indicate some common sounds on the-SPL scale pictorially. As already noted above, the ratio of pressure squared terms was found to obtain the decibel. The reason for this is that we prefer to compare power like quantities such as intensity (which is power/unit area) and we know from equation 13.5 where intensity $\langle I \rangle_t$ as

$$\langle I \rangle_t = P_{\text{rms}}^2 / \rho_c \quad \text{...(13.5)}$$

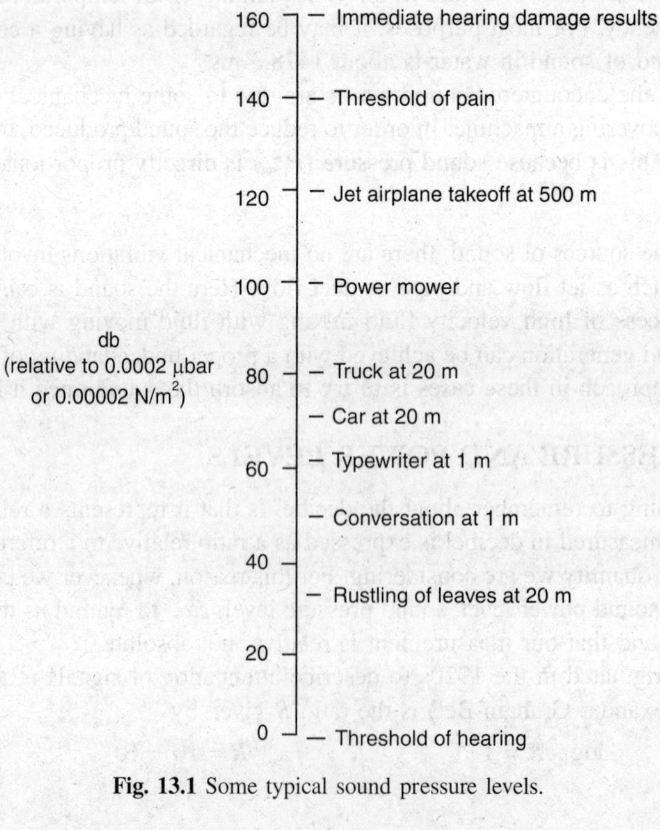

Fig. 13.1 Some typical sound pressure levels.

The quantity ρ_c is known as the characteristic acoustic impedance of the fluid and for air $\rho_c = 428 \text{ kg m}^{-2}\text{s}^{-1}$ at 0°C and $415 \text{ kg m}^{-2}\text{s}^{-1}$ at 20°C.

We see that the ear responds to very small sound pressures and the range of sound pressure amplitudes varies by a factor of about 10^6. The level of noise is expressed in terms of SPL in dB. Sound is often described in terms of loudness but this cannot be measured like sound frequency, sound pressure and sound intensity. Loudness may be described as listener's auditory impression which is synonymous with sound pressure. For example, a sound of 60 dB is louder to the listener if the frequency is 1000 Hz and soft when the frequency is 100 Hz. Similarly, a sound of 20 Hz must have a sound pressure of 80 dB to be audible, but a sound of 2000 Hz is audible at a pressure of 20 dB.

Loudness is expressed in a relative unit, called sone. One sone equals the loudness of 40 dB sound pressure at 1000 cycles per second (Hz), while 40 dB sound at 5000 Hz seems twice as loud and equals the value of 2 sones. Generally, 85 dB with loudness 10–50 sones, depending on the frequency may be critical for ear damage. Normally hearing is most acute in the frequency range 2000–5000 Hz. Perfect hearing lies in the range of about 16 (infra audible) to 20,000 (ultrasonic) Hz.

The quality of unpleasantness of sound depends upon the following important factors:

(*i*) The frequency of sound waves;
(*ii*) The intensity of sound waves;
(*iii*) The time of exposure of sound waves, and
(*iv*) The intermittence of sound waves.

Varies by a factor of about 10^6. The level of noise is expressed in term of SP in dB.

Sound Power Level: Again we note the word level suggests a relative measurement. Sound Power Level is often abbreviated as *PWL* or L_w

$$PWL = 10 \log \frac{W}{W_{\text{ref}}} \text{ dB} \qquad \text{...(13.6)}$$

where W is the acoustic power of the source, and W_{ref} is the reference acoustic power ($W_{\text{ref}} = 10^{-12}$ watt). Some typical acoustic power levels are given in Fig. 13.2.

It is very striking that small sound powers can have such a large effect on the ear. It is estimated that the roar of a football crowd produces only enough energy to heat a cup of coffee. A full symphony orchestra only produces about one acoustic watt. The World Health Organization has recommended 75 dB as explosive limit to industrial noise. Sound beyond 80 dB harms hearing system and so it can be regarded as pollutant. The largest noise a man can hear without much discomfort is thus about 80 dB. A noise of over 90 dB heard for more than 10 milliseconds causes tympanic membrane to contract, because of aural reflex action. A noise of over 140 dB causes the middle ear muscles to change the direction of movement of ear ossicles, which makes a sudden decrease in the noise intensity reaching the inner ear. Such protective reflex actions, however, exist only for a short duration against loud noise. Thus, a noise level up to 65 dB has been considered tolerable for hospital zones.

13.3 EFFECTS OF NOISE AND VIBRATIONS ON MAN

In many practical cases the sound signal cannot be predicted and it is random with time. Such signals for which mathematical description becomes difficult are called noise signals. Noise is known to have other non-auditory effects on people. These other effects may be classed as psychological (*e.g.*, annoyance, fear, effects of noise on actions, mental performance and work performance in industry) and physiological

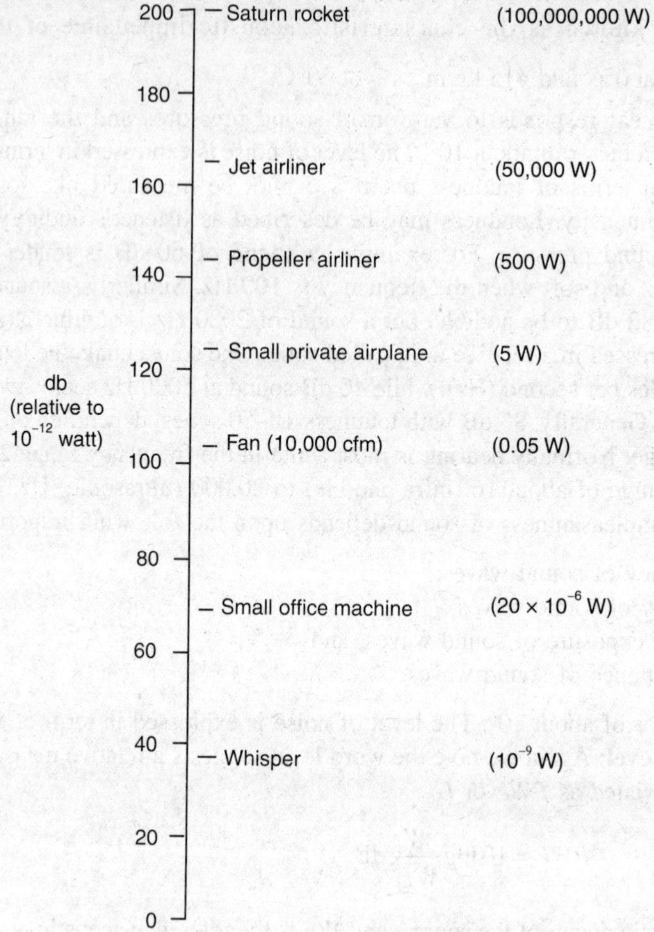

Fig. 13.2 Some typical sound power levels.

(*e.g.*, startle, stress, sleep, vision, pulse rate, blood pressure, and muscle tension). There is a concern that some of these effects mentioned above could be dangerous to health. However, it is difficult to isolate such effects, and research and experimental data have been inconclusive to date. Most acousticians trained in psychoacoustics hold the view that noise induced hearing loss is the major effect of intense noise and that all other observed psychological and physiological effects are of a minor nature at least as far as the health of people is concerned. In the summary of his book, Kryter[4] makes the following statement: "... other than as a damaging agent to the ear and as a masker of auditory information, noise will not directly harm people or interfere with psycho-motor performance. Man should be able, according to this concept, to adapt physiologically to his noise environment, with only transitory interference effects of physiological and mental and motor behavioural activities during this period of adaptation. This concept, or its converse is difficult to substantiate by scientific research and may be hypothetical at this time."

Kryter's view is not universally shared, and in many other European countries, non-auditory effects of noise are considered to be more serious. In Russia, it is believed that noise can lead to increased accidents and absenteeism at work. Some of the non-auditory effects of noise are as follow:

Sleep loss: It is common knowledge that noise interfere with sleep. We are all awakened by intense noise and most will easily be awakened by an alarm clock. However, the phenomenon is more complicated than it might appear at first sight. A person passes from one stage to another during a night sleep. There are several ways of labelling these sleep states. One common way is REM/(rapid eye movement) 1, 2, 3, 4. It is in REM stage most dreams occur. Here the person has internalised his attention but is not in a deep stage of unconsciousness. However, a person is insensitive to sound. The different stages of sleep can be detected by attaching electrodes to the body and monitoring brain waves and other behaviour. It appears that sleep in stage 2 is most easily disturbed, while deep sleep in stage 4 is disturbed least easily.

It has been suggested that while noise undoubtedly produces sleep pattern in people similar to those of a poor sleeper, a person compensates by spending more time in deep sleep, by becoming less responsive to stimuli and by napping. Thus noise may not deprive a person of efficient sleep to produce an adverse effect on health. Thus sleep disturbance caused by noise will reduce a person's feeling of well being, and that if this disturbance is sufficient, it could conceivably have an adverse effect on health.

Noise-Induced Stress: Stress as a state of arousal which can eventually have adverse health effects. The general response to any form of stress is called *adaptation syndrome.* This consists of three stages: alarm, a stage of resistance and a stage of exhaustion. It is suggested by medical authorities that some non-auditory diseases can be produced in the resistance stage of stress, such as ultra and blood pressure. Certainly, it appears that laboratory experiments are inconclusive at present. Although it appears that short-term changes in blood circulation and secretion changes in blood circulation and the secretion of chemicals in the urine are produced, it is uncertain if any long-term effects are produced. The incidence of hypertension in male and female workers at a noisy workshop is more.

Balance: As long ago as 1926, Tullio reported that pigeons could be made to swerve off course, if exposed to intense low frequency noise. From this experiment some have concluded that none has a direct effect on the balance. More recent experiments both at work and in laboratory show that intense noise can produce complaints of vertigo (dizziness), nystagmus (rapid involuntary side to side eye movements) and balance problems in people. However, the levels to produce these effects are quite high about 130 dB.

N-Response: The following effects of noise are called N-Response.

(*a*) Blood circulatory changes caused by vasoconstriction (temporary shrinkage) of the peripheral blood vessels and hence increased blood flow in brain; (*b*) slow breathing; (*c*) changes in electrical resistance of the skin, and (*d*) changes in muscle tension. Such response do begin at quite low sound pressure level (*e.g.*, 70 dB at 1000 Hz). The response becomes progressively greater as the sound pressure level is increased. It is not known at present if these responses decrease with adaptation to noise.

Startle: If people are subjected to sudden impulsive-like sounds they respond with a startle reaction. The reaction is controlled by an auditory-muscular reflex and tends to turn the load and eyes to the source. Presumably this occurs in order to prepare the person for appropriate action in case there is some danger at hand. This reflex action begins at quite low levels and normally causes a change in the rate of heart beat and blood pressure. However, it appears that there is considerable adaptation to startle. It has been reported that babies of mothers, who were pregnant in neighbourhoods exposed to aircraft noise, were much less aroused from sleep by aircraft noise than were the babies of mothers who were pregnant in non-noise exposed neighbourhood.

Effect of Vibrations: Several of the human organs have resonances which can produce sensation of pain if excited by vibration or noise at their resonance frequencies. This resonance frequencies are in the range of 5 to 100 Hz. People seem to be most responsive to vertical vibration. Building response to vibration will be insufficient to cause cracks in plaster finishes or in walls until it becomes unpleasant or painful.

13.4 EFFECT OF NOISE ON PERFORMANCE

The generalised agreed findings are as follows:

(*a*) Steady noise (without meaning) does not interfere with human performance unless the sound level exceeds 90 dB(A) and even then not consistently;

(*b*) Impulsive noises are more disrupting to performance then steady noise. Bursts of such noise can be disruptive even for low level below 90 dB(A);

(*c*) High frequency noise (above 2000 Hz) appears more disrupting than low frequency noise (below 2000 Hz);

(*d*) Noise does not seem to change the rate of work but rather increases the variability of the work;

(*e*) Noise increases the number of errors made at work rather than the total amount of work; and

(*f*) Complex tasks are more affected by noise than simple task.

It has been suggested that people have a limit to the number of features of their surrounding that they can observe in a certain time interval. All the stimuli which are received compete for attention, and the human sensory system receiving all these stimuli containing both useful and useless information has to filter out useless information, such as noise. It is supposed that human mental filter functions as follows:

(*a*) It rejects or ignores signals which do not change with time even though these signals may be important.

(*b*) A person's state of arousal, stress or fatigue may alter the mental filters ability to disorientate.

(*c*) The filter can be overridden or overloaded by useless information which demands attention, such as new, intense or unexpected stimuli. The overloading of the mental filter can cause momentary lack of attention and other specific effects, such as auditory masking and muscular activation. Noise does affect the physiological effects annoyance, fear, effect on action and physiological effect like pulse rate, blood pressures vision but is not understood.

Effect on Blood Pressure: Noise pollution effects human health, comfort and efficiency. It causes contraction of blood vessels, makes the skin pale leads to excessive secretion of adrenalin harmone into blood stream which is responsible for high *blood pressure.* Adrenalin changes hormone content of blood which increases rate of heart beat, constriction of blood vessels, digestive spasms and dilation of pupil of eye.

Auditory Fatigue: It appears in the 90 dB. Temporary deafness occurs at 4000–6000 Hz, while a permanent loss of hearing occurs at 100 dB due to continuous noise exposure. High interior impulsive noise resulting from an explosion or sudden excessive noise of more than 150 dB may cause instantaneous damage to internal structure of ear.

13.5 HEALTH HAZARDS OF NOISE POLLUTION

Although pollution of land, sea and air has been well documented, the least recognised version is the swelling tide of noise which is engulfing urban as well as rural areas. Unlike other pollutants noise lacks visibility, seldom registering on the consciousness, except as a trifling irritant to be dismissed at will and, therefore, less likely to be perceived as a threat. Noise poses a threat to health and is known to cause a number of complications. Declining productivity among workers in certain industries has been directly correlated with noise levels, particularly those under constant exposure to the menace.

Noise not only impairs the physical and psychological functioning of human organism, but also causes nausea, vomiting, pain, hypertension and a host of other complications including cardio-vascular complaints. Experiments with animals have shown that certain intensities (levels) of noise cause convulsions and even death.

Beside affecting health, noise is also a significant factor in economic losses. According to Prof. Glorig, an international expert on sound, "the potential cost of noise induced hearing loss to industry is greater than any occupational disease". In the USA alone, noise reportedly causes losses running into millions of dollars daily, affecting large segments of population.

On the decibel (dB) scale, a whisper is equal to 20 dB, normal speech ranges between 30 to 50 dB, vehicle engines peak at 90 dB while reviving up. A jet engine roars up to 170 dB. Uninterrupted exposure to noise above 80 dB is harmful. 90 dB is the upper tolerance limit presented by the International Standards Union. The effect becomes more pronounced. Eight hours daily exposure causes life-long deafness. At 93 dB the tolerance level drops to four hours and then to two at 96 dB. The higher the intensity of sound, the more vulnerable becomes the hearing organ.

The phenomenal growth in the volume of vehicular traffic has contributed exponentially to demon of decibel in urban areas. Beginning with pressure horns there seems to be no let up on the production of more obnoxious variants to be sounded indiscriminately. The idea is not to warn other motorists or pedestrians, but to frighten them out of the way. On its part, the administration is equally guilty of overlooking defaulting pressure horns users and nonusers of silencers.

Noise levels are particularly acute near railway junctions, traffic roundabouts, bus terminuses, transit area, fish and vegetable marts. Combined with lung power it exceeds 90 dB at peak hours representing a dire threat to patients and pregnant women in the vicinity. Despite "no horn" sign in prohibited zones near hospitals, asylums and elsewhere, indifferent drivers continue with the act. The situation is worse in complexes located close to the airports as they are exposed to shattering blasts by jet aircrafts round the day.

Rock music when played at high volume makes the listeners to rock and dance. The high volume of music increases the listeners' blood pressure and thus listener starts dancing. However, if ordinary song is played at high volume, one does not want to listen, hence no such effect. Also rock music played at low volume will not have that effect.

Recreational noise, another ugly facet, is becoming more widespread in cities and towns. Loudspeakers are turned on with full volume during marriages, festivals, jagrans, musical programmes, particularly at night without the least consideration of others. Even at 50 dB, sound can waken a person from deep slumber. Loudspeakers output is from 60 dB to 80 dB causes the pupils of a slumbering person to dilate, with increasing intake of oxygen, resulting in palpitation. The effect is more pronounced in narrow lanes. TV sets are played at full volume at prime time, invariably disturbing neighbours. Noise making seems to have become latest status symbol in India, be it an election campaign or slogan shouting or advertising ownership of a TV set.

There is no marked vertical escalation in noise levels, but there is a horizontal spread encompassing more areas. In other words, with proliferating numbers of two, four and six wheelers on the roads, noise levels have become uniformly high over large area, doubling every six years for the last 30 years in Delhi. At this rate, the day is not far when hearing losses would be as common as the common cold.

13.6 PREVENTION OF HEARING LOSS IN INDUSTRY

The USA was the first Western country to introduce industrial noise regulations. In 1969, new standards for industrial noise became effective by a revision of the Walsh-Healey Act. These standards set the following permissible noise exposures as are given in Table 13.1. These standards originally affected only 28 million out of the 40 million US non-governmental and non-agricultural workforce. However in 1971, these standards were extended to apply to the employees of all the US companies engaged in interstate commerce when the Occupational Safety and Health Act (OSHA) became law.

Table 13.1 Permissible noise exposures

Duration per day, hours	*Sound level dB(A) slow response*
8	85-90
6	92
4	95
3	97
2	100
1-1/2	102
1	105
1/2	110
114 or less	115

Perhaps, it would be appropriate at this stage to discuss how a plant can comply with the OSHA Noise Regulation Bulletin 334, "Guidelines to the Department of Labour's Occupational Noise Standards"[9] is very useful to consult. These guidelines suggest the following approach:

OSHA Noise Surveys: A noise survey of each area in the plant in which sound levels exceed 90 dB (A) shall be made at least once each year to ensure that sound levels have not increased above the originally existing level. The survey may also establish that noise levels in some areas have been reduced to levels below 90 dB (A) and thereby justify discontinuing application of requirements for administrative controls, ear protection, and audiometric tests of individuals in such areas. A noise survey of an area is recommended whenever a change is made in either equipment or type of operations, so that significant changes is noise level will be acted upon immediately.

Whenever a noise survey shows noise levels in excess of those of listed in Table 13.1, the Bulletin 334 Guidelines state that necessary steps to reduce the noise exposure to or below those levels shall be ascertained and a detailed plan with completion dates for individual steps shall be prepared. Following the original survey which shows existence of over exposures, the steps in a typical compliance plan might include the following, not necessarily in this order and some usually going on simultaneously.

OSHA Compliance Plan

1. A detailed survey of sound levels and sound spectra to determine the excessive sound levels.
2. Initiation of engineering studies to determine methods for reducing sound level at their sources.
3. Planning and initiation of feasible administrative controls, such as modifying production schedules to below the permissible limit or spreading part time noisy operation.
4. Initial audiograms for personnel excessively exposed.
5. Installation of a personal protective equipment program.
6. Follow up audiograms at appropriate intervals to assess effectiveness of the personnel protective equipment program and administrative control.
7. Installation of engineering controls or process changes to reduce noises at their sources.
8. Repeated noise surveys to measure effectiveness of the engineering changes.

The OSHA guidelines state that the orderly completion on schedule, of the various phases of the compliance plan, together with other components of hearing conservation programme.

If employees are working in areas with intense noise levels, then it is very desirable to monitor their hearing regularly. These hearing tests will determine whether or not the administrative and engineering controls and ear protection devices are being adhered to or effectively used. The 90 dB (A) eight-hour workday only protects 23 per cent of workers. Since people's ears are different, 77 per cent of workers will still be suffering appreciable hearing damage in these conditions, with monitoring, it may be possible to identify those persons with "weak" ears who may still be susceptible to hearing damage at the levels set by OSHA.

In some countries (*e.g.*, Japan) it is considered that the most effective method of controlling hearing damage caused by intense industrial noise is the issuance and strict enforcement of the use of ear protective devices. Most of the ear-protectors are uncomfortable to wear. They may cause the ears to get hot and sweaty or tightness of some types of ear protectors can cause pain or headaches. Ear-protectors also create communication difficulties, particularly the hearing of a warning signals. Another disadvantage of the use of hearing protectors is that workers may feel that they can no longer judge the performance of their machine or detect the noise caused by wear or some other malfunction. A final reason is that some ear-protectors, such as ear plugs, can exacerbate diseases of the outer ear such as otitis. However, in these cases the use of ear muffs or medications such as Cortisporin Otic drops or Corticosteroid ointment will normally overcome this problem.

13.7 NOISE POLLUTION IN DELHI, CHENNAI AND LUCKNOW

Noise has been notified under the Air Prevention and Control of Pollution Act, 1981. The main menace of noise in cities come from automobiles, construction equipments, loud-speakers, bursting of crackers. The number of road vehicles have increased enormously and horns with such loud voices are used without understanding what they are doing. The noise level of various metro cities is more than 90 dB. Inhabitants of cities are subjected to this most annoying form of transport noise, which is gradually deafening them. According to a recent study conducted by the Central Road Research Institute in collaboration with the All India Institute of Medical Sciences (AIIMS) and Delhi Police at 14 important intersections in New Delhi have maximum and minimum noise level varying between 93–100 dB and 62–71 dB. In Chennai city, it varies between 119.4 dB and 53 dB, while in Lucknow it is 90 dB and 80 dB respectively. Since the dB scale is logarithmic, each point rise is quite large. The acceptable noise level in residential area is given in Table 13.2. The recommended range for residential area is 40–85 dB and for commercial 50–68 dB. The CRRI report warns that there is a great need to contain higher noise emitting sources, as the higher values of maximum levels—generally above 90 dB indicate that noise besides leading to hearing impairment may create other problems.

Table 13.2 Acceptable noise levels in various areas or cities

Location	*Noise level (dB)*	*Location*	*Noise level (dB)*
City Area	45–55	Music Theatre	30–35
Rural Area	25–30	Radio, TV, Studio	25–35
Residential Area (urban)	35–40	Hospitals	30–35
Suburban Area	30–40	Conference Rooms	35–40
Urban-Business	40–45	Private offices	40–45
Industrial Areas	50–60	Court rooms	35–40
		Bank and Restaurants	50–55

The Delhi Pollution Control Committee has conducted noise survey in Delhi from August to October, 1996 in different parts of Delhi. The analysis of primary and secondary data and the indices have brought out the following noise pollution scenario in Delhi.

- Three wheelers, trucks and motorcycles remain the chief source of noise pollution in Delhi roads followed by generators in residential, commercial and industrial locations.
- 15 of the 46 residential locations have noise level within tolerable limits for over 90% of the time of the day. On the other extreme, 12 locations are noisy round the clock. The rural areas which are away from main National highways have not yet caught up with the noise problem.
- Major commercial areas remain noisier than tolerable limits.
- Industrial areas by and large remained within the prescribed limits.
- All major traffic corridors are highly noisy with peaks even crossing 100 dBA mark.

The DPCC survey report has recommended following steps for control of noise:

- Ban on pressure horns to be effectively implemented.
- Well designed silencers and mufflers be installed on the vehicles, preferably at the manufacturers levels. Three wheeler Autos to be phased out.
- Synchronised traffic signalling to be introduced on all major traffic corridors. A comprehensive traffic management plan should be introduced.
- Use of generators for commercial and residential purposes must be banned.
- Extensive plantation of trees on the road and road side colonies be undertaken in a definite time target.
- Hoarding of noise absorbent material be encouraged.
- As noise induced ailments are preventable but are difficult to control after they have occurred, mass awareness programmes be launched for people's participation in noise abatement programme, remembering that high noise levels result in auditory fatigue and deafness.

13.8 CONTROL OF NOISE POLLUTION

1. Certain codes should be enforced which require sound proofing in the construction of industrial and building apartments. Traders have the option of getting sound deadening tiles put on walls and ceiling. These tiles have high holes which trap sound and prevent it from bouncing back.
2. The government, on its part, could introduce heavy fines on noise polluters. Public should be told to carry out all noisy operations in an open space far off from residential colony. The shopkeepers should use modified technology approaches.
3. Trees absorb and dissipate sound energy and act as a buffer zone. Plants and trees should be planted along the highways, near industrial areas. In metro cities, green belt vegetation should be encouraged so that open space may absorb the sound. For instance in Delhi, road dividers of Daryaganj, Qutab road, tall dense hedges need to be grown.
4. Airport noise can be reduced by reducing the noise at source as far as technically possible.
5. Zoning the area around the airport so that no one is allowed to construct any house or factory within 1 km radius of the airport.
6. Road traffic should be restricted from unnecessary blowing of pressure horns. Decibel metres should be installed along the highway. Any vehicle driver exceeding the noise limit of 75 dB should be fined.
7. Enforce noise limit at once in the city and monitor sites where levels are high.
8. Consult experts on noise control planning and development.
9. Do not allow high pitch pressure horns, vehicles without silencers, random use of crackers and playing of radio, television at high volume. Restrict the use of loudspeakers.
10. Construct flyovers and subways at busy intersections to reduce noise.
11. To reduce noise due to aeroplanes the flight path and the angle of landing and take off may have to be changed.
12. Indians are religious minded and their every occasion, be religious, social marriage and family celebrations are manifested in a noisy manner. A lot of noise nuisance is created during marriage and festival processions which overlooks the health and peace of others. During elections, campaign commences even in midnight. Many countries like USA, UK have enacted laws to control noise menace, why can't we?

REFERENCES

1. Rudnick, I., *Handbook of Noise Control,* McGraw-Hill, New York (1957).
2. Kurze, V. and Bernek, L.L., *Sound Propagation Outdoors Noise and Vibration Control,* Bernek L.L. (Ed.) McGraw-Hill, New York (1971).
3. Welch, B.L. and Welch, A.S., *Physiological Effects of Noise,* Plenum Press, New York (1970).
4. Kryter, K.D., *The Effect of Noise on Man,* Academic Press, New York (1970).
5. Close, W.H., "Surface Transportation" Noise in, Noise and Vibration Control Engineering, Crocker M.J. (Ed.), Purdue Univ., West Lafayelta, Ind.
6. Palmer, J.W., *Anatomy for Speech and Hearing,* Harper and Row, New York (1972).
7. Glong, A., *Noise and Your Ear,* Grane and Stratton, New York (1958).
8. Anon Guidelines to the Department of Labour's Occupational Noise Standards, Bull 334, revised, U.S. Deptt. of Labour, Washington D.C. (1969).
9. Stevens, S.S. and Warshofsky, F., *Sound and Hearing,* Time Life Books, New York (1965).

Chapterwise Problems

CHAPTER 1

Earth's Atmosphere

1. Explain the nebular hypothesis that led to the evolution of earth and universe.
2. Describe the significant processes involved in the evolution of atmospheric oxygen. Explain the significance of the reaction between carbon dioxide and calcium silicate.
3. Show that if the atmosphere is isothermal the temperature change of a parcel of air rising adiabatically is

$$T(z) = T_0\, e^{-\Gamma z/T'}$$

where T_0 and T' are the temperature of the parcel at the surface and of air respectively and Γ is adiabatic lapse rate.
4. What accounts for the numerical difference between the dry lapse rate and wet lapse rate?
5. Describe the prevailing surface wind pattern in both hemisphere.
6. Contrast the essential difference between the Hadley theory and the Eddy theory.
7. Why is water vapour considered to be part of the energy transport system in the Eddy theory?
8. What is temperature inversion? How does an inversion acts like a lid?
9. What is subsidence inversion?
10. (*a*) Calculate the direction and magnitude of the acceleration per unit mass of an air parcel with temperature 10°C that has been compressed isothermally in a dry adiabatic atmosphere from 800 to 1000 m bar.

    (*b*) If the compressed parcel becomes saturated with moisture while assuming the environmental temperature at 1000 m bar is 27°C, what would be its temperature be if it were lifted back to 800 m bar.
11. The morning sounding shows the following

Pressure in m bar	1000	950	900	850	800	750	700
Temperature °C	7	8	8	7	5	3	0

(*a*) If the afternoon maximum temperature at the station (1000 m bar) is 20°C. What will be the maximum mixing height (MMH)?

(*b*) If the mixing ratio below MMH is 8% at the time of maximum temperature, will there be a cloud base in the layer? If so at what CCC?

(*c*) What is the maximum possible mixing ratio in the layer below the MMH without cloud formation in the layer?

12. (*a*) The anemometer at a plant indicates a 10 min mean wind speed of 10 ms^{-1} at the 25 m site where the instrument is exposed. Estimate the mean wind speed at the height of the plant's stack 65 m.

(*b*) List the assumption that were made in the course of making the above estimate.

(*c*) Place upper and lower limit on your estimate representing the extremes of thermal stability.

CHAPTER 2

Principles of Thermodynamics and Gas Kinetics

1. The precise amount of air required for the complete combustion of hydrocarbon fuel can be calculated by considering the stoichiometric conversion of the hydrocarbon to CO_2 and H_2O. Determine the stoichiometric air fuel ratio for combustion of cyclohexane, cyclohexene and benzene.

2. The complete combustion of hydrocarbon C_xH_x can be expressed chemically as

$$C_xH_x + \left(x+\frac{y}{4}\right)O_2 + \left(x+\frac{1}{4}y\right)3.76N_2 \longrightarrow xCO_2 + \frac{y}{2}H_2O + \left(x+\frac{1}{4}y\right)3.76\,N_2$$

Thus the combustion of 1 mole of hydrocarbon produces x moles of CO_2, $y/2$ moles of H_2O in addition to the $\left(x+\frac{1}{4}y\right)3.76$ moles of N_2 assumed to be unadulterated during the course of reaction. Show that if f is the fraction of incomplete combustion in ppm by volume, then unburned fuel, λ in the exhaust is given by

$$\lambda = \frac{10^{\circ} f}{M + f(1+F-M)}$$

where M is the total number of moles of products resulting from 1 mole of fuel and F moles of air.

3. A river with 400 ppm of salt and an upstream flow of 0.025 Ls^{-1} receives an agricultural discharge of 0.005 Ls^{-1} carrying 2000 mg L^{-1} of salt. The salt-become uniformly distributed in the river. A municipality just downstream withdraws water and mixes it with enough pure water from another source to deliver water having no more than 500 ppm salts to its customers. What should be the mixture ratio F of pure water to river water?

4. A lake with constant volume 10×10^6 m^3 is fed by a pollution free stream with flow rate 50.0 m^3 s^{-1}. A factory dumps 5.0 m^3s^{-1} of a non-conservative waste with concentration 100 mg L^{-1} into the lake. The pollution has a reaction rate coefficient k of 0.25/day. Assume that the pollutant is well mixed in the lake. Find the steady state concentration of pollution in the lake.

5. A simple way to model air pollution over a city is a box model that assumes complete mixing and limited ability for the pollution to disperse horizontally or vertically except in the direction of the prevailing winds. Consider a town having an inversion at 250 m a 20.0 km horizontal distance perpendicular to the wind, a wind speed of 2.0m s^{-1} and co-emission rate 60 kg s^{-1}.

Assume CO as conservative and completely mixed in the box. What should be the CO concentration in the box?

6. Consider the air over a city to be a square box 100 km on a side that reaches upto an altitude of 1.0 km. Clean air is blowing into the box along one of its sides with a speed of 4 m s^{-1}. Suppose an air pollutant with reaction rate $k = 5.56 \times 10^{-5}$ s^{-1} is emitted into the box at a total rate of 10.0 kg s^{-1}. Find the steady state concentration of the air assuming it is in a pollutant completely mixed state.
7. If the wind in problem 6 suddenly drops to 1 ms^{-1}, estimate the concentration of pollutants 2 hrs. later.
8. Suppose total world energy consumption of fossil fuels is equal to 3×10^{20} J/year and were to be obtained entirely by combustion of petroleum approximate chemical formula C_2H_3. Petroleum has an energy content of about 43×10^6 J/kg.

   (*a*) Estimate the emission of CO_2 per year.

   (*b*) What is the ratio of grams of C emitted per unit energy for petroleum versus methane.
9. Calculate the work necessary to raise a mass of 1 kg through a height of 1 m on the surface of (*a*) earth where $g = 9.81$ m s^{-2} (*b*) the moon where g is 1.6 ms^{-2}.
10. Use the tables of standard molar enthalpies, find the standard molar enthalpies of the following reaction under standard temperature pressure conditions.

    (*a*) $NO_2 + NO_2(g) \longrightarrow N_2O_4(g)$

    (*b*) $C_2H_6 + O_2 \longrightarrow 2CO_2 + 3H_2O$

    (*c*) $C_2H_4 + H_2 \longrightarrow C_2H_6$
11. Calculate the standard molar enthalpies of the formation of N_2O_5 on the basis of data.

    $2NO(g) + O_2(g) \longrightarrow 2NO_2(g)$ $\quad \Delta H = -114.0$ kJ mol

    $2NO_2(g) + O_2 \longrightarrow N_2O_5(g)$ $\quad \Delta H = -110.2$ kJ mol^{-1}

    $N_2 + O_2 \longrightarrow 2NO$ $\quad \Delta H = +180.5$ kJ mol^{-1}
12. A mass of 1000 g falls freely to a platform from a height of 10 m. What amount of heat will be evolved when it strikes the platform?
13. From the heat of combustion of *n* butane, calculate the standard heat of formation of this compound per mole at 25°C

    $$C_4H_{10} + O_2 \longrightarrow CO_2 + H_2O \qquad \Delta H = 2.876 \text{ kJ/mol}$$
14. From the following series of reaction, find the free energy of formation of N_2O_4(g) at 25°C.

    $\frac{1}{2}N_2 + \frac{1}{2}O_2 \longrightarrow NO$ $\quad \Delta G^\circ_{298} = 85.366$ kJ

    $NO + \frac{1}{2}O_2 \longrightarrow NO_2$ $\quad \Delta G = -34.5$ kJ

    $2NO_2(y) \longrightarrow N_2O_4(g)$ $\quad \Delta G = -5.685$ kJ

15. At 25°C, ΔG° for the reaction $N_2O_4 \rightleftharpoons 2NO_2$ is 5.774 kJ. Show that the degree of dissociation at 25°C when the total pressure is 10 atm is 0.0494.
16. For the reaction $2\ NO_2(g) \longrightarrow 2NO\ (g) + O_2$ at 184°C and atmospheric pressure $K_p = 6.76 \times 10^{-5}$. Find the percentage of dissociation of NO_2 at a total pressure of 1 atm.
17. Azomethane $CH_3N_2CH_3$ decomposes as follow:

$$CH_3N_2CH_3 \longrightarrow C_2H_6 + N_2$$

Reaction rate of this reaction was studied by measuring partial pressure of azomethane as a function of time at 600 K:

t/s	0	1000	2000	3000	4000
p $CH_3N_2CH_3$	8.2×10^{-2}	5.72×10^{-2}	3.99×10^{-2}	2.78×10^{-2}	1.94×10^{-2}

Show that it follows first order reaction hence determine the reaction rate constant.
18. The reaction rate constant for the first order decomposition of N_2O_5 has the value 4.8×10^{-4} s^{-1}. What is half life of reaction, if initial pressure of N_2O_5 was 500 mm Hg. Calculate pressure after (*a*) 10 sec. (*b*) 10 min.
19. At 25°C the half life period for decomposition of N_2O_5 is 5.7 hr and is independent of initial pressure of N_2O_5. Calculate (*a*) the reaction rate constant (*b*) time required for the reaction to go to 90% completion.
20. Consider the following reaction system:

$$A \xrightarrow{1} B$$

$$B + M \xrightarrow{2} C$$

Assume M is present in great excess, so that $[M] \approx$ constant. The concentration of B and C are zero at $t = 0$.

(*a*) Derive analytical expression for the exact dynamic behaviour of this system overtime. Show mathematically under what conditions the pseudo-steady state approximation (PSSA) can be made for [B].

(*b*) Use the PSSA to derive a simple set of equations for the concentration of A, B and C.

(*c*) Draw a graph (on log-log paper) that compares the exact behaviour of the system to calculations based on the PSSA for the following set of conditions. Discuss your results

Set 1		Set 2	Set 3
[M]	10^5	$[M] = 10^5$	$[M] = 10^5$
$[A]_o$	10^2	$[A]_o = 10^2$	$[A]_0 = 10^2$
k_1	10^{-5}	$k_1 = 10^{-3}$	$k_1 = 10^{-1}$
k_2	10^{-6}	$k_1 = 10^{-6}$	$k_1 = 10^{-6}$

21. Can the reaction

$$NO_3 + H_2O \longrightarrow HNO_3 + OH$$

occur spontaneously at normal temperature pressure (NTP). Use the following Gibbs free energies of formation at NTP (all in kJ mol^{-1}) HNO_3, – 74.7; OH, 34.2; NO_3, 115.8; and H_2O_2 – 228.4.

CHAPTER 3

Energy States and Photochemistry

1. What is the heat carried away by evaporating water molecules?
2. Once a molecule is excited into an electronically excited state by absorption of photon. Explain what are the possible fates of this excited molecule.
3. Some of various photochemical processes that can lead to various reactions are as follows. Name the processes associated with these reactions.

$$M_n \xrightarrow{hv} M_m$$

$$M_1 \longrightarrow M_n + \text{heat}$$

$$M_1 \xrightarrow{ISC} M_3$$

$$M_3 \xrightarrow{k_3} M_a + \text{heat}$$

$$M_3 + Q \xrightarrow{k_4} M_m + Q_n$$

$$M_3 \xrightarrow{k_5} \text{Products}$$

4. A swimmer enters a gloomier world (in one sense) as he dives to greater depth. Given the mean molar absorption coefficient for sea water in the visible region is 6.2×10^{-5} dm^{-5} mol^{-1} cm^{-1}. Calculate the depth at which a diver will experience (*a*) half the surface intensity of light (*b*) one tenth of that intensity.
5. When Naphthalene is irradiated with 360 nm light it does not show any absorption. Phosphorescence spectrum of solid solution of naphthalene + benzophenone show component of naphthalene. Account for this observation.
6. A transition of particular importance in oxygen molecule gives rise to Schumann Runge band in the UV spectrum. The wave number (cm^{-1}) of transition from the ground state to vibrational level of upper electronic state $^3\Sigma_u^-$ are as follow:

   50062.6, 50725.4, 51369.0, 51988.6, 52579, 53143.4, 53679.6, 54177.0, 54641.8, 55078.2, 55460.0, 55803.1, 56107.3, 56360.3, 56570.6.

   What is the dissociation energy of the upper electronic state? You should answer this extrapolation to the convergence point of the vibrational sequence.
7. The quantum efficiency for the formation of ethene from heptan-4-one with 313 nm light is 0.21. How many molecules of ethane/second and moles/second are formed when sample is irradiated with 313 nm light at 50 W under the conditions all light get absorbed.

   Note: Quantum efficiency depends on the wavelength of light.
8. A certain substance absorbs 2.0×10^{16} quantum of light per second. At the end of 10 min it is observed that 0.001 moles of the irradiated substance have reacted. What is the quantum efficiency of reaction?
9. The incident radiation from a monochromatic source of 2537Å on a slit 80×10^{-7} J/sec./m^2. The radiation is passed through a slit whose area is 5 square mm and onto a cell whose face transmits 30% of the incident radiation. If the solution in the cell transmits 10% of the radiation

how much energy in Joules is absorbed by the solution per hour? If quantum efficiency of the reaction is one, how many moles of substance must have decomposed?

10. A leaf absorbs energy from a lamp in the red part of spectrum which has an intensity of 1.532 watts with an average wavelength 7576Å. In what interval of time will leaf absorb 10J?
11. Calculate the energy in kJ mole^{-1}, kcal/einstein at which the yellow, green and blue sensitive cones with iodospin pigment of eye have maximum absorption.
12. In Fig. 3.6c assume that the height of the minimum in the curve corresponding to predissociation is 22516 cm^{-1} above the ground level. Estimate the dissociation energy for a molecule in the ground state and in the predissociated state if no transfer to state B occurs.
13. Light of an intensity of 3.2×10^{-6} Einstein sec^{-1} falls on the surface of anthracene crystal. The absorption coefficient is 3×10^{-6} cm^{-1} at the wavelength of this light. Calculate the intensity of light at a depth of 10^{-6} cm below the surface of the crystal.
14. The quantum of CO in photolysis of gaseous acetone ($p < 6$ kPa) at wavelength between 250-320 nm is unity. After 20 minutes irradiation with light of 313 nm, 18.4 cm^3 of CO (measured at 1008 Pa and 22°C) is produced. Calculate the number of quantum absorbed and the absorbed intensity in Js^{-1}.
15. A mechanism for photolysis of acetaldehyde is

$$CH_3CHO + h\nu \longrightarrow CH_3 + CHO$$

$$CH_3 + CH_3CHO \xrightarrow{k_2} CH_4 + CH_3CO$$

$$CH_3CO \xrightarrow{k_4} CO + CH_3$$

$$CH_3 + CH_3 \xrightarrow{k_4} C_2H_6$$

Derive the expression for the rate of formation of CO and quantum yield of CO.

16. A suggested mechanism for photolysis of ozone in low energy light (red light) is

(*i*) $O_3 + h\nu \longrightarrow O_2 + O$

(*ii*) $O + O_3 \longrightarrow 2O_2$

(*iii*) $O + O_2 + M \xrightarrow{k_3} O_3 + M$

The quantum yield of reaction (*i*) is ϕ_1

(*a*) Derive an expression for overall rate of disappearance of ozone

(*b*) Write the expression for the overall quantum yield for disappearance of ozone ϕ_0.

(*c*) At low total pressure $\phi_0 = 2$, what is value of ϕ?

17. The absorption spectrum of O_2 show vibrational structure which becomes a continuum at 56876 cm^{-1}; the upper electronic state dissociates into one ground state and one excited atom. (The excitation energy of which measured from atomic spectrum is 15875. Show that the dissociation energy of ground O_2 molecule is 41000 cm^{-1} or 490.3 kJ/mol.
18. The values of the $\bar{\upsilon}_e$ and x_e in the ground state, $^3\Pi_u$ and a particular excited state $^3\Pi_g$ of C_2 molecule are

	$\overline{\upsilon}_e$	x_e
Ground state	1641.4 cm^{-1}	7.11×10^{-3}
Excited state	1788.2 cm^{-1}	9.19×10^{-3}

Find the number of vibrational energy levels below the dissociation limit and hence the dissociation energy of C_2 molecule in both states.

19. The values of $\overline{\nu}_e$ and x_e in the ground state of HCl and HI are

Molecule	$\overline{\upsilon}_e / cm^{-1}$	x_e
HCl	2991	17.6×10^{-3}
HI	2309	17.16×10^{-3}

Calculate the dissociation energy of these molecules in ground state in kJ/mole.

20. The values of $\overline{\upsilon}_e$ and x_e for some diatomic molecules in the ground state are as follows:

Molecules	$\overline{\upsilon}_e / cm^{-1}$	x_e
ClO	853.8	6.442×10^{-3}
CO	2169.8	6.124×10^{-3}
NO	1904.2	7.39×10^{-3}
OH	3737.76	22.71×10^{-3}
HBr	2648.97	17.07×10^{-3}

Calculate dissociation energy of these molecules in wave numbers as well as in eV.

21. Following are the bond dissociation energies of some of the bonds in various molecules.

Bond	*Bond dissociation energy in eV*
C—H	4.2 eV
H—O	5.113 eV
N—O	1.677 eV
N—NO	4.93 eV
ON—O	3.114 eV
$O_2N—NO_2$	0.5937 eV

Calculate the wavelength of the photon that will break these bonds. Also mention in which region of spectrum this wavelength lie. Whether this light is available on the surface of earth.

22. Write down the term symbol for the ground electronic state for the atoms of nitrogen, oxygen, sodium and aluminium.

23. Following the electronic configuration of some diatomic molecules/ions.

	Electronic configuration	*No. of valence electrons*	*Term symbol*
NO^+	$(1\sigma)^2 (2\sigma)^2 (1\pi)^4 (3\sigma)^2$	10	$^1\Sigma$
CO^+	$(1\sigma)^2 (2\sigma)^2 (1\pi)^4 (3\sigma)^1$	9	$^2\Sigma_1$
O_2^{2+}	$[Be_2][1\pi]^4[3\sigma_g]^2[1\pi_g]^0$	10	$^1\Sigma_g^+$
NO^-	$(1\sigma)^2 (2\sigma)^2 (1\pi)^4 (3\sigma)^1 (2\pi)^2$	12	$^3\Sigma$
N_2^-	$[Be_2][1\pi]^4[3\sigma_g]^2[1\pi_g]^1$	7	$2\Pi_g$

Check if the term symbol for the ground states given are correct or not. What are the first four electronic transition are possible from these states? Write the term symbol for these electronic states.

24. Hydrogen atom show an absorption in Lyman series of lines for an energy of = 2.18×10^{-18} J. Express this value in eV, Joules/mole, wavelength of light associated with it in nm, frequency of the light in Hertz and wave numbers.
25. C—H molecule shown an absorption for the transition from $^2\Pi_r \rightarrow {}^2\Delta$ at a value of 23184 cm^{-1}. Express this value in electron volts, joules/mole, nm and Hertz.
26. The ground electronic state of CO is $^1\Sigma^+$ and first excited electronic state is $^3\Pi_r$. The energy difference between these two states is 48686.70 cm^{-1}. Express these results in Joules, J/mole, Hertz and nm.
27. Following are the excited energy states of some homonuclear molecules.

Molecules	*Excited state 1*	*Excited state 2*	*Energy difference*
N_2	$^1\Sigma_g^+$	$^3\Pi_g$	59619 cm^{-1}
	$^3\Sigma_u^+$	$^3\Sigma_u^-$	16069 cm^{-1}
O_2	$^3\Pi_g$	$^1\Pi_g$	3650 cm^{-1}
	$^1\Sigma_u^-$	$^3\Sigma_u^-$	16736 cm^{-1}
	$^3\Sigma_u^-$	$^3\Sigma_u^-$	30090 cm^{-1}

(*i*) Predict which of these transitions are allowed and calculate the energy difference in eV, Joules, J/mol, nm.

(*ii*) Those transition which are not allowed. Explain why?

CHAPTER 4

Photochemistry of Atmosphere

1. Based on 8 hr measurement, air quality standard for CO is 9 ppm. Express this standard as percentage by volume as well as mg/m^3 at 1 atm and 25°C.

2. The air quality standard for ozone is 0.08 ppm. Express this result in μg/m³ at 1 atm and 25°C. Convert conc. of SO_2 of 0.1 ppm to μgm^{-3} at 298 K and 1 atm. Do the same for 0.1 ppm NO, NO_2 and O_3.
3. What wavelength band of solar radiation leads to photochemical dissociation of NO_2? What determines the lower limit?
4. How many molecules of NO can be oxidized to NO_2 by the reaction of one OH free radical with one proper molecule?
5. What is the most important aspect of photochemistry of ozone to troposphere? Explain.
6. What accompanies horizontal divergence of wind near earth's surface? What effect is this likely to have on thermal stability of this layer?
7. In the primitive earth why whole of water was not photodecomposed to oxygen and hydrogen?
8. Nitrogen molecule is not a photochemically active molecule but a very small part of it gets converted to NO in the stratosphere. Suggest a mechanism for the same.
9. Which substance is the key substance to produce OH radicals on photolysis? Explain.
10. PAN acts as a storage reservoir of NO_x which on decomposition leads to smog formation. Explain.
11. Ordinary oxygen atom is in the $^3P_2$ state. These atoms combine to form oxygen molecule as

$$2O(^3P_2) + M \longrightarrow O_2(X, {}^3\Sigma_g^-) + M.$$

$$\text{Rate} = k[O]^2[M] \qquad k = 3\times10^{-3}\ \text{cm}^6\ \text{molecule}^{-2}\text{s}^{-1}.$$

k is slightly inverse temperature dependent. Based on this information suggest a mechanism of above reaction and explain the result. Also explain why the value of k imply that each collision will lead to product. Above 100 km why primary oxygen species is atomic oxygen?
12. What photochemical reaction lead to the formation of O_3 molecules in stratosphere? Explain, how 0.3 cm layer of ozone is opaque to solar radiation.
13. Discuss the various electronic transition possible in oxygen molecule. If oxygen is said as a strong UV eliminator from solar light how far it will be correct? Justify your answer.
14. What transition lead to Night Glow and why it is so named?
15. Steady state O_3 concentration in the presence of solar radiation, can be given by Leighton relationship. Derive this relationship

$$O_3 = k_{446}[NO_2]/k_{443}[NO].$$

If $k_{446} = 3\times10^{-11}\ \text{s}^{-1}$ and $k_{443} = 1.8\times10^{-14}\ \text{cm}^3\ \text{molecule}^{-1}\ \text{s}^{-1}$.

which set of conditions are consistent with O_3 = 100 molecules cm^{-3}.

[NO] molecule cm^{-3}	16.67	12.5	1.667	8.335	1000
$[NO_2]$ molecule cm^{-3}	1.0	12.5	0.1	5.097	60

What does these answer indicate about functional dependence of O_3 concentration on absolute magnitude of $[NO_x]$ = [NO] + $[NO_2]$?
16. Why Leighton relation holds good in afternoon in a polluted city only?
17. Figure 4.2 gives the spectrum of ozone. Discuss this spectrum in relation to ozone formation and photolysis so as to act a shield to UV and other radiations.
18. It is said that root cause of pollution is NO. If NO had not been there in atmosphere many other pollutant may not have been there. Justify this statement.

19. How NO_3 radical and Nitrous acid (HONO) acts as a sink to NO gas, one for day time other for night time.

CHAPTER 5

Chemistry of Stratospheric Ozone

1. Write the chemical formula for CFC-134 and F-114.
2. What would be the CFC number for $C_2H_4F_2$, CCl_2FCClF_2?
3. The earth's ozone layer is under attack in part by chlorine released when UV radiation makes CFC to decompose. Consider three CFC's F-11, F-12, F-22.
   (*a*) Calculate percentage of chlorine in each of these CFC's.
   (*b*) If F-22 could be replaced by F-11 molecule for molecule by what percentage would the mass of chlorine emission be reduced.
4. Write the chemical formula for F-1211, F-2402, F-114, CFC's. Arrange them in the decreasing order of their sensitivity to ozone layer and justify that.
5. List some items that are threatening the survival of ozone layer.
6. Describe why some type of synthetic rubber are less susceptible to ozone than natural rubber.
7. What constituent released by human being are likely to affect the ozone in the upper stratosphere? Which of these might be expected to have the greatest effect?
8. Why is it that rainbow is observed after a mild shower not otherwise? Why rainbow is always in the bow form?
9. Why is it all the stars appear to be of same colour from the surface of earth?
10. Why is that CFC's are highly inert compounds particularly in troposphere but become highly reactive in stratosphere? How is it that bromoderivatives are still more destructive?
11. Suggest the mechanism by which chlorofloro carbon deplete ozone layer. Suppose we spray lot of methane gas into stratosphere will it retard or accelerate the ozone depletion.
12. Explain why is it that formation of ozone hole has taken place at south pole while it has not been that pronounced in North pole, while CFC's are not released at poles?
13. Ozone has maximum concentration on poles and minimum on equator. If CFC's are pumped from equator into stratosphere, in what way various forms of plant and animal life will be affected?
14. Ozone is one of the most poisonous substance, how its formation on the surface of earth affects us?
15. Even if the Montreal Protocol is completely successful, the long atmospheric lifetimes of CFC's will mean their atmospheric concentration will continue to increase. This can be shown with a simple model. If emission were totally eliminated then the mass of, *m* (*t*) of a given CFC in atmosphere would follow a simple exponential decay process.

    $$m(t) = m(o)e^{-t/\tau}$$

    where *m* (*o*) is initial mass, τ is average atmospheric residence time.

    This relation can be used to find rate of CFC removal, $-\dfrac{dm}{dt}$

$$-\frac{dm}{dt} = m(o)e^{-t/\tau}\left(-\frac{1}{\tau}\right) = \frac{m}{\tau}$$

Applying the mass balance equation.

Rate of increase in CFC man = Rate of CFC addition – rate of CFC removal

$$= P - \frac{m}{\tau}$$

If the emission rate P is constant then we can find the ultimate steady state mass of CFC in the atmosphere by setting giving $\frac{dm}{dt} = 0$.

These equations give a solution

$$m(t) = m(o)e^{-t/\tau} + m(\infty)\left(1 - e^{-t/\tau}\right)$$

or in terms of desired concentration

$$C(t) = C(o)e^{-t/\tau} + C(\infty)\left(1 - e^{-t/\tau}\right)$$

where $C(\infty) = \frac{P(g/y_r)(\tau(y_r))}{\text{Molar mass}} \times 5.63 \times 10^{-21}$

Assuming the following statistics for CFC-12

Atmospheric residence time = 150 year

1985 emission rate = 0.44×10^{12} g/yr

1985 atmospheric conc. = 0.40 ppb.

Suppose emission rate of CFC-12 is simultaneously reduced to 50% of its 1985 value and held constant thereafter.

(*a*) What would be the final, steady state atmospheric concentration of CFC-12?

(*b*) What cut in emission rate would be required for CFC-12 concentration to remain constant at 1985 level of 0.40 ppb?

(*c*) Plot the transient response of CFC concentration versus time for various emission rates corresponding to (*i*) 1985 rates (*ii*) rates cut by 15% (*iii*) rates cut by 50%.

16. Assume the following data for CFC-11 atmospheric residence time = 75 year, atmospheric concentration in 1985 = 0.24 ppb, 1985 emission rate = 342000 tons/year.

(*a*) If emission rate remain constant, what would be the ultimate steady state concentration.

(*b*) What drop in emissions rates would be required to maintain existing concentration?

(*c*) If emission immediately drop to 50 per cent of their 1985 value, sketch the concentration vs. time graph and indicate the ultimate steady state level that would be achieved.

17. Repeat the problem 16 for CFC-13 which has an atmospheric residence time of about 90 years.

(*a*) In 1985 conc. of CFC-13 was 0.032 ppb and emission rate of about 165000 tons/year.

18. Equation 4.59 describe the steady state O_3 concentration in the presence of solar radiation.

$$\frac{d[\text{NO}]}{dt} = 0 \quad \text{and} \quad O_3 = \frac{k_1[\text{NO}_2]}{k_3[\text{NO}]}$$

If k_1 = 0.03 min and k_3 = 30 ppm^{-1} m^{-1} consistent with O_3 = 0.1 ppm

[NO]	0.5	0.125	0.047	0.47	0.001
ppm	0.05	0.125	0.47	0.047	0.01

Calculate NO_2 concentration for each value.

What does the answer indicate about the functional dependence of ozone concentration on the absolute magnitude of NO_x = [NO] + [NO_2]?

19. Explain the Chapmann cycle of oxygen.
20. Explain which chemical species control the natural concentration of ozone in stratosphere? How does ozone stops the UV radiation from reaching earth?
21. Imagine by some process lot of water vapours are sprayed directly into stratosphere, what changes you expect on the surface of earth?
22. Suppose some country decides to fly supersonic transport (SST) at a height of 16-18 km, in what way it will affect the ozone layer?
23. Assuming that O_3 production and loss follow the reaction 4.12, 4.16, 4.25 (with some reaction rate constant k_{425}) and the reaction O + $O_3 \longrightarrow 2O_2$ with reaction rate constant k_{518}, show that ozone concentration is related to oxygen by the relation

$$[O_3]=[O_2]=\left[\frac{k_{-412}\,k_{416}\,[M]}{k_{425}\,k_{518}}\right]^{1/2}$$

Note: This relationship neglects some of the ozone destroying reaction, particularly catalytic cycles, transport of O_3 and other chemicals by atmospheric motion.

24. What happen to O and O_3 according to Chapmann scheme when solar radiation is switched off at sun set? If catalysts are present what happens to X and XO. Assume that k_{416} [O_2] [M] + k_{518} [O_3] ≥ 10^{-1} s^{-1} throughout the stratosphere.
25. A useful general measure of O_3 in the atmosphere is the column of ozone defined in terms of total number of O_3 molecules in a vertical column of atmosphere of unit horizontal cross-section. This number is expressed as Dobsin Unit (DU) defined as the height of the column in hundreth of a millimeter if all the ozone molecules in it were brought to a pressure of 1 atm and a temperature of 0°C. Typical values of column ozone are about 300 DU, meaning that the height of column compressed in this way would be 3 mm. Calculate the number of molecules per square metre in 300 DU.

CHAPTER 6

Carbon Dioxide and Climate

1. Estimate the tons of carbon in atmosphere corresponding to a concentration of 360 ppm of CO_2. Assume the total mass of air equal to 5.1×10^{18} kg. The density of air at standard temp. and pressure (STP) 0°C and 1 atm is 1.29 kg/m^3.
2. The concentration of CO_2 in 1965 was 320 ppm, while in 1990 it was 355 ppm. If we model that growth with a simple exponential function $C = C_0e^{rt}$ (C_0 = initial concentration, *r* growth rate and *t* time), at this exponential rate of growth, in what year would atmospheric concentration

be twice the 280 ppm, pre-industrialized value and year by which 380 ppm value will be reached.

3. Suppose coal based electricity generating unit has 33% efficiency, where 100% heat is converted to water heating but there is coal emission. Suppose coal is replaced by gas for heating water with a 70% conversion efficiency by what factor would carbon emission be reduced? Carbon emission factor for coal is 25.2×10^6 ton C/quad and for natural gas it is 14.5×10^6 tons C/quad.

4. The carbon contents of world resources are as follow

	Natural gas	Oil	Coal
Giga-ton of C	145	230	7100

For each of the fossil fuel estimate the increase in atmospheric CO_2 that would result from complete combustion of the entire resource. Assume an air borne fraction of 50%. Also 1 ppm ≡ 2.12 Gt of Carbon.

5. Extending models beyond the next century is an exercise that may provide some speculation and insight. The total release of carbon during fossil fuel combustion may be calculated with a Gaussian distribution function of the sort

$$\frac{P}{P_m} = \exp\left[-\frac{1}{2}\left(\frac{t - t_m}{\sigma}\right)^2\right]$$

where P and P_m are rate of release of carbon (Gt/year) at time, t and t_m time at which maximum release of carbon is reached and σ = standard deviation of the Gaussian distribution where σ can be found from

$$Q_\infty = \sqrt{2\pi}\sigma P_m$$

where Q_∞ = the total carbon ever released (Gt)

and $t_m = \sigma\sqrt{2 \ln P_m / P_0}$

Suppose the total world wide release of carbon due to fossil fuel combustion is equivalent to burning 200000 quads of coal at 25.2×10^6 tons C/quad. The 1980 rate of release of carbon was 5.2 G ton/year. If peak emission rate is eventually 33 G ton C/year, find the year when peak emission would result.

6. In the reaction 6.1a we considered photoxidation of CO in the presence of NO_x. By invoking PSSA derive rate equation for $[NO_2]$, $[NO]$ and $[CO]$. Integrate these equation numerically for the following three sets of conditions and plot $[NO_2]$, $[NO]$, $[CO]$ and $[O_3]$ as a function of time from $t = 0$ to $t = 3000$ hr. Assume $T = 300$ K and RH = 50%.

	Initial conditions (ppm)		
	1	2	3
$[NO_2]_0$	0.1	0.2	1.0
$[NO]_0$	0.5	0.5	0.5
$[CO]_0$	10	10	10

In each case you should assume the photolysis rate constant for NO_2 as $k_{NO_2} = 0.533\ min^{-1}$.

Discuss your results in term of conversion of NO to NO_2 and the accumulation of O_3.

7. If the earth atmosphere is assumed to be in radiative equilibrium with sun, it is possible to estimate average temperature value T of the earth atmosphere system given by Stefan Boltzmann relation.

$$T = \left[\frac{(1-\alpha)S}{4\sigma}\right]^{1/4}$$

Calculate T for albedo $\alpha = 0.35$. The average temperature is much higher than this value. Why is this so?

8. Suppose that the earth was shaped like penny with one side that faces the sun at all times with uniform temperature all through. Neglect any radiation losses of the rim of the earth and assume there is no albedo or green house effect estimate the temperature of this earth as a black body radiator.

9. The Solar constant and albedo of Venus and Mars are as follow

	Venus	*Mars*	*Earth*
S (Wm^{-2})	2613	589	1372
α	0.75	0.15	0.30
Actual Temp (K)	700	220	288

(*a*) Calculate the equivalent temperature on the surface of Venus and Mars, assuming no green house effect compare this temperature with actual temperature and state relative concentration of Green house gases.

(*b*) If Mars radiates as a black body at 217K, what would be the peak wavelength in its spectrum.

10. The solar flux S striking a planet will inversely proportional to the square of distance from the planet to sun, *i.e.*, $S = k/d^2$ where k is some constant and d is distance.

(*a*) Use the data given in problem 9 to estimate k given the distance from the sun to earth is 150×10^9 m.

(*b*) Use your result from (*a*) to estimate the distance from sun to venus.

(*c*) Mercury is 58×10^9 m from sun and has an albedo equal to 0.06. Estimate its equivalent temperature.

11. The solar flux S arriving at the outer edge of the atmosphere varies by ± 3.3% as earth moves in its orbit (reaching its greatest value in early January). By how many degrees would the effective temperature of earth vary as a result?

12. Global precipitation has been estimated to average about 1 m of water per year across the entire globe. If the surface area of the earth is $5.10 \times 10^{14}\ m^2$, find the energy required to cause that much water to evaporate each year. Compare your answer to the estimated 1987 world energy consumption of 3.3×10^{17} kJ, average sunlight that strikes the earth surface as 167 Wm^{-2} and energy required to vaporize 1 kg of water 2459 kJ/kg.

13. The average rate at which energy is used to evaporate water is given as 94 W/m^2 and latent heat

of vaporization of water 2.61×10^6 J/kg and surface area of earth 5.1×10^{14} m^2 estimate the total world annual precipitation in m^3/year which is equal to world annual precipitation in m^3/year which is equal to the total water evaporated.

14. Why carbon monoxide is considered a dangerous atmospheric pollutant?
15. What factors affect the length of time a pollutant is air borne?
16. Using Wien's displacement law determine the mean effective temperature of the earth's atmospheric system of the resulting long wave radiation peak at 11 μm contrast the magnitude of the radiant flux at the 11 μm with that at 50 μm.
17. What are relative unique radiation properties of gaseous CO_2 that make it a candidate for causing climate change?
18. Is the heating due to increased CO_2 expected to be greater or less at the poles, than through most of the atmosphere. Explain why?
19. Explain why albedo change due to heating or cooling would be expected to produce positive feed back?
20. CO_2 is an inert gas but green house effect warms the earth. On the other hand if somehow CO_2 concentration is increased in stratosphere, it will have cooling effect. Explain.
21. If CFC-11 and CFC-12 emission remains at their current rate, their ultimate concentration in atmosphere would be about 1 ppb and 3.1 ppb respectively. How much global temperature increase would be expected?
22. Verify the following

    (*i*) If there is no quantization of energy, *i.e.* Planck constant is set equal to zero in equation 6.1, it is reduced to Rayleigh-Jean Law

    $$\frac{dU}{dx} = \frac{8\pi kT}{\lambda^5}$$

    (*ii*) Summing the energy in every wavelength range $d\lambda$ from $\lambda = 0$ to $\lambda = \infty$ at a temp. T, Planck relation 6.1 reduces to Stefan Boltzman law

    $$U = aT^4 \text{ where } a = \frac{4\sigma}{c} = \frac{8\pi^5 k^4}{15c^3h^3}$$

    where σ is called Stefan Boltzmann constant.

    (*iii*) At the maximum of Black body emittance (Fig. 6.1) $\frac{dU}{d\lambda} = 0.$

    If in eq. 6.1 $\frac{dU}{d\lambda}$ is equated equal to zero, we get Wien's law *i.e.*

    $$\lambda_{max} T = b = \frac{hc}{4.865k} = 2.898 \times 10^{-3}$$

23. Find the temperature for which $\frac{dU}{d\lambda}$ is maximum at 500 nm and at 10 nm.
24. Assuming that sun behaves as a black body at a temperature, T, calculate $\frac{dU}{d\lambda}$ at 500 nm and percentage increase in these spectral radiance if temperature increases by 100 K.

25. How does the earth manage its radiation balance so as to maintain an average temperature of 290 K?
26. When we are sitting before hot air blower, explain how heat is transferred to our body.

CHAPTER 7

Acid Deposition

1. A relationship exist among corrosion, pollution and SO_2, explain why one expect this interdependence?
2. Explain why soiling and corrosion are the hidden costs of air pollution.
3. Why are oxides of nitrogen and oxides of sulphur usually reported in emission inventory tables rather than that actual oxidation states.
4. What advantage is gained by oxidizing H_2S to SO_2, consider toxicity and odour?
5. Various minerals found on/in earth are sulphides, carbonates or oxides. Arrange in order in which these might have formed and why? Explain why sulphite minerals are very rarely there.
6. Sulphur dioxide in atmosphere gets converted to SO_3 which is then removed from atmosphere. What are the pathways for the SO_2 conversion to SO_3? How the presence of olefins will affect the oxidation?
7. What are the various paths by which SO_2 is oxidized to SO_3 by nitrogen oxides in the presence/absence of sunlight?
8. Studies of combustion reaction depend on knowing the concentration of hydrogen atom and hydroxyl radical. Measurement on a flow system using ESR for the detection gave following information on reaction.

$$H + NO_2 \longrightarrow OH + NO \qquad k_1 = 2.9\times10^{10}\ dm^3mol^{-1}s^{-1}$$

$$OH + OH \longrightarrow H_2O + O \qquad k_2 = 1.55\times10^{9}\ dm^3mol^{-1}s^{-1}$$

$$O + OH \longrightarrow O_2 + H \qquad k_3 = 1.11\times10^{10}\ dm^3mol^{-1}s^{-1}$$

On the basis of initial hydrogen atom and NO_2 concentration of 4.5×10^{-10} mol cm^{-3} and 5.6×10^{-10} mol cm^{-3} respectively, compute and plot curves showing the O, O_2, OH concentration as a function of time in the range 0–10 ns.

9. In a polluted urban atmosphere of alkanes and aldehydes, reaction 7.45 and 7.46 leads to formation of nitric acid. Concentration of various pollutants are as follow:

Pollutant	Alkane	HCHO	CH_3CHO	NO_3	NO_2	OH
Concentration (molecules cm^{-3})	2.5×10^{12}	5×10^{11}	2.5×10^{11}	2.5×10^{12}	1.25×10^{12}	1.6×10^{6}

Calculate the nighttime rate of formation of HNO_3. Also calculate the daytime rate of formation of HNO_3 by equation 4.48 and compare the two results.

10. In Los Angeles, smog formation involve the reaction 4.49a, 4.50 and 7.47 which result in HNO_3 formation

$k_{449a} = 3.2 \times 10^{-17}$ cm^{-3} molecule^{-1}s^{-1} $k_{450} = 4.5 \times 10^{-3}$ cm^6 molecule^{-2}s^{-1}

$k_{747} = 1.3 \times 10^{-21}$ cm^{-3} molecule s^{-1} with typical night concentrations

NO_2 = 7.38 cm^{-3} molecule

NO_3 = 2.47 × 10^7 cm^{-3} molecule

H_2O 50% humidity at 300 K.

Show that rate of formation of HNO_3 is 2.05 × 10^7 cm^{-3} molecule s^{-1}

11. Compare and contrast the behaviour of sulphuric acid and nitric acid as pollutant once formed in atmosphere.
12. In what way acid rains affect the vegetable plants. Explain.
13. Acid rains that acidify the river lakes and ponds makes the fish to disappear from them, explain why?
14. It is said that installation of Mathura Refinery a few miles from Agra will be harmful to Taj because of formation of acid rains. Justify for or against the statement.
15. In what way acid rains accelerates the soil pollutions.

CHAPTER 8

Photochemical Smog

1. What is temperature inversion? How does inversion acts like a lid?
2. What is meant by Lapse rate? Explain why the adiabatic lapse rate for wet air is less than the adiabatic lapse rate for dry air?
3. Los Angeles has the reputation of being a smog capital. The city is situated on the 34° latitude. It has mountains on three sides and fourth side is sea and thus no cross ventilation. Maritimes air that is cooler than surface moves from Pacific sea and covers the city. This air is covered by a warm air and leads to subsidence inversion. Explain how its meteorological conditions and topography feature create an ideal situation for the smog formation.
4. A pollution source is located in a valley with an inversion lid at the ridge top. The valley is 0.5 km wide at the location source and open at a 20° angle downward.

   (*a*) If the mean wind speed is 4 ms^{-1} at 1 km downwind and if concentration of pollutant is uniform downwind, what is the wind speed at 0.5 km downwind?

   (*b*) If instead the concentration is doubled between 1 km and 2 km downwind of the source due to a second source at 1 km. What is the strength of the second source relative to first? Assume all valley cross-section normal to the central line have the same slope on the walls and same depth to ridge top.

5. At what time of day and under what meteorological conditions is maximum ground level pollution likely to occur at location several km inlands from a shore line industrial complex whose pollutants are primarily released from stacks of moderate height of about 40–130 m?
6. What decimal reactions and pollution environment in London type of smog is different from Los Angeles type smog? In what ways smog is harmful to man?
7. As air become more polluted, more of transient radicals formation take place which increase the rate of ozone formation. Justify this statement.

8. How does the change of hydrocarbon/NO_x ratio affects the pollution in the day light? What is the role of PAN in the change?
9. PAN acts as a storage reservoir of NO_x, which on decomposition lead to smog formation. Explain.
10. In what way smog may affect our health and the vegetable plants and many other things that we use in our day to day life?
11. The following is a chemical reaction mechanism for the atmospheric photoxidation of formaldehyde and acetaldehyde

$$NO_2 + h\nu \xrightarrow{1} NO + O$$

$$O + O_2 + M \xrightarrow{2} O_3 + M$$

$$O_3 + M \xrightarrow{3} NO_2 + O_2$$

$$HCHO + h\nu \xrightarrow{4a} 2HO_2^{\bullet} + CO$$

$$\xrightarrow{4b} H_2 + CO$$

$$HCHO + OH \xrightarrow{5} HO_2^{\bullet} + CO + H_2O$$

$$CH_3CHO + h\nu \xrightarrow{6} CH_3O_2^{\bullet} + HO_2^{\bullet} + CO$$

$$CH_3CHO + OH \xrightarrow{7} CH_3C(O)O_2^{\bullet} + H_2O$$

$$HO_2^{\bullet} + NO \xrightarrow{8} NO_2 + OH$$

$$CH_3O_2 + NO \xrightarrow{9} NO_2 + HCHO + HO_2^{\bullet}$$

$$CH_3C(O)O_2^{\bullet} + NO \xrightarrow{10} NO_2 + CH_3O_2^{\bullet} + CO_2$$

$$CH_3C(O)O_2^{\bullet} + NO_2 \underset{12}{\overset{11}{\rightleftharpoons}} CH_3C(O)O_2NO_2$$

$$OH + NO_2 \xrightarrow{13} HNO_3$$

Show that by using PSSA, the mechanism can be written as

$$NO_2 + h\nu \overset{1}{\rightleftharpoons} NO + O_3$$

$$HCHO + h\nu \xrightarrow{4a} 2NO_2 - 2NO + 2OH + CO$$

$$\xrightarrow{4b} H_2 + CO$$

$$HCHO + OH \xrightarrow{5} NO_2 - NO + OH + CO + H_2O$$

$$CH_3CHO + h\nu \xrightarrow{6} 3NO_2 - 3NO + HCHO + 2OH + CO$$

$$CH_3CHO + OH \xrightarrow{7} CH_3C(O)O_2^{\bullet} + H_2O$$

$$CH_3C(O)O_2^{\bullet} + NO \xrightarrow{10} 3NO_2 - 2NO + HCHO + OH + CO_2$$

$$CH_3C(O)O_2^{\bullet} + NO_2 \underset{12}{\overset{11}{\rightleftarrows}} CH_3C(O)O_2NO_2$$

$$OH + NO_2 \xrightarrow{13} HNO_3$$

and the chainlength of the system is given by

$$L_c = \frac{1 + k_5[HCHO] + k_7[CH_3CHO]}{k_{13}[NO_2]}$$

12. Alkyl peroxy nitrates RO_2NO_2, can be presumed to decompose according to the following mechanism.

$$RO_2NO_2 \underset{2}{\overset{1}{\rightleftarrows}} RO_2^{\bullet} + NO_2$$

$$RO_2^{\bullet} + RO_2^{\bullet} \xrightarrow{3} 2RO^{\bullet} + O_2$$

$$RO^{\bullet} + NO_2 \xrightarrow{4} RONO_2$$

Let us assume that a sample of RO_2NO_2 decomposes in a reactor and its decay is observed. We desire to estimate k_1 from the rate of disappearance. To analyze the system we assume that both RO_2 and NO_2 are in pseudo steady state and that $[RO_2] \cong [NO_2]$.
Show that the observed first order rate constant for $RO_2\ NO_2$ decay is related to the fundamental constants of the system by

$$k_{obs} = k_1\left[1 - \frac{k_2}{k_2 + 2k_3}\right],$$

Thus, given k_{obs} and values of k_2 and k_3 and k_1 can be determined.

13. Moshin and O'Brien (1989) have proposed a simplified mechanism for predicting O_3 formation in photochemical smog system.

$HC + OH \xrightarrow{1} aRO_2 + P$	Hydrocarbon loss and product formation
$P + OH \xrightarrow{2} bRO_2 + \text{loss}$	Product loss
$P + h\nu \xrightarrow{p} cRO_2 + \text{loss}$	Product photolysis
$RO_2 + NO \xrightarrow{3} NO_2 + \text{loss}$	No photo oxidation
$NO_2 + h\nu \underset{5}{\overset{4}{\rightleftarrows}} NO + O_3$	Photostationary state
$\text{Everything} \xrightarrow{d} \text{loss}$	Dilution

In this mechanism, HO_2 radicals are lumped with RO_2, and, to a first approximation, all RO_2 are assumed to react with NO. This mechanism does not predict the OH concentration, taking instead the quantity $k_1[OH] = m$ as an input hydrocarbon reactivity parameter.

(*a*) Show that the mechanism can be used to predict the number of net NO photo oxidation product per hydrocarbon consumed.

$$\frac{d([\text{NO}]+[\text{O}_3])}{d[\text{HC}]} = m(a[\text{HC}]+bk_2/k_1[P])+ck_p[P]$$

$$\frac{+k_d([\text{NO}]-[\text{O}_3])}{(m+k_d)[\text{HC}]}$$

Use the quantity [NO] —$[O_3]$ as net NO oxidized effectively cancels out the photostationary state: thus ozone is simply "negative nitric oxide".

(*b*) To solve the equation in part (*a*) product concentration [P] must be expressed as a function of parent hydrocarbon concentration [HC]. Show that

$$P = \frac{mz_1[\text{HC}]}{Z_2 - Z_1^2} + \left(\frac{mz_1[\text{HC}]_0}{Z_1^2 - Z_2}\right)\left(\frac{[\text{HC}]}{[\text{HC}]_0}\right)^{Z_2/Z_1^2}$$

where $Z_1 = m + kd,\ Z_2 = (k_2 m/k1 + kd + kp)Z_1$

The two equation derived express ozone forming ability of any single hydrocarbon in terms of the hydrocarbon's rate parameters, *a*, *b* and k_2/k_1 the overall dilution rate k_d and the effective hydrocarbon reactivity, $m = k_1$ [OH], Moshin and O'Brien illustrate fits of this mechanism to laboratory data by assuming *a*, *b* and k_2/k_1. The value of *m* is calculated by subtracting the dilution rate from a semilog plot of HC versus time.

CHAPTER 9

Particulate Matter in Atmosphere

1. Knowing a particle density and its settling velocity V, show how to determine its diameter.
2. A unit density sphere of diameter 100 μm move through air with a velocity of 25 cms^{-1}. Compute the drag force offered by air.
3. A unit density sphere of diameter 1 μm moves through air with a velocity of 25 cms^{-1}. Compute the drag force offered by air.
4. Calculate the terminal velocity of silica particles (ρ_p = 2.65 g cm^{-3}) of 0.05 μm, 0.1 μm, 0.5 μm, 1 μm diameter.
5. Develop a table of terminal settling velocities of water drops in still air at T = 293 K, P = 1 atm. Consider drop diameter ranging from 1.0 μm to 1000 μm = 1 mm.
6. A bar with volume 500 m^3 has 50 smokers in it, each smoking two cigarettes per hour. An individual cigarette emits, among other things, about 1.40 mg of formaldehyde. Formaldehyde converts to CO_2 with reaction rate constant k = 0.40 h^{-1}. Fresh air enters the bar at the rate of 1000 m^3h^{-1} and stale air leaves at the same rate. Estimate the steady state concentration of formaldehyde in air, assuming complete mixing. At 25°C and 1 atm of pressure how result compare with threshold for eye irritation of about 0.0614 mg m^{-3} or 0.05 ppm.
7. The bar of problem 6 is clean when it opens at 5 p.m. If formaldehyde with reaction rate constant 0.4 h^{-1} is emitted from cigarettes smoke at the constant rate of 140 mgh^{-1} starting at 5 pm, what would be the concentration of formaldehyde at 6 pm?
8. Is an air cleaner capable of removing 3 μm particles likely to be effective on wet process cement kiln dust? Explain.

9. How is particle deposition and removal from the lung influenced by the size of particles?
10. What temporary effect upon climate by particulate matter from a sizable volcanic eruption has the potential to produce?
11. How does particulate matter affect the vegetable plant growth? Explain.
12. A typical coal based power plant make following emission:

S–emission	1.40 gs/kWh
Particulate emission	1.4 g/kWh
Carbon emission	280 gc/kWh

A 15W compact fluorescent light tube produces the same amount of light as a 60W incandescent while using only one fourth of the power. Over the 9000 h life time of these bulbs compute the carbon, SO_2 particulate emission that would be saved if one of these bulbs is replaced by incandescent light and electricity comes from the coal based plant (mentioned above).

CHAPTER 10

Atmospheric Dispersion

1. The following is atmospheric altitude versus temperature data with E as stability category.

Altitude (m)	0	100	200	300	400	500	600
Temp °C	20	18	16	15	16	17	18

(*a*) What would be the mixing depth?
(*b*) How high would you expect a plume to rise if it is emitted at 21°C from a 100 m stack of inner radius 1.5 m if it rises at the dry adiabatic lapse rate? Would you expect the plume to be looping, coning, fanning or fumigating?

2. For the temperature profile given in Problem 1 of maximum surface temperature is 22°C and a weather station ariemometer at 10 m height shows winds averaging 4 ms^{-1}. What would be the ventilation coefficient? Assume stability class C and use the wind at the height half way to the maximum mixing depth.
3. Suppose a bone fire emits CO at the rate of 20 gs^{-1} on a clean night when the wind is blowing at 2 ms^{-1}. If effective stack height is 6 m:
(*a*) What would you expect the ground level CO concentration to be at 400 m downwind?
(*b*) Estimate the maximum ground level concentration.
4. A coal fired power plant with effective stack height of 100 m 1.2 gs^{-1} of SO_2 per megawatt of power delivered. If winds are assumed to be 4 ms^{-1} at that height and just over 3 m s^{-1} at 10 m, how big could the plant be (MW) without having the ground level SO_2 exceed 365 μg m^{-3}? (First decide which stability classification leads to the worst conditions).
5. A 35% efficient coal fired plant with effective heights of 100 m emits SO_2 at the rate of 25.8 g J^{-1} into the plant. If winds are assumed to be 4 ms^{-1} at the stack height and just over 3 ms^{-1} at 10 m, how large could the plant be (MW) without having the ground level SO_2 exceed 365 μg m^{-3}.
6. A stack emits 80 gs^{-1} of NO has an effective stack height of 100 m. The wind speed is 4 ms^{-1} at 10 m and it is a clear summer day with sun nearly overhead. Estimate ground level NO concentration.

(*a*) Directly downwind at a distance of 2 km.

(*b*) At the point downwind where NO is maximum.

(*c*) At a point located 2 km down wind and 0.1 km off the downwind axis.

7. $y = 0$ in equation 10.1 reduces to

$$C(x, 0) = \frac{Q}{\pi u \sigma_y \sigma_z} \left(\exp\left[-\frac{H^2}{2\sigma_z^2} \right] \right)$$

For the stability class C, the ratio σ_y/σ_z is essentially constants and independent of x. Assuming that it is constant

(*a*) Show that distance downwind from a stack at which the maximum concentration occurs corresponds to the point where $\sigma_z = H/\sqrt{2}$.

(*b*) Show that maximum concentration is

$$C_{max} = \frac{Q}{\pi \sigma_y \sigma_z u e} = \frac{0.117\, Q}{\sigma_y \sigma_z\, u}$$

(*c*) Show that C_{max} is inversely proportional to H^2.

8. A copper smelter stands 380 m high and has an inner diameter at top 15.2 m. If 130°C gas exist the stack at 20 ms^{-1} while the ambient temperature is 10°C and the winds at stack height are 8 ms^{-1} use the Brigg's model to estimate effective stack height. Assume a slightly unstable atmosphere of Class C.
9. Repeat problem 8 for a stable isothermal atmosphere with no temperature change with altitude.
10. A 200 MW power plant has a 100 m stack with radius 2.5 m, flue gas exit velocity 13.5 ms^{-1}, exit temperature 145°C, ambient temp. 15°C, wind speed at the stack is 5 ms^{-1} and atmosphere is stable, class E, with lapse rate 5°C/km. If plant emits 300 gs^{-1} of SO_2 estimate the concentration at ground level at a distance 16 km directly downward.
11. A source emits 20 gs^{-1} of some pollutant in winds that average 5 ms^{-1} on a class C day. Use Fig. 10.5 to find the peak concentration for effective stack heights of 50, 100, 200 m. Note whether the concentration is roughly proportional to $1/H^2$ as suggested in problem 7.
12. A long line of agricultural waste emits 0.3 g/m s of particulate matter on a clear fall afternoon with wind blowing 3 ms^{-1} perpendicular to the line. Estimate the ground level particulate concentration 400 m downwind the line.
13. On a high way 10000 vehicles pass per hour in front of a house 200 m away. Each car emits an average of 1.5 g/ms of NO and wind is blowing at 2 ms^{-1} towards the house. Estimate the NO concentration at the house on a clear summer day near noon assuming NO_x is chemically stable.
14. Given the following temperature variation with altitude.

Elevation (m)	0	20	40	60	80	100	120
Temp °C	20	18	16	18	20	18	16

What type of plume would you expect if the exit temperature of the plume was 20°C and smoke stack was (*a*) 20 m tall (*b*) 40 m tall (*c*) 60 m tall (*d*) 80 m tall.

15. Sulphur dioxide is emitted through a stack 50 m high and 2 m diameter with exit temperature 120°C and velocity of 100 ms^{-1} with emission rate 200 gs^{-1}. Show how the ground level

concentration at a downwind distance of 1 km varies with wind speed for stability category D when the pressure is 1000 m bar and ambient temp. 25°C.

16. A plume from a 50 m stack is Gaussian and wind field is uniform and constant.

    (*a*) Under a stability class, how far from the ground level centre-line 0.2 km downwind must go to find a mean contamination equal to 0.1 that of the centre line concentration at the same distance downwind.

    (*b*) In the O stability class what are the values of *x* and *y* that exhibit the same concentration as in (*a*).

CHAPTER 11

Air Pollution Control

1. Small quantities of NO_2 can be formed in combustion exhaust gases by the third order reaction.

$$2NO + O_2 \xrightarrow{k} 2NO_2$$

Estimate the amount of NO_2 that can be formed by this route under conditions typical of those in the exhaust of the automobile. Assume that the exhaust system of a car can be represented as a straight pipe through which the exhaust gases flow in so called plug flow, each element travels through the pipe independently of the other elements. Assume that the concentration of NO at the beginning of the exhaust system is 2000 pm. For initial O_2 concentration of NO_2 formed for a residence time of 2 sec. if temp. of exhaust gases.

(*i*) is constant at 800°C

(*ii*) is constant at 300°C

The reaction rate constant is

$$k = 1.066 \times 10^{-5}\ T^2\ \exp(530/T)(ppm)^{-2}\ min^{-1}.$$

2. Suppose that within a square city, 15 km on a side there are 200000 cars on the road each being driven 30 km between 4 pm and 6 pm and each emitting 3 g/kg of CO. It is a clear winter evening with a radiation inversion that restricts mixing to estimated 20.0 m and the wind is bringing clean air in at a steady rate of 1.0 m s^{-1} along an edge of the city. Use a box model to estimate the CO concentration at 6 pm if there were no CO in air at 4 pm and only source of CO is cars. Assume that CO is conservative and that there is complete and instantaneous mixing in box.
3. An electrostatic precipitator with 6000 m^2 of collector plate area 97% efficient in treating 200 m^3/s of the flue gases for a 200 MW power plant. Assuming in equation 11.6 is equal to 1. How large would the plate area have to be increased so that precipitator becomes 99% efficient.
4. When gasoline burns following reaction take place C_7H_{13}+ $10.25O_2 + 38.54N_2 \longrightarrow 7CO_2 + 6.5\ H_2O + 38.54\ N_2$, where any oxidation of nitrogen to nitrogen oxides has been neglected. Calculate the ratio of air to fuel required for complete combustion of gasoline.
5. On the basis of average temperature gradient in the following situations clarify the degree of stability of atmosphere

Temperature at ground level		Temperature at a height, *h* m	
(*a*)	20°C	500 m	25°C
(*b*)	25°C	800 m	15°C
(*c*)	25°C	2000 m	5°C
(*d*)	30°C	500 m	20°C
(*e*)	25°C	1000 m	20°C

6. A conventional cyclone with diameter 1.0 m handles 3.0 m^3s^{-1} of standard air. Using Ne = 6 determine the cut size and the efficiency as a function of particle diameter of density (*a*) 1500 (*b*) 2500 kg m^{-3}.
7. Solve the problem for high efficiency cyclone.
8. A conventional cyclone with a diameter of 1.0 m is to be used to separate particulate matter of density 1500 kg m^{-3} from standard air at a flow rate of 2.0 m^{-3} s^{-1}. An analysis of the particles yielded the following size distribution.

Size μm	3.0	6.0	10.0	20.0	40.0	80.0	140.0	200
Weight fractionless than the stated size	0.03	0.08	0.15	0.33	0.60	0.85	0.96	0.99

Using Ne = 6 show that the overall separation efficiency (86%).

9. A bag house is to be designed to handle 800 m^3/min of air. The filtration takes place at constant pressure so that the air velocity through each bag decreases during the time between cleanings according the relation

$$V_{bag} = \frac{1}{0.267 + 0.08\,t}$$

where V_{bag} is the m^3/min of air per square metre of cloth and *t* is time in minutes. The bags are shaken in sequence row by row on a 15 min cycle. Each bag is 30 cm in diameter 6 m high. The bag house is hope square in cross-section, with 30 cm spacing between bags and 30 cm clearance from the walls. Calculated (*a*) the number of bags required and (*b*) the width of bag house.
10. Why a bag filter is not suitable for each of the following applications particles from lime slaking, fly ash from coal burning, radioactive fumes, heavy mist from a chemical process?
11. Before the installation of an electrostatic precipitator, the stack gas of a power plant contained 6.0 g particulates per m^3 of gas. The gas flow rate in 350 m^3/min and the new precipitator can remove 2500 kg particulate/day.
    (*a*) What is the emission rate of particulates before and after pollution control in kg/day?
    (*b*) What is the efficiency of the electrostatic precipitator?
    (*c*) Will the new system meet the emission standard of 0.7 gm^{-1}?
12. A cylindrical electrostatic precipitator of diameter 0.3m is used for separating galvenised coal fly ash particles from a furnace gas stream. If the volumetric flow rate of the gas is 0.05 m^3s^{-1}, what will be the length of the precipitator for obtaining a collection efficiency of 99.9%. What per cent and change in electrode collection area is required to increase the collection efficiency from 99.9 to 99.95% (3.67m, 10%)?

13. A venturi scrubber is used to collect open hearth steel furnace fumes. The gas pressure drop is 50 cm of water guage and the scrubber water pressure drop is 1 atm. The gas flow rate is 1500 m^3/min and the water flow rate is 5000 litres/min. Estimate the fume collection efficiency of the scrubber.
14. An automobiles equipped with a catalytic converter to converts CO and hydrocarbon emission in exhaust gas. The catalyst temperature must be controlled at 300°C by adjusting the air intake to the converter. The sensible heat of the product gas after combustion in the convertor is 1.8×10^5 Jm^{-3} at 300°C. During driving, the exhaust gas rate is 2 m^3/min at the gas temperature of 425°C and its molar mass is 30. If the air intake to the converter occur at 25°C. What is the amount of air required? Take the average specific heat of the exhaust gas and air as 1050 J kg^{-1} K^{-1} and neglect the heat loss from the converter to the surrounding (1.5 kg/min).
15. You wish to design a bag house to clean 3000 cubic metre at a filter of ratio 3m^3/m^2 of cloth. The filter bag are 15 cm in diameter and 3 m long. If you design a square bag house with the bags on 30 cm centre what would be the exterior dimension neglecting duct work? An alternate system uses 15 mm diameter porous plastic pipes 1 m long on 25 m centres. For the same filter ratio and flow what would be the extension dimension of a square enclosure?
16. For a given cyclone collector, plot centrifugal force as a function of particle specific gravity 0.5–3.0 gas velocity 175 m/minute to 1750 m/minute and with radius of air volume from 30 to 250 cm.
17. List the advantages and disadvantages of using a bag house wet scrubber or electrostatic precipitator for particulate collection from an aspalt plant drying kiln. The gases are at 250°C and contain 4500 mg/m^3 of rock dust in the 0.1–10 μ size range. Gas flow is 2000 m^3/minute. Consider initial and operation cost space requirement and ultimate disposal of production, etc.
18. Suppose a gaseous effluent of 30 m^3/minute is at 200°C and 50% relative humidity. It is cooled to 65°C by spraying with water that was initially at 20°C. What vol. of saturated gas would you have to design for at 65°C? How much water per cubic meter would the system require? How much water per cubic meter would you have to remove effluents from the system?
19. If an electrostatic precipitator in 90% efficient for particulate removal what overall efficiency would you expect for the two of the precipitator in series. Would the cost of two in series be double the cost of single precipitator. List two specific cases where you might use two precipitator in series.

Appendices

APPENDIX I

SOME VERY GOOD INTERNET SITES ON OUR ATMOSPHERE

1. Go to wikipidia site
   <en.wikipidia.org/wiki/>
   at dash enter
   green house gases, climate change, ozone depletion chloroflorocarbons, smog, acid rain, Environment-issues-in-India or any other subject of your interest
2. Animated diagrams of Green house effect, for teachers and students
   <earthguide/diagram,greenhouse/>
3. Global warming Interactive
   <environment.nationalgeographic.com/.../g.w.overview-interactivehtml>
4. Ten ways to make your house carbon neutral <www:ideahelper:com/10ways to_make_your_house_carbonneutral_55htm>
5. Ozone depletion
   <librarythinkquest.org/26026/.../ozonedepletion>
6. Smog
   <edugreen.teri.res.in/explore/air/smog.htm>
7. Smog photos
   <www.nunukphotos.com/smog_photoshtml>
8. Acid Rain
   <library think quest.org/CRO2154/acid_rain.htm>
9. Acid Rain
   <www.angelfire.com/ks3/acidrainreport/acid.htm>
10. For video clipings on various pollution related topics go to
    <www.youtube.com>
    Enter topic of your interest and see video.
11. Environmental Air Pollution in India, Air quality monitoring
    <www.indiastat.com/environmentandpollution/.../stats.aspx>

APPENDIX II

2009 UNITED NATIONS CLIMATE CHANGE CONFERENCE (COP15) COPENHAGEN

Information
Date 7-18 December 2009
Location Bella Centre, Copenhagen, Denmark
Participant UNFCCC member countries
UN Chronicle: Special Climate Change Issue.

In this write up Annex 1 countries and details of data are not being given. However everyone is requested to go to the internet site

<http:/climatecongress.ku.dk/pdflsynthesisreport>

This site has every aspect of as to how global warming will affect, a wonderful site. The other good sites are

<www.simens.com/answers>
<www.scsentified.com> It is on carbon foot print series.

1. Kyoto Protocol

In the mid nineties, scientific opinion was that human activities has resulted in substantial global warming from mid 20th century with continuous increase in green house gas concentration would generate dangerous climate change.

The Kyoto Protocol is a protocol to the United Nations Framework Convention on Climate Change (UNFCCC) aimed at combating global warming. The UNFCCC is an international treaty with the goal of achieving stabilization of green house gases concentration in the atmosphere on a level that would prevent dangerous anthropogenic interference with the climate system.

The protocol was initially adopted on 11 December 1997 in Kyoto and entered into force on 16 February 2005. 187 states have signed and ratified the protocol.

Under the protocol 37 industrialized countries commit themselves to a reduction of four Green House Gases (GHG) (carbondioxide, methane, nitrous oxide and sulphur hexafloride) and two groups of Gases (hydro floro carbons and per floro-carbons) produced by them. All members countries give general commitments. Annex 1 countries agreed to reduce their collective green house gas emission by 5.2% from the 1990 levels. Emission limits do not include emissions by international aviation and shipping, but are in addition to the industrial gases. Chloroflorocarbons CFC, which are dealt under the 1987 Montreal Protocol on substances that deplete ozone layer.

The benchmark 1990 emission levels were accepted by the conference of the parties of UNFCC, which were the values of "global warming potential" calculated for the IPCC Second Assessment Report. These figures are used for converting the various green house gas emission into comparable CO_2 equivalent when computing overall source and sink.

The protocol allows for several "flexible mechanism" such as "emission trading", the Clean Development Mechanism (CDM) and joint implementation to allow Annex countries to meet their GHG emission limitations by purchasing GHG emission reduction credits from elsewhere, through financial exchanges projects that reduce in emission in non-Annex 1 countries from other Annex 1 countries or from the Annex 1 countries with excess allowances.

Each Annex country is required to submit an annual report of inventories of all anthropogenic green house gas emissions from sources and removal from sinks under UNFCCC and Kyoto protocol. These countries nominate a person (called a "designated national authority") to create and manage its green house gas inventory. Countries including Japan, Canada, Italy, the Netherland, Germany, France, Spain and others are actively promoting government carbon funds, supporting multilateral carbon funds intent on purchasing carbon credits from non-Annex 1 countries and are working closely with their major utility, energy, oil and gas and chemical conglomerates to acquire green house gas certificates as cheaply as possible. Virtually all of the non-Annex 1 countries have also established a designated national authority to manage the Kyoto obligation, specifically the CDM process that determines which GHG projects they wish to propose for accreditation by the CDM Executive Board.

The protocol was adopted COP on 11 December 1997. It was opened on 16th March 1998 for signature by parties to UNFCCC.

Article 25 of the Protocol specifies that the Protocol enters into force on the ninetieth day after the date on which not less than 55 parties to convention, incorporating parties included in Annex 1 which accounted in total for at least 55% of the total carbon dioxide emissions for 1990 of the Annex 1 countries, have deposited their instruments of ratification, acceptance, approval or accession.

The most noteable non-party to the Protocol is the United States which is a party to UNFCCC and was responsible for 36.1% of the 1990 emission levels of Annex 1 countries.

Objectives

The objective is the "stabilization and reconstruction of green house gas concentration in the atmosphere at a level that would prevent dangerous anthropogenic interference with climate system.

The objective of the Kyoto conference was to establish a legally binding international agreement where by all the participating nations commit themselves to tackling the issue of global warming and GHG emissions. The target agreed upon was an average reduction of 5.2% from 1990 level by the year 2012. Contrary to popular belief, the protocol will not expire in 2012. In 2012 Annex 1 countries must have fulfilled their obligations of reduction of GHG emission established for the first commitment period.

Five principal concepts of the Kyoto Protocol are:

- commitment to reduce gases that are legally binding for Annex 1 countries, as well as general commitments for all member countries.
- implementation to meet the Protocol objectives, to prepare policies and measures which reduce green house gases: increasing absorption of these gases and use all mechanisms available such as joint implementation, CDM and emission trading: being rewarded with credits which allow more greenhouse gas emissions at home.
- minimizing impacts on developing countries by establishing an adaptation fund for climate change.
- accounting, reporting and review to ensure the integrity of the Protocol.
- compliance by establishing a compliance committee to enforce commitment to protocol.

The protocol provides for several flexible mechanism which enable Annex 1 countries to meet their GHG emission targets by acquiring GHG emission reductions credits. The flexible mechanism are emission trading, Clean Development Mechanism (CDM) trading and joint implementation trading.

In practice this means that non-Annex 1 countries have no GHG emission restriction, but have financial incentives to develop GHG emission reduction projects to receive carbon credits that can be sold to Annex 1 countries encouraging sustainable development. In addition, the flexible mechanism allow Annex 1 countries with efficient, low GHG emitting industries, and high prevailing environmental standards to purchase carbon credits on the world market instead of reducing GHG emissions domestically, Annex 1 countries typically will want to acquire carbon credits as cheaply as possible, while non-Annex 1 countries will like to sell carbon credits. So carbon credits is a form of currency note.

2. Carbon Credit Systems

Burning of fossil fuels is a major source of industrial **greenhouse gas** emissions, especially for power, cement, steel, textile, fertilizer and many other industries which rely on fossil fuels (coal, electricity derived from coal, natural gas and oil).

The concept of carbon credits came into existence as a result of increasing awareness of the need for controlling emissions. The IPCC observed *that: Policies that provide a real or implicit price of carbon could create incentives for producers and consumers to significantly invest in low-GHG products, technologies and processes. Such policies could include economic instruments, government funding and regulation.* A tradable permit system was accepted as one of the policy instruments that will control GHG emission effectively. The mechanism was formalized in the Kyoto Protocol, and the market mechanisms were agreed through the subsequent Marrakesh Accords.

The Protocol agreed 'caps' or quotas on the maximum amount of Greenhouse gases for developed and developing countries. In turn these countries set quotas on the emissions of installations run by local business and other organizations, generically termed 'operators'. Countries manage this through their own national 'registries', which are required to be validated and monitored for compliance by the UNFCCC. Each operator has an allowance of credits, where each unit gives the owner the right to emit one metric tonne of carbon dioxide or other equivalent greenhouse gas. Operators that have not used up their quotas can sell their unused allowances as carbon credits, while businesses that are about to exceed their quotas can buy the extra allowances as credits, privately or on the open market. As demand for energy grows over time, the total emissions must still stay within the cap, but it allows industry some flexibility and predictability in its planning to accommodate this.

By permitting allowances to be bought and sold, an operator can seek out the most cost-effective way of reducing its emissions, either by investing in 'cleaner' machinery and practices or by purchasing emissions from another operator who already has excess 'capacity'.

Kyoto's 'Flexible Mechanisms'

A credit can be an emissions allowance which was originally allocated or auctioned by the national administrators of a cap-and-trade program, or it can be an offset of emissions. Such offsetting and mitigating activities can occur in any developing country which has ratified the Kyoto Protocol, and has a national agreement in place to validate its carbon project through one of the UNFCCC's approved mechanisms. Once approved, these units are termed Certified Emission Reductions, or CERs. The Protocol allows these projects to be constructed and credited in advance of the Kyoto trading period.

The Kyoto Protocol provides for three mechanisms that enable countries or operators in developed countries to acquire greenhouse gas reduction credits.

- Under **Joint Implementation** (JI) a developed country with relatively high costs of domestic greenhouse reduction would set up a project in another developed country.
- Under the **Clean Development Mechanism** (CDM) a developed country can 'sponsor' a greenhouse gas reduction project in a developing country where the cost of greenhouse gas reduction project activities is usually much lower, but the atmospheric effect is globally equivalent. The

developed country would be given credits for meeting its emission reduction targets, while the developing country would receive the capital investment and **clean technology** or beneficial change in land use.

- Under **International Emissions Trading** (IET) countries can trade in the international carbon credit market to cover their shortfall in allowances. Countries with surplus credits can sell them to countries with capped emission commitments under the Kyoto Protocol.

These carbon projects can be created by a national government or by an operator within the country. In reality, most of the transactions are not performed by national governments directly, but by operators who have set quotas for their country.

There bare two distinct types of Carbon Credits: Carbon Offset Credits (COC's) and Carbon Reduction Credits (CRC's). COC consist of clean forms of energy production, wind, solar, hydro and biofuels. CRC consist of the collection and storage of Carbon from our atmosphere through biosequestration (reforestation, forestation), ocean and soil collection and storage efforts. Both approaches are recognized as effective ways to reduce the Global Carbon Emissions crises.

Emission Markets

For trading purposes, one allowance or CER is considered equivalent to one metric tonne of CO_2 emissions. These allowances can be sold privately or in the international market at the prevailing market price. These trade and settle internationally and hence allow allowances to be transferred between countries. Each international transfer is validated by the **UNFCCC**. Each transfer of ownership within the European Union is additionally validated by the European Commission.

Climate exchanges have been established to provide a spot market in allowances, as well as futures and options market to help discover a market price and maintain liquidity. Carbon prices are normally quoted in Euros per tonne of carbon dioxide or its equivalent (CO_2e). Other greenhouse gasses can also be traded, but are quoted as standard multiples of carbon dioxide with respect to their global warming potential. These features reduce the quota's financial impact on business, while ensuring that the quotas are met at a national and international level.

Currently there are five exchanges trading in carbon allowances: the Chicago Climate Exchange, European Climate Exchange, Nord Pool, PowerNext and the European Energy Exchange. Recently, NordPool listed a contract to trade offsets generated by a CDM carbon project called Certified Emission Reductions (CERs). Many companies now engage in emissions abatement, offsetting, and sequestration programs to generate credits that can be sold on one of the exchanges. At least one private electronic market has been established in 2008: Cantor CO_2e.

Managing emissions is one of the fastest-growing segments in financial services in the City of London with a market now worth about Euro 30 billion, but which could grow to Euro 1 trillion within a decade. Louis Redshaw, head of environmental markets at Barclays Capital predicts that "Carbon will be the world's biggest commodity market, and it could become the world's biggest market overall."

Setting a Market Price for Carbon

Unchecked, energy use and hence emission levels are predicted to keep rising over time. Thus the number of companies needing to buy credits will increase, and the rules of supply and demand will push up the market price, encouraging more groups to undertake environmentally friendly activities that create carbon credits to sell.

An individual allowance, such as a Kyoto Assigned Amount Unit (AAU) or its near-equivalent European Union Allowance (EUA), may have a different market value to an offset such as a CER. This is due to the lack of a developed secondary market for CERs, a lack of homogeneity between projects which causes difficulty in pricing, as well as questions due to the principle of supplementarily and its lifetime. Additionally, offsets generated by a carbon project under the Clean Development Mechanism are potentially limited in value because operators in the EU ETS are restricted as to what percentage of their allowance can be met through these flexible mechanisms.

Raising the price of carbon will achieve four goals. First, it will provide signals to consumers about what goods and services are high-carbon ones and should therefore be used more sparingly. Second, it will provide signals to producers about which inputs use more carbon (such as coal and oil) and which use less or none (such as natural gas or nuclear power), thereby inducing firms to substitute low-carbon inputs. Third, it will give market incentives for inventors and innovators to develop and introduce low-carbon products and processes that can replace the current generation of technologies. Fourth, and most important, a high carbon price will economize on the information that is required to do all three of these tasks. Through the market mechanism, a high carbon price will raise the price of products according to their carbon content. Ethical consumers today, hoping to minimize their "carbon footprint," have little chance of making an accurate calculation of the relative carbon use in, say, driving 250 miles as compared with flying 250 miles. A harmonized carbon tax would raise the price of a good proportionately to exactly the amount of CO_2 that is emitted in all the stages of production that are involved in producing that good. If 0.01 of a ton of carbon emissions results from the wheat growing and the milling and the trucking and the baking of a loaf of bread, then a tax of $30 per ton carbon will raise the price of bread by $0.30. The "carbon footprint" is automatically calculated by the price system. Consumers would still not know how much of the price is due to carbon emissions, but they could make their decisions confident that they are paying for the social cost of their carbon footprint.

The social cost of carbon is the additional damage caused by an additional ton of carbon emissions... The optimal carbon price, or optimal carbon tax, is the market price (or carbon tax) on carbon emissions that balances the incremental costs of reducing carbon emissions with the incremental benefits of reducing climate damages. If a country wished to impose a carbon tax of $30 per ton of carbon, this would involve a tax on gasoline of about 9 cents per gallon. Similarly, the tax on coal-generated electricity would be about 1 cent per kWh, or 10 per cent of the current retail price. At current levels of carbon emissions in the United States, a tax of $30 per ton of carbon would generate $50 billion of revenue per year.

How Buying Carbon Credits can Reduce Emissions

Carbon credits create a market for reducing greenhouse emissions by giving a monetary value to the cost of polluting the air. Emissions become an internal cost of doing business and are visible on the balance sheet alongside raw materials and other liabilities or assets.

For example, consider a business that owns a factory putting out 100,000 tonnes of greenhouse gas emissions in a year. Its government is an Annex I country that enacts a law to limit the emissions that the business can produce. So the factory is given a quota of say 80,000 tonnes per year. The factory either reduces its emissions to 80,000 tonnes or is required to purchase carbon credits to offset the excess. After costing up alternatives the business may decide that it is uneconomical or infeasible to invest in new machinery for that year. Instead it may choose to buy carbon credits on the open market from organizations that have been approved as being able to sell legitimate carbon credits.

We should consider the impact of manufacturing alternative energy sources. For example, the energy consumed and the Carbon emitted in the manufacture and transportation of a large wind turbine would prohibit a credit being issued for a predetermined period of time.

- One seller might be a company that will offer to offset emissions through a project in the developing world, such as recovering methane from a swine farm to feed a power station that previously would use fossil fuel. So although the factory continues to emit gases, it would pay another group to reduce the equivalent of 20,000 tonnes of carbon dioxide emissions from the atmosphere for that year.
- Another seller may have already invested in new low-emission machinery and have a surplus of allowances as a result. The factory could make up for its emissions by buying 20,000 tonnes of allowances from them. The cost of the seller's new machinery would be subsidized by the sale of allowances. Both the buyer and the seller would submit accounts for their emissions to prove that their allowances were met correctly.

Credits versus Taxes

Carbon credits and carbon taxes each have their advantages. Credits were chosen by the signatories to the Kyoto Protocol as an alternative to Carbon taxes. A criticism of tax-raising schemes is that they are frequently not hypothecated, and so some or all of the taxation raised by a government may be applied inefficiently or not used to benefit the environment.

By treating emissions as a market commodity it becomes easier for business to understand and manage their activities, while economists and traders can attempt to predict future pricing using well understood market theories. Thus the main advantages of a tradable carbon credit over a carbon tax are:

- the price is more likely to be perceived as fair by those paying it. Investors in credits have more control over their own costs.
- the flexible mechanisms of the Kyoto Protocol ensure that all investment goes into genuine sustainable carbon reduction schemes, through its internationally-agreed validation process.
- if correctly implemented a target level of emission reductions is achieved with certainty, while under a tax the actual emissions would vary over time.
- it provides a framework for rewarding people or companies who plant trees or otherwise sequester carbon.

The advantages of a carbon tax are:

- possibly less complex, expensive, and time-consuming to implement. This advantage is especially great when applied to markets like gasoline or home heating oil.
- perhaps some reduced risk of certain types of cheating, though under both credits and taxes, emissions must be verified.
- reduced incentives for companies to delay efficiency improvements prior to the establishment of the baseline if credits are distributed in proportion to past emissions.
- when credits are grandfathered, this puts new or growing companies at a disadvantage relative to more established companies.
- it is clear what effect the policy has on the price of energy.

3. Copenhagen Climate Council

The Copenhagen Climate Council was founded in 2007 by the leading independent think tank in Scandinavia, *Monday Morning,* head-quartered in Copenhagen, Denmark. The purpose of this Council was to create global awareness of the importance of the UN Climate Summit (COP15) in Copenhagen, December 2009. The Council was to seek to promote constructive dialogue between government and business, so that when the world's political leaders and negotiators meet in Copenhagen, they will do so armed with the very best arguments for establishing a treaty that can be supported by global business. The strategy was to built upon the following principles:

- **Creating international awareness** of the importance of the Copenhagen UN Climate Summit and the successor treaty to the Kyoto Protocol.
- **Promoting constructive dialogue** between government, business, and science.
- **Inspiring global business leaders** by demonstrating that tackling climate change also has the potential to create huge opportunities for innovation and economic growth.

Bases on these principles the council prepared a manifesto for 2009 meet.

Copenhagen Climate Council comprises 30 global climate leaders representing *business, science,* and *public policy* from all parts of the world.

- **Business leaders** are selected to represent global companies and innovative entrepreneurs, who, through their actions, reveal that sustainable, climate-responsible business is both necessary and profitable.
- **Scientists** are gathered to ensure that the work of the Council is underpinned by rigorous analysis.
- **Policy makers** with experience in public policy are included in the Council to ensure that the work is informed by knowledge of what is required to assist high-level, complex policy negotiations.

The Copenhagen Climate Council created global awareness to the urgency of reaching a global agreement on how to tackle climate change at the UN Climate Conference in Copenhagen, December 2009. To achieve this end, the Copenhagen Climate Council provided a Web 2.0 climate website — **'The Climate Community'** — which features latest climate news, intelligence, solutions and points of view, an online climate community, as well as the rest of the Copenhagen Climate Council activities, such as the **'World Business Summit on Climate Change';** launching the **'Thought Leadership Series';** launching the **'Climate LIFE'** film, book, and digital exhibition; cohosting with CITRIS, the scientific conference **'Unlocking the Climate Code: Innovation in Climate and Energy';** and the Poznan side event **'Business Requirements of a Post-2012 Climate Treaty',** the Council also hosted a Business Round table in Beijing.

4. United Nations Climate Change Conference–2009

The **United Nations Climate Change Conference–2009** was held at the Bella Center in Copenhagen, Denmark, between 7 December and 18 December. The conference included the 15th Conference of the Parties (COP 15) to the United Nations Framework Convention on Climate Change and the 5th Meeting of the Parties (COP/MOP 5) to the Kyoto Protocol. According to the Bali Road Map, a framework for climate change mitigation beyond 2012 was to be agreed there. The conference was preceded by the Climate Change: Global Risks, Challenges and Decisions Scientific Conference, which took place in March 2009 and was also held at the Bella Center.

Connie Hedegaard was president of the conference until December 16, 2009, handing over the chair to Danish Prime Minister Lars Lokke Rasmussen in the final stretch of the conference, during negotiations between heads of state and government. On Friday 18 December, the final day of the conference, international media reported that the climate talks were "in disarray". Media also reported that in lieu of a summit collapse, solely a "weak political statement" was anticipated at the conclusion of the conference.

The Copenhagen Accord was drafted by the China, India, Brazil and South Africa on December 18, and judged a "meaningful agreement" by the United States government. It pulled the negotiations back from the brink of total for leave. It was "recognised", but not "agreed upon", in a debate of all the participating countries the next day, and it was not passed unanimously. The document recognised that climate change is one of the greatest challenges of the present and that actions should be taken to keep any temperature increases to below 2°C. The document is not legally binding and does not contain any legally binding commitments for reducing CO_2 emissions. Leaders of industrialised countries, including Barack Obama and Gordon Brown, were pleased with this agreement but many leaders of other countries and non-governmental organisations were opposed to it. The accord postpone that industrialized nations have to come up with a deeper legally binding emission cuts than the Kyoto protocol, apart from providing substantial amounts of finance and technology to help developing countries, what nations did at Copenhagen was not save humanity but just their national interest.

On December 16, there was speculation that the summit in Copenhagen was in jeopardy. "We have made no progress" said a source close to the talks. "What people don't realise is that we are now not really ready for the leaders. These talks are now 18 hours late.

The Guardian reported that negotiators were openly talking of the best possible outcome being a "weak political agreement that would leave no clear way forward to tackle rising greenhouse gas emissions". This would mean that negotiations would continue into 2010 increasing the damage done by emissions.

On December 18, the head of the United Nations Environmental Program told the BBC that "as of this morning is a summit in crisis" and that only the arrival of heads of state could bring the summit to a successful conclusion.

Early on Saturday 19 December, delegates approved a motion to "take note of the Copenhagen Accord of December 18, 2009". However it was reported that it was not yet clear whether the motion was unanimous, or what its legal implications are. The Copenhagen Accord recognises the scientific case for keeping temperature rises below 2°C, but does not contain commitments for reduced emissions that would be necessary to achieve that aim. An agreement was also reached that would set up a deal to reduce deforestation in return for cash from developed countries. The agreement made was non-binding but U.S. President Obama said that countries could show the world their achievements. He said that if they had waited for a binding agreement, no progress would have been made. In the week following the end of the Copenhagen summit, carbon price in EU dropped to a six month low. The accord asks countries to submit emissions targets by the end of January 2010 and paves the way for further discussions to occur at 2010 UN climate change conference in Maxico and mid year session in Bonn.

In nutshell, the accord reflects a true political willingness to combat climate change, but it remains unclear how this political willingness will translate into concrete steps. As a result Global energy business were disappointed and confused by the deal saying it does not provide enough certainty to justify the large investments needed to carbon emissions.

APPENDIX III

APPROXIMATE BOND DISSOCIATION ENERGIES

Bond broken	*Bond dissociation energy (kJ mol^{-1})*	*Corresponding absorption wavelength in nm*
CH_3—H	439.3	272
C_2H_5 —H	410.03	292
n-C_3H_7—H	410.03	292
i-C_3H_7— H	397.48	300.8
s-C_4H_9—H	401.66	297.7
cyclopropyl—H	443.5	269.7
cyclobutyl—H	405.8	294.7
cyclopentyl—H	397.5	300.9
cyclohexyl—H	401.7	297.7
cycloheptyl—H	389.1	307.4
$HOCH_2$—H	393.3	304.1
$C_6H_5CH_2$—H	368.19	324.8
C=C—H	460.24	259.8
C_6H_3—H	464.4	257.5
C=C—C(C)—H	347.27	344.4
C=C—$C(C)_2$—H	322	371.4
C≡C—C	552	216.6
CH_3—CH_3	376.5	317.6
CH_3—C_2H_5	360	332.2
C_2H_5—C_2H_5	343	348.7
n-C_3H_7—CH_3	364.0	328.6
i-C_3H_7—CH_3	359.8	332.4
s-C_4H_9—CH_3	355.64	336.3
t-C_4H_9—CH_3	351.5	340.2
n-C_3H_7—C_2H_5	343.0	348.7
i-C_3H_7—C_2H_5	339.0	352.8
s-C_4H_9—C_2H_5	335	357
C_6H_5—CH_3	427	280.1
C_6H_5—C_2H_5	406	294.6
C_6H_5—*n*-C_3H_7	410	291.7

(Contd.)

Bond broken	*Bond dissociation energy (kJ mol^{-1})*	*Corresponding absorption wavelength in nm*
$C{=}C—CH_3$	422.6	287.0
$C{=}CC—CH_3$	309.6	386.3
$C{\equiv}C—CH_3$	527	226.8
$C{\equiv}CC—CH_3$	318	376.1
$CH_3—OH$	385	310.6
$C_2H_5—OH$	385	310.6
$C_6H_5CH_2—OH$	339	352.8
$CH_3O—H$	435.1	274.9
$C_2H_5O—H$	435.1	274.9
n-$C_3H_7O—H$	431	277.5
$C_6H_5O—H$	364	328.6
$CH_3O—CH_3$	347	344.7
$CH_3O—C_2H_5$	343	348.7
$C_6H_5O—CH_3$	268	446.2
$CH_3O—OCH_3$	159	752.2
$C_2H_5O—OC_2H_5$	159	752.2
$CH_3\overset{O}{\overset{\Vert}{C}}O—H$	443.5	269.7
$C_2H_5\overset{O}{\overset{\Vert}{C}}O—H$	443.5	269.7
n-$C_3H_7\overset{O}{\overset{\Vert}{C}}O—H$	443.5	269.7
$H\overset{O}{\overset{\Vert}{C}}—H$	364	328.6
$CH_3\overset{O}{\overset{\Vert}{C}}—H$	360	232.2
$C_6H_5\overset{O}{\overset{\Vert}{C}}—H$	364	328.6
$OHC—CH_3$	347	344.7
$OHC—C_6H_5$	401.6	297.8

(Contd.)

Bond broken	*Bond dissociation energy (kJ mol^{-1})*	*Corresponding absorption wavelength in nm*
$CH_3\overset{O}{\overset{\|\|}{C}}—CH_3$	339	352.8
$CH_3\overset{O}{\overset{\|\|}{C}}—C_6H_5$	393.4	304.0
$CH_3\overset{O}{\overset{\|\|}{C}}—\overset{O}{\overset{\|\|}{C}}CH_3$	280	327
$HS—H$	381	313.9
$CH_3S—H$	381	313.9
$RS—H$	381	313.9
$CH_3—SH$	309.6	386.3
$C_2H_5—SH$	297	402.7
t-$C_4H_9—SH$	284.5	420.4
$C_6H_5—SH$	364.0	328.6
$CH_3S—CH_3$	322	371.4
$CH_3S—C_2H_5$	305.5	391.5
$CH_3SO_2—CH_3$	280	427.1
$RS_2—H$	293	408.2
$RS_2—CH_3$	238.5	501.5
$RS—SR$	301	397.5
CH_3NH_2	355.6	336.3
$C_2H_5—NH_2$	343	348.7
$CH_3NH—H$	418.4	285.8
$CH_3NH—CH_3$	343	348.7
$CH_3NH—C_2H_5$	335	357.0
$(CH_3)_2N—N—H$	385	310.6

(Contd.)

Bond broken	*Bond dissociation energy (kJ mo1⁻¹)*	*Corresponding absorption wavelength in nm*
$(CH_3)_2N—CH_3$	318	376.1
$(CH_3)_2N—C_2H_5$	301	397.3
$C_6H_5NH—H$	368.2	324.8
$C_6H_5NH—CH_3$	297.06	402.6
$C_6H_5NH—C_2H_5$	288.7	414.3
$H_2NCH_2—H$	389	307.4
$CH_3—CN$	510.5	234.3
$C_6H_5—CN$	548.1	218.2
$CH_3—F$	460	260.0
$CH_3—Cl$	355.6	336.3
$CH_3—Br$	267	402.7
$CH_3—I$	364	328.6
$CF_2Cl—F$	514.6	232.4
$CF_2Cl—Cl$	318	376.1
$CF_2Cl—Br$	272	439.7
$CFCl_2—Cl$	305.5	391.5
$HO—H$	498	240.1
$HO—OH$	214	558.9
$HOO—H$	364	338.6
$HO_2—NO_2$	96	1246.0
$H—NO_2$	326	366.9
$HO—NO$	205	583.4
$H—ONO_2$	422	283.0
$HO—NO_2$	205	583.4
$HO—ONO_2$	163.2	733.7

(Contd.)

Bond broken	*Bond dissociation energy (kJ mol⁻¹)*	*Corresponding absorption wavelength in nm*
CS — S	431	277.5
OS — O	544	219.8
HS — SH	276	433.3
NH_2 — H	447.7	267.1
CN — H	519.0	230.5
CH_3 — NO	167.4	714.4
i-C_3H_7 — NO	155	771.6
i-C_3H_4	247	484.2
RO — NO	171.5	697.4
RO — NO_2	171.5	697.5

APPENDIX IV

CONCENTRATION UNITS

For gas phase species, the most commonly used units are part per million (ppm), parts per hundred million (pphm), parts per billion (ppb) and parts per trillion (ppt). These units express the number of molecules of pollutant found in a million (10^6), a hundred million (10^8), a billion (10^9) or a trillion (10^{12}) molecules of air. Specification of an American billion is necessary because a British billion is 10^{12} and not 10^9.

Alternatively, because numbers of molecules (or moles) are proportional to their volumes according to ideal gas law ($PV=nRT$), these units may be thought of as the number of volumes of the pollutants found in 10^6, 10^8, 10^9 or 10^{12} volumes of air.

For example, a background concentration of O_3 may be 0.04 ppm; this is 4 pphm, alternatively in every 10^8 volumes of air, 4 volumes are due to ozone. However, it is most convenient to express the concentration in ppm, pphm, ppb or ppt or molecules per cm^3 are used of which ppm is the largest.

Finally, it is important to remember that number of molecules, or volumes of a given gaseous species forms the basis of units in atmospheric chemistry; in water chemistry, mass rather than volume is basis for expressing concentrations in ppm, and so on.

Number per cubic centimeter: A second type of concentration unit is generally used for species such as free radicals (e.g. OH) present at sub-ppt levels. It is the number of molecules, atoms or free radicals present in a given volume of air, usually cm^3. Thus the number of molecules per cm^3 in air at one atmosphere pressure and 298 K is given by

$$\frac{n}{V} = \frac{P}{RT} = \frac{1 \text{ atm}}{0.08206 \dfrac{L \text{ atm}}{k \text{ mole}}} = 0.0409 \frac{\text{moles}}{L}$$

Converting to units of molecules cm^{-3}, one obtains

$$\frac{n}{V} = 0.0409\frac{\text{moles}}{L}\times 10^{-3}\frac{L}{\text{cm}^3}\times 6.02\times 10^{23}\frac{\text{molecules}}{\text{moles}}$$

$$= 2.46\times 10^{19}\text{ molecules cm}^{-3}$$

From the definition of ppm as the number of pollutant molecules per 10^6 molecules of air, 1 ppm corresponds to $2.46 \times 10^{19} = 2.46 \times 10^{13}$ molecules cm^{-3} at 298 K and one atmosphere total pressure. It follows that a concentration of the OH radical in polluted air of 0.1 ppt is $2.46 \times 10^{19} \times 10^{-12} \times 0.1 = 2.46 \times 10^6$ molecules cm^{-3}.

Table AI summarizes the relation between these units at one atmosphere pressure and 298 K. Of course correction must be made if temperature or pressure differs significantly.

Table AI Conversion between units of concentration in ppm, pphm, ppb, ppt and molecules cm^{-3} at one atmosphere pressure at 298 K

Parts per	*Unit*	*Molecules, atoms or radicals per cm³*
10^6	1 ppm =	2.46×10^{13} cm^{-3}
10^8	1 pphm =	2.46×10^{11} cm^{-3}
10^9	1 ppb =	2.46×10^{10} cm^{-3}
10^{12}	1 ppt =	2.46×10^7 cm^{-3}

1 ppm in unit of mass = 4.09 × molar mass mg m^{-3}.

Micrograms per cubic meter: A third unit of gaseous species is mass per unit volume, usually 10^{-6} g per cubic meter (μg m^{-3}. Since one atmosphere at 298 K contain 4.09×10^{-2} moles L^{-1}, 1 ppm must contain $(4.09 \times 10^{-2}) \times 10^{-6}$ or 4.09×10^{-8} moles L^{-1} or 4.09×10^{-5} moles m^{-3}. If molar mass of pollutant is MM grams per mole, then 1 ppm in units of mass is $(4.09 \times 10^{-5}) \times$ M M g m^{-3} or 4.09 × MM μg m^{-3}. (MM stands for molar mass of species)

Returning to the example of O_3 of 0.04 ppm, this concentration in μg m^{-3} is (4.09 × 48) × 0.04 or 7.853 μg m^{-3}. If the pollutant is 0.04 ppm SO_2 (MM = 64) the concentration of SO_2 would be (4.09 × 64) × 0.04 = 10.5 μg m^{-3}.

The conversion between μg m^{-3} and ppm, pphb ppb and ppt can be summarized as follow:

$$4.09 \times 10^{-8}\text{ moles L}^{-1} = 1\text{ ppm}$$

$$\mu\text{ g m}^{-3} = \text{ppm} \times 40.9 \times (\text{MM})$$

$$= \text{pphm} \times 0.409 \times (\text{MM})$$

$$= \text{ppb} \times 0.0409 \times (\text{MM})$$

$$= \text{ppt} \times 4.09 \times 10^{-5} \times (\text{MM})$$

For atmospheric particulate matter, concentration are expressed in mass per unit volume, commonly μg m^{-3} or in number of particles per unit volumes for example, per cm^{-3}.

The strengths of anthropogenic emissions are measured in units of Tetragrams or Metric Tons. Their relation to grams is as follows:

$$1 \text{ Tetra gram} = 1 \text{ Tg} = 1012 \text{ g}$$
$$1 \text{ m ton} = 10^6 \text{ g}$$
$$1012 \times 10^6 \text{ m ton} = 1 \text{ Tg}$$

Thus the global anthropogenic production of SO_2 in 1976 was 104 Tg S Yr^{-1}, which reads as 104 tetragram of sulphur per year.

A number of pollutants have documented effects on people, plants or material at concentrations or approaching those found in polluted air. To protect public health and welfare, Table A2 list seven criteria pollutants concentration observed from remote areas to severe polluted urban areas.

Table A2

Criteria pollutants	*Typical concentration ranges*
CO	≤ 0.2 – 50 ppm.
SO_2	≤ 1 lb – 2 ppm.
O_3	0.1 – 0.5 ppm.
NO_2	≤ 1 ppb – 0.5 ppm.
NMHC	≤ 65 ppbc – 1.5 ppmc.
Total Suspended Particulates	5 – 1500 μ g m^{-3}
Lead	0.0001 – 10 μg m^3

These values are one-hour average except is SO_2 and TSP which are 24-hour average values.

APPENDIX V

CONVERSION FACTOR FOR CHANGING LIGHT ABSORPTION COEFFICIENT FROM ONE UNIT TO ANOTHER

Both units in Either Logarithmic Base *e* or Base 10

$$(\text{cm}^2 \text{ molecule}^{-1}) \times 2.69 \times 10^{19} = (\text{atm at 273 K})^{-1} (\text{cm}^{-1})$$
$$\times 2.46 \times 10^{19} = (\text{atm at 298 K})^{-1} (\text{cm}^{-1})$$
$$\times 3.24 \times 10^{16} = (\text{Torr at 298 K})^{-1} (\text{cm}^{-1})$$
$$\times 6.02 \times 10^{20} = \text{L mole}^{-1} \text{ cm}^{-1}$$
$$(\text{atm at 298 K})^{-1} (\text{cm}^{-1}) \times 4.06 \times 10^{-20} = \text{cm}^2 \text{ molecule}^{-1}$$
$$\times 1.09 = (\text{atm at 273 K})^{-1} \text{ cm}^{-1}$$
$$(\text{L mole}^{-1} \text{ cm}^{-1}) \times 4.46 \times 10^{-2} = (\text{atm at 273 K})^{-1} (\text{cm}^{-1})$$
$$\times 4.09 \times 10^{-2} = (\text{atm at 298 K})^{-1} (\text{cm}^{-1})$$
$$\times 1.66 \times 10^{-21} = \text{cm}^2 \text{ molecules}^{-1}$$

Change of Both Logarithmic Base and Units

$$(\text{cm}^2 \text{ molecule}^{-1}), \text{ base e} \times 1.17 \times 10^{19} = (\text{atm at 273 K})^{-1} (\text{cm}^{-1}), \text{ base 10}$$

$$\times 1.07 \times 10^{19} = (\text{atm at 298 K})^{-1} (\text{cm}^{-1}), \text{ base } 10$$

$$\times 1.41 \times 10^{16} = (\text{Torr at 298 K})^{-1} (\text{cm}^{-1}), \text{ base } 10$$

$$\times 2.62 \times 10^{20} = \text{L mole}^{-1} \text{ cm}^{-1}, \text{ base } 10$$

$$(\text{L mole}^{-1} \text{ cm}^{-1}), \text{ base } 10 \times 3.82 \times 10^{-21} = \text{cm}^2 \text{ molecule}^{-1}, \text{ base } e$$

$$\times 0.103 = (\text{atm at 273 K})^{-1} (\text{cm}^{-1}), \text{ base } e$$

$$\times 9.42 \times 10^{-2} = (\text{atm at 298 K})^{-1} (\text{cm}^{-1}), \text{ base } e$$

$$(\text{atm at 273 K})^{-1} (\text{cm}^{-1}), \text{ base } 10 \times 8.57 \times 10^{-20} = \text{cm}^2 \text{ molecule}^{-1}, \text{ base } e$$

$$\times 5.16 = \text{L mole}^{-1} \text{ cm}^{-1}, \text{ base } e$$

$$(\text{Torr at 298 K})^{-1} (\text{cm}^{-1}), \text{ base } 10 \times 7.11 \times 10^{-17} = \text{cm}^2 \text{ molecule}^{-1}, \text{ base } e$$

$$\times 4.28 \times 10^{4} = \text{L mole}^{-1} \text{ cm}^{-1}, \text{ base } e$$

$$(\text{atm at 298 K})^{-1} (\text{cm}^{-1}), \text{ base } 10 \times 9.35 \times 10^{-20} = \text{cm}^2 \text{ molecule}^{-1}, \text{ base } e$$

$$\times 2.51 = (\text{atm at 273 K})^{-1} (\text{cm}^{-1}) \text{ base}$$

APPENDIX VI

SOME COMMON CONVERSION FACTORS FOR GAS PHASE REACTIONS

Concentration

$$1 \text{ mole L}^{-1} = 6.02 \times 10^{20} \text{ molecules cm}^{-3}$$

$$1 \text{ ppm} = 2.46 \times 10^{13} \text{ molecules cm}^{-3} = 4.09 \text{ (MM) } \mu \text{ g m}^{-3}$$

$$1 \text{ ppb} = 2.46 \times 10^{10} \text{ molecules cm}^{-3} = 4.09 \times 10^{-3} \text{ (MM) } \mu \text{ g m}^{-3}$$

$$1 \text{ ppt} = 2.46 \times 10^{7} \text{ molecules cm}^{-3} = 4.09 \times 10^{-5} \text{ (MM) } \mu \text{ g m}^{-3}$$

$$1 \text{ atm} = 760 \text{ Torr} = 4.09 \times 10^{-2} \text{ mole L}^{-1} = 2.46 \times 10^{19} \text{ molecules cm}^{-3}$$

Second Order Rate Constants

$$\text{cm}^3 \text{ molecules}^{-1} \text{ S}^{-1} \times 6.02 \times 10^{20} = \text{L mol}^{-1} \text{ s}^{-1}$$

$$\text{ppm}^{-1} \text{ min}^{-1} \times 4.08 \times 10^{5} = \text{L mol}^{-1} \text{ s}^{-1}$$

$$\text{ppm}^{-1} \text{ min}^{-1} \times 6.77 \times 10^{-16} = \text{cm}^3 \text{ molecules}^{-1} \text{ s}^{-1}$$

$$\text{atm}^{-1} \text{ s}^{-1} \times 4.06 \times 10^{-20} = \text{cm}^3 \text{ molecules}^{-1} \text{ s}^{-1}$$

Third Order Rate Constants

$$\text{cm}^6 \text{ molecules}^{-2} \text{ s}^{-1} \times 3.63 \times 10^{41} = \text{L}^2 \text{ mol}^{-2} \text{ s}^{-1}$$

$$\text{ppm}^{-2} \text{ min}^{-1} \times 9.47 \times 10^{12} = \text{L}^2 \text{ mol}^{-2} \text{ s}^{-1}$$

$$\text{ppm}^{-2} \text{ min}^{-1} \times 2.75 \times 10^{-29} = \text{cm}^6 \text{ molecules}^{-2} \text{ s}^{-1}$$

PART–II

Water, Earth and Environment

1 Water and Earth

1.1 EVOLUTION OF THE EARTH AND LIFE

It is hypothesized evolution of the earth took place approximately five billion years ago when a cloud in the interstellar gas, known as solar nebula, collapsed, forming protostars and planetary systems. Out of these clouds of cosmic dust emerged the earth, a sphere of molten rocks. Gases bubbled out of the hot interior forming a reducing atmosphere of methane, ammonia, hydrogen cyanide, water vapours, helium and hydrogen, etc., that surrounded the earth. As these gases moved upwards from high to low pressure regions, there occurred a drop in temperature and consequent condensation of gases, falling back on the surface of the infant earth in the form of rain.

The cooling in the upper strata of the atmosphere caused torrents of rains, and the process must have continued for millions of years, for the rains immediately vapourized in the beginning, long before they even reached the surface. Gradually, as the earth cooled and rains could reach the surface shallow oceans resulted. The rains fell not for days or months, not for years but perhaps for centuries, until the earth was totally covered with water. The ceaseless downpour washed away millions of tons of minerals from some mountains and slashed out huge valleys across the surface of the earth. The rains finally slowed and stopped. The lowest regions of the earth's surface had been filled with a primeval broth in which the primitive form of sea and, subsequently, life developed.

The earth at that time appeared to have been rocked by explosions so gigantic and violent, that a thousand hydrogen bombs exploding together could not possibly match them. Mountains arose out of the earth, and deep oceans, were created. In the beginning these first mountains were eroded by the torrential rains, and new ones were subsequently formed.

The composition of the atmosphere has been deduced using spectroscopy and telemetry of allied planets. Space probes have helped a lot in this regard. The earth's primitive atmosphere consisted of methane, ammonia, hydrogen cyanide, and water vapour. Hydrogen and helium had already escaped from the gravitational pull of the earth. These gases were subjected to the strong ionizing radiation and that dissociated these molecules into positive and negative ions. Water vapours condensed more readily on particles carrying negative charge than on those carrying a positive charge. Thus negative electric charges were brought down into the lower part of the cloud, whereas the positive charges were left in the upper part of it, thereby producing sufficiently high differences in the electrical potential for lightning to strike between the clouds. These nearly continuous electric discharges in the atmosphere for many thousands of years made these molecules to unite to form various amino acids.

The bulk of evidence indicates that primitive atmosphere of the earth was reducing and that the present oxidizing atmosphere and hydrosphere are relatively recent, and are in fact, products of biological activity. Bacterial decomposition as well as oxidation was absent in the prebiotic world. This led to reversal of equilibria and consequently structural stabilization and hence polymerization of amino acids. Once polymerization had occurred, the danger of dissolution diminishes, for the polymers may organise into a stable configuration. Hydrophobic bonding comes into play and the polymer is surrounded by a hydration sheath which, in effect, protects it from the solvent powers of the bulk water. These polymers transformed into a highly organised configuration by the electrolytes in the sea water.

Oparin[17, 18] demonstrated in his experiments that suspended proteinoid microspheres assume a coacervate structure upon repeated hydration and dehydration, this might have occurred in rocky crevices along the sea edges. It was at this level that the mechanisms of evolution and natural selection came into operation. Coacervates are more stable and more successful than their fellow molecules in the competition for nutrient material and so survived, while the weaker contemporaries perished and their substance was devoured. One important structural feature separates the coacervate droplet from the cell—the cell has a membrane of lipids which have one hydrophobic and one hydrophilic end and tend to pack in sheets; such sheets or membranes made possible the isolation of the interior of a coacervate from its surroundings.

It is claimed that there are six stages of molecular complexity in the evolution of life. The scientists represent this hierarchy of polymer complexity firstly by a simple polypeptide chain, then the α-helix, followed by folded coil, then linked folded coils as in myoglobin and, finally, the linking of different folded coils together with a ribonucleic acid chain as in the *tobacco mosaic virus*—one of the simplest living things.

The complexity of living molecules venders the probability of their formation extremely small. But there were enormous number of atoms and molecules colliding continuously among themselves for a few hundred million years. In a long span of time, even such a small probability may lead to the formation of a complex organic molecule such as nucleic acid that led to appearance of life. Somehow, at a certain time, or more likely on many occasions, in the distant past, a strange and wonderful thing happened. One of these nucleic acids, perhaps with the aid of a protein, which acted as an enzyme, was able to reproduce itself. Now there were two of these remarkable groups of chemicals. The two floated their separate ways, each gathering in new chemicals, from the soupy sea. Within a short time, each duplicated itself, and the two became four, and the four became eight; soon the seas were filled with many of these "prelife" forms. Could these be the beginning of life?

Ultraviolet and cosmic rays continued to stream down upon the earth. Often these cosmic rays struck one of the "prelife" forms. Every now and then, this caused a change in the nucleic acid—a mutation. Most of the times, the changes were not very important or perhaps they even hurt the tiny "prelife" forms and made it less able to survive in the soupy seas. Once in a while, the change was helpful and made the small "prelife" form, better able to survive. Perhaps it could reproduce more rapidly. Perhaps it was able to capture new chemicals better than its neighbours; perhaps it formed a thicker covering around itself to protect the chemical inside.

Life did not appear suddenly. Instead it came gradually, with one small step at a time. Perhaps it took millions of years, or may be tens of millions, we do not know. Some of the creatures that are known today, such as viruses, seem to be living at times and nonliving at other. Finally a creature developed, which all would recognize as life. It could live by itself, made its own food, or gobble it down from the seas, and reproduce itself. Is this the way life actually began on earth? No one can say for sure, but scientists do know from their experiments in the laboratory, that this is the way it could have happened, but there is no eye-witness account at all, of how actually life began.

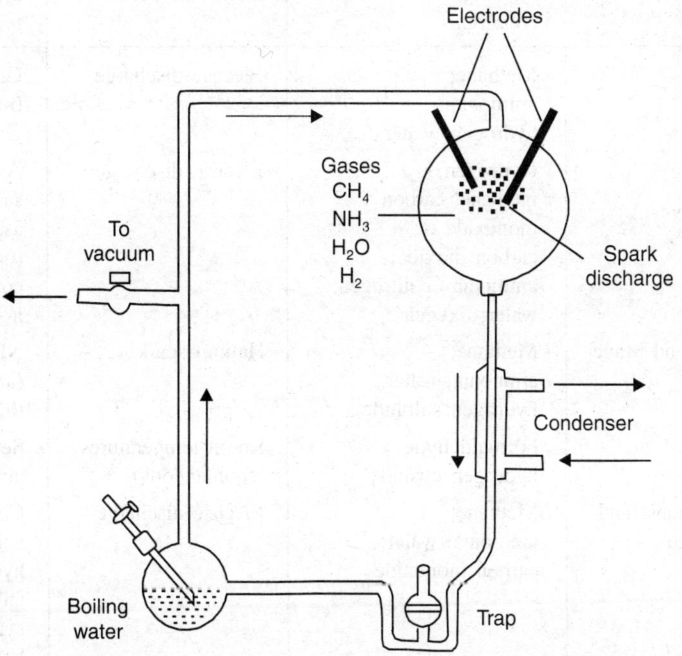

Fig. 1.1 Miller's experimental apparatus (one of several models). Most chemical reactions take place in the spark chamber, and the reaction products are condensed. The boiling water kept the system circulating in the direction shown by the arrows. The amino acids collected at trap were detected by chromatography.

In the 1950's some scientists thought of ways to test ideas on the origin of life in the laboratory. An American researcher, Stanley Miller[11, 12] recreated in their laboratory at the University of Chicago, what was thought to be the atmosphere of the ancient earth. For nearly a week they sent electric sparks through this atmosphere. At the end of this time, they were surprised to find some of the simple chemicals of life in their apparatus. Among them were amino acids, the building blocks of proteins.

A Summary of Some Experiments Similar to Primitive-Earth

Scientists throughout the world became very excited at this discovery. They not only repeated the Miller's experiment, but they made changes of their own. One scientist added a little phosphate—the kind one finds in the rocks of earth, and the amino acids that were formed, joined with each other to form larger compounds. According to all the chemical tests, they were proteins. But, were they like the proteins made by living things? The scientists fed them to bacteria, who took them as though they were real proteins. Still other researchers made chemicals like the nucleic acid residues, porphyrins, which are related to the haemoglobin of our blood, and chlorophyll of the plants. The porphyrins can absorb solar energy and can use it for later reactions.

Another complicated chemical made thus, and which could have been found on the ancient earth was **adenosine triphosphate** (ATP). This is a very important chemical of life, for it is used on a storehouse of energy by almost all living things. ATP is the basic currency for the exchange of energy inside the cell. Without ATP, life could not exist. So, the scientists showed in the laboratory that almost all the

Table 1.1 Synthesis of amino acids and related compounds

Date	*Investigators*	*Starting conditions*	*Energy sources*	*Significant products*
1953	Miller	Methane; ammonia; hydrogen; water	Electric discharge	Glycine, alanine β-alanine, glutamic acid
1955–56	Abelson	Hydrogen, methane, carbon monoxide or carbon dioxide; ammonia or nitrogen; water; oxygen	Electric discharge	Alanine, glycine, sarcosine, serine, aspartic acid (only when conditions are non-oxidizing)
1957	Walter and Mayer	Methane; ammonia; water; hydrogen sulphide	Helium spark	NH_4SCN (ammonium thiocyanate)
1959	Oró	Formaldehyde; hydrogen cyanide	Room temperatures (spontaneous)	Several common amino acids
1959	Pavlovskaya and Pasynskii	Methane; ammonia; water; carbon monoxide	Electric discharge	Common amino acids (depletion of hydrogen from system enhances amino acid synthesis); amino-butyric acid; lysine
1960	Groth and Von Weysenhoff	Methane; ethane; ammonia; water	Ultraviolet light; 407 nm; 219.6 nm	Glycine; alanine; aminobutyric acid
1959–65	Becker, et a1.; Hanafusa et al.; Pcnnaınperuma et al.; Steinman, et al.	*n*-carboxyamino acid anhydrides; aminoacetonitrile sulphate; kaolin; dicyandiamide; HCl and corresponding amino acids	Heat to 373 K	Aspartic acid, threonine, serine, glutamic acid; proline; glycine; alanine; valine; isoleucine; leucine; tyrosine; phenylalanine
1963	Oró	Ethane: methane; ammonia, water	—	Several common amino acids; proline; valine; leucine; asparagine
1965	Grossenbacher; Knight	Methane; ammonia bubbled over ammonium hydroxide	—	Several common amino acids plus threonine; serine; glutamic acid; isoleucine
1965	Oró	Methane; ammonia; water	1300 K heat	Aspartic acid; threonine; serine; alloisoleucine isoleucine; phenyl-alanine; β-alanine; glutamic acid; etc.

Adopted from Margulis, Lynn "Symbiosis in cell evolution" (1985) W.H. Freeman and Company, San Fransisco.

chemicals needed for life could be formed from the nonliving chemicals present in the strange atmosphere and the primeval-broth.

An important question still remained—no doubt these strange chemicals could have been formed, but could they have come together to produce life by themselves? Life is not merely self-reproduction of nucleic acids. Life demands controlled growth, following nutrition' in a closed membrane-bound system, the cell. This means that metabolism is an essential part of life.

The Russian researcher, Oparin[17,18] tried a new experiment. He mixed solutions of two substances—gelatin and gum arabic together, and tiny droplets formed. A mixture of these droplets with enzymes helped chemicals to react much more rapidly than they did in an open system, when the enzymes were just dissolved in water with no droplets present. The findings of this experiments resembled those in case of life.

Oparin's[17,18] droplets were the beginning of an answer. But in many ways they were not like living cells at all, for his droplets were of many different sizes and shapes, and they easily broke apart. Even more important, Oparin's droplets of gelatin and gum arabic were made from chemicals that had been produced by living things. Could nonliving chemicals come together to form such droplets?

Researchers, continued the search. They took the amino acids that were made in a simulated atmosphere, like that of the ancient earth, heated them and formed proteins. Then they added water, as could have occurred on the surface of the ancient earth when it rained or when the tides of the ocean lapped at volcanic shores, and an amazing thing happened. The proteins formed tiny balls. These balls were rather like Oparin's droplets, but there was an important difference, they were all about the same size, namely the size of bacterial cells.

The scientists looked at those tiny protein balls under the powerful microscopes. They appeared rather like bacteria. They had a thickened coat around them just like the bacteria. And this coat, like that of a bacterium, seemed able to let certain chemicals pass through it while it stopped others. More tests were run and it was found that the proteins of these balls, or microspheres as they are called, acted like enzymes. Then it was discovered that these protein microspheres could reproduce if they were kept in the solution in which they were originally formed, for a week or so, they began to form small buds.

These researchers grew excited because there are living things such as yeasts and certain bacteria that reproduce by forming buds. What would happen if the buds were separated from their parent microspheres? It turned out to be easy to separate buds, an electric shock or even a sudden bump would do it. These buds were then left quietly in the solution to grow. They kept growing until they reached the size of their parents. Then they stopped growing. The researchers left the daughter microspheres in the solution and soon these protein balls began to form buds of their own! A new generation of protein microspheres were being formed.

The little protein microspheres, formed entirely from nonliving chemicals, appeared almost like life. This is not life as we know it today, for they did not contain nucleic acids. Scientists believe that the two self-reproducing systems, the nucleic acids and the proteins, could have come together in the seas of ancient earth and could have integrated to form a whole, a primitive cell, somewhat similar to the primitive bacterial cell without a well-organized nucleus. At this stage it seems more logical to assume that life originated as many independent systems, which later got integrated together to evolve into a simple cell.

At the earliest stage of evolution of life on earth, no direct sunlight could penetrate the surface of the earth. The primitive microorganism must have been able to exist in this damp darkness, and in the absence of oxygen must have used dissolved organic substances as their food, whereas others grew accustomed to purely inorganic food. The second class of organisms obtained their energy through the oxidation of sulphur and iron compounds. However, some of the primitive organisms may not have been satisfied with purely vegetable diet. These organisms began to eat one another and the necessity of catching prey or running away from pursuit led to the development of an ability to move.

As time passed, the surface of earth grew cooler and the heavy clouds hiding the sun gradually thinned out. The sun's radiation produced atmospheric oxygen proportionately as a result of the two interrelated processes as follows:

(*i*) dissociation of atmospheric water vapour into hydrogen and oxygen by solar radiation; and
(*ii*) loss of hydrogen, from the earth's gravitation field.

These processes would tend to favour the combination of carbon with oxygen as carbon dioxide in preference to the combination of carbon with hydrogen as methane. With the presence of CO_2 photosynthetic plant life became possible. Photosynthesis, in turn, released much of additional oxygen by metabolic breakdown of water and carbon dioxide. It is this plant—produced oxygen that has slowly accumulated to provide the abundant oxygen in our planet's present atmosphere.

Under the intense radiation present in the upper atmosphere, oxygen formed a layer of ozone, O_3 which absorbs most of the sun's UV radiation and prevent them from reaching the lower atmosphere and the earth's surface. Without this ozone screen, life could have been possible underneath rocks, in deep water or in other places where direct sunlight could not penetrate. The bacteria and blue-green algae of the early oxygen-poor atmosphere may have lived in such restricted environments. Only after the formation of Ozone screen, existence of life became possible, primarily in shallow water and, ultimately, on land. In this way, life on earth began about three billion years ago. Since then a multitude of different types of organisms have emerged, prospered and died out, leaving their fossils on earth for their history to be reconstructed.

The development of man from his not-so-good looking and not-so intelligent ancestors has taken place over the last two million years. The earliest stage of that branch of organic evolution originated in Africa. In 1959, the remains of *Zinjanthropas,* who may have been the great, great, great....grand father of us all, were found in East Africa by the British archaeologist, Dr. Louis Leakey. *Zinjanthropas* the name means East African man, lived 1,750,000 years ago and is the earliest known member of stone age.

1.2 MECHANISMS OF REACTIONS

It may be recalled that modern biochemical pathways are the products of millions of years of slow evolution. It is unlikely that anything resembling a biochemical pathway today, was present or even possible among the mixture of chemicals proposed in the primeval broth, and the atmosphere. With few exceptions the simulated conditions employed by most researchers are harsh, not inimical to living systems today, and destructive to polymers. It is possible, however, that the mechanisms involved in the synthesis of organic compounds on the primitive earth were harsh, that the first prelife forms retained these mechanisms, and that evolution of milder chemistry took place primarily in these structures.

Miller[11, 12] considered two possible mechanisms: (*i*) hydrogen cyanide, aldehydes, nitriles, and amines, and part of the polymers are synthesized in the electric discharge and the amino, hydroxy–aliphatic acids are formed by hydrolysis of the respective nitriles in the solution and, (*ii*) all the products identified are synthesized in gas phase form free radicals and ions, formed in the electric discharges. His tests on both these alternatives, seem to favour the first of these.

HCN is an early product. It may have several origins. It can form directly from the interaction between methane and ammonia, both constituents of the earlier atmosphere:

$$CH_4 + NH_3 \longrightarrow HCN + 3H_2$$

Other hydrocarbons as they form can also give rise to HCN on reacting with ammonia. Carbon monoxide is still another source for HCN:

$$CO + NH_3 \longrightarrow HCN + H_2O$$

Synthesis of organic compounds: Carbon-carbon bonds formation take place on passing electric spark through methane gas yielding two-carbon compounds, e.g., C_2H_2, C_2H_4, C_2H_6. In the presence of HCN, the hydrocarbons yield nitriles :

$$C_2H_4 + HCN \longrightarrow CH_3CH_2CN \text{ (Propionitrile)}$$
$$C_2H_2 + 2HCN \longrightarrow NCCH_2CH_2CN \text{ (Succinodinitrile)}$$

The corresponding acids are obtained when the nitriles are hydrolysed in the aqueous phase:

$$CH_3CH_2CN + 2H_2O \longrightarrow NH_3 + CH_3CH_2COOH \text{ (Propionic acid)}$$
$$NCCH_2CH_2CN + 4H_2O \longrightarrow 2NH_3 + HOOCCH_2CH_2COOH \text{ (Succinic acid)}$$

With HCN, aldehydes form the corresponding hydroxynitriles, which on hydrolysis will yield the hydroxyacids.

$$\underset{\text{Acetaldehyde}}{CH_3CHO} + HCN \rightleftharpoons CH_3CHOHCN \xrightarrow[-NH_3]{2H_2O} \underset{\text{Lactic acid}}{CH_3CHOHCOOH}$$

$$\underset{\text{Propionaldehyde}}{CH_3CH_2CHO} + HCN \rightleftharpoons CH_3CH_2CHOHCN \xrightarrow[-NH_3]{2H_2O} \underset{\alpha\text{-Hydroxybutyric acid}}{CH_3CH_2CHOHCOOH}$$

With ammonia, the hydroxynitriles can form the corresponding aminonitriles from which the amino acids may be obtained by hydrolysis:

$$CH_3CHOHCN \xrightarrow{NH_3} CH_3CHNH_2CN \xrightarrow[-NH_3]{2H_2O} \underset{\text{Alanine}}{CH_3CHNH_2COOH}$$

$$CH_3CH_2CHOHCN \xrightarrow{NH_3} CH_3CH_2CHNH_2CN \xrightarrow[-NH_3]{2H_2O} \underset{\alpha\text{-Aminobutyric acid}}{CH_3CH_2CHNH_2COOH}$$

The use of heat in simulated prebiological synthesis has revealed some interesting facts. Vacillating temperatures can bring about, in sequence, the synthesis of amino acids, the polymerisation of these into protein-like polymers, and the formation from the latter of a microspheres. The ease with which amino acids will polymerise into polypeptide especially in the presence of an excess of aspartic and glutamic acids, makes it possible to visualize the natural conditions under which such events might have taken place.

Ionizing radiations, very high temperatures, and electric sparking create high energy species. Under the impact of sufficient energy, compounds dissociates and give rise to free radicals and ions. The results of the interaction of these can be very complex and unpredictable. Recombination of free radicals of carbon will form predominantly branched chain compounds. It was suggested that the formation of acetylene from carbon radicals as an intermediate in the abiological synthesis of styrene which, in turn, would give rise to the aromatic amino acids phenylalanine and tyrosine.

One of the most interesting mechanisms has to do with the reactions of activated HCN. Hydrogen cyanide appears among the first products of the gases in mixtures used in experiments of the abiological synthesis of organic compounds. Oró detected adenine among the reaction products from a mixture of

HCN, ammonia, and water. This was confirmed in subsequent work and the probable mechanism is given below:

(1) $(H^+) + (N \equiv C)^- + \underset{\displaystyle H}{\underset{|}{C}} \equiv N \longrightarrow N \equiv \underset{\displaystyle H}{\underset{|}{C}} - C \equiv N - H$

(2) $(H^+) + (N \equiv C)^- + N \equiv \underset{\displaystyle H}{\underset{|}{C}} - C \equiv N - H \longrightarrow N \equiv C - \overset{\displaystyle NH_2}{\overset{|}{\underset{\displaystyle H}{\underset{|}{C}}}} - C \equiv N$

(3) $2H - C \equiv N + 2NH_3 \longrightarrow 2NH = \underset{\displaystyle H}{\underset{|}{C}} - NH_2$

(4) 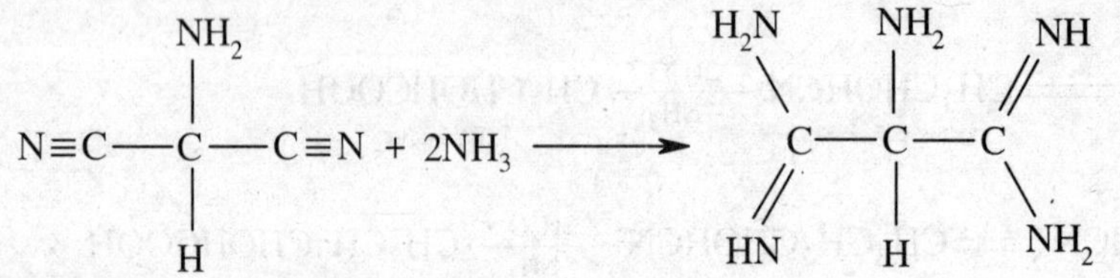

(5) 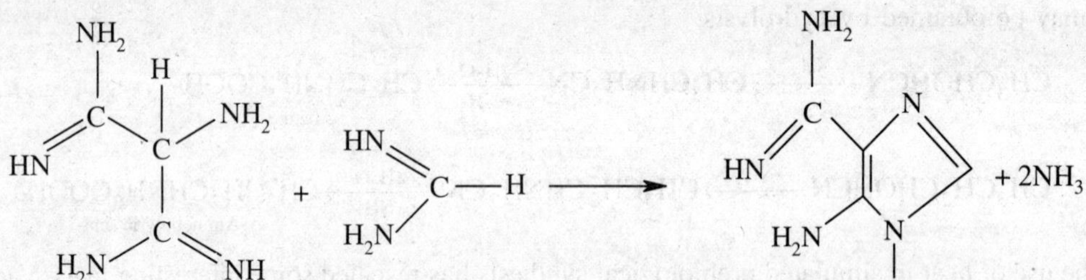

(6) 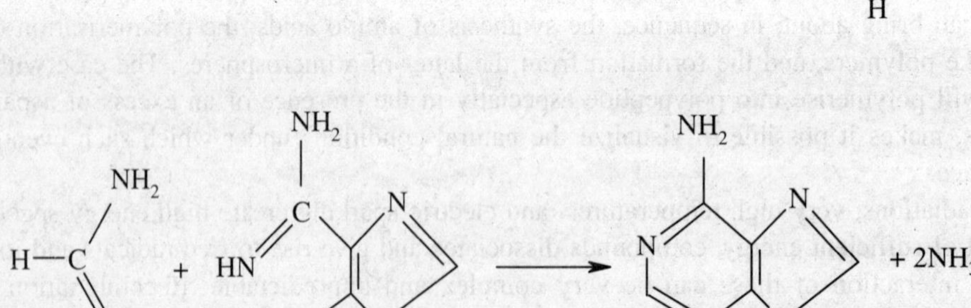

Overall reaction; 5 HCN = Adenine

Similar theories have been suggested for the synthesis of other purines such as guanine, xanthine, and found to be correct, when appropriate experimental conditions prevailed.

It has already been pointed out that available evidence favours that the prebiological atmosphere had a composition comprising mainly carbon monoxide and carbon dioxide, nitrogen and hydrogen, and the seas attained a pH of 8–9. It is often asked 'what kind of prebiological chemistry can occur at ordinary temperatures in dilute solutions at pH 8' (Calvin[5] has doubts, because dehydration reactions are not thermodynamically feasible in excess water, and the primeval broth has lots of water.) At high or low pH HCN will not polymerise, but at these pH values it readily polymerise. Hydrolysis of a tetramer of HCN yields glycine, and still others yield alanine, serine, aspartic acid, and glutamic acid. As yet, so many different views have resulted that any attempt to summarize all would easily fill a treatise, and hence only the most prevalent ones have been presented.

1.3 EXTENT OF ATMOSPHERE

A vast ocean of air surrounds the earth as a thin transparent veil. One quarter of the atmosphere is below 2 km, one half below the altitude of 5 km and three quarters of it is below 9 km. But highly rarefied fringes of atmosphere extend hundreds of kilometers above the earth's surface and it is difficult to say where our atmosphere ends. The total mass of atmosphere is 5.1×10^{18} kg which is only 0.33 per cent of the total mass of water in the oceans. The entire atmosphere is distributed over the 5.1×10^{14} m^2 of the earth's surface. This implies that atmospheric pressure at sea level is about 10^4 kg/m^2 or 10^3 g cm^{-2}. This is approximately equivalent to one atmosphere (1013 m bar). The earth's atmosphere comprises: 78% molecular nitrogen (N_2), 21% molecular oxygen (O_2), 0.93% argon (Ar), 0.03% carbon dioxide (CO_2). These gases are uniformly mixed in the troposphere and the stratosphere except at great heights.

Starting from the earth's surface, the atmosphere can be divided into several distinct regions based on the temperature-altitude profile as shown in Fig. 1.2.

The temperature initially falls up to −63°C with increasing altitude at the rate of 5.3°C per km. This region is called the troposphere, but the temperature change above 15 km is reversed. In the stratosphere, from 15 km to 50 km the temperature increases from −63°C to 15°C. From 50 to ~85 km, is the mesosphere, the temperature drops to −88°C, up to ~85 km where the thermosphere or the ionosphere begins. Here the temperature increases with altitude.

The decrease in temperature with increasing altitude in the troposphere is the result of a strong heating effect at the earth's surface from the absorption of visible and UV solar radiation. Accompanying this, there is strong vertical mixing so that particles and gaseous air pollutants can move from the earth's surface to the top of the troposphere in a few days, depending upon the meteorological conditions. Essentially all the water vapour, clouds, precipitation in the earth's atmosphere, storms, hurricanes, tornadoes and typhoons originate here; as a result, removal of pollutants by precipitators (scavenging) can be an important process.

In the stratosphere, a series of photochemical reactions involving ozone and molecular oxygen occur. Ozone strongly absorbs solar radiation in the range 210–290 nm, whereas oxygen absorbs radiation ≤ 200 nm. The absorption of light, primarily by ozone, is a major factor responsible for the increase in temperature with altitude in the stratosphere. This initiates a series of reactions in which ozone is formed and destroyed, leading to steady state concentration of ozone. This concentration in the stratosphere is essential for life, because it absorbs light wavelength ≤ 290 nm. As a result, only light of wavelength longer than 290 nm can reach the troposphere. This short wavelength cut-off sets a limit on the troposphere photochemistry. Relatively less vertical mixing occurs in the stratosphere, and no precipitation can take place in this region. As a result, massive injection of particles, such as from volcanic eruptions, often produce layers of particles in the stratosphere which persist for long periods of time.

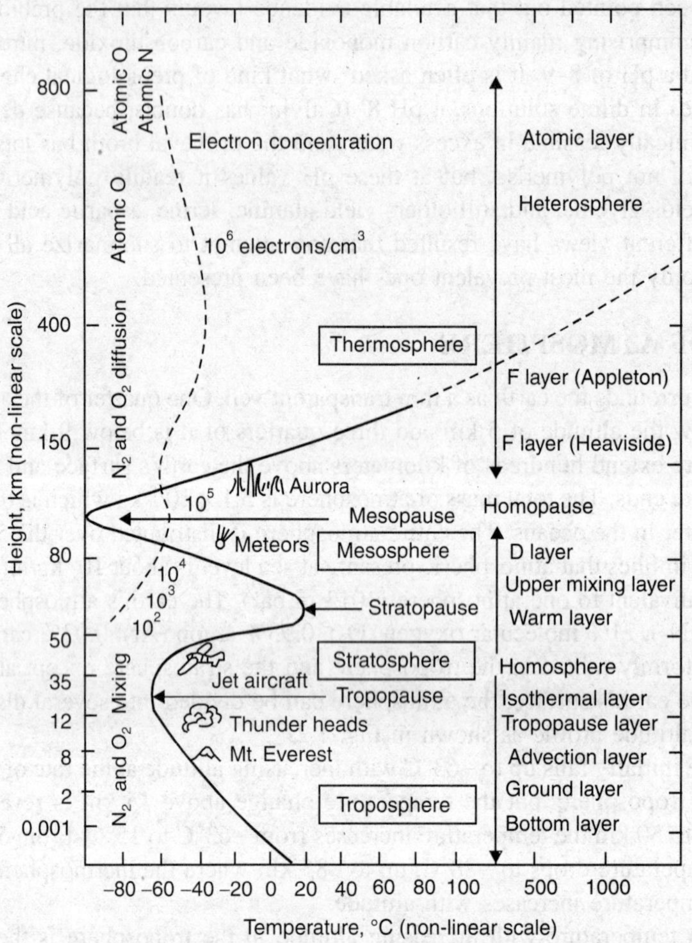

Fig. 1.2 Structure of the earth atmosphere.

In the mesosphere, from ~50 to 85 km, the temperature again falls with altitude and vertical mixing within the region occurs. This temperature trend is due to the decrease in the O_3 concentration with altitude in this region; this, in turn, reduces the heat release through absorption and decomposition of ozone.

At ~ 85 km, the temperature starts rising again because of increased absorption of solar radiation at wavelength < 200 nm by oxygen and nitrogen as well as other atomic species. This region is known as the thermosphere. The stratum of the atmosphere lying between 20 and 110 km is known as the chemosphere. The transition zone between the various shells of the atmosphere are known as tropopause, stratopause and mesopause.

1.4 WATER

Each human being requires 13 kg of air, 2 kg of water and 1.2 kg of food each day for survival. The unique properties of water enable it to play a dynamic role in shaping the landscape and creating special habitats for all organisms. Water is the only substance that exists in all three physical states on earth's surface *i.e.* as liquid, solid and gas. Some of the anomalous properties of water and their environmental and biological significance are shown in Table 1.2. A comparison between certain physical properties of water and *n*-heptane has been made in Table 1.3.

Table 1.2 Some of the anomalous properties of water and their environmental and biological significance

Properties	*Comparison with normal liquids*	*Significance*
State	Liquid rather than gas like H_2S, H_2Se, H_2Te	Provides life medium.
Heat Capacity	Very high	Moderates environmental temperatures, good for heat transportation.
Heat of Fusion	Very high	Stabilises the liquid state.
Heat of Vaporisation	Very high	Moderation effect, important in atmospheric physics, in precipitation—evaporation balance.
Density	Anomalous behaviour as maximum is at 4°C for pure water	Freezing on the surface controls temperature distribution and circulation in bodies of water.
Surface Tension	Very high	Important in surface phenomena *i.e.* droplet formation in atmosphere, transport through biomembranes.
Dielectric Constant	Very high	Thus, good solvent.
Hydration	Very extensive	Thus, good solvent and mobiliser of environmental pollutants, alters the biochemistry of solutes.
Dissociation	Very small	Provides a neutral medium, but with availability of both H^+ and OH^- controls many ionic equilibriums.
Transparency	High	Thickens biologically productive (opaque to infrared) euphatic zone.
Heat conduction	Very high	Can provide an important heat transfer mechanism in stagnant systems such as cell.

Table 1.3 Comparison of physical properties of water with *n*-heptane

	Water $(H_2O)_n$	*n-Heptane* (C_7H_{16})
Molar mass	$(18)_n$	100
Dipole moment	6.237×10^{-30} C m	> $.667 \times 10^{-30}$ C m
Dielectric constant $(\varepsilon/\varepsilon_0)$ at 25°C	78.703	1.97
Density (g/cm^3)	1.0	0.73
Boiling point (K)	373	371.4
Freezing point (K)	273	176
Heat capacity $(J\ g^{-1}K^{-1})$	4.184	2.092

(Contd.)

	Water $(H_2O)_n$	*n-Heptane* (C_7H_{16})
Heat of vaporization ($J\ g^{-1}$)	2436	318
Heat of fusion ($J\ g^{-1}$)	334	142
Surface tension (Nm^{-1}) at 293 K	72.75×10^{-3}	25×10^{-3}
Viscosity at 293K (PaS)	1.002×10^{-3}	0.5×10^{-3}

Pure water, apart from slight ionization, consists of molecules of water with covalent bonds. There is no reason to assume that molecules of pure water, except for small mutual deformation, differ from the H_2O molecules in steam. The water molecule in steam consists of a V-shaped arrangement of nuclei, the OH distances being 96 pm and HOH angle 104°28′, very nearly a tetrahedron of vertex angle 109°.

It is more difficult to picture the electronic distribution, if it were simply spherical and centred on the oxygen nuclei, the molecular dipole moment would have been 18.678×10^{-30} C m instead of 6.237×10^{-30} C m as observed. There must be a considerable screening effect by a concentration of negative charge around the protons. The wave mechanical theory of water molecules gives a qualitative information on this point. According to Mulliken[15] the 10 electron wave function of water molecule can be designated as

$$1s^2\ [2s]^2\ [2p_x]^2\ [2p_y]^2\ [2p_z]^2$$

Of these $[2p_x]^2\ [2p_y]^2$ are responsible for the binding of the protons and when combined give an electron density surrounding the proton positions, whereas $[2p_z]^2$ is not a binding wave function and gives a concentration of negative electricity in two regions at right angle to the plane of 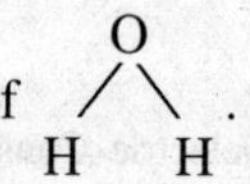. The net electronic density distribution will therefore resemble a tetrahedron with two corners of positive and two negative charges.

The radius of water molecule is more easily determined. The nearest approach of water molecules in ice is 276 pm and very similar values result from the most reliable measurements of water of crystallization. This would make the radius of the water molecule 138 pm, slightly larger than that of isoelectronic O^{2-}, 135 pm; F^-, 133 pm; OH^-, 133 pm ions.

If we attempt to account for the structure of water on the basis of such a molecule with a simple monotonic liquid structure such as mercury, we are faced at the outset by the difficulty of accounting for density. A simple close packed assembly of water molecules of radius 138 pm with face-centred close-packed structure will have a unit cell of length $= r2\sqrt{2} = 138 \times 2 \times \sqrt{2} = 390.3$ pm $= a$.

The density of water in such a close packed structure will be

$$= \frac{4M}{Na^3}$$

$$= \frac{4 \times 18}{6.06 \times 10^{23} \times (390.3 \times 10^{12})^3}$$

$$= 1.998 \text{ g/ml.}$$

Cr conversely, for a density of 1.0 the equivalent radius would be 173.8 pm. We have, therefore, the choice of assuming that either water is a monotonic close-packed liquid in which the effective radius has changed from 138 pm to 173.8 pm in liquid or that the radius is still 138 pm but with the mutual arrangement of the molecules being far from that of a simple liquid. All the well-known abnormal properties of water and extremely asymmetrical and electropolar properties of H_2O molecules would lead us to the latter alternative, but luckily there exists independent evidence from X-ray diffraction study of the internal structure of water.

1.5 THE ARRANGEMENT OF H_2O MOLECULES IN WATER

The best results on X-ray diffraction of water after corrections for absorption, Compton effect, incoherent scattering etc. are furnished by the independent researches of Meyero[15], Steward[24] and Amald[1], all of whom are in almost perfect agreement. The observed diffraction curve is shown in Fig. 1.3 experimental curve 2. This curve shows immediately by contrast with that of mercury which is similar to curve 1, the theoretical curve for close-packed structure. The association of molecules in water is different from that of a simple liquid, as instead of the sequence of maxima in the theoretical sequence at equivalent

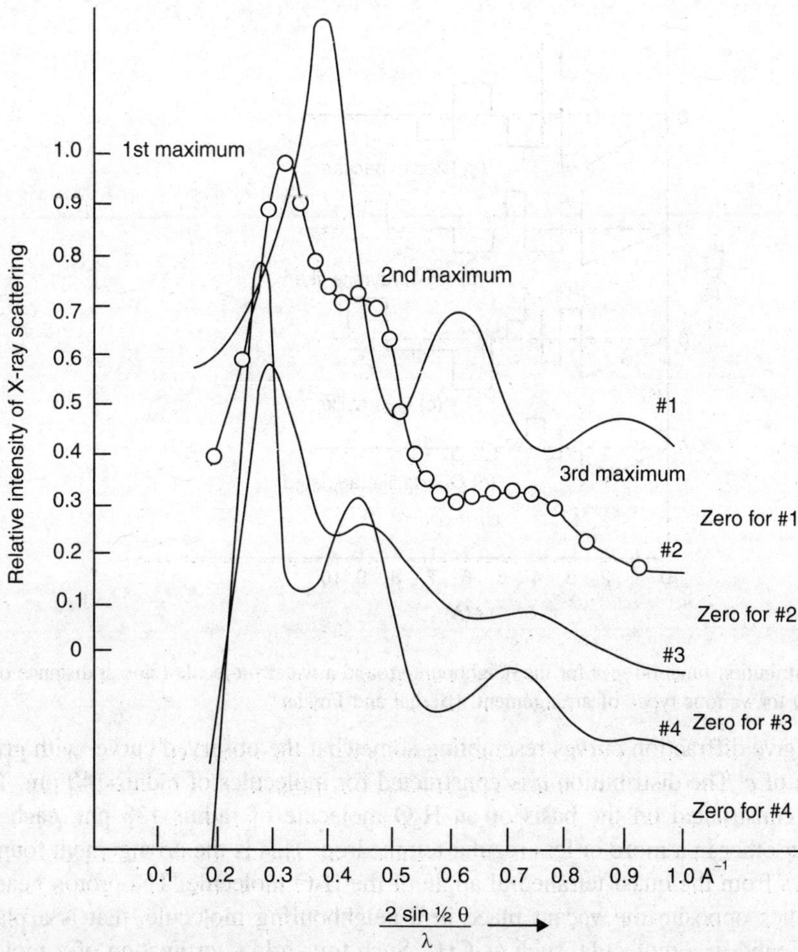

Fig. 1.3 X-ray scattering curves for water. Curve 1, theoretical curve for irregular close packing; curve 2, experimental curve; curve 3, theoretical curve for a quartz-like distribution; curve 4, theoretical curve for an ice-trydymite-like distribution. Bernal and Fowler[4] and (1 Å = 100 pm).

spacing of 260, 165, 110 pm we have a very strong maxima at 327 pm followed by very weaker ones at 230 and 140 pm. It is difficult to determine the variation from the simple liquid arrangement. The best method is to build up models of the distribution functions of the molecules, *i.e.* the function $g(r)$, giving the probability of finding the centre of any molecule at distances between r and $r + dr$ from the centre of a given molecule using different hypothesis, no contract from these, according to the formula

$$I_\theta = \int_0^\infty 4\pi r^2 \left[g(r) - P\right] \frac{\sin Sr}{Sr} dr$$

where $S = (4\pi \sin\theta/2)/\lambda$

the theoretical X-ray diffraction, and to compare it with experiment. Curves relating to various function $g(r)$ are shown in Fig. 1.4, curve (*a*) represents $g(r)$ for a simple close-packed liquid arrangement, curve (*b*) for an ice-like arrangement and curve (*c*) for one somewhat similar to quartz. It can be seen

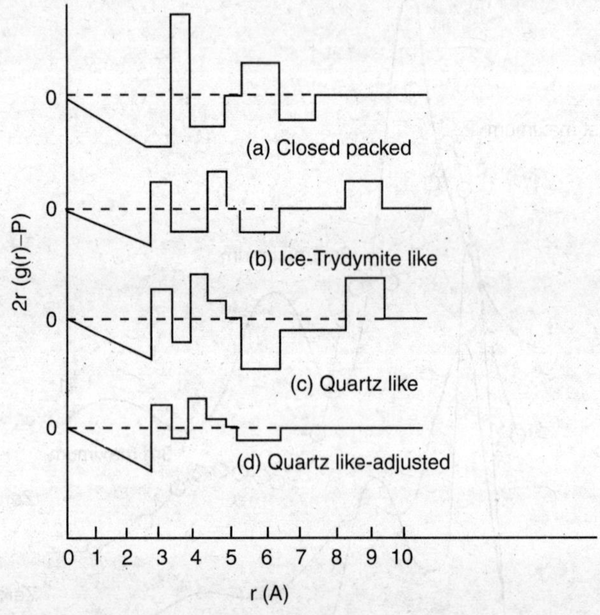

Fig. 1.4 The distribution function $g(r)$ for the neighbours around a water molecule (closest distance of approach 2.76 A) for various types of arrangement. (Bernal and Fowler[4]).

that only *b* and *c* give diffraction curves resembling somewhat the observed curve, with greater resemblance in the case of *c*. The distribution *a* is constructed for molecules of radius 140 pm. The distribution *b* and c are constructed on the basis of an H_2O molecule of radius 138 pm, each of which is surrounded by four other in a more or less regular tetrahedron. This is the arrangement found in ice and necessarily follows from the quasi-tetrahedral angle of the H_2O molecule. The proton near the surface of one molecule lies opposite the vacant place in a neighbouring molecule, that is a place where a proton would be found in a molecule, such as CH_4. Such four-fold coordination of a molecule is not, however, sufficient to determine the entire distribution. In the case of silica, SiO_2, where a similar fourfold coordination holds, there are three main crystalline forms, cristobalite, tridymite and quartz besides amorphous silica glass. Of these the tridymite structure corresponds to that of ordinary ice and it would seem natural to imagine water as simply an irregular version of this (distribution *b*). It is difficult, however in this case to account for the great reduction of volume on passing from ice to water. A regular crystalline arrangement might be expected to increase in volume on becoming irregular. This suggests the possibility that water does not generally correspond to the ice-tridymite pattern (density = 0.92) but to the quartz pattern (density 1.08). In such a pattern, while the intermolecular distance between nearest neighbours is the same at 280 pm, but that between next nearest decreases to 420 pm from the 450 pm in ice-tridymite pattern leading to a 17 per cent decrease in volume. Water might reasonably be expected to have lower density due to irregularities. A close examination of the theoretical curves enables these considerations to be pushed much further. The chief difference between curve 3 of Fig. 1.3 for the quartz like distribution and the observed curve is that the chief maxima of the former are at spacing 350 and 220 pm instead of 330 and 230 pm and intervening minimum at 250 pm relatively much deeper—42 per cent of the chief maximum instead of 72 per cent. Apparently this may be due to tendency towards a close packed water arrangement (curve 1) which has a maximum precisely in this region. It can be seen from the distribution curves for $g(r)$ that beyond r = 270 pm, the density of quartz and close packed structures are out of step for the same radii and, consequently, the effect of adding distribution *a* with *b* is to smooth out differences from that of a uniform fluid; to diminish the regularity of the packing, particularly at greater distances than 400 pm from any molecule. A scattering curve constructed from such an altered distribution function (*d* of Fig. 1.4) is shown in curve 5 of Fig. 1.5. It shows an almost perfect fit with the experimental curve in respect of the positions of the maxima and minima, but it still differs in intensity, the second maximum being undoubtedly low compared with the first. The physical basis of the difference between the ideal quartz-like structure and the actual water structure is the temperature variation of the molecules. At high temperatures, this

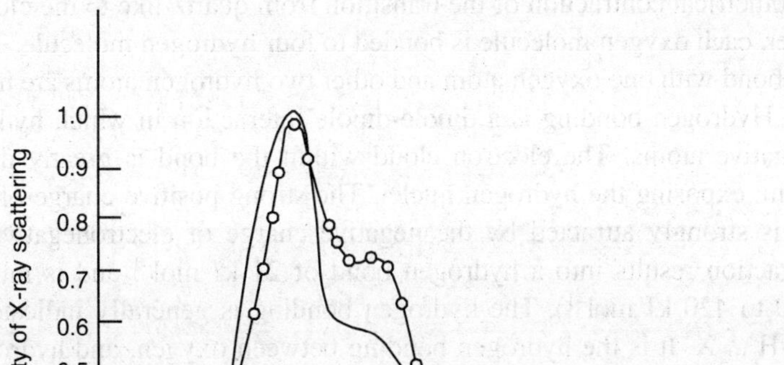

Fig. 1.5 X-ray scattering curves for water, Curve 2, experimental; Curve 5, theoretical curve for quartz-like water modified by some close packing. (Hemal and Fowler[4]).

causes a breakdown of the quartz structure, leading towards the more close-packed structure, and is revealed in the X-ray pattern as a shift of the first and second maxima to opposite directions towards the mean position of 270 pm and of the wiping out of the maximum between them. Thus, between 2°C and 98°C the main maximum moves from 327 to 310 while the second moves from 220 to 249 at 40°C and subsequently disappears.

The inward movement of the chief maximum gives rise to the appearance of a contraction on heating water, which can now be seen to be spurious. At a low temperature of 2°C the deepening of the first minimum suggests a distribution in which the tridymite-ice-like arrangement (Fig. 1.6(a)) is beginning to appear. This arrangement might be expected to be common in super cooled water and its disappearance is the reason for the anomalous contraction of water on heating above 0°C. But as these anomalous changes of volume are small compared with the ice-water change, we will just say that cold water is on the whole has quartz-like with appreciable tridymite-ice-like tendencies.

We may, therefore, consider that there are three chief forms of arrangement of the H_2O molecules in water: Water I, tridymite-ice-like, rather rare, present to a certain extent at temperatures below 4°C; As we go from 0°C to 4°C some of the Tridymite hexagons (Fig. 1.6a) break, there by releasing some free water molecules. These free water molecules enter into the hexagon rings, which makes it a more close packed structure and makes water to have maximum density at 4°C. Water II, quartz-like predominating at ordinary temperatures and Water III, close-packed, ideal liquid predominating at high temperatures for some distance below theoretical point at 374°C. These forms gets continuously converted into each other with change of temperature. It is a sequence of increasing fluidity but not of increasing volume. The immediate effect of break down of the relatively empty ice-like structure of Water I is decrease of volume in Water II, followed by an increase of a more normal kind of water III, where the increase of the mean distance of the neighbouring molecules due to thermal agitation more than compensates for the geometrical contraction of the transition from quartz-like to the close-packed structure.

As stated earlier, each oxygen molecule is bonded to four hydrogen molecule. Two hydrogen atoms form full chemical bond with one oxygen atom and other two hydrogen atoms are held together through hydrogen bonding. Hydrogen bonding is a dipole-dipole interaction in which hydrogen is attached to a highly electronegative atoms. The electron cloud within the bond is greatly distorted towards the electronegative atom, exposing the hydrogen nuclei. The strong positive charge of the poorly shielded hydrogen nucleus is strongly attracted by the negative charge of electronegative atom of a second molecule. This attraction results into a hydrogen bond of 21 kJ mol^{-1} and is much weaker than the covalent bond (330 to 420 kJ mol^{-1}). The hydrogen bonding is generally indicated in formulas by a broken line as X—H ... X. It is the hydrogen bonding between oxygen, and hydrogen that determines the structure of water.

The Frank and Wen[6] model is the most popular model which pictures liquid water as consisting of a mixture, of hydrogen bonded 'clusters' and free or monomeric water. The polymers or clusters are said to be 'flickering' that is, clusters are constantly forming and disintegrating with thermal fluctuations in micro regions of the liquid with an estimated half-life of 10^{-1} to 10^{-12} s. The clusters have been described as ice-like and thus liquid exhibits many of the structural features of a solid.

Viscosity is a good indicator of the extent of H-bonding or structure in liquid water—the more and tighter the molecules bound together, the more difficult their movement and the higher the viscosity. As temperature increases, the viscosity decreases indicating that polymeric clusters are melting and decrease in the number of H-bonds from freezing to boiling point is surprisingly slight. Nemethy and Scherage[17] estimated that there are about 65 water molecules in an average cluster at 0°C, but only 12 are present at 100°C. Since the clusters fragment with increasing temperature as well as 'melt', the

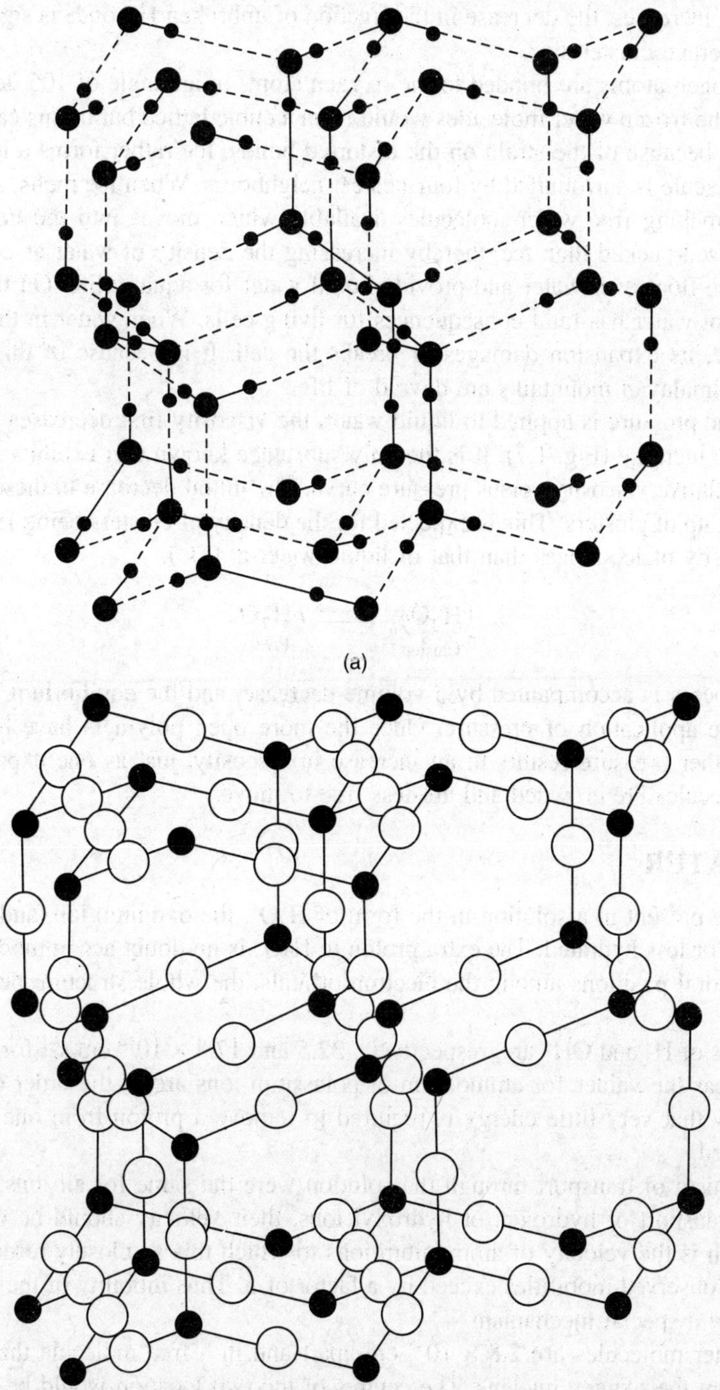

Fig. 1.6 (*a*) Water molecules arranged in Ice-Tridymite structure, ● — H atoms, ● — Oxygen atoms, (*b*) Tridymite structure of SiO_2, ● — Silicon atoms, ○ — Oxygen atoms.

number of clusters increases, the decrease in the fraction of unbroken H-bonds is significantly less than the decrease in average cluster size.

The two hydrogen atoms are bonded to the oxygen atoms at an angle of 105 degrees. If the angle were 109 degree, the frozen water molecules would form a cubic lattice but in this case such a structure would be unstable because of the strain on the distorted bonds. Ice rather forms a hexagonal structure in which each molecule is surrounded by four nearest neighbours. When ice melts, some of the hydrogen bonds break making free water molecules available which moves into the hexagon making the structure more closed-packed than ice, thereby increasing the density of water as compared to that of ice. This makes ice float over water and provide liquid water for aquatic life. On the other hand, this peculiar property of water has fatal consequences for living cells. When water in the cells freezes and become less dense, its expansion damages or breaks the cell. It is because of this reason that polar regions and the Himalayan mountains are devoid of life.

If a hydrostatic pressure is applied to liquid water, the viscosity first decreases and then, at higher pressure, begins to increase (Fig. 1.7). It is the only substance known that exhibits such an anomalous minimum in the relative viscosity versus pressure curve. The initial decrease in these curves, evidently, is due to the break up of clusters. This is expected for the density of clusters being less than that of free water (just as density of ice is less than that of liquid water at 0°C).

$$\underset{\text{Clusters}}{(H_2O)_n} \rightleftharpoons \underset{\text{Free}}{nH_2O} \qquad \text{...(1.1)}$$

Hence the process is accompanied by a volume decrease, and the equilibrium is therefore shifted to the right by the application of pressure. Once the more open polymers have been destroyed, the application of further pressure results in an increase in viscosity, just as one expects for a 'normal' liquid, as the molecules are crowded and are less free to move.

1.6 IONIC WATER

Hydrogen ions are present in a solution in the form of H_3O^+, the oxonium ion, and OH^- and all other ions will be more or less hydrated. The extra proton in H_3O^+ is no doubt accommodated at or near one of the two tetrahedral positions among the electron orbitals, the whole structure being very similar to ammonia.

The mobilities of H^+ and OH^- are, respectively, 32.5 and 17.8×10^{-4} cm s^{-1} for an applied field of 1 volt cm^{-1} whereas the values for ammonium or potassium ions are of the order of 6×10^{-4} cm s^{-1}. Calculations show that very little energy is required to remove a proton from one water molecule to which it is attached.

If the mechanism of transport through the solution were the same for all ions, which necessarily involves bodily transport of hydrogen or hydroxyl ions, their velocity should be of the order of 6×10^{-4} cm s^{-1}, which is the velocity of ammonium ions to which this so closely resembles in electronic structure. But the observed mobilities exceed by a factor of 5. Thus mobility of the hydrogen ion must be taking place by a special mechanism.

A pair of water molecules are 2.8×10^{-8} cm apart and in a free molecule the protons are about 10^{-8} cm away from the oxygen nucleus. The centres of the two location would be about 8×10^{-9} cm apart, which is quite close. In such a situation, the H_3O^+ ion is in sufficiently close contact with water molecules, and if H_3O^+ and H_2O have similar energies, it has been shown that H^+ ion acts as a bond between the two water molecules. The proton jumps to and fro from one molecule to the other when the configuration of two molecule is favourable and this leads to an extra drift down the field.

In the calculations carried out by Bernal and Fowler[4], it was found that if N, the number of favourable configurations which the oxonium ion may be expected to enter into per second, each one

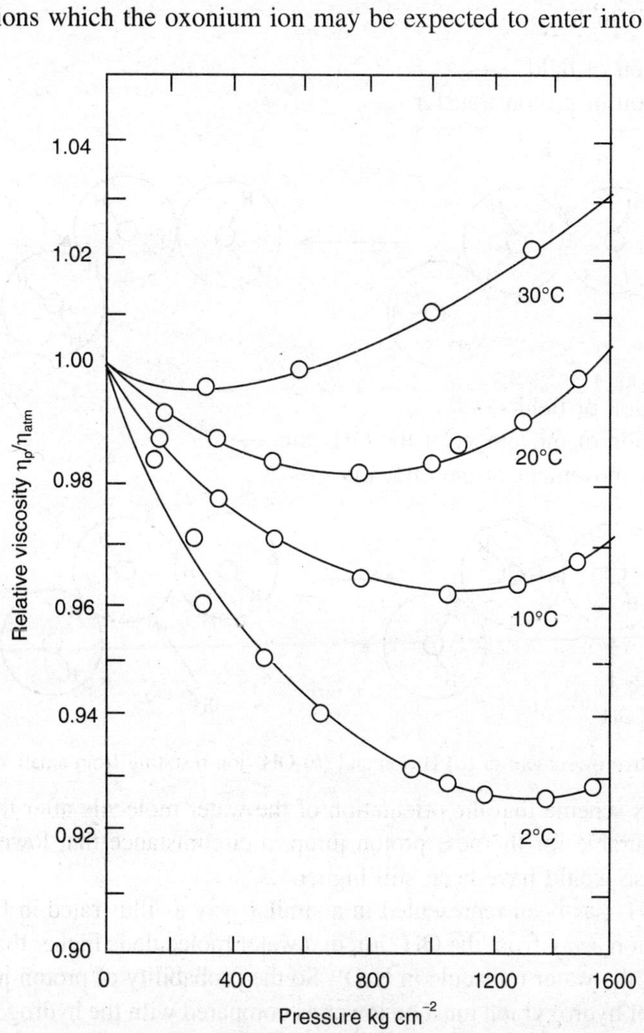

Fig. 1.7 Pressure dependence of the relative viscosity of water (after Stanley and Batten[22])

offering to a proton a chance of jumping in one direction to a neighbouring molecule, is about 2.5×10^{12} and the transition time τ for jumping is given by

$$\tau = 6.5 \times 10^{-13} \text{ s.}$$

This value is practically equal to $1/N$. This speed of migration is so high that it is controlled by τ and not N. Thus, if the probability for jumping from one molecule to the next is infinitely frequent, or, more exactly, the opportunity is presented to the $(OH_3)^+$ ion at least as frequently as it can make use of it, the time required for any single jump will be of the order of τ. In the presence of an external field, the probability of proton migration in the direction of field increases, and the process of proton movement will occur by a chain mechanism, *i.e.* the proton will be passed on from one water molecule to the next and so on—a passing the proton game. Diagrammatically it is shown in Fig. 1.8a resulting in movement of H^+ from *A* to *B*,

(*a*) (*i*) the direction of field $\longrightarrow$

(*ii*) the direction of proton transfer $\longrightarrow$

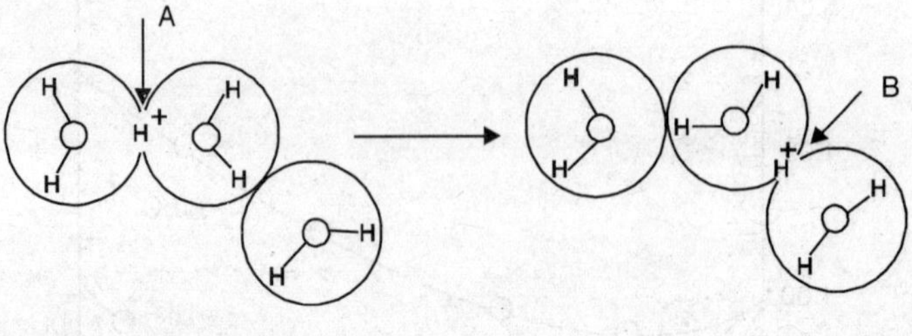

(*b*) (*i*) The direction of field $\longleftarrow$

(*ii*) The direction of movement of the OH^- ion $\longleftarrow$

The direction of movement of the OH^- ion $\longleftarrow$

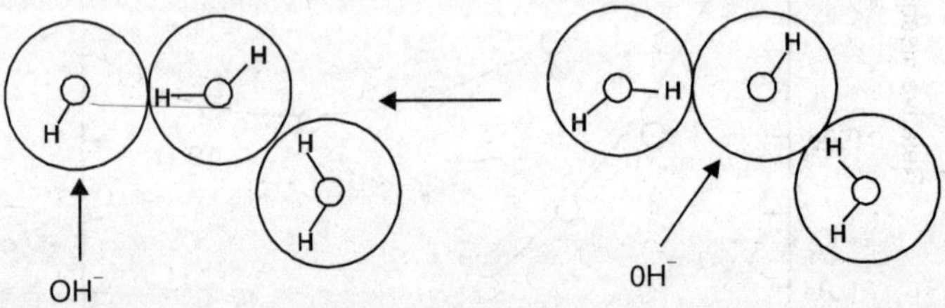

Fig. 1.8 The effective movement of (*a*) H^+ ion and (*b*) OH^- ion resulting from small shifts of protons.

It follows from this scheme that the orientation of the water molecule after the proton has left the oxonium ion is unfavourable for the next proton jump, a circumstance that lowers the rate of proton transfer which otherwise would have been still higher.

The mobility of OH^- has been represented in a similar way as illustrated in Fig. 1.8b. The energy required to pull the proton away from the OH^- ion in a water molecule is higher than that needed to tear the hydrogen ion H^+ of the water molecule in H_3O^+. So the probability of proton jumps and, hence, the velocity of movement of hydroxyl ion must be lower as compared with the hydrogen ion. The unfavourable orientation of the water molecule formed also adds to the spatial difficulties for the next proton jump, as a result of which the rate of transfer of hydroxyl ion is lowered.

Whereas other ions have to move bodily through the water, the H^+ and OH^- ions move by what Bernal and Fowler[4] termed a kind of 'relay ice', small shifts of protons only being necessary.

Water molecules are strongly interacting through the hydrogen bonding and partial positive and negative charges. Such an interaction is a natural preliminary to the outright transfer of proton from the first molecule to the second molecule where it forms a covalent bond

$$H_2O + H_2O \rightleftharpoons H_3O^+ + OH^- \qquad K_0 = a_{H^+}\, a_{OH^-} / a_{H_2O} \qquad ...(1.2)$$

The oxonium species H_3O^+, which is conjugate acid of a base is also called hydronium ion and has a triangular planar structure. Even in pure water, there is a autoprotolytic equilibrium. It is obviously absurd to absorb water activity into K_0 so it is normal to express this equilibrium in terms of constant K_W

$$K_W = a_{H^+} \cdot a_{OH^-} \qquad ...(1.3)$$

At 25°C, $K_W \approx 1 \times 10^{-14}$. This indicates that only a very small portion of water molecules are dissociated. Since the concentration is so low, the activities may be replaced by molality

$$K_W = m_{H^+} \cdot m_{OH^-} \quad ...(1.4)$$

Since water is electrically neutral, $m_{H^+} = m_{OH^-}$

or $$m_{H^+} = K_W^{1/2} = 1 \times 10^{-7}$$

The molality of protons plays a central role in many chemical processes. Its magnitude can vary over a wide range. On the pH scale, this wide span reduces to the range of 0 to 14, where pH is defined as follows.

$$\text{pH} = -\log m_{H^+} \quad ...(1.5)$$

and $$pK_W = -\log K_W \quad ...(1.6)$$

In water, at 25°C, $m_{H^+} = 1 \times 10^{-7}$, the pH is 7.0 so that this value corresponds to neutrality at 25°C.

It may be pointed out that presence of ionic solutes or organic matter changes the ionization of water as well as its activity coefficient tacitly assumed value of 1.0. Yet incorporation of $pK_w = 14$ into calculations on buffer solutions, industrial waste, sea water and other solutions to which it does not apply, leads to answers that may be approximate and should be rounded to first decimal of pH only.

The variations of pK_w for pure water as a function of temperature and pressure are given in Table 1.4 and 1.5 respectively.

Table 1.4 pK_w for pure liquid water at different temperatures

T(°C)	0	5	10	15	20	25
pK_W	14.983	14.727	14.528	14.341	14.165	13.998
T(°C)	30	35	40	45	50	55
pK_W	13.84	13.69	13.549	13.414	13.285	13.165
T(°C)	60	70	80	90	100	125
pK_W	13.049	12.834	12.638	12.462	12.301	11.966

Table 1.5 pK_w for pure water at various pressures

bar	*0°C*	*25°C*	*50°C*	*100°C*	*150°C*	*200°C*
300	14.79	13.68	13.17	12.19	11.58	11.24
1000	14.50	13.66	12.96	11.96	11.32	10.92
3000	13.91	13.18	12.53	11.46	10.19	10.33
10000	-	-	-	10.49	9.78	9.25

The effect of pressure is too small to be important to most workers. A recent article[11] in National Geographic describes water rising from a volcanic vent at the ocean floor at about 250 bar and 400°C has salt content one twelfth of that of regular sea water, suggesting ionic strength substantially less than 0.1. It was highly acidic as a result of dissociation of water. However, at even more extreme conditions, water can become a strong acid. The IUPAC collection[18,19] shows that pK_w is

1.05 at 133 kbar and 8.4°C
3.05 at 93 kbar and 727°C
5.70 at 71 kbar and 373°C

giving pH = 0.5, 1.5 and 2.8 for pure water. Such extremes may perhaps occur in some geologic situations, so that acidity of the water would be a factor in the solution and recrystallization of rocks.

In an acid solution m_{H^+} is greater than pure water and pH is less than 7 and in base solution the pH is greater than 7.

The autoprotolysis equilibrium is maintained in the presence of acids and bases and so if an acid is present, so that m_{H^+} is enhanced, m_{OH^-} concentration must be such, as to maintain $m_{H^+} \times m_{OH^-} = K_W$, hence in an acidic solution the pH can be obtained from $m_{H^+} \equiv K_W / m_{OH^-}$. The measurement of pH is made with an instrument called pH-meter and is of fundamental importance in chemistry. It is the basis of monitoring a wide range of biological and industrial processes, and is of crucial assistance in monitoring acid-base reactions, rates of reactions and strength of various solutions and so on.

1.7 AQUEOUS ELECTROLYTES

In order to account for the high dielectric constant of water, it is necessary to suppose that there exist groups of molecules with a pseudocrystalline structure that is with sufficient orientation of the O—H—O bonds to give an appreciable electric moment. The upper limit of size of these clusters has been estimated from the studies of the infrared absorption spectra is in the 1.1 to 1.3 μ region, as was earlier found by Frank and Wen[6] and these clusters continuously make and break, resulting in high value of the dielectric constant.

The dielectric constant of a liquid is a measure of its capacity to neutralise the attraction between the electrical charges. As for example, in solid NaCl the force of attraction between Na^+ and Cl^- is given by Coulomb's Law

$$F_1 = \frac{q_1 q_2}{4\pi\varepsilon_0 r^2} \qquad ...(1.7)$$

where q_1 and q_2 are charges on sodium and chloride ion separated by some distance r. When it is dissolved, water molecules come between Na^+ and Cl^- ions so the cations and anions are attracted by a force F_2 *i.e.*,

$$F_2 = \frac{q_1 q_2}{4\pi\varepsilon_1 r^2} = \frac{F_1}{\varepsilon/\varepsilon_0} \qquad ...(1.8)$$

where $\varepsilon/\varepsilon_o$ = dielectric constant of water which is 78.7 so that

$$F_2 = \frac{F_1}{78.7} \qquad ...(1.9)$$

It reduces the force of attraction between oppositely charged ions in solution to nearly 1 per cent of original force. This neutralising action is carried out as follows. In an aggregation of water molecules, due to a difference in electronegativity hydrogen and oxygen do not share their bonding electrons equally. This results in a partial negative charge on oxygen atoms and partial positive charge on hydrogen atoms. Liquid water has an equilibrium between cluster and free water. The free water molecule comes in contact with positive ion or negative ion which attracts the oxygen atom or hydrogen atoms of water. This shifts the equilibrium to right side producing more free water. These free water molecules surrounded the cations, oriented with their oxygen next to the ion, whereas molecules around a negative ion turn their hydrogen toward the anion as shown in Fig. 1.9.

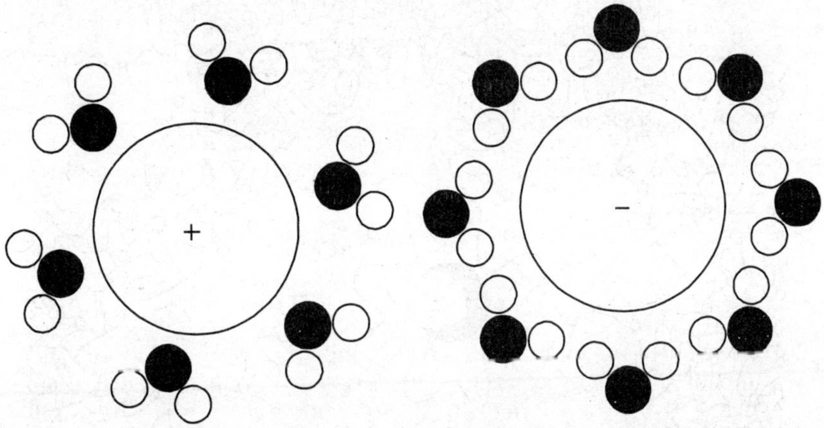

Fig. 1.9 Circles (labelled with plus and minus signs) in water are apart because they polarise the water molecules around them. Since the oxygen atom in the water molecule has more negative charge than the hydrogen atoms, it is attracted to a positive ion (*left*). Because the hydrogen atoms have more positive charge than the oxygen atom, they are attracted to a negative ion (*right*).

In this way water molecules act as cages which separate and neutralise the charges on ions which go into solution. This explains how water acts so effectively as a solvent for electrolytes. This also shows that aqueous electrolytic solution changes its internal structure.

The effect of electrolyte addition on viscosity is complex because certain electrolytes, such as NaCl and $MgSO_4$, increase the viscosity of water and thus appear to be increasing or enforcing the water structure, whereas other electrolytes such as KCl appear to break the water structure since their addition reduces the viscosity.

The thermodynamics of solutions also can provide useful clues to the structural changes. The thermodynamics of electrolytes is dominated by the energetics of transferring a charge from a low (air) to a high (water) dielectric constant medium and any structural effects get swamped out. However changes in entropy on the addition of non-polar solutes to water, such as H_2, O_2, CO, N_2 or the noble gases, indicate that order or structure is being induced in the system, presumably by the formation of a structured iceberg surrounding the solute molecules. This will be discussed in Sec. 1.8 of this chapter.

Figure 1.10 shows a two-zone model for the structure of the hydration atmospheres of ions in solution consisting of an inner zone (A) of strongly bound, electrostricted water molecules surrounded by a region (B) in which the coulombic field of ion is sufficiently strong to disrupt the structure of the bulk water (C) but is too weak to reorient the water molecules into a new pattern. In the case of structure makers, zone A is more important, for structure breakers zone B. Analysis show that zone A can be further divided into subregions, an innermost shell of so called "primary hydrogen", which may be inseparable from the ion even upon evaporation of the solution (water of hydrogen or crystallization), surrounded by what appears to be a Frank-Wen[6] cluster (Fig. 1.10).

With the exception of proton, hydroxyl ions and highly concentrated solutes, all solutes have similar activation energy of viscous flow, diffusion coefficients, equivalent electrical conductivities, dielectric relaxation, strongly suggesting that the same fundamental molecular mechanism form the basis of all these solutes and solvent movement processes.

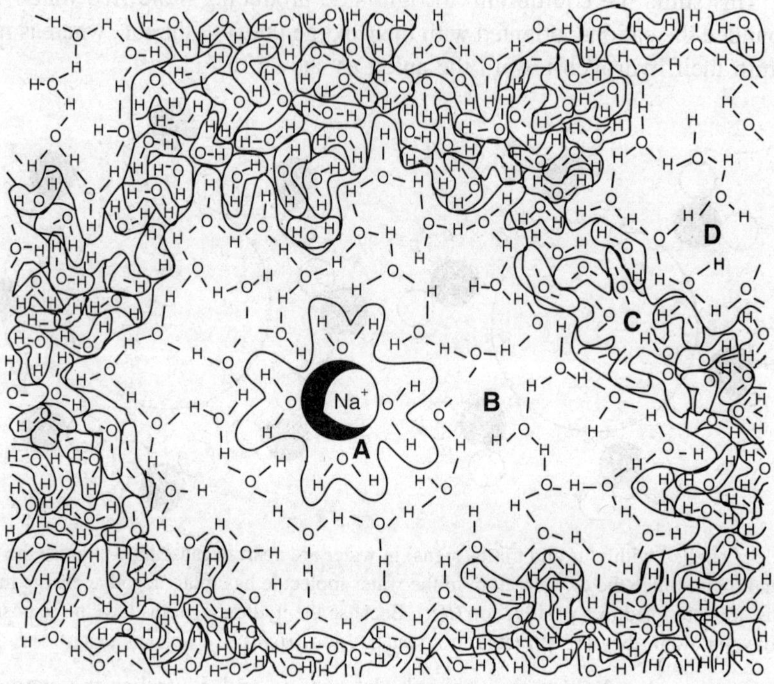

Fig. 1.10 The local water structure near a cation in aqueous solution: A, innermost, dense, tightly bound, electrostricted region (primary hydration); B, rarified, bound, Frank-Wen clusterlike region; C, broken water structure region; D, bulk water—Frank-Wen clusters and "monomeric" water, $D_A > D_D > D_B$. (after Home[10])

In as much as both NaCl and $MgSO_4$ tend to increase the structure of liquid water as diffusion coefficient, equivalent conductivities, fluidity (1/η) of water tends to decrease with increasing salt content. The electrical conductivity is one of the most important properties for the ease with which this property can be measured with relatively great accuracy, even in adverse conditions of study.

Increase of the temperature facilitates transport processes without increase in the conductivity and fluidity. The effect of hydrostatic pressure is not so straightforward; as the pressure increases the relative conductance of an aqueous solution of strong one-to-one electrolytes, first increases, goes through a maxima and then decreases. Analysis of the initial increase reveals unexpected increase in conductance, which is greater than one, except from the pressure induced viscosity and volume changes. Two explanations for this anomalous conductance increment have been offered: (*a*) further dissociation of strong electrolytes under pressure, and (*b*) pressure-induced dehydration of the ions results in a decrease in their effective hydrodynamic radii, which will release more free molecules, thereby shifting the Frank-Wen (*a*) cluster mechanism (eq. 1.10) to the left-hand side. As these clusters are less dense than the free water, the application of pressure tends to favour the more dense free water and destroy the large volume from

$$\underset{\text{Clusters}}{(H_2O)_n} \rightleftharpoons \underset{\text{Free}}{n(H_2O)} \qquad ...(1.10)$$

and this results in a decrease in relative viscosity until all clusters are destroyed. Then relative viscosity increases with increasing pressure, just as expected of a normal liquid. The maximum in conductivity versus pressure curve does not coincide with maximum in the viscosity versus pressure curve, indicating that the situation is not as simple as one might hope.

1.8 CLATHRATE TYPE OF HYDRATES

Two facts are noteworthy: *(i)* Corn sometimes showed frost effects when the temperature was 7–8°C, well above freezing points; *(ii)* The discovery that pipelines carrying natural gases often become clogged with a slushy snow containing water at temperature of 20°C. The plain indication was that these freeze-ups were due to the water.

Methane molecule does not form ions in water, nor does it form hydrogen bonding with oxygen. It is however slightly soluble in water, the dissolving methane molecules form compounds with water known as "hydrates" (Fig. 1.11), in which several water molecules may be joined to one molecule of methane. The dissolution reaction liberates ten times as much heat as when methane is dissolved in hexane in which it is more soluble. Methane molecule occupies more than twice the volume of a water molecule. To form a relatively large cavity for itself on dissolving, a great deal of energy would be required. The force of attraction between methane and water is quite weak and insufficient for this purpose.

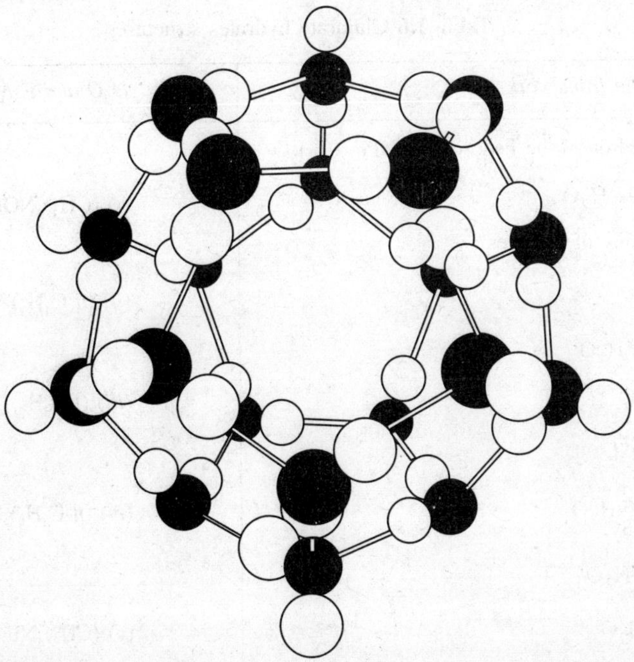

Fig. 1.11 Hydrates is formed when a foreign molecule like methane in water is electrically neutral and just has the right size for the water molecules to collect it around in crystalline cage. It can be a part of a unit of as large as 136 water molecules.

The alternative is that methane may drastically change the water structure itself. Let us assume that dissolved methane is surrounded by a cage of 10 to 20 water molecules (Fig. 1.11). In the space occupied by a methane molecule the attractive force on the water molecules and hence the inward pressure would disappear. Under these conditions, water will freeze at a higher temperature, and the interface between the methane and water may crystallize into ice. This is further supported by the fact that all the non-electrolytes have been found to form crystalline hydrates. Any such hydrate has lower density than ice, consequently its formation can cause a large destructive expansion as in the case of freezing corn at 7–8°C.

Hydrocarbons and other non-polar organic solutes also forms such structure with the surrounding water, and in many instances, it has been possible to isolate a clathrate type of hydrate or cage of water molecules enclosing the quest solute molecule. This water structure possesses clathrate nature, surrounding non-polar solutes as hydrophobic hydration to distinguish it from the coulombic hydration surrounding charged species, such as ions, in solution.

In regard to the latter, the ordering of water in the cerebrospinal fluid into clathrate like structure could form the basis of the anaesthetic properties of certain non-planar solutes by impeding the transmission of neural signals. More recently, Miller (1969) has attributed the disappearance of air bubbles in Antarctic ice at depths greater than 800 to 1200 m to the crystallization of the constituents of air as clathrate hydrates such as $(N_2, O_2)\ 6H_2O$. Miller has also discussed the occurrence of hydrates on other planets, and he has proposed that the Martian polar cap may contain the clathrate hydrate of carbon dioxide.

A list of well-established clathrate hydrate structures are given in Table 1.6. The detailed structures of these compounds are given in A.F. Wells "Structural Inorganic Chemistry"[23] Page 543.

Table 1.6 Clathrate hydrates structures

H_2O molecules only in framework	*H_2O and F, N, S in framework*
(*a*) Truncated Octahedron in the Fedorov net $HPF_6 \cdot 6H_2O$	
(*b*) $(CH_3)_2CNH_2 \cdot 9\frac{3}{4}H_2O$	(*i*) $(CH_3)_4NOH \cdot 5H_2O$
(*c*) Dodecahedral family of hydrates	
(*i*) $Cl_2 \cdot 7\frac{1}{4}H_2O$	(*ii*) $(n\ C_4H_9)\ SF \cdot 20H_2O$
(*ii*) $6.4C_2H_4O \cdot 46H_2O$	
(*iii*) $CHCl_3 \cdot 17H_2O$	(*iii*) $(i\text{-}C_5H_{11})_4\ NF \cdot 38H_2O$
(*iv*) $Br_2 \cdot (8-6)H_2O$	
(*v*) $(C_2H_5)_2NH \cdot 8\frac{2}{3}H_2O$	(*iv*) $(n\text{-}C_4H_9)_3SF \cdot 23H_2O$
(*vi*) $4(CH_3)_3N \cdot 41H_2O$	
(*vii*) $N_4(CH_2)_6 \cdot 6H_2O$	(*v*) $(CH_3)_4NF \cdot 4H_2O$

In this respect, although unimportant in natural water, the tetralkyl ammonium cations form an interesting family, for as n in $(C_nH_{2n+1})_4N^+$ increases, the ion-charge density decreases, and there is a gradual progression from coulombic to hydrophobic character. A complex molecule such as a biomacromolecule, will have a hybrid hydration sheath with coulombic hydration surrounding polar sites and hydrophobic hydration encasing the hydrophobic backbone of the molecule.

Thus, water is unique and a complicated material with distinct and varied chemical properties. It has a definite, though changing, physical structure which depends on the orientation of its molecules with respect to solute molecules and much of the inanimated world.

The oceans cover 70 per cent of the earth's surface. The movement of the water currents follow the global pattern of the winds blowing over the ocean surface. While the general circulation of the atmosphere is largely unrestricted by the surface features of the earth, the circulation of ocean water is hindered by the continents separating various large water basins. Even then the actual ocean current pattern is approximately similar to the wind currents.

1.9 HYDROLOGIC CYCLE

There is no sequence in physical chemistry of our environment more obvious than the hydrologic cycle. It is responsible for our weather, and it makes our rivers run. There are not many processes in our world of comparative importance. The hydrologic cycle as a whole is not essential to life on this planet but

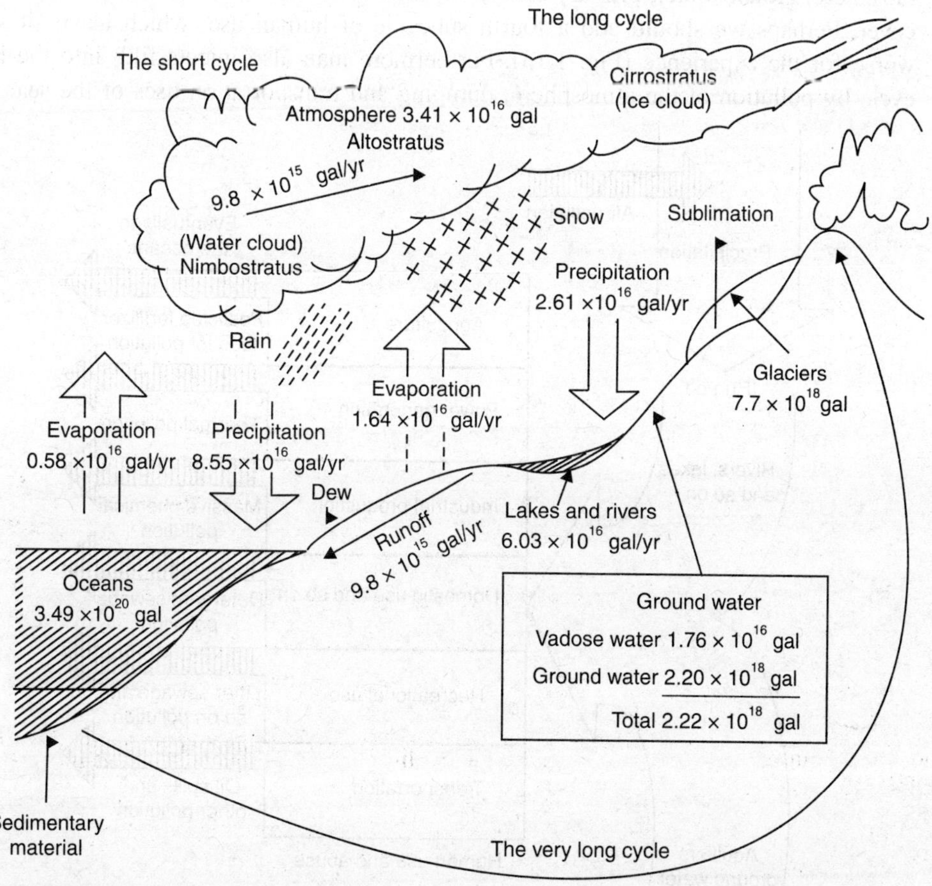

Fig. 1.12 Details of the hydrologic cycle.

the cycle certainly is essential to life forms that have emerged and it is very difficult to conceive what the earth would be like, even the physical appearance of the landscape without the hydrologic cycle.

Of the total water available on earth, 97.2 per cent resides in oceans and 2.5 per cent of the remainder is trapped in polar ice and glaciers—reservoir of little use to man. Man relies on only 0.65 per cent of the hydrosphere, and it is this small yet vital fraction that is renewed through the hydrologic cycle. The hydrologic cycle makes water available to us and replenishes our supplies. Were it not for this solar-powered renewal cycle man would quickly exhaust all his useful supplies of water. In the final analysis, the distribution of the human species on this planet and the success of its enclaves is determined, not by historical events, but by the availability of water.

The hydrologic cycle has three sub cycles (Fig. 1.12):

(*a*) the short cycle of evaporation and precipitation over ocean area;

(*b*) the long cycle of evaporation from ocean areas and evaporation and transpiration from land areas, precipitation overland and eventual partial return to the oceans as river run-off; and

(*c*) the very long cycle, consisting of the degasing of the lithosphere and the recently discovered return of water to the magma by slipping under marine sedimentary deposits at the continental shelves. These cycles exchange water among seven major reservoirs; the atmosphere, the biosphere, ground water, primary water, the oceans, surface water on land masses, and the ice cover. Perhaps we should add a fourth subcycle of human use, which leaves it very much worse for the experience (Fig. 1.13). Furthermore man also injects filth into the hydrologic cycle by pollution of the atmosphere, dumping and transportation uses of the sea.

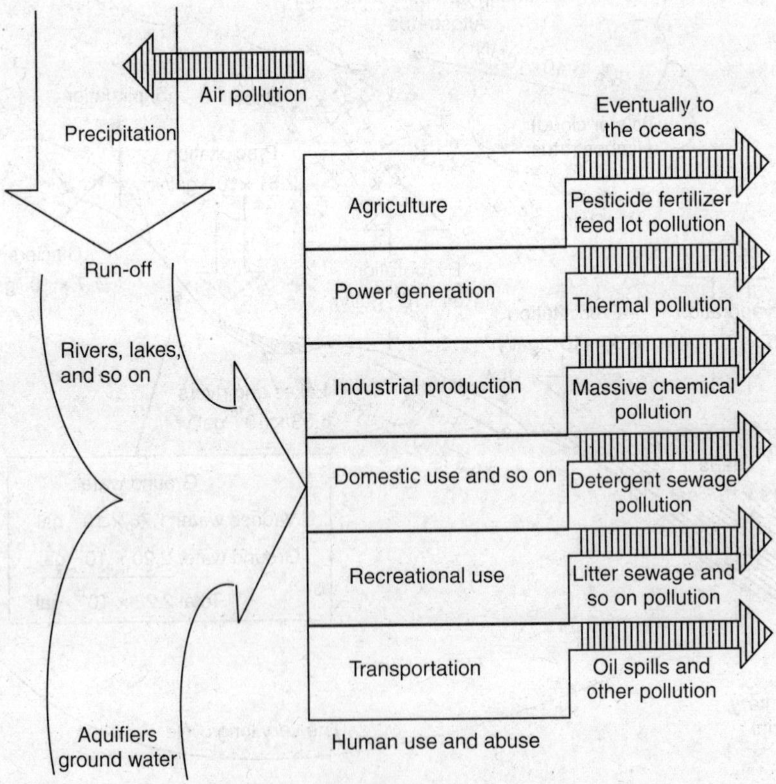

Fig. 1.13 Injection of production into the long cycle.

(a) Short Cycle: Solar energy, some 1.368 kW/m^2, evaporates water from the surface of the oceans and after a period of residence in the atmosphere about 90 per cent of this moisture falls back into seas as precipitation. The oceanic evaporation-precipitation balance is a major factor in determining the salinity of surface waters. Excess evaporation will result in highly saline waters whereas an excess of precipitation leads to low salinity. Because of this dependence on evaporation-precipitation balance, the salinity of surface water exhibits a diurnal cycle, the salinity being highest in the afternoon and also it is affected by seasonal variation.

About 10 per cent of the moisture picked up from the oceans, 19.8×10^{15} gal/year is carried by winds overland. Here, it joins its long-cycle travel by some 1.6×10^{16} gal/year of moisture from land-areas (Fig. 1.12), some evaporation takes place from rivers, lakes, and human impoundments and some from transpiration of leaves of plants. The contribution of transpiration from vegetation, however, is believed to be very larger. Much more water is removed by transpiration than from photosynthesis, in the appropriate ratio of 2000:3. Atmospheric moisture is precipitated as rain, snow, hail and dew. The excess water returns to the seas as run off via two routes—fast moving rivers and streams, and very slow migration of subterranean ground water—to complete the long cycle.

In emphasizing the importance of the hydrologic cycle in the chemistry of our environment, we must not forget that it also largely determines the energy balance of the earth. The high heat capacity of liquid water and high heat of all its phase transitions make it an excellent heat-transfer fluid, ideally suited for the purpose. Coastal inhabitants are very much aware of the moderating effect of the ocean on their climate. The thermal energy transported by the ocean current every hour is equivalent to the burning of 1.75×10^8 tons of coal. All the coal mined in the world in 1 year, would supply warmth at the rate of only one and a half hour.

1.10 THE GEOLOGIC CYCLE

Except during the early history, the material such as minerals, rocks, solid and water on or near the surface have been created, maintained and destroyed by numerous internal and external processes. Collectively, the process is referred to as the Geologic Cycles which is a group of two important cycles:

(i) hydrologic cycle, and *(ii)* tectonic cycle.

The hydrologic cycle as discussed in Sec. 1.9, is the circulation of water from oceans to atmosphere and back to oceans by way of evaporation run off from streams and rivers and ground water flow. Though only a small amount of water is active in the hydrologic cycle at any one time, yet it is tremendously important in facilitating the movement and sorting of chemical elements in solution, weathering of rocks, transporting and depositing sediments and providing water resources.

The tectonic processes are driven by forces deep within the earth. The outermost shell called the lithosphere, is roughly 100 km thick and comprises of the continents and the oceans. It is divided into a series of irregularly shaped plates that move relative to one another. These rigid plates float on another layer of earth called the asthenosphere, on the upper mantle about 200 km below the lithosphere, as shown in Fig. 1.13. The asthenosphere is a continuous layer of little strength. It is solid but not sufficiently strong to creep under stress. This property makes it act as a lubricant, which enables the plates above it to move. Plates move a few centimeters in a year—it would take a plate 50000 years to drift one km—but they do move.

The boundaries between plates are areas where most earthquakes and volcanic activities occur. Three types of boundaries exist, divergent, convergent and transform fault. Divergent boundaries occur

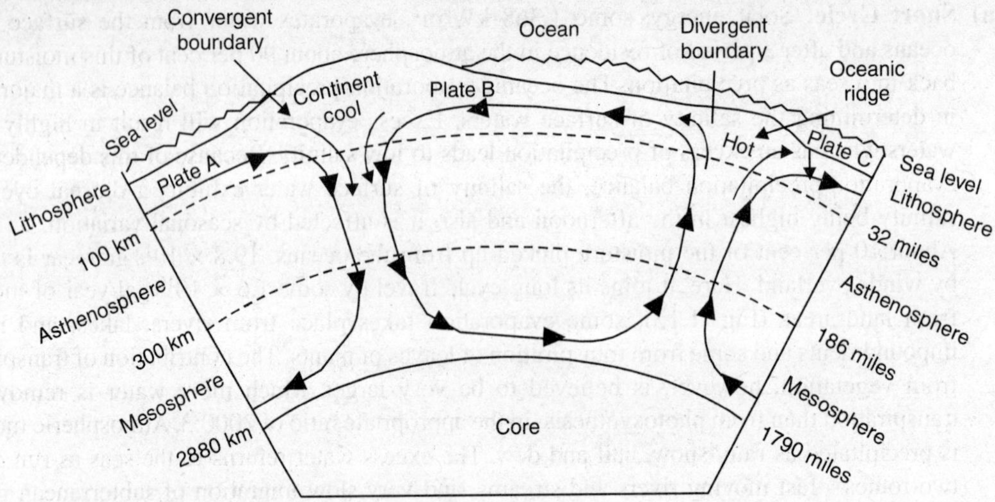

Fig. 1.14 Idealized diagram showing the model of sea-floor spreading which is thought to drive the movement of the lithospheric plates. New lithosphere is being produced at the oceanic ridge (divergent plate boundary). The lithosphere then moves laterally and eventually returns down to the interior of the earth at a convergent plate boundary (subduction zone). This process produces ocean basins and provides a mechanism that moves continents.

at oceanic ridges where plates are moving away from each other, and new lithosphere is produced. Convergent boundaries occur when one plate dives beneath the leading edge of another plate. However, if both leading edges are composed of relatively light continental material it is more difficult for subduction to start, and a special type of convergent plate boundary called a collision boundary, may develop. These produced linear mountain systems such as the Alps and the Himalayas. Transform fault is usually associated with oceanic ridges and boundaries, occur where one plate slides past another, as for example, the San Andreas fault in California.

Beneath the lubricating asthenosphere, are the red hot rocks of two inner mantle that is wrapped around the earth's core, 2900 km below the surface. It is in this mantle that the driving force of earth's mobility originates, hot material rises, cools as it moves along the surface and then sinks again.

Nearly 75 per cent by weight of earth is oxygen and silicon. These two, in combination with aluminium, iron, calcium, sodium, potassium and magnesium account for the minerals that make up 95 per cent of the earth's crust. These minerals, called silicates, are some of the most important rock-forming minerals. The three most important rock-forming silicate minerals are quartz, feldspar and ferromagnesium. Feldspars are the most abundant and perhaps the most important group of rock-forming potassium, sodium or calcium aluminosilicates. Feldspars weather chemically to form clays, micas which are layer lattice aluminosilicates, a major constituent of good soil. Details about these will be discussed in chapter 4.

In general, rocks are classified into three families namely igneous, metamorphic and sedimentary. All these types are involved in the worldwide recycling processes. The internal heat from the tectonic cycle produces igneous rocks from molten materials. These may crystallize beneath or on the earth's surface. Rocks at or near the surface breakdown chemically and physically by weathering processes to form sediments that are transported by wind, water and ice. The sediments accumulate in the depositional basins, such as the oceans, where they are eventually transformed into sedimentary rocks. Once the sedimentary rocks are buried to sufficient depth, they are altered by heat, pressure or chemically active fluids to produce metamorphic rocks which may then melt to restart the cycle.

Igneous rocks are crystallized from a naturally occurring mobile mass of quasi-liquid earth material known as magma. Magma is probably generated in the upper asthenospheric or lithosphere. As it moves up, it displaces the rock it intrudes, breaking a portion of the rock and incorporating them into the mass of moving magma. Magma may crystallise below the surface of earth, but it is exposed at the surface of the earth, where erosion sets in. Magma may be thrown up in a volcanic eruption as pyroclastic debris or lava. The composition varies from place to place.

Sedimentary rocks form when sediments are transported, deposited and then lithified by natural cement, compression or other mechanism. Therefore two types of sedimentary rocks.

(*i*) **Detrital Sedimentary Rocks:** These rocks form from broken parts of previously existing rocks. These rocks include shale, sandstone and conglomerate.

(a) Shale: It is held together by molecular attraction of the fine clay particles. Presence of shale is a danger sign calling for a close look to determine the type of shale-rock content and its physical properties for engineering purpose.

(b) Sandstone and Conglomerate: These are coarse-grained sedimentary rocks that may be very strong and stable for engineering purpose.

(*ii*) **Chemical Sedimentary Rocks:** These rocks are formed by chemical or biochemical processes that remove material carried in chemical solution.

(a) Lime stone: These are most abundant chemical sedimentary rocks. Although it may have sufficient strength, it weathers easily to form subsurface cavern systems and solution pits. In limestone areas, most of the streams may be diverted to subterranean routes because these routes may easily become polluted by calcium that enters the ground water.

(b) Rock salt: It is another important chemical sedimentary rock. It is composed primarily of sodium chloride. Rock salt forms when shallow seas or lakes dry up to precipitate salt. The salt may later be covered up with other types of sedimentary rocks. It is from these sedimentary basins that salt is mined. Because rock salt is abundant, usually completely isolated from circulating ground-water and located in areas with low earthquake risk, it has been suggested that old solution-mining pits may be an acceptable place to dispose off radioactive wastes such as those from fuel reprocessing plants.

(c) Metamorphic rocks: These are changed rocks. Heat, pressure and chemically active fluids produced in tectonic cycle may change the minerology and texture of rocks; in effect, producing new rocks.

Soils may be defined as solid earth material that has been altered by physical, chemical and organic processes such that it can support rooted plant life and can be removed without blasting. Intimate interaction of processes between the rock and hydrologic cycles produce weathered rock material which forms the basic ingredient of soil. Weathering is the physical and chemical breakdown of rocks and is the first step in the soil-forming process. The more insoluble weathered material may remain essentially in place and be further modified by organic processes to form residual soil.

Most of the differences in soils are the effect of climate, topography, the type of parent rock, the organic process and the amount of time required for soil-forming process. The vertical and horizontal movement of material in the soil system produces a distinct soil layering or soil profile divided into zones or horizons (Fig. 1.15).

An important process is leaching, *i.e.* downward movement of soluble material from the upper zone, called the A horizon, to the intermediate zone known as the B horizon, where the leached material is deposited. In different climates, this may produce a hard-pan layer of compacted clay or calcium carbonate, a cemented soil material known as caliche. In addition, material may also move upward in the soil in response to natural variation of water or when extensive irrigation raises the level of ground-water. When accompanied by extensive evaporation, this upward movement causes salts to move up and be deposited in the upper soil layers or on the surface of the ground.

This happened in the Indus River Valley in Pakistan, where about 93000 km were irrigated. The water table rose and extensive evaporation deposited salts, making the soil unsuitable for crops. The problem can be; and has been in some places, rectified, by installing a drainage system for the land and then providing an abundance of water to leach the salts out. This is done in the Great Valley of California, where salts are a problem.

Nobody is yet sure why different oceans spread at vastly different rates or what governs changes in the direction and rate of plate movement or how to explain (apparent) differences in the behaviour of oceanic and continental plates. More fundamentally little is actually known about the mechanisms that drive these surface phenomena, the physics and chemistry of the earth's inner mantle. These are the areas that require further research.

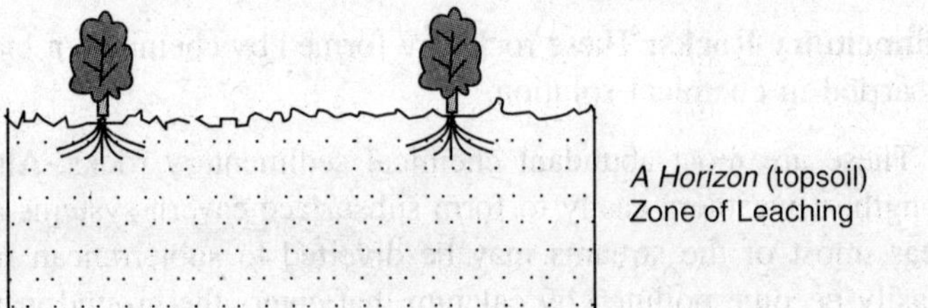

Fig. 1.15 Idealized diagram showing soil horizons and associated zones.

Even so, it is possible to make a guess as to what the surface of the globe might look like 50 million years or so in the future, the Mediterranean sea will disappear, a chunk of north east Africa will break away. Australia would have drifted northward and in the direction of south-east Asia, Los Angeles, would have long since passed; San Francisco will be heading further north for the Aleutian trench. And finally east and west will meet as western edges of North and South America begin to close up to Japan.

REFERENCES

1. Amald, Phys. Zeitt 32,914 (1931).
2. Barbara, J. Finlayson Pitts, J.N. Pitts (Jr), *Atmospheric Chemistry Fundamentals and Experimental Techniques,* John Wiley & Sons, New York, (1986).
3. Arthur, M., Bushwell and Worth H. Rodebush, *Water,* Scientific American, 194,76–89 (1956).
4. Bernal, J.D. and R.H. Fowler, J. Chem. Physics 1, 513–48 (1933).
5. Calvin, M., *Chemical Evolution,* Oxford Univ. Press, London (1969).
6. Frank, H.S. and W.Y. Wen, Trans Faraday Soc., 24, 133 (1951).
7. Humphreys, W.J., *Physics of Air,* 3rd Ed., McGraw Hill, New York (1940), Reprinted by Dover, New York (1964).
8. Hogflit, E., *Stability Constants of Metal-Ion Complex Part A, Inorganic Liquids* (*IUPAC* Data Series 21), Pergammon (1982).
9. Hammon, S.D., M. Linton, Trans Faradey Soc., 65, 2186 (1969).
10. Home R.A., in I.L., Ciaccio, ed. *Chemical and Microbiological Analysis on Water and Water Pollution,* Dekker, New York (1971).
11. Lutz, R.A., R.M. Haymon, National Geographic, 10, 295–304, (1994).
12. Miller, S.L. and L.E. Orgel, *The Origin of Life,* Englewood Cliff's Prentice Hall, New York.
13. Muller, S.L., Production of Some Amino Acids Under Possible Primitive Earth, *J. of Am. Chem Soc.*, 77, 2361-61 (1955).
14. Margulis Lynn, *Symbiosis in Cell Evolution,* W.H. Freeman and Company Sanfransisco (1985).
15. Meyero, A.W., d. Ann Physik S, 701 (1980).
16. Mulliken, Phys. Rev., 41, 756 (1932).
17. Nemethy, G. and H.A. Scheraga, *J. Chem. Phys.*, 36, 3382 (1962).
18. Oparin, A.I., *The Origin of Life and Origin of Enzymes,* Adv. Enzymol, 27, 347–80 (1965).
19. Oparin, A.I., *Genesis and Evolutionary Development of Life,* Academic Press, New York (1968).
20. Orgel, L.E., *The Origin of Life: Molecules and Nature Selection,* John Wiley, New York.
21. Palmer, E. and Newton, C.W., *Atmospheric Circulation Systems, Their Structures and Physical Interpretation,* Academic Press, New York (1969).
22. Stanley, E.M. and Batten R.I., *J. Phys. Chem.,* 73, 1187 (1969).
23. Wells, A.F., *Structural Inorganic Chemistry,* Oxford Univ. Press (1975).
24. Steward, Phys. Rev., 35, 1426 (1930); 37, 9 (1931).
25. "Riches from the Mobile Earth", The Economist, Aug. 16, 1980, page 63-69 (a good article on Plate Tectonics).

2

Ecology

2.1 INTRODUCTION

Ecology is the scientific study of the relationship of living organisms with each other, with their environment and the interrelations of biotic communities with their nonliving environment. As ecology incorporates elements from all spheres of science, it is an extremely wide subject that cannot be covered in a single chapter. The most significant principles of ecology that relate directly to environmental chemistry and, thus, to the use of energy in chemical reactions alone will be taken up here.

Ecosystems: An ecosystem is the biotic community and its nonliving environment as an interacting system. Living organisms and their nonliving environment are inseparably interrelated and interact with one another. Any unit that includes all the organisms that function together in a given area interacting with the physical environment, so that a flow of energy leads to clearly defined biotic structures and cycling of materials between living and nonliving parts constitutes an ecological system or an ecosystem.

An ecosystem consists of three components

(*i*) **Organism:** An organism is an individual unit constituted to carry on activities of life. It comprises species, populations and communities;

(*ii*) **Environmental Factors:** It constitutes soil, climate, and other physical factors and processes such as energy flow, nutrient cycles, water flow, freezing and thawing;

(*iii*) **Ecological Processes:** With the passage of time, organisms have evolved in synergism, the action of individual agencies that together have a greater effect than the sum of their independent effects with their environment; as a result many internal controls keep them in dynamic equilibrium with the world around them.

Some organisms, mainly plants, are rooted to their environment and must be able to survive any climatic variations, whereas others, particularly small animals, can move about and hence escape the vicissitudes of environment.

A distinction should be made between environmental concern and ecological concern as these terms are often confused and not clearly distinguished. The increase of carbon dioxide concentration in the atmosphere is expected to produce a significant climatic change in the future—a warmer climate, which may be wetter world wide, but drier in certain areas, is an environmental concern. However, when the impact of climate change on crop production or ecosystem dynamics is considered, the issue becomes an ecological concern.

Iron or zinc leaching from metal may poison microorganisms in soil, concern about these matters is clearly an ecological issue. It is self-evident that every environmental event has an ultimate ecological consequence, no matter how insignificant it may be. However, a specific event may terminate with the effect on environment or it may involve a multitude of ecological consequences.

2.2 FOOD CHAINS AND FOOD WEBS

Life is maintained by organisms. Plants convert carbon dioxide and water into energy-rich compounds by the use of sunlight, a chemical process known as photosynthesis.

Photosynthesis initially involves the photolytic cleavage of water in which light absorbed by chlorophyll provides the energy for the separation of water which requires some oxidants such as ferric ions. This may be represented as follow:

$$4Fe^{3+} + 2H_2O \longrightarrow 4Fe^{2+} + 4H^+ + O_2$$

This may also be represented by the following reaction in which "A" is any oxidant.

$$2A + 2H_2O \xrightarrow{h\nu} 2AH_2 + O_2$$

This reaction only takes place in the presence of light, and gaseous oxygen and hydrogen are produced early in the photosynthesis process.

The second major stage of photosynthesis is independent of light, and is called the dark reactions. In this ATP or NADH (nicotinamide adenine dinucleotide) act as energy source. The atomic hydrogen combines with carbon dioxide to produce the first chemical compound of carbon, hydrogen and oxygen. This can be represented in simplified form as follows:

$$2AH_2 + CO_2 \longrightarrow 2A + CH_2O + H_2O$$

The carbon subsequently becomes involved in 3-phosphoglyceric acid ($C_3H_5O_3H_2PO_4$) and then in ribulose 1, 5 bisphosphate ($C_5H_{10}O_3(H_2PO_4)_2$). This compound becomes the primary acceptor of CO_2. The reaction proceeds through several stages involving glycerate, glyceraldehyde, fructose and finally glucose-6-phosphate.

This photosynthesis requires the followings:

(*i*) Green plants containing chlorophyll,
(*ii*) Visible light or infrared light in the 800–850 nm range,
(*iii*) Carbon dioxide,
(*iv*) Water,
(*v*) Some oxidant ion such as Fe^{3+} or manganese, and
(*vi*) Phosphorus in the form of phosphate.

Green plants have additional capacity to synthesize higher organic compounds including disaccharide sugars, starches, lipids, proteins and vitamins, which contain five fundamental elements, namely, carbon, oxygen, hydrogen, nitrogen and phosphorus. The complete mechanism can be looked in the Stryer's book[8].

Consumption: Consumers of primary products may be conveniently classified into four categories:

(*i*) **Herbivores:** These are the primary consumers and are those animals which feed directly on plants, examples are meadow mice, deer, seed-eating birds and leaf-eating insects.
(*ii*) **Carnivores:** These organisms are also called secondary consumers and consume animal food.

Thus, the hawk which eats the meadow mouse and birds which consume the leaf-eating insects are secondary consumers.

(*iii*) **Omnivores:** These are multilevel consumers or omnivore refer to those living organisms that consume both plants and animals as food. Man is an omnivore as he consumes vegetables, berries and grains as well as meat. Bears and raccoons are also omnivores since they feed readily upon berries and fruits, but also prey upon fish, clams, crustaceans and other animals. Many animals are omnivores to a limited extent, Baboons normally consumes plant material but will occasionally eat small animals.

Scavengers: A special type of consumption is that of scavengers, who feed upon dead and decaying plant and animal matter. Vultures, sea gulls and even eagles consume extensively on animals that have died of other causes.

Decomposers: If the world only had producers and consumers, it would not work very well because the flow of matter would be in only one direction, that is, from lower to higher order compounds. Something needs to return these compounds to more elemental forms so that they can be recycled and used over again in the food chain. This process requires a group of organisms known as decomposers. Bacteria and fungi play this role. Bacteria generally act on animal tissue and fungi on plant tissue. Plant and animal material is degraded enzymatically and released as basic elements into the environment, where the elements are again available to the producers for reuse.

Some organic materials, such as sugars, lipids and proteins are decomposed rapidly in a series of stages, whereas others, such as cellulose, lignin, hair and bones are decomposed slowly. Decomposition in its most basic form may be represented by the formulae for respiration. Aerobic respiration is essentially an oxidative process represented as follows:

$$C_6H_{12}O_6 + 6O_2 \longrightarrow 6CO_2 + 6H_2O + \text{energy}$$

This is not so simple as it appears, it involves complex set of enzymatic reactions but is essentially the reverse reaction of photosynthesis.

In all such reactions, energy is not recycled. It moves unidirectionally through an ecosystem, being consumed at each step of the food chain.

Food Chains and Webs

The transfer of food energy from the source in plants through a series of organisms that consume and are consumed is called the *Food chain*. In each transfer some energy is lost. Therefore, the shorter the food chain, or the nearer the organism to the beginning of the chain, the greater the energy available to that population. Food chains are of two basic types:

(*i*) The grazing food chain, which starting from a green plant base, goes to grazing herbivores and on to carnivores.

(*ii*) The detritus food chain, which goes from nonliving organic matter into microorganism and then to detritus feeding organisms and their predators.

Food chains are more or less vaguely familiar to everyone, since we eat the big fish that ate the little fish that ate the zooplankton that ate the phytoplankton that fixed the sun's energy. Similar is the example of the cow that ate the grass that fixed the light energy. A much shorter food chain comprises of eating the grains that fixed the sun's energy. In the latter case, human beings function as primary consumers, whereas in the grass-cow-human food chain, humans function as secondary consumers. In general, humans tend to be both primary as well as secondary consumers. Accordingly, energy flow is divided between two or more levels in proportion to the percentages of plant and animal food being consumed. The shorter the food chain, the greater the biomass which can be produced from a given amount of energy.

These considerations help us to understand why the Antarctic sea water are among the most productive in the world during the polar summer. They typically have short, simple food chains; as for example, from phytoplankton to whalebone whales is a 2-link food chain. During the polar summer they also have 24 hour energy input from sunlight and a sharp turnover of water strata with an up welling of nutrients from the bottom to the surface. This favours the growth of plankton. Finally the organisms in the cold polar water have a low respiratory rate, and therefore relatively little energy is lost through respiration. In other words, net productivity is high in relation to gross productivity. Similarly, in the Antarctic Sea, if krill were eliminated by some ecological accident, there would be a catastrophic decline of virtually all marine mammals, birds and fish that depend upon krill for their nutrition.

Food chains are not isolated sequences; they are interconnected. The interlocking pattern is often spoken of as the food web. It is the network through which energy flows in the entire community of plants and animals. Most food webs have a great deal of inherent stability. They will readily repair themselves when disturbed, unless the disturbance is of an extreme nature. When the disturbance is so great that species are wiped out or entire breeding populations are decimated, ecosystem recovery to its previous form never occurs.

The food web of a tropical forest is highly complex, and includes tens of thousands of species. Once the forest is removed, imagine how many food chains are wiped out causing an irreparable damage to the ecosystem or at least liable to cause damage that may require centuries to overcome.

Suppose a forest is cut and the cleared area is planted with some starchy plants. This denudes forest tract rapidly, permits only one or two years of productive agriculture, and leaves a wake of destruction. A forest that may have taken 1000 years to develop can be destroyed in a few years, rapid soil erosion takes place, the entire ecosystem may be irreparably damaged. Either the lessons of ecologic history have been ignored or the economic pressures of life have made people blind to appreciate their importance to us.

2.3 NATURAL SELECTION

Ever since Darwin gave the theory that evolutionary change is a slow, gradual process involving many small mutations and continuous natural selection of those that provide competitive advantage at the individual level, gaps found in the fossil record and transitional forms could not be accounted for. Species that occur in different geographical regions or are separated by a spatial barrier are said to be allopatric; those occurring in the same area are said to be sympatric. Allopatric speciation has been generally assumed to be the primary mechanism by which species arise. According to this conventional view, two segments of a freely interbreeding population become separated spatially. With time, sufficient genetic differences accumulate in isolation so that the segments will no longer interchange genes when they come together again and thereby coexist as distinct species in different niches. Sometimes these differences are further accentuated by *Character displacement.* When two closely related species have overlapping ranges they tend to diverge in one or more morphological, physiological or behavioural characters in the areas of overlap and to converge in parts of their range where each species occurs alone.

British salt marshes provide an example of sympatric speciation resulting from hybridization and polyploidy. When the American salt marsh gram *Spartina alternilora* was introduced into the British

Isles, it crossed with native species *S. mentima,* to produce a new polyploidy species, *S. townsendii,* which has now invaded formerly bare tidal mud flats not occupied by the native species.

Another example of rapid natural selection resulting form human interference is what has come to be known as "Industrial melanism". Dark pigmented tree trunk moths have evolved in industrial areas of England where the barks of trees has become greatly darkened by industrial pollution, which bills the lichens that give normal bark a light appearance. Kettlewell[7] demonstrated experimentally that dark moths survive better in dark woods and light moths in natural woods, presumably because predation by bird is selective for the individuals not protectively coloured. This example leads us naturally onto the subject of direct or purposeful selection of humans.

Selection carried out by people to adapt plants and animals to their needs is known as artificial selection. Domestication of plants and animals involve more than modifying the genetics of a species and the domesticator are required. Accordingly, domestication lead to a special form of mutualism. Because humans are egotistical, we fall into the trap of thinking that domesticating another organism through artificial selection means merely 'bending' the nature to suit our purpose. Actually domestication produces changes in people as well as in the domesticated organisms. Thus our dependence on the corn develops a very different culture from the one that depends on herding cattle. It is a genuine question as to who becomes the slave of whom? The same question can be posed about the relationships of humans with machines. Perhaps the relationship between man and his tractor is not too different from that of man and his draft animal, except that a tractor requires a greater energy input and produces more poisonous wastes.

2.4 RESPONSE TO PHYSICAL FACTORS

Under steady-state conditions, the essential material available in amounts most closely approaching the minimum need tends to be the limiting one, a concept that has come to be known as Liebig's law of the Minimum. A crop yield was often limited, not by nutrients needed in large quantities, such as carbon dioxide and water, as these are often found in abundance in the environment, but by some raw material, such as zinc, needed in minute quantities but are very scarce in soil. The growth of a plant independent of the amount of foodstuff presented to it in a minimum quantity, has come to be known as Liebig's law of the minimum. Extensive work since the time of Liebig has shown that two subsidiary principles must be added to the concept if it has to be useful in practice. The first is a constraint that Liebig law is strictly applicable only under steady state conditions, *i.e.* when inflows balance outflow of energy and materials. The second consideration is a factor of interaction. Thus, a high concentration or availability of some substance, or the action of some factor other than the minimum one, may modify the rate of utilisation of the latter. Sometimes organisms can substitute, at least in part, a chemically closely related substance for one that is deficient in environments. Thus, when strontium is abundant, mollusks can substitute strontium for calcium to a limited extent in their shells. Some plants have been shown to require less zinc when growing in the shade than when growing in full sunlight, therefore a low concentration of zinc in the soil would less likely be limiting to plants in the shade than under the same conditions in sunlight. Thus organisms have an ecological minimum and maximum; arranged in between represents the *limits of tolerance.*

Some subsidiary principles to the law of tolerance may be stated as follows:

1. Organisms may have a wide range of tolerance for one factor and a narrow range of another.
2. Organisms with a wide range of tolerance for all factors are likely to be most widely distributed.

3. When conditions are not optimum for a species with respect to one ecological factor, the limits of tolerance may be reduced for other ecological factors. For example when soil nitrogen is limited, the resistance of grass to drought is reduced.
4. Organisms in nature are not actually living at the optimum range of a particular physical factor. In such cases, some other factor or factors are found to have greater importance. For instance, certain tropical orchids actually grow better in full sunlight than in the shade, provided they are kept cool. In nature however, they grow only in the shade because they cannot tolerate the heat of direct sunlight.
5. Reproduction is usually a critical process when environmental factors are most likely to be limiting. The limits of tolerance for reproductive individuals, seeds, eggs, embryos, seedlings and larvae, are usually narrower than for non-reproducing adult plants or animals.

It may be pointed out that if an organism has a wide limit of tolerance for a relatively constant factor in moderate quantity in the environment, that factor is not likely to be limiting. Conversely if an organism is known to have definite limits of tolerance for a factor, that also is variable in the environment, then that factor merits careful study, since it might be limiting. For instance, oxygen is so abundant, constant and readily available in the terrestrial environment that it is rarely limiting to land organisms, except the parasites of those living in soil or at high altitudes.

To sum up, primary attention should be given to factor that are operationally significant to the organism at some time during its life cycle. The aim of environmental analysis is not to make a long, uncritical lists of possible factors but to achieve these more significant objectives:

1. to discover, by observation, analysis and experiment which factors are operationally significant, and
2. to determine how these factors affect the individual, population or community, as the case may be. In this manner, the effect of disturbances or proposed environmental alterations can be predicted with reasonable accuracy.

Temperature Response

All organisms are sensitive to temperature and have distinct temperature preferences. Some organisms have broad temperature tolerances and some have very narrow. Most cold-blooded animals, known as poikilotherms, have low metabolic rates and body temperatures quite close to the temperature of the air or water in which they are present. In strong sunlight, reptiles may often have body temperatures of 10°C or more, above the air temperature. Warm-blooded animals, known as Homeotherms, have body temperatures that are regulated to remain within narrow limits except for some animals during hibernation, when they essentially go into a cold-blooded mode. The normal body temperature for humans is about 37°C, but body temperatures as low as 22°C and as higher 43°C have survived. Birds have the highest body temperatures of any animals, being normally 41° to 44°C; birds have recovered from 48°C. Plants generally have temperature close to air temperature, except in strong sunlight, when leaves may be 10° to 20°C above air temperature, or under a clear, cold night when they may be 5° to 8°C below the surrounding environment.

All plants and cold-blooded organisms have a behavioural and growth response that increase with increasing temperature above some low temperature tolerance limit: the behavioural and growth response reaches a maximum at some optimum temperature and decreases sharply at higher temperatures, until the thermal maximum is reached. The rate of egg development for most insects follows this temperature response.

Light Response: Light radiation consists of electromagnetic waves over a wide range of wavelengths. Two bands of wavelength readily penetrate the earth's atmosphere, namely, the visible band, together with some parts of adjacent bands, and the low-frequency radio band having wavelength greater than 1 cm. All other radiation are eliminated by the atmosphere. Solar radiations that reach the earth's surface consist of light ranging in wavelengths from about 0.3 micron to 10 microns. Light visible to the human eye lies in the range 3.90 to 7.60 μ (1 micron = 1 μ = 10^{-6} m).

Photosynthesis is a photochemical process requiring light to convert carbon dioxide and water to carbohydrates, starch, and sugars. Most plants respond to increasing light intensity increasing photosynthetic rates.

However, some plants require shade to grow well and must avoid direct exposure to sunlight. These shade-loving plants (skiaphytes) respond photosynthetically with more sensitivity to low levels of solar irradiance than do the heliophytes. However, the sun-plant has a greater photosynthetic rate in full sunlight than does the skiaphytes.

Aquatic plants respond to light in a manner similar to that of land plants. However, there is evidence of light inhibition of photosynthesis in marine phytoplankton at light intensities approaching full sunlight. There is also some evidence that ultraviolet light limits the development of aquatic plants.

Animals respond to light in number of ways. Many organisms have a *biological clock*, which is a physiological mechanism for determining time. For some organisms the biological clock seems to be regulated by external factors such as light, and for other organisms by internal factors. Some plants and animals possess responses having about a 24-hour periodicity. These are known as *Circadian rhythms.* Timing mechanisms regulate endocrine changes, gonadal development, colour changes in birds in spring and autumn, mating behaviour in moths, insect feeding, flowering and many other responses. Light intensity and day light clearly are important in some of these, if an environmental change occurs as for example in the illumination of a construction site; stripmining operation or the industrial activity, it may have direct impact on the biological clocks of many plants and animal nearby.

2.5 THERMAL POLLUTION

The hot effluents are discharged at a temperature 8° to 10°C higher than the temperature of intake water. This unutilised heat, discharged into the river water system, produces pollution known as "thermal pollution". The following sources contribute to thermal pollution:

(*i*) Coal-fired power plants,
(*ii*) Domestic sewage,
(*iii*) Hydroelectric power,
(*iv*) Industrial effluents, and
(*v*) Nuclear power plants

In all these cases water is heated by different means, to be converted to steam under high pressure, say at a temperature of 200°C = 473 K = T_1, which is used in the generation of power. After producing electricity the heated water is thrown into water say at a temperature of 52°C = 325 K = T_2. If this cycle were reversible and of Carnot cycle type, the efficiency can be calculated as follows:

$$\eta_{nev} = \frac{T_1 - T_2}{T_2} = \frac{473 - 325}{325} = 45.0\%$$

Nearly 55 per cent of unutilised heat is lost, goes into water to produce thermal pollution.

Harmful effects of thermal pollution may be summarised as follows:

1. **Reduction in Dissolved Oxygen (DO):** Concentration of dissolved oxygen decreases with increase in temperature of water (see section 3.5), for example, DO concentration is 11.28 mg l^{-1} at 10°C and 7.5 mg l^{-1} at 30°C. Thus cold-water fish swimming at 10°C, which require more oxygen cannot tolerate the hot water of 30°C and may die. Similarly, since the aquatic biota live aerobically, so a healthy steam should have an adequate supply of dissolved oxygen. In oxygen-rich water, bacteria, protozoans and microorganism multiply rapidly and then become food for more advanced creatures.
2. **Change in Water Properties:** A rise in temperature changes the physical and chemical properties. The vapour pressure increases sharply, while viscosity of water, density, solubility of gases decreases but that increases settling rate of suspensions, which seriously affect the food supplies of aquatic organism.
3. **Interference with Biological Activities:** Temperature is considered to be of vital significance to physiology, metabolism and biochemical process in controlling respiratory rates, digestion, excretion and overall development of aquatic organisms. The temperature changes totally disrupt the entire ecosystem. Sharp changes in temperature are often destructive. Because life process of aquatic animals involves several chemical reactions and the rates of these reactions vary according to the changes in temperature.
4. **Variation in Reproductive Rate:** The increase in temperature triggers deposition of eggs by the female, particularly shell-fish, oysters and clams, which spawn within four hours of water temperature reaching the critical level.
   The Atlantic Salmon eggs hatch in 114 days in water at 2°C and in 90 days at 7°C. Herring egg hatch in 47 days at 0°C and in 8 days at 14.6°C. These are the critical temperature for reproduction. The offsprings of Gammarids above 7.8°C may be only females.
5. **Change in Metabolic Rates:** With increase in temperature fishes show a marked rise in basal rate of metabolism with speed. In brown Trout, the rate of oxygen consumption rises steadily till lethal temperature, i.e. 28°C, Sockeyl salmon swim twice as fast in water at 15°C as at 0°C, but above 15°C its speed declines.
6. **Invasion of Destructive Organisms:** Thermal pollutants may permit the invasion of organisms that are tolerant to warm water and highly destructive. Invasion of shipworms into New Jersey's Oyster creek constitute the best example.
7. **Undesirable Change in Algae Population:** The life in an aquatic ecosystem is greatly influenced by the growth of algae. Excess nutrients from the wash out water from farmlands combined with thermal pollution cause an excessive algal growth with consequent accelerations of eutrophic and other undesirable changes. All the major group of algae, like diatoms, green, blue-green algae, have a distinct tolerance ranges for water temperature. High temperature promotes blue-green algal blooms, which disrupts the aquatic food chain.
   A water flea which might be able to tolerate the thermal extreme of 35°C would probably starve to death at that temperature, if the diatoms on which it fed were unable to survive. In turn, fish feeding on water fleas would be similarly hard pressed to survive, regardless of their tolerance or adaptability to the high temperature. In an unpolluted steam diatoms grow best at 16 to 20°C, green algae at 30 to 35°C and blue-green algae at 35 to 40°C.
   Evidently, discharge of hot water to a water course may favour growth of blue-green algae which is a poor food source for aquatic animals. It may prove toxic to fish. Thus, as the thermal pollution load in large water streams increases, the potential for increase of fish disease will

also increase. As blue algae can convert a lake into marsh, it appears to be the indicator of extreme thermal pollution conditions. This unicellular algal bloom inhibits the growth of other aquatic organisms and produces foul odour in the receiving stream.

8. **Longevity:** High temperature of water may induce increase in activity, which exhausts the organisms and shortens its life. Crustacean *Daphnia* survives for 108 days at 7.8°C and 29 days at 21°C. Generally the speed of chemical change is approximately doubled for every 10°C rise in temperature. Mesophylic organisms which can survive in the temperature range of 10 to 40°C severely affect the water quality. Their activities usually reach a maximum between 30 to 38°C and then falls off as the enzymes become less active. However, to assess the total hazard presented by temperature, effects on biochemical systems, degradation and catalytic influences should be considered.
   The increase in temperature, exhausts the microorganism and shortens the life span of fishes. In general, most fresh water fauna population decline with rising temperature of water and a few species can survive at about 40°C.
9. **Effects on Water Quality:** Purification of water streams is an aerobic oxidation process when organic matter is converted into innocuous substances by bacteria. At higher temperature, the higher rate of bacteriae activity causes a burden on the dissolved oxygen resources of the water. When the waste to be disposed in water has a higher demand for oxygen than the supplies, then it may cause putrefaction in water bodies, developing noxious conditions. On the other hand water purification occurs more rapidly and at a lower cost at high temperature.

2.6 CONTROL OF THERMAL POLLUTION

Control of thermal pollution requires immediate attention since in future its detrimental effect on the aquatic ecosystem may be worse. To reduce the magnitude of pollution, the outlet water can be made to impart its heat to the environment and then may be discharged into the water course. The following methods can be adopted to control the high temperature:

1. **Natural Draft Towers:** In these cooling towers, hot water is sprayed down through a rising current of air. The water transfer its heat to the counter current of air by evaporation. The cold water collected at the bottom is returned to the water system. The technique is expensive due to the high installation and operational costs.
2. **Mechanical Draft Cooling Tower:** In this type of cooling towers, air flow is induced by electric fans. Warm water during its passage gets cooled by the action of air. The technique is also not desirable, since it produces noise pollution due to the operation of fans. The cost is a lot higher compared to its utility.
3. **Cooling Ponds:** Reservoir constitutes the simplest method of cooling thermal discharges. Heated effluents on the surface of water in cooling pond requires maximum dissipation of heat to the atmosphere and requires a minimum water area and volume.

By forming a warm water wedge over the cold water rapid heat loss to the atmosphere is encouraged by increasing the energy transfer due to evaporation, convection and back radiation. This warm water wedge acts as a cooling pond.

REFERENCES

1. Odum, Eugene P., *Basic Ecology,* Sundens College Publishing.
2. Southwick, C.H., *Ecology and the Quality of Our Environment,* D Van Nostrand Company.
3. Barnett, S.A., *Instinct and Intelligence: Behaviour of Animals and Man,* Prentice Hall (1967).
4. Ehrlich, P.R., A.H. Ehrlich and J.P. Holdren, *Human Ecology: Problem and Solution,* W.H. Freeman & Co. (1973).
5. Smith, W.F. and Smith, A.M., *Minamata,* Renehart and Winston, New York (1975).
6. Wallbank, W.T., *A Short History of India and Pakistan,* Scott Foreshman (1958).
7. Kettlewell, H.B.D. *Further Selection Experiments on Industrial Melanism in the Lepidoptera*", Heredity, 10,287-301 (1956).
8. Lubert Stryer, *Biochemistry,* W.H. Freeman and Company, New York, (page 517-543 for Photo-synthesis) (1988).
9. Smith, R.L., *Elements of Ecology*, 3rd Edition, Harper Collins Publishers Inc., New York (1992).

3

Water Pollution

3.1 INTRODUCTION

Ever since man progressed from hunting to an agriculture, with corresponding development of stable communities, the phenomenon of water pollution has been his constant companion. As agricultural methods improved, a smaller percentage of population produced all the food needed; larger communities and diverse secondary industries developed and grew into the present modern society. Concurrent with this growth was the increase in waste material and the problem of its disposal. When the waste from a community was relatively small the, easiest method of disposal was the throw it away, usually into the nearest receptacle. Since man cannot exist without water, community development and city growth centered in areas where the water supplies were adequate and continuous. Initially, this meant development in river valleys, and thus the nearest receptacle for wastes was the river or stream nearest to inhabitation.

As civilization progressed and trade became more important, cities were developed on trade routes and along navigable waterways. The result being that, at present, narrow coastal strips contain the major proportion of population and industry. This development, of course, immediately opened another disposal area. What better disposal system could be found than estuaries, bays and the vast wide ocean. The major fallacy is that disposal to the ocean is a misnomer, in that oceanic conditions are not found in the inshore waters. What is seen, however, is a coastal water system that is completely unique to each coastal section, with characteristics that depend in part on the given topography and wind conditions.

It is in this particular area that the question of pollution takes on a new meaning. Is a body of water polluted when it directly affects man, or should it be classified as polluted when the ecological structure is first upset? The hydrosphere is a dynamic system in physiochemical and biological equilibria, and there is no doubt that a normally active waterway has a large capacity to assimilate wastes. However, in many areas this capacity has now reached its limit or even been exceeded, so that many water ways have become increasingly contaminated. Some examples of water systems, where the effects of pollution have become apparent, are the Adriatic, Baltic, Mediterranean seas, the Ganga, Yamuna, Gomti, Godavari and Thames rivers, the Great lakes in America and Canada, the Dal Lake in India.

However, any dynamic system has a remarkable capacity for regeneration and with intelligent planning even the most seriously polluted waterways can be brought back into a normal state. An example of river regeneration on a large scale is the ongoing attempt to clean the Ganges.

A common ground for pollutants is their mode of entry into the water system. Broadly speaking, waterways are subject to entry of pollutants by

1. Direct discharge into the river-water system
2. Run off and/or seepage with subsequent transport
3. Reactions and transport across the air-water interface
4. Reactions and transport across the water-sediment interface

Under all conditions, entry of the pollutant is either in dissolved or particulate form, or may be converted to these forms by chemical reactions.

3.2 POLLUTION OF LAKES

Large lakes support fishes and supply water to the cities. Lakes are essentially transitory; these may, ultimately fill up and as they age, they may even vanish. Very few are more than a few tens of thousands of years old, and those that are, such as Lake Baikal in Sibena and Lake Tanaganyka, have survived only because of their vast size, but with time even they may vanish. Every particle of silt brought into the lakes settles to the bottom, and every creature which grows in them has to die, and thus help, to fill up their basin. In shallow waters around the lake edges, this process is further accelerated by the growth of rooted aquatic plants, weeds. These accumulate silt and cause the steady encroachment of the shores.

Lakes are formed by some geological accident such as large-scale subsidence faulting or damming of valleys by rock falls or the gouging action of glaciers. Initial properties of each lake depend on local conditions. If the drainage basin of a lake is rocky and infertile, very small amount of nutrient salts needed for plant growth, namely potassium, nitrate, phosphates are carried into the lake. Its water is therefore infertile with low biological productivity. On the other hand, if the drainage basin is fertile and is composed of easily weathered rocks and rich soil, relatively large quantities of nutrients find their way into the water and the lake becomes productive. These two examples are the extremes of a continuous variable series.

An unproductive lake also known as oligotrophic lake, does not remain indefinitely in the same condition: not only does its basin change as erosion occurs and more fertile soils are built up, but it also acts as a trap for fertility. As nutrients reach a lake, they tend to become fixed in the bodies of plants and animals, and when these die and decay they form a subaquatic soil which slowly releases the nutrients for the lake dwellers. So, with time, the length of which depends on local circumstances, the fertility and productivity increases. Similarly, a lake which was initially fairly infertile becomes richer with the passage of time. This has important effects on plant and animals, the total quantity of living matter increases and different species succeed one another. These changes are of considerable scientific interest.

First, the change from unproductivity to productivity involves a great increase in amount of plankton, the small plants and animals which live in the open water, and these affect the usefulness of water to man. Unproductive lakes have a sparse plant-plankton, composed very largely of desmids and, correspondingly, few planktonic animals. As fertility increases, desmids are replaced by diatoms and then by various flagellates and other green algae; finally, blue-green algae appear and these can become so abundant at some seasons as to form a water bloom. Such an outbreak of plantation growth can give water the appearance of pea soup. The increase in tiny plants is followed by an increase in small animals, mostly crustaceans among plankton. The resulting large amounts of algae and animal matter pose great problems for the marine engineer. Planktons can also affect the use of water for other purposes for example, industrial cooling, considerable enrichment of lake water, or eutrophication, which is, a symptom of aging and is, therefore, definitely a disadvantageous process from man's point of view.

Pollution, even by highly purified effluents increases the rate of aging of lakes, but it is questionable whether an increase in the rate at which a perfect normal process occurs leads to an alteration in the natural quality of water. Unless alterations produced by pollution are very striking, or detailed rewards of previous conditions and rate of change of a particular lake are available, it is very difficult to assess how much of the alternation is natural and how much has been caused by human activities. Nagin Lake in Kashmir and Nainital Lake, depth was reduced from 27 m to 18 m during the last 80 years due to siltation of the lake bed. Dal lake will soon turn into a marsh. In 1991, a strong red layer of algae surfaced all along the bank and later spread to over one fourth of the lake surface. This is the first sign of its transition to a marsh. Srinagar's Dal lake, a principle tourist attraction in Kashmir, is getting choked by the spread of aquatic plants and algae in it. Its area has shrunk from 33 sq. km to 10 sq. km in 50 years.

3.3 POLLUTION IN RIVERS

Rivers are classified according to certain properties as shown by its water composition. These are as follows:

The dissolved salts in water, the most important of which is calcium bicarbonate, leached out by lime-bearing rocks. When the water flows from noncalcarous rocks, such as slates, shales or granite, very little calcium is present in solution, the water is therefore soft and may be acidic. Acidity occurs particularly where the landscape is peaty because the water is quite unbuffered, the acid that comes from peats is not neutralised, rather it lowers the pH. Hard waters, on the other hand, are alkaline, and their pH does not change readily. Suppose, for instance, one adds some hydrochloric acid to hard water, the following reaction occurs.

$$Ca(HCO_3)_2 + 2HCl \longrightarrow CaCl_2 + 2H_2O + 2CO_2$$

The HCl is replaced by an equivalent amount of CO_2 which is a very weak acid and the pH is little changed. Thus hard water, quite apart from the fact that it contains, calcium salts are needed by animals for shells or other skeletal structures is more stable chemically than in soft water. Other important dissolved salt ions, such as potassium nitrate and phosphates, are needed for plant growth are leached out of the rocks and oils and more fertile drainage basins. On the whole, the softer lowland rocks and soils produce more than the mountain-forming rocks.

The rate of flow of water is indirectly important as it influences the type of river bed and amount of silt deposition. A turbulent river will replenish dissolved oxygen and hence fishes like trouts and invertebrates will survive in such rivers.

The temperature of the water is dependent on the altitude and type of source. All rivers in northern India are cold in winter, however, sluggish lowland rivers may become very warm, occasionally reaching temperatures as high as 25°C, while a high altitude stream remains cool.

The dissolved oxygen content of water is measured in term of Biochemical Oxygen Demand (BOD). It is a measure of the amount of oxygen necessary to decompose organic matter in a unit volume of water. As the amount of organic waste in water increases, more oxygen is used, resulting in a higher BOD. Many animals, notably trout and many invertebrates, can live only in well-aerated water. At a pressure of one atmosphere and taking into account the fact that air contains only 20.9 per cent O_2,

100 per cent saturation value of dissolved oxygen varies from about 14 ppm (10.30 ml/ lt) at 0°C, 8.67 ppm (6.39 ml/lt) at 20°C, 7.85 ppm (5.78 ml/lt) at 25°C and zero at 100°C. The saturation values for water in contact with pure oxygen are nearly five times high. Living organisms and their decay, use up oxygen which can only be replaced from the atmosphere or by the photosynthetic activities of a green plant. In a swift upland stream, the turbulence of water rapidly replenishes any oxygen used and the water remains saturated. If it tend to become over-saturated, either because of increase in temperature or because of photosynthesis by plants, excess oxygen is lost to the atmosphere; thus the percentage of saturation remains around 100% all the times. In a sluggish river, there is less turbulence and oxygen is not readily absorbed from or given up to air. Thus, the amount in solution may fall considerably, particularly at night. It is prevented however under ordinary circumstances from falling to very low levels by the fact that the rate of oxygenation increase as rate of deoxygenation by microbes falls. This rate of oxygenation is proportional to the saturation deficit; if the rate of oxygen uptake under given conditions is x when the saturation deficit is 10% (*i.e.* percentage saturation = 90%), it becomes $2x$ when the deficit is 20% (details in section 3.9). The relative inability of sluggish water to exchange oxygen with the atmosphere can also result in the phenomenon of supersaturation. Under conditions of bright sunshine the plants produce a large amount of oxygen and this being pure oxygen can theoretically result in nearly five times the concentration, which could be dissolved from the air at the same temperature. In practice, 500% is never attained but values up to 200% have been recorded. In a turbulent stream, where easy exchange with the atmosphere tends to keep the balance, the percentage saturation with oxygen is therefore always higher than in a sluggish stream. Values tend to fluctuate widely and the lowest values are reached at night. The warmer the water, the greater the rate at which oxygen is used up and there is less in solution to use, so that very low values tend to occur mainly during summer nights.

These then are the physical and chemical factors, which determine the type of habitat which any river offers to living things. River and stream waters may be hard or soft, fertile or infertile, swift-flowing or sluggish, warm or cold, all combinations and intermediates may occur, and any individual river may vary in respect of these characteristics along its length.

At the base of all living communities lie different plant. The structure of the plants is to some extent correlated with the rate of flow of water, those which grow in turbulent habitats being either cushion-shaped or provided with strong flexible stems.

The algal community of rivers is essentially sessile, it grows on solid bodies and can develop only where these are present, namely in places where the sub-stratum is soft mud it can grow only on weeds or along the hard parts of the bank. In muddy reaches also the steady deposition of silt on to the bottom smother tiny plants; silt also greatly lowers the light intensity and renders algal growth less possible. In such reaches the algae which do occur, other than those living on weeds, live in the open water and are therefore, planktonic, as are the algae of the open lake.

But river plankton is not indigenous, it cannot maintain itself in the way that lake plankton does, as, except in very large rivers like the Nile and the Danube, it is carried away so fast by the water that no individuals remains for long in the river.

Figure 3.1 demonstrates the general effects diagrammatically. In section A and B the purely physical and chemical phenomena, and the way in which they change as one passes downstream, are shown. The example shown is a case of fairly severe pollution, but not so severe as to cause complete deoxygenation. Section C shows the corresponding changes in microorganisms. The bacteria are at first abundant and then decline as the organic matter is used up; sewage fungus appears, increases and then declines in the same way. When oxygen conditions permit, large numbers of protozoa are built up and most of them belong to species which feed on the bacteria. The algae at first decrease in numbers, but then, as nutrient salts are released from the organic matter and conditions improve, they increase greatly in numbers and then declines like the other algae. Section D shows the effects on the larger animals. Fishes and clean water invertebrates decline rapidly in numbers and usually disappear as they are

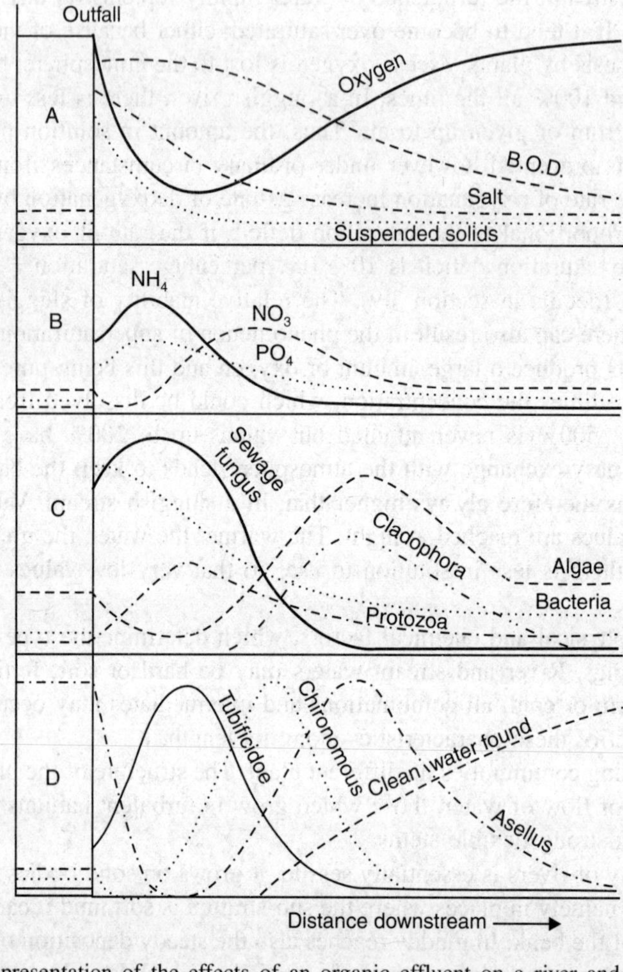

Fig. 3.1 Diagrammatic presentation of the effects of an organic effluent on a river and the changes as one passes downstream from the outfall. A & B, physical and chemical changes; C, Changes in microorganisms; D, Changes in larger animals. HBN Hynes[10].

overcome by lack of oxygen, blanketing effects of suspended solids and sewage fungus or release of poisons. These bodies then reappear in small numbers and little variety. Their numbers and variety increase steadily as recovery takes place. In the badly polluted zone, the clean water animals are replaced by a very abundant "pollution fauna" consisting largely of sludge worms (*Tubi ficidame*), blood worms (*Chironomous*) and the water louse (*Asellus*). These three species succeed one another in importance as one proceeds downstream from the outfall. A point that should be stressed here is that although it is legitimate in this content to speak of a pollution fauna, none of the animals which belong to it are confined to polluted water. All are indeed, normal inhabitants of river muds, ponds, and they merely happen to be favoured by organic pollution.

3.4 EFFLUENTS

Broadly speaking, effluents fall into six groups:

1. *Inert suspension* of finely divided matter result from many types of mining and quarrying, such as that of China clay and from washing processes, such as those of coal and root crops.
2. *Poisons in solution* occur in waste waters from many industries. They include acids and alkalis, chromium salts from tanning, electroplating, phenols and cyanides from chemical industries and gas works, tar acids, ammonia chiefly from gas work and ice factories; Cu, Pb, Zn from various industries including tool for thermal power stations, mines, insecticides from animal dips and agriculture, detergents and soaps and many others including the future threat of radioactive isotopes. Some of these substances are fairly rapidly precipitated in waters into which they flow, as for example Cu and Pb are precipitated when they run into hard water but others are persistent. A few, such as ammonia are destroyed fairly rapidly by oxidation, while phenols, cyanides, are similarly destroyed but much slowly.
3. *Inorganic reducing agents,* such as sulphides and sulphites occur as constituents of the effluents of several types of industries. These substances use up the oxygen in river water. Important among them are ferrous salts, which are present in many underground waters. These often reach the rivers quite naturally, but large volumes are pumped by mines and can produce serious pollution. Ferrous salts are usually acidic in nature, but as the acid is neutralised usually by the loss of carbon dioxide to air, the ferrous salts become oxidised, often by bacterial action, and ferric hydroxide is precipitated. This then takes on the role of an inert suspension.
4. *Oil* is washed into rivers from spills on roads and factory floors, from workshops and garages, and often by the thoughtless motorist who drains it to the nearest stream. This last is a particularly reprehensible habit, as a small quantity of oil can cover an enormous area of water, of the order of four acres per gallon.
5. *Organic residues* include the effluents from a great variety of activities including dairies, ensilage, manure heaps, and cattle yards, slaughter houses, sugar mills, textile manufacture, canning plants, laundries, breweries, tanneries, fish meal factories, paper mills and most important of all domestic sewage. These effluents vary a great deal but they have much in common. They all contain complex organic compounds in solution and suspension, often together with toxic substances and various salts. Sewage for instance, unless it is from a very small town, usually contains some gas work effluents and probably some other industrial wastes; the drainage from silage is usually acidic, as is that from sugar factories; paper mill wastes often contain sulphites. Fundamentally, however, the basic property of there effluents is that they contain unstable compounds which are readily oxidised and so use up the dissolved oxygen in water. Some of these compounds are more readily decomposed than others, for example, dairy wastes oxidizes rapidly but wood pulp is comparatively stable.
6. *Hot water:* Vast amounts of water find use for cooling purposes by steam electric power plants and other industries. This water is discharged at a raised temperature, to the rivers directly. Increasing the water temperature, decreases dissolved oxygen in water which is harmful as it alters the chemical, physical and biological characteristics of that system. High temperature decrease the density and viscosity of water, causing an increased settling rate of suspended solids. Evaporation rate is increased and in marine systems, this may lead to localized areas of abnormally high salinity.

A rough rule of thumb has been that the rate of any chemical reaction, including those of respiration and oxidation, approximately doubles with every 10°C increase in temperature. In thermally polluted water, fish need more oxygen due to an increased respiration rate. However, the available oxygen in such water gets decreased. Thus, thermal pollution affects fish in a double barreled fashion.

The addition of heated water to a cooler body of water may accelerate the lowering of DO levels due to density differences between the two. The less dense warm water tends to form a layer on top of the cooler, more dense water. This occurs particularly when the receiving body of cool water lies deeper. The resulting blanket of 'hot' water cannot dissolve as much atmospheric oxygen as the underlying cold water which is denied contact with the atmosphere. Normal biological reduction of the DO level of the atmospherically unreplenished lower layer may give rise to anaerobic conditions.

Other reactions are also influenced. Trout eggs hatch in 165 days when incubated at 1°C. When water temperatures have been 11°C only 32 days are needed and no hatching takes place at water temperatures above 15°C. Such a result can be disastrous to fish populations. If the fish hatch early and find no natural food organisms available, they ofcourse do not survive. The natural food of such hatchlings depends on the food chain originating with plants whose abundance is a function of day length as well as temperature.

Pollution of river water by hot water leads to growth of bacteria and it is common practice to chlorinate the water. The dose of chlorine should be carefully regulated so that none remains as free chlorine when the water returns to the river. This is necessary because chlorine combines with organic matter in water, thiocyanates of gas effluents to form very toxic compounds. Thus, chlorination may produce poisonous effluents even if no free chlorine remains; moreover chlorine kills microorganisms important in certain food chains and may in this way exert a negative influence on fish population.

Various remedial solutions to thermal pollution have been in use. Installation of a cooling tower as part of industry, to remove heat from water before returning it to natural water supply is one way. Two types of towers, wet and dry, have been used. Water is run over baffler in a thin layer in the wet tower. Cool air, entering at the bottom, moves upwards and absorbs heat from the water. In dry towers, air is forced over water-containing pipes by huge fans. Heat get exchanged by convection from the pipes. The use of cooling ponds or lakes has been another alternative. Both of these have borne positive results to the thermal pollution problem.

3.5 OXYGEN IN WATER

Under normal conditions, surface waters are saturated with a gas in relation to its atmospheric concentration and its pressure, and in most cases the water will hold the maximum permissible concentration of gases, *i.e.* 100% saturation. The total dissolved concentration of a gas in a liquid is given by Henry's law. It states that partial pressure of solute in vapour phase (P_j) to the mole fraction of solute (x_j) is solution related by

$$x_j = \frac{1}{K_j} P_j \qquad \text{...(3.1)}$$

where K_j is the proportionality constant. It is used to correlate the data on solubility of gases in a liquid.

The solubility of gases is often expressed as the Bunsen absorption coefficient, α_j, which is the volume of gas measured at 0°C and 1 atm. pressure dissolved by unit volume of solvent, if the partial pressure of gas is one atmosphere

$$\alpha = \frac{V_j^{\circ}(g)}{V(l)} \qquad ...(3.2)$$

but $V_j^{\circ} = n_j^{\circ} RT^{\circ}/p^{\circ}$ while volume of the solvent $V(l) = \frac{nM}{\rho}$ where n is the number of moles of solvent, M its molar mass, ρ the density of liquid. Thus

$$\alpha = \frac{n_j^{\circ} RT^{\circ}/p^{\circ}}{nM/\rho} \qquad ...(3.3)$$

when the partial pressure of gas $P_j = p^{\circ} = 1$ atm, the solubility by Henry's law, x_j, is given by

$$x_j = \frac{n_j^{\circ}}{n + n_j^{\circ}} \approx \frac{n_j^{\circ}}{n} = \frac{1}{K_j} \qquad ...(3.4)$$

Substituting this value of n_j°/n in equation (3.3) gives

$$\alpha_j K_j = \frac{RT^{\circ}/p^{\circ}}{M/\rho} = 0.022414 \text{ m}^3/\text{mol} \frac{\rho}{M} \qquad ...(3.5)$$

which is the relation between the constant K_j and Bunsen absorption coefficient of gases in moles/unit volume of solvent, $\frac{n_j^{\circ}}{n}\frac{M}{\rho}$ is directly proportional to α_j (eq. 3.3). This makes α_j more convenient parameter than K_j for a discussion on solubilities. Some of the values of α for certain gases in water are given in Table 3.1. Note the increase in α with increase in boiling point of gas.

Table 3.1 Bunsen absorption coefficient, α, of some gases in water at 298K

Gas	t_b/°C	α_j	$K_j = 1/\alpha_j$	*Solubilities* moles/vol= $\alpha_j M/\rho$
Hydrogen	−252.8	0.0175	57.14	0.316
Nitrogen	−195.8	0.0143	69.93	0.2582
Oxygen	−182.96	0.0283	35.33	0.5109
Methane	−161.5	0.03	33.3	0.5162
Ethane	−88.3	0.041	24.4	0.7402

Oxygen is the most investigated, because of its importance in the life cycle. The overall concentration of oxygen ranging from 6–12 ppm found in the hydrosphere is not in dispute, but the rate of replacement or replenishment is still very much a unknown quantities. In the long term, over a period of a year, for example, in an uncontaminated system the enhanced capacity of cold water to dissolve oxygen during autumn and winter is balanced by photosynthetic activity and low solubility of oxygen in warmer water during spring and summer. This implies that even though concentration may change, the percentage saturation remains at a relatively constant level. The percentage saturation, given by

$$\frac{100(CO_2)}{(CO_2)'} \qquad ...(3.6)$$

where (CO_2) is the observed concentration and $(CO_2)'$ is the equilibrium concentration at that temperature and salinity of the observed sample is generally found to be between 97–105% for surface water. This tends to increase as the depth increases, but local conditions have a dominating effect on temperature distribution.

Calculated values of oxygen saturation in water at different temperatures are reproduced in Table 3.2. Figure 3.2 is a nomogram for converting O_2 concentration to percentage saturation in fresh-water. In this figure there are two scales for temperature as well as for dissolved oxygen concentration as found experimentally. The A-scale of temperature should be used with the A-scale of dissolved oxygen.

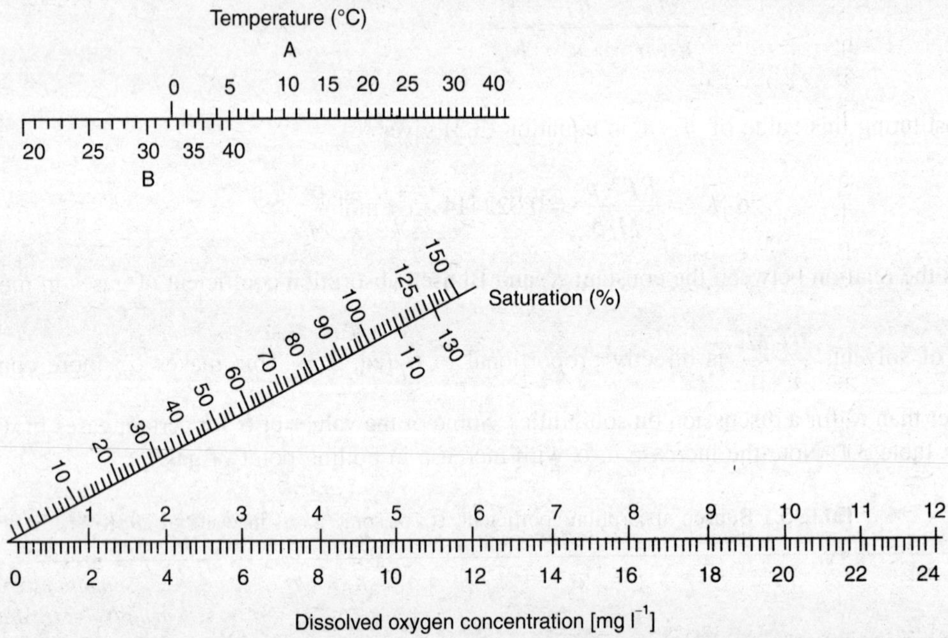

Fig. 3.2 Nomogram for interconverting O_2 concentration and percentage saturation in fresh water.

Temperature t °C	*Dissolved oxygen conc. x, mg l^{-1}*	*Joining t °C with x mg l^{-1} intercept of oblique line*	*Calculated from Table 3.2*
10	9 mg l^{-1}	80%	$\frac{9}{11.28} = 79.8$
20	8 mg l^{-1}	85%	$\frac{8}{9.08} = 88.1$
25	5 mg l^{-1}	60%	$\frac{5}{8.27} = 60.45$

By joining the temperature value with the value of dissolved oxygen, where this line intercepts the oblique line, that reading should be taken as the percentage saturation of oxygen in fresh water. This can also be calculated with the help of Table 3.2 as shown in the previous page.

Oxygen solubility decreases with rise in temperature and salinity, so that cold fresh water can dissolve more oxygen than warm salty water. Since most legislations regarding oxygen level are based on percentage saturation, a standardised set of tables should be formulated and adhered to by all concerned.

As discussed in sec. 1.3–1.5, water has a peculiar property to adopt different structures for different types of solutes. Dissolved gases give it a clathrate structures, ionic compounds impart in hydration structure, where predominant forces are ion-dipole interactions, for protein and sugars get hydrated through dipole-hydrogen bonding interactions. These structures, as discussed in sec. 1.6, have been identified for pure water. However, in rivers, lakes, sea and oceans, the presence of all three types of substances simultaneously in water puts strains on those structures, consequently, the one with lowest energy gets depleted. Hence, addition of any salt or sugar to an oxygen-saturated solution should immediately decrease its solubility. As the enthalpy of clathrate structures may be of the order of thermal energy (= 2.5 kJ deg^{-1} mol^{-1}) with low activation energy (≈10 kJ deg^{-1} mol^{-1}), the rate of oxygen release is quite fast as compared to rate of replenishment of oxygen in water at any temperature. This also explains why the rate of diffusion of dissolved oxygen is not instantaneous and does not take place uniformly throughout the hydrosphere. Even in a relatively small localized area uniform distribution is the exception rather than the rule.

Table 3.2 Solubility or O_2 in water in equilibrium with air at 760 mm Hg pressure and 100% relative humidity (from EAWAG 1973) Units : mg l^{-1}

°C	*0.0*	*0.1*	*0.2*	*0.3*	*0.4*	*0.5*	*0.6*	*0.7*	*0.8*	*0.9*
0	14.60	14.56	14.52	14.48	14.44	14.40	14.36	14.33	14.29	14.25
1	14.21	14.17	14.13	14.09	14.05	14.02	13.98	13.94	13.90	13.87
2	13.83	13.79	13.75	13.72	13.68	13.64	13.61	13.57	13.54	13.50
3	13.46	13.43	13.39	13.36	13.32	13.29	13.25	13.22	13.18	13.15
4	13.11	13.08	13.04	13.01	12.98	12.94	12.91	12.88	12.84	12.81
5	12.78	12.74	12.71	12.68	12.64	12.61	12.58	12.55	12.52	12.48
6	12.45	12.42	12.39	12.36	12.33	12.29	12.26	12.23	12.20	12.17
7	12.14	12.11	12.08	12.05	12.02	11.99	11.96	11.93	11.90	11.87
8	11.84	11.81	11.78	11.76	11.73	11.70	11.67	11.64	11.61	11.58
9	11.56	11.53	11.50	11.47	11.44	11.42	11.39	11.36	11.34	11.31
10	11.28	11.25	11.23	11.20	11.17	11.15	11.12	11.10	11.07	11.04
11	11.02	10.99	10.97	10.94	10.91	10.89	10.86	10.84	10.81	10.79
12	10.76	10.74	10.72	10.69	10.67	10.64	10.62	10.59	10.57	10.55
13	10.52	10.50	10.47	10.45	10.43	10.40	10.38	10.36	10.34	10.31
14	10.29	10.27	10.24	10.22	10.20	10.18	10.15	10.13	10.11	10.09
15	10.07	10.04	10.02	10.00	9.98	9.96	9.94	9.92	9.89	9.87

(Contd.)

16	9.85	9.83	9.81	9.79	9.77	9.75	9.73	9.71	9.69	9.67
17	9.65	9.63	9.61	9.59	9.57	9.55	9.53	9.51	9.49	9.47
18	9.45	9.43	9.41	9.39	9.37	9.36	9.3	9.32	9.30	9.28
19	9.26	9.24	9.23	9.31	9.19	9.17	9.15	9.13	9.12	9.10
20	9.08	9.06	9.05	9.03	9.01	8.99	8.98	8.96	8.94	8.92
21	8.91	8.89	8.87	8.86	8.84	8.82	8.81	8.79	8.77	8.76
22	8.74	8.72	8.71	8.69	8.67	8.66	8.64	8.63	8.61	8.59
23	8.58	8.56	8.55	8.53	8.51	8.50	8.48	8.47	8.45	8.44
24	8.42	8.40	8.39	8.8	8.36	8.35	8.33	8.32	8.30	8.29
25	8.27	8.26	8.24	8.23	8.21	8.20	8.18	8.17	8.16	8.14
26	8.13	8.11	8.10	8.08	8.07	8.06	8.04	8.03	8.01	8.00
27	7.99	7.97	7.96	7.94	7.93	7.92	7.90	7.89	7.88	7.86
28	7.85	7.84	7.82	7.81	7.80	7.78	7.77	7.76	7.74	7.73
29	7.72	7.70	7.69	7.68	7.66	7.65	7.64	7.63	7.61	7.60
30	7.59	7.57	7.56	7.55	7.54	7.52	7.51	7.50	7.49	7.47

When wastes with a high BOD are discharged, the dissolved oxygen in the receiving water is consumed by aerobic enzymes at a specific rate and will not be replaced for some finite length of time. If the resulting mixture remains on the surface, the rate of renewal depends largely on the surface film, but if the mixture is denser and sinks, the rates of mixing, diffusion, depend on the difference in temperature, salinity, density and dissolved oxygen. Whenever consumption of oxygen exceeds renewal, a steadily declining state is achieved which has a marked effect on various organisms.

When the dissolved oxygen concentration falls as much as about 2 ppm, various anaerobic bacteria will begin obtaining their oxygen requirement by reducing the nitrate (NO_3^-) ion or sulphate (SO_4^{2-}) ion, and the water body may be considered as anoxic. Organic matter utilise nitrates as an oxidant first, (denitrification). When all the nitrates and nitrites are consumed, sulphate reduction by bacteria can occur. If organic matter of biosphere is having the composition $(CH_2O)_{106}$ $(NH_3)_{16}H_3PO_4$, these three type of oxidative metabolism can be represented by

$$(CH_2O)_{106}(NH_3)_{16}\,H_3PO_4 + 138O_2 \longrightarrow 106CO_2 + 122H_2O + 16HNO_3 + H_3PO_4 \qquad ...(3.7)$$

$$(CH_2O)_{106}(NH_3)_{16}\,H_3PO_4 + 84.8\ HNO_3 \longrightarrow 106\ CO_2 + 148.4\ H_2O + 42.4N_2 + 16NH_3 + H_3PO_4 \qquad ...(3.8)$$

$$(CH_2O)_{106}(NH_3)_{16}\,H_3PO_4 + 53SO_4^{-2} \longrightarrow 106\ CO_2 + 106H_2O + 16\ NH_3 + 53S^{2-} + H_3PO_4 \qquad ...(3.9)$$

Apparently under these circumstances a fundamentally different chemistry is involved in the system other than normally oxygenated water. This induces many changes and effects. Numerous micro- and macroscopic form of life have been identified in even the most depleted water systems but this life cycle is foreign to what is generally accepted as being normal, as in former case sulphur and nitrogen have taken over the role of oxygen. A third condition which is difficult to duplicate, lies intermediate between the two extremes. This anoxic and oxygenated water has a complicated set of equilibria that affects all the constituents present, such as metal and organic compounds on one hand and the biological structure of the system on the other. These changes also depend on the frequency rate and the degree of change and can fluctuate in a very complicated manner. The confusion that is generated in such a water may decrease the extent of life in it.

In cells of organisms, the oxidative system of enzymes and electron carrier is located in the special organelles called mitochondria. These organelles can be regarded as efficient low temperature furnaces where organic molecules are burned with oxygen. Most of the released energy is converted to high energy bonds of ATP. However, in an anoxic situation, equilibria and confusion starts prevailing in mitochondria which stops working and that leads to death of cell and thus of life.

Thus the question of dissolved oxygen concentration, percentage saturation, distribution, diffusion and associated effects are not a simple one and adequate training in the field of water chemistry is an absolute necessity before one can safely assume the responsibility of allowing oxygen consuming discharges.

3.6 CARBON DIOXIDE IN WATER

The carbon dioxide-carbonate system, though complex, is an important system in the hydrosphere. It is involved in the atmosphere-surface water interchange, the chemistry of the water system, the biological structure of organisms, and the deposition of carbonaceous sediments. In the short term, it controls the pH which directly affects some of the chemical equilibria in any local system and the ions can act as ligends for a number of complexing metals. It is also of primary concern in the biological cycle of all organisms with a carbonaceous skeletal structure, where carbon is utilized in the growth, death and decay patterns.

Its distribution is not uniform and depends on the biological activity in any area. Similar to oxygen, the carbon dioxide concentration in surface water is a function of partial pressure of the gas in air. However, the pattern of distribution differs markedly from oxygen. As for example, the surface water in the Pacific ocean is unsaturated, whereas the surface water of Indian ocean near the equator is supersaturated. Thus in the Indian Ocean, dissolved carbon dioxide concentration increases with increase in temperature. The distribution of total carbon tends to increase with depth due to the settling of the decay fragments of dead organisms from the biologically rich surface.

The solubility of carbon dioxide in various classes of water systems and the thermodynamic equilibria involved in its reaction are not well understood. As carbon dioxide reacts with water, so that its total solubility does not obey Henry's law, it makes the system complex. Estimates of uncombined carbon dioxide can be correlated to atmospheric conditions by a consideration of Henry's law. The concentration of gas shows deviation from the predicted course.

In solution, carbon dioxide will form carbonic acid at a low pH.

$$CO_2\,(g) \rightleftharpoons CO_2\,(aq) \qquad ...(3.10)$$

$$CO_2\,(aq) + H_2O \rightleftharpoons H_2CO_3 \qquad ...(3.11)$$

This species CO_2 (aq) (molecular CO_2) associated with water of hydration to form a clatharate structure and H_2CO_3 (aq) although different, are not normally distinguished, so that overall reaction can be written as

$$CO_2(g)+H_2O \rightleftharpoons H_2CO_3^* \qquad ...(3.12)$$

where
$$H_2CO_3^* = CO_2(aq)+H_2CO_3(aq) \qquad ...(3.13)$$

$$K^*_{H_2CO_3} = [H_2CO_3^*]/P_{CO_2} = 2.8\times10^{-2} \qquad ...(3.14)$$

while
$$K_{H_2CO_3} = [H_2CO_3]/P_{CO_2} = 1.58\times10^{-3} \qquad ...(3.15)$$

The ionization equilibria, which involves fast reaction are

$$H_2CO_3^* + H_2O \rightleftharpoons H_3O^+ + HCO_3^- \qquad ...(3.16)$$

$$K_{a_1} = [H_3O^+][HCO_3^-]/[H_2CO_3^*] = 4.45\times10^{-7} \qquad ...(3.17)$$

the ionization constant for H_2CO_3 is about 3.1×10^{-4} but the value in (eq. 3.17) is smaller, because only a part of dissolved CO_2 is in the form of H_2CO_3

$$HCO_3^- + H_2O \rightleftharpoons H_3O^+ + CO_3^{2-} \qquad ...(3.18)$$

$$K_{a_2} = [H_3O^+][CO_3^{2-}]/[HCO_3^-] = 4.69\times10^{-11} \qquad ...(3.19)$$

The variation of concentration of H_2CO_3, HCO_3^- and CO_3^{2-} solution as a function of pH is illustrated in Fig. 3.3 for a fixed total carbonate concentration, $S = 1 \times 10^{-2}$ moles/lt, where

$$S = [H_2CO_3^*] + [HCO_3^-] + [CO_3^{2-}] \qquad ...(3.20)$$

We obtain the following expression from Eq. 3.20

$$[H_2CO_3^*] = \frac{S}{1+K_{a_1}/[H_3O^+] + K_{a_1}K_{a_2}/[H_3O^+]^2} \qquad ...(3.21)$$

$$[H_2CO_3^-] = \frac{S}{1+[H_3O^+]/K_{a_1} + K_{a_2}/[H_3O^+]} \qquad ...(3.22)$$

$$[CO_3^{2-}] = \frac{S}{1+[H_3O^+]/K_{a_2} + [H_3O^+]^2/K_{a_1}K_{a_2}} \qquad ...(3.23)$$

Figure 3.3 is plotted from such a consideration. The various equilibrium constants on which this figure is based are for pure water containing $H_2CO_3^*$ and thus CO_2.

Under normal circumstances the pH of a natural system is found to range between 7.0–8.5 and in this area the main species in solution is bicarbonate ion. In this form the carbon is not easily assimilated by aquatic organisms and fluctuation in its availability are kept at a minimum.

The dissociation constants K_{a_1} and K_{a_2} inversely proportional to the ionic strength of the solution, which controls the concentration of the ions in question. At the same time, however the constants are directly proportional to temperature. This means that the increase in ionic strength shifts various species concentrations to the left in Fig. 3.3. As the temperature increased the curves are shifted to the right, but the total amount of carbon decreases since the solubility of carbon dioxide is expected to decrease in the warmer water. The magnitude of the shift due to both considerations is once again a local phenomenon.

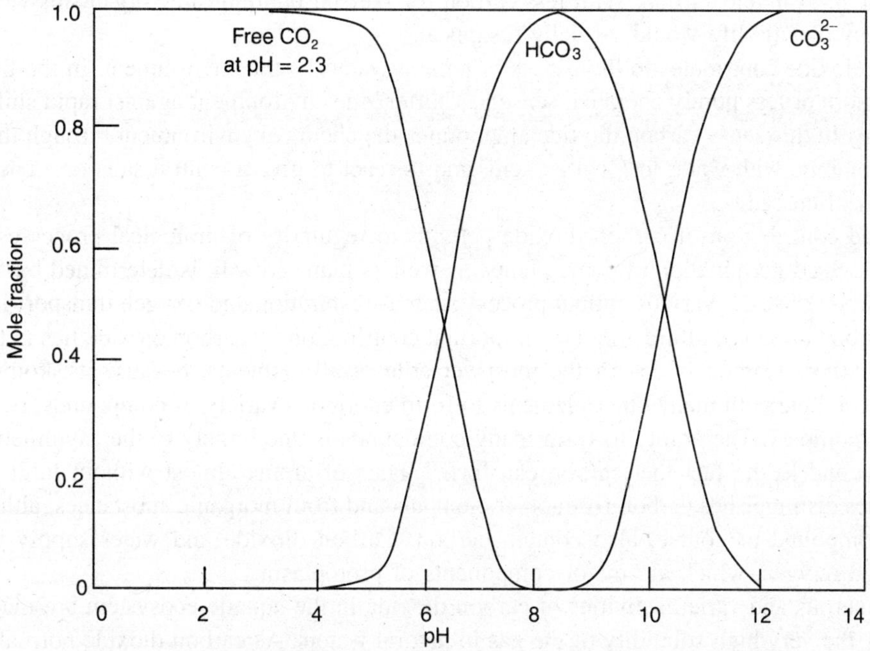

Fig. 3.3 Distribution of H_2CO_3 – HCO_3^- – CO_3^{2-} as a function of pH at 298 K.

The present accepted value for the total concentration of carbon dioxide in the hydrosphere is about 4×10^{-3} M with more than 80% of this in the H_2CO_3 form. A fully saturated solution, however, will sustain a concentration of the order of 5×10^{-2} M, with an increase by a factor of more than 10. So it appears that the increase in atmospheric concentration can be adequately handled by the hydrosphere. The solubility of the gas is dependent on temperature, salinity and pressure, but it is only over surface water, which will have the excess concentration. This mean that absorption of the gas will not be uniform but instead a highly localised phenomena. Under these circumstances, some areas can become saturated or supersaturated relatively quickly while other areas will remain unsaturated.

A further complication is arising because of the green-house effect of carbon dioxide, which means the temperature of the earth is increasing with increase in concentration of carbon dioxide in air. This has been discussed in chapter 6 in Part-I. This should result in a reduced capacity of water to absorb the gas which will eventually further increase the atmospheric concentration of carbon dioxide. However it may be slowed down to some extent if the heating effect melts the polar ice caps. This would mean that more water would become available to absorb the gas but with a small change in surface area, which would increase the solubility of the gas in water till water is warmed up sufficiently to begin losing the gas again which in turn, would obviously revitalize the cycle. However, melting of the polar ice caps will raise the sea level and thus changes in coast line and disappearance of many small islands like the Maldives, Andaman and Nicobar islands.

The reactions induced by these changes would tend to be minimal as the bicarbonate ion (HCO_3^-) concentration would increase, but because of the buffering capacity of water, the pH change would be negligible. In this case, the carbon availability would increase which would result in an increase in the population of various organisms. This means a significant bloom of phytoplankton and nutrients as well as algae would be available in water on a grand scale. On the other hand, if the pH changes it will result in a decrease in available carbon. With less carbon for skeletal building, few organisms would mature and eventually aquatic life would virtually disappear.

Carbon dioxide contributes to the fitness of natural waters as the environment. In the first place, it serves in a more or less purely chemical sense to "buffer" the environment against rapid shift in acidity and alkalinity. In this sense carbon dioxide ameliorates the chemical environment through the ability of the gas to combine with water to form an acid, and to react to give a neutral salts or a base. This has already been discussed.

A second contribution of carbon dioxide pertains to regularity of biological processes in aquatic communities. Seed germination of some plants, as well as plant growth, is determined by the concentration of carbon dioxide. Various animal process such as respiration and oxygen transport in blood are related to carbon dioxide. A third and most important contribution by carbon dioxide lies in the fact that it contains carbon. Carbon is one of the most versatile of all elements, because its atomic structure allows it to combine with many other elements to form enormous variety of compounds, some of them exceedingly complex. The ability to form many compounds is due largely to the asymmetrical nature of the bonds and to the fact that carbon can form chains of atoms almost without limit. This latter characteristics distinguishes carbon from other elements and from inorganic substances, although many inorganic compound of course, also contain carbon. Carbon dioxide and water supply the carbon, hydrogen and oxygen, which are major components of protoplasm.

The numerous and varied activities of carbon dioxide in the aquatic ecosystem are made possible primarily by the very high solubility of the gas in natural waters. As carbon dioxide normally exists in a delicate equilibrium with several other compounds, forming a complex buffer system. The solubility of carbon dioxide varies considerably in nature. Under laboratory condition, however, the solubility of the gas varies inversely with temperature. It can safely be said that at normal natural temperatures, carbon dioxide is much more soluble than oxygen in water. At 20°C and atmosphere pressure of 760 mm Hg, water in equilibrium with atmosphere carbon dioxide contains about 0.88 vol of the gas; only 0.03 vol of oxygen are contained in water under similar condition. Although air contains some 700 times more oxygen than carbon dioxide (by volume), the proportion in water at equilibrium is more nearly equal, about 4 ml of carbon dioxide per litre and 6 ml of oxygen per litre. If we consider the amount of carbon dioxide "located up" in various combined forms, the total amount of the compound in water, particularly in the oceans, is much greater than in air.

Carbon dioxide in natural water is derived from a number of sources. Bacterial decomposition of organic matter in the tropholytic zones, and respiration by animals and plants, contribute to the store of carbon dioxide. In the case of plants, the greater net contribution is at night when photosynthesis is not taking place. Ground waters flowing or seeping into lakes and streams may carry carbon dioxide, the amount being determined by the extent of decomposition in the top soil, and we shall see presently, by the chemical nature of the underlying rocks. Within the body of water, certain chemical reactions between acids and various carbonates release carbon dioxide. In order that carbon dioxide be released from soil and water, both become acidic at night. Finally, the atmosphere directly furnishes some carbon dioxide to natural waters, and rain, as it falls through the atmosphere dissolves some of the gas and delivers it to lakes and other water sources. In order to increase the concentrations of CO_2 and HCO_3^-

in water, the solution has to become alkaline. The system become alkaline at day time when precipitation has to occur by absorbing CO_2 through photosynthesis.

The importance of rain in supplying carbon dioxide to inland water lies in reactions wherein carbon dioxide is maintained and transported in forms other than a gas. As rain percolates through soil containing carbonates, due to its decomposition, some more carbon dioxide gas become dissolved in rain water. The reaction between CO_2 and water results in the formation of carbonic acid (H_2CO_3). If this weak acid encounters carbonate holding rocks, limestone ($CaCO_3$) for example, the latter dissolves as calcium bicarbonate, or $Ca(HCO_3)_2$. The solution of $Ca(HCO_3)_2$ remains stable only in the presence of certain amount of free, or equilibrium carbon dioxide. Free carbon dioxide represents the CO_2 in H_2CO_3 plus that in simple solution. The formation of H_2CO_3 and its dissociations shown in the reactions (eq. 3.12, 3.16 and 3.18) can be put together as

$$CO_2 + H_2O \rightleftharpoons H_2CO_3 \rightleftharpoons H^+ + HCO_3^- \rightleftharpoons 2H^+ + CO_3^{2-}$$

Note that in the above reactions, carbon dioxide is contained in two states: bicarbonate and carbonate radials. This carbon dioxide is called *combined carbon dioxide.* Solution of $CaCO_3$ is dependent on the addition of CO_2 in an amount greater than that of free carbon dioxide. This additional CO_2 is known as *aggressive carbon dioxide.* The major points to be noted from these considerations include (*i*) the relations between rain and soil in supplying compounds containing carbon dioxide in natural water, (*ii*) the chemical reaction by which these compounds are formed, and (*iii*) the occurrence of carbon dioxide in its three forms : free (CO_2 in solution plus that in H_2CO_3), half bound (HCO_3^-), and bound (CO_3^{2-}). The direction of reactions involving the forms of carbon dioxide and the very occurrence of these in natural waters are reciprocally related to the acid base relationships in the medium. When carbon dioxide dissolves, the reaction and end-products depend on the nature of solvent, particularly with relation to the hydrogen ion concentration. Under acid conditions, the combination is as shown above; H_2CO_3 is formed, followed by dissociation to $H^+ + HCO_3^-$. Under highly basic conditions the reaction is

$$\text{Base}(OH^-) + H_2CO_3 \longrightarrow H_2O + \text{base}^+ + HCO_3^-$$

Such equilibria works in mild acidic (pH 7 to 8.5, when HCO_3^- is the major form) and alkaline conditions when nature can withstand it due to the buffering action. However, continuous addition of alkali or acid will not allow the system to become acidic at night or alkaline at day time and consequently germination of plant in soil or water will be hindered. This is a consequence of pollution of water and soil.

Table 3.3 Solubility product of some carbonates

Carbonates	$CaCO_3$	$MgCO_3$	$SrCO_3$	$BaCO_3$	$FeCO_3$	$PbCO_3$
$K_{sp} = [M^{2+}][CO_3^{2-}]$	2.88×10^{-9}	3.46×10^{-8}	1.096×10^{-10}	5.13×10^{-9}	2.089×10^{-11}	1.656×10^{-11}
Conc. of Metal ion at pH 4	4.25×10^{3}	1.73×10^{3}	5.48×10^{2}	2.56×10^{2}	10.45×10^{2}	82.8
Conc. of Metal ion at pH 6	4.48×10^{-2}	7.77×10^{-1}	1.70×10^{-3}	7.97×10^{-2}	3.25×10^{-4}	2.57×10^{-4}
Conc. of Metal ion at pH 8	1.33×10^{-6}	2.0×10^{-5}	6.37×10^{-8}	2.78×10^{-6}	1.21×10^{-8}	9.6×10^{-9}

Equilibria with insoluble carbonates may contribute to CO_3^{2-} control as in the case of $CaCO_3$. Solubility product of some of the carbonates are given in Table 3.3. The maximum concentration of various carbonates which may be present at different pH, assuming that total carbonate concentration $S = [CO_3^{2-}] + [HCO_3^-] + [H_2CO_3^*] \equiv 1 \times 10^{-2}$ M. Apparently, carbonates of Sr^{2+}, Ba^{2+}, Fe^{2+}, Pb^{2+}, get precipitated before calcium or magnesium start precipitating. Thus, these ions do not interfere, with the precipitation of Ca or Mg ions.

As is evident from Fig. 3.3, at pH 8, when the concentration of $[CO_3^{2-}]$ is as low as 1.72×10^{-5} M, if 9×10^{-4} M Ca^{2+} is present, spontaneous precipitation takes place. The precipitate removes CO_3^{2-} ions from the solution and equilibria shift to right-hand side to produce more and more carbonate ion to be removed as precipitate till the concentration of Ca^{2+} does not fall below 1.33×10^{-6} M. Making use of equation 3.23

$$\left[CO_3^{2-}\right] = \frac{S}{1 + \frac{\left[H_3O^+\right]}{K_{a_1}} + \frac{\left[H_3O^+\right]^2}{K_{a_1} K_{a_2}}} \quad ...(3.23)$$

Substituting the values of K_{a_1} and K_{a_2}, we have

$$\left[CO_3^{2-}\right] = \frac{S}{\left[1 + \frac{\left[H_3O^+\right]}{4.45 \times 10^{-5}} + \frac{\left[H_3O^+\right]^2}{20.87 \times 10^{-18}}\right]} \quad ...(3.24)$$

Thus, if $[H_3O^+]$ concentration is known (by pH measurement) and if solubility (eq. 3.20) S be known, concentration of carbonate can be calculated. Using the solubility product value given in Table 3.4, concentration of that metal ion can be estimated.

$$\left[M^{2+}\right] = K_{sp} / \left[CO_3^{2-}\right] \quad ...(3.25)$$

There are a number of complication in the treatment of metal carbonate equilibria, as in calcium carbonate. It has different crystalline forms (calcite and aragonite) but with slightly different solubility constants, the problem of supersaturation, *i.e.* the solubility product concentration may exceed without precipitation taking place.

Another important activity of CO_2 is in the eggs to be laid down by fish. The egg shell is made of $CaCO_3$. If river water becomes acidic this shell will dissolve and hence egg will not fertilize which will lead to depletion of aquatic life. This is a consequence of pollution of water. If $CaCO_3$ precipitation controls the carbonate ion equilibria through the Ca^{2+} concentration of the water, it in turn controls the pH and act as a buffering system for changes in carbon dioxide pressure. There is evidence that supersaturation may be encouraged by organic materials in the oceans, which hinders the nucleation process, that is essential for the initiation of the solid phase. Many soil waters have carbon dioxide concentrations higher than that dissolved is normal water since soil water absorbs carbon dioxide as soon as it is formed from various reactions of microorganisms.

As $CaCO_3$ is more soluble under pressure, but when the water come to the surface, the pressure is released, excess CO_2 leaves the solution, and $CaCO_3$ gets deposited. An important property of water found useful in industrially or in the domestical situation is its alkalinity, which is a measure of water quality. Titration of water with acid to a pH of 8.3 (the phenolphthalein end point say V_1 ml = P) converts all carbonates to bicarbonates and the amount of acid required for this is a measure of carbonate and hydroxy content. Titration with acid to pH 5 (methyl orange end point) (say V_2 ml = T) converts carbonates and hydroxides to salts, the difference between the two titre values ($T - P$) gives volume used for conversion of bicarbonate to CO_2 and is taken to be a measure of bicarbonate content.

Large amounts of carbonate alkalinity and any caustic alkalinity are undesirable since such water will be corrosive and unpalatable. Caustic alkalinity is present whenever the amount of acid required to reach phenolphthalein end point exceeds that needed to go from the former end point to methyl orange. Various reactions that take place are

$$\left.\begin{array}{l} CO_3^{2-} + HCl \longrightarrow HCO_3^- + Cl^- \\ OH^- + HCl \longrightarrow H_2O + Cl^- \end{array}\right] \quad V_1 = P \text{ with phenolphthalein}$$

$$HCO_3^- + HCl \longrightarrow H_2O + CO_2 + Cl^- \qquad V_2 = T \text{ with Methyl orange}$$

Two cases arise. Suppose that a solution contains caustic alkali, so that it will not have any free bicarbonate in it. How to carry out the titration: Take a known volume of water (say 20 ml) and add a drop of phenolphthalein as indicator and add a standard HCl (0.02N) solution drop-wise till the colour of solution changes from pink to colourless. Note the volume of HCl used (say P ml). Take another 20 ml of water and add a drop of methyl orange as indicator. Now add standard HCl (0.02N) drop-wise till the solution change, colour from yellow to orange. Note the volume of acid added (say T ml)

$$\text{Concentration of caustic alkali} = 2P - T$$

$$\text{Concentration of } CO_3^{2-} = 2(T - P)$$

If water contains carbonate and bicarbonate ions, caustic alkali cannot be present. In such cases

$$\text{Concentration of } CO_3^{2-} = 2P$$

$$\text{Concentration of } HCO_3^- = (T - 2P)$$

Thus, knowing the concentration of carbonate and bicarbonate ion in water S (from Eq. 3.20) can be found out.

Table 3.4 Calculation of alkalinity relationship

Results of titration	*OH^- alkalinity as alkali*	*CO_3^{2-} alkalinity as $CaCO_3$*	*HCO_3^- alkalinity as $CaCO_3$*
P	0	0	T
$P < \frac{1}{2}T$	0	$2P$	$T - 2P$
$P = \frac{1}{2}T$	0	$2P$	0
$P > \frac{1}{2}T$	$2P - T$	$2(T - P)$	0
$P = T$	T	0	0

For the purpose of calculation of alkalinity relationship, Table 3.4 can be used where *P* and *T* stands for the litre value of water in the presence of phenolphthalein and methyl orange as indicators.

3.7 NUTRIENTS

The term nutrient covers a number of elements that are essential for the proper health and growth of all organisms. Some of these are required only in trace amounts but if any fraction is constrained or absent, the growth pattern of the subject organisms are virtually affected. The most common elements utilized on the small scale are transition metals such as Cu, Fe, Co and Mn and are referred as trace metals. The better known nutrients, on the other hand, are compounds obtained from the chemical reactions of nitrogen, phosphorus, oxygen, carbon and sulphur.

In most cases, the element is involved in a nutrient cycle throughout the biosphere, including the atmosphere, the hydrosphere and the lithosphere. In other cases, such as phosphorus, the cycle is short-circuited and the element is restricted to the hydrosphere and lithosphere. In all cases, however, the hydrosphere is never removed from the circuit and the necessary elements are normally present there in varying concentrations.

Fluctuations in these concentration and distribution of the nutrients both depend on a number of variables, the most important of which is the biological activity of the area, and can show both seasonal and other effects. These short-term variations are normally restricted to surface water, where the penetration of sunlight is sufficient for photosynthesis. This implies that the primary consumption of the major nutrients is by plants which form the fundamental or lowest step or first trophic level of the food chain and are, in fact, referred to as the plant nutrient.

These nutrients are initially present in an inorganic form and are converted by the metabolic actions which obtain energy by the consumption of inorganic compounds to produce organic forms. After the death of an organism, decay involves further metabolic reaction which return the element to the inorganic form, and in this way the nutrient cycle goes on. When consumption of nutrient exceeds supply, the biological activity is at its peak (in summer and spring) and a steady decrease in concentration occurs. In this way, the biota is restricted by the available food supply. Similarly, in autumn and winter, when activity is decreased, the nutrient concentration will build up and be available for the following spring and summer.

Similar to land forms, the plants of the hydrosphere can be induced to grow more prolifically by the use of extra nutrients in the form of fertilizers. The use of excess concentration of one or both of these species will lead to an increased production of phytoplankton with accompanying consumption of other essential nutrients such as the transition metals. Under these circumstances the trace nutrients are exhausted first and may not be resupplied as readily as nitrates and phosphates, so that the area in question may become less able to sustain peak growth. If all nutrients are present in sufficient quantities, a condition similar to a population explosion can occur that is referred as a plant, or algae bloom, and can have very serious consequences. When algae are present in sufficient amounts, all the oxygen will be consumed for the decomposition of algae. It is said to induce an neutrophic effect, an effect that accelerates the life cycle of the water body in question. This latter process of enrichment and aging in system is called eutrophication.

There is still much to be learned about elements and nutrients for healthy biological growth. Some nitrogen, phosphorus and sulphur are termed as non-specific elements as all organisms require them for the production of protein while others tend to be more specific, such as silica, for growth of diatom, copper requirement of oysters or magnesium as an integral part of chlorophyll. Most of the specific elements are absorbed by specific organisms in a particular form only.

It is not the intention in this chapter to investigate any further the role of metal as nutrients but rather to take a better look at the role of the two specific elements; nitrogen and phosphorus.

Nitrogen

In one form or another, nitrogen is found throughout the biosphere. In its gaseous form, nitrogen, which is inert, forms 79% of the atmosphere and a major reservoir for the total cycle. In the hydrosphere, its total chemical forms have a concentration of approximately 5×10^5 M. This latter value is generally an accepted average, but it must be remembered that it is subject to local variations to a marked degree. A complete discussion of the Nitrogen cycle will be given in Chapter 5. Here, emphasis is on the fate of nitrogen-nutrients in the hydrosphere.

The aqueous redox chemistry of nitrogen involves primarily NO_2^-, NO_3^- and NH_4^+ and take part in oxidation-reduction processes as follows:

$$NO_3^- + 2H^+ + 2e \longrightarrow NO_2^- + H_2O \quad ...(3.26)$$

$$NO_2^- + 8H^+ + 6e \longrightarrow NH_4^+ + 2H_2O \quad ...(3.27)$$

$$NO_3^- + 10H^+ + 8e \longrightarrow NH_4^+ + 3H_2O \quad ...(3.28)$$

These reactions depend on the pH value and the equilibrium composition of nitrogenous system depends on this as well as on the pE of the system. Nitrate has a corporatively narrow range of *pE* (or E_h) over which it can exist at significant concentrations; the most common form being NH_4^+ or ammonia. At neutral pH ranges the NO_2^- stability region lies in the vicinity of $pE = 6.5$.

Before the nitrogen can be utilised it has to be fixed by a biological system, which in the hydrosphere appears to be done by the blue-green algae. The reactions involved are

$$N_2 \longrightarrow 2N\,(\text{fixation}) \quad ...(3.29)$$

followed by hydrogenation

$$2N + 3H_2 \longrightarrow 2NH_3 \quad ...(3.30)$$

The NH_4^+ in water is subject to nitrification reaction in two steps. The reactions in both steps are oxidation and are carried out by two species of autotropic bacteria Nitrosomonas and Nitrobacter (Nitrite & Nitrate bacteria)

$$4NH_4^+ + 7O_2 \longrightarrow 4H^+ + 4NO_2^- + 6H_2O \quad ...(3.31)$$

$$2NH_3 + 3O_2 \longrightarrow 2H^+ + 2NO_2^- + 2H_2O \quad \text{Nitrosomonas} \quad ...(3.32)$$

$$2NO_2^- + O_2 \longrightarrow 2NO_3^- \quad \text{Nitrobacter} \quad ...(3.33)$$

so that overall reaction is

$$NH_3 \text{ or } NH_4^+ \longrightarrow 2NO_2^- \longrightarrow NO_3^- \quad ...(3.34)$$

and the end product of the oxidation is a nitrate ion which is the form in which the nitrogen is assimilated to form amino acids and proteins.

On the death of an organism, other type of bacteria can convert the nitrogen in amino acids by a process called ammonification.

$$\underset{\text{Amino Acid}}{NH_2-(CH_2)_x\ COOH} \longrightarrow NH_3 + CO_2 + H_2O \quad ...(3.35)$$

which reintroduces ammonia back into the cycle and helps to restore the balance of nature. To complete the cycle, NO_2^- and NO_3^- ions are converted through bacterial action by denitrification to the gaseous states as either N_2 or N_2O. In this way, under normal circumstances, the total amount of nitrogen fixed equals the total amount returned to the environment as gas.

As can be seen, the major role in the conversion of inorganic nitrogen to a usable form in biological scheme is played by various types of bacteria. These are subject to environmental conditions in much the same way as other organisms and disruptions in these conditions can result in changes in bacterial activity. Landner and Larson[2] have investigated unusually high concentrations of nitrite ion in the bottom water in the Bay of Koping in Sweden and have concluded that there is an over abundance of nitrosomonas bacteria as compared with Nitro bacteria. This has apparently resulted from excess fertilization of the water by runoffs from fertilized land areas, thus stimulating reaction 3.30. But relative inhibition of Nitrobacter and reaction 3.31 by pollution cannot be ignored, since it is known that excess nitrates, cyanates, some heavy metals and other chemical species will inhibit the efficiency of the Nitro bacteria.

It now becomes possible for some environmentally dangerous conditions to arise. It is known that excess concentrations of nitrite ions (NO_2^-) are toxic to many varieties of aquatic organisms, and that excess nitrate ions (NO_3^-) can constitute a human health hazard as well as stimulate excess algae growth. Initially the nitrogen cycle was relatively stable, and the concentrations of all the species of nitrogen compounds were such that denitrifying reaction was capable of returning all the nitrogen fixed into the cycle.

Manufacture of nitrogenous fertilizers from zero ton in 1910 has increased to 100 million tons in 2000. India in 1959 used 23470 tons of urea but in 2000-01 it consumed 12 million tons. This is extremely soluble and so easily goes from soil to ground water reservoir, lakes, rivers and streams, thereby damaging the nitrogen cycle, changing the natural nitrogen cycle has given us benefits. It has increased the lifespan of human beings, thanks to the substantial increase in nitrogen through fertilizers and cultivation-induced biological nitrogen fixation. The flip side: the run away increase in nitrogen in the ecosystem has several serious environmental consequences. The end result is an inability of denitrifying bacterial to keep up with the pace. This should increase nitrates and nitrates concentration producing corresponding deleterious effect. The excess nitrogen also interfere with the trees ability to produce chlorophyll, lower their resistance to frost and cause conifers to loose needles.

Throughout the world, lakes, bays and coastal water systems are major receptacle for large quantities of human and industrial nitrogenous wastes. The effect of these discharges are usually only considered if a noticeable short-term problem arises, such as the eutrophication of a body of water or contamination of drinking water. It is believed that eutrophication of coastal waters is controlled by the nutrient nitrogen concentration. Of equal importance, however, indeed in some cases even more so are the long-term effects engendered by changes in the nitrogen cycle. If the reactions considered earlier are disrupted, the overall result can be quite devastating and exceedingly difficult to alleviate. It is therefore apparent that before too many nitrogenous wastes are allowed into the water system the various rates of fixation, nitrification and denitrification should be comprehensively appreciated.

Ideally, waste nitrogen should be recycled back to soil as fertilizer, where it can be utilised more efficiently than in the water system.

Phosphorus

In conjunction with nitrogen, the presence of phosphorus in a system satisfies the needs for major nutrients. The phosphorus movement is simpler than nitrogen but is relatively restricted. Since most of the phosphorus compounds of concern are solid, the movement extends between land and water, with the water system forming the eventual sink, *i.e.* a one-way flow from rock to water.

The phosphate cycle in the hydrosphere is much the same as the nitrogen cycle, in that it involves microbiological reactions. The inorganic phosphate is converted primarily in the biota to form adenosine triphosphate (ATP) and adenosine diphosphate (ADP) which are generally considered to supply, by their reactions, the energy requirements for all biochemical reactions in the organism. Although proteins do not contain phosphorus, yet no protein can be made without it. Similarly if the phosporylation of sugar is hindered, photosynthesis of carbohydrates from carbon dioxide would be inhibited. A very general reaction scheme for the consumption and utilisation of phosphorus in regard with sugar is

$$\begin{array}{ccc} \text{ATP} & \xrightarrow{\text{glucose}} & \text{glucose phosphate } + \text{ADP} \\ \uparrow & & \downarrow \text{fructose} \\ \text{Inorganic phosphate} & & \text{Inorganic phosphate + sucrose} \end{array} \quad ...(3.36)$$

In the life cycle of an organism, phosphorus as inorganic phosphate is returned to the system to close the cycle and initiate further biological activity as conditions permit.

In the hydrosphere the forms of naturally occurring phosphorus compounds are pH dependent, similar to the carbon dioxide system

$$H_2O + H_3PO_4 \underset{}{\overset{K_1}{\rightleftharpoons}} H_3O^+ + H_2PO_4^{2-} \qquad K_1 = 6.73\times10^{-3} \quad ...(3.37)$$

$$H_2O + H_2PO_4^- \overset{K_2}{\rightleftharpoons} H_3O^+ + HPO_4^{2-} \qquad K_2 = 6.152\times10^{-8} \quad ...(3.38)$$

$$H_2O + HPO_4^{2-} \overset{K_3}{\rightleftharpoons} H_3O^+ + PO_4^{3-} \qquad K_3 = 4.365\times10^{-13} \quad ...(3.39)$$

At pH less than 2, the predominant form is H_3PO_4; between 3 and 7 it is $H_2PO_4^-$; between 8 and 11, HPO_4^{2-} and above 13 it is PO_4^{3-}. Thus, at normal pH of water, *i.e.* 6.8, the prominent form is HPO_4^{2-} (about 90%) with $H_2PO_4^-$ (about 10%).

3.8 ORGANIC AND INORGANIC COMPOUNDS

It may be recalled that most pollutants are present in water under normal condition, in concentrations low enough so as to present no problem. It is only when these concentrations increase beyond threshold levels that a problem ensues. Introduction of organics, in solution and as particulate matter, into the water system is a new phenomenon, and for many the necessary degradative processes are either conspicuous by their absence or relatively slow. These generate a wide varieties of species. In a polluted system, the concentrations depend on the type of discharge and on the physical and chemical properties of the receiving water. The variety of species identified are:

1. Insecticides
2. Pesticides
3. Petroleum
4. Detergents

5. Polymers
6. Protein fragments
7. Carbohydrates
8. Carboxylic and Humic acids
9. Phenolics
10. Vitamins

and all of them have a detectable effect. Out of these the first six will be dealt with in details in chapters 8, 5 and 4. These are like foreign bodies and frequently have an effect out of proportion to their concentrations. The species in the latter half of the group constitute necessary requirements in any system and can frequently play the role of limiting nutrients.

Pesticides and industrial organics are usually introduced indirectly through runoffs from agricultural areas and through precipitation from the atmosphere or directly from organic industrial processes. In all these areas, reactions occur between the available species present, but most of the reactions are not well understood. It is known that metal ions are complexed, but the rates of formation, dissociation and transfer between species have yet to be comprehensively determined. How these complexes inhibit the precipitation of various phosphates and carbonates is also not known. This also include the degradation and oxidation rates of various compounds that have to be worked out.

Chemical fertilizers and potent pesticides came into use in a bid to expand food production while the traditional practice of using human waste as fertilizers declined. The upshot of developments like these resulted in a steady deterioration of the ecosystem. Such uncontrolled use of water led to pollution diseases in Japan and India, that gained wide attention.

1. **The Minamata Disease:** Minamata, a city in the southern island of Kyushu, in Japan, received reports on an outbreak of a disease that attacks the central nervous system. The symptoms were sickness, madness, paralysis, loss of speech and vision, loss of emotional controls, crippling of arms and legs and finally death, all brought on because of damage to the central nervous system. Forty years ago 100,000 people died in agony at Minamata. The cause was organic mercury which had entered the water via the factory's effluents and had built up in the fish and shell fish consumed by the local residents. The victims of the disease brought a suit against the company which used to release mercury in the effluent of its acetaldehyde facility. They asked for full compensation and it was settled in their favour in March 1973.
2. **Yokkaichi Asthma:** In the second half of the 1950's, the petrochemical industry began to grow rapidly. In 1959, Yokkaichi's first industrial complex for oil refining petrochemical production and power generation began its operation. The local residents then began complaining of foul odours and irritating fumes, and they also protested that dirt in the air was soiling their laundry. Soon there was a marked increase of asthma in the city which came to be known as "Yokkaichi asthma". In September 1967, after studies by Mie university it became clear that the principal culprit was sulphur dioxide released from smoke stacks, and some of the victims took the responsible companies to court. The case was decided in their favour in July 1972 and six companies agreed to make compensation payments.
3. **Itai-itai disease:** In the late 1950's, strange disease in the Jinzu river basin in Toyarms prefecture was detected. The victims suffered bone deformation and fractures and their bones became so brittle that even a single sneeze could sometimes cause a fracture. This extremely painful affiliation was named the 'itai-itai' (ouch-ouch) disease by the local physicians.
   It was discovered that the disease was caused by cadmium poisoning. The source of the cadmium was the Kamioka Mining Station, a lead and zinc mine located on a stream that flows into the Jinzu River. The cadmium had entered the patients' bodies when they drank water from this river, the cadmium content of these water also contaminated the local farmland. A group

of victims filed a damage suit against the mine authorities and the district court ruled in their favour in June 1971.

4. **Bombay Minamata:** Similar incidents were identified in many countries but *Bombay Minamata* must be mentioned here. In 1972 almost 300 persons were officially listed as its victims. A series of investigations by the Institute of Science, Bombay, revealed that industries on the banks of the Kalu River are creating another pollution disaster. Kalu is a small river, about 50 km long which originates in the Western Ghats. It flows through the industrial suburbs of Ambarnath, Ulhas Nagar and Kalyan in north east Bombay and finally enters into the Arabian sea. It receives effluents from over 150 industrial units.

   The institute investigated a stretch of 10 km of the river water from Ambivale to Titwala. They found that the river, in addition to other pollutants, contained mercury and lead above the permitted level. These elements are insoluble in water and sink to the bottom. Certain bacteria in the river bed convert the insoluble mercury into soluble methyl mercury. The fish get poisoned and the cattle that graze on the plants in shallow water also accumulate it. Methyl mercury content of pyereus plants was the most abundant at Ambivali, was 3 to 110 ppm. There was 5 ppm of mercury in the milk of milch cattle that graze on the pyereus plants. A child drinking one litre of milk everyday consumes 35 ppm of mercury in a week.

   Over a period of several month this can lead to accumulation of over 0.3 mg of mercury which is well above the safe level, thereby leading to Minamata disease.

   The fish that are poisoned by mercury are killed by other toxic wastes. Thus dead fish, dying plants, poisoned live stock, paralysed human being and infirm children make up a deadly cycle in the village of Ambivali.

5. **Fluorosis:** In India 25 million people spread over 150 districts in 15 states are leading a painful, crippled life due to fluorosis. Fluorosis destroys the human body, in which teeth become black and rotten. Parts of body become stiff and disabled. Youth of 25-30 years have to take the support of a walking stick. Fluoride is ingested through water, food, cosmetics and fluoridated tooth paste, and drugs, though the highest intake is through drinking water. Fluorosis affects bones, teeth, our tissues and other organs of the body, leading to death after prolonged illness. It leads to dental decolouration and deformation of bones causing knock-knees, bowlegs and stiffening of the joints, joint pains, stiffness of back, back pain, difficulty in getting up and so on are common complaints.

   In endemic areas, a large percentage of the people suffer from gastro-intestinal complaints, loss of appetite, nausea, vomiting, pain in stomach, constipation and intermittent diarrhoea and flatulence. The expectant and lactating mothers are one of the most vulnerable groups of the community. There is a high incidence of still births/abortions. Besides fluoride is also known to influence aging. Many complain of muscular weakness, pain and rigidity of joints, anaemia and inability to carry on even routine domestic activities. Provision of safe drinking water and creating awareness among people of the dangers of excess fluoride in drinking water have not gained importance in India.

   The maximum level of fluoride which the human body may tolerate is 1.5 ppm which is the permissible WHO limit for drinking water. However, the average Indian diet-compares very poorly with the West, so what applies there may not be true here. Keeping in view the various sources through which fluoride finds entry into the body, 1 ppm for water is considered as the upper limit, but still lower limit will be better.

The states of Andhra Pradesh, Bihar, Delhi, Gujarat, Haryana, Karnataka, Maharashtra, Madhya Pradesh, Orissa, Punjab, Rajasthan, Tamil Nadu, Uttar Pradesh, Jammu & Kashmir and Kerala are endemic for fluorosis through underground water. However this can be easily controlled if reverse osmosis plant for water purification is attached to water taps.

6. **Arsenic Poisoning:** So far arsenic contamination in ground water has been discovered in Gangetic plains and Padma-Brahmaputra basin in India, Taiwan, Antofagasta in Chile, Laguna Region in North Mexico and Monte Quesmado in Northern Argentina.

   Dr. Dipankar Chakroborti, Director, School of Environmental Studies and his team have carried out analysis of drinking water from thousands of tube-wells from six districts of West Bengal. He believes that contamination is due to the over exploitation of ground water for irrigation that has in turn led to a fall in the water table. Now when there is rain, water percolates through soil to such a low level, reacts with pyrites-containing arsinite at a lower level. The water from the West Bengal wells are contaminated with arsenic by more than 30 time the WHO safe level of 0.05 mg/litre. The presence of as little as 100 mg of arsenic in water may cause acute arsenic poisoning and thus accumulation in the body.

3.9 ANALYTICAL EXAMINATION OF SEWAGE

Sewage can be classified according to the following criteria:

1. *Turbidity* expressed in international turbidity units.
2. *Suspended solids* (*SS*) measured in parts per million.
3. *Total dissolved solids* (TDS) measured in parts per million.
4. *Acidity* in terms of hydrogen ion concentration or pH.
5. *Dissolved oxygen concentration in parts per million of O_2:* This is determined by Modified Winklers method. In this method dissolved oxygen determined in the same way as the BOD value, by reacting the oxygen dissolved in the sample with manganous hydroxide and iodide and titrating the iodine liberated with sodium thiosulphate (See below).
6. *The potassium permanganate value* (*PV*), *Tidy value:* This value is found by adding a certain volume of acidified N/100 potassium permanganate solution to a given volume of the sample followed by excess ferrous ammonia sulphate and back titrating with N/100 $KMnO_4$ solution. The permanganate value is given in parts per million of O_2 and indicates the concentration of unstable organic compounds in the sample of sewage.
7. *The biochemical oxygen demand* (*BOD*): These are of two types depending upon environmental temperature. At 27°C, BOD is found out after 3 days and is called $(BOD)_3$ and after 5 days at 20°C it is called $(BOD)_5$. This gives the amount of oxygen used up when the stoppered sample is allowed to stay in the dark at room temperature for 3/5 days. Light must be kept out of the bottle to keep algae from adding oxygen by photosynthesis and the stopper is used to keep air from replenishing DO that has been removed by biodegradation. The oxygen required is determined by iodometric determination $(BOD)_5$.

$$MnSO_4 + 2KOH \longrightarrow Mn(OH)_2 + K_2SO_4$$

$$2Mn(OH)_2 + O_2 \longrightarrow 2H_2MnO_3$$

$$H_2MnO_3 + 2KI + H_2SO_4 \longrightarrow K_2SO_4 + MnSO_4 + 3H_2O + I_2$$

$$I_2 + 2Na_2S_2O_3 \longrightarrow Na_2S_4O_6 + 2NaI$$

Both the sample and a blank are shaken up with air and stored for 5 days. At the end of that time, a mixture of manganese sulphate, potassium iodide and potassium hydroxide are added to both, and the iodine evolved is titrated against standard sodium thiosulphate solution. The BOD value, which is also quoted in parts per million of O_2, gives a measure of the amount of material in the sewage which is readily biodegradable.

8. *The chemical oxygen demand (COD):* It is the total oxygen required to oxidize the entire organic matter, chemically. In this determination the sewage is reacted with strong boiling potassium dichromate in the acid state. The sample is boiled for 2 hours. After the reaction has been completed, the excess potassium dichromate is titrated against ferrous ammonium sulphate solution. This value, which is also quoted in terms of ppm of O_2, gives the total amount of organic matter present in the sewage. The reaction can be expressed as

   $$C_nH_mO_x\,(\text{organic matter}) + (n + m/4 - x/2)\,O_2 \longrightarrow nCO_2 + m/2\,H_2O$$

   For any given sample of sewage, the following always applies

   $$\text{COD} > \text{BOD} > \text{PV}$$

9. *The methylene blue value:* This gives the time required for a given sample to start fermenting. It is carried out by adding methylene blue to a sample of sewage and measuring the time in hours required for the blue colour of this dye to disappear.

For each of these parameters scientists specify a range of acceptable values for different activities. As given in Table 3.5, the specified values are however often dependent on economic as well as technical conditions.

The rate of decomposition of organic wastes is assumed to be proportional to the waste available. If L_t represent the oxygen demand left after time t, then assuming a first order reaction, we can write

$$\frac{dL_t}{dt} = -kL_t \qquad ...(3.40)$$

where k = BOD reaction rate constant (time^{-1}). The solution to equation 3.40 is

$$L_t = L_0 e^{-kt} \qquad ...(3.41)$$

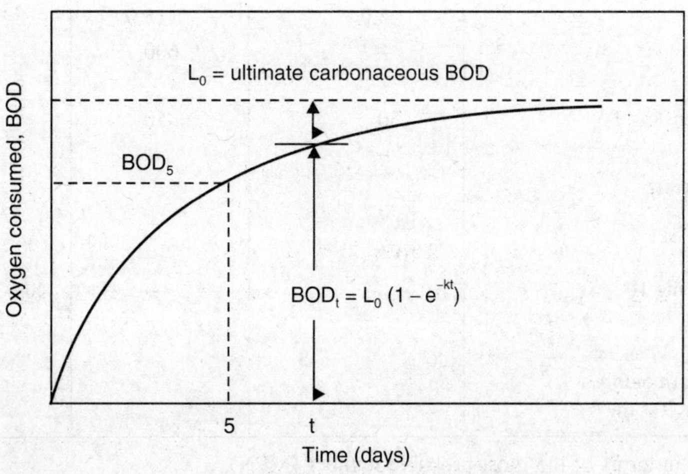

Fig. 3.4 Idealized carbonaceous oxygen demand: the oxygen consumed.

where L_0 = ultimate oxygen demand. It is the total amount of oxygen required by microorganisms to oxidize waste organic matter to carbon dioxide and water. This oxygen demand is sum of the amount of oxygen consumed by the waste in the first '*t*' days (BOD_t), plus the amount of oxygen remaining to be consumed after time *t*. That is

$$L_0 = (\text{BOD})_t + L_t$$

or $$(\text{BOD})_t = L_0 - L_t.$$

or $$(\text{BOD})_t / L_0 = (1 - e^{-kt}) \qquad ...(3.42)$$

A graph of this equation is shown in Fig. 3.4 which shows meaning of various terms.

Table 3.5 Pollution control standards for water

Characteristics	*Industrial effluents*		*Inland surface water*
	Inland surface water	*Public sewers*	
BOD (5 day 20°C), mg l^{-1}	30.0	500.0	3
COD, mg l^{-1}	250	–	–
pH	5.5-9.0	5.5-9.0	6.0-9.0
Total Suspended Solid, mg l^{-1}	100	600	–
Temperature °C	40	45	–
Oil and Grease, mg l^{-1}	10	100	0.1
Phenolic compounds, mg l^{-1}	1	5.0	0.005
Cyanides, mg l^{-1}	0.2	2.0	0.01
Sulphides, mg l^{-1}	2.0	–	–
Fluorides, mg l^{-1} (total residuals)	2.0	–1.5	–
Chlorine, mg l^{-1}	1.0	–	–
Insecticides, mg l^{-1}	zero	–	zero
Arsenic, mg l^{-1}	0.2	–	0.2
Zinc, mg l^{-1}	5.0	15.0	–
Chloride, mg l^{-1}	–	600	600
Sulphate, mg l^{-1}	–	–	1000
Ammonical nitrogen, mg l^{-1}	50	50	–
Nitrates, mg l^{-1}	–	–	00
Radioactive materials			
Alpha emitter uc/ml	10^{-7}	–	10^{-9}
Beta emitter uc/ml	10^{-6}	–	10^{-8}
Dissolved oxygen, mg l^{-1}	–	–	40% of saturation value or 3 mg/l whichever is higher.
Coliform organism (monthly average) -MPN per 100 ml	–	–	under 5000*

*These are measured in terms of the most probable number (MPN).

Source: Deodhar and Nath, *Water pollution and its control in India*, Indian Science Congress, 1981.

This model of a first order reaction for BOD oxidation is an over simplification since the model does not include bacterial cell concentration as a parameter and does not correspond to established microbial growth relationship. However, the model is adequate for most practical purposes. The importance of the concept of BOD in water quality must be emphasized. Drinking water usually has a BOD of less than 1 mg/lt and water is considered fairly pure with a BOD of 3 mg/lt, but when the BOD value reaches 5 mg/lt the water is of doubtful purity. A typical BOD curve carbonaceous oxidation and nitrification phases for waste water is shown in Fig. 3.5.

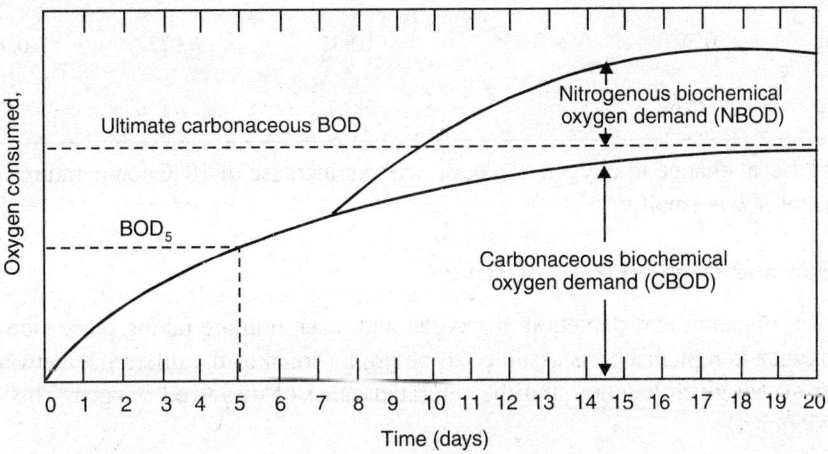

Fig. 3.5 Illustrating the carbonaceous and nitrogenous biochemical oxygen demand.

During the carbonaceous stage, ammonia is produced by the breakdown of organic nitrogen materials but this process is so slow that nitrifiers do not predominate until nearly the end of the carbonaceous stage. Ammonia is oxidized to nitrate by Nitrosomonas and Nitrobacter by reaction 3.32 and 3.33. The oxidation process speeds up toward the end of the first stage and slows down again as the ammonia is oxidized. Ammonia exerts a very high oxygen demand, requiring over $4^1/_2$ times its own weight of oxygen for complete oxidation. Thus if nitrification is allowed to occur in the receiving stream, second stage BOD exerts its presence and a further decrease in the oxygen resource will be experienced.

The reaction rate and thus k depends upon a number of factors including the nature of waste, the ability of the available microorganisms to degrade the wastes and the temperature. The rate of biodegradation increases with increasing temperature. To account for these changes, the reaction rate constant is k, is calculated from modified Arrhenius equation

$$k_t = k_{20}\,\theta^{(T-20)} \qquad ...(3.43)$$

where k_{20} is the reaction rate constant at 20°C, k reaction rate constant at T°C. θ is temperature coefficient for most purposes it has a value of 1.047.

Example: What proportion of total BOD of a waste water would be utilized in five days with k values at 20°C as 0.115, 0.23, 0.345, 0.46, 0.576? Also calculate value of k at 30°C and corresponding BOD five day values.

Solution: t = 5 days

Different k values at 20°C	0.115	0.23	0.345	0.46	0.576
	From equation 3.42 $(BOD)_t/L_0 = 1 - e^{-kt}$				
$(BOD)_t/L_0$	0.437	0.683	0.822	0.90	0.944
	From equation 3.43 $k_t = k_{20}\ \theta^{(T-20)}$				
Corresponding k values at 30°C	0.182	0.364	0.546	0.728	0.912
Corresponding $(BOD)_t/L_0$ at 30°C	0.597	0.838	0.935	0.973	0.989
% increase	36.6	22.8	13.7	8.1	4.7

See the significant change in oxygen depletion with an increase of 10°C temperature. This increase is more significant if k is small.

Deoxygenation and Oxygenation of Rivers

From Fig. 3.1 it is apparent that depletion of oxygen and reaeration are taking place side by side. The rate at which oxygen is replenished is assumed to be proportional to the difference between the actual DO in the river at any given location and the saturated value of dissolved oxygen. This difference is called oxygen deficit D.

$$\text{Rate of aeration} = k_r D \qquad ...(3.44)$$

$$= k_r\,(\text{Saturated dissolved oxygen} - \text{actual dissolved oxygen})$$

$$= k_r\,(DO_s - DO) \qquad ...(3.45)$$

The reaeration constant k_r is much dependent on the particular conditions in the river. A fast moving shallow stream will have a much higher reaeration constant than a sluggish stream or a pond. O'Connor and Dobbin's empirical relation is the most accepted one as

$$k_r = \frac{3.9u^{1/2}}{H^{3/2}} \qquad ...(3.46)$$

where k_r = aeration coefficient

u = average stream velocity

H = average stream depth.

The deoxygenation caused by microbial decomposition of wastes and oxygenation by aeration are competing processes that are simultaneously removing and adding oxygen to a stream. Combining the two relation 3.40 and 3.44 yields the 3.47 expression for the rate of increase of the oxygen deficit.

Rate of increase of deficit = Rate of deoxygenation – Rate of oxygenation

$$\frac{dD}{dt} = kL_t - k_r D$$

$$= kL_0 e^{-kt} - k_r D \qquad ...(3.47)$$

The solution to this equation is as follow:

$$D = \frac{kL_0}{k_r - k}\left(e^{-kt} - e^{-k_r t}\right) + D_0 e^{-k_r t} \qquad \text{...(3.48)}$$

Equation 3.48 is the classic *Steerer-Phelps oxygen sag equation.* In this equation t represents the time of travel in the stream from the point of discharge to the point in question downstream.

For the special case $k_r = k$, the solution becomes

$$D = \left(k L_0 t + D_0\right)e^{-kt} \qquad \text{...(3.49)}$$

If the stream has constant cross-sectional area, and it is travelling at a speed u, then time and distance downstream are related by

$$x = ut \qquad \text{...(3.50)}$$

and

$$D = \frac{kL_0}{k_r - k}\left(e^{-kx/u} - e^{-k_r x/u}\right) + D_0 e^{-k_r x/u} \qquad \text{...(3.51)}$$

A plot of this dissolved oxygen is given in Fig. 3.6

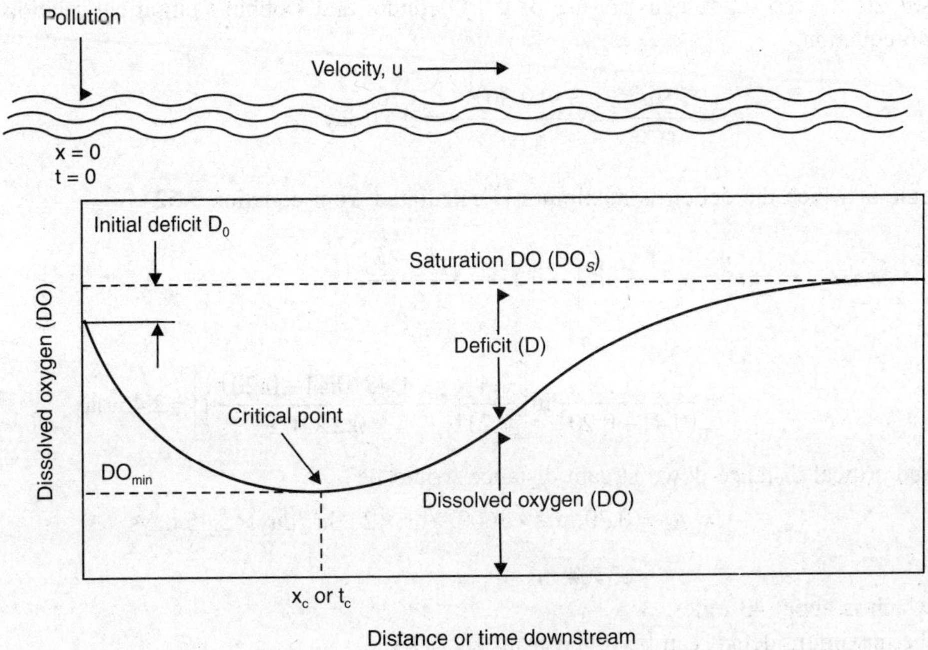

Fig. 3.6 Streeter-Phelps oxygen-sag curve.

As can be seen, there is a stretch of river immediately downstream of the discharge point $x = 0$ where oxygen depletion rate is faster than reaeration causing dissolved oxygen to drop. It reaches minimum at a particular location of time called critical point. At this point

rate of deoxygenation = rate of aeration

Beyond the critical point, reaeration exceeds deoxygenation and stream naturally recovers as shown in Fig. 3.1A.

The location of critical point and the corresponding minimum value of DO are of obvious importance. It is at this point stream condition is worst. Setting the derivative of oxygen deficit equal to zero and solving for critical time yields

$$t_c = \frac{1}{k_r - k} \ln\left[\frac{k_r}{k}\left\{1 - \frac{D_0(k_r - k)}{k L_0}\right\}\right] \qquad ...(3.52)$$

The maximum deficit can then be found by substituting the value obtained for the critical time into 3.48.

Example 3.1: Just below the point where a continuous discharge of pollution mixes with a river, the BOD is 10.9 mg/lt and DO is 7.6 mg/lt. The river and waste mixture has a temperature of 20°C, deoxygenation constant of 0.20/day, an average speed of 0.30 m/s and an average depth of 3.0 m.

(*a*) Find the time and distance downstream at which the oxygen deficit is maximum.

(*b*) Find the minimum value of DO.

Solution: From Table 3.3 the saturation value of DO at 20°C is 9.08 mg/lt, so the initial deficit is

$$D_0 = 9.08 - 7.6 = 1.48 \text{ mg/L}$$

To estimate the reaeration constant use of the O'connor and Dobbin's empirical relation is made use of 3.46 equation.

$$k_r = \frac{3.9u^{1/2}}{H^{3/2}} = \frac{3.9(0.30)^{1/2}}{(3.0)^{3/2}} = 0.41/\text{day}$$

(*a*) time at which the deficit is maximum is calculated from equation 3.52

$$t_c = \frac{1}{(k_r - k)} \ln\left[\frac{k_r}{k}\left(1 - \frac{D_0(k_r - k)}{k L_0}\right)\right]$$

$$= \frac{1}{(0.41 - 0.20)} \ln\left[\frac{0.41}{0.21}\left(1 - \frac{1.48(0.41 - 0.20)}{0.2 \times 10.9}\right)\right] = 2.45 \text{ day}$$

and critical distance down stream distance would be

$$x_c = ut_c = 0.30 \text{ m/s} \times 3600 \text{ s/hr.} \times 24 \text{ hr./day} \times 2.45 \text{ days}$$

$$= 63504 \text{ m}$$

which is about 43 miles.

(*b*) the maximum deficit can be found from eq. 3.48

$$D = \frac{kL_0}{k_r - k}\left(e^{-kt} - e^{-k_r t}\right) + D_0 e^{-k_r t}$$

$$= \frac{0.20 \times 10.9}{0.41 - 0.20}\left(e^{-0.2 \times 2.45} - e^{-0.41 \times 2.45}\right) + 1.5\, e^{-0.41 + 2.45}$$

$$= 3.11 \text{ mg/lt.}$$

So the minimum value of DO will be the saturation value minus this maximum deficit

$$DO_{min} = (9.08 - 3.11) = 5.97 \text{ mg/lt}.$$

3.10 INDIAN RIVERS

The state of rivers is largely an expression of their terrestrial environments, and the magnitude of human activity. In India all the 14 major rivers and their numerous tributaries have become increasingly polluted. During the past few years, however, attention has been paid to the Ganges, the largest and most important Indian river. Many scientific investigations of the river from the Himalayas to its delta on the Bay of Bengal side were made under the "Ganga Action Plan". This was initiated by the Government of India in Feb., 1985, to restore the river water quality by providing treatment of domestic and industrial wastes along the river course, Sinnarkar *et al.*[4] Krishnamurthy *et al.*[5] These studies and other similar studies on other Indian rivers are focusing on a few physiochemical and biological water quality parameters.

The total available water for use in India is estimated as 1900 billion cubic metres per year. About 86% of this is in the form of surface runoff. A staggering 70% of the available water is polluted with community wastes from human settlements, and accounts for four times as much waste water than in the industrial effluents. Most of these wastes are discharged untreated into the water courses. Two-third of all illnesses in India are related to water-borne diseases as given in Table 3.6.

Oxygen and water are essential for life. Water is used everywhere for drinking, bathing, irrigation, cooling and in the industry. When the quantum of wastes goes up, the amount of oxygen required to consume it by the aerobes and the protozoa increases. But rivers can supply only a limited amount of oxygen. As this oxygen begins to deplete, the aerobes and protozoa begin to die off. Finally the oxygen level falling in the upper layer, the aerobes rise and begin attacking the waste matter. This they do by extracting hydrogen. The hydrogen combines with sulphur from the wastes to produce the fouls smelling hydrogen sulphide gas. The water becomes turbid and sunlight cannot penetrate the surface. The algae in the absence of sunlight begin to die and so do the fish. What remain is a polluted river, smelling and sluggish, devoid of life and a menace for the living. This can be seen happening over and over again.

How are these processes disturbed when other ecological balances enter the picture? How does the building of bridges affect turbidity of water? How does industrial cooling, which raises the temperature of the river, affect the aerobes and protozoa? How trace metals enter the food chain when the sewage water is used for immigration? How do flood prevention measures and hydroelectric projects influence the breeding pattern of fish. On an oxygen-water planet, what all eventually disturbs the delicate balance between the two? We do not know the complete answers to these question.

Scientists have devised various parameters for measuring the pollution and purity of water.

Before Yamuna enters Delhi, 100 ml contain about 7500 coliform organism. But further downstream, at Okhla, after it has received Delhi's waste water, the MPN skyrockets to 24000,000 per 100 ml, according to tests conducted by the Central Board for Prevention and Control of Water Pollution. In Kanpur the MPN value of Ganga water was as high as 10^5 per 100 ml in 1966. Today the situation is only likely to be worse. BOD value is expected to be less than 5 ppm or 3 mg. But municipal wastes regularly exceed the figure even after treatment. Suspended solids (SS) should be less than 100 mg/l but this limit is observed only in its violation. Large-scale pollution of the Damodar River near Asansol results in BOD values as high as 270 mg l^{-1} and SS exceeds a concentration of 100,000 mg l^{-1}. Thus while standards exists, they are not enforced.

Table 3.6 Examples of human infections

A. Animal infections that are of public health importance because they are transmissible to man

1. Tetanus from horses and cattle transmitted by inoculation or contact with animal faeces.
2. Bubonic plague from wild rodents by insect (flea) bite.
3. Anthrax from herbivorous animals by direct contact.
4. Rabies from dogs, hats, etc., by bites.
5. Bovine tuberculosis from cattle through ingestion or air-borne transmission.
6. Jungle yellow fever from monkeys through mosquito bites.
7. Several types of encephalitis from birds and fowl through mosquito bites.
8. Trichinosis from swine through ingestion.

B. Primarily human infections in which the infective agent has a certain period of extrahuman residence before transmission

1. Schistosomiasis ("snail fever") from snails.
2. Urban yellow fever from mosquitoes.
3. Hookworm from soil by skin penetration.
4. Malaria (also a mosquito infection) from mosquitoes.
5. Typhus from lice.

C. Infections that persist or multiply in the external environment and are transmissible from man to man

1. Cholera, typhoid fever, bacillary dysentery, poliomyelitis, and infectious hepatitis from water and food through ingestion.
2. Staphylococcal and streptococcal diseases from food, air, and the proximate environment through contact and inhalation.
3. Smallpox from air, dust, and the proximate environment through inhalation.
4. Coxsackie and ECHO virus diseases from water through ingestion.

Ganga Action Plan (GAP): It is estimated that nearly 1.2 million cubic meters of water are released in the Ganges from about 27 major cities, situated along its banks. At Varanasi it receives the largest amount of biological and chemical filths from millions of people taking bath in its water, every year. Moreover, about 0.06 million cubic meters of untreated sewage are dumped into it every day. It also receives ashes of about 40000 human bodies, about 10,000 incompletely burnt dead human and animal bodies dumped into it at Varanasi. Ganga is polluted at Kanpur by jute, chemical, metal and surgical industries, tanneries, textile mills and domestic sewage as pollutants. Ganges water near Garh Mukteshwar, Meerut is heavily contaminated with nitrates, fluorides and metals.

The main objective of the Ganga Action Plan was to improve the quality of the Ganges by reducing the pollution load and by establishing self-sustaining sewage treatment plant systems. The Plan by December 1991, completed the following important tasks:

(*i*) Renovation of existing trunk sewers in order to prevent the outflow of sewage into the river.
(*ii*) Construction of interceptors, drains and sewers to divert the flow of sewage and other liquid wastes away from the river.
(*iii*) Renovation and installation of sewage treatment plants with special attention on recovery of resources such as methane gas for the generation of power and treated water for the purpose of irrigation and agriculture.

(*iv*) Low cost sanitation schemes in adjoining areas of rivers.

(*v*) Biological conservation measures.

(*vi*) Development of river front and afforestation on the banks of river.

(*vii*) Prevention and control of industrial pollution, in cooperation with regulatory agencies and providing some other schemes.

(*viii*) Regular and continuous monitoring of water quality at several places.

A total of 261 schemes necessary for completion of work under Ganga Action Plan were sanctioned. The state wise schemes were as follows.

S.No.	*Scheme*	*Uttar Pradesh*	*Bihar*	*W. Bengal*
1.	Interceptions and diversions	40	17	31
2.	Sewage treatment plants	13	7	15
3.	Low cost sanitation	14	7	22
4.	Electric crematoria	3	8	17
5.	River front facilities	8	3	24
6.	Other schemes	28	3	1
	Total	106	45	110

For sewage treatment plants, new technologies, in collaboration with the Dutch Government, the Up flow-Anaerobic Sludge Blanket Technology were developed. This technology was in use at Kanpur, Mirzapur and Chhapra. Phase II of GAP had the following main objectives:

(*i*) Pollution abatement schemes in 25 class I towns along the main stream of Ganga which were not considered earlier.

(*ii*) Schemes in Class II and Class III towns along main stream of the Ganges.

(*iii*) Schemes of pollution abatement in tributaries particularly Yamuna, Gomti and of the Ganges.

As the GAP went into its second phase under the 8th Five-Year Plan, the whip was to be cracked on over 600 polluting industries, sugar units, pesticides manufacturing factories, fertilizer plants, pharmaceutical distilleries, tanneries, paper and pulp plants and refineries were told to set their house in order or face prosecution and closure.

Drawbacks of GAP: A vast stretch of 600 km along the Ganga continue to remain highly polluted as the Ganga Action Plan tried to clean it but ran into obstacles. The Plan identified 27 cities which were dumping millions of tons of untreated sewage into the river. Problems, such as delay in the construction of sewage treatment plant and lack of electricity supply constantly plague the GAP. Meanwhile 120 industries along its banks pump their effluents into it, as a result pollution level in many sections of the river continues to remain high.

To arrest the growing pollution of the Yamuna, the Government of India had launched a Yamuna Action Plan on 5th July 1993. The important components of the plan are:

1. Interception and diversion of waste water.
2. Establishment of sewage treatment plants.
3. Low cost sanitation.
4. Improved wood crematoria.

5. River front facilities.
6. Afforestation along the river and schemes for biological conservation of aquatic species.

The standard of cleanliness set for Yamuna were the same as those for the Ganges. The Yamuna will be cleaned up to have maximum BOD of 3 mg l^{-1}, minimum DO of 5 mg l^{-1} and pH value of 6.5–8.5. After cleaning the Yamuna, the Central Ganga Authority will take up cleaning of Gomti in UP and Damodar in West Bengal. The programme will then move on to the national river action plan.

3.11 INDUSTRIAL MANAGEMENT OF WATER

The state of rivers, lakes in a country is an expression of their environments and magnitude of human activities. In India all the 14 major rivers and their tributaries have become increasingly polluted. Nearly 70% of the water is polluted with community wastes from human settlements. For centuries rivers and lakes have been vegetating itself of the community wastes, a biodegradable organic matter and there would not be any pollution even today. However 30% of industrial effluents make the 70% biodegradable matter to nonbiodegradable and hence all rivers are highly polluted.

The majority of mills and factories utilise large volumes of water for the process and other works, thus resulting in pollution of water, to an extent dependent on the particular industry and the nature of the process. The large water requirements of many industries and the utilisation of water for power generation, and cooling purposes, as a solvent, were indeed some of the chief reason for the establishment of so many factories on the banks of rivers, lakes and the cause of so many pollution problems.

An in-depth study of Industrial Pollution has been published by us in a report "Control Water, Control Industrial Pollution". The report published by Vatavaran[14], an NGO, deals with how to control the industrial pollution in a lucid way.

Water is one of the natural resources that has been consistently overused by our industries. The industries consume it, waste it and pollute it without worrying about consequences. The industrial demand for water has increased steadily in absolute terms, as well as in proportion to agricultural and domestic consumption. The overall demand for water by industries was 6 bkl in 1970, 150 bkl in 1990, and 180 bkl in 2000 AD and is expected to be 270 bkl by 2025 AD. The water discharged by industries is 100 times more polluted than discharges of community waste of human settlements. Such discharged water contain effluents, acids and metal ion which, if not removed at source, create more problems to sewage disposal.

Industries using similar technologies and process consume widely varying quantities of water. Steel plant use 12-21 cubic metre of water per ton of steel. The range for fertilizer plants is 9 to 40 cubic metres per ton of urea, paper manufacturing units consume anything between 230–450 cubic metre of water per ton of paper. The wide variation in specific consumption of water among the 19 paper plants and a fertilizer plants adopting similar technologies effect differences in operational efficiency and use of obsolete technologies. The lowest water consumption figures of our plants are far higher than the international level, showing Indian industry in a poor light. As in the case of steel, manufacturers in the USA, Europe and South East Asia, require as little as 10% of the water consumed by an Indian plant of the same size.

The staking of Indian industry's thrust could arise because of the following reasons:

(*i*) A lack of awareness about conservation of water;
(*ii*) Water priced at an unreasonable low value;
(*iii*) The industrial units not being charged for cleaning up the polluted water they discharge.

A more sustainable water deal in India calls for water prices high enough to induce industries to use it wisely. For such rationality, water rates to industries should not only reflect the real cost but also cover the expenditure of cleaning up and adequately upgrading the water for the entire range of downstream as usage and disposal of sewage. Industries may be charged at a uniform rate for every unit of pollution that they add to the water or differentially according to the various kinds of pollution they cause to the entire environment. Some permutation of the two paths should bring out an even stricter regime.

Any pollution abatement programme should be acceptable even if it impedes economic development. A basic law for water pollution, or rather to environmental pollution, should be the protection of the health of the people. Economic development cannot override the health of its citizens. In fact, the argument that economic development is being impeded suppresses any proper water management policy.

An industrialist may place a product in the market at a lower price at the cost of pollution he creates. Who is going to pay for the cost of pollution. As long as a mineral is not being used it is not a pollutant because it does not disturb our environment. Once it is processed the entire mineral and thus its contents are distributed over the entire environment. It is then that certain parts of it become pollutants, particularly in water. Water may flow through its cycle of river-to-sea and then evaporate to rain again but pollutants get distributed all along the river path and stay to be there which affects animals, their milk, milk product, fruits, and everything that we eat. It is these polluted eatables that we will take and hence will affect our health. Who is going to pay for this? This could be the cause of somebody's death also. What is the meaning of such an economic development that spoils the health of the concerned parties? For whom these developments are being brought? Even industrialist cannot live in isolation and will be the affected party. So, industries should be forced to make water free from pollutants before discharging.

Alternatively there should be a proper water management policy concerning water for industries. The industry that is polluting water or environment should entirely be responsible for it and should pay the price. This is the so-called Polluter Pay Principle. Unless this is enforced, industry may bring pressure from time-to-time to get the kind of interpretation that serves its own ends. If industrialist are made to pay half the price, they will show increased expenditure and obtain a 100% grant from the government and, as a result, it will never be possible to enforce the law. So, industries must bear the entire burden of pollution.

Faced with levies of this kind, industries will look for enhanced water-efficient plants. Japan, Germany, the USA have attained striking gains in industrial water productivity. They made water wastage and consumption more expensive, deliberately, and it worked well. In 1965, Japan's industrial output was worth $21 per cubic metre of water which has increased in 25 years to $77. This lead to water conservation and recycling. In the former West Germany similar strategies between 1975-90 increased industrial output by 44%. In the USA industrial water consumption in 2007 AD is expected to be only one third of what it was in 1977, mainly because of the increased cost of disposing off the waste water imposed by environmental legislation. In India two states of Gujarat and Tamil Nadu introduced water levels and it worked very well. It shows that industries here are perfectly capable of using water more efficiently without affecting production provided laws are enforced. In both these states the government became popular with such strict measures on this account.

Indian industrialists should not be allowed to pollute rivers, lakes and even sea with disdain. Industries must pay for this because water too is an input in its profit; without proper legislation, nature's first resource will be squandered away. But if priced properly, industries and others will treasure water as commodity.

3.12 THE CHEMISTRY OF SEWAGE PURIFICATION

Waste effluents from human communities are of two types: (*i*) domestic sewage is largely composed of metabolic wastes, as well as animal and vegetable wastes from the kitchen, which are almost totally biodegradable, and (*ii*) Industrial wastes may be non-biodegradable but most of them are over poisonous to the organisms which degrade domestic waste and as a consequence, must be treated separately. Sewage from modern large towns and cities must be adequately treated before discharge.

The actual composition of sewage varies greatly. An analysis of typical domestic sewage showed it contained the constituents shown in Table 3.7.

Table 3.7 Approximate analysis of sewage

Constituent	%
Lipids (ether soluble fraction)	30%
Amino acids, starch and pectins	8%
Hemicellulose	3%
Cellulose	4%
Lignin	6%
Protein	25%
Alcohol soluble fraction	3%
Detergents	2.5%
Ash	18.5%

In addition, sewage contained suspended mineral matter, bacteria, viruses and dissolved salts. Sewage treatment can suitably be classified into three distinct types. The three stages of sewage treatment is shown in Fig. 3.7.

(*i*) *Primary treatment:* removes larger suspended particles from sewage and produces a liquid which contains only colloidal and dissolved matter.

(*ii*) *Secondary treatment:* removes the bulk of the organic matter present in the fluid, using biochemical oxidation processes, and provides an effluent which can be discharged into the sea and rivers, making use of natural biochemical processes to complete the purification.

(*iii*) *Tertiary treatment* processes use the effluent from secondary treatment as feed and produce water which is of drinking water standard. Most sewage plants only go upto secondary treatment, but in recent years the demand for tertiary treatment has stepped up as the need for complete recycling of water has increased.

Primary and Secondary Treatment of Sewage: The crude sewage from the sewers is first stored for about one week before treatment. Normal sterilization treatment with chlorine, ozone or similar substances is extremely effective in killing off bacteria but has little effect on the much smaller viruses. By allowing bacteria to multiply, the viruses are killed off. Bacteria can then be separated, by filtration system and sterilization without any difficulty.

The crude sewage next passes through a rough screen that separates off large lumps, disintegrates them and resuspends them in the main sewage flow. Grit and oil removal tanks follow, where floating insoluble materials are taken off, as are heavy stones and grit particles which sink to the bottom. Invariably such tanks are circular in shape and are equipped with top and bottom scraping devices to eliminate sludge and scum. The bottom scraper is connected to a rotating spider network suspended either at the centre of the tank or form a radical bridge above the water surface. In former case, drive is from the centre of tank and in the latter case at the periphery. In the radial bridge assembly there is considerable flexibility in force. The sludge is drawn off at the bottom of the steeply sloping tank by means of telescopic valves by which extraction flow rate can be adjusted and also prevent valve of lockage, fully automatic diaphragm extraction valves are also widely used. The sludge from these primary settling tanks can then be passed on to a sludge digester if sewage methane is required for combustion purposes. Alternatively the sludge can be passed on directly to driers and incinerators. The liquid from the primary settling tank passes on to a biological purification plant.

The typical bacterial purification beds, usually have a diameter of 40 m or more and head loss in the flow of water usually exceeds 250–300 mm of water. Sprinkled joints which cause little friction and are not sensitive to variations in inlet pressure are used. These beds need very large quantities of filtering material, yet the purification efficiency of a single bed seldom exceeds 80%. To achieve better purification, either several bacterial beds in series or recirculation flow are necessary.

The tanks are either rectangular or circular in shape. Depths of water varying approximately between 3.5 m and 7 m are employed and air is supplied by rotary blowers. This air is distributed by mean of nonclog diffusers fitted with flaps or through porous media. The amount of air which passes through is automatically adjusted on the basis of the concentration of impurities in the effluent. This is done by means of a redox potential regulator or a dissolved oxygen probe. In general, the treatment time varies between 1.5 and 3 hours, which is roughly half the time usually required in the more traditional types of bacterial beds. Excess sludge is removed by means of telescope or automatic valve and water is returned to the primary settling tank. Water, which is drawn from the biological purification plant is commonly considered clean enough to be pumped into the rivers. If further purification is needed—as for example, when the water is to be recycled directly into the town's water system—clarification, using such materials as activated carbon, followed by sterilization with chlorine, ozone or ultraviolet radiation is necessary.

Digesting of Sludge: During methane fermentation, sludge loses between 40% and 60% of its volatile matter and its volume is, reduced by about 60%. In order to increase the capacity of a certain sludge digester one of the following technique is commonly employed:

1. The mixture of sludge and effluent is heated, using steam or hot water, in order to increase the rate of fermentation; and
2. The sludge is thoroughly recirculated inside the digester in order to eliminate void space which also improves its homogeneity.

In many cases, multistage digestion systems are used to improve efficiency. In general, when planning digestion plants for the sewage outflow of a city, it is reasonable to allow a digester volume of about 30 litre per head of population served. Thus a total digester plant volume of around 30 million litre is desirable for a city of 1 million inhabitants.

The methane produced during the digestion of sewage sludge is burned primarily for steam production to heat the sludge fed into the digester. In very large plants, sewage methane can also be used for operation of gas engines. These operate on the total energy principle, to produce electricity for work purposes, and any excess heat is used for sludge heating. However, it is generally uneconomical to install gas engines unless the sewage plant serves at least 300,000 inhabitants, for smaller plants the investment costs tends to be far too high.

Much of the digested sludge produced is then dried on special drying beds. Modern practice tends to frown on the use of these for several reasons. Drying beds occupy a large ground area and offensive smells are created which lowers the amenity value of the surrounding district. The sludge, even when dry, is difficult to handle and has an unpleasant appearance. New sewage works therefore prefer a system of thickening followed by filtration.

After emerging from a thickener, the sludge is passed into one or more vacuum filters where most of the water is drawn off. The filtered sludge has a dry matter content of more than 30% and is often pelletized and stored, to be used either directly as an agricultural fertilizer, or mixing with house hold refuse prior to composting.

In general, the digested mixture of primary and secondary sludge is filtered directly. However, in the case of a multistage plant, and where biological treatment is carried out later, filtration may be used at the beginning of the process.

Filtration and Removal of Colloidal Matter: The water which emerges from the second filter contain colloidal particles, insect larva, algae, as well as dangerous amoeboid organisms, all of which cannot be eliminated by standard sterilisation procedures even.

Filtration is carried out by passing the water through porous material. In most cases filtration mediums serve only to remove particles larger than the pores of the medium but also as a matrix for endearing microorganisms which secrete enzymes. These enzymes coagulate most of the colloidal material present in the water and facilitate their removal. Often surface active filter mediums are used. These serve to retain colloidal material in the adsorbed state on the filter medium itself.

The main filter mediums for water are sand and kieselguhr. The advantage of these materials, compared with more static filter mediums like cloth or porous porcelains is that granular filter mediums are much easier to wash. Kieselgur is employed for smaller filters, like those used in the cleaning of swimming pools. For large scale industrial practice, sand and gravel beds are preferred between 600 mm and 1500 mm thick. Once the filtration medium is fully loaded, it is agitated by a mixture of compressed air and water but the pure sand is reclaimed by sedimentation.

Both slow and fast filtration are employed in the purification of water obtained from sewage effluents. Slow filtration is three stage process. The first consists of coarse filters which remove large particles. These operate at a rate of about 1 m/hr. The second consists of prefilters which remove every thing but colloidal particles. The water passes through these at a rate of 500 mm/hr. Finally, there are proper filters, through which water passes at the rate of only 200 mm/hr. The purpose of these is to coagulate colloidal particles susceptible to bacterial action and to remove them, and to act as surface active agents for adsorption of other colloids. An essential feature of this form of filtration is the formation of bacterial membrane, which can take several days after the medium has been washed.

Fast filtration takes place at a rate of up to 20 m/hr. This is incapable of removing colloid particles which have to be coagulated prior to passage of the water into the filter plant.

The filter medium used for fast industrial water filtration consists of 98% SiO_2 , with effective size being between 0.3 mm and 1 mm diameter for standard filters, and 2 mm diameter for prefilters. Many different types of filter plants are in the market, each having its own operational characteristics. Sewage treated by primary and secondary methods has, in the past, been discharged directly into rivers without further treatment and even now in the vast majority of cases. Organic wastes separated in this way most of which is biodegradable to yield methane gas and the solids may be used as organic fertilizer. The biochemistry of this decomposition will be discussed in detail in Section 3.13.

Tertiary Methods of Water Treatment

These are summarised as follows:

1. Activated carbon in the form of either granules or powder
2. Micro-straining;
3. Special selective coagulants;
4. High density aeration processes;
5. Removal of phosphate by precipitation;
6. Removal of nitrogen compounds;
7. Electro-chemical methods; and
8. Sterilisation of effluents.

1. **Activated Carbon:** One of the most efficient agents for removal of dissolved and suspended impurities is activated carbon that can be activated by heating to 930°C, when adsorbed material is driven off from its surfaces.

Carbon surfaces are, in general, more active with respect to aromatic than aliphatic compounds and reveal a greater adsorption for branched chain compound than for straight chain types. Amines, carbonyl group, double bonds, sulphonic groups and nitro groups all tend to make the compounds concerned adsorb more rapidly.

Adsorption on carbon is prevented if the pH of the solution is above 9. An activated carbon plant generally uses continuous withdrawal of the spent carbon charge at the bottom of each column. This is then passed into the regeneration plant where the carbon is heated to 930°C in a atmosphere of steam. The steam desorbs the impurities adsorbed on the carbon surface. The furnace gases carrying these impurities are passed through after-burners a wet scrubber before being released into the atmosphere, some 5% of the carbon, representing about 5 kg/1000 m^3 water treated, has to be discarded. The rest in recycled. The water which emerges from the carbon columns is passed into a chlorinator and is then discharged as quality drinking water.

2. **Micro-straining:** This is a form of simple filtration which uses very finely woven stainless steel wire fabrics drawn to diameters that are barely visible. The wire is woven on special high precession looms to produce these so-called microfabrics. Thin mats of intercepted solids are formed on their surfaces which, together with the fabric support, have relatively high flow ratings at low hydraulic resistance and can intercept solids which are even smaller than the apertures in the woven fabric. The highest grade of such fabrics has 250 apertures per mm^2 and each opening is just 65 μ across. After partial blockage of the apertures by intercepted solids has occurred, such fabrics have been found adequate for removing organisms with a diameter of as little as 7 to 12 μ. Such micro-straining techniques can be used for the classification of potable water and even for filtering finished drinking water because of their efficiency in removing both inorganic particles such as fine dust, metallic scales, etc. and living organisms such as algae, diatoms, amoebae, and some of the larger bacteria. The use of two-stage micro-straining has been found effective in reducing suspended solids at a cost lower than that of competing tertiary purification system.

3. **Coagulants:** The terminal settling velocity of particles in water is given by Stokes law, which can be written as follow:

$$\text{Terminal velocity} = \frac{g d^2 (D_s - D_w)}{18\eta} \text{ms}^{-1}$$

where g is the acceleration due to gravity, about 9.8 m s^{-2}, d the diameter of the particles in metres, D_s the density of the solid particle in kg m^{-3}, D_w, the density of water in kg m^{-3} (about 1000) and η, the viscosity of water.

Let us assume that the density of certain impurities is 1500 kg m^{-3}, and the diameter 10^{-4} m. The viscosity of water 0.001 kg m^{-1} s^{-1} (at 20°C), the terminal velocity of such a particle at 20°C is given as

$$\frac{9.8\times10^{-8}\,(1500-1000)}{18\times0.001}=2.7\times10^{-3}\ \text{ms}^{-1}$$

The particle thus settles at a rate of just 2.7 mm s^{-1} and assuming that the sedimentation bed is 1 metre deep, precipitation is complete within $=\dfrac{1}{2.7\times10^{-3}\times60}=6.17$ min. Applying the same formula one can see that particles with a diameter of 0.1 μ need no less than 102900 hr or around 12 years to precipitate out. Coagulation is the process of co-precipitation of tiny particles of matter with larger pieces of solid matter so that they sink to the bottom much more rapidly. A typical precipitant is aluminium sulphate, which reacts with dissolved calcium bicarbonate as follows:

$$Al_2(SO_4)_3+Ca(HCO_3)_2\longrightarrow Al(OH)_3\ \text{floc}+3CaSO_4$$

The aluminium hydroxide flocculent engulfs colloidal suspended particles and carries them down. The most widely used flocculents are aluminium sulphate, ferric sulphate, ferrous sulphate and sodium aluminate. A mix of potassium permanganate and ferrous sulphate reacts as follows:

$$FeSO_4+KMnO_4+2H_2O\longrightarrow Fe(OH)_3+MnO_2+KHSO_4$$

The ferric hydroxide constitutes the flocculent which precipitates colloidal particles, while excess potassium permanganate is useful for the destruction of organic matter. Small particles, particularly when their diameters are less than 2 μ carry electric charges which keep them suspended. Alum coagulants, when dissolved in water, contain positively charged aluminium ions, which help dissipate this charge. Other polyvalent ions, such as the ferric ions do the same. This enables the particles to coalesce to form larger units which, of course, precipitate out more rapidly than the original colloidal particles.

Metal salts are rather expansive, are at times have the disadvantage as they themselves may form colloidal particles which cause a carryover of fine turbidity. A large variety of powder substances are added to the coagulants for easier settling. These are called coagulant aids, the most important of which are activated silica, bentonite clay, kieselguhr, clay, precipitated calcium carbonate.

4. **High Intensity Aeration Process:** These consist of forced aeration of activated sludge and constitute extremely effective methods of accelerating the reaction process because (1) sufficient circulation of sewage and activated sludge is known to achieve constant and intimate intermixing, and (2) the supply of excess oxygen supports the biological process which are involved.

An example of such a process is the Simplex Cone, made by the Ames Crosta Mills Company. The cone which is driven by a totally enclosed electric motor, rotates at 30–60 rpm, draws liquid through the uptake tube, and discharges it in a heavy torrent over the lip of the cone to strike the liquid in the tank with some force, thereby mixing, enables the biological action to proceed rapidly and readily.

In another technique, air is dissolved in the liquid under pressure. Once the pressure in the unit is released, the dissolved air comes out in fine bubbles that separate solid particles as a sludge blanket which rises to the surface and can be easily separated by straining. A high intensity aeration system can eliminate as much as 2.1 kg m^{-3} of biological oxygen demand per 24 hours.

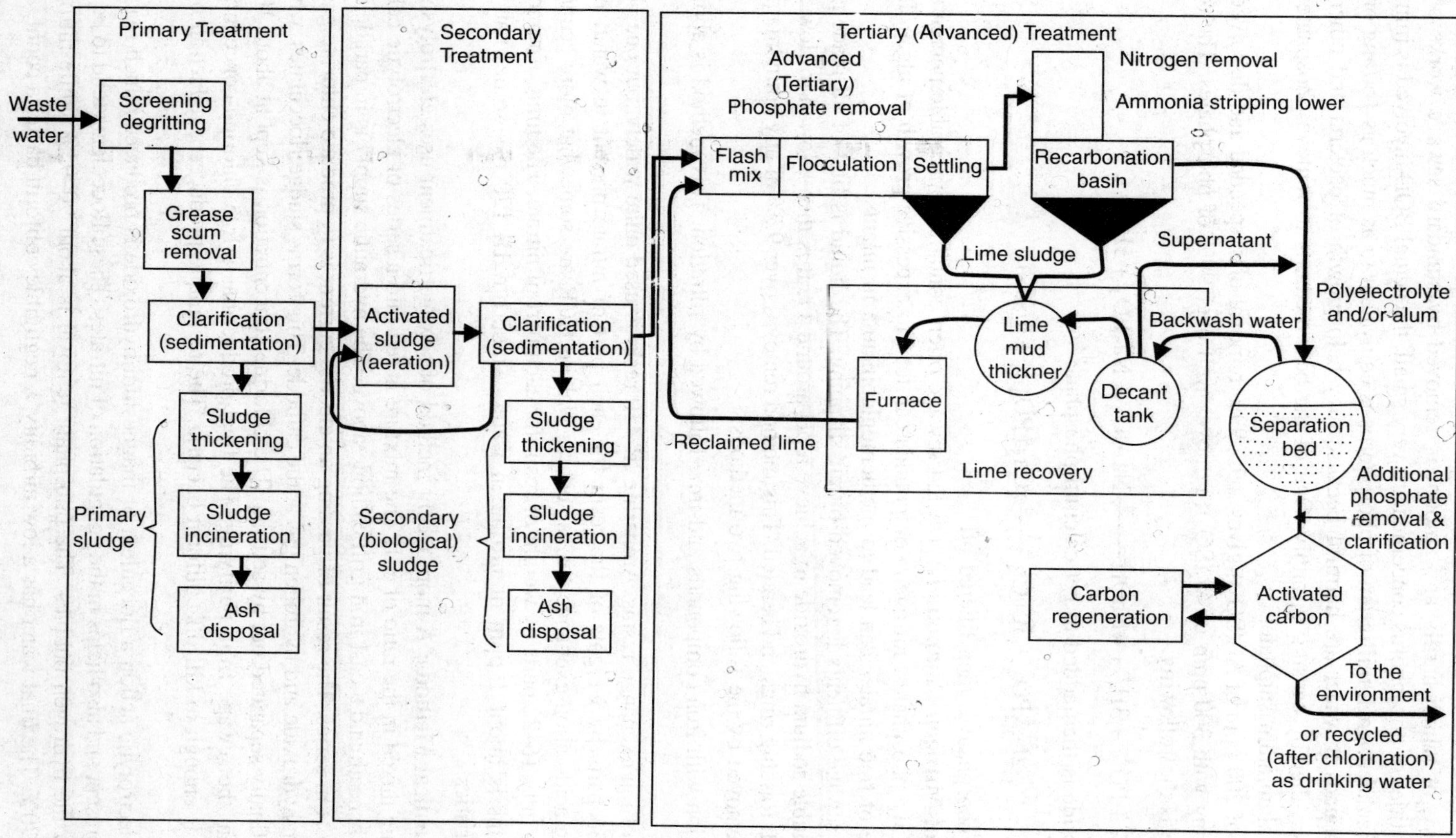

Fig. 3.7 The three stages of treatment for sewage and similar organic wastes. Many metropolitan and small city areas have yet to complete even primary treatment facilities, although most have plans to achieve secondary treatment in the next few years. Tertiary treatment must be added as rapidly as possible or cities will either choke in their own wastes, ruin the quality of the countryside, or both.

5. **Removal of Phosphate by Precipitation:** Sewage contains about 1.5 g to 3.7 g of phosphorus per person per day, of which hardly any amount is removed by standard sewage works practice. Although the quantities of phosphorus concerned are very small, the actual BOD involved is immensely high. It can be calculated: each milligram of phosphorus is equivalent to as much as 160 mg of BOD. Phosphorus compounds in water are harmful because they help growth of bacteria and other living organisms in water, thereby causing turbidity. One of the best methods to remove phosphates is by coagulation with aluminium sulphate.

The addition of 200 ppm of $Al_2(SO_4)_3$ gives a 95–99% removal of suspended and dissolved phosphates, while dosing with 300 ppm of $Al_2(SO_4)_3$ removes well over 99% of phosphates. The reaction which takes place is the following:

$$Al(OH)_3 + PO_4^{3-} + 3H^+ + 4NaOH \longrightarrow NaAlO_2 + Na_3PO_4 + 5H_2O$$

Calcium chloride is then added to get calcium phosphate

$$2Na_3PO_4 + 3CaCl_2 \longrightarrow Ca_3(PO_4)_2 + 6NaCl$$

The calcium phosphate is then filtered off.

6. **Removal of Nitrogen Compounds:** Treated sewage often contains appreciable percentages of bound nitrogen. Like phosphates, nitrogen compounds can lead to excessive growth of algae. In addition, the presence of free nitrates can lead to methemoglobinemia in infants.

Ammonia in sewage effluents is removed by air stripping. The liquid is first of all made alkaline by addition of soluble sodium hydroxide to achieve a pH around 11. It is then allowed to flow down a tower packed with Raschig rings and aerated. This method removes over 92% of any ammonia present. Nitrates can be removed by the following two methods:

(*i*) Coagulation with iron compounds and lime followed by filtration. This method is about 90% effective.

(*ii*) The use of ion exchange resins. Amberlite lR 410 resin is used quite widely, but one needs to use 30–55 parts by weights of the resin for each part of nitrate present in water. Nuclear sulphonic acid-cation exchange resins such as Nalcite HCR are somewhat more economical, requiring only 16–22 parts by weight of resin for each part of nitrate. The limit of removal for both resins is about 1 ppm of nitrate in water. Amberlite 1R-120 has somewhat similar characteristics.

7. **Electrochemical Methods:** A more recent method of sewage treatment uses electrolysis. Sewage and seawar are mixed in the ratio of 3:1. The mixture is fed into a series of electrolytic cells with a graphite anode beneath and a cast iron grill cathode above. The hydraulic supply is in parallel but the electricity supply is in series. The result of passing an electric current is to produce chlorine gas at the anode and sodium hydroxide and hydrogen gas at the cathode. This forms sludge flocculates, which are recirculated and finally separated out by settling. The chlorine concentration is kept at about 100 ppm in order to prevent the sewage from purifying. The final product has a free chlorine concentration of about 5% which is enough to kill off coliform bacteria. The technique produces a reduction of about 60% in BOD.

During the electrolytic action a flocculent of magnesium hydroxide is produced which reacts with any phosphates present and also helps in the coagulation of nitrates. These flocculents tend to be floated to the surface by the hydrogen bubbles. The phosphate removal is about 90–95% while nitrogen removal is around 70%. The final water has a low turbidity, a negligible coliform bacteria count, and can be discharged without any ills effects into the sea close to recreational areas. Power consumption is of the order of 0.4 kWh m^{-3} of sewage treated. The flow sheet of electrochemical method is given in Fig. 3.8.

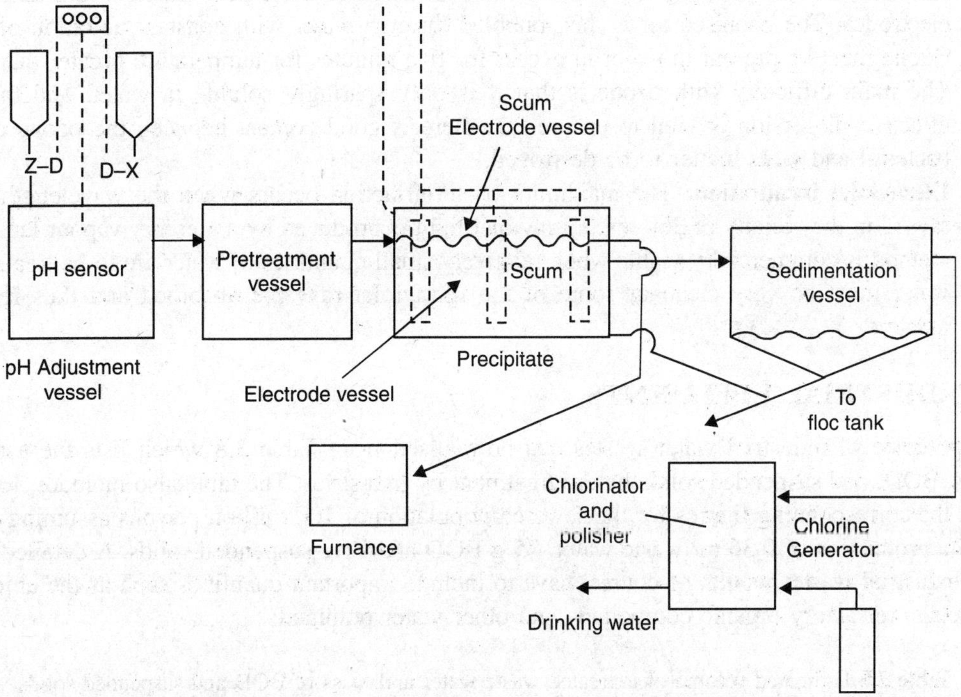

Fig. 3.8 Flow sheet of "Elechem" process for the electrolytical purification of sewage wastes[15].

8. Sterilisation of Effluent

The main methods of sterilising water are as follows

(*i*) Chlorine and its derivative,

(*ii*) Ozone dosing, and

(*iii*) Ultraviolet irradiation

(*i*) **Chlorine and its derivatives:** For the sterilisation of water either chlorine gas or one of the two salts, sodium hypochlorite or calcium hypochlorite, is used. Chlorine in small concentrations effectively kills most bacteria. Enough chlorine is added so that some free chlorine remains in the water after 2 hours of contact. This free chlorine-even in the most minute concentration is readily detected by the use of potassium iodide and a starch indicator. For slow sterilisation, a slightly acid medium is preferred. If the temperature of water is less than 10°C, an excess of chlorine compensate for a slow reaction rate.

For rapid sterilisation, an excess of chlorine is used in water and allowed to destroy bacterial and other harmful organic matter for a period of about 10 minutes. At the end of that time ammonia may be added to the water, which converts excess chlorine into the chloramines; NH_2Cl, $NHCl_2$, and NCl_3. These chloramines are quite odourless and have no specific tastes and they are bactericides in their own way.

(*ii*) **Ozone dosing:** Ozone cannot be stored because it disintegrates spontaneously; it must be produced on site from air. There are several commercial ozonisers in the market designed to operate at about 12,000 V obtained by means of a standard transformer. The air entering the ozoniser must be perfectly dry to avoid the formation of nitric acid which would destroy the electrodes. The ozonised air is then bubbled through water with constant agitation of water. Ozone must be present in water in excess for five minutes for temperature grcater than 10°C. The main difficulty with ozone is that it is only sparingly soluble in water, and therefore effective dispersion is vital to ensure that there is good contact between the ozone and the bacterial and virus matter to be destroyed.

(*iii*) **Ultraviolet irradiation:** The maximum bacterial action occurs when the wavelength of the rays is in the vicinity of 260 nm. Ultraviolet light is produced by a mercury vapour lamp. This method is commercially viable when relatively small quantities of water are to be treated. The water must be very clean or some of the ultraviolet rays are absorbed and thus rendered inactive.

3.13 INDUSTRIAL EFFLUENTS

The importance of industrial water wastes can be realised from Table 3.8 which lists the estimated volumes, BOD, and suspended solids before treatment by industries. The table also includes, for comparison, the corresponding figures for the sewered population of 100 million persons assuming a daily per capita production of 0.36 m^3 waste water, 75 g BOD and 91 g suspended solids. A detailed inventory of industrial wastes would, of course, have to include important quantities such as the amount of toxic metals, refractory organic compounds and other water pollutants.

Table 3.8 Estimated volume of untreated waste water and mass of BOD and suspended solids for industries and sewered population (of 100 million persons)

	Waste water (billion m^3)	*BOD (10^6 kg of O_2)*	*Suspended solid (10^6 kg)*
Primary metals	13.58	183.3	1775
Chemicals	11.67	3667.0	716.5
Paper and allied products	6.0	2225.0	1134.0
Petroleum and coal	4.08	191.5	173.0
Food and kindred products	2.175	1625.0	2500.0
Transportation equipment	0.76	45.0	–
Rubber and plastics	0.51	15.0	19.0
Machinery	0.47	22.5	19.0
Textile mill products	0.44	333.5	–
Electrical machinery	0.28	25.7	8.5
All other manufacturer	1.415	147.5	350
Total manufacturer	41.38	8481	6695
Domestic (sewered)	16.66	2750	3333

In areas with heavy industrial concentration, industrial wastes are many times as great as domestic wastes.

1. **Food and Kindred Products:** Wastes produced by the food processing industry-meat and dairy products, beet sugar refining, brewing and distilling, canning, etc. tend to be troublesome mainly due to their high content of decomposable organic matter, which can result in oxygen depletion and water supply impairment in much the same way as domestic sewage. The meat (cattle, hogs, poultry) processing wastes, come from stock yards, and slaughter houses, packing plants and rendering plants, and contain blood, fats, proteins, feathers and other organic wastes. In many countries these wastes are being dumped into rivers. The dairy industry produces organic wastes high in protein, fat, and lactose from milk and cheese processing whey from the production of cheese is an important BOD source. The beet sugar refining industry produces wastes of high BOD content, including sugar and protein. Breweries and distilleries produce organic solids containing nitrogen and fermented starches from grain processing and alcohol distillation. The processing of food to produce canned or frozen products leads to enormous amounts of wet solid wastes. Food processing often involves some danger of infectious disease.
2. **Textile Products:** Textile mill wastes, generated by cooking the fibres and desizing the fabrics, have high BOD and are quite alkaline, requiring neutralisation and other treatment. The wastes arise from impurities in the fibre and from the chemicals used in processing. The production of 1000 kg of wool leads typically to 1500 kg of impurities (wool fibres, sand, grease, burrs etc.) and 300 to 600 kg of process chemicals with a total of 200 to 250 kg BOD. Cotton processing leads to relatively less total BOD but the waste water may have 200 to 600 mg/ BOD.
3. **Paper and Allied Products:** Paper and pulp mills, in addition to being notorious air polluters, produce a vast amount of water pollution. The effluent, is a complex mixture of the chemicals used in the kraft-process stray-wood chips, bits of bark, cellulose fibres, and dissolved lignin (woody tissue carbohydrates). About 50% of the wood used as input is eventually discarded as waste material. Mill effluent tends to be deep brown in colour and can thus interfere with aquatic photosynthesis. It also contains compounds toxic to fish such as methyl mercaptan, paper and wood pulp preservatives such as penta-chlorophenol and sodium pentachlorophenate. Sulphate liquor is also toxic to shellfish.
4. **Chemical Industry:** A wide variety of water pollutants are produced by chemical plants manufacturing acids, bases, synthetic fibres, pesticides, detergents, and many other compounds, both organic and inorganic. Acid wastes result not only from acid manufacturing plants but also from almost all other chemical plants as well and are customarily neutralised to give a pH of at least 6. Wastes from DDT and rayon manufacturing are especially acidic because a large amount of sulphuric acid is used. The production of the herbicide 2, 4, –D leads to dichlorophenol in water and effluent may contain 25 ppm dichlorophenol and some dioxan despite treatment that removes 95 to 98 per cent of the compound. The waste waters of the phosphate industry contain elemental phosphorus, fluorine, silica and large amounts of suspended solids. In recent years mercury wastes from chemical plants (as well as other types of factories) have caused great concern.
5. **Petroleum Industry:** Oil drilling wastes include drilling muds, salt water, brines pumped out of the well with the crude oil and some oil as well. Oil refineries and petrochemical plants produce an astounding number of different pollutants. These include hydrocarbons, acids, alkalis, cyanides, numerous sodium salts, phenolic compounds numerous, inorganic and organic sulphur compounds and halogenated and nitrogenated hydrocarbons. Many of these compounds

has detectable tastes and odours at concentrations in the ppb range; ethyl mercaptan at 0.00019 mg l^{-1}, isoamyl acetate at 0.0006 mg l^{-1}, hydrogen cyanide at 0.001 mg l^{-1}, etc. Others cause fish flesh to acquire adverse tastes at concentrations of less than 1 ppm. *o*-chlorophenol at 0.015 mg l^{-1}, ethyl benzene at 0.25 mg l^{-1}, kerosene at 0.1 mg l^{-1} etc.

6. **Coal Industry:** Coal mines lead to acid mine drainage, including typically 100 to 6000 ppm sulphuric acid. Coal preparation wastes from coal wasteries contain very large amounts of suspended solids—coal, shale, clay, sandstone, etc.
7. **Rubber and Plastics:** Wastes from rubber production have a high BOD, taste and odour. Synthetic rubber is prepared from butadiene and styrene in a soap solution and coagulated with an acid-brine solution, and the wastes have some of all the materials, used. Odour from rubber plants can give rise to unpalatable water for several hundred kilometres.
   Manufacture of plastics produces wastes with hydrocarbons and other organic compounds, as well as various reagents.
8. **Metal Industries:** Steel mills are producing water wastes from the cooking of coal, the washing of blast furnace flue gases, and the pickling of steel. These wastes are acidic and are having cyanogen, phenol, ore, coke, limestone, alkali, oils mill scale, fine suspended solids. Larger mills may recover by-products but smaller mills do not find this economical and may simply neutralise the acidity with lime, which leads to a large volume of sludge. Other metal industries are facing similar problems, and their wastes generally have the metal being produced or that used such as chromium, lead, nickel, cadmium, zinc, copper, silver, etc. as well as acids, alkaline cleaners, grease and oil.
9. **Other Industries:** Other industries are also producing a wide variety of water pollutants. Leather tannery wastes have high total solids, hardness, salt, sulphides, chromium, alkalinity, lime and BOD. Polishing of optical glass produce wastes containing detergents and suspended solids that do not settle readily. Laundries having turbid wastes are alkaline and have organic solids. Radioactive wastes result from nuclear fuel processing plants, and hospitals and research laboratories using radio isotopes. Soft drink bottling plants are also producing highly alkaline wastes with high BOD from the washing of bottles, which involves removal of cigarette butts, paper and other debris left in the bottles by previous users.

3.14 TREATMENT AND DISPOSAL OF INDUSTRIAL WASTES

The water treatment methods used by industries are generally the same primary, secondary, and tertiary treatment methods which are used by municipalities. Many industries are releasing water wastes of fairly uniform or predictable quality and as the exact pollutants are known, it is often possible to choose the best treatment path for those particular pollutants. Biological treatment has been found to be well suited for wastes from dairy and food industries, as for example, chemical treatments have worked well for wastes which are being released from the metal plating industries. As new methods are invented or old methods improved, an industry may find that it can change to a better treatment.

The vast demand for water by many large industrial users have led to an emphasis on water reuse and recovery of valuable by-products from their waste water. Some of our Industries major effluents have been complied in Table 3.9.

One method of disposal of effluent that is gaining popularity among industries producing concentrated unusable inorganic wastes is deep-well injection. The petroleum drilling industry has made great use of brine injection wells to dispose of oil field brines. A deep well for injection of toxic wastes needs certain geological features that do not exist in every part of the country. There must be impermeable a porous strata of rocks such as shell sedimentary rock stratum, usually sandstone, limestone or dolomite above and below the injection stratum so as to accept large amounts of the liquid wastes.

Table 3.9 Major effluents of some industries

S.No.	*Industries*	*Physico-chemical characteristics*
1.	Dairy products	High BOD ≈ 816 mg l^{-1}, High COD ≈ 1400 mg l^{-1} alkalies, acids, greases, fats, lactose, colloidal solids mainly protein, pH ≈ 8.0
2.	Sugar	High BOD ≈ 650 – 800 mg l^{-1}, COD ≈ 60 – 95 mg l^{-1}, High suspended matter ≈ 1500 – 1800 mg l^{-1} dissolved organic compounds, pH 7.8–8.0.
3.	Fruits, vegetables and yeast	Dissolved and suspended organic matter.
4.	Beer	High dissolved solids containing N-and fermented starches and allied products.
5.	Distilleries	Very High BOD ≈ 29000 mg l^{-1} and COD 65000 mg l^{-1}, low pH ≈ 4.3, high organic matter, high suspended solids 40000 mg l^{-1}, containing nitrogen.
6.	Soft drinks	High BOD, high pH and high suspended solids.
7.	Pickles	High suspended solids, colour and organic matter.
8.	Slaughter house	High suspended solids, dissolved organic matter, protein and blood.
9.	Synthetic mills	Zn, toxic substances, S^{-2} and high pH.
10.	Cotton industries	Alkali pH (8–11). organic matter BOD 200–600 mg l^{-1} dyes, detergents, chromium, fibres, suspended solids 30–50 mg l^{-1}.
11.	Tanneries	High pH ≈ 9.5, suspended matter ≈ 3200 mg l^{-1}, dissolved. organic matter BOD ≈ 7000 mg l^{-1} colour, Ca and Cr salt.
12.	Paper and pulp mills	Suspended solids, high or low pH, colour, fibres, high BOD, COD, dissolved substances.
13.	Antibiotics	Toxic organics, high acidity or alkalinity;
14.	Synthetic drugs	High suspended solids, dissolved organic matter, including, vitamins, high acidity or alkalinity.
15.	Organic chemical industries	Toxic compounds, Phenols, high acidity or alkalinity.
16.	Explosives	Alcohol, metals, TNT and organic acids.
17.	Pesticides	Aromatic compounds, acidity, high organic matter.
18.	Fertilizers	High pH, high NH_3, high F^- content, acidic or alkaline, suspended solids ≈ 37000 mg l^{-1}, BOD ≈ 30 mg l^{-1}, COD ≈ 330 mg l^{-1}.
19.	Rubber industries	Variable pH high BOD, chlorides, suspended and dissolved solids.
20.	Steel mills and iron foundries	Low pH, BOD ≈ 630 mg l^{-1}, acids, phenols, 13000 mg l^{-1}, cyanides, alkali, limestone oil, suspended solids, of coal and clay.
21.	Metal plating	Low pH, metals, toxic cyanides of Cd, Cr, Zn, Cu, Al.
22.	Photography	Organic and inorganic reducing agents, silver salts and products alkalies.

The injection stratum must lie below ground water and have no commercial brine or mineral deposits. The method however may not be a completely satisfactory solution to industrial water pollution for all times to come.

3.15 SOME HAVOC CAUSING CASES OF UNTREATED INDUSTRIAL EFFLUENTS

Some of the industrial pollution that caused havoc in India are given as sample cases.

1. **Bicchri, Rajasthan:** Mr. O.P. Agarwal set up a Jyoti Chemicals and Silver chemicals factory in Bicchri, Rajasthan in 1987 to manufacture H-acid. H-acid (or 1 napthol –8 amino –3, 6 disulphonic acid) is an intermediate in the preparation of azo dye. The effluents of the acid are highly acidic that decomposed soil to sand. The gases it released caused extensive damage to standing crops. Drinking water contaminated with H-acid effluents caused liver damage.
   Till 1978, the processed water from such units in western countries was collected and dumped into sea. This is not allowed any longer that forced the closure of several units there. The highly acidic dark effluents (Photograph 3) were either stored in unlined pits or released through open drains. The stored effluents seeped down into aquifier in four months to cause damage. Till 2001 authorities did not know how to clean this water. The effluents flowed down from Silver Chemicals to pollute 90 wells enroute. The water changed colour to red, apart from the following harmful consequences.
   The delicate root hair associated with plant roots which absorb nutrients are totally destroyed. In short land has turned baren.
   Cattle skins have peeled off after drinking contaminated water. Many cows and buffaloes died; those that survived cannot conceive.
   There is no fodder for cattle to eat because gases have destroyed most of the trees (See photograph 3).
   Sometimes ago, every unit of 10 families used to contribute 150 kg of milk per day, now there is nothing to sell. Fertile land has become barren so no one wants to purchase.
   People have given up farming and live on thrashing stone. No water to drink, ten years after factories were shut down and ground water remain polluted. After creating such pollution havoc, O. P. Agarwal had no remorse and shifted to Vapi, Gujarat to produce H-acid under the name Golden Colour Chem—probably to create the same havoc again.
2. **Medak, Andhra Pradesh:** Medak is an industrial district just off the Hyderabad on the Pune Highway. Here untreated effluents from about 100 industrial units, have been transferred to Nakkavague drain that has poisoned the drain and ground water. It has affected hundreds of acres of the agricultural land. It has made underground water of about 15 hamlet undrinkable. Caused many deaths due to brain humour. Cattles have perished after drinking the water. The crop yield has decreased. Children who bath in this water develop rashes and skin ailment. The quality of effluents is such that they cannot be treated anywhere. At night, tankers are used to carry effluents to the neighbouring state Karnataka which is about 25 km away or to far off places for dumping. The same tankers supply drinking water at day time. Why such a situation arose, no state official has an answer but factories are still releasing effluents.
3. **Gujarat:** Gujarat has more than 90000 industrial units but according to state government about 8000 are polluting units. Major polluting industries are located in Vadodara Petrochemical Complex, Naridesari, Ankleshwar, Vapi, Vatra and Hazira near Surat. In this corridor a number of potential disaster similar to the Bhopal gas tragedy have taken place. All the major river, streams of Gujarat are in a bad state due to effluents discharged by industry be it the Kolak, the Mahi, the Daman Ganga or Amalakhadi.

Take the case of farmers from 11 villages between Lali and Navagram who used to irrigate their fields, untreated effluents released into Khari river with pH = 2, BOD and COD 14 and 16 times the permissible limits. Nearly 100 tube wells and bore wells have now distinct red colour. Paddy yields have gone down to half, milk quality has deteriorated. Water contained 210 times the permissible limits of mercury.

Vapi industrial estate lies in Southern Gujarat. Factories here produce pesticides, agrochemicals, organochloro dyes and dye intermediates, have been dumping untreated effluents in the river. As per a CPCB action plan for Vapi, factories were not allowed to dump their effluents into the river Bhil Bhadi and were asked to set up a common effluent treatment plant (CETP) but even now are releasing untreated effluents into Bhil Bhadi which ultimately meet and pollutes the Kolak river and the sea. Sixty people have died of cancer (Photographs 4 & 5) in past ten years while 20 are fighting a losing battle. Nearly 20 years ago cancer cases were not so rampant.

Blame for such irregularities lies with regulatory agencies—state pollution control board and state development corporation. Vapi CETP was commissioned in Jan. 1997 and has a capacity to handle 55000 cubic metres of effluents per day. There are 625 members of CETP of the 1400 industrial units in Vapi. But CPCB found that the design parameters of CETP were wrong, units were expected to send their effluents to the plant only after primary treatment which they do not do. So whenever effluents reach the inlets with higher concentration or quantity, the officers are forced to let the effluents into Daman Ganga without treatment (Photograph 6).

The 1650 hectare Ankleshwar Industrial Estate (AIE), Bharuch has 1500 industrial units which manufacture dyes, paints, pigments, pharmaceutical, chemicals and pesticides. The water it releases has 100 times permissible mercury level. Effluents have severely contaminated the underground aquifiers (Photograph 7). Villagers of this region are forced to use such water for past six years to cultivate wheat and cotton because they have no option. Water here in past seven years have become so polluted that one cannot wash hands.

3.16 SOME METHODS OF PURIFYING INDUSTRIAL EFFLUENTS

Industrial effluents are far more difficult to purify than domestic wastes because they contain such a wide variety of different pollutants, many of which are harmful to the bacteria, which carry out biodegradation processes in normal sewage works. It is necessary to deal with the purification of industrial effluents separately within the factory complex, and to discharge the effluent only after it has shed any components which would interfere with normal sewage purification operation.

The purification of industrial effluents approximately proceeds along the same paths as normal sewage purification, sedimentation, coagulation, microfiltration techniques, adsorption and electrochemical methods are widely used as are a number of biochemical methods using bacterial organisms immune to specific organic compounds. Certain materials however offer special problems.

Metals: The most toxic metals are cadmium, lead, mercury, beryllium and arsenic, whereas several others are also exceedingly harmful. It is essential that every trace of such metals be removed. The three main techniques employed are precipitation with reagents, cation exchange techniques, and solvent extraction methods.

Precipitation: Precipitation involves the choice of a reagent that can produce an insoluble compound with the noxious metal. The solubility product of the two ions concerned is then of vital importance. Table 3.10 gives the solubility product of a certain number of selected insoluble salts.

In order to precipitate the maximum quantity of noxious metal one adds excess of the anion concerned. For example, if one wished to precipitate as much as possible of the Hg_2^{+2} ion from solution, a large excess of chloride ions would be used. Often the precipitates obtained are colloidal and are therefore difficult to filter. Precipitation combined with coagulation often helps in efficient filtering.

Cation Exchange: Most metal ions can be removed almost totally by the use of cation exchange. Lead, copper, zinc, mercury, chromium, nickel, and arsenic all can be removed from the effluent in this way. Ion-exchange methods used in conjunction with precipitation methods work best particularly when the concentration of the metal ion is low. The concentration of most metal ions in effluents can be reduced by a combination of precipitation and ion exchange to less than 0.001 ppm.

Table 3.10 Solubility product of selected salts

Salts	*Solubility product, K_{sp}*
HgS	2×10^{-49}
Hg_2Cl_2	2×10^{-18}
$Pb(COO)_2$	2.7×10^{-8}
PbS	3.4×10^{-28}
CdS	3.6×10^{-29}
CuS	2×10^{-47}
ZnS	1.2×10^{-23}

Dissolved Salts and Suspended Matter: Industrial effluents often contain dissolved salts which are harmful to the environment. The most frequently employed methods for removal of such salts are ion exchange and electrolysis.

Many organic solutions and suspensions are best treated by the use of strong oxidation methods designed to break down the noxious material into carbon dioxide, water and oxides of sulphur, nitrogen, etc. Even partial oxidation sometimes helps. Complex materials with active groups are separated out by the use of adsorbent such as activated carbon and activated silica.

Electrochemical methods are often used for the removal of suspension such as oils and paints in water. Hydrogen is produced electrolytically and carries such materials upwards to produce a froth which is easy to separate out. If sea water is added to the effluent, electrolysis produces chlorine, which is capable of oxidizing organic matter.

3.17 CHEMISTRY AND BIOCHEMISTRY OF SEWAGE

The major constituents of organic waste are polysaccharide, polypeptides, fats and nucleic acids with approximate composition given in Table 3.7. Most waste treatment methods in use today are biological processes taking advantage of the catalytic and metabolic activities of microorganisms which hydrolyse potential pollutants. The products of these reactions are ultimately converted to low molar mass fatty acid or alcohols from which many key biochemical reactions originate. Figure 3.9 summarises the major reactions involved in the biological waste treatment while detailed examples are presented in the following paragraphs.

Polysaccharides: The carbohydrate content of living tissues is primarily in the form of polysaccharides.

Cellulose: Forms the bulk of the cell wall material in higher plants and, as such, is the most widespread and abundant naturally occurring organic polymer. It is also a minor component of animals fungi and a few bacteria. Cellulose is an insoluble, linear polymer of at least 3000 β – D (1–4) linked glucose residues. Parallel chains of the polymer form crystalline zones and amorphous or paracrystalline zones. Microbial decay of this stable structure is the function of at least three groups of enzymes. The first group of celluloses, (C_1) may attack the cellulose fibres causing a loss of tensile strength and the production of reactive cellulose, C_x celluloses perform the subsequent depolymerisation to low molar mass disaccharides and oligosaccharides. The third group of enzymes catalyses the hydrolysis of these di-and oligosaccharides to glucose and are described as β-glucosidases. Where cellobiose is the substrate, cellulobiase is the enzyme responsible.

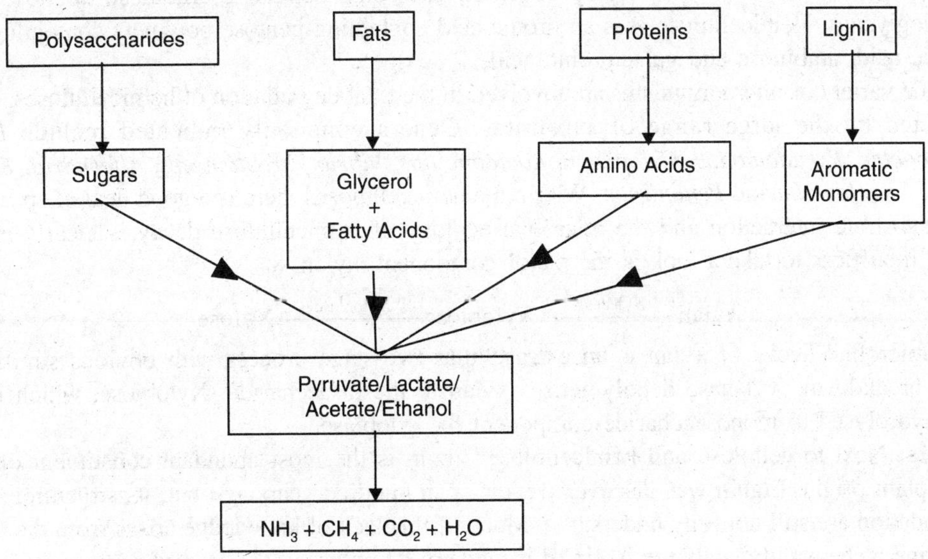

Fig. 3.9 The breakdown of organic matter during sewage treatment.

Enzymatic breakdown of cellulose

$$\text{Cellulose} \xrightarrow{C_1} \text{reactive cellulose} \xrightarrow{C_x} \text{cellulobiase} \xrightarrow{\text{cellobiase}} \beta\text{-glucose}$$

Starch: It is composed of amylose, a linear polymer in which glucose units are linked by the $\alpha - D$ (1 – 4) linkages and branched chain glucose polymer, amylopectin with $\alpha - D$ (1 – 4) as well as $\alpha - D$ (1 – 6) linkages. The hydrolysis of starch is catalysed by two principal enzyme systems α and β amylase. The α- amylase, found in a variety of bacteria and fungi, are believed to cleave the glycosidic linkage in $\alpha - D$ (1 – 4) glucans to produce a mixture of maltose and glucose. This, in itself, may be a two-step reaction with the intermediate production maltotriose. It is possible that an α-glucosidase (maltase) is responsible for subsequent monosaccharide production. β-Amylases perform a step-wise hydrolysis of alternate linkage and are clearly important in amylopectin decay.

Pectin: Pectins are predominantly polymers, of galacturonic acid (I), methylated galacturonic acid (II), arabinose (III), and galactose (IV).

Protopectinases solubilise the protopectin, the parent material is hydrolysed to polygalacturonic acids (pectin) in which high proportion of the carboxylic groups are present as methyl ester. Other enzymes mediate in the fissure of glycosidic linkages, partially depolymerising the substrate to form pectin. These enzymes are variously described as pectin polygalacturanase LPG, pectinase, pectolase, and polygalacturonase; names which may be synonymous (or representative) of a multienzyme system. Esterases, which hydrolyse pectin and pectinic acids by splitting off methyl ester groups are designated pectin methyl esterase (PM or PME), pectase, pectin methoxylase or pectin esterase. Finally, poly galacturonase breaks down the remaining substrate to galacturonic acids.

Hemicellulose: It is an imprecise term for a group of plant polysaccharide materials having no structural relationship to cellulose. Hydrolysis of the resulting material yields a complex mixture of hexose sugars, pentose sugars and frequently uronic acids. Hemicellulose is composed of hemicelluiose A and hemicellulose B. Hemicellulose A is predominantly composed of β – *D* (1– 4) linked xylose. Some molecules are linear, some highly branched and some contain L-arabinose-furanose residues arbinoxyloglycans. Hemicellulose B is an uronic acid containing pentose polymers containing xylose, glucuronic acid, arabinose and galacturonic acid.

A wide variety of microorganisms are involved in the total degradation of hemicelluloses, as would be expected by the large range of substrates. Genera commonly indicated include *Bacillus, Achromobacter; Pseudomonas, Cylophaga, Lactobacillus, Vibrio, Streptomyces, Alternaria, Fusarium, Rhizopus*, *Chaelomium* and *Penicillium.* Within the hemicelluloses there is a good deal of specificity of microbe-substrate interaction and the microbial ecology of hemicellulose decay is clearly extremely complex. It suffices to take a look at the major component, xylan.

$$\text{Xylan} \xrightarrow{\text{Xylanase}} \text{Xylobiose} \xrightarrow{\text{Xylobiase}} \text{Xylose}$$

The microbial decay of xylan is an extracellular two-stage process with obvious similarities to cellulose breakdown. Xylanase depolymerises xylan to the disaccharide, Xylobiose, which is subsequently hydrolysed to monosaccharide component by xylobiase.

Lignin: Next to cellulose and hemicellulose; lignin is the most abundant constituent of organic waste of plant origin. Lignin was discovered more than hundred years ago but, its structure, synthesis and degradation are still not fully understood. Much of this lack of knowledge arises from the difficulty of extracting a chemically unaltered lignin. It is made of a mixture of the aromatic monomer protocatechuic acid (*v*), p-hydroxybenzoic acid (*vi*) vanillic acid (*vii*) vanillin (*viii*) α-ethoxy propion guaiacone, vanilloyl methylketone, etc.

Softwood lignin is made up of β-aryl ester linked α-guaiacyl glycerol-β-coniferyl ester units. The total decay of lignin is clearly the function of several groups of microorganisms-fungi, actinomycetes, bacteria. Much research however has involved the Basidiomycetes and especially the so-called "white rot" fungi. Therefore, biochemistry of lignin degradation by fungi such as Fomes, Armillaria, Polyporus, Pleusotus is frequently (though not satisfactorily) described. Notwithstanding, the decay of lignin in the high carbohydrate-high protein micro-environment found in sewage works must be attributed to bacteria, actinomy and micro-fungi.

The initial steps of lignin breakdown involve the cleavage of ether linkages to depolymerise and solubilise the phenyl-propane macromolecule. Demethylation of the products (initially guaiacyl-glycerol-coniferyl ether) produces water-soluble monomers and dimers suitable for intracellular breakdown, where ring fission can occur. A number of coenzymes are involved in the initial steps with α-phenoloxidase and peroxidase most frequently coming into play.

Lipids: The fats and fatty acids content, which escapes digestion, absorption or deposition contains a variety of fatty substances (triglycerides, phospholipids, cholesterol, waxes and fixed oils). Fats and fatty acids also include acetic acid, propionic acid, butyric acid, lauric acid and a range of branched chain acids.

Lipase enzymes, common in plants and animals, are found in both aerobic and anaerobic micro-organisms. Enzymatic hydrolysis of fats yields a mixture of glycerol and fatty acids.

$$\begin{array}{l} CH_2OCOR_1 \\ | \\ CHOCOR_2 \\ | \\ CH_2OCOR_3 \end{array} \xrightarrow{\text{lipase}} \begin{array}{l} CH_2OH + R_1COOH \\ | \\ CHOH \ \ + R_2COOH \\ | \\ CH_2OH \ + R_3COOH \end{array}$$

Long-chain fatty acids are subsequently metabolised by β-oxidation in which two carbon units, at a time are split from the chain.

Proteins: Proteins are long-chain polymers of 1-amino acids with the general structure NH_2—CH(R)—COOH, where *R* may be a hydrogen atom, a *R* single methyl group, a short carbon chain (with or without functional groups such as SH, NH_2 and COOH) or acyclic structure. The amino acids joined by the peptide bonds (CO—NH) give rise to proteins.

Exo-enzymes hydrolyse the peptide bonds are collectively described as proteases and appear to be of two major types: (*i*) exopeptidases which cleave peptide bonds towards the end of the terminal amino acids. (*ii*) Proteolytic activity produces a mixture of peptones, peptides and amino acids. There is evidence that some proteases are specific for linkage between certain amino acids.

Amino acids serve as carbon and nitrogen sources for numerous heterotrophs and their intercellular deamination and decarboxylation is well documented. Some of them are metabolised in reality (arginine, tryptophan) whilst others are more resistant (lysine, methionine). Proteins and amino acids may also be attacked through fermentation to produce a number of unpleasant smelling amines, mercaptans and hydrogen sulphides.

Methanogenic Bacteria

One of the major groups of micro-organisms engaged in the aerobic digestion of sludge are the methanogenic bacteria. These organisms couple organic oxidation with carbon dioxide reduction to produce methane. The substrates used for methanogenic bacteria fall into three distinct categories.

(*a*) The lower fatty acids containing six or less carbon atom;
(*b*) The normal and iso-alcohols containing from one to five carbon atoms;
(*c*) Three inorganic gases (hydrogen, carbon monoxide and carbon dioxide).

In addition, microbial species may be highly specific for one or more of these substrates (Table 3.11).

Table 3.11 Methanogenic Bacteria species and their carbon substrates

Species	*Substrate*
M. formicicum	CO, CO_2, H_2
M. omelianskii	alcohols.
M. suboxydans	butyrate, valerate, caproate
M. vanielli	formate, H_2
M. harberi	methanol, acetate, formate.

Methanogenic bacteria are intolerant of pH extremes (optimum 6.4–7.2). As for example, if the input of carbohydrate is high compared to protein there is relatively little ammonia available and more carbon dioxide than nitrogen is produced. This will cause a drop in the pH value. By contrast high protein input results in too much ammonia production, leading to alkalinity. The present state of knowledge indicates that methane can be formed in two different ways:

A. With some substrates, like (*i*) ethanol, (*ii*) butyrate, or (*iii*) hydrogen, methane is produced by atmospheric carbon dioxide reduction. Removal of carbon dioxide stops the reaction.

B. Methane is formed by carbon dioxide reduction, when that gas is formed during substrate oxidation (*iii*) carbon monoxide, (*iv*) propionate, (*v*) acetate.

(*i*) $$2C_2H_5OH + CO_2 \longrightarrow 2CH_3COOH + CH_4$$

(*ii*) $$4H_2 + CO_2 \longrightarrow CH_4 + 2H_2O$$

(*iii*) $$CO + H_2O \longrightarrow CO_2 + H_2$$

$$CO_2 + 4H_2 \longrightarrow CH_4 + 2H_2O$$

$$CO_2 + 3H_2 \longrightarrow CH_4 + 2H_2O$$

(*iv*) $$4C_2H_5COOH + 8H_2O \longrightarrow 4CH_3COOH + 4CO_2 + 24H$$

$$3CO_2 + 24H \longrightarrow 3CH_4 + 6H_2O$$

$$4C_2H_5COOH + 2H_2O \longrightarrow 4CH_3COOH + CO_2 + 3CH_4$$

(*v*) $$CH_3COOH \longrightarrow CH_4 + CO_2$$

Both mechanisms appear to involve the reduction of carbon dioxide to N^5, N^{10} methylene-tetrahydrofolate and hence via methycobalamine and co enzyme-M to methane. The electrons involved in this reaction are derived from either the organic substrate or the hydrogen.

Karnal Technology: When water is not the real problem, sewage can be treated by Karnal technology. It is a technology developed at the Central Saline Soil Research Institute in Karnal, Haryana for treating sewage safely by using certain species of trees such as eucalyptus, *leucaena* and poplar.

The technology consist essentially of growing trees in specially prepared plots and irrigating them with solid waste free sewage. The sewage is let into the plots in such a way that there is no water logging nor does the sewage contaminate the ground water. It uses the entire biosystem as a living filter for supplying nutrients to soil and plants. It reduces the soil pH from highly alkaline to neutral without building up salinity. In addition to reducing pollution and improving soil fertility, it also pays dividends in the form of wood of the tree grown. It is basically ideal for the plains, where most of the sewage pollution, especially of rivers, takes place.

REFERENCES

1. Green, E.J. and Carritt, D.E., J. Mar. Res, 25(2), 140–47 (1967).
2. Landner, L. and Larsson Ambio T., 2 (5), 154–157 (1973).

3. Anon (1985), *An Action Plan for Prevention of Pollution of the Ganga*, Department of Environment, Govt. of India, New Delhi, India 24 pp. illustr.
4. Sinnarkar, S.N., Rosarwani, S.K., Bhatt, G.G., (Eds), *River Ganga: An Overview of Environment Research;* National Environmental Engineering Research Institute, Nagpur, India 120 pp. (1987).
5. Krishnamurty, C.R.; Bilgrami, K.S., TMDas, RP Mathur (Eds), *The Ganga—A Scientific Study,* Northern Book Center, New Delhi, India 246 pp; illus. (1991).
6. Water Treatment Handbook, Degremont Laing Ltd. (1965).
7. Metcalf & Eddy. Inc., *Waste Water Engineering,* McGraw Hill, New York (1972).
8. Klein, L., *River Pollution*, Butterworths, London (1962).
9. Coaccio, L.L., *Water and Water Pollution Handbook,* Marcel Dekker, New York (1971).
10. Hynes, H.B.N., *The Biology of Polluted Waters,* A Liverpool Univ. Press; U.K. (1960).
11. Joint Ed. Arnold E Green Burg APHA, Joseph J Corinors AWWA, David Jenkins WPCF, *Standard Methods for the Examination of Water and Waste Water,* Prepared and published by American Public Health Association, 15th Ed.
12. Reid, Geord K, Wood. Ricard D., *Ecology of Inland Waters and Estauries,* D. Van Nostrand Company (1976).
13. Goterman H.L., Clymo R.S., Ohnstad MAM, *Methods for Physical and Chemical Analysis of Fresh Water,* Blackwell Scientific Pub., Oxford 2nd Ed (1978).
14. Sindhu, P.S. and Iqbal Malik, *Control Water, Control Industrial Pollution,* Published by Vatavaran (1996).
15. O'M Bockris, J., Ed., *Environmental Chemistry,* Chapter 5 by R.M.E. Diamant, The Chemistry of Sewage Purification, Plenum Press, New York.
16. Sindhu, P.S., *Chemistry of Atmosphere,* New Age International Publishers, New Delhi (1997).

4

Soils and the Environment

4.1 INTRODUCTION

SOILS: Soils consist of minerals, organic matter, water and air arranged in a complicated physico-chemical system that provides the mechanical foothold for plants in addition to supplying their nutritive requirements. Vertical profiles through soil reveal horizontal layers (horizons) which differ in their physical characteristics and which, in part, determine the kind and amounts of vegetation which a soil will support. Broadly speaking, three types of horizons may be identified. The uppermost layer, which may be from 30 cm to almost 60 cm in thickness, is the surface soil in which most of the life processes take place. This is called the A-horizons. Water and acid remove the finer particles of soils, salts of sodium, potassium, calcium and magnesium, soluble organic matter and oxides of iron and aluminium. Little or no carbonate remain in this horizon for any mature soil. The second horizon called the B-horizon, is the subsoil extending to about 90 cm below the surface. This is a zone of accumulation in which part of the material leached out of the upper horizon has been deposited. Still further below the surface, to a depth of about 180 cm is a layer of loose and partly decayed rock which is the parent material of the soils. This in brief is a description of the basic soil profiles which are conventionally designated the A, B and C horizon and can be further differentiated as shown in Fig. 4.1. These depth of layers are not fixed, rather they depend on rainfall. As the rainfall increases, calcium carbonate deposits disappear entirely from the soil profile A and move to the B-horizon. However with very low rainfall, it may occur in surface soil. Thus in desert region the soil is high in carbonates not only of calcium and magnesium but of sodium and potassium as well. On addition of limited amounts of irrigation water these and other soluble salts form a crust on the surface of soil. The inorganic fraction of surface soils may fall into any one of a number of textural classes, depending on the percentage of sand, silt and clay. Sand contents largely of primary minerals, such as quartz, and has a particle in the size from 2 mm to 0.02 mm. The silt consists of particles in the range 0.02 mm to 0.002 mm and clay particles are smaller than 2 micron (0.002 mm) in diameter. Figure 4.2 illustrates the textural classes into which soils can be differentiated, depending on the percentages of each of the three principal constituent.

The important physio-chemical process by which soils provide the nourishment to plants are controlled largely within the clay fractions of soil. The clay fractions are basically layer-lattice aluminosilicates. An essential characteristic of this plate-like particles of secondary aluminium silicates comprising clay is the substance of negative surface charges. As a result clay particles attract ions

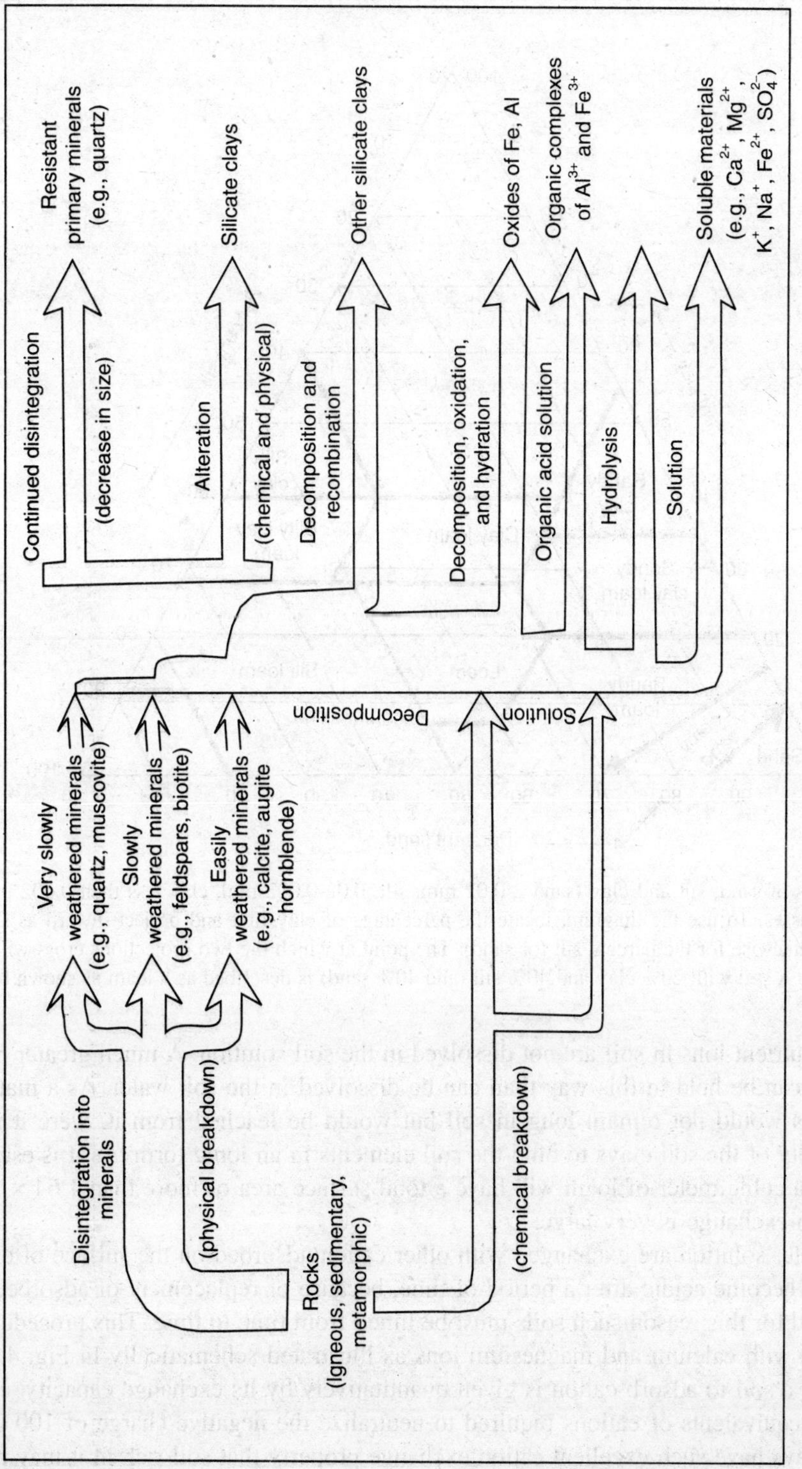

Fig. 4.1 Pathways of weathering that occur under moderately acid conditions common in humid temperate regions.

especially cations to their surfaces thereby showing ion exchange, is one of the most important properties of soils.

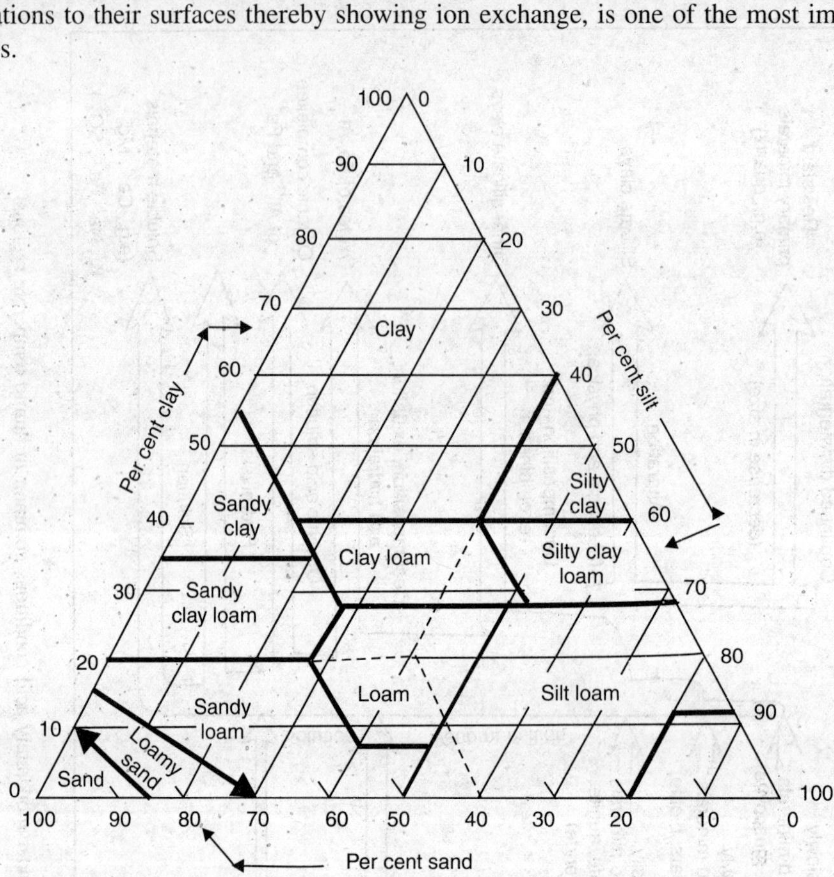

Fig. 4.2 Percentages of sand, silt and clay (sand 2-0.02 mm, silt, 0.02-0.002 mm, clay less than 0.002 mm) in major textural classes. To use the diagram, locate the percentage of clay first and project inward as shown by the arrow. Do likewise for the percent silt (or sand). The point at which the two projections cross will identify the class name. A soil with 20% clay and 40% silt (and 40% sand) is described as a loam as shown by the broken lines.

Most of the nutrient ions in soil are not dissolved in the soil solution. A much greater reservoir of nutrient elements can be held in this way than can be dissolved in the soil water. As a matter of fact, dissolved nutrients would not remain long in soil but would be leached from it, were it not for the extraordinary ability of the soil clays to bind the soil elements in an ionic form. As it is estimated that a total surface of a cubic meter of loam will have a total surface area of more than 1.64×10^7 m^2, the probability for ion exchange is very large.

Cations in water solution are exchanged with other cation adsorbed on the surface of clays. Thus, most soils tend to become acidic after a period of time, because of replacement of adsorbed cation by hydrogen ions, and for this reason such soils must be limed from time to time. This procedure replaces the hydrogen ions with calcium and magnesium ions as illustrated schematically in Fig. 4.3.

The ability of a soil to adsorb cation is given quantitatively by its exchange capacity, customarily expressed as milliequivalents of cations required to neutralize the negative charge of 100 g of soil at pH 7. Smectite clays have such excellent cation exchange property that soil rich in it may have cation

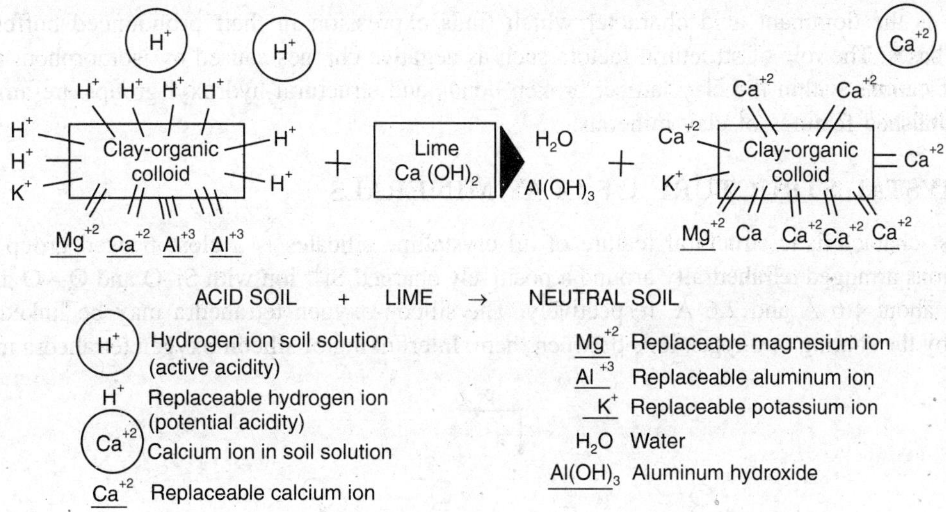

Fig. 4.3 The cation-exchange reactions when an acid soil is limed.

exchange capacity of about 100 meq per 100 g compared to less than 10 meq per 100 g for predominantly kaolin-type of soil. Sandy soils, as can be expected, tend to have low exchange capacity, which can be increased by adding organic matter.

4.2 CLAY MINERAL CONTENT

The clay minerals are essentially, crystalline silicates having a layer lattice structure. Our knowledge about their structural scheme and various aspects of their chemical behaviour has considerably advanced in recent years and a new discipline which might be appropriately described as the structural chemistry of clay minerals, has taken shape. This has resulted from the rapidly expanding knowledge of the manner in which the structure of clay crystals determine their chemical behaviour.

The chemistry of clays has a number of special features which have been kept in view in trying to correlate it with their structure. Firstly, the clay minerals being, by and large, insoluble substances, their reactions under not too drastic conditions, are practically confined to their surfaces. Clay chemistry in this sense, is more or less synonymous with the surface chemistry of clay crystals, consequently, the surface structure, rather than the bulk structure of clay crystals has a more direct bearing on their chemical behaviour. It will be shown later on that the surface structure of clay minerals, which can be fairly unequivocally inferred from their bulk structure, bears a close relationship to their chemical behaviour.

A second distinct feature of surface chemistry of clay is their extensive surface area on which reaction with reagents can take place. The large specific surface of naturally occurring clay arise from their small particle size which imparts the character of disperse or colloidal systems, with well-understood behaviour pattern. The realisation that they are crystalline in nature came much later. Presently their structure is well correlated to their colloidal properties.

A third and, perhaps, the most important feature of the chemistry of clay lies in the fact that clay crystals are predominantly ionic in nature. This endows them with a pronounced electrochemical character which is strikingly brought out in characteristic ionic properties of clay such as a well-known capacity for ion exchange and migration in an electrical field. The main feature of the electrochemistry of clays is the dominant acid character which finds expression in their pronounced buffer action towards bases. The role of structural factors such as negative charges caused by isomorphous replacements of cations within the clay lattice, broken bonds and structural hydroxyl groups are among the well-established features of clay minerals.

4.3 CRYSTAL STRUCTURE OF CLAY MINERALS

The most characteristic structural feature of all crystalline silicates is a close-packed group of four oxygen ions arranged tetrahedrally around a positively charged Si^{4+} ion with Si–O and O—O distances equal to about 1.6 Å and 2.6 Å, respectively. The silicon-oxygen tetrahedra may be linked to one another by the sharing of oxygen ions between them. Interlinking of silicon oxygen tetrahedra may give

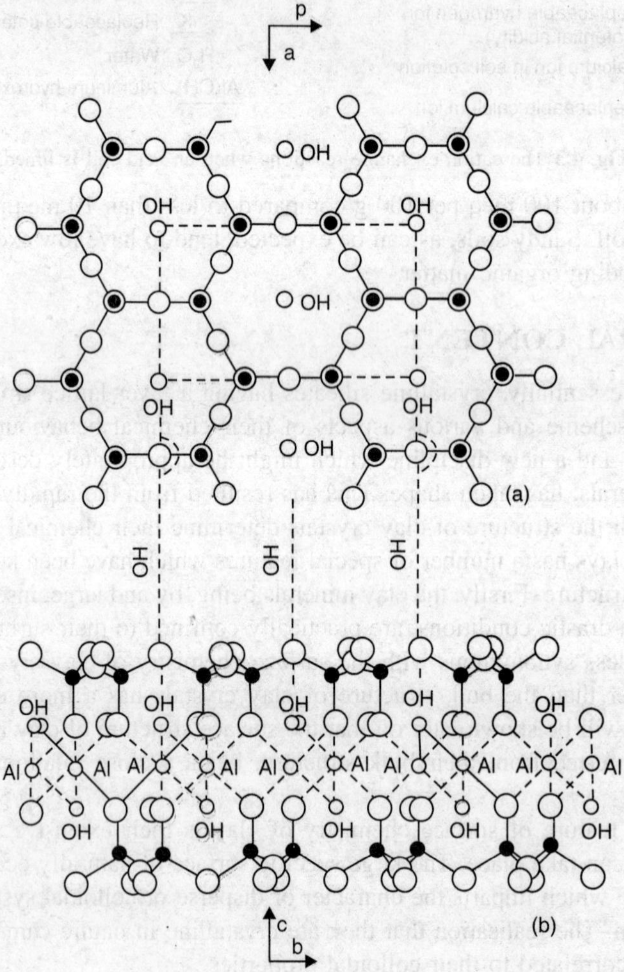

Fig. 4.4 (*a*) The hexagonal network of linked silicon-oxygen tetrahedral groups, with hydroxyl at the centre of each ring of six vertices. (*b*) Two of the above sheets seen edgewise with their vertices turned inwards. The layers of vertices and hydroxyls pack into each other, providing situations for Al or Mg atoms. The whole forms a firmly bound double sheet. Certain O and OH atoms which would otherwise be superimposed have been slightly displaced.

rise to complex silicon-oxygen anions which can be recognised as such in the structure; the anions, $(Si_2O_7)^{6-}$, $(Si_3O_9)^{6-}$ and $(Si_5O_{16})^{12-}$, for instance, occur respectively in minerals such as hemiomorphite, bentonite and beryl. The silicon-oxygen linkages may also extend indefinitely in one, two or three dimensions in space, giving in turn, chain, sheet and framework structures represented respectively by $[(SiO_3)^{2-}]_n$, $[(Si_2O_5)^{2-}]_n$ and $[(SiO_2)]_n$.

Occurrence of the silicon-oxygen sheet of composition $[(Si_2O_5)^{2-}]_n$ is a characteristic structural feature of layer-lattice silicates. Elucidation of the general scheme of their structure was given by Pauling[47,48], who recognised that several Si—O tetrahedra form a hexagonal net (Fig. 4.4a) produced by the sharing of each of the three coplanar oxygen of the tetrahedral between neighbouring silicons. The vertices of all tetrahedra in the net, point in the same direction. One negative valency of each vertical oxygen remains unsatisfied and the layer of the linked Si—O tetrahedra has the composition $[(Si_2O_5)^{2-}]_n$.

The basic unit in the second kind of layer in clay minerals is the close-packed group of six OH ions arranged octahedrally around a central Al^{3+} ion as in minerals like hydrargillite or gibbsite. Several such Al—OH octahedra are linked to give an octahedral layer in which the negative charge of each OH ion is balanced by the positive charges, each 0.5 unit strong, of the two neighbouring Al^{3+} ions present at the octahedral centres. In accordance with Pauling's[48] electrostatic valence rule, only two thirds of the possible aluminium positions will be occupied in the octahedral layer giving the composition $[Al_2(OH)_6]_n$.

The dimensions of the Si—O tetrahedra and the Al—OH octahedra are such that the distance between the vertices of two neighbouring octahedra is above the same as that between the centres of two adjacent hydroxyl groups. When the hydroxyl groups at the two adjacent carriers of each aluminium octahedron are removed, a charged layer with the composition $[Al_2(OH)_4]_n^{2+}$ results. The layer can be fused with the silica layer having the composition, thus giving rise to a neutral two-layer crystal lattices of composition $[Si_2O_5Al_2(OH)_4]$ in which the vertical oxygen ions of the Si—O tetrahedra

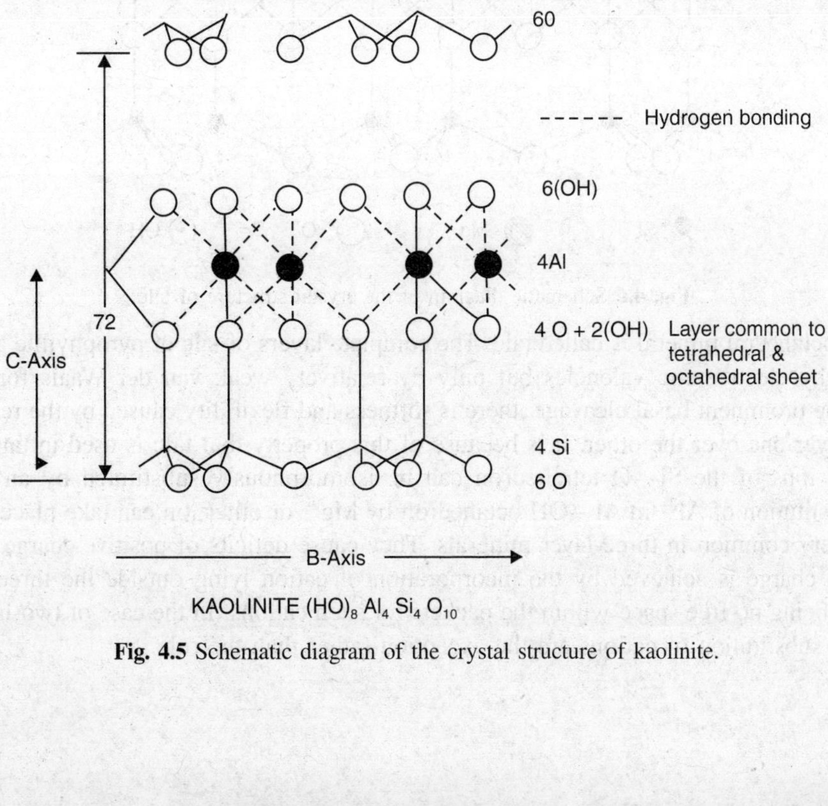

Fig. 4.5 Schematic diagram of the crystal structure of kaolinite.

occupy the places left vacant on the removal of the OH groups from the Al—OH octahedra. Such a neutral packet is present in kaolin clay minerals: kaolinite, halloysite, dickite (Fig. 4.5).

If from each Al—OH octahedron of the octahedral layer, four hydroxyl groups are removed, two from each side, the layer will have the composition $[Al_2(OH)_2]_n^{4+}$. Such depleted layer can be sandwiched between two layers of linked Si—O tetrahedra, each of composition $[(Si_2O_5)_n^{2-}]$, giving rise to a neutral three-layer packet (Fig. 4.4b) having the composition $[Si_2O_5.HO.Al_2.OH.Si_2O_5]$, which is identical with the mineral pyrophyllite. Actually, the octahedra layer is fused between the two tetrahedral layers, a feature of their fusion being the replacement of two-thirds of the OH groups in one of the two sheets of OH groups in the Al—OH layer by the vertical oxygen of the Si—O layer.

When aluminium is present in octahedra layer, only two thirds of possible positions are filled to balance the structure and minerals like pyrophyllite are called dioctahedral minerals (see octahedral layer in Fig. 4.4b). When magnesium is present, all the positions are filled (see Fig. 4.6) to balance the structure, which is the brucite structure and has the formula $Mg_3(OH)_6$. If, from each $Mg(OH)_2$ octahedral layer, four hydroxyl groups are removed, two from each side, the layer will have the composition $[Mg_3(OH)_2]^{4+}$. Such a depleted layer can also be sandwiched between two layers of Si—O tetrahedra, each of composition $[(Si_2O_5)_n^{2-}]$, giving rise to another neutral three-layer mineral having the composition $[Si_2O_5.HO.Mg_3.OH.Si_2O_5]$.

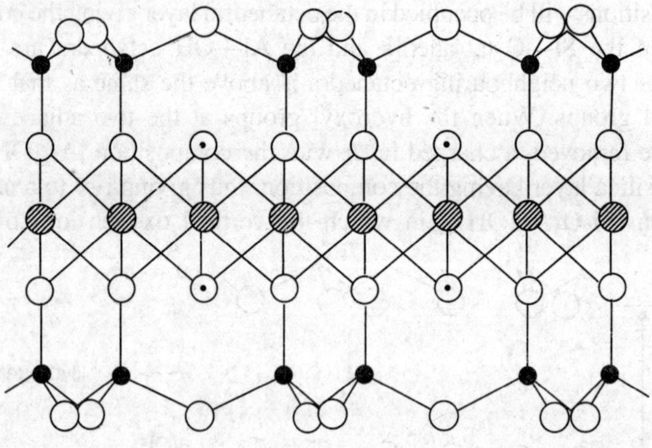

Fig. 4.6 Schematic diagram of the crystal structure of talc.

This tri-octahedral mineral is called talc. The complete layers of talc or pyrophyllite are not chemically bound by electrostatic valencies but only by relatively weak van der Waals forces. Thus, in addition to the prominent basal cleavage, there is softness and flexibility caused by the ready sliding of successive layer one over the other. It is because of this property that talc is used in talcum powder.

The Si^{4+} ions of the Si—O tetrahedron can be isomorphously substituted by an Al^{3+} ion and similarly substitution of Al^{3+} in Al—OH octahedron by Mg^{2+} or other ion can take place. Such substitutions are very common in three-layer minerals. They cause deficits of positive charge in the layers. Balancing of charge is achieved by the incorporation of cation lying outside the three-layer crystal lattice, there being no free space within the packets for such cations. In the case of two-layer minerals, isomorphous substitution of cations are the exception rather than the rule.

Clay minerals are, generally speaking, the result of stacking, one above the other, of either the two-layer or three-layer crystal lattice. The clay minerals of kaolin group, kaolinite, dikite, nacrite and hallosite are made up natural two-layer crystal lattices $[Si_2O_5.HO\ Al_2(OH)_3]_n$. The individual member of this group, therefore, have the same composition and the only way in which they differ is the exact manner in which the two-layer crystal lattices are stacked, one above the other, in their crystals. The oxygen plane of one of the two-layer lattice touches the OH plane of the one adjacent to it. This gives rise to hydrogen bonding which keeps the neutral layers of the crystal together, as shown in Fig. 4.5.

The three-layer crystal lattice $[Si_2O_5.HO.Al_2.OH.Si_2O_5]_n$ with isomorphous substitution, enters into the composition of clay minerals of the montmorillonite (now called smectite) and illite groups. In illite or hydrous mica, Al^{3+} replaces Si^{4+} in tetrahedral layer, is the principal isomorphous substitution. When every fourth Si^{4+} in tetrahedral layer is replaced by Al^{3+} and the resulting charge deficiency is balanced by a K^+ ion, one obtains muscovite mica. Hydrous mica or illite differs from muscovite in that it shows a lesser amount of isomorphous replacement and has a greater number of OH groups.

The montmorillonite clays consist of distinct clay minerals such as montomorillonite, beidellite, nantronite, and saponite, isomorphous replacements are known to occur both in tetrahedral and octahedral layers. Mg^{2+} for Al^{3+} in octahedral layer is the major replacement in montmorillonite. In beidallite, some tetrahedral Si^{4+} is replaced by Al^{3+}. Deficiencies of positive charges caused by these replacement are balanced by the incorporation of uni- and bivalent cations, which remains external to the three-layer crystal lattice. These cations are exchangeable with other ions in solution and give a definite cation exchange capacity to the minerals. Replacement of all Al^{3+} in montmorillonite by Fe^{3+} and some Si^{4+} by Al^{3+} give rise to nantronite. Magnesium in this mineral is found to be low, particularly when iron was high Ferrous iron was in insignificant amounts. Replacement of some of the Mg^{2+} by Li^+ in talc give rise to clay mineral hectorite. There is no charge in silicon layer, but the exchange capacity is high (75–80 meq per 100 g). There is also some proxying of F for OH but this will not effect the charge distribution or exchange capacity. Replacement of some Si^{4+} by Al^{3+} in talc gives rise to clay mineral saponite, replacement of Si^{4+} by Al^{3+} and some Mg^{2+} by Al^{3+} in octahedral layer of talc also gives the saponite (aluminium). Replacement of all Mg^{2+} in saponite (aluminium) by Zn^{+2} gives mineral sauconite. The list of various montmorillonite minerals are given in Table 4.1 and those of mica in Table 4.2. All montmorillonite (smectite) minerals have a high cation exchange capacity and readily disperse in water.

The Micas: The classification of the micas divided them into two main groups according to the nature of central octahedral layer. The heptaphyllite or dioctahedral group resemble pyrophyllite in which two-thirds of the total octahedral spaces are occupied by trivalent metals. Muscovite, $K_2Al_4\ (Al_2Si_6)O_{20}(OH)_4$, is an example. In the octaphyllite or trioctahedral group all octahedral positions are occupied by divalent metals chiefly magnesium and ferrous ions as in Biotite $K_2(MgFe)_6\ (Al_2Si_6)\ (OH)_4\ O_{20}$.

As noted earlier, in pyrophyllite and talc the units are chemically independent, being held together only by van der Waals forces between successive layers. In vermiculites, as will be seen later polar water molecules and balancing cations of considerable mobility form a cementing bond between successive layers. In muscovites and biotites potassium cations, which balance the charge caused by replacement of one quarter of the silicon atoms by aluminium find themselves in twelvefold coordination between rings of oxygen atoms from successive Si_2O_5 layers. Hence, they provide an electrostatic bond of moderate strength shown in Fig. 4.7. In brittle micas half the silicon atoms are proxied by Al and the layers are still more firmly held together by divalent calcium.

Table 4.1 Formulae for some members of smectite group with cation exchange capacity in the range 67-100 meq/100 g of clay

Dioctahedral Smectite	
Montmorillonite	$(HO)_4 Si_8 (Al_{3.34} Mg_{0.66}) O_{20}$ ↓ $Na_{0.66}$
Beidellite	$(HO)_4 (Si_{6.34} Al_{1.66}) Al_{4.34} O_{20}$ ↓ $Na_{0.66}$
Beidellite	$(HO)_4 (Si_6 Al_2) Al_{4.34} O_{20}$ ↓ $Na_{0.66}$
or	
Nantronite	$(HO)_4 (Si_{7.34} Al_{0.66}) Fe^{3+}_4 O_{20}$ ↓ $Na_{0.64}$
or	
Nantronite (aluminian)	$(HO)_4 (Si_{6.34} Al_{1.66}) Fe^{3+}_{4.34} O_{20}$ ↓ $Na_{0.66}$
Trioctahedral Smectite	
Hectorite	$(HO)_4 (Si_8 (Mg_{5.34} Li_{0.66})) O_{20}$ ↓ $Na_{0.66}$
Saponite	$(HO)_4 (Si_{7.34} Al_{0.66}) Mg_6 O_{20}$ ↓ $Na_{0.66}$
or	
Saponite (aluminium)	$(HO)_4 ((Si_{6.66} Al_{1.34}) Mg_{5.34} Al_{0.77}) O_{20}$ ↓ $Na_{0.66}$
Sauconite	$(HO)_4 (Si_{6.54} Al_{1.46}) (Zn_{5.28} Mg_{0.22} Al_{0.24} Fe^{+3}_{0.26}) O_{20}$ ↓ $Na_{0.66}$

Table 4.2 Formulae and name of some micas

Dioctahedral (heptaphyllite) Micas	
Muscovite	$(HO)_4(Si_6Al_2)Al_4O_{20}$ ↓ K_2
Margarite or Brittle Mica	$(HO)_4(Si_4Al_4)(Al_4)O_{20}$ ↓ Ca_2
Parogonite	$(HO)_4(Si_6Al_2)Al_4O_{20}$ ↓ Na_2
Trioctahedral (Octaphyllite) Micas	
Biotite	$(HO)_4(Si_6Al_2)(MgFe^{2+})_3O_{20}$ ↓ K_2
Biotite is called Phlagophite if content of Fe^{2+} in octahedral layer is zero.	
Illites	$(HO)_4(Si_{8-y}Al_y)(Al_4Fe_4Mg_4Mg_6)O_{20}K_y$
Illites include trioctahedral as well as dioctahedral types	Frequently y varies in the range 1-1.5.
Vermiculite	$(HO)_4(Si_{8-x}Al_x)(Mg.Fe)O_{20}yH_2O$ ↓ $(Mg, Ca)_x$ where $x = 1$ to 1.4 and $y = 8$

The negative charge in micas lie in the tetrahedral layer whereas in the case of montmorillonites it lie in the octahedral layer. As a consequence, each three-layer crystal lattice of mica are held by a stronger coulombic force than that of the three-layer lattice of montmorillonites. This is reflected in the swelling up of montmorillonites when water dipoles can separate the three-layer lattices, hydrates the clay surface and exchangeable cation, thereby leading to swelling of the mineral, something not observed in the case of micas. The extent to which swelling occurs depends on the exchangeable cations. In other words, the extent of hydration of the planar surface is dependent upon exchangeable cations, which is a cooperative phenomenon. This is the reason why smectites have a higher cation exchange capacity than mica though isomorphous replacement is more in mica.

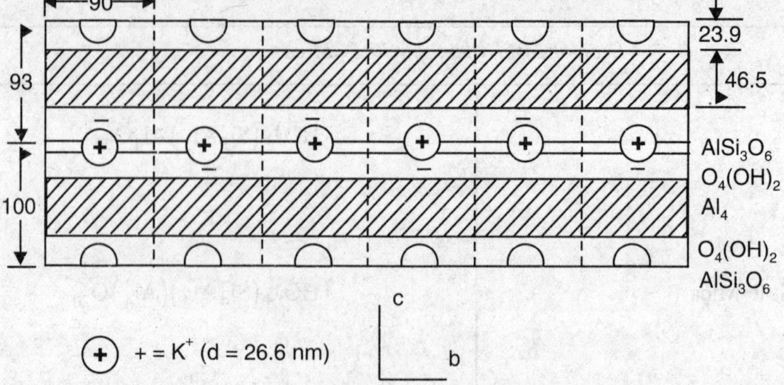

Fig. 4.7 Diagrammatic edge view of muscovite, showing the manner in which the silicate units are held together by potassium.

The Vermiculites: Vermiculites consist of sheets of trioctahedral mica separated by layers of water molecules occupying a definite space, 49.8 pm, which is about the thickness of two water molecules.

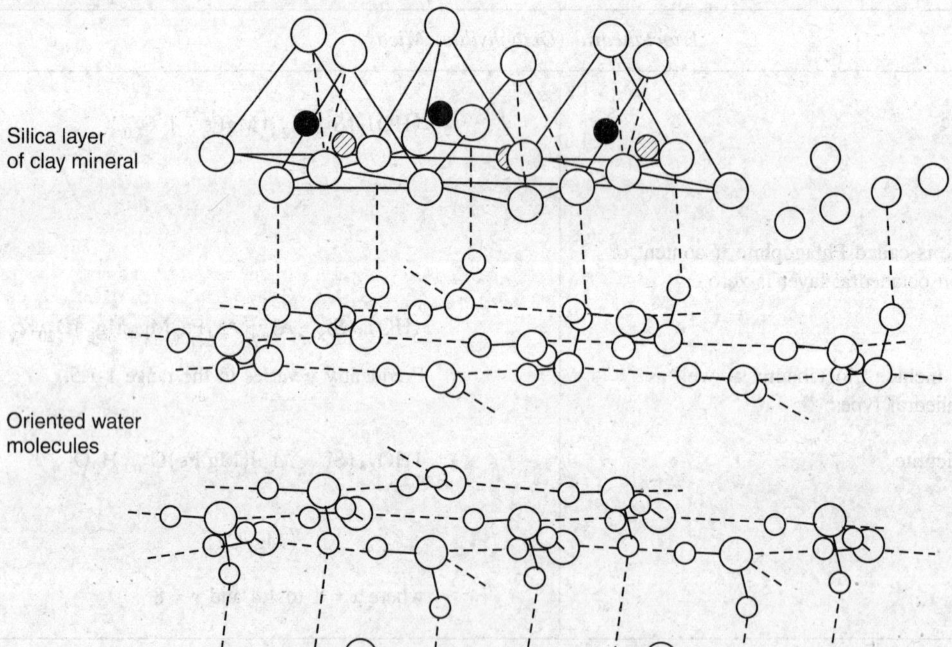

Fig. 4.8 Diagrammatic sketch of the vermiculite structure, showing layers of water Hendricks and Jefferson[20].

In its natural state, therefore, the mineral consist of an alternation of mica and a double water layer. The structure is unbalanced chiefly by substitution of Al^{3+} for Si^{4+}. These substitutions may be partially balanced by other substitutions within the mica lattice. The charge deficiency is satisfied by cations, which occur chiefly between the mica layers and are largely exchangeable. In the natural mineral which has the same cation exchange capacity as smectite, or somewhat higher, the balancing cation is Mg^{2+}, sometimes with a small amount of Ca^{2+} is also present. Hendricks and Jefferson[20] postulated an extended hexagonal network of water molecules as shown in Fig. 4.8. In the weathering of micas very delicate relationships between the cementing cations, water dipoles and the Si_2O_5 planes govern the early stages of decomposition. The work of Barshad[3] on the transformation of vermiculites to biotites and vice-versa clearly illustrates this. Vermiculites may be regarded as hydrated biotites in which cementing potassium has been replaced by magnesium or calcium. When calcium and magnesium is replaced by potassium, dehydration occurs readily and biotite results. Similarly, biotite on exposure to solutions containing magnesium and calcium can slowly be transformed into vermiculite. This penetration of water molecules evidently depends upon two factors. The first is the net charge on structural units which must not be too high or the dipoles will not be able to force the layers apart. On the other hand, when it is zero, as in pyrophyllite or talc, there is not sufficient polarization of dipoles to make an effective entry and hydrous forms of these minerals are unknown. The second is the nature of cementing cation. When this tends to assume a considerable envelop of water molecules, entry is facilitated. Barshad found that ammonium, like potassium, tends to be hydrated and hence favours the biotite structure. Magnesium, calcium and sodium favour the hydrated vermiculite structure.

The Chlorites (2:2 layer Clays): An interesting mineral chlorite has a *layer lattice silicate* and the structure consist of alternate mica-like and brucite-like layers. The layers are continuous in the '*a*' and '*b*' dimension and are stacked in the '*c*' direction with basal cleavage between the layers. The mica-like layers are trioctahedral with a general composition $[(HO)_4(Si_6Al_2)\ (Mg,Fe)_4,O_{20}]^{2-}$, but instead of a potassium atom balancing the charge on the silica layer a positive charge is carried by interheaved brucite layer of composition $(MgAl)_6(OH)_{12}$. The bonding between the layers is partly electrostatic in character as a consequence of substitution within the lattice. Although common in the early stages of weathering of basic igneous rocks, typical chlorites are of weak persistence compared with muscovite. This might be expected from the high content of magnesium in relation to silicon. However, more aluminous and resistant members are now known to occur in soils. A second important bonding mechanism arises from adjacent sheets of oxygen and hydroxyl ions where principal bonding mechanism arises from pairs of adjacent oxygen and hydroxyl ions. This mechanism is similar to that found in kaolin. Substitution within tetrahedral sheets vary from Si_3Al to Si_2Al_2 and substitution in octahedral layer is from Mg_5Al to Mg_4Al_2 with Fe^{2+}, Mn^{2+} partially replacing Mg^{2+} and Fe^{3+} or Cr^{3+} partially replacing Al^{3+}. The substitution of all Al^{3+} for Si^{4+} in tetrahedral sheets expand them sufficiently to accommodate the somewhat larger octahedral layers between them.

Sepiolite-palygorskite-attapulgite: The structure of attapulgite as worked out by Bradlley[4] is now generally accepted and is as follows. Attapulgite consists of double silica chains running parallel to the *c*-axis with the chains linked together through oxygen at their longitudinal edges. The apexes of the tetrahedron in successive chains point in opposite directions. The linked chains, therefore, form a kind of double ribbed sheet with two rows of tetrahedral apexes at alternate intervals in the top and bottom of the sheets. The ribbed sheets are arranged so that the apexes of successive sheets point together, and the sheets are held together by the Al and/or Mg in octahedral coordination between the apex oxygen of successive sheets as shown in Fig. 4.9. The octahedral layer is similar to that in the layer-lattice clay minerals, but it is continuous only in one direction. The octahedral layer completed by the central OH groups and by hydroxyls at open sides. The mineral has good cleavage parallel to 110, caused by the weak link through oxygen of the silica chains in the ribbed layer, chains of water molecules running parallel to the *c*-axis, fill the interstices between the amphibole chains. The cavities will accommodate four molecules of water per unit cell and this water would account for the dehydration loss at low temperatures. The dimensions of a cell, as suggested by Bradley, are $a_0 \sin \alpha = 129$ pm, $b_0 = 180$ pm, and $c_0 = 52$ pm. The balanced structure and composition of the ideal cell is $(OH_2)_4(HO)_2Mg_5Si_8O_{20}$ in which trivalent cation are considered equivalent to 1.5 Mg. Some substitution of Al^{3+} for Si^{4+} is considered probable. Bradley pointed out that substitution of Al for Mg or Si or both should weaken the structure, so that it appears doubtful that extensive replacement takes place and an aluminium end

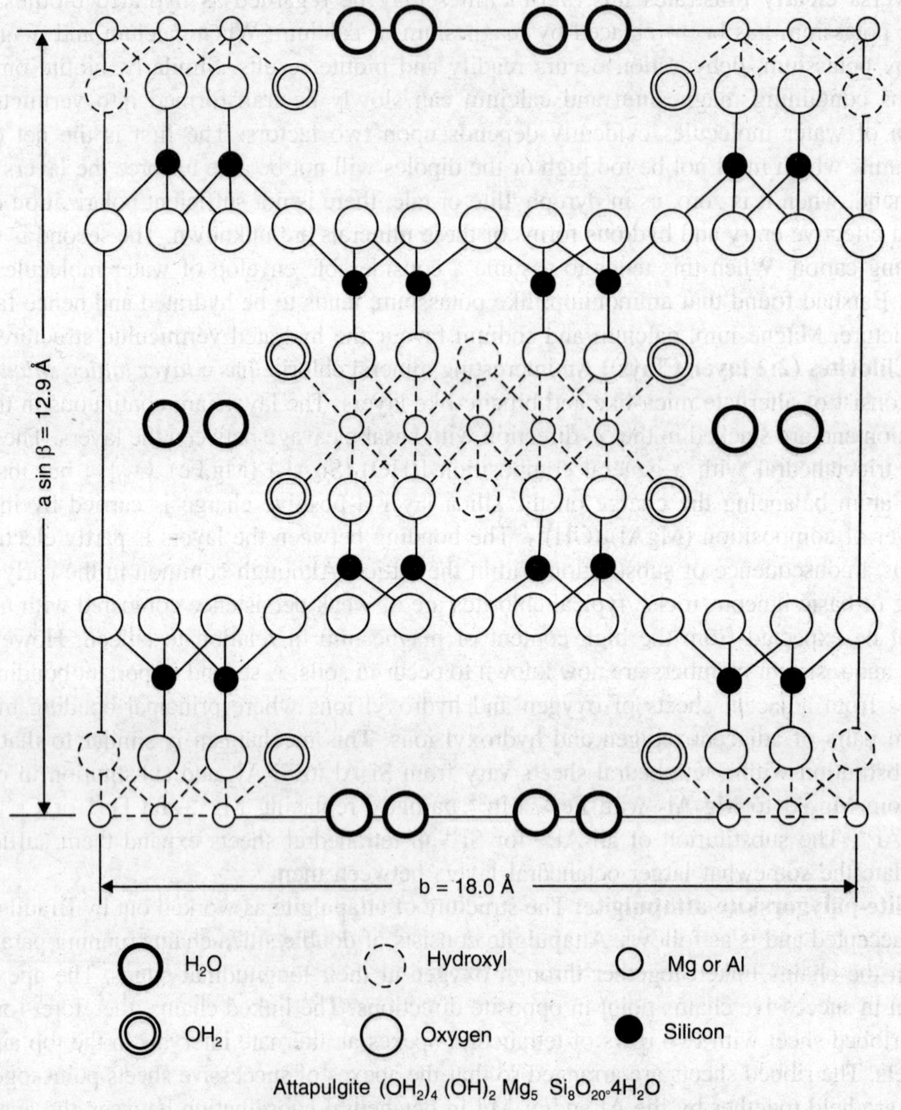

Fig. 4.9 Schematic presentation of the structure of attapulgite. Bradley[4].

member would not be expected. Attapulgite and palygorskite are structurally the same; attapulgite is simply short fibered palygorskite.

Sepiolite differs structurally from attapulgite as in former case in triple silicon-oxygen chain are present. Thus the octahedral strips containing magnesium occupy eight position, whereas in the case of attapulgite it was five. This makes channels to be larger and thus accommodate more water molecules, the formula of which is (Si_{12}) (Mg_8) O_{30} $(OH)_6(OH_2)_4.8H_2O$. Water in this has four different energies namely—(1) molecular water, (2) Water bound to brucite octahedral edges; (3) hydroxyls analogous with those of various layer mineral structure; and (4) balancing protons analogous with exchange ions. These categories are in the ratio of 6 : 4 : 3 : 2.

The unit cell dimensions are $a \sin \beta = 134$ pm, $b = 270$ pm, and $c = 53$ pm. The exchange properties of sepiolite are similar to those of attapulgite. It is extremely difficult to obtain relatively pure sepiolite and consequently structure is more dependent on chemical data.

Mixed-Layer Clay Minerals: Mixed-layer clay minerals are those in which different kinds of layers alternate with each other. M. C. Ewan[36] showed that mixed layer structure or what is called interstratification, are of three types.

(*a*) Regular Layers of different types called, for instance, *A* and *B* alternate according to a specific law. The most simple law will be of the form *ABABAB* ... Some such regular mixed layers have been given mineral names;

(*b*) Irregular Layers. No law governs the alternation, the layers are randomly interstratified;

(*c*) Segregation of alternating crystal lattices, being themselves in some cases interstratified and results into some irregular layers. It is a very complex case and is difficult to study.

In soils and sediments, the majority of layers are the four following types: illite, montmorillonite, vermiculite and chlorite. These four types of layers can indeed interstratify simultaneously but are difficult to analyse. The interstratification of these four minerals two-by-two give six types of mixed minerals.

Illite-Montorillonite, I-M; Montmorillonite-Vermicullate, M-V; Illite-Vermicullite, I-V; Montmorillonite-Chlorite, M-C; Illite-Chlorite, I-C; Vermicullite-Chlorite, V-C.

4.4 IDENTIFICATION OF CLAY MINERALS IN SOILS

Soil clay exist in crystalline form. In soil science a substance is considered to be amorphous, if it is amorphous to X-ray, because that show lack of regularity in internal atomic arrangement. In crystals, the atomic arrangement may be repeated in a three dimensional pattern. This repeating unit forms crystal lattice by translation in three dimensions. The size or length of the edges of unit cell in *x*, *y* and *z* directions are expressed in terms of *a*, *b* and *c* each of which represents a discrete unit length for a specific crystal.

The atomic groups in a crystal lattice can be arranged in planes at equal spacings along the crystallographic direction. Several types of atomic planes can be drawn in the crystal with interplanar spacings called *d* spacings. According to the Miller indices system, this plane is given the number 001. Since most of layer lattice silicates have same *a* and *b* values the basal 001 spacing plays a fundamental role in identification of clay mineral species by X-ray diffraction analysis. Schematic representation structure of some clay minerals are shown in Fig. 4.12 along with their *d*-spacing. They are usually identified by an integral sequence of the basal (001) diffraction peaks which corresponds to the sum of the thickness of the component layers. For example, two layers of 14 Å vermicullite clay will yield a basal d spacing $2 \times 14 = 28$Å. Interstratified chlorite-montmorillonite clay is either $14 + 17 = 31.0$ Å for an expanded montmorillonite component or $14 + 12.4 = 26.4$Å for air dry montmorillonite component.

The X-ray diffraction is perhaps the most widely used technique in the identification of clay. X-ray diffraction analysis is a nondestructive method but is not applicable to analysis of amorphous or noncrystalline material. Most of the time X-ray (Cu-Kα Radiation) with $\lambda = 1.5405$ Å is used. The diffraction follows the Bragg's relation

$$n\lambda = 2d \sin\theta$$

where d = spacing between atomic planes in the crystal
λ = wavelength
θ = glancing angle of diffraction
n = order of diffraction.

The samples for X-ray diffraction analysis can be prepared as a random oriented samples or as an oriented samples. In a random powder sample, the crystals lie in a random position to each other. With the aid of glycerol and gum tragacanth, the clay sample is made into a paste and rolled into a rod of 0.3 to 0.5 mm thickness. This can also be prepared by passing the paste into specially designed wedge holders.

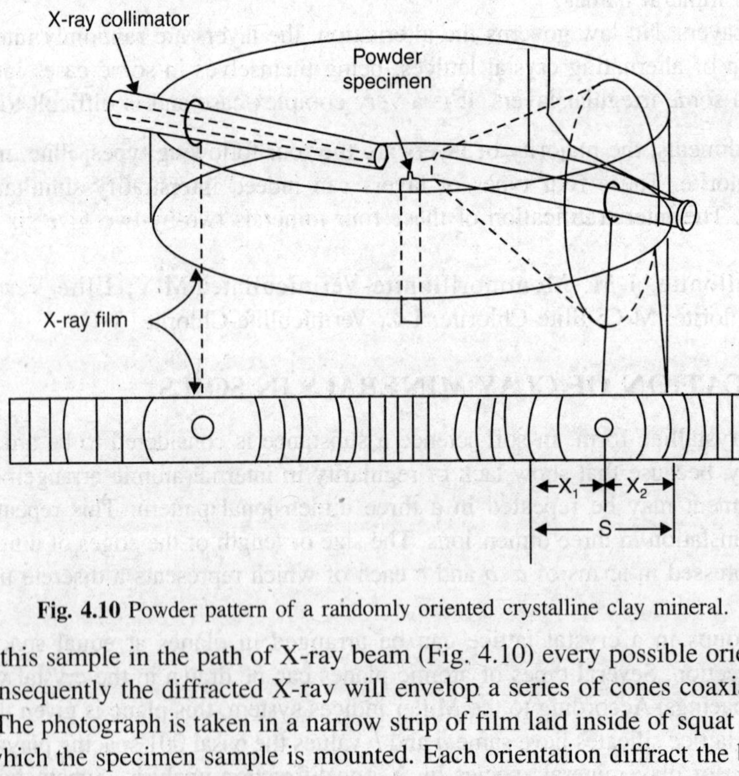

Fig. 4.10 Powder pattern of a randomly oriented crystalline clay mineral.

On placing this sample in the path of X-ray beam (Fig. 4.10) every possible orientation to crystals are available consequently the diffracted X-ray will envelop a series of cones coaxial with the path of incident beam. The photograph is taken in a narrow strip of film laid inside of squat cylindrical camera on the axis of which the specimen sample is mounted. Each orientation diffract the beam to yield dark lines on the photographic plate. On laying it flat, powder pattern in the form we see it is obtained. Measure the distance X_1 and X_2. Add the two values X_1 and X_2 for the same cone to get the value of S from which appropriate θ value can be calculated. The θ value in radiations for the set of reflecting planes is given by the equation

$$4\theta = S/r$$

where r is the radius of camera. Initially when powder pattern was developed sin θ and cos θ tables in degrees were more conveniently available, so θ was converted to degrees. This method is followed even now though sin θ and cos θ tables in radians are now available. The angle is converted to degrees as

$$\theta = \frac{S}{4r} \cdot \frac{180}{\pi}$$

The constant factor 180/4π can be eliminated by designing the camera which be a multiple of it. So radius of camera was selected as equal to 180/π = 5.73 so that θ = 2.5 S and

$$d = \frac{n\lambda}{2 \sin 2.5S}$$

Since all the quantities on the right hand side are known, d for various lines can be calculated. The various lines so obtained are assigned to various lattice planes d values.

A more popular method of mounting samples is by orienting the samples on microscopic glass slide or on porous ceramic plates. A clay suspension is made properly and pipetted on to the slide, so that approximately 15–25 mg of clay is transferred per 10 cm^2. After the sample has been allowed to dry at room temperature, it is ready for analysis with a direct recording X ray spectrophotometer, in which the X-ray pattern are printed on charts (Fig. 4.11). The results are normally shown in terms of 2θ values and from the θ values d is calculated from the Table 4.3.

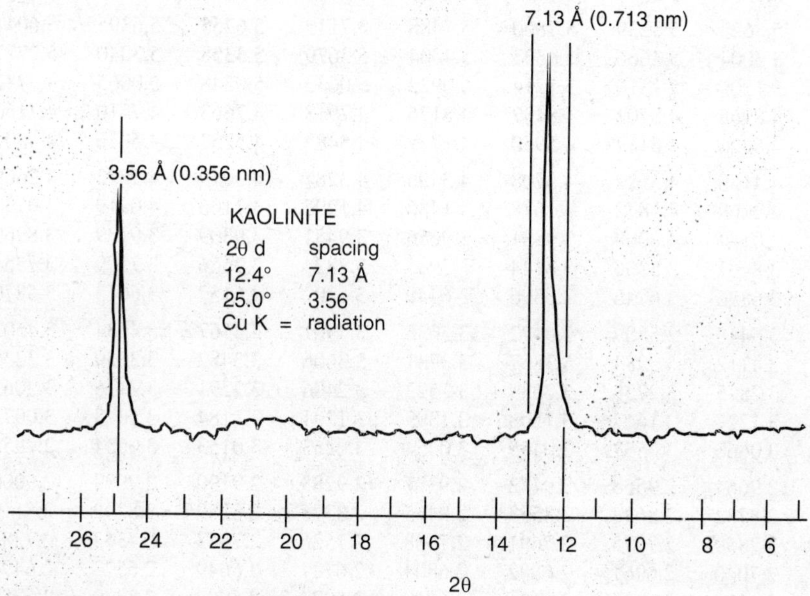

Fig. 4.11 Interpretation and identification of kaolinite using its X-ray diffraction pattern. At 12.4°, 2θ reading corresponds to a d spacing of 7.13 Å (0.713 nm), which is the diagnostic d spacing of kaolinite. See Table 4.3 for 2θ d spacing values.

Identification of clay minerals in soils is carried out by X-ray diffraction. The diffraction characteristics of many of the clay minerals have considerable similarity so that identification based solely on *d*-spacing cannot be taken safely. The diffraction data presented here are taken from the G. Brown "The X-ray Identification and Crystal Structure of Clay Minerals"[16].

Table 4.3 X-ray diffraction 2 θ-*d* spacing conversion table

2θ*d* Spacing Values for Cu Kα Radiation with λ = 1.5405 Å (0.1540 nm)										
2θ	0.0	0.1	0.2	0.3	0.4	0.5	0.6	0.7	0.8	0.9
0	∞	882.63	441.32	294.21	220.66	176.53	147.11	126.09	110.33	98.076
1	88.263	80.245	73.555	67.897	63.047	58.845	55.167	51.922	49.038	46.457
2	44.135	42.033	40.122	38.378	36.779	35.308	33.950	32.693	31.526	30.440
3	29.425	28.476	27.587	26.751	25.964	25.223	24.522	23.859	23.232	22.636
4	22.071	21.532	21.020	20.531	20.065	19.619	19.193	18.785	18.394	18.018
5	17.659	17.312	16.979	16.660	16.352	16.054	15.768	15.491	15.225	14.967
6	14.717	14.476	14.243	14.017	13.798	13.586	13.381	13.181	12.988	12.800
7	12.617	12.440	12.267	12.099	11.936	11.777	11.622	11.471	11.325	11.182
8	11.042	10.906	10.773	10.644	10.517	10.394	10.273	10.155	10.040	9.9270
9	9.8168	9.7098	9.6042	9.5010	9.4001	9.3015	9.2053	9.1105	9.0173	8.9264
10	8.8378	8.7500	8.6645	8.5506	8.4989	8.4181	8.3387	8.2609	8.1847	8.1100
11	8.0360	7.9644	7.8935	7.8234	7.7549	7.6880	7.6220	7.5571	7.4932	7.4305
12	7.3688	7.3081	7.2484	7.1897	7.1320	7.0751	7.0192	6.9642	6.9100	6.8567
13	6.8042	6.7524	6.7015	6.6513	6.6019	6.5532	6.5053	6.4550	6.4114	6.3655
14	6.3203	6.2757	6.2317	6.1883	6.1456	6.1035	6.0619	6.0209	5.9804	5.9405
15	5.9011	5.8623	5.8239	5.7860	5.7488	5.7119	5.6755	5.6395	5.6041	5.5691
16	5.5345	5.5004	5.4666	5.4333	5.4004	5.3679	5.3358	5.3040	5.2727	5.2417
17	5.2111	5.1809	5.1510	5.1214	5.0922	5.0633	5.0348	5.0065	4.9787	4.9511
18	4.9238	4.8968	4.8701	4.8437	4.8176	4.7918	4.7663	4.7410	4.7160	4.6913
19	4.6669	4.6426	4.6187	4.5950	4.5715	4.5482	4.5253	4.5026	4.4801	4.4577
20	4.4357	4.4138	4.3922	4.3708	4.3496	4.3287	4.3079	4.2872	4.2669	4.2467
21	4.2267	4.2069	4.1872	4.1678	4.1486	4.1295	4.1106	4.0919	4.0733	4.0550
22	4.0367	4.0187	4.0008	3.9831	3.9656	3.9481	3.9309	3.9139	3.8969	3.8801
23	3.8635	3.8467	3.8306	3.8144	3.7983	3.7824	3.7666	3.7509	3.7354	3.7200
24	3.7047	3.6896	3.6746	3.6596	3.6449	3.6302	3.6157	3.6013	3.5870	3.5728
25	3.5587	3.5448	3.5309	3.5172	3.5036	3.4901	3.4767	3.4634	3.4502	3.4371
26	3.4241	3.4112	3.3984	3.3857	3.3731	3.3606	3.3482	3.3359	3.3236	3.3115
27	3.2995	3.2875	3.2758	3.2639	3.2522	3.2406	3.2291	3.2176	3.2063	3.1951
28	3.1839	3.1727	3.1617	3.1508	3.1399	3.1291	3.1184	3.1078	3.0973	3.0868
29	3.0763	3.0660	3.0557	3.0455	3.0354	3.0253	3.0153	3.0054	2.9955	2.9857
30	2.9760	2.9664	2.9568	2.9472	2.9377	2.9283	2.9190	2.9098	2.9005	2.8914
31	2.8823	2.8732	2.8643	2.8553	2.8465	2.8376	2.8289	2.8202	2.8116	2.8029
32	2.7945	2.7859	2.7775	2.7691	2.7608	2.7526	2.7443	2.7362	2.7281	2.7200
33	2.7120	2.7040	2.6961	2.6882	2.6804	2.6727	2.6649	2.6573	2.6496	2.6420
34	2.6345	2.6270	2.6195	2.6121	2.6048	2.5974	2.5902	2.5830	2.5757	2.5686
35	2.5615	2.5541	2.5474	2.5404	2.5334	2.5295	2.5196	2.5129	2.5060	2.4993
36	2.4926	2.4859	2.4793	2.4727	2.4661	2.4596	2.4531	2.4466	2.4402	2.4338
37	2.4274	2.4211	2.4149	2.4086	2.4024	2.3962	2.3901	2.3840	2.3779	2.3719
38	2.3659	2.3599	2.5340	2.3480	2.3421	2.3362	2.3305	2.3247	2.3189	2.3131
39	2.3074	2.3018	2.2962	2.2905	2.2849	2.2794	2.2739	2.2684	2.2629	2.2574

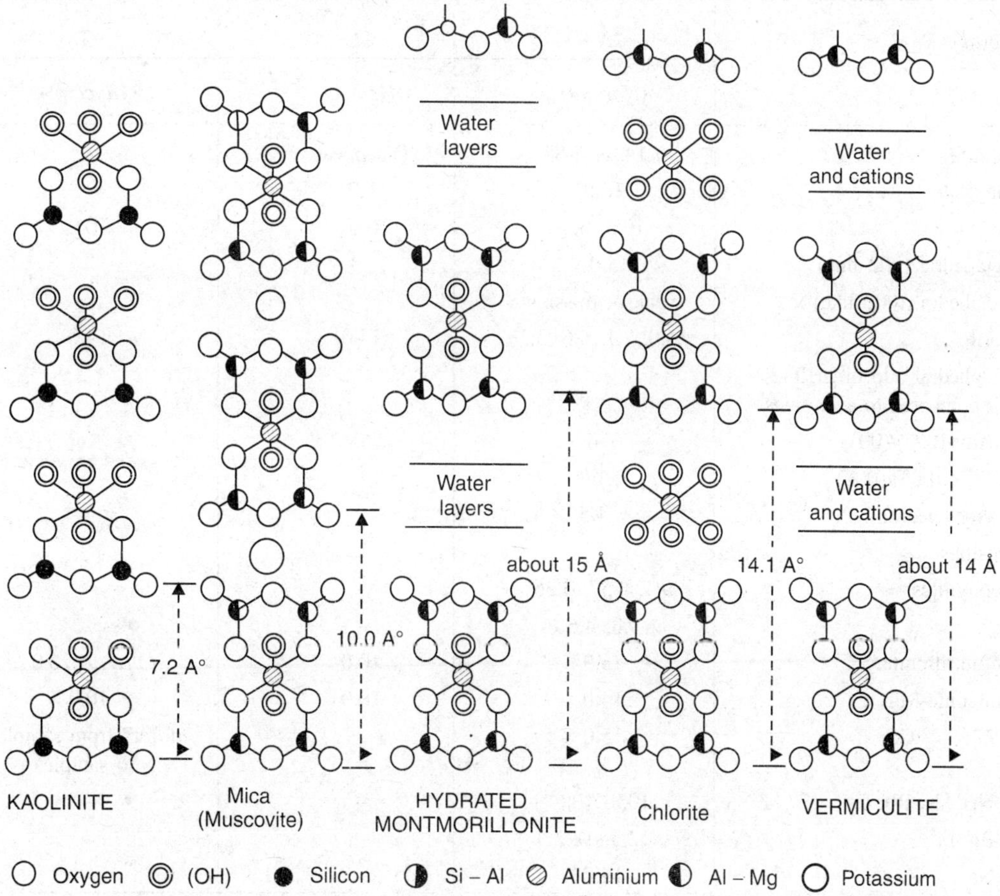

Fig. 4.12 Schematic representation of the structure of the principal clay minerals.

The usual tests are carried out on X-ray diffraction of well oriented flakes of clays:

Tests of natural or normal clay	*N*
Tests after action of glycerol on flakes	*G*
Test after heating to 490°C	Ch

The glycerol treated sample is prepared by placing the sample clay or soil over glycerol for at least 48 hours at room temperature in a desiccator. The sample is taken out and X-ray diffraction pattern is recorded. After recording the pattern the specimen is placed in muffle furnace at 490°C for three hours and X-ray diffraction pattern is recorded as soon as possible to avoid any rehydration.

"The interpretation is based on basal spacing obtained for the pure single mineral. These spacings given in Å are the following:

Mineral	*N*	*Ch*	*G*
	(*Normal*)	(*Heated*)	(*Glycerol*)
Kaolinite	7.14, & 3.58	Disappear	–
Halloyssite	10.8	7.2	–
Illite	10.0, 3.3	10.0	10.0
Dioctahedral (002 line)	5.0		
Trioctahedral (002 line)	weak or missing		
Chlorite	14.0, 7.1, 4.6, 3.5	14.0	14.0
Trioctahedral odd order lines weak compared to even lines			
Intensity (002)/(001)	3–4		
(003)/(002)	< 1.0		
For dioctahedral,	> 1.5		
Intensities			
Pyrophyllites	9.1, 4 57, 3.36	–	–
Talc	9.3, 4.6	–	–
Montmorillonite	13.9	10.0	17.7 to 20.0
Vermiculite-Mg	14.0	10.0	10.0
- Ca	15.0	–	differs from sample to sample
For NH_4, K, Rb, Cs	10.8, 10.6	–	–
For Ba, Li	12.3, 12.2	–	–
For Na	14.8	–	–
Attapulgite	Main lines are 10.5, 6.44, 5.42, 4.48, 3.24, 2.15		
Sepiolite	Main lines 12.1, 7.6, 5.05, 4.5, 4.3, 3.75, 3.33		

If these minerals are mechanically mixed, the pattern would give diffraction pattern of the two minerals present at the same time but not interstratified. Regular interstratified mixed layers do not behave as mixtures but as true minerals, whose basal spacing is equal to the sum of those of the two interstratified layers. A new periodicity is introduced which was not there in parent minerals. For instance, chlorite-montmorillonite mixed layers show a basal spacing equal to 14.0 + 13.9 = 27.9A. The sequence of other diffraction lines are as follow:

Normal	27.9		14.0			9.3		7.0		
Heating	24.0			12.0			8.0		6.0	
Glycerol	32.0	16.0			10.6		8.0			5.3

The irregular mixed layers have no periodicity in their lattice. Their sequence is not regular. In fact, if widely separated reflections of each of the two component minerals are not affected, the reflections which are close to each other will interfere. A single reflection is obtained by interference of diffracted rays, very close to one another; this is called a maximum. Every structure shows a succession of maxima peculiar to it resulting in a nonlinear series of reflections.

A first approach to identification of mixed layer structures is thus possible. As a result, guides to identification have been set up such as in table (p. 120). The simplified schematic representation is presented in Fig. 4.13.

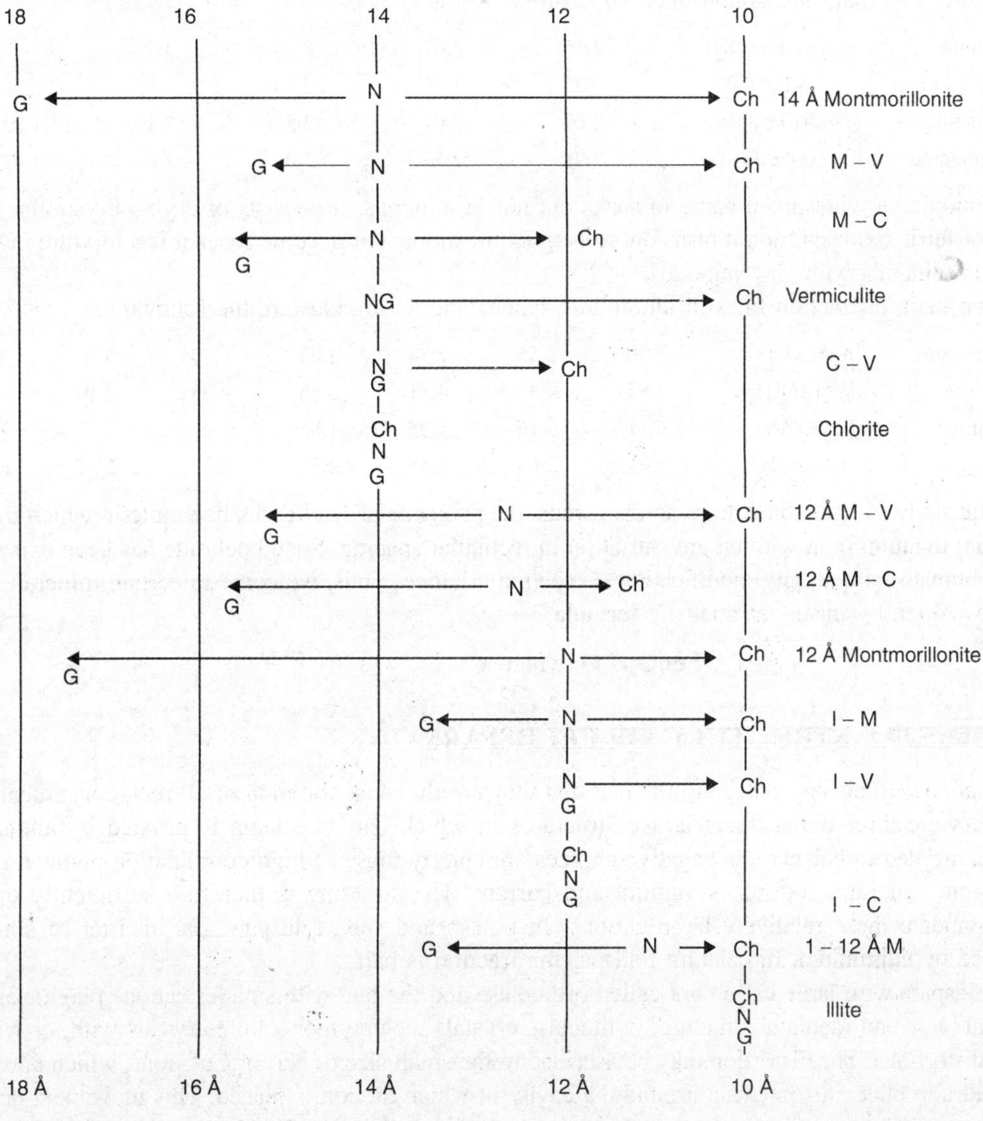

Fig. 4.13 Schematic representation of the behaviour of the reflections corresponding to basal spacing during X-ray diffraction after different treatments (after Lucas, Camez and Millot[14], G: Treatment with glycerol, N: Natural sample, Ch: Heating at 490°.

4.5 IRON AND ALUMINIUM HYDROXIDES

Iron and aluminium oxides and hydroxides occur in a great number of weathering products in soils, and even in rocks rich in clays. However magnesium is precipitated in alkaline region. As the solubility product of magnesium carbonate is less than magnesium hydroxide, magnesium is found as carbonate and not as hydroxide or oxide in alkaline soils. The oxide and hydroxide matter of soil is amorphous in nature and has very high surface area and thus are quite reactive. Knowledge of these minerals will be particularly useful in the study of lakerites and bauxites. There are four iron mineral species besides magnetite. The main diffraction lines, in Å are as follows:

Goethite	(α-FeOOH)	4.15	2.67	2.43	1.7	
Lepidocrocite	(γ-FeOOH)	6.25	3.28	2.46	1.93	
Hematite	(α-Fe_2O_3)	3.67	2.67	2.5	2.19	1.83
Maghamite	(γ-Fe_2O_3)	2.93	2.50	2.07	1.60	1.47

Limonite, a widespread name in rocks but not in minerals. It consists of crypto crystalline aggregates of ferric oxides. Often it turns out to be goethite alone, but in some cases it is a mixture involving formula minerals with clay minerals.

The main diffraction lines of aluminium oxides and hydroxides are the following

Corundum	(α-Al_2O_3)	3.48	2.55	2.38	2.08	1.74	1.6	1.4
Gibbsite	(γ-$Al_2(OH)_6$)	4.85	4.37	4.31	2.45	2.38	2.04	
Bohmite	(γ-AlOOH)	6.11	3.16	2.35	1.86			
Diaspore	(α-AlOOH)	4.71	3.98	2.55	2.31	2.13	2.07	1.63

The study of some bauxite minerals reveals the presence of ferriferous boehmite in which iron can be equal to aluminium without any variation in recticular spacing. Such boehmite has been experimentally obtained without any modification of boehmite lattice. Thus, we come to certain minerals which can be written by means of dualistic formula

$$(Al_{2-x}\ Fe)_xO_3H_2O, \text{ where } x \leq 1.$$

4.6 CRYSTAL STRUCTURE OF FELDSPARS

Feldspars owe their importance to the fact that they are the most common of all rock constituents. The feldspars are three-dimensional lattice structures in which part of silicon is proxied by aluminium. Cations needed to balance the negative charges are chiefly those of high coordination number, chiefly potassium, calcium, sodium, strontium and barium. The structure is therefore sufficiently open to accommodate these relatively large cations. In potash and soda feldspars, one quarter of silicon is replaced by aluminium. In calcium feldspar the fraction is half.

Feldspars with large cation are called orthoclase and the one with smaller cations plagioclase. All feldspars have an identical structure. Orthoclase crystals include monoclinic crystals with, or without, some distortions. The distortion may be ascribed to the small size or Na^+ or Ca^{+2} ions, which allows the tetrahedra to close in somewhat around the cavity in which the ion is placed. This movement destroys the symmetry planes and rotation axis present in monoclinic orthoclase. Table 4.4 gives the crytallographic data of some of the various feldspars.

Orthoclase have a peculiar type of chains, which link rings of four tetrahedra. Figure 4.14a shows two rings of four tetrahedra, the heights of the atoms in pm units above the plane of the diagram being marked. On the left, the two upper tetrahedra are pointing upward and the two lower downwards. This is reversed on the right. It will be readily understood that these rings can be superimposed in such a way that the oxygen atoms marked 42 are common to both. If this process is continued, a chain of rings as shown in Fig. 4.14b results. The pattern repeats at twice the height of a tetrahedron (42 pm), thus accounting for the *a*-axis (84 pm) of all feldspar. The process of linking produces additional rings of four tetrahedra. One quarter of the atom at the centre of the tetrahedron is aluminium and the remainder is silicon. X-ray analysis fails to distinguish them, and they will be considered as identical though the possible influence of this substitution upon the symmetry must not be forgotten.

Table 4.4 Feldspar group crystallographic data

(a) Monoclinic or nearly monoclinic							
		a	:	*b*	:	*c*	β
Orthoclase	$KAlSi_3O_8$	0.65865	:	1	:	0.554	116°3′
Soda orthoclase	$(K, Na)AlSi_3O_8$						
Hyalophane	$BaAl_2Si_2O_8$	0.6584	:	1	:	0.5512	115°34
Microcline	$KAlSi_2O_8$						
Anorthoclase	$(NaK)AlSi_3O_8$	0.657	:	1	:	0.554	115°2′

Triclinic, the Plagioclase Feldspar												
		a	:	*b*	:	*c*	α	:	β	:	γ
Albite (Ab)	(Ab) $NaAlSi_3O_8$	0.6353	:	1	:	0.5517	94°.3′	:	116°3′	:	88°9′
Oligoclase	Ab_6An to Ab_3An	0.6335	:	1	:	0.5554	93°4′	:	116°23′	:	90°5′
Andesine	Ab_3An to Ab An	0.6357	:	1	:	0.5521	93°.23′	:	116°29′	:	89°59′
Labradorite	AbAn to $AbAn_3$	0.6377	:	1	:	0.5547	93°31′	:	116°3′	:	89°54′
Bytownite	$AbAn_3$ to $AbAn_6$						93°22′	:	116°	:	90°41′
Anorthite (An)	(Art) $CaAl_2Si_2O_8$	0.6347	:	1	:	0.5501	93°13'	:	115°55'	:	91°12

The structure of orthoclase is shown in Fig. 4.15 projected on (010). A chain can be seen running parallel to '*a*' with alternate rings, approximately perpendicular and parallel to the plane of the diagram. At the same time, it can be seen how the chains are linked to each other. Oxygen atoms marked AAA, lie on two fold axes as shown, and by the operations of these axes they form the connecting links to chains to right and left. The oxygen atoms marked RR lie on the reflection planes at heights 0 and 50 *i.e.* b/2 in the diagram. They link the chains to mirrored neighbouring chains above and below. The atoms in the upper half of unit cell simply repeat those in the lower half by reflection across (010) at height b/2, therefore, only the b coordinates of the atoms in the lower half have been given. The potassium atoms lie on the reflection planes fitting into cavities. Each potassium atom has six oxygen neighbours at 28.5 pm and four at 31 pm.

The framework of linked Si and Al tetrahedra being similar for the monoclinic and triclinic form and is continuous throughout the whole structure. At high temperatures, the K and Na atoms are evenly diffused in the framework producing a uniform crystal. At lower temperature there is segregation by diffusion into potash rich and soda rich lamellae producing alternate sheets with monoclinic or pseudo-monoclinic and triclinic symmetry, respectively. It has been found that the *a*-axis of potash feldspar and albite are markedly different (84.5 pm and 81.4 pm), whereas the *b* and *c*-axes are almost identical (129 nm, 71.5 pm and 128.6 pm, 71.7 pm). This accounts for the lamellae being parallel to (100), since the *b* and *c*-axes of the two types of crystal coincide in the (100) plane, whereas the *a*-axis running through the lamellae becomes shorter in the albite regions and longer in the potash feldspar.

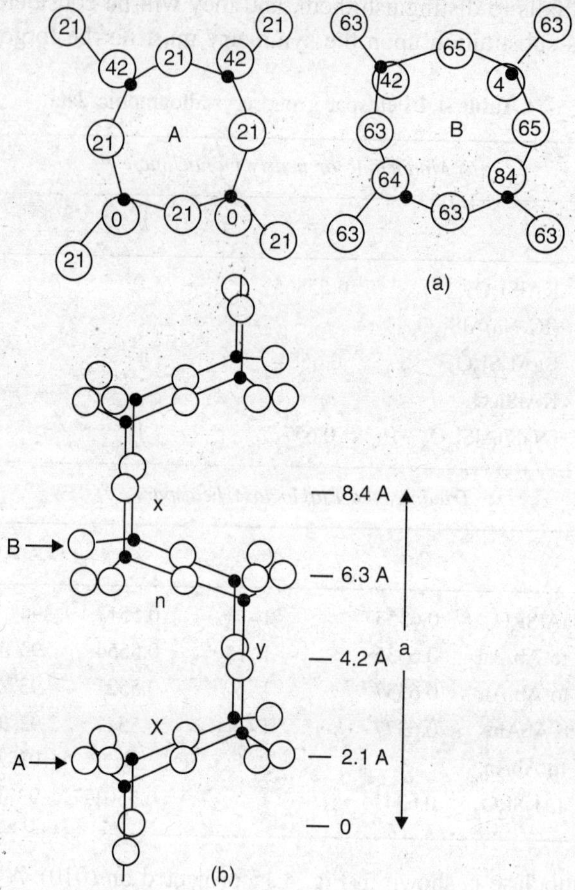

Fig. 4.14 The essential structural feature of all feldspars. (*a*) Rings of four tetrahedral groups, with two *adjacent* vertices pointing up and two down. (*b*) The same rings joined to form a chain by placing B over A. The oxygen atoms projecting to right and left form the means of linking this chain to its neighbours.

In all the feldspar structures, the only places for cations in the silicon-aluminium-oxygen framework are large cavities described above. It is not surprising to find that the feldspars contain only large ions such as Na, K, Ca and Ba. They do not, for instance, contain the series of small ions Mg, Fe, Cr, Mn, etc., which enter so freely into the structure of pyroxenes, amphiboles or the mica minerals. The structures have no suitable place for ions which require a situation within an octahedral group.

4.7 HYDROLYSIS OF FELDSPARS

Billions of years ago, the earth is said to have evolved from a red hot ball. Slowly it cooled down and, in the process, the formation of feldspars took place. These silicate minerals may have been stable at that temperature but not at room temperature that too in contact with water. Thus change from feldspars to clay minerals is a forgone conclusion.

These silicates were found in the composition of the elements available in a particular environment. As earth cooled down, there came new formation of rivers, seas and the oceans. For millions of years

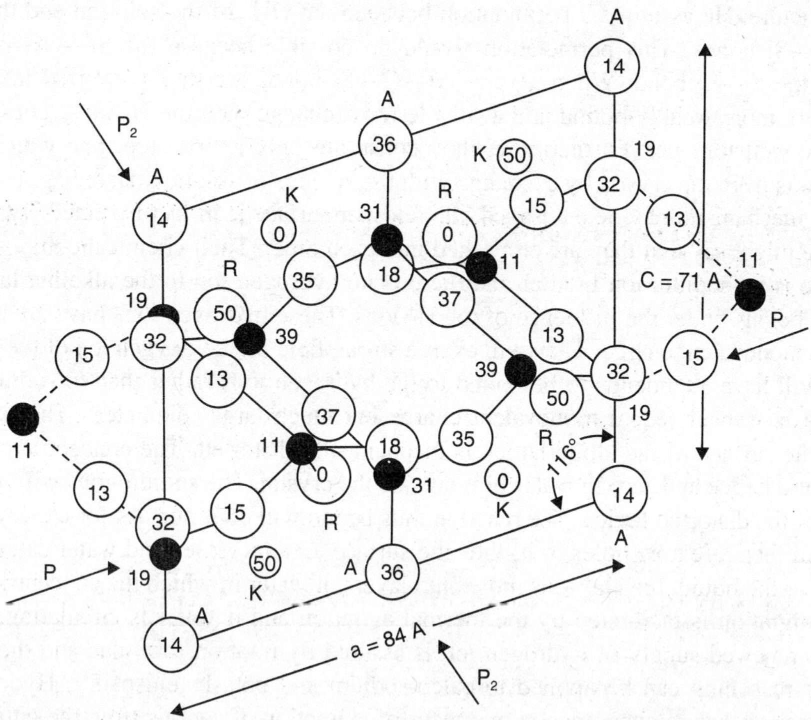

Fig. 4.15 The structure of orthoclase, $KAlSi_3O_8$. Only atoms in the lower half of the unit cell are shown. Those in the upper half being their reflections in a plane at height 50. Note the chain of Fig. 4.14 running parallel to *a*.

it appears, it had been raining. In this process feldspar started breaking. This resulted in the disappearance of feldspar and the formation of layer lattice silicate mineral took place. The exact mechanism is not known but an attempt is being made to present a thorough review of this process of change.

No crystal is completely insoluble. An element which is called insoluble in common language is simply very slightly soluble. No matter what crystal is put into water, there will be ions migrating from the crystal surface into the aqueous medium. This includes the feldspar silicates, when in contact with water they are altered by hydrolysis. The reactions which take place between the crystal surface and water, have been approached in two schematic ways.

Fredesickson[29] made comparison between the structure for water such as it was imagined by the Bernal and Fowler[5], and the structure of albite, as an example, when he considered a kind of arrangement of the adjacent layer of water on to the crystal. On the surface, albite are exposed oxygen ions whose negative valencies are not entirely satisfied. Water adheres to the crystal surface like a film, and the positive poles of water molecules saturate the negative surface offered at the contact. The negative poles of water face outward, where they will attract the positive poles of the following layer, and so on. As one moves farther away from the crystal, the organisation of water layers decreases. On the contrary, in contact with albite, the first layer of water would be arranged with respect to the albite surface in a manner similar to an epitaxy. In this way, the exchange between the Na^+ cations of the albite and the H^+ ions of the solution take place.

Devore[8] did not think that hydration of the feldspar surface by means of the association between water molecules and unsaturated oxygen ions could weaken the bonds between the alkaline ions and the crystal structure. He assumed a permutation between the OH^- of the solution and the oxygen ions of the Si—O—Si bonds. This permutation would be possible because the Si—OH bond would be stronger than the Si—O bond. When several Al—O—Si bonds are thus converted into Si—OH, the alkaline ions are more weakly bound and available for exchange with the H^+ ions. These two schemes can work very well together. Furthermore, they constitute only a first step, one which precedes the release of cations from the crystal by exchange with the H^+ ions of solution. Here begins the true attack.

Whatever mechanism may be envisaged, the feldspar surface is in fact hydrated. Two mediums are in contact, one might say that they are enmeshed into each other. Their chemical compositions are very different, there is no sodium ion in after and there is no hydrogen ion in the alkaline lattice. It is here that alteration begins under the influence of the H^+ ions. These hydrogen ions have an extremely small diameter for a monovalent charge. They will exert a strong field on the oxygen ion of the crystals. These oxygen ions will have a tendency to be bound to the hydrogen ions rather than to sodium ions, whose ionic potential is weaker (equal monovalent charge but much larger diameter). The sodium ions are released and the surface of the albite lattice is mechanically distorted. The concentration of sodium is high in the albite lattice and zero in water film outside the crystal. The sodium ions will migrate through the channels of the distorted lattice. The reaction thus begun will continue. As for the crystal, the lattice is distorted and there are now holes in it; thus the surface area increases and water can easily enter the structure. As for the liquid, the Na^+ ions move into layers of water in which the concentration of sodium is low. This migration is facilitated by the thermal agitation and if water is circulating, sodium ion is exported. The renewed supply of hydrogen ion is assured by rotation formulae and thermal agitation.

The same reasoning can be applied to calco-sodium and potash feldspars[47]. Hydrolysis thus accounts for a base exchange phenomenon by means of extraction of cations from the lattice. The Na, K, and Ca ions are released because of their weak ionic potential.

To alter a feldspar or other silicates, it doesn't suffice simply to release alkaline or alkaline earth cations from the lattice. It has been found that continuous leaching of feldspars with water releases silica and alumina, but in proportions which are variable accounting to the different pH conditions. There is more alumina than silica in an acid medium, more silica than alumina in a neutral medium and there are equal quantities of both in an alkaline medium. These experiments also conformed that during weathering of feldspar, alumina accumulates in the solid state because of its low solubility between pH 4 and 9. Further, when the Si—O—Al bonds are numerous, the weaker part than the Si—O—Si will be extracted first, the rate of which will depend upon the pH.

'Non-layered silicates are hydrolysed and lose substance under the influence of excess water. Feldspars are impoverished in cations and silica. The conditions are rapidly assembled for the neoformations of smectite. Ferromagnetic minerals are hydrolysed and the conditions become appropriate for the neoformation of chlorite. Volcanic glass is hydrolysed and with or without any obvious colloidal phase, the conditions for the neoformation of montmorillonite are realised. All these neoformations of clay minerals are formed from the silicates subjected to subtraction. If the flow of solution is accelerated, the leaching of cations is speeded up, the neo formed clay minerals are hydrolysed in their turn, and silica and alumina alone remain and organise into kaolin. If leaching is intense right from the beginning, the preceding stages are bypassed and kaolinisation prevails, with the direct release of iron. If drainage is still more intense, and if no water level subsists to retain silica in solution at the disposal of alumina, gibbsite is formed. Thus we pass from laterization to bauxitization by subtraction.

Layered silicates once formed are not insensitive to these environment of leaching. Although stable in the hydrosphere than other silicates, they will not resist intense leaching. Subtractions that were favourable to them when they were discrete will be fatal if they are intensified. In brief, in this environment we do not see progressive transformation by degradation as the layered minerals themselves are subjected to a even more degraded leaching. This also explains how, from feldspars surface we attained the present soil surface which is made up of clay minerals of metal oxides and organic matter (to be taken up in the latter section). This evolution by subtraction can stop any time if equilibrium is established and this gives rise to a variety of weathered products and soils. At any moment this evolution can definitely be stopped by erosion and this is the termination of weathering and the slowing down of sedimentation. All these facts and mechanisms are independent of place but solely dependent on environmental conditions. The mechanisms of transformation from feldspars to clayminerals must have been very slow. It may have taken many centuries to form soil in an undisturbed situation.

4.8 ACID CHARACTER OF CLAY MINERALS

Cation exchange is a surface property of clay crystals. With the exception of montmorillonite which shows inner crystalline swelling, cation exchange is confined to the outer surfaces of the individual clay crystals. The smaller the size of the crystal, the larger is the outer surface and the larger the cation exchange power. Exhaustive grinding in a ball mill, for example, considerably reduces the crystal size with a corresponding marked increase in the cation exchange capacity. These cations hold the three layer crystal lattice and all these exchangeable cations are not available for exchange in micas. Montmorillonites, on the other hand, are dispersed as single flakes and all the exchangeable cations can be replaced. Marshal[35], as well as Mitra and Sindhu[40] claims to have obtained satisfactory agreement between cation exchange power of montmorillonite and sauconite, with the negative isomorphous charge calculated from the results of chemical analysis.

There is little or no isomorphous replacement in the silicate minerals namely pyrophyllite and kaolinite but they do show a cation exchange capacity[21, 17, 44] which can be quite large if particle size is small. Broken bonds[22] on the lateral surface of the crystal bind the cations from the solution. This is amply reflected by the amphoteric nature of clay minerals which can combine with cations as well as anions.

Montmorillonite, freshly percolated through beds of an H-resin, contains exchangeable H^+ ions which liberate a strong clay acid[22, 23, 50, 54, 38]. Some exchangeable Al^{3+} ions are also present[1, 28, 31]. When allowed to age in water, the H-Al montmorillonite tends to become Al-montmorillonite[7, 13, 39]; pH and conductivity titration curves of freshly prepared acid montmorillonite obtained by passing Na-clay through H-resin reveal two and three clear inflections/breaks respectively. The pH titration curve is shown in Fig. 4.16a. Strong H^+ ions are neutralised up to the first inflection/break; between the first and second breaks, Al^{3+} ions react with the base and are precipitated as $Al_2(OH)_6$. A weak acid is neutralised, having its origin in the broken bonds, between the second and third breaks. In the case of acid sanconite[40], four inflections/breaks have been observed. Of these, the stronger hydrogen is neutralised at the first break; between the first and second inflections/breaks, Al^{3+}, and between second and third inflections/breaks Zn^{2+} react with base and are precipitated as $Al_2(OH)_6$ and $Zn_3(OH)_6$, respectively. A weak acid is neutralised between the third and fourth breaks. All acids except the weak part can be extracted with NaCl solution.

In the acid environment provided by H-resin, Al^{3+} ions are mobilised from the edges of crystals, which then proceed to occupy the exchange sites. In the case of sauconite, Zn and Al occupy the trioctahedral layer with some replacement of Si^{4+} by Al^{3+} in the tetrahedral layer. By the action of H-resin both Zn^{2+} and Al^{3+} are dislodged from the lateral surfaces and then occupy the exchange sites. The weak acid component which could not be extracted by neutral salt extract has been observed by Mitra *et al.*[37], Mitra and Kapoor[38] in montmorillonite, Mitra and Sindhu[40] in sauconite, and by Sindhu[43] in attapulgite.

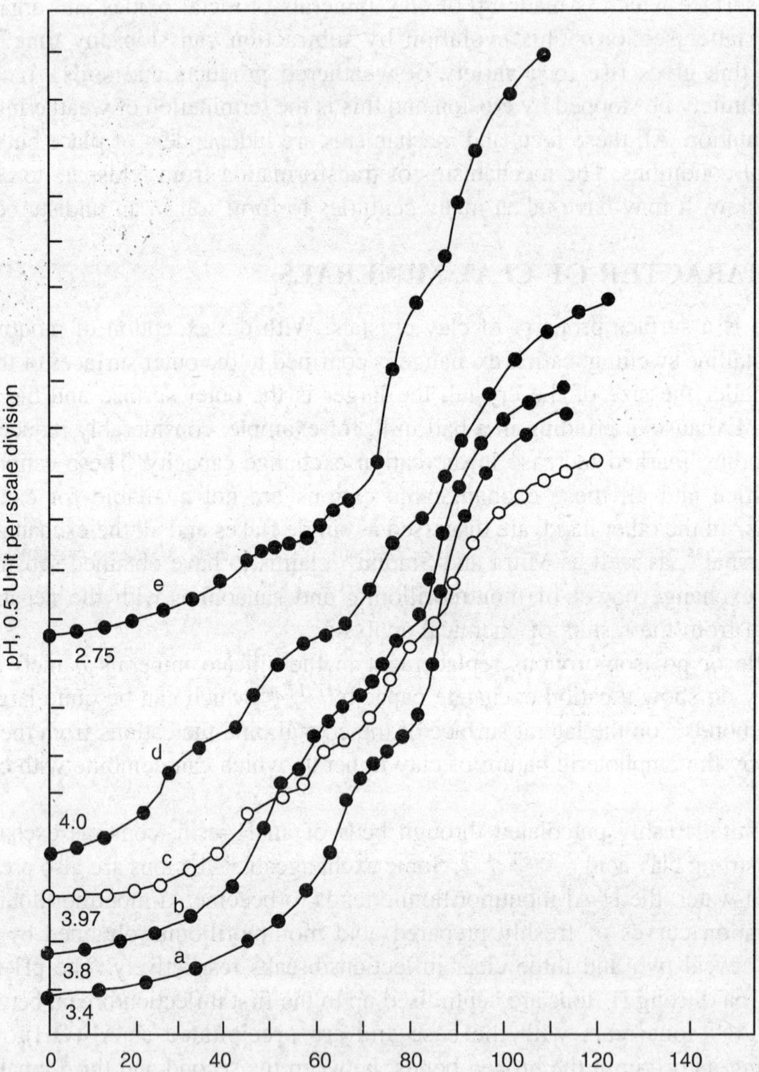

Fig. 4.16 Potentiometric titration curves of acid montmorillonite: (*a*) freshly prepared; (*b*) aged for one week; (*c*) aged for four weeks; (*d*) aged for nine weeks; (*e*) aged for four weeks and titrated in presence of salt (NaCl). Initial pH is indicated on the curves (Mitra and Kapoor[38]).

The octahedral cations exposed on the lateral surfaces carry some residual positive charge. Due to the lattice termination these surfaces will have both positive and negative centres on them. Acid clay suspension has low pH and the exposed positively charged cations of the octahedral layer will complete their octahedral shell with water molecules. In the titration of acid clay, cations have reacted with base and when more of the latter is added OH^- ions displace the water molecules from the octahedral shell of octahedral cations at the edges. The cations of the base thus balance the negative centres present on the lateral surfaces. The exposed cations, in effect, act as a lewis acid and they are responsible for the weak acid function for acid clays.

The pH and conductometric titration curves of acid montmorillonite which was allowed to age in water for different periods of time before titration were published by Mitra and Kapoor[38], the pH-titration curves of which are shown in Fig. 4.16. No change in base exchange capacity was observed. The titration curves showed four inflection/breaks and as more time was allowed a new acidic species appeared. During aging, strong H^+ ions migrate from exchange sites to the edges where they attack the Si—O—Al bonds, the Al^{3+} ions thus released go to occupy the sites vacated by the strong H^+ ions and the protons after attacking the Si—O—Al bonds are converted into very weakly acidic Si—OH groups. The H^+ ions which migrate to the edges can combine with the OH groups attached to edge aluminium ions besides attacking the Si—O—Al bonds. Both these processes cause the pH to rise. As a result Al^{3+} ions which have already moved from the edges and exchanged sites with strong H^+ ions hydrolyse to give basic Al ions. These basic Al ions are not only formed by hydrolysis of exchangeable Al^{3+} ions but they are also mobilised from the edges more so when the aging has sufficiently advanced and the pH is already in the neighbourhood of 4.0. Under these conditions $Al(OH)^{2+}$ or $Al(OH)_2^+$ ions move from the edge, may survive and migrate to the exchange sites displaying a strong H^+. Thus, species neutralised upto first inflection is strong H^+, between the first and second inflections Al^{3+} is precipitated, between the second and third inflections it is $Al(OH)^{2+}$ or $Al(OH)_2^+$ ion that precipitated as aluminium hydroxide whereas between the third and fourth inflections it was the weak acid, which was not extractable with NaCl, that was neutralised. However, when acid clay was allowed to age for one year, there was total disappearance of H^+ ions, but no change in total cation exchange capacity of montmorillonite was observed.

The pH and conductometric titrations of acid sauconite, which was allowed to age in water for different periods of time, were carried out by Mitra and Sindhu[40]. Figure 4.17 shows the pH titration curves. The titration curves show that strong hydrogen ions disappear after a day of aging. On further aging exchangeable Al^{3+} ions disappear. The H^+ and Al^{3+} are replaced by Zn^{2+} after 15 days of aging. When clay is allowed to age, the latter category of protons are available for mobilising Al^{3+} and Zn^{2+} ions from the edges. These protons combine with OH groups attached to the edge Al's and Zn's besides attacking the Si—O—Al and Si—O—Zn bonds. All these processes cause the pH to rise. Al^{3+} ions would not be stable in solution above pH 4.5. Hence as the pH rises and approaches this values, Al^{3+} ions which have already moved from the edges and exchanged sites with strong H^+ are precipitated as $Al(OH)_3$. Zn^{2+} (and $ZnOH^+$) ions are however stable in solution up to a pH of about 6.5. Accordingly Zn^{2+} ($ZnOH^+$) ions continue to move from the edges and take the place of exchangeable H^+/Al^{3+} ions. If only exchangeable H^+ and proton produced by the hydrolysis of exchangeable Al^{3+} ions were responsible for the mobilisation of Zn^{2+} ions would be the same irrespective of the age of clay as in montmorillonite, but in sauconite it has increased by 60%. This increase in cation exchange capacity signifies that there is an increase in negative charge. This is explained by the isomorphous replacement of Zn^{2+} by Al^{3+} in the octahedral layer, providing it excess positive charge which is offset and balanced by the negative charge in the tetrahedral layer. As Al^{3+} ions from octahedral layer move out, negative charge will increase which increases the exchange capacity of sauconite.

The acid character of attapulgite as studied by Sindhu[42] is similar to montmorillonite except that there is no formation of $Al(OH)^{2+}$ and $Al(OH)_2^+$ but there is no change in cation exchange capacity. The strange thing about attapulgite was that it is a magnesium mineral but there was no exchangeable magnesium ion in acid or aged acid clay.

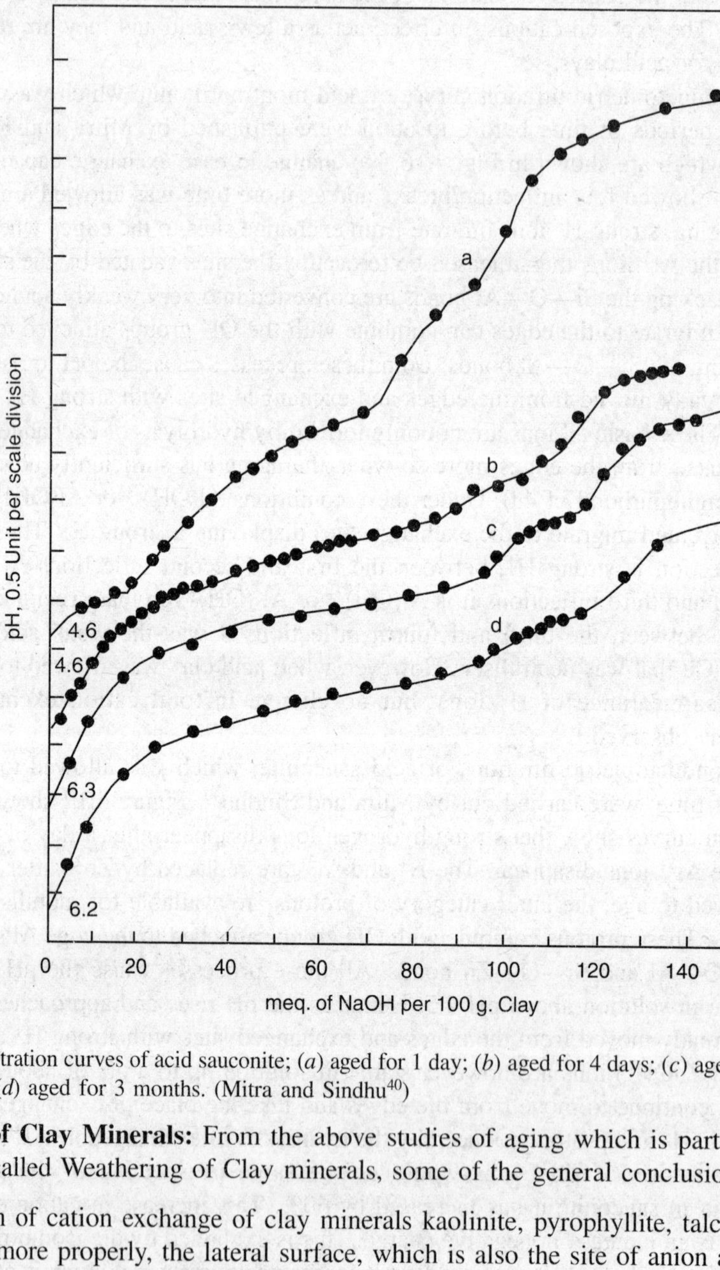

Fig. 4.17 pH titration curves of acid sauconite; (*a*) aged for 1 day; (*b*) aged for 4 days; (*c*) aged for 15 days; and (*d*) aged for 3 months. (Mitra and Sindhu[40])

Weathering of Clay Minerals: From the above studies of aging which is part of weathering of soils can also be called Weathering of Clay minerals, some of the general conclusion can be drawn.

1. The origin of cation exchange of clay minerals kaolinite, pyrophyllite, talc is due to broken bonds or more properly, the lateral surface, which is also the site of anion adsorption as will be discussed section 4.9. The CEC in this case increases as the particle size decreases.
2. Cation exchange capacity (CEC) of all monomorillonite clays does not depend on the particle size. The origin of CEC is due to the isomorphous replacement of Si^{4+} by Al^{3+} or Al^{3+} by Mg^{2+} in tetrahedral or octahedral layers or both and due to the lateral surfaces.

3. All montmorillonites give strong acid, when passed through H-resin. The-H-clays immediately start breaking till a homoionic suspension is obtained. Those minerals in which isomorphous replacement in the octahedral layer give positive charge to it. The exchange should increase on aging as in the case of sauconite, saponite (aluminium), nantronite (aluminium) and beidellite.
4. Rates of decomposition of acid clays under similar conditions may be as follows: Aluminium minerals < Fe minerals < Mg-minerals < Zn minerals.
5. In micas, where the isomorphous substitution is very high but tetrahedral layer shows low CEC and it depends on the particle size which also increase the CEC due to the lateral surfaces.
6. Rate of decomposition of muscovite is faster than that of montmorillonites or may be rate of decomposition of mica is faster than that of corresponding montmorillonites.

4.9 ADSORPTION OF ANIONS BY CLAY MINERALS

As has been mentioned earlier, the cleavage planes of clays and micas are very different in surface comparison to the lateral surfaces. The author believes that a clear distinction should be drawn between surfaces produced by violent mechanical cleavage in the (hkO) planes, and the normal edges of crystals which have simply ceased to grow. One of the most vivid demonstration of differences in chemistry between kaolinite surface and edges was first presented by Thiessen[55] in an electron microscopy study of adsorption of negatively charged colloidal gold by clay particles. Colloidal gold were preferentially taken up at edges as shown in Fig. 4.18. Polyphosphates as studied by Michael[45] are also strongly adsorbed on edges. Such modified edges no longer take up gold particles preferentially, but show that clay edges are positively charged.

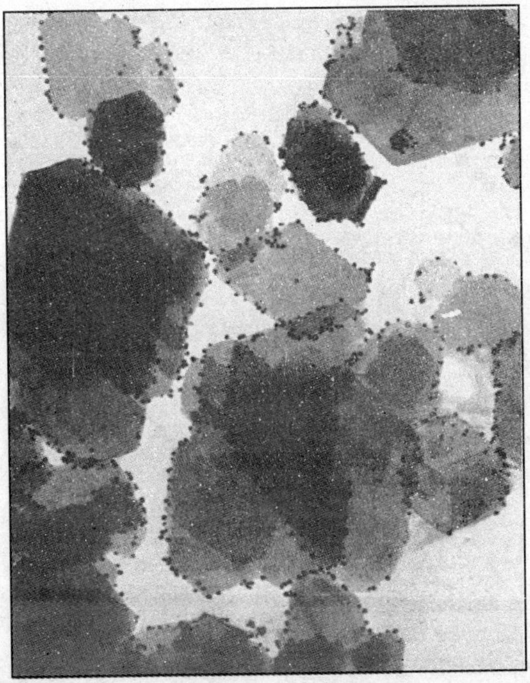

Fig. 4.18 Electron micrograph of a mixture of a kaolinite and a gold sol.

Clay suspensions show the phenomenon of Thixotrophy in which the fluid system is under high shearing force but solid under low shearing force. Such transformation is defined as gel-liquid-gel and is very common with montmorillonite clays aqueous suspensions. The highly hydrated clay suspension forms a structure in which edges are attracted by the planar surface and form a what is called "house of cards structure". The vacant space is occupied by water when gel like structure is obtained. On applying high shearing force, the house of cards as well as water structure is demolished. Now clay is just a suspension in water and thus behaves as a liquid. On allowing it to stand, edges again stand over the planar surface to become gel. Such behaviour show that edges carry a positive charge.

The ability of silicate minerals to combine with anions as well as cations points to their basically amphoteric nature. However, compared with its acid character the reactivity of clay as a bases, as a rule, are subdued. Anions of many strong acids, like Cl^- and SO_4^{2-} are often seen to be negatively adsorbed[10] instead of being taken up from solution. Phosphate fixation by clay minerals is considered another important manifestation of a fundamental basic character of clay minerals for which structural mechanism was investigated by Mitra and Sindhu[41] for four silicate minerals namely pyrophyllite, kaolinite,

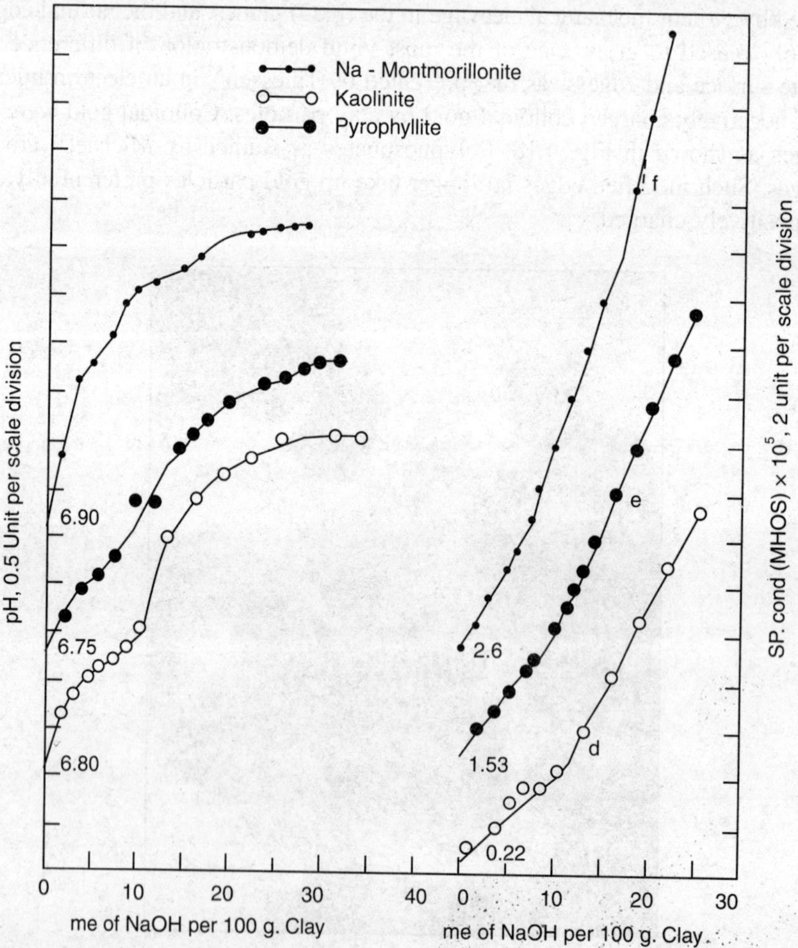

Fig. 4.19 pH and conductometric titration curves of Na-montmorillonite, pyrophyllite and kaolinite with NaOH (Initial pH and conductance are indicated on the curves) (Mitra and Sindhu[41]).

montmorillonite and mica. Na clays were titrated with acid (HCl, H_3PO_4) and base (NaOH, Na_3PO_4) and adsorption of phosphate was correlated. The pH and conductometric titration curves, shown in Fig. 4.19, indicate definite zones of buffer action against the added alkali followed by inflection proves that reaction between four minerals and alkali does take place and that this reaction has a definite end point. On titrating these minerals against HCl, pH falls slowly initially but much more rapidly subsequently, after which the pH continues to fall at an increasingly lower rate. The fact that all these minerals had a pH greater than 6.0 shows that H^+ and Al^{3+} are not present. Exposed planar surfaces of clay minerals have hexagonal openings through which no hydrated cation can pass. Under such conditions lateral surface of silicate minerals are the only possible seat of the reaction responsible for their ability to combine with acid and base. For this it is necessary to consider the atomic structure of this surface in some details. The first point to note is that due to rupture of bonds there will be incomplete tetrahedral and, octahedra on this surface. The four silicate minerals have the following kinds of bonds:

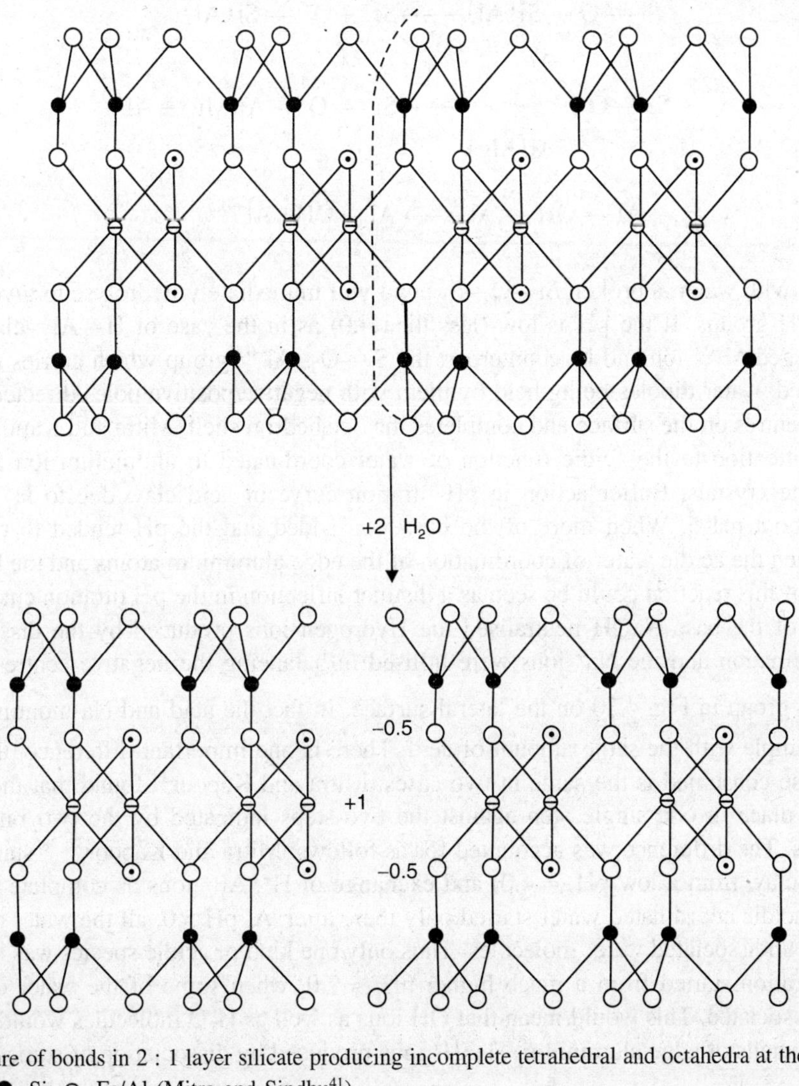

Fig. 4.20 Rupture of bonds in 2 : 1 layer silicate producing incomplete tetrahedral and octahedra at the edges. O, O; ⊙, OH; ●, Si; ⊖, Fe/Al (Mitra and Sindhu[41]).

$$\text{Si}-\text{O}-\text{Si(Al)}$$

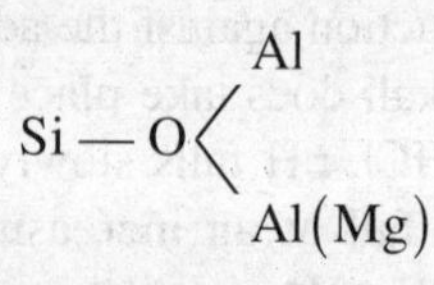

$$\text{Al(Mg)}-\text{OH}-\text{Al}$$

The dashed line in Fig. 4.20 shows one way in which the bonds may rupture in a three-layer lattice. Assuming that the bonds are ionic or predominantly so, their rupture will expose equivalent amounts of positive and negative centre on the lateral surfaces

$$\text{Si}-\text{O}-\text{Si(Al)} \longrightarrow \text{Si}^{+} + \text{O}^{-} - \text{Si(Al)}$$

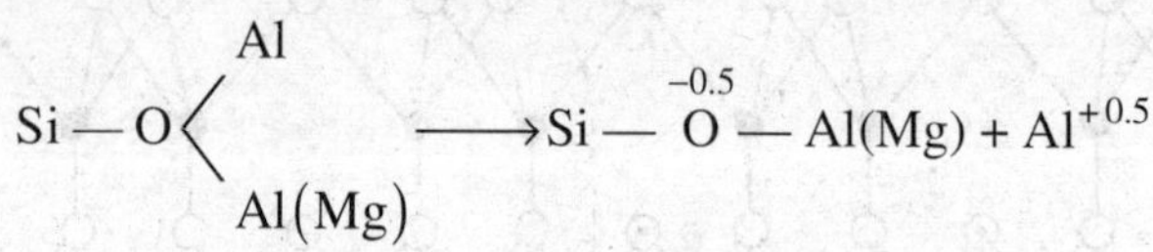

$$\text{Al}-\text{OH}-\text{Al} \longrightarrow \text{Al}-\overset{-0.5}{\text{OH}} + \text{Al}^{+0.5}$$

In contact with water, a broken Si—O—Si bond will immediately hydrolyse to give a pair of very weakly Si—OH groups. If the pH is low (less than 4.0) as in the case of H—Al—clay, the exposed positively charged $Al^{+0.5}$ ion and its counterpart the Si—O—$Al^{-0.5}$ group which carries negative charge will be hydrated, water dipoles being held by them with negative/positive poles directed towards positive/negative centres on the surface and completes the octahedron shell. Mitra and Kapoor[37, 38] were the first to draw attention to the acidic function of water coordinated to aluminium ion at the edges of montmorillonite crystals. Buffer action in pH titration curve of acid clay, due to H^+ and Al^{3+} were complete at about pH 5. When more of the base was added and the pH tended to rise further, the reaction between the acidic water of coordination of the edge aluminium atoms and the base set in, and the end point of this reaction could be seen as a distinct inflection in the pH titration curves (Fig. 4.16). The OH ions of the base NaOH neutralised the Hydrogen ions produced by the dissociation of the water of coordination and the Na^+ ions were utilised in balancing the negative centres (e.g., the two Si—$\overset{-0.5}{\text{O}}$—Al) group in Fig. 4.20 on the lateral surface. In fact the acid and Na-montmorillonite were of the same sample with the same amount of acid. There is one important difference, though the total quantity of base consumed is the same in two cases, Mitra and Kapoor[38] found that the reaction with the base took place in one single step against the two steps indicated by the two inflections in the titration curves. The difference was accounted for as follows: Mitra and Kapoor[37, 38] started titration of their H—Al—clay; from a low pH (≈ 4.0) and exchange of H^+, Al^{3+} ions as complete at pH 5.0. The reaction with acidic coordinated water started only there after. At pH 5.0, all the water of coordination would exist as undissociated water molecules. Thus only one kind of acidic species was involved. In Na clay where titration started from a much higher pH, ≈ 7.0, when some of the water of coordination would have dissociated. This would mean that OH ions as well as H_2O molecules would now complete the octahedral shells of edge aluminium. The H^+ ions produced by dissociation of some of the water of coordination are used up in balancing some of the negative centres, other such centres still have oriented water dipoles as their near neighbours. As a result two kinds of acidic species on the lateral surface accounting for the reaction with the base will take place in two stages. Since the H^+ ions were derived from the water of coordination, the same total quantities of the base (about 15 meq/100 g) is used up irrespective of whether the reaction takes place in one step or in two steps. In fact one could hardly distinguish between the titration curves of Na-montmorillonite and Na-mica, except for the fact that more of the base is consumed by mica than montomorillonite.

Kaolinite and pyrophyllite show that reaction takes place in one step, where the initial pH was 6.0. One would be tempted to think that at a comparatively low pH, most if not all, of the coordinated water is in an undissociated state.

Titration curves with HCl have only one inflection/break which corresponds with the alkali titration values. The acid neutralises the edge OH after which pH sharply falls. However, in the titration with a base, the latter neutralises not only H^+ produced by the progressive dissociation during the titration of the H_2O molecule initially in a state of coordination with edge aluminiums, but also the H^+ which is produced by the dissociation of some of the coordinating water molecules and had come to balance some negative centres ($Si—\overset{0.5}{O}—Al$). Since these H^+ ions are the counterparts of the OH^- ions covering some of the positive Al-centres, their neutralisation, in addition to the neutralisation of H^+ dissociated from the coordinating water molecules during the titration with NaOH, will obviously give a higher titre value with base than with the acid. The Cl^- serving as counter ions of the positive centres may also replace some water molecules. Uptake of phosphate and collateral variations in pH and specific conductance: Fig. 4.21 shows the variations in the pH and specific conductance observed when increasing amounts of H_3PO_4 were added to aqueous suspension minerals. The results were corraborated with the adsorption of phosphates. The pH values corresponding to the three stages of dissociation of phosphoric acid are 2.1, 7.2 and 12.3 (at 25 °C). In the pH range 7.0 and 2.5, for all practical purposes $H_2PO_4^-$ will be significant. Like HCl, the H_3PO_4 is used up in two ways. First it neutralised OH groups attached to the edge aluminiums.

$$Al—OH + H^+ + H_2PO_4^- \longrightarrow AlH_2PO_4 + H_2O$$

Secondly, $H_2PO_4^-$ are taken up against those positive Al-centres that have water in their coordination shells. The $H_2PO_4^-$ may form moderately strong coordinate bonds with the edge aluminiums as suggested by Ksu and Lotse[27], in this case they displace water molecules from the coordination shell. The $H_2PO_4^-$ held by a strong coordination bond to an Al ion carrying 0.5 unit of positive charge will give the moiety $\left[\begin{array}{c} \mathrm{O} \\ | \\ \mathrm{Al—O—OH} \\ | \\ \mathrm{OH} \end{array}\right]^{-0.5}$ carrying 0.5 unit of negative charge. The H^+ ion of the added H_3PO_4 will be electrostatically bonded to two such moieties in addition to being held against negative centres like $Si—O—Al^{-0.5}$ already existing on the surface. The negative centres as cited above draws the H^+ of the added H_3PO_4 and this tends to immobilise and subsequently, $H_2PO_4^-$ replace OH^- ions. What is more important is the fact that maximum adsorption value in terms of meq/100g agree with the amount of the H_3PO_4 reacting as an acid with lateral aluminiums, signifying a one-to-one correspondence. What was more important is the fact that the acid combining capacity, the base combining capacity and the capacity to take up phosphate are all the same. Further, the reaction with H_3PO_4 is a surface reaction and there is no precipitation in the sense that a separate phosphate phase is formed[24, 26] in which case more of phosphate would be taken up.

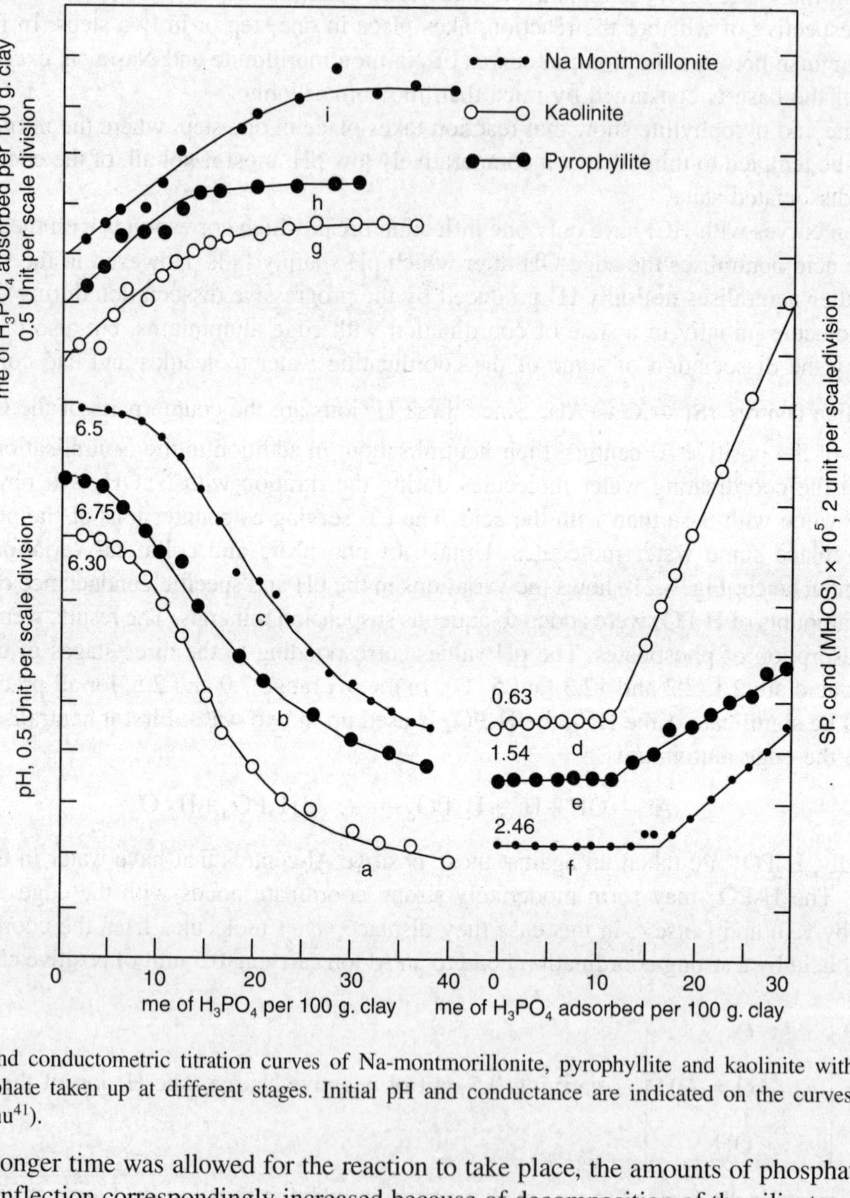

Fig. 4.21 pH and conductometric titration curves of Na-montmorillonite, pyrophyllite and kaolinite with H_3PO_4 and phosphate taken up at different stages. Initial pH and conductance are indicated on the curves. (Mitra and Sindhu[41]).

When a longer time was allowed for the reaction to take place, the amounts of phosphate taken up and point of inflection correspondingly increased because of decomposition of the silicates as was also noted by Low and Black[32, 33] but a one-to-one correspondence between the amount of the H_3PO_4 reacting as an acid in terms of titre value and that of phosphate taken up is maintained. This is also in line with Hsu's[24] thinking who distinguished between two stages of reaction in phosphate fixation, one progressing at a rapid rate and reaching completion in a matter of hours or minutes and the other at a much slower rate. The rapid fixation is surface adsorption which under acidic conditions is followed by precipitation.

Interaction with sodium phosphate: Figure 4.22 shows curves of pH, specific conductance and uptake of phosphate on addition of Na_3PO_4. Titration curve resemble that with NaOH. Uptake of phosphate increases linearly with addition of Na_3PO_4 till inflection is reached, after which there is a decrease. However, in terms of meq/100 g about three times more Na_3PO_4 has to be added than NaOH to reach the inflection/break. Na_3PO_4 at pH 10.5 is in a hydrolysed state as Na_2HPO_4 and NaOH. The NaOH produced neutralises the H^+ ion form the dissociation of water of coordination. Since one molecule of NaOH is available from the hydrolysis of one molecule of Na_3PO_4, three times more Na_3PO_4 in terms of meq of the latter per 100 g of clay have to be added to effect the neutralisation than if a solution of NaOH is used.

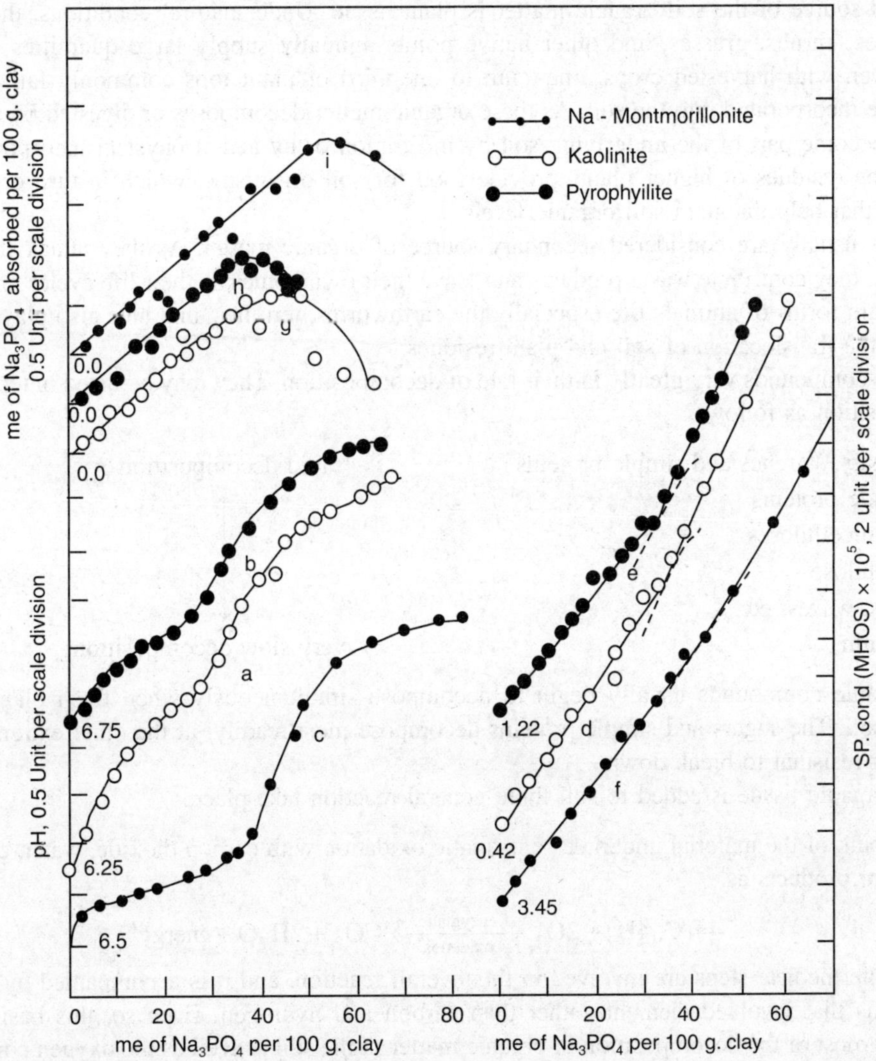

Fig. 4.22 pH and conductometric titration curves of Na-montmorillonite, pyrophyllite and kaolinite with Na_3PO_4 and phosphate take up at different stages. Initial pH and conductance are indicated on the curves. (Mitra and Sindhu[41]).

It can be seen from Fig. 4.22 that maximum Na_3PO_4 adsorption by the minerals is in the pH range 9.5 to 11. In this pH range, the hydrogen of the Na_2HPO_4 can be expected to function as an acid towards the basic OH groups attached to the edge aluminium atoms. Each such OH groups is thus replaced by a PO_4^{3-} ion. In effect each lateral Al centre is covered by $Na_2PO_4^-$. Accordingly three times more Na_3PO_4 is taken up than H_3PO_4. The decrease in phosphate uptake beyond the inflection/break in pH/conductometric curves is evidently due to desorption of PO_4^{3-} by OH^- ions which now accumulate in the system[41].

4.10 SOIL ORGANIC MATTER

The original source of the soil organic matter is plant tissue. Under natural conditions, the tops and roots of trees, shrubs, grasses, and other native points annually supply large quantities of organic residues. Even with harvested crops, one-tenth to one-third of plant tops commonly fall to the soil surface to be incorporated into the soil. As these organic matter decomposes or digested by soil organisms, they become part of the underlying soil by infiltration or by actual physical incorporation. Accordingly, the residues of higher plants provide food for soil organisms, which in turn create stable compounds that help maintain soil organic levels.

Animals usually are considered secondary source of organic matter. As they attack the original plant tissues, they contribute waste products and leave their own bodies as their life cycles are consummated. Certain forms of animals life especially the earthworms, termites, and ants also play an important role in the translocation of soil and plant residues.

Organic compounds vary greatly in their rate of decomposition. They may be listed in terms of ease of decomposition as follow:

1. Sugars, starches and simple proteins — rapid decomposition
2. Crude proteins
3. Hemicelluloses
4. Cellulose
5. Fats, waxes, etc.
6. Lignins — very slow decomposition

All organic compounds usually begin to decompose simultaneously when fresh plant tissue is added to a soil. The sugars and simple proteins decompose most readily; at the other extreme, lignins are the most resistant to break down.

When organic tissue is added to soil three general reaction take place.

1. The bulk of the material undergoes enzymatic oxidation with carbon dioxide, water, energy and heat as major products as

$$-(\text{C, 4H}) + 2O_2 \xrightarrow[\text{oxidation}]{\text{enzymatic}} CO_2 + 2H_2O + \text{energy}$$

Many intermediate steps are involved in this overall reaction, and it is accompanied by important side reactions that involved elements other than carbon and hydrogen. Even so, this basic reaction accounts for most of the decomposition of organic matter in the soil as well as the oxygen consumption and release of CO_2.

2. The essential elements such as nitrogen, phosphorus and sulphur are released and/or immobilized by a series of specific reactions relatively unique for each element. As enzymatic changes of soil organic matter proceed, simple organic products emerge. For example, as proteins are attacked by

microbes, amino acids appear. These in turn are broken to produce first amino compound and sulphides and finally nitrates and sulphates. Similar breakdown of other organic compounds release inorganic phosphates as well as Ca^{2+}, Mg^{2+} and K^+ ions.

3. Compounds very resistant to microbial action are formed either through modification of compound in the original plant tissue or by microbial synthesis. Collectively, these resistant compounds comprise of soil humus. The bulk of these compounds have resulted from two general type of biochemical reactions—decomposition and synthesis. Decomposition occur as chemicals in plant residues are broken down drastically modified by soil organisms. Even lignin is split and degraded in many of its structural units being destroyed. Synthetic reaction involves breakdown products of lignins as the phenols and quinones. These products present initially as separate molecules are enzymatically stimulated to join together as polyphenols and polyquinones as polymers which are quite complex, with no sharply defined physical or chemical properties, like melting point.

Humic substances are dark coloured, acidic predominantly aromatic, hydrophylic, chemically complex polyelectrolyte like materials, that range in molar mass from a few hundreds to several thousands. A definite structure to humus cannot be given at the moment and require more research before a satisfactory account may be given.

Most of the humus can be extracted with alkaline solution as shown by Schinitzer and Khan[52]. On acidification of this extract to pH 2, the precipitate formed, the precipitate formed is known as humic acid (HA) and the fraction remaining in solution is known as fulvic acid (FA), which is soluble in both acid and alkali. The 0.4M to 0.5 NaOH is the most suitable extractant for isolating humic acid materials, low in ash content and is called Humin. It appears that structurally these three fractions are similar but they differ in molar mass and functional groups. FA has a lower molar mass although they contain more oxygen, less of carbon and nitrogen with higher content of oxygen containing functional groups (—COOH, —OH, — C=O) per unit mass than the other two humic acid fractions. The chemical structure and properties of humic fractions appear to be similar to those of HA. The insolubility of humin seams to arise from it being firmly adsorbed on soil and sediment constituents. Fractions of HA into Hymatomelanic acid or grey HA and brown HA is done with alcohol though it is not done very often.

It appears that up to 50% of the alphabetic structures in HA's and FA consist of fatty acids entailed to phenolic OH groups. The remaining aliphatics are made up of more loosely held fatty acids and alkanes that seem to be physically adsorbed on humic material and which are not structurally humic components and possibly composed of aliphatic chains joining aromatic rings. As shown by a variety of chemical degradation experiments on HA's and FA's extracted from soils differing in locations and pedologial histories, the major HA and FA degradation products are phenols, benzene and carboxylic acids.

Some clues as to what type of structural arrangement may be the most probable one are provided by physico-chemical methods. Scanning electron microscopy and viscosity measurements show that both the shape and size of HA and FA particles are strongly affected by pH. At low pH, HA and FA tend to aggregate as elongated fibers and bundles of fibers. The aggregation appears to be brought about by hydrogen bonding, van der Waals forces and interaction between the π-electrons of adjacent aromatic rings as well as by homolytic reactions between free radicals. As the pH increases these forces become weaker and because of increasing ionization of —COOH and —OH group, particles separate and begin to repel each other electrostatically so that molecular arrangements become smaller and smaller but better oriented. Thus, one witnesses aggregation at low pH and dispersion at high pH, which is also the basis for separating humic substances into HA's and FA's. Both HA's and FA's behave like flexible, linear synthetic polyelectrolytes so that one deals here not with chemical structures largely or exclusively composed of condensed rings, but there must be present numerous linkages about which relatively free rotation can occur. It becomes, in this way even more apparent that humic substances are not single molecules.

4. The colloidal surfaces of humus are negatively charged, the source of the charge being hydroxy (OH), carboxylic (—COOH), or phenolic (Ph—OH) groups. The extent of the negative charge is pH dependent (high at high pH values).

5. At high pH values the cation exchange capacity of humus may on mass basis be 150–300 meq/100 g which far exceeds the cation exchange capacity of most of the clay minerals.

6. The water holding capacity of humus on mass basis is 4 to 5 times that of the clay minerals.

7. Humus has a very favourable effect on aggregate formation and stability.

8. The black colour of humus tends to distinguish it from most of the other colloidal constituents of soils.

9. Cation exchange reactions with humus are qualitatively similar to those occurring with silicate clays.

The humic micelles, like clay particles carry a swarm of adsorbed cations (Ca^{2+}, H^+, Mg^{2+}, K^+, Na^+, etc.). Thus colloidal humus may be represented by the similar illustrative formula used for clay

$$H^{+}\ \boxed{\text{Micelle}}\ \begin{matrix} Ca^{2+} \\ Al^{3+} \end{matrix}$$

and similar cation exchange reaction take place but rather an association of molecules of microbiologist, polyphenolic, listening and condensed lignin origin. It is noteworthy that *X*-ray analysis, electron microscopy and viscosity measurements of FA point to a relatively open flexible structure perforated by voids of varying dimensions and various cations may fit into the voids, provided the charges are complementary. A chemical structure that is in harmony with many of the requirements was proposed by Schnitzer[51] Fig. 4.23.

This molecular arrangement may account for a significant part of the FA structure, each of these compounds that make up the structure has been isolated from FA prior to and following chemical degradation.

The colloidal nature of humus has following major characteristics:

1. The tiny colloidal humus particles (micelles) are composed of carbon, hydrogen, and oxygen (probably in the form of polyphenols, polyquinones, polyuronides and polysaccharides.
2. The surface area of humus colloids per unit mass is very high, generally exceeding that of clay minerals. Humus enhances minerals breakdown and, in turn, nutrient availability in two ways. First humic acid can attach the minerals and bring about their decomposition, thereby releasing essential base forming cations.

$$\underset{\text{Microline feldspar}}{KAlSi_3O_8} + \underset{\text{Humic acid}}{H^{+}\ \text{Micelle}} \longrightarrow \underset{\text{Acid feldspar}}{HAlSi_3O_8} + \underset{\text{Adsorbed } K^{+}}{K^{+}\ \text{Micelle}}$$

The potassium is changed from a molecular to adsorbed state, which is more readily available to higher plants.

The second mechanism for increasing the availability of some cations is through the formation of stable organomineral complexes with these ions. For example polysaccharides and fulvic acids form such complexes with metal ions such as Fe^{3+}, Cu^{2+}, Zn^{2+}, and Mn^{2+}. The cations are attracted from the minerals in which they are found and held in complex form by the organic molecules. Later they may be taken up by plants or may take part in synthesis of clay and other inorganic constituents.

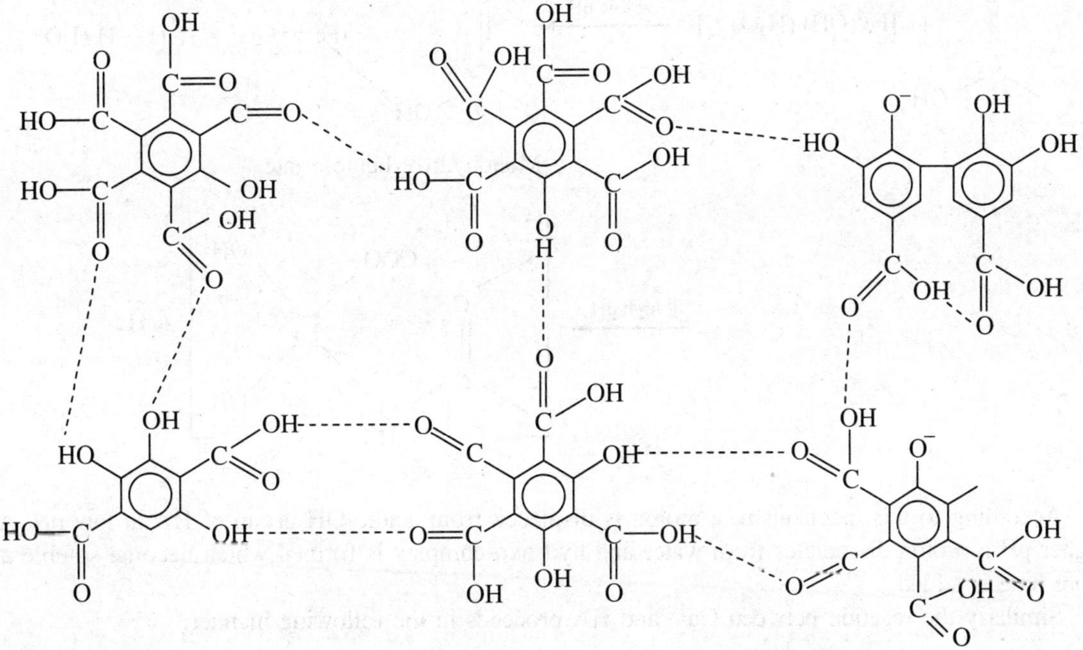

Fig. 4.23 A partial chemical structure for Fulvic Acid (FA).

These examples are reminders of the beneficial interactions of clay and humus. It is most fortunate that these two soil constituents tend to reinforce each other in many soil-plant relationships.

4.11 REACTIONS OF HA's AND FA's WITH METALS AND MINERALS

Humic substances are capable of interacting with metal ions, metal oxides, metal hydroxides and more complex minerals to form metal organic association of widely different chemical and biological stabilities. Interactions between humic substances and metal ions have been described as ion exchange, surface adsorption, chelation, coagulation and peptisation reactions[51]. The reactions of organic matter with Zn^{2+} hinders the oxidation of organic matter by H_2O_2 and destroy the ability of soil to complex with Zn^{+2}. A more adequate knowledge of metal – HA and metal – FA interactions is desirable from the point of view of soil genesis, soil structure formations, nutrient availability and especially the chemistry of micro and toxic elements and their mobilisation in environment.

Potentiometeric titrations of humic substances in the absence and presence of various metal ions have been carried out by various researchers. Metals of first transition series of the periodic table form complexes with humic substances, and that the order of stabilities of different metal complexes follows that of Irving-Williams series[25]

$(Pb^{2+} > Cu^{2+} > Ni^{2+} > Co^{2+} > Zn^{2+} > Cd^{2+} > Fe^{2+} > Mn^{2+} > Mg^{2+})$

Similar results were obtained by Khan[30], Khanna and Stevenson[34].

Van Dijk[58] rationalises the titration of HA + Fe^{3+} with base in the following manner:

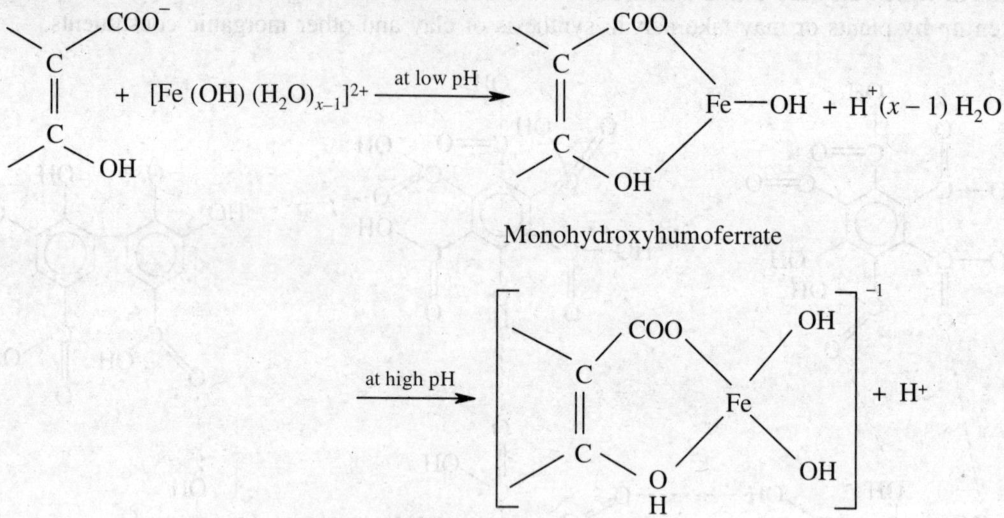

According to this mechanism, a proton is displaced from acidic OH group of HA at low pH. At higher pH a proton dissociates from water and hydroxo-complex is formed which become soluble as more base is added.

Similarly the reaction between Cu^{2+} and HA proceeds in the following manner:

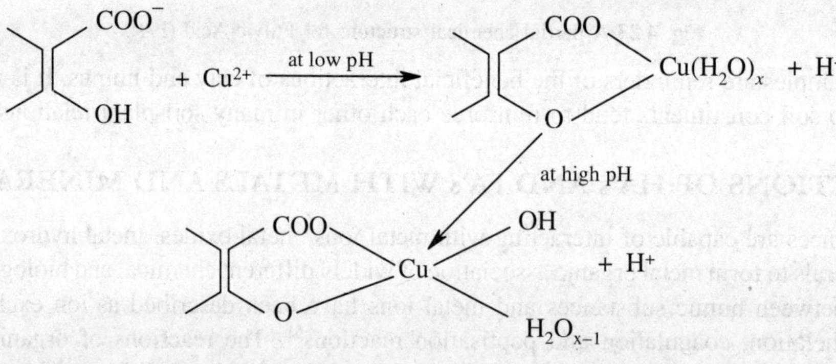

The reaction between Cu^{2+} and FA at pH 5.0 proceeds in the following manner:

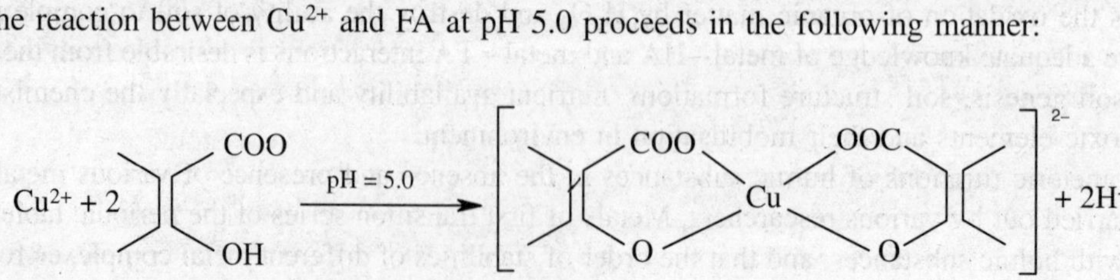

Thus at low pH a proton is displaced from an acidic OH group of the HA, whereas at higher pH protons dissociate from water molecules covalently bound to Cu^{2+}, so that hydroxo-complex is formed. Van Dijk also notes that the capacity of HA to bind metal ions equals approximately the number of titrable H^+ ions divided by the valency of the interacting metal ions.

Metal –HA and metal –FA stability constants have been determined but results vary widely. There is still considerable uncertainty about how to define the problem, which method to use and how useful such constants are once they have been determined.

4.12 DECOMPOSITION OF SOIL MINERALS BY ORGANIC MATTER

HA is capable of extracting ions by chelation process from otherwise insoluble solids and enable the transfer of these ions in chemical environments where they would normally be precipitated. In this way, HA and FA induce the decomposition of soil minerals. The available data indicate that dissolution occurs initially at a rapid rate, subsequently this rate decreased and become very slow or constant with time (50-50). Of the two types of humic matter, HA appears to be more effective at $pH > 6.5$ than FA in mineral dissolution by chelating the metals from the crystal structure[11, 58]. It appears that cations in octahedral sheets dissolve at higher rates than those in tetrahedral sheets.

If metal/HA or metal/FA ratios are low, the complexes are water soluble, but if the ratios are high, the metal –HA or –FA complexes become water insoluble. Strong solvent activity have also been reported for HA's for hematite, pyrolusite, feldspar, biotite, enstatie, actinolite and epidote, Mn^{3+} and Mn^{4+} oxides and Mn^{2+} hydroxides and basic salts, which are also gradually dissolved. Similarly FA have been shown to be efficient for dissolving metals form goethite, gibbsite, clay minerals, chlorites and mica[52].

A point of considerable interest is that minerals rich in iron are most susceptible to attack by humic acid materials[58]. The high affinity of HA and FA for iron and the low crystal field stabilisation energies of Fe^{3+} and Fe^{2+} provide possible explanations for these findings. The complexing of iron by humic materials appears to have an adverse effect on the structural stability of Fe-rich minerals, along with Fe and other major constituent elements such as Mg, Al, K, and Si are also brought into solution.

At least two, possibly three, different binding sites for iron were found to occur in humic materials. (*a*) Fe^{3+} was strongly bound and protected by tetrahedral and octahedral coordination. This form of iron exhibited considerable resistance to complexation by known sequesting agents and reduction; and (*b*) Fe^{3+} was adsorbed on external surfaces of humic materials weakly bound octahedrally and easily complexed and reduced. These observations were made on laboratory prepared Fe—HA and Fe—FA complexes.

Another process of decompositions of soil minerals can also occur by ferrolysis, defined as the breakdown of clay minerals due to alternate reduction and oxidation of iron oxides. Under oxidising conditions ferrous ion is converted into ferric ion and H^+ is produced

$$Fe^{2+} + H_2O \longrightarrow FeOH^{2+} + H^+ + e$$

which will attack the clay mineral. This process is more chemical in nature and can take place in the absence of humic matter.

Mineral weathering is an essential part in process of soil formation. It produces clay and releases the elements used for differentiation of soil horizons, and those required for plant growth. The dissolution products can either be immobilised by changing physico-chemical conditions, or else are free to move in the soil or will be leached to enrich rivers, lakes and oceans, causing eutrophication of the environment. In the presence of humic matter, the weathering process is enhanced but with the formation of metal-humic acid complexes and chelates, the harmful effect of the dissolution products is reduced considerably. Accumulation of excessive amounts of free ions in water bodies and soils is prevented because of adsorption and bonding by humic substances. In acid soils, mineral weathering may release large amounts of free Al, Fe and Mg ions creating toxic conditions for plant growth. However, when humic matter is present, the elements released during the decompositions process are in the form of metal-humic acid complexes. As chelates, their activities are decreased substantially. Depending on saturation and denaturation of the complexing sites, the complexes can be immobilised or remain mobile in the soil pedon. Current concepts on the formation of spodic, mollic and andic horizons are based on the formation and mobility of metal-humic chelates although a mechanism for formation and movement of inorganic Si—Al and Fe (III) —Al hydroxy complexes has been reported recently. Taylor[56] proposes that the formation of spodosol (Podzol) is attributed to migration of soluble Fe(III)—Al hydroxy complexes as opposed to Fe-organic complexes[12]. From the agricultural standpoint, it can be concluded that humic matter is capable of dissolving more rapidly, than generally possible in the prevailing condition, the essential elements required for plant growth. The dissolution product in the form of a chelate is believed to provide the carrier mechanism by which depleted nutrients at root surfaces can be replenished. Minerals that are otherwise insoluble and useless for plant nutrition become rapid suppliers of plant nutrients by decomposing in the presence of humic matter.

4.13 THE ROLE OF CLAY FRACTION IN THE PARTICLE ARRANGEMENT AND STABILITY OF SOIL AGGREGATES

The electrochemistry of ionic solids, clay minerals suspensions can be discussed in terms of electrical double layer which is assumed to exist at the solid-liquid interface. The inner sheet of double layer is believed to be made up of ions which are firmly built into the surface of the solid by chemical or valence forces and it endow the latter with a surface charge. This charge is balanced by an equivalent amount of ions of the opposite sign, the counter ions stay in the liquid phase and form the outer sheet of the double layer. If an electrolyte is now added to the liquid its ions enter into an exchange equilibrium with those of the already existing double layer. This is the main feature of the electrochemistry ionogenic surface.

As the concentration of electrolyte increases, there will be tendency of the cations to accumulate on the solution side of fixed double layer which decrease the thickness of double layer, thereby leading to flocculation of colloid. The higher the valence of cation, the lower the concentration of solute required to bring about sufficient change in the double layer.

The physico-chemical properties of valence the clay fraction of soils exerts a profound influence on many soil properties. The stability of soil aggregates *e.g.* to wetting, drying or mechanical stress applied during cultivations depends upon the forces operating between individual clay particles as well as between groups of particles. Swelling and dispersion are controlled by the same forces but the distinction lies on the distance between the particles. An understanding of swelling and dispersion of clays in soils is essential for a proper appreciation of physical soil characteristic such as retention and movement water in soils and the mechanisms controlling the stability of aggregates.

Clay Particle Interactions: The forces of attraction and repulsion which affect the interactions between clay particles and thus their behaviour in an aqueous suspension are largely dependent on the type and concentration of the electrolyte in the system. However, clay mineralogy and the composition of the exchangeable cations are also important. The total (attractive and repulsive) potential energy curve for the interaction of two clay particles (DLVO theory) has been described by Olphen[57].

The total potential energy curve for highly dispersed clay suspension exhibit a deep minimum at close separation (*e.g.* 5 nm). The existence of a minimum shows that, in addition to repulsive forces, attractive forces are operative between the clay particles. The repulsive forces origin is the isomorphos replacement in octahedral or tetrahedral layer which gives a negative charge to the planar surfaces. Ordinarily, the planar surfaces should repel one another. The origin of attractive forces is the positive charge carried by lateral as discussed in section 4.8 and 4.9, which are attracted by the planar surfaces. However, at high electrolytic concentrations (say 5M) attractive forces are suppressed and repulsive forces only operate, and thus potential energy minima disappear. These short range repulsive forces are thought to reflect the work required to dehydrate interacting surfaces containing ionic or polar species. They have thus been termed hydration forces.

As the electrolytic concentration in a stable suspension is increased, the attractive forces are suppressed; the particles coalesce to form loose flocs. This process of flocculation is usually reversed on lowering the electrolytic concentration and is supposed to involve a secondary minimum[59]. In some areas, however, clay particles may interact strongly so that the coalescence of the particles is not readily reversible with respect to electrolyte concentration. These particles are described as being in a primary potential energy minimum.

The Na- and Ca- montmorillonite provide examples of particle coalescence into secondary and primary minima respectively. When a suspension of Na- montmorillonite is treated with NaCl, floccules are formed which have an open (or "Card-house") type structure. On lowering the electrolyte concentration, the flocculation is reversed. In the case of Na-montmorillonite, the second minimum occurs at a double layer thickness of 5 to 10 nm. When $CaCl_2$ is added to a suspension of Ca-montmorillonite, flocculation into a weak secondary minimum initially occurs (Greene *et al.*[18]). However, if the suspension is shaken, the individual clay particles in the floccs undergo rearrangement into a parallel alignment. This induces them to condense into a deep primary minimum and thus form quasi crystals[2]. This condensation is irreversible so that even when the electrolyte concentration is reduced the clay particles remain in primary minimum. These quasi crystals can only be dispersed by applying a considerable amount of mechanised energy such as ultrasonics[19].

In addition to effects of electrolyte and exchangeable cations, the flocculation behaviour of clay particles is affected by their shape, size and charge density. The different shape of particles of montmorillonites and illites provides an explanation for the observation that when freeze-dried Ca-montmorillonite was placed in water, the quasicrystals failed to disperse, whereas Ca-illite redispersed, whereas the addition of $CaCl_2$ to a suspension of Ca-montmorillonite followed by shaking resulted in the formation of loose floccs, the Ca-illite suspension was flocculated by $CaCl_2$[18]. Hence, the flocculation behaviour of montmorillonite and illite clays are different. Although the flocculation values for Na- and Ca-montmorillonites are 12 meq l^{-1} NaCl and 25 meq l^{-1} $CaCl_2$, respectively, compared with the corresponding values of 45 meq l^{-1} NaCl and 0.25 meq l^{-1}. $CaCl_2$ for Na- and Ca-illites, kaolinite particles are flocculated even by very low electrolyte concentrations when the suspensions are near or below natural pH values. However, at pH values above 8, water adsorbed on lateral surface dissociate and thus the increase in negative charge occurs at the lateral surface as was discussed in section 6.9.

The association of clay particles and their organisation are mainly dependent on their mineralogy and composition of exchangeable cation. In Na-montmorillonite, the bond between the platility and sodium ions are weak and easily broken by water dipoles. By contrast, in Ca-montmorillonite, the force of attraction between stack layers, one above calcium ion and one below, is so strong that each clay crystal consist of a stack of several unit layers to form quasi-crystals. A large mica crystal would show poor surface development programme and its electrochemical reactivity would be very weak. Mechanical combination of the crystal would open up new faces on which ionic reaction would occur. The cleavage plane of mica on which K^+ ions are distributed can serve as the seat of important ionic reactions (Fig. 4.7). In the interior of the mica crystal one finds, oxygen ions whose centres form two hexagonal rings, one above and other, below the K^+ ions. The negative charge balancing the K^+ ions resides in the underlined tetrahedral layer of the lattice being caused by the isomorphous replacement of every fourth Si^{4+} in this layer by Al^{3+} ions. The coulombic force of attraction between the stacking layers is so strong that each clay crystal consisting of a stack of several unit layers, which cannot be penetrated by water molecules and thus micas do not disperse. In each crystal of kaolinite, usually larger than mica, several unit layer are linked by hydrogen bonding between hydrogen of hydroxy layer and hexagonal layer of the oxygen atom of another unit. As kaolinite crystal do not carry any charge, they do not disperse easily.

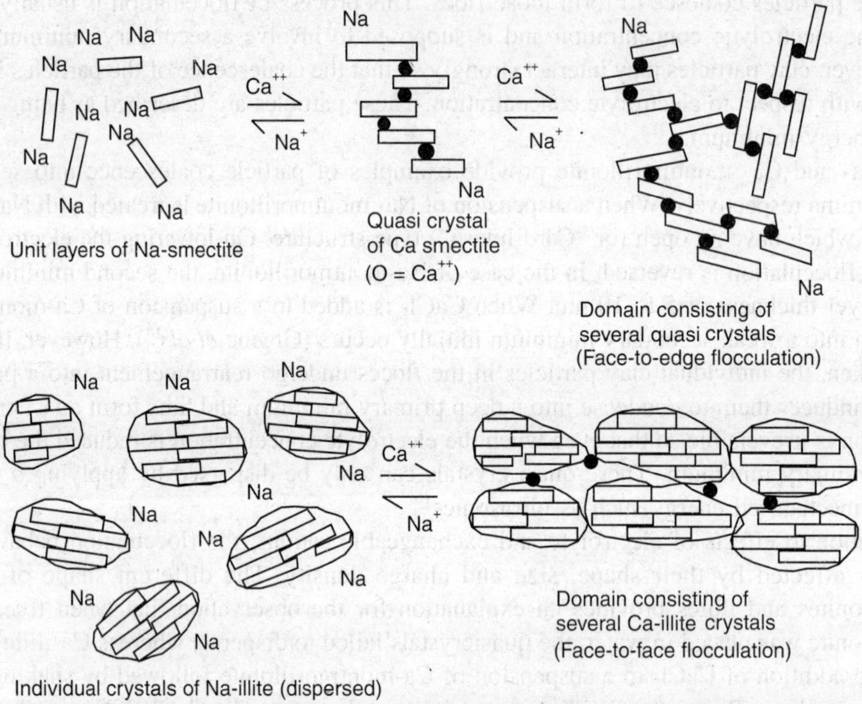

Fig. 4.24 Schematic representation of the formation of quasi-crystals and domains.

The individual crystals of clay minerals can further be linked into a groups known as domains due to interaction between their external surfaces (Fig. 4.24). These domains can be of colloidal size.

Association of clay particles can occur via face-to-face and edge-to-edge interactions (van Olphen[59]). Consequently, a number of structures are possible. Face-to-face association can produce stacks whose major dimensions are similar to those of the unassociated unit layers, with the thickness of the resultant particles having been increased. When the clay particles are associated in an edge-to-face fashion, a cardhouse structure results in voluminous floccs. Edge-to-edge association can either result in card house floccs or a ribbon-line structure. At high solid concentrations, both these association leads to the formation of gels. This system gives rise to the phenomenon of "Thixotropy", involving transition in the system

$$\text{Gel} \rightleftharpoons \text{Sol} \rightleftharpoons \text{Gel}$$

is fluid under high shearing force but solid like under low shearing forces. Such association of clay particles and the resulting strength of inter particle bonding have been investigated using optical measurements[53]. (*e.g.* optical density, birefringence, light scattering, *X*-ray diffraction, swelling pressure, electron microscopy, Rheology and porosity).

4.14 WIND EROSION OF SOIL

Wind erosion is the process by which loose surface material is picked up and transported by the wind, thereby changing the microtopography of soils and any agricultural activity associated with them. Fertilizers and Pesticides may be lost or redistributed. It operates in an environment that lacks a protective cover of vegetation and it is particularly significant in both hot and cold deserts, exposed mountain regions, regions of high wind velocity as in case of semiasic areas, as well as some of the more humid regions that experience periodic draughts. Fine material is deposited in houses, and in farm machinery which may be clogged. The cost of removal for particles and cleaning will lead to loss of working hours and a decline in productivity.

Vast quantities of water, be it the high speed flowing river water or monsoon rains, remove the fine particles of soil from the surface, particularly areas without vegetation and thus erode the soil. The fertile portion of soil is thus washed away, leaving behind a thick layer of sandy-loam and gravel. Huge quantities of such eroded soil lie in the sub-mountain tract. The slow movement of rivers in the plains decrease the silt carrying capacity of water. Only the fine sediments are transported to the central portion. Six centimeters or top fertile layer of the Shivalik hills of Punjab required 2400 years for its formation but may disappear in a single year of strong flooding.

Those areas where, the average annual rainfall is less than 300 mm, wind erosion of soil takes place particularly on unirrigated surfaces in the hot months of summer. The average wind velocity is 20 km/hr but, in an extreme situation, it may increase to 90–100 km/hr. During the summer months, the top soil behaves as a dry powder. Scorching dry, strong wind and dust storms damage the roads, streets and agricultured soil, which changes to barren land every year. One cubic kilometer of air can take away 1000 tons of soil particles from one place and move a distance of about 250 km in 12 hours.

In soils having high cation exchange capacity, the soil particles are held together by strong forces. Such well-structured soil has strong mechanical strength and particles are difficult to move. On the other hand, soils with weak structures may be rapidly abraided. The following are structural units stability factors.

1. **Water:** Soils with low CEC water is easily removed by surface drying and the cohesive bond is broken. On the other hand where finer materials predominate, water molecules are adsorbed on clay surfaces by electrostatic forces and are held there despite the high suctions caused by drying. Such wet soils tend to bind the mass together and wind erosion decreases.

   Silt and clay form surface crusts by the impact of raindrops. Whereas coarser particles are left loosely on the surface, whereas easily dry particles may be removed by the wind. Once the soil movement has begun, the crust may be fairly rapidly broken down and further drying is enhanced. The crust is part of the general surface smoothening caused by raindrop impact, a trend that increases the avalanching effect.
2. **Texture:** The influence of texture on erodibility is largely related to the soil moisture retention properties. The higher the proportion of silt and clay in a soil, the greater is the production of clods and lower is soil erodibility conversely, a high proportion of sand produces few clods and a highly erodible soil.

3. **Organic Cements:** Continuous addition of organic matter is necessary to improve cohesion, and that material left on the surface breaks down at a slower rate and is more useful than material mixed in by ploughing. This can be done by planting more trees and plants.
4. **Disintegrating Process:** Calcium carbonate weakens soil structure and increases erodibility. Sandy soils are an exception and have little structure in the first place. Addition of lime is beneficial, for it is mostly silt-sized and acts as a weak cement.
5. **Soil Erosion Rates:** Bagnold[65] derived a relationship between the rate of sand flow q (the weight of sand which moved along a lane of unit width past a fixed point in unit time) and the wind velocity over an eroding surface, u, such that

$$q = C\sqrt{\frac{d}{D}}\frac{\rho}{g}u^3$$

In this equation, D = standard grain diameter of 0.25 mm,

d = mean grain diameter in mm of the sand in question,

ρ = density of air, and

g = the gravitational constant (ρ/g has a value of 0.125 in SI units) and

C = an empirical constant.

The values of C were found with respect to the size distribution of the material involved, and the following values were used : 1.5 for uniform sand; 1.8 for naturally graded sand; 2.8 for sand with a wide range of grain sizes. When sand is driven over a surface containing a large number of hard, immobile elements, $C = 3.5$. On substituting the constants into the equation, it reduces to $a = 0.875\sqrt{d}\,u^3$.

Several problems arise in respect to the above relationships. Firstly, it predicts sand flow at velocities below the threshold of movement and, therefore, threshold must be independently determined. Secondly, the equation assumes that the mode of transport is both saltation and creep but not suspension, and it may therefore not describe fully the movement of very fine materials and, thirdly, the correct C value estimation may prove difficult, particularly if the equation is extended in its use for predicting soil flow.

4.15 SOIL ACIDIFICATION

The behaviour of clays described in the foregoing sections is also applicable to soils. Soil pH can be measured in 0.01 M $CaCl_2$ solution extract or in a suspension of soil in water. However, the pH of suspension in water is preferred. The buffering of soil pH is due to several soil properties, the effect of which is to make the drop in pH less than in an equivalent volume of water when acid is added. The reactions that occur with added acid are summarised as follow:

1. Carbonates are dissolved by carbonic and other acids.

$$CO_2 + H_2O \rightleftharpoons H_2CO_3 \rightleftharpoons H^+ + HCO_3^-$$

$$CaCO_3 + H^+ + HCO_3^- \rightleftharpoons Ca(HCO_3)_2$$

The product, calcium bicarbonate, is soluble and leached out of the soil by excess rain.

2. Cation exchange, which removes protons from the solution

$$M^+soil + H^+ \rightleftharpoons H^+soil + M^+$$

where M⁺ is an exchangeable cation. The acid soil so formed further breaks up to give Al-soil, or depending upon the nature of clay minerals and metal oxides present in the soil.

3. Acid dissolves folic acid but precipitates humic acid as

$$—COO^- + H^+ \rightleftharpoons —COOH$$

and at a high pH

$$\rangle CO^- + H^+ \rightleftharpoons \rangle C—OH$$

The effect of these reactions is to reduce the negative charge on ionised carboxyl and phenolic hydroxyl groups, respectively thereby making humic acid less soluble.

4. Aluminium buffering

$$AlOH(H_2O)_5^{2+} + H^+ \rightleftharpoons Al(H_2O)_6^{3+}$$

5. Proton absorption by clays and oxides of Al and Fe

$$\rangle AlO + H^+ \rightleftharpoons AlOH$$

$$\rangle AlOH + H^+ \rightleftharpoons AlOH;$$

These reaction lower the pH dependence on negative charge and are the source of positive charge that erodes the soil.

6. Hydrolysis of minerals

$$\underset{\text{Feldspar}}{4KAlSi_3O_8} + 4H^+ + 18H_2O \longrightarrow \underset{\text{Kaolinite}}{Si_4Al_4O_{10}(OH)_8} + 4K^+ + 8Si(OH)_4$$

In this reaction orthoclase feldspar decomposes and kaolinite and silicic acid are formed. Hydrolysis of other silicates also removes H^+ from solution. The two paths of conversion of one mineral to another are as follows:

1. Mica ⟶ Intermediate stages ⟶ Vermiculite ⟶ Semectite
2. Biotite ⟶ Mixed layer mineral ⟶ Vermiculite ⟶ Hydrobiotite ↓ Semectite ⟵ Vermiculite

In all these reactions, soil components remove protons from acidified solutions and pH decreases to an extent that depends on the buffer capacity of the soil.

In a neutral soil the exchangeable cations that dominate are Ca^{2+}, Mg^{2+}, K^+ and Na^+. As a result of acidification, these basic cations are replaced by protons and aluminiums. The loss of basic cations is permanent if they are leached out of the root zone or are removed in a harvested crop. The loss is temporary if they are taken up by plants and returned to the soil as litter or on the death of the plants. Some of the basic cations taken up by trees are held in woody tissues where they are effectively removed from the soil for many years. A distinction therefore needs to be made between acidification of the soil and acidification of the soil-plant ecosystem. Source of acidity (and alkalinity) are listed in Table 4.5.

Table 4.5 Source or acidity and alkalinity in soil

Acidity	
1.	Addition of acids and acid forming chemicals from the atmosphere.
2.	Uptake of (cations) more than anions by plants increases acidity of soils.
3.	Removal of basic cations in plant harvest and by leaching.
4.	Oxidation process such as nitrification.
5.	Microbial production of organic acids.
6.	Increase in soil organic matter content.
7.	Volatilisation of ammonia from ammonium compounds.
Alkalinity	
1.	Addition from the atmosphere of carbonates and weatherable minerals in dust.
2.	Uptake of anion > uptake of cations by plants.
3.	Reduction process.
4.	Weathering of primary minerals.
5.	Hydrolysis of exchangeable sodium ions, which may be present in high concentration after irrigation with saline water.

Effect of the Atmosphere

Deposition from the atmosphere can cause soil acidification in several ways:

(*i*) Acids and acid forming chemicals on precipitation (rainfall, snow and fog) falling on plants and soils;

(*ii*) Absorption of gases from the atmosphere by plants and soil (dry decomposition);

(*iii*) Deposition of aerosols (occult deposition) on to plants and soils; and

(*iv*) Organic acid washed by rain from the leaf canopy through fall as well as down the stems of plants (stem flow).

The individual components that cause acidification fall into the three following groups:

(*i*) **acids in rain water:** H_2CO_3, HNO_3, H_2SO_4, and HCl.

(*ii*) **acid gases:** CO_2, SO_2, SO_3, and HNO_3. In soil water they form H_2CO_3, H_2SO_3 and H_2SO_4 respectively, and

(*iii*) **acid forming chemicals:** NH_3 from NH_4^+ in soil and inorganic fertilizers.

$$2NH_3 + 3O_2 \xrightarrow[\text{oxidation}]{\text{Microbial}} 2NH_2^- + 2H^+ + 2H_2O$$

$$2NH_4^+ + 3O_2 \xrightarrow[\text{oxidation}]{\text{Microbial}} 2NH_2^- + 4H^+ + 2H_2O$$

Followed by $$2HNO_2 + O_2 \xrightarrow[\text{oxidation}]{\text{Microbial}} 2HNO_3$$

The first is production of NO_2^- ions by Nitrosomonas is as follow:

$$NH_4^+ \xrightarrow{[O]} \underset{\text{Hydroxyl amine}}{HONH_2} \xrightarrow{-2H} \frac{1}{2}\underset{\text{Hyporitrite}}{HONNOH} \xrightarrow{[O]} NO_2^- + H^+ + \text{energy}$$

The conversion of NO_2^- to NO_3^- is carried but by Nitrobacter and is so quick that there is no accumulation of nitrite ions in soil.

These reactions described as nitrification, which causes acidification of soils will be referred to later on.

Hydrochloric, nitric and sulphuric acids are strong acids and lower the pH of rain water below 5.6, the value of pure water in equilibrium with an atmosphere containing 340 ppmv of CO_2.

Effect of Plants: If plants take up more cations than anions, they synthesise more organic acids and release protons from their roots. This lowers the pH. Thus, to balance the uptake of one calcium ion Ca^{2+}, two nitrate ions are needed. An imbalance of this uptake is one cause of soil acidification. Uptake of excess anions raises the soil pH. The elements required in largest amounts from soil are nitrogen and potassium. Nitrogen can be taken up by plants as NH_4^+ or NO_3^- and potassium is taken up as K^+. If NH_4^+ rather than NO_3^- is taken up by plants, the uptake ratio of cations charge: anion charge will usually be greater and acidification will occur in the root zone. Acidification is also caused by legumes which obtain their nitrogen by biological fixation and take up excess cations.

Soil acidification caused by an uptake ratio greater than 1 will be temporary if all the plant material is returned to the soil. This is because the organic acids decompose to carbon dioxide and water, and the basic cations are released into the soil. Release of the basic cations counter balances the acidification that occurs during plant uptake. If plants or trees are harvested, acidification results from the removal of the basic cations from the site, whether cation uptake exceeded anion uptake or not.

Oxidation-Reduction Process: Oxidation and reduction processes usually increase and decrease corresponding acidity. The processes are catalysed by enzymes from microorganisms. Soil microorganism carry out many more reactions of these two categories. An important reaction is the oxidation of plant residues which can be illustrated with glucose as the starting substance. The oxidation occurs stepwise using a series of enzymes, with pyruvic acid as the product of the first step. The reaction is

$$C_6H_{12}O_6 \longrightarrow CH_3COCOOH + 4H^+ + 4e^-$$

where e^- represents one electron. There is a further reaction;

$$O_2 + 4H^+ + 4e^- \longrightarrow 2H_2O$$

in which oxygen gains electrons and is reduced to water.

In the absence of O_2, other substances gain electrons and are reduced

$$Fe^{3+} + e^- \longrightarrow Fe^{2+}$$

or for the reduction of nitrate

$$NO_3^- + 2H^+ + 2e^- \longrightarrow NO_2^- + H_2O$$

and of nitrite

$$2NO_2^- + 8H^+ + 6e^- \longrightarrow N_2 + 4H_2O$$

This denitrification process is a step-wise reduction as

$$\begin{array}{ccccccccc} NO_3^- & \longrightarrow & NO_2^- & \longrightarrow & NO & \longrightarrow & N_2O & \longrightarrow & N_2 \\ X = +5 & & +3 & & +2 & & +1 & & 0 \end{array}$$

where *X* is the oxidation state of nitrogen compound. Each step in the reaction is triggered by a specific enzyme.

One of the important reactions that causes acidification is

$$NH_4^+ + 4O \longrightarrow 2H^+ + NO_3^- + H_2O$$

which represents the process of nitrification. It occurs with urea and ammonium fertilizers.

$$CO(NH_2)_2 + 2H_2O \longrightarrow (NH_4)_2CO_3$$

and
$$(NH_4)_2CO_3 + 8O \longrightarrow 2HNO_3 + H_2CO_3 + 2H_2O$$

The acidifying effect of these fertilizers in soil depends largely on whether crop plants are present, which they normally would be, unless the crop has failed. In the absence of a crop, the proton equivalence of 28 kg N (2 kg atoms of N) is 4 kg for $(NH_4)_2SO_4$ and 2 kg for urea, assuming that H_2CO_3 dissociates to CO_2 and water. If the nitrate and sulphate so produced, are leached out of the soil they would carry with them 80 and 40 kg of calcium from the addition of the ammonium sulphate and urea, respectively. In presence of a crop, the acidifying effect depends on the extent of nitrification and on the relative uptake of cations and anions. This makes it difficult to calculate the extent of acidification in the presence of a crop, but it appears generally to be about half of that in its absence, as calculated above. However, depletion of Ca^{2+} from soil will definitely take place in every case.

There are more oxidation-reduction resections

$$4FeS_2 + 15O_2 + 8H_2O \rightleftharpoons 2Fe_2O_3 + 16H^+ + 8SO_4^{2-}$$

Oxidation of ion pyrites produces highly acidic soils.

Oxidation of glucose to pyruvic acid occurs during the aerobic respiration of microorganisms. Oxygen acts as electron acceptors in these condition. In an aerobic conditions, there are other electron acceptors such as Fe^{3+} which becomes reduced to Fe^{2+} with the consumption of H^+. Hence, when acid soils are flooded they become anaerobic and the pH rises.

Increase of Humus Content: Humus has pH-dependent negative charges, which give it a cation exchange capacity at pH 7 of about 3 moles of charges per kilogram; the value differs from soil to soil and depends on the method of measurement. The charge at pH 7 arises mainly from the dissociations of protons from carboxylate groups. The acid groups are the result of oxidation by microorganisms of the plant material which is incorporated in the soil. If the plant material has a high content of basic cations, they balance the negative charge on the humus. Introductions of legumes, such as subterranean clover, can cause acidification. This is probably because there is a shortage of basic cations to balance the acidic groups on the increased content of soil organic matter, although fixation of nitrogen by the clover might be a partial cause.

Weathering of Minerals: Many primary minerals contain the basic cations Ca^{2+}, Mg^{2+}, K^+ and Na^+ which pass into solution when the minerals hydrolyse by process of chemical weathering. The process can be expressed in terms of consumption of H^+, or production of OH^- expressed generally as

$$(\text{Ca, Mg, K, Na, Al})\ \text{Silicate} + H_2O \longrightarrow Ca^{2+} + Mg^{2+} + Na^+ + K^+$$

$$+Al_2(OH)_6 + Si(OH)_4 + OH^-$$

The products may include silicic acid, hydrous oxides of aluminium and iron, kaolinite or other clay minerals and the basic cations pass into solution. Because of the production of basic cations and OH^- (or consumption of H^+), the process is known as hydrolysis of minerals, as for example ferromagnesium minerals, are present in the soils. The higher the temperature, the faster it proceeds. The rate is also increased by addition of acids, but it is not known whether acidification, as it occurs under field conditions, has any effect.

The effect of weathering of most minerals is to increase alkalinity, although oxidation of iron pyrites releases protons as mentioned above. Iron pyrites are formed by reduction of iron sulphate by microorganism when organic matter is present. It also occurs in the waste material from coal mining. Reclaimation of this waste and the drainage of marine and estuarine clays and pests can give very acidic soils. They are known as acid sulphate soils and can have a pH below 3.

On one hand, during acidification the concentration of protons in soil is increased and basic cation are displaced from exchange sites on clays and humus. The basic cations are lost permanently if they are leached out of the soil or removed in a harvested crop. However, in forests, because of cycling of basic nutrients and their accumulation, the distinction needs to be made between acidification of soils and that of the ecosystems.

4.16 EFFECTS OF SOIL ACIDITY ON PLANTS

Acidity creates chemical and biological conditions in soils which are harmful to most plants although there are exceptions. The effects are summarised in Table 4.6.

When soils become acidic their capacity to absorb cations is reduced so that nutrient cations, especially Ca^{2+} any Mg^{2+}, pass into solution and are leached in drain water. Acidity also increases the solubility of most metals. As pH decreases below 5.4, the soil solution contains increasing concentration of aluminium ions which displace other cations. Acid soils therefore usually have low contents of calcium and magnesium, and in extreme conditions, the supply to plants may be deficient.

Table 4.6 The main effects of soil acidification on plants

1.	Reduced supply of nutrients; Ca^{2+} and other cations lost in drainage water: N, P and S remain immobilised for longer periods in soil organic matter;
2.	Increased concentrations of metal ions in solution, especially those of Al, Mn, Cr, Cu, Ni, Zn, which may become toxic;
3.	The uptake of phosphates and molybdates is reduced;
4.	The form of nitrogen taken up by plant roots may be NH_4^+ instead of NO_3^- because nitrification is inhibited; and
5.	Nitrogen fixation by legumes may be reduced unless the Rhizobium strain is acid tolerant.

The most common problems in acid soils is, however, the toxicity of aluminium to plants, and for some species the toxicity of manganese. Solution concentrations of 10^{-6} M Al can harm sensitive plants. The concentration of other metals, including Cr, Cu, Ni and Zn are increased in acid soils but their toxicity is almost entirely restricted to polluted soils.

Acid soils can affect plants in several ways. Plant species and varieties differ, however, in their sensitivities to the conditions in acid soils. Among the crop plants, barley, cotton and luceme (alfalfa) are sensitive, whereas the tea plant actually requires acid conditions. The usual agricultural practice for most crops, at least in temperate regions is to maintain a soil pH of 6.0-6.5 by the addition of lime, applied as calcium carbonate. The selection of breeding of varieties of crops and pasture plants may decrease soil pollution due to acidity.

4.17 LIMING ACID SOILS

Materials used for neutralisation of acid soils are restricted to the carbonates, hydroxide, oxide or silicate of calcium or magnesium or both. The reaction of each of these materials with acid soils results in the formation of weakly dissociated compounds which, in effect, destroy the hydrogen ions. Neutral salts such as gypsum ($CaSO_4$) are not used, since they would not correct the acid condition. Sodium and potassium compounds are not used because of their relatively higher costs compared with similar calcium compounds and because of the deleterious effects of excess sodium and potassium may have on soil characteristics and plant growth. A typical reaction of liming materials within acid soil is as follows:

$$2HX + CaCO_3 \longrightarrow CaX + H_2O + CO_2$$

The liming value of various pure compounds is as follows:

$CaCO_3$	: Dolomite	: CaO	: $Ca(OH)_2$	: $CaSiO_3$ slag
1	: 1.09	: 1.78	: 1.35	: ≈ 0.7

Since montmorillonite clay acids are highly dissociated and the calcium activity does not change appreciably over a moderate range in percentage saturation, limestone reacts more rapidly with them than with kaolinite clay acids soils with high cation exchange capacities, are more buffered than soils with low capacities and therefore possess more potential acidity at a given pH value. This increases the rate of lime reaction.

What-Lime Does to the Soils: Highly acidic soils are not productive soils. To increase productivity of acid soils, lime is the first step for the following reasons:

1. Lime reacts with phosphorus to form the more soluble phosphates in limed soil than the relatively insoluble iron and aluminium phosphates formed under acid conditions.
2. Lime makes potassium more efficient in plant nutrition. All plants absorb more potassium than they need when it is plentiful. Lime reduces the excessive uptake of potassium which is a sound practice. When lime is abundant, plants take up more calcium and less potassium which is passed on to the animals that require more calcium than potassium. It is therefore desirable to increase the percentage of calcium in the plant. As the plant absorbs more of calcium and less of potassium, liming the soils is desirable.
3. Microbiological transformations take place more rapidly under limed than under acid conditions. This affects the availability of nitrogen, sulphur and phosphorus.
4. The solubilities of trace elements, such as boron, manganese, copper, zinc and molybdenum as well as iron, are related to factors that control or vary with soil pH.
5. Other basic ions may be made more available by reduced leaching losses accompanying liming. As liming raises the pH of soils, it also reduces the plant uptake of other basic ions by competing with them for absorption into the plant.

6. Liming affects the solubility of toxic substances in soil. Aluminium and manganese are more soluble under acidic conditions. The presence of these elements in solutions appreciably reduces the plant uptake of calcium, aluminium and seems to have a greater depressing effect than manganese. The presence of these inhibiting substances in soil solution may explain why good growth is not obtained from the application of gypsum to acid soil even though calcium is being supplied by these conditions.
7. Liming influences the characteristics of the soil (Fig. 4.25). The formation of stable aggregates in the soil depends upon cementing agents in the soil that will hold individual particles in aggregates. Humus colloidal particles, with calcium as the principal adsorption ion, fulfill this requirement. Calcium humate remains flocculated even under excess moisture. It is this characteristic of calcium-saturated humus that gives rise to be favourable structure found in the black soils of sub humid regions.
8. Liming influences the distribution of roots in soil. Root distribution studies show that plant roots develop more fully in limed than in unlimed zones. Liming the soil, therefore, will encourage an increased root system with the added advantage of nutrient and water availability that are associated with greater root development.
9. Over a period of years, a good liming programme improves the physical conditions of the soil by decreasing its bulk density, increasing its infiltration capacity and increasing its rate of percolation of water.
10. There is less soil erosion following an adequate liming programme. This results in increased vigour and density of plants following the lime application.

Physical Guarantees of Lime: The chemical activity of liming material is determined by the solubility of the chemical compounds in the lime. For example, calcium oxide is more soluble than calcium carbonate, while limestone is more soluble than dolomite. Calcium silicate is the least soluble of the liming materials. It is evident that, finer the lime particles, the faster they react in the soil to become available to plants. To arrive at a satisfactory solution to the problem of how much lime to apply, the requirement of the crop is a good starting place. Since soil acidity and lime level have a fairly good correlation in humid region, the pH of the soil is used as an index of the lime needs of various crops. The relative tolerance of selected field crops to soil acidity can be studied from Fig. 4.25. This figure indicates that alfalfa, sugar beets, sweet clover and barley require the highest pH while blue berries require the lowest.

After the pH requirement of the crop is known, it is then necessary to test the pH of the soil. The soil texture and humus content must also be estimated to arrive at the relative buffer capacity. The relationship between texture and the buffer capacity are shown in Fig. 4.26. The more clay and humus is there in a soil, the more lime is needed to change the pH.

Sandy soils require approximately one tone of limestone to effect a change of soil pH by one unit, say from 5.5 to 6.5. Loams require approximately two tons, clay loams three tons, muck soils five tons.

There is also a variation in lime needs to change the pH, depending upon the type of clay present as well as the range of the pH change desired. As the pH change is closer to 7.0, the amount of lime required to effect the same pH change is greater as for example, it takes perhaps 20 per cent more lime to change the soil pH from 5.5 to 6.5 than it does to change it from 4.5 to 5.5.

Soil pH and Nutrient Availability: The general relationship between soil pH and plant nutrient availability is shown in Fig. 4.27. From this chart it is obvious that primary nutrients such as nitrogen, phosphorus and potassium, as well as the secondary nutrients namely sulphur, calcium and magnesium, are as available or more available at a pH of 6.5 than at any other pH. Molybdenum, manganese and boron availability is also similar to that of the primary and secondary nutrients.

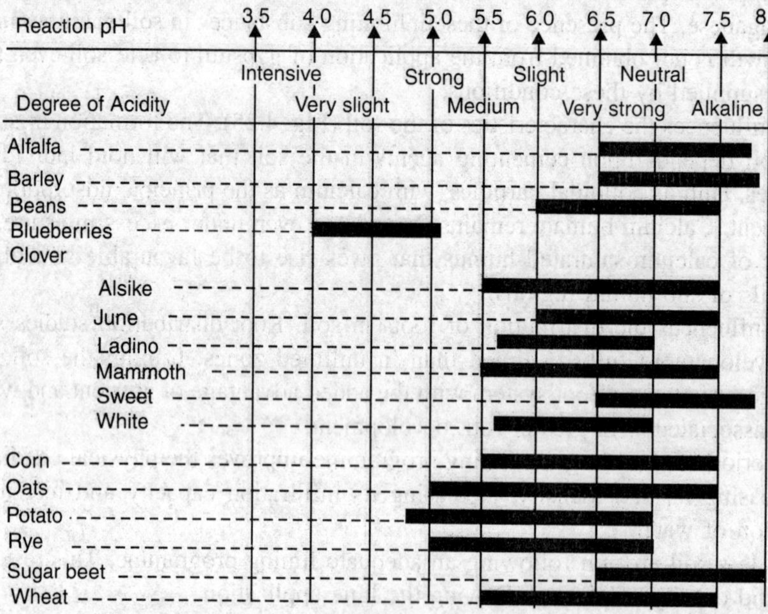

Fig. 4.25 The relative tolerance of selected field crops to soil acidity.

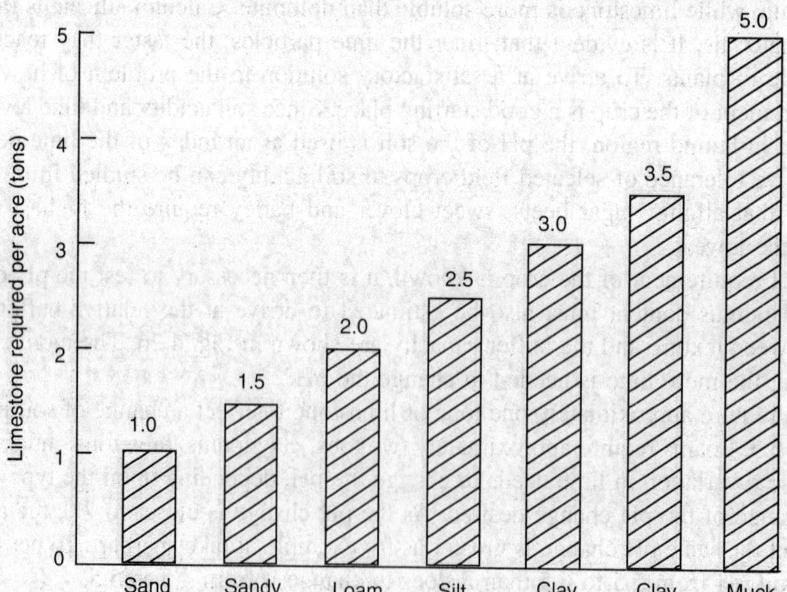

Fig. 4.26 The limestone required to change the pH of the common soil textures and of muck by one pH unit. (Data from various sources).

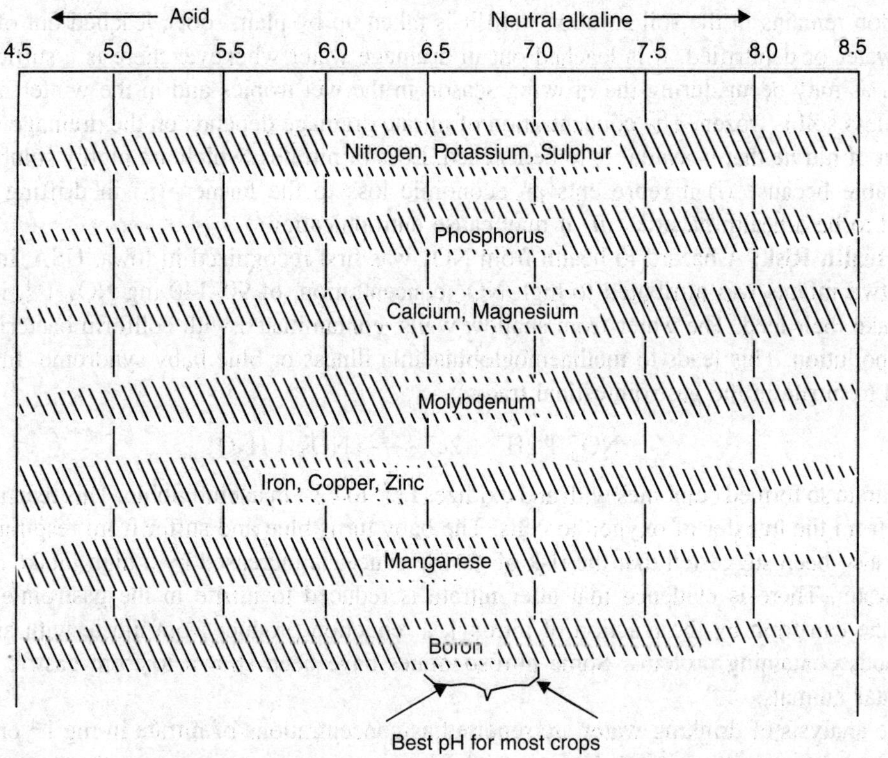

Fig. 4.27 The relationship between soil pH and the relative availability of plant nutrients. The wider the bar, the greater the availability.

The minor nutrients, namely—iron, manganese, boron, copper, chlorine and zinc are less available at a pH of 6.5 than at more acid reactions. In general, minor elements are sufficiently available so that plant growth is not limited at a pH of 6.5. A pH value between 6.5 and 7.0 is therefore considered to be the pH range in which most nutrients are desirably available to plants.

Overlimiting: When excessively large amounts of lime are applied to sandy soils low in humus, there can be injury to plant growth, which may be due to anyone or a combination of these causes:

1. Boron deficiency;
2. Iron, manganese, copper or zinc deficiency;
3. Phosphorous availability may be reduced to an critically low level. This can be reduced by application of large amounts of manure, green manure crops, compost, phosphorous fertilizers, boron or a mixture of minor elements. Overliming injury is not so common as was once thought.

4.18 NITRATE, ITS SOURCES AND AS A HEALTH HAZARD

Nitrate is present in most soils comes from the three sources:

(*i*) Microbial breakdown of soil organic matter, organic manure and plant residues,

(*ii*) From nitrate fertilizers and nitrate formed by microbial oxidation of NH_4^+ from ammonium fertilizers or urea,

(*iii*) Addition from the atmosphere.

The NO_3^- ion remains in the soil solution until it is taken up by plant roots, leached out of the soil in drainage water or denitrified. It is leached out in drainage water whenever there is a sufficient excess of rainfall, as may occur during the growing season in the wet tropics and in the winter in temperate regions unless soil is frozen. The concentration of nitrate drainage depends on the drainage volume and the amount of nitrate that is available to be leached. Loss of nitrates, which are highly soluble in water, is undesirable because (*i*) it represents an economic loss to the farmer, (*ii*) in drifting water it is considered to be a health hazard, (*iii*) it may cause entrophication.

The Health Risk: A hazard to health from NO_3^- was first recognized in Iowa, USA, in 1945. The illness of two infants was attributed to high NO_3^- concentration, of 90–140 mg NO_3^- l^{-1}, in the water used to make their feed. The water from shallow wells, contaminated with coliform bacteria, indicator of nitrate pollution. This leads to methaemoglobinaemia illness or blue baby syndrome. In this nitrate is reduced to nitrite in the gastrointestinal tracts

$$NO_3^- + 2H^+ + 2e^- \longrightarrow NO_2^- + H_2O$$

The nitrate so formed combines with and oxidizes Fe^{2+} to Fe^{3+} in haemoglobin. This reaction prevents the blood from the transfer of oxygen to cells. The baby turns blue and suffer from respiratory failure.

It has also been suggested that the risk of gastric cancer is increased by a high intake of nitrate in drinking water. There is evidence that after nitrate is reduced to nitrite in the gastrointestinal tract, nitrosoamine can form by the reaction of nitrosyl, a reduction product from nitrite with amines from cooked foods containing proteins. Some nitrosoamines have been shown to form gastric tumours in experimental animals.

Nitrate analysis of drinking water are reported as concentrations of nitrate in mg l^{-1} or amount of nitrogen present in as nitrate (NO_3^-–N) in mg l^{-1}. The European commission set the maximum admissible level at 50 mg NO_3^- l^{-1}(11.3 mg NO_3^-–Nl^{-1}).

Nitrate is a particular hazard to cattle and sheep because the lumen of these animals provide an environment conducive to its reduction to nitrite. Nitrate intake from water and feeds has caused nitrite poisoning of live stock in the USA for many years.

4.19 PESTICIDES IN SOILS

Pesticides are chemicals that are used to control the population of harmful organisms. The use of pesticides is commonly criticized but they contribute substantially to the increased production of food. Their manufacture and use does however entail hazard to the environment and will be discussed in Chapter 7.

Some of the ways through which pesticide reach soil are:

(*i*) intentional application for destroying soil borne fungal and bacterial pathogens, nematodes and soil dwelling pests.

(*ii*) drift and spillage of pesticide.

(*iii*) Washing of the pesticide deposited on plants and plant leaves. All this make about 50% of the pesticide to come down to soils and thus the principle source of pesticide and thus part of soil system. Most of the pesticides in soil is controlled by five process. (*i*) Volatilization into atmosphere (*ii*) adsorption-desorption processes by soil components (*iii*) biodegradation (*iv*) chemical decomposition (*v*) plant absorption.

(*i*) **Volatilization:** Pesticides have a wide range of volatility as shown by the values of vapour pressure in Table 4.7. In soil the pesticide molecules are partitioned between the vapour and liquid phases. The partition coefficient is also known as Ostwald solubility coefficient, *B*, and equals

$$B = \frac{\text{Concentration of pesticide in liquid phase}}{\text{Concentration of pesticide in gas phase}}$$

As diffusion is about 10^4 times faster in gases than in water, diffusion is predominately in the vapour phase if B is less than 10^4, and in the liquid phase if $B > 10^4$.

From the partition coefficients given in Table 4.7, it can be seen that the soil sterilising agent ethylene dibromide, diffuses through soil as a vapour whereas simazine diffuses in the liquid phase.

Table 4.7 Properties of pesticides that affect their volatility

	Vapour pressure (mm in Hg)		*Solubility in water (μgl^{-1})*		*Partition coefficient, B*
Ethylene-dibromide	11.0	(25°C)	4.3×10^3	(30°C)	36
Trifluralin	1.0×10^{-4}	(25°C)	0.58	(25°C)	3.2×10^2
Disulfoton	1.8×10^{-4}	(20°C)	15	(20°C)	5.5×10^3
Dimethoate	8.5×10^{-6}	(20°C)	3×10^4	(20°C)	2.5×10^8
Simazine	6.1×10^{-9}	(20°C)	5	(20°C)	7.4×10^8

The approximate value of B can be calculated as follows: one mole of ethylene dibromide ($C_2H_2Br_2$) = 186 in gram and occupies 22.4 l at STP, at a pressure of 11 mm Hg 1 litre of air contains

$$= \frac{11}{760} \times \frac{1}{22.4} \times 186 = 0.12 \text{ g}$$

$$= 120 \ \mu g \, l^{-1} \text{ of } C_2H_2Br_2$$

$$\text{Partition coefficient } B = \frac{4.3 \times 10^3}{120} = 36$$

Disappearance of pesticides from soils does not mean they breakdown. Rather some chemicals are lost to the atmosphere only to return to soil or surface water through rain is now known.

(*ii*) **Adsorption and Desorption:** Pesticides are mostly organic compounds which are generally hydrophobic and have limited water solubility and low leachability. On account of this and also since pesticides are applied at fairly low concentration, often below 2.5 kg/hectare, their sphere of interaction with soils is generally limited to about 15 cm of soil only. The application of 1 kg/ha corresponds to roughly 1 ppm of soil.

The sorption of pesticide directly controls its concentration in soil solution and thus influences its bio-availability, transport, persistence and degradation. Since surface area and reactivity are directly related to sorption, montmorillonite and vermicullite are the most compulsive adsorbents followed by illite and kaolinite.

Nonpolar pesticides like chlorinated hydrocarbon are weakly adsorbed through van der Waals forces and hydrophobic bonds, may be on organic matter of soils. Polar pesticides are bound to clay mainly through exchange cations. The bonding is predominantly electrostatic and consist of ionic bond in the case of cationic pesticides like diquat and paraquat. Basic pesticides like triazines which can be protonated at low solution pH or on the surface of the clay by an exchangeable proton. The dissociation of water of hydration of exchangeable cation like Al^{3+} or by another protonated base already adsorbed also releases proton. Protonation of basic pesticides make them cation which is adsorbed over negatively charged planner surface of clays and organic matter

$$\underset{\text{Basic Pesticide}}{\boxed{\text{Pesticide}}} + H^+ \longrightarrow \underset{\text{Cation}}{\boxed{\text{Pesticide}}\ H^+}$$

$$\boxed{\text{Pesticide}}\ H^+ + M^+\ \boxed{\text{Clay}} \longrightarrow \boxed{\text{Pesticide}}\ H^+\ \boxed{\text{Clay}} + M^+$$

In case of polar pesticides, dipole-dipole interaction between the exchangeable cation and the pesticide mediated through the water of hydration is main type of this interaction.

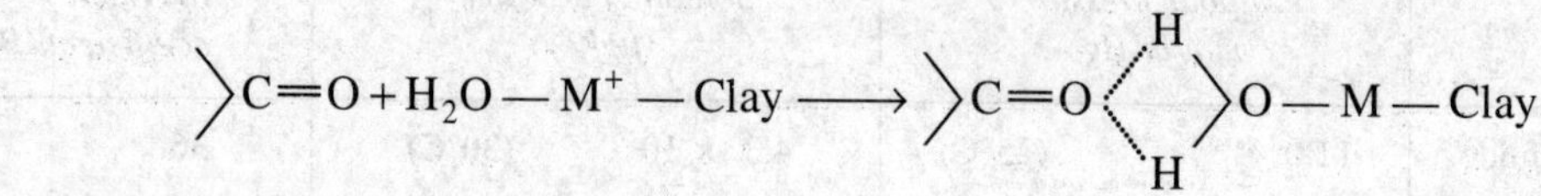

Hydrogen bonding and hydration of exchangeable cation also permits inter layer adsorption of some pesticides.

In dehydrated soils, when the water content is very low, the cation can interact directly with the pesticides without the mediation of water molecules. In case of transition metals like Cu^{2+}, Ni^{2+}, Zn^{2+} coordinate bonds involving stronger orbital overlap have been shown to be formed.

Persistence of pesticides in soils is very much dependent on adsorption-desorption processes which is dependent on the various factors as follows:

1. **Solubility:** There is inverse relationship between solubility and adsorption and is more likely to be true within a group of chemically similar compounds.
2. **Water content of soil:** Most pesticides are adsorbed when soil is dry than when they are moist which leads to their greater adsorption. Water molecules are themselves polar, thus when more water is added to pesticide-soil complex, the water molecules begin to compete with pesticide molecules for adsorption sites on the soil-colloid and force more of pesticide into solution. Addition of moisture releases more of adsorbed pesticide to solution.
3. **Soil pH:** With the change of pH, the form of pesticide changes. Thus as the pH become acidic, nitrogen containing pesticides are converted to cationic form which is strongly adsorbed soil complex, or pesticides with carboxylic acid group will be strongly adsorbed in its acid form than in the salt form. So we can say

	Cationic form	Neutral	Anionic form
pH	Acidic	Neutral	Alkaline

Adsorption decrease ⟶

Solubility inverse ⟶

4. **Soil temperature:** Adsorption is an exothermic process. When hydrogen bond or ionic bonds are formed heat is given off. Thus when soil temperature increases, some bonds break and desorption of some molecules take place. Also increase in temperature may increase solubility of pesticide and thus make more pesticides available in soil solution.
5. **Leaching:** It is dependent on solubility of pesticide. However leaching of pesticide down, as far as the water table is concerned, there is currently little evidence that this may occur.

6. **Volatilization:** Neutral pesticides are more volatile than their cationic or anionic form. Since in that form solubility is very low, volatilization contribution comes from some limited pesticides.
7. **Photodegradation:** Few organic pesticides are completely resistant to photolysis, but this is probably not a major means of pesticide in activation or disappearance in soil.

Adsorption is generally measured by allowing the soil to react with aqueous solution of pesticides over a range of concentrations. At equilibrium, the amount adsorbed is obtained as the difference between the amount added and the amount remaining in solution. On plotting amount adsorbed, x against equilibrium concentration, C, invariably a straight line is obtained (Fig. 4.28). The slope of this line is known as adsorption coefficient, K or distribution adsorption coefficient. Average values of K for some commonly used pesticides (Table 4.8) vary greatly between weakly adsorbed 2, 4, *D* and strongly adsorbed, DDT and paraquat, pesticides. Clay minerals adsorb pesticide in the order, montmorillonite > vermiculites > illite > kaolinite, which is also the order of decreasing surface area.

Table 4.8 Values of adsorption coefficient, K of some pesticides

Pesticide	*K*	*Pesticide*	*K*
DDT	$(1–10) \times 10^4$	Atrazine	26
Lindane	7–50	Simazine	1–7
2,4, *D*	2	Paraquat	200–500

Weak adsorption, that is a small value of K indicates a high mobility of a water soluble pesticide in soil and a large value indicates low mobility. The values can be used to calculate the rate of transport of the pesticide to plant roots and to the drainage water. Paraquat, for example, is strongly adsorbed by clays and so it is not likely to be leached out of soil, nor will it move to plant roots; its effect is therefore by contact with plant leaves. Water soluble pesticides can, however escape adsorption by being washed through cracks and pores in soil into ground water and then into aquifers similarly, surface run off can carry water soluble pesticides into surface waters, e.g. rivers and lakes.

(*iii*) **Microbial Degradation:** Biodegradation is the major operative process leading to degradation of chemicals in the environment mainly in soil. The process involves the interaction of chemicals with the microorganism present. The microorganism present in soil are algae, fungi, bacteria and actinomycetes. This process is considered the major one because most pesticides end up in soil and soil is a reservoir of microorganism. The microorganisms utilize the organic molecules present in soil for the production of energy for their growth and reproduction. Most of the organic compounds can serve as source of energy to these organisms. The pesticides ending up in the system establish an equilibrium between soil colloid fraction and soil solution. Any microbe that can adopt to this chemical as an energy source, immediately attacks the molecules present in soil solution and rapidly increase in number and degrade the chemical at a much faster rate. The pesticides can also be microbially degraded by co-metabolism without supplying energy for the microorganisms. Certain polar groups such as —OH, —COO^-, and —NH_2 on pesticide molecule provide points of attack for the organisms.

Among the practices suggested to reduce pesticide levels in soils is the addition of easily decomposed organic matter. The growth of high nitrogen cover crop or the addition of large quantities of animal manures also helps. In either case microbes in soil will increase and some of them may be good enough to decompose the pesticide.

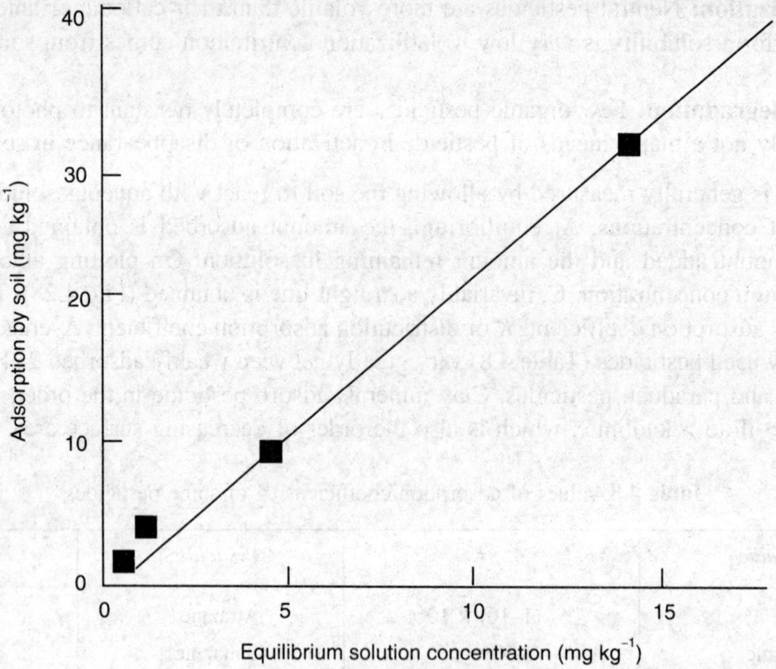

Fig. 4.28 The adsorption of a pesticide, *e.g.* atrazine, by a soil. The slope of the line gives the adsorption coefficient, *K*. In this experiment *K* is approximately 2 mg kg^{-1}/mg l^{-1} (= 21 kg^{-1} or 2 cm^{-1} g^{-1})

Ideally, a pesticide should remain long enough to kill the target organism and then decompose to a harmless composition. 2.4 D as well as 2, 4, 5 T is degraded completely to CO_2, H_2O and Cl_2 in soil within one month, whereas DDT can persist for several years, its principal degradation product DDE is also persistent, though much less effective. Microbial degradation of most triazine occur very slowly in soil. Adsorption to soil colloid may slowly catalyze the chemical degradation of atrazine to its 2-hydroxy and N-dealkylated derivatives and eventual ring cleavage and liberation of C and N for microbial and plant growth. The persistence time of some pesticides are given in Table 4.9.

Table 4.9 Common range of persistence or some herbicides and pesticides

Pesticide	*Persistence time*
Chlorinated hydrocarbon in pesticides (e.g. DDT, Chlordane, dieldrin)	2-5 yrs.
Triazine herbicides (e.g. atrazine, simazine)	3-4 months
Benzoic acid herbicides (e.g. amiben, dicamba)	2-12 months
Phenoxy herbicides (2, 4-D, 2, 4, 5- T)	3-4 weeks
Organophosphate insecticides (e.g., malathion, diazinon)	1-12 weeks
Carbonates	1-8 weeks

(*iv*) **Chemical Degradation:** Chemical reactions occurring in soil can destroy the activity of pesticide or can activate the pesticide. Normally the adsorption of pesticides enhance the chemical degradation, whereas on the other hand it prevent the attack of microbes to the pesticides. pH plays an important role in the chemical degradation of pesticides. The major routes of transformation of organophosphorous pesticides in soil are hydrolysis, oxidation and reduction. Most of the organophosphorus are resistant to hydrolysis in acidic and neutral pH.

The methyl-parathion undergoes degradation in crops by hydrolytic cleavage of *p*-nitrophenol followed by oxidation of the dimethyl thiophosphoric acid. Methyl thiophosphate and dimethyl thiophosphate were recognized as the main degradation products within the first 3 days, which were subsequently transformed to phosphoric acid, whereas in soil the major metabolites were O, O-dimethyl-O-*p*-aminophenyl thiophosphate, O, O-dimethyl-O-*p*-aminophenyl phosphate, O-methyl-O-*p*-nitromethyl thiophosphate and smaller amounts of O-methyl-O-*p*-nitrophenyl phosphate. Phosphoric acid, O, O-dimethyl thiophosphate; O-methyl-thiophosphate and O, O-dimethyl phosphate were also found in soil.

The scheme of degradation of methyl-parathion in soil is given on next page.

Plant Absorption: Pesticides are commonly absorbed by minor plants, a necessary process for the effectiveness of most herbicides and of some insecticide. The absorbed chemicals may remain intact inside the plant or they may be degraded. Some degradation products are quite harmless to humans and other creatures, but others are toxic, in some cases even more toxic than the original pesticide. Pesticide residues, especially those found in edible portion of plants are of critical concern to humans.

Pesticides in the Environment: The three main uses of pesticides, which are of benefit to man are:

(*i*) To control weeds, fungal diseases, insects and other pests which would otherwise reduce the yield of agriculture crops.

(*ii*) To stop the spoilage of food crops during storage by controlling insect pests such as grain weevils;

(*iii*) To kill disease carrying organisms.

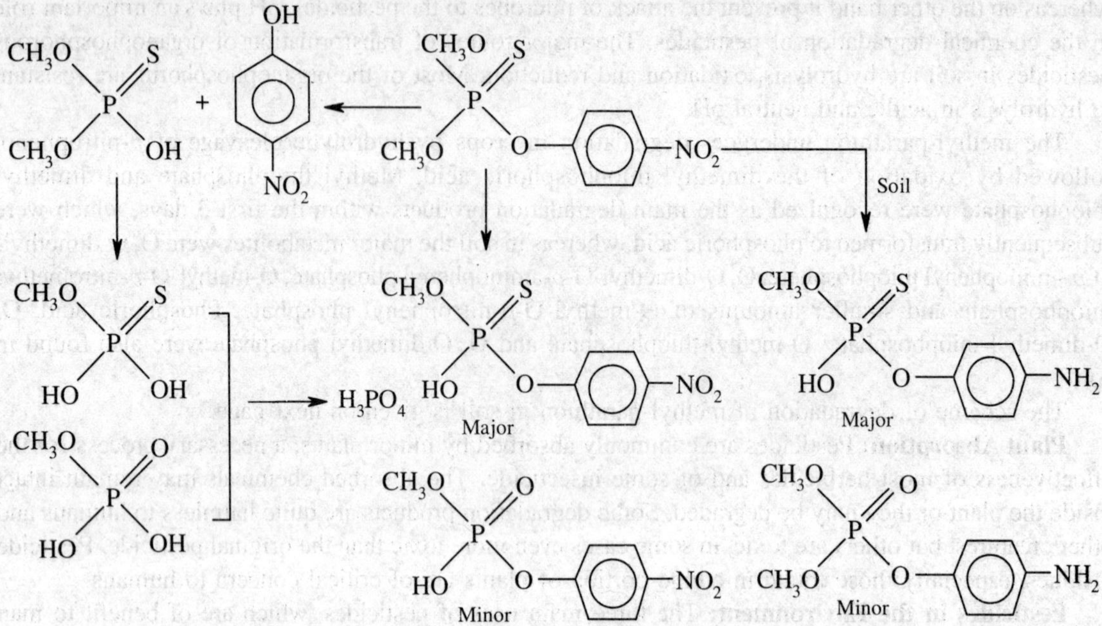

Environmental hazards were first recognised with DDT. Three properties of DDT cause problems, namely:

(*i*) DDT and its principal decomposition product DDE are highly persistent in soils.

(*ii*) It is insoluble in fat and therefore accumulates in fatty and oil tissues.

(*iii*) It is toxic to fish and birds.

Application of DDT, which reach the soil, are taken up by earthworms and insects. Concentration increase in birds eating the earthworms and insects, and increase further in predatory birds such as the eagle. Similarly in water, DDT is taken up by plankton which are consumed by small fish, which in turn are consumed by bigger fish and birds, the concentration of the chemical in tissues of the animals increase with each step and cause their death. For this reason DDT is now prohibited in India and in many countries.

Because of the experience with DDT, a thorough testing of the environmental impact of a chemical is required before it can be marketed. Based on the results of these tests, the pesticides now in use are generally considered safe as long as the guidelines governing their use are followed. There is, however, public concern that they might cause cancer or act as poisons, although there is little evidence to justify these fears.

Environmental checks are made by measuring the concentrations of pesticides in water and food. In the European Community, the maximum permissible concentrations in drinking water are:

1. $0.1\ \mu\ gl^{-1}$ for any individual pesticide and related produce.
2. $0.5\ \mu\ gl^{-1}$ for total pesticides and related products.

These concentrations are based on detection limits rather than on concentrations that are a risk to health. They are nevertheless mandatory. From the above discussion it should be apparent that use of pesticides will change the character of soil and thus pollute the soils.

4.20 HEAVY METALS IN SOIL (See also Section 5.6)

Heavy metals are natural components of the environment, but are of concern because they are being added to soil, water and air in increasing amounts. They can however be harmful if taken up by plants or animals in large amounts, as can be other heavy metals which are not known to be essential nutrients. Often heavy metals discharged by industry of which As, Cd, Cr, Cu, Pb, Hg, Ni and Zn are listed by European commission directives as representing the greatest hazards. Soil also receives chemical elements from active volcanos including As, Hg and Se and these can be a hazard. A third source of hazard to soil is fly ash, i.e. the ash left after the combustion of coal for the production of electricity. However, the most hazardous of the all these are Cd, Pb and Zn. The concentration of these elements in earth's crust and soils are given in Table 4.10.

Table 4.10 Average concentrations of some hazardous metals

Metal	*Earth's crust* $\mu g\ g^{-1}$	*Rocks with highest concentration*	*Soils* $\mu g\ g^{-1}$	*Soils* $kg\ ha^{-1}$
As	1.5	shales and clays	0.1 – 50	0.2 – 100
Cd	0.1	shales and clays	0.01 – 24	0.02 – 4.8
Cr	100	ultra basic	5 – 1500	10 – 3000
Cu	50	basic	2 – 250	4 – 500
Hg	0.05	sandstone	0.01 – 0.3	0.02 – 0.6
Ni	80	ultra basic	2 – 1000	4 – 2000
Pb	14	granite	2 – 300	4 – 600
Zn	75	shales and clays	10 – 300	20 – 600

Amount of metals per hectare calculated for a soil depth of 15 cm and a bulk density 1.3.

The effects of emissions from industry depend on their nature. Solid waste from the mining of metal ores has a local effect whereas emissions into the atmosphere from metal refineries can contaminate large areas of land. Similarly metals in sewage sludge are more widely distributed on agricultural land than these disposed off in landfill sites.

Cadmium: Uptake of cadmium by plants is greatest from soils high in cadmium but is decreased by raising the pH by liming. Uptake varies with the soil properties, especially those that control adsorption, and also with the plant species, cultivator and source of cadmium. It has been shown, for example, that uptake is less from soils to which sewage sludge or organic manure has been added than from soil to which an equal amount of cadmium salt has been added. Much of the cadmium that enters the plant is retained in the roots.

Lead: The concentration of lead in soil solution is low and the uptake by plants from soils even near highways is also low. Addition of lead salts to soils or nutrients solution produce only small increases in uptake, and most of this is retained in the roots. Lead is excluded from cereal grains, fruits and edible roots. The lead deposited on leafy vegetables in urban gardens is a greater hazard though about one half is removed by washing.

Zinc: In the Earth's crust and soil, zinc is the most abundant metal, is an essential micronutrient for plant, animals and man. However, at higher concentrations in soil, zinc can be phytotoxic, reducing the growth of plants or killing them at tissue concentrations that may not be toxic to animals. The principal hazard with zinc is therefore with plants. Where zinc toxicity occurs, liming the soils to increase the pH lowers the solution concentration, thus provides a remedy.

Radionuclides: Radionuclides are a danger, rarely because of the elements themselves but due to the radiation they emit. Some occur naturally, exposing all living organisms to small and usually harmless doses of radiation. Particularly dangerous are some man made nuclides that accumulate as they pass through a biocycle.

These nuclides, with their half lifes, are Cesium-137 30 years; Iodine-131 8 days; Strontium-89 52 days; Strontium-90 28 years. Details have been discussed in section 9.12.

4.21 WATER MANAGEMENT FOR SOILS IN INDIA

The total precipitation received by the country as a whole is around 350 mhm per annum—which is theoretically capable of placing its entire land surface under approximately 115 cm of water—this resource must be treated as a scarce resource because of firstly, its highly uneven spread in space as well as in time, and secondly, the steadily increasing demands that are being made on it not only by agriculture but also by the industrial and domestic sectors. Since the soil, however well endowed it may be, is incapable of any biotic production in the absence of moisture, the management of water lies at the very heart of land management. An overview of water management scenario in India reveals that our traditional policies should be reviewed urgently.

The traditional policy on the conservation of water had been to rely on the creation of surface storages, whether big, medium or small. Between 1950 and 1995 we have spent over Rs. 50,000 crores on such projects and created a storage capacity of around 20 mhm. During eighth plan, government spend around Rs. 27500 crores on this sector or Rs. 5500 crores per annum.

However, we have come to a dead end on this route as investments in surface projects have later shown unmistakable signs of becoming unproductive. Thus, an investment of Rs. 11,107 crores on major and minor projects during Seventh plan, according to Ministry of Water Resources statistics, not only failed in the creation of any additional potential but resulted in the loss of 0.6 mh of even the potential that existed at the end of sixth plan. Seventh Plan had envisaged the creation of an additional potential of 4.3 mh, the net loss of planned potential between 1985 and 1990 amounted to 4.9 mh, the replacement value of which at current prices would be well in excess of Rs. 30,000 crores. This is a serious development and no explanation has so far been offered.

The track record of small surface projects has, if anything been worse. According to the Land Use Statistics (L.U.S.) the net area served by such projects declined from 8.2 mh in 1961 to 6.8 mh in 1989 inspite of the fact that around Rs. 6000 crores were invested in them during the intervening period of 28 years. However, in this case the reason behind the debacle is known—it lies in the premature siltation of reservoirs, which being much smaller than those of the major and medium sector, are rendered inoperative much more quickly.

In view of these developments, the time has come to reconsider the traditional approach to the problem of water conservation and adopt the only other available option, namely, the storage of water mainly in the form of soil moisture and ground water within, rather than on the surface of the country's land mass. Not only is this option incomparably cheaper but it also involves no high technology and has proved to be a great success wherever it has been adopted. It consist of basically reducing the run off of water to the sea by creating biotic as well as engineering impediments to the free flow of water along slopes so that it may get a better chance to percolate into the soil and sub soil strata. The restoration of permanent vegetal cover, whether of grasses or trees, on all denuded lands—as far as possible through natural regeneration, the construction of innumerable small weirs, check dams and small tanks across all drainage lines in all micro-catchments and the treatment of all erosion prone agricultural lands for the conservation of both soil and water.

It is imperative that a nation wide programme for soil and water conservation be given the highest possible priority for a variety of reasons. It would, in the first place, help to reduce run-off losses and increase the availability of water in the form of soil moisture and ground water or water stored in countless small tanks and ponds. The enhancement of soil moisture would be particularly beneficial for the nation's rain fed agricultural lands which account for nearly two-thirds of the total land under cultivation. The storage of water in village tanks and ponds would not only benefit local communities, but would also help in the replenishment of ground water. It may be noted here that ground water not only accounts for more than 50% of the total area under irrigation but is also around 100% more efficient than canal water in terms of productivity per hectare. This is indeed the reason of productivity per hectare, and its demand by farmers who can develop it quickly and easily with their own resources, assisted by bank loans, wherever necessary.

In the second place, such a programme would help in controlling the premature siltation of reservoirs and tanks—which in most cases are irreplaceable—in moderating floods by reducing the quantity of water and top soil that rivers have to carry at peak periods, and by protecting their carrying capacity against siltation. In the third place, the return flow of water that takes place from fully charged ground water aquifers into springs and rivers during lean season would also help to migrate draughts.

The wisdom of conserving water mainly within the land mass and in Nature's own way rather than in man-made reservoirs becomes apparent when we consider the great difficulties that are encountered in putting to actual use the irrigation potential created by surface storages. The total potential created between 1950 to 1990 was 20.2 mn., of which according to the Ministry's own claim, only 15.8 mh had been utilised by 1990, thus revealing an unutilized gap of 4.4 mh.

Traditional policies on water has damaged land in two ways.

(*i*) Due to land denudation and erosion, large quantities of the top soil had been displaced year after year as a result of the action of rain waterway. The quantity of top soil so eroded was estimated to be 6000 million tons per annum in 1912 — today the loss must be at least double this amount. This results in progressive reduction in fertility level and degradation of land.

(*ii*) Complete neglect of drainage problems, water logging and salinisation of the soil, large tracts of once fertile lands have become unproductive. According to the latest available estimates, the area affected by these maladies increased from 14 mh in 1981 to 17.6 mh in 1985, thus registering a growth rate of 0.9 mh per annum. Assuming the same growth rate has prevailed during the last decade, the area affected in 1995 should be around 27 mh. Control of erosion on an average cost around Rs. 4000 per hectare, the amelioration of water logged and saline lands requires much larger outlays of the order of Rs. 30,000 to 40000 per hectare—as drainage, especially under ground drainage is an expensive proposition.

In view of the above analysis, a thorough revamping of existing policies in water management is unavoidable. Such a review must take particular note of the following considerations.

(*a*) Further investment in the creation of fresh potential in the surface irrigation sector must be suspended and all available resources should be diverted towards the utilization of the very substantial potential that has been already created but not used.

(*b*) The working of existing canal systems must be improved so as to increase their productivity in terms of yields per hectare. Once this is done, irrigation rates which are at present inordinately low, can be raised so as to prevent the loss of some Rs. 3000 crores per annum on maintenance and operational costs alone.

(*c*) Special attention must be paid to the problem of water logging and salinisation which has received little attention so far.

(*d*) Existing flood protection policies—which aim at dealing with the symptoms of the disease rather than its real cause viz., poor land management must be revised in favour of policies aimed at the prevention rather than the control of floods, through dykes and bunds.

(*e*) The predominant position attained by ground water in the field of irrigation must be recognised and the management of this resource strengthened. The replenishment of this resource must be assisted through natural means such as better land management as well as through artificial recharge, and it must not be treated as a source of only minor irrigation.

It needs to be stressed that existing policies of land management as well as water management are highly unsatisfactory and are the cause of very substantial losses to the government and damage to the economy. Key to better water management lies essentially in improved land management, through a country wide control of denudation and soil erosion.

Check Dams and Water Harvesting (Photographs 8, 9, 10, 11)

Alwar district in Rajasthan which was once sustained by ecosystem of Aravalli had come to a state that it does not have a bucket of water for irrigation and wells were dry. Rain here is quite low 600 mm of which 500 mm falls during monsoon. The denuded hills could not hold rain water. Runoff wasted away the fertile top soil and brought pebbles to agriculture lands. It also meant that ground water was not getting recharged. At this time in 1985, Tarun Bharat Singh and his five friends restored a check dam in Gopalpura. In next monsoon the dry wells of Gopalpura had water. The news spread like a wild fire. 3000 water harvesting structures in 650 villages of Alwar were build by the local people and in the year 2000 when there was draught in Rajasthan, Alwar was not affected.

Check dams are small dams that are build to stop free flow of water. The sites are so selected that they have a catchment area. When monsoon comes these check dams are filled with water. Dry wells get charged with water.

Similarly Hardev Singh, Balwant Singh Jadeja a Post Graduate in English who was selected for the post of deputy superintendent of police, was offered a post of sarpanch. He opted for latter. About 10 years ago Rajsmadhiyala (Rajkot) village in Gujarat was declared desert. In 1986-88 Jadeja constructed 12 check dams and planted thousand of trees. Since then villages have implemented many more projects. In the year 2000 even though there was only 316 mm rainfall against average of 500 mm and Gujarat had draught all round, most wells have water at 3 m depth. All 280 wells, five hand pumps and 35 bore wells have water. Surface water was also available for 10 months. The village earned 2.5 crore more than the neighbouring villages. This is all because of water management and water harvesting. There are many more instances where people have gone for water harvesting.

REFERENCES

1. Aldrich, D.G. and Buchanan J.R., Anomalies in Techniques for Preparing H-bentonites, *Soil, Sci. Soc. Am. Proc.,* 22, 281-85 (1958).
2. Aylmore, L.A.G. and Quirk J.P., Swelling of Clay Water System, *Nature* 183: 1752-53.
3. Barshad, I., Vermiculite and Its Relation to Biotite as Revealed by Base Exchange Reactions, X-ray Analysis, Differential Thermal Curves and Water Content, *Am. Mineralogist,* 33,655 (1948).
4. Bradley, W.F., The Structural Scheme of Attapulgite, *Am. Mineralogist* 25: 405-410 (1940).
5. Bernal, J.D. and Fowler, R.H., A Theory of Water and Ionic Solution with Particular Reference to Hydrogen and Hydroxyl Ions., *J. Chem. Phys.,* Vol. 1, 515-48 (1933).
6. Brown, G., The X-ray Identification and Crystal Structure of Clay Minerals, *Min. Soc.* (Clay Mineral Group), London (1961).
7. Colman, N.T. and Craig. D., The Spontaneous Alternation of Hydrogen clay, *Soil Sci.,* V 91, P-14 (1961).
8. Devore, G.W., Surface Chemistry as a Chemical Control on Mineral Association, *Journal Geol,* 64, 31–57 (1956).
9. Dean, L.A & Rubins, E.J., *Soil Sci.,* V. 63, p. 377 (1947).
10. de Haan, F.A.M. & Bolt, G.H., Soil *Sci. Soc. of Am.,* V. 27, 390 (1963).
11. de Borger R., *Rev. Agric.,* V. 20, 555–565 (1967).
12. De Coninck, F., Major Mechanisms in Formation of Spodic Horizons, *Geoderna,* V. 24, 101–1283 (1980).
13. Eeckman J.P. and Laudelout H., Chemical Stability of Hydrogen Montmorillonite Suspension, *Kolloid Z.,* V. 178, 991–107 (1961).
14. Lucas, I. Cemez Tet., Millot G., Determination Pratique Aux Rayon X-des Mineraux Argileux Simples et Inter stratifires, *Bull Sero Carte Geol Als, Lor,* 12, p 21–33 (1959).
15. Fripiat, J.J., *Trans 9th int. Cong. Soil Sci.,* Surface Fields and Transformation of Adsorbed Molecules in Soil Colloids, 1,679–689 (1968).
16. Grim, R.E., *Clay Mineralogy,* McGraw Hill Book, Company, New York (1968).
17. Gruner, J.W., *The Crystal Structure of Kaolinite,* Z. Krist, 83,75–88 (1932).
18. Greene, R.S.B., Posner AM and Quirk J.P., A Study of the Coagulation of Montmorillonite and Illite Suspensions by $CaCl_2$ Using the Electron Microscope. In "*Modification of Soil Structure*", Eds. W. W. Emerson, R.D. Bond and A.R. Dexter. pp. 35-40, John Wiley & Sons, Chichester (1978).
19. Greene R.S.B., Posner A.M. and Quirk J.P., Factors Affecting the Formation of Quasi-Crystals of Montmorillonite, *Soil Sci. Soc. Am:. Proc.,* V. 37, 357-60 (1973).
20. Hendricks, S.B. and Jeffersons, M.E., Structure of Kaolin and Talc-Pyrophyllite Hydrates and Their Bearing on Water Sorption of Clays, *Am Mineralogist,* 23 : 863-75 (1938).
21. Hendricks, S.B., *Ind. Eng. Chems.,* 37,625 (1945).
22. Harward, M.E. and Coleman, N.T., Some Properties of H- and Al-Clays and Exchange Resins, *Soil Sci.,* 78, 181–88 (1954).
23. Higdon, W.T. and Marshall, C.E., Electrochemical Properties in Relation to Two Methods of Preparation of Colloidal Clays, *J. Phys., Chem.,* 62, 1204 (1958).
24. Hsu, Pa. Ho, *Soil Sci.,* V. 97, 398 (1965).
25. Irving, H. and Williams, R. PJ, *Nature :* 162,746 (1948).
26. Kittric; J.A. and Jackson, M.L. *J. Soil Sci.,* V. 7 : 81 (1957).
27. Ksu. S and Lotse, E.G., *Soil Sci. Soc. Am. Proc.,* V. 36, 725 (1972).
28. Lai, T.M., Mortland, M.M. and Timnich, A. High Frequency Titration of Clay Minerals, *Soil Sci.,* 83, 359-68 (1957).

29. Fredsickson, A.F., Mechanismede lalte ration. *Bul. Geol. Soc. Amer:* 62, 221-231 (1951).
30. Khan, S.U., *Soil Sci. Soc of Am Proc.,* 33, 851-54 (1969).
31. Low, P.F., The Role of Aluminium in the Titration of Bentonite, *Soil Sci. Soc. Am. Proc.,* 19,135-39 (1955).
32. Low, P.F. and Black, C.A., *Soil Sci., Soc. Am. Proc.*, V. 12, 180 (1947).
33. Low, P.F. and Black, C.A., *Soil Sci.*, V. 70, 273 (1950).
34. Khanna, S.S. and Stevenson, F.J., *Soil Sci.,* 93, 298-305 (1962).
35. Marshall C.E., *The Physical Chemistry and Mineralogy of Soils,* John Wiley, New York (1964).
36. Mc Ewan D.M.C. and A Ruiz, *Amil Interstratified Clay Minerals in Soil Components,* Vol. 2, Inorganic Components, J.E. Gieseking (ed.) Springer-Verlag, New York, pp. 265-334.
37. Mitra, R.P., Sharma, B.K., Kapoor, B.S., Three stages, in the Titration of Montmorillonite in Water and in Acetonitrile-benzene Mixture, *Ind. J. Chem.,* 1, 225-26 (1963).
38. Mitra, R.P. and Kapoor B.S., Acid Character of Montmorillonite Titration Curves in Water and Some Non-aqueous Solvents, *Soil Sci.,* V. 108, 11-23 (1968).
39. Mitra, R.P. and Singh, H., On the Strong Acid Character of Montmorillonite Clays and its Disappearance on Aging, *Naturwissenschaften,* 9, 319-20 (1959).
40. Mitra, R.P. and Sindhu, P.S., Acid Character of Sauconite: Increase in Cation Exchange Capacity on Aging in Water and the Role of Zn^{2+} and Al^{3+} Ions, Clays and Clay Minerals., V. 19 : 391-97 (1971).
41. Mitra, R.P. and Sindhu, P.S., Uptake of Phosphate by Some Layered Silicate Minerals in Relation to Their Titration Curves with Acid and Base, *J. Indian Soc.* Soil Sci., V. 24, 29-47 (1976).
42. Sindhu, P.S., The Acid Character of Attapulgite in Terms of Its Structure, *Clay Research,* V. 3, 75-80 (1984).
43. Sindhu, P.S., Some Special Features of the Reaction between Aluminium and Orthophosphate Ions., *Ind. J. of* Chem., V. 14A, 611-13 (1976).
44. Nagelschemidt, G. *Miner: Mag, 28,* 140 (1935).
45. Michael, A.S., Deflocculation of Kaolinite by Alkali Polyphosphate, *Ind. Eng. Chem.*, 50, 951 (1958).
46. Nash, V.E. and Marshall, C.E., Cationic Reactions of Feldspar Surfaces, *Soil Sci Soc. Am. Proc.*, 21, 149 (1957).
47. Pauling, L., *Proc. Natl Acid Sci.,* U.S., 16,123 (1930).
48. Pauling, L., *JACS,* 51, 1010 (1929).
49. Paver, H. and Marshall, C.E., The Role of Aluminium in the Reactions of Clays, *J. Soc. Chem. Ind.,* 53, 750-60 (1934).
50. Slabough, W.H., The Heat of Neutralization of Hydrogen-Bentonite, *JACS,* 74, 4462-64 (1952).
51. Schnitzer, M., Humic Substances: Chemistry and Reactions, P1-64 in *Soil Organic Matter*, Eds. M. Schnitzer and S.V. Khan, Elsevier Scientific Publication, New York (1978).
52. Schnitzer, M and Khan, S.U., *Humic Substances in the Environment,* Dekker, New York (1972).
53. Swartzen-Allen, S.L. and Matijevic, E., Surface and Colloid Chemistry of Clays, *Chem. Rev.,* 74, 385-400 (1974).
54. Thompson, A.C. and Culbertson, J.L., Acidic Properties of Bentonite, *J. Phys. Chem.,* 63, 1917-20 (1959).
55. Thiessen, P.A., Wechselseitge Adsorption won Kolloden Z. *Electrochem.,* 48, 675 (1942).
56. Taylor, R.M., Proposed Mechanism for the Formation of Soluble Si-Al and Fe(III)-Al Hydroxy Complexes in Soils, Geodarma, 42, 65–77 (1988).
57. Van Olphen, H., *An Introduction to Clay Colloid Chemistry,* Wiley-Interscience, New York (1977).
58. Van Dijk, H., *Geoderma*, 5, 53-67 (1971).
59. Quirk, J.P., *Some Physico-chemical Aspects of Soil Structure Stability–A Review. Modification of Soil Structure* Eds. W. W. Emerson, R.D. Bond and AR Dexter, pp. 3-16, John Wiley and Sons, Chickester (1978).

60. London, J.R., (Ed.) *Booker Tropical Soil Manual* Longman, Harlow (1991).

61. Morgan, R.P.C., *Soil Erosion and Conservation,* Longman, Harlow (1986).

62. Harrison, R.M. (Ed.), *Pollution, Causes, Effects and Control* Royal Soc. of Chemistry, Cambridge, U.K. (1990).

63. Schlesinger, W.H., *Biogeochemistry. An Analysis of Global Change,* Academic Press (1991).

64. McLaren, R.G. and Cameron, *K. Soil Scie.* An Introduction in the Properties and Management of New Zealand Soils, Gafred Univ. Press, Auckland, NZ (1990).

65. Barngnold, R.A., *The Physics of Brown Sand and Desert Danes.* London, Chapman and Hall (1941).

66. Vasilev, N.G., Oucharenko, F.D., Chemistry of Surface Acidic Forms of Natural Laminar Silicates, USP, Khim-46(8), 1488–1511 (1973).

67. Vasilev, N.G., Buntova, M.A., Oucharenko F.D., Mechanism of Breakdown. Process of Aging of H-Forms of Clay Minerals, Colloid J. USSR, 37(1), 11–15 (1975).

68. Mukerjee, S.K., Electrochemistry of Clays and Clay Minerals in Bulletin No. 9, page 92 on Mineralogy of Soil Clays and Clay Minerals, Edited by S.K. Mukerjee and T.D. Biswas. The Bulletin was published by India Society of Soil Science, New Delhi (1974).

5

Environmental Chemistry of Some Elements

5.1 INTRODUCTION

Chemical elements tend to circulate in the biosphere in characteristic paths from environment to organisms and back to the environment. These paths, known as biogeochemical cycles, are essential to life. Each cycle has two compartments or pools (*i*) the large, slow moving generally nonbiological portion or the reservoir, and (*ii*) a smaller but more active portion that is exchanging rapidly between organisms and their immediate environment is called labile or cycling pool. Each cycle has two groups: (*i*) gaseous type in which the reservoir is in the atmosphere or the hydrosphere; and (*ii*) sedimentary, in which the reservoir is the earth's crust.

Of the ninety two elements known to occur in natural between 30–40 are required by living organisms. Some elements such as C, H, O, N are needed in large quantities, others are needed in small or even in minute quantities and are referred to as trace elements. Essential elements exhibit biogeochemical cycles. The non essential elements, not required for life, although less closely coupled with organisms also exhibit, cycles, often flow along with essential elements because of their chemical affinity.

The rationale for classifying biogeochemical cycles into gaseous and sedimentary types is that some cycles, such as those involving C, N and O adjust rather quickly to perturbation because of the large atmospheric or oceanic reservoirs or both. Local increase in CO_2 production by oxidation or combustion, for example, tend to be quickly dissipated by the movement of air and the increased output compensated for by increased plant uptake and carbonate formation in the seas. Gaseous type cycles with large atmospheric reservoirs can be considered well buffered globally because of a large capacity to adjust to change. Sedimentary cycles which involve elements such as P and Fe, tend to be more easily disrupted by local perturbations because the great bulk of material is present in a relatively inactive and immobile reservoir in the earth's crust. Consequently, some portion of the exchangeable material tends to get lost over long periods of time. Return or recycle mechanisms in many cases are chiefly biotic. The aim of conservation of natural resources in the broadest sense is to perpetuate processes.

The elements that are believed to be essential for the growth and development of organisms are listed in Table 5.1. The chemistry of these elements is best understood in terms of their position in

Table 5.1 Biologically essential elements

Elements	*General biological function or occurrence*
Hydrogen	Necessary constituent of virtually all organic compounds
Boron	Required by algae and plants; function not known
Carbon	The basic element for organic compounds
Nitrogen	Found in amino acids and derivatives, and other organic compounds
Oxygen	Present in many organic materials and needed for respiration
Fluorine	Small amounts seem to be essential; strengthens teeth
Sodium	A major component of biological electrolytes; nerve action
Magnesium	Enzymes, chlorophyll
Silicon	Structural material in diatoms
Phosphorus	Used in energy transfer processes, DNA and RNA, bones and teeth
Sulphur	Some amino acids, proteins
Chlorine	A major component of biological electrolytes
Potassium	Component of biological electrolytes; nerve action
Calcium	Bones, enzymes, biological electrolyte component
Vanadium	Necessary in plants, marine animals
Chromium	Enzymes
Manganese	Enzymes
Iron	Enzymes, haemoglobin (O_2 transport); biological redox processes
Cobalt	Enzymes
Copper	Enzymes
Zinc	Enzymes
Arsenic	Function uncertain; evidence for need in rats
Selenium	Liver action, some plants
Bromine	May be necessary for animals
Molybdenum	Enzymes
Tin	Functions uncertain, but seems to be needed in rats
Iodine	Thyroid hormones

periodic table (Fig. 5.1). As can be seen, the essential elements occur chiefly among the lighter members of the table. Toxicity, with no known biological use, is encountered among many of their heavier analogues. The property trends in periodic table are discussed in many textbooks of inorganic chemistry[1, 2]. The properties of a family of representative elements show a general similarity, but with trends governed by the smaller size and higher electronegativity of the lighter elements. Among the representative metals, the stronger and more covalent bonding is shown lighter members of a family in their typical compounds, but higher coordination numbers and more ionic character are typical to the heavier members. Across a period, the number of valence electrons and the electronegativity increases with a change from a metallic to nonmetallic character. The transition elements owe much of their chemistry to their incompletely filled *d*-orbitals. Complex formation is particularly important among

1	2	3	4	5	6	7	8	9	10	11	12	13	14	15	16	17	18
$_{1}$H 1.008																	$_{2}$He 4.003
$_{3}$Li 6.941	$_{4}$Be 9.012											$_{5}$B 10.81	$_{6}$C 12.01	$_{7}$N 14.01	$_{8}$O 16.00	$_{9}$F 19.00	$_{10}$Ne 20.18
$_{11}$Na 22.99	$_{12}$Mg 24.31											$_{13}$Al 26.98	$_{14}$Si 28.09	$_{15}$P 30.97	$_{16}$S 32.06	$_{17}$Cl 35.45	$_{18}$Ar 39.95
$_{19}$K 39.10	$_{20}$Ca 40.08	$_{21}$Sc 44.96	$_{22}$Ti 47.90	$_{23}$V 50.94	$_{24}$Cr 52	$_{25}$Mn 54.94	$_{26}$Fe 55.85	$_{27}$Co 58.93	$_{28}$Ni 58.71	$_{29}$Cu 63.55	$_{30}$Zn 65.37	$_{31}$Ga 69.72	$_{32}$Ge 72.60	$_{33}$As 74.92	$_{34}$Se 78.96	$_{35}$Br 79.90	$_{36}$Kr 83.80
$_{37}$Rb 85.47	$_{38}$Sr 87.62	$_{39}$Y 88.91	$_{40}$Zr 91.22	$_{41}$Nb 92.91	$_{42}$Mo 95.94	$_{43}$Tc (99)	$_{44}$Ru 101.1	$_{45}$Rh 102.9	$_{46}$Pd 106.4	$_{47}$Ag 107.9	$_{48}$Cd 112.4	$_{49}$In 114.8	$_{50}$Sn 118.7	$_{51}$Sb 121.8	$_{52}$Te 127.6	$_{53}$I 126.9	$_{54}$Xe 131.3
$_{55}$Cs 132.9	$_{56}$Ba 137.3	$_{71}$Lu 175.9	$_{72}$Hf 178.5	$_{73}$Ta 180.9	$_{74}$W 183.9	$_{75}$Re 186.9	$_{76}$Os 190.2	$_{77}$Ir 192.2	$_{78}$Pt 195.1	$_{79}$Au 197.0	$_{80}$Hg 200.6	$_{81}$Tl 204.4	$_{82}$Pb 207.2	$_{83}$Bi 209.0	$_{84}$Po (210)	$_{85}$At (210)	$_{86}$Rn (222)
$_{87}$Fr (223)	$_{88}$Ra (226)	$_{103}$Lr (257)	$_{104}$Rf 262	$_{105}$Db 262.1	$_{106}$Sg (263)	$_{107}$Bh (262)	$_{108}$Hs (265)	$_{109}$Mt (266)	$_{110}$Uun (281)	$_{111}$Uuu (272)	$_{112}$Uub (285)	$_{113}$Uut	$_{114}$Uuq	$_{115}$Uup	$_{116}$Uuh	$_{117}$Uus	$_{118}$Uuo

Lanthanides	$_{57}$La 138.9	$_{58}$Ce 140.1	$_{59}$Pr 140.9	$_{60}$Nd 144.2	$_{61}$Pm (147)	$_{62}$Sm 150.4	$_{63}$Eu 152.0	$_{64}$Gd 157.3	$_{65}$Tb 158.9	$_{66}$Dy 162.5	$_{67}$Ho 164.9	$_{68}$Er 167.3	$_{69}$Tm 168.9	$_{70}$Yb 173.0
Actinides	$_{89}$Ac (227)	$_{90}$Th 232.0	$_{91}$Pa (231)	$_{92}$U 238.0	$_{93}$Np (237)	$_{94}$Pu (242)	$_{95}$Am (243)	$_{96}$Cm (248)	$_{97}$Bk (247)	$_{98}$Cf (251)	$_{99}$Es (254)	$_{100}$Fm (253)	$_{101}$Md (256)	$_{102}$No (254)

Fig. 5.1 A standard form of the periodic table with atomic numbers, atomic weight. Elements known to be essential in organisms are shown in grey.

these elements. The heavier transition elements and some of the representative metals immediately following them in the periodic table form highly covalent interactions with many of the nonmetals. They tend to be soft acids.

When a biogeochemical cycle, is broken at any point, it will disturb the entire universe. Thus, in the carbon cycle, as will be seen in Section 5.2, carbon burns to produce carbon dioxide, which is photochemically converted to plant biomass. As long as we do not disturb the cycle, namely if percentage of CO_2 in the atmosphere does not change, it is not a pollutant. However, when more of CO_2 is being produced than is being converted to plant biomass, its percentage will increase in the atmosphere, and the same CO_2 has become a pollutant.

5.2 THE CARBON CYCLE, CO_2 AND CARBONATE[3]

For the carbon cycle, it is seen that the atmospheric pool is very small compared to that for carbon in the ocean and in fossil fuels and other storages in the earth's crust. Until the onset of the industrial age, it was believed, flow between the atmosphere, the continents and the oceans were balanced. During the past century, however, the CO_2 content has been steadily rising because of new anthropogenic inputs, as shown by the dotted lines in Fig. 5.2. Although the burning of fossil fuel is believed to be the main source of the new input, agriculture and deforestation also contribute. All agree that forests are important carbon 'sinks' because forests biomass is estimated to contain 1.5 times and forest humus 4 times the amount of carbon in the atmosphere. The rapid oxidation of humus, and the release of gaseous CO_2 normally held in the soil has effects on the cycling of other nutrients. The flux between the reservoir and the exchangeable pool has been fundamentally altered.

The build up of CO_2 and depletion of oxygen over geological time have resulted because production has an average, exceeded respiration. Also, geological and purely chemical processes such as release of oxygen from iron oxides, formation of reduced nitrogen compounds and release of oxygen from water by ultraviolet action, have probably contributed to oxygen accumulation. Both the low concentration of CO_2 and the high concentration of O_2 are now restricting photosynthesis. Most plants increase their rate of photosynthesis if either the CO_2 concentration is increased or O_2 concentration is decreased. This makes green plants very responsive regulators.

It is believed that at the beginning of the industrial revolution (about 1800), the CO_2 concentration in the atmosphere was about 290 ppm = 0.29%; in 1958 when accurate measurement were made, it was about 315 ppm; in 1980 it had risen to 335 ppm. If the concentration double the preindustrial level, which would happen by the 2050, a warming of global climate is likely to occur, with a mean rise in temperature of 1.5 to 4.5°C along with a rise in the sea level as polar ice melts, and changes in rainfall pattern that could disrupt agricultural production. Recently documented, mean sea level already has shown a rising trend, it has risen by about 12 cm in this century. This is making climate to change and flooding of coastal regions. Such a threat should not be ignored.

In addition to CO_2, two other forms of carbon are also present in the atmosphere, though in small amounts; carbon monoxide (CO) at about 0.1 ppm and methane (CH_4) at about 1.6 ppm. As with CO_2, these two rapidly and thus, have short residence time, about 0.1 year for CO, 3.6 years for CH_4 and 4 years for CO_2. Both CO and CH_4 arise from incomplete anaerobic decomposition of organic matter, but in the atmosphere both are oxidized CO_2. An amount of CO equal to that from natural decomposition is now injected into the air by incomplete burning of fossil fuels, especially automobile exhaust.

Very little is known about the flow of carbon in sedimentary cycle in qualitative terms. The movement of solid matter through the air as dust, indicated as fall out, may return to earth as rain or as a dry fall out. To this man is adding more material fall out, relatively small in amount but biologically significant

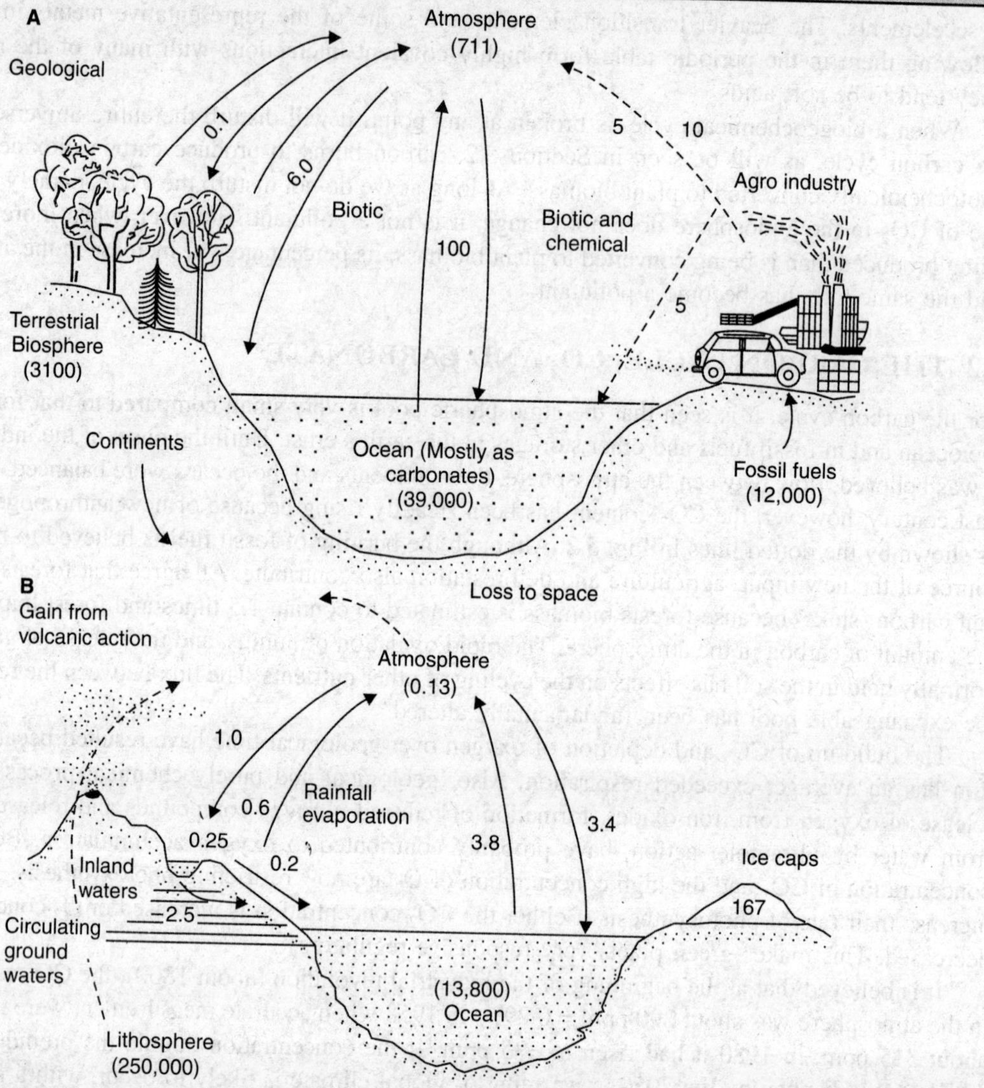

Fig. 5.2 **A.** The CO_2 cycle. Figures are 10^9 tons CO_2 in major biosphere compartments and circulating between compartments (arrows). **B.** The hydrological cycle, Figures are geograms (10^{20} grams) H_2O in major biosphere compartments and circulating between compartments (arrows).

because of their poisonous effects. Annual flow of sediments varies from continent to continent. Lowlands and oceans tend to gain soluble mineral nutrients at the expense of the uplands during the periods of minimum geological activity. Under such conditions local biological recycling mechanisms are extremely important. In other words, the longer the vital elements such as carbon can be kept within an area and used over and over again by successive generation of organisms, the less new material will be needed from the outside.

In evaluating atmospheric CO_2 effects, gas-solution equilibria and acid-base equilibria of carbonate (Eqs. 3.10 to 3.23) have to be considered, both of which influence one another. The rate at which equilibrium can be achieved is of equal importance. This may take a few years for surface water (*i.e.*

to depth of less than 100 m) but a period of the order of 1000 years for deeper ocean layers which make up over 90% of the total. Carbon dioxide–Carbonate equilibria in water has already been discussed in chapter 3, section 3.6.

In solution, carbon dioxide forms carbonic acid (H_2CO_3)

$$\underset{}{CO_2 + H_2O} \longrightarrow \underset{\text{Acid}}{H_2CO_3} \rightleftharpoons \underset{\text{pH 7–8.5}}{HCO_3^- + H_3O^+} \overset{H_2O}{\rightleftharpoons} \underset{\text{pH>9}}{CO_3^{2-} + H_3O^+}$$

Under normal circumstances, the pH of a natural system is found to be in the range 7.0–8.5, where main species is bicarbonate ion. In this form, carbon is not easily assimilated by aqueous organism and fluctuations in its availability are kept to a minimum. As the solution becomes alkaline, the equilibria shifts to right hand side producing more of carbonate ions. The concentration of carbonate ion is related to pH as follow:

$$CO_3^{2-} = \frac{S}{1 + \dfrac{H_3O^+}{4.8 \times 10^{-5}}} \qquad ...(3.24)$$

and metal ion concentration = $M^{2+} = K_{sp}/[\,CO_3^{2-}]$ ...(3.25)

As in the case of $CaCO_3$, with calcium concentration 6.52×10^{-4} M at a pH 7, in order that it may precipitate, the concentration of CO_3^{2-} should be > $2.88 \times 10^{-9}/6.52 \times 10^{-4} = 4.2 \times 10^{-6}$ M. However according to eq. 3.24, carbonate concentration at pH = 7 is 9.97×10^{-3} M, hence $CaCO_3$ will be precipitated. As precipitate removes CO_3^{2-} ions, equilibria will shift to the right-hand side to produce increasing amounts of CO_3^{2-} to be removed as precipitate till carbonate concentration drops as low as 4.2×10^{-6} M. It should also be apparent from Eqs. 3.24 and 3.25 that pH also fixes the concentration of metal ions. Another important activity of CO_2 involves fish eggs. The egg shell is made of $CaCO_3$. If river becomes acidic, this shell will dissolve and hence eggs will not fertilize, resulting in a depletion of aquatic life.

Carbon dioxide in natural water is derived from a number of sources. Bacterial decomposition of organic matter in the tropholytic zone, and respiration by animals and plants, contribute to the store of carbon dioxide. In the case of plants, the greater net contribution is at night when photosynthesis is not occurring. Ground water flowing or seeping into lakes and streams, may carry CO_2, the amount being determined by the extent of decomposition in the top soil. In water, chemical reaction between acid and various compounds of carbonates release CO_2 for which it is essential that both should be acidic at night. The system has to become alkaline at day time when precipitation has to occur by absorbing CO_2 through photosynthesis.

The importance of rain in supplying CO_2 to inland water lies in a reaction in which it is transported to solution as H_2CO_3. This acidic water encounters carbonate holding rocks, lime stone, etc. when the latter dissolves as bicarbonate or $Ca(HCO_3)_2$. This solution remains stable only in the presence of free CO_2. However, continuous addition alkali or acid will not allow the system to become acidic at night or alkaline at day time and consequently, germination of plant in soil or water will be hindered. This is a consequence of pollution of water.

5.3 THE NITROGEN CYCLE[4, 5]

In one form or another nitrogen is found all through the biosphere. In its gaseous form, it is present as diatomic molecule N_2 which is rather inert and comprises 79% of the atmosphere and forms the major reservoir for the total cycle. In the hydrosphere, its total chemical forms have an ion concentration of approximately 5×10^{-5} M. This latter value is generally accepted average, but it must be remembered that it is subject to local variations to a marked degree.

The aqueous reduction/oxidation (redox) chemistry of nitrogen involves primarily NO_2^-, NO_3^- and NH_4^+ which take part in oxidation-reduction process expressed in the following equations.

$$NO_3^- + 2H^+ + 2e^- \longrightarrow NO_2^- + H_2O \quad ...(3.26)$$

$$NO_2^- + 8H^+ + 6e^- \longrightarrow NH_4^+ + 2H_2O \quad ...(3.27)$$

$$NO_3^- + 10H^+ + 8e^- \longrightarrow NH_4^+ + 3H_2O \quad ...(3.28)$$

These reactions depend on pH value whereas equilibrium in compounds of the nitrogenous system depends on this as well as on the E_h (or *pE*) of the system. Nitrate has a comparatively narrow range of E_h over which it can exist at significant concentrations; the most common form is NH_4^+. At neutral pH ranges, the NO_2^- stabilizes in the vicinity of $E_h = 6.5$. The essential features of the nitrogen cycle are shown in Fig. 5.3. Table 5.2 gives an estimate of amount of material involved.

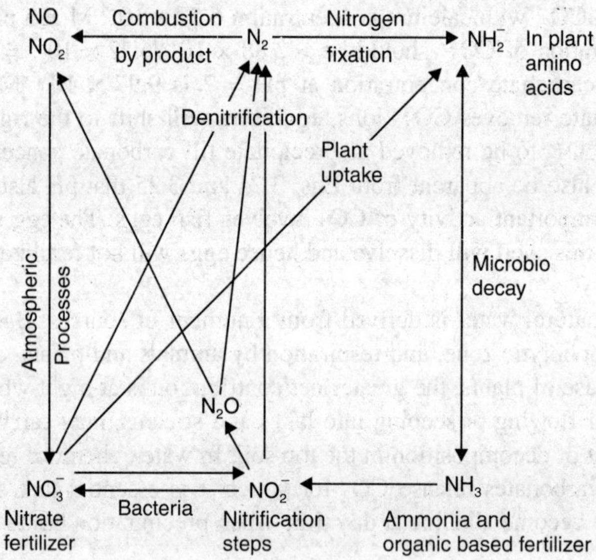

Fig. 5.3 The nitrogen cycle.

Before nitrogen can be utilised it has to be fixed by a biological system, which in the hydrosphere appears as blue-green algae. The reactions involved are

$$N_2 \longrightarrow 2N(\text{fixation}) \quad ...(3.29)$$

conversion of atmospheric N_2 to N (II) is also microbial process, most important bacteria associated with it are found in leguminous plants (*e.g.* clover). The reaction is followed by hydrogenation.

$$2N + 3H_2 \longrightarrow 2NH_3 \quad ...(3.30)$$

Table 5.2 Quantities involved in the nitrogen cycle

Nitrogen inventory (10^9 metric tons)		
Atmospheric N_2		3,800,000
Plants		13
Animals		1
Dead organic matter:	land	760
	ocean	900
Dissolved N_2		20,000
Inorganic nitrogen (not N_2):	land	140
	ocean	100
Rocks, sediments		18,000,000
Rates of processes in the nitrogen cycle (10^6 metric tons per year):		
Fixation: Biological	(land)	44
	(oceans)	10
	(atmosphere)	7.6
Industrial		30
Total		91.6
Denitrification:	(land)	43
	(oceans)	40
Total		83
River run-off		30

In both cases the process is quite complex biochemically, but evidently depends on a substance containing enzyme in which both Fe and Mo take part and lead to the formation of ammonia which is immediately incorporated into amino acid in plants. Plants however cannot generally absorb nitrogen in any form other than the nitrate ion. Other natural fixation processes are insignificant although much nitrogen is now fixed industrially.

Ammonia produced by fixation incorporates the form NH_4^+ in water which is subject to nitrification reaction in two steps. The reactions in both steps are oxidative and are carried out by two species of autotropic bacteria nitrosomonas and nitrobacter

Nitrosomonas

$$4NH_4^+ + 7O_2 \longrightarrow 4H^+ + NO_2^- + 6H_2O \quad \Delta G = -351.1 \text{ kJ mol}^{-1} \quad ...(3.31)$$

$$2NH_3 + 3O_2 \longrightarrow 2NO_3^- + 2H_2O \quad ...(3.32)$$

$$2NO_2^- + O_2 \longrightarrow 2NO_3^- \quad \text{Nitrobacter,} \quad \Delta G = -74.47 \text{ kJ mol}^{-1} \quad ...(3.33)$$

so that overall reaction is

$$NH_3 \text{ or } NH_4^+ \longrightarrow NO_2^- \longrightarrow NO_3^- \quad ...(3.34)$$

and the end product of oxidation is nitrate, the form in which nitrogen is assimilated to form amino acids and proteins. The detailed chemistry of these reactions is not fully understood. It appears that the first stage in nitrification is the oxidation of ammonia to hydroxylamine (5.1). Following this, hydroxylamine may react with nitrite already formed (as nitrous acid) to produce nitro hydroxylamine (5.2) which is subsequently oxidized to nitrite. Nitrobacter oxidizes nitrite to nitrate directly (3.33).

$$2NH_4^+ + O_2 \longrightarrow 2NH_2OH + 2H^+ \quad ...(5.1)$$

$$NH_2OH + HNO_2 \longrightarrow NO_2NHOH + 2H \quad ...(5.2)$$

$$2NO_2NHOH + O_2 \longrightarrow 4HNO_2 \quad ...(5.3)$$

$$HNO_2 + H_2O \longrightarrow 4HNO_3 + 2H$$

There are a large number of physical and chemical factors which influence the nitrification process, some of which are indirect.

If organic matter in soil is low in nitrogen, the nitrate available to plant will be limited. In addition, If microbial oxygen consumption is high then the resulting anaerobic conditions will prevent oxidation of ammonia. The same retardation occurs at deep water levels where oxygen is excluded from the soil pores.

It is true that during nitrification process hydrogen ion production tends to acidify the environment in which the soil buffering capacity normally counter acts. This is just as well because the nitrifying organisms unlike the ammonifer eqs. 5.5 & 5.6 are very sensitive to acidity. *Nitrobacter* species have an optimum pH between 5 and 8 whilst *Nitrosomonas* function best between 7 and 9. Therefore, in slightly acid environment *Nitrobacter* soon runs out of nitrite substrate, while in alkaline reaction there may be a build up of the biological toxic nitrite. At a pH less than 5, nitrification stops all together. Soil containing colloidal clays may strongly adsorb ammonium ion and temporarily restrict their input into the nitrification pathways.

Nitrification is a reversible reaction under anaerobic conditions, since reduction of nitrate ion can provide entry of bacterial source through the net change

$$NO_3^- + \frac{1}{2}C\,\text{organic} \longrightarrow NO_2^- + \frac{1}{2}CO_2 \quad ...(5.4)$$

On the death of an organism other type of bacteria can convert the nitrogen in amino acids by a process called "ammonification". Some nitrogen containing organic substances are more easily decomposed than others. As for example, those residues containing an abundance of non-nitrogenous organic matter (say cellulose) may encourage the development of a high proportion of tricropes capable of utilising glucose polymers but not able to ammonify any nitrogen containing matter. The presence of naturally occurring bacteriostatic phenolic substances and lignoprotein complexes may also retard ammonification.

Ammonification is both an aerobic as well as an anaerobic process. So that the displacement of oxygen by water does not inhibit ammonia production to any great extent. In other words, various microorganisms become involved and the end product (in addition to CO_2 and NH_3) are different (5.5-5.6). However, as the previous step in the nitrogen cycle—nitrification—is necessarily aerobic, anaerobic environment will inhibit the overall cycle.

The environment pH also influences the rate of ammonification which appears most efficiently in alkaline/neutral environment

$$\text{N} - \text{Organic matter} \xrightarrow{\text{Aerobic}} NH_4^+ + CO_2 + SO_4^{2-} + H_2O \qquad ...(5.5)$$

$$\text{N} - \text{Organic matter} \xrightarrow{\text{Anaerobic}} NH_4^+ + CO_2 + \text{amines and organic acids} \qquad ...(5.6)$$

However, even in an acid environment, fungal proteases ensure some splitting of protein material.

The nitrite ion is relatively toxic because of its interaction with haemoglobin. Nitrites in the blood results in the oxidation of the Fe(II) of haemoglobin to form methemoglobin, which has no oxygen carrying ability; this disease is called methemoglobinemia. While cases of directs nitrite poisoning are rare, the nitrate ion can be reduced to nitrite in the stomach, and for this reason food and water with high nitrate content are dangerous. This reduction is especially possible in the stomachs of infants, where the low acidity allows the growth of nitrates in the body can react with organic amines possibly to form carcinogenic nitrosoamines. Nitrites (often produced from KNO_3 in site) is a common additive incurred meat, where it acts to prevent the growth of bacteria, but is also added to produce flavour and an attractive colour, the latter is through the formation of nitrosylmyoglobin. Denitrification is the reduction of nitrate to nitrite, nitrous oxide, nitric oxide, ammonia and gaseous nitrogen. Some of these products are reincorporated into soil environment (NO_2, NH_3) and as a result are only temporarily removed from the immediate nitrogen cycle. Others are released in a gaseous form (N_2, NO, N_2O) into the atmosphere. The former is often referred to as nitrate reduction the latter as true denitrification.

A few autotrophic bacteria are capable of reducing nitrate by using it as an electron acceptor. As example, *Micrococcus* denitrification and sulphur bacterium *Thiobacillus* denitrification

$$5S + 6KNO_3 + 2CaCO_3 \longrightarrow 3K_2SO_4 + 2CaSO_4 + 2CO_2 + 3N_2 \qquad ...(5.7)$$

However, the great majority of nitrate-reducing bacteria are facultatively anaerobic heterotrophs and include species of the genera *Pseudomonas, Achromobacter; Micrococcus* and *Bacillus.* The reducing power comes from glucose (5.8 & 5.9) and for every two molecules of nitrate one molecule of nitrous oxide or nitrogen is produced.

$$C_6H_{12}O_6 + 6KNO_3 \longrightarrow 6CO_2 + 3H_2O + 6KOH + 3N_2O \qquad ...(5.8)$$

or $$5C_6H_{12}O_6 + 24KNO_3 \longrightarrow 30CO_2 + 18H_2O + 24KOH + 12N_2 \qquad ...(5.9)$$

Heterotrophic denitrifiers are often found in great abundance in soil ($1 \times 10^{4-6}\ g^{-1}$), performing aerobic activities until oxygen becomes scarce. Thus, the potential for denitrification is enormous.

The most likely pathways for nitrate reduction and denitrification appears as follow:

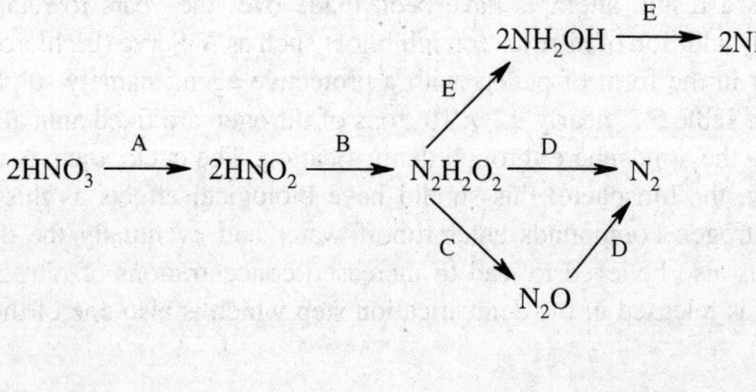

...(5.10)

The reaction are composed of the following steps:

A. Reduction involving the addition of two electrons and nitrate reductase:

$$HNO_3 + 2H \longrightarrow HNO_2 + H_2O \qquad ...(5.11)$$

B. Formation of an unstable intermediate, hyponitrite, with the help of nitrite reductase

$$NO_2^- + 6H \longrightarrow N_2O_2H_2 + 2H_2O \qquad ...(5.12)$$

C. At high temperatures, hyponitrite is reduced by hyponitrite reductase to nitrous oxide

$$N_2O_2H_2 \longrightarrow N_2O + H_2O \qquad ...(5.13)$$

D. At low temperatures, nitrogen gas is formed

$$N_2O_2H_2 + 2H \longrightarrow N_2 + H_2O \qquad ...(5.14)$$

This effect of temperature on the end products of denitrification is illustrated by denitrifier (bacteria) which produces approximately equal quantities of nitrogen and nitrous oxide at 30°C but 63% nitrous oxide at 57°C.

E. Reduction of hyponitrite to ammonia proceeds under micro-aerophilic conditions and recognises hydrazine as an intermediate product. Micro-organisms involved in this reaction include *Bacillus subtilis* and some *Azotobacter* species. Very little is known concerning the enzyme systems of denitrification.

Fertilizers make up an important artificial source of nitrogen input to the environment. Because nitrogen is an essential nutrient for plants, and limited in quantity, large scale use of nitrogen-containing fertilizers has become common place. These may take the form of nitrates, which can be made available immediately to plant. Alternatively, fertilizers based on ammonia or organic nitrogen can be used. All these must be converted to inorganic nitrate ions so as to be used by plants, and since this takes place over a period of time, they can provide a longer lasting source of nitrogen. Nitrates tend to be quite soluble and weakly held by ion-exchange force, and hence may easily be leached from the soil and wasted. Organic nitrogen fertilizers may or may not be soluble, but soluble materials, can be formulated to resist dissolution (*e.g.* pellets may be coated with sulphur and wax). However, it should be emphasised that movement of materials through the soil is highly complicated by absorption and ion-exchange processes. Even readily soluble materials can be retained for long periods in some types of soils. Thus ammonia itself, although a highly soluble gas, is an effective fertilizer. It is held in the soil as the ammonium cation by ion-exchange. Other widely used compounds are NH_4NO_3 and urea. It is noteworthy that since nitrogen is assimilated by plants as the nitrate ion, the original source of the form a fertilizer should be calculated to the rate of release of NO_3^-, extent of leaching (which results in waste and ultimately water pollution), and other economic factors. This also leads to *eutrophication* of water and related problems. As a result, attempts have been made over the years to retard the release of nitrogen from fertilizers by addition of nitrification inhibitors such as *N*-Serve (trichloromethyl pyridine) or preparing the fertilizer in the form of pellets with a protective agent, namely sulphur.

From the estimates in Table 5.2, nearly 9.1×10^7 tons of nitrogen are fixed annually while only 8.3×10^7 tons are released to the atmosphere through denitrification. The cycle, then, is not balanced and nitrogen is building up in the biosphere; this should have biological effects as this material acts as fertilizer. The bulk of nitrogen compounds enter runoff water and eventually the oceans. However, increased fertilizer used is also believed to lead to increased concentrations of nitrogen oxides in the atmosphere as more N_2O is released in the denitrification step which is also one of the cause of ozone hole at poles[30].

Oxides of nitrogen in the atmosphere are a cause of concern with respect to air pollution problems. The main compounds causing concern are NO and NO_2, which are by-products of combustion processes. They are formed during the reaction of N_2 in the air with O_2. The primary product is NO, but some NO_2 is produced as well. Nitrogen oxide (often called NO_x) production is directly related to the temperature in the combustion zone of the furnace or engine, and can be reduced by operating at lower temperatures. This however reduces the efficiency of the device. Reduction of the amount of excess air in the combustion chamber will reduce NO_x emission, but at the expense of an increase in the amount of incompletely burned fuel. A two-stage combustion, the first at high temperature with an air deficiency, followed by completion of the combustion at a lower temperature, can be quite effective in reducing NO_x emission.

Absorption of NO_x from the combustion gases is more difficult than SO_2 removal, although various methods have been proposed both for acid and alkaline scrubbing of stack gases. Alkaline scrubber can remove SO_x and NO_x simultaneously, but neither NO nor NO_2 is effectively absorbed by a simple alkaline solution. However, a mixture of NO and NO_2 enters into an equilibrium reaction.

$$NO + NO_2 \rightleftharpoons N_2O_3 \qquad ...(5.15)$$

The equilibria lies to the left under normal conditions. Since N_2O_3 is anhydride of nitrous acid, if it is absorbed in a basic solution, the equilibria is shifted to the right.

$$N_2O_3 + H_2O \longrightarrow 2HNO_2 \qquad ...(5.16)$$

The nitrite salts formed in a basic solution can be oxidised to nitrate. This process however requires equimolar amounts of both NO and NO_2 in the gas stream or recycling of the excess. Other processes have been proposed, *e.g.* absorption of NO by alkaline sulphite solution which results in the production of a compound from SO_3^{2-} and NO_2 with the formula $O_3SN_2O_2^{2-}$ but practical applications are best with problems of oxidation and stability.

Catalytic reduction of NO_x is possible, reaction with CO, NO_3 or CH_4, for example, can produce N_2

$$NO + CO \xrightarrow{\text{Catalyst}} CO_2 + \frac{1}{2}N_2 \qquad ...(5.17)$$

This reaction is favoured thermodynamically, but is very slow under normal circumstances. Various metals, *e.g.* Ag, Cu, Ni and Pd are effective catalysts, but are subject to poisoning in practical use and in some cases are expensive. Some oxides, *e.g.* copper chromite and Fe_2O_3 are also effective catalyst for this purpose. Systems of this kind may be applicable to automobile exhaust, but would have to be built on a very large scale to cope with power generator gases.

5.4 THE PHOSPHORUS CYCLE[6]

There are about 18000 million metric tonnes of phosphate rocks, China has 37 per cent of reserves, Morocco/Western Sahara 32 percent and USA 5–6 per cent. Export is dominated by Morocco alone as China is not a major exporter. The geopolitical importance of Western Sahara independence or integration with Morocco continues to be minimal because of two reasons the world is unaware of this conflict and phosphorus limitation has not become a global concern. But this conflict could have major impacts on world economy through Western Sahara/Morocco monopoly of over one third of the global supply of rich phosphate rocks. Presently, the Moroccan government controls all trade and other activities in

Western Sahara. India is the largest country in the world that is most dependent on foreign sources of phosphate.

Approximately 90% of phosphate rock is used for super phosphate fertilizer and animal feed supplement and rest for industrial chemicals. Very little phosphorus is recycled. Excess runoff and discharge of phosphorus in surface water has resulted in eutrophication and algal overgrowth, which creates oxygen poor conditions, reduces water quality and damage ecosystem. Phosphate fertilizers transform to less available forms which remain bound to upper soil layer and most of it cannot be used by plants. Apparently phosphate become unavailable to plants due to combination of pH mediated insolubilised as shown by Sindhu[9]. This also changes the soil character.

Phosphorus is a component of nucleic acids, phospholipids, inositol phosphates, sugar phosphates, coenzymes, etc. Plant residues may contain up to 0.5% phosphorus and phosphorus of organic origin contributes up to 70% of that element in soil. Plants are, for the most part unable to take up organic phosphorus and only utilize it in an inorganic soluble form such as orthophosphate ($H_2PO_4^-$).

Soil is made up of a mixture of clay minerals, metal oxides and carbonates, salts (soluble and insoluble) and organic matter. A good soil contains a fair amount of clay minerals and phosphate in it. Phosphorus is rendered insoluble in two different ways:

(*i*) by adsorption and (*ii*) by precipitation.

The ability of clay mineral and various oxides in soil to combine with anion as well as cations points to the amphoteric nature of broken bonds and it has been discussed in great detail in the studies of Mitra and Sindhu[7]. Broken bonds exposed on lateral surfaces would create equivalent amounts of negative as well as positive centres and negative centres immobilise cations, the positive centres would hold anions.

If pH is low, the exposed positively charged $Al^{+0.5}$ ion and its counterpart the $Si—\overset{-0.5}{O}—Al$ groups will be hydrated, water dipoles being held by them with negative/positive poles directed towards positive/negative centres on the surface. The Al ion at the edge whose coordination octahedra has been broken completes its octahedral shell with water dipoles held to it by coordination bonds.

Around pH 7.0 when some of the water of coordination are in a dissociated condition, these edge aluminium atoms will be covered partly, by water and partly by OH ion. In the acidic medium these OH ions are progressively neutralized by hydrogen ions which is subsequently replaced by $H_2PO_4^-$ as

$$Al—OH + H^+ + H_2PO_4^- \longrightarrow AlH_2PO_4 + H_2O \quad ...(5.18)$$

In alkaline medium, water dipoles and OH ions are replaced by the $Na_2PO_4^-$ and quantitative results were obtained by Mitra and Sindhu[7].

The second way of preparing phosphates is by precipitation. As discussed in chapter 4 the clay minerals decompose to give exchangeable cations (*e.g.* Fe^{3+}, Al^{3+} etc). Sindhu[8] has found that these cations precipitate in acidic medium, at low pH of 4, as variscite ($AlPO_4 . 2H_2O$) and strengite $FePO_4 2H_2O$.

Unlike nitrogen, phosphorus does not undergo any valency changes during assimilation and enters into organic combinations as phosphate. Nor is it likely that changes occur during degradation of organic matter. This lack of volatility and reactivity of phosphorus produces somewhat limited and less spectacular biological cycle (Fig. 5.4) than that of nitrogen. The involvement of soil micro-organism-phosphorus transformation may include, (*a*) changing the solubility of inorganic phosphorus (*b*) mineralising organic material to release orthophosphate, and (*c*) converting available phosphorus to microbial protoplasm (immobilisation), and (*d*) oxidation and reduction reactions.

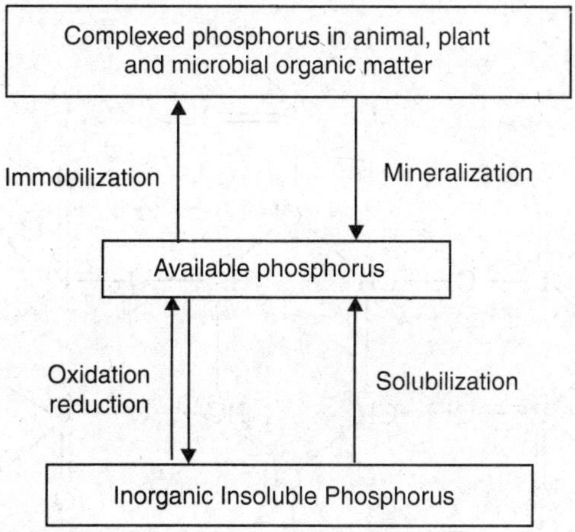

Fig. 5.4 The phosphorus cycle.

(a) Solubilisation: A great many microbes have the potential to solubilise inorganic phosphorus compounds. They may achieve this end in two way by acid production or by the simulation of a reducing environment. Acids may be generated by chemolithotrophs producing nitrous and nitric acids or sulphuric acid, by all respiring bacteria which produce carbonic acid and those which produce organic acids as metabolic by-products.

As has been discussed in section 3.2, the plants growing in soils do not undergo photosynthesis at night rather CO_2 is released. In order that CO_2 be released from soil and water, both become acidic at night. This makes the phosphates soluble. In other words phosphate is released to plants for growth at night. The reverse is the case at day time, when plants absorbs CO_2 and photolysis takes place with the release of O_2. In order to absorb CO_2 at day time soil and water become alkaline which leads to precipitation of phosphate in soil and thus less phosphate is available at day time.

Similarly when there is rain, water dissolve CO_2 from air and it becomes acidic which also leads to release of phosphate and calcium from soil. This acid production increases phosphate availability in the form of three calcium phosphates-monocalcium phosphate ($Ca(H_2PO_4)_2$), dicalcium phosphate ($CaHPO_4$) and tricalcium phosphate $Ca_3(PO_4)_2$. Of these, monocalcium phosphate has the highest solubility, while tricalcium phosphate is practically insoluble. The reaction moves from right to left as pH decreases:

$$Ca(H_2PO_4)_2 \rightleftharpoons CaHPO_4 \rightleftharpoons Ca_3(PO_4) \quad ...(5.19)$$

Under an aerobic conditions, ferric phosphate (strengite $FePO_4.2H_2O$) may be reduced to ferrous phosphate with an increased solubility. Tricalcium and aluminium phosphate are not dramatically affected. In addition, hydrogen sulphide, produced by such bacteria as *Desulfavibrio,* reacts with ferric phosphate to form ferrous sulphide and hypophosphate. The evolution of carbon dioxide and secretion of organic acids by plant roots may also have an important influence on phosphate availability.

(b) Mineralisation: The most abundant class of phosphorus-containing organic matter in soil are esters such as inositol phosphate which in itself may account for some 50% of the total organic phosphorus. These hexaphosphate of myoinositol (myo-IHP or phytic acid) Fig. 5.5 is a common component of soil humus. These organic phosphates are slowly decomposed in soil despite the abundance of microbes to produce phosphate. The degradation of nucleic acids is a substantial extracellular reaction.

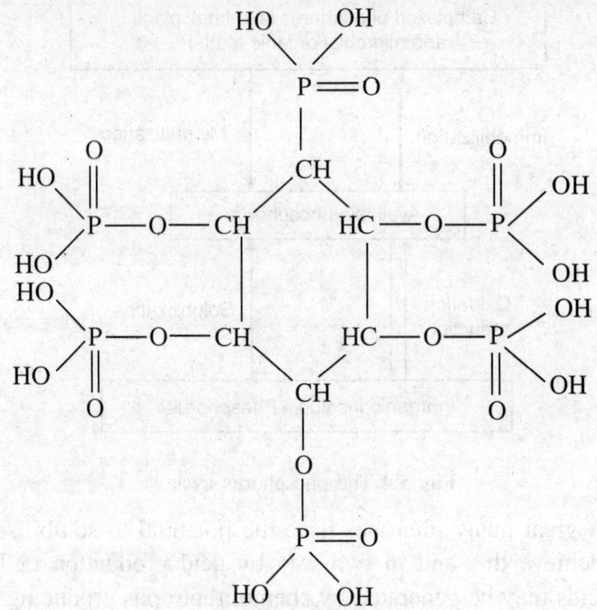

Fig. 5.5 Phytic acid.

Only a small proportion of soil phosphorus is present as phospholipids or phosphatides. A number of animal, plant and microbial enzymes catalyse the hydrolysis to release orthophosphate.

(c) Immobilisation: Microbial immobilisation occurs when relatively large amounts of carbon and nitrogen are available (C:P > 2-300 : 1 ; N : P = 10:1). Microbes are comparatively rich in phosphorus – their demand for this element is set against the requirements of higher plants.

(d) Oxidation and Reduction: Phosphorus, like nitrogen, may exist in a number of oxidation states from the −3 of phosphine (PH_3) to the +5 of PO_4^{-3}. Microbiological intercellular oxidation of orthophosphite to orthophosphate $\left(HPO_3^{2-} \longrightarrow HPO_4^{2-}\right)$ has been recorded for a number of soil heterotrophs although there is no evidence for this being an energy yielding reaction.

It is interesting to note that nitrogen fixation by *Azotobacte vinelandic* is inhibited by orthophosphite. The microbial reduction of phosphate in the environment has not been conclusively demonstrated. However, it has been suggested that under anaerobic conditions the reduction of phosphate to phosphite or even phosphine gas can occur.

$$H_3PO_4 \longrightarrow H_3PO_3 \longrightarrow H_3PO_2 \longrightarrow PH_3$$

Clostridium butyricum and *Escherichia coli* may mediate in this change which is analogous to denitrification.

5.5 SULPHUR AND THE SULPHUR CYCLE[10]

Sulphur is a relatively abundant, essential element of considerable importance. Sulphur exists in environment as mineral sulphate and sulphides, inorganic matter, as fertilizer and soil pH treatments, as gaseous sulphur dioxide and as sulphuric acid in rain water. It is a major component of air pollution, particularly in industrialised area, although natural sources of sulphur also contribute.

As a plant component, animal and microbial organic matter, sulphur (as sulphydryl, *R—SH* groups) provides a linkage between the polypeptide chains in the protein molecule helping to maintain its three dimensional structure. Aminoacids, cystine, cysteine, methionine, vitamin-B as in thiamine, biotin, and trictic acid and a variety of excretory products.

Oxidation state of sulphur in its compounds varies from –2 to +6. The variety of oxidation states means that a range of enzymic and chemical systems are involved in the transformation of sulphur.

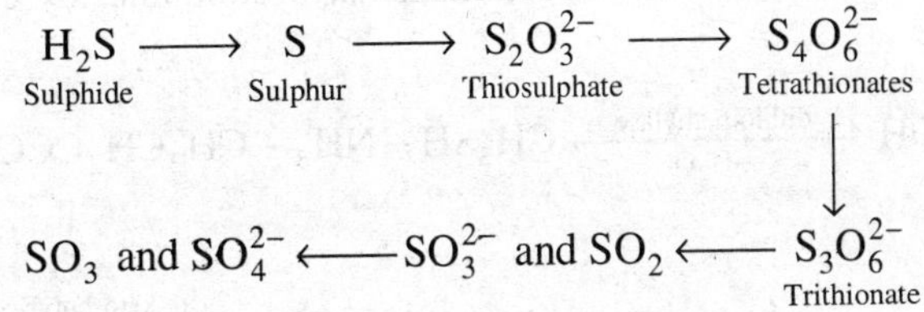

Oxidation of sulphides produces chiefly SO_2 as the intermediate product. The most important inter-relationships of these states are shown in the sulphur cycle Fig. 5.6.

A wide variety of heterotrophic microorganisms are capable of decomposing sulphur-containing organic residues. A portion of this sulphur is used in cell synthesis whereas, in many instances, the remaining is released as hydrogen sulphide which is unstable in an aerobic environment and is rapidly oxidised by chemical and biological mechanisms. A sulphur-containing amino acid released during

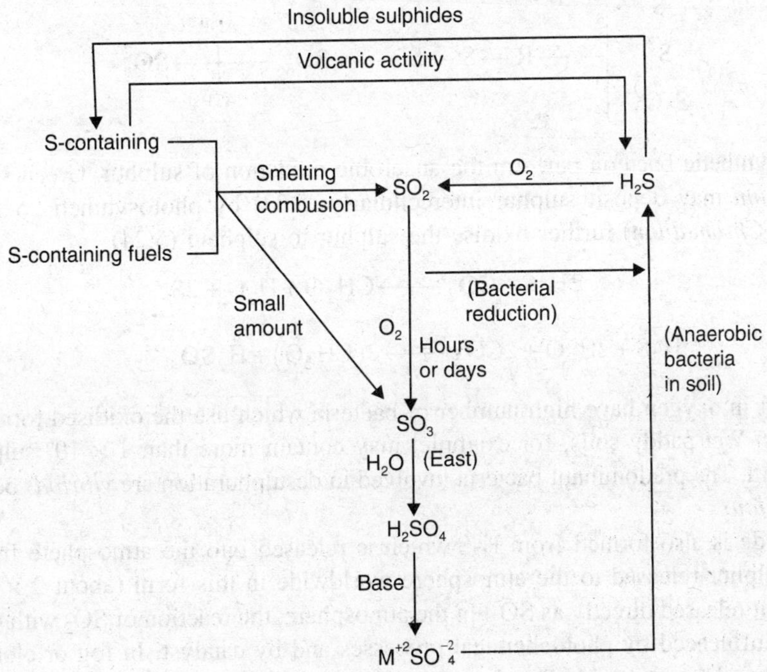

Fig. 5.6 The sulphur cycle.

protein hydrolysis may undergo desulphydration; *e.g.*, cysteine is degraded by the enzyme *desulphydrase*

$$HSCH_2CH_2CHCOOH \xrightarrow{+H_2O} H_2S + NH_3 + H_3C-\underset{\underset{O}{||}}{C}-CH_2COOH \quad ...(5.20)$$

(Cysteine: $HSCH_2CH_2CH(NH_2)COOH$; Pyruvic acid)

On other occasions, cysteine-sulphur may be oxidised all the way to sulphate without involving sulphide by a combination of chemical and biological mechanisms. Methionine, by contrast may be oxidised to mercaptan.

$$CH_3SCH_2CH_2CHCOOH \xrightarrow[+H_2O]{\text{dithiomethylase}} CH_3SH + NH_3 + CH_3CH_2\underset{\underset{O}{||}}{C}COOH \quad ...(5.21)$$

(Methionine: $CH_3SCH_2CH_2CH(NH_2)COOH$; α-keto-butyric acid)

Sulphur, often as sulphate, may be assimilated by microorganisms to produce sulphur amino acids. Members of autotrophic genera are typical pseudomonads (gram negative rods with a single polar flagellum) to oxidise sulphur and/or ferrous iron. The oxidation of sulphur or sulphide by *Thiobacillus* species involves their reaction with sulphydryl group of glutathione to form glutatione polysulphide. Sulphide oxidase then converts sulphide to sulphate. Thiosulphate substrates are split into sulphite and sulphur. The sulphur is oxidised to sulphate.

$$\left.\begin{matrix} S^0 \\ S^{2-} \\ S_2O_3^{2-} \end{matrix}\right] \longrightarrow R-S-S \longrightarrow SO_3^{2-} \xrightarrow[ATP]{ADP} SO_4^{2-} \quad ...(5.22)$$

Many photosynthetic bacteria perform the anaerobic oxidation of sulphur. Green sulphur bacteria such as *Chlorobium* may deposit sulphur intercellularly (5.23) by photosynthetic oxidation. Purple sulphur bacteria (*Chromatium*) further oxidise the sulphur to sulphate (5.24)

$$2H_2S + CO_2 \longrightarrow CH_2O + H_2O + 2S \quad ...(5.23)$$

$$H_2S + 2H_2O + 2CO_2 \longrightarrow 2(CH_2O) + H_2SO_4 \quad ...(5.24)$$

Soils deficient in oxygen have high number of bacteria which use the oxidised form of sulphur as electron acceptors. Wet paddy soils, for example, may contain more than 1×10^6 sulphate reducing organisms per gram. The predominant bacteria involved in desulphuration are *vibriods* belonging to the genus *Desulphovibno.*

Sulphur dioxide is also formed from H_2S which is released into the atmosphere from volcanoes. The amount of sulphur released to the atmosphere worldwide in this form (about 2×10^9 tons/year) exceeds the amount released directly as SO_2. In the atmosphere, the reaction of SO_2 with O_2 is relatively slow. The rate is influenced by photochemical processes and by catalyst. In fog or cloud, SO_2 reacts with water to form sulphurous acid. This is followed by oxidation

$$H_3SO_3 + \frac{1}{2}O_2 \longrightarrow H_2SO_4 \quad ...(5.25)$$

Sulphuric acid is responsible for the corrosive properties of air in industrial localities. The effects can be noticed in the deterioration of some construction materials, as well as the metals, paints, etc.

Rain water containing H_2SO_4 is acidic and contributes to the acidity of lakes and streams. Rain-water pH values, near 4, are quite low for water which is in equilibrium with atmospheric CO_2. Of course, oxides of nitrogen also contribute to this acidity. The chemical transformations relevant to acid water production are as follows (reactions 5.26 to 5.31).

$$FeS_2 + 3O_2 \longrightarrow FeSO_4 + SO_2 \text{ (absence of moisture)} \quad ...(5.26)$$

$$2FeS_2 + 7O_2 + H_2O \longrightarrow 2FeSO_4 + 2H_2SO_4 \quad ...(5.27)$$

$$4FeSO_4 + 2H_2SO_4 + O_2 \longrightarrow 2Fe_2(SO_4)_3 + 2H_2O \quad ...(5.28)$$

$$2Fe_2(SO_4)_3 + 12H_2O \longrightarrow 2Fe_2(OH)_6 + 6H_2SO_4 \quad ...(5.29)$$

$$FeS_2 + Fe_2(SO_4)_3 \longrightarrow 3FeSO_4 + 2S \quad ...(5.30)$$

$$2S + O_2 + 2H_2O \longrightarrow 2H_2SO_4 \quad ...(5.31)$$

One of the consequence of acidic effluents is to render the water courses into which they drain unsuitable for growth of aquatic plants, fish and many organisms; moreover this water becomes unsatisfactory for domestic consumption or irrigation purposes.

As emission of SO_2 presents a major pollution problem, considerable attention has been given to reduce this emission. Worldwide combustion of fossil fuels produces SO_2, the order of 10^8 tons annually, so that an increasing use of high sulphur fuels requires some means of reducing the amount of SO_2 evolved in order to control pollution problems. Pretreatment of fuels represents one approach. Methods are available for removing sulphur from petroleum. So gasoline and diesels should be made sulphur-free before marketing.

Removal of SO_2 from exhaust gases is another approach to the pollution problem and a variety of techniques have been proposed. It may be pointed out that the removal of SO_2 from industrial exhaust gases also removes the acid gases such as HCl completely as also NO_x to some extent.

Sulphur dioxide can be removed with a base such as $CaCO_3$ conveniently.

$$2CaCO_3 + 2SO_2 + O_2 \longrightarrow 2CaSO_4 + 2CO_2$$

Calcium carbonate or calcium magnesium carbonate are most widely used in such processes, since they are available cheaply. In another approach, the crushed carbonate is added with the fuel and the sulphate salts are precipitated by electrostatic precipitator or washed from the gas stream by a scrubber. Alternatively the stack gases may be scrubbed by a slurry of $CaCO_3$.

This method can be very effective, but produces large amounts of sludge containing $CaSO_4$, $CaSO_3$, unreacted carbonate and fly ash which raises severe disposal problems.

Another process that has been suggested involved catalytic conversion to sulphur.

$$SO_2 + H_2O \longrightarrow HSO_3^- + H^+ \quad ...(5.32)$$

$$HSO_3^- + \text{Citric acid} \longrightarrow HSO_3^- \text{ Citrate complex} \quad ...(5.33)$$

$$HSO_3^- \text{ Citrate complex} + H_2S \longrightarrow 3S + \text{Citric acid} + 3H_2O \quad ...(5.34)$$

The sulphur can be recovered for industrial use by physical methods.

In yet another method, known as alkylated alumina process, the stack gas comes in contact with alkalinised alumina (Na_2O + Al_2O_3), thereby converting SO_2 into sodium sulphate Na_2SO_4. It is then treated with synthetic gas (A mixture of H_2, CO and CH_4)

$$Na_2SO_4 + (H_2, CO, CH_4) \xrightarrow{\text{high temp.}} Na_2O + H_2S + CO_2 + H_2O \quad ...(5.35)$$

Sulphur is then recovered from H_2S by the Claus reaction.

$$H_2S + SO_2 \longrightarrow 2H_2O + 3S \quad ...(5.36)$$

SO_2 so required can be obtained by burning H_2S in air

$$H_2S + \frac{3}{2}O_2 \longrightarrow H_2O + SO_2 \quad ...(5.37)$$

Recently researches at the Argonne National Laboratory in Illinois, USA used microwaves to split toxic H_2S into hydrogen and sulphur. H_2S is passed through a microwave reactor when 98% of the gas breaks down and sulphur vapours are condensed and collected as liquid. Any untreated H_2S is mixed with incoming gases and sent back through the reactor.

For economic viability, these processes have to operate on a very large scale and must do so reliably with a minimum of maintenance requirement. The volumes of gases to be handled are enormous say for example, of the order of 10^6 cubic metres per hour from a 750 MW coal-fired generating plant. Despite these problems, stack gas scrubbers are expected to work at reasonable operational costs and solve waste disposal problems.

Yokkaichi Asthama: In 1959 the Yokkaichi petrochemical and power complex went into operation in Japan. The local residents then began complaining of foul smell and irritating smell. Soon there was increase in asthma in the city called 'Yokkaichi Asthama , which was due to the sulphur dioxide released from smoke stacks. Six companies had to make compensation to all the victims.

5.6 BIOLOGICALLY IMPORTANT METALS AND TRACE ELEMENT

5.6*a* Introduction

In addition to the elements already discussed, a variety of others have important biological effects. The biological availability of an element, and its ultimate effect on an organism, may depend on the chemical form in which the element is encountered. Thus, solubility properties, the oxidation state, or the bonding characteristics in a particular compound may determine whether it will be inert, poisonous or detoxified by normal metabolic pathway. Toxic compounds occurring naturally are very often degraded by organisms to nontoxic compounds, but excessive concentrations or new compounds may overcome these metabolic capacities.

Table 5.3 lists some of the metallic elements released in trace amounts to the environment by industrial and other processes which may pose concern as health hazards. Strong evidence of toxicity is not present for all of them. There is evidence that a few of these are harmful in an indirect way for the environmental forms. Many of the heavy elements not listed here are potentially fairly toxic but either are very rare or highly insoluble in forms encountered in the environment.

Antimony, bismuth, tellurium and zirconium are also encountered in atmospheric and aqueous industrial emissions. Health hazards are not established but the elements are toxic.

The importance of the chemical form of an element is illustrated by nickel sub sulphide Ni_2S_3 and nickel are highly toxic. Nickel carbonyl is decomposed to produce nickel. Most other nickel compounds, if not carcinogenic entirely, certainly have weaker effects. Both Ni_2S_3 and $Ni(CO)_4$ are possibly by-products from the use of nickel catalyst as well as from some of the sources given in Table 5.3. Such catalyst may find very widespread use in coal gasification processes.

Table 5.3 Trace metals that may pose health hazards

Element	*Atomic No.*	*Source*	*Health effects*
Beryllium	4	Coal, nuclear power industry, rocket fuel mine tailings.	Acute and chronic system poison, cancer.
Boron	5	Coal, cleaning agents, medicinal, glass making and other industry	Micro nutrient to plant but excess is toxic to plants, slightly toxic to animals.
Vanadium	23	Petroleum (Venezuela, Iran), chemicals and catalysts, steel and other alloys	Probably no hazard at current levels.
Chromium	24	Metal electroplating	A micronutrient to plant, chromates carcinogenic.
Manganese	25	Mining and industrial wastes	Relatively nontoxic, micro-nutrient.
Nickel	28	Fuel Oil, coal, tobacco smoke, chemicals and catalysts, steel and nonferrous alloys	Lung cancer (as carbonyl).
Copper	29	Waste pipes, algae control, industrial smoke	Possible liver damage with prolonged exposure, toxic to plants.
Zinc	30	Metal electroplating, mining, industrial smoke	Possible lung effects, low toxicity in solution.
Germanium	32	Coal	Little innate toxicity
Arsenic	33	Coal, petroleum, detergents, pesticides and mine tailings	Arsenic poisoning
Selenium	34	Coal, Sulphur	May cause dental caries, carcinogenic to rats, essential to mammals in low doses.
Yttrium	39	Coal, petroleum	Carcinogenic in mice on long term exposure
Molybdenun	42	Industrial smoke	Possible lung effects.
Silver	47	Electroplating, photographic waste	Discolouration of skin.
Cadmium	48	Coal, zinc mining, water mains and pipes, tobacco smoke	Cardiovascular disease and hypertension in humans suspects interferes with zinc and copper metabolism
Antimony	51	Industry	Shortened life span in rats.
Mercury	80	Coal, electrical batteries and industry	Nerve damage and fatality
Lead	82	Auto exhaust (from gasoline), paints and in water through lead joint of pipes.	Brain damage, convulsions, behavioural disorder, fatality.

The function of metal ions in biochemistry can be correlated with their coordination tendencies. As for example, typical ligands, chiefly through electrostatic effects are involved with control processes involving ion transfer. Mg (II) and Ca (II) also are largely electrostatic in their interactions, but readily form complex. They exhibit rapid ligand exchange, and often are components of enzymes involved to trigger reactions and hydrolysis processes. Electronegative ligands such as oxygen donors are favoured by both these groups. Stronger bonding preference for nitrogen donors, and stronger interactions with sulphur donors are found with the ions of Zn, Co, Mn, Fe and Cu. These are also common in enzyme systems which control a variety of processes. Those that readily form two different oxidation products especially Cu, Fe, are often involved in redox processes.

The contribution of metals and their ions towards pollution problems is of very serious concern, largely because their physiological effects may be substantial even at extremely low concentrations. In some cases, effects on behaviour and on the nervous system begin with subtle changes and are difficult to detect in their initial stages, some metals may be accumulated in the body so that dangerous levels can build up from long term ingestion of very small amounts. There is little real knowledge about their effects and what constitutes a safe level of concentration. Levels that have been set are often based as much on intuition than on facts. At low concentrations, metal pollution is very wide spread, with many industrial processes contributing to it. Natural water normally contain considerable mineral content, in particular cases, levels of harmful inorganic components may be significant quite independent of any man-made pollutants. Finally, because concentration that are significant may be so small, unexpected sources may be important. As for example, although the amount of metals contained in coal and oil are small, the vast amounts of these materials, that are burned result in the liberation of appreciable total quantities of mercury, cadmium, and other metals. Generally, the elements of major concern are found in the environment as compounds of low volatility, nevertheless their contribution to atmospheric problems must not be ignored as they may be present in fine particulate material. In water, they must be considered in suspended forms as well as in true solution.

The actual role played by trace metals vis-a-vis human health is complex and by no means fully understood, and this makes the determination of safe levels very difficult. For instance, zinc is known to be an essential element, and a comparatively nontoxic metal by conventional tests. However, there is some evidence that suggests that amounts of zinc very much below present acceptable levels can destroy some of the natural anti-tumour mechanism of the body. On the other hand, iron deficiency seems to encourage tumour development.

5.6*b* Beryllium

It is the lightest alkaline earth element and its properties are quite different from the elements of this group. This is because beryllium has low atomic number, high ionization energy, greater electronegativity, and thus favours covalent bonding. It has found a number of high level technological uses such as in structural materials, in nuclear reactors, and in missiles, as an additive to the rocket fuel, and as an alloy with Cu for providing high corrosion resistance. While these usages offer at least the potential for industrial pollution, another source is the amount that may be released from burning of coal, which usually contains up to 3 ppm Be. It also differs from other alkaline earths as it is highly toxic, as has been recognized long before.

5.6*c* Boron

Boron is the most electronegative element of this group and thus it broadly resembles the non metals, particularly Si. Boron in nature is encountered as boric acid or its salt. Boric acid, $B(OH)_3$ is a very weak acid and this is the form found in aqueous systems. Boron compounds are widely used industrially. Sodium tetraborate $Na_2B_4O_7$ is used in detergents, and consequently enters the water system. Although it is necessary as a micronutrient to plants, including algae, it is known to be toxic to plants at higher concentrations.

Toxicity of boron is not very well understood, but a maximum of 1 mgl^{-1} has been set for drinking water.

5.6*d* Magnesium and Calcium

Magnesium and Calcium are extremely common in natural water systems, with lime-stone (calcium carbonate) $CaCO_3$ and dolomite $CaMg(CO_3)_2$ being two widespread natural sources. Solubility in water is influenced by pH and CO_2 content. The two salts are the chief cause of hardness in water, which manifests itself by precipitation with soaps. Both calcium and magnesium are essential elements needed in significant amounts by living organism. Except in the context of these problems, they are not harmful either in solution or as particulate material.

Chemically magnesium behaves in a similar way to calcium but generally is found in much lower concentrations in water, both exist solely as the 2 + ions in solution with interactions that are predominately coulombic. The ions favour coordination by oxygen donors, simple complexes are weak, but chelation can give moderate stability. Magnesium being lighter, shows stronger bonding with solvent or ligand molecules and forms more stable complexes. Both are important in a variety of complexing and rapid exchange determines their function.

In the overall photosynthetic process in plants, carbohydrates are synthesised from CO_2 (from the atmosphere) and hydrogen donor, H_2A such as H_2O, H_2S or even H_2

$$nCO_2 + 2nH_2A \longrightarrow (CH_2O)_n + nH_2O + 2nA \qquad ...(5.38)$$

Photosynthesis is enormously facilitated by catalyst such as chlorophyll in which the catalytic function of magnesium is enhanced by inclusion in porphyrin. The structure of chlorophyll is shown in Fig. 5.7.

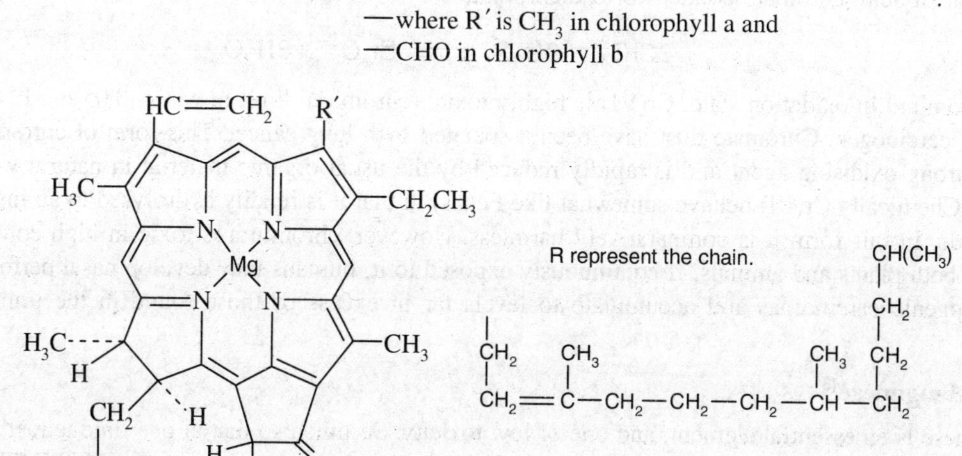

Fig. 5.7 Chlorophyll structure, Green plant pigment catalyst for photosynthesis.

The primary photosynthetic step is not carbohydrate production from water and CO_2 but the conversion of photoenergy into chemical energy which is then utilised in a number of ways. The electrons of chlorophyll are excited by the absorption of photons. This excitation enables the green pigment to act as an electron source for the production of adenosine triphosphate (ATP). With water serving as a proton source the triphosphopyridine nucleotide (TPN) is reduced to form $TPNH_2$, which in turn is responsible for the reduction of CO_2 and production of glucose, carbohydrates are far from being the only products of photosynthesis; fatty acids, fats, carboxylic acids and amino acids are also formed.

Calcium is the essential cation in structural features such as bones as hydroxyapatite $Ca_{10}(PO_4)_6(OH)_2$ and is highly insoluble in water. Protonation of the phosphate ion, however, increases the effective solubility. In spite of chemical similarities calcium is not interchangeable with magnesium as phosphate because of significant size differences. Calcium metal is extremely reactive and has little industrial use as such; magnesium metal, on the other hand, is considerably less reactive and is important structurally since it combines lightness and strength.

The heavier members of the alkaline-earth family, namely strontium and barium, are comparatively rare. Both resemble calcium more than magnesium. Since one of the radioactive isotopes of strontium is found in abundance and as Ca^{2+} and Sr^{2+} have the similar atomic mass, the replacement of Ca^{2+} in bones by Sr^{2+} is of importance and will be discussed in chapter 9.

5.6*e* Chromium[12]

Although familiar as an inert metal used for decorative and protective plating, large amounts of chromium compounds are used industrially for other purposes, especially in treating steel to resist-corrosion. Solutions containing CrO_4^{2-} are effective in passivating steel, presumably by reacting to form a protective oxide layer on the surface. Wastes from such operations are a prime source of chromium pollution. In acid, chromate condenses to dichromate

$$2CrO_4^{2-} + 2H_3O^+ \rightleftharpoons Cr_2O_7^{2-} + 3H_2O$$

Chromium in oxidation state, Cr(VI) is highly toxic, (chromium limit in water, 0.05 mg l^{-1}) and a possible carcinogen. Chromate dust have been associated with lung cancer. This form of chromium is also a strong oxidising agent and is rapidly reduced by the usual organic material in natural water to Cr(III). Chemically Cr(III) behave somewhat like Fe(III) in that it is readily hydrolysed to an insoluble hydroxide. In this form it is comparatively harmless. However, chromium is toxic in high concentrations to both plants and animals. If continuously exposed to it, humans may develop nasal perforation, bronchiogenic carcinomas and accumulate to levels far in excess of those found in the immediate habitat.

5.6*f* Manganese[13]

Manganese is an essential element, and one of low toxicity. So much so that at one time tetraethyllead (TEL) was thought to be replaced by methylcyclo-pentadienyl manganese tricarbonyl (MMT). This compound alone may be used in lead-free gasoline in concentrations up to 0.13 gm Mn/gal. In its highest oxidation state as permanganate ion MnO_4^- it is a strong oxidising agent. In its more stable state Mn(IV) it is encountered in natural systems, chiefly as insoluble MnO_2 or as compounds of Mn(II). The chemistry of these states resembles that of Fe(II) and Fe(III); Mn(II) salts are soluble in water but MnO_2 is insoluble, although often producing a colloidal or gelatinous and highly absorbant precipitate. Mn(II) is encountered only under reducing conditions in natural waters, as is Fe(II). Sources of manganese are industrial pollution and acid mine drainage. According to the International standards for drinking water maximum allowable limit of manganese is 0.5 mg per litre, based largely on staining rather than toxicity, has been prescribed by WHO.

5.6.7*g* Iron[17, 18, 19]

Iron is a common and environmentally important element. It forms two oxidation states in nature, Fe(II) and Fe(III). In water Fe(III) forms a strongly hydrated ion. This pale-purple hydrated ion is stable in a very strongly acid medium and tend to hydrolyse to yellow solution at pH 2-3.

$$\left[Fe(H_2O)_6\right]^{3+} + H_2O \rightleftharpoons \left[Fe(H_2O)_5OH\right]^{2+} + H_3O^+ \quad ...(5.39)$$

$$K_1 = 8.91 \times 10^{-4}$$

$$\left[Fe(H_2O)_5OH\right]^{2+} + H_2O \rightleftharpoons \left[Fe(H_2O)_4(OH)_2\right]^{+} + H_3O^+ \quad ...(5.40)$$

$$K_2 = 5.5 \times 10^{-4}$$

At pH 4-5 hydroxo species polymerize to form a brownish solid dimer. Polymerization of aqua ions is quite common especially if they have a high charge.

$$2H_2O + 2\left[Fe(H_2O)_6\right]^{3+} \longrightarrow \left[(H_2O)_4Fe\begin{matrix}OH\\ \lt \quad \gt \\ OH\end{matrix}Fe(H_2O)\right]^{4+} + 2H_3O^+ \quad ...(5.41)$$

$$K_3 = 1.23 \times 10^{-3}$$

At even higher pH value a reddish brown precipitate of hydrous oxide is formed. There is no evidence that $Fe(OH)_3$ exists, though we write it. Hydrolysis of Fe(III) does not yield hydroxide but gives a reddish brown gelatinous precipitate of hydroxide $Fe_2O_3(H_2O)_n$. At least part of the precipitate consists of FeOOH. Heating the hydrous oxide to 200°C yield reddish brown α-Fe_2O_3, The structure comprises a hexagonal close packed lattice of O^{2-} ion with Fe^{3+} ion in two-thirds of octahedral voids. However if Fe_3O_4 is oxidised, γ-Fe_2O_3 is formed which has a cubic close-packed arrangement of O^{2-} with Fe^{3+} ions randomly distributed in both octahedral and tetrahedral voids. As a result of hydrolysis reactions such as the above, Fe(III) salts are virtually insoluble in water with a neutral pH, indeed, iron is present mostly as hydroxo — complexes at pH = 4.

Fe(II) on the other hand, does not form these complexes significantly until pH = 10 and is soluble near neutrality. Hydrated Fe(II) is a pale blue-green ion $[Fe(H_2O)_6]^{2+}$. The potential of the Fe^{3+} and Fe^{2+} couple, 0.771 is such that molecular oxygen can convert ferrous into ferric ions in an acid medium. The relationship, if solid phase is taken to be FeO(OH), then equilibrium reaction is

$$FeO(OH) + 3H_3O^+ \rightleftharpoons Fe(III)(aq) + 5H_2O \quad ...(5.42)$$

$$K_4 = 10^4$$

The redox equilibrium reaction is

$$Fe(III)(aq) + e^- \rightleftharpoons Fe(II)(aq) \quad ...(5.43)$$

$$K_5 = 4.5 \times 10^{13}$$

or
$$2Fe^{2+} + \frac{1}{2}O_2 + 2H^+ \rightleftharpoons 2Fe^{3+} + H_2O \quad \text{...(5.44)}$$

$$E^0 = 0.46$$

Combining Eqs. (5.42) and (5.43)

$$FeO(OH)(s) + 3H_3O^+ + e \rightleftharpoons Fe(II)(aq) + 5H_2O \quad \text{...(5.45)}$$

$$K_6 = K_4K_5 = 4.5\times10^{17} \text{ and } \log K_6 = 17.653$$

Since
$$K_6 = [Fe(II)]\big/\left[H_3O^+\right]^2\left[e^-\right] \quad \text{...(5.46)}$$

It is desirable to express the oxidising or reducing properties of a system in terms of the concept of pE. In a formal sense, pE(= –log Ce^-) is entirely analogous to pH. Although there are no free electrons in chemical systems, the value of pE provides a measure of oxidising or reducing properties of a solution. Conditions that exhibit low pE values will favour reduction and high pE values will favour oxidation. Taking log in eq. 5.46

$$\log K_6 = \log Fe(II) - 2\log\left[H_3O^+\right] - \log Ce$$

or
$$\log Fe(II) = \log K_6 - 2pH - pE \quad \text{...(5.47)}$$

At pE value of air-saturated water 12.5 and pH 7

$$\log Fe(II) = 17.653 - 14 - 12.5 = -8.847.$$

In this way the concentration of Fe(II) is insignificant. Indeed, this is the pE value required for any realistic concentration of Fe(II) to be stable in the presence of solid FeO(OH) and the oxygen partial pressure that would be in equilibrium with this. It will be found that Fe(II) can be formed only under an aerobic conditions.

For the solution reaction of FeO(OH), $[Fe(H_2O)_4(OH)_2]^+$ is the most readily formed soluble species.

$$FeO(OH)(s) + 3H_2O + H_3O^+ \rightleftharpoons \left[Fe(H_2O)_4(OH)_2\right]^+ \quad \text{...(5.48)}$$

$$K_7 = 4.9\times10^{-4}$$

At pH = 8.1, the concentration of soluble ferric iron is about 3×10^{-11} M. The true dissolved Fe(III) concentration in neutral water such as ocean is difficult to evaluate due to the presence of colloidal material, however it seems to be about 10^{-6}–10^{-8} M. This value is much higher than the calculated value from solubility equilibria cannot be accounted for with certainty, but may involve the formation of soluble complex of an unknown kind, perhaps with natural organic ligands.

Ferric hydroxide precipitates interact strongly with water and get hydrated through hydrogen bonding. This, coupled with the regular nature of the chains, and cross linking formed as the material precipitates, leads to a gelatinous flocculation.

On the other hand, if very small aggregates form, as might be the case in extremely dilute solutions, the individual particles will be strongly hydrated and in favourable cases further stabilised against coagulation by adsorbing ions or a particular charge on their surfaces. Under these conditions a more or less stable suspension or colloid results, which is easily distinguishable from dilute solutions.

Iron is by far the most widespread and important metal with a functional role in living systems. Iron-containing proteins participate in two main process: oxygen transport and electron transfer. There are other molecules whose function is to store and transport iron itself. Iron also occurs in conjunction with molybdenum in enzymes that catalyse nitrogen fixation (see reaction 3.30).

The chief heme proteins are the haemoglobin (Hb), myoglobin (Mb), cytochromes and some enzymes such as catalase and peroxidase. These are characterised by the presence of peroxidase. These are characterised by the presence of the heme group (Fig. 5.8) as iron-containing unit.

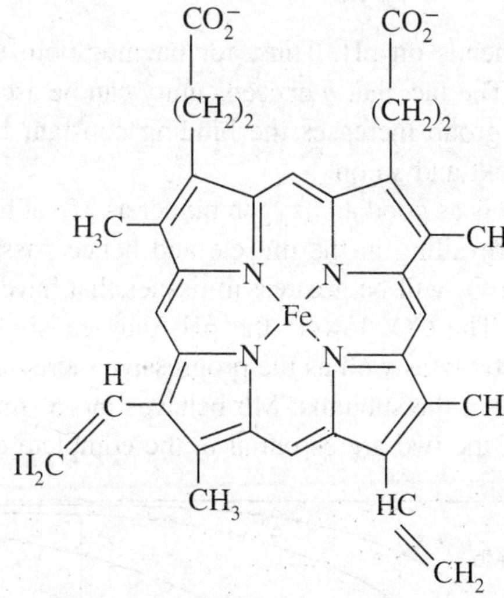

Fig. 5.8 Structure of Heme.

Haemoglobin and myoglobin are closely related. The former has a molar mass of 64500 and consists of four subunits, each containing one heme group; myoglobin is very similar to one of the subunits of haemoglobin: A detailed discussion of its working can be found in R.E. Dickerson and I. Geis *The Structure and Action of Proteins*[17]. Haemoglobin has two functions: (*i*) binds oxygen molecules to its iron atoms and transports them from the lungs to muscles, where they are delivered to myoglobin molecules which store the oxygen molecule until it is required for metabolic action and (*ii*) the haemoglobin then uses certain amino groups to bind carbon dioxide and carry it back to lungs.

The heme group is attached protein in both haemoglobin and myoglobin through a coordinated histidine-nitrogen atoms. In the position trans to the histidine-nitrogen atom there is a water molecule in the deoxy-species and an oxygen molecule in the oxygenation species. The structure of Fe–O_2 grouping is still unknown, but changes in the oxidation state of iron and introduction of O_2 changes the structure of Fe-heme complex.

Haemoglobin is not a simply passive container for oxygen but an intricate molecular machine. This may be appreciated by comparing its affinity for O_2 to that of myoglobin. For myoglobin (Mb) we have the following simple equilibrium.

$$\text{Mb} + \text{O}_2 \rightleftharpoons \text{MbO}_2 \qquad K = \frac{[\text{MbO}_2]}{[\text{Mb}][\text{O}_2]} \qquad ...(5.49)$$

If f represents the fraction of myoglobin molecules bearing oxygen and p represents the equilibrium partial pressure of oxygen, it follows that

$$k = \frac{f}{(1-f)p} \text{ or } f = \frac{Kp}{1+Kp} \qquad ...(5.50)$$

This is the equation for the hyperbolic curve labelled Mb in Fig. 5.9. Haemoglobin with its four units has more complex behaviour; it approximately follows the equation.

$$f' = \frac{Kp^n}{1+Kp^n} \text{ with } n \approx 2.8 \qquad ...(5.51)$$

where the exact value of n depends on pH. Thus, for haemoglobin the oxygen binding curves are sigmoidal as shown in Fig. 5.9. The fact that n exceeds unity can be ascribed physically to the fact that attachment of O_2 to one heme group increases the binding constant for the next O_2 which, in turn, increases the constant for the next and so on.

It will be seen that while Hb is as good an oxygen binder as Mb at high oxygen pressure, it is much poorer at the lower pressures prevailing in the muscle and hence passes its oxygen on to the Mb as required. Moreover, the need for O_2 will be greatest in tissues that have already consumed oxygen and simultaneously, produced CO_2. The CO_2 lowers the pH, thus causing the Hb to release even more oxygen to the Mb. The pH sensitivity as well as the progressive increase of the O_2 binding constants in Hb are due to interactions between the subunits; Mb behaves more simply because it consists of only one unit. It is clear that each of the two are essential in the complete oxygen transport process.

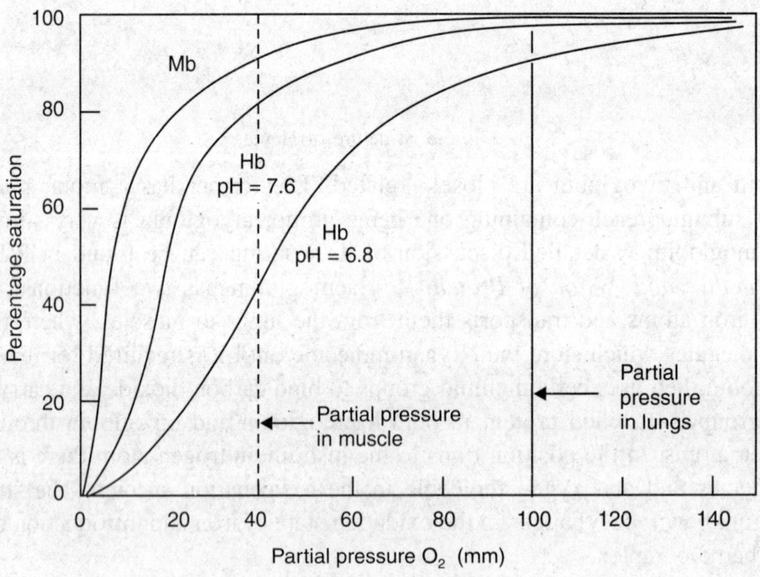

Fig. 5.9 The oxygen-binding curves for myoglobin (Mb) and haemoglobin (Hb), showing also the pH-dependence (Bohr effect) for the latter.

Carbon monoxide, PF_3 and a few other substances are toxic because they become bound to the iron atoms of Hb more strongly than O_2. Their effect is one of complete inhibition for Hb to carry oxygen which ultimately leads to the breaking up of Hb unit.

Cytochrome *c* can be characterized as a shell one residue thick protein surrounding the heme which serves as an electron carrier. Its minimum molar mass is 13000 with one Fe and six S atoms. The heme group here is covalently attached through several side chains to amino acid residues of the protein and, in addition, an octahedron about the iron atom is completed by coordination of a histidine-nitrogen atom and one other ligand, not yet identified. The cytochrome *c* molecule in a great range of organisms are nearly identical, indicating that this type of electron carrier developed very early in the evolutionary process and has served its purpose well, not requiring much change.

Ferredoxin are relatively small proteins (molar mass 6000–12000) which contain water soluble protein containing a 2Fe—2S cluster with cysteine-sulphur bonds. The so-called inorganic sulphur have redox potentials close to that of the standard hydrogen electrode. They appear to occur in all green plants including algae, in all photosynthetic bacteria and protozoa and in some fermentative anaerobic bacteria. These molecules play an essential role as electron-transfer agents at the low potential end of the photosynthetic process. Light causes electrons to flow from H_2O to NADPH and leads to generation of a proton-motive force. Protons are pumped by this complex as electrons return to the reaction centre[31]. While having similar redox properties, the several characteristic ferredoxins are diverse chemically in that iron content varies from two to eight atoms/molecules.

Under anaerobic conditions, considerable amounts of iron pyrites, FeS_2 are formed in coal mines. Exposure of pyrites to air and moisture leads to the formation of the so-called acid mine drainage (eq. 5.26–5.31). The high acid water cause considerable damage to natural water systems.

Iron in water is not toxic, but it does have a number of harmful effects. It can cause rusting stains on clothing, porcelain material, etc. as well as giving the water itself an undesirable colour and taste. The United States Public Health standard for maximum iron content in drinking water is 0.3 mg l^{-1} and this is based on colour and taste effects rather than on toxicity.

5.6*h* Copper[16]

Copper is an essential metal for many organisms. It is used in enzymes that modify redox reactions and also in some oxygen-carrying systems. Its function in these is associated frequently with its ability to exist as both Cu(I) and Cu(II). The most common state is Cu(II). Like many essential metals, large amounts are toxic and the USPH limit in drinking water is set at 1 mg l^{-1}. It is particularly toxic to lower organisms and has been used as algicide in lakes. Other sources are erosion of copper pipes, industrial wastes and weathering of rocks. The carbonate and hydroxide are of low solubility so that the natural level in water is low in the absence of complex-forming substance.

5.6*i* Zinc

Zinc appears to be one of the most important metals biologically. It is probably second only to iron among the heavy metals (i.e. excluding Na^+, Ca^{2+} and Mg^{2+}). In the past 50 years more than 200 zinc-containing proteins have been identified, most of them enzymes. The Zn^{2+} ion is contained in several

Table 5.4 Some metalloenzymes of zinc

Name	*Molar mass*	*Zn atom per molecule*	*Source*
Carbonic anhydrases	~30, 000	1	Erymrocytes
Carboxypeptidases	~35, 000	1	Pancreas
Dehydrogenases	~85,000	≥ 2	Yeast, liver
Alkaline phosphatases	89,000	4	E. coli.

dehydrogenases, aldolases, peptidases, phosphatases, an isomerase of yeast, a transphosphorylase and a phospholipase, attesting to its importance in carbohydrate, lipid, and protein harmones, neurotransmitters, haem and nucleic acid metabolism in virtually all organisms. Table 5.4 lists a few of the enzymes known to contain zinc as an essential constituent.

Among the enzymes most thoroughly studied are the following:

(*i*) *Aldol dehydrogenases,* from diverse species including yeast, horses and humans, catalyse the oxidation of ethanol or the reduction of acetaldehyde, using diphosphopyridine nucleotide (DPN) as a cofactor. Crystalline yeast alcohol dehydrogenase has a molar mass of 150,000 and contains four Zn^{2+} ions and binds DPN molecules per mole. Its structure and chemistry are not known in detail.

(*ii*) *Alkaline phosphatases of bacteria* catalyse hydrolysis of a variety of phosphomonoesters and also act as phosphotransferases. The one from *E. Coli* contains four Zn^{2+} ions per unit of molar mass 89000. Much still remains to be learned of the chemistry, whereas the structure is unknown.

(*iii*) *Carboxy peptidase* is specific for hydrolysis of *c*-terminal amino acids of proteins, peptides, and corresponding esters. Carboxy peptidase from beef pancreas has a molar mass of 34,300 and contains one zinc atom per molecule. Chemical and X-ray crystallography studies show that the Zn atom is at the active site and give a moderately detailed idea of how the enzyme functions and what role is played by the metal: the Zn^{2+} ion is bound to the enzyme by two N and one O atoms, a distorted tetrahedron being completed by a water molecule. A peptide carbonyl group displaces the water molecule when the substrate is bound, and hydrolysis of the peptide bond containing the bound carbonyl group then takes place by a series of proton transfers.

(*iv*) *Metallothioneies* are curious proteins whose function is still unknown. They have been found in kidneys, liver, and gut of horses, rabbits and humans. They have a molar mass of ≈ 10,000 and contain ≈ 9% by weight of metal, and nearly 1/3 of their amino acids residues are cysteines

$$\begin{array}{c} -NH-\underset{\displaystyle CH_2SH}{\underset{|}{CH}}-CO- \end{array}$$

The metal content is mainly zinc but an appreciable fraction is cadmium. Metallothinoeine is the only biological material known to contain cadmium.

The following factors responsible for the importance of zinc are:

(*i*) Zinc is the most widely used Lewis acid in life processes, having readily variable coordination number of six, five and four in corresponding order of increasing Lewis acidity and is indispensable as an acid catalyst.

(*ii*) Zinc readily forms complexes with N, O and S donors. The stronger the donor, the weaker the Lewis acidity. Imidazole acts as a σ-donor and π-acceptor.

(*iii*) Ligand exchange processes of Zn^{2+} are usually rapid. The zinc metalloenzyme carbonic anhydrase is, in terms of turnover the most active enzyme known.

(*iv*) The coordination geometry of Zn^{2+} is easily deformed by the ligands in its coordination sphere, normally proteins. Nitrogen is often involved in the imidazole moiety of histidine unit. This deformation produces a type of stress (entasis) which is believed to play the role of a catalyst.

(*v*) In coordination with water it can generate OH^- ion and thereby also function in base catalysed enzymatic processes. The coordination of water on to complexed zinc sites in enzymes is particularly important because it can provide H^+ and OH^- for donation to a suitably positioned substrate at near neutral pH. The species is a stabilised hydroxide ion that can exist at a pH where free OH concentrations are normally low.

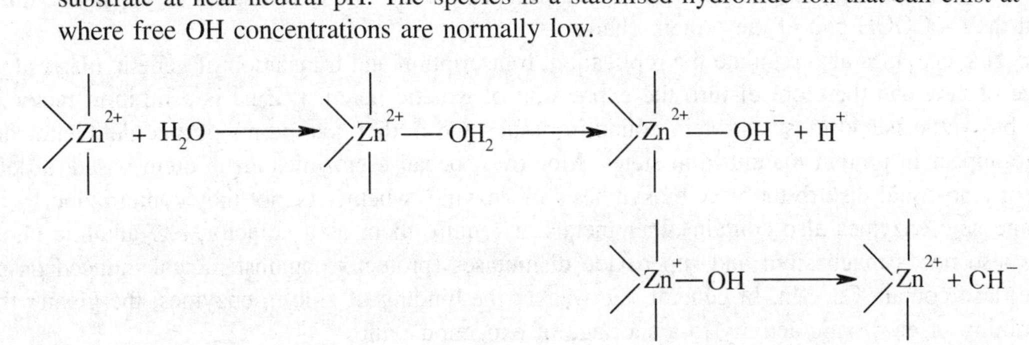

Thus zinc complex exhibit a useful variation of degree of Lewis acidity in conjunction with an accommodating flexibility of coordination geometry.

It appears to be the active centre of the enzyme carbonic anhydrase which is widely distributed in tissues and is associated with such diverse processes as respiration with the parathyroid hormone, the mobilisation of Ca^{2+} from bone. This enzyme donates OH^- to CO_2

$$CO_2 + OH^- \xrightarrow[\text{Zn bound OH}^-]{} HCO_3^- \xrightarrow{H_2O} H_2CO_3 + OH^-$$

The catalysation of this reaction by zinc is by a rate factor of 10^7 relative to uncatalysed reaction by generating nucleophile. The scheme of reaction is as follows:

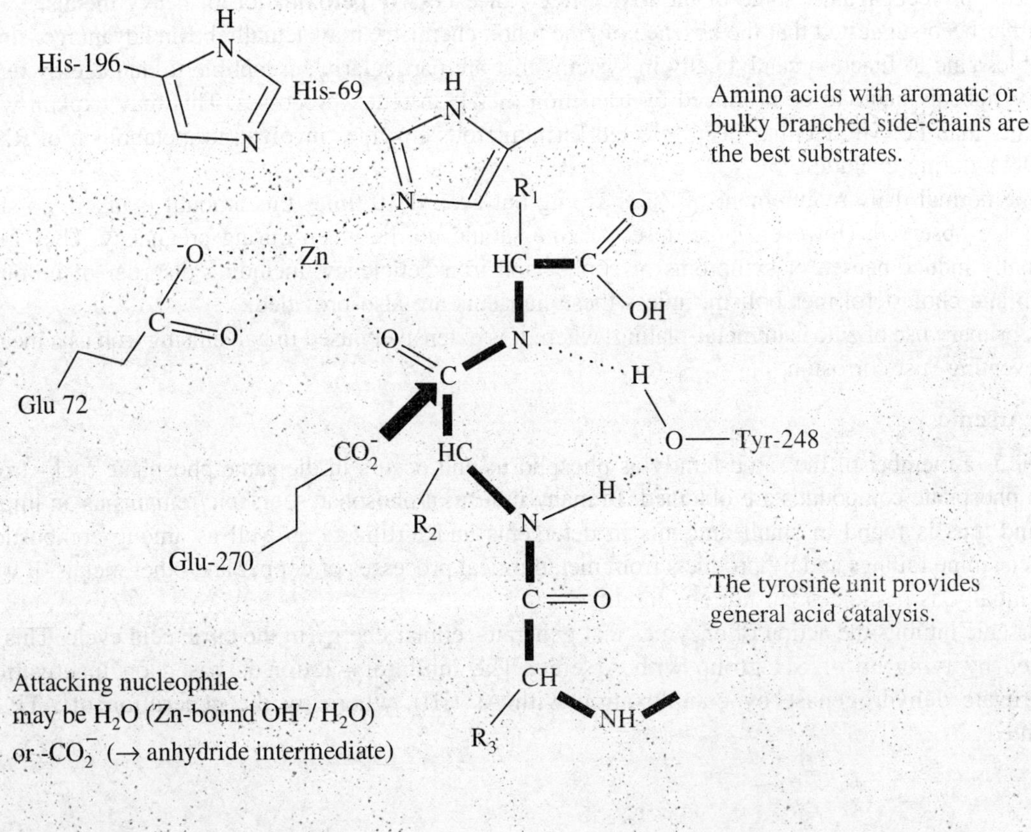

The coordination of Zn^{2+} on to carbonyl oxygen polarises the carbonyl group catalysing nucleophilic attack on carbon, protein hydrolysis by carboxypeptidase — A, which selectively cleaves amino acids at the — COOH end of the protein chains.

The zinc enzymes also regulate the replication, transcription and translation of genetic material. A shortage of zinc can therefore disturb the expression of genetic heredity. Zinc is a limiting factor in protein biosynthesis and mostly protein bound in tissues and in diet; symptoms of zinc deficiency are liable to appear in protein malnutrition states. Moreover, because enzymes are protein based, a deficiency of zinc could disturb the very biosynthesis of enzyme, whether or not they contain zinc.

Some zinc enzymes also contain other metals or require them as a cofactor, *e.g.* alkaline phosphatase also needs magnesium and superoxide dismutases (protective against radical induced tissue damage) also contain Cu, Mn. In general, the weaker the binding of zinc in enzymes, the greater the susceptibility of enzymatic activity to a shortage of exogenous zinc.

Furthermore, coordination of Zn^{2+} ion to O, N, and S can stabilise the tertiary structure of proteins by providing cross links, which unlike —S—S— bridges from cysteine units are stable under reducing conditions. These cross links may well be involved in a stabilising action of zinc on membrane structures. Zinc deficiency labilises cell membranes and increases membrane permeability partly through the increased peroxidation of membrane phospholipid. This last effect may result from the necessary presence of zinc in superoxide dismutases. Zinc binds to —SH groups in cysteine units and is one of the metals, including Cu and Cd that induce biosynthesis of high cysteine metal-binding protein metallothioneine present in most cells. This protein is important in the transport and short-term storage of zinc, and it protects tissues from the toxic effects of Cd, Hg and possibly Pb by coordinating these elements in forms that reduce their availability. In this way, and also by direct competition for active sites, zinc protects against some of the toxic effects of excessive peroxidation of heavy metals.

It has been suggested that the absence of zinc redox chemistry may actually be an advantage, since it enables zinc to function catalytically in systems that are particularly susceptible to damage by reactive oxyradicals liable to be produced by transition metals in redox processes. This may explain why Zn rather than Fe, Mn, etc. has been selected for numerous enzymes involved in metabolism of RNA and DNA during evolution.

The normal daily requirement of Zn is 15 mg but even if 10 times this amount is taken, no side effects are observed. However, large doses of zinc should not be given during pregnancy. They may eventually induce nausea or symptoms of copper and iron deficiency, including disorder of immune system and cholesterol metabolism, unless these nutrients are also provided.

A primary use of zinc is in metal-plating, where it is extensively used in galvanising iron as a means of preventing rust corrosion.

5.6*j* Arsenic

Arsenic is a member of the same family as phosphorus and occurs in the same phosphate rocks from which phosphate compounds are obtained. In many industrial phosphates, arsenic remains as an impurity, and thus is found in small amounts in detergents and fertilizers as well as among combustion products, mine railings and by-products from metallurgical processes of copper and other metals. It was used initially as pesticides but not any more.

Arsenic inhibits the action of enzymes that generate cellular energy in the citric acid cycle. This is affected by tying up of SH group with Arsenic. The inhibitory action is based on inactivation of pyruvate dehydrogenase by complexation with As (III), where by the generation of ATP is prevented.

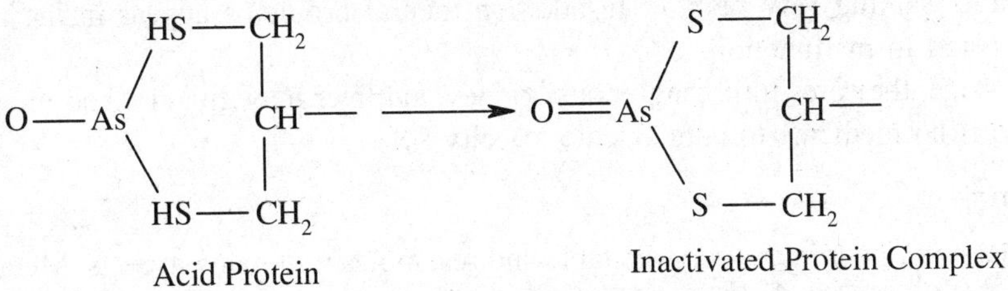

As phosphorus and arsenic are chemically similar, some of the biochemical processes involving phosphorus are inhibited. In the generation of ATP, the enzymatic synthesis of 1, 3-diphosphoglycerate 3-phosphate is an important step. Arsenite interferes by forming 1 arsena-3-phosphoglycerate instead of 1, 3-diphosphoglycerate which hydrolyzes to form 3-phosphoglycerate and arsenate.

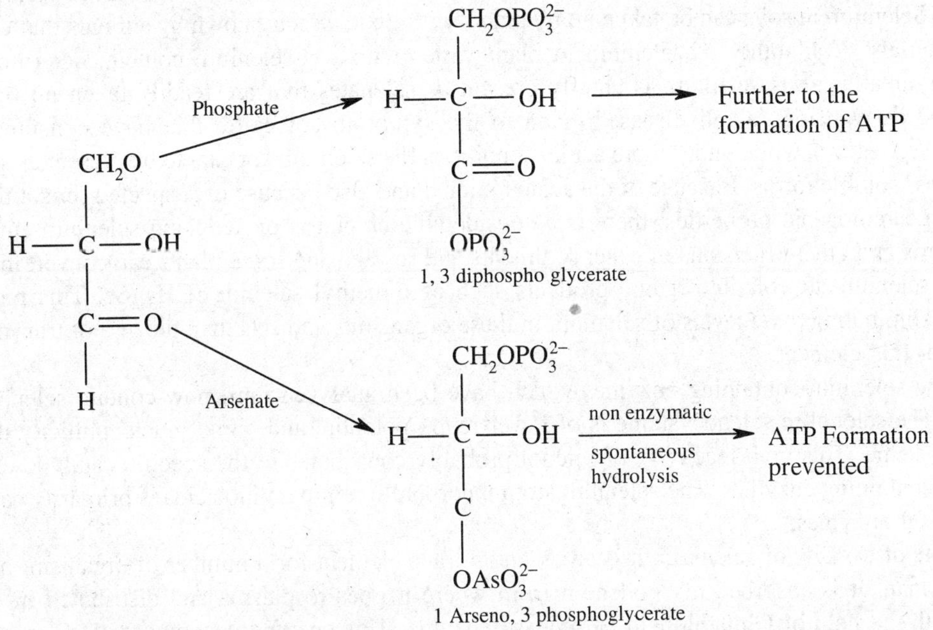

Arsenic poisoning in underground water is quite prominent in the Gangetic plains and Padma-Meghna-Brahmaputra basin along the Himalayas. Here sediments are iron pyrites where arsenic got adsorbed. This arsenic leached into aquifiers as follows. The arsenic is adsorbed as arsenites on iron pyrites rocks in the aquifier sediments. Over exploitation of underground water for irrigation led to fall in water table which has introduced oxygen at the surface of water. This oxygen oxidises arsenite to arsenate which is desorbed from pyrite surface and thus released to water thereby poisoning water.

Arsenic affects the human body in many ways. The first sign of poisoning, which emerge as long as 10 years after someone starts drinking arsenic laden water, is appearance of black spots on the upper chest, back (Fig. 5.10 – see colour plate) and arms a condition termed melanosis in medical parlance. Palms of hands or soles become hard and lose sensation (keratosis) (Fig. 5.11 – see colour plate). A patient may also suffer from conjunctivitis, bronchitis, diarrhoea and abdominal pain.

In the second stage, white spots appear mixed with the black ones, legs swell and the palms and soles crack and sometimes bleed (Fig. 5.12 – see colour plate). These sores can become infected and make working and walking very painful. In addition, neural problems appear in the arms and legs, kidney and liver start to malfunction.

In the third stage, the sores turn gangrenous, kidney and liver may give up, and in 20 years cancer shows up. There is no medicine to cure arsenic toxicity.

5.6*k* Selenium[29]

Selenium falls below sulphur in the periodic table and resembles it in many aspects. Metal selenides are found frequently with sulphides and, consequently, selenium compounds may appear in the wastes from processing of sulphide ores. Selenates resemble sulphates and are among the most common oxidised form of the element, although selenites are also comparatively stable. Selenium and its compounds are employed in electronic and photoelectric devices, in glass for colour, and in photocopying equipment.

The toxic nature of selenium is well known, even very small amount can produce sickness and irritation. Selenium in soils can be taken up by plants with toxic effects shown by animals that consume these materials. Availability of selenium to plant varies with soil-selenium content, determines soil acidity. In alkaline soils selenium is chiefly present as selenates that are readily taken up by plants (hence the early name "alkali disease" given to the symptoms of cattle that have consumed high selenium vegetation), while under more acidic conditions the selenium appears to be present as selenites or other less soluble forms. Because of the acute toxicity, and also because of suspected (but at this time unproven) carcinogenic properties, there is a considerable emphasis on reducing selenium emissions.

Bacteria and other organisms in water sediments and sewage and some plants can convert inorganic forms of selenium to volatile organic products such as dimethyl selenide $(CH_3)_2Se$. This may be a means of eliminating toxic levels of selenium in these organisms, and sets up a process of transport and cycling for this element.

All the selenium-containing enzymes which have been analyzed until now contain selenocystein residues. The selenol in selenocysteine is of lower redox potential and higher nucleophilicity than the thiol in cysteine. This high reactivity of selenol probably contributes to the specific catalytic effect of selenium-containing enzymes. The selenium atom incorporation into amino acids is primarily catalysed by pyridoxal enzymes.

In spite of toxicity of selenium, it is an essential trace element for a number of organism, possibly including man. It is absorbed in duodenum from where it goes to plasma and distributed in various tissues with the help of Glutathionein peroxidase enzyme. This enzyme decomposes H_2O_2 or organic peroxides with glutathione (GSH) as a reductant according to the following equation:

$$2GSH + H_2O_2 \text{ (or ROOH)} \rightleftharpoons GSSG + 2H_2O \text{ (or } ROH + H_2O)$$

The molar mass of this enzyme varies according to the origin of enzyme but lies in the range 76,000 and 92000. In any case, the enzyme is composed of four subunits, each of which has molar mass varying from 19,000 to 23,000 and involves one atom of selenium as selenocysteine residue. In the reduced substrate glutathione, the selenol moiety (—SeH) is formed. The moiety is either carboxy methylated by monoidoacetate or monochloro acetate in activating enzyme. Incubating the oxidised enzyme with CN^- in alkaline solutions for a long time liberates selenium also, resulting in activation of the enzyme. In this way, this enzyme prevents the formation of peroxides and free radicals from the oxidation of unsaturated fats.

Recently evidence indicates that selenium is effective in preventing the onset of cancer, that is to say, the incidence of cancer can be correlated, inversely, with selenium levels in the blood. Interestingly, this effect is counteracted by high levels of zinc. If the above observation is valid, it illustrates the difficulty in dealing with trace elements through elimination of selenium from food and water supplies but as too high an intake certainly will do harm.

It is well known that excess inhalation of oxygen creates hole in the heart. This problem is observed in the labourers of Himachal Pradesh but not in Kashmir. On looking into the anomaly, it was found Kashmiri people use Kesar *saffron* in their drinks, where it is used as a flavour in beverages and other eatables. Kesar was found to contain selenium which show antioxidant properties and thus prevents the hole formation in heart. The mechanism is not understood yet but fact is confirmed. This established fact led to place multivitamin tablets on shelves of two types, one with selenium other without selenium all over the world.

5.61 Cadmium[23, 27, 28]

Cadmium and zinc belong to the same group in periodic table, with similar properties. In addition, they are usually mined and refined together, for cadmium essentially as an impurity (in the ratio 1:200, Cd : Zn), only rarely does cadmium appear as an ore in its own right.

Cadmium is virtually absent from the human body at birth and accumulates with age up to 50 years. At this age the average person not exposed to abnormal amounts of cadmium has a total body burden of 20-30 mg Cd. Cadmium metabolism involves: (*a*) lack of homeostatic control mechanism, and (*b*) tenacious retention in the body with a long half-life estimated as 16-33 years in man.

Cadmium[28] enters the biosphere through its increasing use in electroplating, in plastics as stabilisers, in paints as pigments, in cadmium batteries, and as a contaminant in phosphate fertilizers and sewage sludges. In the absence of Cd-emitting factories, such as zinc refineries, the levels in the air approximate 0.001 μg Cd/m^3, which would lead to a maximum inhaled amount of 0.02 μg/person/day. In large cities higher levels approaching 0.03 μg/m^3 may be found. Most municipal waters contain less than 1–3 μg Cd/lt which is well below the upper limit of 10 μ g/lt for drinking water set by the World Health Organization. Even at this upper level the consumption of 2.51, would provide only 25 μg Cd/day. Food is thus normally the major source of cadmium to animals and nonsmoking humans. For man a "provisional tolerable weekly intake" of 400-500 μg Cd/person has been proposed by WHO. This approximates 1 μ g/kg body weight for most individuals or 55–70 μg Cd/day supplies may do harm, just as too high an intake certainly will. Unlike zinc, cadmium is a highly toxic element. The reason for the high toxicity evidently lies in its similarity to zinc; it can replace the latter in enzymes but because of stronger bonding and perhaps stereochemical differences, the function of the enzyme is disrupted. Worldwide industrial use of cadmium is expanding and is used as an antifriction agent (bearings), a rust proofer (especially in paints and varnishes), for alloying (with lead and copper) and many other functions.

The ability of plants to absorb and concentrate cadmium means that it is a real danger. Fortunately, retention is very low (1–2%) and accumulation very slow. Effects related by some to environmental cadmium levels, include cirrhosis of liver, lung damage and destruction of testicular function are given in Table 5.5.

In 1955 there was an outbreak of cadmium poisoning in Jinzu River North Japan with high metal content in river water, rice and soyabean (0.37–3.36 ppm dry wt) called Itai-Itai-Byo disease, the symptoms were typical of extreme lumbago, followed with painful lack of mobility, in some patients, by skeletal collapse. The victims suffered bone deformation and fractures. Their bones became so brittle that even a single sneeze could cause fracture. Cadmium had made the bones so brittle that inhibited bone repair mechanisms. It occurred mostly in post menopausae, multiparous women consuming poor diets low in protein and calcium.

Table 5.5 Pathological effects of cadmium

Pathology	*Daily intake (in μg) for 1–3 years*
Hypertension	175
Anaemia	530
Retardation of growth	1300
Abnormalities of pancreas and spleens	1300
Heart abnormalities	2125
Liver and kidney damage	5250

The cause of the disease was cadmium poisoning. The source of cadmium was the Kamioka mining station, a lead and zinc mine located on a stream that flows into river Jinzu. The cadmium had entered the patients bodies when they drank water from this river; the cadmium content of water from this river also contaminated the local farm land.

5.6*m* Mercury[24, 25, 26, 27]

Mercury is the third member of the zinc, cadmium family. It is a very toxic, cumulative poison, having its chief effects on the nervous system. It favours still more covalent bonds than cadmium, especially with sulphur which coordinates very strongly with Hg(II). Indeed, mercury binds strongly to the sulphur atoms of sulphur containing amino acids (Cystine, Methionine) present in many proteins. Such binding is able to inactivate enzyme and upset the chemical reactions, which they otherwise catalyse in the body The membrane properties of the cell wall get disrupted and normal cellular activities are inhibited. The damage to the body is usually permanent, there being no known effective treatment. The Hg—C bond has comparatively low polarity and so organo-mercury compounds are quite stable in aqueous medium, in contrast to the more polar organo-zinc and cadmium compounds that are easily hydrolysed. Inorganic mercury compounds tend to accumulate in the liver and kidney tissues. This causes damage but can be readily eliminated through the urinary tract and hence may not allow it to accumulate in the body to dangerous levels. Elementary mercury finds use as a catalyst in some chemical manufacturing processes, especially in the manufacture of vinyl chloride, the base of many plastics. It was mercury effluent from a vinyl chloride plant that caused the largest known epidemic of mercury poisoning in Minamata Bay, Japan (1953–60).

The Chloro-alkali industry has been the largest industrial users of mercury. It produces chlorine and caustic soda by electrolysis of NaCl solution. The function of mercury in this process is to serve as cathode of the electrolysis cell. Theoretically no mercury is consumed during the process of producing Cl_2 or caustic soda but there is an actual loss of mercury. Some of the mercury gets lost to the products, some leaves with effluent water and some gets lost through air ventilation system. The operating conditions of the electrolysis cell result in a loss of 0.05–0.25 kg of mercury for every ton of Cl_2 produced. Most of this loss ultimately shows up in waste-water discharges from the plants.

The second largest consumption of mercury has been in the production of electrical appliances. The use of mercury and its compounds as fungicides constitutes the third largest category of mercury consumption. In this category use is made of the toxicity of mercury compounds, towards living organisms. These substances have been used to destroy fungi in the paint, pulp and paper, and agriculture industries. Mercury compounds have been added to paints for use in humid regions to act as a latex preservative and a mildencide. Paints for use on ships must contain "antifouling" agents (usually HgO) to prevent barnacle growth. For Milddew-proofing of marine paints, phenylmercuric acetate (PMA) or its derivatives are used. PMA has been found to be a most important organo-mercury commercial compound, with a formula as follows:

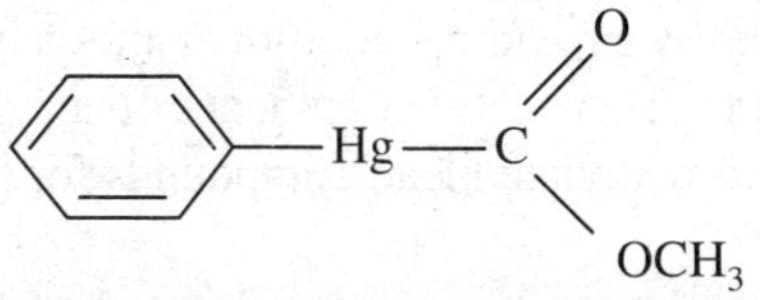

Phenyl mercury acetate (PMA)

The use of PMA by the pulp and paper industry for in-process slime control has resulted in PMA discharges in waste water. The incineration of paper produced when PMA was used constitutes a possible source of air borne mercury pollution, because trace amounts of mercury remains in such papers.

A current area of great concern involves the use of mercury compounds as seed dressings in agriculture. The amounts of mercury in seed-dressing formulations are small, but the effect is large, because the seed dressing has been applied to a large volume of seeds that have been subsequently sowed over millions of acres. Which results in an extremely widespread dissemination of the mercury compounds. Also there has been enough evidence suggesting that mercury undergoes translocation in plants namely transfer of mercury from seed to crop. It is found that mercury gets translocated from grain seed to the harvested grain, then to a hen, and finally to eggs. Mercury also finds it way into animal tissue by way of food chains.

Metallic mercury in its compounds, upon settling into the beds of lakes, gets converted by an aerobic microorganisms into CH_3Hg^+ or $(CH_3)_2Hg$. This latter compound is highly volatile and escapes from the bed into the water. $(CH_3)_2Hg$ is stable in alkaline solutions but when exposed to acidic surroundings it changes to CH_3Hg^+. This ion is soluble in water and concentrates in living organisms, usually appearing in body lipids. Such an accumulation occurs due to direct absorption of ion or an amplification by a food chain process.

The natural volatility of elemental mercury implies that there is a mercury cycle in the environment. Mercury-ore deposits normally produce a considerable mercury vapour pressure that results in a low level, worldwide distribution of mercury. The element has been found in ice samples that predate industrial use of the metal, although such use has increased the amount available to enter into this cycle. The inorganic forms of mercury undergo conversions that are mediated by microorganisms, as for example, insoluble HgS is converted to soluble Hg(II) by bacterial oxidation and soluble Hg(II) is reduced by some bacterial enzymes to Hg(0). If Hg(II) and Hg are present together, they will enter into an equilibrium reaction to produce Hg(I), the latter exists as the diatomic ion Hg_2^{2+}.

An alternative method by which Hg(II) is eliminated by some bacteria is through methylation. This appears to proceed through transfer of a CH_3^- group to Vitamin B_{12}, a cobalt-containing coenzyme. Vitamin B_{12} contains cobalt in a porpyrin-like environment of four nitrogen (as in haemoglobin) atoms with a fifth nitrogen atom from a dimethylbenzimidazole group also attached. A methyl group on the active site of the Co(I) may be transferred as CH_3^- to a suitable acceptor and replaced by H_2O. The cobalt is oxidised from the less common Co(I) to the more common Co(II) state in this process. A small amount of methyl mercury ion CH_3Hg^+ can be further methylated to $(CH_3)_2Hg$ in similar process. It can also be reduced bacterially to Hg(0).

The methyl mercury ion CH_3Hg^+ produced in sediments is partially extracted into the water phase and assimilated into the food chain. Dimethyl mercury can volatilize into the atmosphere, where it is photolysed to Hg(0) and methyl radicals. The reactions are summarised in Fig. 5.13.

These reactions of mercury undoubtedly occur naturally although at quite low levels. Industrial input can upset the natural balance by providing additional inorganic starting material, or by direct input of organo-mercury compounds. There is also evidence that Hg(II) can be methylated by water soluble methylsilicon, methyltin and methylated lead compounds for these to be introduced from industrial wastes.

The environmental hazards of mercury are well documented in scientific and popular literature in David Seltzer's "*Prophecy*"[25]. These include the *Minamata Disease.* Minamata, a city in the southern Japan received reports of an outbreak of a disease that attacks the central nervous system. The symptoms were sickness, madness, paralysis, loss of speech and vision, loss of emotional control, crippling of arms and legs and finally death, all because of damage to the central nervous system. Nearly 100,000 people died in agony at Minamata. The cause was detected as methyl mercury, which had entered the water via the factory effluents and had built up in fish and shell fish consumed by the local residents.

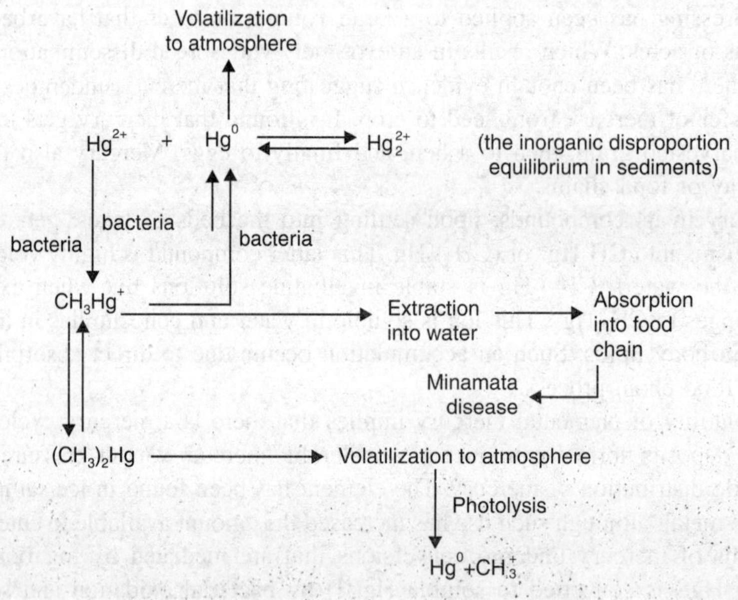

Fig. 5.13 Reactions of mercury in the biosphere.

Bombay Minamata: In 1972, 300 persons were officially listed as its victims. A series of investigations by the Institute of Science, Bombay revealed that industries on the banks of Kalu river are creating another pollution disaster. Kalu is a 50 km long river which originates in the Western Ghats. It flows through the industrial suburbs of Ambarnath, Ulhas Nagar and Kalyan in north east Bombay and finally enters the Arabian Sea. It receives effluents from over 150 industrial units. The effluents that enter into the river contain a lot of mercury and lead. These elements are insoluble in water and sink to the bottom. Certain bacteria in the river bed convert the insoluble mercury into soluble methyl mercury. The fish get poisoned and cattle that graze on the plants in the shallow water also accumulate it. Methyl mercury content of pycrus plants, the most abundant at Ambivali was, 3 to 110 ppm. There was 5 ppm of mercury in milk of milch cattle that graze on pycrus plants. A child drinking one litre of milk everyday consumes 35 ppm of mercury in a week. Over a period of several months this can lead to an accumulation of over 0.3 mg of mercury which is well above the safe level thereby leading to Minamata disease.

The fish that are not poisoned by mercury are killed by other toxic wastes. Thus dead fish and dying plants poison the livestock, paralyse human being and cause infirmity among children, making up a deadly cycle in the village of Ambivali.

5.6*n* Lead[20, 21, 22]

Lead is the last member of the carbon family and exhibits metallic properties. However, lead has weak electropositive character and electron-acceptor property. Bonding with nonmetals such as carbon, provides a considerable covalent character. Lead compounds have the oxidation state (IV) and (II). The former is found in organometallic compounds which are highly toxic and have been cause of concern in the recent years. The USPH limit for lead in water is 0.05 mg/l and like mercury, Pb(II) forms comparatively covalent bonding with appropriate donor groups generally favouring nitrogen and sulphur over oxygen donors, and it may owe some of its physiological reaction to replacement of other metals in certain enzymes. Previously 99.8% of lead used in manufacturing chemicals, was used in the production of the popular antiknock agent. The addition of tetraethyl lead to gasoline raises its octane member, but on combustion, 25 to 75% of it in the exhaust is particulate as elemental lead, lead oxide and lead halide. These may affect plant and animal life directly and even affect weather conditions down wind of emission zones due to formation of the so-called ice nuclei. Many common lead salts such as carbonate, hydroxide, and sulphide are highly insoluble in water, this normally limits the dissolved lead content in water. However, use of lead as antiknock agent in gasoline, in India is almost removed.

Lead compounds find use as paint pigments because they have low water solubilities, good covering power and they exist in a variety of colours. The most widely used compound is 'white lead' or $Pb(OH)_2.2PbCO_3$. 'Red lead' Pb_3O_4 is a bright red powder used as a pigment in corrosion-resistant paints. An attractive yellow colour has been produced by the inclusion of "chrome yellow", $PbCrO_4$.

An interesting use of lead has been the production of pottery glaze. A glaze is a thin layer of glass fused on to the surface of clay objects. The main constituent of ceramic glazes is silica, which reacts with other oxides to produce complex silicates (glasses). The lead compound, PbO, is added because it gives the final glaze attractive properties unattainable with other oxides.

In the last century, the mining and smelting of lead, its dispersal in air, water and soil has increased fivefold above the prehistoric level. This is poisoning water, vegetation, air and thereby perturbing the lead cycle on earth. All this has led to mental disturbances as reflected by terrorism, barbarism, crime, cultural degradation. Visible every where are irritability, emotional instability, general restlessness and reduced acuity.

The current standard of safe lead level in blood is 10–15 μ d dlt^{-1}. The blood does not need lead, yet everyone absorb it in varying amounts. Lead is not excreted but is stored for many years in tissue, chiefly in the bone marrow from which it is released back into the blood stream to wreak cellular havoc. For middle aged men, the risk of hypertension rise for pregnant woman it means as greater chance of miscarriage, premature delivery and still birth and children are susceptible to a range of psychological, neurological, kidney and blood abnormalities, including partial hearing loss, hyperactivity, lower IQ, impaired ability to absorb iron and calcium, reduced ability to metabolise vitamin D, disturbances in the formation and maintenance of red blood cells, decreased muscle and bone growth. Like mercury, lead prefers bonding with nitrogen and sulphur rather than with oxygen. It forms bonds with various amino acids and enzymes, thus disturbing the functioning of enzymes, which may even cause cancer.

Tap water contains quite a large amount of lead. The problem does not lie in municipal treatment plants or public distribution lines but in residential plumbing systems, where water collects small particles from corroded pipes. A stop gap remedy that is the best is to run the tap water for sometime before using water. This must be done for every drinking water faucet.

Another source of lead poisoning is auto-emissions, flaking paints and smelters that have lead in dusty soil and sediments. In cities, children, particularly those coming from low income section, encounter it in the backyards. Powder-fine outdoor dirt shifts inside to become household dust, presenting a danger to children who crawl on the flour and gnaw at furniture. Such dust can also contaminate fruits and vegetables, but a washing removes most of the lead content.

Soldering wire of electronic circuits contain more of lead, quite a lot of which enter into the atmosphere during soldering of circuits.

However, a diet with adequate proteins, low in saturated fats and rich in minerals such as iron, zinc and calcium helps in protecting against lead poisoning.

5.6*o* The Halogens[28]

Chlorine is the most abundant halogen, being widely distributed in water systems as the chloride ion. Sea water is near 0.5 molar in chloride, but a certain amount is also present in fresh-water. Chlorides of certain metals are insoluble but those commonly encountered in nature are soluble. Chlorine is widely used as oxidising agent, and is added to drinking water to oxidise organic matter. In water, chlorine is hydrolysed

$$Cl_2 + 2H_2O \longrightarrow Cl^- + ClOH + H_3O^+$$

The hypochlorite ion ClO^- is commonly encountered as sodium hypochlorite used as an oxidising agent or bleach. Chloride is not harmful, although the total salinity with which it is associated will, if high enough, render water unfit for use. NaCl and to lesser extent KCl are among the final salts to crystallise when fresh water evaporates. This gives rise to natural salt deports in arid regions and in some cases a gradual build-up of salt in irrigated soils, when irrigations as waters containing small amount of chloride evaporate in the absence of some periodic flushing process. A good deal of chloride as solid NaCl or $CaCl_2$ is used to melt snow and ice, and this also contributes to locally high growth of vegetation, with many plants having very limited tolerance toward this. The Roman practice of sowing salt on the ruins of Earth age recognised that a high level of salinity would destroy soil productivity and at least symbolically was intended to prevent regrowth. Chloride ions may accelerate corrosion, for example by taking part in the mechanism of metal attack.

The heavier halogens, bromine and iodine, are much less common than chlorine in natural water, but they are accumulated from the sea water by sea weed. Iodine is particularly essential to health and often is added in common salt, because in many areas water and foods are deficient in it.

The first member of the halogen family is fluorine[32]. Human populations obtain widely varying quantities of fluorine, from the food, the water supply, and the atmosphere depending on location. The air does not constitute a significant source of fluorine for man, except in areas of industrial pollution. Surface waters, such as are used for drinking and working in most communities, generally contain less than 1 ppm, or even 0.1 ppm or less. In some parts of the world, the population is dependent on water from deep wells or artesian bores which are naturally high in fluoride content. Normal dietary intake lies in the range of 0.78–1.03 mg F/day from food for family, nonfluoridated cities and 1.73–3.44 mg F/day in fluoridated environment.

Fluorine does not concentrate in any tissue, other than the bones and teeth. Soluble fluorides are rapidly and almost completely absorbed from the gastrointestinal tract, even at high intakes and are rapidly distributed throughout the body as fluoride' ion, in a pattern similar to that of a chloride. It readily breaks, the erythrogates. The disappearance of fluoride from blood is quite fast.

Uptake of fluoride by bone depends on its vascularity and growth activity, so that in fully mature bones the rate of F^- deposition and retention is greatly reduced and can occur by two processes. The first involves a rapid exchange of F^- ion in the tissue fluid with OH^- or CO_3^{2-} on the mineral crystal surface. The second process involves slow bone formation leading to storage, in which fluoride ion is incorporated into the hydroxyapatite lattice to form fluoroapatite. Since there is no change in the Ca:P ratio, the phosphate group cannot be replaced by F^- ions. However, the carbonate content of fluoroapatite is decreased and magnesium content increased. This suggest a replacement of $CaCO_3$ by the MgF_2.

The overall reaction is

$$\underset{\text{Hydroxyapatite}}{Ca_{10}(PO_4)_6(OH)_2} + 2F^- \longrightarrow \underset{\text{Fluoroapatite}}{Ca_{10}(PO_4)_6F_2} + 2OH^-$$

The hydrogen is embedded in the oxygen in OH^- and the group is effectively spherically symmetric with the radius of OH^- being equal to that of F^-. The sizes of the OH^- and F^- ions are quite similar so that overall structure is not changed, by such a limited replacement; but the hardness and crystallinity of the substances are improved. On the other hand, excess floride can be harmful.

The caries experienced in the permanent teeth was found to decrease as F^- concentration in drinking water increases. The decrease in dental caries is a linear function of fluoride expressed on a log scale. When considered in relation to "incidence of mottling of the enamel, and plotted on the same graph, the two lines intersect at close to 1 ppm F in the water. If concentration is $\geq$ 1 ppm, enamel of teeth is eroded, mottled teeth are characterised by chalky white patches distributed irregularly over the surface of tooth with a secondary infiltration of yellow to brown staining these sites give a corroded appearance to teeth. Mottled enamel is almost entirely confine to permanent teeth and develops only during their period of formation. The fully formed enamel of adult teeth is unaffected by fluorine and the deciduous teeth are only affected by a high fluoride intake where other sign of fluorosis are evident. However, fluorosis leads to deformation of bones resulting in knock-knees, bowlegs and stiffening of joints, joint pains, stiffeners of back, backpain, difficulty in getting up are some common complaints. The dangerous factor is that not only the old but also the youth and children are also the increasing victims of this disease have to take the support of a walking stick.

In endemic areas, a large percentage of people suffer from gastro-intestinal diseases complains, loss of appetite, constipation and intermittent diarrhoea and flatulence. It leads to a high incidence of still births and abortion.

In India, 25 million people spread over 100 districts in 15 states are suffering from fluorosis. Provision of safe drinking water and creating awareness among people of the dangers of excess fluoride in drinking water have now gained an important place in the Indian health programme. Andhra Pradesh, Bihar, Delhi, Gujarat, Haryana, Karnataka, Maharashtra, Madhya Pradesh, Orissa, Punjab, Rajasthan, Tamil Nadu, Uttar Pradesh, Kerala are partly affected by Fluorosis.

REFERENCES

1. Cotton, F.A. and Wilkinson, G., *Advanced Inorganic Chemistry*, Interscience, New York.
2. Manku, G.S., *Theoretical Principles of Inorganic Chemistry*, Tata McGraw-Hill, New Delhi (1985).
3. Bolin, B., The Carbon Cycle, in Chemistry in Entrainment, *Scientific Am.*, p. 53 (1973).
4. Delwiche, G.C., The Nitrogen Cycle, ibid, p. 43 (1973).
5. Benemann, J.R. and Valentire, R.C., *The Pathways of Nitrogen Fixation*, Advances in Microbial Physiology.
6. Halsted, R.L. and Mekercher R.B., Biochemistry and Cycling of Phosphorus, Chap. 2 in *Soil Biochemistry* (EA Paul & A.D. Mclaren Eds), Vol. 4, Marcel Dekker, New York.
7. Mitra, R.P. and Sindhu, P.S., Uptake of Phosphate by Some Layered Silicate Minerals in Relation to Their Titration Curves with Acid and Base, *J. Indian Soc. of Soil Science*, Vol. 24(1), 29-47 (1976).
8. Sindhu, P.S., Some Special Features of the Reaction between Aluminium and Orthophosphate, *Indian J. Chem*, Vol. 14A, 611–13 (1976).
9. Sindhu, P.S., Interaction of Some Silicate Minerals with Acid, Base and Phosphate, Ph.D. Thesis, Delhi University, Delhi (1974).
10. Kellog, W.W., Cade, R.D. Allen, E.R., Lazrus, A.L. and Martell A.E., The Sulphur Cycle, *Science*, 176, 573 (1972).
11. Pfeiffer, J.B. ed., Sulphur Removal and Recovery from Industrial Processes, *Advances in Chemistry Series*, No. 139, American Chemical Society, Washington DC (1975).
12. Committee on Medical and Biological Effects of Environmental Pollution "Chromium", National Academy of Science, Washington DC.
13. *Ibid* Manganese (1973).
14. *Ibid* Lead (1972).
15. *Ibid* Nickel (1975).
16. *Ibid* Copper (1977).
17. Dickerson, R.E., and Geis, I., *The Structure and Action of Protein* Harper and Row (1969) (on Hemoglobin Structural Details and Chemical Data).
18. Wang, J.H., Accounts Chem Res. For Synthetic Biochemical Models. V. 3, 90 (1970).
19. Huisman, T.H.J. and Schroeder, WA., eds. *New Aspects of the Structure, Function and Synthesis of Homoglobin*, Butterworths (1971).
20. Ewing, B.B. and Pearson, J.E., Lead in the Environment, *Environmental Science & Tech.*, J.N. Pitts (Jr), R.L. Metcalf and A.C. Lloyd eds), Vol. 3, P, Wiley, New York (1974).
21. Sindhu, Jagjit Kaur, Lead Poisoning, *Chemistry Education*, Jan-March 47-48 (1988).
22. Needleman, H. and Maher, C., A Critical Appraisal of Air Quality Criteria for Atmospheric Lead, United States Environmental Pollution Agency (Nov. 1976).
23. Page, A.L. and Bingham, F.T., Cadmium Residues in the Environment, *Residue Reviews*, 48, P. 11 (1913).
24. Saha, J.G., Significance of Mercury in the Environment, Residue Reviews, 42, P. 103 (1972).
25. Seltzer, David, *Prophecy*, A Study of Unrelenting Terror, Granada Publishing, New York (1979).
26. Smith, W.E. and Smith, A.M., *Minimata. A Case Study of Mercury Poisoning*, Holt, Rinehart and Winston, New York (1975).
27. Huckabee, J.W. and Blaylock B.G., Transfer of Mercury and Cadmium from Terrestrial to Aquatic Ecosystem, in *Metal Ions in Biological System*, p. 125 (S.K. Dhared) Plemn, New York (1973).
28. Underwood Eric *Trace Elements in Human and Animal Nutrition* Fourth Edn. Chapter 9: Cadmium; Chapter 13: Fluorine, Academic Press (1977).
29. *Selenium in Biology and Medicine* Eds., Julian E. Sphallholz, John L. Martin, Howard E. Ganther, AVI Publishing Company Inc; West Port, Connecticut.
30. Sindhu, P.S., *Chemistry of Atmosphere*, New Age International Pub. Ltd., New Delhi (1997).
31. Lubert Stryer, *Biochemistry*, W.H. Freeman and Company, New York (1988).

6 Polymers and Plastics

6.1 INTRODUCTION

Synthetic polymers are macromolecules, that is, molecules containing many atoms and are of very large size and high molar mass (ranging from 10,000 to 10,000,000). They are made up of a small number of repetitive groups. A complex array of polymeric substance are produced in nature in vast quantities. However, most of the naturally produced polymeric substances do not accumulate since they are degraded microbiologically and their components are returned eventually to the carbon, nitrogen and other elemental cycles. It is obvious that if there was no natural agency for the degradation, substances such as chitin (found in insect cuticle, butterfly wings, lobster shell, etc., or keratin (the sclero-protein comprising the bulk of wool, hair, feathers and hooves), which had been deposited over the world for centuries, we would, today, be knee deep in feathers and butterfly wings.

Many macromolecules are components of living systems; examples include proteins, which are polymers of amino acids; starch and cellulose, which are polymers of sugars. Since biological polymers are built-up in living systems, they can be broken down in these systems and are called biodegradable. Biodegradable substances will not cause long-term environmental problems, although the decomposition of large amounts of such materials can be a problem of immediate concern. In aqueous solution, their degradation will contribute to the biological oxygen demand and utilisation of the dissolved oxygen. Under anaerobic conditions, decomposition of proteins will produce undesirable H_2S, NH_3 and CH_4.

Synthetic polymers, on the other hand, are often not biodegradable. In fact, they are chemically tailored for long life in the environment. Although environmental stability is desirable, there are many instances where a polymer material is used once and then discarded; *e.g.*, as plastic bottles, food wrappers, and plastic cups. Those polymers designed for as long life have often been used to manufacture these disposable items. As these 'throw-away' items accumulate in the environment, it has become apparent that biodegradable polymers are needed to prevent accumulation of plastic litter on the road side, in parks, and on beaches. As a consequence, a few polymer materials have been recently designed that are subject to rapid environmental decay, and are being used in the manufacture of disposable plastic items.

Synthetic polymers are made up of a great number of monomers joined together in a regular fashion. Some examples are given in Table 6.1. Most plastic items are not formulated from polymeric material alone. They usually contain lower molar mass compounds called plasticisers, which increase their flowability and processibility. Esters of phthalic acid are examples of plasticisers. Fillers, inert materials such as wood flour, grot mica, asbestos or glass fibre, are often added to increase the strength of the plastic or to provide bulk at low cost. Plasticisers may be leached or evaporated from plastic materials in use. The film that deposits on the inside of the windows of new cars in warm weather is caused by plasticiser distilled from the plastic of the car interior. Both plasticiser and fillers contribute to the environmental disposal and degradation problems. Fillers may be inert, but if released as dust during incineration of wastes, they can contribute to atmospheric particulate material and, in the case of asbestos particles become health hazards in view of the known connection between asbestos dust and lung cancer. The bulk of monomers used in polymers synthesis are obtained from petroleum. Therefore, our present dependence on plastic materials is ultimately a dependence on a source of crude oil.

6.2 POLYMER SYNTHESIS

Polymers may be formed from monomers either in chain-reaction or step-reaction polymerization. The details of polymer synthesis can be found in a good textbook on polymers[1, 2, 3, 4]. The chain reaction process is illustrated by the free radical polymerization of ethylene or its derivatives.

The specific example of the polymerization of acrylonitrile to Orlon is given in reactions 6.1 to 6.5. In free radical polymerization, a small amount of initiator, such as a peroxide is added to the monomer to start the polymerization. The initiator decomposes to form the radical (6.1), which adds to the monomer in the chain initiating step (6.2). Further addition of monomer takes place in the chain propagating step (6.3). Polymer growth stops in the chain terminating steps which require reaction of two radicals such as the disproportionation reaction shown in (6.6).

Initiation

$$ROOR \longrightarrow 2RO^{\bullet} \qquad ...(6.1)$$

$$RO^{\bullet} + CH_2{=}\underset{CN}{\underset{|}{CH}} \longrightarrow ROCH_2-\underset{CN}{\underset{|}{\dot{C}H}} \qquad ...(6.2)$$

Propagation

$$ROCH_2-\underset{CN}{\underset{|}{\dot{C}H}} + CH_2{=}\underset{CN}{\underset{|}{C}} \longrightarrow ROCH_2\underset{CN}{\underset{|}{CH}}CH_2\underset{CN}{\underset{|}{\dot{C}H}} \qquad ...(6.3)$$

Termination

$$2ROCH_2\underset{CN}{\underset{|}{CH}}(CH_2\underset{CN}{\underset{|}{CH}})_nCH_2\underset{CN}{\underset{|}{\dot{C}H}} \xrightarrow{\text{combination}}$$

$$ROCH_2\underset{CN}{\underset{|}{CH}}-(CH_2\underset{CN}{\underset{|}{CH}})_nCH_2-\underset{CN}{\underset{|}{CH}}-\underset{CN}{\underset{|}{CH}}-CH_2-(\underset{CN}{\underset{|}{CH}}CH_2)_n\underset{CN}{\underset{|}{CH}}CH_2-OR \qquad ...(6.4)$$

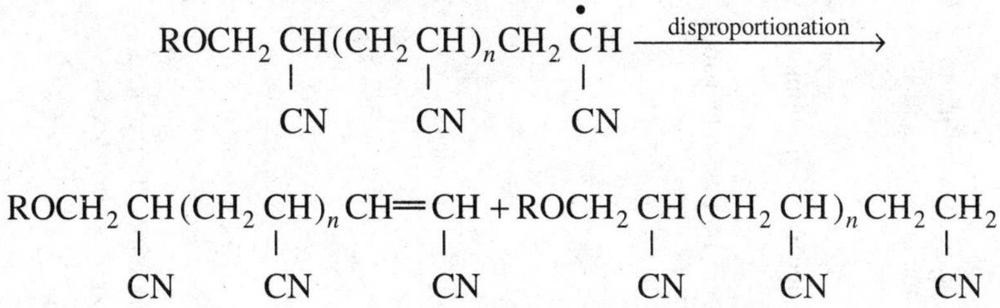

...(6.5)

Ionic processes may also be involved in chain reaction polymerization. Cationic polymerization of vinyl monomers is initiated by acids. The molar mass of the oligomer or polymer will depend on the reaction conditions and the ratio of acid to olefin. Anionic polymerization is initiated by strong bases such as butyl lithium $CH_3CH_2\,CH_2CH_2^-Li^+$, abbreviated as *n*-Bu Li. A carbanion intermediate is formed (6.6) that continues to add monomer (6.7) until the carbanion is destroyed by the addition of a proton or the elimination of a negatively charged group (6.8)

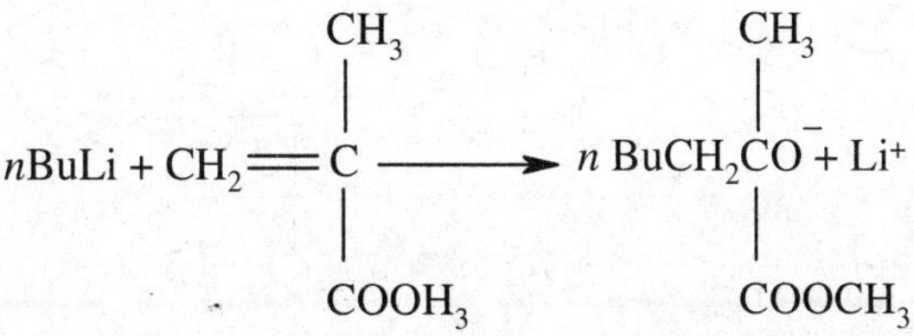

...(6.6)

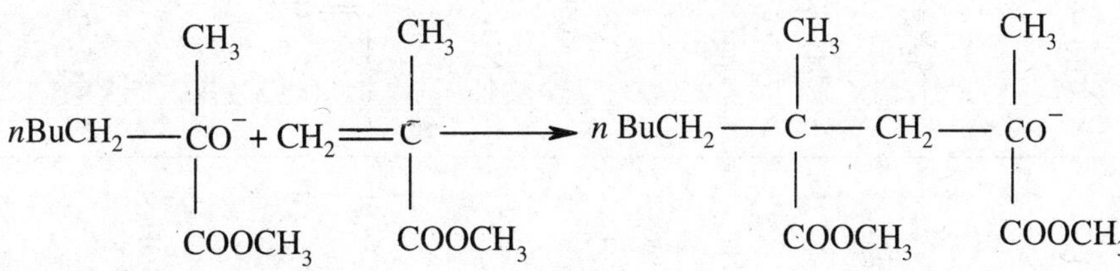

...(6.7)

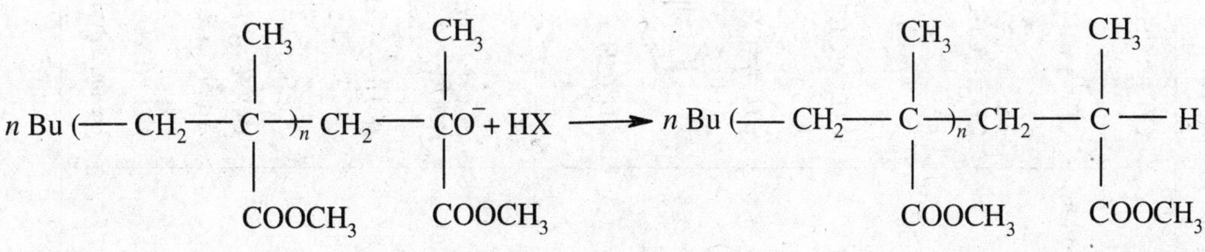

...(6.8)

In the chain polymerization reaction, the molar mass of the final polymer is dependent on the maintenance of the active species, be it the radical, cation or anion formed initially. In step-reaction polymerization, each reaction that adds a new monomer unit is independent of the preceding one. The formation of nylon 6, 6 illustrates this process by reaction (6.9).

Table 6.1 Some typical monomers and their polymers

Polymer	*Monomer*	*Repeating unit*	*Uses*
Polyethylene (marlex)	$CH_2= CH_2$	$-CH_2-CH_2-$	House wares like buckets, carpets, cable insulation and packing material.
Polytetrafluoroethylene (Teflon)	$CF_2= CF_2$	$-CF_2-CF_2-$	Non sticking coating of utensils, insulations for electric items.
Polypropylene	$CH_2= CHCH_3$	$-CH_2-\overset{CH_3}{\overset{\vert}{CH}}-$	Pans, drinking glasses, musical equipment electronics equipment, tubes, pipes and fibre.
Poly (vinyl chloride)	$CH_2= CHCl$	$-CH_2-CHCl-$	Floor covering, corrugated roof, gramophone record.
Poly styrene	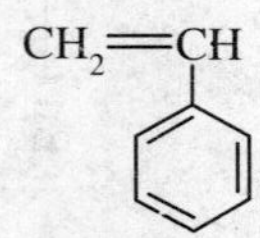	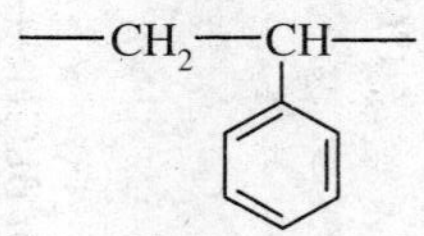	Food containers, cosmetic bottles, television cabinets, plastic cups, toys.
Polycaprolactam (Nylon 6) (perlon, caprolon)	$NH_2(CH_2)_5$ COOH	$-NH(CH_2)_5-\overset{O}{\overset{\Vert}{C}}-$	Fibres and clothes.
Poly (ethylenetetraphthalate) (Dacron, Terylene, Trevira, Mylar)	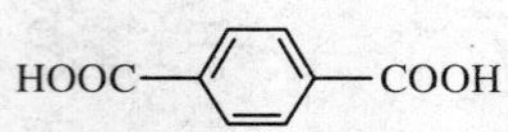 & $HO-CH_2-CH_2-OH$	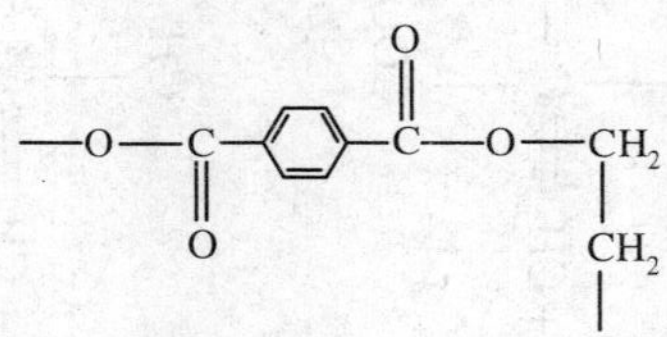	Fibres and textiles.
Nylon 6, 6	HOOC $(CH_2)_4$ COOH Adipic Acid NH_2 $(CH_2)_6NH_2$ Hexamethylene-diamine.	$-C(CH_2)_4\overset{O}{\overset{\Vert}{C}}-NH(CH_2)_6-NH-$	Synthesis of fibre, production of stockings and weaning materials, ropes, bristles for brushes

(Contd.)

Poly (cis-1, 4, isoprene) Natural rubber	$CH_2=C(CH_3)-CH=CH_2$	$-CH_2-C(CH_3)=C(H)-CH_2-$	Used for making tyres, and many rubber items.
Polychloroprene (Neoprene)	$CH_2=C(Cl)-CH=CH_2$	$-CH_2-C(Cl)=CHCH_2-$	Synthetic tyres, hoses for petrol and oil.
Polymethylmethacrylate (Lucite, Plexiglass)	$CH_2=C(CH_3)-C(=O)OCH_3$ (printed as CH_3 / $CH_2=C-H$ / $C=O$ / OCH_3)	$-CH_2-C(CH_3)(COOCH_3)-$	Signs, sky lights, aircraft canopies, optical equipment, dentures.
Lexan (a poly carbonate)	$COCl_2$ & 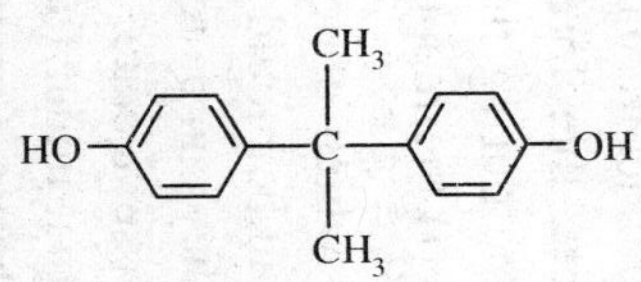	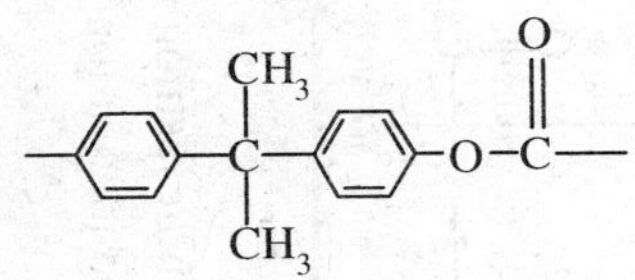	Moulding type of resin.
Phenol-formaldehyde resin (Bakelite)	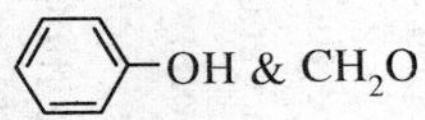	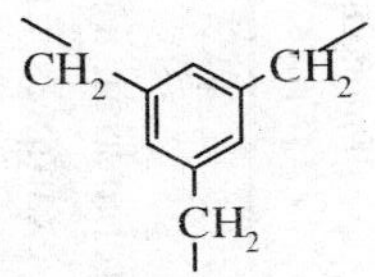	Cooking utensils, electric plates, switches and electrical appliances.

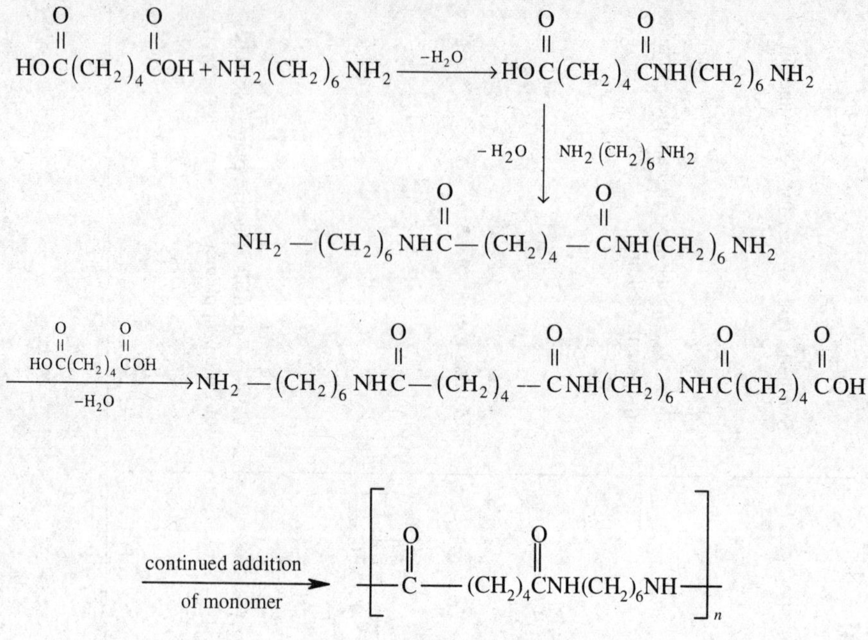

...(6.9)

Nylon 6, 6

Polyester Dacron is prepared in a similar fashion from ethylene glycol and terephthalic acid. These stepwise polarization are examples of condensation polymerization. A small molecule, in this case water, is eliminated in each step.

If a substance during its stage of manufacture has passed through a plastic condition (is capable of flowing) it is called a plastic. Plastics are, therefore, an arbitrary group of artificially prepared substances of organic origin which sometime during the stage of manufacture could be caused to flow. Plastics can be divided into two classes (*i*) thermoplastic and (*ii*) thermosetting.

(*i*) Thermoplastic resins are those which under the influence of heat and sometimes pressure become plastic (can flow) and can be moulded into different shapes under this condition. They retain their shapes on cooling. They can be remoulded into different shapes on heating as in the case of poly(styrene) and cellulose acetate.

(*ii*) Thermosetting resins are those that under the influence of heat and pressure become soft and can be moulded into different shapes. On further heating, they become hard and infusible on account of chemical change. They retain their shapes and cannot be remoulded. Examples include phenol-formaldehyde resin and urea-formaldehyde resins.

6.3 POLYMERS DEGRADATION

All polymers can be degraded but not necessarily regenerates the material from which the polymer was developed. Polymerization involves the addition of monomer to the end of a growing chain and depolymerization, namely removal of the monomers one by one. Thus depolymerization is considered to be reverse of polymerization and monomer and polymer may be considered to be in equilibrium as in styrene:

$$nCH_2{=}CHC_6H_5 \rightleftharpoons {+}CH_2-CHC_6H_5{+}_n$$

The energy of activation for the polymerization of styrene is about 109–117 kJ and the value of depolymerization lies between 91.6–96 kJ. The equilibrium reaction that differ by activation energies of 29–33 kJ are considered reversible. Apparently, all polymerization reactions are not reversible.

Thus heating polystyrene leads to depolymerization, because monomers have to be removed one by one, degradation process is slow. Polyethylene glycol produces acetaldehyde, divinyl ether gives dioxane and a number of other products, polyvinyl chloride and polyvinylidene chlorides liberates hydrogen chlorides on charring. Obviously in the latter case secondary reaction predominates in the thermal degradation. Even polystyrene shows benzene and toluene as secondary products. The thermal degradation of various important polymers are summarized in Table 6.2.

Table 6.2 Thermal degradation of polymers

Polymer	*Monomer*	*% Monomer*	*Degradation products*
Poly (methyl methacrylate)	$-CH_2-C(CH_3)(COOCH_3)-$	100	Monomer
Poly (methyl acrylate)	$-CH_2-CH(COOCH_3)-$	Traces (less than 1%)	Chain fragments
Poly (styrene)	$-CH_2-CH(C_6H_5)-$	42	Monomer, dimer, trimer, tetramer
Poly (α methyl styrene)	$-CH_2-C(CH_3)(C_6H_5)-$	100	Monomer
Poly(acrylonitrile)	$-CH_2-CH(CN)-$	Traces (less than 1%)	Hydrocyanic acid, ammonia
Poly (methyl acrylonitrile)	$-CH_2C(CH_3)(CN)-$	100	Monomer

(Contd.)

Polymer	*Monomer*	*% Monomer*	*Degradation products*
Poly (ethylene)	$-CH_2-CH_2-$	Less than 1%	Chain fragments.
Poly (vinyl chloride)	$-CH_2-CH(Cl)-$	Nil	Hydrogen chloride and conjugated products and dioxan.
Poly (vinylidene chloride)	$-CH_2-C(Cl)_2-$	Nil	—
Poly (tetrafluoro ethylene)	$-CF_2-CF_2-$	100	Monomer
Poly (isoprene)	$-CH_2-C(CH_3)=CH-CH_2-$	12	Monomer
Poly (butadiene)	$CH_2-CH=CH-CH_2-$	1.5	—

Chemical depolymerization could also be studied for condensation reactions. As the polymer was built up by elimination of simple molecule, in a condensation polymerization, the addition of this molecule would cause degradation. Typical examples of reactions yielding monomers include acid hydrolysis of polyesters, gelatin, polyamide and amino caproic acid.

Depolymerization can also be brought about by the addition of free radicals, involving both chemical and thermal degradation. Thus, if polyisobutylene, polyethylene, polystyrene are heated in the presence of benzoyl peroxide, the degradation is more rapid and monomers are obtained. Depolymerization can also be brought about in solutions, where polymers and monomers in toluene are in equilibrium. On continuous addition of benzoyl peroxide to a solution of polystyrene in toluene, equilibria shifts and monomers are obtained at a particular temperature and concentration.

6.4 PHOTOCHEMICAL DEGRADATION

Solar radiation is only effective in degrading those polymers that absorb at wavelengths greater than 290 nm since the more energetic UV radiation with shorter wavelength is absorbed by the ozone layer. Consequently, those photochemical reaction are possible which involve absorption at wavelength greater than 290 nm. The energy of some of the common bonds encountered in polymers are given in Table 6.3.

It is therefore evident from Table 6.3 that if an unsaturation, aromatic or carbonyl group, or oxygen, is present in any form in a polymer will show absorption at wavelength greater than 290 nm, hence homolytic bond dissociation will readily occur to give a free radical. These radicals can react with any oxygen present, leading to oxidation of the polymer chain. In other words polymer material can suffer deterioration by thermal, photo and oxidative degradations leading to aging, weathering and fatigue.

Table 6.3 Bond dissociation energies of some common bonds

Bond	*Bond dissociation energy kJ/mol*	*Wavelength absorbed in nm*	*Bond*	*Bond dissociation energy kJ/mol*	*Wavelength absorbed in nm*
C—H	414	288	C—N	301	397
C—C	347	344	C=N	314*	381
C=C*	260	460	C≡N	272*	439
C—Cl	334	357	C—S	272	439
C—O	356	334	C—P	259	462
C=O*	376	318	P=S	338	354

*These bond dissociation energy correspond to upper most π-bond.

Oxygen in the ground state occurs as a diradical O—O. Oxygen normally reacts with oxygen-containing organic compounds in a radical chain reaction involving the ground state diradicals with following reactions (6.10–6.15)

$$R^{\bullet} + O_2 \longrightarrow RO_2^{\bullet} \quad ...(6.10)$$

$$RO_2^{\bullet} + RH \longrightarrow RO_2H + R^{\bullet} \quad ...(6.11)$$

$$RO_2H \xrightarrow[\text{Cleavage products}]{} RO^{\bullet} + \dot{O}H \quad ...(6.12)$$

$$RO^{\bullet} + RH \longrightarrow ROH + R^{\bullet} \quad ...(6.13)$$

$$HO^{\bullet} + RH \longrightarrow HOH + R^{\bullet} \quad ...(6.14)$$

$$2RO_2^{\bullet} \longrightarrow \text{Cleavage products} \quad ...(6.15)$$

These reactions may also be initiated by the mechanical shearing of the polymer during processing. The oxidative processes initiated by shearing may result in the loss in strength of polymer but the extent of the oxidation is not usually large enough to give cleavage products that are sufficiently small for biodegradation.

Photodecomposition triggered by peroxides: Peroxides are formed during the processing of polymers as defined in 6.10 and 6.11. The UV absorption of peroxides extends to the long wavelength side of 300 nm so they absorb sunlight. The peroxides are cleaved photochemically with a quantum yield close to one that forms alkoxyl and hydroxyl radicals (6.12). The radicals thus formed react further as shown in reactions (6.13–6.15).

Decomposition triggered by carbonyl groups: Carbonyl groups, present as a result of decomposition of polymer peroxides, will absorb sunlight in the vicinity of 300 nm. The excited carbonyl groups can initiate reactions leading to cleavage of the polymer. This process does not lead to complete degradation of polymers, at least over a reasonable length of time as in polystyrene and other plastics in which degradation take place very slowly. However, incorporation of the carbonyl group in polymer leads to comparatively rapid degradation of polymers and plastics.

Although many polymers contain small amounts of carbonyl groups that may be formed by the slow decomposition of the peroxides produced during the processing of the polymer, in general there is a small number of carbonyl groups that results in photodegradation. It has been found, however, that the co-polymerization of small amounts (~1%) of carbonyl-containing monomers, together with the normal monomer, results in a photo-degradable polymer. As for example, phenyl-vinyl ketone can be polymerized together with the styrene monomer to produce a polystyrene which contains 1% phenyl ketone residues.

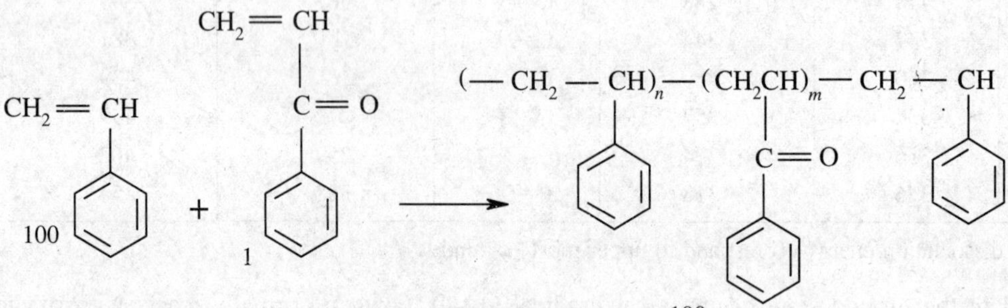

...(6.16)

n + m = 100

The small number of phenyl ketone groups does not have much of an effect on any of the physical properties of the polymer, except on its ability to absorb light; the polymer can be moulded, foamed, etc., by the same techniques used for fabrication of conventional polystyrene. Some photodecomposition may take place before the plastic item is discarded, but this is not usually a problem because fairly extensive chain breaking is required for the physical properties of the polymer to change significantly.

The carbonyl group was chosen to effect the polymer degradation because it absorbs light in the 300–325 nm region. This is a critical region for photodegradation, because the polymer will not be degraded by sunlight passing through window glass since the latter only transmits light on the long wavelength side of 325 nm. The polymer will be degraded in the environment, because the atmosphere transmits light of wavelength greater than 290 nm. Consequently, materials made from these polymers will be stable before use, as for instance when placed in display windows, but will decompose when discarded. Cups prepared from polystyrene which contains "built in" carbonyl groups essentially disappear when exposed to sunlight for two weeks. A wettable powder is produced which gets dispersed in the environment. These small particles are eventually degraded to sufficiently small chain length to be attacked by micro-organisms and hence become biodegradable.

Many compounds have been developed that inhibit the spontaneous decomposition of polymers by reacting with the peroxides formed during processing. Recently, compounds have been devised that provide the same initial stabilizations but they accelerate decomposition. The additives are iron (III), or copper (II) complexes of dinonyl dithiocarbamates (reaction 6.17) which stabilises the polymer by reacting with the peroxide groups generated during processing. However, once the sulphur of the complex is oxidised, the metal ion is released (6.18). The free metal accelerates photochemical decomposition of the peroxides. The following mechanisms of protection and decomposition have been suggested:

Protection of Polymer

$$(R_2NCS_2)_2M \xrightarrow{ROOH} \left(R_2N\overset{\overset{\displaystyle S}{\|}}{C}-\overset{\overset{\displaystyle O}{\|}}{\underset{\underset{\displaystyle O}{\|}}{S}}-O \right)_2 M \qquad ...(6.17)$$

$$\left(R_2NC(=S)-S(=O)_2-O \right)_2 M \xrightarrow{H_2O} 2RNCS + M^{2+} + SO_4^{2-} + SO_2 + 2ROH \qquad ...(6.18)$$

Accelerated decomposition to peroxy radicals

$$ROOH + M^{2+} \xrightarrow{h\nu} ROO^{\bullet} + H^{+} + M^{+} \qquad ...(6.19)$$

$$ROOH + M^{+} \xrightarrow{h\nu} RO^{\bullet} + H^{-} + M^{2+} \qquad ...(6.20)$$

The free radicals so produce may accelerate the decomposition of polymer.

It is possible to formulate polymers of specific life times by using varying amounts of the metal complex of the dithiocarbamates. For instance, polyethylene containing 0.05 of this metal complex showed no evidence of degradation after outdoor exposure for seven months, but after ten months exhibited a 75% decrease in elongation on stretching. Stretching properties are related to chain length, the observed behaviour demonstrate extensive chain breaking between the end of the seventh and tenth months of the test. Decomposition was not observed when none of the metal complex was used.

From the molecular orbital theory we know that conjugation in molecules brings the energy levels more near one another. In other words, more is conjugation more the π-electron delocalization, consequently lesser is the energy difference between consecutive energy levels. In other words such polymer show absorption at still lower wavelength, as compared to monomer. Thus styrene-phenylvinyl ketone polymer show photo-decomposition because phenyl vinyl ketone make more delocalised π-electrons. Conjugated polymers in general should satisfy the condition and may be degraded, photochemically. The conjugation length of a polymer is the average distance between two defects which interrupts conjugation. The degeneracy of the ground state give rise to more defects as in polyacetylene (PA). Some of conjugated polymer with nondegenerate ground state are poly (p-phenylene sulphide) PPS, polypyrrole PPY, polythiophene PTP and polyfuran PFU.

PA

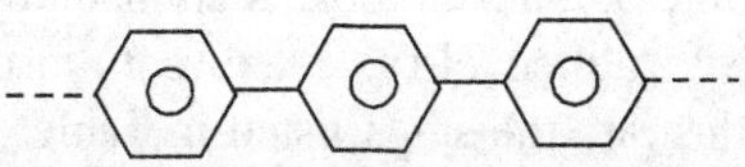

PPP

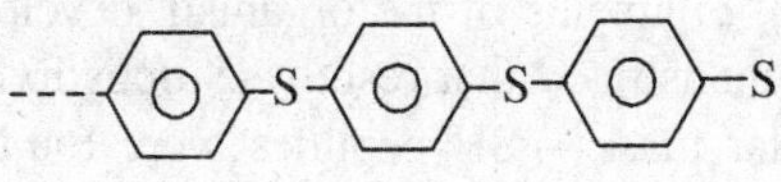

PPS

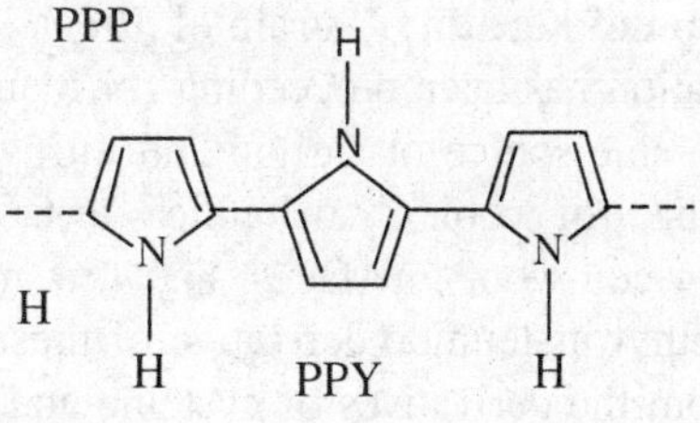

PPY

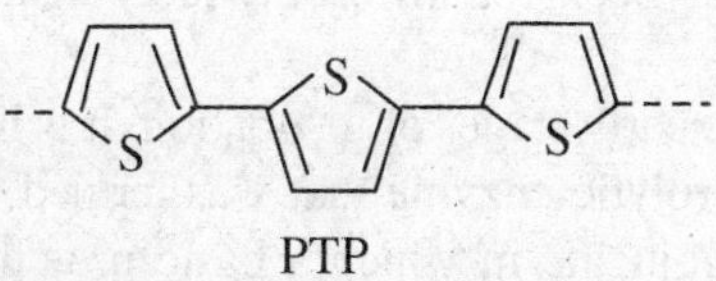

PTP

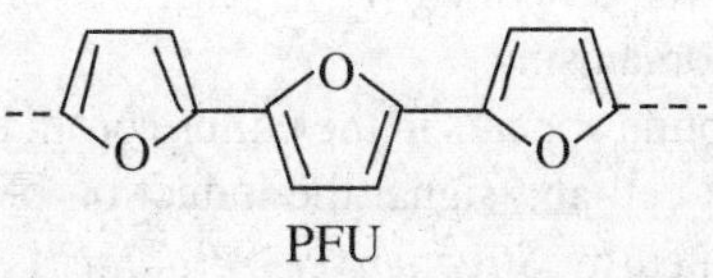

PFU

6.5 BIODEGRADATION OF NATURALLY OCCURRING POLYMERIC SUBSTANCES

Although the naturally produced polymer substances are quite different in their chemical composition, there are many points in common in the manner in which they are degraded by the microorganisms. At first glance it seems that insoluble polymer substances as cellulose, chitin, keratin, lignin and rubber cannot enter a microbial cell, hence cannot be degraded at all. Had the situation been so, these polymers should have deposited over the world for centuries and we would be knee deep in feathers and butterfly wings today. Apparently, there is some natural agency that has been degrading such substances. The question then arises how does the cell detect the presence of usable substance and secrete the adaptive enzyme that can hydrolyse the substrate? A microorganism that can attack, an insoluble substance first establishes intimate contact with the surface of the polymer. Then activity at the cell surface causes an initial attack on the polymer whereby some oligomeric material is solubilised. This soluble oligomeric material can enter the cell and serves as an inducer for the synthesis of a specific hydrolytic enzyme that is secreted into the medium. The secreted enzyme binds to the insoluble substrate and catalyses hydrolytic cleavage of specific linkages. The substrate is thus solubilized and provides a nutrient source for growth of the microorganism (Table 6.4).

Table 6.4 Stages in microbial attack on an insoluble substrate

(*a*)	Intimate contact established between microbial surface and polymer surface.
(*b*)	Activity at microbial surface, initial attack on polymer.
(*c*)	Oligomeric substances released by initial attack.
(*d*)	Oligomeric substances enter cell, induced elaboration of specific hydrolytic enzyme.
(*e*)	Hydrolytic enzyme binds to insoluble substrate.
(*f*)	Substrate is hydrolysed and solubilised.
(*g*)	Solubilised products enter cell and support growth of cell.

This analysis of how a microorganism attacks an insoluble substrate is based on studies conducted on degradation of keratin[5, 6], rubber, citrus, peel, cellulose[7, 8, 9] and chitin[10,11]. Studies on keratin degradation may serve to illustrate the steps listed in Table 6.4.

Degradation of keratin: A strain of *Streptomyces Fradiae* was isolated from soil in an area where keratin degradation has been proceeding for many years in chicken yards. This organism grew readily on wool as the sole source of carbon and nitrogen. Filaments of the organism developed an intimate contact with fibres of wool. Cytochemical tests demonstrated that reactive Sulphydryl (—SH) groups were present in cell-wool interfaces, and low molar mass —SH peptides were found in the culture broth. The N-ethyl maleimide derivative of these —SH peptides were chromatographed and found to be different from the derivatives of cysteine and glutathione[12]. Thus, the activity at the cell surface of *Streptomyces Fradiae* in contact with wool represents a type of protein disulphide reductase activity which originally was described from yeast[13, 14] and has been found to be widely distributed in plant tissues and microorganisms.

The —SH peptide appears in the culture broth in the early stages of incubation. It is likely, therefore, that they enter the cell and signal the induction of hydrolytic enzyme that was termed keratinase. The enzyme is elaborated by *Streptomyces Fradiae* as a conjugate in which a basic protein is bound to a highly acidic polymer. Keratinase conjugate binds to wool according the Freundlich adsorption equation, and peptides are rapidly released from wool by the hydrolytic action of enzyme. The rate of hydrolysis of the insoluble substrate is proportional to the amount of enzyme adsorbed on the surface of the polymer.

We see that here the manner in which an ecological relationship has evolved whereby polymer reduction in nature is balanced by polymer degradation, brought about largely by microbial agencies. But now it is apparent that man must stop dumping huge amounts of waste into rivers, must stop burning discarded rubber tyres, and in short, must stop poisoning the environment in which we all live.

Let us look at some examples of how microbial process can be employed to achieve this goal (Table 6.5). As a first principle it may be stated that if the waste material to be converted, transformed or utilized, then the microbial population employed to conduct the process must be concentrated. This condition can be achieved by carrying out the process in a fermentor in which conditions for microbial growth can be optimised.

One of the first successful efforts to combine waste disposal problem with the production of a worthwhile product was brought into operation about forty-four years ago at Rhinelander, Wisconsin. Torula yeast was produced in a continuous formation of spent liquor from the sulphite pulping process. Of the sulphite liquor solids, about 20% consists of sugars, principally pentoses, and organic acids readily utilisable by *Candida utilis*.

Table 6.5 Microbial utilization of waste materials

Waste material	*Organisms (Chemical Latin name)*	*Products*	*References*
Sulphite pulp liquor	*Candida Utilis*	Torula Yeast	16, 17
Whey	*Saccharamyces Fragilis*	Yeast lactalbumin (Wheast)	18
Potato culls and peelings	*Endocnycopsis fibuliger* and *C. Utilis*	Torula yeast	19
Feathers (Wool, hair, feathers, Hoover)	*Streptomyces fradiae*	Water soluble peptide (scleroprotein)	20
Citrus pulp and peels	*Scalerotium rolfoil*	Oxalic acid and water soluble polysaccharides	10, 11, 15
Cellulosic wastes (plants)	*Trichoderma viride* *Cellulomonase* *Cellvibrio, Clostridia* and many fungi	Glucose	7, 8, 9
Rubber	*Streptomycesopi* a filamentous fungi	Isoprene	5, 6
Lignin	Higher *basidio-mycetous* fungi	Methoxylated aromatic propane network	5, 6

The problem of disposal of whey from the manufacture of cottage cheese has now been solved. The lactose-utilising yeast *Saccharomyces Fragilis* grown on sterilised whey is recovered centrifugally and most of the lactalbumin of whey is removed with the yeast, thus providing a high-protein feed supplement. In both these case the carbon sources were water soluble sugars.

In 1956, a process was developed whereby the amylase-secreting yeasts, *Endomycopsis fibuliger* or *E. Chodati,* grew on starch waste material together with *Candida utilis* which grew at the expense of sugar solubilised by amylase secreting species. This system has also been applied on wastes obtained from peeling potatoes. In practice, urea is supplied as a nitrogen source and the fermentor is inoculated with equal numbers of each species. After incubation for 24 hours, the cell population consists typically of 1425×10^6 *C. utilis* cell per ml and 175×10^6 *E. fibuliger* cells per ml. Since the process of peeling potatoes entails 20–30% loss by weight, it is obviously necessary that an efficient and constructive solution to the waste disposal problem must be employed.

As a by-product of the poultry industry, vast quantities of chicken feathers are produced annually. Feathers consist largely of keratinaceous protein. As the feathers are largely wasted, they present a waste disposal problem. A large batch of chopped feathers was added to mineral salt medium in a 1000 litre fermentor. After autoclaving, the fermentor was inoculated with a string of *streptomyces fradiae.* After incubation for 36 hours, the feathers were solubilised completely. The broth then was filtered and spray-dried; the product was a light coloured powder (60% yield) consisting principally of water soluble peptides. It proved satisfactory in chicken diets (supplied ~25%) of the total protein in diet. The powder also proved suitable for chemical use as a cheap source of peptides.

Another solid waste disposal problem is created by modern food technology because of the fondness of many people for orange juice. At present, the only use for the vast tonnage of orange peels is cattle fodder. For some time we have been interested in a plant pathogen, *Sclerotium rolfoil,* that is able to solubilise a wide variety of plant materials. When inoculated into a suspension of orange peels in tap water, the fungus grew rapidly and the peels were solubilised within 36 hours and gave us a number of worthwhile products like oxalic acid and polysaccharides. From the foregoing examples of microbial utilization of organic wastes and from the principles in Table 6.4 on the stages involved in microbial attack on an insoluble substrate, the message emerges clearly—**keep waste organic matter away from aquatic environment**.

Let us consider what happens when certain organic wastes are not subject to microbial utilization, but are dumped into rivers. *Saccharomyces fragilis* is not commonly distributed in nature and, in any event, could not compete with enteric bacteria for utilization of lactose. Therefore, if large quantities of whey are dumped into a river, the growth of enteric bacteria present through faecal contamination is encouraged greatly. It is also worth noting that lactose-positive enteric pathogens are being isolated with increasing frequency. If sulphite pulp liquors are dumped into streams, a variety of bacteria including *Escherichia coli, Aerobacter aerogenes* and species of *Pseudomonas* utilises the pentose components of the waste. The two examples just given illustrate the fact that dumping concentrated metabolisable organic wastes into an aquatic environment serves to increase the frequently undesirable microbial population.

6.6 DISPOSABLE SYNTHETIC POLYMERS

Highly branched hydrocarbons were especially resistant to microbial attack, and substances containing terminal neopentyl groups were almost completely resistant. Such studies indicated that resistance of alkyl benzene sulphonate detergents to microbial degradation resided in the alkyl chain. Detergents with straight, unbranched alkyl groups proved to be biodegradable.

Although the total output of high-polymer synthesis by the chemical industry of the world may be small in comparison with the total synthesis of polymers by plants on land and in the sea, nevertheless, vast tonnage of synthetic polymers are intended for disposal each year. Nylon, polyethylene, polypropylene, and most other synthetic polymers are not subject to microbial degradation and accumulate when disposed as waste. It is possible that a strain of microorganism may evolve that can attack such materials. But we have only to consider the millions of years that have elapsed after the appearance of lignin before a group of fungi evolved that was capable of attacking the polymer is, from which our fossil reserves of coal developed. Several years ago, (Late) W.J. Nickerson drew attention to the prospect of the world being covered at some future date with a Saran Wrapping[15]. The use of disposable plastics has increased greatly in recent years and is a growing cause of concern. To make matters worse, we do not have the least notion of the ground rules that govern susceptibility or resistance of a synthetic polymer to microbial attack. For example, the resistance of nylon to microbial attack inexplicable at present. It is a linear polyamide without substituents (see Table 6.1). Although many bacteria can elaborate amidases, apparently none can attack nylon. At the opposite extreme, the amide linkage in penicillin a most improbable molecule that gave rise to fermentation technology and that probably is never produced under natural conditions is hydrolysed readily by amidases elaborated by a variety of microorganism (Fig. 6.1).

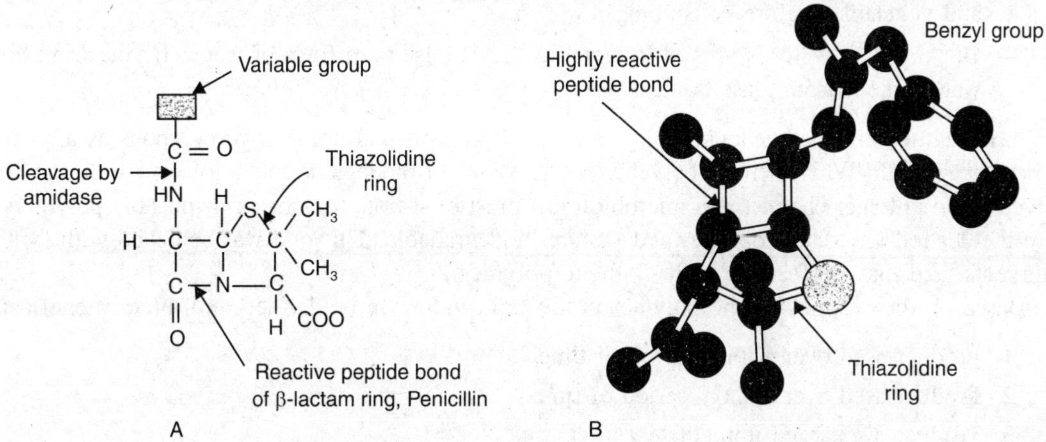

Fig. 6.1 Cleavage of the penicillin molecule by amidase.

Just having a particular linkage does not imply that a molecule may be biodegradable. Biodegradation of polymer-molecules is a configuration based process. This will be explained with the help of polymer amylase, a linear polymer made up of glucose as repeating unit with α-D1-4 linkages. In aqueous amylose solution, glucose has chain configuration and polymer has random coil configuration. To this if we add enzyme α-amylase, amylose is degraded to maltose and glucose. To amylose solution if we add KI_3, a blue amylose-I_3^- complex is obtained. In this complex, glucose has boat configuration that form a helical structure around I_3^- ions. Now to this if we add amylase enzyme, biodegradation does not take place. Presence of I_3^- is not the cause, because if we heat the amylose-I_3^- complex, near about 60–70°C it decolourises, when glucose goes to chain form and polymer in random coil confirmation. Now to this solution if we add amylase, biodegradation take place even in the presence of I_3^-. It shows that biodegradation is configuration based process. This may amount to that polymer and enzyme should be in a very ultimate contact with one another, only then biodegradation can take place.

It is noteworthy that chemicals and plastics in common use may be having a insidious effect on the human body—not just causing cancer but also affecting the hormonal system, damage the reproductive system, affecting foetuses and, thus the future generation, increasing childhood hyperactivity and leading perhaps to a decline in intelligence. Some of these chemicals termed as endocrine disrupters are:

Bis phenol A—leaches from polycarbonate water jug and plastic linings of food cans.

Nony phenol—a compound added to PVC plastic and is also a break-down product of industrial detergents and pesticides.

Polychlorinated biphenyls (PCB)—found in electric transformer.

Dioxins—a by-product of near ubiquitous combustion including waste incineration; vinclozolin which is a common fungicide that reach human being as a pesticide residue.

Laboratory experiments indicate that breakdown products of certain plastics including some used for storing food, can speed up the growth of cultured breast cells much like estrogen.

So what should we do: Stop microwaving food in plastic containers.

— Think twice before breast feeding the new born children, because pesticides tend to accumulate in breast tissues. Our food has a high pesticide residue. Despite the low consumption of pesticides on a perceptual basis, people in Delhi have, nonetheless, revealed world's highest level of DDT and other pesticides in human body fat and in mother's milk. So wash the fruits and vegetables before consuming it.

— Our drinking water is also of low quality. Try to use pure form of water. If you drink bottled water take it from glass bottles.

In the same vein, tremendous number of microorganisms can utilise hydrocarbons as a source of carbon for growth. Why, then, is polyethylene resistant to biodegradation? However time has surely arrived when polymer chemist and microbiologist must cooperate to achieve disposable polymers that have the desired physical properties and yet are biodegradable. This end was achieved with synthetic detergents, and may be feasible with synthetic polymers.

Based on their longevity, the polymer products in India can be divided into three categories:

1. Product used over a long period of time
2. Product used over a short period of time
3. Product discharged immediately after use.

8 per cent polymer products have a long life span as they stay in use for years. Some examples include plastic panels, pipes, room dividers, furniture boxes, crates, buckets, tubs, water tanks, crockery suitcases, briefcases, minibarrels, doors, windows, sandles, baskets and P-traps for rural sanitation, plastics used in radio, T.V. sets, cars, blood transfusion sets and counting vials used by atomic energy departments lasts for almost a life time.

82 per cent polymer products, lasting for a week to three months, fall into this category. Goods like oil, acid, acetone, alkaline cans, cosmetic jars, squeezable tubes, detergent and sanifresh bottles, etc., are some of the examples of this category. 10 per cent polymer items are discarded immediately after use. Examples are the multicoloured plastic carry bags, soft drink glasses, coffee cups, mineral water bottles, some soft drink bottles, syringes, blood vials, and explosive shells. Thousands of such items form a major proportion of the garbage of residential colonies, educational institutes, bazars, hospitals and industrial areas.

6.7 POLYMER RECYCLING

All synthetic products affect the environment over their life cycle from the point of manufacturing to disposal of wastes. Plastics are no exception. Plastics has infiltrated all sectors of our economy. From 1.88 million tones (1995–96) the domestic demand for plastics was nearly 4 million tonnes by year 2001–02 and by 2004–05 it rose to 10 million tonnes. India's plastic wastes recycling industry presently handle plastics valued around Rs. 2550 crores, Rs. 6400 crores by 2001 and Rs. 14000 crores, 2004–05.

Phenomenal increase in the use of plastic by the masses in our country started in 1980. There were no cling films to wrap sandwiches before this time. There were no disposable ball pens. Milk was not sold in sachets. Noodles, potato chips or other fancy looking things were not available. At present, there would not be a single household in an urban area which is not discarding some plastic trash every day.

In most colonies the ragpickers collect most of these articles. These ragpickers either are employees of some big *kabari* or they work on their own. Ultimately the plastic trash from rag picker and local *kabari* reaches the big *kabari* who in turn sells all this to a wholesaler. From the wholesaler the plastic trash reaches the plastic recycling units. This whole cycle is shown in Fig. 6.2.

The rate of profit increases with finer segregation, quantity and quality from item to item at each level of the chain.

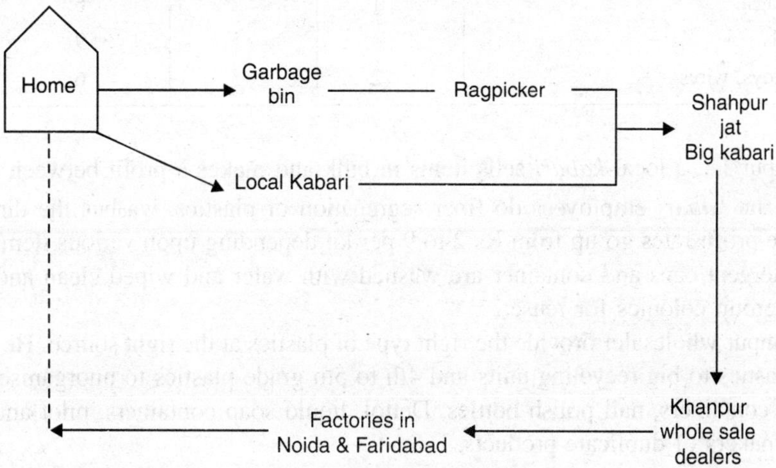

Fig. 6.2 Plastic Chain in Delhi.

At the household level, the local *kabari* takes them without weighing and pay unreasonably low. The rates given in Table 6.6 are applicable for items sold in bulk.

Table 6.6 Purchase of sale of plastic waste in the chain (of 1997)

(Rs. per kg)

Nos.	*Items*	*Local Kabari*		*Shahpur Jat Kabari*	*Khanpur wholesale dealer*
		Buys	*Sells*	*Sells*	*Sells*
1.	Pepsi glasses, coffee cups				
2.	Soft drink and Mineral water bottles	5	7	10	18
3.	Fevicol can				
4.	Jugs, mugs, buckets (white)	30	32	38	45
5.	Jugs, mugs, buckets (coloured)	12	14	18	22

(Contd.)

Nos.	Items	Local Kabari		Shahpur Jat Kabari	Khanpur wholesale dealer
		Buys	Sells	Sells	Sells
6.	Plastic hangers	10	12	15	20
7.	Shampoo bottles	11	12	18*	—
8.	Plastic suitcase	12	14	18	25
9.	Milk sachets	4	5	14	22
10.	Cooking oil, acid cans in good condition	6–8	8–14	—	
11.	Detergent cans				
12.	Carry bags	1	2	6	6
13.	Chappals	4	5	9	10
14.	Plastic toys, wires	2	3	6	8

*Resold

(*i*) At Shahpur Jat, a local *kabari* sells items in bulk and makes a profit between Rs 1-2 per kg.

(*ii*) Shahpur Jat *kabari* employees do finer segregation of plastics, washes the dirty items. As a result the profit rates go up from Rs 2 to 9 per kg depending upon various items. Cooking oil, acid, detergent cans and container are washed with water and wiped clean and sold to lower income group colonies for reuse.

(*iii*) The Khanpur wholesaler provide the right type of plastics at the right source. He sells 1st to 3rd grade plastics to big recycling units and 4th to 5th grade plastics to unorganised illegal units. Lipstick containers, nail polish bottles, Dettol, liquid soap containers, pilot and ball pens are sold to makers of duplicate products.

Table 6.7 The plastic business

Category	*Total No.*	*Area of operation*	*Earnings/day*
Rag pickers	3.5 lakhs	Garbage dumps	Rs. 20–40
Unskilled personnels	1.8 lakhs	Mini units	Rs. 50–70
Semi-skilled staff	60 thousands	Middles and big units	Rs. 70–100
Skilled staff	20 thousands	Middle, big and mega units	Rs. 200–400
Managerial staff	2500	Big and mega units	Rs. 500–1000

In Delhi at a small colony Tri Nagar, recycling of plastic is being done in one room house, where a family of 5 or 6 people live. They buy 4th grade plastic trash of PVC put it in a small machine. The colours added are invariably carcinogenic. The machine is operated for 12 to 15 hours a day and the family members work on it in shifts. From this machine the plastic grains pop out like pop corns. During the entire process noxious fumes of HCl, Cl_2, HCN, CO, Dioxins are emitted, which are inhaled by new borns, pregnant women, young and elderly people of the family all the time. After all these hardship, the family's monthly income is not more than Rs. 4000. In 1991 Delhi alone had around 53400 such plastic recycling units.

At Rampura, these grains are used to make carry bags. Plastic grain are bought from Sadar Bazar at the rate of Rs. 100 to Rs. 180 per kg depending on the grade. From one kilogram of low-grade grains, 400–500 thin plastic carry bags are made and from one kilogram of high grade grains about 200 thick plastic bags are made.

Small machines are fixed in one-room houses. The grains are fed and plastic bags are extracted. The thickness of the bag can be controlled by a knob. As the grains melt in the machine, noxious fumes fill the house. These families have a monthly income Rs. 4000 only. The workers and their families suffer from skin allergies, bronchial diseases and eye infections due to inhalation of fumes.

On heating PVC releases HCl which reacts with oxygen and produces gaseous chlorine. Chlorine reacts with carbon to release different kinds of polychlorinated hydrocarbons including polychlorinated dibenzofurans and dibenzodioxins. The most toxic of all these is 2, 3, 7, 8 tetrachlorodibenzo-*p*-dioxin (2,3,7,8 TCDD). Death from acute exposure to dioxin has never been observed. Global human exposure has been estimated to be about 120 pg/day expressed as 2, 3, 7, 8 TCDD toxin equivalents. Inhalation of PVC smoke leads to irritation, choking of respiratory tract leading to hypoxia. Therefore the most important initial treatment goal is to ensure adequate tissue oxygenation. The less exposed person generally suffers from a transient effects, malaise, myalgia that can last a few days and minor respiratory irritation. People working in PVC smoke should run at least two kilometres after their shift in open ground. This will provide excess oxygen required to the tissues of respiratory tracts and may save them from some of the ill effects of dioxins.

Dioxin is one of the most toxic chemical known to human kind. The vulnerable daily intake (TDI) has not been set as yet but lies between 5–10 picogram per killogram of body weight. Dioxin emission in air of most Japan urban area is said to be highest in the world. It leads to cancer, reproductive disorder like birth defects, reduced immunity and inhibits the formation of foetal genibals. Cancer was the biggest killer in Japan in 1998, all because 80% emission come from PVC polymer garbage.

Our field visits to recycling/reprocessing units and waste dealers have brought to light the need for upgradation of the working conditions of operations as also the recycling technology. Also it is important to pay some attention to the social status of ragpickers and waste collectors who contribute towards clearance of plastics waste from public places and thus play a key role in the environmental management of plastics waste. **Is it not the duty of the society, social system to care for such silent workers.**

Plastics have become a symbol of change of our society to throwaway society. They are nonbiodegradable but recyclable, to some extent. With the technological advances, plastics recycling is economically feasible for plastics packaging. Recycled materials compete with original materials in terms of price and performance. It is an established fact without reuse, total/ absolute diversion of material from waste stream is impossible.

To enhance the demand for recyclable materials, various mechanisms and options are to be assessed these include.

(*i*) Plastic products of which various materials are made up of should be either recyclable or biodegradable, no other polymer should be used.

(*ii*) User charge a tax to ensure that companies bear the cost of solid plastic waste they produce.

(*iii*) Certain percentage of government purchased products be made of recyclable polymer or made of recycled material or price preference on items containing recycled material be encouraged.

(*iv*) Maintenance of recycling standard: Recycling standard will provide mechanism for coordinating recycling activities and for establishing broad based recycling infrastructure. This recyclable material must be labelled appropriately and must reach a specified recycling site.

(*v*) The recycled material of first grade should be used and in respect of second grade material, it should be a blend of 50:50 (recycled: virgin). However third grade recycled material use should be discouraged for the manufacture of carry bags. To sustain the use of plastic through carry bags, certain gradation and acceptable consumer quality must be insisted and adhered to by the manufacturer.

(*vi*) Carry bags and containers made of recycled plastics conform to the specifications mentioned in the prevention of Food and Adulteration Act, 1954 and the rules made there under (*a*) such carry bags and containers are not pigmented (*b*) the minimum thickness of carry bags made of recycled plastics shall not be less than 25 microns (*c*) reprocessing or recycling of plastics is undertaken in accordance with the Indian Standards (IS) 14534 (*d*) the minimum thickness of carry bags made of virgin plastics should not be less than 20 microns.

(*vii*) Passenger airlines, hotel and restaurants generate lot of plastic wastes. This waste be identified, graded source and sent for appropriate use or recycling.

(*viii*) The third grade PE or PP plastics recycling should be banned. Such plastics so generated should be resorted to energy recovery, a technology that is encouraged world over. This plastics can be burnt to produce electricity with no residue ash or particulate matter emission. This waste can also be used for composites and volume applications like lumber, etc. These can act as a substitutes for wood and concrete products.

(*ix*) PVC resin is consumed for bottles and used bottles are collected back for recycling. However PVC polymer should be completely banned because of two reasons (*i*) every stage of recycling is associated with the emission of highly toxic gases like dioxins, HCl, (*ii*) this polymer cannot be burnt and thus cannot be destroyed.

(*x*) Polyethylene tetraphthalate (PET) is extensively used for making mineral water/liquor bottles. Because of durability and glass like clarity PET bottles and containers are longterm assets for consumers. However 50% of these bottles are available for recycling. An organised collection system of PET bottles is required. The recycling capacity be fully utilized and appropriate product application are to be identified and promoted in India.

6.8 CARRY BAGS—A MENACE

On an average in Delhi, eight such bags reach a middle class household everyday. Each bag weigh 2 g and constitute only 2% of the litter, but of all the nonbiodegradable products these bags are most visible. As most of these bags (nearly 90%) are made of 4th to 5th grade plastics scrap, the capacity for recycling is only 0.5%. Clean bags are sold at only one hundred rupees per kilogram or almost 450 carry bags. The extreme light mass of these bags demands that a huge quantity has to be collected for a paltry sum, makes them least attractive for the ragpickers. That is why they keep lying on ground, thus choking the grass and land mass. Many of them find their way in the open sewers, where they block the waste water flow. As the sewers get blocked, methane gas produced in the sewage underneath cannot escape and this resulted in series of explosions in the under ground sewage network on March 23, 1995 in New Delhi.

Most of the carry bags are used for eatables. The pigments used for colouring these plastic bags are mostly cadmium based. Invariably these dyes leach on to the food stuff. Curds become green in colour and Rasmalai (sweets made from milk) turns pink by the time consumers bring them home. This is slowly poisoning the masses.

While consumers of the developed countries are reducing their plastic consumption, we like gullible fools are lapping up plastic waste and literally choking ourselves with it. Unless we the consumers, start saying 'no' to the use of plastics, the producers would continue to entice us and flood the market with carry bags. Today it blocked sewers; who knows, tomorrow we might be completely buried into nonbiodegrable heaps. The government is well advised to ban the manufacture of plastic carry bags and ideally we all must go back to the traditional Indian cotton bags.

REFERENCES

1. Flory, P.J., *Principles of Polymer Chemistry,* Cornell Univ. Press, New York (1953).
2. Odian, G., *Principles of Polymerization,* 2nd Edition, John Wiley, New York (1981).
3. Billmeyer, Jr. F. W., *Textbook of Polymer Science,* 3rd Edition, John Wiley, New York (1984).
4. Misra, G.S., *Introductory Polymer Chemistry,* Wiley Eastern Limited, New Delhi (1993).
5. Noval, J.J. and Nickerson, W.J., *J. Bacteriol,* 77,251 (1959).
6. Nickerson, W.J. and Noval, J.J., *U.S. Pat*., 2,988,487 (1961).
7. Selky, K. and Maitland, C.C., *Biochem, J.* 104716 (1967).
8. Mandels, M. and Rease, E.T., Dev. *Indust. Microbiol,* 5, 5 (1964).
9. Norkrans, B., *Adv. Appl Microbial.*, 9, 91 (1967).
10. Berger, L.R. and Reynolds, D.M., *Biochim: Biophysic Acta*, 29522 (1959).
11. Jeuniause, C., *Arch. Intern Physiol, Biochim,* 65,135 (1951).
12. Noval, J.J. and Nickerson W.J., in Symposium in Sulphur in Proteins, p. 55 (R. Benesch ed.) Academic Press, New York (1959).
13. Nickerson, W.J., Falcone *G., Science*, 124,318 (1956).
14. Nickerson, W.J. and Falcone G., in Symposium on Sulphur in Proteins, p. 409 (R. Benesch ed.) Academic Press, New York (1959).
15. Nickerson, W.J., *Ind. Eng. Chem.,* 48, 1411 (1956).
16. Inskeep, G.C., Wiley A.J., Holderby J.M. and Highes L.P., *Ind. Eng. Chem.,* 43, 1702 (1951).
17. Wiley, A.J., *Industrial Fermentations,* (L.A. Under bofler & R.J. Hickey. eds.) Chemical Publishing Co., New York 1954, Vol. 1, p. 307.
18. Wickerham, L.J. and Kuehner C.C., U.S. Pat., 2,764, p. 487 (1956).
19. Tveit, M.T., U.S. Pat., 3.105, p. 799 (1963).
20. Nickerson, W.J., *Fermentation Advances,* p. 64 (D. Perlman ed.), Academic Press, New York (1969).

7

Fossil Fuels and Petroleum

7.1 NATURE OF PETROLEUM

Crude petroleum is a complex mixture of hydrocarbon that are separated into fractions of differing boiling ranges by the refining process. The various fractions either are used directly as such or are modified chemically to yield more efficient energy source. In addition, they may be converted to petrochemical compounds such as polymers and detergents. Aluminium, calcium, copper, iron, chromium, sodium, silicon and vanadium are found in almost all samples of petroleum in the 0.1–100 ppm range either as carboxylic acid salts or as porphyrin chelates. Lead, manganese, barium, boron, cobalt and molybdenum are observed occasionally.

The origin of petroleum lies in plants and animals which lived on earth and in the sea many million years ago. These organism died and their remains became buried under the earth. Due to bacterial action and under the action of earth's heat and high pressure, these remains were converted to liquid hydrocarbons, namely petroleum. The gaseous hydrocarbons that were produced, constituted natural gas. The evolution of life and its conversion to petroleum is depicted in Fig. 7.1.

7.2 THE AUTOMOBILE FUEL

The low-boiling fraction of crude petroleum obtained by distillation is called gasoline and is a source of energy in the automobile internal combustion engine. The fuel mixture of gasoline vapour and air burns within the working cylinder of most automobiles driven by a four-stroke engine Fig. 7.2. Gasoline is mixed with air in the carburettor and drawn into the cylinder during the first stroke. The air-gasoline vapour mixture is compressed on the second stroke and ignited by the sparking plug. The expanding gases formed by combustion then drive the piston down on the third stroke that provides power to drive the car and the combustion product are forced out of the cylinder during the fourth stroke. The engine gives maximum efficiency when gasoline-air mixture is highly compressed before firing in the cylinder. The modern engines squeeze this mixture to one seventh or even to one eighth of its normal volume.

As the compression ratio increase, a point is reached when knocking is observed. On sparking, all the fuel does not burn smoothly and the end portion burns with explosive violence giving rise to metallic rattle called knocking. This results in loss of power. The loss of power defines the rating of a fuel by an octane number. Higher the octane number, less is the rattling. Thus pure linear alkane gasoline has an octane number of 68 only.

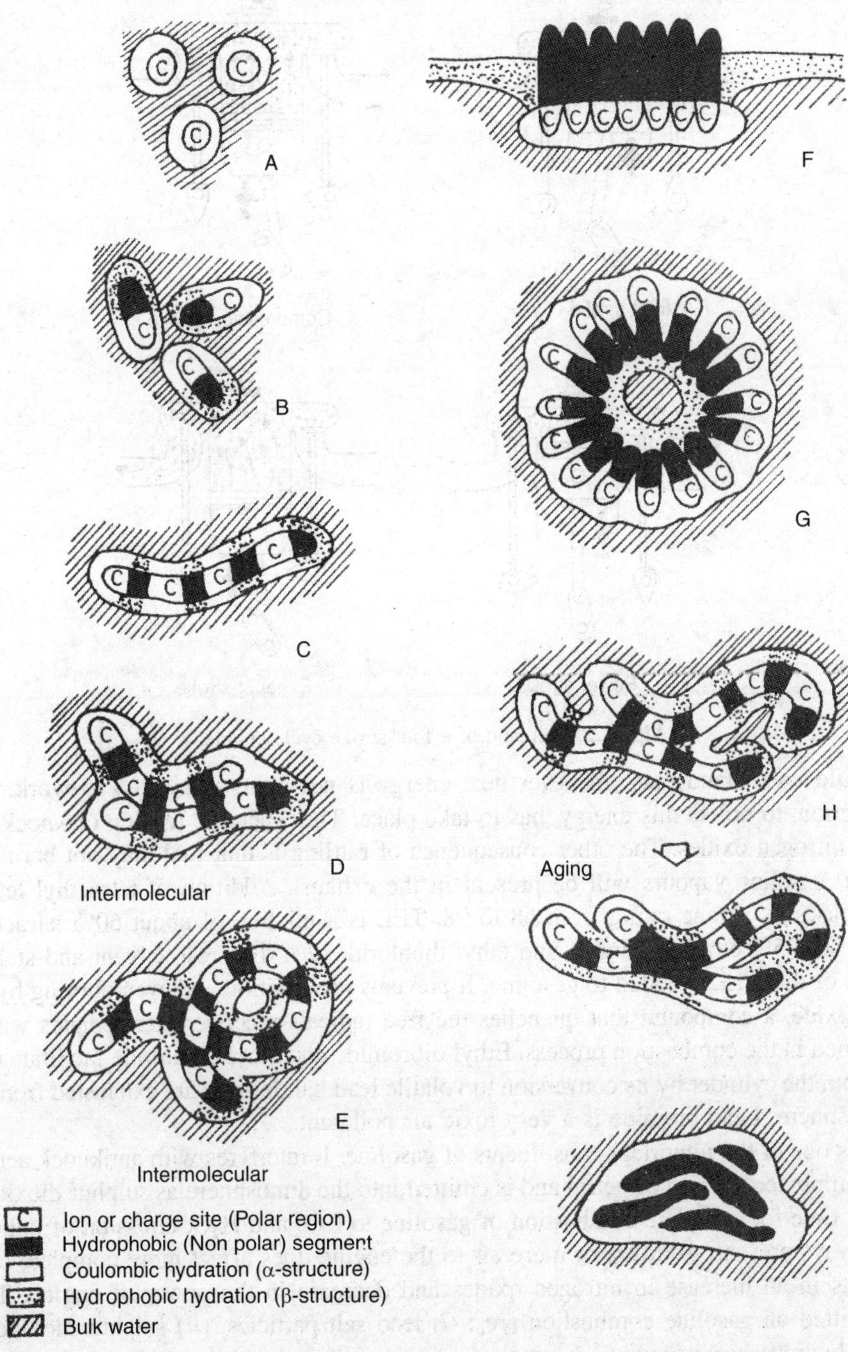

Fig. 7.1 Evolution and hydration types. A. Hydrophilically hydrated ionic constituents of the primordial ocean. B. The simple amino acid building blocks with their hybrid hydration envelops. C, D, E. Building up of macromolecules. F, G. Formation of films and enclosed structure. H. The aging process, entailing the destruction of hydrophilic sites and subsequent cross linking and dehydration of the biopolymeric material. I. The end product of it all hydrophobic petroleum.

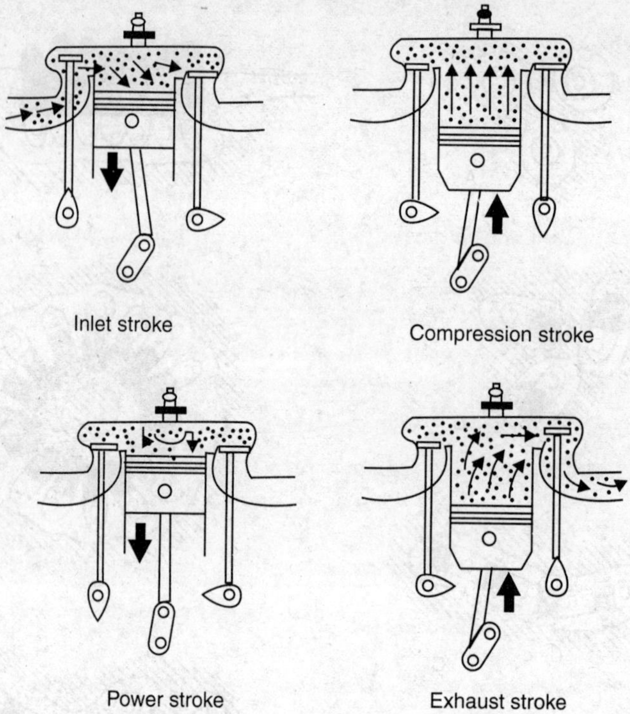

Fig. 7.2 Schematic of a four-stroke cycle engine.

One should understand that whenever heat energy is not being converted to work, some other chemical reaction, to utilise this energy, has to take place. Thus metallic rattling or knocking leads to formation of nitrogen oxides. The other consequence of rattling is that fuel may not burn completely, hence carbon, gasoline vapours will be present in the exhaust. Addition of tetraethyl lead, TEL, to gasoline increases its octane rating from 68 to 98. TEL is a mixture of about 60% tetraethyl lead or tetramethyl lead, 35% ethyl dibromide and ethyl dichloride, 2% dye, rest solvent and stabiliser, 0.66 gram per litre of this TEL is added to gasoline. It prevents pre-ignition and even burning by its conversion to lead oxide, a compound that quenches the free radicals reactive intermediates with impaired electrons formed in the combustion process. Ethyl dibromide and ethyl dichloride facilitate the removal of the lead from the cylinder by its conversion to volatile lead halides that are exhausted from the engine into the atmosphere. Lead bromide is a very toxic air pollutant.

Sulphur is one of the important constituents of gasoline. It interferes with antiknock action of TEL or even with improved version of petrol and is emitted into the atmosphere as sulphur dioxide. As there is insufficient time for complete combustion of gasoline to CO_2 and H_2O, emission of unburnt hydrocarbons is always present. Addition of more air to the engine does affect more complete combustion, but this results in an increase in nitrogen oxides and decrease in the power of engine. The various pollutants emitted on gasoline combustion are : (*i*) lead salt particles, (*ii*) sulphur dioxide, (*iii*) TEL vapours, (*iv*) hydrocarbon vapours, (*v*) nitrogen oxides, (*vi*) carbon monoxide, (*vii*) carbon dioxide, (*viii*) water vapours and (*ix*) carbon particles. (*i*), (*ii*) and (*iii*) are removed with the reformation of fuel while the others are removed by introducing a catalytic convertor and modification in the engine.

Lead Pollution: These will be explained with Delhi as a specific example. Gasoline consumption in Delhi in 1985–86 was 2.5844×10^9 litres, in 1989–90 it went up to 3.9657×10^9 litres and in

1994–95 it went up further to 4.10664×10^9 litres. Furthermore, 0.15 g l^{-1} of lead is assumed to be added (which is equivalent to 0.66 g l^{-1} of TEL) though in the earlier year it was more than double this amount. Assume that 50% of this gasoline is burnt outside Delhi and the rest in Delhi. The population of Delhi is taken to be 1,00,00,000.

The safe limit of lead in our system is 25 micrograms per litre. Our system does not need lead yet everyone absorbs it in varying amount. It is not excreted from our system but is stored over many years in our tissues, chiefly in the bone marrow from where it released into the blood stream to wreck the cellular system. It can damage the nervous system, kidneys of adults and increase in hypertension. In pregnant women it enhances chances of miscarriage, premature delivery and stillbirth. Children are susceptible to a range of physiological, neurological, kidney and blood disorders including partial hearing loss, loss of learning abilities.

On burning 1000 litre of petrol, 0.1198 kg of lead chlorides are thrown into the air as particulate matter. For the three years under consideration the figures are as follows:

1985–86	531 kg/day	=	53.1 mg/day/person.
1989–90	815 kg/day	=	81.5 mg/day/person.
1994–95	843 kg/day	=	84.38 mg/day/person.

The amount for 1994–95 is equivalent to 84.38 mg/day/person. Even if 99% of this lead salt were to just fall over plants or soil and 1% lead in air at all times must be 8.43 kg.

The concentration of lead in ambient air in research journals are expressed in micrograms per cubic metre (μ g m^{-3}). The area of Delhi state is 1483 km^2. Particles dispersion in air are found up a height of 20 metres. So the total volume of air containing lead = $1483 \times 20 \times 10^6 = 29660 \times 10^6$ m^3.

So the lead concentration in the μ g m^{-3}

$$= \frac{8.435 \times 1000}{29660 \times 10^6} = 0.2841 \mu g m^{-3}$$

The concentration of lead in ambient air is found by a special arrangement which amounts to collecting a sample of air and be allowed to stand for 4 hours. The concentration of lead in dispersed form in this air is estimated. According to a January 1996 World Bank report, Delhi had a lower lead level at 0.2775 μ g m^{-3} which agree with other calculated value of lead.

Since a man inhales 13.5 kg of air or a volume of 10.586 m^3, every day lead taken into lungs will be $0.2841 \times 10.587 = 3.008$ μ g. As lungs extract no more than a fifth of oxygen in air that enters them, the proportion cannot be higher for lead even. So lead retained by lungs = 3.008/5 = 0.6016 μ g/day. Since a healthy man has 40 to 45 dl of blood, concentration of lead in terms of microgram per decilitre (μ g dlt^{-1}) = 0.6016/40 = 0.01504 μ g/dlt/day. This lead that enters our blood or body does not leave it till we die.

The average international level of lead allowed in blood is between 10 to 15 μ g/dlt, the lower limit being valid for children. At 30 times that limit, we would all be quite dead. The safe limit may be reached in 10/0.01504 = 665 days = 1.82 years. Since the concentration mentioned above had been prevailing for the last many years, every Delhi citizen must have a very high concentration of lead.

However, DTC bus drivers, Red line and Blue line bus drivers who work for 18 hours a day, bus conductors and shopkeepers in highly congested area, one may be inhaling more lead all the times. In other words, these people must have crossed the safe limit in 66 day. This is frightening for its inexorable corollary is that rising accumulation of lead will significantly reduce life spans in the city and will severely damage the intelligence, health, and reproductive capacity of those who are born today. Since the rise in vehicle population in the city only began in earnest in 1992, these effects have still to show themselves. But by the time they did, it would be too late to do anything to prevent the havoc they will wreak.

Grass, vegetables grown around Delhi, and the soil, must be having a high lead concentration which also enters our systems. As lead salts are mostly insoluble, Delhi cannot get rid of the lead that had been falling over it for at least four to five decades. In such a situation, addition of lead to petrol had to be stopped.

Sulphur Pollution: In 1959, Yokkaichi, a Japanese petrochemical industry, went into operation. The local residents then began complaining of foul odours and irritating fumes. They also protested that dirty air was soiling their laundry. Soon there was a marked increase of asthma in the city, which the citizen started to term Yokkaichi Asthma. Later it was confirmed that the principal cause of this asthma was the sulphur dioxide released from the smoke stacks of petroleum. The victims took those responsible to the court, where the case was decided in their favour and substantial compensation payments were an order. Sulphur dioxide also cause acid rains and may lead to formation of smog and haze.

Reformation of Gasoline: Following steps may be taken in the order of priority.

(*i*) Sulphur should be removed completely from gasoline for which several processes are available.

(*ii*) On passing crude naphtha vapours over Pt-Alumina as catalyst at 770°C, aromatic hydrocarbon formation take place. This sulphur free gasoline is blended with benzene, toluene and xylene (BTX) with octane number 104 which is more than octane rating of lead gasoline. Such gasoline can be used without TEL and does not show knocking effect. However, exhaust of such a gasoline contains benzene which is air toxic.

(*iii*) In USA, unleaded gasoline was introduced in 1974 to cater for cars equipped with catalysts. Leaded gasoline has been banned in the USA from January 1995. The final version of modified gasoline used in the USA is given in Table 7.1.

(*iv*) To gasoline 2% Methyl tertiary butyl ether (MTBE) is added. It enhances the octane number and enhance combustion quality.

Table 7.1 The final version of modified gasoline used in the USA

Parameters	*Producer limit*	*Averaging limit*	*All gasoline limit*
Volatility (Psi)	7.0 psi/48.3 kPa	–	7.0 psi/48.3 kPa
Sulphur (ppm)	40	30	80
Aromatics (Vol %)	25	22	30
Benzene(Vol%)	1.0	0.8	1.2
Olefin (Vol %)	6.0	4.0	10.0
Oxygen (Wt%)	1.8–2.2	–	2.7 Max 1.8 Min
T50°F(°C)	300(149)	290(143)	320(166)
T90°F(°C)	210(99)	200(93)	220(109)

The effect of different properties of modified gasoline performance are given in Table 7.2.

Presence of oxygenates, e.g., ethanol, methanol, MTBE, ETBE, TAME, etc., in gasoline are advantageous as the oxygenates help in reducing carbon monoxide emissions especially from older vehicles and they are excellent octane improvers.

In USA, aliphatic alcohols, glycerols, ethers and polyethers are added to the fuel provided that the amount of oxygen in the finished fuel does not exceed 2.7% weight. Ethanol and ETBE are excluded from this regulation. These fuels are as follow:

- Gasohol, consisting of gasoline with 10% volume of ethanol is permitted. This contains 3.7% wt of oxygen.
- A mixture of TBA and methanol up to a maximum concentration of 3.7% oxygen provided that methanol does not form more than 5.0% of the mixture.
- Methanol up to 5% volume, plus at least 2.5% of volume co-solvent (ethanol, propanol or butanols) plus corrosion inhibitor with maximum oxygen contents 3.7% weight.

Table 7.2 Effect of properties of reformed gasoline on engine performance emission and deposits

Property	*Effect on engine performance,emission and deposits*
Reid Vapour Pressure (RVP) (psi)	Lower RVP reduces vehicular evaporative running loss and refueling emission.
Aromaticity	Lower aromatics reduce the reactivity of emissions, produce less engine deposits and contribute to lower benzene emissions.
Benzene	It is air toxic.
Olefins	Lower olefins reduce the reactivity of the emissions and produce less engine deposits.
Oxygenates	Oxygenates help reduce carbon monoxide emissions and they are excellent octane improvers.
Sulphur	Lower sulphur improves catalytic converters performance.
90% Distillation Temperature	Lowering the 90% point should help reduce hydrocarbon emissions and engine deposits.
Derivability Index (Function, T_{10}, T_{50} T_{90})	T_{10}, T_{50}, T_{90} must all be balance to provide a good derivability index.
Deposit Control Additives	Deposit control additives are necessary to prevent intake system and combustion chamber deposits which increases emissions and impair vehicle driveability.

Since 1979, the use of ethanol has been mandatory in Brazil through their government *Proalcohol* programe. Specially designed vehicles were produced in Brazil to operate on ethanol fuel. However, such a fuel produced formaldehyde in the exhaust which in turn, is highly toxic. Also the addition of alcohols to gasoline leads to emission of formaldehyde which can be removed by using catalytic converters. So, alcohol fuel should be used only if a catalytic converter is used.

7.3 EMISSIONS FROM TWO-STROKE ENGINE

Mostly the crankcase scavenged two-stroke engine is used to power the Indian two- and three-wheelers and is shown in Fig. 7.3. In conventional two-stroke engines the fuel-air mixture from the carburettor enters the engine crankcase, where it is compressed by a factor of 1.1 to 1.3 and during scavenging period enters the engine cylinder via transfer ports. The fresh fuel-air mixture in the crankcase which is slightly compressed pushes the burned gases out of the cylinder. During this process, a portion of the fresh mixture also finds its way to the exhaust port and is lost without being burned. This phenomenon of fresh mixture loss is sometimes termed as mixture short-circuiting. It has been estimated that depending upon the engine operating conditions, 15 to 40% of total fuel supplied to the engine is short-circuited. This is the principle cause of high hydrocarbon emissions observed from these engines. This results in a poor fuel economy of the vehicles powered by these engines.

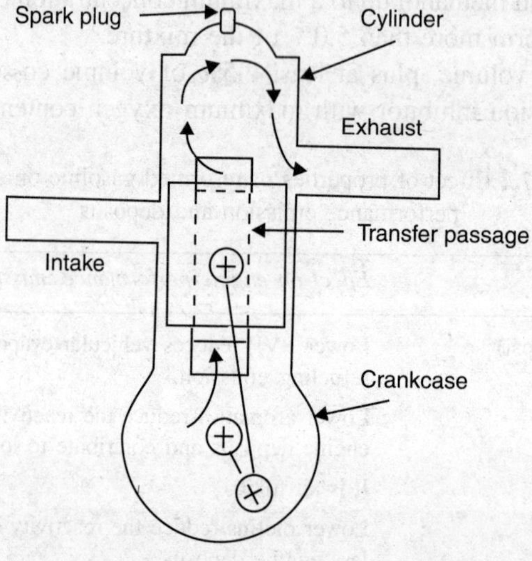

Fig. 7.3 Schematic diagram of a conventional crankcase. Scavenged two-stroke engine.

The presence of a high amount of residual burnt gases in the cylinder results in slow flame propagation and poor combustion. As a result the emission has high hydrocarbon content while NO_x emissions are rather low. The emission depends upon the engine load, speed and may vary considerably. The CO concentration varies in the range of 1.1 to 4.8, while hydrocarbon emission varies from 12000 to 36500 ppm, and C (2000 to 6090 ppm hexane equivalent). The concentration of NO_x is less than 200 ppm. Depending upon the design there is considerable variation in emissions. Research and development efforts to reduce scavenging through fuel losses may be divided into following categories.

(*i*) Improvements in the engine with premixed fuel-air scavenging.

(*ii*) Scavenging with stratified charging.

(*iii*) Fuel injections and scavenging by air unconventional and more complex design have been made to reduce the hydrocarbon emissions. Such available emission controlled technologies have been well documented in the Indian Institute of Petroleum, Dehradun, Report on "Vehicular Emission and Control Perspectives in India".[13]

Now, scooter with four stroke engine are also available in India. Emission of hydrocarbons and smoke is considerably reduced but concentration of NO_x increases, which can be removed by using catalytic converter.

Emission from Four-stroke Engines

All the Indian cars and jeeps are powered by four-stroke, spark ignited, gasoline fuelled, carburetted engines. Most of the new cars are equipped with positive crankcase ventilation (PCV) system. Some of the Indian models employ exhaust gas recirculation, evaporation emission control technologies.

Gases and vapours that escape from the combustion chamber past the engine piston and rings under pressure and enter the crankcase to be discharged into the atmosphere through draught tube and other vents are known as crankcase blowby. The blowby gases emitted from the crankcase are rich in hydrocarbons and contribute about 20% of total hydrocarbon emission on uncontrolled gasoline vehicles.

The hydrocarbon-rich blowby gases are prevented from escaping into the atmosphere and are recycled to intake air system through a PCV valve. This valve restricts the flow of blowby gases during idling to prevent rough idling and high flow at lighter loads.

A schematic diagram of the system is shown in Fig. 7.4. This is a low cost hardware that has been adopted in India by car producers. PCV can knock off a substantial portion of the hydrocarbon emissions.

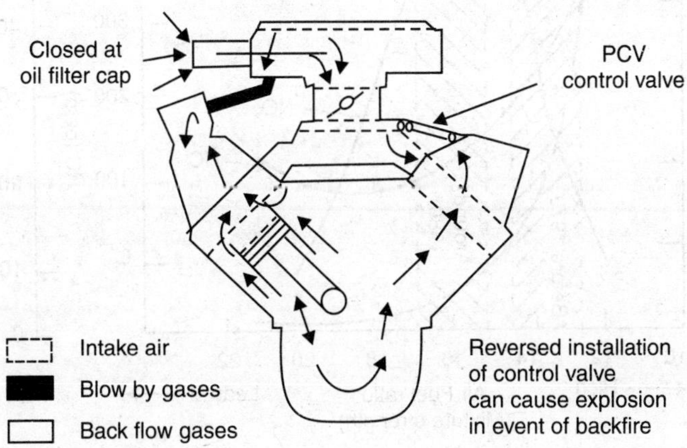

Fig. 7.4 Positive crankcase ventilation system (PCV).

Evaporative emissions are estimated to contribute another 20% of total hydrocarbons. The two major sources of evaporative emission are the carburettor bowl and fuel tank. There is no standard for evaporative emissions and no evaporative emission control system is presently in use on any of the automotive vehicles. Hydrocarbon emissions in the exhaust are primarily caused due to the "Wall-quenching" phenomenon, *i.e.*, quenching or extinction of the flame at the cold walls of the engine combustion chamber due to heat transfer, leaving behind a layer or volume of unburned and partially burned fuel-air mixture. High combustion pressure and temperatures promote reactions, leading to the formation of oxides of nitrogen. In addition to these, there are several complex reactions resulting in particulate emission. The air-fuel ratio is one of the most important parameters which greatly influences the exhaust emissions. Figure 7.5 shows the trends in leaning of air-fuel ratio and its effect on emissions and fuel consumptions. Reduced compression ratio, improved combustion chamber design, lean burn engines, stratified charge engines, etc., help greatly in reducing exhaust emissions.

7.4 EMISSION FROM DIESEL VEHICLES

Almost all the light and heavy duty commercial vehicles in India are powered by four-stroke, compression ignited diesel engines. Most of these vehicles are powered by direct injection diesel engines

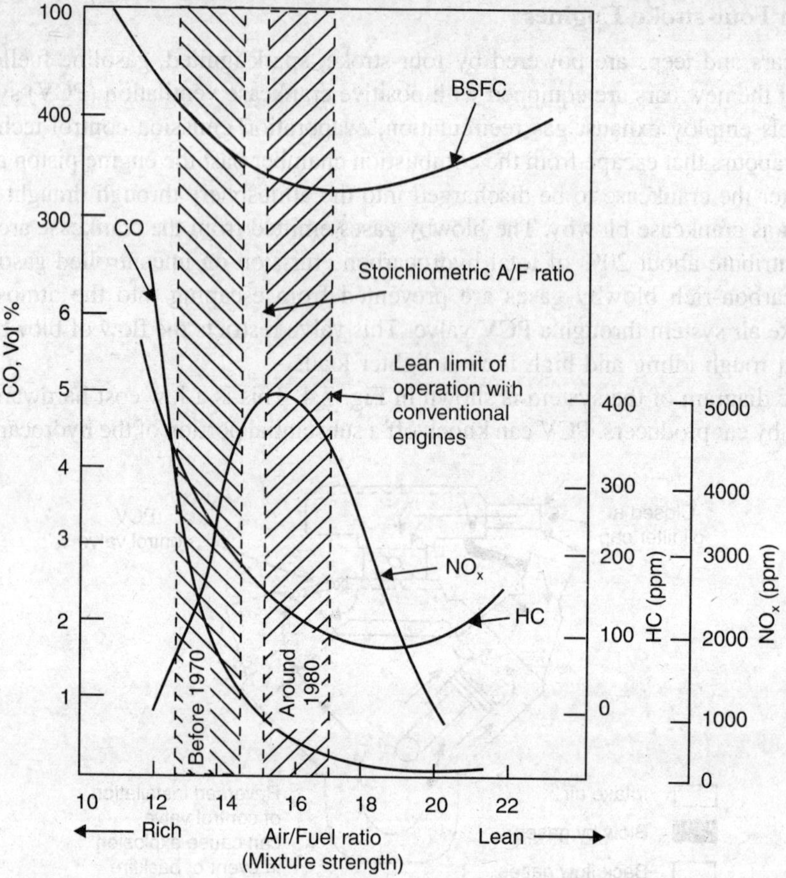

Fig. 7.5 Trends in leaning of air-fuel ratio and its effect on emissions and fuel consumption.

except one make of light commercial vehicles and some passenger cars. Most diesel vehicles are naturally aspirated. Turbo charging is being employed on same models. Pollutants emitted by diesel engines are CO, NO_x, unburned hydrocarbons, SO_2, particulates, aldehydes, smoke and odour. The exhaust smoke from the diesel is made up of soot particles. As the volatility of diesel is very low, not much evaporative losses occur from the fuel tank. The concentration of carbon dioxide and unburned hydrocarbon in diesel exhaust are relatively low compared that in gasoline engine but nitrogen oxides are present in high concentration. Typical emission from a diesel engine under different operation conditions are given in Table 7.3.

Table 7.3 Typical emissions from a four-stroke diesel engines under different operating conditions

Operating conditions	*CO* %	*HC* ppm	NO_x ppm	*Formaldehyde*
Full load rated speed	0.20	29	921	4.3
Half load rated speed	0.03	70	493	6.8
No load rated speed	0.03	90	109	1.8
Idle	0.03	106	119	6.8

Particulate Emission: In addition to pollutants, large particulate emissions include polynuclear hydrocarbons, some of which are known carcinogens. Such air inhalation of borne particulates poses health hazards.

Particulates from diesel engines are composed mainly of soluble organic fraction (SOF), soot, sulphates and other oxides. Particulates smaller than 2.5 microns are respirable and constitute over 90% of the total mass of particulates. Not only the mass emitted but also the chemical composition has been observed to vary significantly with the engine, driving cycle and fuel composition, particularly the sulphur content. Diesel particulates structure is shown below in Fig. 7.6.

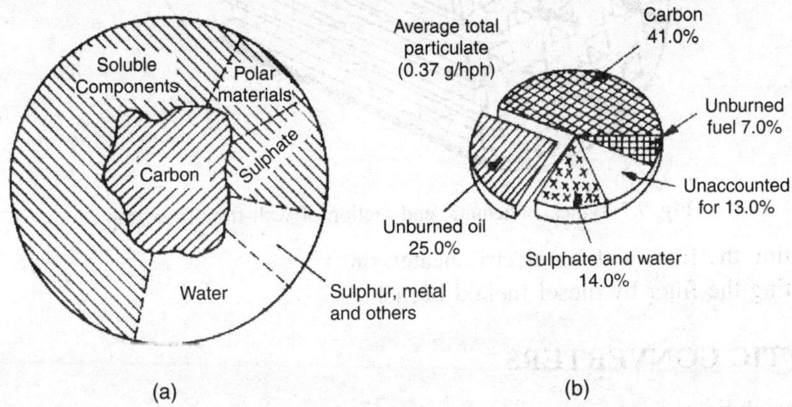

Fig. 7.6 (*a*) Schematic structure of diesel particulates, (*b*) Typical particulate composition.

Its main constituents are defined below:

1. **SOF:** The SOF is the portion of the particulates which is extractable using a solvent such as methylene chloride. It consists of fuel and lubricant derived from hydrocarbons and polynuclear aromatics.
2. **SOOT:** Soot is a carbonaceous matter which results from incomplete combustion of fuel and lube oil, sulphates and oxides. It includes SO_4^{2-}, H_2O, NO_3^-, PO_4^{3-} etc. Other elements like Fe, Ca, Zn, etc., are also present in traces.
3. **Polynuclear aromatic hydrocarbons (PAH):** Diesel particulate contain several PAH's, some of which are known carcinogens. PAH's are found in the range of 152 to 723 *g/g* of total particulates. Fuel composition has no effect on PAH.

The particulate trap utilises a filter media with very small pores, Fig. 7.7, which is effe...e in capturing the particles. As the trapped mass of particulates increases, resistance to flow increases and at a certain level pressure drop across the trap then its regeneration is required. Otherwise, the fuel consumption of the engine increases considerably.

Corning has developed a ceramic wall flow traps by taking cellular ceramic substrates and plugging alternative channels at opposite ends. This forces the exhaust gases to flow through the porous ceramic channel walls and hence referred to as 'Wall Flow' filters or traps. The porosity of these filters is around 50% with pore size ranging from 12 to 22 μm. A typical diesel trap structure is shown in Fig. 7.7. Depending upon the pore size the filtration efficiency varies in the range 60–95%. In sophisticated vehicles the regeneration is automatic. In less sophisticated vehicles the filter is removed from vehicles periodically and regenerated either.

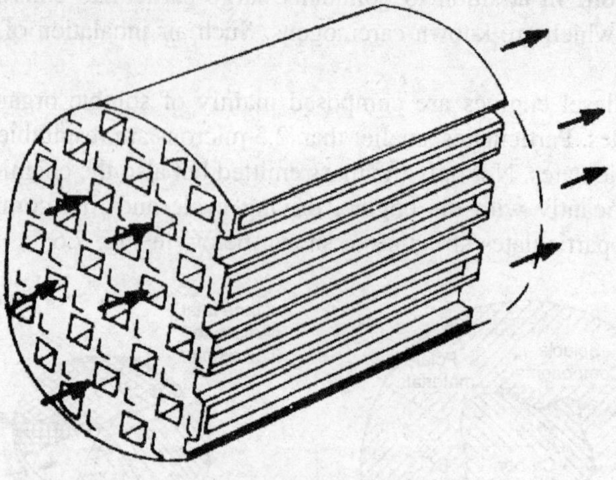

Fig. 7.7 Typical schematic and section of wall-flow filter trap.

(*i*) by heating the filter with an electric heater, or

(*ii*) by heating the filter by diesel fuelled burner.

7.5 CATALYTIC CONVERTERS

All successful catalytic converters developed so far employ precious metals as catalyst, such as platinum/rhodium. Hydrocarbons are oxidised over the catalyst in a two-stroke engine, but CO conversion is low as part of the hydrocarbon is not completely oxidised to CO_2. Their partial oxidation produces CO. Emission tests on Moped and Motor cycle are given in Table 7.4.

From Table 7.4 it may be noted that engine design improvements resulted in a decrease in CO and HC emissions. Use of catalytic converters reduced the emission considerably further. Results with fresh catalyst and those aged for 15000 km reveal that there is a negligible decrease in its efficiency. However, lead is extremely poisonous for catalytic converter and thus gasoline has to be free of lead. If air to fuel ratio is higher, say more than 2%, catalyst cell blockage may occur because of carbon and other residues.

Table 7.4 Emission test results

Emissions g/km	*1.2 hp (moped)*		*2.7 hp (Moped)*		*15 CC Motorcycles*		
	Without catalyst	*With catalyst*	*Without catalyst*	*With catalysts*	*Series produce, without cat.*	*Improved versions*	
						Without cat.	*With cat.*
CO	1.013	0.015	0.771	0.022	21.7	1.7	0.8
HC	2.716	0.090	3.205	0.037	16.9	10.4	1.9
NO_x	0.049	0.049	0.090	0.067	0.01	0.03	0.02

Oxidation catalyst for cars were introduced first in the US market 1975 to meet the CO and HC emission standards. It oxidizes the CO and HC to CO_2 and H_2O and, hence, is called oxidation catalyst.

The catalyst is made up of Pt-Pd with active catalyst material amount to 1.6 g per car of platinum and palladium in a ratio 5:2. Rhodium acts as a promoter and further increases its efficiency.

Further research led to the development of a variety of catalytic converters that can control all the three emissions that of HC, CO and NO_x simultaneously, and hence the name three-way catalysts. This system uses a catalyst system which operates best near the stoichiometric air-fuel ratio. To maintain the exhaust constituents at this precise value of oxygen content, closed-loop air-fuel ratio control systems are required. If air-fuel ratio is maintained correctly more than 90% reduction in all the three pollutants is achieved. Various automotive manufacturers have developed closed-loop controlled carburettor and fuel injection systems to meet this requirement and also to improve vehicle fuel economy and performance.

The advent of Zircona and/or Titania exhaust gas oxygen sensor (Lambda-sensor) was a great help in designing closed-loop feed back A/F control system. It measures the amount of excess oxygen in the exhaust relative to the oxygen in air. When the exhaust goes from rich to lean or vice versa the oxygen sensor produces a signal. This signal is used to control the amount of fuel supplied to the engine. Figure 7.8 shows the Ford three-way catalyst/feed back control system in the form of a block diagram. The components comprising the control-system include.

(*a*) Exhaust gas oxygen (EGO) sensor (Lambda sensor).
(*b*) Electronic control unit (ECU)
(*c*) Vacuum regulator solenoid (VRS)

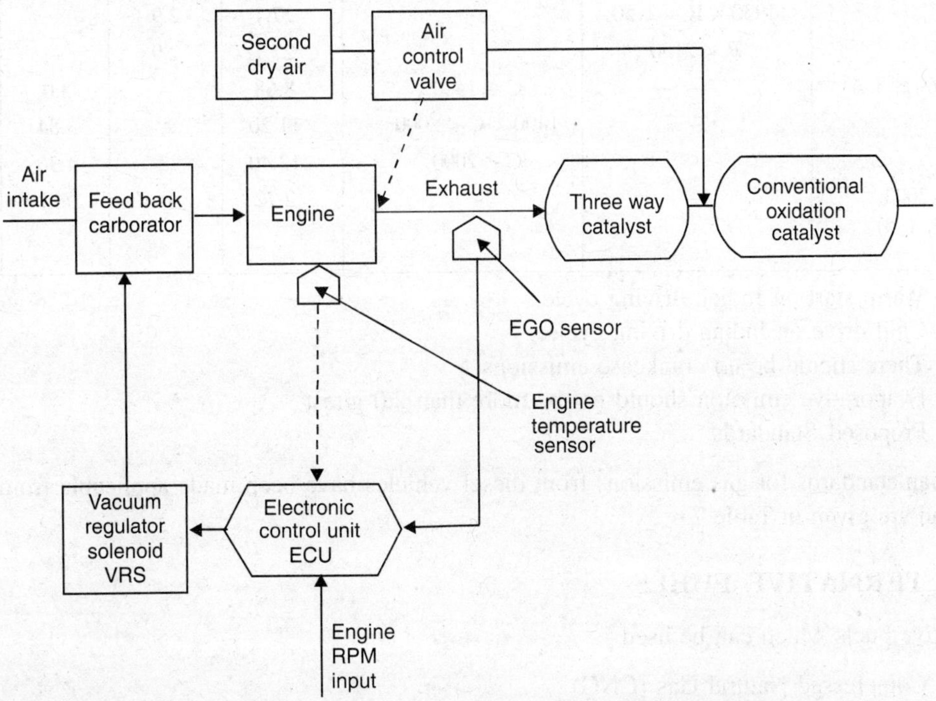

Fig. 7.8 Ford three-way catalyst/feedback control system block diagram.

The three-way catalyst system, while reducing the levels NO_x, hydrocarbons and CO have the drawback that it increases the concentrations of reduced species such as ammonia, cyanide and organic amines, hydrogen sulphide and organic sulphides.

Diesel Oxidation Catalysts (DOC) are noble catalyst similar to those used in gasoline. These use monolithic honey comb as support. Depending on the type of engine and its exhaust, DOC oxidises 30 to 80% of HC and 40 to 90% of CO. They do not reduce NO_x as there is always an excess of oxygen present in diesel exhaust. That is why, three-way catalyst are not effective in diesel engines. DOC shows very little oxidation of dry soot. These however remove 30 to 50% total organic fractions (SOF).

Indian standards on mass emissions of gasoline vehicles have been made applicable from April 1, 1991 and are given in Table 7.5.

Table 7.5 Indian emission standards for gasoline vehicles

Effective date	*Reference mass R (kg)*	*Cubic capacity C (cm^3)*	*CO g/km*	*HC g/km*	*HC + NO_x*	*CO idle*
1-3-1990	–	–	–	–	–	3%
1-4-1991(1)	R < 1020	–	14.3	2.0	–	3%
	1020 < R < 1250	–	16.5	2.1	–	3%
	1250 < R < 1470	–	18.8	2.1	–	3%
	1470 < R < 1700	–	20.7	2.3	–	3%
	1700 < R < 1930	–	22.9	2..5	–	3%
	1930 < R < 2150	–	27.1	2.9	–	3%
	R < 2150	–	27.1	2.9	–	3%
1-4-1996 (2, 3, 4)	–	C ≤ 1400	8.68	–	3.0	3%
	–	1400 < C < 2000	11.20	–	3.84	3%
	–	C < 2000	12.40	–	4.36	3%
1-4-2000 (2, 3, 4, 5)	–	–	2.72	–	0.97	3%

1. Warm start on Indian driving cycle.
2. Cold drive on Indian driving cycle.
3. There should be no crankcase emissions
4. Evaporative emission should not be more than 2.0 g/test
5. Proposed Standards.

Indian standards for gas emissions from diesel vehicles have been made applicable from April 1, 1992 and are given in Table 7.6

7.6 ALTERNATIVE FUELS

Alternative Fuels which can be used

(*i*) Compressed Natural Gas (CNG)
(*ii*) Propane
(*iii*) Methanol and Ethanol
(*iv*) Biofuel
(*v*) Hydrogen

Table 7.6 Indian gaseous exhaust emission standards for diesel vehicles

A. Gross Vehicle Weight > 3.5 Ton

Effective date	*Test cycle*	*Emission limit, g/kW hr*			
		CO	*HC*	NO_x	*PM*
01.04.1992	ECE 13 Mode	14.0	3.5	18.0	–
01.04.1996	ECE 13 Mode	11.2	2.4	14.4	
01.04.2000	ECE 13 Mode	4.5	1.1	8.0	0.36

B. Gross Vehicle Weight 3.5 Tons

Reference mass R (kg)	*Emission limits, g/km.*			
	CO HC + NO$_x$ *Effective from 1.04.1992**		*CO HC + NO$_x$* *Effective from 1.04.1996***	
R < 1020	14.3	2.7	5.0	2.0
1020 < R 1250	16.5	5.1	5.7	2.2
1250 < R < 1470	18.8	5.4	6.4	2.5
1470 < R < 1700	20.7	5.8	7.0	2.7
1700 < R < 1930	23.00	6.2	7.7	2.9
1930 < R < 2150	24.9	6.5	8.2	3.5
R > 2150	27.1	6.9	9.0	4.0

*The test should be as per Indian driving cycle with warm start

**The test should be as per Indian driving cycle with cold start.

There are 3.6 million vehicles in the world using propane, 4 million vehicles uses ethanol, 7 thousand uses natural gas, sixteen thousand uses methanol and nearly 1 million uses CNG.

The popularity of ethanol is going down as it produces 8% less energy than was envisaged, has high cost and has the highest emission of green-house gases amongst the alternative fuels. All alternative fuels produce lower emissions than gasoline but propane and the compressed natural gas are most economical, have the lowest emissions and, thus, are likely to be the fuel of the future.

Natural oil and gas may be present together or separately. When together, the gas may be present on top in the form of gas cap or may be dissolved in oil. This is known as associated gas as it is produced along with the crude oil. When gas alone is present, it is called free gas. When crude oil reaches the surface from an oil well, natural gas (methane, ethane, propane, ...) is separated from oil. From the natural gas certain hydrocarbons can be obtained in the liquid form, called "Natural Gas liquids" (NGL). These compounds rank among the world's most valuable energy resources. NGL compounds are propane, butane, pentane, hexane and heptane.

Natural gas in India is found in Andhra Pradesh, Gujarat, Nagaland, Rajasthan, Tamil Nadu, Tripura western and eastern offshore coastal area. At present 3.85 million cubic metres of gas is being flared every day by ONGC. Oil India Limited, (OIL); another public sector undertaking, had produced

3.45 billion cubic metres per day in 1994–95 out of which 3.3 billion cubic meter of gas was flared every day.

By the year 2015, demand for petroleum products would be around 132 million tons of which our indigenous production is estimated to be around 60–65 million tons only. Thus, if our natural gas reserves were to be properly prospected and exploited, we would not be required to import any petroleum products.

7.7 COMPRESSED NATURAL GAS (CNG)

CNG (Compressed Natural Gas) is a mixture of hydrocarbons consisting of approximately 80 to 90 per cent methane in gaseous form. Due to its low energy density, it is compressed to a pressure of 200 to 250 kg/cm^2 (to enhance the vehicle on board storage in a cylinder) and hence the name Compressed Natural Gas. Natural Gas is colourless, odourless, non toxic but inflammable and lighter than air. CNG as a vehicle fuel was introduced under a Pilot Project in 1992 in the cities of Delhi, Mumbai and Baroda. It is also available in the cities of Surat and Ankaleshwar. Almost any petrol vehicle with or without catalytic converter, can be converted to operate on CNG. Diesel engines can also be converted to CNG, either by installing dual fuel kit or converting the existing diesel into spark ignition engine. CNG option for diesel vehicle can be achieved by replacing existing diesel engines with an equivalent rated petrol engine of same torque and installing bifuel petrol CNG kit. CNG buses are available from Ashok Leyland and TELCO.

In India only Ministry of Surface Transport (MOST) approved CNG kits are allowed to be fitted on cars. Based upon the authority letter of MOST, regional transport authority includes the CNG fuel modes in the vehicles when the vehicle is presented for inspection along with the installation certificate. Installation certificate is issued by respective installer workshop authorised by kit supplier in writing to the Regional Transport Authority. A fuel switch on dash board is fitted to enable the vehicle to run on petrol, in case CNG runs out.

CNG is the safest gaseous fuel. It is less likely to auto-ignite on hot surface than petrol and require high energy for ignition. In case of fuel spill, natural gas is lighter than air and in case of vehicle fire, the fuel will dissipate into air and burn above the ground level.

CNG has a much higher octane number and lower cetane number, which makes CNG, a good alternative for spark ignition but unsuitable for compression ignition engines. Higher octane number makes it superior fuel than petrol. As the octane no. of CNG is high, no antiknock additives are required. Due to absence of any lead or benzene content in CNG, the lead fouling of spark plug and lead or benzene pollution is eliminated. Being a gaseous fuel, CNG mixes with air easily and evenly.

The energy equivalence and other properties of natural gas, unleaded gasoline and diesel are given in Table 7.7. CNG is currently approximately 73% cheaper than petrol and 44% cheaper than diesel. Use of CNG reduces CO emission by 70–90%. Because CNG has more energy, CO_2 emission is reduced by 15–20% thus decreasing green house effect. As it is stored in cylinder, ground water pollution is eliminated.

Since CNG is a gaseous fuel, storage capacity for CNG in a vehicle is comparatively less than that of petrol. The quantity of CNG filled by dispenser during refuelling also depends upon pressure at the dispensing station. At maximum permitted filling pressure (200 bar g) an amount of 8/9/10 kg CNG is stored in 40/50/60 litre of size cylinder respectively, which is equivalent to approximately 11.2/12.5/13.9 litres of petrol equivalent. However, the gas quantity depends on ambient conditions and actual fill pressure. The onboard cylinder in CNG vehicle is refuelled from the dispensing unit under pressure, in a manner similar to petrol filling.

Table 7.7 Energy equivalence and other properties of CNG, petrol and diesel

Property	*Unit*	*Petrol*	*Diesel*	*CNG*
Energy equivalence	–	1.39 lt	1.18	1 kg
Cost	Rs.	43.53	30.48	19.20 /kg
Cost on energy equivalence	Rs.	60.5	35.96	19.20
CNG Cheaper	–	68%	46.6%	–
Density relative to Air	Air 1	3.4	3.9	0.6
Auto ignition Temp.	°C	232–282	225	540
Flammability Limit	% in Air	1–8	0.6–5.5	5–15
Flame temperature	°C	1977	2055	1790
Octane No.	–	97	–	127
Cetane No.	–	10	45	10

7.8 PROPANE

Another automobile fuel commonly known as propane, HD_5, but different from liquefied petroleum gas (LPG), contains 95 to 98% propane and 5 to 2% other hydrocarbons. Of the more than 8.3 million alternative fuel vehicles in the world, 3.6 million use propane. Like LPG, it liquefies under pressure at room temperature. Propane is chemically simpler than gasoline that burns in the engine under heat. Propane gas exhaust contains fewer pollutant and almost no particulate matter. It can be easily liquefied and as soon as pressure is reduced it evaporates. It has higher octane rating and better antiknock properties. Propane gas tank is much lighter than CNG tank. Any leakage can be detected by adding 5 ml methyl mercaptan.

Propane has the lowest greenhouse gas emissions and low air toxicity. Its exhaust has a very low concentration of nitrogen oxides, hence ozone formation potential decreases considerably. Smog and acid rain formation almost completely eliminated. Propane is positioned to be the first clean fuel to achieve market acceptance and proven reliability. Public interest and demand will grow as government and industry would use such a clean fuel.

Propane is a by-product of oil refineries and is produced during light refining and at present it is not being tapped. In Masana, Gujarat field ONGC is flaring 200 metric tons of propane every day. To tap propane no change is needed in refining process in refineries.

The total cost of a car operating on propane fuel is Rs. 14000/- (excluding cylinder)

Petrol Price per litre	–	Rs. 43.53
CNG Price per kg	–	Rs. 19.20
HD_5 Price per kg	–	Rs. 10.60
Running cost approximately	–	25 per cent of petrol

Cost of a propane dispensing stations = a single CNG dispensing station

In the last few years, in India, many hobbyists have converted their cars with LPG cylinders but could not make it fool proof. Cars thus fitted, have low pick-up, slow speed, starting troubles, leakages, frequent tuning/adjustments, breakdowns, missing, slow running, etc. The reason being the crude process/material, lack of technical expertise, non-professionalism.

7.9 METHANOL AND ETHANOL

Natural gas as well as biogas can be the most common sources for methanol production as vehicular fuel. Methanol is used most often in vehicles as a fuel blend to provide more oxygen to fuel for burning but can be used as a fuel on its own. Methanol energy content is approximately one half that of gasoline, although advanced technology is expected to make it more efficient with up to 10% to 20% more power that is comparable to gasoline vehicles.

Methanol has a relatively low vapour pressure and absorbs a relatively large amount of heat when it vapourises, it tends to cool the fuel intake area, which delays vehicle warm up. As a result, cold weather engine start-ups in methanol-fuelled vehicles can be troublesome relative to propane- or gasoline-powered cars. Start up difficulty in methanol-fuelled vehicles can begin at temperatures below 9°C and becomes severe below freezing.

Unlike gasoline, when stored in a tank, methanol vapours are explosive at normal atmospheric temperatures. Because the flames are invisible a colourant must be added to enable fire detection. Saturated vapours are explosive at temperatures between –1°C and 5°C when 100% methanol vapours are enclosed in a tank.

Another difficulty relating to the use of liquid methanol is that it corrodes metal and synthetic materials in the conventional gasoline engines. As a result, flexible-fuel engines that use gasoline and methanol interchangeably must be reinforced with stainless steel or specially treated material to prevent rapid corrosion.

Another disadvantage is that there should be no emission of methanol by evaporation or through exhaust pipe. The reason being that methanol is highly toxic in nature and man or animal who inhale it will eventually become blind.

Ethanol: Ethanol is derived primarily from fermenting sugars in corn and cane sugar and other organic materials. While it produces fewer harmful emissions, it can contribute to environmental stress through the large amounts of energy fertilizers and pesticides used to produce the extensive feed stock required to manufacture ethanol.

Large amounts of energy are required to apply fertilizers and pesticides and to plough and harvest crops used to make ethanol. The US Department of Energy calculates that the production of 100 BTU of ethanol from corn requires a total energy investment of 109 BTU (44 BTU to grow the corn and 65BTU's to process it into ethanol). As a result, the end product, ethanol, produces about 8% less energy than was invested in its production.

Despite the cost barrier and, from a life cycle perspective, the negative environmental impact, the options provided to North America by the growth of food for fuel suggests that the use of ethanol is expected to grow as an environment friendly octane enhancer for gasoline.

7.10 BIOFUEL

(a) Biofuel: Addition of oxygenate as fuel additive (alcohol and ethers) that contain oxygen, enhances octane number, enhance combustion, reduce exhaust emission. Indian government decided to mandate by 2008, 10% blending ethanol with gasoline but it may not be good news for Indians. To meet its biofuel target India need an additional 110 million tons of sugarcane to produce ethanol, which in turn will need 30 cubic kilometers of additional irrigation. This water will come at the expense of other irrigated crop. In India which houses four per cent of its fresh water, biofuel crop will eat into the resources needed to meet a growing food demand. It may be beneficial to Brazil, USA but not India. The alternative that is being tried is Jatropha curcas and Pongamia yield ore that, after processing, makes biodiesel. It is a hardy plant that can grow in abundance in infertile soils, is not grazed upon by animals. The standard seeding method of Jatropha propagation takes two years for the plant to yield. TERI's mycorrhiza[14] application speeds up the process to yield Jatropa in seven months. These plants can grow on poor soils and need very little water or manure. Oil is extracted and put through a process called transesterification (conversion of fatty acid esters by incubation with alcohol and alkali), making it a suitable blend for petroleum derived diesel. Growing the crop and processing them will create fresh employment. The Jatropha blend reduces greenhouse gas emission by half.

(b) Hydrogen: Hydrogen is being aggressively explored as a fuel. It can be used in fuel cell to power electric motors or burned in internal combustion engine (ICE). It is an environmentally friendly fuel that has potential to dramatically reduce our dependence of foreign oil, but several significant challenges must be overcome before it can be widely used. Such cars will run on hydrogen and electricity, emit only water and none of noxious fumes. It will be three times more energy efficient than a standard petrol powered car. It will produce NO_x when burned in ICE. Recently, in June 2008 Honda Motors has started producing fuel cell vehicle that propels zero emission car running on hydrogen. This car can reach speeds up to 160 km per hour.

(c) Nuclear Fuel: Whilst many in the traditional "environmentalist" are skeptical of the use of nuclear power, citing concerns over issues such as the long-term disposal of radioactive waste, the majority of scientific opinion is that the use of nuclear energy is one important option which must be kept on the table, given its large potential to displace highly greenhouse-gas intensive fossil fuel based energy generation systems, given the potential for grave consequences as a result of anthropogenic forcing of climate change. France and Sweden, for example, whose electricity generation is 78% and 48% from nuclear energy respectively, and have some of the lowest per capita greenhouse gas emissions in the developed world, providing evidence for the effectiveness of this approach for displacing the use of carbon dioxide-intensive fuels. Sweden and France have been able to eliminate the use of fossil fuels almost entirely through the use of nuclear energy and renewable energy systems, such as hydroelectricity. In recent years, several prominent environmentalists, including James Lovelock, Tim Flannery and Stewart Brand have spoken out in support of the use of nuclear energy, given the grave concerns over the potential damage of anthropogenic forcing of climate change.

7.11 HYDROCARBONS IN OUR ENVIRONMENT

(a) Hydrocarbons in Atmosphere: Most of the atmospheric methane is produced by anaerobic bacteria in swamps and bogs, a much smaller percentage is released by natural gas wells. Automobile is the main natural hydrocarbon releaser in air. It is estimated that it is responsible for 50% hydrocarbon, industry contributes 14%, solid waste disposal 5%, gasoline marketing 4%, organic solvent evaporation 10% and fires 11%.

Emission from automobile are 20% of the emissions from the crankcase ventilators (blowing), 60% from exhaust pipe and 20% from gasoline evaporation from gas tank and carburettors. Automobile emissions can be reduced by 20% when these hydrocarbons are recycled back through carburettor and into the engine. This process is called positive crankcase ventilation (PCV) (see Fig. 7.4). The emission resulting from evaporation can be controlled by proper design of carburettor and gas tank to minimize escape of hydrocarbons. The addition of these changes will further reduce the hydrocarbon emissions by another 20%.

The most difficult source of emission to control is that from the exhaust pipe. The main reason for this is that the internal combustion engine as it is presently designed will always emit some hydrocarbon. The main source of these hydrocarbons is due to "wall-quenching". The burning of gasoline close to the walls of the cylinder is quenched because of the zone next to the wall being much cooler than the combustion zone in the centre of the cylinder. This relatively cold area slows the rate of combustion so that the hydrocarbons in this region are swept out by the piston before combustion is complete. The crevice above the piston rings between the cylinder wall and the piston is also a zone where wall quenching occurs (Fig. 7.9).

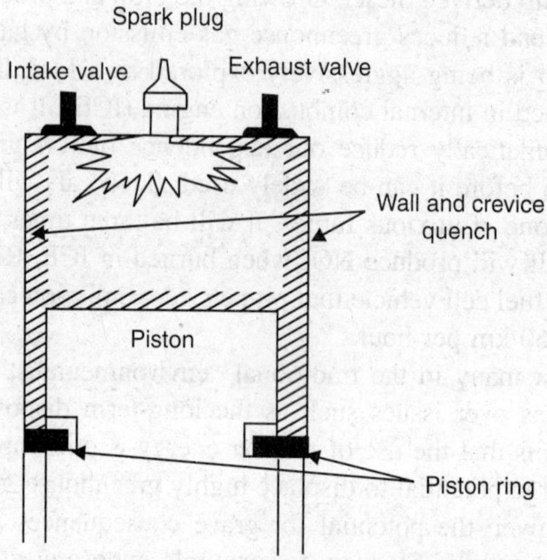

Fig. 7.9 Areas of wall and crevice quench in the cylinder of a four-stroke cycle internal combustion engine.

There would still be hydrocarbon emission from the automobile even if it were possible to eliminate wall quenching. This is because there is insufficient time for complete combustion of gasoline to CO_2 and H_2O. Addition of more air to the engine does affect more complete combustion but this results in an increase in the NO_x emission and a decrease in the power as shown in Fig. (7.5). Since the present design of the internal combustion engines cannot be built with satisfactory low pollution levels, some types of emission controls have to be added to the automobile if both NO_x and hydrocarbons emission are to be controlled.

(b) Petroleum in Environment: The major environmental effects of petroleum are due to either crude oil itself or to the automobile hydrocarbon emissions from gasoline. It is estimated that 6 million metric tons of petroleum hydrocarbon enters the environment each year from both these sources. Oil spills generally are associated with the release of hydrocarbons to oceans or other water bodies which spread over large areas of the water surface. Slow degradation of petroleum residues affects the aquatic life and thus is an environmental problem. Oil spills on land are confined to small areas. Automobile emissions, on the other hand, cause atmospheric problems. Much spilled oil also enters the atmosphere by evaporation, but problem associated with oil spill and automobile emissions are sufficiently different to warrant separate discussions.

Natural Oil Seepage: Natural oil seeps led to the discovery of most of the major oil fields in the world. The rate of seepage is different for different sites but often the rate of seepage is so slow that the oil is effectively dispersed by environmental forces. Most seepages produce an extensive oil slick on the ocean surface to produce less than 40 gal/day and the environmental effect appears to be limited. It has been estimated on the basis of very incomplete data that 160 – 62000 × 10^6 litres of oil per year are released into environment.

Oil Spills: The hundreds of oil tankers that deliver crude oil are designed to ride low in the water when carrying a full load. After they unload their cargo they must take on sea water as ballast for their return trip to the oil fields. This ballast, along with the residual crude oil in their tanks is pumped into ocean before the next load of crude oil is loaded. It is estimated that 800×10^6 gal of oil per year are introduced into the sea from this source and another 130×10^6 gal are added from the bilges of all other ocean vessels. This oil is released close to the shore where it may be carried on to the beaches.

Off-Shore Wells: With the exception of cases when the wells that blew off as in January 1969 in Santa Barbara, California, North Sea Blowout in 1977 or the blow out of rig in Allawaram village in Andhra Pradesh in India at the Krishna-Godawari basin on January 8, 1995, the danger of leakage from off-shore oil wells appear to be minimal. In these blow-outs, natural gas had rushed out at pressures much beyond the calculations, blowing out the well head, and having ignited. The various lapses at Krishna-Godawari basin were as follow:

1. Pressure of the gases at area 3000 meters below surface was not estimated properly. This was possible to be measure by using seismic data of a particular area with sand and shell pattern studies, rock densities and structural analysis. The propagation of sound waves in the "interval velocity method" after these studies would have indicated the pressure.
2. The well was being created over an uplifted fault block. This clearly indicates lack of subsurface data by way of analysis, correlation and identification of these structures.

The very nature of oil exploration does create environmental problems. However, ONGC has a hostile structure whose sole objective is taking out, with the least cost to itself and no regard for the people or the environment around, the natural resources of the regions. Its effluent treatment plants let the water into fields. These once lush and paddy fields now have been transformed into alkalised wastelands, streams and ponds once teeming with fish now reduced to muck and slime. ONGC should be made to follow the minimal Environment and Pollution Control Measurement as they have laid down for themselves in their publication. "Environment and Pollution Control Measures taken by ONGC".

Lubricating and Crankcase Oil: About 2400×10^6 gal of lubricating oil are produced each year. About half of this is consumed. Of the remaining, about 600×10^6 gal are recovered and burned as fuel in incinerators and boilers and 110×10^6 gal are purified for reuse. The remaining 500×10^6 gal are a major source of pollution. Perhaps most waste oil will eventually be re-refined. Direct combustion of it is not desirable because such combustion results in the introduction of heavy metals. It is estimated that 1–1.5% by weight of crankcase oil is lead so that its recovery can be an important source of this metal also.

7.12 ENVIRONMENTAL REACTIONS OF PETROLEUM AND PETROLEUM PRODUCTS

Hydrocarbons are relatively inert. Alkanes and alkenes react with oxygen and ozone but these reactions are not effective in removing hydrocarbons from environment. They are significant in other ways. However, ozone reactions, are important at ambient temperatures in smog formation whereas the formation of CO and CO_2 is important in combustion.

5 to 10 per cent of the total organic emissions from automobiles contain oxygen substituents. These compounds are formed by the incomplete oxidation of hydrocarbons in the internal combustion engines. Aldehydes make up the major portion of these oxygenated compounds but ketones, alcohols, ethers and nitro alkanes are also formed.

7.12*a* Reaction with Ground State Molecular Oxygen $^3\Sigma_g$

Oxygen molecule has two unpaired electrons in the ground state. So gas phase oxidation products of hydrocarbons with oxygen may be expected to exist as free radicals in nature. These reactions proceed at a very low rate at 25–100°C. Reactions 7.3–7.5 are much faster than 7.6 to 7.10 so that appreciable concentration of peroxides develop slowly below 100°C. Reaction 7.1 will take place at an appreciable rate only if radicals initially formed are stable. At temperatures about 100°C, reactions 7.1 and 7.6 proceed at a significant rate so that a chain process involving reactions 7.3 and 7.6 predominates. As the reaction temperature reaches 250°C, oxidation becomes autocatalytic. This is because of the increased importance of reaction 7.7 in which accumulated peroxides decompose to form two radicals. Temperature in the 345–500°C range favour hydrogen abstraction and radical decomposition to olefines 7.8–7.9.

$$O_2 + RCH_2CH_2R \longrightarrow R\dot{C}HCH_2R + HO_2^{\bullet} \qquad ...(7.1)$$

$$R\dot{C}HCH_2R + O_2 \longrightarrow RCH{=}CHR + HO_2^{\bullet} \qquad ...(7.2)$$

$$\xrightarrow{+O_2} RCH(\dot{O}_2)CH_2R \longrightarrow R{-}\underset{H}{C}(\dot{O}){-}CH_2R + \tfrac{1}{2}O_2 \qquad ...(7.3)$$

$$\xrightarrow{HO_2^{\bullet}} R{-}CH(O_2H)CH_2R + O_2 \qquad ...(7.4)$$

$$RCH(\dot{O}_2)CH_2R + RCH(\dot{O})CH_2R \longrightarrow R{-}CH(O_2H)CH_2R + RC({=}O){-}CH_2R \qquad ...(7.5)$$

$$RCH(\dot{O}_2)CH_2R + RCH_2CH_2R \longrightarrow R{-}CH(O_2H)CH_2R + RCH_2\dot{C}HR \qquad ...(7.6)$$

$$RCH(O_2H)CH_2R \longrightarrow RCH(\dot{O})CHR + H\dot{O} \qquad ...(7.7)$$

$$R\dot{C}HCH_2R \longrightarrow RCH{=}CH_2 + \dot{R} \qquad ...(7.8)$$

$$\longrightarrow RCH{=}CHR + \dot{H} \qquad ...(7.9)$$

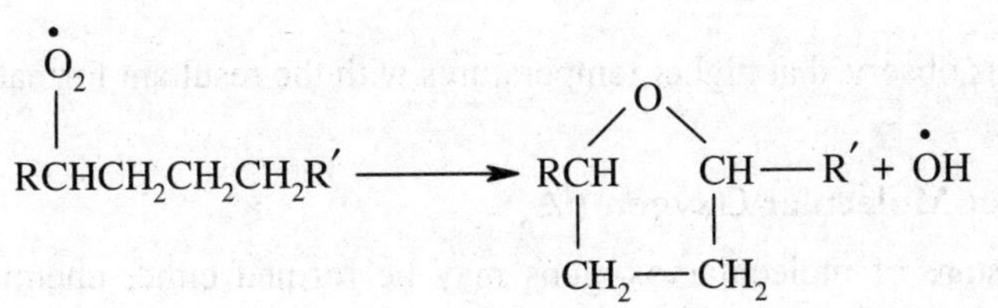

...(7.10)

The actual products observed when neopentane (2, 2 dimethyl propane) and *n* pentane are oxidised with an equimolar amount of oxygen in the gas phase are shown in 7.11, 7.12 and 7.13. More extensive oxidative degradation takes place in the internal combustion engine, where a much larger proportion of oxygen is used.

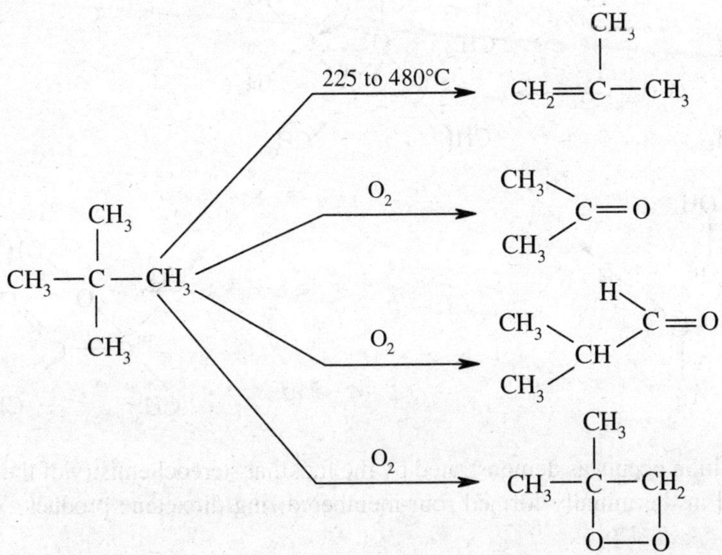

...(7.11)

...(7.12)

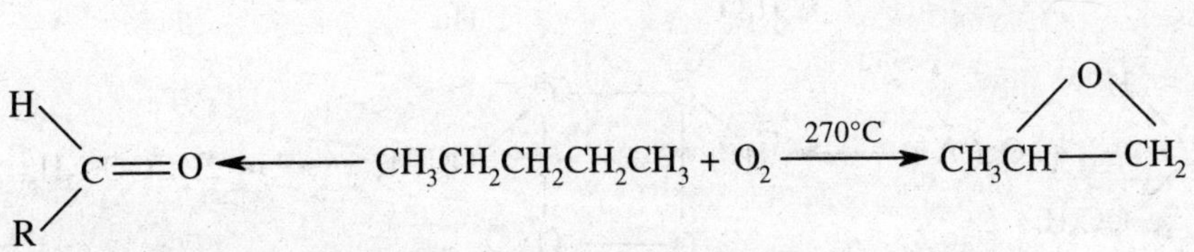

...(7.13)

Where R = CH_3, C_2H_5, C_3H_7

The following reaction mechanism is consistent with these reaction products

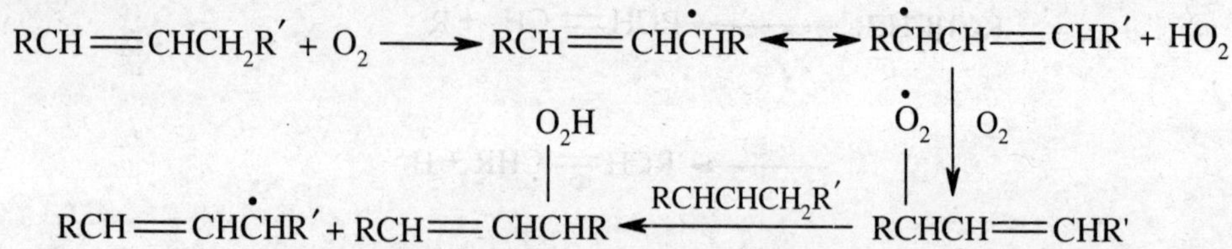

...(7.14)

Peroxide decomposition is observed at higher temperatures with the resultant formation of alcohols, aldehydes and ketones.

7.12*b* Reaction with Singlet Molecular Oxygen ($^1\Delta_g$)

Singlet oxygen, an excited state of molecular oxygens may be formed either chemically or by the photosensitised reaction of oxygen. The photochemical formation is probably the most significant from the environmental viewpoint but not in smog formation. The reaction of Singlet oxygen with olefin with the most probable mechanism is as follow:

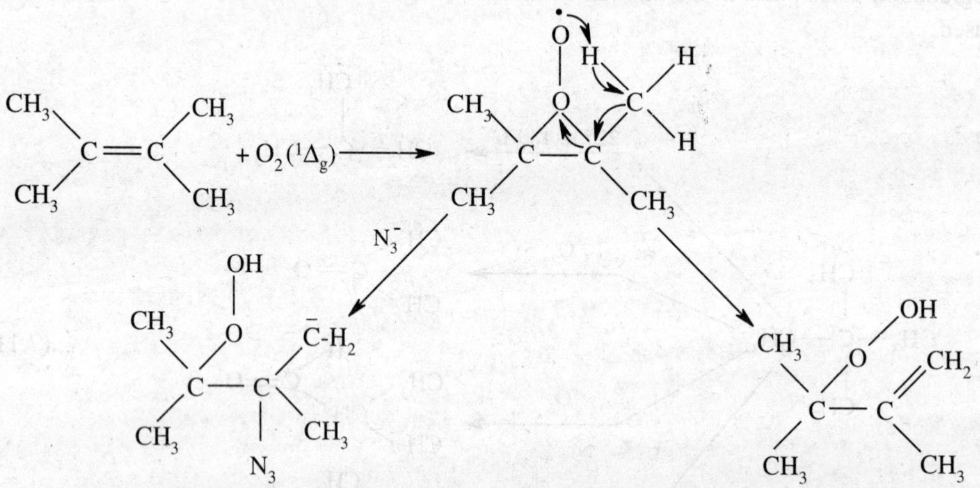

...(7.15)

Stereospecific addition occurs as demonstrated by the fact that stereochemistry of the 1, 2 diethoxy-ethylene is maintained in the initially formed four-membered ring dioxetane product

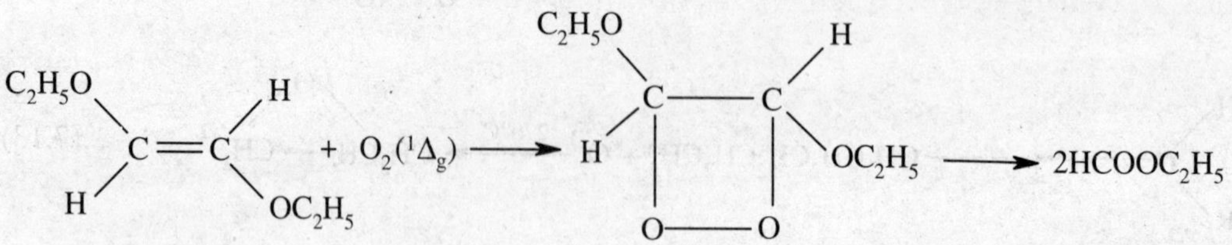

...(7.16)

The reaction of singlet with oxygen rich olefins follow a different reaction pathway to yield a carbonyl product

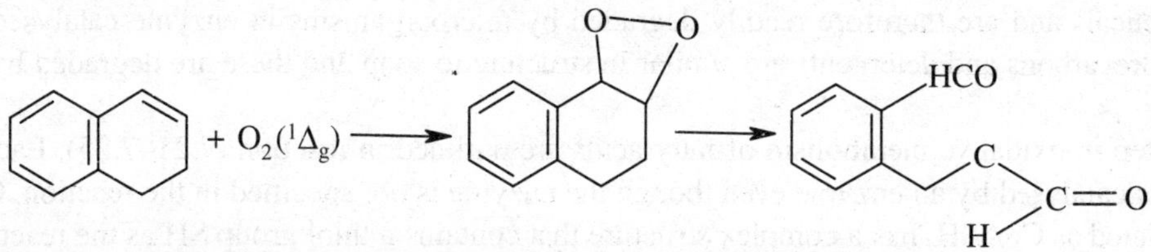

...(7.17)

7.12*c* Reaction with Ozone

Ozone is present in the atmosphere in trace amount. Levels of ozone and peroxide greater than 0.15 ppm cause eye irritation. Ozone is very reactive substance. It forms free radical intermediates that can react with bio-molecules of the body. Ozone makes 1, 3 dipolar addition to the olefin bonds in gas phase to form molozonite which dissociates into a diradical which in turn, decomposes to aldehydes or ketones and another diradical. This diradical reacts with oxygen to give radicals that are similar to those postulated previously in the reaction of oxygen with hydrocarbon 7.18-7.20.

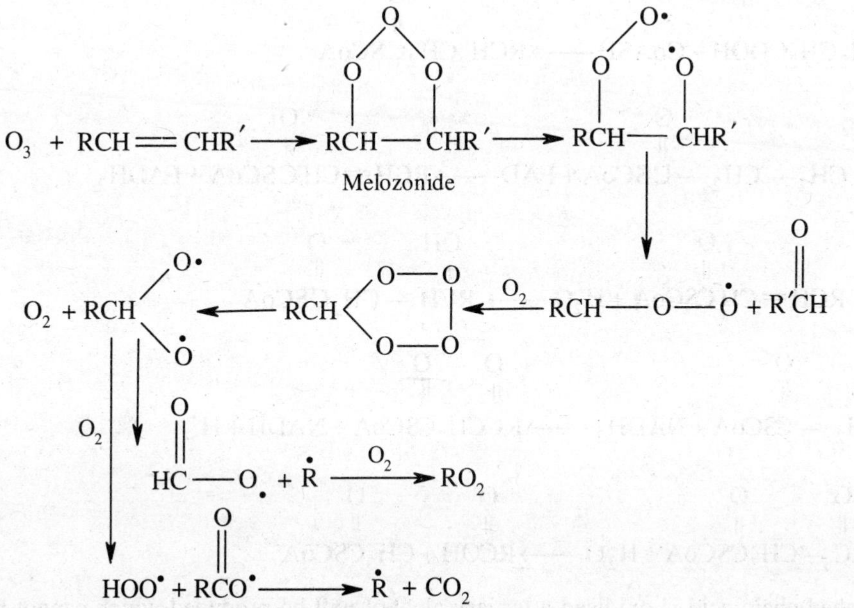

...(7.18)

...(7.19)

...(7.20)

Ozone reacts with hydrocarbons at 25–50°C to yield products that are similar to those obtained by reaction of hydrocarbons and molecular oxygen at 125–250°C.

7.13 BIODEGRADATION OF HYDROCARBONS

Microorganisms are the most effective agents for the total degradation of such molecules. Microorganisms degrade hydrocarbons, soaps and detergents because they can use energy and organic fragments released for the synthesis of bio-molecules essentially for their growth. The metabolism of these compounds is affected by enzymes, high molar mass proteins, which catalyses a broad range of chemical transformation in living systems. Enzymes are characterized either by their extreme catalytic efficiency or specificity. Many enzymes require a small, nonprotein prosthetic group called coenzyme for the catalytic activity. These prosthetic groups are usually bound to the enzyme during the course of reactions and sometimes they combine with one or more of reacting chemical species. Fatty acids (protonated soap) occur widely in biochemicals and are therefore readily degraded by microorganisms in enzyme catalysed reaction. Linear hydrocarbons and detergents are similar in structure to soap and these are degraded by the same pathways.

The step in oxidative metabolism of fatty acids are outlined in reaction (7.21–7.25). Each of these reactions is catalysed by an enzyme even though the enzyme is not specified in the reaction. Coenzyme A, abbreviated as CoASH, has a complex structure that contains a thiol group SH as the reactive centre. The acetyl coenzyme A $\left(\begin{matrix} O \\ \| \\ CH_3CSCoA \end{matrix}\right)$ produced in reaction 7.25 may be used by microorganism for synthesis of the specific fatty acids needed for growth by a pathway that is essentially the reverse of the one shown here, FAD and NAD^+ are the oxidised forms of hydrogen transfer coenzymes while $FADH_2$ and $NADH_2$ are the respective reduced forms:

$$RCH_2CH_2COOH + CoASH \longrightarrow RCH_2CH_2\overset{\overset{O}{\|}}{C}SCoA \qquad ...(7.21)$$

$$R—CH_2—CH_2—\overset{\overset{O}{\|}}{C}SCoA + FAD \longrightarrow RCH{=}CH\overset{\overset{O}{\|}}{C}SCoA + FADH_2 \qquad ...(7.22)$$

$$RCH{=}CH\overset{\overset{O}{\|}}{C}SCoA + H_2O \longrightarrow \underset{}{R\overset{\overset{OH}{|}}{C}H}—CH_2\overset{\overset{O}{\|}}{C}SCoA \qquad ...(7.23)$$

$$R\overset{\overset{OH}{|}}{C}H—CH_2—\overset{\overset{O}{\|}}{C}SCoA + NADH_2 \longrightarrow R\overset{\overset{O}{\|}}{C}CH_2\overset{\overset{O}{\|}}{C}SCoA + NADH + H^+ \qquad ...(7.24)$$

$$R\overset{\overset{O}{\|}}{C}—CH_2\overset{\overset{O}{\|}}{C}SCoA + H_2O \longrightarrow R\overset{\overset{O}{\|}}{C}OH + CH_3\overset{\overset{O}{\|}}{C}SCoA \qquad ...(7.25)$$

If a branched chain acid is oxidised a tertiary alcohol will be produced which cannot be oxidised to ketone thus further reaction is inhibited. This scheme shows why the highly branched ABS detergents are degraded very slowly and ranked as bio-non-degradable detergents and thus pollutant.

The microbial oxidation-breakdown of detergents and petroleum requires the terminal oxidation of the hydrocarbon chains to a carboxyl acid group. This terminal oxidation is well documented. Molecular oxygen is the oxidizing agent and the oxygen is activated by binding to iron containing enzyme cytochrome P-450. The reaction follows the sequence shown below:

$$RCH_2CH_2CH_3 \longrightarrow RCH_2CH_2CH_2OH \longrightarrow RCH_2CH_2COOH$$
$$\longrightarrow RCOOH + CH_3COOH \qquad ...(7.26)$$

These reactions are carried out by microorganism but are slow especially with high molar mass materials.

The microbial oxidation of aromatic compounds present in detergents and petroleum is also catalysed by enzyme cytochrome P-450 with sequence given in 7.27 for the oxidation of benzene.

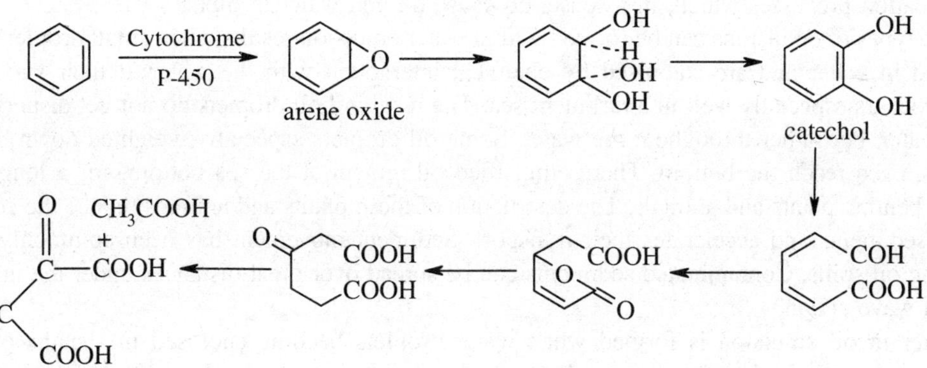

...(7.27)

The more highly condensed aromatics such as naphthalene or of the alkyl substituted benzenes proceed via salicylic acid which is then oxidised to catechol (reaction 7.28) and finally to oxalo acetate;

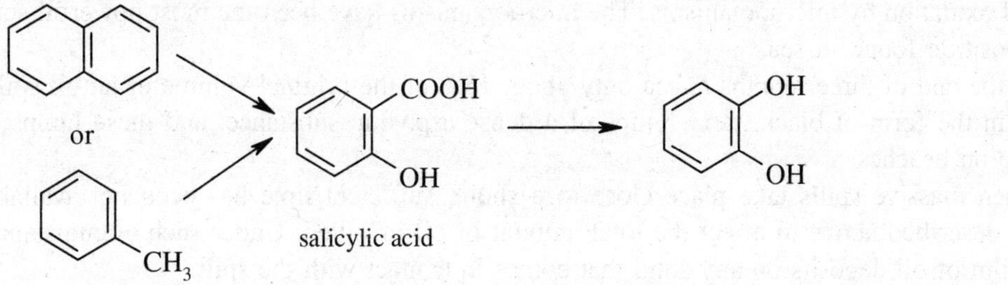

...(7.28)

The sulphonic acid residues present in sulphonic acid detergents are subject to hydrolysis by a variety of aquatic bacteria. Therefore, these sulphonic acid groups do not slow the environmental degradation of these detergent. Nature's technique involves microbial decomposition of petroleum, but it suffers from four disadvantages;

(*a*) It is slow;
(*b*) It imposes an enormous requirement of precious oxygen. 1 gal of crude oil requires all the oxygen dissolved in 320000 gal of air saturated sea water;
(*c*) It can be adversely affected by environmental conditions such as low temperature and low oxygen concentration and hindered by the presence of other contaminants as heavy metals;
(*d*) It is a selective process and tends to be least effective on the most toxic petroleum constituents. An oil spill will persist and continue to spread for an extended period of time. The more volatile and soluble lighter hydrocarbons evaporate and heavy hydrocarbon remain.

7.14 CRUDE OIL AND MARINE LIFE

Since crude oil is composed of chemically inert hydrocarbon, spilled oil may persist on the ocean surface for many years. One gallon of oil will cover an area of four acres of water when completely spread out. During the time it has been spreading many changes that take place in the oil. All oils regardless of type, have volatile components that evaporate readily. Within a few days 25% of the volume of a typical oil spill is lost through evaporation. The remaining oil has been subjected to emulsification processes which, in a sense, do cause oil and water to mix.

Two types of emulsions can be formed. Oil in water emulsion results when small droplets of oil are dispersed in water and are stabilised by chemical interaction with the water at their surfaces. This process occurs especially well in a turbulent sea. The resulting oil droplets do not get dispersed on top of the water, but rather throughout the water. Some oil droplets especially weighted down by mineral particles, even reach the bottom. There emulsified oil remain at the sea bottom, for a long time and damage benthic plants and animals. The destruction of these plants and animals reduce the cohesion of bottom sediments and accelerates their transport. Sediment movement has been commonly observed following oil spills. Contaminated sediments can be spread over great distances under the influence of tide and wave action.

Water in oil emulsion is formed when water droplets become enclosed in sheaths of oil. The emulsion gets stabilised by interactions between the enclosed water droplets and various resinous and asphaltic materials found in most crude oil. This type of emulsion shows up as a floating, sticky, viscous mass that, on occasion, contains enough water to cause the total volume to be greater than that of the original oil.

Most oil that survives the emulsification processes is then degraded by spontaneous photo-oxidation and oxidation by microorganisms. The microorganisms have been the most powerful source of oil decomposition found in sea.

By the end of three months at sea only about 15% of the original volume of an oil spill remains. This is in the form of black, tarry lumps of a dense asphaltic substance, and these lumps frequently wash up on beaches.

When massive spills take place close to a shore, sufficient time has been not available for the process described above to affect the total amount of oil involved. Under such circumstances a thick sticky film of oil deposits on any solid that comes in contact with the spill.

Biological and Physical Effects

The effects fall into two categories; those caused by coating and asphyxiation, and those that result from oil's toxicity. The effects caused by coating and asphyxiation are:

1. **Reduction of Light Transmission:** Measurements have shown that ambient light intensity 2 metres below an oil slick was 90% lower than at the same depth in clear water. Photosynthesis by marine plant life may be hampered by extended periods of such conditions.
2. **Dissolved Oxygen Reduction:** Oil films are able to retard the rate of oxygen uptake by water. Measurements have shown a significantly lower dissolved oxygen level in water beneath a slick than in clear water. The final effects of this could not be determined.
3. **Damage to Water Birds:** Swimming and diving birds get covered with oil. This causes their feathers to mat together, reducing their buoyancy and preventing flight. The insulative value of the feathers also decreases and some birds quickly die of cold, caused by evaporation of water or by its lower temperature. It has been estimated that 25,000 sea birds died because of the Torrey Canyon incident.
4. **Smothering:** The effects of oil on plant life along shore lines are becoming a matter of concern. Intertidal algae and lichens are known to have been killed by smothering coats of oil which washed ashore in some areas.

The toxic effects of oil are the result of some of the many compounds that make up this very complex mixture. These compounds have been found to vary widely in their molar masses and structures.

Until recently, low boiling saturated hydrocarbons were regarded to be harmless to a marine environment. It has now been demonstrated that these substances, in low, water concentrations, cause anaesthesia and necrosis in a wide variety of lower animals. Cell damage and death result in animals exposed to high concentrations. It has been indicated that larval and other young forms of marine life may be particularly affected.

Low-boiling aromatic hydrocarbons have been abundant in oil and represent its most dangerous fraction. These compounds have been acute poisons to life. These compounds are more soluble than saturated hydrocarbon in water. The aromatics are able to kill marine organisms by direct contact with oil or through contact with dilute solutions.

Generally, one way of getting rid of spilled oil is burning. It has been effective only if it can begin immediately after a spill. The lighter fraction of oil evaporates immediately after a spill, and the remaining oil is having a higher ignition temperature and can be difficult to ignite. This factor coupled with the ability of water to conduct heat away from the burning oil, has caused some attempts to be unsuccessful. One drawback has been that water pollution has been converted to air pollution, since dense black smoke is generated. The residue after burning has been a black solid that can be easily collected and removed.

Straw and saw dust are used as sorbents. After becoming saturated with oil they might be removed and disposed of. Shredded polyurethane foam, an absorbent, is used and has the advantage that when it is saturated, it can be removed and the oil can be squeezed out. The foam can be reused. The recovered liquid contains 80% oil which can be recovered for use. A volume of 1.5 cubic metre of foam will reportedly soak upto one ton of crude oil. The method is prohibitive with extremely large slicks due to the amounts of material involved. The soaked foam becomes so heavy, to remove it, is also a problem.

7.15 COAL

Coal was formed from the debris of giant tree ferns and other vegetation that was growing in swamps and bogs some 300 million years ago. When plants died they were covered with sediment that prevented their aerial oxidation. Further build-up of sediments and other geologic processes subjected these materials to high pressure that resulted in their conversion to coal. The carbonisation of wood and woodlike matter, to various extents, produce different forms of coal as follow:

Peat *57%C*	*Lignite* *60-70%C*	*Bituminous coal* *75-90%C*	*Anthracite* *93%C*
Found in Nilgiri Hills, South India	Kashmir, Neyvelli (Madras) used as boiler fuel	Rajasthan, Bokaro, Raniganj (Bengal), Madhya Pradesh, Orissa	Kashmir & Eastern Himalyas.

The carbonisation of plant carbohydrates result in the formation of highly condensed carbon. These forms of coal on heating gives out many important organic compounds. Coal also contains a variety of inorganic ions that ultimately end up in ash residue resulting from bituminous coal with major constituents being Na, K, Ca, Al, Si, Fe and Tc.

The gasification of coal was carried out extensively in 1920 in the United States but the discovery of natural gas wells resulted in the shift to the cleaner and cheaper natural gas as a fuel. As a consequence only small advances have been made since 1920 in gasification of coal. Coal liquefaction technology is at a more primitive stage of development than gasification. At present the only liquefaction plant is located in South Africa. In this plant the coal is first converted to CO and H_2 which is hydrogenated at 200°C in the presence of iron as catalyst to get a mixture of hydrocarbons. Methanol can be produced instead of hydrocarbons, if copper is used as a catalyst.

7.16 DETERGENTS

Anything that behaves as a cleaning agent is called detergent. This would include soaps, alkaline dish washer and surfactant. Surfactants, or surface-active agents, increase the wetting and cleaning power of water and are principal active ingredients in both soap, tooth paste, shaving cream, shampoo, dish washing compounds, detergents and cleaning agents of various kinds. The largest use of surfactants is in household detergents but there are also many industrial applications. Most detergents are formed by the chemical transformation of petroleum. Consequently, the environmental degradation of many of these compounds is similar to that of petroleum. Soaps, the sodium and potassium salts of fatty acids are prepared by alkaline hydrolysis of plant and animal fat. Since fat is a bio-product, soaps are completely biodegradable.

Surfactants generally consist of a polar or hydrophilic group (such COO^-, SO_3^{2-}, or NH_4^+) that are soluble in water and oily or lipophilic groups that are soluble in lipids. They fall generally into three groups:

1. Anionic surfactants in which the negative ion is surface active. These are the most commonly used and soaps can be classified as anionic detergents. The most common surfactants are alkyl benzene sulphonates (ABS) and linear alkyl sulphonates (LAS).
2. Cationic surfactants in which the positive ion is surface active. These have usually quaternary ammonium group $RN(CH_3)_3^+$ and have poor cleaning properties but are good germicides and frequently encountered when this property is important.
3. Nonionic surfactants in which the entire molecule is surface active. A popular type is polyoxethylene nonylphenol.

Properties of Soaps and Detergents

Soaps and detergents have similar structure and their mechanism of cleaning action are similar. Both contain a hydrophobic portion, a long chain hydrocarbon chain to which a hydrophilic (polar) group is attached. Such surface active material concentrates at the surface of a aqueous solution. This lowers the surface tension of water so that it penetrates the surface and interstices of the object being cleaned. If the binding of the detergent to the substance being cleaned is greater than the binding of the soiling agent, the later will then be displaced into the water phase. Oil and greases will also adsorb the hydrophobic end of the surface-active agent. The soiling agent can be washed away because of the affinity of the polar group of adsorbed soap or detergent for water keeps the dirt suspended in water. The main disadvantage in the use of salts of carboxyl acids as cleansing agents is the insoluble precipitates that form with Ca^{2+} and Mg^{2+}. This curdy grey precipitate deposits on clothes and leaves a ring in the bath tub. The deposits are not only undesirable but also greater amounts of soap must be used to make up for that loss in precipitation. Another disadvantage of soap is that they are weak acids that protonate in mild acidic or neutral solution. The polarity of the head of detergent decreases when protonated and, as a consequence, it does not solubilise the oil and grease as rapidly.

The main advantage of detergents is that they are not precipitated by hard water. However, their cleansing power is decreased markedly by Ca^{2+} and Mg^{2+} ions. So that it is necessary to add "builders" commonly polyphosphate, to chelate these ions and to prevent "water spiting" on dishes caused by insoluble Ca^{2+} and Mg^{2+} salts when automatic dishwashers are used. Similar quantities of brighteners, soil deposition preventives, foam control agents and other materials also may be present.

Synthesis of Soaps and Detergents

Soaps are prepared by the saponification of fats hydrolysis to convert the triglycerides in the fats to the sodium salts of carboxylic acids:

$$\begin{array}{l} RCOO-CH_2 \\ \quad | \\ RCOO-CH \\ \quad | \\ RCOO-CH_2 \end{array} + 3NaOH \longrightarrow \underset{\text{Soap}}{3ROONa} + \underset{\text{Glycerol}}{\begin{array}{c} CH_2OH \\ | \\ CHOH \\ | \\ CH_2OH \end{array}}$$

Some long chain fatty acids occur naturally, it is not surprising to find some foam and suds at "water falls" and other places of turbulence in river streams.

Detergents: Polar group in most detergents is a sulphonic acid group RSO_3^- and consequently such detergents are anionic in nature. Some alkyl sulphates R—OSO_3^- are also used. A synthetic detergent is not a pure chemical entity but is a mixture of olefins in the feed stock and to variation in the extent of polymerization of the olefins.

Alkyl benzene sulphonate (ABS) is prepared from propane obtained by cracking the hydrocarbons and benzene made by catalytic reforming of petroleum. ABS synthesis:

1. Propene is polymerised

$$4CH_3CHCH_2 \xrightarrow[300-500°C]{H_3PO_4} (CH_3)_2CH(CH_2CH_2)_2\overset{\overset{CH_3}{|}}{C}H{=}CHCH_3$$

2. This polymer is abbreviated as $RCH{=}CHCH_3$ reacts with benzene

$$RCH{=}CHCH_3 + C_6H_6 \xrightarrow[\text{or } H_2SO_4]{AlCl_3 \text{ or HF}} RCH_2\underset{\underset{CH_3}{|}}{C}H-C_6H_5$$

The sulphonate group is then introduced in the aromatic ring

$$RCH_2-\underset{\underset{CH_3}{|}}{C}H-C_6H_5 + SO_3 \longrightarrow RCH_2-\underset{\underset{CH_3}{|}}{C}H-C_6H_4-SO_3H \xrightarrow{NaOH} ABS$$

These reactions are an over simplification, without going into the mechanism. Since this detergent has numerous side chains, ABS detergent is not biodegradable detergent.

Over a 180–200 km trip down a waterway only 30 to 50 per cent of the ABS might biodegrade showing ABS surfactant resistance to biological agents. It also inhibits, the oxidation of certain phenols by enveloping them and thus isolating them for dissolved oxygen in water. It also interferes with the waste treatment processes by stabilising small particles in colloidal suspension and reducing the activity of biological filter beds and activated sludge. The most notorious effect was their tendency to produce stable foam in rivers sometimes several metres high, over an extent of several hundred metres.

People who worked on the foamy water problem reasoned that if something could be done to make ABS attractive to bacteria, its foaming properties should be destroyed. This was accomplished by a modification in the structure of ABS that produced biodegradable detergents called LAS, –linear alkyl sulphonate. Why this makes a difference to microbes no one knows for sure, but it is a good sign that the microorganisms can easily digest the linear chain and use it as a source of carbon for food, and the foam, odour and taste formerly imparted to water by ABS has been virtually eliminated. Hence, there is emphasis on LAS detergent, that is used everywhere now.

It is not possible to synthesize linear hydrocarbon molecules by acid catalysed polymerisation of olefins, because more substituted carbonium ion is formed which results in branched chain formation. However, linear hydrocarbon can be separated or prepared by other methods.

Urea crystallises in a unique fashion to form hydrocarbon-urea inclusion complex from solution. This is the basis of the method of separating linear and branched hydrocarbon. The urea crystallises as helices and the linear hydrocarbon are encased in the central channel of the fibers formed by these helices. Branched chain do not fit into the 5Å diameter tube and so the linear hydrocarbons can be separated.

The use of molecular sieves has proved to be cheaper method for the separation of linear and branched hydrocarbons. Molecular sieves are semipermeable inorganic zeolites (a clay mineral) that have a channel structure that is similar to that of urea when crystallised from hydrocarbons. The molecular sieves with a cross section of 5Å are used to screen linear hydrocarbons from petroleum selectively.

Linear hydrocarbon can be synthesised from olefin by use of catalyst such as the complex of triethyl aluminum and $TiCl_4$. The product can be alkane or a derivative. In one process, the "Alfor process" linear alcohols are formed by the air oxidation and subsequent hydrolysis of the polymer formed initially (7.29–7.31). Olefins can be formed by the thermolysis of the adducts of the organometallic derivative

$$-\overset{|}{\text{M}}-CH_2CH_3 + nCH_2{=}CH_2 \longrightarrow -\overset{|}{\text{M}}-(CH_2CH_2)_n\,CH_2CH_3 \qquad ...(7.29)$$

$$CH_3CH_2(CH_2CH_2)_nOH \xleftarrow{H^+} -\overset{|}{\text{M}}-O(CH_2CH_2)_n\,CH_2CH_3 \xleftarrow{O_2} \qquad ...(7.30)$$

$$CH_3CH_2(CH_2CH_2)_{n-1}\,CH{=}CH_2 \xleftarrow[\text{Ni}]{\Delta} \qquad ...(7.31)$$

The synthesis of biodegradable linear detergent are completed by the addition of a polar head to linear hydrocarbon. The alkane can be chlorinated as in 7.32 and converted to LAS (reactions 7.33–7.35) by the procedure similar to those used in synthesis of the ABS detergent. The chain branching to the benzene ring does not impede the microbial degradation of these compounds.

$$nRCH_2CH_3 + Cl_2 \longrightarrow R-\underset{}{\overset{CH_3}{\overset{|}{C}H}}-Cl \qquad ...(7.32)$$

$$R-\overset{CH_3}{\overset{|}{C}H}-Cl + C_6H_6 \longrightarrow R-\overset{CH_3}{\overset{|}{C}H}-C_6H_5 \qquad ...(7.33)$$

$$R-\overset{CH_3}{\overset{|}{C}H}-Cl \xrightarrow{-HCl} RCH{=}CH_2 \xrightarrow[C_6H_6]{HF} R-\overset{CH_3}{\overset{|}{C}H}-C_6H_5 \qquad ...(7.34)$$

$$R-\overset{CH_3}{\overset{|}{C}H}-C_6H_5 \xrightarrow[2NaOH]{1SO_3} R\overset{CH_3}{\overset{|}{C}H}-C_6H_4SO_3Na \qquad ...(7.35)$$

LAS

Alternatively the alkanes can be oxidised to a mixture of alcohols as in (7.36) and the compound can be converted sulphate ester (7.37) or non ionic detergent with ethylene oxide (7.38).

$$nRCH_2CH_3 \xrightarrow[\Delta]{O_2} RCHOH \text{ (mixture of acohols)} \qquad ...(7.36)$$

$$R-\overset{CH_3}{\overset{|}{C}H}-OH + H_2SO_4 \longrightarrow R\overset{CH_3}{\overset{|}{C}}HOSO_3H \xrightarrow{NaOH} R\overset{CH_3}{\overset{|}{C}}HOSO_3Na \qquad ...(7.37)$$

$$R-\overset{CH_3}{\overset{|}{C}}-OH + 8\,CH_2\overset{O}{\frown}CH_2 \xrightarrow{NaOH} R\overset{CH_3}{\overset{|}{C}}HO(CH_2CH_2O)_7CH_2CH_2OH \qquad ...(7.38)$$

High level of LAS, however can increase fishfry mortality.

7.17 THE BUILDERS

To the detergents generally are added compounds called 'builders' which are sodium salts of phosphoric, carbonic, sulphuric, silicic and boric or similar acids. In the absence of a builder, unsequestered hardness ions such as Ca^{2+} and Mg^{2+} can poison the surfactant molecules to remove dirt. One of the common builder used is sodium tripolyphosphate ($Na_5P_3O_{10}$). Most of the detergents contain 35 to 50% of $Na_5P_3O_{10}$ but the amount used has been steadily decreasing as substitute preparations are placed on the market. Sodium tripolyphosphate has the following structure.

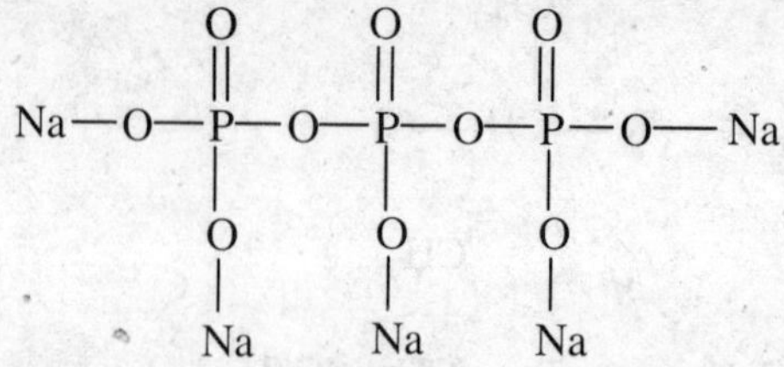

The builder tie-up with hard water ions Ca^{2+} and Mg^{2+} to form a large water soluble $Ca_{10}(P_3O_{10})_2$ ion. In this form, metal ions cannot interfere with the action of detergent. Builder also undergo hydrolysis which cause the water to become alkaline. This alkalinity is necessary for effective removal of dirt. $Na_5P_3O_{10}$ undergoes a slow hydrolysis reaction to produce orthophosphates that are completely non-toxic. The reaction is

$$P_3O_{10}^{5-} + 2H_2O \longrightarrow 2HPO_4^{2-} + H_2PO_4^-$$

These phosphate ions contribute to the eutrophication of waterway.

Literally, thousands of compounds have been screened and tested as possible partial or full replacement for phosphate builders but no replacement is possible because of the following reasons.

1. Polyphosphates as well as phosphates are non-toxic to aquatic life and pose no health hazard to man.
2. Phosphates are safe to use on colours, fibres and fabrics of all types.
3. Phosphates are non-corrosive and non-flammable and hence to use it in washing machine is quite safe.
4. They do not interfere with our waste treatment procedures.
5. Phosphates are the most economical material available for builders.

Another cheap builder is nitrilotriacetic acid or NTA. It has been found to be effective as a water softener that disperse dirt and help maintain the proper alkalinity of water, which is also necessary for loosening of dirt. The water softening action of NTA resembles that of tripolyphosphate. It absorbs moisture from the air which cause the detergent to cake. Its anaerobic degradation is difficult, and show some teratogenic effects (effects that damage the foetus *in utero*) in combination with heavy metals. NTA can also tie up with some biological essential elements and thus harm natural ecosystem.

NTA is believed to be converted mainly to nitrite and nitrate, which are toxicants under some circumstances and also furnish nutrients to microorganisms and higher plants. These nitrites led to formation of nitrosamines, a group of compounds which are highly carcinogenic. Hence, NTA was removed from market depending further investigations of its biological effects.

A few builders in use are:

(*i*) Sodium carbonate — a mildly alkaline material. It disperses the dirt particles. But can be injurious when ingested, especially by children.

(*ii*) Borax — a good cleansing agent but removes dyes from fabrics. Sensitive trees and shrubs have been damaged by waste water from industrial cleaning operations in which borates were used;

(*iii*) Sodium metasilicate — and silicates are effective water softeners but are strongly alkaline which irritates the skin and eyes.

These builders pose some disadvantage. The build-up of calcium and magnesium salt residue is one of the problem. These solids may be deposited on articles to be washed which gives dull looks. The precipitating builders give much more alkalinity to the wash water than do phosphates. Phosphates produce wash water with a pH of 9 to 10.5, non-phosphate builders produce wash water with a pH of 10.5 to 11. A solution with a pH of 11 and above is corrosive to skin, causes serious eye irritation, and injests toxicity. At these high pH values the solutions form gels with protein tissues and become very difficult to remove before serious damage has been done. All this also signifies that phosphate-free detergents are no better for the environment either. However, phosphates may partly be replaced by enzymes.

The enzymes are used with the intension to attacking stains that other detergents cannot remove, especially stains from proteogenous materials; blood, urine, milk, meat juice, and chocolate, etc. The enzymes have been produced by cultures of microorganism (*Bacillus subtilis*) which secretes the enzymes during its digestive process.

Soil is absorbed with sweat and sebum (secretion of sebaceous glands which contain fatty acids, cholesterol and dead cells) in the network of cotton fabrics of the fabrics which are made of cellulose. Enzyme alkaline celluloses partly modify the cellulose and subsequently the solid is removed by detergent. Similarly alkaline lipases act on oil spots as well as on the soil adsorbed on them.

Alkaline proteases remove stains like blood and proteinaceous materials which come through handling food items. In the absence of proteases, proteogenous dirt coagulates on the fabric on account of washing conditions. This precipitate also retains other dirt components, subsequently difficult to remove. It results in the fabric appearing grey and unclean after several washings. The enzymes should have the property to be able to function in alkaline pH and high temperature.

Detergent also contain oxidising, bleaching and chelating agents which can hinder the activity of certain enzymes. Some surfactants even denature the enzyme. Hence, an ideal detergent enzyme must be stable in alkaline conditions, have stability over a wide temperature range, be compatible with the oxidizing, bleaching and chelating agents and perfumes, and should not require metal ions for their own activity.

Detergents also contain miscellaneous ingredients such as brighteners, perfumes, antire deposition agent and in some products, enzymes. Brighteners are dyes which absorb invisible ultraviolet light and emit white or blue light. Antiredeposition agents keep dirt in suspension once it has been removed from a fabric. Carboxy-methylcellulose (a product of wood) is commonly used for this purpose.

Unfortunately the enzymes dust in manufacturing plants also attacks the skin of workers hands. When inhaled the enzymes are able to sensitize the lungs, causing allergic reactions resembling hay fever or asthma. The enzymes where found to have several potentially toxic substances that caused disintegration of blood constituents and agglutination of red blood cells when injected into the body cavity of mice. The potential effects of enzymes on aquatic organisms are not fully recognized.

7.18 ENVIRONMENTAL EFFECTS OF BIODEGRADABLE ORGANIC COMPOUNDS

The presence of biodegradable organic compound in lakes and river can result in an environmental problem. Oxygen required for the microbial degradation of these compounds and relatively small quantities of organic material can deplete the supply of dissolved oxygen. The amount of oxygen consumed in microbial degradation of organic compound can be estimated if the crude assumption is made that the organic compound is pure carbon and is converted to carbon dioxide.

$$C + O_2 \longrightarrow CO_2 \quad ...(7.39)$$

For this reaction and the ratio of molar mass of oxygen to atomic mass of carbon it can be seen that 32/12 or three parts of oxygen are required to oxidize 1 part of carbon. Since water can dissolve only about 10 ppm of O_2, this means in the absence of further dissolution of atmospheric oxygen that only 4 ppm of carbon compound can deplete all the dissolved oxygen. This calculation is an over simplification since the microbial oxidation does not proceed instantly and oxygen is always replenished. However, the oxygen near the bottom of deep lakes can be depleted by dissolved organics because of the slow rate of oxygen transport to these low level. Fish have difficulty breathing when the oxygen level drops to 5 ppm.

The biological oxygen demand (BOD) for a mixture of the biodegradable organic material in water, is determined by the amount of oxygen required for the microbial oxidation of the dissolved organic content of a water sample. The test measures the amount of dissolved oxygen lost from a water sample kept in a sealed container at 20°C over a five-day period. Optically about 75% of the organic material is oxidised under these conditions. Chemical Oxygen Demand (COD) is another measure of dissolved organic compound. It is determined by measuring the equivalents of acidic MnO_4^{2-} or $Cr_2O_7^-$ necessary for the oxidation of organic constituents. Neither BOD nor COD measures the total oxidisable carbon content. Total oxidisable carbon is measured by the amount of CO_2 formed on combustion after the water and the carbonate in the sample is removed. Obviously even biodegradable compounds such as detergents reduce the oxygen level in lakes and rivers. However, the major sources of dissolved organics in lakes and rivers are the effluents from sewage disposal plants, manure from animal feed lots, industrial wastes and decomposing plants and algae.

REFERENCES

1. Brenner, T.R., Biodegradable Detergents and Water Pollution in *Advances in Environmental Science,* Vol. 1 (J.N. Pitts and R.L. Metcalf, Eds.) p.147, Wiley-Interscience, New York (1969).
2. "Catalysts for the Control of Automotive Pollutants," *Advances in Chemistry Series,* No. 143, American Chemical Society, Washington DC (1975).
3. Ebel, R.H., Catalytic Removal of Potential Air Pollutants from Auto Exhaust in *Advances in Environmental Sciences,* Vol. 1 J.N. Pitts & R.L. Metcalf) p. 237, Wiley-Interscience, New York (1969).
4. Gibson, P. T., "Initial Reaction in the Degradation of Aromatic Hydrocarbons" in *Degradation Synthetic Organic Molecules in the Biosphere,* National Academy of Science, Washington DC (1972).
5. Kearns, D.R., Physical and Chemical Properties of Singlet Molecules Oxygen, *Chem. Rev.,* 71, p. 395 (1971).
6. Leedy, H.A., "Ozone Chemistry and Technology", *Advances in Chemistry Series*" # 21 ACS, Washington DC (1959).
7. Mayo, F.R., "Oxidation of Organic Compound, I, II, III", *Advance in Chemistry Series* #, 75, 76, 77 ACS, Washington, DC (1968).
8. Mc Kenna, E.J., Microbial Metabolism of Normal and Branched Chain Alkanes in *Degradation of Synthetic Organic Molecules in the Biosphere*", National Academy of Science, Wash. DC (1972).
9. Niven, W.W., *Industrial Detergency,* Reinhold, New York (1995).
10. Pennert, S.S. and Licerman, Eds. "Energy".
Vol. 1 "Demands Resources, Impact Technology and Policy"

Vol. 2. "Non Nuclear Technologies"

Vol. 3. "Nuclear Energy and Energy Policies", Addison Wesley, Reading, Massachusetts (1974, 1975, 1976).

11. Ruedisilo, L.C., and Firebaugh M. W. Eds., *Perspectives on Energy,* Oxford Univ. Press, New York (1975).
12. Wigg, E.E., Campion, R.J and Petersop, W.L., "The Effect of Fuel Hydrocarbon Composition on Exhaust Hydrocarbon and Oxygenated Emission", Society of Automotive Engineers, Automotive Engineering Congress, paper 72025 (1972).
13. "Vehicular Emission and Control Perspectives in India" A State of Art Report, Indian Institute of Petroleum, Dehradun.
14. TERI's Technologies for Sustainable Development, TERI, Darbari Seth Block, IHC Complex, Lodhi Rd., New Delhi (2005).

8 Pesticides

8.1 INTRODUCTION

Through the ages insects have been a major competitor for food. Every cultivated crop may be attracted by insects from the day it is planted until harvested. Pesticides are chemical compounds specifically designed to destroy harmful insects. They have proved tremendously beneficial to human population reducing or eliminating insects and other pests which transmit disease, destroy agricultural crops, damage homes and stored products, and directly or indirectly affect human health and welfare. Hence, the value of chemical pesticides in controlling mosquitos, termites, cockroaches, weevils and bovers, snails, rats and a multitude of other animals have been great, and it is difficult to imagine modern disease control and agricultural programmes without some forms of chemical control. If modern agriculture attempted to operate without chemical pesticides of some sort, crop production would probably decline, precipitously in many areas, food prices would soar far higher and food shortages would become even more severe.

Before going ahead, it is necessary to make a distinction between various terms being used in literature as under:

Insects: Any arthropod of the class Insecta, have a head, thorax, abdomen, two antennae, three pair of thoracic legs and, usually, one or two pair of thoracic wings.

Pests: A destructive animal, especially insect, attracted towards crops, livestock.

Fungicide: A fungus-destroying substance.

Herbicide: Substance toxic to plants and used to destroy unwanted vegetation.

Insecticide: Chemical substance used for killing insects.

Pesticide: A chemical substance used for destroying insects and other organism harmful to cultivated plants or animals.

Large quantities of these pesticides have been manufactured and distributed in the environment to control plant and animal pests. The environmental stability of many of these compounds is appreciable. However, the toxicity and environmental problems associated with the commercial use of halo-organics that are released in the environment has only recently been appreciated. For instance, polychlorinated biphenyls (PCB's) and chloroethenes are extremely stable industrial compounds with an environmental half life of about 15 years.

It has been estimated that 90 million kg of chloroethenes are lost to the environment each year in the process of manufacturing 2.3 billion kg of polyvinyl chloride plastic. Since vinyl chloride is a gas at room temperature this compound probably occurs mainly as an atmospheric contaminant.

A potential environmental problem of a different kind may result from the extensive use of fluorochlorocarbons, as propellants in aerosol cans and as refrigerants in refrigeration systems. These compounds are colourless, odourless, and essentially stable in the biosphere. An illustration of their stability is demonstrated by observing that dichlorodifluoromethane decomposes at the rate of 1% a year at 500°C. Carbon tetrachloride, an important industrial solvent, is almost as stable as the Freons. Freons and carbon tetrachloride are not susceptible to environmental degradation until they reach the stratosphere, where they are photodissociated by ultraviolet light. This process may lead to partial destruction of the ozone layer (as discussed in Part 1 of this book). Even a partial loss of ozone layer could result in a dangerous increase in the amount of UV-radiation reaching the surface of the earth.

8.2 CHEMISTRY OF CHLORINATED ORGANIC COMPOUNDS

Organochlorine derivatives are prepared industrially by direct chlorination of hydrocarbons. Chlorination is useful in the preparation of organochlorine compounds, such as 1,3 dichloro -1 propene, which are effective in killing nematodes (soil worms). The overall reaction is

$$CH_2{=}CHCH_3 + 2Cl_2 \longrightarrow ClCHCH{=}CH_2Cl + 2HCl$$

Aromatic compounds are generally more difficult to chlorinate than aliphatic compounds. The aromatic chlorination reaction proceeds by an ionic pathway. Polychlorinated biphenyls are prepared from biphenyl by direct chlorination using ferric chloride as a catalyst. The mechanism involves the initial formation of a chloronium ion that undergoes an electrophilic addition to the aromatic ring system

$$Cl_2 + FeCl_3 \rightleftharpoons FeCl_4^- + Cl^+$$

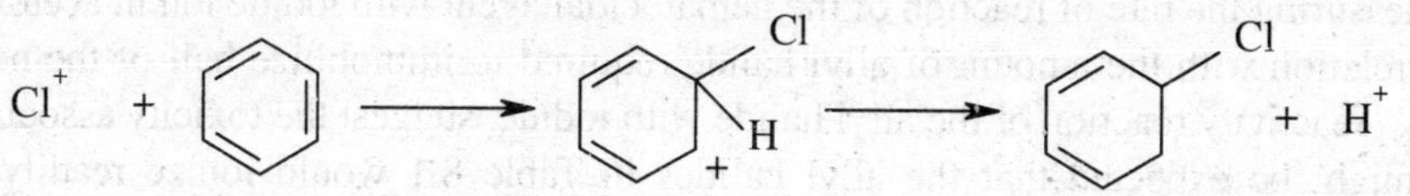

Formation of polychlorinated biphenyls may take place on chlorination of wastes in sewage disposal plants from textile mills which may be dangerous because it is environmentally stable. Hence, chlorination of waste should be avoided.

Reactions: The aliphatic organochlorine compounds are generally more reactive than the corresponding aromatic derivatives. Two limiting reaction pathway (substitution and elimination) are observed for aliphatic chloroderivatives in the presence of nucleophilic reagents. Substitution may proceed by a unimolecular (S_{N1}) (Reactions 8.1 + 8.2) or bimolecular (S_{N2}) (Reaction 8.3) pathway. The rate-limiting step of the S_{N1} reaction is ionization of the alkyl halide to form the planar carbonium ion (8.1). Whereas the rate of the S_{N2} reaction is proportional to the concentrations of both alkyl halide and nucleophile (8.3)

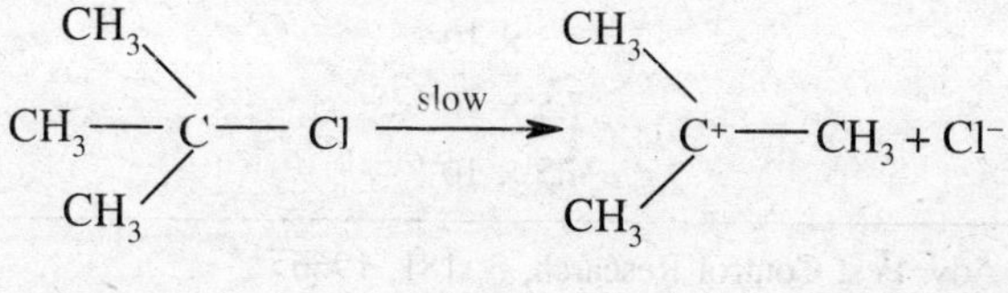

...(8.1)

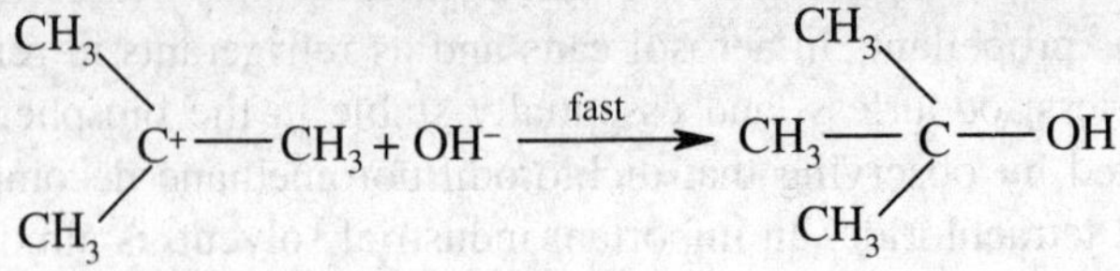

...(8.2)

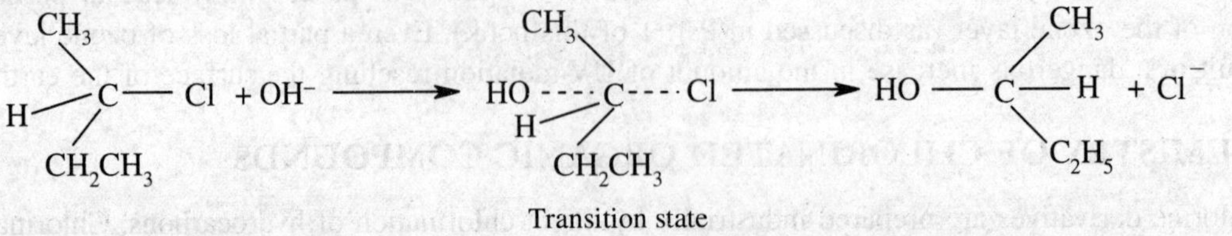

...(8.3)

The mechanism of the substitution reaction is determined by the solvent, substituting reagent and structure of derivative. The S_{N1} (carbonium ion) mechanisms is favoured by polar solvents that stabilize the ionic intermediates or with poor nucleophile such as water is the displacing agent. Carbon compounds, which are sterically hindered, halogenation are resistant to direct (S_{N2}) displacement or which contain tertiary, allylic, benzylic halogens ionize to stabilised carbonium ions, usually react by S_{N1}, pathway. The formation of an intermediate carbonium ion is the most likely pathway under environmental conditions. Usually the environmental reactions occur in a polar solvent (water) near pH 7 with a poor nucleophile (water) or with low concentration of nucleophile substituting agent, *i.e.* S_{N1} conditions.

It has been observed that the toxicity of nematocides is proportional to their S_{N2} reactivity, is determined by measuring the rate of reaction of the nematocidal agent with iodide ion in acetone (Table 8.1). A good correlation with the amount of allyl halide required to immobilize half of the nematodes (ED_{50}) and the S_{N2} reactivity reaction of the allyl halide with iodide, suggest the toxicity associated with a nematode. It might be expected that the allyl halides in Table 8.1 would ionize readily to form

Table 8.1 S_{N2} Reactivity and nematocidal action[a]

Compound	ED_{50}^{b}	S_{N2} *Reactivity (I^- in acetone)*
$CH_2{=}CHCH_2Cl$	1.5×10^{-3}	1.0
$CH_2{=}CCl\ CH_2Cl$	2.6×10^{-4}	0.72
$CH_2{=}C(CH_3)CH_2Cl$	2.8×10^{-4}	1.58
$HC{\equiv}CCH_2Cl$	2×10^{-4}	1.78
trans–$ClCH{=}CHCH_2Cl$	7.7×10^{-5}	2.9
cis $ClCH{=}CHCH_2Cl$	3.3×10^{-5}	8.6
$CH_2{=}CHCH_2Br$	7.5×10^{-5}	506
$HC{\equiv}CCH_2Br$	4.5×10^{-6}	909

(*a*) Adapted from W. Moje, Adv. Pest Control Research, 3, 181, 1966.

(*b*) Molar concentration required to produce 50% inhibition of the mobility of the citrus nematode *T. semipenetrans.*

carbonium ions, but the S_{N2} reaction is not favoured under these conditions because acetone is a weakly polar solvent and iodide is a very good nucleophile.

Two limiting reaction mechanisms are available for the elimination reaction; a unimolecular (El) (reaction (8.1 and 8.4)) and biomolecular (E2) (reaction 8.5) pathway

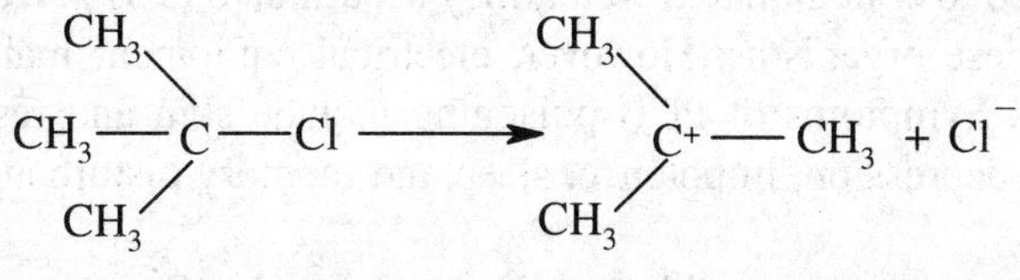

...(8.1)

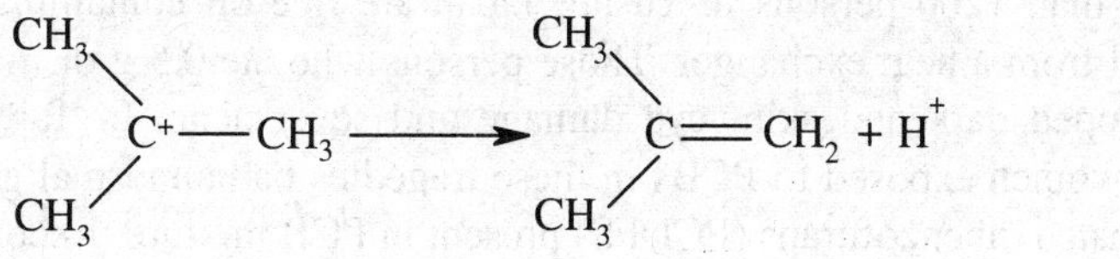

...(8.4)

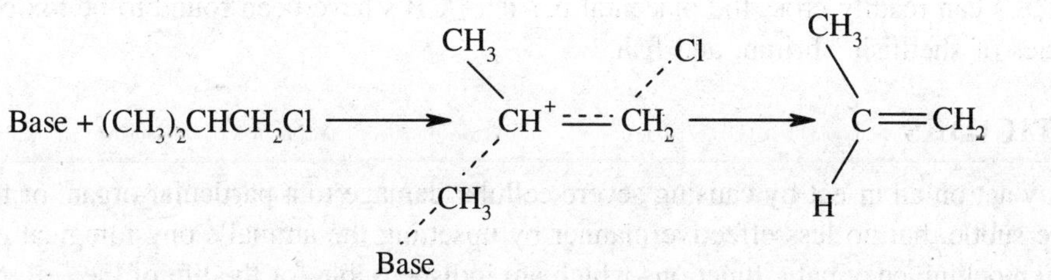

...(8.5)

Those factors favouring the S_{N1} reaction also favour El elimination because the same carbonium ion intermediate is involved, (reaction 8.1). The relative proportion of elimination to displacement is determined by the relative nucleophilicity and basicity of the group attacking the carbonium ion. The E2 elimination is observed with strong bases in nonpolar solvents and is not likely to take place under environmental conditions. As for instances, the dehydrochlorination of DDT and DDT analogues with sodium hydroxide proceeds by an E2 mechanism but it may not take place in the environment.

8.3 POLYCHLORINATED BIPHENYLS (PCB)

The polychlorinated biphenyls have been available commercially for over five decades. The synthesis from chlorine and diphenyl yields a mixture of derivatives with the extent of chlorination being dependent on the chlorine-biphenyl ratio. These compounds were sold in the USA by Monsanto Co. Under the trade name of *Aroclor: Aroclor* products containing various amounts of chlorine are differentiated by a four digit number; the last two digits indicate the percentage of chlorine.

PCBs are resistant to acid, base, heat and oxygen. This extreme stability makes them especially useful as a dielectric material in capacitors and transformers. They have also been used as plasticizers and solvents in plastics and printing inks. However, since 1970, Mansanto has refused to sell PCB's to customers in plasticizer operations, where disposal of the end product is not controlled. As a consequence they are now only being used in 'Captive' applications, such as transformers where they can be recovered. However in Delhi when a transformer is replaced, the old one either discarded or sent for repair in DUSU workshop courtyard with pool of oil. Nobody is there to tell the workers that they are handling the poison. However PCB's are not used as insecticides. Its manufacture in Europe, has been totally stopped. Closely related compounds, to be polybrominated biphenyles are used as flame retardants.

PCBs are not very water soluble, but mix easily with oil or fat. When organisms like algae and phytoplankton are exposed to contaminated water, they accumulate PCB's. Humans can be poisoned by eating fish feeding on these organisms. However, electrical equipment maintenance workers are directly exposed to PCB's. Symptoms of PCB poisoning include skin and respiratory problems, headaches, fatigue, dizziness, depression, impotence, sleep and memory disturbance. PCB poisoning is also linked to cancer.

Toxicity of PCB's has not been established. However, toxic effects were amply demonstrated in 1968 when more than 1200 persons in Yusho, Japan ate rice oil contaminated with PCB's that had leaked into the oil from a heat exchanger. Those persons who ate 0.5 g or more (average consumption was 2 gm) developed darkened skin, eye damage and sever acne. In Taiwan, 2,061 died in 1978. Children born to women exposed to PCBs in these tragedies had abnormal growth. These deaths were due to polychlorinated dibenzofurans (PCDFS) present in PCB mixture. Recovery was slow with symptoms still present even after three years. Several infants were born with the same synptoms, demonstrating that PCB's can readily cross the placental barrier. PCB's have been found to be toxic to mink and some species of shellfish, shrimp, and fish.

8.4 INSECTICIDES

Insecticides may act on an insect by causing severe cellular damage to a particular organ, or they may work in a more subtle, but no less effective manner by upsetting the animal's physiological organisation, usually by blocking enzymatic functions which are indispensable for the life of the cell. Nicotine, for example, acts on the functioning of heart and nerves, cyanides block cellular respiration by acting on the cytochrome oxidase enzyme, phosphoric ester inhibits the enzyme cholinesterase and thus blocks the transmission of impulses along the nerves. However, many insecticides act simultaneously on several organs and affect several important and vital functions. In 1973, C.M. Williams divided pesticides into three types:

First Generation Pesticides

These are as follow:

(*i*) **Fumigant Insecticides:** These include sulphur dioxide, ethylene oxide, carbon disulphide, hydrocyanic acid, methyl bromide, carbon tetrachloride, trichloro ethylene, paradichloro benzene.

(*ii*) **Mineral Insecticides:** Arsenic and derivatives of arsenic like As_2O_3, Paris Green $(MeCOO)_2\ Cu_3Cu(AsO_2)_2$, Sodium arsenite, calcium arsenate, lead arsenate, fluorides and fluorosilicates (barium fluorosilicate).

(*iii*) **Organic Insecticides of Plant Origin:** Nicotine, pyrethrum and pyrethrin, rotenone.

Second Generation Pesticides

Synthetic Organic Insecticides: These are of three types (*i*) Organic halogen compounds (*ii*) Organic phosphorous compounds (*iii*) Carbamates.

DDT, BHC, Aldrin, dieldrin, endrin, toxophore, etc., are called persistent chemicals with half lives of several years and perhaps, even decades. The other two groups have relatively short half lives and are easily biodegradable into harmless substances, so do not pose any serious environmental problems when compared to organochlorines.

Third Generation Pesticides

Pheromones and Attractants.

Although the First Generation insecticides are not used all around the world but these will be discussed here.

8.5 PESTICIDES INTERFERING WITH RESPIRATION

All higher organisms rely on respiration to breakdown organic molecules thereby providing energy and raw materials necessary for their continued survival. Consequently, all are, in principle, vulnerable to the action of inhibitors of this process.

Respiration falls into a number of phases. Organic food stuffs contain carbohydrates, fats or proteins and these are first broken down to their constituent monomers which are subsequently oxidized through glycolysis.

Splitting of sugar and/or the tricarboxylic acid (Kreb's) cycle in which two carbon units of acetyl CoA (a thioester) with sulphur-containing coenzyme A takes part in a variety of acetyl transfer reactions and are degraded to CO_2. For glucose the process is described by

$$C_6H_{12}O_6 + 6O_2 \longrightarrow 6CO_2 + 6H_2O \quad ...(8.6)$$

During the oxidative reactions of these procedures, nicotinamide adenine dinucleotide (NAD^+) is reduced to NADH, thus conserving the reducing power of substrates. The reducing equivalents are in turn, relinquished through a chain of electron carriers during respiration chain electron transport, so that the NAD^+ is regenerated, and electron flow through the chain reactions that culminates in the reduction of oxygen to water. This electron transport is coupled to the synthesis of the high energy intermediate, adenosine 5-triphosphate (ATP) from diphosphate (ADP) and in organic phosphate by a process called oxidative phosphorylation. A certain amount of ATP is also formed directly during glycolysis.

The reaction of glycolysis takes place in the soluble part of the cell, the cytoplasm. The pyruvate is further metabolised inside the mitochondria by a reductive decarboxylation which yields acetyl CoA. The conversion of pyruvate to acetyl CoA is accomplished by pyruvate dehydrogenase complex on which arsenic containing pesticide are thought to act. The overall reaction is

$$\underset{\text{pyruvate}}{CH_3COCOOH} + NAD^+ + CoASH \longrightarrow \underset{\text{acetyl CoA}}{CH_3CoSCoA} + NaDH + H^+ + CO_2 \quad ...(8.7)$$

Kreb's cycle has an enzyme α-oxo glutarate dehydrogenase which also proceeds by an analogous mechanism.

The treatment of insects with arsenic-containing compounds causes inhibition of respiration. Arsenite ion inhibits keto-acid oxidation by combining with two thiol groups in the lipoamide moiety of the dihydrolipoamide transacetylase.

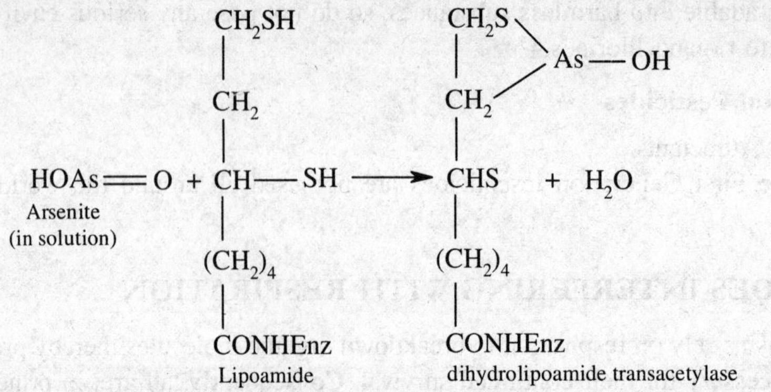
...(8.8)

Arsenite ions do not react with any group of biological importance other than thiols. Reaction 8.8 will inhibit the occurring of reaction 8.7 and thus this insect will die.

The biggest disadvantage of first generation pesticide lies in the fact that chemicals may remain in or on the crop, or in the soil, drift to other nearby crop areas, flow into streams and drainage, and inhibit respiration in man or animals by a similar mechanism as (8.8), thereby create hazard for man or his animals and his environment. This was the reason that inorganic pesticides had to be given up.

8.6 SECOND GENERATION PESTICIDES

DDT (Dichloro Diphenyl Trichloroethane) also called (1, 1, 1, trichloro - 2,2 bis (p-chlorophenyl) ethane. DDT is a broad-spectrum, persistent organochlorine pesticide. It has been used against crop predators such as apple, maggots and also against the vectors of human diseases such as malaria, yellow fever and plague. It is particularly effective against malarial mosquitoes.

Synthesis: It is synthesised by the interaction of chlorobenzene and chloral in an acidic medium by a Friedel-Crafts type reaction. The mechanism of the reaction involves the electrophilic addition of a carbonium ion species to aromatic ring as shown by reactions 8.9 to 8.11.

$$CCl_3CHO \overset{H^+}{\rightleftharpoons} CCl_3\overset{+}{C}H{-}OH \qquad ...(8.9)$$

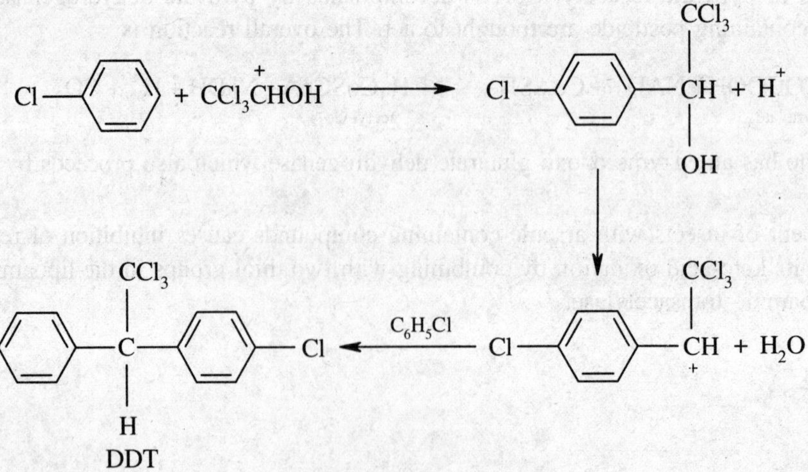
...(8.10)

...(8.11)

The *p, p*-isomer shown is not the only reaction product. Appreciable amounts of *o, p′* and *o o′*-DDT are also obtained. This synthetic approach has been applied for the preparation of a great variety of DDT analogues. One important analogue is methoxy chlor, currently being used in place of DDT in many applications because it is biodegradable. However, it is less effective as a pesticide and about three times as costly to prepare because anisole is a more expensive starting material than chlorobenzene.

The half life of DDT in the environment is about eight years. It undergoes a relatively rapid elimination of hydrochloric acid to yield DDE (1, 1, dichloro 2, 2 bis (*p* chloro phenyl) ethene (II). However, the subsequent hydrolysis of DDE to DDA [1, 1, dichloro 2,2′ bis (*p*-chlorophenyl) acetic acid] (III) is extremely slow since the structure of DDE involves the relatively inert aryl chloride and vinyl groups. The degradation of DDT is shown in reaction 8.12 but complete degradation of DDT in soils and in cultures of microorganisms is given in Table 8.2.

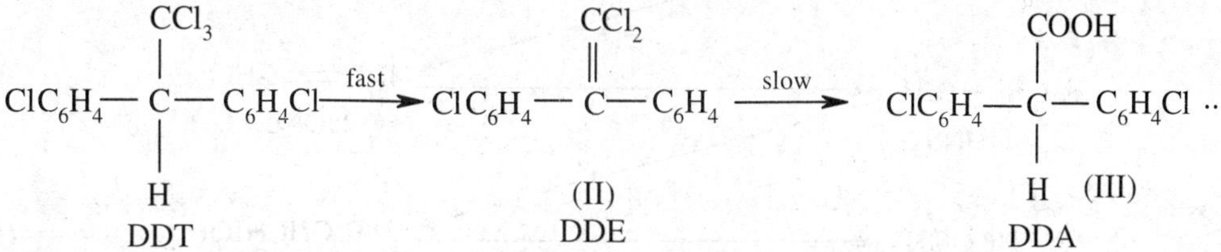

...(8.12)

Mechanism of action of DDT and DDT analogue: Insects sprayed with DDT exhibit hyperactivity and convulsions consistent with the action of the DDT on the nervous system. Many theories have been suggested for the toxic effect of DDT, but the exact mechanism is not known with certainty. A plausible theory with experimental basis is that the DDT molecules are of the correct size to be trapped in the pores of the nerve membranes.

This distorts the membrane and sodium leaks through and depolarizes the nerve cell, so that it can no longer transmit impulses. This theory states that the toxicity of DDT is not due to its chemical activity but to its size and geometry that allows for the blockage of the pores of the nerve membranes. The theory is supported by the observation that a variety of quite different compounds, that are stereochemically similar to DDT, exhibit DDT-like activity.

A further evidence for this theory is provided by the fact that the biological activity of methoxy chloro analogues decreases rapidly when the alkyl group, contains the longer chain. Methoxy chloro analogous are not able to fit in the nerve pores.

Insects are especially susceptible to DDT because it is readily absorbed through insect cuticle. DDT is not appreciably absorbed through the protein skin of mammals. If the polarity of the DDT analogue is increased by introducing —NO_2, —COOH, $COCH_3$ and OH substituents into the aromatic ring, the toxicity is lost. This loss is probably due, in part, to decreased absorption through the cuticle and, in part decreased absorption on the nonplanar nerve membrane.

Insect resistance to DDT: When DDT came into widespread use after world war II, it appeared to be effective against a number of insect species. However by 1950 its effectiveness seemed to decline. The reason was that the insects that survived DDT, produced succeeding generations that were increasingly resistant to DDT. Therefore, it became necessary to use large quantities of DDT to control resistant strains. This resulted in a rapid build-up of DDT and DDE in the environment. It was observed that DDT resistant flies were knocked down by normal doses of DDT, but after some buzzing around on the floor they eventually recovered. This observation suggested that the resistant strains have a mechanism for rapidly detoxifying the DDT. Subsequent investigations revealed that DDT-resistant strains contained an enzyme (DDT dehydro chlorinase) that was present in those flies which were susceptible to DDT and that it catalyzed the conversion of DDT to DDE.

Table 8.2 Degradation of DDT

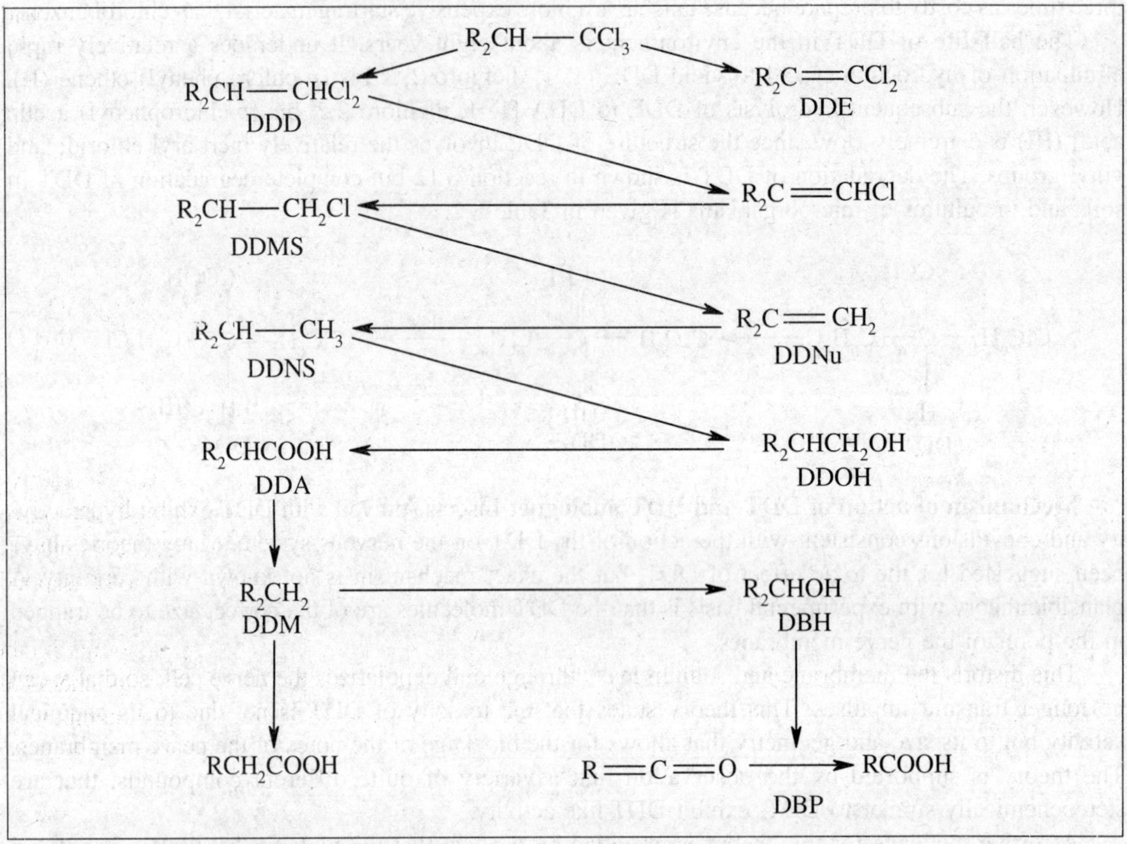

An interesting result that is consistent with the proposed action of DDT dehydrochlorinase is the observation that a deuterium substituted DDT derivative is more effective than DDT against resistant strains. Yet it exhibits the same level of toxic effect against nonresistant strains as DDT. The enzyme reaction of DDT probably involves the cleavage of C—H (or C—D) and C—Cl bonds in the rate determining step. However, the removal of deuterium by base is slower than proton removal, because of greater strength of the C—D bond (5 kJ mol^{-1}) as compared to the C—H bond. Since the deuterated analogue is converted to DDE more slowly by DDT dehyrdrochlorinase, it has a greater time to extend its toxic effect on the resistant strains.

Resistance of insects to insecticide is a commonly observed phenomena. Resistance develops where 100 per cent insect mortality does not occur and where sublethal doses of the insecticides remain in the vicinity for long periods of time. The surviving insects breed under these conditions and only those progeny that are resistant manage to survive in the presence of the insecticide. Resistance is not observed as frequently with those insecticides that are destroyed rapidly in the environment.

A non-resistant population of insect may build-up again when the insecticide is no longer used. This occurs if the nonresistant strain (wild type) adapts more successfully to the insecticide-free environment than the resistant strain. However, a population of resistant insects form very rapidly if the insecticide is reintroduced.

DDT is poisonous to human beings and carcinogenic in nature. The smallest fatal dose is estimated to be 500 mg/kg. Nevertheless, human diet containing 1 ppm or higher levels of DDT enhances the activity of enzymes in the liver. The activated enzymes enable the liver to attack poisons and drugs. If the individual happens to be taking medications these may be broken down more easily than is desirable. Moreover, the enzymes activated by DDT attack hormones. It is also speculated that regular consumption of diet containing higher levels of DDT may cause birth defects and cancer. As a result the FAO has prescribed that the daily intake of DDT through diet should not exceed 0.01 mg/kg.

Bioamplification: DDT has proved to be hazardous to several species of fish and bird because it is not easily metabolised by the animals. Rather, it is known to concentrate in the fatty tissues. Its concentration progressively increases in those species which occupy a higher trophic level in the food chain. This phenomena is called bioamplification.

As an example, spraying a marsh to control mosquitoes will result in the accumulation of traces of DDT in the cells of microscopic aquatic organisms such as plankton. The DDT then becomes concentrated in the tissue of the fish and shell fish which feed on planktons. This may lead in turn to residues in small fish then to larger fish and finally to such fish-eating birds as gulls, ospreys, bald eagles, pelicans, and falcons. The level of accumulated DDT are often so high that the consequences may be disastrous.

The birds of prey, which occupy a significantly high position in the food chain concentrate enormous quantities of DDT in their tissue. Their body attempts to metabolize large amounts of DDT by an alteration of the normal metabolic pattern. This alteration involves the use of hormones which normally regulate the calcium metabolism of the bird and are vital to its ability to lay eggs having thick shells. When these hormones are used to metabolise DDT, they are chemically modified and are no longer available for the construction of eggshells. As a result the eggs are easily damaged and survival rate of birds decreases.

As in human system, the presence of DDT in the bird's body, increases the activity of enzymes in the liver. The activated enzymes attack the bird's sex hormones. This leads to three consequences: firstly, breeding takes place abnormally late. Secondly, new eggs are not laid to replace early losses, and thirdly, fewer eggs are laid. All these factors result in a decline of bird population.

It has been shown that cows fed on forage previously treated with DDT accumulate the insecticide in their milk. The cream, in particular, is rich in DDT and so is butter made from it. Calves fed with this milk can exhibit fairly serious signs of poisoning.

Initially, the effectiveness with which DDT eliminated the insect-pests was so spectacular that Paul Mueller who discovered the pesticidal properties of DDT was awarded the Nobel Prize in 1948. However, gradual ecological problems began to emerge as a result of which the application of DDT was perforce restricted if not completely prohibited.

8.7 LINDANE

Lindane is synthesised by the free radical addition of chlorine atoms to benzene. Of the eight possible isomers, only four are formed in appreciable amounts (reaction 8.14). These isomers differ in the relative orientation of the chloro-substituents. Axial (*a*) groups are perpendicular to the cyclohexane ring and are labelled in the α-isomer in 8.13. Equatorial groups are in equatorial plane of the ring as shown in the β-isomer. Only one of these isomers, the γ-isomer known as Gamma-HCH, exhibits significant insecticidal action. Initially the total reaction mixture was used as the insecticide.

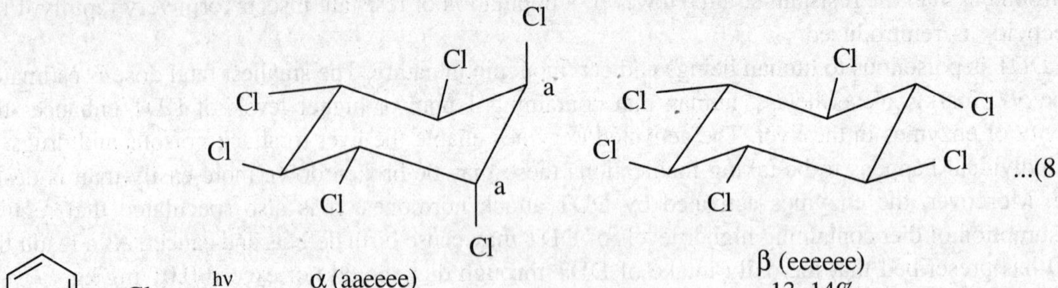

α (aaeeee) 53–70%   β (eeeeee) 13–14% ...(8.13)

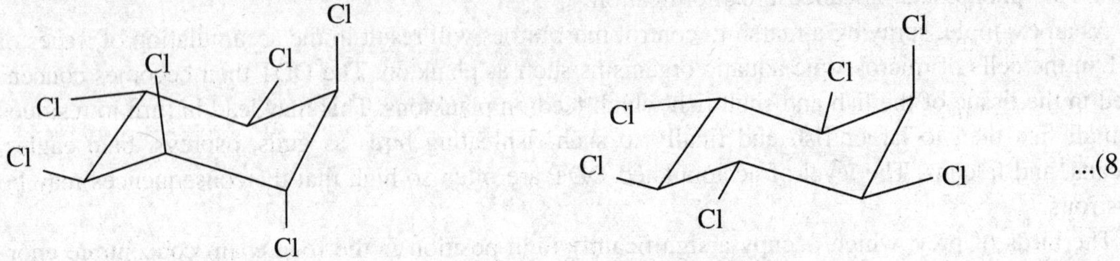

γ (aaaeee) 11–18%   δ (aeeeee) 6–10% ...(8.14)

However, that practice has been discontinued because it results in the distribution of large amounts of inactive chlorocarbons in the environment.

Lindane tends to accumulate in the fatty tissues of mammals, thus it is no longer used in animal husbandry. It is used to control limited insect infestations such as borers, beetles, and hornets and it has been used in flea collars. Insects treated with gamma-HCH develop violet quivering of the body, followed by convulsions and paralysis. It causes excessive release of acetylcholine and hence, leads to hyper activity at cholinergic synapses.

Environmental Degradation: Lindane is a very inert chloro-organic compound, which undergoes a slow elimination of HCl in aqueous solution (reaction 8.15). Presumably this is an E_1 reaction

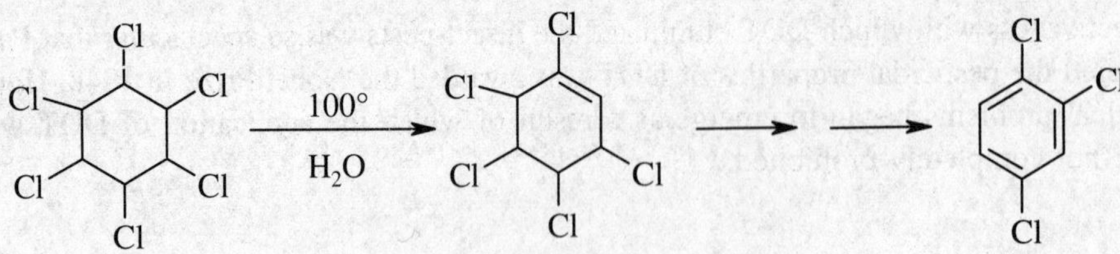

and other isomer ...(8.15)

A series of what are probably E2 elimination reactions (8.15) occurs in an aqueous base to give the same trichloroaromatic compounds that are observed in neutral solution. From these limited data, it is clear why lindane and its aromatic degradation products persist for a long time in the environment.

8.8 POLYCHLORINATED CYCLOPENTADIENE DERIVATIVES

A series of bicyclic insecticides such as chlordane, heptachlor, aldrin and dieldrin can be synthesised by Diels-Alder addition of hexachloro cyclopentadiene to olefins. The preparation of these compounds is shown in reactions 8.16–8.18. Obviously, it is possible to synthesize a whole series of potential insecticides by this route by simply varying the diene and olefin.

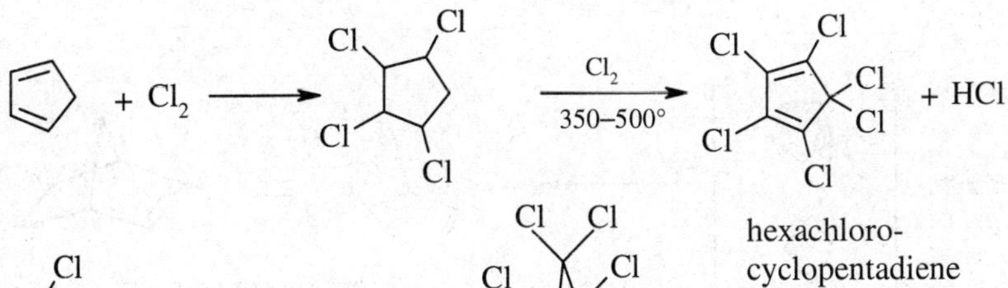

...(8.16)

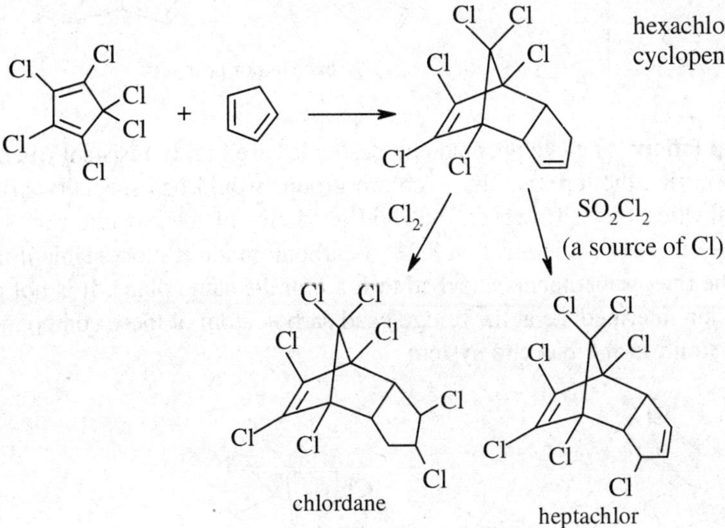

...(8.17)

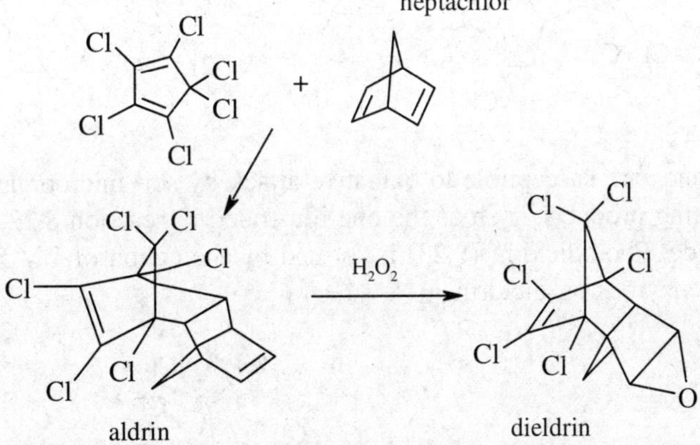

...(8.18)

The endo adduct (see the structures) is the major product resulting from the Diels-Alder addition of a diene and an olefin. No intermediate has been detected, but the reaction product can be rationalised as originating from a transition state in which there is maximum interaction between the pi-bonds of the olefins and a diene. This is illustrated in reactions 8.19–8.20, in which the upper cyclopentadiene may be regarded as the olefin and the lower cyclopentadiene, the diene.

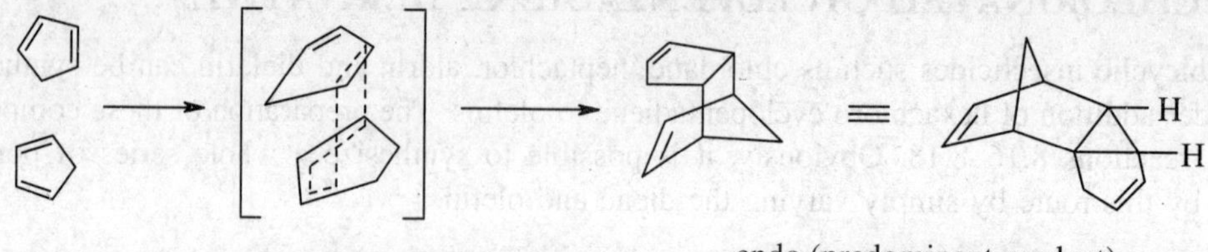

...(8.19)

endo (predominant product)

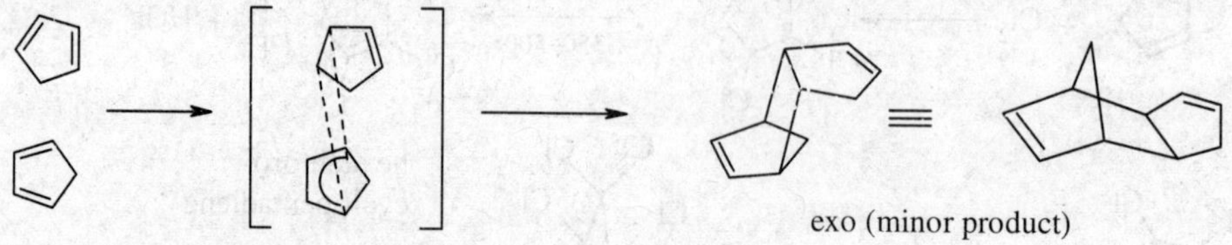

...(8.20)

exo (minor product)

Environmental Degradation: The cyclopentadiene pesticides are highly resistant to environmental forces. One might expect that the tertiary, allylic chloro groups would be especially susceptible to S_N1 hydrolysis or E1 elimination (8.1). However, the rigid geometry of the bicyclic ring system prevents carbonium ion formation as shown in reaction 8.21. A carbonium ion is more stable if the positive central carbon atom and the three substituents attached to it are in the same plane. It is not possible to obtain a planar carbonium ion intermediate at the bridge head carbon atom of these compounds without introducing extraordinary strain in the bicyclic system.

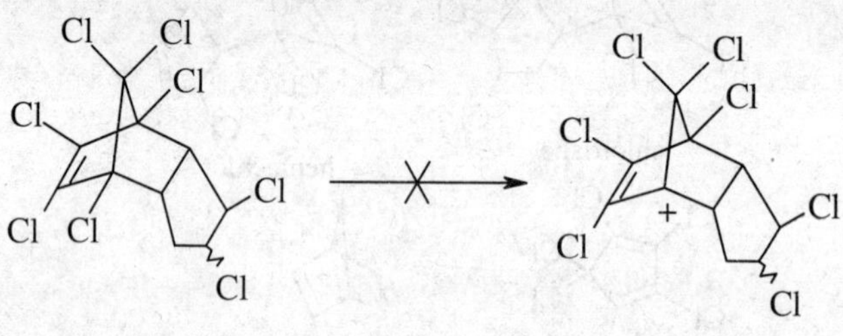

...(8.21)

These compounds are susceptible to oxidative attack by soil microorganisms, plants and animals. However the resulting products, such as the one illustrated in reaction 8.22, are often more toxic than the starting pesticide. Photodieldrin (8.23) is formed by the action of UV radiation as well as by the action of oxidase enzymes on dieldrin.

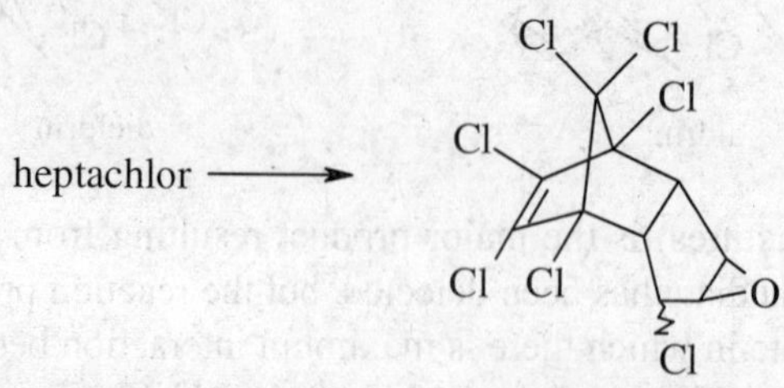

...(8.22)

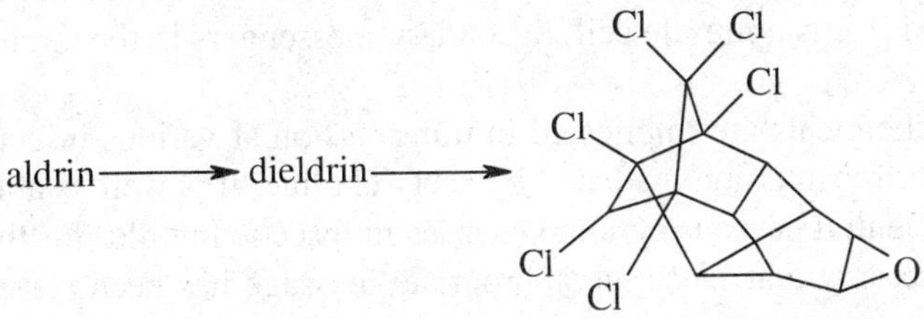

...(8.23)

Most of the pesticides in current use act by interfering with the passage of impulses in the pest's nervous systems. Insects depend on an integrated nervous system which enables external stimuli to be translated into effective action. The nervous system contains specialised cells called neurons, which consist of a cell body and a long process called an axon (Fig. 8.1). The axon terminates near a receiving cell of some sort and the gap between the cells is called synapse. Axonal transmission in insects is mediated by an ionic mechanism and that in the case of most synapses, when the impulse reaches the end of the axon it triggers the release of a transmitter substance into the synapse. The exact mechanism of transmitter release is not fully understood. It is known, however, that arrival of an impulse at the nerve ending leads to a flow of calcium into the cell, and, in some way, this brings about release of the transmitter perhaps from vesicles in which it may normally be stored. The transmitter diffuses across the gap and combines with its receptor on the post-synaptic membrane. The binding alters the ionic

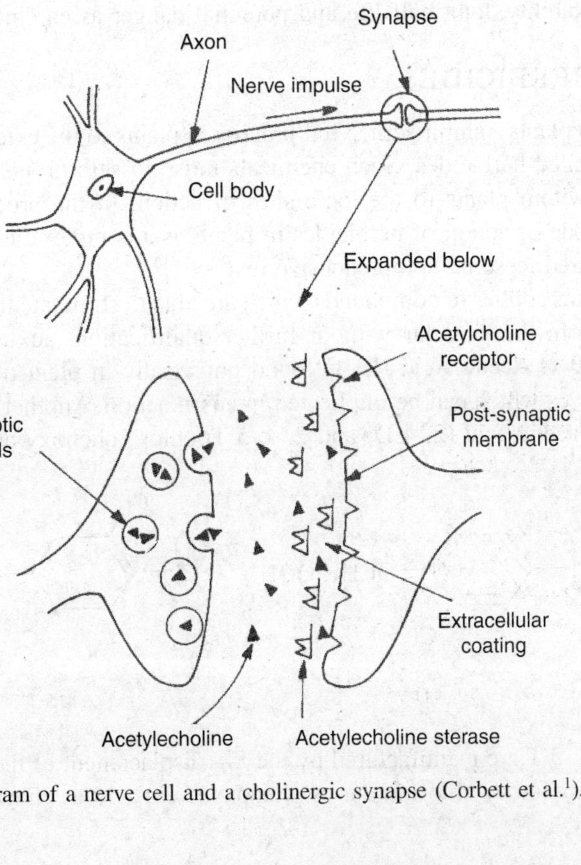

Fig. 8.1 Diagram of a nerve cell and a cholinergic synapse (Corbett et al.[1]).

permeability of the membrane and this results in the post-synaptic cell passing an impulse if it is a nerve cell, or in muscle contraction if it is a muscle cell. Secondary messengers in the target cell may also be activated.

A number of different chemicals are implicated in transmission at various insect synapses but we shall be concerned only with transmission mediated by acetylcholine. It is well established that acetylcholine is the transmitter at central nervous system synapses in insects. In order for the nervous system to operate properly it is necessary that, once the appropriate message has been passed, excess acetylcholine should be removed from the synapse, both to prevent repetitive firing and to allow a succeeding message to be transmitted. This removal is affected by the enzyme acetylcholinesterase, which catalyses hydrolysis of the ester bond as follows:

$$\underset{\text{Acetylcholine}}{(CH_3)_3N^+CH_2CH_2OCOCH_3} \longrightarrow \underset{\text{Choline}}{(CH_3)_3N^+CH_2CH_2OH} + CH_3COOH \quad ...(8.24)$$

The result of this inhibition is that acetylcholine accumulates in synapses so that proper nerve function is impaired. This leads to degeneration of vital tissues following loss of water from the insect integument which leads to death of insect. Insect treated with cyclodiene pesticide, related compounds and gamma HCH develop violent quivering of the body followed by convulsions and paralysis. This is due to the excessive release of acetylcholine and hence leads to hyperactivity at cholinergic synapses.

Dieldrin is very slow to act whereas aldrin exhibits its effect quickly. It has been found that dieldrin, although slow acting, was much more toxic than aldrin.

The manufacture of aldrin and dieldrin and other related compound were banned in the USA in 1974 because of stability, long half life and potential danger as carcinogens.

8.9 AUXINS HERBICIDES

Chemicals toxic to plants, inhibit some vital process in plants to the extent that the plants die or can no longer grow are called herbicides. Such chemicals must possess properties that enables them to enter plants and move within plants to the toxic sites of action. In the process, growth of weeds etc. are controlled. The mode of action of herbicides in plants is a complex field and cannot be covered here. The reader is referred to some comprehensive text[12, 13, 14].

The terms auxins applies to compounds which are able to stimulate the growth of excised coleophile tissue. When used in the singular without further qualification, auxin refers to naturally occurring compound Indole-3-yl Acetic Acid (IAA) found universally in plant tissue with closely related molecules and derivatives which can be implicated in auxin action. Another two important auxins are 2, 4, Dichloro phenoxy acetic acid (2, 4 D) and 2, 4, 5 Trichloro phenoxy acetic acid (2, 4, 5 T)

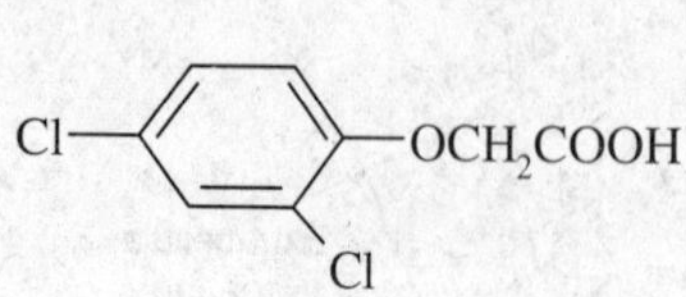

2, 4 D

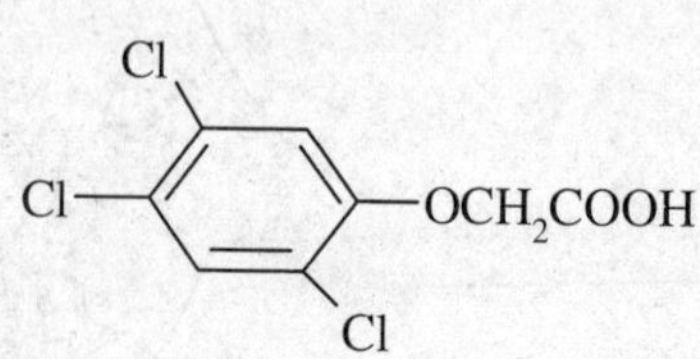

2, 4, 5 T

2, 4 D and 2, 4, 5 T are manufactured by the S_{N2} displacement of the chlorogroup of chloroacetate with phenoxide anion or by chlorination of phenoxyacetic acid as shown below:

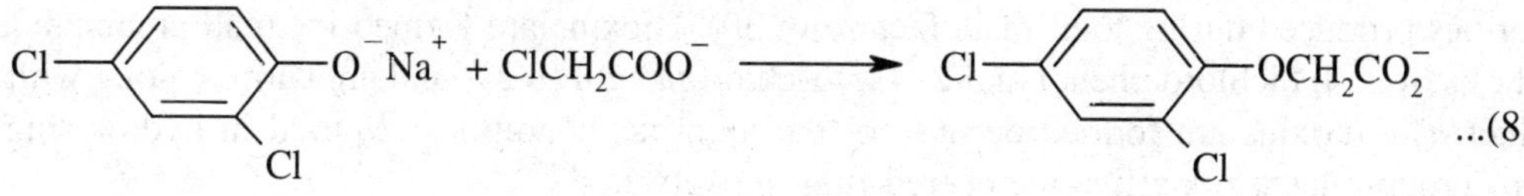

...(8.25)

2, 4 D

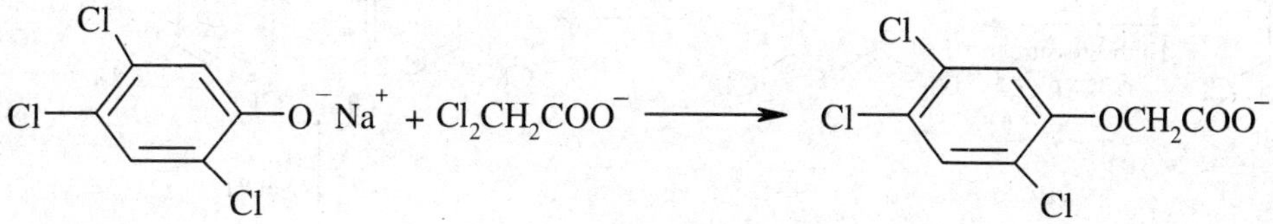

...(8.26)

2, 4, 5 T

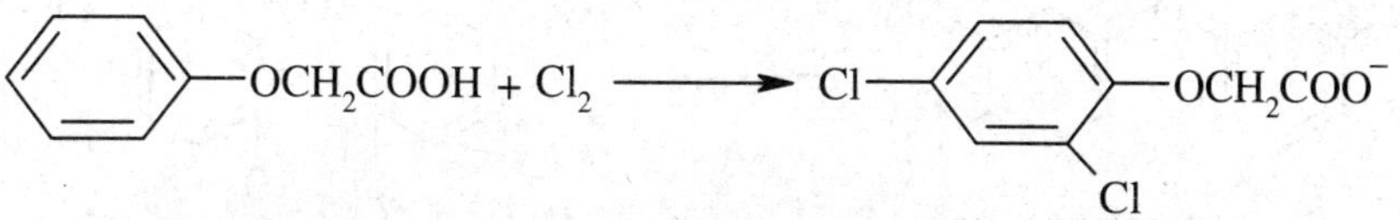

...(8.27)

2, 4 D

A variety of closely related 2, 4 D analogues also shows herbicidal activity. The carboxyl can be replaced by the amide, nitrile or ester groups. In addition, a longer side chain can be used in place of the acetic acid moiety. However, this side chain must contain an even number of carbon atoms to be active. This result is consistent with β-oxidation of the side chain in the plant to form 2, 4 D.

Environmental Degradation: In recent years a good deal of attention has been focussed on auxin-binding components of plant tissues. Assuming that there must be a site to which auxin can bind in order to trigger the correct cell responses. Both soluble and membranous auxin binding components have been identified and there may well be binding sites with different properties located on the various membranes of the cell.

The secondary question of what changes are brought about when auxin binds to a site has not been answered. It is recognised that the growth stimulation shown by excised stem section falls into two phases. First there is a rapid response (within minutes), simulated by low pH and perhaps due to auxin stimulating the pumping of protons into the cell wall and loosening it. The second phase of the response occurs 35–45 min after treatment. Inhibitor work shows it to be dependent on macromolecule synthesis and ribonucleic acid synthesis in particular is implicated.

Pure 2, 4 D and 2, 4, 5 T are subject to rapid environmental degradation. These compounds are cleaved by microorganism to the corresponding phenol and oxalic acid which are readily degraded further, and subsequently. In addition, the photochemical cleavage of the acetic acid side chain, replacement of ring chlorines with hydroxyls followed by polymerization to humic acids. This suggest that the long term environmental impact of these compounds is small:

$$\underset{\text{2, 4 D}}{Cl_2C_6H_3-OCH_2COOH} \longrightarrow Cl_2C_6H_3-OH + \begin{matrix} COOH \\ | \\ COOH \end{matrix} \qquad ...(8.28)$$

The environmental problem which has been associated with these compounds is due to dioxin impurities produced during their manufacture (8.30). Dioxins are formed in small amounts during the synthesis of 2, 4, dichloro phenol and 2, 4, 5 trichlorophenol (8.29) and are carried along with the final product. The dioxins are formed because of the vigorous conditions required in hydrolysing the aryl chloro group-related toxic five-membered ring derivatives :

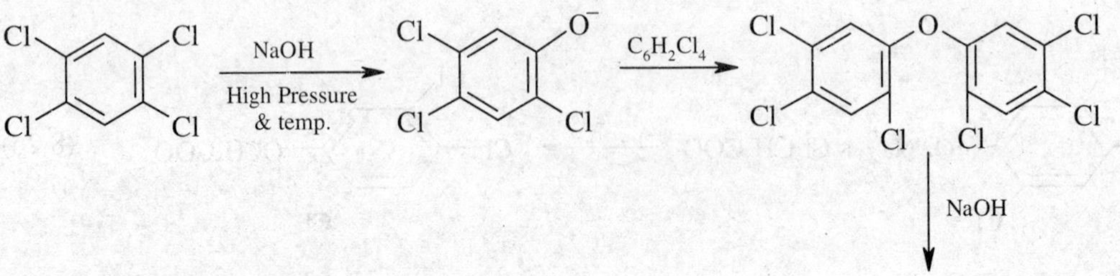

...(8.29)

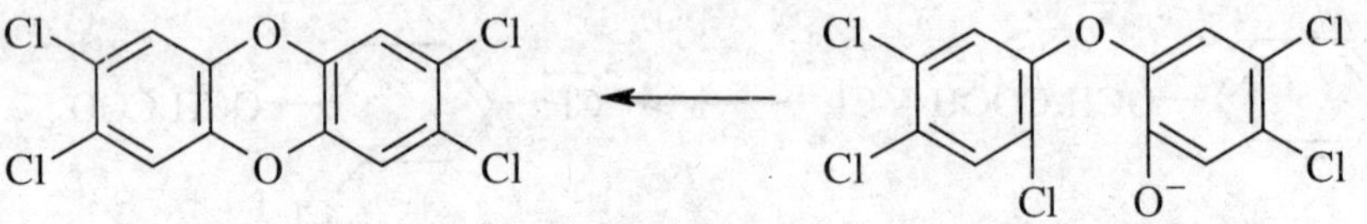

...(8.30)

2, 3, 7, 8 tetrachlorodioxine

Dibenzo furans formed in the manufacture of PCB's may be responsible for the toxicity of these compounds.

The dioxins produced during the manufacture of 2, 4, 5 T have been shown to be toxic, tratogenic (cause birth defects) and acnegenic in laboratory animals.

The toxicity to humans and a wide array of mammals was demonstrated when dioxins were accidentally applied to the soil in a horse arena. Contact with only 32 micrograms of dioxins per gram of soil was sufficient to kill the birds, cats, dogs and horses that used the arena. Humans were less sensitive but one child developed hemorrhagic cystitis and several other skin lesions. The acnegenic effect has been observed in workers involved in the manufacture of 2, 4, 5 T. Similar toxic effects were noted in 1976 when 1–5 kg of tetrachloro dibenzodioxin was released over a 123 acre area in northern Italy from a plant which manufactures 2, 4, 5 trichlorophenol. High levels of dioxins have been found in fish in Vietnam where large amounts of 2, 4, 5 T were used as defoliants. Dioxins is highly carcinogenic and mutagenic. Several generations of post war Vietnam population are suffering from reproductive abnormalities. More than two decades after spraying 2, 4, 5 T dioxins is still present in the food chain though its concentration is gradually diminishing.

The biochemical cause of the toxicity of dioxins is not known, but two different acnegenic effects have been correlated with the induction of two different enzymes. Higher levels of cytochrome P-450 are induced by dioxins and structurally related tetrachloroazobenzene derivatives. A direct correlation between the ability to produce chloracne and the affinity for the binding sites specific for P-450 induction was observed. Acne, associated with enhanced levels of porphyrins, is also caused by dioxins because of increased level of δ-aminolevulinic acid synthesis, an enzyme responsible for one of the steps in porphyrin biosynthesis.

8.10 LONG-TERM EFFECTS OF ORGANOCHLORO COMPOUNDS

The chronic (long-term) toxicity of chlorinated hydrocarbons is the main problem arising from their use. This toxic effect was first observed on the populations of eagles, hawks, falcons pelicans, and other birds at the end of a food chain, were decreasing. Closer examination revealed that the birds were mating later in the season and were producing eggs with much thinner shells than normal. Analysis of the birds and their eggs revealed that they contained large amount of organochlorine compounds. Laboratory studies confirmed that when birds were fed DDT and other PCB's they produced thin shells.

The cause of the delayed mating has been traced to the enhanced levels of the cytochrome P-450 oxidase enzymes in the livers of those birds containing the organochlorine residues. Enhanced level of oxidase enzymes are known to be induced by barbiturates and polycyclic aromatic compounds. It was found that DDT, DDE, Dieldrin, and PCB's also cause such enhancement. The enhancement of oxidase enzymes affects mating behaviour because these enzymes catalyse the oxidation of steroid hormones. Specifically it was found that an increase in oxidase enzymes resulted in a decrease in the level of estradiol in the birds. It is known that the birds will mate only if a certain level of estradiol is present in the birds. If breeding is delayed then the offspring hatch at a time when there is less food available for their growth. This is an especially important factor with large birds that require a full season for the young to develop.

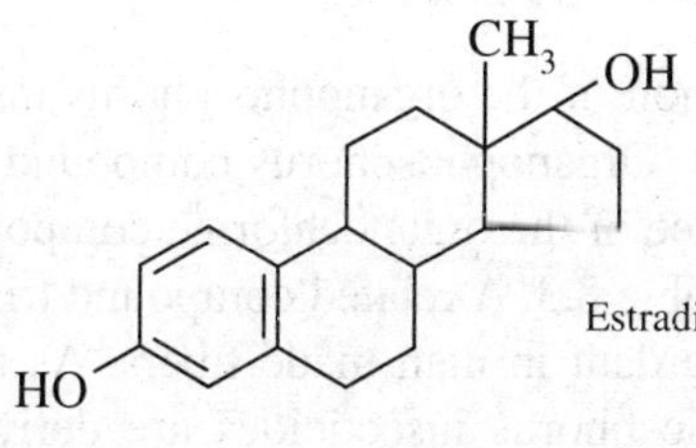

Estradiol

The production of egg shells is also influenced by estradiol. The shell of an egg is only formed in the last day before laying. About 60% of the calcium needed to form egg is obtained from the bird's food intake and the remainder comes from calcium stored in the bone marrow. A low level of estradiol results in a low level of stored calcium. The absence of stored calcium could be a significant factor in egg survival since a 20% decrease in shell thickness results in extensive egg breakage. A 20% decrease occurs with as little as 25 ppm of chloro-organics in the egg.

Finally, it has been observed that some pesticides inhibit the formation of eggshells in the presence of an adequate supply of calcium. Presumably these pesticides interfere with the enzyme carbonic anhydrase that is responsible for the conversion of carbon dioxide to carbonate. The latter is required to combine with calcium to form the calcium carbonate of the shell. So far it has been established that DDE, and not dieldrin, interferes with the formation of the shell.

The chronic toxicity of chloro-organics is not limited to birds. Mink are highly susceptible to PCB's with levels of 5 ppb which is sufficient to halt their reproduction. This is an economic problem because mink ranchers have been using Great Lakes coho salmon for mink feed. However this can no longer be used since the salmon contains 5 ppb of PCB's.

At present, there are no reports of chronic toxicity of PCB's and chlorinated insecticides in man, but these are some indication that it may exist. The first is the observation that pesticides induce high levels of liver enzymes in a wide array of vertebrates, including the monkey. It seems very likely that such induction will be observed in man as well. Secondly, it has been observed that a dose of 46 mg/kg DDT can produce a fourfold increase in tumors in animals. These data suggest a correlation between the ingestion of DDT and cancer. In addition, there is a growing list of chlorinated organic compounds that are known to produce tumors. Obviously man is in as vulnerable position as the predatory birds in the food chain so that it may be essential for his survival that the level of chlorinated hydrocarbons in the environment be reduced.

Chlorinated hydrocarbons are ubiquitous in the environment. Residues have been found in locations and in species where chloro-organics have never been used; for example they have been found in the polar ice pack, in animals in arctic regions, and in the air over the centre of the ocean. The chlorinated hydrocarbons must be carried through the atmosphere to achieve such wide distribution. The detection of chlorinated hydrocarbons in rain water is consistent with this observation. Further, DDT is sufficiently volatile so that the entire stock manufactured to date could be volatilized and still not saturate the atmosphere.

The global circulation of chlorinated hydrocarbons demonstrates that all nations have a stake in the reduction in the manufacture of these compounds. Banning the use of DDT, aldrin and dieldrin in India will be of little help if other countries continue to manufacture them for use. Obviously some means must be devised for an international control of chloro-organics.

8.11 ORGANOPHOSPHORUS PESTICIDES

The chemistry and mechanism of action of the organophosphorus insecticides are quite different from those of the organochlorine materials. Organophosphorus compound merit consideration because they are being used more frequently in place of the organochlorine compounds. Some common phosphorus-containing insecticides are listed in Table 8.3. A related compound tris (2,3 dibromo propyl) phosphate, has been widely used as a flame retardant in man made fibers. As illustrated by the examples in the table, the majority of the organo phosphorus insecticides are derivatives of phosphoric acid or the sulphur analogues of phosphoric acid. The phosphoric acid esters used in the synthesis of these insecticides are prepared industrially by the reaction of phosphoryl chloride with alcohols (8.31). Under suitable conditions, di-ester chlorides (dialkyl phosphorochloridates) may be prepared. The same reaction pathway may be used for the preparation of corresponding sulphur analogues (dialkyl phosphorothio chloridates) (8.32). Both the phosphoric acid and thiophosphoric acid, derivatives may also be prepared by the action of chlorine on the corresponding derivative of phosphorous acid (8.33).

$$Cl{-}\overset{\overset{\displaystyle O}{\|}}{\underset{\underset{\displaystyle Cl}{|}}{P}}{-}Cl + 2ROH + 2R_3N \longrightarrow RO{-}\overset{\overset{\displaystyle O}{\|}}{\underset{\underset{\displaystyle OR}{|}}{P}}{-}Cl + 2R_3NHCl \qquad ...(8.31)$$

$$Cl{-}\overset{\overset{\displaystyle S}{\|}}{\underset{\underset{\displaystyle Cl}{|}}{P}}{-}Cl + 2ROH + 2R_3N \longrightarrow RO{-}\overset{\overset{\displaystyle S}{\|}}{\underset{\underset{\displaystyle OR}{|}}{P}}{-}Cl + 2R_3NHCl \qquad ...(8.32)$$

$$\text{RO}-\overset{\overset{\text{O(S)}}{\|}}{\underset{\underset{\text{RO}}{|}}{\text{P}}}-\text{H} + \text{Cl}_2 \longrightarrow \text{RO}-\overset{\overset{\text{O(S)}}{\|}}{\underset{\underset{\text{OR}}{|}}{\text{P}}}-\text{Cl} \qquad ...(8.33)$$

Table 8.3 Some common organophosphorus insecticides

A. Organophosphorus compounds	**Structure Formula**
Acephate	$(CH_3O)_2-\overset{\overset{O}{\|}}{P}NH\overset{\overset{O}{\|}}{C}CH_3$
Azinphosmethyl	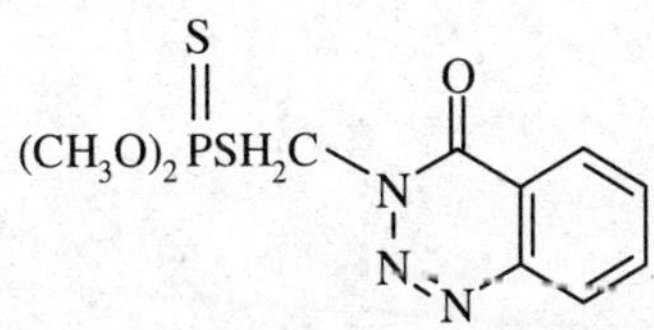
Diazinon	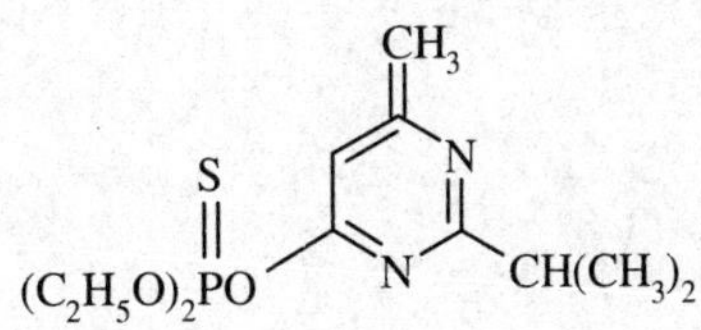
Dichlorvos	$(CH_3O)_2-\overset{\overset{O}{\|}}{P}OCH{=}CCl_2$
Dimethoate	$(CH_3O)_2-\overset{\overset{S}{\|}}{P}SCH_2CONHCH_3$
Dioxathion	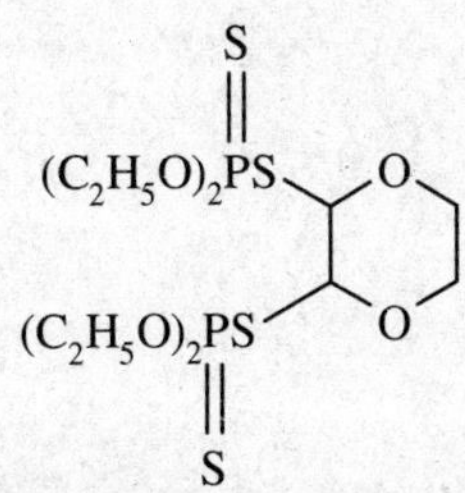
Disulfoton	$(C_2H_5O)_2-\overset{\overset{S}{\|}}{P}SC_2H_4SC_2H_5$

(*Contd.*)

Ethion	$(C_2H_5O)_2—P(=S)SCH_2SP(=S)(OC_2H_5)_2$
Fenhion	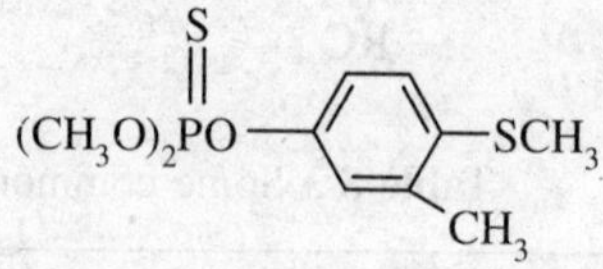
Fonofos	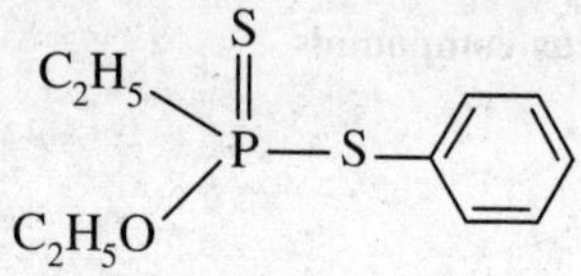
Malathion	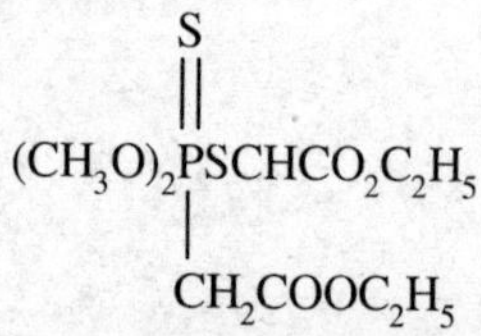
Parathion	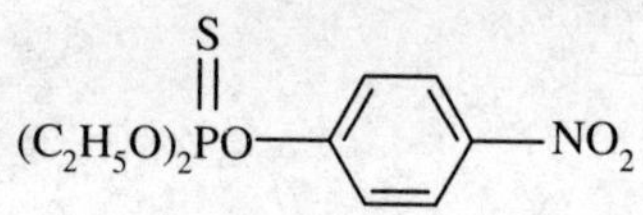
Phorate	$(C_2H_5O)_2P(=S)SCH_2SC_2H_5$
Phosphamidon	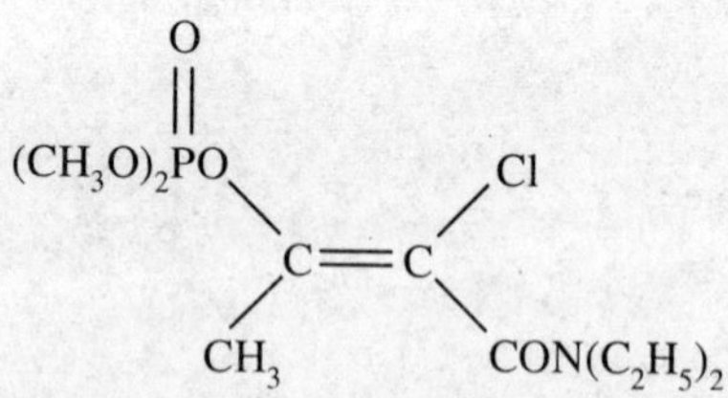
Tetrachorvinphos	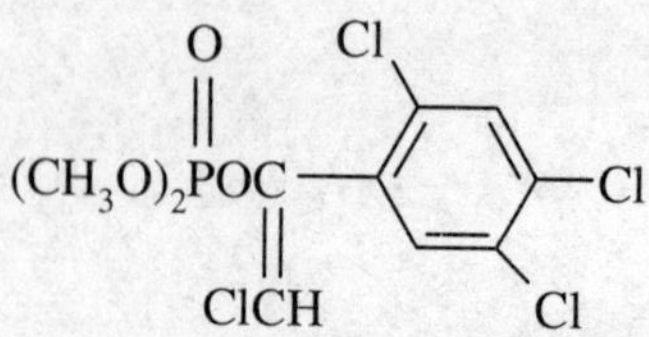
Tetra ethyl pyrophosphate TEPP	$(C_2H_5O)_2—P(=O)—O—P(=O)—(OC_2H_5)_2$

(*Contd.*)

Sahradana (OMPA) octa methyl pyrophosphamide 

Trichlorfon 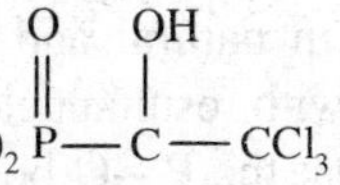

B. Carbamates

Carbary 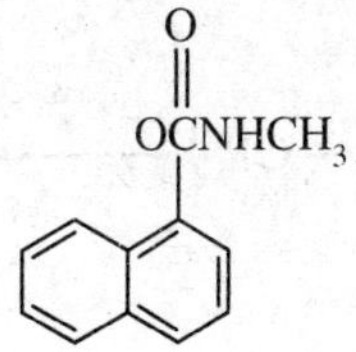

Carbo furan 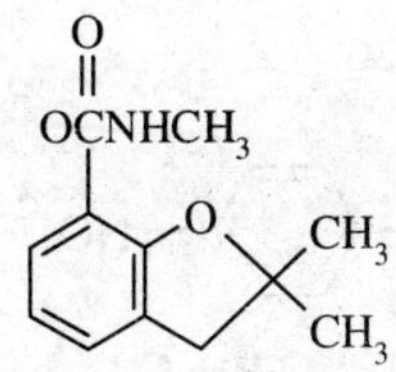

m Tolyl methyl carbamate 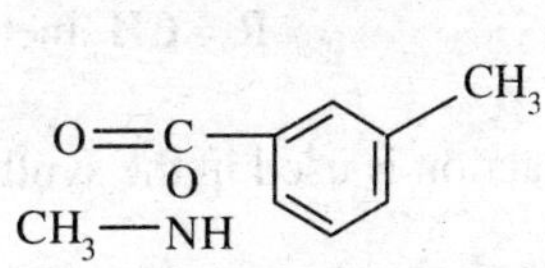

Xylyl carb 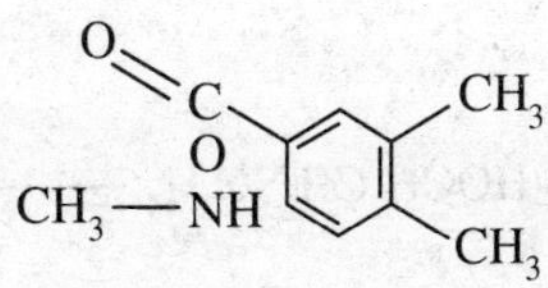

Methomyl 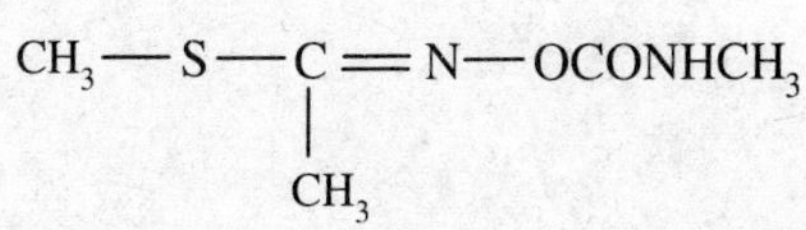

Most nucleophiles attack phosphorochloridates at the phosphorus atom and displace the chloro group (8.34). This reaction is analogous to the S_{N2} displacement at carbon. The transition state shown in 8.34 may actually be a stable intermediate although this has not been established with certainty.

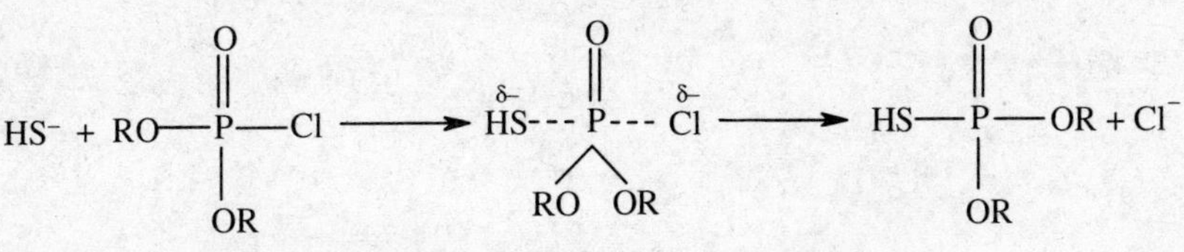

...(8.34)

Most of the organophosphorus insecticides are triesters of phosphoric or thiophosphoric acid, prepared by the attack of a nucleophile on the phosphochloridate or phosphothiochloridate as shown in the examples which follow. Since the ester group is more difficult to displace than the chlorogroup, the triesters exhibit greater stability than the phosphorochlorodiates. As a consequence of the slow displacement of the ester, a reaction pathway different from that illustrated in (8.34) is sometimes observed. As for example, in natural and acid solutions, water can attack the α-carbon atom of the phosphate ester grouping with resultant cleavage of the C—O bond (8.35). It has been shown by isotope tracer studies using ^{18}O that the P—O bond is not broken. This pathway is more likely under environmental conditions than the displacement on phosphorus (8.34)

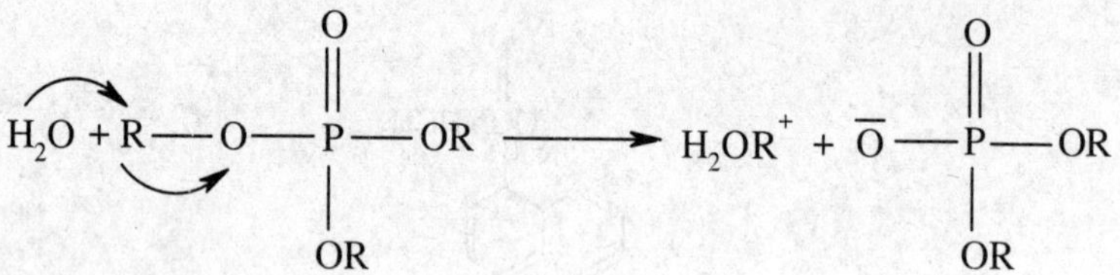

...(8.35)

The synthesis of parathion or methyl parathion involves an S_{N2} like displacement reaction (8.36).

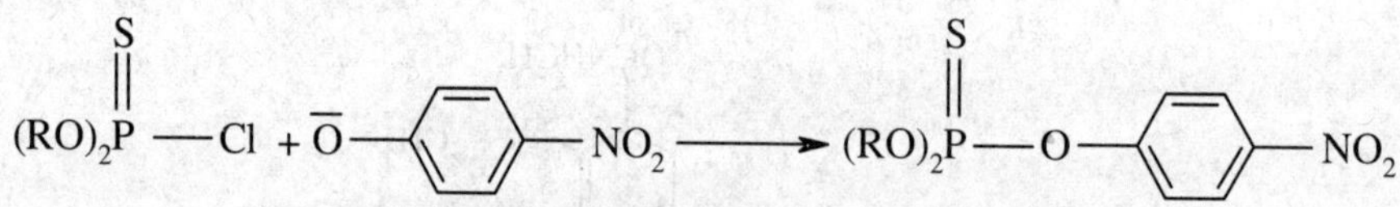

...(8.36)

$R = C_2H_5$ parathion.

$R = CH_3$ methyl parathion.

A similar type of reaction is used in the synthesis of demeton. However a rearranged product is also formed.

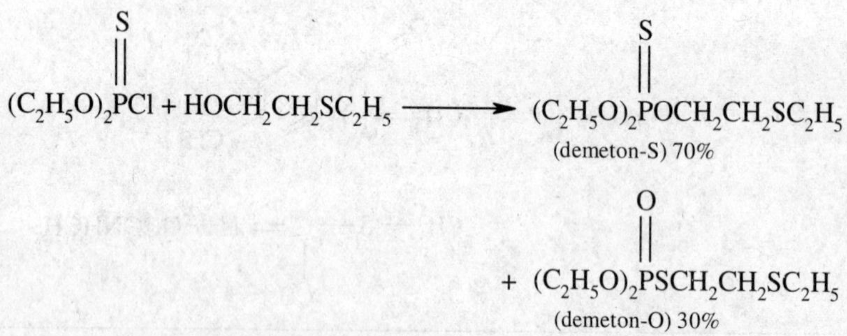

...(8.37)

The formation of the rearranged product probably involves the participation of the sulphur of the thioether since the oxygen analog does not rearrange. The following mechanism has been suggested:

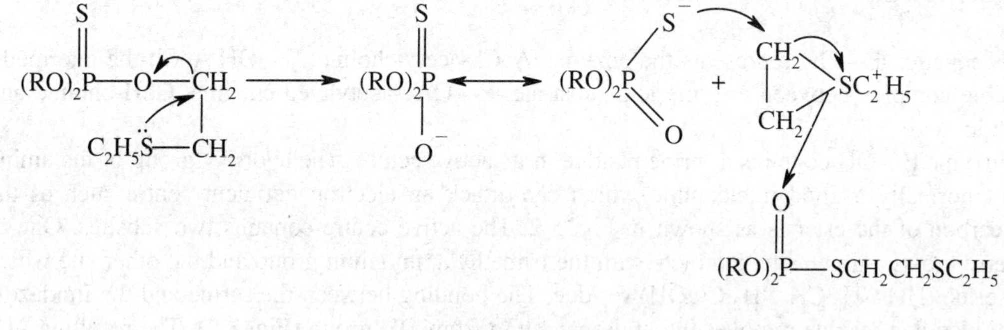

...(8.38)

The synthesis of malathion involves a nucleophilic addition of a thioacid to a double bond that is conjugated to a carbonyl group (Michael addition):

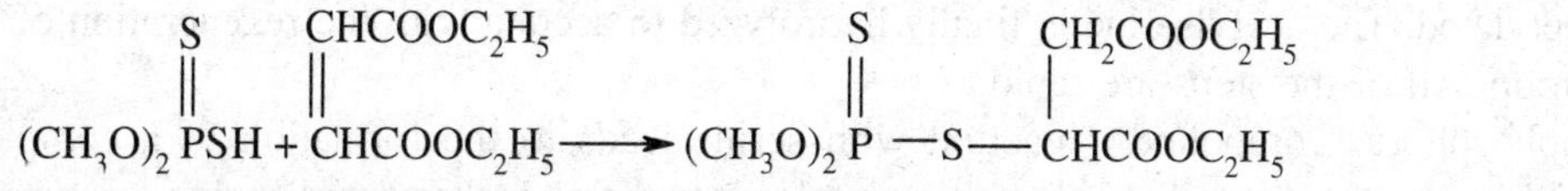

Malathion ...(8.39)

Application as Pesticides: These organophosphorous compounds exhibit high insecticidal action against a wide variety of species. This is not really an asset because they kill beneficial as well as harmful insects. These compounds are highly toxic to vertebrates as well as to insects and as a consequence, are much more dangerous to the workers using them than are chloro-organics. Demeton is unique in that it shows systematic action, that is, it is absorbed by the plant and its toxic effect is retained in the plant for 1–1.5 months. Obviously this insecticide is mainly useful for ornamental plants and shrubs that are not consumed as animal or human food. Malathion can be used safely by home gardeners because it is readily detoxified by the vertebrates. The ester groups are readily hydrolyzed by the esterase enzymes present in mammals but absent in insects. The lethal dose for killing 50% of the mammals ingesting it (LD_{50}) is 500–1500mg/kg of body weight for malathion as compared to an LD_{50} of 6–12 mg/kg of parathion and 25–50 mg/kg for methyl parathion.

The toxic effect of organophosphorus as well as carbamates insecticide is similar to polychlorinated cyclopentadiene derivatives. The toxic effect is due to their interference in the transfer of nerve impulses from one nerve cell to the next. When a nerve impulse reaches the end of a nerve cell, it triggers the release of a minute amount of the acetylcholine. The acetylcholine activates a receptor on an adjacent nerve cell causing it to carry the impulse to the next nerve cell. The acetylcholine is then hydrolyzed to choline and acetic acid by the enzyme acetylcholinesterase which catalyses hydrolysis

$$\underset{\text{Acetylcholine}}{(CH_3)_3\overset{+}{N}CH_2CH_2OOCCH_3} \longrightarrow \underset{\text{Choline}}{(CH_3)_3\overset{+}{N}CH_2CH_2OH} + CH_3COOH \quad ...(8.24)$$

The pesticide bind chemically by forming a charge transfer bond to the acetylcholinesterase so that it can no longer catalyse the hydrolysis of acetylocholine. The resulting excess of acetylcholine hyperstimulates nerves to produce convulsions, irregular heartbeats, and choking in vertebrates and insects.

It is now generally agreed that organophosphorous compounds and carbamates react with acetylcholine esterase by a mechanism as follow

$$E - OH + AcCh \underset{k_{-1}}{\overset{k_2}{\rightleftharpoons}} E - OHAcCh \xrightarrow[ChH.]{k_2} E - OA \xrightarrow[k_3]{H_2O} E - OH + AcOH \quad ...(8.40)$$

In this equation E—OH represents the enzyme, AcCh-acetylcholine, E—OH AcCh-the intermediate reversible complex between enzyme and substrate, E—OAc-acetylated enzyme, ChH-choline and AcOH-acetic acid.

The enzyme E—OH contains a serine residue in its active centre. The hydroxy group of the amino acid is not normally a good nucleophile which can attack an electron deficient centre such as the carbonyl carbon of the ester is as shown in Fig. 8.2. The active centre contains two subsites. One of these, referred to as anionic site, interacts with the trimethylammonium group and the other site which contains serine (HO CH_2 CH NH_2 COOH) residue. The bonding between the serine and the imidazole group of a histidine residue removes the hydrogen atom from OH group (Fig. 8.2). The resulting $-O^-$ ion is much more reactive than the OH group, so that, following the initial binding of acetylcholine to the enzyme, the serine attacks the carbonyl group of the ester. Choline is liberated and the serine becomes acetylated. The acetylserine is finally hydrolyzed to acetate with the regeneration of the free hydroxyl group. All of the steps are rapid.

Organophosphorus compound's reaction with serine leads to the formation of a complex with enzyme which is not an intermediate of reaction but leads to the phosphorylated enzyme, an irreversible reaction that leads to inhibition of reaction 8.24 and insects die.

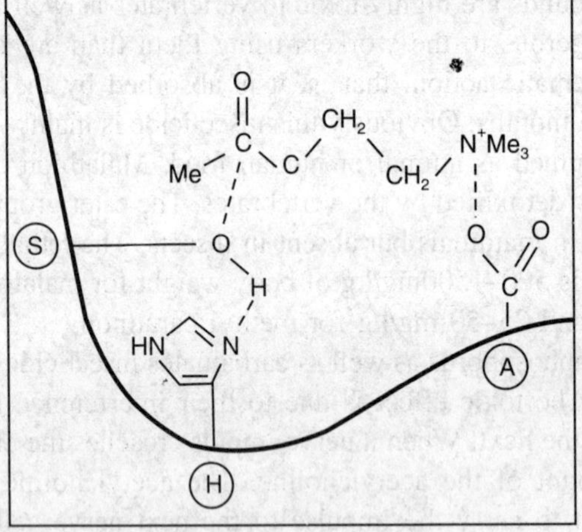

Fig. 8.2 Diagram of the active site of acetylcholinesterase. S, H and A represent serine, histidine and an acidic amino acid respectively[1].

It may be pointed out that in an organophosphorus compound, the nucleophile attacks along the line of the bond of phosphorus and leaving group. Thus if nucleophile (*i.e.* the serine—OH) is relatively fixed in enzyme, it is not possible for the leaving group in an organo phosphorous compound to occupy the same site on the enzyme as the leaving group in acetyl choline.

Environmental Degradation: The organo-phosphorus insecticides undergo a very rapid hydrolysis in the environment. This is a mixed blessing. There is no build-up of residues; however frequent applications of the pesticide are required. Some typical hydrolytic reactions include reactions (8.41–8.43). The nitro group of parathion is reduced by soil micro-organisms to give the nontoxic aminoparathion, which is stable and remains bound to the soil. It is not known if this inactive form of parathions constitutes an environmental problem.

$$(RO)_2P(=S)-O-C_6H_4-NO_2 \xrightarrow{H_2O} (RO)(OH)P(=S)-O-C_6H_4-NO_2 + (RO)_2P(=S)OH + HO-C_6H_4-NO_2 \quad ...(8.41)$$

$$(CH_3O)_2P(=S)SCH(CH_2COOC_2H_5)COOC_2H_5 \xrightarrow[H^+]{H_2O} (CH_3O)_2P(=S)-OH + HS-CH(CH_2COOC_2H_5)COOC_2H_5 \quad ...(8.42)$$

$$\xrightarrow[OH^-]{H_2O} (CH_3O)_2P(=S)-SH + CHCOOC_2H_5{=}CHCOOC_2H_5 \quad ...(8.43)$$

A problem associated with these organophosphorous insecticides is their environmental conversion to more toxic substances. For example, Demeton-O, the by-product formed in the synthesis of Demeton-S (8.37) has an LD_{50} of 2 mg/kg for vertebrates as compared to the an LD_{50}, of 100 mg/kg for Demeton-S. The Demeton- rearrangement (8.38) takes place readily in plant tissues and may explain the long-term systematic action of Demeton. If the rearranged product is truly the toxic agent, then trace amounts could still be effective. Parathion is readily oxidized by the atmospheric oxygen or enzymatically to a derivative that is four times as toxic as parathion itself. Since parathion exhibits no anticholinesterase activity itself, its insecticidal action may be due to its enzymatic conversion to the oxygen analogue in insect.

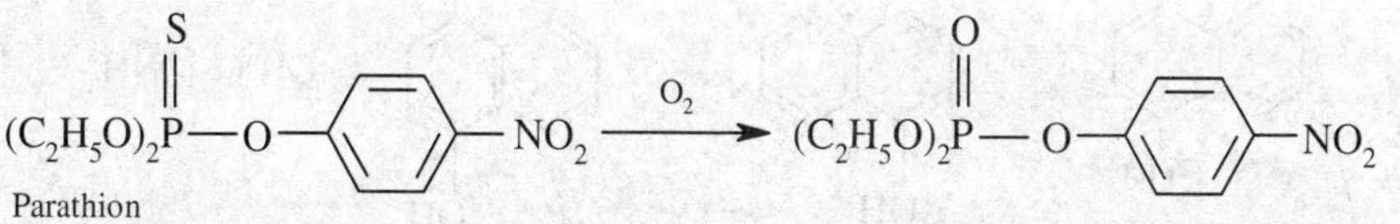

Parathion ...(8.44)

8.12 THE CARBAMATES

Most carbamates are phenyl N-methyl carbamates *i.e.* carbamate esters derived from substituted phenols. The carbamate insecticides have been used extensively as a replacement for DDT to control the Gypsy moth as well as other insects. Its toxicity to insects is due to its antiacetylcholinesterase action as in organo phosphorus insecticide though the mechanism is a little different. The polarization of carbonyl carbon gives a positive charge to carbon. This leads to protonation of this oxygen thereby leading to the formation of a reversible complex as a tetrahedral intermediate as shown in Fig. 8.3.

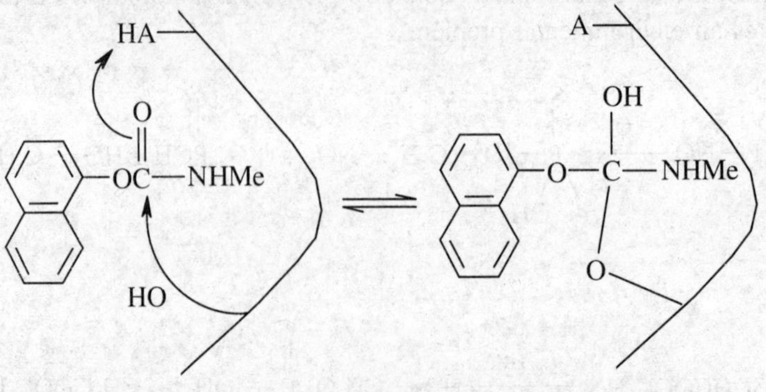

Fig. 8.3 Proposed route of formation of the reversible complex between acetylcholine sterase and carbamates. OH represent the serine and HA represent a hydrogen donor.

In this complex there is a covalent bond between carbamate and enzyme, rather than noncovalent interactions. The formation of such a complex engages the site of enzyme and hydrogen, reaction 8.24 is inhibited. This leads to accumulation of acetylcholine and eventually to the death of insect.

It may be pointed out that the geometry of nucleophilic attack at carbonyl carbon differs from that of nucleophilic attack at tetrahedral phosphorous. Unlike, phosphorus compounds the planar carbonyl group is attacked from either side of plane in which its atoms lie. This means that if the nucleophile (*i.e.* the serine OH) is relatively fixed in the enzyme, it is not possible for the leaving group an organo phosphorus compound to occupy the same site on the enzyme as the leaving group in acetyl choline and in a carbamate. In other words the leaving group in phosphorus ester binds to a region of enzyme which is different from that which binds the leaving group of acetylcholine and by implication of carbamates.

Carbaryl is not very toxic to mammals. LD_{50} for rats is 0.5 gm/kg, presumably is not because it is readily oxidized and excreted. Carbaryl is not persistent in environment because it is rapidly destroyed

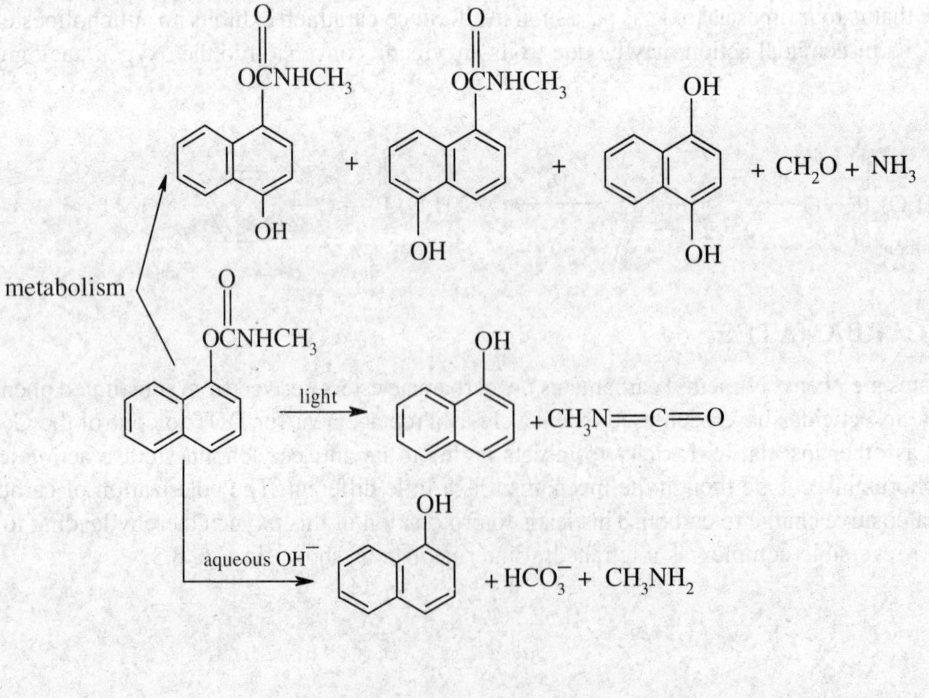

...(8.45)

...(8.46)

...(8.47)

by a number of pathways and, as a consequence, it must be applied frequently for insect control. Some of the key reactions of this material are given in reactions (8.45–8.47).

Insects treated with organophosphorus and carbamates insecticides show initial hyperactivity followed by convulsions and then paralysis through the release of acetylcholinesterase. It leads to degeneration of vital tissues following loss of water from insect integument and ultimately death. Development of resistance to carbamates and organophosphorus pesticides has turned out to be due to alteration in acetylcholinesterase rendering it less susceptible to inhibition.

8.13 THIRD GENERATION CONTROLLER: PHEROMONES

Chemicals released externally by insects and which elicit immediate behavioural responses from members of the same species are called pheromones. These compound messengers do not exert toxic effect with in a species, and as such cannot be called pesticides. However, it has been demonstrated that interference with natural communication mechanisms can provide effective population control of pest species.

Among the many types of pheromones, sex attractants received the greatest attraction. Pheromones (Gr. *phero* meaning 'convey' and harmony) are released by one sex of the species (usually the female) to attract the opposite species for the purpose of mating. Other types of pheromones are involved in aggregation and colonisation of a host (population attractants), egg-laying, and foraging and maintenance of a social order in colonies of insects such as bees and ants.

The insect usually detect pheromones by sweeping the air with their antennae which possess thousands of olfactory hairs. These hairs contain sensory nerve cells which have receptor sites for the chemicals. The presence of a sufficient number of molecules at their receptors leads to depolarization of the membrane of the sensory cell and this generates a nerve impulse which is transmitted to the central nervous system, the greater the number of molecules arriving at the receptor, the greater is the frequency of impulses transmitted. It has been shown that a sensory cell will respond to one molecule of pheromone and it has been estimated that the presence of some **two hundred molecules** is sufficient to induce a behavioural change. Thus three hundred grams of a pheromone with molar mass 300, will kill the following number of insects

$$6.06 \times 10^{23} \text{ molecule are present in 300 g}$$

$$\text{200 molecules will be present in } \frac{300 \times 200}{6.06 \times 10^{23}}$$

or No. of insects which will change behaviour by

$$300 \text{ g of } i = \frac{6.06 \times 10^{23}}{300 \times 200} = 1.01 \times 10^{21} \text{ insects}$$

Isolation of Sex Pheromones: Pheromones are present in trace amounts in insects so that special techniques coupled with considerable skill, ingenuity, and intuition are required to decipher the structures of these compounds. For example only 500 μg (5×10^{-4} g) of pure male attractant was isolated from 75000 virgin females fall armyworm moths. Obviously the usual techniques of organic chemistry cannot be used for structure analysis of these pheromones. A sensitive bioassay is required, which is usually based on the observation of the response of the insect to the pheromone. A common technique is to allow the effluent from a gas chromatograph to pass through a cage containing the insects. Compounds are present in too small an amount to be detected by the detectors in the gas.

Chromatograph: These can be detected by the insects as evidenced by their agitated behaviour. Once a sensitive assay has been developed, large numbers of the insects can be bred or collected to provide a source of pheromones. The whole insect or the glands that secret the pheromones are extracted and the pure pheromone is separated from the hundreds of other compounds in the extract. The structure assignment of the trace amount of purified pheromone is usually based mainly on spectroscopic measurements. As a consequence, the postulated structure must then be confirmed by an unambiguous synthesis. The final objective of the project is to device a synthesis for the preparation of large amounts of the pheromones to use for insect control.

This description is an oversimplification that assumes pheromone to be a single compound. However the sex pheromone of the boll weevil, for example, is a mixture of four compounds. No attractive power is observed if one of the components is missing. Occasionally, masking agents are present that inhibit the effectiveness of the pheromones. A major problem is that the insect produces other compounds that are structurally similar and difficult to separate from the biologically active compounds. Sometimes only after the completion of the total synthesis of a biologically inactive compound is it discovered that pheromone was present as a lesser contaminant in the compound whose structure was determined. Pheromones literature is replete with controversy because of these chemical difficulties and in many instances incorrect structures have been assigned to the pheromones.

Examples of Pheromones: The sex pheromone of the boll weevil has been studied extensively. This insect is important because of the extensive cotton crop lapses that it causes each year. The four compounds that make up this pheromone were isolated from fecal material in the proportion indicated below:

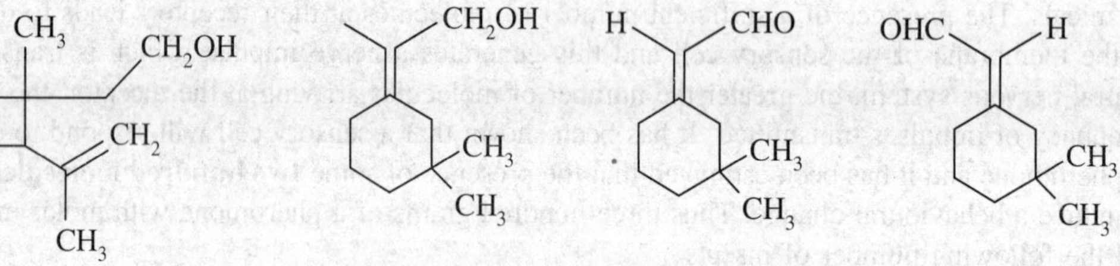

13 10 1 1

Boll weevil pheromone 'grandlure'

The sex pheromone of the pink bollworm moth, another predator of cotton, has been identified as a mixture of *cis-cis* and *cis-trans* isomers of 7, 11, hexadecadienyl acetate. *Cis*-7-hexadecyl acetate also attracts the male and was originally believed to be a pheromone, and it is currently being used as an attractant under the commercial name Hexalure. Recent attempts to detect Hexalure in the extract of 1.2 million pink bollworm moths proved unsuccessful.

$$CH_3\overset{O}{\overset{\|}{C}}OCH_2(CH_2)_5\underset{cis}{CH{=}CH}(CH_2)_2\underset{cis\text{ or }trans}{CH{=}CH}(CH_2)_3CH_3$$

Pink bollworm moth pheromone

$$CH_3\overset{O}{\overset{\|}{C}}OCH_2(CH_2)_5\underset{cis}{CH{=}CH}(CH_2)_7CH_3$$

hexalure

The structure of a number of other pheromones are listed in Table 8.4. It is remarkable that their structures are so similar yet the attractants are species specific.

Not all insect attractants are pheromones. For example, it has been established that *Hexalure* is not a pheromone, but it exerts a strong attraction to the pink bollworm moth. Some other attractants that have been used in insect control are listed in Table 8.5. *Siglure* and *Medlure* were especially useful in the program for the eradication of the Mediterranean fruit fly in Florida.

Table 8.4 Examples of insect sex pheromones

Insect		*Pheromones*
Cabbage looper		$CH_3(CH_2)_3CH{=}CH(CH_2)_5CH_2O\underset{\underset{O}{\|}}{C}CH_3$
Gypsy moth	*cis*	$CH_3\overset{\overset{CH_3}{\vert}}{C}H(CH_2)_4CH\overset{O}{—}CH(CH_2)_9CH_3$
Honey bee queen		$CH_3\underset{\underset{O}{\|}}{C}(CH_2)_5CH{=}CHCOOH$
Fall armyworm moth	*cis*	$CH_3CO\overset{\overset{O}{\|}}{}CH_2(CH_2)_7CH{=}CH(CH_2)_3CH_3$
Cabbage looper moth	*cis*	$CH_3CO\overset{\overset{O}{\|}}{}CH_2(CH_2)_5CH{=}CH(CH_2)_3CH_3$
Oriental fruit moth (8% trans isomer is required for maximum activity)	*cis*	$CH_3\overset{\overset{O}{\|}}{C}CH_2(CH_2)_5CH{=}CH(CH_2)CH_3$
Silkworm		$CH_3(CH_2)_2CH{=}CHCH{=}CH(CH_2)_8CH_2O$
Pine beetle		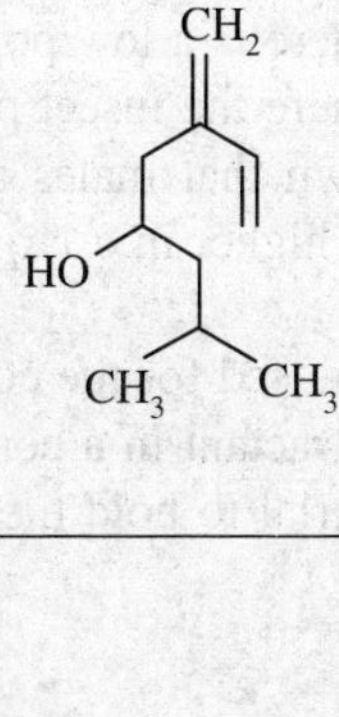 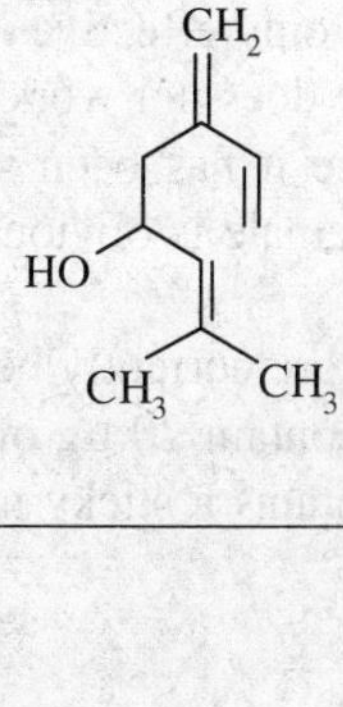 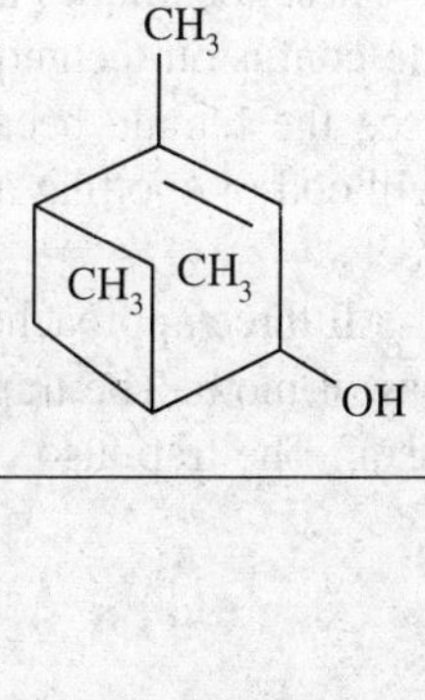

Table 8.5 Insect attractant

Insect	*Attractant*
Mediterranean fruit fly	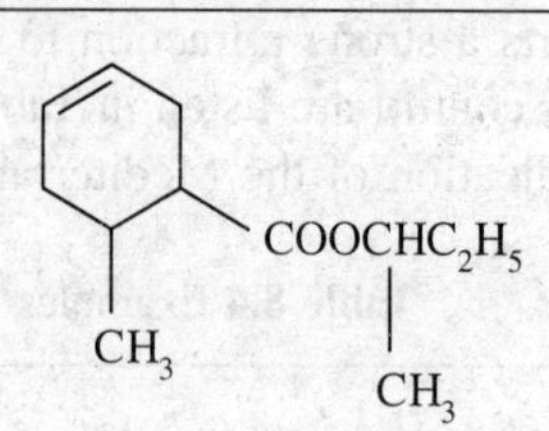
Melon fly	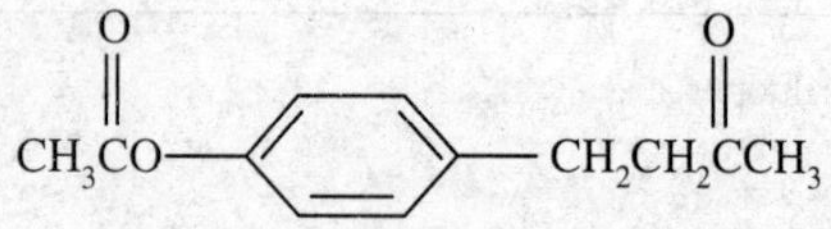
Ants	$CH_3\overset{\overset{O}{\|}}{C}OCH_2CH_2CH_2CH_2CH_3$
Sugar beet wire worm	$CH_3CH_2CH_2CH_2COOH$
June beetle	$CH_3CH_2CH_2\overset{\overset{CH_3}{\vert}}{C}HNH_2$

8.14 USE OF PHEROMONES AND ATTRACTANTS IN INSECT CONTROL

Pheromones have been used in three different approaches to insect control;

(*i*) To monitor the population of a specific insect the pheromone is placed in traps and the number of insects trapped is recorded. The population density of the insect can be monitored in this way to measure the effectiveness of the insecticides being used to control or to monitor insect migration into new areas;

(*ii*) To trap males, a large number of traps containing the pheromone are used to trap all the males in an areas so that they are not available for mating

(*iii*) For male confusion large amounts of the pheromone are distributed so the air is permeated with the female sex pheromone. The male is surrounded by attractant and is unable to locate the female.

These techniques are only effective if there is a low population of the insect to be controlled. The male confusion technique does not work where the insect population is sufficiently large for the male to see the female because it has been shown that males are attracted to any females in sight. The likelihood of spotting a female is obviously higher in a large population density region of a particular insects.

All three approaches are currently being used for the control of the Gypsy moth in low infestation areas of moth. The trap contain 20 μg of attractant in a control release agent that releases it over one season. The trap also contains a sticky material to hold the moth once it flies into the trap. The male confusion technique has been tested on a small scale with encouraging results. Significantly, fewer females laid fertile eggs in the area where disparlure was sprayed. It should be emphasized that other methods of control will probably be required in areas of heavy infestation.

Pheromones have the advantage of being very specific for one insect and in most cases only very small amount are required. Some scientists have expressed the opinion that the insects will not develop resistant strain because the pheromone is produced naturally by the insect. However, others note insects have rapidly developed resistance to the synthetic pesticides and see no reason why they should not develop varieties that do not respond to those pheromones. However, even the pessimists agree that it will probably take longer for the insects to develop resistance to pheromones than it took for them to develop resistance to the second-generation insecticides.

8.15 JUVENILE AND MOULTING HORMONES

The juvenile hormones are secreted by the insect Corpora allata, two glands at the base of the brain. The hormone's major function is to maintain insect larva in a juvenile stage at a moult, so that they prevent a progressive change from one developmental stage to the next. The structures of the hormones, which are found in different amounts in different tissues, are give below:

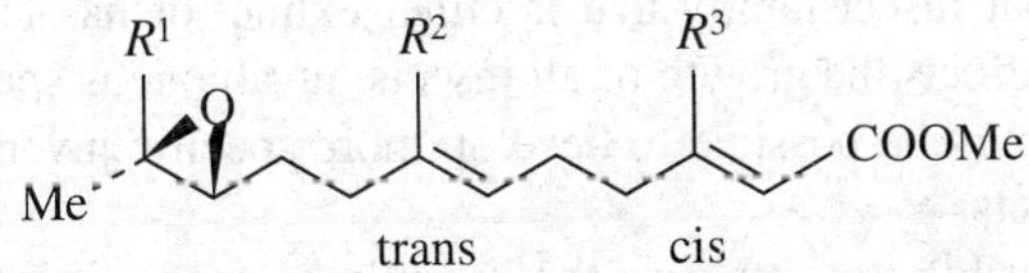

	R^1	R^2	R^3
JH O	Et	Et	Et
JH I	Et	Et	Me
JH II	Et	Me	Me
JH III	Me	Me	Me

A plausible working hypothesis is that juvenile hormone enter a cell, combines with a cytoplasmic receptor protein, and is admitted to the nucleus, where it influences the pattern of messengers being transcribed from the DNA. This would involve, in conjunction with other hormones, maintaining the expression of genes required in the juvenile phase while ensuring that the adult genes are not expressed. The resultant insect combine juvenile and adult characteristics, cannot feed or reproduce and soon dies.

In other words, insect hormones regulate growth and maturation of the animal from the larva to the adult. The juvenile hormone controls the development of the immature larva through successive growth states. However, it must be absent, for the metamorphosis of the larva to the adult to take place or a deformed larva or adult is formed, that dies. The moulting hormone is required for the differentiation to the adult stage. If the immature larva is injected with the moulting hormone, it passes through its life cycle at a rapid pace and dies prematurely. Insect growth and development could be controlled by the use of these hormones.

The general approach for the isolation of the insect hormones is similar to that of the pheromones. A large supply of insects is required for the isolation of a very small amount of the purified hormone. The juvenile hormone was fixed isolated from the silkworm because silk worms are available in large quantities from breeders who supply them to the silk industry. α-Ecdysane has been isolated as the moulting hormone from the tobacco hornworm.

A juvenile hormone mimic is even produced by plants. This was discovered accidentally when it was observed that the European linden bug did not mature when grown in the presence of paper prepared from North American trees, however it did grow in the presence of paper manufactured in Europe. The compound juvabione, responsible for the growth retardation, was traced to the Balsam fir and its structure has been determined

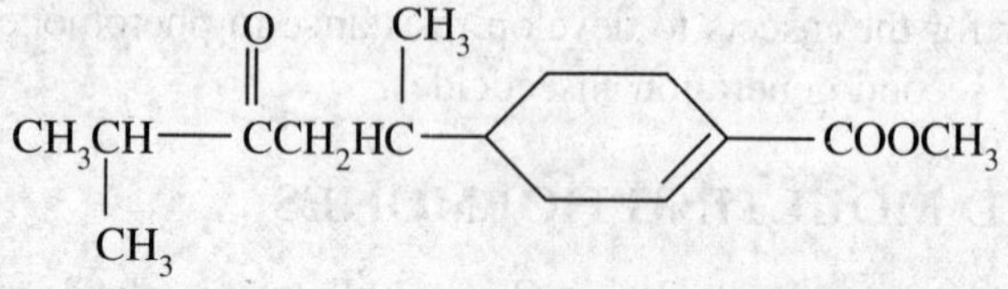

Juvabione

It is surprising that Balsam fir contains a juvenile hormone, which affects the growth of a family of insects that are not natural predators. It has been suggested that juvabione is the juvenile hormone of a former predator from that insect family that is either extinct or has learned to avoid Balsam fir. While the juvenile hormone affects the growth of all insects, juvabione is specific for the Pyrrhocoridae family. This suggests that it may be possible to develop more specific juvenile hormone analogue that will affect only harmful insects.

Although it has been possible to synthesise the juvenile hormone, it is too unstable (at the epoxy ring) and too difficult to prepare in large amounts to be used commercially. However, a number of analogues have been prepared that show commercial promise surprisingly, some are considerably more active than the natural hormone. Methoprene was the most potential of a series of juvenile hormones.

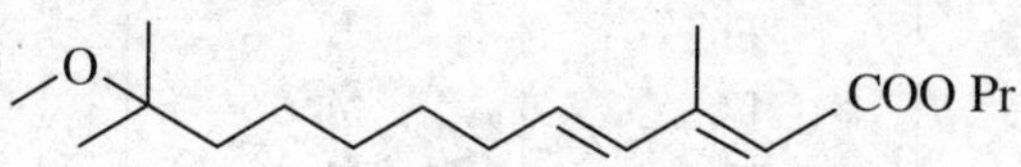

Methoprene

Activity was measured in synchronised sensitive stages of the greater wax moth, yellow mealworm and yellow fever mosquito, being, respectively, 0.055 ×, 130 × and, 1070 ×, as active than JH I. The compound is used a larvicide against flood-water mosquitoes, where even 1 ppb is effective. The fact that methoprene is not rapidly degraded by esterase preparation which metabolizes juvenile hormones presumably contributes to its

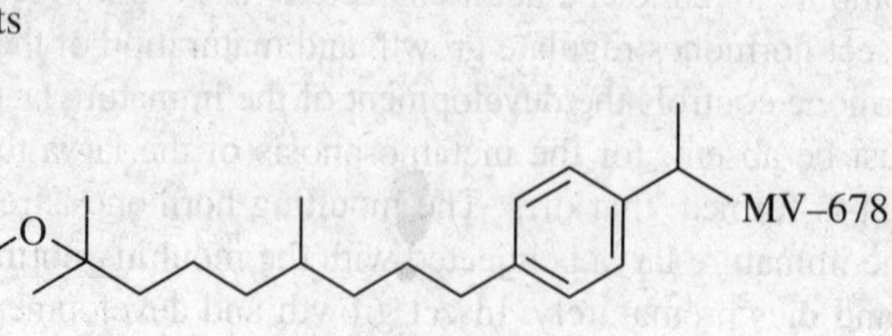

Juvenile hormone analogue MV 678 can be used to exterminate fire ants colonies by incorporation in a bait which is taken to the nest, where it disrupts the development of new workers activity.

Use of hormones in insect control

Insect hormones must be applied at a specific stage in the insect's life to influence its development. The effective time periods are often very brief so that the hormones must be present for a long time enough to ensure that significant proportion of the insect populations is exposed during the sensitive period. One application of juvenile hormone analogs is for the control of flood-water mosquitoes. The eggs of this mosquito are dormant until they are in water. Since the farmer controls the time at which he irrigates the field, he can apply the juvenile hormone analog at the precise time when all the mosquitoes hatch and are in their susceptible grown up state. The hormone analogue is also formulated in controlled-release polymer spheres that slowly release the compound over the life cycle of the mosquito (seven to ten days). Precise timing of the application of the pesticide is not required with this formulation.

The observation that some plants produce juvenile hormone analogues suggested that some of these composed, might be stable when absorbed in plant tissues (systemic). This has been observed with the juvenile hormone analogue ethyl pivalyl-alanyl-p aminobenzo. It is absorbed and retains its juvenile hormone activity in the sunflower plant for 1-2 weeks.

As in the case of pheromones, it is not known whether insects will develop resistance to these hormones. Those who argue that no resistance will develop note that insects need these hormones at a specific time in their own development. If resistance developed, it would result in the death of the insect. Those who argue that resistance will develop note that as the insect develops to an adult, it must be able to inactivate these hormones. The insect could develop, more efficient ways of doing this and in this way negate the effect of the added hormone.

There are several problems associated with the use of insect hormones. One is the lack of specificity. This will not be a problem if the hormones are applied only at certain times during the year when the target insect is going through its metamorphosis. Another problem is that hormones are only effective when the insect is changing from the final larval stage into an adult. However, insects are usually most destructive in their larval stage so that the hormone would not curtail the most destructive period. This means that the hormones are best suited for the treatment of light infestations where the damage is tolerable or against flies and mosquitoes, where the adult is the pest. A third problem associated with the use of hormone is their effect on the development of arthropods (a class of phylum Insecta) in the sea. As for example, the moulting hormone also affects the moulting the crabs and cray fish and a juvenile hormone analogue affects the development of a corn banacle. Obviously, these hormones will have to be used judiciously to avoid destruction beneficial arthropods.

8.16 FOURTH GENERATION INSECT CONTROLLER, ANTIJUVENILE HORMONE FROM PLANTS

An excess of juvenile hormone or a mimic will only disrupt development at limited stages in the insects life cycle. Bower *et al.* (1976) reasoned that, since juvenile hormone levels are high for most of the life cycle, a compound interfering with the synthesis, movement or action of juvenile hormone might be more generally disruptive.

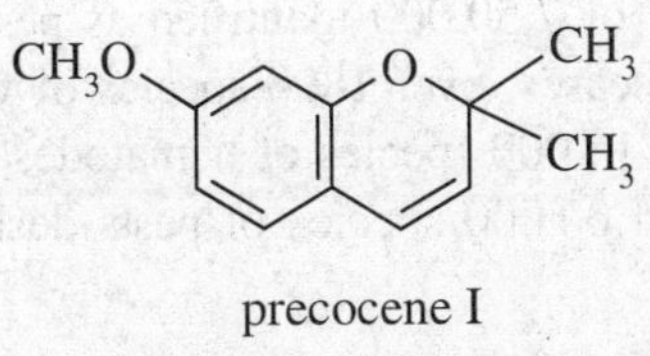

precocene I

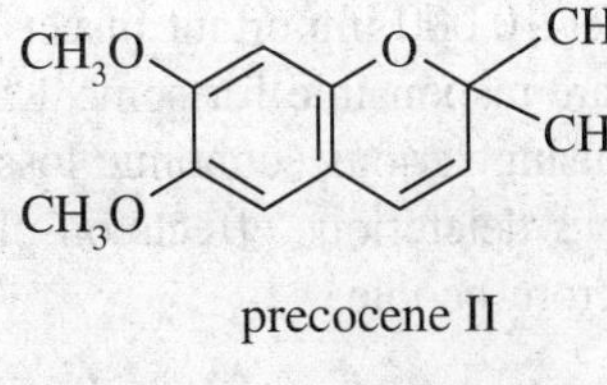

precocene II

This was followed by a screening programme to find plant extracts with antijuvenile hormone activity culminating in the discovery, that extracts of *Ageratum houstonianum* induced precocious (hence the name) metamorphosis in immature hemipteran and prevented development of ovaries in normal adult insects. Subsequent purification yields the precocenes I and II.

The adult females that develop after treatment with the precocenes are sterile and only a few of the males are capable of successful mating with normal females. Moreover, treatment of newly developed adult females with precocenes prevents the development of ovaries and in the case of the cockroach, prevents the secretion of sex pheromones. Treatment of the adult potato beetles with precocenes caused it to prematurely stop feeding and go into a dormant stage. Precocene II also prevented the development of the eggs of the milkwood bug and the Mexican bean beetle.

The wide range of biological activity exhibited by the precocenes suggest that antiallatrotropins should be very effective agents for insect control. They have the decided advantage over juvenile hormones in that their toxic properties are effective on both the adult and egg as well as on the larva. In addition, their effect on the larval stage is to accelerate it development to the adult and thereby shorten the larval stage which is generally the most destructive.

8.17 INSECT STERILISATION

The insect itself can become the insecticide when sterilisation is used for pest control. Large numbers of viable but sterile males are released into the infected area in this approach and the sterile males compete with nonsterilised males for the females. If the number of sterile males is significantly greater, it is then possible to control or in some cases eradicate an insect population. This approach is obviously species-specific and it appears unlikely that resistant strains will develop.

For a sterilisation programme to be successful, one must be able to raise large numbers of healthy, vigorous insects that can be sterilised without affecting their vigour and mating competitiveness. The sterilisation is usually accomplished with X-rays or gamma rays from a Cobalt-60 source. The biological effect of both type of irradiation is similar. A dosage of 2500 roentgens (R) was effective for the sterilisation of the male screw-worm fly although the fly can tolerate 20,000 R without any other apparent side effects. Then these sterilised insects must be released in suitable numbers to be effective but not destructive.

It seems likely that this approach will not be used to eliminate major insect infestations because of the large numbers of sterile insects that it will generate. However, it can be used to eliminate or manage insect populations that have been suppressed by other means. It could also be used to prevent the spread of insects into a new area. It has been especially effective in peninsulas and islands where there is a natural barrier to prevent the influx of nonsterile insects once control has been established.

8.18 CURRENT POSITION OF PESTICIDES IN INDIA

Currently, in India, the use of pesticides have generated much interest and the debate is not only relevant but also long due. It is difficult to visualise agriculture being pesticide-free, though there is a growing awareness about the hazards of pesticide use.

There are over 10,000 important insect species, out of 7,50,000 identified as pests. Over 50,000 species of fungi are responsible for some 1,500 plant diseases, over 1800 species of weeds out of the known 30,000 causing serious economic loss, and about 15,000 species of nematodes producing more than 15,000 serious deleterious effects on plants. Over 1,00,000 species of pests destroy food which could feed 13.5 crore people.

In India, the consumption of pesticides went up from 80,000 tons to 1,00,000 tons between 1990 and 1995 and increased by 19,000 tons between 1995 and 2005. The country's average pesticide consumption is around 450 g per hectare.

In an unprecedent pest attack on cotton crops in Warangal district of Andhra Pradesh in 1997–98 led to crop failure that led many farmers to commit suicide. Highly poisonous pesticide Methomyl was used to kill this pest. Season after season the consumption of this pesticide had been increasing there by eliminating all helpful pest from soil. This cash crop they have been growing for several years in continuation, but indiscriminate use of pesticide was the cause of this. There are many other such incidents that had taken place.

Nearly 67% of the total pesticide used are for cotton (52 to 55%) and rice (17 to 18%) cultivation in five states—Andhra Pradesh, Karnataka, Gujarat, Punjab and Maharashtra. About 95% of the total herbicides used in the country today are for wheat, rice and tea. Cereals, oil seeds and pulses consume only 6 to 7%. The cropwise consumption scenario presents a picture of imbalanced and indiscriminate use of pesticides.

Pros and Cons

The environmental hazards associated with effluents and chemical residues in crops and agricultural runoff is alarming. About three lakh Indian farmers lost their lives due to pesticide poisoning.

Pests damage farm-products worth roughly Rs. 6,000 crore a year namely about 10 to 20% of the total products, which India can ill-afford. Maintaining self sufficiency in food grains for an exponentially growing population have made chemical pesticides today an inseparable part of a farmer's regime.

However, the harmful effects of pesticides seem to out weigh the benefits. These harmful effects lead to salinization of the soil, contamination of ground water, depletion of wild life population and encouragement to resistant pest strains. The hazards caused by pesticides can be broadly categorised as occupational, health and environmental.

Occupational hazards are direct in effect, *i.e.*, they affect the farmers and the labourers who use them. Ignorance on the usage and safety measures is the main reason why 10,000 people die every year and another 40,000 suffer from various effects of pesticide poisoning.

Use of protective clothing, including gloves, aprons and masks while spraying pesticides is more of an exception rather than the rule in our country. The indirect hazards of pesticides to the common man are pesticide residues in food, produce ill-effects, when consumed without washing or when the concentration exceeds the safe tolerance levels. Food commodities from wheat, rice, milk and milk products are found to contain high level of DDT, BHC, Lindane, Dieldrin and Malathion. Even though the Prevention of Food Adulteration Act, 1954, provides the teeth to tackle excessive limits of these, it is seldom enforced. Hence, an average Indian's body has the highest DDT level in the world, varying from 12.8 to 3.1 ppm.

Seepage into the soil of the pesticide residues affects groundwater sources. As a consequence, ponds, streams and rivers get polluted making this water unpotable, besides lowering fish production. Pollution due to the effluents discharged by the pesticide manufacturing plants also get into the "food-chain" thereby accentuating the already existing problem.

Another dimension of excessive use of chemical pesticides is the resistance development, sudden resurgence of target pests and induced outbreaks of secondary pests. The number of pests resistant to insecticides shot up from 12 in 1950 to more than 500 in 1990 (Table 8.6).

In India, insect resistance to the pyrethroid group of insecticides has created problems in controlling crop pests such as *Heliothis armigera, Plutella* and *Spodoptera litura.* DDT, BHC, and Malathion

Table 8.6 Increase in number of insect pests resistant to insecticides

Year	*Species*	*Year*	*Species*	*Year*	*Species*
1910	0	1954	25	1976	364
1920	0	1957	26	1980	> 400
1930	0	1960	137	1990	> 500
1940	0	1963	159		
1950	12	1968	224		
1951	16	1970	324		

which used to control malaria mosquito-*Anopheles culicifacies* have now become ineffective. Furthermore, pesticides have decimated a host of natural predators which kept pests in check. By destroying the opposition, pesticides have made it easier for the pests to have a field day.

Even though India's share of global pesticide use is only 3.75%, WHO has graded most of the pesticides used by Indian farmers as hazardous.

As a consequence the Banerjee Committee which submitted its report on 31 pesticides, the government banned the use of 12 pesticides, restricted the use of 18, including DDT, and reviewed 17 others. The Agriculture Ministry amended the Insecticides Act 1968, in June 1993, to allow use of nine pesticides after introducing new and safer compounds.

Now 25 pesticides are banned in India, pesticides are under review (Table 8.7).

Table 8.7 Curbs on pesticides

Pesticides banned in India	*Pesticides under review in India*
1. Dibromochloro para (DBCP)	1. Penta chloronitro benzene
2. Endrin	2. Aluminium Phosphate
3. Pentachloro nitro benzene (PCNB)	3. Capta fol.
4. Penta chloro phenol (PCP)	4. DDT*
5. Texaphene	5. Methyl Bromide
6. Ethyl parathion	6. Sodium cyanide
7. Chlorodane	7. Lindane*
8. Heptachlor	8. Methyl parathion
9. Aldrin	9. Nicotine Sulphate
10. Pasaquat-di-methyl sulphate	10. Alachlor
11. Nitrofen	11. Benomyl
12. Tetradifon	12. Diuron
13. BHC	13. Penarimal
14. Chlorobenzilate	14. Methomyl
15. Dieldrin	15. Oxyfluorfen
16. Ethylene-di-bromide	16. Phosphamidon

(*Contd.*)

17. Phenyl Mercury acetate	17. Thiometon
18. Copper acetoarsenate	18. Triazophon
19. Ethyl mercury chloride	19. Tridenorph
20. Menazon	20. Ziram
21. Sodium methane arsenate	21. Monochlorophos
22. Calcium cyanide	
23. Aldricerb	
24. Maleic hydrazide	
25. Trichloroacetic acid	

Viable Alternatives

1. All the pesticides being used in India are on an adhoc basis, because we do not know what are our problems. Considering the nature of the problems, the prevailing illiteracy, long-term as well as short-term objectives should be framed. These can only be done if a comprehensive national pesticide policy for our country is formulated.
2. Chemical pesticides should be replaced by pheromones, juvenile and moulting Hormones, the 3rd and 4th generation pesticides. However, these can be used by an educated farmer only. So farmers should be educated to the use of these pesticides. In other words, replacement of chemical pesticides by 3rd or 4th generation pesticides is a human resource problem which involves a time factor of nearly 10 years.
3. In the mean time, farmers should be trained in handling and use of pesticides, telling them the right time and concentration of pesticide to be used. At the same time, the sale of such hazardous pesticides as are banned in western countries, should also be banned in developing countries, instead their sale is being promoted.
4. Organic farming, biodegradable pesticides, pesticides of natural origin, biopesticides, vermiculture and microbial insecticides should be offered as substitutes for intensive farming.
5. While there is no doubt that these alternatives make sound environmental sense, the cost and time factor seem to be the biggest impediment producing these commercially.
6. Organic farming can give best results preferably when the holdings are large. But a reduction in yield to the tune of 10 to 30% is inevitable. However, this can be compensated for by the fact that organically grown commodities fetch premium prices, resulting in increased net returns. For example, the production and export marketing of Eco- Tea by Bombay- Bunnah Co. fetched it 30–150% higher returns in the West.
7. On the other hand, continued use of pesticides will dampen our current export of agri-commodities, live stock and aquaculture opportunities and will effectively block any future export potential (as a result of the successful conclusion of Uruguay Round of GATT), as overseas customers become more and more stringent with respect to chemical residues in agriculture imports. The maximum residue limit in key export markets like Japan and the USA is zero for most chemicals.

Conclusion

Although the movement towards the alternatives is appropriate, its progress is poor. While there is no doubt that pesticides have contributed to a spectacular increase in food grain production, it is also true that if such chemical-intensive farming is allowed to continue, it may shake the very foundation of agriculture.

Neither various effects of pesticides nor their alternatives figure in the national agriculture policy. The task of convincing today's farmers of viable alternatives to the high yield-high return economics of pesticides has to begin at the grassroots with the help of NGOs and field workers of all agencies.

Today, pesticides is more or less a "dirty" word despite its growing importance in saving precious grains. The ground reality is that it is not easy to replace or substitute pesticide use, keeping in view the peculiarities of our nation. However, no efforts must be spared by the government to promote viable alternatives.

The State governments must also do their bit by properly enforcing acts or laws, at their disposal. There is a growing awareness among the general public about the ill effects of pesticides. A huge export market both domestic and international, waits for organically grown food. People are ready to pay more, in the interest of their health and indirectly, to conserve their environment also, Government intervention in this regard would be a very positive step.

The million dollar question is will the viable alternatives replace the chemical intensive way of farming in a big way, in the near future? This is a question which can be answered only by time.

Organic Farming

Before ending this chapter a short note on Organic Farming has to be mentioned Organic farming is a form of agriculture, which avoids or largely excludes the use of synthetic fertilizer, pesticides, plant growth regulators and live stock feed additives. As far as possible organic farmers rely on crop rotation, crop residues, animal manure and mechanical cultivation, to maintain soil productivity, it has to supply plant nutrients, control weeds, insects and other pests. The role of organic agriculture whether in farming, processing distribution or consumption is to sustain and enhance the health of ecosystem and organism from smallest in the soil to human beings. Organic farming is highly labour and knowledge intensive, where as conventional farming is capital intensive requiring more energy and manufactured inputs. Organic farming generally produces somewhat lower yields but sustain better yields during drought years allowing it to reap high yields. Studies so far has shown that organic farming requires less water, uses few and always natural pesticides, prevent soil erosion, leaches dramatically few nitrates. It has shown to have improved nutrient qualities including as much as double the flavonoids, an important antioxidant.

REFERENCES

1. Corbett, J.R., Wright, K., Bailie, A.C., *The Biochemical Mode of Action of Pesticides,* Academic Press, London (1984).
2. "Insect, Neurobiology and Pesticide Action" (1980), Society of Chemical Industry, London.
3. Bowers, W.S., Ohta, T. Cleere, J.S. and Marsella, P.A., "Discovery of Insect Antijuvenile Hormones in Plant" Science, 193, 542–47 (1976).
4. Fest C., and Schmidt, K.J., *The Chemistry of Organophorous Pesticide,* Springer Verlag, Berlin (1973).
5. Jacobsons, M., *Insect Sex Pheromones,* Academic Press, New York (1972).
6. Kuhr, R.J., Dorough, H.W., *Carbamate Insecticides, Chemistry, Biochemistry and Toxicology,* CRC Press, Cleveland (1976).
7. Melnikov, N.R., *Chemistry of Pesticides,* Springer Verlag, New York (1971).
8. Shanklan, D.L., *In Insecticides, Biochemistry and Physiology* (C.P. Wilkinsoned), pp. 229–270, Plenum Press, New York (1976).

9. Wood, D.L., Silvester, R.M. and Nakayima, M., *Control of Insect Behaviour by Natural Products,* Academic Press, New York (1970).
10. Carson, R., *Silent Spring,* Houghton Mifflin Co, Boston Massachusetts (1962).
11. Sindhu, P.S., *Chemistry of Atmosphere,* New Age International Publishers, New Delhi (1997).
12. Ashton P.M. and Crafts, A.S., *Mode of Action of Herbisides,* John Wiley & Sons, New York (1973).
13. Audus, L.J., Herbicide Behaviour in Soil. II. Interactions with Soil Microorganisms, In *The Physiology and Biochemistry of Herbicides,* Audus L.J. (Ed., Academic Press, New York, 5: 163–206 (1964).
14. Kearney, P.C. and Kaufman, D.D. (Eds.), *Herbicides, Chemistry, Degradation and Mode of Action,* 2nd ed., Vol. 1 (1976), 500, Vol. 2 (1976), p. 501–1036, Marcel Dekker Inc., New York.
15. Hayes, W., Lawn E., *Handbook of Pesticide Toxicology*, Vol. 1, San Diego, CA: Academic Press (1991).

9 Nuclear Chemistry and the Environment

9.1 INTRODUCTION

The phenomenon of natural radioactivity was first discovered by Becquerel in 1896. He observed a blackening of photographic emulsion in the vicinity of uranium compounds. This was subsequently attributed to radiation emanating from uranium compounds. In the following years, the outstanding researches of Madam Curie established the fact that certain nuclei are not completely stable. These nuclei emit radiations of three main types called alpha, beta and gamma radiations. The nuclei emitting these radiations are called radionuclides and the phenomenon is termed radioactivity.

It soon became apparent that radioactivity was a useful tool for studying the structure and properties of matter, both on a microscopic and macroscopic scale. The phenomenon of radioactivity has been used to reveal the structure of the atomic nucleus, to estimate the age of the earth, and to measure the rates of sediment formation on ocean bottoms.

With the advent of man's utilization of nuclear energy, studies of the natural levels of radioactivity have become necessary to understand fully the environmental influences of radioactivity produced by the atomic energy industry. Only by having understanding of natural radioactivity and the manner in which it varies, can an intelligent interpretation of monitoring data be ensured, whether in the vicinity of a reactor site, the oceans, the atmosphere, or in the tissues of man.

Traces of naturally occurring radioactivity can be demonstrated in all living and non-living substances. In addition, ionizing radiations originating beyond the earth are continually bombarding its surface and exposing all inhabitants to another source of radiation. Natural radioactivity, to which man is exposed, thus is both of terrestrial and extra-terrestrial origin.

About 340 nuclides have been found in nature, of which about 70 are radioactive and are found mainly among the heavy elements. At least 36 of the naturally occurring elements have radioactive isotopes, and all isotopes of elements with atomic number greater than 83 are radioactive isotopes.

The relative abundances of the isotopes now present on earth are derived from the isotopic ratios produced when the universe was formed. Alpher and Herman[8] called attention to the surprising uniformity of the ratios of elements, not only on earth, but in the atmospheres and on the surfaces of the planets, in meteorites, in interstellar matter, and even in extra galactic nebulae. This observation, together with the observed constancy in the relative abundance of the isotopes everywhere on earth, has led to the inference that a constant isotopic distribution is a fundamental property of the universe. Alpher and Herman[8], and Ranbama[9] have reviewed the relationships between this inference and the several hypothesis by which the origin of the universe is explained. However, minor shifts in isotopic ratios do occur gradually in nature as a result of geophysiochemical processes that are sensitive to slight differences in mass.

When matter, of which the universe is now formed, first came into being, many of the isotopes may be presumed to have been radioactive, but in the several billion years during which the world is believed to have existed, all the shorter half-life isotopes must have disappeared. The radionuclides which now remain are those that have half lives comparable to the age of the universe. Radioisotopes with half lives of less than about 10^8 years would have disappeared whereas radionuclides having half lives greater than 10^{10} years have decayed very little up to the present time. Kohman[10] divides the residual radioactivity of the earth into the primary radionuclides, whose half lives are sufficiently long that they have survived the interval since their creation, and the secondary radionuclides which are derived by radioactive decay from parent.

In addition to naturally occurring radionuclides of terrestrial origin, a few are produced by interactions of cosmic rays with the gases of the earth's atmosphere and with the elements on the earth's crust. Of these the most notable are carbon-14 and hydrogen-3.

Following a general introduction to the subject of radioactivity, the properties of the three kinds of rays will now be considered. All the three kinds of rays, α-rays, β-rays and γ-rays ionize air to a different extent; in this respect α-rays are the most effective, being of the order of 100 times as powerful as β-rays, and the latter approximately 100 times as effective as γ-rays. α-particles constitute helium nuclei with +2 charge on it, being denoted as ${}^4_2He^{2+}$. The initial velocity V_0 cm per sec of an α particle is related to its range R cm in air by the equation.

$$V_0^3 = aR$$

where the constant a has a value of about 1.0×10^{17} cm^2 s^{-3}. Since the α-particle has appreciable mass and moves with high velocity it has considerable kinetic energy; it is thus very effective in removing electrons from molecules close to its path and so produce considerable ionization. The α-particle is a massive particle, travelling through matter slowly but with high momentum. It has a high probability of colliding and thus ionizing atoms and molecules along its path, giving up its energy during each impact. As a result, α-particles lose energy very rapidly and can only travel short distances in a dense medium.

β-rays are associated with electrons emitted from nucleus of atom. The velocity of β-particle, as it is called, varies from nuclei to nuclei. However, because of low mass the kinetic energy of β-particles is small, consequently the ionization power to gases is 100 times less than in case of α-particle. As β particles are much smaller than α-particles, they travel much faster, they undergo fewer interactions per unit length of track to give up their, energy more slowly. This means that β-particles travel further in a dense medium.

Gamma rays (γ-rays) do not have any particle associated with it but demonstrated exceptional penetrating properties, suggest that they consist of electromagnetic radiation analogous to X-rays.

During the period 1900–1940, artificial radioactivity was discovered and particle accelerators for producing man-made radioactivity were devised. In 1939, Hahn and Strassmann detected neutron induced nuclear fission of uranium. Potential uses were immediately recognised and within a few years, during World War II, the technology for large scale utilisation of the energy released in nuclear fission was developed. Whether the energy is released explosively in a nuclear weapon or more slowly in a nuclear power plant, hazardous radioactive fission products and other radioactive substances are formed.

The extent to which these sources of ionizing radiation enter the environment and add to the natural background level of ionizing radiation, and thereby become a threat to the health of this and future generations is a matter of serious concern now and will continue to be so in the future.

With the signing of international agreements limiting open-air testing of nuclear weapons by nations and in the absence of nuclear warfare, this source of atmospheric contamination has been greatly reduced but has not been eliminated. Atmospheric weapons tests and underground nuclear detonation or tests, are still being conducted, but attention has been focussed on the environmental impact of nuclear power plants. The possibility of a direct impact arises because the plants produce large quantities of radioactive material of the same type that is produced in nuclear weapon plants. Within a nuclear power-plant and in subsequent handling step, the radioactive material is confined to containers that are designed to prevent its release to the environment. The problem of storage of radioactive waste from the nuclear power plants is an unsolved but by no means a new problem. The possibility of indirect environmental impact stems from use and production in the plants of materials needed to fabricate nuclear weapons, even though the materials used or produced, would require extensive purification and processing.

The potential uses that will meet the needs of mankind may or may not be obvious at the time of a discovery. The development of applications is expedited as and then potential benefit are recognised. Too often the cost has been taken to be simply the development cost. The true cost may not be recognised until many years have passed. Harmful effects involving new phenomena as well as familiar ones may slowly emerge. The harmful effects of ionising radiation have been characterised and studied in depth for several decades. Although much is shown, much remains to be learned. Some of the effects do not become evident until many years after exposure and even then one cannot always be certain that an observed effect was caused by ionising radiation rather than some other agent capable of producing precisely the same effect.

We shall begin the chapter with a brief summary of properties of radioactive materials. Such properties include the radioactive half life and the type and energy of the radiation emitted. We shall then consider the interaction of ionizing radiation with matter, dosimetry, and the effects of such radiation on the human body. Next, we shall examine the environment with respect to both natural and man made sources of ionizing radiation. Finally, current levels of pollution exposure will be summarised.

During the last decade or so, the public, industry, and government have increasingly recognized that cost benefit analysis is an essential part of the decision making process. This represents a drastic change from the situation that prevailed during the industrial revolution. It is now common place for environmental impact to be a key factor or even a critical factor in making or delaying of decisions that affect agriculture, automobile industry the power industry, etc. Hopefully, this chapter will provide the reader with a basis for thinking objectively when considering the pros and cons of issues that involve ionizing radiation.

9.2 THE KINETICS OF NUCLEAR DISINTEGRATION

Kinetics of radioactive decay has been studied extensively. Thus thorium-232 decays to give radium-228 and bismuth-210 disintegrates to give polonium-210 as

$$^{232}_{90}\text{Th} \rightleftharpoons {}^{228}_{88}\text{Ra} + \alpha\text{-rays} \qquad T_{1/2} = 1.4\times10^{10}\ \text{Yr}$$

$$^{210}_{83}\text{Bi} \rightleftharpoons {}^{210}_{84}\text{Po} + \beta\text{-rays} \qquad T_{1/2} = 5.01\ \text{days}$$

Such rates of nuclear emissions have been found to be first order processes. For a radioactive substance, the number of atoms decaying per unit time in a sample is directly proportional to the number of atoms of radioactive substance present. A given radionuclide is designated by a chemical symbol, a subscript corresponding to the atomic number, and a superscript indicating the mass number. Thus, carbon-14 is written as $^{14}_{6}C$, or simply ^{14}C since the atomic number is implicit in the chemical symbol.

If N is the number of atoms of a single radionuclide present at time t, the rate of change of N is given by

$$\frac{dN}{dt} = -\lambda N \qquad ...(9.1)$$

The proportionality constant, λ is the disintegration constant, a characteristics constant for each radionuclide and has units sec^{-1}. The disintegration rate or the activity for a single substance A, at any instant is equal to (λ) N. The unit of activity is curie, where one curie, 1 Ci, = 3700×10^{10} disintegrations per sec (dps). Commonly used subunits are mCi (10^{-3} Ci), μCi (10^{-6} Ci) and pCi (10^{-12} Ci). For most purposes, it is useful to know the specific activity of a radionuclide. This is the activity per unit weight of the element, *e.g.*, mCi g^{-1} of carbon. Another useful quantity is activity concentration, *e.g.* pCi l^{-1} of solution.

Integration of Eq. (9.1) between two limits t_1 and t_2 leads to

$$N_2 = N_1 \exp\left(-\lambda(t_2 - t_1)\right) \qquad ...(9.2)$$

where N_1 and N_2 are number of nuclide at times t_1 and t_2 respectively. Commonly, the time at the beginning of the decay period of interest, is set equal to zero which gives the familiar expression.

$$N = N_0 \exp(-\lambda t) \qquad ...(9.3)$$

where t is time lapsed and N_0 is the number of atoms of the radionuclide present in the sample when $t_1 = 0$.

When N is equal to $N_0/2$, t is equal to $T_{1/2}$ the half life of radionuclide. Then expression,

$$\frac{N_0}{2} = N_0 e^{-\lambda T_{1/2}}$$

$$\log_e \frac{1}{2} = -\lambda T_{1/2} \text{ or } T_{1/2} = \frac{0.693}{\lambda} \qquad ...(9.4)$$

Equation (9.3) may then be written in the form

$$N = N_0 \exp\left(-0.693 t / T_{1/2}\right) \qquad ...(9.5)$$

which is more useful than Eq. (9.3), because it is $T_{1/2}$ rather than λ, which is tabulated for radionuclide. When Eq. (9.5) is rewritten as

$$N = N_0 (0.5)^{t/T_{1/2}} \qquad ...(9.6)$$

the resulting equation is useful for quick exact calculations or working approximations. The mean life τ is of interest in certain calculations and is equal to $1/\lambda$.

By multiplying Eqs. (9.3), (9.5) and (9.6) by λ and replacing λN by A, the activity, equations showing the time dependence of activity of a sample is obtained. Thus Eqs. (9.5) and (9.6) become

$$A = A_0 \exp\left(-0.693t/T_{1/2}\right) \qquad ...(9.7)$$

and

$$A = A_0 (0.5)^{t/T_{1/2}} \qquad ...(9.8)$$

where A_0 is the initial activity. A typical decay curve corresponding to these equation is shown in Fig. 9.1. To illustrate the use of Eq. (9.8), we might ask how long it would take for a sample of radionuclide to decay to about one thousandth of some specified activity. Recognising that $(0.5)^{10}$ is equal to 1/1024, about ten half lives will be required to provide the specified reduction in A/A_0 for any radionuclide. Thus for the fission products $^{90}_{38}Sr$ ($T_{1/2}$ = 28.5 years), $^{22}_{11}Na$ ($T_{1/2}$ = 2.6 years and ^{32}P ($T_{1/2}$ =14.6 days) about 285 years, 26 years and 146 days respectively would be required for a 1 Ci sample of each to decay to about 1 mCi.

In the laboratory or in the field, we measure a counting rate *R* using a detector. The counting rate is proportional to *A* and depends on the response efficiency of the detector. Assuming that the efficiency remains constant during measurement, a plot of R/R_0 or of A/A_0 would give a decay curve like that shown in Fig. 9.1.

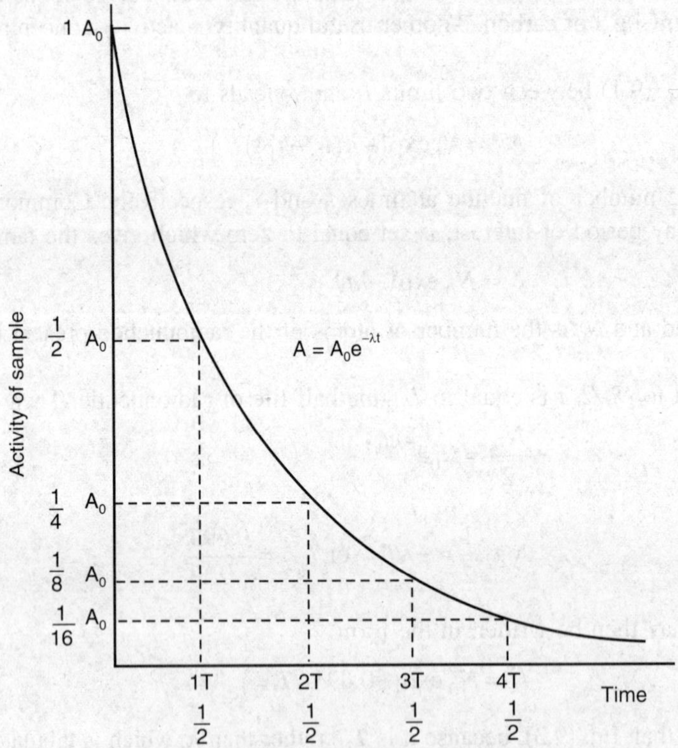

Fig. 9.1 Variation of activity with time.

A few examples of half lives of naturally occurring radionuclides are given in Table 9.1.

The primordial radionuclides have half-life values comparable to or longer than the age of the earth, *i.e.* about 4.5 × 10^9 Yr. One of the most important single sources of background radiation is ^{40}K. Other important radionuclides are the isotopes $^{232}_{90}Th$, $^{228}_{88}Ra$ and $^{235}_{92}U$. Most short-lived radionuclides

Table 9.1 Half-lives of some radionuclides in nature

Radionuclide	*Radioactive decay*	*Half-life $T_{1/2}$*	*10 $T_{1/2}$*
^{3}H	β-radiation	12.33 Yr.	123.3 Yr.
^{14}C	β-radiation	5570 Yr.	55700 Yr.
^{22}Na	β^{+}-radiation (positron)	< 2.6 Yr.	26 Yr.
^{32}P		14.5 days	145 days
^{33}P		< 25 days	25 days
^{35}S		< 87 days	87 days
^{40}K	β-radiation (90%) + γ-ray (10%)	1.3 × 10^{9} Yr.	13 × 10^{9} Yr.
^{50}V	β-radiation	4 × 10^{14} Yr.	40 × 10^{14} Yr.
^{129}I	β-radiation	1.72 × 10^{7} Yr.	17.2 × 10^{7} Yr.
^{215}Po	α-rays	18.3 × 10^{-3} sec.	18.3 × 10^{-2} sec.
^{226}Ra	α-rays	1622 Yr.	16220 Yr.
^{232}Th	α-rays	1.4 × 10^{10} Yr.	14 × 10^{10} Yr.
^{238}U	α-rays	4.49 × 10^{9} Yr.	44.9 × 10^{9} Yr.

in nature are formed as daughters of long-lived parents. About 20 short-lived radionuclides are also being produced continuously by cosmic ray interaction in the atmosphere. Two such cosmogenic radionuclides of interest in terms of environment are ^{3}H and ^{14}C. These are produced in the atmosphere by nuclear reactions of fast neutrons (energy in the MeV range) with ^{14}N. The reactions are

$$^{14}_{7}N + ^{1}_{0}n \longrightarrow ^{14}_{6}C + ^{1}_{1}H$$

$$^{14}_{7}N + ^{1}_{0}n \longrightarrow ^{12}_{6}C + ^{3}_{1}H$$

The relative yields for the production of ^{3}H and ^{14}C from ^{14}N are about 1:100. Among the other radionuclides of cosmogenic origin in the atmosphere, soil and sea are ^{7}Be, ^{22}Na, ^{26}Al, ^{32}Si, ^{32}P, ^{33}P, ^{35}S, ^{36}Cl, ^{39}Cl, ^{37}Ar, ^{53}Mn and ^{60}Cs.

These are present in smaller amounts and according to the current thinking they are not of much important particularly to environmental pollution.

9.3 KINETICS OF DISINTEGRATION OF MIXTURES OF RADIONUCLIDES

In nature and in nuclear power reactors there are complex mixtures of radionuclides. Mixtures are of three types: (1) independent (2) genetically related (parent decays to radioactive daughter), and (3) a combination of (1) and (2). For a nongenetically related mixture of only two radionuclide *A* and *C*, such that

$$A \longrightarrow B \text{ (stable)}$$

and

$$C \longrightarrow D \text{ (stable)}$$

and *C* has the longer half life, the composite decay curve might be as shown in Fig. 9.2, where the component decay curves are also shown.

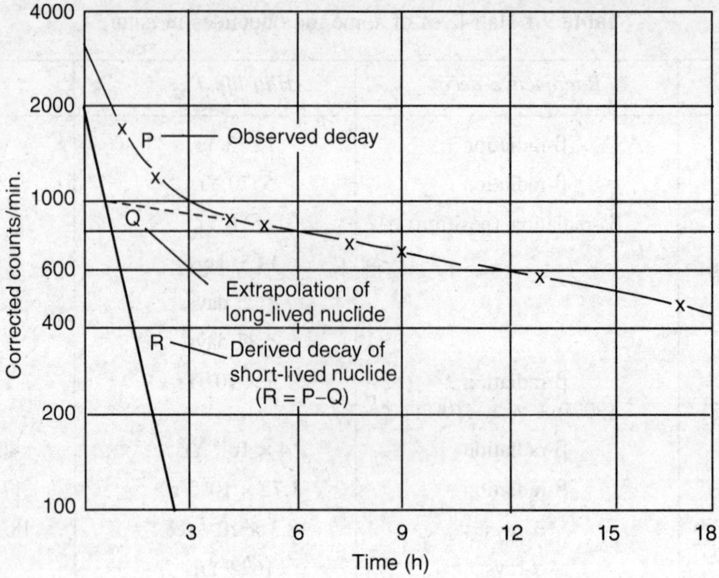

Fig. 9.2 Decay curve of mixed Cl-38 and Na-24 sample.

Most of the naturally occurring radionuclides in the rocks and soil of the earth are members of one of the three series known as the uranium, thorium and actinium series. The series are named after the longest lived member, the parent of the series. Similarly, fission products are commonly members of chain typically having three or four members.

For a given series one can write *A—B—C—D—X* to which is a stable end product. The parent daughter relationships in the three naturally occurring series are represented in Table 9.2–9.4 and contain characteristics data.

Table 9.2 The Uranium (4*n* + 2) series

Nuclide	*Name used in older literature*	*At wts.*	*At no.*	*Half life time*	*Major radiation*
U	uranium I	238	92	4.47×10^9 yr.	α-rays
Th	uranium X_1	234	90	24.2 days	α-rays (γ-rays)
Pa	uranium X_2	234	91	1.17 min	β-rays
Pa	uranium Z	234	91	6.67 hr	β-rays
U	uranium II	234	92	2.44×10^5 Yr.	α-rays
Th	(Ionium)	230	90	7.7×10^4 Yr.	α-rays
Ra	Radium	226	88	1600 Yr.	α-rays (γ-rays)
Rn	Radon	222	86	3.824 days	α-rays
Po	Radium A	218	84	3.05 min.	α-rays (99.98) β-rays (0.02%)
Pb	Radium B	214	82	26.8 min.	β-rays

(*Contd.*)

Bi	Radium C	214	83	19.8 min.	α-rays (0.02%) β-rays (99.8%)
Po	Radium C′	214	84	164 μ sec.	α-rays
Ti	Radium C″	210	81	1.30 min	β-rays
Pb	Radium D	210	82	22.3 Yr.	β-rays
Bi	Radium E	210	83	5.01 days	β-rays
Po	Radium F	210	84	138.4 days	α-rays
Pb	Radium G	206	82	stable	

Table 9.3 The Thorium (4*n*) series

Nuclide	*Name used in older literature*	*At wts.*	*At no.*	*Half life time*	*Major radiation*
Th	Thorium	232	90	1.4×10^{10} Yr.	α-rays
Ra	Mesothorium I	228	88	5.75 Yr.	β-rays
Ac	Mesothorium II	228	89	6.13 Yr.	β-rays
Th	Radiothorium	228	90	1.91 Yr.	α-rays
Ra	Thorium X	224	88	3.64 ray	α-rays
Rn	Thoron	220	86	55.6 sec	α-rays
Po	Thorium A	216	84	0.15 sec	α-rays
Pb	Thorium B	212	82	10.64 hr.	β-rays
Bi	Thorium C	212	83	60.6 min	α (36%) β (64%)
Po	Thorium C′	212	84	304 nsec	α-rays
Tl	Thorium C″	208	81	3.054 min	α-rays
Pb	Thorium D	208	82	Stable	

Table 9.4 The Actinium (4*n* + 3) series

Nuclide	*Name used in older literature*	*At wts.*	*At no.*	*Half life time*	*Major radiation*
U	Actinouranium	235	92	7.04×10^{8} Yr.	α-rays
Th	Uranium Y	231	90	25.5 Yr	β-rays
Pa	Protactinium	231	91	3.25×10^{4} Yr.	α-rays
Ac	Actinium	227	89	21.77 Yr.	α-(1.4%)
Th	Radioactinium	227	90	18.72 days	α-(98.6%)
Fr	Actinium K	223	87	22 min.	β-rays
Ra	Actinium X	223	88	11.43 day	α-rays

(Contd.)

Nuclide	*Name used in older literature*	*At wts.*	*At no.*	*Half life time*	*Major radiation*
Rn	Pctinon	219	86	3.96 sec.	α-rays
Po	Actinium A	215	84	1.78 m sec.	α-rays
Pb	Actinium B	211	82	36.1 min	β-rays
Bi	Actinium C	211	83	2.14 min	α-rays (99.7%) β-rays (0.28%)
Po	Actinium C′	211	84	0.53 sec	α-rays
Ti	Actinium C″	207	81	4.77 min	β-rays
Pb	Actinium D′	207	82	Stable	

The uranium series, for example, begins with $^{238}_{92}U$ and ends with $^{206}_{82}Pb$. Because all members of the series have mass number represented by four times an integer plus two, the series is also known as the $4n + 2$ series.

To determine the amount of daughter *B*-present in these or any series at a time *t,* one starts with the rate equation for the net rate of change of N_B atoms.

$$\frac{dN_B}{dt} = \text{rate of formation } B \text{ by decay of } A - \text{rate of decay of } B$$

$$= \text{activity of } A - \text{activity of } B$$

or
$$\frac{dN_B}{dt} = \lambda_A N_A - \lambda_B N_B \qquad ...(9.9)$$

In this equation it is assumed that substance *A* can decay in only one mode, namely, that in which daughter *B* is formed. There are many cases in which branching (decay by more than one mode) occurs. When branching occurs, the $\lambda_A N_A$ term must be multiplied by the fraction of decay in which the substance is formed.

The value of A_B, the activity of the daughter at any time can be obtained in terms of $(A_A)_0$ and $(A_B)_0$, the activities of the parent and daughter at some reference time $t = 0$ by integrating, Eq. (9.9) to obtain $\lambda_B N_B$, and A_B. The resulting equation is then

$$A_B = \frac{(A_A)_0 \lambda_B}{(\lambda_B - \lambda_A)}\left[\exp(-\lambda_A t) - \exp(-\lambda_B t)\right] + (A_B)_0 \exp(-\lambda_B t) \qquad ...(9.10)$$

The last term represents the decay of the daughter radionuclide present in a sample of *A* when $t = 0$. Three types of parent-daughter system can be recognised on the basis of the relative values of λ_A and λ_B. When $\lambda_A <<< \lambda_B$ that is $T_A >>> T_B$, the system can attain a steady state called secular equilibrium. When $\lambda_A < \lambda_B$ or $T_A > T_B$ the system can reach a steady state called "transient equilibrium". Finally, when $\lambda_A > \lambda_B$ or $T_A < T_B$ a steady state cannot be attained.

For systems that can attain secular equilibrium Eq. (9.10) reduces to the following when only pure A is present at $t = 0$.

$$A_B = (A_A)_0\left[l - \exp(\lambda_B t)\right] \qquad ...(9.11)$$

In this equation, t, the lapsed time, is short compared to the half life of the parent, hence $(A_A)_0$ remains essentially constant during time t. The growth of the daughter activity according to Eq. (9.11) is exponential in nature. Recalling that exp $(-\lambda_B t)$ can be replaced by $(0.5)\ t/T_{1/2}$, we see that in one half life of the daughter, its activity is equal to 50% of that of the parent, after two half lives, 75% and after ten half lives 99.9%. When a system has attained secular equilibrium, the activity of the daughter is equal to that of the parent. The number of atoms and therefore the weight of each is, however, very different.

Two examples of parent-daughter combinations that can attain secular equilibrium are

$$^{226}_{88}\text{Ra} \xrightarrow{1600\text{ yr.}} {}^{222}_{86}\text{Rn} \xrightarrow{8.83\text{ days}} {}^{218}_{84}\text{Po} \longrightarrow$$

and

$$^{106}_{44}\text{Ru} \xrightarrow{369\text{ days}} {}^{145}_{45}\text{Rh} \xrightarrow{30\text{ sec.}} {}^{106}_{46}\text{Pd}\,(\text{stable})$$

The former is important in the uranium series and the latter is an important contributor to fission product wastes. In the uranium series, the parent of each series is a primordial radioactive nuclide. The primary rocks that have been disturbed by weathering, leaching or the escape of the radon isotope that contributes the noble gas member of each series, all members except the last are in secular equilibrium with the parent. This is important to keep in mind when considering the radioactivity background associated with uranium and thorium content of rocks, soils and water.

9.4 RADIATION UNITS

As already mentioned, the ionization of a gas provides means of detecting radiation and the first widely used radiation unit, the roentgen, was based on the ionizing effect on air of γ radiation.

Exposure is a quantity expressing the amount of ionizing caused in air by X-rays or γ-rays. The special unit is roentgen (abbreviated to R) which corresponds to the production of ions (of one sign) carrying a charge of 2.58×10^{-4} coulombs per kg (C kg^{-1}) of air. Now an ion (of either sign) carries a charge of 1.602×10^{-19} C and so 1R corresponds to the production of

$$= \frac{2.58\times10^{-4}}{1.602\times10^{-19}} \text{ ion pairs per kg of air.}$$

$$= 1.61\times10^{15} \text{ ion pairs kg}^{-1}$$

The average energy required to produce an ion pair in air is 5.4×10^{-18} J and so the energy absorbed by air corresponding to an exposure of 1R is

$$1.61\times10^{15}\times5.4\times10^{-18}\text{ J kg}^{-1} \text{ of air}$$

$$= 8.69\times10^{-3} = 0.00869 \text{ J kg}^{-1} \text{ of air}$$

The roentgen is inadequate as a radiation unit since it applies only to X- and γ-rays and their effect on air, even then it is still in use. Usually human tissue is the medium of interest and the energy deposition is often higher in tissue than in air. In the case of γ-rays, over the commonly encountered range of photon energy, the energy deposition in tissue for an exposure of 1R is about 0.0096 J kg^{-1}. The concept of absorbed dose of radiation was introduced to overcome these difficulties.

Absorbed dose is a measure of energy deposition in any medium by all types of ionizing radiation. The unit of absorbed dose is the *rad* which is defined quantity of absorbed radiation of any type which deposits 100 ergs g^{-1} or 0.01 J kg^{-1} of material. Since it is a unit which measures the absorbed energy independent of stopping power of the medium, it is particularly suitable for the study of biological effect. The roentgen, on the other hand is basically a unit which measures radiations dose, regardless of the extent to which it is absorbed.

Since an exposure of 1R results in an energy deposition of 0.00869 J kg^{-1} in air and 0.0096 J kg^{-1} in tissue, it follows that 1R gives an absorbed dose of

$$= \frac{0.00869}{0.01} = 0.869 \text{ rad in air}$$

and

$$\frac{0.0096}{0.01} = 0.96 \text{ rad in tissue}$$

In many cases, then, exposure in R and absorbed dose in rads have almost the same value. Note that the medium is always specified.

Although the rad is a very useful physical unit, it transpires that in biological systems the same degree of damage is not necessarily produced by the same absorbed dose of different types of radiation. It is found, for example, that 1 rad of alpha radiation can do as much biological damage as 10 rad of gamma radiation. This difference in the radiobiological effectiveness must be taken into account if we wish to add doses of different radiation to obtain the total biological effective dose. To do this we must multiply the absorbed dose of each type of radiation by a quality factor (*QF*) which reflects the ability of the particular type of radiation to cause damage. The quantity obtained when an absorbed dose is multiplied by the quality factor is known as the dose equivalent, the unit of which is the rem.

Dose equivalent (rem) = Absorbed dose (rad) × *QF*.

The value of the quality factor is found to depend on the density of ionization caused by the radiation. An alpha particle produces about 10 million ion pairs per centimetre of track in tissue, whereas a beta particle produces about one hundred thousand per centimetre. γ-rays cause ionization of a similar density as β-ray and since all others are comparable to that of gamma rays, the *QF* for gamma rays is taken as unity and the *QF* for beta rays is also one. For alpha particles, the *QF* is taken as 10. The *QF* for neutrons is dependent on the neutron energy, but a value of 3 is usually taken for thermal neutrons (10 MeV) and 6.5 for fast neutrons. The values of *QF* are summarised in Table 9.5.

Table 9.5 Summary of QF values

Types of radiation	*QF*
X-ray	1
γ-ray	1
β-ray (β^-, β^+ and e^-)	
< 0.03 E_{Max}	1.7
> 0.03 E_{Max}	1.0
α-ray	10.0
Recoil nuclei	20.0
Thermal neutron (10 MeV)	3
Fast neutron > 10 MeV	6.5

Example 9.1. In one year a worker receives a gamma dose of 2 rad, a thermal neutron (N_S) dose of 0.5 rad and a fast neutron dose (N_F) of 0.1 rad. What is the total dose equivalent?

Solution:

$$\text{Dose equivalent} = \text{Absorbed dose} \times QF$$

$$\therefore \quad \gamma\text{-ray dose equivalent} = 2 \times 1 = 2 \text{ rem.}$$

$$N_S \text{ dose equivalent} = 0.5 \times 3 = 1.5 \text{ rem.}$$

$$N_F \text{ dose equivalent} = 0.1 \times 6.5 = 0.65 \text{ rem.}$$

$$\text{Total dose equivalent} = 4.15 \text{ rem.}$$

While calculating the dose equivalent due to radioactivity material inside the human body, it is necessary in certain rare cases to apply a further factor which takes account of non-uniform distribution of the radioactivity. In such cases the dose equivalent is expressed as follow:

$$\text{Dose equivalent} = \text{Absorbed dose} \times QF \times DF.$$

where DF is a distribution factor. Normally the value of DF, is one, but in rare cases mentioned, it is given a value of 5.

In terms of occupational radiation exposure the roentgen, rad and rem are rather large units. It is often convenient to have smaller units and this is done by using the prefixes milli (m) and micro (μ). Thus

$$1 \text{ R} = 1000 \text{ mR} = 1000000 \text{ } \mu\text{R.}$$

$$1 \text{ rad} = 1000 \text{ mrad} = 1000000 \text{ } \mu \text{ rad.}$$

$$1 \text{ rem} = 1000 \text{ m rem} = 1000000 \text{ } \mu \text{ rem.}$$

Dose rate: The roentgen, rad and rem are units expressing an amount of radiation, which may have been received over any period of time. In controlling radiation hazard, it is usually necessary to know the rate at which the radiation is being received. Thus, if a man works in an area for 2 hours and receives a dose equivalent of 0.4 rem, then the dose equivalent rate is a 0.2 rem per hour (rem h^{-1}). Similarly, we have exposure rates ($R\ h^{-1}$) and absorbed dose rates (rad h^{-1}).

The relationship between, dose, dose rate and time is

$$\text{Dose} = \text{Dose rate} \times \text{time.}$$

Example 9.2. If a man is permitted to receive a dose equivalent 100 m. rem in a week, for how many hours may he work in a area in which the dose equivalent rate is 5 m rem h^{-1}?

Solution:

$$\text{Dose} = \text{Dose rate} \times \text{time}$$

$$\text{Time} = \frac{\text{Dose}}{\text{Dose rate}} = \frac{100}{5}$$

$$= 20 \text{ hours}$$

Flux: It is often convenient to express a radiation field as the number of particles or photons crossing an area of 1 cm^2 in 1 sec. This is strictly fluence rate but is commonly referred to as flux (denoted by ϕ). The concept is best illustrated by a practical example.

Consider a point source which emits neutrons at the rate of Q per sec. Now the flux at distance r is the number of neutrons passing through an area of 1 cm^2/s^{-1}. Since the neutrons are being emitted uniformly in all directions, the flux at distance r is the number of neutrons emitted per second divided by the area of the sphere of radius r. This area is $4\pi r^2$ and so the flux. ϕ is given by Fig. 9.3.

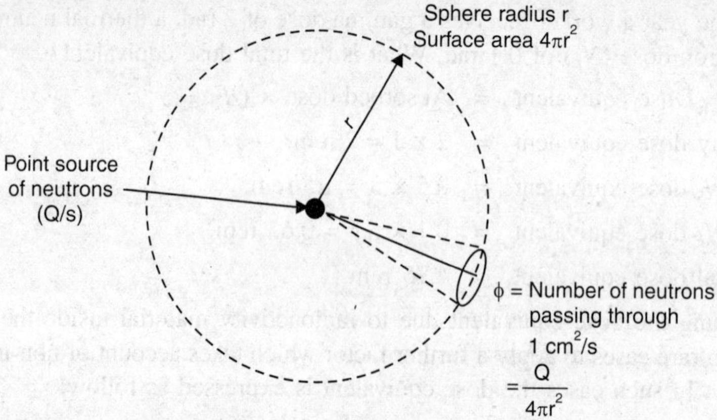

Fig. 9.3 Flux from a point source.

$$\phi = \frac{Q}{4\pi r^2} \text{ neutrons per cm}^2 \text{ per sec } \left(n \text{ cm}^{-2} \text{ s}^{-1}\right) \quad ...(9.12)$$

Note if r is doubled, r^2 increases fourfold and ϕ reduces by four times. This relationship is the inverse square law.

Example 9.3. Calculate the flux at a distance of 50 cm from a source which emits 2×10^7 ns^{-1}.

Solution: $$\phi = \frac{Q}{4\pi r^2} = \frac{2 \times 10^7}{4\pi \times 50 \times 50} = 6.4 \times 10^2 \text{ cm}^{-2}\text{s}^{-1}$$

Example 9.4. Calculate the gamma photon flux at 1 m from a 1 Ci cobalt-60 source. (Cobalt-60 emits 2 gamma rays per disintegration).

Solution: We know $1 \text{ Ci} = 3.7 \times 10^{10} \text{ dis s}^{-1}$, but for ^{60}Co there are two photons per disintegration.

$$\therefore \quad Q = 2 \times 3.7 \times 10^{10} = 7.4 \times 10^{10} \text{ photons s}^{-1}$$

$$\phi = \frac{Q}{4\pi r^2} = \frac{7.4 \times 10^{10}}{4\pi \times 100^2}$$

$$= 5.9 \times 10^5 \text{ photons cm}^{-2}\text{s}^{-1}$$

The units which have been introduced are summarised in Fig. 9.4.

Thus flux at a distance, r, is inversely proportional to the square of the distance. Since the radiation dose rate is directly related to flux, it follows that the dose rate (D) also follows the inverse square law. This relationship holds good for a point source, a point detector and negligible absorption of radiation between source and detector.

Thus inverse square law may be written as

$$D \propto \frac{1}{r^2}$$

or $$D_1 r_1^2 = D_2 r_2^2 \qquad ...(9.13)$$

where D_1 and D_2 are dose rates at distances r_1 and r_2 from the source.

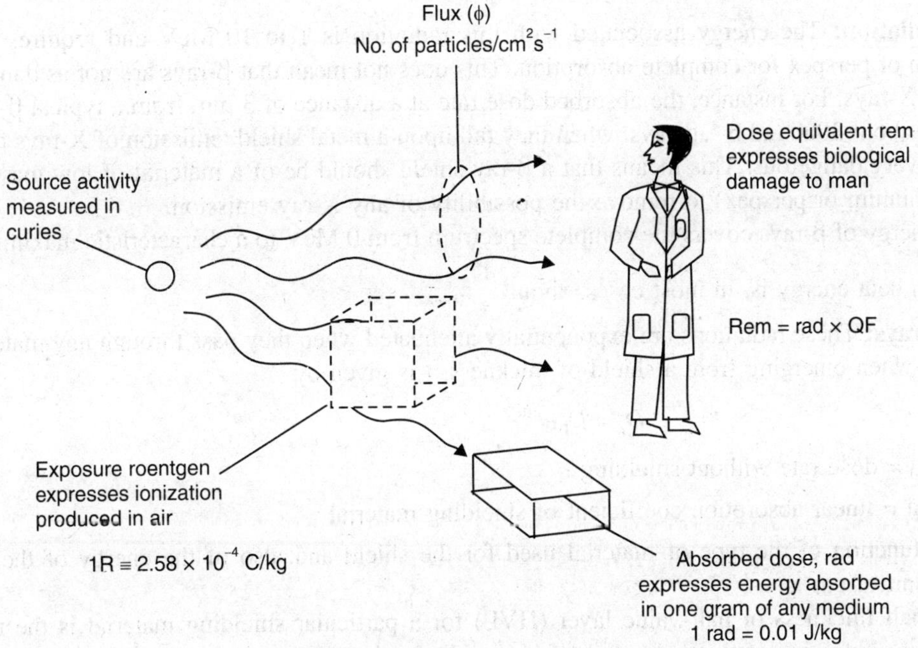

Fig. 9.4 The relationship of units.

A useful expression for calculating the approximate dose rate from a γ-source of C curies, emitting photons of total energy E MeV per disintegration is

$$D = 0.5\frac{CE}{r^2} \qquad ...(9.14)$$

where C is activity of source in curies, E is the total γ energy/disintegration in MeV and r is the distance from the other source in metres.

Example 9.5. Calculate the activity of sodium-22 source which gives a dose rate of 3 rem h^{-1} at 1m, ^{22}Na emits one gamma photon of energy 1.28 MeV per disintegration.

Solution: From equation 9.14

$$C = \frac{Dr^2}{0.5E}$$

Here D = 3 rem h^{-1}
r = 1 m
E = 1.28 MeV

$$C = \frac{3 \times 1}{0.5 \times 1.28} = 4.68 \text{ curies.}$$

On account of the different penetrating powers of the various atomic radiations, widely different amounts and types of shielding are required to deal with them.

α-particles: A thin sheet of paper is sufficient to stop an α-particle so it never presents a shielding problem.

β-radiation: The energy associated with this radiation is 1 to 10 MeV and requires shielding upto 1 cm of perspex for complete absorption. This does not mean that β-rays are not as dangerous as γ-rays or X-rays. For instance, the absorbed dose rate at a distance of 3 mm from a typical β-source of 1 m Ci is about 3000 rad h^{-1}. β-rays, when they fall upon a metal shield, emission of X-rays take place which is very dangerous. This means that a β-ray shield should be of a material of low mass number (e.g. aluminium or perspex) to remove the possibility of any X-ray emission.

As energy of β-rays covers the complete spectrum from 0 MeV to a characteristic maximum, E_{max}. The mean beta energy is, in most cases, about $\frac{1}{3}E_{max}$.

γ and X-rays: These radiations are exponentially attenuated when they pass through any material. The dose rate when emerging from a shield of thickness, t is given by

$$D_t = D_0 e^{-\mu t} \qquad \text{...(9.15)}$$

where D_0 = dose rate without shielding

μ = linear absorption coefficient of shielding material

and is a function of the type of material used for the shield and also of the energy of the incident photons unit of cm^{-1}.

The half thickness or half-value layer (HVL) for a particular shielding material is the thickness required to reduce its intensity to one half its incident value. Writing the half-value layer as $t_{1/2}$, the previous equation becomes

$$\frac{D_0}{D_t} = 0.5 = e^{(-\mu t_{1/2})}$$

or

$$t_{1/2} = \frac{0.639}{\mu} \qquad \text{...(9.16)}$$

The concept of half-value layer is very useful in doing rapid approximate shielding calculation. One HVL reduces the intensity to one half, two HVL requires to the intensity to one quarter, and three HVL to one eighth.

Another value sometime used in shielding work is the one-tenth value layer, $t_{1/10}$. By a calculation similar to that carried out above, it can be shown that

$$t_{1/10} = \frac{\ln 10}{\mu} = \frac{2.303}{\mu}$$

Some typical values of $t_{1/2}$ and $t_{1/10}$ for lead and water are given in Table 9.6. The values for μ and $t_{1/2}$ are available in handbooks in tabular or graphical form as a function of energy for a given absorber material.

Table 9.6 Approximate values of $t_{1/2}$ and $t_{1/10}$

γ-radiation energy (MeV)	Centimetres of lead		Centimetres of water	
	$t_{1/2}$	$t_{1/10}$	$t_{1/2}$	$t_{1/10}$
0.5	0.4	1.25	15.0	50.0
1.0	1.1	3.5	19.0	62.5
1.5	1.5	5.0	20.0	70.0
2.0	1.9	6.0	22.5	75.0

Neutron Shielding: It is complicated by the very wide range of energies encountered. The most important reactions are

(a) Elastic Scattering: When a neutron collides with the target nucleus and bounces it in a manner similar to collision of two billiard balls, the neutron loses some of its initial energy and this energy is transferred to the target nucleus. All of this transferred energy appears as kinetic energy of the target nucleus. Light elements are best for slowing down the neutrons by an elastic scattering and so materials with a high hydrogen contents (such as parafin, water, concrete) are used.

(b) Inelastic Scattering: Here, the incoming neutrons impart some of their energy to the scattering material and excite the target nucleus. These target nuclei usually emit γ-rays later, when they return to their ground state. The inelastic scattering is important for heavy nuclei.

(c) Neutron Capture Reactions: In these reactions neutrons are captured by nuclei which then disintegrates to another particle or photon. One very important neutron capture reaction is

$$^{10}_{5}B + ^{1}_{0}n \longrightarrow ^{7}_{3}Li + ^{4}_{2}He^{2+}$$

The α-particle so emitted is easily absorbed. Thus the incorporation of boron-10 in shields means that neutrons are absorbed and the resulting α-particles cause no further shielding problems.

9.5 RADIATION CHEMISTRY

It is perhaps appropriate at this point to compare briefly the interaction and the effects of the interaction of ionizing radiation with matter. It should be recalled that when light in the wavelength range of about 200–1000 nm is absorbed, one photon transfers its energy to one molecule in the interaction or primary step. When absorption occurs, the energy of the photon corresponds to the energy needed to excite the molecule from a lower discrete level to a higher discrete level. For light in the visible or UV region, sufficient energy is available to cause electronic transitions. Ionization requires more energy, *e.g.*, 15.5 eV to form N_2^+ and 12.5 eV to form O_2^+ corresponding to light having a wavelength of about 100 nm.

The electronically excited molecules will undergo one or more of a variety of process or secondary steps. For example, a molecule may undergo deexcitation by the processes of flourescence or phosphorescence, it may undergo photolysis, *i.e.*, dissociate into chemically reactive species, or it may collide with other molecules thereby transferring energy and possibly undergoing chemical reaction.

Charged particles and γ-rays with energy in the range of thousands to million of electron volts can readily produce ionization in the primary interaction step. In fact, a single α-particle may, depending on its energy, undergo 10^5 or so ionization events in passing through air before becoming a normal helium atom in thermal equilibrium with its surrounding. Although a gamma ray may disappear in a single primary ionization event, the electrons or the positron-electron pair released into the absorbing medium will produce many additional ion pairs.

The chemical changes that occur as a consequence of ionization are studied in the branch of chemistry called radiation chemistry. Electronically excited molecules are also produced directly by ionizing radiation and indirectly when an ion is neutralised. Many of the observed chemical changes are the same as those involved in photochemistry. Dissociation into fragments is called radiolysis rather than photolysis. As a point of contrast, ionizing radiation, which eventually transfers its energy by intense force can populate a wider range of excited states than light can. Furthermore, the reactions of ionic species are dominant in radiation chemistry that are uncommon in photochemistry.

9.6 BIOLOGICAL EFFECTS OF RADIATIONS

The interaction of ionizing radiation with the human body arising either from external sources outside the body or from internal contamination of the body by radioactive substances leads to biological effects which may later show up as clinical effects which may later show up as clinical symptoms. The nature and severity of these symptoms and the rate at which they appear depends on the amount of radiation absorbed and the rate at which it is received. Radiation injuries can be divided into two

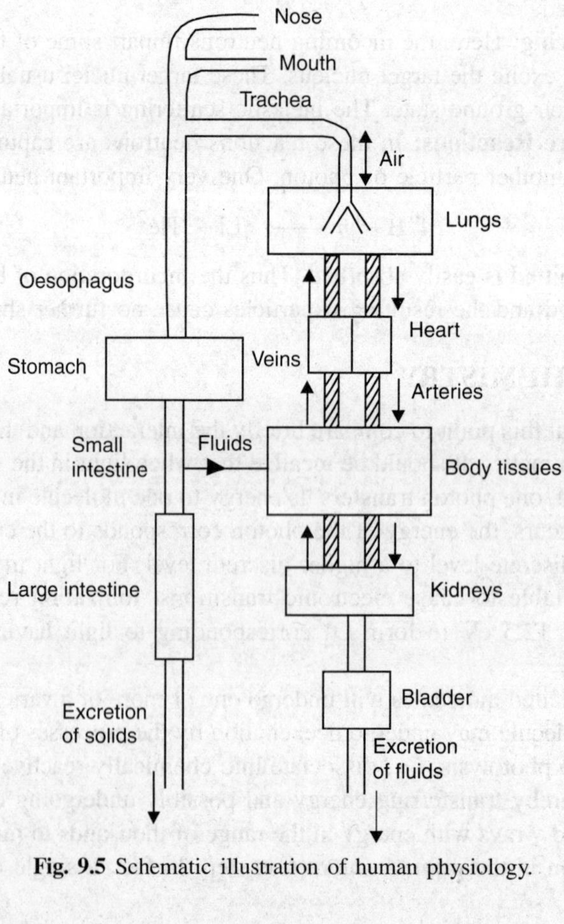

Fig. 9.5 Schematic illustration of human physiology.

classes: somatic effects in which the damage appears in the irradiated person himself; genetic effects which arise only in the offspring of the irradiated person as a result of radiation damage to germ cells in the reproductive organs, the gonads.

Man can be regarded as a machine consisting of various interrelated systems each performing some important functions. The systems which are most relevant to an understanding of the behaviour of radioactive substances entering the body are the circulatory, respiratory and digestive system as shown in Fig. 9.5.

The Circulatory System: The circulatory system is a closed circuit of tubes around which blood is pumped by the action of the heart. Blood circulates to almost every region of the body carrying food and oxygen to every cell. It also picks up waste produced and carbon dioxide and transfers them to the excretory organs. The heart has two pumps: the left side pumps the blood through the arteries to the tissues. The nourishment is transferred from the tissues to the cells via the tissue fluid. The blood, after passing through tissues, returns to the right side of the heart via veins. The blood is then pumped to the lungs where it becomes oxygenated before returning to the left side of the heart again. The blood in the arteries contain a lot of oxygen and is bright red in colour while the blood returning from the tissues contain very little oxygen and is dark red in colour. The body contain about 5 litres of blood which circulates on the average once a minute.

The Respiratory System: Respiration involves transportation of oxygen to the lungs and the elimination of carbon dioxide. Oxygen is carried to the tissues as described above to produce CO_2 as a gaseous waste product and this is carried back by the blood to the lungs and breathed out. The volume of air breathed per day is approximately 10.5 m^3 of which half is considered to be breathed during the eight hours at work.

In the process of respiration, man inhales gaseous and particulate matter. Gases pass freely into the lungs and enter the blood stream to a greater or lesser extent depending on their solubility. In the case of particulate matter, only a fraction of the inhaled material is deposited in the upper respiratory passages and subsequently swallowed. The behaviour of the material deposited in the lungs depends mainly on its solubility. Highly soluble materials are absorbed rapidly into the blood stream, whereas insoluble material may persist in the lungs for many months, clearly, then the respiratory system represents a route of entry for radioactive substances which can then be transported by the blood stream to the other parts of the body.

The Digestive System: It consist of the oesophagus, the stomach, the duodenum and the small intestine which is connected to the large intestine. Food ingested via mouth is converted into a form suitable for the production of heat and energy, and the molecules necessary for the growth and repair by enzymes in the digestive tract before being absorbed into the blood stream and passed via liver to the tissues. The unabsorbed food together with bacteria and cells shed from the intestine wall are passed out as solid waste. Liquid waste is excreted from the body via the kidney and bladder as urine. Soluble radioactive contamination, when swallowed, may pass through the digestive tract and become absorbed into the blood stream which carries it to all parts of the body. Insoluble contamination passes through the digestive tract and will be eventually excreted in the faeces. During its passage through the body it will irradiate the tract and the large intestine.

All living creatures and organisms consist of tiny structures known as cells. The basic components of a cell are the *nucleus* surrounded by liquid called cytoplasm and a membrane which forms the cell wall. Figure 9.6 shows the structure of a typical human cell.

The simplest picture of the cell is that the cytoplasm is the 'factory, of the cell, while the nucleus contains all the information which the cell needs to carry out its function and reproduce itself. The cytoplasm breaks down food and converts it into energy and small molecules. These small molecules are later converted into complex molecules needed by the cell either for maintenance or duplication.

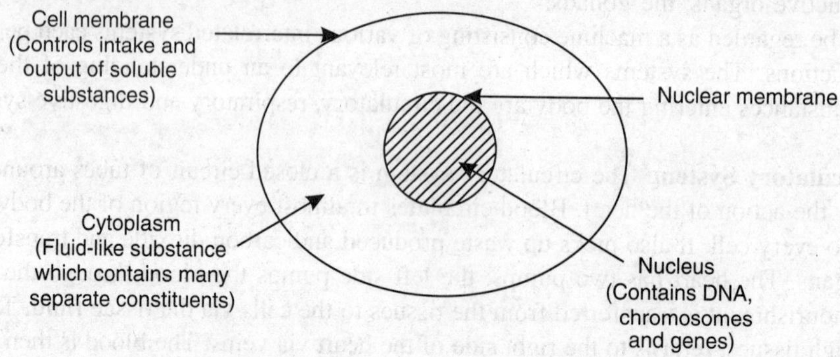

Fig. 9.6 Structure of human cell (schematic).

The nucleus contains the chromosomes which are tiny thread like structures made up of genes. Human cells normally contain 46 chromosomes. The genes consist of DNA and protein molecules and carry the information which determines the characteristics of the daughter cell.

Cells are able to reproduce to compensate for cells which die. The life of different types of human cell and hence the rate of reproduction varies from a few hours to many year. Reproduction of cells occurs in two ways, known as mitosis and in mitosis the chromospheres duplicate by splitting lengthwise. The original cell divides into two new cells which are each identical to the original cell.

Meiosis is a special kind of division which occurs during the formation of the sexual reproduction cells, namely the sperm in the male and the ovum in the female. It occurs only once in the cell's life-cycle and only in the reproductive cells. In sexual reproduction a sperm and an ovum unites and the chromosomes are added to form a new cell containing genetic material (*i.e.* genes) from each of the parents. The embryo and subsequently the offering develops from this single cell—the fertilised ovum.

9.7 THE INTERACTION OF IONIZING RADIATIONS WITH CELLS

The basic difference between nuclear radiations and heat and light is that the former have sufficient energy to cause ionization. In water, of which cells are largely composed, ionization can lead to molecular changes and to the formation of chemical species of a type which are damaging to chromosome material. The damage takes the form of changes in the construction and the function of the cell. In the human body, these changes may manifest themselves as clinical symptoms such as radiation sickness, cataracts or, in the longer term cancer.

For an acute exposure say 25 rem: Approximate lower limit of detection by blood analysis. There is a decrease in the number of leukocytes (white blood cells). In order to determine accurately the change, a base blood count must be obtained before exposure to ionizing radiation. The count will rise again after a few days although full recovery may take several months. Lymphocytes count is a useful dose indicator over the whole body range 25 to 400 rem. Above 400 rem the number of cells surviving is too small for accurate counting.

Significant symptoms are not likely to be seen in individuals who have been exposed to less than 100 R $\approx$ 96 rad, but as the dose rises above this figure, vomiting and nausea occur in increasing frequency and will be seen in almost all exposure of about 300 R $\approx$ 288 rad. Nausea and vomiting may be followed by a latent period of as much as 2 weeks when the dose is 100 to 250 R, but less than a day when the dose is greater than 700 R. The signs and symptoms which then develop usually include epilation, sore throat, haemorrhage, purpura, *petechiae* and diarrhoea. Among the most striking changes are those which occur in the composition of blood. Injury to the blood forming organs reduces the rate at which the blood elements are produced and has a dramatic effect on the composition of circulating blood.

In addition to the effects that become apparent immediately following heavy irradiation, some of the consequences may not appear for many years. Some of the delayed effects, like changes in the texture of pigmentation of hair, will be apparent relatively soon after the initial injury. Cataract, leukemia and cancer may not appear for 5 or more years. Some long term effects are complied in Table 9.7. The life expectancy of animals exposed to near lethal doses have been shown to be 25 to 50% lower than the normal.

Table 9.7 Some long-term effects of radiation

Effect	*Mean latent period*	*Evidence*
Leukemia	8–10 years	Atomic bomb casualties, Medical X-ray treatment
Bone cancer	15 years	Radium luminous dial painters.
Lung cancer	10–20 years	Mine workers.
Thyroid cancer	15–30 years	Atomic bomb casualties, Medical treatment.
Cataract formation	5–10 years	Atomic bomb casualties.
Life shortening	—	Experiments with mice.

At lower doses (200 to 500 rads), the reduction is from 2 to 4% per 100 rads. The effect of acute exposure on life shortening in people is not yet known quantitatively, but this information may emerge from extensive follow-up studies of the wartime bombing in Japan.

The processes leading to radiation damage may be easily divided into four stages:

(*i*) The initial physical stage, lasting only a minute fraction ($\sim 10^{-16}$) of a second in which energy is deposited in the cell and causes ionization. In water, the process may be written as

$$H_2O \xrightarrow{\text{radiation}} H_2O^+ + e^-$$

(*ii*) The physiochemical stage, lasting about 10^{-6} seconds in which the ions interact with other water molecules resulting in a number of new products. For example, the positive ion dissociates

$$H_2O^+ \longrightarrow H^+ + OH$$

the negative ion, that is the electron, attaches to a neutral water molecule which then dissociates.

$$H_2O + e^- \longrightarrow H_2O^-$$

$$H_2O^- \longrightarrow H + OH^-$$

Thus the products of the reaction are H^+, OH^-, H and OH. The two former ions, which are present in quite a large extent in ordinary water, takes no part in subsequent reactions. The other two products, H and OH are called free radicals that have an unpaired electrons and are chemically highly reactive. Another reaction product is hydrogen peroxide which is a strong oxidising agent

$$OH + OH \longrightarrow H_2O_2$$

(*iii*) The chemical change, lasting a few seconds in which the reaction products interact with the important organic molecules of the cell. The free radicals and oxidising agents may attack the complex molecules which form the chromosomes. They may, for example, attack themselves to a molecule or cause links in long-chain molecules to be broken.

(*iv*) The biological stage in which the time scale varies from tens of minutes to terms of years depending on the particular symptoms. The chemical changes discussed above can affect an individual cell in a number of ways. For examples, they may result in

(*a*) early death of the cell
(*b*) prevention or delay of cell division or
(*c*) a permanent modification which is passed on to the daughter cells.

The effects of radiation on the human body are the result of damage to the individual cells. The somatic effects arise from damage to the ordinary cells of the body and affect only the irradiated person. The genetic effects, on the other hand, are due to the cells in the reproductive organs, the gonads. The important difference is that in this case, the damage may be passed on to the person's children and to subsequent generations.

Recessive genes are only recognised when by chance two of the recessive type genes come together. A considerable number of diseases are associated with recessive genes and will, therefore, only manifest themselves when both parents have the same recessive genes. Spontaneous mutation account for the fact that an appreciable fraction of the world's population suffers from one of the 500 or more defects or diseases attributable to heredity.

Radiation can induce gene mutations which are indistinguishable from naturally occurring mutations. It is noteworthy that heat and chemicals can also cause mutations. Mutated genes are generally recessive and so it is generally assumed that all mutations are harmful. This cannot be strictly true since man is believed to have attained his present advanced state via a series of mutation. However, this has occurred over an enormous time span and the number of harmful mutations which have had to be eliminated from the species over this time is incalculably large.

Since ionizing radiation can cause an increase in the mutation rate therefore its use will increase the number of genetically abnormal people present in future generations. Clearly the consequences of excessive genetic damage would be very serious indeed and strict control must be exercised over the radiation exposure of the general population.

9.8 MANMADE AND NATURAL RADIATION

Cosmic radiation reaches the earth from interstellar space and from the sun. It is composed of a wide range of penetrating radiations which undergo many types of reactions with the elements they encounter in the atmosphere. The atmosphere acts as a shield and reduces very considerably the amount of cosmic radiation reaching the earth's surface. This filtering action means that dose rate at sea level is less than at high altitudes. As for example, the mean dose rate to the gonads at sea level, at the equator, has been measured as 23 m rem/year, while the equivalent dose rate at 3000 m, was ≈ 56 m rem/year. The average dose rate in British Isles from cosmic radiation is about 50 m rem/year.

One very important radionuclide arises from the interaction of neutrons in cosmic radiation with nitrogen in the upper atmosphere to form carbon-14 as follow.

$$^{14}_{7}N + ^{1}_{0}n \longrightarrow ^{14}_{6}C + ^{1}_{1}p$$

The carbon-14 which has a half life of 5570 years diffuses to the lower atmosphere, where it may become incorporated in living matter. Similarly small concentrations of other radionuclide such as ^3H ($T_{1/2}$ = 12.26 years), ^{36}Ca ($T_{1/2}$ = 3.08 × 10^5 years), ^{41}Ca ($T_{1/2}$ = 1.10 × 10^5 years) are maintained in the lower atmosphere by cosmic radiation. They are much less important than ^{14}C.

When a radionuclide is produced in an accelerator or a nuclear reactor, its rate of formation is dependent upon its half life. If t is the irradiation period for a target containing atoms of a specific nuclide A which undergoes nuclear reaction to form radionuclide B, the activity of B at the end of irradiation is given by

$$A_B\,(dps) = (R_f)_B\left[1 - \exp\left(-0.693t/(T_{1/2})_B\right)\right] \quad ...(9.17)$$

In this equation it is assumed that $(R_f)_B$, the rate of formation of B, is constant during the irradiations. The magnitude of the term $(R_f)_B$ is directly proportional to

(*i*) number of atoms of A in the material being irradiated

(*ii*) the number of reactant projectile particles, *e.g.*, neutrons of protons striking the target every second.

(*iii*) the probability that the product B will be formed when the projectile particle reacts with a nucleus of A.

Equation (9.17) is similar to (9.11) for the growth of the daughter nuclei that can attain secular equilibrium with its parent. After an irradiation time of ten half lives of B, A_B, is essentially equal to R_f and the product maches a "saturation level" and the nuclei B is decaying as rapidly as it is being formed. The decay time, after the termination of irradiation, is sometimes referred to as the cooling time because the heat released by a given radioactive source is proportional to the heat released by a given radioactive source which, in turn, is proportional to the disintegration rate. The appropriate equation becomes

$$A_B = (R_f)_B\left[1 - \exp\left[-0.693t/(T_{1/2})_B\right]\right]\left(\exp\left(-0.693\,t'/(T_{1/2})_B\right)\right) \quad ...(9.18)$$

The rocks and soil of the earth contain small quantities of the radioactive elements uranium, thorium and their disintegration products. The concentration of these elements varies considerably depending on the type of rock formation. In sandstone and limestone regions, the concentration is much lower than in granite. Thus the dose rate from this source depends on the geographical location. In the British Isles the dose rate to the whole body varies between about 30 and 60 m rem per year. Other long-lived isotopes such as ^{48}Ca ($T_{1/2}$ > 7 × 10^8 years), ^{50}V ($T_{1/2}$ = 4 × 10^{14} years), etc. do not contribute significantly to the human dose because they occur in nature in small concentration.

The human body contains small amounts of ^{14}C and ^{40}K radionuclide and contributes about 20 m rem per year to the gonads. A significant contribution to radioactivity in the body comes from gaseous decay products of the uranium and thorium radioactive series, namely ^{222}Rn, ^{220}Rn. These gases effuse from the rocks and soil and are present in easily measurable concentration in the atmosphere. They are also inhaled by humans, animals and plants with the result that most food stuff contain measurable amounts of natural radioactivity. Of ordinary foods, cereals have a high radioactive content while milk products, fruits and vegetables have a lower content. The intake of this type of radioactivity varies greatly with diet and is thought to contribute an average gonadal dose of about 4 m rem every year in the British Isles.

Manmade Sources of Radiation

The early experiences of man made sources of radiation involved X-rays and various uses of radium. As early 1896, a letter appeared in *Nature* describing the effects of repeated exposure of the hands to X-rays, and during the next 15 years many more cases were reported. These cases arose both from experiments with X-ray sets and also from their use in various treatments. By 1911, Hesse had studied the histories of 94 cases of tumors induced in man by X-rays of which 50 cases were among radiologists.

These studies illustrated the early types of damage produced by X-rays but gave no indication of the long-term effects. For certain types of damage, such as skin cancer, there is a latent period between 10–30 years and some radiologists observed malignant skin changes as late as 25 years after discontinuing fluoroscopic examinations. By 1922, it was estimated that more than 100 radiologists had died from occupationally produced cancer. Similarly, it has been estimated that the death rate from leukemia among early radiologists was about nine times that among other physicians.

Other studies have shown that the average life expectancy of the pioneer radiologists was reduced by approximately 2–3 years compared with physicians in general practice. Using results from animals experiments on the relationship between life shortening and the amount of irradiation, it is estimated that the total dose received by the average radiologist from 1935–1958 ranged between 400–800 R. The results of this study have been used by the International Commission on Radiological Protection (I.C.R.P.) to set the occupational exposure limit of 15 rems per year for most organs of the body.

Experience was also gained early in this century of the effect of internal dose from various nuclides such as radium ^{226}Ra ($T_{1/2}$ = 1622 years), ^{228}Ra ($T_{1/2}$ = 6.7 years), ^{228}Th ($T_{1/2}$ = 1.91 years) and their disintegration products. Even earlier it has been recognised that there was a remarkable high incidence of lung cancer among the miners in the Schneaberg cobalt mines of Saxony and the Joachimsthal pitchblende mines in Bohemia. Eventually, this high rate of lung cancer was shown to be due to radiation from the disintegration products of uranium namely ^{226}Ra, ^{222}Rn, ^{218}Po, etc. These mines contain large concentrations of uranium.

During the second and third decades of the twentieth century there were many cases of overexposure to radium. A considerable number of these arose from the use of radium as a therapeutic agent. It was administered for a large variety of diseases ranging from arthritis to insanity.

The most serious overexposures to radium occurred in the radium dial painting industry in the United States, most of the persons employed were women and they had the habit of pointing their paint brushes with their lips. Many of these women probably ingested tens or even hundreds of microcuries of radium. It is not known how many radium dial painters actually died from the effect of radiation damage.

The maximum permissible body burden of ^{220}Ra set by International Commission on Radiological Protection (I.C.R.P.) is 0.1 μ Ci, which corresponds to an average dose to the skeleton of about 30 rem/year.

Current Sources of Manmade Radiation

In addition to the natural background radiation, there are several other sources of human exposure which are peculiar to the last few decades. Some of them are as follow:

Diagnostic Radiology: It has been estimated that 75–90% of the total exposure of the population from medical uses of radiation comes from the diagnostic use of X-rays. The most critical regions of the body are bone marrow, the gonads and foetus. The bone marrow is the site of the primitive blood forming cells and so irradiation of this region could lead to the induction of leukaemia. Irradiation of the gonads is significant in view of the possibility of genetic damage. Irradiation of pregnant women should be very severely limited because it can lead to physical and mental deformities of the child.

Radioisotopes are used in medicine as a means of tracing the path and location of specific chemicals in the body. Since radioactive isotopes are chemically identical to stable isotopes of the same element, they will follow the same path and be concentrated to the same degree as the nonactive isotopes in the body. By counting methods the location of the active, and hence of the ordinary nonactive isotopes of the element may be determined.

The increasing use of radioisotopes, more particularly the development of the nuclear power industry results in an evergrowing stores of radioactive waste.

Continued dispersal of low and intermediate levels of radioactive waste to the environment means that members of the general population will receive an increasing exposure from this source. At present the contribution to the total exposure of the population from waste disposal is negligible but it seems certain to increase as more power stations are built. For this reason a very strict control should be exercised over the dispersal of radioactive waste.

The nuclides of concern in radioactive fallout from nuclear weapons testing are similar to those arising from the operation of nuclear power stations. The two most important radionuclides are strontium-90 ($T_{1/2}$ = 28.8 years) and Cesium-137 ($T_{1/2}$ = 30.0 years). ^{90}Sr concentrates in the skeleton and ^{137}Cs is distributed uniformly throughout the body. Some of the radionuclides created during a nuclear test are injected into the troposphere (10–20 km) and are carried around the earth several times. They gradually return to the earth over a period of a few years and consequently give appreciable doses to the world population. The dose from nuclides injected into the troposphere reaches a peak shortly after each weapon test.

The dose from all occupational exposure, both research and production, is very small when averaged over the whole population. The estimated dose in the United Kingdom is about 0.4m rem/year of which atomic energy workers contribute about 40%. The remainder is due to uses of radiation in industry and medicine.

9.9 MAXIMUM PERMISSIBLE DOSES

The International Commission on Radiological Protection (I.C.R.P.) was established by the Second International Congress of Radiology (I.C.R.) in 1928. Since its inception, the ICRP has been the one internationally recognised body responsible for recommending values of maximum permissible exposure to the ionizing radiation. Most of the legislation of various countries relating to the exposure to radiation are based on the recommendation of I.C.R.P.

The early recommendations of I.C.R.P. were concerned with protection against X-rays and radium. Some of the earliest recommendations dealt with the length of time that a worker should be engaged in radiation work; these were

(*a*) Not more than 7 h/day

(*b*) Not more than 5 days/ week

(*c*) Not less than 1 month's holiday per year

(*d*) Off days to be spent, as much as possible, out of doors.

The maximum permissible doses were defined very loosely.

In 1950, the I.C.R.P. extended its scope to deal with the many new problems resulting from the introduction of nuclear reactors. The maximum permissible exposure levels recommended by ICRP have changed considerably over the four decades as shown in Table 9.8.

All of these changes have led to a decrease in the occupational exposure values. This continuous decrease in the maximum permissible exposure levels has not been due to any positive evidence of damage to persons working within the earlier permissible dose levels, but rather to an increasing awareness of the uncertainties in much of the available experimental data.

Table 9.8 Values or maximum permissible occupational exposures

Dose rate	*Date*	*Comments*
0.1 of an erythema dose per year	1925	Proposed by A. Mutscheller and R.M. Sievert. This corresponds to an exposure of ~ 30 R/year from 100 kV X-rays or ~ 70 R/year from 200 kV X-rays.
0.2 R/day or 1 R per working week	1934	Recommended by ICRP
15 rem/year or ≈ 0.3 rem/week	1950	Recommended by ICRP
5 rem/year or ≈ 0.1 rem/week	1956	Recommended by ICRP

Over the years several different expressions have been used to describe acceptable levels of exposure to radiation. The most important of these are tolerance dose and maximum permissible dose. The term tolerance dose had the unfortunate connotation that it seemed to imply a threshold dose below which no radiation damage would occur. On the basis of presently available evidence there is no proof of the existence of a threshold dose for certain types of somatic dose. Consequently, the term tolerance dose is not used nowadays but has been replaced by the term maximum permissible dose (MPD). In 1958, the ICRP gave the following definition of MPD.

"The permissible dose for an individual is that dose, accumulated over a long period of time or resulting from a single exposure, which in the light of present knowledge carries a negligible probability of several somatic injuries; furthermore, it is such a dose that any effects that ensure more frequently are limited to those of a minor nature that would not be considered unacceptable by the exposed individual and by competent medical authorities. Any severe somatic injuries (*e.g.* leukemia), that might result from exposure of individuals to the permissible dose would be limited to an exceedingly small fraction of exposed group, effects such as shortening of life span, which might be expected to occur more frequently, would be very slight and would likely be hidden by normal biological variations. The permissible doses can therefore be expected to produce effects that could be detected only by statistical methods applied to large groups."

The problem in setting MPD values is to balance the expected benefits to be derived from the use of radiation and atomic energy against the risk from radiation exposure. In the absence of any definite evidence of a threshold dose for certain types of somatic damage it must be assumed that there is some risk associated with a dose of any size. The proper balance between risk and expected benefit may be very difficult to define and it may vary considerably from country to country depending on the social and economic factors. Thus, the MPD values given by ICRP are only recommended values and each country must accepts or modify them in accordance with its own particular risk and benefit conditions.

The responsibility for setting levels that will ensure a correct balance between risks and benefit is very considerable. If the MPD's are set at too high a level an unacceptable degree of injury may be sustained. On the other hand, if the MPD's are unnecessarily low, many of the benefits of radiation and atomic energy will be lost.

The ICRP recommendation are intended to limit somatic effects in individuals and hereditary effects in the population as a whole. It is therefore necessary to consider not only doses to individuals but also the average dose to the population. Table 9.9 lists the approximate dose limit for adults exposed in the course of their work and also for members of the public. Since some organs are more sensitive than others to the effect of radiation, different limits are specified for various organs.

Table 9.9 Summary of dose limits for individual

Organ	*MPD for adults exposed occupationally*		*Member of the public*
	Rem/year	*Rem/13 week*	*Rem/year*
Gonads, red bone marrow	5	1.7	0.5
Skin, bone, thyroid	30	10	11.5
Hands, forearms feet and ankles	75	25	7.5
Other single organs	15	5	1.5

The following points should be stressed in applying these dose limits:

(*i*) All unnecessary dose should be avoided.

(*ii*) Adults exposed in the course of their work who might exceed three-lengths of the MPD should be subject to medical supervision and personal monitoring. Such persons are often termed Classified Radiation Workers.

(*iii*) The permissible quarterly (13 week) doses for occupational exposure are based on one half-third of the annual dose but are rounded up. The object of the 13 week MPD's is to limit somewhat the rate of accumulation of dose. For instance, if only annual limits were specified it would be possible for a worker to receive 5 rem (whole body) on the 31st December and another 5 rem on 1 January, giving a total of 10 rem in two days. This is obviously undesirable and is only permitted in very special circumstances (see below).

(*iv*) In special circumstances, it may be justifiable to permit the quarterly quota to be repeated in successive quarter provided that the total whole body dose accumulated does not exceed 5 (N–18) rem, where N is the age of the person in years. Thus, it is permissible under special circumstance for a person to receive in a single year a dose of 12 rem to the gonads and red bone marrow, provided that the $O = 5\ (N–18)$ rule is obeyed.

(*v*) Women of reproductive capacity should not be exposed at the increased whole body rate of 3 rem/13 week, but should be limited to 1.25 rem/13 week.

(*vi*) The suggested limits for members of the public are in all cases a factor of one-tenth of the equivalent limits for adults exposed in the course of their work.

(*vii*) In the case of uniform whole body irradiation, the gonads and red bone marrow are considered to be the critical organs, thus for uniform irradiation the whole body dose limits are the same as those for the gonads and red bone marrow.

In very special circumstances, it is permissible to receive doses of up to twice the annual limit provided that the 5 (N–18) rem limit is not exceeded and the quarterly limit has not exceeded in the last twelve months. Planned special exposures should not be permitted to a worker who has previously received abnormal exposures in excess of five times the permissible annual dose as also is to women of reproductive age. It is generally felt that planned special exposures should only be permitted under very unusual circumstances, where no alternative solution is possible.

Abnormal Exposures in Emergencies or Accidents

It is obvious that no dose limit can be specified for accidental exposures. However, ICRP have recognised the need to protect the livelihood of a worker involved in such an accident. They recommended that no special limitations should be placed on his future radiation dose following a "Once-in-a-lifetime" accidental exposure of less than 25 rem.

It is worth noting that when a worker has used up his once in a lifetime special exposure allowance he is severely limited as far as future radiation work goes. After such an exposure, he must be employed on work which has an extremely low probability of leading to another high dose.

In an emergency, male volunteers may receive large doses for the purpose of rescue or of saving valuable installations. It is difficult to specify limits to cover such an event since each situation will be unique. It is often recommended that a whole body dose limit of 10 rem should be applied if saving life easily is not involved. When rescue operations have to be undertaken it may be possible to work to specified dose limits. Each situation must be assessed by the competent person or persons and a decision reached on the basis of this assessment.

It is probable that the entire body dose of up to 100 rem might be acceptable in saving a life. If the operation would require doses much in excess of this level, then the risks and possible result of the operation would have to be judged very carefully. One important consideration would be the accuracy of the information regarding the likely dose in the accident area and a second; the condition of casualties and their likelihood of survival. For instance, if the dose rate estimate were lower by a factor of two, a rescuer might unwillingly receive a sufficiently large dose to produce severe early somatic effects.

Examples of the maximum permissible concentration (MPC) of radionuclides in water and air are given in Table 9.10. Note that the values are for occupational exposures in a controlled area where normal operations could involve sources of ionizing radiation. Values for 40 hour exposure are also given in the reference for Table 9.10. For person outside a controlled area one-tenth of MPC and MPBB values given in the table are recommended.

Table 9.10 Maximum Permissible Body Burden (MPBB) and Maximum Permissible Concentrations (MPC) in Water and Air for Selected Radio nuclides for occupational exposures (a)

Radionuclide	*Form (b)*	*Location (c)*	*MPBB (μ Ci) Total body*	*MPCC (μ Ci/ml) 168 hr-week*	
				Water[d]	*Air*[d]
^{3}H	3	Body tissue	103	0.03	2×10^{-6}
	S	Entire body	2×10^{3}	0.05	3×10^{-6}
^{14}C	S	Fat	300	8×10^{-3}	4×10^{-6}
	S	Entire body	400	0.01	2×10^{-6}
^{85}Kr	See note b	Entire body	–	–	3×10^{-6}
^{90}Sr	S	Bone	2	10^{-6}	10^{-10}
	S	Entire body	20	4×10^{-6}	3×10^{-10}
	I	Lung	20	–	2×10^{-9}
	I	GI	10^{-3}	4×10^{-4}	6×10^{-8}

(*Contd.*)

^{131}I	S	Thyroid	0.7	2×10^{-5}	3×10^{-9}
	S	Entire body	50	2×10^{-3}	3×10^{-7}
	I	GI	–	6×10^{-4}	10^{-7}
	I	Lung	–	–	10^{-7}
^{137}Cs	S	Entire body	30	2×10^{-4}	2×10^{-8}
	S	Liver	40	2×10^{-4}	3×10^{-8}
	S	Spleen	50	2×10^{-4}	3×10^{-8}
	S	Muscle	50	2×10^{-4}	4×10^{-8}
	I	Lung	–	–	5×10^{-9}
	I	GI	–	4×10^{-4}	8×10^{-8}
	S	Spleen	0.03	7×10^{-6}	2×10^{-8}
	S	Kidney	0.04	8×10^{-6}	2×10^{-10}
	S	Entire body	0.4	8×10^{-5}	2×10^{-9}
^{222}Rn	–	Lung	–	–	10^{-8}
^{226}Ra	S	Bone	0.1	10^{-7}	10^{-11}
	S	Entire Body	0.2	2×10^{-7}	2×10^{-11}
	I	Lung	–	–	2×10^{-11}
	I	GI	–	3×10^{-4}	6×10^{-8}
^{238}U	S	GI	–	4×10^{-4}	3×10^{-11}
	S	Kidney	5×10^{-3}	6×10^{-4}	3×10^{-11}
	S	Entire Body	0.5	0.01	6×10^{-11}
	I	Lung	–	–	5×10^{-11}
	I	GI	–	4×10^{-4}	6×10^{-8}
^{239}Pu	S	Bone	0.04	5×10^{-4}	6×10^{-13}
	S	Entire Body	0.4	3×10^{-4}	5×10^{-12}
	I	Lung	–	–	10^{-11}
	I	GI	–	3×10^{-4}	5×10^{-8}

(*a*) From "Maximum Permissible Body Burden and Maximum Permissible Loads of Radionuclides in Air and in water for Occupational Exposure" National Bureau of Standards Handbook 69, 1963 USA.

(*b*) S stands for soluble form; I for insoluble form, for Kr, immersion in air is assumed.

(*c*) Where a specific organ is named it is considered to be a critical organs. GI refers to the gastrointestinal tract and the entries in this table refer specifically to the lower large intestine. For Cs the entire body is in critical organ category.

(*d*) Values are for continuous exposure for 50 y, 50 weeks/year.

9.10 NUCLEAR FISSION AND NUCLEAR REACTOR

The discovery of nuclear fission in 1938 proved to be the basis of a new source of energy, potentially greater than the entire world reserves of fossil fuels. The development of nuclear energy has been very rapid. The first fission reactor was operated in a converted squash court in Chicago in 1942 by Enrico Fermi. The rate of subsequent development has been such that nuclear energy now supplies a significant fraction of the power requirements of most of the advanced countries of the world. The advantages of the nuclear reactor as a source of power are offset to some extent by a number of special problems. These include:

(*i*) the protection of the operator and maintainer:

(*ii*) the safe treatment and disposal or storage of the enormous quantities of radioactivity produced; and

(*iii*) achieving an acceptably low risk of injury to the public due to potential large releases of radioactivity in the event of a reactor accident.

None of these problems is insuperable and it is generally agreed that the nuclear energy industry has an excellent safety record. In order to understand the hazards associated with reactors, a basic knowledge of nuclear fission and reactor technology is necessary.

The Fission Process: The splitting of a nucleus into approximately two equal parts is known as nuclear fission. Certain types of heavy nuclei notably uranium and thorium, are found to undergo spontaneous fission at a low rate. Fission can be induced, by the addition of energy as for example, by bombardments with neutrons. For some nuclides fast neutrons are required whilst for others, *e.g.* uranium-235, thermal neutrons can cause fission. Such materials, are said to be fissile. The process of fission results in the release of energy, mainly in the form of kinetic energy of the fission fragments. This is rapidly converted into thermal energy and raises the temperature of the fuel material.

A fundamentally important feature of fission is that the fission fragments are so unstable that they give off neutrons, usually between 2 and 4 neutrons per fission (Fig. 9.7).

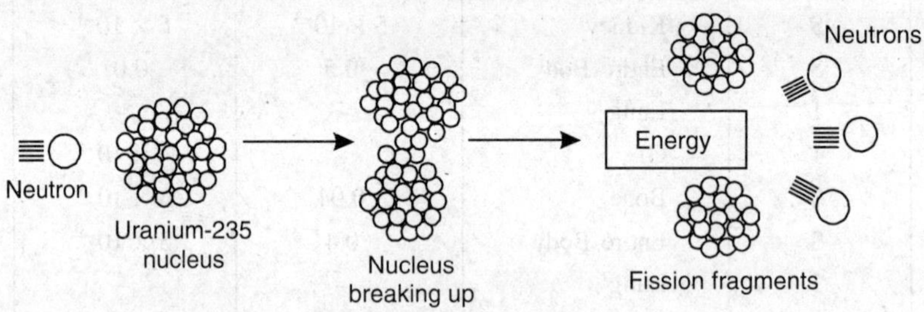

Fig. 9.7 Fission.

Most of the neutrons are emitted almost instantaneously, and are called prompt neutrons, but a small fraction is released some seconds or even minutes after fission—these are delayed neutrons.

The neutrons produced may themselves cause further fission with the emission of more neutrons, so making possible a chain reaction. In practice, some of the neutrons escape from the system and others are captured in a non fission reaction. In a reactor, wastage of neutrons by capture reactions is minimised by taking care to exclude material of high capture cross-section from the core. Some loss is inevitable since even the fuel material can capture neutrons without undergoing fission. By increasing the size of the core the fraction of neutrons escaping can be reduced to a sufficiently low level to permit a chain reaction. If a core is very small, neutrons can easily escape but in larger cores they would require to travel further and are, therefore, more likely to cause fission. Another method of reducing leakage is to put a reflector of some light material around the core to reflect escaping neutrons back into the core.

Consider an assembly of fuel material in which, on average, 2–3 neutrons are produced by each fission. If on average, 1.3 neutrons are lost by leakage or capture, one neutrons is available to cause a further fission. This results in a self sustaining chain reaction in which the fission rate is constant. Such a system is said to be critical. An average loss of 1.31 neutrons for every fission means that only 2.3–1.31 = 0.99 neutrons are available for fission. The system is then sub-critical and the fission rate will decrease. Conversely, for a loss of 1.29 neutrons per fission, 1.01 neutrons remains. This results in a super-critical system in which the fission rate will increase.

Thus for any given type of reactor there is a minimum size of core below which the system cannot go critical. Alternatively, a certain minimum mass of fuel is required for a chain reaction to be possible, this is the minimum *critical mass*; for instance, a nuclear weapon can, in principle, be made from two pieces of fissile material each of which is slightly more than half of the critical mass. The weapon is detonated by bringing the two pieces rapidly together to give a supercritical mass. This result in the fission, and hence the rate of energy released rises very rapidly to enormous values. Fortunately this is rather difficult to achieve in practice.

The critical mass concept has a very important implication. In plants such as reactor fuel manufacturing or processing plants, it is absolutely vital to ensure that a critical mass of material is not assembled in anyone place since a chain reaction and a very serious radiological hazard would result.

In fission, the splitting of the atoms of a fissionable material can occur in many different ways. The most likely division of the heavy atom is into fragments of mass about 97 and 135. For example

$$^{235}_{92}U + ^{1}_{0}n \longrightarrow ^{135}_{52}Te + ^{97}_{40}Zr + 4\,^{1}_{0}n$$

The distribution of mass number of fission products is of the form shown in Fig. 9.8. The fission process produces about 300 different nuclides in the mass number range 72 to 160, most of which are neutron rich and decay by a series of beta emission through a chain of radioactive nuclides. In the example of fission shown above tellurium-135 and zirconium-97 both undergo a series of beta decays until a stable nuclide is produced.

$$^{135}_{52}Te \longrightarrow ^{135}_{53}I \longrightarrow ^{135}_{54}Xe \longrightarrow ^{135}_{55}Cs \longrightarrow ^{135}_{56}Ba$$

and $$^{97}_{40}Zr \longrightarrow ^{97}_{41}Nb \longrightarrow ^{97}_{42}Mo$$

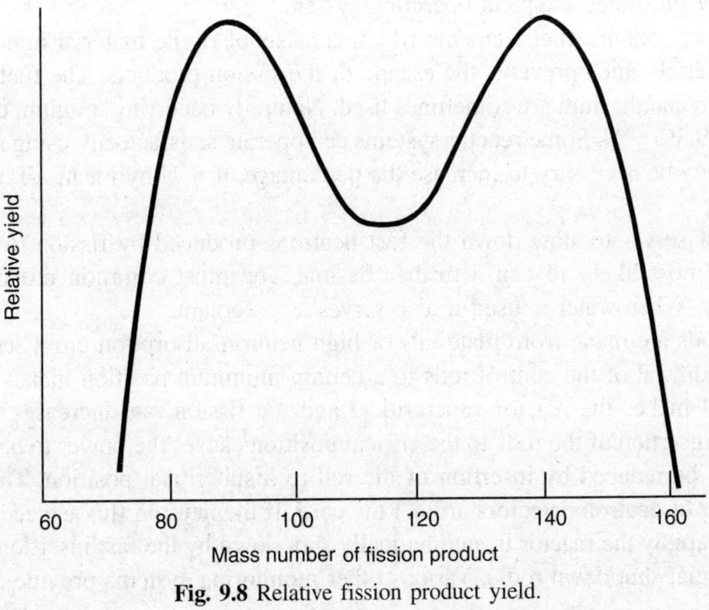

Fig. 9.8 Relative fission product yield.

It should be noted that the yields of the mass numbers which are plotted on a logarithmic scale range from about 10^{-5} to more than 6×10^{-2} (6%). There are 75 known fusion chains in the mass number range 72–161. On examining the periodic table we see that fission product elements cover a wide range of chemical behaviour. The total number of fission products is over 200 with half lives between a fraction of a second and 2×10^5 years. The large number of chain reactions (75) reflect the fact that symmetrical fission into two fragments of equal mass number is relatively uncommon when fission is induced by thermal neutrons. Instead, a light and a heavy nuclei are usually produced. This gives rise to the doubly peaked fission yield curve shown in Fig. 9.8. The fission yield for a given mass number is the total (maximum) yield for the chain of that mass number. It is expressed as the percentage of fission in which fission products of the given mass number are formed. Individual fission products may or may not have the same yield as the chain. One reason for the latter is that there may be direct formation of members later in the chain.

When fission of ${}^{236}_{92}U$ occurs, the sum of masses of the pair of one light and one heavy fragment must be equal to 236 since mass number is conserved. Similarly, the resulting fission product masses plus the masses of the neutrons evaporated by the fragments must also be equal to 236. In about two out of 1000 fission events a rare fission known as tertiary fission occurs, ${}^4He$ or ${}^3H$ is released. The utilisation of nuclear fission to produce energy results in the formation, within the fuel, of hundreds of different types of radioactive fission products with half lives varying from a fraction of a second to many years. The inventory of fission product in the fuel builds up over the period of irradiation. The various nuclides present in fission products have a wide range of physical and chemical properties and so the radiotoxicity of different nuclides varies enormously. In order to prevent the escape of the fission products, the fuel is usually contained within another material of suitable properties known as the *cladding*.

Reactor System: The central feature of a nuclear reactor is the core, which contains the fissile material, a means of controlling the fission rate and a moderator in the case of thermal reactor but not in the case of fast reactors. The other main features are a cooling system to remove the heat generated by fission and a radiation shield, often called the biological shield. These various aspects are described below and Fig. 9.9 illustrates a typical operating system.

The reactor core contains fuel elements which consists of fissile material in a fuel can or covered by a cladding material which prevents the escape of the fission products. The fuel is usually uranium although plutonium and thorium are sometimes used. Naturally occurring uranium contains 0.7% of the isotope ${}^{235}U$ and 99.3% ${}^{238}U$. Some reactor systems can operate satisfactorily using natural uranium, but in other cases it may be necessary to increase the percentage of ${}^{235}U$ by means of an expensive enrichment process.

The moderator serves to slow down the fast neutrons produced by fission to the thermal energy range, which are most likely to cause further fission. The most common moderator materials are graphite and water. When water is used it also serves as a coolant.

The control rods are made from materials of high neutron absorption cross section such as boron or cadmium. Withdrawal of the control rods to a certain minimum position makes the reactor critical. Further, withdrawl makes the reactor supercritical and the fission rate increases. When the required power is reached, insertion of the rods to the critical position causes the power to become constant. The reactor power can be reduced by insertion of the rod to a subcritical position. The control system is based on a number of neutron detectors around the core. If the neutron flux exceeds a present value or if it is increasing rapidly the reactor is automatically shut down by the fast insertion of the control rods (or sometimes special shut down rods). Various other monitoring systems provide data which can shut down the reactor if something goes wrong, for example, high gamma radiation levels outside the biological shield, excessive core or coolant temperatures, or loss of coolant flow.

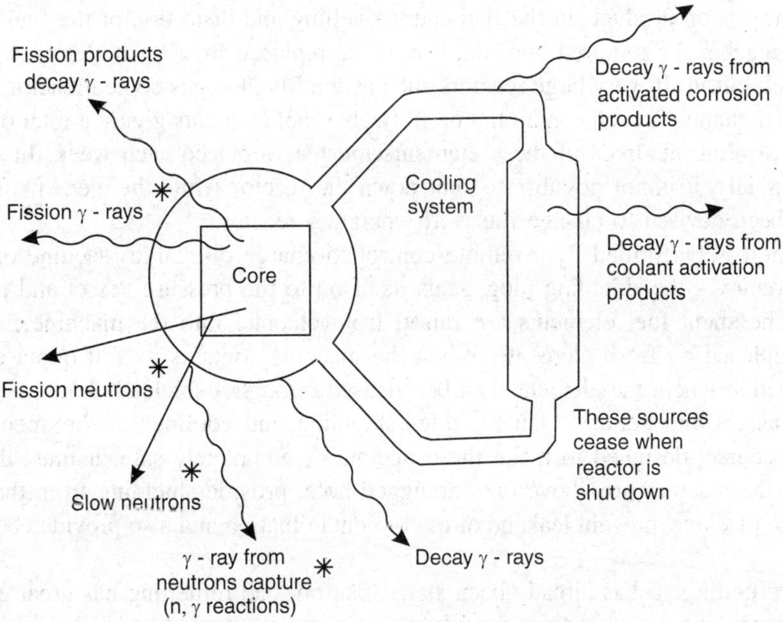

Fig. 9.9 Sources of radiation in a nuclear reactor system.

Fission causes energy release within the fuel and consequent temperature rise of the fuel and cladding material. The heat is removed by a coolant which is circulated in a closed loop and transfers the heat to water in a boiler. Steam from the the boiler is then used to drive turbines. Various subsidiary systems are provided for coolant purification, sampling and so on.

Coolants in common use include carbon dioxide, helium, water, heavy water, liquid metal sodium. The integrity of the cooling system must be very high. Most coolants become radioactive to some extent and even minor leaks may pose radioactive contamination problems. Major leaks could limit the heat removal from the core and affect the safety of the reactor. It is important to realise that in a power reactor the decay of the fission products generates a substantial amount of heat and so cooling must be guaranteed even when the reactor is shut down.

The biological shield serves to attenuate the neutron and gamma radiation from the reactor core and cooling system so that the operators and maintenance staff do not receive excessive doses. The most commonly used materials are lead, concrete, iron, water and polythene. In the case of power reactors the factor of reduction required may be of the order 10^8 to 10^9. Weaknesses can arise in the shield due to streaming radiation down coolant ducts and other penetrations. The shield is often constructed in two parts—the primary shield around the reactor core and the secondary shield around the cooling system. This arrangement permits access to the coolant system (circulatory, heat exchangers, etc.) when the reactor is shut down since the gamma radiator from the decay of fission products is attenuated by the primary shield.

Refuelling Reactors: In a nuclear reactor, one pound of natural uranium (about 16.4 cm^3) can produce as much energy as 5 tons of coal burned in a conventional power station. This, inspite of the fact that only 0.3% of the uranium atoms are burned up, which represents about 40% of the fissile

U-235 atoms. The percentage of uranium which can be burned up is limited, since a stage would be reached when insufficient fissile material is present to sustain a chain reaction. In addition, pressures generated by the fission products in the fuel cause swelling and distortion of the fuel elements. When a fuel element reaches the required burn up, it must be replaced by a fresh elements.

On-load refuelling: In very large reactors such as the Magnox gas cooled reactor there are several thousands of fuel channels, each containing up to twelve fuel elements giving a total of 30000 or more fuel elements. Several hundreds of these elements must be replaced each week. In order to run the reactor economically, it is not possible to shut down the reactor while the elements are replaced. So methods have been devised to change fuel with operating reactor.

The operation is performed by a remote controlled charge discharge machine on the top of the reactor which removes the shielding plug, seals itself on to the pressure vessel and removes the fuel channel cap. The spent fuel elements are raised from channel into the machine, which is heavily shielded and replaced by fresh elements. When the machine contains its full quota of elements it is moved to a position where the elements can be released through a shielded chute into a large tank of water, known as cooling pond, which provides shielding and cooling for the spent elements. The machine is, of course, designed to make these operations completely safe. It must therefore provide shielding from the reactor core above the unplugged hole, provide shielding from the spent fuel elements inside the machine, prevent leakage of gas or contamination and also provide cooling for the fuel elements.

Off-load refuelling: It has already been stated that on-load refuelling has great economic advantage. In some types of reactor systems the advantage may be outweighed by the difficulties as

(*i*) The problems associated with controlling the reactor power whilst less than fifty per cent of one of the elements are removed;

(*ii*) In high pressure reactor, there are great difficulties in achieving leak height penetrations of the pressure vessel;

(*iii*) In case of reactors for marine propulsion the additional weight of on-load refuelling equipment may be unacceptable.

If, for any of these reasons, it is not possible to refuel on-load, the core is designed to give as long a life as possible between refuelling operations and the entire core may be replaced in the same operation. This may entail shutting down the reactor for several weeks or even months depending on the type of reactor.

9.11 RADIATION HAZARD FROM REACTORS

In general reactors are a lesser radiation hazard when they are operating than when they are shutdown. The shielding is designed to give acceptable radiation levels at positions, where men are required to work and thorough surveys are made during the commissioning of the plant and, subsequently, at regular intervals. Systems are provided which permit safe means of sampling coolants and other radioactive effluents. During shut down periods, a great variety of non-routine jobs may be undertaken, some of them on highly radioactive systems. It is during such periods that exposure of personnel to radiation and radioactive contamination must be carefully controlled.

A serious fault or maloperation could cause considerable damage to the plant and give rise to dangerously high levels of radiation or radioactive contamination. If the hazard is confined to the reactor site, it is often called a site emergency but if it extends off site it may become a public emergency.

The main source of radiation from a reactor at power are the core and the coolant. The radiation from the core include fission neutrons, fission gamma rays and activation product decay gamma rays. The last two arise predominantly in the core structure and shield. Radiations from the coolant, activation of impurities and fission product contamination of the coolant are illustrated in Fig. 9.9.

The neutrons produced in a reactor as a result of fission process are fast neutrons in the 0.1 to 15 MeV range, with an average energy ≈ 2 MeV. Those emerging from the surface of the biological shield have undergone varying degrees of moderation and so neutrons of all energies from thermal to fast may be present.

Fission gamma rays are emitted immediately after the fission fragments and vary in energy from 0.25 to ≈ 7 MeV. The gamma radiation resulting from the decay of fission products in the fuel elements is small compared to the fission gamma radiation but, whereas the latter ceases on shutdown of the reactor, the fission products continue to emit radiation for many years after the fuel has been withdrawn from the reactor.

Neutron capture in the structural material of the reactor and in the shield results in the emission of so-called capture gamma rays and also makes these materials radioactive. The radiation from the decay of the radioactivity, as in the case of fission products, continues to be emitted when the reactor is shutdown. Neutron scattering also leads to gamma ray emission but does not, in general, reduce radioactivity.

An important factor in the choice of a reactor coolant is the extent to which it becomes radioactive. The most common coolants are water, air, carbon dioxide and sodium. In the first three of these materials the following reaction on oxygen is of great significance in shield design.

$$^{16}_{8}O + ^{1}_{0}n \longrightarrow ^{16}_{7}N + ^{1}_{1}p$$

Nitrogen-16 decays with a half life of 7.2 sec and emits very penetrating (6.1 MeV) gamma rays.

Air is used as a coolant in a number of reactors of early design and is used in the Magnox reactor to cool the biological shield. Argon, which is present to the extent of 1.3% in air, is activated by the reaction.

$$^{40}Ar + ^{1}_{0}n \longrightarrow ^{41}Ar + \gamma\text{-rays}$$

Argon-41 decays with a half-life of 1.8 hour and emits γ-rays of energy 1.29 MeV.

Fast reactors use liquid metal coolants such as sodium or sodium-potassium alloy. The important radioactive nuclides produced are sodium-24 and potassium-42.

Although reactor coolants are very pure by normal standards there are always some impurities. When subject to the very high neutron flux in the reactor core, these impurities become radioactive to an appreciable extent. Impurities also arise as a result of corrosion or erosion of the core and the coolant system. For example, in Magnox and advanced gas-cooled reactors, graphite dust collects around the cooling system. In water-cooled reactors, iron, cobalt and manganese are common impurities due to corrosion of the coolant system. Deposition of this corrosion material in the core causes the buildup of a film of corrosion products on fuel surfaces. As a result of irradiation by neutrons, these films become highly radioactive. The continuous release of material from the core and subsequent deposition in the out-of-core regions causes a build up of radioactivity in the coolant system.

Reactor coolants usually contain readily detectable levels of fission contamination. This can occur in three ways:

(*i*) uranium contamination on the fuel element surface;

(*ii*) uranium impurity in the fuel cladding material, and
(*iii*) release from a damaged fuel elements.

The coolant is being continually cleaned up by a treatment system and so that long-lived fission products do not build-up appreciably. The predominant fission products are usually krypton-88 and xenon-138, their particulate disintegration daughter products rubidium-88, cesium-138 and isotopes of iodine, ^{133}I, and ^{135}I. A seriously damaged fuel element could lead to a considerable fission product activity being spread around the cooling system. Most gas cooled reactors are fitted with a system known as 'burst can detection' which continuously and automatically 'sniffs' each channel in turn and gives warning and thus to shut down with any increase in the fission, product activity. The channel containing the damaged fuel can is then unloaded and the faulty element replaced.

Almost all of the radioactive nuclides mentioned in the preceding paragraph decay by beta emission. Beta radiation is so easily absorbed that the shielding designer does not even need to consider if, he concentrates on the associated γ-remission.

A radionuclide of considerable importance which is produced in all reactor systems is tritium. This nuclide has a half-life of 12.3 years and decays by low energy beta emission only and is produced by the following reactions

$$^{2}_{1}H + ^{1}_{0}n \longrightarrow ^{3}_{1}H + \gamma\text{-rays}$$

$$^{6}_{3}Li + ^{1}_{0}n \longrightarrow ^{3}_{1}H + ^{4}_{2}He$$

$$^{10}_{5}B + ^{1}_{0}n \longrightarrow ^{7}_{3}Li + ^{4}_{2}He^{2+}$$

$$^{7}_{3}Li + ^{1}_{0}n \longrightarrow ^{3}_{1}H + ^{4}_{2}He + ^{1}_{0}n$$

In reactors cooled or moderated by heavy water, large amounts of tritium build up. In light water; because of much lower concentration of deuterium and frequency of water change, tritium buildup is usually less significant.

Graphite contains lithium as an impurity and, consequently, tritium is produced in the Magnox and Advanced Gas Cooled Reactor system.

Contamination can occur because of a coolant leak. In pressurised water systems the leak may be direct to the atmosphere or via a heat exchanger into the secondary system in which case radioactivity, mainly ^{88}Kr, ^{138}Xe, and ^{41}Ar in the gaseous forms would be carried over with the stream into the turbines and then to the atmosphere via the condenser air ejector. The fission product gases ^{88}Kr, ^{138}Xe decay to their particulate daughters ^{88}Rb and ^{138}Cs. A leak in the heat exchanger of a gas cooled reactor would normally cause steam to leak into the primary system because of higher secondary pressure.

The reactor core contains about 5 curies of fission products per watt of operating power. Thus a 100 MW reactor, contains about 500 mega curies (MCi) of fission products. This vast inventory of radioactivity is contained within the fuel can or cladding which is often called the first level of containment. The second level of containment is the boundary of the primary systems, that is the pressure vessel and the coolant system. This boundary also contains the radioactivity of the coolant which, in a 100 MW reactor, may amount to some thousands of curies. Provided that the primary and secondary containments remains intact there is little risk of serious contamination. In practice some contamination does occur during the operation of most reactors. It is therefore necessary to sample the coolant periodically. A contamination hazard could arise in both cases but the risk is minimised by good design of facilities.

In its most common usage, the term 'reactor containment' refers to the structure within which the system is housed. An example of such a containment is the familiar sphere of the Dounreay Fast Reactor. The function of this containment is to protect the general public by limiting the release of fission products in the event of a serious reactor accident.

Maintenance: A reactor represents a large capital investment and shut-down periods, whether scheduled or not, are costly. Reactor systems are comparatively simple well-engineered and normally very reliable. A major overhaul may include decontamination and refilling of coolant circulators, control rod mechanisms, (inspection of heat exchangers and various other jobs on radioactive systems if followed by a tight time schedule can lead to a general standards of safety both radiation and conventional, though it may slow down the work. All major jobs should be planned in consultation with all involved parties and sometimes it is desirable to rehearse particularly difficult jobs. Personnel should receive instruction on general safety matters and be familiar with the particular hazards associated with their own work.

When the reactor is shutdown, the primary shield gives adequate protection from the fission products in the core. The radiation hazard to personnel working on the primary system is due to radioactivity within the system. The dose rate in the vicinity of the primary system tends to decay rapidly in the first 24 hours after shutdown mainly due to decay of coolant activities or their clean up by the treatment system. Thereafter the levels do not change significantly from day to day. The half-lives of most of the radioactive corrosion products are of the order of one month to about five years. The dose rates vary considerably from reactor to reactor but in systems with corrosion problems, levels of several rem/h are frequently encountered on certain components. If the dose rate is excessive; additional shielding on hot spots should be provided.

Some degree of control is required over personnel working in areas of high dose rate. This often takes the form of a manner control point at the entrance to the area. Personnel entering the area are given a working time based on a radiation survey to the area. In addition to their normal film badger they are required to wear some form of a direct reading. The time of entry and exit and diameter reading should be lodged.

9.12 RADIOACTIVE WASTE MANAGEMENT

The radioactive waste products that become available from fission atomic energy originate either from detonation of nuclear weapons or from the various stages in the civilian exploitation of the atom. The latter wastes may in turn, be divided into (*i*) low level wastes, which may be discharged into the environment under controlled conditions without violating the various regulations pertaining to the disposal of radioactive waste and (*ii*) high level wastes which are too intensively radioactive be discharged safely into the environment and therefore be treated for storage under controlled conditions. Intermediate wastes may be defined as those requiring some degree of decontamination before being discharged to the environment.

It has been shown that mining, milling, refining and fuel fabrication are not likely to produce widespread radioactive contamination although radium-bearing wastes discharged into the rivers surrounding environment by the uranium mills had elevated the radium content of drinking water. Thus waste disposal from the mines and refineries pose problems similar to those of chemical industries and significant environmental contamination, can be avoided by following commonsense procedures.

In the case of the fuel fabrication plants, the extent of environmental contamination by enriched uranium is limited by economic considerations. The value of enriched uranium is great so as to make it inconceivable that widespread environmental contamination could be produced with this substance. In the future, it is possible that plutonium and other alpha emitters, more toxic than uranium, may be utilised for reactors and fuel fabrication operations may then be required to exercise additional care in the release of these materials to the environment because of their toxicity.

Any release of radioactivity is a potential source of radiation exposure to the population at large. Radiation exposure can occur via many different exposure pathways. Consider for example, a release of activity from a chimney stack. This would be dispersed by air movements and could result in radiation exposure of the population in a number of ways.

(*i*) direct external β-or γ-radiation from the plume;
(*ii*) inhalation of radioactivity resulting in internal dose;
(*iii*) direct external β- or γ-radiation from deposition (fallout) of radioactivity;
(*iv*) consumption of food stuffs contaminated by depositions; and
(*v*) consumption of meat or milk from animals that have grazed on contaminated ground.

Similarly, radioactivity discharged into the river, lake or sea could result in a number of exposure pathways;

(*i*) contamination of drinking water supplies;
(*ii*) external dose to swimmers or to people using contaminated beaches; and
(*iii*) consumption of contaminated fish, shellfish or plants. Some of these exposure pathways result from complex route known as food chains. For example, activity discharged to the sea may be taken up by planktons which are eaten by insects which in turn are eaten by fish. The fishes are subsequently eaten by man. An example of a massive food chain is illustrated in Fig. 9.10.

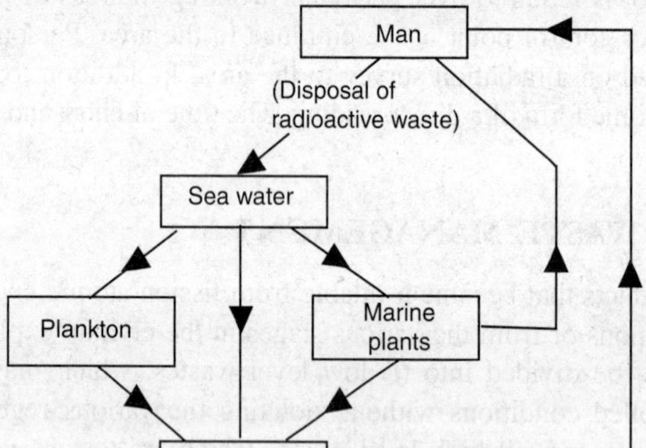

Fig. 9.10 A simple marine food chain.

The three most important fission products are ^{90}Sr, ^{137}Cs and ^{131}I. ^{90}Sr is potentially the most dangerous from the standpoint of its long-range public health significance because (*i*) it is formed abundantly in the fission product (about 5% of the total atoms produced in fission), (*ii*) its half-life (about 28 years) is quite long; and (*iii*) it is chemically similar to calcium. Because of its similar chemical properties, it is readily absorbed by living things and thus may pass through the food chains to man. With calcium, it is deposited in the skeleton, thus presenting a potential hazard of bone cancer or, possibly injury to the blood-forming tissue in the bone marrow.

^{137}Cs is formed in the fission process in slightly greater amounts than ^{90}Sr, with a half life of about 30 years and like potassium, is highly soluble. However, unlike ^{90}Sr it does not get fixed up in the skeleton but is eliminated with a biological half-life of about 100 days. As it is a gamma ray emitter, in 100 days, it is considered a potential hazard to genetic code form irradiation of gonadal tissue.

^{131}I-radioiodine with a half-life of 8.1 days. When absorbed into the body, radioiodine tends to concentrate in the thyroid, which in the case of a child can weigh less than a gram. Because of its volatility, radioiodine to escape to the atmosphere to an extent a greater than most other radionuclides. The radioiodine may be inhaled if an individual is enwrapped in a passing cloud, but most extensive exposure result from deposition of ^{131}I on pasture land. The cow grazes large areas of grass surface each day, from which the radioiodine is absorbed and transmitted to human via milk. The public health risk can be minimised by diverting the contaminated milk to powdered or condensed milk, thereby permitting time for the 8-days decay.

Figure 9.11 shows the major pathways for radioactive material released to the atmosphere, to surface or ground-water, and to soil. There is experimental evidence that radioactivity may be one of the carcinogens in smoke that cause higher incidence of lung cancer among cigarette smokers. The radionuclide directly involved is the α-emitter, ^{210}Po, which is the last radioactive member of the uranium series. This isotope is contained in cigarette smoke and it is formed by decay of its long-lived (22 yr) grand parent ^{206}Pb which is also present in the smoke. The ^{210}Po that is carried directly to water soluble form and very likely eliminated from the lungs. Atoms of ^{206}Pb, however, are incorporated in insoluble particles in the smoke and remain in the lung. The tissue near the particle is irradiated with α-particles from ^{210}Po in tobacco starts with ^{226}Ra in the soil or in the phosphate fertilizer produced from uranium-bearing phosphate rock. The decay product ^{222}Rn is released as a gas below the tobacco plant, atoms of the radon daughters attach themselves to particles in air and some of these are collected on the surface of the leaves of the plant. They are retained by the leaf and in this way ^{210}Pb is incorporated in the cigarette. Unfortunately for the smokers, the specific activity *e.g.* pCi/g of ^{210}Pb is about 10^4 times greater in smoke particles than in dried tobacco.

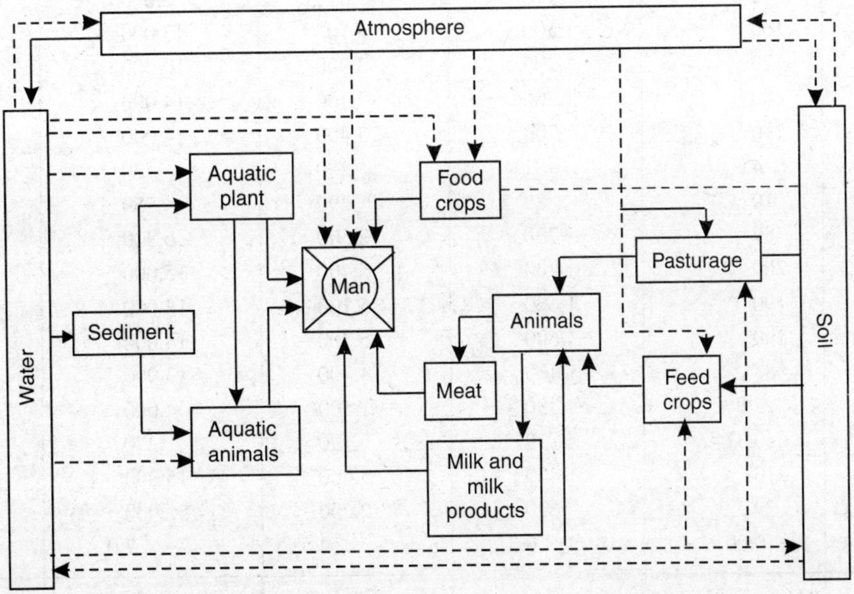

Fig. 9.11 Major pathway for the uptake by man of radioactive material from the environment.

One hundred megawatt days of reactor operation, experienced at a uniform rate over a period of 150 days, will result in accumulation of fission products equivalent to approximately 1.5 × 10^5 Ci of miscellaneous beta activity after 30 days of cooling. Among the long-lived radionuclides will be 380 Ci of strontium-90 (760 Ci or ^{90}Sr–^{91}Y), about 290 Ci of cesium-137, 6500 Ci of barium-140, 36 Ci of krypton-85. Using these basic figures, the estimate of the amount of stable radionuclides that will accumulate in wastes in future years is solely dependent on estimates of the rate of development of the nuclear power industry. The amounts of individual fission products that would be associated with these production rates for the United States are listed in Table 9.11. If one assumes that the total accumulation at the end of the given period in Table 9.11 are stored for various periods up to 30 years, the accumulation will be as shown in Table 9.12.

Among the various techniques for solving the problem of permanent storage of high-level wastes are methods for converting the radioactive fission products into solid salt cakes. This may be accomplished by several methods, including adsorption of fission products on clays which can then be vitrified, incorporation of wastes in glasses and other related techniques. All these techniques have as their object the fixation of fission products into relatively insoluble form.

The possibility that relatively large amounts of strontium-90, cesium-137 and possibly other fission products may be used as sources of thermoelectric power offers a partial solution to the long-term waste disposal problem. A necessary condition is that the isotopes when separated for use as isotopic power sources must be packaged in such a way that they do not cause environmental contamination.

Table 9.11 Effect of Period of Accumulation on Total Activity in Waste System (Fission-product Activity, MC)

Nuclide	*1960*	*1970*	*1980*	*1990*	*2000*
^{137}Cs	2.6	140	710	2,300	6,000
^{90}Sr	2.8	170	810	2,600	6,900
^{91}Y	160	1,500	5,100	13,000	33,000
^{89}Sr	120	1,200	3,900	10,000	25,000
^{95}Zr	180	1,700	5,400	14,000	35,000
^{95}Nb	190	1,700	5,400	14,000	36,000
^{106}Ru	4.7	81	280	730	1,800
^{129}Te	10	89	290	750	2,000
^{131}I	89	730	2,300	6,300	16,000
^{140}Ba	200	1,700	5,400	15,000	35,000
^{141}Ce	190	1,600	5,100	14,000	34,000
^{143}Pr	190	1,600	5,300	14,000	35,000
^{144}Ce	88	1,400	4,700	13,000	31,000
^{147}Pm	7.8	390	1,500	4,000	10,000
^{85}Kr	0.37	19	90	270	700
^{133}Xe	210	1,800	5,600	15,000	37,000
^{147}Nd	.79	690	2,200	5,9000	15,000
^{151}Sm	0.00	0.3	2.0	7.0	20
Total	1,810	17,200	56,700	152,000	376,000

Table 9.12 Effect of Decay time on Accumulate Activity

Period during which activity is accumulated in waste system, yr	*Fission-product activity, MC at decay time, yr*					
	0	3	5	10	20	30
1960	1,810	15.57	8.32	5.07	3.59	2.72
1970	17,200	593	412	284	204	158
1980	56,700	2,523	1,879	1,351	998	763
1990	152,000	7,470	5,690	4,250	3,170	2,430
2000	376,000	19,370	15,080	11,330	8,410	6,440

A seemingly ideal solution to the long-range problem of high level waste disposal would be to convert the waste to a solid inert form which would then be stored in specially selected geologic formation such as salt beds of south-eastern New Mexico, in shales, and crystalline rocks such as granite in the hope that acceptable repositories may be developed (in different parts of the USA).

In exceptional cases where sufficient land is available and there is thorough knowledge of the geology and hydrology of an area, the ground may be used for the disposal of liquid radioactive wastes. This practice has been followed on a large scale at Hanford. It occupies 1476 square km adjacent to Columbia river with an average rainfall of 16.5 cm. Here, a total of 2.5 million curies of fission products have been discharged to cribs and trenches up to June 1959. This radioactive material is contained in 17.7×10^9 litres of liquid, the Hanford studies of the fate of these fission products constitute a classic experiment in the field of waste management.

Wastes may also be produced in solid forms consisting of contaminated equipment, miscellaneous trash machine hurnings, processed filter cakes and an infinite variety of other debris ranging in activity from almost undetectable levels of short-lived activity to multicurie amounts of long-lived isotopes. These solid wastes may be compressed under high pressure into solid cakes and may be disposed by either land burial or sea disposal.

The Atlantic and Pacific oceans have been used since 1946 for the disposal of packaged low-level wastes originating mostly from research and development facilities such as Brookhaven, where 250 lt 18 gauge steel drums are used to contain mixtures of low-level wastes and centre which were permitted to harden before disposal into the Atlantic Ocean. From 1954 to 1958, a total of 3290 Ci of activity was dumped into the Atlantic Ocean in a total of 3164 drums. The total number of drums of low-level wastes buried at sea by American Atomic Energy Commission since 1946 is over 45000, containing about 22000 Ci at the time of burial.

The British have for some years followed the practice of discharging fission products into offshore coastal waters by means of a pipeline about 4 km. Hydrological and radiobiological studies of the region have shown that it is safe to discharge from 20,000 to 100,000 Cis per month. Although the capacities of the oceans and seas are certainly not infinite. One finds upon study that they are capable of receiving considerable quantities of wastes, provided the radioecology of the particular marine environment is understood.

The relatively short-lived low-level wastes associated practice may be disposed of in a variety of ways including land burial, sea disposal, or even storage in open dumps in the case of the such rapidly decaying isotopes as ^{32}P, ^{131}I, ^{24}Na, etc. which find frequent use in many research laboratories. However, the magnitude of the problem increases enormously as one moves to the storage and ultimate disposal of the high-level wastes developed in the processing of irradiated fuel.

REFERENCES

1. Friedlander, G., Kennedy, J.W. and Miller, J.M., *Nuclear and Radiochemistry,* 2nd Ed. Wiley, New York (1964).
2. Harvey, B.G., *Introduction to Nuclear Physics and Chemistry,* 2nd Ed. Prentice Hall, Englewood Cliffs, New Jersey (1969).
3. O'Donnell, J. H. and Sangster, D.P., *Principles of Radiation Chemistry,* American Elsevier, New York (1970).
4. Peter Alexander, *Atomic Radiation and Life,* 2nd Ed., Penguin Books.
5. Morgan, K.Z. and Turner., J.E., *Principles of Radiation Protection,* Ed. Wiley (1962).
6. Eisenbud, *Environmental Radioactivity,* McGraw-Hill, New York (1963).
7. Glasstone, S. and Sesonste, A., *Nuclear Reactor Engineering,* Van Nostrand (1963).
8. Alpher, R.A. and Herman, R.C., Origin, Abundance, and Distribution of Elements, *Am. Rev. Nuclear Sci.* 2 : 1 (1953).
9. Rankama, K., *Geology,* McGraw-Hill Book Company, New York (1954).
10. Kohman, T., Section 6 in *Radiation Hygiene Handbook,* H. Blaty (Ed.), McGraw-Hill Book Company, New York (1959).
11. Cuthbert, F.L., *Thorium Production Technology,* Addison Wesley Publishing Company Inc. Reading, Mass. (1958).
12. United Nations Comprehensive Report of Scientific Committee "Effects of Atomic Radiation" (1958).
13. United Nations Second Comprehensive Report of Scientific Committee "Effects of Atomic Radiation" (1962).
14. Lea, D.E., '*Actions of Radiations on Living Cells*', Cambridge University Press (1956).
15. Rydberg, J., Liljenzin J.O., Choppin Gregory R., *Radiochemistry and Nuclear Chemistry,* Butterworth Heineman, 3rd Ed. (2001).
16. For further information on Nuclear Environmental Services, E-mail: <allan.ashworth@atkinglobal.com>.

10 Garbage Management

10.1 GARBAGE

In most urban localities there is an overflow of open garbage dumps in the bylanes, parks and on the road side. Besides being eyesores, these garbage dumps are health and environmental hazards for the millions of people living there. Crowds of ragpickers, pigs, cows, dogs and cats are seen wallowing in the garbage. This rotting biodegradable matter is a breeding ground for flies and mosquitoes, germs responsible for plague, tuberculosis, cholera, dysentery, diarrhoea, leprosy, eye infections, skin allergies and breathing problems multiply here. This is also where people often slip and fall, where even a perfumed hanky on the nose cannot keep the stench away. Garbage can be segregated into five types; some of these can be treated or disposed off separately. These are:

(*i*) Hospital wastes
(*ii*) Non-biodegradable wastes that can be recycled
(*iii*) Non-biodegradable wastes that cannot be recycled
(*iv*) Toxic chemical wastes
(*v*) Biodegradable mass

The quantities of garbage generated and garbage cleared in 1995 in some major Indian cities are given in Table 10.1.

The approximate composition of city garbage is 50% organic matter, 25% plastic and 25% other matter. The uncleared garbage stays in the city and year after year the heap increases and hence cities are becoming more and more inhygienic. It may also be pointed out that toxic effluents which are dumped into the rivers or over the biodegradable garbage only render the degradability of the biomass difficult. That is why effluents create problems.

How do the residents of colonies treat their garbage? Unfortunately, rather casually, everyday the houses are swept clean and garbage tripped over the boundary wall onto the back lane. There, it piles up and slowly becomes a permanent heap for the neighbours, who only add on to it. It rots, stinks, breeds flies, mosquitoes and bacteria, the boundary wall no longer provide protection against diseases and pollution. Our attitude towards places which we litter with fruit peels, non-biodegradable cups and other sundry wastes, remains the same.

There are too many open eating stalls on our road sides. Stall owners rarely have a bin for leaf plates or wrappers. Most people are so used to this situation that they do not even bother to search for civic amenities' dustbins—they just throw the wastes on the road or in a corner. Untreated organic matter, if left in the heat and humidity even for a short time, start decomposing and then it becomes impossible to scrap off and becomes a stinking permanent mass. The onus to keep our neighbourhood clean must lie with us. It is our garbage, we produced it, we must dispose it in a fitting manner.

Table 10.1

City	*Garbage generated tons/day*	*Garbage cleared tons/day*	*Annual municipal budget Rs. in crores*	*Expenditure on garbage cleared per ton per day in Rs.*
Ahmedabad	1500	1200	270	6000
Bangalore	2130	1800	237	3607
Mumbai	5800	5000	2436	13348
Kolkata	3500	3150	250	2175
Delhi	7880	4850	1016.38	5742
Lucknow	1500	1000	48	1315
Chennai	2675	2140	145	1856
Patna	1000	300	15	1370
Surat	1250	1000	170	4657

The three steps of garbage management of a city corporation are: collection, transportation and disposal. However, in most colonies dumping bins are absent. In many places where formal systems of garbage disposal have been introduced garbage is transported to landfill sites. At these landfill sites, the garbage decays and settles down to fill up the sites. The maximum expenditure in garbage management is on transportation. If expenditure on transportation is not met fully, the garbage decays in the cities. That is what is happening in most of the Indian cities (see Table 10.1).

How to Manage Garbage? Article 21 of our constitution gives us the right to a clean and wholesome environment. To live in a garbage free city, the residents should collectively sue the municipal corporations in courts and thus force them to remove garbage from cities.

The other alternative is to phase out the heaps of open stinking garbage from their localities by starting a scientific, eco-friendly decentralised garbage management scheme. Community participation and cooperation of the public are vital for efficient waste management. Dr. Iqbal Malik gave a lead in setting up a community brigade called "Cleaning Brigade" to manage the garbage in some colonies of Delhi. They recommend that garbage should be managed by private contractors with the following procedure:

1. Garbage should be collected from door-to-door and taken to the community bin.
2. Sorting of recyclable matter should be undertaken by the employees of the contractor. These recyclable are at present, picked out and biodegradable are left strewn in and around the bin.
3. Recyclable are sold on a day-to-day basis initially. Then a monthly contract can be fixed with most wholesale trash dealers of the area, out of which the contractors make a lot of money. This reduces the cost of management of garbage to negligible amounts.
4. The local ragpickers still exist and they pick those inexpensive recyclable which have been left behind by the employees of the contractors.

5. The contractor should provide his workers with a uniform, gloves and boots for segregation of garbage. These boys become an extended family of the residents for whom they work. The total money earned by contractors should be distributed as part of pay to his employees.
6. Biodegradable matter should be converted into compost by vermicasting. This compost can be sold and will be another source of income.
7. When residents start understanding garbage management, the residents may be asked to collect their garbage in two separate bins say green bin for biodegradable garbage and another, yellow bin for non-biodegradable garbage.

If people become more responsible about the amount of garbage they produce and about segregating it, perhaps we can avoid creating new generations of ragpickers. If the children could be freed from such activity, NGO's could focus more on education and other alternatives. However, within the existing situation, we need to ensure that waste pickers are accepted as respectable citizens which can be done by providing more hygienic methods of waste handling and they be provided with simple technologies which they can use to improve their working conditions and earning capacity. It is desirable to focus primarily on the social implication of garbage generation and disposal and the issue of social responsibility among citizens, especially towards such underprivileged sections of our society. Alternatives to garbage management strategies cannot at present, replace the formal system; at best there can be parallel and complementary systems which can take some pressure off the formal one and, at the same time, improve and give respect to the working and economic conditions of those engaged in the informal system. Further, it is preferable that children should not be working, especially doing this task, however, at a practical level their work should be recognised and their working conditions as well as earnings should improve.

10.2 HOSPITAL WASTES

Though garbage dumps are all around us, some dumps are more dangerous than others. Hospital wastes is a case in point. The amount of waste produced in a hospital depends more on in house than outdoor patients; about one and a half kg of waste is produced per bed per day. Of the total waste produced in hospitals $\approx 47\%$ is biomedical waste which is hazardous as it is contaminated with disease-causing pathogens. The human and animal wastes along with items saturated or dripping with blood and body fluids, discarded medical equipment, soiled cotton, plasters, dressings and surgical and autopsy wastes can become a major health hazard as they provide fertile environs for bacteria, viruses and other microorganisms to multiply. Pathogens like *Escherichia, Salmonella, Vibrio, Hepatitis Shigella* are prevalent and active till these wastes are incinerated and can be carried far away through various agents. Surgical and autopsy waste, if untreated, can be most infectious. The list of diseases caused due to improper disposal and treatment of hospital wastes endless, but the majority of them are deadly, such as AIDS, viral hepatitis, tuberculosis, bronchitis, gastroenteritis and other skin and eye related disorder.

Hospital waste should be segregated and should be incinerated. According to a notification issued by the Ministry of Environment and Forests dated 24th April 1995. The following rules are to be followed by all hospitals and nursing homes:

(*i*) Every hospital, nursing home, clinic having more than thirty beds or catering to more than a thousand patients per month shall install an incinerator in its premises;
(*ii*) No biomedical waste shall be stored beyond a period of three days;
(*iii*) No person shall recycle or reuse any biomedical waste, except glassware;

(*iv*) All biomedical wastes which are not incinerable shall be pretreated, disinfected and shall be disposed off in an environmentally sound manner; and

(*v*) Schedule II of the notification also mentions the type, colour, coding, labelling for waste containers.

As of today only 5% of Delhi's 2500 nursing homes are registered; 80% of health care is provided by the private sector, 70% of these are mismanaged and in an appalling conditions, 0.0001% of our private Hospital, like ESCORTS, follow all the rules laid down by the Ministry. Different coloured bins have been provided for different types of wastes. Incinerators here are always in working conditions. Although 8% of the so called private hospitals provide better facilities than our 6% of the good Government hospitals, but where waste management is concerned the two barely differ. Most dispensaries throw their wastes behind the boundary walls. Almost all small medical practitioners throw disposable syringes either in their garbage or outside their shops. *Vatavaran,* a NGO, that is working for a pollution-free environment has found that 20% of our hospitals have *bhatties* in place of incinerators, 50% do not even have that. They either take the waste to a landfill site burn it in the open, 10% hospitals cart their waste to another hospital where it is incinerated, waste goes to for incineration, 20% hospitals do have proper incinerators but they are out of order very often.

Controlled air incinerators are the best incinerators to get rid of all types of hospital wastes. In these incinerators the heat and air for combustion is regulated in such a way as to first volatilise/gasify the waste in conditions of inadequate air and then totally destroy the waste by adequate heat and excess air. These are twin-chambered incinerators—where refractory-lined chambers are mounted on top of each other. Volatilisation of wastes is achieved in the primary chamber with supply of air in controlled quantities through the twin air ducts. The Thermax forced draft fan is the source of supply for combustion air. The heat source is the high efficiency, automatic On/Off Thermax fuel oil burner.

The volatilised/gasified waste passes to the secondary chamber through a specially designed opening to achieve sufficient velocity. The combustion is carefully regulated for a minimum turbulence, thus avoiding emission of pollutants and making use of a dust collector or bag-filter redundant.

An educator system at the outlet brings down the temperature of the exit gases from 950°C to 300°C before releasing them into the atmosphere. This is a safe limit for emissions, confirming to strict pollution control norms. Hospital wastes, on the basis of moisture content, can be segregated into two types. Moisture content of left over food is 70%, Pathological wastes like experimental dead animals amputated body parts and sputum, etc., 85%, blood bags, saline bottles, syringes have as low moisture content as 20%.

When a hospital does not segregate the waste and puts everything together, the incinerator gets spoiled. Hospitals which have *bhatties,* in place of incinerators, can put anything in it and it burns. However, in these bhatties enough heat is not available to destroy all the pathogens and this results high air pollution.

Incineration is an expensive proposition. A hospital which runs two units of its incinerator for a few hours everyday can dispose of about 70 kg of infectious waste. The cost involved is Rs. 1400 a day or Rs. 20 per kg on the other hand, if the hospital takes its waste to the landfill site as it would spend Rs. 7 kg, and if hospital buries its waste in its backyard it does not spend any money at all.

But the question arises that why even the big private hospitals with a so-called good image, who charge so much from the patients to provide an infection-free environment and other such facilities to their patients avoid spending any money on proper disposal of their hospital waste? Are they not cheating the public. When there is no incinerators present even in our big cities where small nursing homes can deposit their waste and other infections waste to be incinerated on a commercial basis.

10.3 NON-BIODEGRADABLE BUT RECYCLABLE WASTES

The predominant non-biodegradable but recyclable waste is plastics. The raw material for the manufacture of plastic items constitutes resins. These resins consist of thousand of particles which melt into a syrupy liquid when heated and can be shaped into almost any form. Although various types of plastics may look the same, they are in fact quite distinct groups of materials of different molecular structures.

Phenomenal use of plastic by masses in our country started in 1980. There were no cling films to wrap our sandwitches before that. There were no discardable ball pens, Maggi Noodles, Uncle Chips or other fancy looking things were not available. At present, there would not be a single household in the urban areas of the country which is not discarding some plastic trash everyday. House wives either sell these materials to local *kabari* who come in the colony or throw them with the garbage.

The quantity of garbage plastic produced in 1995 in some of the major Indian cities are given in Table 10.2.

Table 10.2

City	*Plastic garbage generated in tonnes*	*If this plastic is spread on a ground of 100 m × 100 m the height of hill in meters*
Ahmedabad	68437.5	5.47
Bangalore	97181.25	7.77
Mumbai	264625	21.17
Kolkata	151687.5	12.135
Delhi	35925	28.76
Lucknow	68437.5	5.475
Chennai	122047	9.76
Patna	45625	3.65
Surat	57031	4.562

If this plastic is spread in a ground of 100 m × 100 m, the approximate height of such a plastic hill is given in column 3 of table 10.2.

In most of colonies the ragpickers collect most of these discarded items. The rag pickers are either employees of some big kabari or they work on their own. In almost all eventuality the plastic trash reaches the big *kabari* who, in turn sells all this to a wholesaler and finally to plastic recycling units. More details about it have been given in section 7 and 9 of Chapter 6.

The other recyclable but non-biodegradable garbage are non-ferrous metals. Most of this garbage is imported from other countries and recycled here. The metallurgical process used in these metals are highly inferior, producing a lot of pollution and toxic wastes.

Lead is one of the toxic wastes. All over the world, lead storage batteries are used to provide power for electrical systems of automobiles, electronic instruments, computer etc. The waste cells are sent to places where disposal and processing costs are the lowest and affected population lacks access to information and political, educational and financial resources to oppose imported wastes—India being one of them. Here cost of recycling scrap is much less because human resource is in plenty and cheap.

From the lead scrap, lead plates are regenerated by heating with coal. It does not form slag with silicates. Other metals do not melt at 600–700°C and thus do not interfere. These new lead plates so produced are used to make new storage cells. As the labour is quite cheap in India, lead smelting is quite profitable and popular. Around Delhi there are nearly 260 illegal lead smelting units. The environmental problem with lead is as follow: Lead has a low melting point (around 630°C), Red hot coal has temperature of 950°C and that keeps lead in molten state. This high temperature makes a lot of lead vapours to escape to air particularly in the illegal smelter units which are of very inferior quality. So that places near lead smelter air has a very high lead and then lead level of water, soil and fodder rises by 175 times than the normal level.

Phasing out is the best solution to eliminate the risk of lead pollution in the environment and to the poor smelters. The alternative storage cells with alkaline medium of KOH with Nickel-Cadmium electrodes, or Silver-Zinc electrodes or Ni/Nickel hydroxide as anode and finely divided iron as cathode called Ni-Fe cell be introduced. These cells are rather more efficient than lead storage cells. Silver-Zinc cell and Ni-Fe cells are being used in aeroplanes all over the world. However till government does not decide, this cannot be implemented.

Till then the smelters should be updated as has been done in USA and UK. The technology being used in these countries is pollution free and lead vapours may not escape into air. The lead acid batteries makers use lead plates to make lead storage cell but such cells may not work for six months: However if negative electrode is given a paste of mixture containing.

Tri basic lead sulphate ($3PbOPbSO_4H_2O$) 85%.

Tetra basic lead sulphate ($4PbO, PbSO_4$) 15% along with some lignin sulphate (expander) and $BaSO_4$ as binder. With this paste applied over lead plate will increase the life span to 2 years. The lead basic sulphate and tetra basic sulphate are readily available in India. For different purposes the rate of withdrawl of current is different and different plates have to be developed. This amounts to that for different purposes different lead storage cell should be used.

Also legally (*i*) lead trash cannot be imported (*ii*) lead smelting is not allowed. So what is taking place is illegal; how, that needs no mention. The import of lead trash and its smelting should be allowed legally. The latest lead storage cell technologies should be imported and popularised. That will compensate for the cost of latest technology to some extend. If this problem be properly tackled it can be highly economic preposition because every country needs lead storage cells in large numbers, but very few countries are producing them.

10.4 NON-BIODEGRADABLE NON-RECYCLABLE GARBAGE

House construction is one of the very big industries in India. This leaves a lot of silicates as garbage which have to be disposed off at a landfill site. The landfill site is a courtyard for the thousand odd families who live around it. The requirements for suitable landfills are as follows:

(*i*) Avoiding environmentally sensitive areas e.g. plains, permafrost area, critical habitats of endangered species and recharge zones which supply the sole source of culinary water;

(*ii*) Avoiding limestone-containing areas with caves and sink holes;

(*iii*) Avoiding sites traversed by pipes or conduits as sewer pipe and water lines;

(*iv*) Evaluating such soil site which have low-to-moderate permeability and can support vehicular traffic in adverse weather;

(*v*) Knowing the kinds and amounts of wastes expected;

(*vi*) Rainfall pattern, water movement through soil and soil permeability should be known;
(*vii*) To find any future alternative uses of site, *i.e.* would the landfill site make a good location for a park or farming site; and
(*viii*) Construction of multistorey buildings at landfill sites should be avoided.

Though landfill sites should only be used for disposing of non-biodegradable garbage, but biodegradable garbage is also disposed off at such sites. So separate landfills must be earmarked for biodegradable and crub and subsequently land can be reclaimed accordingly. Rather, innovative schemes should be initiated under which citizens can carry a truck load of garbage to their respective sites and get money in return.

Solid wastes deposited in landfills degrade chemically and biologically to produce solid, liquid and gaseous substances. Biological activity within a landfill first results in aerobic decomposition. An aerobic decompositions sets in once the oxygen supply in the pit is exhausted. This produces organic acids, carbon dioxide, sulphates of iron, manganese and a substantial amount of methane.

To tap the gas being released from the dump, large perforated coiled pipes can be inserted at different locations in the area for drawing out the gas. The size of the coil, the diameter of the coiled pipe, the number of perforations can be selected on the basis of the gas potential. The entire area can then be covered with polyethylene with only the end of the coils striding out. These ends can be further linked to the distribution pipes. This can become profitable energy producing centre.

10.5 SOME TOXIC CHEMICAL GARBAGE

The garbage from market places which have dry cleaning units, dyeing units, furniture shops, leather processing units, acid cleaning units, paints and varnish shops, tyre and vehicle workshops and LPG gas cylinder stores was qualitatively analysed. The list of chemicals present, and their probable effects on health and environment, is presented below:

S. no.	*Chemicals in garbage*	*Probable effects on health and environment*
1.	Chromium	Carcinogenic
2.	Cadmium	Heavy metal, can cause free radical damage to cells, accumulates in liver and kidney, hair loss, osteoporosis.
3.	Ammonia	Light headedness, disorientation.
4.	Caustic Soda	Skin rashes
5.	Lead	Gets deposited in the bone marrow and leads to numerous harmful effects
6.	Metals like Copper and Nickel, etc.	Metals when washed in acid to clean rust and dross release fumes in air. Waste water which goes in drains contains these heavy metals
7.	Potassium dichromate	Water soluble but gets converted to chromium sulphate which is most difficult to remove from soil and environment
8.	Sulphur dichloride	Pungent smell
9.	Amyl acetone	Highly inflammable, can lead to fires, to eye, nose and throat irritation.
10.	Benzene, Benzedine	Carcinogenic and highly flammable
11.	Carbon tetrachloride	Liver toxicant, damages optic and cranial nerves.

(*Contd.*)

S. no.	Chemicals in garbage	Probable effects on health and environment
12.	Formaldehyde	Irritant to mucus membrane. Leads to hardening of tissues.
13.	Formic acid	Eye irritant, allergen
14.	Methylene chloride	Moderately toxic causes dizziness, lethargy, skin burns
15.	Napthalene	Excessive perspiration, confusion
16.	Perchlorophenol	Breathing problems, convulsions
17.	Trichloro ethane	Heart attack, miscarriage
18.	Toluene	Long-term exposure affects central nervous system, mutogenic.

10.6 BIO-DEGRADABLE GARBAGE

In nature, all dead matter is gradually acted upon by the forces of nature such as the sun, wind, rain and microbes which serve to break down complex materials into simpler molecules. If such material is left to decay on road sides it begins to decompose and stink and invites many diseases. Instead, the process of decomposition can be used to convert organic wastes which we generate everyday to produce compost which can help us keep our surroundings clean and green. The various methods of compositing are as follows:

Anaerobic (Absence of air)	*Aerobic (presence of air)*	*Vermicomposting (presence of air and earthworms)*
Can be used for generating biogas which can be a source of fuel	(*a*) more complete than anaerobic in 2–3 weeks (*b*) very little odour	(*a*) it is complete and odourless (*b*) rich in macro and micronutrients (*c*) can be dried and stored easily
Disadvantages		
It takes long time (4–8 months)	(*a*) attracts animals unless properly covered and protected	(*a*) earthworm needed, safe to handle
	(*b*) maintainance of 55°C temperature.	(*b*) controlled conditions like moisture, temperature necessary.
Needs a patch of land	Requires land and sheds	Land, water, sheds and earthworms

Vermicasting: 'Vermi' stands for earthworms. Earthworms eat soil and various kinds of organic matter which undergo complex biochemical changes in the intestines which are then excreted out in the form of granular casts of earthy smell. Earthworm excreta together with their cocoons and undigested food are called vermicastings. Some earthworm species which live close to the surface and have greater preference for organic matter than soil can be isolated and used to produce vermicastings. Species found best-suited for the generation of manure by this method are *Endrilus engeniae*, *Eisenia Fetida* and *Perionyx excavatus.* These are all surface feeders which feed on organic matter and enhance organic decomposition by microbes. Earthworm culturing techniques and production of vermicastings are as follow:

1. Earthworm culturing can be done in shallow cement tanks, wooden boxes and stone lined pits or plastic 1 m × 1 m × 0.3 m tubs which can accommodate around 2000 worms.

2. Saw dust, coconut husk or any other hard material can be used as base. Dung of domestic animals such as cattle, horse, pigs or poultry droppings mixed with kitchen wastes forms an ideal feed for the worms. Previously decomposed matter helps faster formation of vermicastings.
3. The worms that feed actively assimilate only 5–10 % and the rest is excreted as loose granular mounds of vermicastings on the surface, generally away from the food source. These have to be brushed aside and collected into separate trays. The collected castings have to be left overnight in conical heaps for the worms to move to the bottom. The tops of the cones which are free of worms are then collected and lightly air dried. The dried vermicasting are sieved through a 3 mm mesh to separate cocoons and young ones from the vermicast. The dried castings are ready for use as manure.

Application of vermicastings to plants is similar as in the case of compost or organic manure.

Based on controlled aerobic fermentation of city waste, Excel Industries, Mumbai has developed a process known as CELRICH Bioorganic soil enricher semi-mechanical processing plants in four models of 150, 300, 500 and 700 tons per day waste processing with capital cost for 2, 4, 6 and 7.20 crores borne by company. Municipal corporation have to lease out 1/5 to 1/6 size of their dumpyards and assure daily delivery for a period of 20 to 30 years. In return Municipality will get a lease rent 1 to 2.5%, royalty ex-factory sales values of recovered products. However the treatment cost is to be borne by the corporation on actual treatment substrates consumed. A 500 tons of organic garbage will produce 45625 MT of organic manure equivalent to 9000 MT of chemical fertilizer.

In the process waste is treated with herbal extracts and other treatments for defouling the smell and accelerated bioconversion. Due to exothermic reaction temperature of waste increases to 75°C within 36 hours and that kills pathogens and accelerate the fermentation. As the process proceeds, mass changes colour to dark brown humus like substance and volume reduces by 35 to 40%. Well decomposed biomass is separated with grading and sieving machines.

10.7 OIL SLUDGE AND OIL ZAPPER

One of the biggest and most frequent threats for any oil refinery is oil sludge. There are 16 refineries in India all create toxic waste, 20000 tons of petroleum sludge every year. A refinery has to construct polymer lined pit with special leachate collection system to prevent the dumped sludge from leaking into the earth and ground water. A pit costs about 10 million rupees and each refinery needs many such pits. A pit gets filled up in three to four years. Two hundred tons of Oilzapper, a friendly bacteria, clears 20000 tons of oil waste. Oilzapper is essentially a cocktail of five different bacterial strains that are immobilised and mixed with a carrier material (powdered corn cob). Oilzapper feeds on hydrocarbon compounds present in crude oil and oil sludge. It is sprayed over oil sludge, about 200 g of it converts a kilogram of oil into carbon dioxide and water. Oilzapper is neatly packed into sterile polythene bags for safe transportation. The shelf life of this product is three months at ambient temperature. Oilzapper has elicited tremendous response in India and abroad. This technology is shared by BPCL, Mumbai for their internal use but is marketed by TERI, New Delhi. Another member of this family is Oilivorous-S, a little different from of Oilzapper, and has additional bacterial strain that makes former more effective against sludge and crude oil with high sulphur content. This is also marketed by TERI, New Delhi.

10.8 BIOGAS

Development of biogas technology in India began almost half a century ago. It was in 1937 that a sewage treatment plant employing anaerobic digestion process was set up in Bombay. This generated the interest of scientist and researchers in the field of anaerobic digestion of cattle dung for the production of combustible gas, called biogas and rich organic manure in the form of digested slurry. Pilot experimental plants were designed in the 1940's and the prototype (moveable gas holder plants) was developed in the 1950's. A few plants were installed for actual use in field evaluation-cum-demonstration, during the next ten years. However, systematic extension of moveable gasholder plants began in 1962 when Khadi and Village Industries Commission (KVIC) included biogas technology as one of their programmes. A low cost fixed dome biogas plant named "Janta" plant was developed by the Planning Research and Action Division (PRAD), a division of UP State Planning Institute and was released for extension in 1978. Thus through the efforts of KVIC state government agencies, like Action for Food Production (AFPRO), about 100,000 biogas plants were set upto 1980. Thus, the biogas programme in India has reached a stage from where it has started making an impact on the rural energy scenario.

The raw material for the gas is biodegradable organic matter which is subjected to fermentation in the absence of air caused by anaerobic bacteria. The process is carried out in concrete tanks made under ground called a digester. Organic matter slurry is pored into this digester. A gas holder made up of mild steel, covers the digester. An anaerobic fermentation of organic matter leads to formation of biogas with a typical composition, methane 55%, hydrogen 7.5%, carbon dioxide 35%, nitrogen 2.6%. This gas has a calorific value of 35 kJ g^{-1}. The optimum temperature for fermentation is 34–48°C. The biogas thus collected can be used as fuel. 1 kg of dry animal dung gives about 160 *l* of biogas.

Many good designs of biogas plants have been suggested as majority of men are not prepared to work regularly with organic materials to run any of these plants, not much success has been attained. Even in urban area many biogas plants have not been set up. It is not difficult to set up a good biogas plant but one has to lay down the pipe lines so that gas could reach people's home. This makes it a very costly preposition. The alternative is to give up laying down pipes. The entire gas should be compressed and then made to expand when carbon dioxide will liquify and thus can be removed. The rest of the gas can be cooled with liquid carbon dioxide and compressed and thus be filled in gas cylinders like compressed natural gas (CNG) which can be supplied at home on demand. Removal of carbon dioxide will increase maximum flame temperature to nearly 1100°C. Biogas plant if installed adjacent to electric generator plant can be used as fuel for the production of electricity.

Compressed biogas can also be used as a fuel for running city bus service with a non-polluting exhaust. Such a management of organic garbage will give particulate matter free, cheap fuel, organic fertilizer with no flyash problem. With time entire organic matter may be managed to be converted to biogas and pollution of garbage be substantially tackled.

10.9 THERMAL POWER PRODUCTION AND FLY ASH

Man needs energy in devise forms and for various purposes. As his economic condition improves he tries to utilise a cleaner source of energy. Electricity is the cleanest source of energy available at homes. Water is converted to steam by heating with coal or natural gas and it is the steam that runs the turbines to produce electric power. However thermal power production is associated with a great deal of pollution.

Most of the Delhi industrialists do not pay any money for the power they consume. It is the ordinary man who has to pay the price in the name of progress. In such a situation coal may never be replaced by natural gas as a cleaner source for power production.

Indian coal has high ash content (about 30–45%). High ash coal means more wear and tear of plant and machinery, low thermal efficiency of boiler, slogging, choking and scaling of furnace and most serious of all generation of large amount of fly ash.

As a rule of thumb for every MW of installed capacity of power approximately one acre of land is required for disposal of the ash generated. Indian coal is associated with 30 to 45% of ash in addition to non-carbonaceous material like stones associated with coal. Fortunately, except the coal from Assam, Indian coal is reasonably low in sulphur (≈ 0.5%).

The Talcher thermal plant in Orissa generates 3000 tons of ash per day. The approximate content of various toxic metal (in kg) in 3000 tons of ash is as follow:

Chromium	*Iron*	*Nickel*	*Cobalt*	*Zinc*
404	1387	373	74.66	373
Copper	Cadmium	Thorium	Uranium	—
300	33.3	110	30.6	—

However, if exposed to natural leaching like acid rains only 15% of these metals are leached out, the long term impact of such an addition is little understood.

Combustion of coal in the furnace of a thermal plant at 1200 to 1500° celsius under a jet of air partially melts the inorganic constituents. The result is formation of a mass of spherical and cryptocrystalline particulate, silicious in composition and dominantly of high temperature quartz. The trace metals are redistributed in the fly ash. The cryptocrystalline grains and the hollow and spherical ones known as cenosphere and phenosphere have large specific and internal surfaces and provide ideal sites for condensation of the volatilised metals. High temperature silica, the dominant mineral of the ash has no lattice site to accommodate any of the toxic metals. Thus the entire mass of toxic metals is confined to the surface of the newly formed ash particulate, and is easily released into the environment. Even a mild acidic rainfall can remove sufficiently large quantity of toxic metals.

A noteworthy feature of fly ash is the increased concentration of the metals with a decrease in the particle size. When inhaled the 10 to 1 micron size particles are caught in the nasal mucus, 75% of which are swallowed. On reaching the stomach most of the metals are extracted by gastric juices and could easily enter the body fluid. The submicron particles enter deeper into the lungs and are deposited on alveolar walls where the metals could be transferred to blood plasma across the cell membrane. The residual particles being silica or mullite are refractory and inert and cause silicosis.

Thus thermal power plants increase suspended particle matter in air, rains despoil the land, choke the drainage system. These are amenable to technological solution by way of installation of precipitators and scrubbers and provision of tall stacks, ash ponds etc. However. experience has shown that many of the primary aims of good technological efforts become 'secondary with the passage of time and with 'other factors' taking over the primary aims, the end result is back to square one. For example, the problem of disposal of fly ash or the lime sludge of the gas scrubber, the decreased efficiency of the precipitator with continuous accumulation of the less conductive ash particles on the precipitator blades; leakage or overflow of the ash ponds due to local aquifer conditions and monsoon cloud bursts, nonavailability of open land for disposal of fly ash in the most of Indian power plants are some of the hazards associated with the curative measures in thermal plant projects.

The per annum generation of fly ash in the coal furnaces of about 75 thermal plants is around 60 million tonne. Slurry disposal lagoons/setting tanks become breeding grounds for mosquitoes and bacteria. In addition fly ash holds the potential to contaminate the underground water resources with traces of toxic metals present in fly ash.

Huge investment/expenditures are made just to get the fly ash out from thermal power plants and dump it in the ponds. Existing 75 thermal power plants alone need about 50,000 acres of precious land for disposal of fly ash during their life span of 30 years.

The requirement of land, for disposal of fly ash slurry, would increase with the addition of more thermal plants. The annual expenditure on road transportation just for dumping the fly ash is about Rs. 50 crores. Considering 10% annual growth in power generation through power plants, the annual fly ash generation is expected to exceed 100 million tonnes by 2000 AD. Appreciating the overall concern for environmental problems created by fly ash which otherwise is a very useful product of thermal plants, the Technology Information Forecasting and Assessment Council (TIFAC), of the Department of Science and Technology (DST) identified safe disposal and gainful utilization of Fly ash as a trust area and commissioned a Techno-Market survey: "Technologies for disposal of Thermal Power Stations"[4] Technology demonstrate that it can be used in overbridges (Okhla Fly Overbridge), road embankment, for construction of hydraulic structures, and dwelling units.

The main problems associated with thermal power production can be tackled.

1. Attaching an electric precipitator to the power production unit will decrease the particulate matter of fly ash in air and thus a nuisance can be eliminated. It may be pointed out that a precipitator does not decrease the total ash content but helps to control fly ash which is very essential.
2. The other problem is fly ash. Some of the ways to use fly ash are as follow:
   (*i*) Since toxic metal content is 0.1% which is quite small, and rest is like fine sand. It can be used in any construction work in place of sand.
   (*ii*) Fly ash and cement if mixed in the ratio of 4 : 1, add some water and make bricks without heating. Such bricks will be more stronger than the ordinary bricks, and if properly made may not require any plastering. This can also be used to make blocks of different size according to the requirement of corporations. Unglazed tiles for use on footpaths can also be made from it.
   (*iii*) Ordinary bricks are made from clay sorts which has a layer lattice structure as shown in Fig. 4.4. On heating water molecules are eliminated which may lead to Si—O—Al bond and thus in a three dimensional extended structure some thing of feldspar like structure. This three dimensional extended structure and thus bricks are so strong. If fly ash is mixed with aluminium oxide or even bauxite and heated to 1200°C, there will be formation of feldspar and thus to bricks. These bricks will be stronger than ordinary bricks and we also save the clay soil.
   (*iv*) Mix the fly ash with aluminium oxide. Add aqueous sodium hydroxide solution to above mixture and leave it as such for a week. Now make bricks from such a chemically treated fly ash.
3. Another way to tackle fly ash problem is given by TERI as follows:

Naturally occurring mycorrhizal fungi is sprayed over fly ash, where through its mycelial network, accomulate heavy metals from fly ash and retain them in their cells. These mycelial threads along with dense root biomass, assist binding to ash particles. Mycorrhizal fungi form a reciprocating relationship with the living roots of higher plants, they provide nutrition to plants and in turn receive carbon from them. Mycorrhizal association with small quantity of compost added, benefits plant tremendously in terms of their growth rate tolerance to biological and environmental stresses.

TERI scientist collected different strains of mycorrhizal fungi from diverse regions of India and abroad. These were then isolated, selected, multiplied and tested under greenhouse/nursery conditions to know their growth pattern on fly ash dumps. Strains that offer high tolerance, assist in survival, and provide nutritional support to plants were selected for the purpose. With additional doses of organic and mycorrhizal fertiliser to optimise the impact, the mycorrhizal strains were then applied to the plants on fly ash dumps. In some time, like sprouted on grey, degraded, toxic wastelands in the form of green vegetation. Tree species, sheesham, eucalyptus, meethineem and like; floriculture and aromatic species marigold, sunflower, lemongrass, tuberose, gladiolus, lily and species for biofuel Jatropha and vetiver have been tried. Mycorrhiza biofertilizer is promising enough to metamorphose the 30000-odd hectares of fly ash dumps in India into a huge commercial proposition.

4. The best solution is stop making use of coal. In its place use should be made of natural gas. Almost all pollution problems are eliminated and power plant becomes more efficient and thus power become cheaper than in a coal based thermal plant.
5. Biogas plant be installed just adjacent to power plants so that biogas so produced is used for production of electricity, liquid matter a source of water and solid residue can be used as organic fertilizer. Net result is no garbage.

REFERENCES

1. "Garbage Management, My Experience" by Dr. Iqbal Malik, A *Vatavaran* Report (1996).
2. "Hospital Waste: A Dangerous Infusion" by Iqbal Malik, Bajinder Singh, Ajay Kumar, Meenu Ratnani, Alka Tomar, Vijay Bhan, Dr. Jagjit Kaur, A *Vatavaran* Report (1996).
3. "Plastic and Our Lives" by Iqbal Malik, Nandini Raghuphy, Alka Tomar, Bajinder Singh, Ajay Kumar, Meenu Ratnani, Dr. P.S. Sindhu, A *Vatavaran* Report (March 1996).
4. "Fly Ash Mission" Technology Information, Forecasting & Assessment Council (TIFAC), Department of Science & Technology, Ministry of Science & Technology, Government of India, Technology Bhawan, New Mehrauli Road, New Delhi-110016 (1994).
5. "TERI's Technologies for Sustainable Development", Published by TERI, New Delhi (2005).

Chapterwise Problems

CHAPTER 1

Water and Earth

1. The ocean stores a vast amount of heat but except for climate moderation, this huge store of energy is of little use to human. Explain the difference between high quality and low quality energy.
2. Ecosystem requires energy to function. From where does this energy come and where does it go. How does the flow of energy conform to the laws of thermodynamics?
3. Photosynthesis and cellular respiration are complimentary processes. Explain how they exemplify the law of conservation of matter and thermodynamics?
4. Water cycles work at highly variable rates. Describe a water cycle that occurs in a matter of several days and one that would take more than a year to complete.
5. Describe the path a molecule of water might follow through the hydrological cycle from ocean to land and back again.
6. It is believed that hydrogen bonding is one of the cause of the development of life on earth. Suppose from tomorrow onward, water molecule had no hydrogen bonding with other molecule. Give ten ways that would have affected the life on earth.
7. Water under a pressure of 71 kbar and at 373°C has a pKw of 5.70. What will be the fate of CO_3^{2-}, S^{2-}, SO_3^{2-}, Cl^{-1} salts under such condition? Explain.
8. Gases like O_2, N_2, CH_4, NO and inert molecules like CCl_4, $CHCl_3$ dissolve into water. Explain the nature of interaction that makes these substance to go into aqueous solution.
9. Explain what is Tectonic Cycle? What is the source of energy for this cycle? Explain.
10. What are sedimentary rocks and how their formation has taken place?

CHAPTER 2

Ecology

1. In the biosphere, matter flows in a circular pathways while energy flows a linear pathways. Explain.
2. Every year fires burn large tracts of forest land. Describe the process of succession that occur after a forest fire destroys the existing biological community. In the composition of the final

successional community likely to be the same as that which existed before the fire. What factors might alter the final outcome of succession process?

3. What is the difference between saying that a duck has webbed feet because it needs them to swim and saying duck is able to swim because it has webbed feet?
4. It is often argued that complexity in a biological community creates stability and justifies preservation of native communities. What would be the effect on our government environmental policy of complexity was not related to stability?
5. Try some role playing with your family friends or classmate. Imagine that you are from a family of farmer who lived by the side of a river for several generation but now finds difficult to get water from taps. Another might be a son of gardener who find no more trees or plants in Metro cities. What issues would you want to discuss?
6. Suppose by some mechanism intensity of visible light reaching surface of earth (*i*) increases (*ii*) decreases by 10%, what drastic ecological changes in food production, insects, animal and human life do you expect?
7. If temperature of Indian ocean water increases by from 20°C to 22°C what kind of change in fish population is expected in it and why?
8. It is said that Antarctic sea water is among the most productive in the world during polar summer. Explain the behaviour.
9. How food chains and food web are related to energy in the entire ecological system of earth?
10. Explain what is the Liebig's law of the Minimum by taking two examples. Is this in some way related to the limits of tolerance.
11. In what way plant and animal life is divided into cold and warm blooded animals? Is that related to their movement to from one regions to another.
12. What is the cause of thermal pollution and explain its harmful effects? How to control thermal pollution?

CHAPTER 3

Water Pollution

1. What are some of the most serious infectious disease in the world? How are they transmitted?
2. How do stress, diet and lifestyle affect environment health? What diseases are not clearly related to these factors?
3. How would you feel or act if your child were dying of diarrhoea? Why do we spend more money on AIDS or cancer research than childhood illness?
4. What is subsidence? What are its result?
5. Describe some problems associated with dam building and water diversion projects.
6. Water pollution may be defined as anything that decrease water quality in ways that adversely affect organisms or make it unsuitable for desired human uses. Suppose that part of the silt in a river is from natural sources making fish to die and part is human cause. Is death of fishes a pollution problem but effect on human being not.
7. Suppose that a farmer litters that causes silting in the river because it fertilizes the field and thus likes it but a fishing enthusiast dislikes it because it kills same fish. How can we decide whether it is pollution or not?
8. The dissolved oxygen in an unseeded sample of diluted waste having an initial DO of

9.0 mg/L is measured to be 3.0 mg/L after 5 days. The dilution factor P is 0.03 and reaction rate constant k is 0.22/day.

(*a*) What is $(BOD)_5$, CBOD?
(*b*) What would be the remaining oxygen demand after 5 days?

9. In example 8 the wastes had an ultimate BOD equal to 300 mg/L. At 20°C, the 5 day BOD was 200 mg/L and reaction rate constant 0.22/ day. What would be the 5 day BOD of this waste at 25°C.
10. Just below the point where a continuous discharge of pollution mixes with a river, the BOD is 10.9 mg/L and DO is 7.6 mg/L. The river and waste mixture has a temperature of 20°C, a deoxygenation constant of 0.20/day an average flow speed of 0.30 m/s and an average depth of 3.0 m.

    (*a*) Find the time and distance down stream at which the oxygen deficit is a maximum.

11. In a standard 5 day BOD test.

    (*a*) Why is the BOD bottle stoppered?
    (*b*) Why is the test run in the dark or in black bottle?
    (*c*) Why is it usually necessary to dilute the sample?
    (*d*) Why is it sometimes necessary to seed the sample?
    (*e*) Why is not ultimate BOD measured?

12. A BOD test is run using a 100 ml of treated waste water mixed with 200 ml of pure water. The initial DO of the mixed is 9.0 mg/l. After 5 days, DO is 4.0 mg/l. After a long period of time, the DO is 2.0 mg/lt and it no longer seems to be dropping. Assuming that notification has been inhibited so that the only, BOD being measured is carbonaceous.

    (*a*) What is the 5 day BOD of the waste water?
    (*b*) What would be remaining BOD after 5 days have elapsed?
    (*c*) Estimate the reaction rate constant k (day^{-1}).

13. In chlorinated water, that is used for drinking, chloroform is routinely and inadevertly, created in concentrations of approximately 3.0–70 µg/l. Suppose a municipal drinking water supply has an average chloroform concentration of 70 µg/lt. The potency factor for chloroform 6.1 × 10^{-3} (mg/kg/day) and an average adult weighs 70 kg and drinks 2 litres of water everyday.

    (*a*) Everyday intake of chloroform by an adult and thus calculated maximum lifetime risk of cancer for an adult associated with chloroform (Risk = Daily intake × Potency)
    (*b*) If a city of 500000 is served by this supply what numbers of extra cancer per year would be expected.
    (*c*) Compare the extra cancers per year caused by chloroform in drinking water with the expected number of cancer deaths from all causes.

14. The chronic daily intake (CDI) of a toxic substance can be calculated

$$\text{CDI}(\text{mg/kg/day}) = \frac{\text{Total dose (mg)}}{\text{Body weight} \times \text{Lifetime (days)}}$$

where (Total dose) = (Containment concentration) × (Intake rate) × (Exposure duration) × (Absorption fraction)

Estimate the cancer risk for a 60 kg worker exposed to a particular carcinogen under the following circumstances. Exposure time is 5 days per week, 50 weeks per year and over a 20 year period of time. The worker is assumed to breath heavily for 2 hr per working day at the rate of 9 m^3/day. The carcinogen has a potency factor of 0.02 (mg/kg/ day)$^{-1}$. The absorption factor is estimated at 80 per cent and its average concentration in air is thought to be about 0.05 mg/m^3.

15. A sample of ground water has 100 mg/lt of Ca^{2+} and 10 mg/l of Mg^{2+}. Express its hardness in units of meq/litre and mg/lt of $CaCO_3$.
16. A sample of water at pH 10.0 has 32.0 mg/lt of CO_3^{2-} and 56.0 mg/lt of HCO_3. Find the alkalinity as $CaCO_3$.
17. An analysis of a sample of water with pH 7.5 has produced the following concentration (mg/lt):

Cations		Anions	
Ca^{2+}	80	Cl^-	100
Mg^{2+}	30	SO_4^{2-}	201
Na^+	72	HCO_3^-	165
K^+	6		

Find the total hardness (TH), the carbonate hardness (CH), the noncarbonate hardness (NCH), and the alkalinity, all expressed as $CaCO_3$. Find the total dissolved solids in mg/lt. Compare your results with TH 323, CH 135.2, NCH 187.8, alkalinity = 135.2, TDS 654 mg/lt.

18. A sample of ground water has 150 mg/lt of Ca^{2+} and 60 mg/lt of Mg^{2+}. Find the total hardness expressed in milliequivalents per litre and mg/litre of $CaCO_3$. How would this water be classified soft, hard.
19. A sample of water at pH 10.5 has 39.0 mg/litre of CO_3^{2-} and 24.5 mg/ litre of HCO_3.

(*a*) Ignoring the combination of [H^+] and [OH^-] to alkalinity, what is the alkalinity as $CaCO_3$.
(*b*) Including the combination of [H^+] and [OH^-] find the alkalinity as $CaCO_3$.

20. A sample of water has the following concentrations of ions (and the pH is near neutral)

Cations	mg/lt	Anions	mg/lt
Ca^{2+}	95.0	HCO_3^{2+}	160.0
Mg^{2+}	26.0	SO_4^{2-}	135.0
Na^+	15.0	Cl^-	73.0

(*a*) What is the total hardness (TH)?
(*b*) What is the carbonate hardness (CH)?
(*c*) What is the noncarbonate hardness (NCH)?
(*d*) What is the alkalinity?
(*e*) What is the total dissolved solid concentration?

21. A sample of water has the following concentration of ions

Cations	mg/lt	Anions	mg/lt
Ca^{2+}	40.0	HCO_3^-	110.0

Mg^{2+}	10.0	SO_4^{2-}	67.2
Na^+	?	Cl^-	11.0
K^+	7.0		

(*a*) Assume no other constituent ions are missing, use an anion–cation balance to estimate the concentration of Na^+.

(*b*) What is the total hardness?

22. Solubility of CO_2 in water can be given by Henery relation

$$[CO_2]_{aq} = 55.56\ K_H PO_2 = 0.03395\ P_{CO_2} \text{ at } 25°C$$

Also equilibria of type exist

$$[CO_2]_{aq} + H_2O \rightleftharpoons H^+ + HCO_3^-$$

with
$$K = \frac{[H^+]+[HCO_3^-]}{[CO_2]_{aq}} = 4.47 \times 10^{-3}$$

Estimate the pH of natural rain water assuming that the only substance affecting it is CO_2 that is absorbed from atmosphere. Assume that the concentration of CO_2 is 350 ppm at 25°C and 1 atm pressure.

CHAPTER 4

Soils and the Environment

1. What is a soil? Why are soil organisms so important? Describe four kinds of erosion. Why is erosion a problem?
2. Explain some possible effects of over-irrigation.
3. What can farmers do to increase agricultural production without increasing land use?
4. Explain what is meant by soil pollution. Soil erosion is limiting our capacity to feed ourselves, why do we allow it to continue?
5. How does overgrazing encourage undesirable forage species to flourish?
6. Why can grazing animal generate food on land that would otherwise be unusable by human?
7. Draw and describe the structure of clay minerals montmorillonite, vermicullite, muscosite, biotite and attapulgite.
8. Why feldspars decomposed in contact with water to form various clay minerals? Explain the mechanism of their decomposition.
9. In a soil, how will you identify the clay minerals? Explain.
10. When a suspension of montmorillonite is passed through H-Resin, Acid form of it obtained. Explain the pH, and conductivity curves so obtained. What is the relevance of these curve to the erosion of soils?
11. In Fig. 4.18 electron micrograph photograph shows that lateral surfaces of kaolinite and thus other clay minerals carry positive charge. How this charge has originated there?
12. What is the cause of origin of positive and negative charges on clay platelets? How that leads to adsorption of various cations, anions, pesticides and thus other chemicals polluting the soils.

13. Addition of metal ions pollutants to any soil may form different complexes with Humic acid and Fulvic acid present. In what way that leads to pollution of soils and retard or accelerated the plant growth?
14. It is essential that pH of soil be near about 7 and should be neutral, because an acid and alkaline soil lead to soil pollution. Explain with reaction involved.
15. Discuss the various types of reaction that can take place in acid soil and how they affect plant growth. Why most of the plants cannot grows in alkaline soils?
16. In what ways regular addition of lime to soil stops it from getting polluted? Explain.
17. How the nutrient availability is related to soil pH?
18. Regular addition of Inorganic pesticides pollute the soil. Explain how?
19. Explain with example the hazards and benefits of associated with the use of pesticides to plants.
20. How does presence of heavy metal like Cd^{2+}, Hg^{2+}, Zn^{2+} pollute the soils? Explain with examples.
21. If you are asked to play a favourable role for an NGO for water harvesting, explain how you will go about it.
22. Nitrogen is added in soil in various forms as fertilizer but it pollutes soil and also health hazard. Explain.
23. In what way wind erodes the soil. Is this significant?

CHAPTER 5

Environmental Chemistry of Some Elements

1. Define LD-50. Why is it more accurate than simply reporting toxic dose?
2. What is the difference between acute and chronic toxicity?
3. In what way various cycles help to decide if an element is pollutant or not?
4. Describe the carbon cycle. For the growth of plant it is essential that soil becomes acidic at night and basic in the day. This can only take place in a neutral soil not in an acidic or alkaline soil. Explain why?
5. Total carbonate concentration in a water is 1×10^{-2} M and metal ions $Ca^{+2} = 6.52 \times 10^{-4}$ M, $Ba^{2+} = 6.5 \times 10^{-2}$ M individually. Using the Table 3.4 calculate CO_3^{2-} concentration in water at pH 4, 5, 6, 7, 8 and suggest which ion will get precipitated as carbonate at different pH values?
6. Describe the most likely pathways for the nitrate reduction. An imaginary situation arise that each step can exist in isolation. Based on the information available which steps will lead to pollution and why?
7. How soil microorganisms transform phosphorus to change its (*i*) solubility (*ii*) mineralizing organic material to release orthophosphate (*iii*) Converting available phosphorus to protoplasma.
8. Sulphur, often is assimilated by four categories of microorganisms to produce sulphur amino acids. Explain with reactions.
9. Worldwide combustion of fossil fuels produce 10^8 tons of SO_2 which is very high. One way is pretreatment of fuels. The other is removal of SO_2 from exhaust gases. Explain some of the methods used for its removal.
10. What are the chief function of Haemoglobin? What is similarity and difference between myoglobin and Haemoglobin. Inhalation of CO leads to break up of Haemoglobin. Explain why?

11. Next to iron, zinc is one of the most important metal biologically. Explain how?
12. In the citric acid cycle that generates cellular energy arsenic inhibits it and thus acts as a poison. Explain.
13. How the formation of organic arsenic compounds formation take place? Explain why they are poisonous in nature.
14. Cadmium and mercury had led to Itai-Itai Byo, Minimata poisoning and Bombay minimata. Explain how these disease were caused?
15. What actions of man lead to lead poisoning and in what way it affect our system?
16. In water, containing fluoride ions, is used for drinking it leads to fluorosis. How does it affect our teeth and bones?

CHAPTER 6

Polymers and Plastics

1. Can you suggest a reason why polymerization should take place in a (head to tail) that yields a polymer with regularly alternating groups.
2. Material similar to foam rubber can be made by the following sequences

   adipic acid + excess 1,2 ethanediol $\longrightarrow$ C

   C + excess – *p* OCN—C_6H_4— C_6H_4—NCO *p* $\longrightarrow$ D

   D + limited $H_2O \longrightarrow$ E

   write equations for all steps and show structures for C, D, E. Be sure to account for the cross linking in the final polymer, and its foamy character. Remember A foam is dispersion of a gas in solid.
3. Treatment of benzyl alcohol, $C_6H_5CH_2OH$ with cold conc. H_2SO_4 yields a high boiling resinous material. What is the likely structure for this material, and how is it formed.
4. Linseed oil and tung oil are two important constituent of paints. These are esters derived from acids that contain two or three double bonds per molecule, 9, 12 octadecadienoic acid for example. On exposure to air, paint forms a tough protective film; oddly enough, after the initial rapid evaporation of solvent, this drying of paint is accompanied by a gain in weight. What kind of process do you think is involved? Be as specific as you can be.
5. Give the mechanism of the thermal degradation of poly (styrene) into monomer and oligomers.
6. (*i*) In the cyclization of polymethacrylonitrile, a colour develops gradually from yellow through orange to red and black. Explain.

   (*ii*) On UV irradiation of polystyrene develop yellow colour. Explain.

   (*iii*) On exposing to UV radiation, nylon undergoes both chain scission and cross linking. How can this be explained?
7. Three samples of flexible tubing A (Vulcanized rubber), B (polyvinyl chloride), C (polyethylene) are subjected to the following tests.

   (*a*) Heated on a Bunsen burner Tube A and C burn.

   (*b*) Soaked overnight in chloroform Tube A swells, B become very stiff and somewhat smaller on drying out, C is unaffected.

   (*c*) Immersed in lead nitrate solution, A turns black while B and C are unaffected.
   Explain this behaviour of the three different polymers.

8. A cycle for polymers can be given as follow:

   Air ⟵ Biological action

   ↙ ↗

   Polymer ⟶ Soil

   If any one part of these reactions do not go to completion it may lead to pollution. Explain.
9. Science has advanced enough and can synthesize even enzyme which can degrade any polymer. Today the biggest polymers which are polluters are polyvinyl chloride (PVC), polyethylene. So if enzymes for there degradation are synthesised, will the pollution problem will be solved or not. Discuss.
10. In table 6.4 are given the stages in microbial attack on pian insoluble substrate like polymers. Which step is not feasible that stops biodegradation of polymers?
11. Biodegradation of any substance provides a source of energy for their growth. If that is not there, biodegradation will not take place. Discuss and explain the statement.
12. Eighty two per cent of polymer products are disposed of after a week to three months. Most of these products are made of PVC and PE and are the biggest menace of the day. Suppose you have to address at two different seminars:

    (*a*) From industrialist side: What can be done with these disposable polymer products?

    (*b*) From environmentalist side: Why these polymer products should be banned? Write down you talks.

CHAPTER 7

Fossil Fuels and Petroleum

1. A few years ago laundry detergents commonly contained enzymes for added cleaning power. What do you suppose these enzymes were supposed to do? Can you imagine any disadvantage to adding enzymes to household products?
2. What is composition of TEL that was added to gasoline? What was it purpose and in what way it was a pollutant?
3. Every Delhi citizen who has been living in this Metro city must have crossed the safe limits of lead in his system. How far this statement is true and what is the alternative?
4. What is PCV and how does it decrease pollutants emission from petrol driven vehicles?
5. What is CNG and how its use will decrease pollution?
6. What are pathways by which hydrocarbons are removed from atmosphere? Removal of hydrocarbon mean less or more pollution in atmosphere. Explain.
7. How is hydrocarbons are degraded by microorganisms in nature? Explain. Why is it difficult to degrade an alkane with side chains?
8. Some years ago there was explosion in Gulf area that lead to spilling over of lot of oil over sea. This lead to the death of large number of fishes and birds. Explain why?
9. Compare and contrast between soaps and detergents. How detergent alkyl benzene sulphonate (ABS) is synthesized? Why ABS is non-biodegradable, while detergent LAS is biodegradable?
10. Even if a detergent is biodegradably, it acts as a pollutant. Explain how?
11. Name three builders that are added to detergent. What are their function and how they retard the precipitation of Ca^{2+}, Mg^{2+}. Which one is the best builder and why?

12. Explain the cleaning action of soaps.
13. Name and explain what are role of brighteners, perfumes, anti redeposition agent, enzyme that are added to detergents.

CHAPTER 8

Pesticides

1. Estimate the chronic daily intake (CDI) of DDT for a 70 kg individual consuming 2 g of fish per day for a lifetime from a stream with 20 ppb of DDT. What would be the maximum lifetime cancer risk to this individual due to DDT?
2. Suppose 100 g of the pesticide heptachlor is accidently leaked into a 30000 m^3 pond. What would be the concentration in the pond if it immediately mixes uniformly throughout the pond? If the half life is 2 days, what would be concentration after 1 week?
3. Suppose 1.0 g/day of Heptachlor leaks into a 30000 m^3 pond. If heptachlor has a half life of 2 days and complete mixing oceans.

   (*a*) What would be the steady state concentration in pond?

   (*b*) Suppose a 70 kg individual drank 2 lit/day of that water for 5 years. Estimate the maximum risk factor of cancer due to that exposure to heptachlor. Potency factor of heptachlor is 3.4 $(mg/kg/day)^{-1}$.

4. Arsenites were the first generation pesticide. Discuss the mechanism of the arsenite action that helped to kill pests. Since it did not react with any group biological importance yet it is not used. Explain.
5. Explain the mechanism of action of DDT and DDT analoges. Also explain why deuterated DDT exhibit same level of toxic effects as DDT?
6. How bioamplification of DDT in fishes and birds led to decline in their population? Explain their long term effects.
7. It is said that level of DDT in mothers milk living in Delhi is quite high. Write down a food chain that explains this increase in DDT level in women. What precaution every person living in Delhi should take to overcome this pollution problem?
8. On chlorination of benzene one of the eight isomer is an active pesticide. Identify it. How does it help to kill insects? How it is degraded in environment?
9. Discuss the environmental degradation of the various cyclopentadiene pesticides. How this pesticides leads to violent quivering, convulsions, followed by paralysis in the pests.
10. 2, 4D and 2, 4, 5 T series of herbicides are completely biodegradable. Explain how? Even then these herbicides are associated with environmental problems. Explain.
11. The toxic effect of organophosphorus as well as carbamates insecticides is similar to polychlorinated cyclopentadiene. Explain the similarity and difference between the two types of pesticides.
12. Explain the environmental degradation of organophosphorus insecticides in nature.
13. The third generation pesticides are called pheromones. Give name and structure of two pheromones. Give the mechanism by which pheromones kills pests. In what way pheromones are better than organochloro pesticides like DDT.
14. Explain how pheromones and attractants are used for insect control?
15. How the fourth generation pesticides, anti-juvenile hormones are used to kill pests?

CHAPTER 9

Nuclear Chemistry and the Environment

1. What happened at Chernobyl?
2. Suppose that a nuclear power plant has been proposed for your neighbourhood. What safeguard would you demand. Is there anyway to be sure you are safe—or safe enough. What constitutes safe enough?
3. The cancer caused by exposure to radiation is thought to be approximately 1 lethal cancer per 8000 person-rems of exposure. Exposure to cosmic radiation increases with increasing altitude at sea level it is about 40 m rem/year and at the elevation of Denver (about one mile high) it is about 65 m rem/year.

   (*a*) Compare the lifetime probability of dying of cancer induced by cosmic radiation for a person living at sea level and that a person living in Denver.
   (*b*) Estimate the number of cancer deaths per year in Denver, with a population of 0.5 million, caused by cosmic radiation. Compare it to total cancer deaths in Denver, assuming the typical cancer death rate of 193 per 100000 per year.
   (*c*) If all Indians, 1 billion are supposed to be living at sea level, estimate the total cancer deaths per year caused by cosmic radiation.

4. The dose to residents living near Chernobyl when it exploded has been estimated at about 5 rems. Using the data from problem 3, estimate the individual cancer risk faced by someone exposed to that amount of radiation.
5. It has been estimated that about 75 million people in the Ukraine and Byelorussia were exposed to an average of 0.4 rem of radiation as a result of the chernobyl nuclear accident.

   (*a*) How many extra cancer death might eventually be expected from the exposure? (see problem 3).
   (*b*) If the normal probability of dying of cancer from all causes is 0.22, how many cancer deaths would you normally expect among those 75 million people? What is the percentage increase in cancer deaths that would be expected due to chernobyl?

6. The average radiation exposure to someone living within 20 miles of a nuclear power plant has been estimated at about 0.02 m rem/year. How many extra cancer deaths would be expected per year in India (population 1 billion) due to nuclear power if that were the average exposure? (Use potency data given in problem 3).
7. Living in an average US home with 1.5 pci/L of radon is thought to cause a cancer risk equivalent to that caused by approximately 400 m rem/year of radiation. Using the radiation potency factor and population given in problem 3.
   (*a*) Estimate the annual cancers in USA caused by radon in homes.
   (*b*) Estimate the lifetime risk to an individual living in a home with that amount of radon.

8. Suppose a soil under a single story house emits 1.0 pci/m^{2}s of radon gas equivalent to 270 m rem/year. As a worst case assume that all of this gas finds way through the floor and into the house. The house has 250 m^2 of floor space, an average ceiling height of 2.6 m and an air change rate 0.9 ach. Estimate the steady state concentration of radon in the house assuming ambient concentration is negligible. The decay constant of randon is 7.6×10^{-3} hr^{-1}. Also

estimate probability of dying due to cancer in that house. Potency data may be taken from problem 3.

9. What weight of Ra^{225} whose $T_{1/2}$14.8 sec is required to yield 1 millicurie of radiation?
10. A sample of radioactive substance show an intensity of 230 millicuries at a time *t* and an intensity of 162 millicuries 600 sec latter. What is the half life period of radioactive material?
11. In a naturally occurring thorium decay series $^{230}_{90}Th$ decays in succession by α, β, β, α, α, α, β. Write an equation to show the mass number, atomic no. and symbol of the nuclide formed by the decay.
12. (*a*) What is the difference between a chain reaction and branch chain reaction?
    (*b*) What is a moderator in a nuclear reactor? Give example.
    (*c*) Explain what is the critical mass of a fissile material and why does the critical mass vary.
13. Calculate the radiation intensity in roentgens per hour due to the γ-rays from a 1 curie Co^{60} source at a distance of 1 metre. Co^{60} decays to the 2.505 MeV level of Ni^{60} which in turn decays to the 1.333 MeV level and then to ground state. There is no transition between the 2.505 MeV level and the ground state. Mass absorption coefficients for Al is 8×10^{-2} cm²/gm. They are close to the value for air.
14. Using the mass absorption coefficient say 9×10^{-2} cm²/gm calculate the radiation intensity at a distance of 1 metre from a 1 curie Co^{60} source which is surrounded by 1 cm of lead shielding.
15. Calculate the exposure (rem per 13 week period) of bone containing 1 μμC of Pu^{239} per gram. Assume that all the particles and recoil nuclei are absorbed in the bone. (Ref. Table 9.10).
16. Calculate the thickness of lead required to shield a worker from a 10 curie emitting 1 quantum of 1 MeV γ-radiation per disintegration. The worker is to be at a distance of 3 metre from the source for 1 hour per day, 5 hour per day week.
17. In disassembling cyclotron targets, a member of the crew is required to remove two small screws. This operation takes 5 minutes. On the average, he opens 20 targets per week. The targets average 0.5 curie of 1 MeV γ-rays and very large quantities of β-radiation. Design a simple equipment to permit this operation to be carried out safely. Make reasonable assumptions about the working distance.
18. The absorption of 48 MeV helium ions in cobalt produces about 4γ-quanta of average energy 1.5 MeV per helium ion absorbed. The photons are approximately isotropic. Estimate the thickness of lead shielding which must be placed around a 5 mg per cm² cobalt target exposed to 10μ amp of helium ions in order to give a radiation level of 5 mr per hour at a distance of 10 feet. Assume a reasonable value for the cobalt total cross-section.
19. Confirm that the energy absorbed from roentgen is 1 gm of air is 87×10^{-7} J (assuming 34 eV per ion pair).

CHAPTER 10

Garbage Management

1. What is the difference between toxic and hazardous wastes? Give some examples of material in each case.
2. List some benefits and drawbacks of recycling wastes. What are the major types of material recycled from municipal wastes and how are they used?
3. What is composting? How does it fit into solid disposal?

4. List ten toxic substances in your home and how you would dispose of them?
5. One way to remove ammonia during waste treatment is by ammonia stripping. When organic matter decomposes, ammonia is released when an equilibria sets in

$$NH_3 + H_2O \rightleftharpoons NH_4^+ + OH^- \quad K = 1.82 \times 10^{-5}$$

By doing the equilibria to left less soluble ammonia is formed which can be removed in a gas striping tower. In such a tower water is allowed to brickle downward over slates and air is blown from bottom. The NH_3 fraction in gaseous state can be calculated as

$$[NH_3] \text{ fraction} = \frac{1}{1 + K[H^+]/K_w} = \frac{1}{1 + 1.82 \times 10^{9-pH}}$$

Calculate NH_3 fraction for pH range 4–9. Plot a graph of NH_3 fraction Vs pH.
Also dissociation constant of NH_4OH to NH_4^+ and OH^- is 1.78×10^{-5}.
Plot a pH titration curve of NH_4OH Vs HCl.
Do the two curve provide similar information or different, discuss.
Note: Gas stripping is also used for removal of volatile organic chemicals.
6. The reaction and solubility product of CaF_2 are as follow:

$$CaF_2 \rightleftharpoons Ca^{2+} + 2F^- \quad K_{sp} = 3.9 \times 10^{-4}$$

Find the equilibrium concentration of fluoride in pure water. Express the answer in units mol/lt, mg/lt compare with recommend level of fluoride concentration of 1.8 mg/lt.

Appendices

APPENDIX I

SOME VERY GOOD INTERNET SITES ON ENVIRONMENTAL POLLUTION

1. Go to wikipidia site
   <en.wikipidia.org/wiki/>
   at dash enter
   Soil pollution, water pollution, pestiside pollution, radiation pollution, plastic pollution or any site of your interest.
2. Water Pollution
   <edu.green.teri.res.in/explore/water/pollution.htm>
3. India Ganges
   <www.planetark.org/dailynewsstorycfm?newsid=9513>
4. Water pollution in India
   <www.india.stat.com>
5. Pestisides in drinking water
   <www.waterbornediseases.org/factssheets/pestisides.htm>
6. Soil Pollution
   (*i*) <www.steadyhealth.com/encylopedia/soil_pollution>
   (*ii*) <10answers.yahoo.com/questions/index>
   (*iii*) <botany.uwc.ac.za/.../grade10/ecology/conservation/poll.htm>
7. Pesticide Pollution Control
   <httpi//www.sciencemag.org/cgi/content/abstract/158/3803/871>
8. For video clipings on various pollution topics go to
   <www.youtube.com>
   Enter topic of your interest and see video

APPENDIX II

SOME RELATIONSHIPS BETWEEN COMMONLY USED UNITS

UNITS

Density kg m^{-3}

1 Pound per cubic foot	=	16.018 kg m^{-3}

Concentration

1 molal solution = 1 mol/litre	=	1 mol dm^{-3}
1 milligram per litre in water	=	1.0 ppm
(Sp Gravity of water = 1)	=	1000 ppb
	=	1.0 g m^{-3}

Pressure P_a = Nm^{-2}

1 atmospheric pressure	=	760 mm of Hg
	=	760 torr
	=	101325 N m^{-2}
	=	1.01325 bar

Viscosity N s m^{-2} = 10 Poise (P)

1 kg/ms	=	1 Ns/m^2
1 cP	=	10^{-3} N s m^{-2}

Energy J = Nm

1 British Thermal Unit (Btu)	=	1055 J
1 kilowatt hour	=	3600 kJ
1eV	=	1.602×10^{-13} J
1 MeV	=	1.602×10^{-19} J
1 J	=	6.242×10^{12} MeV
Power W	=	J s^{-1}
kilowatt	=	1000 J s^{-1}
	=	1.34 horse power
1 horse power	=	746 W
(kcal mol^{-1}) × 4.184	=	kJ mol^{-1}
× 0.04336	=	eV
× 349.8	=	cm^{-1}
$\times 489 \times 10^9$	=	K
(cm^{-1}) $\times 1.196 \times 10^{-2}$	=	kJ mol^{-1}
$\times 1.24 \times 10^{-4}$	=	eV

$$\times 2.859 \times 10^{-3} = \text{kcal mol}^{-1}$$
$$\times 1.4387 = \text{K}$$
$$(\text{kJ mol}^{-1}) \times 0.2390 = \text{kcal mol}^{-1}$$
$$\times 0.0104 = \text{eV}$$
$$\times 83.59 = \text{cm}^{-1}$$
$$\times 116.87 \times 10^{9} = \text{K}$$
$$(\text{eV}) \times 96.49 = \text{kJ mol}^{-1}$$
$$\times 23.06 = \text{kcal mol}^{-1}$$
$$\times 8.066 \times 10^{3} = \text{cm}^{-1}$$
$$\times 1.4887 = \text{K}$$
$$\text{Energy (kcal mol}^{-1}) = 2.859 \times 10^{4}/\lambda \text{ (nm)}$$
$$\text{Energy (kJ mol}^{-1}) = 1.196 \times 10^{5}/\lambda \text{ (nm)}$$

1. Roentgen = 1R = Quantity of X-ray or γ-ray which will produce

2.1×10^{9} ion pair in 0.001293 gm of air

= 1.624×10^{12} ion pair/gm of air

1 Rad = 100 erg absorbed/g

= 10 μJ g^{-1} = 0.01 Jkg^{-1}

= 6.24×10^{-13} eV g^{-1}

= 1.83×10^{12} ion pair/g of air

1 Curie = 3.7×10^{10} events or disintegration/second

Water exposed to 1 R of 1 MeV γ-ray dose = 0.965 rad

air received a dose of 0.869 rad.

Multiples of the base units are illustrated by length.

Fraction	10^{9}	10^{6}	10^{3}	1	10^{-2}	10^{-3}	10^{-6}	10^{-8}	10^{-9}	10^{-12}
Prefix	giga	mega	kilo	metre	centi	milli	micro	angstrom	nano	pico
Unit	Gm	Mm	km	m	cm	mm	μm	Å	nm	pm

Physical Constants

Constants	*Symbol*	*Value*
Velocity of light	c	2.9979×10^{8} m s^{-1}
Avogadro number	N	6.02×10^{23} mol^{-1}
Planck's constant	h	6.626×10^{-34} J s
	$\frac{h}{2\pi} = \hbar$	1.602×10^{-34} J s
Electron rest mass	m_e	9.1083×10^{-28} g
Proton rest mass	m_p	1.67239×10^{-24} g
Neutron rest mass	m_n	1.67470×10^{-24} g
Charge of proton	e	1.602×10^{-19} C
Gas constant	R	8.3144 J K^{-1} mole
Boltzmann Constant	$k = \frac{R}{N}$	1.38×10^{-23} J K^{-1}
Faraday constant	$F = Ne$	9.64846×10^{4} C mol^{-1}
Permeability of vacuum	μ_0	$4\pi \times 10^{-7}$ NA^{-2}
Permittivity of vacuum	ε_0	8.854×10^{-12} F m^{-1}
Note: $\varepsilon_0 \mu_0 = 1/c^2$	$4\pi\ \varepsilon_0$	1.112×10^{-10} J^{-1} C^2 m^{-1}

PERIODIC TABLE

1	2	3	4	5	6	7	8	9	10	11	12	13	14	15	16	17	18
${}_{1}$H 1.008																	${}_{2}$He 4.003
${}_{3}$Li 6.941	${}_{4}$Be 9.012											${}_{5}$B 10.81	${}_{6}$C 12.01	${}_{7}$N 14.01	${}_{8}$O 16.00	${}_{9}$F 19.00	${}_{10}$Ne 20.18
${}_{11}$Na 22.99	${}_{12}$Mg 24.31											${}_{13}$Al 26.98	${}_{14}$Si 28.09	${}_{15}$P 30.97	${}_{16}$S 32.06	${}_{17}$Cl 35.45	${}_{18}$Ar 39.95
${}_{19}$K 39.10	${}_{20}$Ca 40.08	${}_{21}$Sc 44.96	${}_{22}$Ti 47.90	${}_{23}$V 50.94	${}_{24}$Cr 52	${}_{25}$Mn 54.94	${}_{26}$Fe 55.85	${}_{27}$Co 58.93	${}_{28}$Ni 58.71	${}_{29}$Cu 63.55	${}_{30}$Zn 63.37	${}_{31}$Ga 69.72	${}_{32}$Ge 72.60	${}_{33}$As 74.92	${}_{34}$Se 78.96	${}_{35}$Br 79.90	${}_{36}$Kr 83.80
${}_{37}$Rb 85.47	${}_{38}$Sr 87.62	${}_{39}$Y 88.91	${}_{40}$Zr 91.22	${}_{41}$Nb 92.91	${}_{42}$Mo 95.94	${}_{43}$Tc (99)	${}_{44}$Ru 101.1	${}_{45}$Rh 102.9	${}_{46}$Pd 106.4	${}_{47}$Ag 107.9	${}_{48}$Cd 112.4	${}_{49}$In 114.8	${}_{50}$Sn 118.7	${}_{51}$Sb 121.8	${}_{52}$Te 127.6	${}_{53}$I 126.9	${}_{54}$Xe 131.3
${}_{55}$Cs 132.9	${}_{56}$Ba 137.3	${}_{71}$Lu 175.9	${}_{72}$Hf 178.5	${}_{73}$Ta 180.9	${}_{74}$W 183.9	${}_{75}$Re 186.9	${}_{76}$Os 190.2	${}_{77}$Ir 192.2	${}_{78}$Pt 195.1	${}_{79}$Au 197.0	${}_{80}$Hg 200.6	${}_{81}$Tl 204.4	${}_{82}$Pb 207.2	${}_{83}$Bi 209.0	${}_{84}$Po (210)	${}_{85}$At (210)	${}_{86}$Rn (222)
${}_{87}$Fr (223)	${}_{88}$Ra (226)	${}_{88}$Lr (257)	${}_{104}$Rf 262.1	${}_{105}$Db 262.1	${}_{106}$Sg (263)	${}_{107}$Bh (262)	${}_{108}$Hs (265)	${}_{109}$Mt (266)	${}_{110}$Uun (281)	${}_{111}$Uuu (272)	${}_{112}$Uub (285)	${}_{113}$Uut	${}_{114}$Uuq	${}_{115}$Uup	${}_{116}$Uuh	${}_{117}$Uus	${}_{118}$Uuo

Lanthanides	${}_{57}$La 138.9	${}_{58}$Ce 140.1	${}_{59}$Pr 140.9	${}_{60}$Nd 144.2	${}_{61}$Pm (147)	${}_{62}$Sm 150.4	${}_{63}$Eu 152.0	${}_{64}$Gd 157.3	${}_{65}$Tb 158.9	${}_{66}$Dy 162.5	${}_{67}$Ho 164.9	${}_{68}$Er 167.3	${}_{69}$Tm 168.9	${}_{70}$Yb 173.0
Actinides	${}_{89}$Ac (227)	${}_{90}$Th 232.0	${}_{91}$Pr (231)	${}_{92}$U 238.0	${}_{93}$Np (237)	${}_{94}$Pu (242)	${}_{95}$Am (243)	${}_{96}$Cm (248)	${}_{97}$Bk (247)	${}_{98}$Cf (251)	${}_{99}$Es (254)	${}_{100}$Fm (253)	${}_{101}$Md (256)	${}_{102}$No (254)

A standard form of the periodic table with atomic numbers, atomic weight.

Glossary

A soil horizon Uppermost soil horizon sometimes referred to as the **zone of leaching.**

Absorption Taking up, incorporation, or assimilation of a material.

Adsorption Attachment of molecules of gas or molecules in solution to the surface of solid materials with which they come in contact.

Acid deposition Combination of wet deposition from the sulphuric acid and nitric acid droplets in air over a solid surface.

Aerobiosis Bacterial decomposition in the presence of air.

Aerosol Particles of diameter range 1/100 pm 101 pm suspended in air.

Age element Elements that characteristically accumulate in tissue with age.

Aggregate Any hard material such as crushed rock, sand, gravel, or other material that is added to cement to make concrete.

Air pollutant Chemical whose concentration in air builds up to cause harm to life or material things

Air pollution Addition of impurities that deteriorates the quality of air.

Algae Simple one celled or many celled plants, usually aquatic capable of carrying on photosynthesis.

Algal bloom Population explosion of algae on water surface due to increase in plant nutrients in water.

Alveoli Tiny sacks at the end of the bronchiole tubes in the lungs where oxygen was inhaled, air is transferred to haemoglobin in the blood and carbon dioxide in the blood is transferred to lungs and exhaled.

Alkaline soils Soils in arid regions that contain a large amount of soluble mineral salts (Primarily of sodium) which in the dry season may appear on the surface as a crust or powder.

Alluvium Unconsolidated sediments, including sand, gravel, and silt, deposited by streams.

Ambient Existing conditions, encompassing atmosphere.

Antibody A chemical produced in a living organism in response to a foreign substance entering the organism.

Antigen A foreign substance that produces an antibody in response to an organism.

Anaerobic Characterized by the absence of free oxygen.

Anhydrite Evaporated mineral ($CaSO_4$) calcium sulphate.

Anthracite A type of coal characterized by a high percentage of carbon and low percentage of volatiles; providing a high heat value. Anthracite often forms as a result of metamorphism of bituminous coal.

Anticline A type of fold characterised by an upfold or an arch. The oldest rocks are found in the center of the fold.

Applied ecology The branch of ecology that predicts as to how there will be minimum adverse impacts of technology and development on environment.

Appropriation doctrine, water law Holds that prior usage of water is a significant factor. The first to use the water for beneficial purposes is prior in right.

Aquiclude Earth material that retards the flow of groundwater.

Aquifer Earth material containing sufficient groundwater that the water can be pumped out. Highly fractured rocks and unconsolidated sands and gravels make good aquifers.

Arable land Land that is capable of being cultivated and supporting agricultural production.

Area (strip) mining Type of strip mining practised on relatively flat areas.

Artesian Referring to a groundwater system in which the groundwater is isolated from the surface by a confining layer and the water is under pressure. Groundwater that is under sufficient pressure that it will flow freely at the surface from a spring or a well.

Asbestos Fibrous mineral material used as insulation. It is suspected of being either a true carcinogen or a carrier of carcinogenic trace elements.

Ash fall Volcanic ash eruption that blows up into the atmosphere and then rains down on the landscape.

Ash flow Mixture of volcanic ash, hot gases, and fragments of rock and glass that flows rapidly down the flank of a volcano. May be an extremely hazardous event.

Ash, volcanic Unconsolidated volcanic debris, less than 44 mm in diameter, physically blown out of a volcano during an eruption.

Avalanche A type of landslide involving a large mass of snow, ice and rock debris that slides, flows, or falls rapidly down a mountainside.

Azonal soil Recent surface material such as flood plain deposits that do not have distinctive soil layering.

Bacteria Smallest living organism with fungi, they comprise the decomposed level of food chains and webs.

Balancing doctrine Asserts that public benefits and the importance of a particular action should be balanced with the potential injury to certain individuals. The method of balancing is changing now, and courts are considering possible long-range injury caused by certain activities to large numbers of citizens other than the immediate complainants.

Barrier island Island separated from the mainland by a salt marsh. It generally consists of a multiple system of beach ridges and is separated from other barrier islands by inlets that allow the exchange of seawater with lagoon water.

Basalt A fine-grained extrusive igneous rock. It is one of the most common igneous rock types.

Basaltic Engineering geology term for all fine-grained igneous rocks.

Bauxite A rock composed almost entirely of hydrous aluminium oxides. It is a common ore of aluminium.

Bed material Sediment transported and deposited along the bed of a stream channel.

Bedding plane The plane that delineates the layers of sedimentary rocks.

Bentonite A type of clay that expands to many times its original volume.

Biosphere The portion of the earth and its atmosphere on wetting. Capable of supporting life.

Biotite A common ferromagnetic mineral, a member of the mica family.

Biomass The total weight of living organisms per unit area.

Biodegradable Material that can be broken down into simpler substances by bacteria or other decomposer.

Biological amplification Increase in concentration of certain fat soluble chemicals such as DDT in successively higher tropical levels of a food chain or web.

Bituminous coal Very common type of coal characterized by relatively high carbon content and low volatiles. Sometimes called soft coal.

Blackbody A material that is able to absorb 100% of the radiation it receives.

Blowout Failure of an oil, gas, or disposal well resulting from adverse pressures that may physically blow part of the well casing upward. May be associated with leaks of oil, gas, or, in the case of disposal wells, harmful chemicals.

BOD Biological oxygen demand, a measure of the amount of oxygen necessary to decompose organic materials in a unit volume of water. As the amount of organic waste in water increases, more oxygen is used, resulting in a higher BOD.

Braided river A river channel characterized by an abundance of islands that continually divide and subdivide the flow of the river.

Breccia A rock or zone within a rock composed of angular fragments. Sedimentary, volcanic, and tectonic breccias are recognized.

Breeder reactor A type of nuclear reactor that actually produces more fissionable (fuel) material than it uses.

Brine Water that has a high concentration of salt.

Breeder nuclear fission reactor Nuclear reactor that produces more fuel than it consumed.

Brittle Material that ruptures before any plastic deformation.

C soil horizon Intermediate soil horizon sometimes known as the zone of accumulation. Lowest soil horizon sometimes known as the zone of partially altered parent material.

Cadmium A metallic element having an atomic number of 48 and an atomic weight of 112.4. As a trace element, it has been associated with serious health problems.

Calcite Calcium carbonate ($CaCO_3$). Common carbonate mineral that is the major constituent of the rock limestone. It weathers readily by solutional processes, and large cavities and open weathered fractures are common in rocks containing the mineral calcite.

Caliche A white to gray irregular accumulation of calcium carbonate in soil of arid regions.

Carnivore Animal that obtains its food by feeding only on other animals.

Capillary action The rise of water along narrow passages, facilitated and caused by surface tension.

Carbonate A compound or mineral containing the radical (CO_3^{2-}). The common carbonate is calcite.

Carcinogen Any material that is known to produce cancer in man or other animals.

Cesium-137 A fission product produced from nuclear reactors and having a half-life of 33 years.

Centrifugal force The tendency of a particle to move in a straight line when rotated creates this imaginary outward force.

Centripetal acceleration The inward turning of air resulting from a force acting to keep it in circular motion about a fixed centre.

Channelization An engineering technique to straighten, widen, deepen, or otherwise modify a natural stream channel.

Circum-Pacific belt One of the three major zones where earthquakes occur. This belt is essentially the Pacific plate. It is also known as the ring of fire, as many active volcanoes are found on the edge of the Pacific plate.

Clay May refer to a mineral family or to a very fine-grained sediment. It is associated with a good deal of environmental problems such as shrinking and swelling of soils and sediment pollution.

Climate A description of aggregate weather conditions; the sum of all statistical weather information that helps describe a place or region.

Climate change A study dealing with variations in climate on many different time scales, from decades to millions of year and the possible causes of such variation.

Coal A sedimentary rock formed from plant material that has been buried, compressed, and changed.

COD (Chemical Oxygen Demand)

The measure of the total depletion of dissolved oxygen in polluted water.

Coliform Bacteria A normal harmless type of bacteria that resides in the intestinal tract of human beings and water may be contaminated with other diseases causing organism found in untreated human and animal waste. Coliform Bacteria count number of colonies of fecal coliform bacterial present in a 100 million free of water.

Columnar jointing System of fractures (joints) that break rock into polygons typically with five or six sides. These polygons form columns. This type of fracturing is common in basalt and is most likely caused by shrinking during cooling of the lava.

Composite volcano Steep-sided volcanic cone produced by alternating layers of pryoelastic debris and lava flows. It characteristically has magma that is intermediate in silica content (60 per cent).

Conchoidal fracture A shell-like or fan-shaped fracture characteristic of the mineral quartz and natural glass.

Cone of depression A cone-shaped depression in the water table caused by withdrawal of water table at rates greater than the rates at which the water can be replenished by natural groundwater flow.

Conglomerate A detrital sedimentary rock composed of rounded fragments, 10 per cent of which are larger than 2 mm in diameter.

Connate water Water that is no longer in circulation or in contact, with the present water cycle. Generally saline water trapped during the deposition of sediments. Contact metamorphism type of metamorphism produced when country rocks are in close contact with a cooling body of magma below the surface of the earth.

Continental drift The movement of continents in response to sea-floor spreading. The most recent episode of continental drift supposedly started about 200 million years ago with the break up of the supercontinent Pangea.

Continental shelf Relatively shallow ocean area between the shoreline and the continental slope that extends to approximately a 600-foot water depth surrounding a continent.

Contour (strip) mining Type of strip mining used in hilly terrain.

Convection The transfer of heat involving the movement of particles; for example, the boiling of water in which hot water rises to the surface and displaces cooler water which moves towards the bottom.

Convergent plate boundary Boundary between two lithospheric plates in which one plate descends below the other (subduction).

Coriolis Effect-(Force) The deflective force of the earth's rotation on all free moving objects, including the atmosphere and ocean. Deflection is to the right in Northern Hemisphere and to the left in the Southern Hemisphere.

Corrosion A slow chemical weathering or chemical decomposition that proceeds from the surface in objects such as pipes experience corrosion when buried in soil.

Creep A type of downslope movement characterized by slow flowing, sliding, or slipping of soil and other earth materials.

Crystal setting A sinking of earlier-formed crystals to the bottom of a magma chamber.

Crystalline A material with a definite internal structure such that the atoms are in an orderly, repeating arrangement.

Crystallization Processes of crystal formation.

Cybernatics The study of communication and control system in living organism and machines.

Cyclone An area of low atmospheric pressure characterized by rotating and converging winds and ascending air. Cyclone Flow Winds blowing in and counter clockwise direction.

Debris flow Rapid downslope movement of earth material often involving saturated, unconsolidated material that has become unstable due to torrential rainfall.

Deep-well disposal Method of waste disposal that involves pumping waste into sub-surface disposal sites such as fractured or otherwise porous rocks.

Decibel (dB) unit used to measure sound power or sound pressure.

Demography the mathematical study of population.

Detrital Mineral and rock fragments derived from preexisting rocks.

Diamond A very hard mineral composed of the element carbon.

Dilatancy of rocks Inelastic increase in volume of a rock that starts after the stress on the rock has reached one-half of the breaking strength of the rock.

Discharge The quantity of water flowing past a **particular** point on a stream. Generally is measured in cubic feet per second (cfs).

Disseminated mineral deposit A mineral deposit in which ore is scattered throughout the rocks. Examples include diamonds in kimberlite and many copper deposits.

Divergent plate boundary Boundary between lithospheric plates characterized by the production of new lithosphere. Found along oceanic ridges.

Doldrums A region of the earth near the equator in which hot humid air is moving upward forming a vast low pressure region. Local squalls and rainstorms are common and steady winds are rare.

Dose dependence Relates to the fact that effects of a certain trace element on a particular organism depend upon the dose or concentration of the element.

Drainage basin The area that contributes surface water to a particular stream network.

Drainage net The system of stream channels that coalesce to form a stream system.

Dredge spoils Solid material, such as sand, silt, clay, rock or other material deposited from industrial and municipal discharges, that are removed from the bottom of water bodies to improve navigation.

Driving forces Those forces that tend to make earth material slide.

Ductile Material that ruptures following elastic and plastic deformation.

Earthquake Natural shaking or vibrating of the earth in response to the breaking of rocks along faults. The earthquake zones of the earth generally correlate very well with lithospheric plate boundaries.

Ecology Branch of science that deals with relationships between organisms and their environments and each other.

Ecosystem the biotic community and its nonliving environment as an interacting system.

Ecotone a transition zone between two adjacent and differing communities.

Ecotype A group of living organisms adapted to a set of environmental conditions.

Economic geology The application of geology to locating and evaluating mineral materials.

Effluent Any material that flows outward from something. Examples include waste water from hydroelectric plants and water discharged into streams from waste-disposal sites.

Effluent stream Type of stream where flow is maintained during the dry season by ground water seepage into the channel.

Elastic deformation Type of deformation where the material returns to its original shape after the stress is removed.

Engineering geology The application of geologic information to engineering problems.

Entropy A measure of that portion of heat energy not available for work.

Environment That which surrounds an individual or a community both physical and cultural surroundings. Environments is also sometimes used to denote a certain set of circumstances surrounding a particular occurrence, for example, environments of deposition.

Environmental geology The application of geologic information to environmental problems.

Environmental geology map A written statement that assesses and explores possible impacts associated with a particular project which may affect the human environment. The statement is required by the National Environmental Policy Act of 1969.

Environmental law A field of law that is growing rapidly and becoming a significant part of our jurisprudence.

Environmental resource unit (ERU) A portion of the environment with a similar set of physical and biological characteristics. Supposedly a natural division characterized by specific patterns or assemblages of structural components such as rocks, soils, vegetation, etc., and natural processes such as erosion, runoff, soil processes, etc.

Ephemeral Temporary or very short-lived. Characteristic of beaches, lakes, and some stream channels that change rapidly (geologically).

Epicenter The point on the surface of the earth directly above the focus (area of first motion) of an earthquake.

Erosion The wearing away of earth's surface by wind and water.

Eutrophication The process of enrichment and aging in an ecosystem.

Evaporite Sediments deposited from water as a result of extensive evaporation of sea water or lake water. Dissolved materials left behind following evaporation.

Extrusive igneous rocks Igneous rock that forms when magma reaches the surface of the earth. A volcanic rock.

Fauna a collective term for all the animals in a given region or geological period.

Fault A fracture or fractured system that has experienced movement along opposite sides of the fracture.

Feldspar The most abundant family of minerals in the crust of the earth. They are silicates of calcium, sodium, and potassium.

Fissile materials Materials such as uranium-238, which is not naturally fissionable but upon bombardment by neutrons is converted to plutonium-239, which is fissionable.

Fecundeh The production of sex cells.

Fission The splitting of an atom into smaller fragments with the release of energy.

Floodway district That portion of a channel and floodplain of a stream designated to provide passage of the 100-year regulatory flood without increasing the elevation of the flood by more than one foot.

Floodway fringe district Land located between the floodway district and the maximum elevation subject to flooding by the 100-year regulatory flood.

Fluvial Concerning or pertaining to rivers. An earthquake originates.

Folds Bends that develop in stratified rocks due to tectonic forces.

Foliation Property of metamorphic rock characterized by parallel alignment of the platy or elongated mineral grains. Foliation is environmentally important because it can affect the strength and hydrologic properties of rock.

Fog A cloud with its base at or very near the earth's surface.

Food chain a series of organisms dependent upon one another for food.

Formation Any rock unit that can be mapped.

Fossil fuels Fuels such as coal, oil, and gas that have formed by the alteration and decomposition of plants and animals from a previous geologic time.

Fracture zone A fracture system that may or may not be active and may or may not have an alteration zone along the fracture planes. Fracture zones are environmentally important because they greatly affect the strength of rocks.

Fungicide Substance or mixture of substances used to prevent or kill fungi.

Fungus Simple or complex organism without chlorophyll. The similar forms are unicellular, higher forms have branched filaments and complicated life cycles, *i.e.,* molds, yeasts and mushrooms.

Frontal Fog Fog formed when rain evaporates as it falls through a layer of cool air.

Fumarole A natural vent from which fumes or vapours are emitted; for example, geysers and hot springs characteristic of volcanic areas.

Fusion Combining of light elements nuclei to form heavy elements with the release of energy.

Gaging station Location at a stream channel where the discharge of water is measured.

Geochemical cycle Migratory paths of elements during geologic changes and processes.

Geologic cycle A group of interrelated cycles known as the hydrologic, rock, tectonic, and geochemical cycles.

Geomorphology The study of landforms and surface processes.

Geostationary Satellite A satellite that remains over a fixed point because its rate of travel corresponds to the earth's rate of rotation. Because the satellite must orbit at distances of about 35,000 km, photographs from this type of satellite are not as detailed as those from polar satellites.

Geothermal energy The useful conversion of natural heat from the interior of the earth.

Glacier A landbound mass of moving ice.

Gneiss A coarse-grained, foliated metamorphic rock in which there is banding of the light and dark minerals.

Gravel Unconsolidated, generally rounded fragments of rocks and minerals greater that 2 mm in diameter.

Groin A structure designed to protect shorelines and trap sediment in the zone of littoral drift. The structure is generally constructed perpendicular to the shoreline.

Groundwater Water found beneath the surface of the earth within the zone of saturation.

Grout A mixture of cement and sediment that is sufficiently fluid to be pumped into open fissures or cracks in rocks, thereby increasing the strength of a foundation for an engineering structure.

Habitat The natural abode of an organism including its total environment.

Herbicide Any chemical that kills or injures plant life.

Herbivore Organism that feeds on plants.

Hematite An important ore of iron, a mineral (Fe_2O_3).

Hetrosphere A zone of the atmosphere beyond about 80 km where gases are arranged into four roughly spherical shells, each with a distinctive characteristics.

High-value resource Materials such as diamonds, copper, gold, and aluminium. These materials are extracted wherever they are found and transported around the world to numerous markets.

Homeostasis The maintenance of equilibria in an organism or biological system.

Humus Black organic material in soil.

Hydrocarbon Organic compounds consisting of carbon and hydrogen.

Hydro consolidation Consolidation of earth materials upon wetting.

Hydro fracturing Pumping of water under high pressure into subsurface rocks to fracture the rocks, thereby increasing their permeability.

Hydrograph A graph of the discharge of a stream with time.

Hydrologic cycle Circulation of water from the oceans to the atmosphere and back to the oceans by way of evaporation, runoff from streams and rivers, and ground-water flow.

Hydrology The study of surface and sub-surface water.

Hydrothermal ore deposit A mineral deposit derived from hot water solutions of magmatic origin.

Igneous rock Rocks formed from the solidification of magma. They are extrusive if they crystallize on the surface of the earth and intrusive if they crystallize beneath the surface.

Impermeable Earth materials that retard greatly or prevent the movement of fluids through them.

Infiltration The movement of surface water into rocks or soil.

Influent stream Type of stream that is everywhere above the groundwater table and flows in direct response to precipitation. Water from the channel moves down to the watertable, forming a recharge mound.

Insecticide Substance or mixture of substance intended to prevent, destroy, or repel insect.

Intrazonal soil Soil characteristically determined by local conditions such as excess water due to flooding or high groundwater table.

Ionosphere A complex atmospheric zone of ionized gases that extends between 80 and 400 km, thus coinciding with the lower thermosphere and hetrosphere.

Island arc A curved group of volcanic islands associated with a deep-oceanic trench and subduction zone (convergent plant boundary).

Itai-itai disease Extremely painful disease that attacks bones, causing them to become very brittle so that they break easily. Associated with heavy metals, especially cadmium, in concentrations of a few parts per million in the soil or in food consumed by the victims of the disease.

Joule the absolute MKS unit of work or energy equal to 10^7 ergs.

Juvenile water Water derived from the interior of the earth that has not previously existed as atmospheric or surface water.

Karst topography A type of topography characterized by the presence of sinkholes, caverns, and diversion of surface water to subterranean routes.

Kimberlite pipe An igneous intrusive body that may contain diamond crystals disseminated (scattered) throughout the rock type.

Land ethic Ethic that affirms the right of all resources, including plants, animals, and earth materials, to continued existence and, at least in some locations, continued existence in a natural state.

Landslide Specifically, rapid downslope movement of rock and/or soil. Also used as a general term for all types of downslope movement.

Land-use planning Complex process involving development of a land-use plant to include a statement of land-use plan to include a statement of land-use issues, goals, and objectives; a summary of data collection and analysis; a land-classification map; and a report that describes and indicates appropriate development in areas of special environmental concern. A extremely controversial issue.

Laterite A soil formed from intense chemical weathering in tropical or savanna regions.

Lava Molten material that is produced from a volcanic eruption, or a rock that forms from solidification of molten material.

Leachate Obnoxious liquid material capable of carrying bacteria produced as surface water or ground water comes into contact with solid waste.

Leaching Process of dissolving, washing, or draining earth material by percolation of groundwater or other liquids.

Lignite A type of low-grade coal.

Lithosphere Outer layer of the earth approximately 100 kilometers thick of which the plates that contain the ocean basins and the continents are composed.

Littoral Pertaining to the near-shore and beach environments.

Loam medium porosity soil consisting of almost equal amounts of sand & silt and somewhat less clay considered to be best oil for growing crops.

Loess Angular deposits of windblown silt.

Low-value resource Resources such as sand and gravel that have primarily a place value. They are extracted economically because they are located close to where they are to be used.

Magma A naturally occurring silica melt, a good deal of which is in a liquid state.

Magma tap The attempt to recover geothermal heat directly from magma. Feasibility of such heat extraction is unknown.

Magnitude, Earthquake A number on a logarithmic scale which refers to the amount of energy released by an earthquake.

Marl Unconsolidated clays, sands, or mixtures of these materials that contain a variable content of calcareous material.

Mesotropic lake Lake with a moderate supply of nutrient.

Mesosphere The layer of the atmosphere above the stratosphere where temperature drops fairly rapidly with increasing height.

Meteoric water Water derived from the atmosphere.

Mica A common rock-forming silicate mineral.

Mutagen A substance that produces changes in genetic material.

Monsoon Seasonal reversal of wind direction associated with large continent, especially Asia. In winter the wind blows from land to sea; in summer it blows from sea to land.

Mudflow A mixture of unconsolidated materials and water that flows rapidly downslope or down a channel.

National amiotent air quality standard (NAAQS) Federal standard that specifies the maximum allowable level, averaged over a specific time period for a certain pollutant in ambient air.

National Environmental Policy Act of 1969 (NEPA) Act declaring that it is national policy that harmony between man and his physical environment be encouraged. Established the Council on Environmental Quality. Established requirements that an environmental-impact statement be completed prior to major federal actions significantly affecting the quality of the human environment.

Neutron A subatomic particle having no electric charge and found in the nuclei of atoms. Neutrons are crucial in sustaining nuclear fission in a reactor.

Niche The role of an organism in a habitat.

Nonrenewable resource A resource that is cycled so slowly by natural earth processes, that once used, it is essentially not going to be made available in any useful time framework.

Normal lapse rate The average drop in temperature with increasing height in the troposphere; about 6.5°C per km.

Nuclear reactor A device in which controlled nuclear fission is maintained. The major component of nuclear power plant.

Oligotrophic lake A lake with low supply of nutrient.

Oil shale An organic-rich shale containing substantial qualities of oil that may be extracted by conventional methods of destructive distillation.

Omnivorous A living organism that consumes both plant and animal food.

Ore An earth material from which a useful commodity can be extracted profitably.

Osteoporosis Disease characterized by a reduction in bone mass.

Outcrop A naturally occurring or man-caused exposure of rock at the surface of the earth.

Overburden Earth material (spoil) that overlie an ore deposit. Most frequently refers to material overlying or extracted from a surface (strip) mine.

Oxidation Chemical process of combining with oxygen.

Parasite An organism that derives its nutrition by feeding upon another living organism.

P wave One of the seismic waves produced by an earthquake. The fastest of the seismic-waves, it can move through liquid and solid materials.

Pathogen Any material that may cause disease; for example, microorganisms, including bacteria and fungi.

Pebble A rock fragment between 4 and 64 mm in diameter.

Pedology The study of soils.

Pegmatite A coarse-grained igneous rock that may contain rare minerals rich in elements such as lithium, boron, fluorine, uranium, and others.

Percolation test A standard test used to determine the rate at which water will infiltrate into the soil. Primarily used to determine the feasibility of the septic-tank disposal system.

Permafrost Permanently frozen ground.

Permeability A measure of the ability of an earth material to transmit fluids such as water or oil.

Pesticide An agent that kills pests; e.g., insecticide.

Petrology The study of rocks and minerals.

Physiographic determinism A type of site selection that utilizes the philosophy of designing with nature.

Physiographic province A region characterized by a particular assemblage of landforms, climate, and geomorphic history.

Placer deposit A type of ore deposit found in material transported and deposited by agents such as running water, ice, or wind; for example, gold and diamonds found in stream deposits.

Plastic deformation A type of deformation involving a permanent change of shape without rupture.

Plate tectonics A model of global tectonics which suggests that the outer layer of the earth known as the lithosphere is composed of several large plates which moves relative to one another. Continents and ocean basins are passive riders on these plates.

Plutonium-239 A radioactive element produced in a nuclear reactor. It has a half-life of approximately 24,000 years.

Point bar Accumulation of sand and other sediments on the inside of meander bends in stream channels.

Pollution Any substance, biological or chemical, in which an identified excess is known to be detrimental to desirable living organism.

Pool Common bed form produced by scour in meandering and straight channels with a relatively low channel slope. Characterized at low flow by slow-moving, deep water. Generally but not exclusively, found on the outside of meander bends.

Porosity The percentage of void (empty space) in earth material such as soil or rock.

Predation The process of one organism killing another for food.

Pyrite Iron sulphide, a mineral, commonly known as fool's gold. Environmentally important because in contact with oxygen-rich water, it produces a weak acid that may pollute water or dissolve other minerals.

Pyroclastic activity Type of volcanic activity characterized by eruptive or explosive activity in which all types of volcanic debris, from ash to very large particles, are physically blown from a volcanic vent.

Quartz Silicon oxide, a very common rock-forming mineral.

Quick clay Type of clay which when disturbed, such as seismic shaking, may experience a spontaneous liquefaction and lose all shear strength.

Rad (Radiation Absorbed Dose) The amount of radiation absorbed by an object or organism, expressed as the amount of energy per gram of absorbing issue or material.

Radioactive The property of some elements to spontaneously emit radiation from unstable atomic nuclei.

Radioactive waste Type of waste produced in the nuclear fuel cycle. Generally classified as high-level or low-level.

Radon A colourless, radioactive, gaseous element.

Reclamation, mining Restoring of land used for mining to other useful purposes, such as agriculture ore recreation, after mining operations are concluded.

Recycling The reuse of resources reclaimed from waste.

Regional metamorphism Wide-scale metamorphism of deeply buried rocks by regional stress accompanied by elevated temperatures and pressures.

Rem Radiation dose equivalent of any type of radiation.

Renewable resource A resource such as timber, water, or air that is naturally recycled or recycled by man-induced processes within a time framework useful for man.

Reserves Known and identified deposits of earth materials from which useful materials can be extracted profitably with existing technology and under present economic and legal conditions.

Resistivity A measure of an earth material's ability to retard the flow of electricity. The opposite of conductivity.

Resources Includes reserves plus other deposits of useful earth materials that may eventually become available.

Riffle A section of stream channel characterized at low flow by fast, shallow flow. Generally contains relatively coarse bed-load particles.

Riparian rights, water law Each land owner has the right to make reasonable use of water on his land provided the water is returned to the natural stream channel before it leaves his property. Furthermore, the property owner has the right to receive the full flow of the stream undiminished in quantity and quality.

Rippable excavation Type of excavation that requires breaking up the soil before it can be removed.

Riprap A layer or an assemblage of broken stones emplaced to protect an embankment against erosion by running water or breaking waves.

Riverine environment Land area adjacent to and influenced by a river.

Rock Geologic An aggregate of a mineral or minerals. Engineering-any earth material that has to be blasted in order to be removed.

Rock cycle A group of processes that produce igneous, metamorphic, and sedimentary rocks.

Rocksalt Rock composed of the mineral halite.

Rotational landslide Type of landslide that develops in homogeneous material. The movement is likely to be rotational along a potential slide plane.

S wave Secondary wave, one of the waves produced by earthquakes.

Saline Salty; characterized by a high salinity.

Salinity A measure of the total amount of dissolved solids in water.

Salt dome A structure produced by the upward movement of a mass of salt. Frequently associated with oil and gas deposits on the flanks of a dome.

Sand Grains of sediment having a size between 1/16 and 2 mm in diameter: Often sediment composed of quartz particles of the above size.

Sand dune A ridge or hill of sand formed by wind action.

Sandstone A detrital sedimentary rock composed of sand grains that have been cemented together.

Sanitary landfill A method of solid-waste disposal not producing a public health problem or nuisance. Confines and compresses the waste and covers it at the end of each day with a layer of compacted, relatively impermeable material such as clay.

Scenic resources The visual protein of an aesthetic experience. Scenery is now recognized as a natural resource with varying values.

Schist A coarse-grained metamorphic rock characterized by a foliated texture of the platy or elongated mineral gains.

Schistosomiasis Snail fever, a debilitating and sometimes fatal tropical disease.

Sea wall An engineering structure constructed at the water's edge to minimize coastal erosion by wave activity.

Secondary enrichment A weathering process of sulphide ore deposits which may concentrate the desired minerals.

Sedimentology The study of environments of deposition of sediments.

Seismic Referring to vibrations in the earth produced by earthquakes.

Seismograph An instrument that records earthquakes.

Septic tank A tank that receives and temporarily holds solid and liquid waste. Anaerobic bacterial activity breaks down the waste. The solid wastes are separated out, and liquid waste from the tank overflows into a drainage system.

Sewage sludge Solid material that remains after municipal wastewater treatment.

Shale A sedimentary rock composed of silt-and clay-sized particles. The most common sedimentary rock.

Shield volcano A broad, convex volcano build up by successive lava flows. The largest of the volcanoes.

Silt Sediment between 1/16 and 1/256 mm in diameter.

Smog Polluted air and water vapour.

Soil Earth material modified by biological, chemical, and physical processes such that the material will support rooted plants.

Soil erosion Movement of topsoil component from one place to another by exposure to wind, flowing water or both.

Soil horizons Layers in soil (A, B, C) that differ from one another in chemical, physical, and biological properties.

Soil survey A survey consisting of a detailed soil map and descriptions of soils and land-use limitations. Generally prepared by the Soil Conservation Service in co-operation with local government.

Solar energy Collecting and using energy from the sun directly.

Solid waste Material such as refuse, garbage, and trash.

Spoils Mining Banks or piles that are accumulations of overburden removed during mining processes and discarded on the surface.

Stratopause The boundary between the stratosphere and the mesosphere.

Stratosphere The zone of the atmosphere above the troposphere that is characterised at first by isothermal conditions and then a gradual temperature increase, the earth's ozone is concentrated here.

Strain Change in shape or size of a material as a result of applied stress. The result of stress.

Stress Force per unit area may be compression, tension, or shear.

Subduction A process in which one lithospheric plate descends beneath another.

Subsidence A sinking, settling, or otherwise lowering of parts of the crust of the earth.

Subsurface water All of the waters within the lithosphere.

Surface water Waters above the surface of the lithosphere.

Surface wave One of the types of waves produced by earthquakes. These waves generally cause most of the damage to structures on the surface of the earth.

Suspended load Sediment in a stream or river carried off the bottom by the fluid.

Syncline A type of fold in which younger rocks are found in the core of the fold. Rocks in the limbs of the fold dip inward toward a common axis.

System Any part of the universe that is isolated in thought or in fact for the purpose of studying or observing changes that take place under various imposed conditions.

Symbiosis The living together of two or more species.

Tar sands Naturally occurring sand, sandstone, or limestone that contains a very viscous petroleum.

Thermocline The subsurface layer of a lake characterised by a significant temperature change.

Thermal pollution Increase in water temperature that has harmful ecological effects on an aquatic ecosystem.

Tectonic Referring to rock deformation.

Tectonic creep Slow, more or less continuous movement along a fault.

Tectonic cycle A group of processes which collectively produce external forms on the earth, such as ocean basins, continents, and mountains.

Temperature inversion A layer in the atmosphere of limited depth where the temperature increases rather than decrease with height.

Tephra Any material ejected and physically blown out of a volcano; mostly ash.

Texture, rock The size, shape, and arrangement of mineral grains in rocks.

Thermosphere The zone of the atmosphere beyond the mesosphere in which there is a rapid rise in temperature with height.

Tidal energy Electricity generated by tidal power

Till Unstratified, heterogeneous material deposited directly by glacial ice.

Transform fault Type of fault associated with oceanic ridges. May form a plate boundary such as the San Andreas fault in California.

Translation (slab) landslide Type of landslide in which the movement takes place along a definite fracture plane such as a weak clay layer or bedding plane.

Tropical cyclone Severe storm generated from a tropical disturbance. Called typhoons in most of the Pacific Ocean and hurricanes in the Western Hemisphere.

Tropopause The boundary between troposphere and stratosphere.

Troposphere The lowest layer of the atmosphere marked by considerable turbulence and generally a decrease in temperature with increasing height.

Tsunami Seismic sea wave generated by submarine volcanic or earthquake activity. Characteristically has very long wavelength and moves very rapidly in the open sea. Incorrectly referred to as *tidal wave.*

Tuff Volcanic ash that is compacted, cemented, or welded together.

Unconfined aquifier Type of aquifier in which there is no impermeable layer restricting the upper surface of the zone of saturation.

Unconformity A buried surface of erosion representing a time of non-deposition. A gap in the geologic record.

Uniformitarianism Concept which states that the present is the key to the past. That is, we can read the geologic record by studying present processes.

Virus A microscopic living particle, usually smaller than bacteria, that is parasitic on living cells and incapable of reproducing outside of living cells.

Volcanic breccia, agglomerate Large rock fragments mixed with ash and other volcanic materials cemented together.

Volcanic dome Type of volcano characterized by a very viscous magma with high silica content. Activity is generally explosive.

Water table The surface that divides the zone of aeration from the zone of saturation. The surface below which all the pore space in rocks is saturated with water.

Weathering Changes that take place in rocks and minerals at or near the surface for the earth in response to physical, chemical, and biological changes. The physical, chemical, and biological breakdown of rocks and minerals.

Zonal soil Soil in which the profile is in adjustment to the climatic zone.

Zone of aeration The zone or layer above the water table in which some water may be suspended or moving in a downward migration toward the water table or laterally towards a discharge point.

Zone of saturation Zone or layer below the water table in which all of the pore space of rock or soil is saturated.

Index of Part-I

A

Acctaldehyde 7, 171
Acid deposition 173, 163-73
– vegetation injury 174
– Al^{3+}, Hg^{2+} poisoning 174
Acid fog 176
Adenine 9
Adenosine triphosphate (ATP) 3
Adiabatic lapse rate 178-182
Advective inversion 13
Aerosols 206
Air circulation 13, 15
Air glow 79
Air pollution control 241-60
Air pollution in India 271-84, 261-70
Air quality 195-97
Alanine 7
Albedo 133
Alkyl peroxy radical 186, 188
Ammonia 3, 84, 260
α-Amino butyric acid 7
Angular momentum, atoms 45-47
– atmosphere 16
Anharmonicity constant 51-55
Atkin nuclei range 203
Atmosphere 11, 12, 13, 15
– Dispersion 222, 235-36
– Photochemistry 75-97
– Stability 11-13, 226, 229
Automobile fuel 262

B

Beer-Lambert law 75
Benzopyrene, BaP 212
BeP 212
Bhopal gas tragedy 276-81
Black body radiation 129
Breezes 21
Brownian diffusion 200
Buys Ballot law 19

C

Cancer 110
Carbon credit system 326
Carbon monoxide
– oxidation 125
– spectra 124
Carbon dioxide spectra 57, 127
Carnot cycle 37
Catalytic converters 267
Centrifugal force 20
Chappius bands 80
Chapman cycle 105
Chloro-Bromo carbons 119, 121
Chlorofluoro carbons 111-115, 116
Climate change 139-148, 160
ClO_x role in stratosphere 108
COP 15 324, 331
Copenhagen climate council 330
Coriolis effect 17-20, 21
Cunnigham factor 199
Cyclone separators 247-50

D

Decibels 285-86

Deposition process 173, 236
Dessert 145
Diethyl hydroxyl amine (DEHA) 193
Diethyl nitrosoamine 193
Dispersion 235
Dissociation energy 55, 65
Dissociation products 64
Doldrums 15
Doppler broadening 51

E

Earth atmosphere 10
– circulation 13
– energy balance 129
– evolution 2, 4
Ecological consequences 153-54
Electromagnetic radiation 48-49
Electronic states 58, 59, 68
Electronic transitions 60
– Chlorofluoro carbons 112-13
– CO 62, 64, 124, 125
– CO_2 127
– DNA 121
– HNO_3 93-94
– HONO 92
– H_2S 176
– I_2 64
– N_2 84
– NO 66-67, 85-87
– N_2O 57
– NO_2 87-91
– NH_3 84
– O_2 61-63, 65, 76-80
– O_3 80-82, 106, 139
– Proteins 121
– SO_2 163-64
Electrostatic precipitator 245-47
El Nino: southern oscillation 148-51
Emission controls 239
Energy law 33
– balance 129-133, 138
– levels 45
Engine – two strokes 264
– four strokes 262-64
Entropy 36
Excited molecules fate 71-73

F

Florescence 73
Flow around structure 24
Flue gas 240
Fluoride 260
Fly ash 216, 273
Fog 182
Formaldehyde 101-4
Four stroke engine 262, 265
Frank-Condon principle 64
Frictional forces 18, 20

G

Gas turbines 269
Gaussian plume 223, 225
Geostrophic wind 18
Glutathione 278
Gradient wind 19
Green house effect 133-36, 137
Grottus-Draper law 69

H

Hadley cell 15
Haemoglobin 124, 278
Half lives 42
Hartley continuum 81
Herzberg continuum 77
Hot bands 53
Huggin's band 82
Hydantoin 281
Hydrocyanicacid 8
Hydrogen peroxide radical 125
Hydrogen sulphide 259
Hydro peroxide HO_2 39, 86, 98, 126
Hydroxy radical 39, 86, 97, 102, 114-17, 139, 165-66

I

Intensity of spectral lines 49, 63
Inter system crossing (ISC) 49-50, 63
Isopleth plot 188

J

Jablonski diagram 72
Jet engines 269

K

Kinetics, unimolecular 39
– bi and tri molecular 39

Kyoto protocol 325

L

La Nino 151
Lapse rate 11-12, 178
Lead 214, 263
Leighton relation 91
Leporte selection rule 48
Life times of reactions 42, 98
Light intensity 203-06
Light scattering 203
Little ice age 143
Local wind systems 23
Lofting and looping 223
London type smog 183-85

M

Markonikov rule 99
Material balance 2
Mercury poisoning 114
Mesopause 11
Mesosphere 11
Methane 58, 95
 – oxidation 115
Mie scattering 204-06
Milankovitch hypothesis 160
Molar extinction coefficient 70
Molecular vibrations 51
Morse potential 51-57
Mount and valley winds 21
Myoglobin 2

N

Night glow 79
Nitrate radical NO_3 91, 98
Nitric acid, HNO_3 93, 171, 173
Nitric oxide NO 85-87, 125-127, 184, 187
Nitrogen dioxide NO_2 57, 171, 187
Nitrogen, N_2 84
Nitrous acid, HONO 92
Nitrous oxide N_2O 57, 171, 187
Noise and health 291
Noise and hearing loss 290
Noise induced stress 289
Noise performance 290
Noise pollution 291, 293
 – control 295

O

Odours 261
Oparin's droplets 2
Organic nitrogenous compounds 95
OSHA-compliance plan 293
Oxygen, O_2 65, 76-80
Ozone 80-82, 98, 106, 169, 197
Ozone hole 120
Ozone loss-cooling 139
Ozone-olifin reaction 166
Ozone shield 105

P

Particle pollutants 211
Particulate matter 198-221
 – chemical composition 208-210
 – diesel 212
 – plant growth 194, 217
 – size 201
Particulate organics 190-93
Perhydroxy radical 95-97
Peroxy acetyle nitrate (PAN) 188-89
Peroxybenzoyl nitrate (PBzN) 189
Peroxynitric acid (HO_2NO_2) 94
Phosphorescence 73
Photochemical smog 178, 183-86
 – adverse effects 193
 – inhibition 192
Photochemistry of atm. 75-104
Photo dissociation 71, 79
Phytochrome 217
Plume rise 230
Polycyclic aromatic hydrocarbons (PAH) 212
Positive crankcase ventilation (PCV) 265
Predissociation 66
Pressure broadening 51
Pressure drop 244, 249
Prevention of hearing loss 292
Prionitrile 7
Propionaldehyde 7
Propionic acid 7

Q

Quantum yield 71, 74

R

Radiation inversion 13, 182
Rate laws 41

Reaction rate constant 41
Reaction rates 40
Reformation of gasoline 216, 263
Reverse pulse baghouse 243
Reyleigh scattering and particle size 204
Russell-Saunders coupling 46

S

Schumann-Runge 77, 110
Scrubbers 253-55
Seabreezes 21
Skin cancer 110-11
SMOG 182
- adverse effect 193
- inhibition 192
- Photochemical 183-185

Smoker-polluter 218-20
Solar constant 132
Space garbage 220
SST 109
Stark-Einstein law 69
Stratosphere 10, 105-122
Stefan-Boltzmann law 137
Subsidence inversion 13, 182
Succinic acid 7
Sulphur dioxide 139, 256, 165-70
- oxidation 163-68

Sulphur monoxide 163-65
Sulphuric acid 173
Sun 121
- flare up 122
- SOHO 122

T

Temperature inversion 12, 178-81, 232
Term symbol, atoms 46
- molecules 58

Tetraethyl lead (TEL) 263
Thermodynamics laws
- first law 33
- second law 36
- third law 38
- zeroth law 33

Thermosphere 11
TOGO 151
Tornadoes 12
Tropopause 10-12
Troposphere 10-12
Two stroke engine 264

U

Urban dome 152
Urban-rural circulation 23-24

V

Valine 278
Vector model of atom 46
Ventilation time 239
Vertical motion divergence 21
Vibrational frequencies 130
Vibrational spectra 51, 52

W

Water 56
- photochemistry 82-84, 108
- resources 145

Width of spectral lines 50-51
Wien's displacement law 129

Y

Younger-Dryas 141

Index of Part-II

A

Acetylcholine 275, 283, 293-94
Acid character of clay mineral 127-131
- attapulgite 130
- montmorillonite 127
- sauconite 130
Acid rain 71, 174
Activated carbons 85
Aeration process 86
Aerobic decomposition 360
Aldrin 281
Anaerobic decomposition 36
Anion adsorption 136
Antijuvenile hormone 303
- insect sterilization 304
Aqueous electrolytes 27
Arsenic 70, 202-04
Arsenite inhibition 276
ATP 5
Atmosphere 5, 11, 13, 16
Attapulgite 113
Attractant 300
Aurin herbisides 284
- degradation 285
- & dioxins 286
Automobile fuel 234
Autoprotolysis 22

B

Beidallite 107
Beryllium 192
Bicchri 94
Biochemical oxygen demand (BOD) 70, 73, 90
Biodegradable garbage 355
Biogas 360
Biotite 111
Bombay minamata 69, 208
Boron 192
Brightners 267
Brucite 108, 115
Buffering action of CO_3^{2-} 61
Builders 265
Bunsen absorption coefficient 53

C

Cadmium 165, 205
Calcium 193
Carbamates 222, 295-97
Carbon cycle 175-77
Carbon dioxide 66
- in water 57
Carbonic acid & carbonates 58-62
- estimation 63, 177
Carrybags menace 232
Catalytic converter 244
Check dams 168
Chemical oxygen demand (COD) 71
Chlorite 113
Chlorophyll 193
Choline 284, 293
Chromium 194
Circulatory system 327
Clathrate hydrates 27-29
Clay minerals 106-115
- acid character 127-31

– anion adsorption 131, 136-38
Clay minerals and X-ray identification 115-21
Close packed structure H_2O 18
– open structure, H_2O 19
Coacervate 4
Coagulants 85
Coal 261
Compressed natural gas (CNG) 248
Copper 199
Crude oil marine life 257

D

Dark reaction 37
Decomposer 38
Demeton 292
Denitrification 65, 213-14
Detergents 262
Dichloro diphenyl trichloroethane (DDT) 164-276
– bioamplification 279
– degradation 278
– long term effects 279
– mechanism of action 277-78
Dickite 108
Dieldrin 281
Diesel emission 242
Dioxins 228
Dissolved oxygen (DO) 54, 56, 74
Distribution factor (DF) 371

E

Ecology 36
Ecosystem 36
Effluents 51, 89, 93, 95
Electrochemical methods 88
Evolution of earth 3-5
Evolution of life 5-8
Eutrophication 182

F

Feldspars structure 122-24
– hydrolysis 124-27
Fluoroapatite 211
Fluorosis 69, 211
Fly ash 360-63
Food chains and webs 37, 38
Four stroke engine 236, 241
Fuels – biofuels 250
– CNG 248
– fossils 234
– hydrogen 251
– methanol and ethanol 250
– nuclear fuel 251
– propane 249
Fulvic acid (FA) 141
– reaction with metal ions & minerals 141-43

G

Ganga action plan (GAP) 78-80
Garbage 351
Garbage management 352
Gasohol 239
Gasoline reformation 238
Geologic Cycle 31-35
Gibbsite 107, 115, 122
Goethite 122

H

Haemoglobin 197-99
Halloysite 108
Halogens 210
Hectorite 110
Heme 197
Herbisides 284
– degradation 285
– & dioxins 286
Hospital Waste 353
Humic acid (HA), 139-44
– Colloidal nature 140
– Reactions 141
Humus 139
Hydrargillite 107, 115
Hydrogen bonding 19
Hydrologic cycle 29
Hydrolysis of minerals 172
Hydroxyapatite 211

I

Illite 111
Industrial effluents 90
Industrial pollution of water 80, 92
Insecticides 274
Insect sterilization 304
Ionic product of water 23
– temp & pressure effects 23
Ionic water 20-24
Ionizing radiation 4
Iron 195
Itai-itai disease 68

J

Juvabione 303
Juvenile hormones 301
  – insect control 303

K

Kaolinite 107, 136-38
Karnal technology 100
Keratin biodegradation 224

L

Lakes pollution 47
Lateral surface structure 133
Lead 165, 209
Lead pollution 236
Lignin 98, 138, 225
Liming acid soils 154-57
Lindane 279
  – degradation 280
Lipids 98

M

Magnesium 193
Malathion 293
Manganese 194
Marine life 259
Mercury 206
Metamorphic rocks 33
Methanogenic bacteria 99
Methylene blue value 71
Methyl parathion 290
Microstraining 85
Miller experiment 5
Mineral insecticide 274
Minimata disease 68, 206
Montmorillonite 109, 128-29, 136-39
Muscovite 109-11

N

Nantronite 110
Natural oil seepage 252
Nematocidal action 272
Nitrate health hazard 157
Nitrobacter 65, 179
Nitrogen cycle 177-183
Nitrosomonas 65, 179
Non-biodegradable but recycleable wastes 355
Non-biodegradable non-recycleable wastes 356
Nuclear chemistry 310
Nuclear fission 361
  – Kinetics 312, 315
Nuclear reactors 337-42
Nutrients 64

O

Omnivores 38
Open structure water 19-20
Orange peels biodegradation 225
Organic insecticide 308
Organic matter 159
Organophosphorus pesticide 288
  – degradation 294
  – toxic effect 293

P

Palygorskite 113
Parathion 290
Paris green 274
Pectin 97
Penicillin 227
Periodic table 174, 380
Pesticides 270-308
  – viable alternatives 304, 307
Pesticides, in soils 158-64
  – in environment 163
Petroleum biodegradation 257
Petroleum reaction with singlet O_2 256
Petroleum reactions with triplet O_2 254
Pheromones 297-301
  – isolation 297
  – & insect control 300
Phlagophite 110
Phosphate precipitation 88
Phosphorus cycle 183-86
Phytic acid 186
Polychlorinated cyclopentadiene 281
  – degradation 282
Polychlorobiphenyls (PCB) 228, 277
  – long term effects 286
Polymers 213
  – biodegradation 224
  – chemical degradation 218
  – photo degradation 220
  – protection 223
  – synthesis 214-18
Positive crankcase ventilation (PCV) 241

Precocene 352
Pyrophyllite 106, 136-39

Q

Quality factor (QF) 320
Quartz 16

R

Radiation 310-49
- biological effect 326
- chemistry 325
- flux (ϕ), 322
- intraction with cells 328
- man made & natural 330
- maximum permissible doses 333-37
- & respiratory system 327
- rad 320
- rem 321
- rotengens 32
- units 319

Radiation hazards-reactors 342, 333-37
Radioactive waste management 345
Radio nuclides 166, 317
- disintegration 317
- Kinetics 312
- Kinetics of mixtures 315

Reactor maintenance 345
Rivers pollution 48-50
- Indian 77-80

S

Sandstone and conglomerate 33
Saponite 109
Sauconite 109, 133
Sedimentary rocks 33
Selenium 204
Sewage purification 70
- primary & secondary treatment 82
- tertiary treatment 85

Sepiolite 115
Shale 33
Sewage chemistry 86
- chemistry and biochemistry 96

Soil horizons 34
- acidification 148, 151
- aggregates 144
- & liming 154
- & crop tolerance 180
- decomposition by organic matter 143
- organic matter 138, 152
- & heavy metal ion 157, 165
- & pesticides 158-64

Solubility product of & pH 61-62
Steerer-Phelps eqn 75
Sterilization of effluents 89
Sulphur cycle 186-90
Sulphur pollution 238

T

Talc 108
Tectonic cycle 31
TEL 238
Thermal pollution 42-44
Thixotrophy 146
Thorium (4n) series 317
Tolerance limit 40
Trimidite 18-19

U

Uranium (4n + 2) series 316

V

Vermicullite 111-13

W

Water 12-14
- clathrates 27-29
- closepacked structure 18
- ionic water 22, 24-27
- open structure 19
- pollution 46-100
- structure 15-18

Water management of soils 166
Weathering of clay minerals 127-31
Whey biodegradation 225
Wind erosion of soils 147

Y

Yamuna action plan 79
Yokkaichi asthma 68, 238

Z

Zinc 165, 199-202
Zinjanthrophas 18